Functional
Human Anatomy

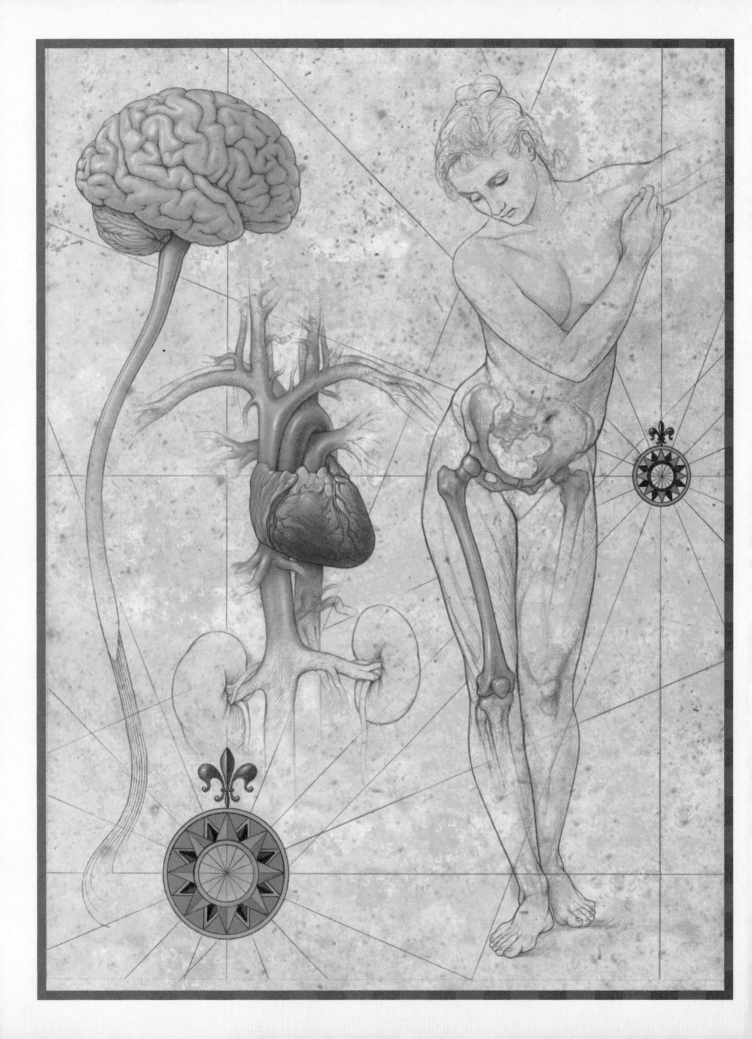

FUNCTIONAL HUMAN ANATOMY

DAVID T. LINDSAY, Ph.D.

University of Georgia,
Athens, Georgia

With 635 *illustrations*

St. Louis Baltimore Berlin Boston Carlsbad Chicago London Madrid
Naples New York Philadelphia Sydney Tokyo Toronto

Editor-in-Chief:	James M. Smith
Editor:	Robert J. Callanan
Senior Developmental Editor:	Jean Babrick
Project Manager:	Carol Sullivan Weis
Production Editor:	Christine Carroll
Designer:	Nancy J. McDonald
Manufacturing Supervisor:	Karen Lewis
Cover illustration:	John Daugherty

Printed in the United States of America

Composition by Clarinda Company
Color separation by Color Associates, Inc.
Printing/binding by Rand McNally

Mosby–Year Book, Inc.
11830 Westline Industrial Drive
St. Louis, Missouri 63146

Library of Congress Cataloging in Publication Data
Lindsay, David T. (David Taylor), 1935-
Functional human anatomy / David T. Lindsay.
p. cm.
Includes index.
ISBN 0-8016-6471-3
1. Human anatomy.
[DNLM: 1. Anatomy. QS 4 L7514 1995]
QM23.2.L56 1995
611--dc20
DNLM/DLC
for Library of Congress   94-38262
                                                    CIP

To Suzie, Anne, and Sara

PREFACE

THIS TEXT

Functional Human Anatomy is written for students who are preparing for careers in the allied health sciences. The text is most appropriate for programs that teach anatomy and physiology as separate, rather than combined, subjects. The goal, however, is universal: to know the human body. *Functional Human Anatomy* emphasizes the relationships between anatomy and function at each level of organization that students will need to understand in their professional programs. The text assumes basic knowledge of cell chemistry, cellular structure, and the functions of organ systems but reviews these foundations in introductory chapters.

THEMES

There are as many goals in learning and teaching anatomy as there are students and instructors. From nursing and physical therapy to scientific illustration, dance, and occasionally premedical studies, most students are interested in what their knowledge of human anatomy will enable them to do, rather than in anatomy for its own sake. As instructors, our challenge is to make the immense details of anatomy accessible and understandable, while organizing this material to our students' advantage.

It is said that anatomy never changes. As molecular biology reveals the action of more and more genes and medical imaging techniques display functions at cellular and tissue levels, however, our understanding of the relationships between structures and functions in the human body will become more intimate and useful. Our students will administer tests and treatments based on this information.

SPECIAL FEATURES

NARRATIVE STYLE

This text uses an active style to bring the subject to life. For example, the straightforward statement, "Muscles are organs of the skeletal muscle system; they move the body, hold it in position, give it shape, and heat it," introduces Chapter 9 on skeletal muscle. The goal is to teach with the students rather than at them. Rather than compiling an encyclopedia of separate topics, this text emphasizes relationships among organs and systems.

ACCESSIBILITY

Anatomy should be accessible, not impenetrable. Students complain that it is hard to organize the formidable amount of information in anatomy. Needing background information on epithelium, one student complained that his text enumerated 11 kinds of epithelium but did not relate them to each other.

Not knowing what is considered important, students often attempt the impossible: to learn every term. *Functional Human Anatomy* supplies "Perspectives," "Brief Tour," and "Overview" discussions to help students set priorities in studying. Perspectives are short, leading sections that outline the significance of the chapter. Brief Tours and Overviews state the central features of the section ahead. For example, the Brief Tour of the Skull (Chapter 6) outlines the major features of the skull before readers are expected to fit in the individual bones. By leading students into the material, this text advances beyond the task-oriented objectives and study questions that all anatomy texts supply. To help students understand epithelium, for example, Chapter 3 outlines its major features before describing different varieties. Throughout the text the outline and significance of each subject precede its details.

SIMPLICITY

Simplicity is power. Simplicity is earned. As teachers, we seek the simplicity that integrates parts into useful wholes, not the simplicity that erodes content. Simplicity begins with the details of anatomy and finds ways to assemble them into mental pictures of structures and functions. Anatomical perspective and experience emerge from these images, and when subsequently applied in a profession, they become real power to earn.

Each chapter employs several techniques to climb this ladder of understanding.

Student objectives list learning goals for the material ahead. Students need be able to trace blood through the cardiovascular system or impulses through tracts of the central nervous system, for example, to understand how drugs reach target tissues or how stimuli reach the brain.

Vocabulary toolboxes list the chapter's hard-working terms, as well as their derivations and pronunciations. It is gratifying how unfamiliar terms make sense when broken down into Latin and Greek roots and become powerful when pronounced professionally. Other important terms appear in boldface throughout the text.

Boldfaced topic sentences begin major sections of the text, stating as clearly as possible the central idea ahead. Recorded in notebook and in mind, these sentences reappear in the chapter summary to enable students to organize details successfully.

Prompt questions after each section alert students to items they should memorize. Rereading the text quickly reveals the answers.

Study questions at the end of each chapter convert objectives into questions that force students to organize the material they have studied. Instructors can use these questions in written exams.

Ten-pointer questions (in the *Instructor's Resource Guide*) are short problems that can be assigned a week before a quiz, to develop a feel for anatomy. Students write answers, discuss them, critique them with the professor, and commit the final version to memory for the quiz. In the author's course at the University of Georgia, these projects are like money in the bank, worth 10 out of 30 weekly points if the work is done in a professional manner. In his course, critiques of the quizzes are published to help improve everyone's work by showing the differences between successful and less successful answers. Critiques also appear in the *Resource Guide*.

CASE STUDIES AND ANATOMY

This text integrates clinical case studies into the chapters to illustrate how structures function. Chapter 18 begins with an account of a spinal injury and then shows how the anatomy of the spinal cord and spinal nerves accounts for the effects of the injury. Such disorders are integral with functional anatomy because they disclose how structures function. Indeed, medical and biological research progress by learning how defects interfere with normal processes. It is important to integrate clinical matters with anatomy in introductory courses because many students will deal with patients and clients who are seeking treatment for disorders or ways to improve their physical performance.

ILLUSTRATIONS

This text is illustrated profusely, with more than 600 figures. Look especially for narrative descriptions of physiological processes given directly in the figures themselves (see Figure 9.15). Summary figures (see Figure 9.16) located near the ends of chapters tie the discussions of disorders, processes, and events to the anatomy of the system. Other figures present practice quizzes that ask for names and functions of structures (Figure 13.26).

CHAPTER HIGHLIGHTS

Part I of this text deals with anatomical fundamentals of the body, cells, and tissues, as well as embryonic development. The color-coded pages of this section identify it as preliminary or optional material separate from the organ systems described in the main text. Part II discusses the anatomy of support, motion, and shape, including bone, skeleton, muscle, and skin. Part III considers the cardiovascular system and vascular integration. Neural integration follows in Part IV. All these discussions converge in the descriptions of the respiratory, digestive, urinary, and reproductive systems given in Part V.

ANCILLARIES

INSTRUCTOR'S RESOURCE GUIDE
This guide includes 10-pointer questions, quiz critiques, and answers to study questions. Also included are review questions and a study program for final exams, as well as source lists of video, CD-ROM, and computer simulations.

TESTBANK
Elizabeth Harper has expanded the author's stock of multiple-choice questions into an extensive testbank, also available in computerized format.

LABORATORY MANUAL
This manual by Drs. Carol D. Rodgers and Jaci L. VanHeest carries the text's functional approach into the laboratory with exercises that identify structures and relate them to their functions.

STUDY GUIDE
Dr. DeLoris Wenzel has crafted a student study guide that combines the best of educational theory with the author's approach to anatomy.

TRANSPARENCIES
One hundred and fifty transparencies selected from the text are available for lecture and lab. Large labels make the names easy to read.

VIDEOS, CD-ROMS, AND SIMULATIONS
The Anatomy and Physiology CD-ROM Tutorial uses human dissection photos, diagrams, x-rays, CT scans, animations, and an interactive mode to provide visual review.

Human Cadaver Videotape shows the dissection of human musculature, as well as the internal organs of the thorax and abdomen.

Human Body Videodisc includes over 1500 still images, as well as 2 hours of video segments and animations.

ACKNOWLEDGMENTS

Although the author's name goes on the cover, inside lies the work of many contributors. Deborah Allen, of Mosby, saw value in this project at the beginning and got it under way. Gordhan Patel, Larry Pomeroy, Judy Willis, and Ray Damian, my colleagues and department heads, granted me time to write. Senior Developmental Editor, Jean Babrick, exercised a steady, firm, and occasionally desperate hand in bringing the manuscript to production. Our respect for each other has grown at every step. Kristin Shahane produced the ancillaries and Christine Carroll brought the figures and text together. Bob Callanan, Editor, oversaw the steps of assembly, and Anne McKeough, Director of Marketing, placed this book in your hands.

I am especially indebted to Patrick Jackson, whose expert editing sharpened the neural integration chapters. Cynthia J. Moore provided an essay on genetics of cancer that is the basis for the cancer sections. All of the reviewers, especially Mary Ellen St. John and the late John Short, improved this text with their criticisms, suggestions, and experiences in their own labs and classrooms. The following people helped find material and illustrations: Harvey Cabaniss, Victor Crosby, Victor Eroschenko, Mary Evans, Mark Farmer, Jen Feuerstein, Samuel Griffin, William Hamilton, Ron and Greta Hendon, Pat Jettie, Farris Johnson, Cathy Kelloes, Hillary Newland, Lynne Radford, and Jill Trefz.

Sonya Hooks, Christina Hunt, Genia King, Marian Thomas, and Kathy Vinson took endless messages for me while I wrote. David Gann graciously accepted my late copying charges. Mary Ellen and Bob Taylor offered recovery at the lake and the mountains. Capt. Karl R. Whitney, MC, USN, Ret., greeted each new chapter with enthusiasm that made us both happier while the next chapter got under way. I am indebted to all of you, but especially to Suzie Lindsay, whose love is always there and who bled zealous red ink on all drafts of the manuscript.

David T. Lindsay

REVIEWERS

William A. Brothers
San Diego Mesa College

David Byman
Penn State University, Worthington-Scranton
Campus

Elizabeth Harper
San Diego City College

Patrick Jackson
Canadian Memorial Chiropractic College

Carol Rodgers
University of Toronto

Roscoe Root
Lansing Community College

John Short
DuQuesne University

Mary Ellen St. John
Central Ohio Technical College

PREFACE TO STUDENTS

This preface describes how to study anatomy and how to use this text. The allied health sciences begin with anatomy, and anatomy begins with names and terms that identify the many details of the human body. From these words, your course and text will construct images of the body's structures, and from them, you will learn your way around the human body and how it works.

The process is like climbing a mountain. As you begin in a valley, surrounded by trees and forest, your perspective is rather limited. But as you climb up to the ridges and summits, the view widens and the meaning of your trip clarifies. Details of kidney structure make considerable sense when viewed from the summit of the cardiovascular mountain.

HOW TO STUDY ANATOMY

GOALS

The primary goal of your anatomy course is to know your way around and through the human body. As is each of the steps in learning anatomy, this course is one of your steps toward a career. Set your own personal goal for this course and then do what is necessary to attain it.

LANGUAGE OF ANATOMY

Begin by learning the vocabulary of anatomy and then learn to think in this new language. Vocabulary terms name body structures and describe their functions. A good way to learn these words is to write each of them on a small index card with a description of the body part and its function on the back. Then practice with the cards. Riffle through the words and describe the functions, and turn them over and recite the names when you see the functions. Practice will quickly put you in command.

KEEP IT SIMPLE

Fat chance, you say. With so many details to learn, how am I possibly going to organize them? The answer is simplicity. Simplicity is power! Simplicity is earned. When you look for simplicity you will find it in concepts that relate structures to each other. You will find that many apparently different structures are made of similar components and do similar things. Look constantly for similarities and differences among body parts. What you find will transform that card file list into working images of organs and tissues that will guide your way around the human body. Most especially, (1) make mental images of the structures and (2) learn anatomy one layer at a time.

IMAGES

Make mental images of body structures and their functions. These pictures will help you remember anatomical names by giving you something to hang them on, a network to hold them together. Describe your image of the esophagus, for example, to a friend to ensure that you are accurate and complete, or draw the structure from memory.

LAYERS

Learn anatomy in layers, a little at a time. You will find that each study session reinforces what you have already learned and makes you more receptive to new material. Repetition is essential! Mental images usually do not emerge all at once, especially not the night before a quiz. Be good to yourself; schedule several hours each day to review old material and to add new.

DO NOT TRY TO LEARN IT ALL

We do not expect you to learn everything this course and text offer; there is not enough time even in three courses. You will do much better if you learn to recognize what material is important and know it thoroughly. Everything seems important to beginners, but as you learn more, you will begin to discriminate. Until then, ask your instructor what the quizzes and exams will cover. Remember the course's goal, as well as your own, and decide accordingly.

BE HAPPY WITH WHAT YOU HAVE LEARNED

The body is an impressive place, far more clever than you and I can imagine. Welcome to anatomy!

HOW TO USE THIS TEXT

What follows is a list of special features found in each chapter of this text. Use these features to help with anatomical terms, mental images, and organization of material.

OBJECTIVES

Each chapter begins with a list of objectives, describing what you should be able to do after you have studied the text. Objectives are the immediate goals and priorities of the chapter. Study questions at the end of the chapter ask you questions about the objectives.

VOCABULARY TOOLBOX

This is a short list of hard-working terms in each chapter, along with their pronunciations and derivations. Pronunciations give you the professional sound of the words, and derivations show you that unfamiliar terms make sense when broken down into their Latin and Greek roots. You will find other important **boldface terms** in the text itself, often with pronunciations and derivations.

RELATED TOPICS

This is a list of topics and chapter references that you should know about as you study each chapter. Review them if necessary.

PERSPECTIVE

A short paragraph describes each chapter's significance to anatomy in general and to health science.

BRIEF TOURS AND OVERVIEWS

Your text contains more details than you need to know about each subject, but instructors choose this text because it gives them room to teach. To help you ferret out the details, major sections of chapters begin with descriptive overviews on which to hang details. Read these "Brief Tours" before you plunge into the main material—they help you form mental images of the topic, and they simplify the discussion by suggesting how to organize the deeper material ahead. If the Brief Tour discusses it, it is important. For example, the Brief Tour of the Skull in Chapter 6 lays out the major spaces and partitions of the skull so that you can recognize later how each bone contributes to the overall structure.

BOLDFACE TOPIC SENTENCES

Leading paragraphs start with boldface sentences that declare the main point of the section. These explicit sentences are excellent tools for organizing your reading and lecture notes because they usually state a unifying principle that organizes the material. These sentences reappear in the chapter summary.

PROMPT QUESTIONS

Details come thick and fast in any anatomy text, and this text is no exception. Prompt questions after each section of text help you keep up with the flow. These short questions alert you to terms and facts you will need to remember.

CASE HISTORIES

Many chapters begin with clinical cases that illustrate anatomy and show how it relates to situations that you may encounter. For example, Chapter 18 begins with an account of a spinal injury; the text shows how the anatomy of the spinal cord and spinal nerves accounts for the paralysis that resulted. Special triangular icons identify these case studies and references to them throughout the text.

SUMMARY

The summary at the end of each chapter gives you an overview of the entire chapter with brief, powerful sentences keyed to pages in the text. Many of these sentences are taken from boldfaced topic sentences in the text.

STUDY QUESTIONS

Study questions ask you to do what the objectives at the beginning of each chapter said you should be able to do at the end of the chapter. Each question refers to an objective. Your instructor will give you the answers. Tests may be based on these questions, or perhaps they will be graded as homework. Other instructors may leave the study questions up to you. **Ask your instructor.**

TEN-POINTER QUESTIONS

At the University of Georgia, our course helps students develop a feel for anatomy by assigning short, investigative problems during the week before a quiz. Ten-pointers are very popular because students can write answers ahead of time and have the instructor critique them. These questions are worth 10 out of 30 points on our weekly quizzes. Your instructor has a supply of 10-pointers; ask for them.

GLOSSARY

Definitions of unfamiliar terms can be found in the glossary at the back of the text.

INDEX

Looking for a topic in the text? Find its page number in the index.

YOUR COMMENTS

Because you are using this text, I would like to know how to improve later editions for other students like yourself. All comments and suggestions are welcome. Reach me at:

Department of Cellular Biology
University of Georgia
Athens, GA 30602
(706) 542-3310
fax (706) 542-4271
e-mail: lindsay @ zookeeper.zoo.uga.edu

David T. Lindsay

CONTENTS IN BRIEF

CONTENTS

Functional
Human Anatomy

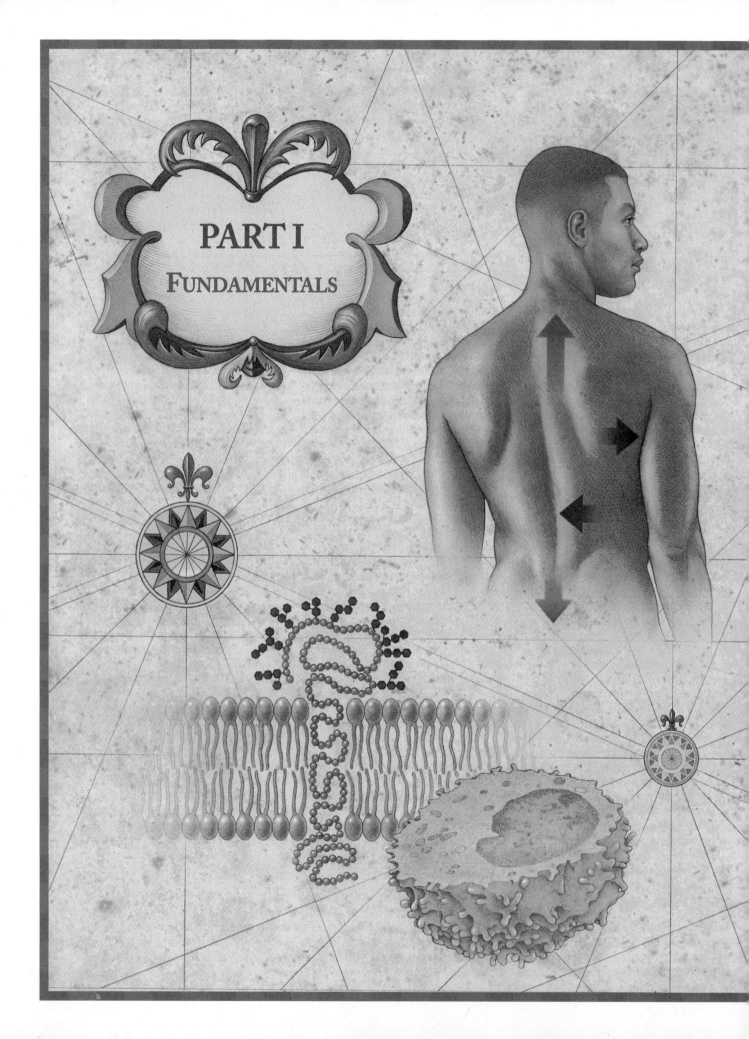

PART I

FUNDAMENTALS

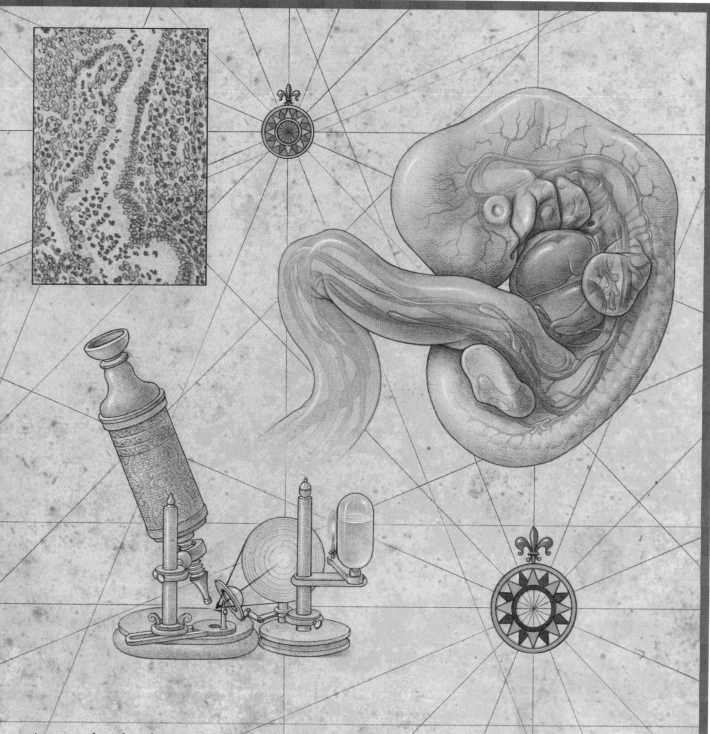

Anatomy describes the parts of the body and how they go together. The word *anatomy*, from the Greek, means "to cut or dissect" to reveal these parts. It takes more than a sharp scalpel to study anatomy, however. This section introduces some of the fundamental concepts that organize the seemingly endless complexity of the human body. First of all, to get an overall picture of human anatomy you need to identify individual organs by their anatomy and the functions they perform, as well as by their location in the body. **Chapter 1,** Anatomical Basics, deals with such matters and presents several unifying principles to help you organize the great variety of structures you will find. **Chapter 2,** Cell Basics, discusses the properties of cells on which organs depend. It will help you identify cells and organelles by their appearance and functions. Eventually, with that information, you will be able to visualize the internal structures of organs that, for example, enable muscles to contract and sweat glands to secrete. **Chapter 3,** Tissues, presents the four fundamental tissues that make up the organs of the body and explains their functions. Finally, **Chapter 4,** Developmental Basics, outlines the processes and events that convert fertilized eggs into humans, some of whom study anatomy.

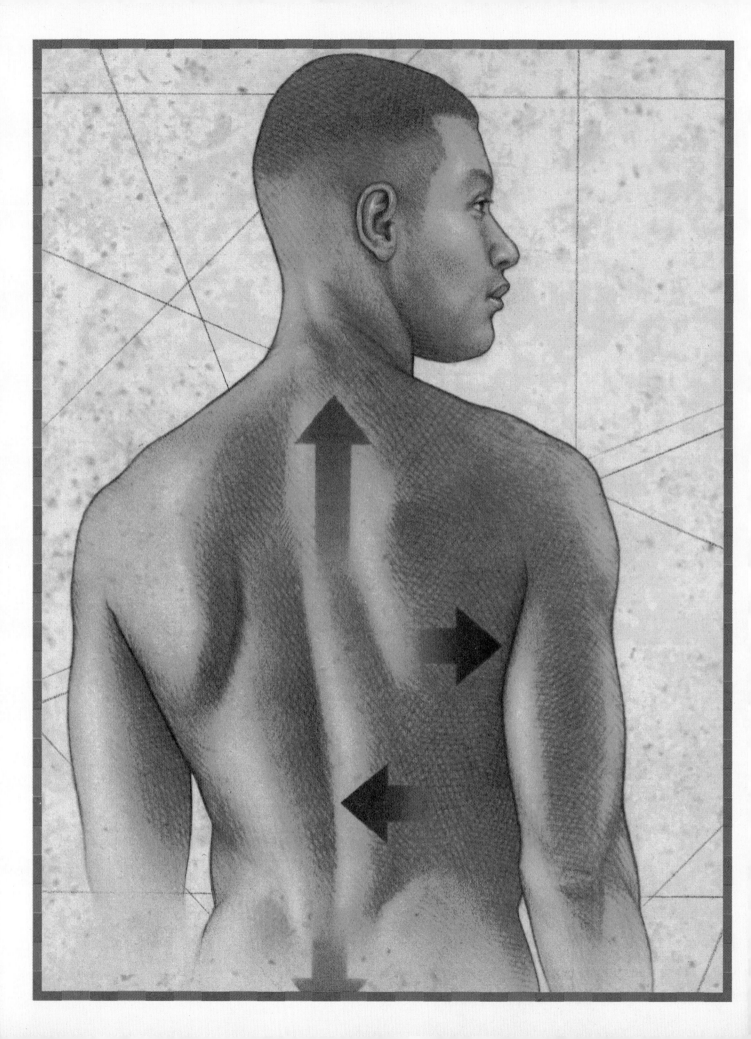

CHAPTER 1

Anatomical Basics

OBJECTIVES

After studying this chapter, you should be able to do the following:

1. Identify the major regions of the body.
2. Name the different levels of organization in the body, and give an example of each.
3. Name the body's organ systems, and identify their functions.
4. Identify the internal cavities of the body.
5. Describe the membranes that line body cavities.
6. Name several structural units that carry out the functions of different organs.
7. Explain how the surface-volume rule favors small structures.
8. Give an example of a distributive tree, and explain how such trees function.
9. Describe an example of negative feedback, and explain how it contributes to homeostasis.
10. Describe the features of the human body plan.
11. Recognize the anatomical position, and use positional terms to describe the location of body parts.
12. Identify sagittal, frontal, and transverse sections of the body.
13. Describe some aspects of internal structure revealed by medical imaging techniques.

VOCABULARY TOOLBOX

The following terms refer to the body (*L*, Latin; *G*, Greek):

anatomy	ana, *G* up, throughout; tom-, *G* cut, to cut up, dissect.	**dorsal**	dors-, *G* the back.
		somatic	soma-, *G* the body.
caudal	cauda, *L* the tail.	**ventral**	ventr-, *L* the underside, the belly.
cephalic	cephal-, *G* the head.	**visceral**	viscer-, *G* the organs of the body cavity.

The following prefixes indicate positions and directions:

ante-	rior, *L* before.	**inter-**	nal, *L* among, between.
extern-	al, *L* outside, outer.	**post-**	erior, *L* after.
infer-	ior, *L* low, underneath.	**super-**	ior, *L* above.

RELATED TOPICS

- **Coelome, Chapter 4**
- **Female reproductive system, Chapter 26**

- **Neural development, Chapter 18**
- **Organelles, Chapter 2**

PERSPECTIVE

The health sciences begin with anatomy. You must know the structure of the human body to understand how the body works and how its disorders are treated. Your overall objective in this text is to learn your way around and through the human body. When you have finished your work, you will be able to describe the body's organ systems and most of its organs. Then you will begin to understand how disease can interfere with the body and how treatment can restore its functions. Beyond these practicalities, you will begin to see that the structure that carries us through life is cleverly built.

Anatomy is the science that studies the structure of the body. The word *anatomy* comes from a Greek word that means to cut up or to dissect. You will learn the structure of the kidney, for example, by studying what the science of anatomy has disclosed about this organ.

Anatomy organizes the body according to regions, levels, and organ systems. Physicians emphasize **regional anatomy** because patients present symptoms in various regions of the body. Pain in the abdomen may indicate intestinal disorder, whereas a head cold implies that the nasal tissues are inflamed. The causes of these conditions may lie at other levels of organization, however. Thus it also is necessary to understand how the various cellular, tissue, and organ **levels of organization** work together to make the body function. Abdominal pain may be traced to damaged cells that line the intestine, and a viral infection in the cells that line the nasal cavity is probably causing the cold. **Systemic anatomy** organizes the body into organ systems. The digestive system and the respiratory system are the sites of the abdominal pain and the cold, respectively. All three approaches attempt to understand the structure of the body and how its different regions, levels, and systems work together as a whole organism. This text emphasizes systemic anatomy, but it uses all three approaches to learning your way through the body.

REGIONS OF THE BODY

Clinicians diagnose a patient's condition by examining the various regions of the body where symptoms occur. Each region usually presents several levels of organization and organ systems for examination. Do you remember your last physical exam? Your physician was watching your facial expressions and eye movements for signs of abnormal cranial nerve function. Saying, "Ah," told your doctor whether or not the cranial nerves of your pharynx and tongue were working normally, and the lining of your mouth also disclosed signs of disease. She also could draw conclusions about the condition of your heart from the carotid arteries pulsing in your neck. By palpating (PAL-pate-ing; examining by touch) the lymph nodes of the neck, the physician learned from their size whether or not your immune system was combating an infection.

Identify the principal regions of the body in Figure 1.1, and then inspect the organ system diagrams shown later in Figures 1.3 through 1.14 to see what organ systems each region contains.

The body consists of head, neck, trunk, and the upper and lower extremities. The head (cephalic region) shelters the brain in the cranium (the portion of the

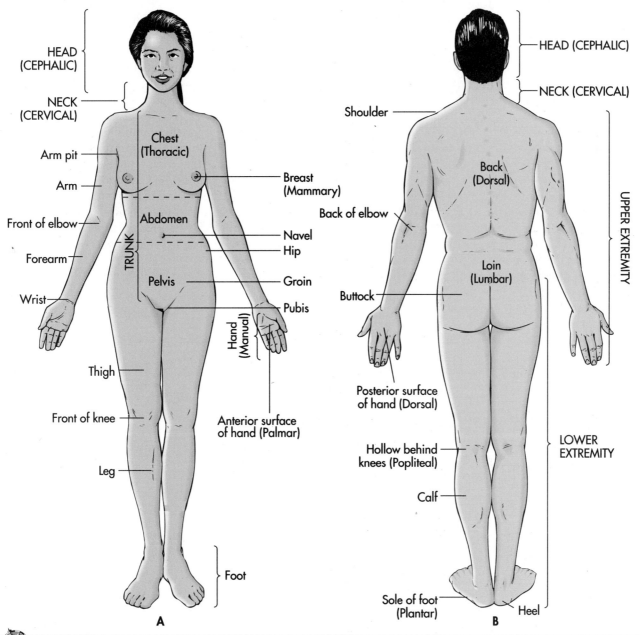

Figure 1.1

Regions of the human body. **A**, Female, anterior. **B**, Male, posterior.

skull surrounding the brain), and the face (facial region) houses the eyes, ears, and nose. The nose opens to the respiratory system, and the mouth is the entrance to the digestive system. The neck, or cervical region, joins the head to the trunk, which contains the **visceral** organs and anchors the extremities.

The trunk is divided into several regions. The thorax (chest) is the portion supported by the rib cage. The abdomen, marked by the navel (umbilical region), lies between the thorax and the pelvic region. In the rear the back extends from the neck to the sacral region, and the lumbar region lies between the thorax and the sacral region.

The shoulders join the upper extremities to the thorax. Marked in front by the muscles of the pectoral region and by the scapular region of the back, the shoulders (acromial region) connect the arm (brachium) to the body. The armpit (axilla) lies beneath this junction, and the elbow (cubitus) links the arm to the forearm (antebrachium). The wrist (carpus) allows the hand (manus) and fingers (phalanges or digits) considerable motion. The palm of the hand is the palmar region, and the back of the hand is the dorsum.

Regions of the lower extremities correspond to those of the upper extremities. The pelvic region joins the thigh (femoral region) to the trunk at the hip (coxa). The inguinal and pubic regions mark this junction in front and the buttocks (gluteal region) behind. The perineal region marks the location of the anus and external genitalia. The thigh articulates with the leg (crural region) at the knee (genu), and the ankle (tarsus) links the foot (pes) to the leg. The sole of the foot is the plantar region, and the upper surface of the foot is the dorsum. The toes are phalanges or digits.

- Name regions within the trunk and extremities.
- What are the anatomical terms for hip, arm, leg, foot, and fingers?

LEVELS OF STRUCTURE

Anatomists organize the body into a hierarchy of levels. Anatomy emerges as each level adds structures and functions to the levels below. **Molecules and atoms** represent the lowest rank (Figure 1.2). Water, ions, and many other molecules participate in metabolic reactions inside cells. Larger macromolecules, represented by enzymes and DNA molecules in the figure, catalyze these metabolic reactions and store genes that specify the structure of the enzymes. **Organelles** (small organs) are cytoplasmic structures, such as the nucleus and plasma membrane shown in the figure, that perform specific services for cells (see Organelles, Chapter 2). The nucleus stores genes in DNA molecules, and the plasma membrane governs what substances enter and leave the cell.

Cells are the smallest units of life capable of reproduction and metabolism. Cells are invaluable because they carry out specific functions. Figure 1.2 illustrates muscle fibers whose contractions enable the heart to pump blood, but cells specialize in hundreds of other activities. For example, tough, flat cells cover the surface of your skin with an impervious protective coating and cone cells detect light in the retinas of your eyes.

Tissues are groups of similar cells that perform particular functions. They specialize much like cells. A lone muscle fiber would not pump much blood, but with thousands of fibers organized as a tissue in the wall of the heart, it is easy to see how blood can flow. This tissue, the myocardium (myo- *G*, muscle), pumps blood because it combines the small contractions of many fibers into large forces that propel blood through the heart. Only at the organ level, however, does the ability to pump truly emerge.

Organs carry out unique functions by combining two or more tissues. The ability of the heart to pump blood emerges from the combined actions of several tissues. First of all, the heart pumps because its chambers have walls that press on the blood when the walls contract. Valves direct blood through the chambers. Each chamber has a thin internal lining that prevents blood from clotting on the walls as it rushes past, and a slippery exterior covering allows the heart to move in the chest. Muscle fibers, myocardium, and

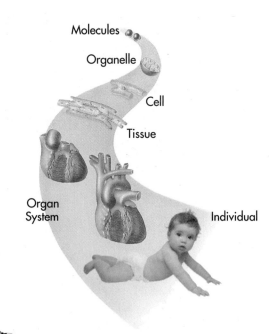

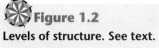
Figure 1.2
Levels of structure. See text.

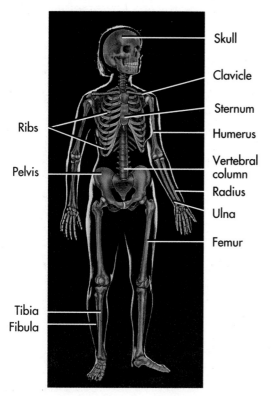

Figure 1.3
Skeletal system. Some 206 bones support the body.

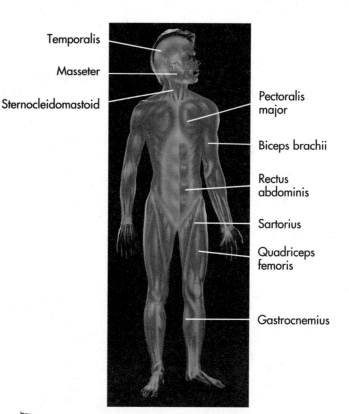

Figure 1.4
Muscular system. Approximately 600 muscles move the skeleton.

the heart itself have even greater significance at the organ system level.

Organ systems are associations among organs that serve broad functions. The heart itself is part of the cardiovascular system, which delivers blood through vessels to virtually every tissue in the body. This communication carries oxygen from the lungs to the tissues, nutrients from the digestive system, and hormones that regulate tissues.

Finally, from such integral actions emerge **individuals,** at the top of the hierarchy. Individuals, like each other level of the hierarchy, are greater than the sum of their parts. As you go up the ranks of the hierarchy, properties emerge from interactions among the ranks below, properties that no cell, tissue, organ, or organ system alone can claim.

This text moves freely up and down the organizational hierarchy to describe anatomy and explain functions. You should be able to do the same. Although abdominal pain presented itself at the organ system level, you should learn to visualize the cause of the pain at lower levels. When shown a low-level function, you also should be able to follow the consequences that emerge at higher levels. Like the properties of organisms, anatomy requires you to understand all levels of organization and how they interact.

- Which are the highest and lowest levels of structure?
- Why are lower levels important in anatomy?
- State the definitions for cell, tissue, and organ.

ORGAN SYSTEMS

The body depends on different organ systems. Figures 1.3 to 1.14 display the organ systems outlined below. The skeleton, muscles, and skin support, move, and shape the body. Other organ systems integrate the functions of the body. The cardiovascular and lymphatic systems communicate with other systems through vessels and fluids. The nervous and endocrine systems conduct stimuli and secrete hormones. Some organ systems are concerned with maintenance. The digestive and respiratory systems deliver nutrients for the tissues to the blood, and the excretory system removes wastes. The reproductive system produces new individuals.

SUPPORT, MOTION, AND SHAPE
SKELETAL SYSTEM

The bones support, protect, and shape the head, trunk, and extremities, and they provide rigid structures for muscles to pull upon to move the body (Figure 1.3). Joints between bones make the skeleton flexible or rigid as needed.

MUSCULAR SYSTEM

Three different types of muscles provide for movement. The skeletal muscles move the skeleton because most of them are attached to bones (Figure 1.4). The cardiac muscle of the heart pumps blood, and visceral muscles twist and squeeze the digestive tract to propel food and fecal matter. Visceral muscles constrict the walls of arteries to control blood pressure, and they also erect hairs and "goose bumps" in the skin.

INTEGUMENTARY SYSTEM

The skin is the external surface of the body (Figure 1.5). It blocks infectious organisms and excessive wa-

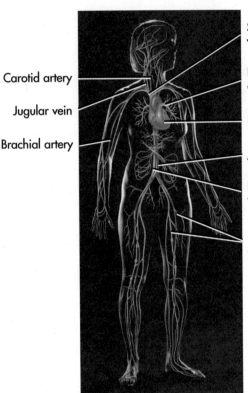

Superior
vena cava

Carotid artery

Pulmonary
artery

Jugular vein

Heart

Brachial artery

Aorta

Inferior
vena cava

Femoral artery
and vein

Figure 1.6
Cardiovascular system. The heart and vessels distribute blood to organs throughout the body.

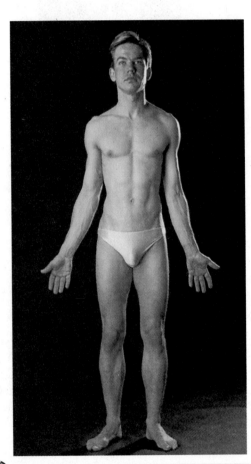

Figure 1.5
Integumentary system. The skin protects, regulates, and communicates.

CIRCULATORY SYSTEMS AND VASCULAR INTEGRATION
CARDIOVASCULAR SYSTEM

The cardiovascular system is a fundamental fluid-bearing communication network that links virtually all other organ systems throughout the body. The heart, vessels, and blood distribute oxygen from the lungs and nutrients from the digestive system (Figure 1.6). The cardiovascular system also distributes waste products to the kidneys and delivers hormones throughout the body.

LYMPHATIC AND IMMUNE SYSTEMS

The vessels of the lymphatic system return tissue fluids (lymph) to the cardiovascular system, and the immune system protects against foreign cells (Figure 1.7). The principal organs are the thymus, spleen, and lymph nodes. Lymphocytes, the principal cells derived from these organs, patrol nearly all tissues, recognizing and destroying materials different from the body's own.

NEURAL AND HORMONAL INTEGRATION
NERVOUS SYSTEM

The brain, spinal cord, and peripheral nerves conduct impulses that coordinate most systems of the body (Figure 1.8). The brain and spinal cord occupy the skull and vertebral column, and the peripheral

ter loss and regulates body temperature through perspiration. Its sensory receptors are sensitive to stimuli that the brain interprets as sensations of touch, temperature, and pain. Hairs, nails, breasts, and sweat glands are specialized structures of the skin. Skin also can disclose to other individuals information about the owner's age, sex, and physical or emotional condition.

nerves extend laterally into the trunk and limbs to conduct impulses that register touch, taste, pain, temperature, vision, sound, and muscular motion. The brain processes these streams of impulses into more streams that reveal themselves as thoughts or flow out peripheral nerves, causing muscles to contract and glands to secrete.

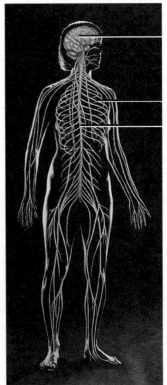

Figure 1.7
Lymphatic and immune systems. The immune system removes foreign substances in lymph fluid and lymph organs.

Right lymphatic duct

Thoracic duct

Thymus gland

Spleen

Lymph node

Figure 1.8
Nervous system. Information flowing through the brain, spinal cord, and peripheral nerves coordinates the body.

Brain

Peripheral nerves

Spinal cord

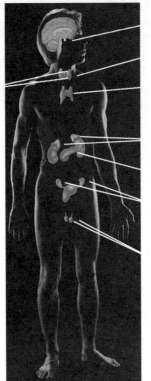

Parathyroid glands (behind thyroid)

Pituitary

Thyroid gland

Thymus gland

Adrenal glands

Pancreas

Ovaries (in females)

Testes (in males)

ENDOCRINE SYSTEM

The pituitary, pancreas, and adrenals, gonads, and other endocrine glands regulate internal processes by secreting hormones through the cardiovascular system to various target tissues. The endocrine system (Figure 1.9) is as broadly distributed as the nervous and immune systems; cells of the gut, blood vessel walls, and certain nerve endings also freely secrete hormones.

Figure 1.9
Endocrine system. Ductless glands regulate body functions by secreting hormones.

BODY MAINTENANCE
DIGESTIVE SYSTEM

The gut is a tube that opens at the mouth, passes through the thoracic and abdominopelvic cavities, and exits at the anus (Figure 1.10). Along its length, special organs such as the stomach hold food for the intestine to digest and absorb into the blood. Glands such as the salivaries, liver, and pancreas assist digestion.

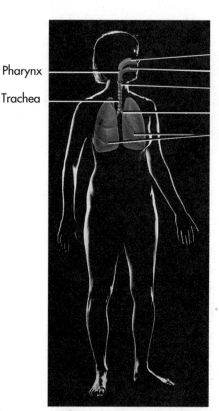

Figure 1.11
Respiratory system. The lungs absorb oxygen and release carbon dioxide.

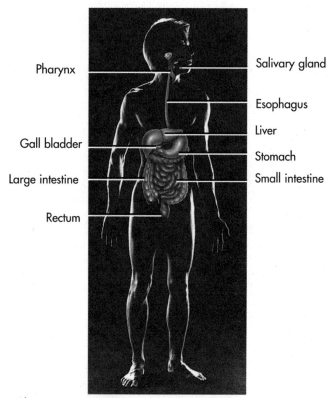

Figure 1.10

Digestive system. Essentially a tube with associated glands, the digestive system supplies nutrients to the body.

RESPIRATORY SYSTEM

The nasal passages of the head, larynx, and trachea deliver air to the lungs (Figure 1.11). The lungs deliver oxygen to the blood for cellular respiration in the tissues and remove waste carbon dioxide. Air flowing through the larynx enables us to speak.

URINARY SYSTEM

The urinary system regulates the composition of blood and other bodily fluids by filtering the blood and excreting the waste substances as urine (Figure 1.12). The kidneys deliver urine via ureters to the bladder for discharge outside. The cardiovascular system delivers blood to be filtered, hormones to regulate excretion, and of course nutrition.

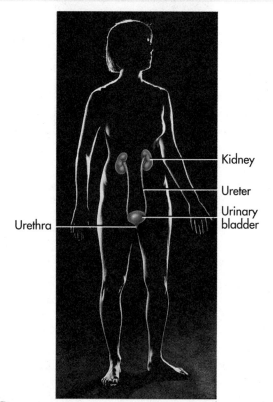

Figure 1.12
Urinary system. The kidneys regulate the composition of body fluids by excreting excess substances.

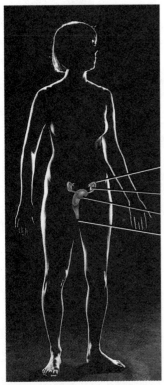

Figure 1.13

Female reproductive system. Ovaries produce eggs, and the uterus and placenta nourish the fetus.

Uterine tube
Ovary
Uterus
Vagina

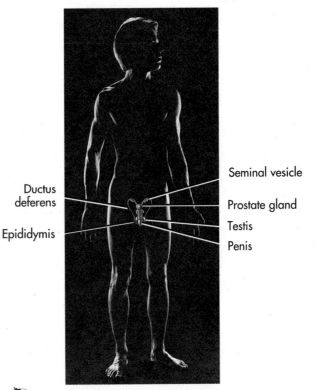

Figure 1.14

Male reproductive system. Testes and penis produce and deliver sperm.

Ductus deferens
Epididymis
Seminal vesicle
Prostate gland
Testis
Penis

REPRODUCTIVE SYSTEM

Both male and female reproductive systems depend on the cardiovascular system for nutrients, but they depend on hormones to control egg and sperm production. The gonads produce eggs and sperm, and the genitalia enable them to unite at mating (Figures 1.13 and 1.14). After conception the female reproductive tract nourishes and delivers the fetus.

- Which organ system or systems give(s) the body mechanical support?
- Absorbs nutrients?
- Protects the body?
- Coordinates its motions and other actions?

BODY CAVITIES AND MEMBRANES

The body contains a dorsal and a ventral cavity. The dorsal cavity, shown in Figure 1.15A, contains the brain and spinal cord in two subdivisions of the cavity. The **cranial cavity** houses the brain within the skull, and the spinal cord occupies the **vertebral canal** within the vertebral column. The heart, lungs, and digestive organs reside in the ventral cavity.

The ventral cavity consists of the thoracic and abdominopelvic cavities seen in Figures 1.15B and

1.15C. The **thoracic cavity** (thore-ASS-ik) occupies the chest and is separated from the **abdominopelvic cavity** (ab-DAH-min-oh-PEL-vik) by a muscular wall, the **diaphragm** (DIE-a-fram). The thoracic cavity contains the lungs, heart, and a central partition called the **mediastinum** (MEE-dee-ah-STY-num). The abdominopelvic cavity is further subdivided into the **abdominal cavity,** which extends from the diaphragm downward through the abdomen to the brim of the pelvis and the **pelvic cavity,** which lies entirely within the bony bowl of the pelvis. Unlike the diaphragm, no partition separates the abdominal and pelvic cavities; they are continuous, but the rim of the pelvis marks the boundary between them. The abdominopelvic cavity contains the digestive organs and the spleen, as well as the ovaries and uterus in females.

Thoracic and abdominopelvic organs fill their respective cavities except for a narrow space around each organ and the walls of the cavities that allows the organs freedom to move. A thin **serous membrane** (SEER-us) lines this space and adheres tightly to the walls and organs; it secretes watery serous fluid that lubricates the slick, slippery surfaces.

The serous membrane encloses a space within the abdominopelvic cavity known as the **peritoneal cavity,** and the membrane is accordingly known also as the **peritoneum.** The portion that lines the wall of the abdominopelvic cavity is called the **parietal peri-**

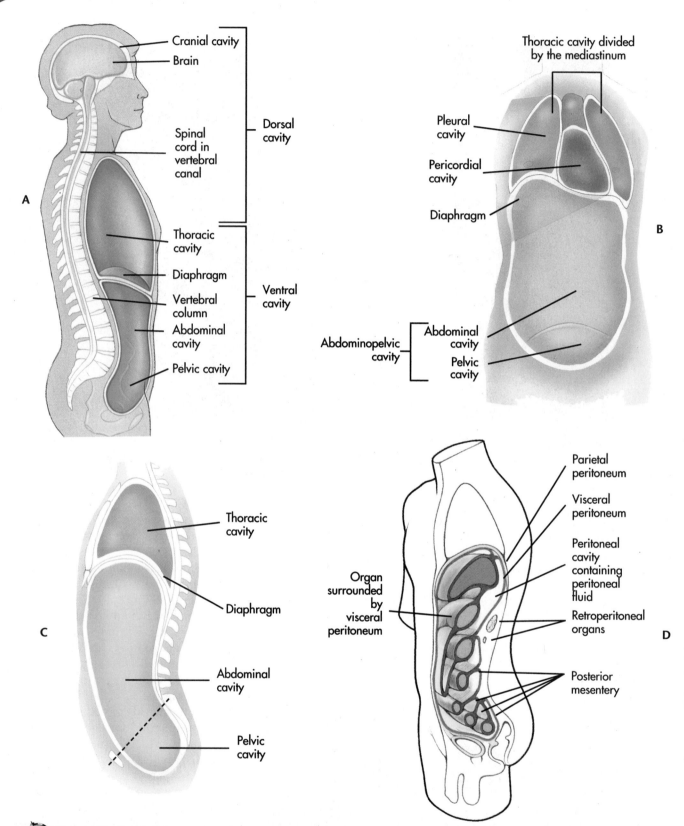

Figure 1.15

The body cavities and serous membranes. **A,** Midsagittal section through the dorsal (*yellow*) and ventral (*blue*) cavities. The ventral cavity is subdivided into thoracic (pleural and pericardial) and abdominopelvic (abdominal and pelvic) cavities. **B,** Frontal view of **A. C,** Lateral view of **A. D,** The peritoneum encloses the peritoneal cavity within the abdominopelvic cavity in this sagittal section. Parietal peritoneum covers the abdominal walls, and the visceral peritoneum covers the organs and posterior mesentery, which suspends these organs from the posterior wall of the cavity. Some organs such as the kidneys are behind the peritoneum (retroperitoneal) in the abdominal wall.

toneum; it is continuous with the **visceral peritoneum** that adheres to the surfaces of the visceral organs. Most of the organs are between the visceral peritoneum and the wall of the abdominopelvic cavity. Other organs such as the kidneys are located in the body wall, behind the parietal peritoneum. They are **retroperitoneal** (retro, *L,* behind). The peritoneal cavity has no opening, except for two potential openings through the female reproductive tract (see Chapter 26, Reproductive System).

Cavities and serous membranes also surround thoracic organs. **Pleural cavities** (PLOOR-al) surround the lungs and the **pericardial cavity** envelopes the heart. Each is a separate space, without openings. **Parietal pleura** lines the walls of the thoracic cavity and the mediastinum, while **visceral pleura** covers the lungs themselves. **Visceral pericardium** covers the heart, and **parietal pericardium** lines the rest of the pericardial cavity. Pleura and pericardium are serous membranes.

Mesenteries are curtains of peritoneum that suspend organs in the abdominopelvic cavity. In Figure 1.15D a section has been sliced through the body to reveal the intestine and the **posterior mesentery** that attaches it to the posterior wall. There is only one mesentery and intestine; the section has cut them several times as they fold back and forth across the abdominal cavity. Careful inspection of the mesentery shows that it is a sandwich composed of two sheets of visceral peritoneum, with vessels and nerves between that deliver to the abdominal organs.

Other internal cavities are lined with **mucous membranes,** known collectively as **mucosa,** that secrete various fluids, including mucus. Oral mucosa leads to the lining of the digestive tube, and nasal mucosa to the trachea and lungs. The bladder and reproductive organs also have mucosal linings. Other cavities, such as the orbital cavities that house the eyes and synovial cavities of the knee and other joints, have nonmucosal linings.

- What are the divisions of the ventral cavity?
- Which of these cavities contains the peritoneal cavity?
- Do any organs occupy the peritoneal and pericardial cavities?

ANATOMICAL PRINCIPLES

FORM AND FUNCTION
You can often tell what an object does by its appearance. The heft and shape of bricks make them ideal for walls—add some mortar and they fit solidly together. Anatomy uses the same kind of thinking. The skeleton supports and protects the body because its bones are rigid and strong. Skeletal muscles enable the body to move because they pull on the skeleton.

You would not expect support and motion from a soft, immobile organ such as the brain, any more than you would expect a wall to be made of balloons.

Form and function are different aspects of the same processes. Form represents the slower processes in which large structures persist for days or years, but we usually interpret faster interactions as functions. Within minutes the glucose molecules you absorbed from your breakfast were converted into carbon dioxide, water, and other molecules. You do not see these substances or the reactions themselves ("Hey, Mom, look at my glucose!"), but you do see structures, which emerge from such interactions, that are large and stable enough to be recognized ("Hey, Mom, look at me!").

Anatomy describes larger, relatively long-lived structures of the body that can be seen directly or with microscopes, whereas molecular biology and physiology are concerned with the more rapid processes that maintain and renew more permanent structures. Accordingly, anatomists attempt to describe the regions, levels, and systems of the body, but they leave the underlying, short-term mechanisms and functions to cell biologists and physiologists.

- Why is the flow of blood through arteries considered a function rather than a structure, and why are the extremities considered structures instead of functions?

FUNCTIONAL UNITS
Organs consist of many small functional units, like pieces in a jigsaw puzzle, that collectively perform each organ's functions. To understand anatomy, you must learn the structure and function of each type of unit (Figure 1.16). For example, lungs contain about 300 million sacs called alveoli, which deliver oxygen from the air to the blood and receive carbon dioxide in exchange (Figure 1.16A). At least 20 million intestinal villi (rhymes with "willy") absorb nutrients in the intestines (Figure 1.16B). Thousands of muscle fibers enable individual skeletal muscles to contract (Figure 1.16C). The nervous system contains billions of neurons that conduct impulses throughout the body and store information in the brain. The cardiovascular system has perhaps the most units (Figure 1.16D), with some 150 km of capillaries per kg of tissue (200 miles per lb).

EXCHANGE SURFACES
Most functional units exchange molecules across a surface, from one side of a surface to the other, absorbing some and giving off others, as you can see in Figure 1.16. In the lungs, air travels through branching passages to alveoli, where oxygen diffuses from the air across the wall of each alveolus into capillaries, while carbon dioxide returns from the blood to the alveolar air (Figure 1.16A). A similar process

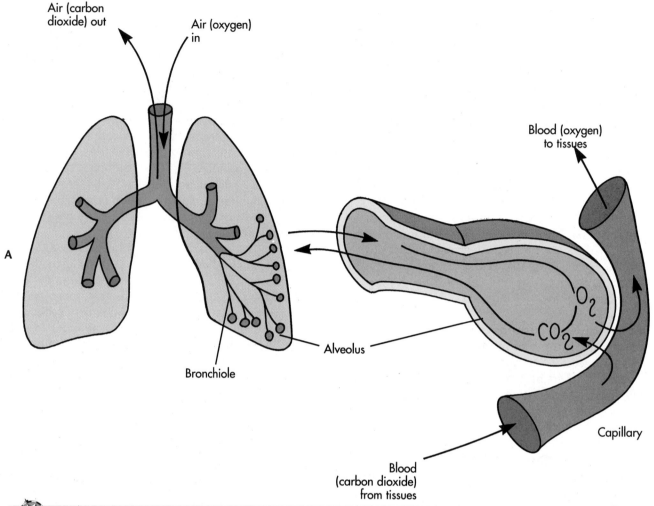

Air (carbon dioxide) out

Air (oxygen) in

Blood (oxygen) to tissues

Alveolus

Bronchiole

A

O_2

CO_2

Capillary

Blood (carbon dioxide) from tissues

✸ Figure 1.16

A, Alveoli are the functional units of the lungs. Air enters the lungs through the trachea and bronchi, and bronchioles distribute the air to the alveoli. As the air circulates inside, it gives oxygen to the blood and receives carbon dioxide from it. The next breath flushes out the carbon dioxide–laden air and brings in fresh, oxygenated air. Meanwhile, blood charged with oxygen exits the lungs and transports the oxygen to the body tissues.

occurs in villi (Figure 1.16B). These intestinal structures release digestive enzymes into the gut and absorb nutrients from digested food. Streams of impulses that stimulate muscles to contract cross from a neuron (Figure 1.16C) to the muscle over connections called motor endplates. Virtually all organs employ capillaries to provide nutrients and waste removal (Figure 1.16D). Consequently, organs can be thought of as internal surfaces that secrete and absorb essential molecules.

SURFACE-VOLUME RULE

Small units are more efficient than large ones, because more surface area can be packed into a smaller overall space (Figure 1.17A). For example, the surface area of all the alveoli (air chambers) in one lung adds up to about 32.5 m^2 (350 ft^2). If this area

were combined into one huge alveolus, each lung would be 1.5 m (5 ft) in diameter, a very improbable size. Indeed it cannot exist, for reasons that you shall see.

The surface-volume rule forces structures to miniaturize. The larger an alveolus or any functional unit becomes, the smaller its surface area gets relative to the volume it encloses. To be sure, both surface and volume increase, but volume increases more rapidly. Mathematically speaking, the surface area increases as the square of the radius (r^2), while the volume increases much faster as the cube (r^3).

Organs observe the surface-volume rule by developing many small, efficient units rather than a few large, inefficient ones. Smaller structures are more efficient because more molecules can cross the relatively larger surface. The large, inefficient hypothetical lungs in

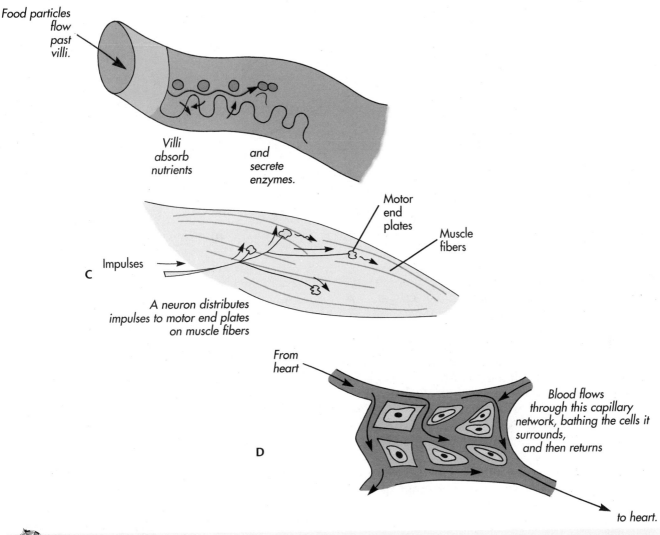

Food particles flow past villi.

Villi absorb nutrients   and secrete enzymes.

Motor end plates

Muscle fibers

Impulses

C

A neuron distributes impulses to motor end plates on muscle fibers

From heart

Blood flows through this capillary network, bathing the cells it surrounds, and then returns

D

to heart.

Figure 1.16, cont'd

B, Intestinal villi absorb nutrient molecules digested from food particles. They release mucus and enzymes into the gut contents. **C,** The branches of a nerve cell distribute stimuli to muscle fibers of a skeletal muscle across motor endplates. **D,** Capillaries branch among the cells of a tissue, delivering nutrients and oxygen and removing wastes.

Figure 1.17A become more and more effective as their walls fold and fill the internal space with more surface in Figures 1.17B and 1.17C. Because the organ now contains three branches, its wall area has increased approximately three-fold without increasing the volume it occupies. How large an organ actually becomes depends on the volume or weight of the body it serves; the internal surface area must be large enough to sustain life through the particular process it performs. At 32.5 m² (350 ft²), each lung satisfies the adult body's respiratory demands, but when asthma, emphysema, or other diseases diminish this capacity, the individual's physical ability declines accordingly.

DISTRIBUTIVE TREES

Branching passages deliver materials to the functional units of organs. These pathways are known as distributive trees. Air reaches the alveoli of the lungs through the trachea, which branches into numerous narrow bronchioles that lead ultimately to each alveolus (see Figure 1.16A). This tree distributes air from a small entrance in the nose to the vast interior surfaces of the lungs. The rest of this figure shows distributive trees that distribute stimuli to skeletal muscles (see Figure 1.16C) and blood to cells of a tissue (see Figure 1.16D).

HOMEOSTASIS AND NEGATIVE FEEDBACK

Your body is not the same one it was a month ago, yet it is the same. A new epidermis has silently replaced the cells that covered your skin last month. A fading suntan and longer hair may be your only clue to its newness. Today your temperature is probably 37.5 degrees Celsius (98.6 degrees Fahrenheit) and

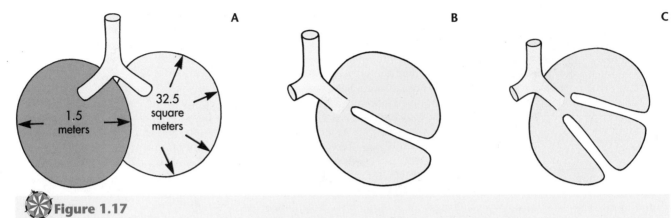

Figure 1.17

Smaller units are more efficient. **A,** The surface area of a lung with one large alveolus is small compared to its volume. **B,** Surface increases, but not volume, as folds fill the space. **C,** Surface area has nearly tripled.

your heart beats about 72 times per minute. These numbers most likely will fall within the same narrow range tomorrow and next week.

The body tends to maintain constant conditions internally despite external change, and this constancy is called **homeostasis. Negative feedback** processes are responsible for homeostasis. Negative feedback balances the size of the pupil of the eye (Figure 1.18), for example, by opposing increases and decreases of light. Strong light on the retina stimulates muscles of the iris to constrict the pupil and admit less light, and dim light relaxes these muscles, which opens the pupil wider and allows more light to enter. Consequently a relatively constant amount of light

flows into the eye because the intensity of light regulates how much light enters. The system is self-regulating. The process is considered negative because it opposes imbalance. By contrast, positive feedback would imbalance light still further by opening the pupil wider in bright light and constricting it in dim.

- What requirement dictates that the lungs contain many small alveoli instead of a few large ones?
- Name an exchange surface.
- Name a distributive tree.
- How does the pupillary opening react to dim light? To bright light?
- What condition and what process do these reactions illustrate?

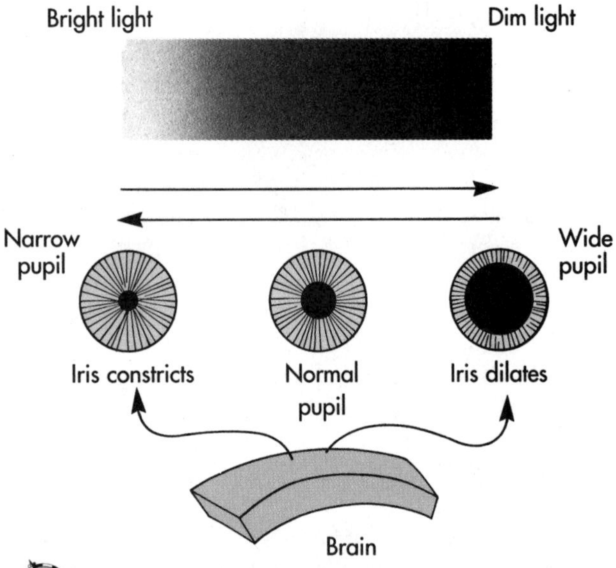

HUMAN BODY PLAN

Everyone recognizes that engines, tires, and steering wheels are fundamental structures that enable automobiles of all types to run. The human body also shares certain fundamental, anatomical features with its relatives that possess backbones—fish, reptiles, amphibians, birds, and other mammals. Knowing these basics is useful when learning the anatomy of the organ systems and regions of the body.

VERTEBRAL COLUMN AND NOTOCHORD

Our vertebrate ancestors have bequeathed us a notochord (NO-toe-kord) **and vertebral column** (VER-teh-bral) seen in Figure 1.19. The notochord is a semirigid rod that appears in the embryo and is replaced during development by the vertebral column, which supports the trunk and marks the central axis of the body.

TWO BODY CAVITIES

The vertebral column supports two great body cavities. The **dorsal cavity** contains the delicate brain inside the skull and the spinal cord in the verte-

Bright light   Dim light

Narrow pupil   Wide pupil

Iris constricts   Normal pupil   Iris dilates

Brain

Figure 1.18

Pupillary light reflex. The eye attempts to admit a constant amount of light. Brighter light causes the brain to constrict the iris and dimmer light dilates it, continually adjusting the pupil to the intensity of light that enters.

BILATERAL SYMMETRY

The human body plan has right and left sides. Left and right are not duplicates, however; they are mirror images of each other. No one understands fully how this useful property called **bilateral symmetry** comes to be, although its advantages are evident daily. The digestive organs are not bilaterally symmetric. Nevertheless, there is a form of symmetry in which organs are occasionally reversed, with the liver on the left and stomach at right, a condition called situs inversus.

AXIAL SEGMENTATION

The body is segmented. Segments are most obvious in the vertebral column and ribs. The spinal cord also is segmented; it sends pairs of spinal nerves left and right to the skin and skeletal muscles. Axial segmentation is obscured in the skull and extremities, and it is largely missing in the visceral organs.

TUBE-WITHIN-A-TUBE

The digestive system is a tube that occupies a greater tube—the trunk. The digestive tube lies within the ventral cavity. It opens at the mouth; passes through the neck, thoracic, and abdominal cavities; and exits through the anus in the floor of the pelvic cavity.

- Cite two features of the human body plan.
- Name a segmented body structure.
- Cite an example of bilateral symmetry.

ANATOMICAL POSITION

Figure 1.20 shows the body in **anatomical position,** the standard position anatomists use to describe the location of body regions and parts. The subject stands erect, facing the observer. The feet are flat on the floor, with the palms of the hands directed forward and the thumbs outward. A **central axis of symmetry** passes through the head and trunk. The subject's left and right sides appear reversed because the subject is facing you.

POSITIONAL TERMS

Positional terms locate body parts in respect to each other and in respect to the central axis. Follow the examples in Figure 1.20 and practice with study questions 14 and 15.

SUPERIOR/INFERIOR

These terms distinguish what is above from what is below. The head is **superior** to the neck, and the lower extremities are **inferior** to the upper extremities (see Vocabulary Toolbox). The hands are inferior

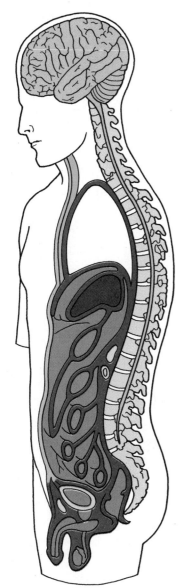

✹ Figure 1.19
Human body plan (see text).

bral column. The **ventral cavity** is the combined thoracic and abdominopelvic cavities. It is suspended from the vertebral column by the ribs and muscles of the body wall. The ventral cavity derives from the embryonic coelomic cavity (see Coelome, Chapter 4).

DORSAL HOLLOW NERVE TUBE

The brain and spinal cord develop from a hollow tube. Part of this **neural tube** (NOOR-al) enlarges and becomes the brain, whereas the spinal cord remains tubular. The space within the tube becomes the ventricles of the brain and the central canal of the spinal cord (see Neural Development, Chapter 18).

to the shoulders in the anatomical position, but when the upper extremities are raised above the head, the hands are superior.

ANTERIOR/POSTERIOR

The front of the body in Figure 1.20A is **anterior**—the face, palms of the hands, toes, knees, and genitalia can be seen from this view. The **posterior** side (Figure 1.20B) reveals the back of the head, hands, and trunk, along with the buttocks, and the heels. These terms are essentially synonymous with ventral and dorsal. In its strict sense, the word *ventral* refers to the ab-domen, but chest and chin are also considered ven-tral. The term *dorsal* refers to the back.

LEFT/RIGHT

Although these terms are obvious, remember that they refer to the *subject's* **left** and **right,** not the ob-server's. In anterior views, left and right are reversed because the subject is facing you. What is on the sub-ject's left is at your right as you view the figure. Poste-rior views do not reverse the sides, however, because you and the subject are facing the same direction; the subject's left is also at your left.

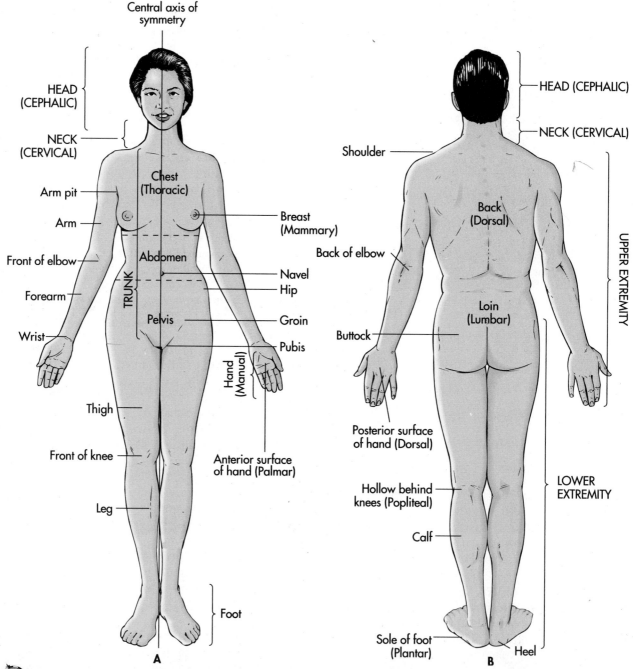

Figure 1.20

The anatomical position. **A,** Female, anterior. **B,** Male, posterior. Identify body parts and use positional terms to locate them.

MEDIAL/LATERAL

On either side of the central axis, structures closer to the axis such as the eyes and nipples are **medial.** **Lateral** structures such as shoulders and hips are farther toward the sides. **Median** structures such as the nose and breastbone are midway between the sides. The arms are lateral to the trunk, and a median structure such as the nose is medial to the eyes.

IPSILATERAL/CONTRALATERAL

These terms locate the same or opposite sides. Traditionally reserved for the nervous system, these terms are seldom used for other structures. Neurosurgeons and neuroanatomists say that each side of the brain controls **contralateral** muscles, meaning that each side controls the other side of the body. **Ipsilateral** structures are on the same side of the body. Each eye, for example, sends impulses ipsilaterally and contralaterally to the brain, that is, to both left and right sides.

PROXIMAL/DISTAL

These terms indicate what is near and far from some specified region, especially in the extremities. With respect to the trunk, the thigh is **proximal,** whereas the feet are **distal.** Similarly the knee is distal to the thigh and proximal to the leg, and the wrist is proximal to the hand. Outside the extremities, you can refer to the colon as a distal portion of the digestive tube, farther along the tube from the stomach than is the small intestine. Viewed from the stomach, however, the small intestine is proximal to the colon.

AXIAL/APPENDICULAR

The head, neck, and trunk are **axial** because the central axis of the body passes through them. The brain, spinal cord, and abdominal organs also are axial. Limbs are **appendicular;** they are joined to the body's sides as lateral appendages.

CEPHALIC/CAUDAL

These terms distinguish between head and tail and are similar to superior and inferior. The coccyx (KOKS-iks) is the remains of the fetal tail and marks the inferior end of the vertebral column. It is the most **caudal** landmark in the body. Even though the feet are inferior to the coccyx in the anatomical position, they are not more caudal because they are part of the lower extremities that join the trunk at the hips, which are cranial to the coccyx. The brain is strictly **cephalic** (se-FAL-ik), and the spinal cord extends from the brain caudally down the vertebral column.

INTERNAL/EXTERNAL

These words refer to structures inside and outside a given area. The heart and abdominal cavity are **internal,** and the skin and hair are **external.** These same terms also apply to individual body parts. The clear transparent cornea of the eyeball is external, and the iris and pupil are internal.

SUPERFICIAL/DEEP

This is another aspect of the difference between external and internal. The external surface of the skin is **superficial,** and muscles and bones are **deep** to the skin.

SOMATIC/VISCERAL

Somatic (so-MAT-ik) organs and tissues are associated with the skin and skeleton (bone and skeletal muscles, extremities, body wall), and the internal organs of the thoracic and abdominopelvic cavities (lungs, heart, liver, intestines, spleen) are considered **visceral** (VISS-er-al). These terms also make functional distinctions. Somatic organs are controlled voluntarily by the nervous system, but visceral organs are primarily under involuntary control.

PARIETAL/VISCERAL

These terms distinguish between the linings of body cavities. The lining of the body wall is **parietal** peritoneum, while the **visceral** peritoneum covers the visceral organs. This same distinction applies to the pleural and pericardial cavities. If you are ever diagnosed with peritonitis, you can be certain that both the parietal and visceral peritoneum are infected.

- How would a man sitting with his hands on his lap move his body into the anatomical position?
- What parts of the body are distal to the trunk? Superior to the shoulders? On the anterior surface? Medial to the hips? Caudal to the abdomen?

PLANES AND SECTIONS

Sections through the body reveal the structure of internal organs. Three mutually perpendicular sections—sagittal, frontal, and transverse—serve this purpose. **Sagittal sections** (G, arrow, SA-jit-tal) cut parallel to the central body axis and separate the body into left and right portions. Midsagittal (median) sections divide the body along the **median plane** into equivalent halves, as Figure 1.21A shows, and they expose such median organs as the liver and heart. Lateral organs such as the lungs and kidneys on either side of the median plane are exposed by sagittal sections (called parasagittal by some) that pass to one side of the median plane (Figure 1.21A). **Frontal** (or coronal) **sections** are perpendicular to sagittal sections; they separate anterior portions of the body from posterior. Frontal and sagittal sections of the body are cumbersome to use because they are so large, but transverse sections are smaller. Also known as **horizontal sections,** transverse sections cut perpendicular to the body axis and separate its superior regions

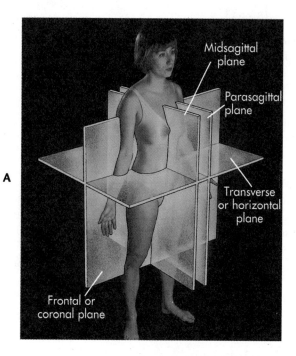

A

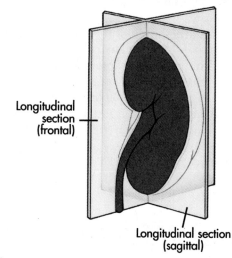

B

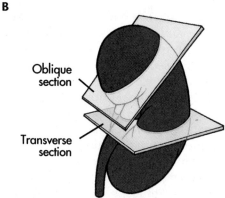

**Figure 1.21**

Planes and sections. **A,** Sagittal sections separate right from left sides. Frontal sections separate anterior parts from posterior. Transverse sections separate superior from inferior. **B,** Longitudinal and transverse sections are useful in individual structures such as the kidney.

from inferior. Transverse sections can expose the internal anatomy of virtually the entire body by slicing a series of sections across the central axis, much as a butcher slices bologna at the deli.

It is not, however, necessary to section the entire body to reveal smaller parts such as the kidney (Figure 1.21B), because they can be sectioned along their **longitudinal axes,** which pass from one end to the other, as the central body axis does in the trunk.

- Which type of section separates left from right?
- What is another name for transverse sections?

MEDICAL IMAGING

Chances are that you will never see your own internal organs. Although they are only about a centimeter beneath your skin, neither you nor your physician can see your organs directly without expensive surgery to expose them. Medical science uses several visualization techniques that minimize or entirely avoid surgical intrusion. Some of these techniques enable investigators to build three-dimensional computer models of internal organs that rival the models in a laboratory. These rapidly growing sources of information are making it possible to dissect the body electronically and to draw computer maps that follow thoughts through the circuitry of the brain.

ENDOSCOPY

An endoscope is a narrow tube inserted through a small incision to see and perform surgery on internal organs (Figure 1.22). Endoscopes are really bundles of smaller rods and tubes inside a larger one. The physician guides the endoscope by pulling or pushing flexible rods in the tube that bend the tip of the endoscope up, down, or sideways. Fiber optics show him where the endoscope is going by conducting light into the endoscope and returning images to a video screen.

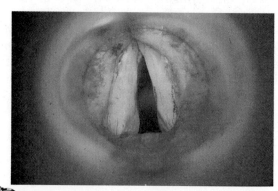

Figure 1.22

Endoscopy. View of the vocal cords.

Various tubes carry fluids in and out or accept surgical instruments. Endoscopes are frequently used to remove cartilage from damaged joints or for abdominal surgery. To remove the gall bladder, for instance, the peritoneal cavity is inflated with air to provide space for maneuvering and the endoscope is inserted through a small incision at the navel. Surgery and removal of the gall bladder is done entirely through the endoscope.

X-RAY

X-radiographs are a long-standing, relatively inexpensive technique used to reveal internal structures as shadows on X-ray film. A radiation source sends a beam of X-rays through the body onto a photographic film. Dense structures such as bone and fibrous tissues leave shadows that appear light when the film is developed. It is easy to recognize teeth and location of the brain in Figure 1.23, but it takes an experienced

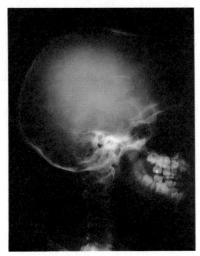

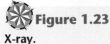
Figure 1.23
X-ray.

eye to define the softer tissues of the head. Because X-radiation may kill cells and cause cancer, modern X-ray instruments use low doses and very sensitive films to minimize the risk of exposure to patients. Fetal X-radiation is avoided whenever possible because of the danger to the developing fetus. By contrast, very sensitive films can reveal considerable fine detail in mammograms at very little risk to the patient, so that the benefit of detecting breast cancer by X-ray far exceeds the risk of inducing it.

ULTRASONIC SCANNING

Ultrasonic scanning reveals motion. Ultrasonic scanning instruments send thin beams of high-frequency sound waves into the body from a sound head

placed on the body by a clinician. The instrument converts echos from denser tissues into a video screen sonogram that can show heart valves in motion, detect blood flow, or reveal a fetus moving in the uterus, among many other applications (Figure 1.24). Unlike

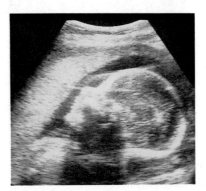
Figure 1.24
Ultrasonic scanning.

X-rays, ultrasound is considered harmless, thus it is especially useful in diagnosing fetal disorders. Because bone and air-filled spaces dissipate sound waves, sonography is of little help in visualizing the brain and lungs. Sonograms (echograms) are stored on videotape as permanent records of the examination.

COMPUTED TOMOGRAPHY

Otherwise known as CT or CAT scanning, this technique constructs transverse sections of the body by computing images from narrow beams of X-rays. The scanner circles the patient as he lies in a tray, and the X-ray detector and computer construct an image of the organs one section at a time. The scanner can build a three-dimensional view from a series of sections as the tray advances the patient through the scanner. Internal detail is considerably better than on an X-ray film because the circling X-ray source allows the computer to eliminate shadows from the image. Figure 1.25 shows a transverse section through the head and neck; you can readily see the brain, eyes, spinal cord, and other structures. CT scanning can enhance certain details by subtracting unwanted background from the image. One version of this technique, known as digital subtraction angiography (DSA), displays vessels of the heart as if they had been isolated from the wall of the heart.

MAGNETIC RESONANCE IMAGING

This technique (MRI) provides better detail than CT scanning by detecting signals from hydrogen atoms in the soft tissues. The patient rides a tray into a powerful magnet that causes the nuclei of hydrogen

Figure 1.25
Computed tomography.

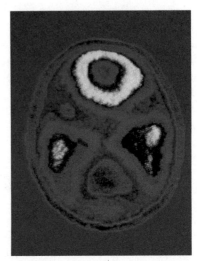

Figure 1.27
Positron emission tomography.

atoms to line up in the magnetic field. Detectors and a computer assemble an image similar to Figure 1.26 from minute amounts of electromagnetic radiation emitted from the energy absorbed by the hydrogen atoms. The computer adds color to indicate the density of hydrogen atoms in different tissues. As computer speed and memory increase in the future, it will be possible to construct highly detailed three-dimensional, movable, even electronically dissectible images of body parts. Already it is possible to map the location of nerve impulses in a functioning brain—in effect, perhaps, to map an image of a thought.

POSITRON EMISSION TOMOGRAPHY

The main advantage of PET scanning is its ability to follow metabolic changes in tissues, especially in the brain, where brain or tumor metabolism consumes glucose more rapidly or increases blood flow. Investigators inject labeled isotopes of glucose or other metabolic markers into the blood, and the patient, again resting on a tray, is scanned for radiation released by the isotope. A computer constructs the type of image shown in Figure 1.27. These maps of metabolic activity come at the cost of lower resolution, however, because the radiation detectors cannot resolve detail as fine as MRI.

- Why are noninvasive imaging techniques so useful?
- Which imaging technique displays shadows of internal structure?
- How does ultrasonic scanning produce an image?
- Which shows structural detail more precisely, MRI or PET scanning?

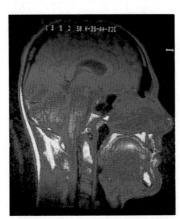

Figure 1.26
Magnetic resonance imaging.

SUMMARY

Anatomy organizes the body into regions, levels, and organ systems. The major regions of the body are head and neck, trunk, and the extremities. Body structure emerges from different levels of organization, ranging from simple molecules up to the entire complex individual. Twelve interacting organ systems perform the various functions of the body. *(Page 5.)*

The body contains two great internal cavities. The dorsal cavity contains the brain and spinal cord, and the visceral organs occupy the ventral cavity. The ventral cavity is subdivided into thoracic, abdominal, and pelvic cavities. Serous membranes enclose cavities within these cavities. The pleural and pericardial cavities surround lungs and heart within the thoracic cavity, and the peritoneal cavity resides in the abdominopelvic cavity. Mucous membranes line cavities within the digestive, respiratory, urinary, and reproductive systems. *(Page 11.)*

Particular structures usually have particular functions. Form and function are long- and short-term aspects of processes that form cells, tissues, and organs. Anatomy concentrates on the longer-lived structures, and physiology and cell biology deal with faster interactions. *(Page 13.)*

Structure and function follow certain principles. Organs contain many small functional units that take up and release materials or signals across exchange surfaces. The surface-volume rule maximizes transport by packing organs with many miniature functional units. Distributive trees connect these units to central sources of supply. Negative feedback mechanisms maintain homeostasis in the body by continually adjusting to changing conditions. *(Page 13.)*

The human body shares at least six fundamental features with its vertebrate relatives. The vertebral column and notochord support the trunk. Dorsal and ventral cavities contain the brain and spinal cord, and the visceral organs, respectively. The brain and spinal cord develop from a hollow tube. The body is bilaterally symmetric and axially segmented. The digestive tube is a tube within the trunk. *(Page 16.)*

Anatomists locate body regions and systems by referring to the anatomical position. Anatomy uses positional terms to describe locations that may be above or below, in front or back, or in other locations. *(Page 17.)*

Sagittal, frontal, and transverse planes and sections reveal internal anatomy by cutting the body along the central axis (sagittal and frontal) or across it (transverse). Medical imaging techniques use these sections to reveal internal structures. *(Page 19.)*

STUDY QUESTIONS
OBJECTIVES

1. Name the regions of the body and identify its major cavities. **1**

2. List the levels of anatomic organization in order of decreasing complexity. Give a definition and an example of each level. **2**

3. List the functions of the organ systems. Some systems have several functions, and you also will find several systems involved in the same function. **3**

4. Which organ systems are located in: the head, thorax, abdominal cavity, or lower extremities? **1, 2**

5. Name the subdivisions of the ventral cavity. What organs does each subdivision contain? Which organs do the pericardial and pleural cavities contain? Which type of membrane lines the peritoneal cavity? **3, 4, 5**

6. Name the functional units of several organ systems and describe what these units do.

7. Why are many small units more efficient than a few large ones? **6**

8. How do materials reach the functional units of various organs? **7**

9. Why is homeostasis important? **8**

10. How does the pupillary reflex regulate the size of the pupil? What conditions narrow the pupil and what conditions enlarge it? Theoretically, what size could the pupil attain if positive feedback controlled it? **9**

11. Describe the fundamental features of the human body plan. **9**

12. What evidence of segmentation does the body display? **10**

13. In what regard is the body bilaterally symmetrical? Which organ systems are bilaterally symmetrical? Consult Figures 1.3 to 1.14. **10**

14. Using the anatomical position, describe the location of the left thumb with respect to the little finger of the left hand, elbow of the left arm, central axis of the body, right thumb, and first toe of the left foot. **10**

15. What structures of the body are inferior to the navel, proximal to the knee, caudal to the thorax? **11**

16. Refer to Figures 1.3 to 1.14. Which system would a transverse section through the lumbar region of the back expose? A transverse section through the leg? A median section through the skull? **11**

17. What internal structures do X-rays, ultrasonic scanning, CT, PET, and MRI reveal? **12**

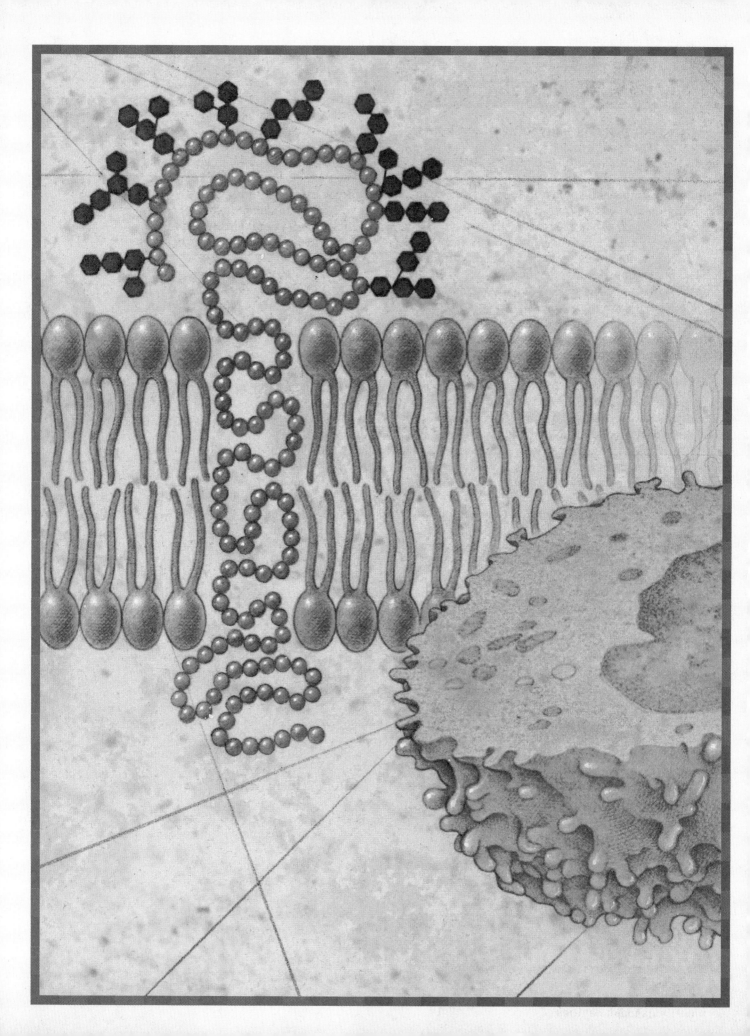

CHAPTER 2

Cell Basics

OBJECTIVES

After studying this chapter, you should be able to do the following:

1. Name and describe four fundamental activities of cells.
2. Describe the structure and functions of several specialized cells.
3. List the organelles of cells, and describe their functions.
4. Explain what features of the plasma membrane make it fluid and mosaic.
5. Name the organelles of the endomembrane system and their functions.
6. Name three components of the cytoskeleton, and describe their functions.
7. Identify various junctional complexes according to their structures and functions.
8. Describe the 9 + 2 axoneme of cilia and flagella and how it is thought to work.
9. Compare and contrast the variety of structures found in skeletal muscle fibers, acinar cells, hepatocytes, and fibroblasts.
10. Distinguish among the mechanisms and structures that enable substances to cross the plasma membrane.
11. Describe two signal transduction pathways that enable cells to respond to external stimuli.
12. Describe and give examples of the role of the cell cycle in various tissues and organs.
13. Describe the processes of mitosis and meiosis and how they differ.
14. Describe differences between cancerous and normal cells, and outline the process that transforms normal cells into cancer cells.
15. Describe the role cells may play in the aging process.

 VOCABULARY TOOLBOX

Cellular terms:

-blast	*G* a bud or sprout. Fibroblasts proliferate and produce connective tissue.	**nucleus**	*L* a little nut, a kernel.
		soma	*G* body. Chromosomes, ribosomes, and lysosomes are cytoplasmic organelles.
-cyte or **cyto-**	*G* a hollow place, a cell. A melanocyte is a pigment cell, and cytoplasm is the substance inside cells.	**vesicle, vesicul-**	*L* a little bladder, blister.

RELATED TOPICS

- **Cancerous tissues, Chapter 3**
- **Epithelial tissues, Chapter 3**
- **Homeostasis, Chapter 1**
- **Neurons and synapses, Chapter 17**
- **Skeletal muscle, Chapter 9**

PERSPECTIVE

Most of us know that the body consists of cells, the fundamental units of life on which the form and function of the body rely. Various specialized cells make up the body's organ systems. Skeletal muscle fibers move the body, for example, and cells that line the intestine secrete digestive enzymes and absorb nutrients. Fewer people know, however, that cellular organelles, structures within the cell, are both the basis for the fundamental similarities among cells and for the various specialized features of cells. This chapter describes the structure and function of cellular organelles. It also illustrates several examples of specialized cells that enable organ systems to function.

This chapter discusses the structure and function of cells. It provides the background for understanding the higher levels of organization among tissues and organ systems discussed in the rest of the text. Organ systems rely on specialized cells whose unique characteristics are based on common cellular structures and functions that all cells share, but which specialized cells modify in various ways. The first section, *Fundamental Properties of Cells,* outlines four fundamental cellular activities that cells modify for different functions. The following section, *Cell Sketches,* presents examples of specialized cells that enable different organ systems to function. You should memorize what these cells do and how they express the basic four activities, because the rest of the chapter describes internal structures and processes that these cells and many others have modified.

FUNDAMENTAL PROPERTIES OF CELLS

Adults have at least 100 trillion cells. Some 200 varieties of cells in 12 organ systems and 50 different tissues are responsible for the different structures and functions of our bodies. Nearly all these cells share four fundamental properties.

1. **Cells produce energy and structure from metabolic reactions**. A network of approximately 10,000 reactions catalyzed by enzymes obtains energy by oxidizing glucose molecules in the presence of oxygen to carbon dioxide and water. This energy-capturing process is called cellular respiration. Certain pathways of reactions in the network produce small, building-block molecules that other pathways assemble into macromolecules and the cellular structures that are the ultimate basis of structure and function. Genes govern the metabolism of different cells and organs by specifying which enzymes and pathways a cell exhibits.

2. **Cells reproduce themselves through growth and cell division,** processes to which cells devote much energy and structure. When large enough, cells divide into two and distribute copies of their genes to both daughter cells. Continuous proliferation in this manner permits the body to grow, organs to develop, wounds to heal, and cells to replace themselves.

3. **Cells interact with the environment surrounding them.** What a cell does depends on the genes that govern it from inside, as well as the conditions outside. Neurons, the cells that conduct impulses in the nervous system, transmit information by stimulating other neurons, and skeletal muscles contract when nerves stimulate them. Numerous hormones that circulate through the blood regulate reproduction and sexual development. Such interactions enable organs to function.

4. **Cells are self-regulating;** that is, they continually adjust to changes by attempting to remain constant. Feedback loops maintain this homeostasis (see Homeostasis and Negative Feedback, Chapter 1) by regulating internal cellular processes. Negative feedback regulates the supply of building-block molecules, for example, by adjusting to cellular demands. Excess molecules inhibit production until the supply declines, then production resumes. Similarly, entire cells may respond by dividing when stimulated and ceasing division when the stimulus stops.

   - Which process acquires energy and produces cellular structures?
   - Which process permits growth, wound healing, and cell replacement?
   - Cite an example of a cell's response to its environment.
   - Name the process that self-regulates cells.

CELL SKETCHES

The cells described in this section illustrate how cells adapt the four fundamental functions to different organ systems.

SKELETAL MUSCLE FIBER
Skeletal muscle fibers enable muscles to move the skeleton. These cells devote their metabolism to assembling, maintaining, and operating the contractile **myofibrils** (Figure 2.1). Nerves stimulate thousands of skeletal muscle fibers in a muscle, causing the myofibrils to shorten and the muscle fibers and the entire muscle to contract. These long, narrow, cylindrical cells often reach many centimeters from one end of a muscle to another. Muscle fibers attain this huge size when immature, dividing cells fuse together. Although muscle fibers can grow in size, they do not divide.

ACINAR CELLS
The pancreas (Figure 2.2A) promotes digestion by secreting into the intestine enzymes that cleave proteins and other macromolecules into smaller fragments. **Acinar cells** (*L,* like a berry; ah-SINE-ar) produce these enzymes. Shaped like cubes, these cells form several million small cups, called acini (singular, acinus; Figure 2.2B), that secrete enzymes into the intestine. Inside each acinar cell, as Figure 2.2C shows, **zymogen granules** (ZIME-o-jen; zym-, *G,* ferment, -gen, *G,* produce) filled with digestive enzymes move toward the surface of the cell, which releases them into the acini. Intestinal hormones accelerate secretion of zymogen granules when undigested food is present in the gut, and they slow production after food has been processed. Acinar cells seldom divide, but they probably are capable of proliferating when replacements are needed.

HEPATOCYTES
The liver (Figure 2.3A) takes up substances such as glucose and cholesterol from the blood and releases others, including blood proteins and urea, into it. **Hepatocytes** (heh-PAT-o-sites; hepat-, *G,* liver) are the principal liver cells (Figure 2.3B) that perform these many functions. Hepatocytes are polygonal (many-sided) cells past which blood flows on its way to the heart. When there is excessive glucose in the blood, hepatocytes store the glucose as molecules of glycogen, but when blood glucose levels fall, negative feedback restores the glucose by breaking down glycogen. Hepatocytes rarely divide, but injury to the liver will stimulate them to replace lost tissue.

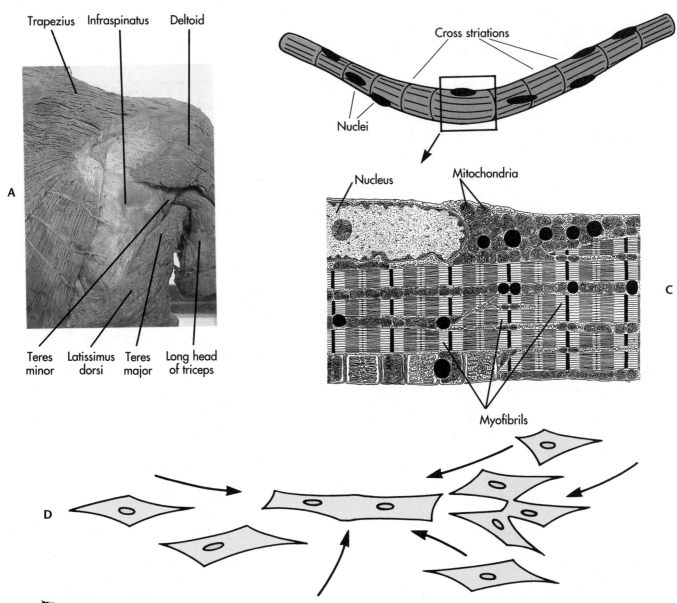

Trapezius   Infraspinatus   Deltoid

Teres minor   Latissimus dorsi   Teres major   Long head of triceps

Cross striations

Nuclei

Nucleus   Mitochondria

Myofibrils

![Figure 2.1 icon] **Figure 2.1**

Skeletal muscles **(A)** are composed of skeletal muscle fibers **(B)**, which are long cells filled with myofibrils **(C)** that shorten when the nervous system stimulates muscle fibers to contract. Muscle fibers **(D)** form when immature cells fuse together.

FIBROBLASTS

Most tissues contain fibroblasts (Figure 2.4A) that synthesize collagen fibers responsible for the toughness and resilience of bone, tendons, and skin, for example. **Fibroblasts** (FIE-bro-blasts; fibr-, *L,* fiber) are stellete (star-shaped) cells (Figure 2.4B). Collagen genes govern collagen production in fibroblasts. Fibroblasts routinely divide, and the process of wound healing especially stimulates them to deposit collagen into damaged tissues.

Governed by their own genes and local environment and regulated by negative feedback, skeletal muscle fibers, hepatocytes, acinar cells, and fibro-

blasts are examples of specialized cells that enable organ systems to perform their various functions.

- Which cell contains myofibrils?
- Name the cell that secretes collagen.
- Which cell secretes zymogen granules?
- Which of these cells are capable of dividing?

CELL ORGANELLES

Organelles are cytoplasmic structures that carry out particular functions. Most cells have in common a set of organelles (small organs) that perform the

Text continued on p. 31.

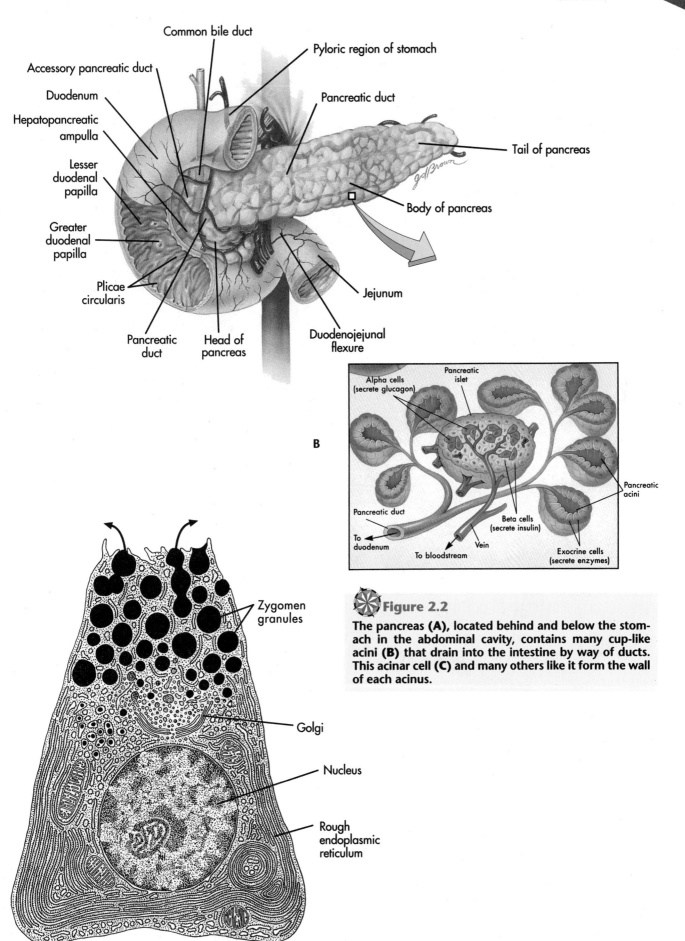

Common bile duct
Pyloric region of stomach
Accessory pancreatic duct
Pancreatic duct
Duodenum
Hepatopancreatic ampulla
Tail of pancreas
Lesser duodenal papilla
Body of pancreas
Greater duodenal papilla
Plicae circularis
Jejunum
Pancreatic duct
Head of pancreas
Duodenojejunal flexure

A

B

Alpha cells (secrete glucagon)
Pancreatic islet
Pancreatic duct
To duodenum
Beta cells (secrete insulin)
To bloodstream
Vein
Pancreatic acini
Exocrine cells (secrete enzymes)

Zygomen granules

Golgi

Nucleus

Rough endoplasmic reticulum

Figure 2.2

The pancreas **(A)**, located behind and below the stomach in the abdominal cavity, contains many cup-like acini **(B)** that drain into the intestine by way of ducts. This acinar cell **(C)** and many others like it form the wall of each acinus.

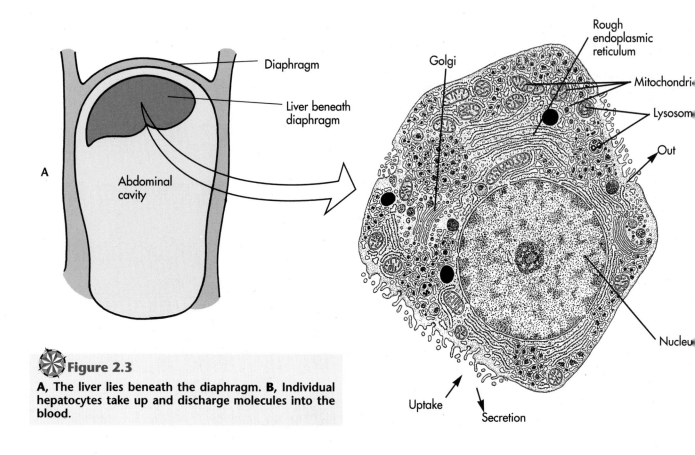

Figure 2.3

A, The liver lies beneath the diaphragm. **B,** Individual hepatocytes take up and discharge molecules into the blood.

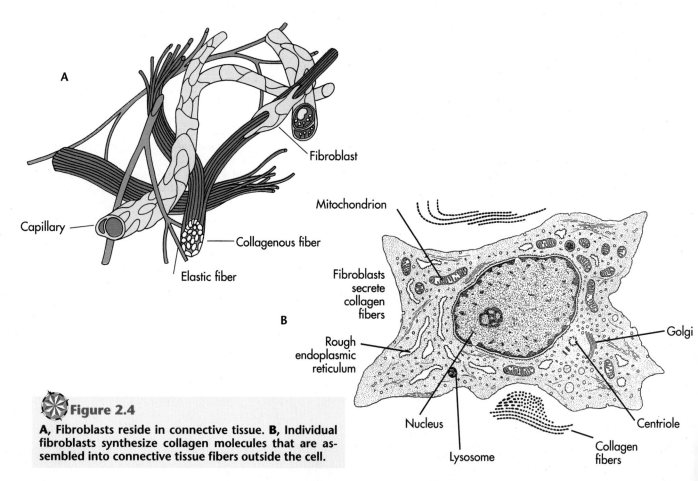

Figure 2.4

A, Fibroblasts reside in connective tissue. **B,** Individual fibroblasts synthesize collagen molecules that are assembled into connective tissue fibers outside the cell.

standard functions all cells require. Plasma membranes, nuclei, and mitochondria, described here and shown in Figure 2.5, carry out housekeeping functions such as containing the cytoplasm, storing genes, and providing energy, respectively. Other, more specialized organelles such as flagella, the whip-like structures that propel sperm, appear only in cells that make use of them. Most organelles are small bodies suspended in the **cytoplasm** (SITE-o-plazm), the complex fluid within the cell. Metabolic reactions that enable organelles to function take place on the surface and the interior of organelles. Other reactions take

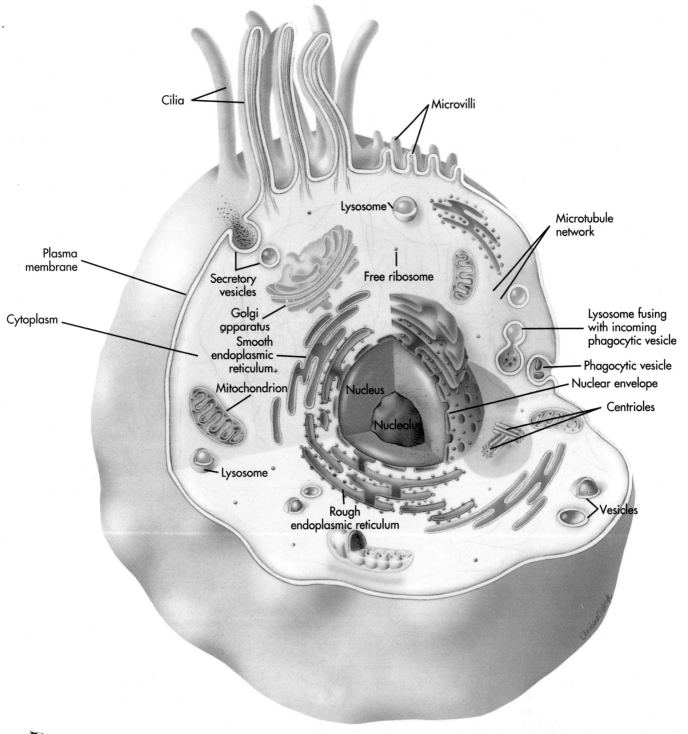

Cilia

Microvilli

Lysosome

Microtubule network

Plasma membrane

Secretory vesicles

Free ribosome

Cytoplasm

Golgi apparatus

Smooth endoplasmic reticulum

Lysosome fusing with incoming phagocytic vesicle

Phagocytic vesicle

Nuclear envelope

Mitochondrion

Nucleus

Centrioles

Nucleolu

Lysosome

Vesicles

Rough endoplasmic reticulum

Figure 2.5

Generalized cells like this one do not exist, but if they did, this cell is what you probably would see. Identify the plasma membrane, cytoplasm, nucleus, and mitochondria.

place among enzymes and molecules dissolved in the **cytosol,** the soluble portion of the cytoplasm.

Cell anatomists and pathologists know organelles so well that they usually can decide what a cell does by observing the organelles it contains. Make a list of the organelles described below and their functions so that you can identify the function of other cells described in study questions 9 and 10 at the end of this chapter.

PLASMA MEMBRANE

A plasma membrane covers the exterior of all cells. This membrane is a thin, flexible, dynamic sheet of phospholipid molecules and proteins that separates the cytoplasm from the external environment (Figure 2.6). Cells communicate with the external environment and other cells through this highly specialized covering. Although water and different small molecules such as oxygen diffuse readily through it, the cell membrane is quite specific about other substances. Cell membranes routinely take up glucose molecules into the cytoplasm and transport potassium and sodium ions in and out of the cell. Membranes also secrete proteins and polysaccharides to the outside. Receptor molecules on the external surface of the membrane detect chemical stimuli, and specialized junctions attach the membranes of adjacent cells together.

A fluid mosaic of phospholipid molecules and proteins accomplishes all these functions. The **phospholipids** (FOSS-fo-LIP-ids) make up two layers shown in Figure 2.6; one layer faces the exterior of the membrane, and the other faces the cytoplasm. In each layer the phospholipid molecules align tail to tail, with their hydrophilic heads (water-loving ends of the phospholipid molecules that bind to water molecules) on each surface of the bilayer, where they readily associate with water molecules, and their hy-

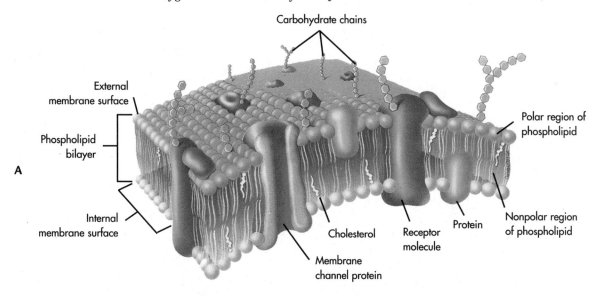

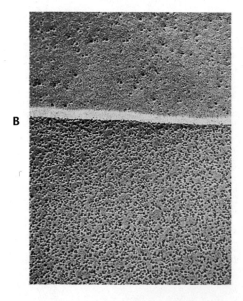

Figure 2.6

The plasma membrane. A, The phospholipid bilayer consists of two layers of phospholipid molecules with integral membrane proteins protruding through both layers onto the external and internal surfaces. **B,** Scanning electronmicrograph of the external surface reveals closely scattered membrane proteins.

drophobic tails (water-fearing portions that bind water molecules poorly) sandwiched between. Together both layers form the **phospholipid bilayer.** Various integral proteins penetrate through both faces of the membrane like icebergs in a sea. Although the phospholipids fit as precisely as tiles in a mosaic, the arrangement is nevertheless fluid because the phospholipids and integral proteins diffuse from place to place on the membrane.

NUCLEUS

The nucleus, the largest organelle, is the genetic information center of the cell. A **nuclear envelope** surrounds the nucleus (Figure 2.7) with a double mem-

the nuclear envelope. Human nuclei contain 23 chromosome pairs that store approximately 100,000 different genes that govern cell structure and metabolism. Figure 2.8 shows each pair and regions where genes are located on them. In Figure 2.7 these chromosomes have uncoiled (decondensed) into long, thin **chromatin fibers** that extend throughout the nucleus and

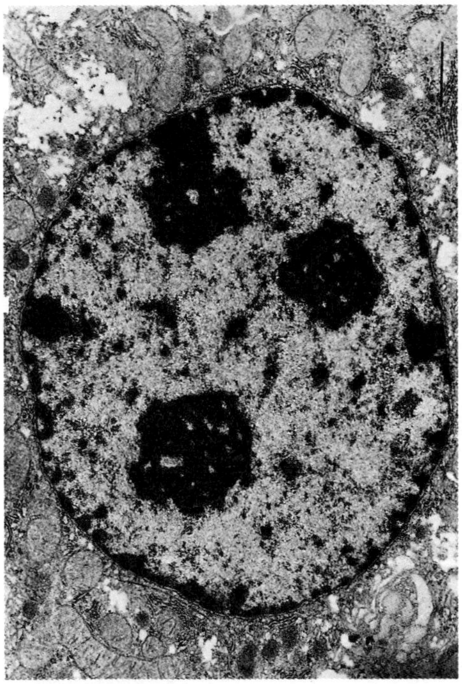

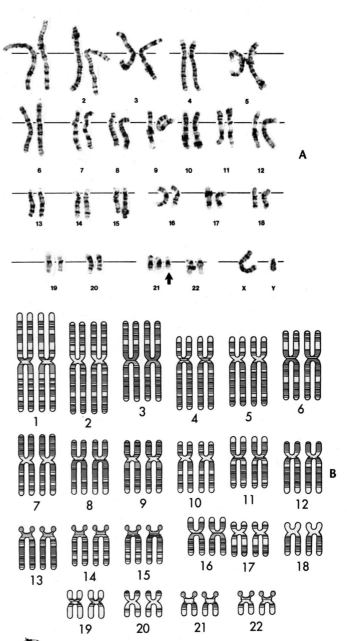

Figure 2.7

A nuclear envelope encloses chromatin threads and nuclear particles. Nuclear pores communicate with the cytoplasm. (Electronmicrograph courtesy of C. Kelloes and M. Farmer, Center for Advanced Ultrastructural Research, University of Georgia.)

brane containing numerous **nuclear pores** that evidently communicate with the cytoplasm. The external membrane frequently is continuous with the endoplasmic reticulum (described later). **Lamin protein fibers** attach chromosomes to the inner membrane of

Figure 2.8

A, Human chromosomes. **B,** Twenty-three pairs of human chromosomes contain genes that govern cell metabolism and anatomical structures. There are 22 homologous pairs and 2 sex chromosomes, the X and Y. Members of a homologous pair are the same length and shape and store homologous genes. The Y chromosome is shorter than the X and contains different genes. The locations of many genes on the chromosomes are known. X and Y chromosomes are not shown.

attach to the nuclear envelope. Chromatin fibers coil tightly and **condense** into visible chromosomes when the cell divides. Nuclei usually contain one or more **nucleoli** (*L,* a little nucleus) that assemble ribosomes, the organelles that synthesize protein molecules in the cytoplasm (see Ribosomes, next section).

Chromosomes store genes in immensely long deoxyribonucleic acid (DNA) molecules, which are the backbone of chromatin fibers. Cells duplicate these DNA molecules when they divide. Genes store information that governs the structure and shape of protein molecules. The DNA of different genes encodes this information in different sequences of nucleotides; the cell transcribes this information and translates (Figure 2.9) it into the various sequences of amino acids that make up proteins and enzymes. What makes

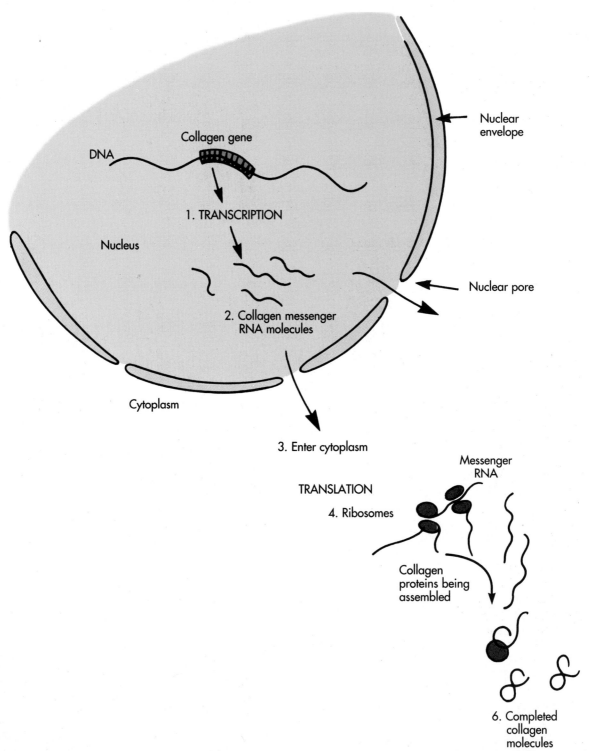

Figure 2.9

Transcription and translation. Follow description in text.

skeletal muscle fibers so valuable, for example, is their ability to transcribe and translate the genes that govern muscle proteins. Fibroblasts transcribe and translate collagen genes and acinar cells, the genes for digestive enzymes. In **transcription** of the collagen genes, for example, the nucleus synthesizes molecular copies, called messenger RNA, of the information encoded in the genes concerned with collagen. The messenger RNA molecules then enter the cytoplasm, probably through nuclear pores, and bind to ribosomes. Ribosomes **translate** the information in the messenger RNA molecules into sequences of amino acids that become collagen molecules. Transcription of genes into messenger RNA and translation of this RNA into proteins are universal cellular processes

that assemble all cell proteins; thus nuclei, chromosomes, and ribosomes are constituents of all cells.

RIBOSOMES

Ribosomes are small, nonmembranous organelles that assemble proteins from amino acids. The cytoplasm of most cells contains several million **ribosomes** (RYE-bo-somes, rhymes with homes); each one in Figure 2.10 consists of a large and a small subunit assembled in the nucleolus. Both subunits enter the cytoplasm, possibly through nuclear pores, and bind to molecules of messenger RNA. Ribosomes that also bind to the endoplasmic reticulum (described later) assemble proteins for export or storage in membranes; whereas in the cytosol, free ribosomes that do not

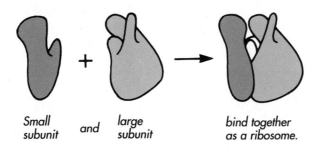

Small subunit   and   large subunit   bind together as a ribosome.

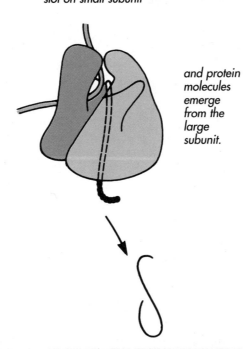

Messenger RNA fits into slot on small subunit

and protein molecules emerge from the large subunit.

Figure 2.10

The large and small subunits of ribosomes bind together during translation. The messenger RNA molecule fits into a slot in the small subunit, and newly translated molecules of protein emerge from the large subunit. Ribosomes appear as small dots in the illustrations of skeletal muscle fibers, acinar cells, hepatocytes, and fibroblasts shown in Figures 2.1 through 2.4.

bind to the endoplasmic reticulum assemble protein molecules that remain in the cell for its growth and maintenance. An initial leader sequence of amino acids in the proteins acts as a name tag that determines whether ribosomes bind to the endoplasmic reticulum or remain free in the cytoplasm.

MITOCHONDRIA

Mitochondria are cellular respiratory organelles that oxidize carbohydrates in the presence of oxygen to carbon dioxide and water and trap some of the energy liberated in energy-rich ATP molecules (adenosine triphosphate). The **mitochondria** (my-toe-KON-dree-ah; singular, mitochondrion) in Figure 2.11 are flexible cylindrical or spherical organelles that accumulate where ATP is needed. Skeletal muscle fibers contain numerous mitochondria that power contraction, and hepatocytes contain approximately 1700 mitochondria that provide the energy to synthesize molecules and transport them to the blood. Like

nuclei, mitochondria are enclosed by a double membrane. The **outer membrane** in Figure 2.11B is smooth, but numerous folds or **cristae** (KRIS-tee; *L,* a crest on a helmet; singular, crista) crease the **inner membrane,** providing additional surface for integral membrane proteins that serve as respiratory enzymes. Substrate molecules and other respiratory enzymes occupy the **inner chamber.** During respiration, hydrogen ions accumulate between the membranes in the **outer chamber** and diffuse back across the inner membrane into the inner chamber, releasing energy that is captured in ATP.

Mitochondria contain DNA and genes for mitochondrial proteins. **Mitochondrial DNA** molecules are circular in form; they store genes for respiratory enzymes and RNA molecules needed during translation. Mitochondrial DNA duplicates rather like nuclear DNA. Mitochondria reproduce by dividing but cannot do so independently of the cell nucleus, which also contains genes for mitochondrial proteins.

- What is the function of the plasma membrane?
- Why is the nucleus considered the information center of the cell?
- Where are ribosomes found, and what is their function?
- Why are mitochondria called respiratory organelles?

ENDOMEMBRANE SYSTEM

Almost all cells contain internal membranes that cluster and fold throughout the cytoplasm; there they appear as sheets and spherical **vesicles.** At any

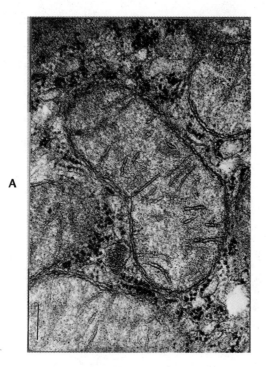

A

**Figure 2.11**

Mitochondria supply ATP. A, An electronmicrograph of a mitochondrion. (Electronmicrograph courtesy of C. Kelloes and M. Farmer, Center for Advanced Ultrastructural Research, University of Georgia.) **B,** Diagram of entire organelle. Find the mitochondria in the skeletal muscle fiber, acinar cell, hepatocyte, and fibroblast shown in Figures 2.1 through 2.4.

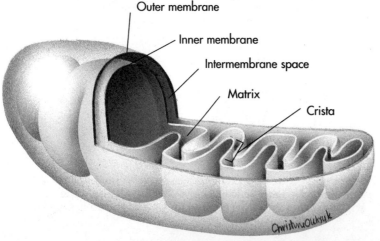

Outer membrane

Inner membrane

Intermembrane space

Matrix

Crista

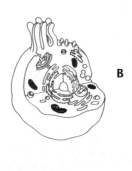

B

moment, sheets may become vesicles as one component flows into another. For example, minutes later portions of the nuclear envelope can become part of the plasma membrane, and patches of the cell membrane may migrate into the internal membranes of the cytoplasm. The principal components of this dynamic, internal **endomembrane system** are the nuclear envelope, the endoplasmic reticulum, Golgi complex, lysosomes, peroxisomes, and various transport vesicles. The fluid nature of the plasma membrane is central to the interplay among all these organelles. The membrane provides for synthesis and recycling of membranes, supplies surfaces for metabolic reactions, and encloses certain enzymes that would otherwise digest the cytoplasm.

ENDOPLASMIC RETICULUM

The endoplasmic reticulum is a cytoplasmic network of membranous sacs that is continuous with the nuclear envelope and the plasma membrane. Figure 2.2C shows the **endoplasmic reticulum** (ER) molded around the nucleus in an acinar cell. In hepatocytes (Figure 2.3B) the ER is a stack of flattened sacs near the nucleus. Each sac encloses an internal space (Figure 2.12) called a **cisterna** (sis-TERN-a; *Med Eng*, a water tank), which resembles the space between the membranes of the nuclear envelope and is continuous with it. The ER synthesizes proteins, polysaccharides, and phospholipids, which are stored in the cisternae and in the membranes themselves for later transport elsewhere in the cell. The amount of membrane surface area in the ER indicates the synthetic activity of cells; clearly, pancreatic acinar cells and hepatocytes are vigorous synthesizers, devoting approximately 60% of their total membrane surface to ER. Quiescent cells such as eggs have much less endoplasmic reticulum.

Endoplasmic reticulum may be rough or smooth based on the binding of ribosomes to the cytoplasmic

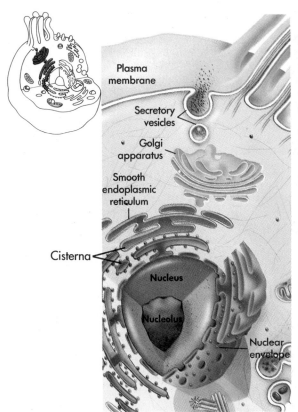
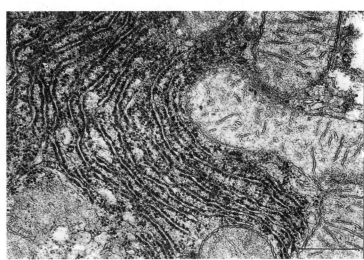

B

Figure 2.12

A, The endoplasmic reticulum (ER) is a series of flattened sacs containing cisternae. To see the relationship of the ER with the nucleus and mitochondria, find it in the cells shown in Figures 2.1 through 2.4. **B,** Electronmicrograph of rough and smooth ER. (Electronmicrograph courtesy of C. Kelloes and M. Farmer, Center for Advanced Ultrastructural Research, University of Georgia).

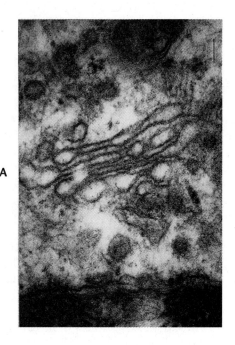

A

**Figure 2.13**

The Golgi complex sorts the products of the endoplasmic reticulum and packages them into transport vesicles to be secreted outside or used internally. The Golgi complex usually is associated with the endoplasmic reticulum; see Figures 2.2 through 2.4. (Electronmicrograph courtesy of C. Kelloes and M. Farmer, Center for Advanced Ultrastructural Research, University of Georgia.)

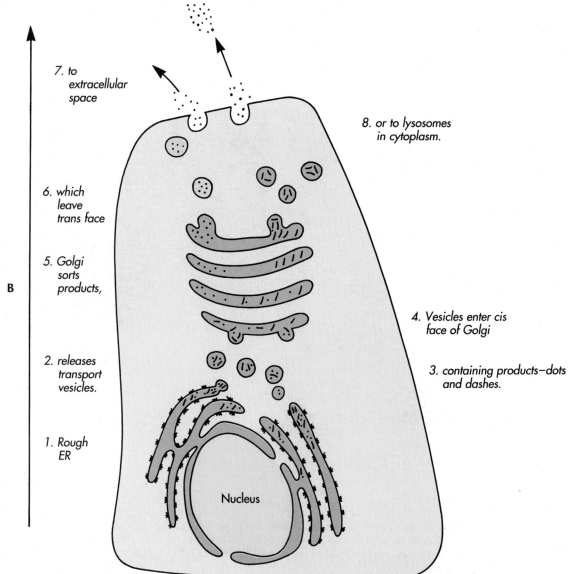

7. to extracellular space

8. or to lysosomes in cytoplasm.

6. which leave trans face

5. Golgi sorts products,

B

4. Vesicles enter cis face of Golgi

3. containing products–dots and dashes.

2. releases transport vesicles.

1. Rough ER

Nucleus

face of sacs (rough ER) or their absence (smooth ER). **Rough ER** in Figure 2.12B synthesizes proteins that will be incorporated into the plasma membrane or secreted outside the cell. Proteins assembled on the ribosomes are inserted through the endoplasmic reticulum membrane into the cisternae, where they are transported elsewhere in the cytoplasm (see Golgi Complex, next section). In hepatocytes, **smooth endoplasmic reticulum** (Figure 2.3B) also assembles polysaccharides and lipid molecules. The interstitial cells of the testis have especially large quantities of smooth ER, which synthesize the steroid hormone testosterone, but there is virtually no smooth ER in acinar cells.

GOLGI COMPLEX

This stack of membranes and cisternae is a cytoplasmic traffic cop (Figure 2.13). The **Golgi complex** (GOAL-jee) is a cluster of membranous sacs found in acinar cells, hepatocytes, and fibroblasts (Figures 2.2C, 2.3B, and 2.4B) that resembles a stack of soup bowls. The complex is usually located near the rough endoplasmic reticulum, but no ribosomes stud its membranes. The Golgi complex (GC) sorts newly assembled proteins from the rough ER and directs them toward destinations inside and outside the cell. Sorting takes place as materials move from the side nearest the endoplasmic reticulum (**cis face;** *L,* the near side) to the opposite side (**trans face**). In Figure 2.13 the contents of cisternae in the ER (1) pinch off the endoplasmic reticulum as small **transport vesicles** (2, 3, and 4) fuse with the nearest Golgi membranes on the cis face. Sorted proteins emerge from the trans face (5) in other transport vesicles (6) marked with identification molecules that specify their destination. As a result, certain transport vesicles become zymogen granules (7) to be secreted outside, and others (8) are delivered to the lysosomes.

LYSOSOMES

These cytoplasmic vesicles help recycle spent macromolecules and organelles. Lysosomes (LYE-so-somes) are spherical membranous vesicles (Figure 2.14) with at least 50 different enzymes that cleave all varieties of macromolecules into their constituent subunits. Cellular metabolism then recycles these products into new macromolecules, thus permitting cells to adjust to changing conditions. Most cells contain lysosomes, and some cells such as hepatocytes have as many as 200 of them (Figure 2.3B).

Lysosomal enzymes are assembled in the endoplasmic reticulum and sorted in the Golgi complex. Lysosomes fuse with cytoplasmic vesicles that contain material to be degraded. These vesicles may come from outside the cell (see Endocytosis) or from **autophagic vesicles** (aw-toe-FAY-jik; self-eating), which capture

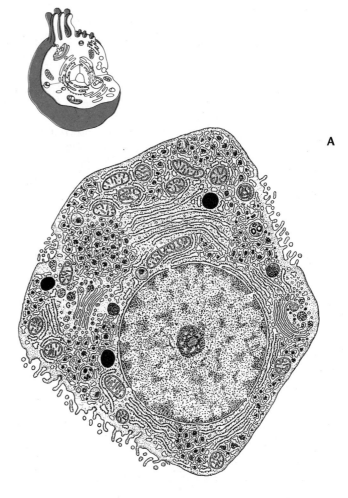

A

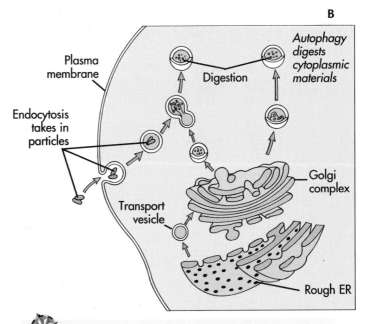

B

Figure 2.14

Lysosomes derive from the Golgi complex and endoplasmic reticulum. A, Lysosomes in heptocyte. **B,** Lysosomes digest materials from outside and inside the cell).

spent ribosomes and mitochondria from the cytoplasm. In either case, digestion is isolated within the lysosomal membrane, which prevents the digestive process from attacking the rest of the cytoplasm. After digestion, the membrane releases the products into the cytoplasm.

Several inherited diseases involve lysosomes. In **Tay-Sachs disease** phospholipids accumulate in neurons because the affected lysosomes lack an enzyme, β-N-hexosaminidase, needed to digest brain phospholipids. This defect is lethal within several years of birth; this gene is located on chromosome 15 in Figure 2.8.

PEROXISOMES

Peroxisomes are oxidizing organelles noted for their ability to produce and degrade the highly toxic substance, hydrogen peroxide. Peroxisomes (per-OX-ih-soam, rhymes with homes) are spherical, membranous organelles, resembling lysosomes that contain enzymes concerned with oxidative degradation. Peroxisomes oxidize ethyl alcohol and fatty acids to carbon dioxide and water. They also convert amino acids and fatty acids into carbohydrates, from which mitochondria produce ATP. Unlike mitochondria, however, peroxisomes do not produce ATP. Instead, they release heat that contributes to the body's temperature.

Peroxisomes evidently bud from the ER, not from Golgi membranes. Most peroxisomal enzymes are synthesized in the cytosol and imported into peroxisomes. Most cells contain peroxisomes, and they are especially abundant (about 500 per cell) in hepatocytes.

- Which organelle of the endomembrane system acts as a cytoplasmic traffic cop?
- Which organelle contains cisternae?
- Which organelle degrades cytoplasmic materials for metabolic recycling?
- Which organelle produces heat?

CYTOSKELETON

Cell shape and movement depend on the cytoskeleton, an internal network of protein fibers shown in Figure 2.15. Three types of cytoskeletal filaments allow organelles to move within the cytoplasm or cells, to move about, to change shape, or to attach to other cells. The filaments are classified by diameter and composition. The largest are **microtubules,** 24 nm in diameter; **intermediate filaments** (11 nm) are next in size, and **microfilaments,** at 7 nm, are the narrowest. Microtubules are assembled from **tubulin** protein molecules, intermediate filaments from several varieties of proteins, and microfilaments from **actin.**

MICROTUBULES

Microtubules form a cytoplasmic network, shown in Figure 2.16, that attaches to the cell membrane (see Junctional Complexes), mitochondria, and other organelles. Pigment granules in pigment cells ride throughout the cytoplasm or cluster around the nucleus on this microtubule network. Neurons disperse transport vesicles along microtubules (Figure 2.16B) to distant parts of the cell. Microtubules form the mitotic spindle and the axonemes of cilia and flagella (see Cilia and Flagella).

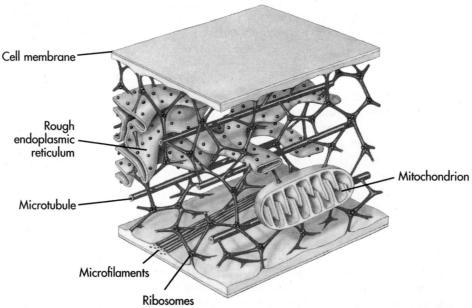

Cell membrane
Rough endoplasmic reticulum
Microtubule
Mitochondrion
Microfilaments
Ribosomes

✱Figure 2.15

The fiber network of the cytoskeleton attaches to organelles and the cell membrane.

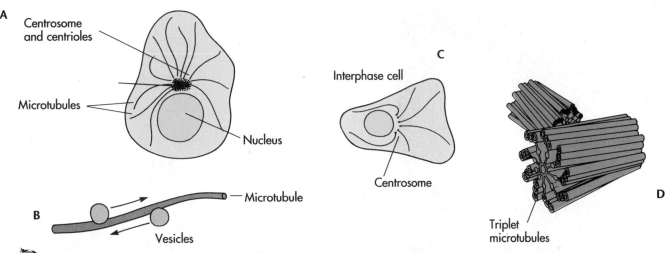

A
Centrosome and centrioles
Microtubules
Nucleus

B
Microtubule
Vesicles

C
Interphase cell
Centrosome

D
Triplet microtubules

Figure 2.16

A, Microtubules extend throughout the cytoplasm. **B,** Transport vesicles can move in either direction on a single microtubule. **C,** A cell's microtubules point outward from the centrosome and centriole. **D,** A pair of centrioles; one centriole is perpendicular to the other.

Centrioles and centrosomes organize the microtubules of the cytoskeleton and mitotic spindle. **Centrioles** are short cylinders composed of nine outer, triplet microtubules and a central, singlet microtubule, as you can see in Figure 2.16D. In most cells a pair of centrioles associates with a cluster of proteins, called the **centrosome,** near the nucleus. Centrioles duplicate in dividing cells. How duplication takes place is unknown, because there is no clear evidence that centrioles contain DNA for this purpose.

INTERMEDIATE FILAMENTS

These filaments form a strong internal network that supports many types of cells. Intermediate filaments in Figure 2.17A and B are not as well known as microtubules and microfilaments, because their protein constituents (IF proteins) are difficult to characterize. The best-known intermediate filaments are constructed of **cytoplasmic (soft) keratins,** protein molecules related to the hard keratins (KARE-a-tins) found in hair, fingernails, and toenails. Epidermal cells of the skin also contain a tough, filamentous keratin network (Figure 2.17A) that connects neighboring cells. The **terminal web** of acinar cells and many others is a mat of intermediate filaments that lies beneath the apex of the cell. Lamins of the nucleus and chromosomes also are intermediate filaments.

ACTIN MICROFILAMENTS

Actin microfilaments are contractile; they are found in nearly all cells (Figure 2.18A) and are responsible for changing the cells' shape and for a variety of motions. In dividing cells (Figure 2.18B) a band of microfilaments known as the **contractile ring** constricts the cytoplasm into separate cells. Bundles of actin microfilaments move the microvilli of intestinal cells and eggs (Figure 2.18C). The myofibrils of skeletal muscle fibers are modified cytoskeletal fibers. Other bundles, called **stress fibers** (Figure 2.18D), distribute forces from the plasma membrane into the cytoplasm when

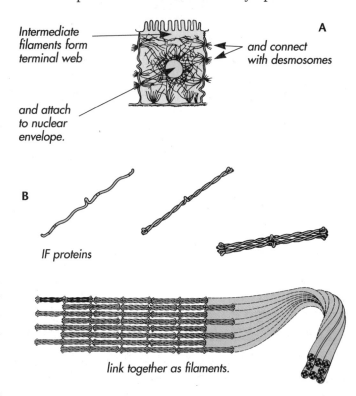

A
Intermediate filaments form terminal web
and connect with desmosomes
and attach to nuclear envelope.

B
IF proteins
link together as filaments.

Figure 2.17

A, Intermediate filaments of epidermal cells are linked together as networks. **B,** Intermediate filament, *IF* proteins link together as filaments.

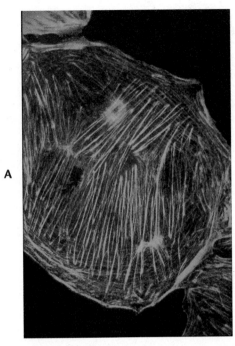

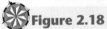

Figure 2.18

Actin microfilaments occur **(A)** throughout the cytoplasm. **(B)**, Contractile rings cleave dividing cells. **(C)**, Actin microfilaments participate in the terminal web and microvilli of intestinal cells and **(D)** form stress fibers.

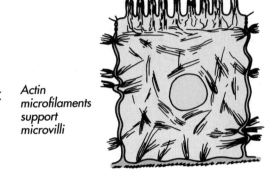

B   Contractile rings constrict dividing cells.

C   Actin microfilaments support microvilli

D   and form stress fibers in stretched cells.

cells stretch. Mats of actin microfilaments bind the cytoskeleton to intregral proteins of the cell membrane. Similar mats are involved in cell movement. When fibroblasts move over a surface they extend **lamellipodia** (footplates, la-MELL-ih-POE-dee-ah; lamelli-, *L,* a small plate; pod, *G,* a foot) that attach to the substrate and draw the cell forward as cytoplasm flows into the lamellipodia, while old attachments release their hold behind.

- Microtubules, intermediate filaments, and microfilaments are elements of the cytoskeleton. Which element has the largest diameter?
- Which is capable of contraction?
- Which group contains cytoplasmic keratin proteins?

JUNCTIONAL COMPLEXES

Junctional complexes join cells to other cells and to external structures (Figure 2.19). Three main types of complexes connect cells together, and two others attach cells to external structures. **Tight junctions** fuse the membranes of adjacent cells together in waterproof bands that prevent leakage between cells. These connections also act as boundaries between apical and basal domains of the plasma membrane (the portions of the membrane above and below the junctions, respectively) because the integral proteins cannot cross the tight junction. **Desmosomes** link cells together with patches of fibers that cross the narrow space between the plasma membranes and extend into the cytoskeleton on each side. In **belt desmosomes** the connecting patches become belts of fibers around entire cells. Cells communicate through **gap junctions** that allow ions and small molecules to pass freely between cells.

Hemidesmosomes and adhesive plaques are junctional complexes that join cells to external substrates. Each of these complexes anchors the coverings of organs to basement membranes and underlying tissue (see Chapter 3, Epithelial Tissues). As the name indicates, **hemidesmosomes** are half-desmosomes in which the fibers attach to the basement membrane instead of another cell. **Adhesive plaques** resemble hemidesmosomes, but no cytoplasmic plaque backs the undersurface of the plasma membrane. How the plaque adheres externally to the basement membrane is unknown.

- Which junctional complex separates the plasma membrane into apical and basal domains?

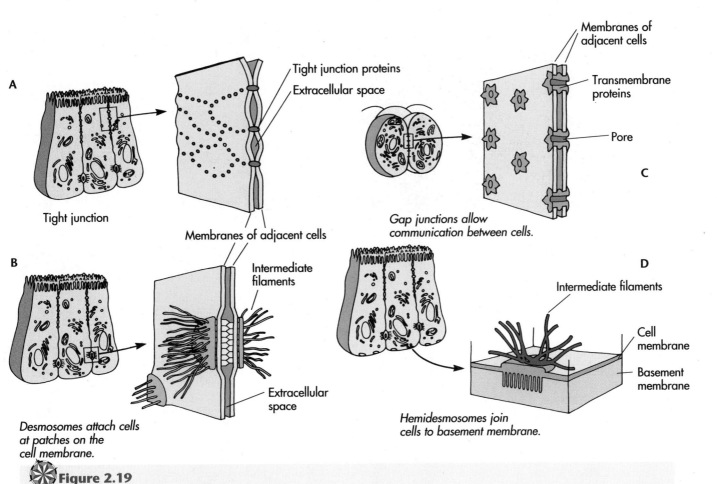

Figure 2.19

Junctional complexes attach cells to other cells and to the basement membrane. **A,** Tight junctions. **B,** Desmosomes. **C,** Gap junctions. **D,** Hemidesmosomes.

- Which complexes attach to basement membranes?
- What are gap junctions?

CILIA AND FLAGELLA

Long, slender, rhythmically beating flagella send sperm through a female's reproductive fluids toward the egg, and shorter cilia of the trachea (Figure 2.20A and B) drive mucus up the lung passages and out. A central **axoneme** (AKS-o-neem), shown in Figure 2.20C, is responsible for all ciliary and flagellar motion. All axonemes consist of the standard **"9 + 2" arrangement of microtubules,** nine outer, doublet microtubules linked to each other and to two central, singlet microtubules. **Radial spoke proteins** connect the central singlets to the outer tubules like spokes in a wagon wheel. Other proteins form the arms and links that join the outer doublets to each other around the rim. A plasma membrane surrounds the axoneme, and a **basal body** anchors the axoneme to the cytoskeleton. Basal bodies resemble centrioles and axonemes; they consist of nine outer, triplet microtubules connected to a central singlet.

Axonemes bend when doublets and singlets slide along themselves (Figure 2.20D). Cross-bridges between arms and doublets provide the motive forces that slide the microtubules past one another and make the axoneme bend. This mechanism resembles the mechanism of muscular contraction described in Chapter 9, Skeletal Muscle. Cilia and flagella develop from basal bodies, which duplicate by unknown means when cells divide.

Kartagener's syndrome (kar-TAJ-en-ers) illustrates the role of genes in the structure of axonemes. This relatively rare inherited disorder immobilizes cilia and flagella because the radial spoke proteins and arms are missing. As you might expect, patients suffer frequent respiratory infections, and males are infertile. Kartagener genes also affect anatomy on a larger scale by reversing the symmetry of abdominal organs, a condition called **situs inversus,** in about half the patients who suffer from the disease. How symmetry and axonemes are related is a mystery.

- What is the 9 + 2 arrangement of microtubules in the axoneme?
- What effects does Kartagener's syndrome have on the structure of axonemes?

A

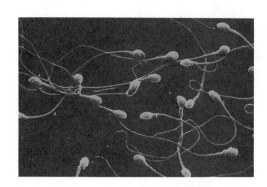

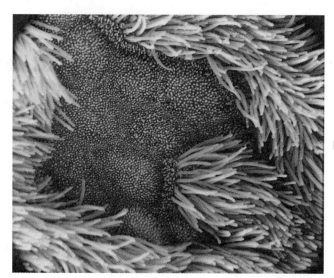

B

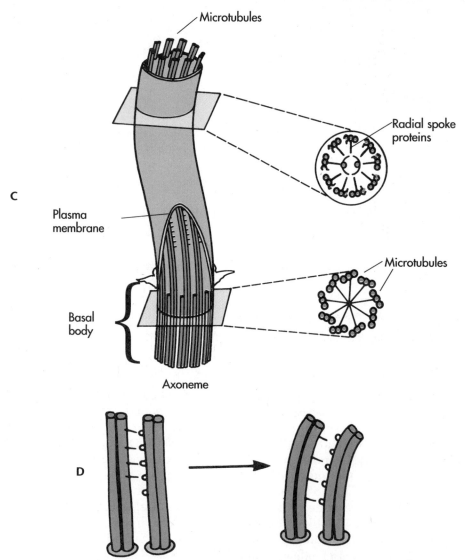

Microtubules

Radial spoke
proteins

C

Plasma
membrane

Microtubules

Basal
body

Axoneme

D

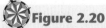 **Figure 2.20**

Cilia and flagella. **A,** A sperm's flagellum. **B,** Ciliated cells move mucus in the lining of the trachea. (Scanning electron-micrograph by permission of R.G. Kessel and R.H. Kardon, *Tissues and organs: A text-atlas of scanning electron microscopy,* 1979, W.H. Freeman and Co.) **C,** The axoneme supports cilia and flagella with microtubules. **D,** Microtubules slide along each other and bend the axoneme.

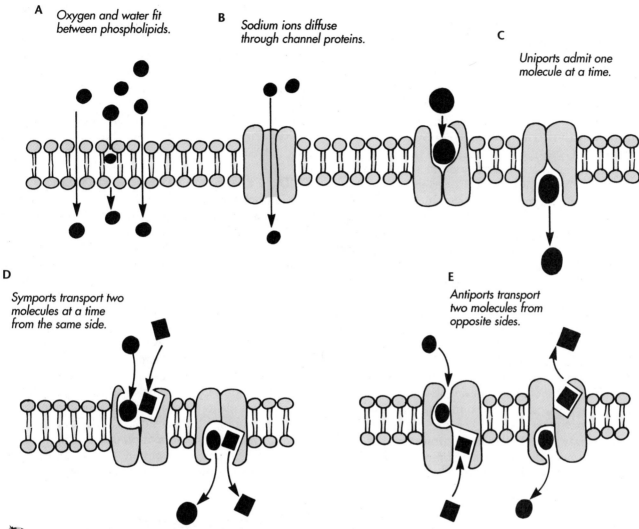

A, Oxygen and water fit between phospholipids.

B, Sodium ions diffuse through channel proteins.

C, Uniports admit one molecule at a time.

D, Symports transport two molecules at a time from the same side.

E, Antiports transport two molecules from opposite sides.

Figure 2.21

A, Oxygen, water, and carbon dioxide molecules can diffuse across the membrane because they fit between the phospholipids. **B,** Sodium ions diffuse through pores in channel proteins. **C,** A uniport carrier protein transfers molecules across the membrane one at a time. **D,** Symports are carrier proteins that transfer two different molecules at a time from the same side. **E,** Antiports transfer two different molecules from opposite sides of the membrane.

ACTIVITIES OF THE CELL MEMBRANE

The plasma membrane governs what enters and leaves cells, and it also relays external stimuli to the cytoplasm.

MEMBRANE TRANSPORT

The plasma membrane is quite selective about what molecules and particles will pass through it. For example, neurons allow sodium ions to rush into the cytoplasm and acinar cells export digestive enzymes. Precisely which substances enter and leave a given cell depends on the structure of the phospholipid bilayer and integral proteins of the membrane. Small molecules diffuse between the phospholipids, and larger molecules are transported by integral proteins, while still larger particles are engulfed by the membrane.

DIFFUSION

Small molecules such as water, carbon dioxide, and oxygen readily diffuse across the membrane because they fit between the phospholipids. Random thermal energy causes these molecules to move from regions of high concentration on one side of the membrane (Figure 2.21A) toward lower concentrations on the other side. Molecules that are too large or too polar (differently charged ends) do not pass. Overall the membrane is selectively permeable or **semipermeable.**

TRANSPORT PROTEINS

Transport proteins are integral proteins that conduct ions and larger molecules across the cell

membrane. These proteins enable molecules that would otherwise be blocked by the phospholipid bilayer to cross. **Channel proteins** are integral proteins that admit substances through a channel, or pore, in the protein molecule (Figure 2.21B). The size and shape of the channel determine what enters or leaves. Thus sodium ion channel proteins preferentially admit sodium ions but not calcium ions. Glucose enters cells by way of **carrier proteins** (Figure 2.21C, D, and E) that bind specifically to glucose molecules and transport them to the cytoplasmic face of the membrane for release inside the cell. Carrier proteins have no channels; they merely transfer molecules across the membrane. Carrier proteins that transport one molecule at a time are called **uniports** (uni-, *L,* one; Figure 2.21C). **Symports** (sym-, *G,* together) transfer two different molecules in the same direction (Figure 2.21D), and **antiports** (anti-, *G,* against) move them in opposite directions (Figure 2.21E). All of these mechanisms are **passive transport** processes that expend no cellular energy in moving the molecules across the membrane.

Most cell membranes continually pump sodium ions out of and potassium ions into the cytoplasm, so that sodium ions accumulate outside and potassium ions inside the cell. Such pumping processes are known as **active transport,** because these processes require energy from ATP to concentrate ions and molecules on one side of the membrane or the other. **Sodium-potassium pumps** are integral proteins that

are sprinkled universally among the membranes of human cells (Figure 2.22A and B). These pumps, like many others in the cell, contain an enzyme that liberates energy from ATP to transport three sodium ions out of the cell for every two potassium ions that enter. Sodium-potassium pumps enable neurons to conduct impulses (see Chapter 17, Neurons).

- Which of the processes listed are passive and which require energy—symport, antiport, diffusion, sodium-potassium pump, sodium ion channels?

ENDOCYTOSIS AND EXOCYTOSIS

Membranes take up dissolved proteins and larger particles by endocytosis, and exocytosis releases particles from the cell, as you can see in Figure 2.23A. Hepatocytes take up cholesterol molecules from the blood by engulfing them in **endocytotic vesicles,** and acinar cells secrete zymogen granules by exocytosis. Membrane receptors are integral proteins that allow **receptor-mediated endocytosis** to select what enters the cell (Figure 2.23B). In this process, cell membranes continually form **coated pits** that take into the cytoplasm molecules such as cholesterol that have bound to receptors on the external face of the pit. The coating on the cytoplasmic face enables the pit to pinch off of the membrane and become an endocytotic vesicle. **Clathrin,** one of the best-known coat proteins, promotes endocytosis with a basket-like network around the pit. Endocytotic vesicles fuse with the **endosome,** a large, central sac

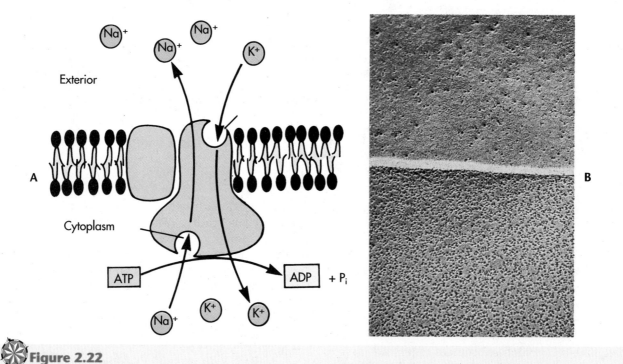

🏵 Figure 2.22

A, Active transport by the sodium-potassium pump exchanges three sodium ions for two potassium ions. **B,** The plasma membrane contains many transport proteins, as this scanning electronmicrograph shows.

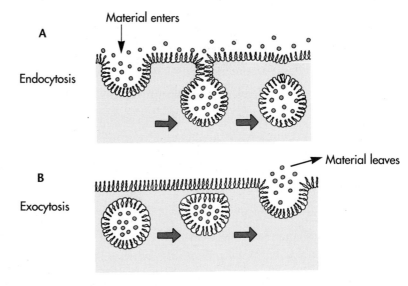

A

Material enters

Endocytosis

B

Material leaves

Exocytosis

C

Granule   Plasma membrane

Figure 2.23

A, Receptor mediated endocytosis takes up extracel-lular materials into vesicles that pinch off from the membrane and enter the cytoplasm. **B,** In exocytosis cytoplasmic vesicles fuse with the membrane and re-lease their contents to the outside. **C,** Endomem-brane system recycles transport vesicles back to the plasma membrane. Careful examination of acinar cells and hepatocytes in Figures 2.2C and 2.3B will disclose such vesicles, although you will not be able to tell whether the contents are entering or leaving.

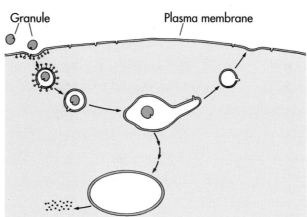

that distributes contents to the lysosomes and recy-cles membranes back to the plasma membrane.

Exocytosis is essentially the reverse of endocyto-sis. As a zymogen granule approaches the membrane of an acinar cell in Figures 2.23A and 2.2C, the mem-brane of the zymogen granule fuses with the plasma membrane and opens the granule's contents to the ex-terior of the acinar cell, where they flow to the intes-tine. The phospholipids and proteins of the zymogen granule membrane become part of the plasma mem-brane and can recycle through the endomembrane system into new zymogen granules.

SIGNAL TRANSDUCTION

Cells respond to signals received by the plasma membrane. Neurons receive stimuli from other neu-rons, and ovaries and testes respond to hormones that promote egg and sperm development. Whether sig-nals originate locally or long distance, cells respond to stimuli when receptors on their plasma membranes signal the cytoplasm that specific molecules have landed. This process is called **signal transduction** (trans-, *L,* across; -duct, *L,* lead) because integral mem-brane proteins transfer the signal across the mem-brane into the cytoplasm.

Membrane receptors detect signals among the molecular traffic flowing past. These receptors are integral proteins that bind to certain external mole-cules but not to others, and then they transfer their re-sponse through the membrane to the cytoplasmic face. For example, neuronal receptors bind to acetyl-choline but not to gonadotropic hormones, and recep-tors on testis cells preferentially bind gonadotropin. These receptors detect molecules the cell has tuned in to, somewhat as satellite antennas receive radio and TV signals. Most signal transduction is accomplished by two pathways—the cyclic AMP and calcium ion pathways—shown in Figures 2.24 and 2.25.

CYCLIC AMP SIGNALING

Epinephrine (adrenalin) stimulates muscle cells, through the cyclic AMP signal transduction pathway, to break down stored glycogen into glucose, which fuels muscle contraction. As shown in Figure 2.24, epinephrine binds to receptors (1) in the cell mem-brane that cause the level of **cyclic AMP** (cAMP; de-rived from ATP, the energy-storing molecule) to rise (2). This increase in turn causes **protein kinases** to activate enzymes (3) that break down glycogen into glucose (4). Protein kinases are enzymes that can ac-

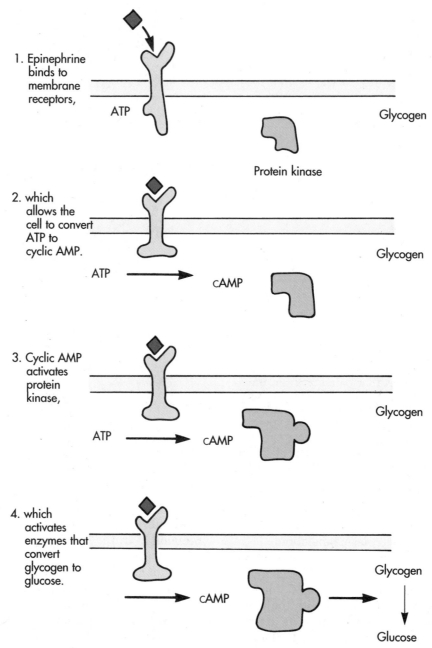

1. Epinephrine binds to membrane receptors,

ATP

Glycogen

Protein kinase

2. which allows the cell to convert ATP to cyclic AMP.

ATP → cAMP

Glycogen

3. Cyclic AMP activates protein kinase,

ATP → cAMP

Glycogen

4. which activates enzymes that convert glycogen to glucose.

cAMP →

Glycogen

Glucose

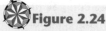 **Figure 2.24**

The cyclic AMP signal transduction pathway.

tivate many different kinds of enzymes by attaching phosphate molecules to them. This pathway continues to break down glycogen as long as epinephrine is present.

CALCIUM ION SIGNALING

Skeletal muscle fibers activate contraction through brief pulses of calcium (Ca++) ions, and neurons stimulate other neurons through similar pulses (see Contraction, Chapter 9, and Synapse, Chapter 17). The **Ca++ signaling pathway** (Figure 2.25) employs membrane receptors (1) that admit small quantities of calcium ions (2) into the cytosol from cisternae of the en-

doplasmic reticulum in muscle fibers. There these ions activate protein kinases (3) that bring about contraction. Not all calcium is released from internal reservoirs such as the endoplasmic reticulum. Neurons admit calcium ions through the plasma membrane from outside. Calcium ion signaling is based on the fact that the level of free calcium ions inside the cytosol is 10,000-fold lower than elsewhere. Calcium-ion pumps in the endoplasmic reticulum and plasma membrane continually evacuate Ca++ ions from the cytosol. Thus a brief pulse of calcium ions can serve as an intracellular signal until the calcium pumps remove the ions, usually within less than a second.

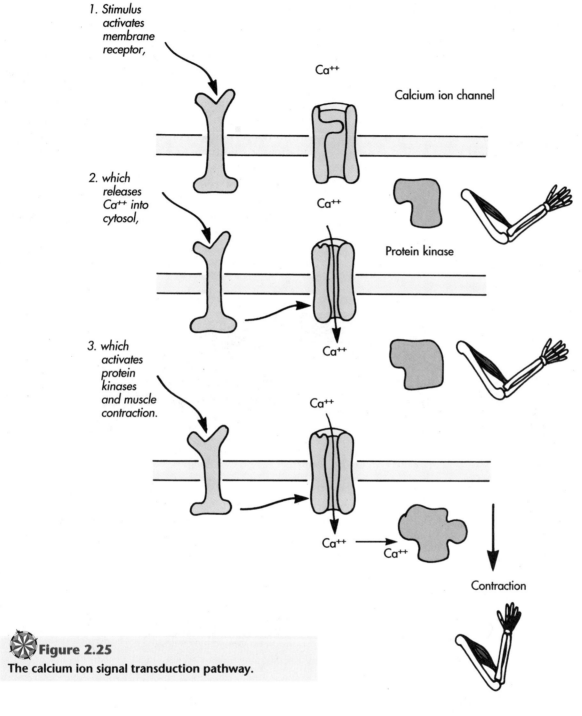

1. *Stimulus activates membrane receptor,*

Ca⁺⁺

Calcium ion channel

2. *which releases Ca⁺⁺ into cytosol,*

Ca⁺⁺

Protein kinase

3. *which activates protein kinases and muscle contraction.*

Ca⁺⁺

Ca⁺⁺

Ca⁺⁺ → Ca⁺⁺

Contraction

✳ Figure 2.25
The calcium ion signal transduction pathway.

- What membrane structures allow endocytosis to take up specific molecules into the cytoplasm?
- How do high cAMP levels affect protein kinase?
- How do calcium ions affect the calcium signaling pathway?

CELL CYCLE

Although your body may not change visibly from day to day, its cells are continually being replaced by new cells. Approximately 1 trillion cells, 1% of the total in your body, die every day, but most are replaced by new cells. The skin, the blood, and the linings of the gut and other tissues continually renew themselves. At this rate, you can expect to produce and discard about 250 times your body weight of cells over your lifetime. Relatively few of your cells will last as long as you do. You can count on many nerve cells, skeletal muscle fibers, and the core of your lenses to survive into crusty old age, but almost all other cells will be replaced many times. Maintaining this overall constancy while changing individual cells is one of the fundamental roles of the cell cycle.

The cell cycle is the name given to the life cycle of cells. A cell passes through two major periods in each generation of the cycle, shown in Figure 2.26. It spends most of its time and energy during **interphase,** growing in size by transcribing genes and synthesizing macromolecules. When a cell has duplicated its DNA and chromosomes and has doubled in size, the processes of mitosis and cytokinesis quickly divide the cell into two smaller, equivalent daughter cells. **Mitosis** delivers identical sets of chromosomes to each of the new cells, and **cytokinesis** cleaves the cytoplasm in two. The new daughter cells resume interphase and continue another cycle of growth and division.

CLONES

You are a **clone** (*G,* a branch) of cells that descended from a fertilized egg. Beginning with a single cell, one round of the cell cycle produced two cells, as you can see in Figure 2.27A. The next round doubled this number to four, and a third generation yielded eight cells while you were still a very young embryo. A group of cells that descended from a common progenitor cell is called a clone. All cells in the clone are genetically identical to each other and to their progenitor (a fertilized egg in your case) because mitosis delivered them identical sets of genes. Provided all cells continue to grow and divide, the number of cells will increase exponentially according to the number

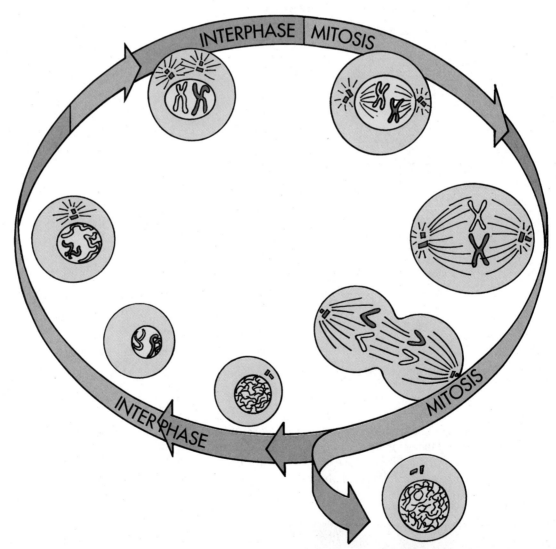

**Figure 2.26**
Cell cycle. Cells grow larger and duplicate their chromosomes during interphase. Mitosis separates the duplicated chromosomes, and cytokinesis cleaves the cytoplasm into two equivalent cells that then resume interphase.

A

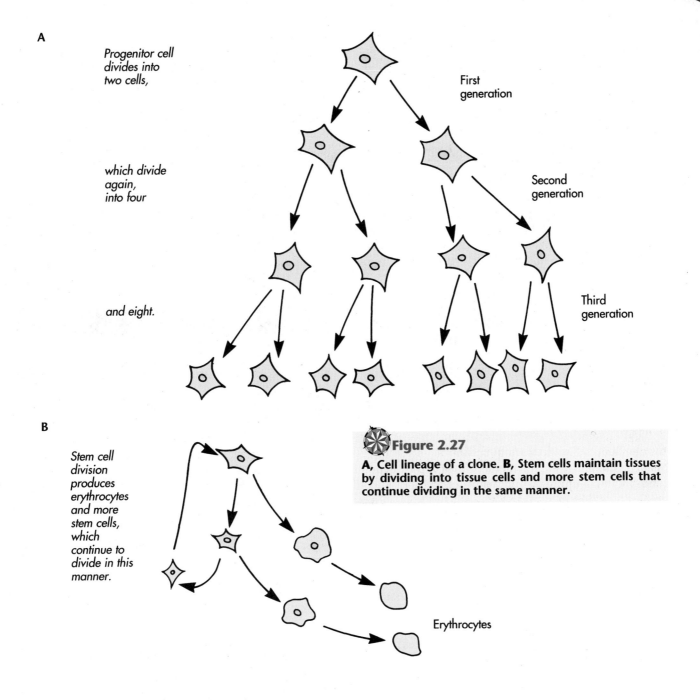

Progenitor cell divides into two cells,

First generation

which divide again, into four

Second generation

and eight.

Third generation

B

Stem cell division produces erythrocytes and more stem cells, which continue to divide in this manner.

✷ Figure 2.27

A, Cell lineage of a clone. **B,** Stem cells maintain tissues by dividing into tissue cells and more stem cells that continue dividing in the same manner.

Erythrocytes

of generations that have passed since the original cell existed.

STEM CELLS

Proliferating stem cells are responsible for replacing worn-out cells. Stem cells are progenitors from which clones of cells (such as erythrocytes) and more stem cells develop. For example, erythrocytes develop from stem cells in red bone marrow (Figure 2.27B). In the simplest case, each stem cell produces another stem cell and a cell that becomes an erythrocyte. This assures a continuous supply of erythrocytes because the stem cells continue producing erythrocytes and more stem cells. Other stem cells replenish

worn-out skin and the lining of the gut. Organs develop from populations of stem cells.

• Name the phases of the cell cycle.
• Why are the cells of a clone identical?
• How do stem cells differ from nonstem cells?

MITOSIS

Mitosis delivers identical sets of chromosomes to the nuclei of daughter cells, as shown in Figure 2.28. Mitosis (mito-, *G,* a thread; -sis, *G,* the act of) separates each chromosome in the nucleus into two identical copies called sister chromatids, so that each daughter nucleus receives one sister chromatid from each chromosome (46 chromosomes in all). For convenience, mitosis is divided into four phases that blend imperceptibly with each other into a continuous process.

PROPHASE

Mitosis begins with **prophase** (pro-, *G*, before) when the chromosomes condense and sister chromatids can be seen in each chromosome (Figure 2.28 A and B), joined together by a **centromere** (SEN-tromeer). The nuclear envelope begins to disperse, and the nucleolus begins to break down, signaling a temporary stop to macromolecule synthesis. Centrioles have duplicated during interphase and moved to opposite ends of the nucleus; they now organize the **mitotic spindle.** Microtubules assemble the spindle around the centrosome and centrioles, and the chromosomes begin attaching to the spindle fibers through connections called **kinetochores** (kin-NET-o-kores).

METAPHASE

The chromosomes jostle to the **metaphase plate** (met-, *G*, between, change) located halfway between the opposite ends of the mitotic spindle (Figure 2.28C). The spindle now contains two types of microtubule fibers. Two **kinetochore fibers,** one for each sister chromatid, extend toward the opposite poles (ends) of the spindle from each chromosome, but these fibers do not completely reach the poles. **Polar spindle fibers** extend the rest of the distance, linking the kinetochore fibers to the poles of the spindle.

ANAPHASE

The sister chromatids separate (Figure 2.28D) and move toward the poles of the spindle, thereby sending identical sets of 46 single chromatids, now called chromosomes, to opposite ends of the spindle. **Anaphase** (ana-, *G*, again) ends when the chromosomes reach the poles.

TELOPHASE

Telophase is the end of mitosis (telo-, *G*, end), the converse of prophase. In **telophase** (Figure 2.28E) the chromosomes cluster at the poles of the spindle, and the nuclear envelope reconstitutes from the endoplasmic reticulum and lamins of every chromosome. The spindle microtubules disassemble, and **cytokinesis** divides the cell into two daughter cells as the chromosomes begin to decondense into interphase chromatin threads. The nucleoli reappear and the daughter cells resume transcription and growth as they enter interphase.

- In which phase of the cell cycle does a cell duplicate its chromosomes? Separate sister chromatids? Reestablish the nucleus? Condense its chromosomes? Transcribe genes?

MEIOSIS

Eggs and sperm carry out a special version of mitosis, known as meiosis, that reduces the number of chromosomes by half, from 23 pairs (**diploid number**) to 23 single chromosomes (**haploid number**), in two maturation division cycles illustrated in Figure 2.29. The first meiosis (mei-, *G*, less) halves the number of chromosomes by separating pairs of chromosomes rather than sister chromatids. The second division more closely resembles mitosis because sister chromatids do separate from each other. The overall process delivers four cells whose nuclei each contain 23 single chromosomes. Follow the discussion in Figure 2.29.

MEIOSIS I

The first maturation division begins when chromosomes condense during prophase I. Each of these 46 chromosomes contains two sister chromatids duplicated in the previous interphase. As the chromosomes jostle about the spindle, **homologous pairs,**

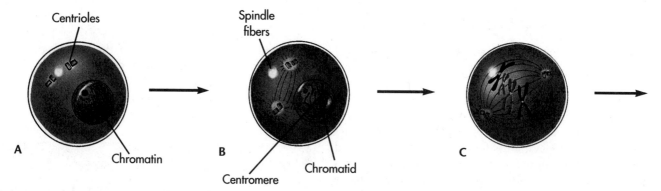

A, Chromatin · Centrioles · **B,** Spindle fibers · Centromere · Chromatid · **C,**

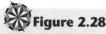
Figure 2.28

Mitosis produces daughter cells that contain identical sets of chromosomes. A, Early prophase. **B,** Late prophase. **C,** Metaphase.

pairs of chromosomes that contain comparable genes and have the same shape and length, match up—homolog to homolog. This process of homologous pairing is called **synapsis,** and the structure containing the four chromatids of each homologous pair is called a **tetrad.** A **synaptonemal complex** intimately binds the homologs together—gene to gene. Chromatids frequently lie across each other during synapsis, and these points of crossover are called **chiasmata** (kye-AS-mah-tah; chias-, *G,* cross). At chiasmata chromosomes can break and exchange pieces of homologous chromatids. The process of **crossing over** recombines sections from one homolog with those of the other, thus adding variety to the genetic makeup of new individuals.

Prophase I comes to an end when the nuclear envelope disperses. **Metaphase I** brings the tetrads to the metaphase plate, and **anaphase I** separates all 23 homologous pairs of chromosomes, sending one homolog from each pair to each pole of the spindle. **Telophase I** encloses the homologs that have reached the ends of the spindle in a nuclear envelope. Each nucleus is now haploid, and each chromosome contains two sister chromatids.

MEIOSIS II

Meiosis II separates the sister chromatids from each other and delivers four haploid nuclei, each with 23 single chromosomes from the original prophase I nucleus. **Meiosis II** begins without further chromosome duplication, although the chromosomes may decondense during a brief interphase. **Prophase II** is brief. The chromosomes condense and attach to the spindle, and the nuclear envelope breaks down. Metaphase II, anaphase II, and telophase II closely resemble mitosis, except that the nuclei are already haploid. **Metaphase II** arranges the chromosomes upon the metaphase plate; **anaphase II** separates sister chromatids from each other and sends them to op-

posite poles of the spindle. **Telophase II** reconstitutes the nuclear envelope around each set of chromosomes.

- Which phase of meiosis reduces the number of chromosomes to haploid?
- In which phase does crossing over occur?
- When do tetrads form?
- Which meiotic division most closely resembles mitosis?

CANCER CELLS

Cancer cells illuminate the structure and function of normal cells. Cancer is the second major cause of death in the United States, claiming 506,000 lives in 1990 out of a total 2,162,000 deaths. In that year, only heart disease killed more people, 725,000. Cancer of the lungs, colon, and breast are the top three killers. Normal cells become cancer cells by **transformation** into continually dividing cells that **metastasize** (meh-TAS-ta-size; change position; meta-, *G,* change; -stasis, *G,* standing). Metastatic cells shed connections with their home tissue and enter the vascular system, where they invade other organs and reduce the resistance of the patient, who eventually dies (see Chapter 3, Cancerous Tissues). Colon cancer illustrates the steps that lead to cancer in many tissues (Figure 2.30). Presumably, transformation begins in the wall of the colon when a single cell proliferates more rapidly than normal and forms a small clone of rapidly growing cells (step 1). This stage of transformation is very difficult to detect among the millions of colon cells because the clone is so small. As the rapidly dividing cells accumulate, they form tumors that may disturb colon metabolism or be detected by medical imaging. Tumors are excessive tissue growths, usually with abundant blood vessels, that can occur in virtually all organs (but only rarely ap-

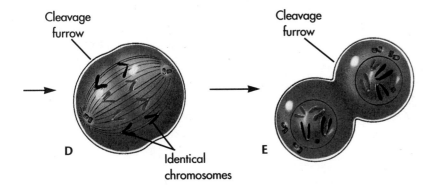

Figure 2.28, cont'd

D, Anaphase. E, Telophase and cytokinesis.

FIRST DIVISION (MEIOSIS I)

SECOND DIVISION (MEIOSIS II)

Early prophase I
The duplicated chromosomes become visible (shown separated for emphasis, they actually are so close together that they appear as a single strand)

Chromosome

Nucleus

Centrioles

Chromatids

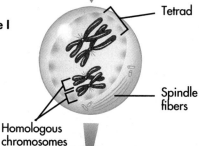

Prophase II
Each chromosome consists of two chromatids

Middle prophase I
Homologous chromosomes synapse to form tetrads

Tetrad

Spindle fibers

Homologous chromosomes

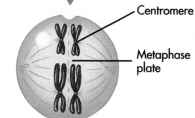

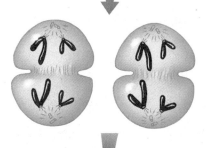

Metaphase II
Chromosomes align at the metaphase plate

Metaphase I
Tetrads align at the metaphase plate

Centromere

Metaphase plate

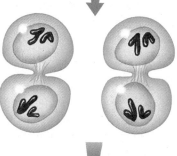

Anaphase II
Chromatids separate and each is now called a chromosome

Anaphase I
Homologous Chromosomes moves apart to opposite sides of the cell

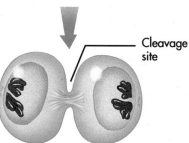

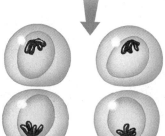

Telophase II
New nuclei form around the chromosomes

Telophase I
New nuclei form, and the cell divides; during interphase (not shown) there is no duplication of chromosomes

Cleavage site

Haploid cells
The chromosomes are about to unravel and become less distinct chromatin

Figure 2.29

Meiosis halves the number of chromosomes by separating homologous pairs of chromosomes.

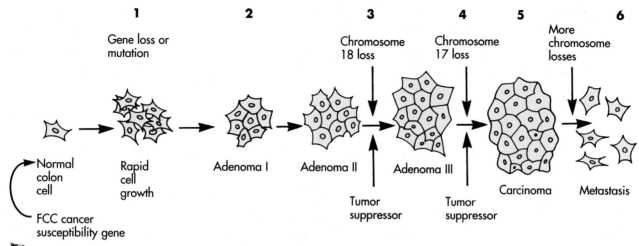

Figure 2.30

Steps in transformation of colon cancer cells (described in text).

pear in skeletal muscle). They may appear as polyps on the lining of the colon, as fibrous growths in the lungs, or as lumps in the breast. Only after the tumors have progressed further (steps 2 through 5) do metastatic cells begin to invade other tissues (step 6). Many tumors of the colon and other organs remain benign (not metastatic) when transformation stops short of metastasis.

Cancer cells have large nuclei and relatively little cytoplasm. Some retain the specializations and organelles of their precursor cells. **Fibrosarcomas** are tumors containing malignant fibro-blasts that produce excessive collagen, and **melanomas** are malignant pigment cells recognized by dark pigment nevuses (patches) in the skin and internal organs. Benign tumors are usually named with the suffix -oma (*G,* tumor) to recognize their origins; a **fibroma,** for example, is derived from fibroblasts, and **adenomas** arise from glandular tissues. In contrast, malignancies may be called **sarcomas** or **carcinomas.** A fibroma that becomes malignant is a fibrosarcoma.

Cancer is a genetic disease of the cell cycle and signal transduction pathways. Presumably, transformation begins when one or more genes that participate in these regulatory pathways is lost or mutates. Mutation is spontaneous genetic damage caused by errors in DNA duplication or environmental factors such as radiation, viral infection, sunlight (skin cancer, melanoma), or tobacco (cancers of lung and larynx). At least four different varieties of genes are known to be involved in transformation, and two of them are concerned with the cell cycle and signal transduction. **Tumor suppressor genes** are *normal* genes that are known to suppress tumor growth and transformation, but in cancer, mutations relax this restraint and cell division proceeds rapidly. A widely studied tumor suppressor known as p53 regulates transcription in the nucleus of colon and other cancer cells. **Oncogenes** are *mutant* genes that also relax cell

cycle controls. One of these genes, ras, governs links between the cytoskeleton and cell membrane in its normal form, but mutation disrupts these connections in cancer cells. Figure 2.30 illustrates several tumor suppressors and oncogenes in colon cancer. About 100 oncogenes and at least a half-dozen tumor suppressors currently are known from all forms of cancer. Recently a third variety involved in metastasis has been identified.

All of these genes come into play after the initial mutations have occurred. The recent discovery of a new colon **cancer susceptibility gene,** however, promises to identify individuals who are particularly susceptible to colon cancer, before transformation begins. This mutant gene known as FCC (familial colon cancer) appears to encourage mutations that lead to transformation. Located on chromosome 2 and found in about 15% of colon cancer cases, this gene apparently interferes with replication of DNA in colon cancer cells. Although relatively few cancer patients have this gene, those who have FCC face an 80% to 90% risk of colon cancer. Another recently discovered mutant gene that fails to correct errors in DNA replication is associated with at least 70% of colon cancers. A breast cancer susceptibility gene, called BRCA1 (breast cancer 1) and found on chromosome 17, carries a comparable risk to those who inherit it. These genes offer researchers the chance to identify cancer risks before the disease is underway, and ultimately to prevent its progress. The genetics of cancer cells promises treatments for cancer, as well as greater understanding of the internal workings of all cells.

- What changes do mutations and metastasis produce in cancer cells?
- Why is it difficult to detect transformation in its early stages?
- Name four types of genes concerned with transformation.
- Why is cancer considered a disease of the cell cycle?

CELLS AND AGING

Aging is the general deterioration that gradually affects the structures and functions of our bodies as time passes. A woman born in the United States in 1976 can expect to live 77 years, and a man has a 69-year life expectancy. Children born in 1995 are estimated to add 3 more years to each of these predicted life spans. This increase is unlikely to run indefinitely; the maximum documented human life span is considered to be 115 years. Death can occur in older people because of degenerative diseases such as Alzheimer's (which affects the nervous system) and atherosclerosis (cardiovascular), but these diseases are the result of the aging process, not its cause.

Almost every organ system ages. Skin becomes wrinkled and loses its elasticity, and hair becomes gray. Bones begin to lose mass and break more readily than the resilient bones of younger people. Muscle mass declines, and skeletal muscle fibers atrophy. The cardiovascular system itself is remarkably free of anatomical degeneration, but cardiac output slows, and atherosclerosis can thicken arterial walls and block circulation. Lungs lose their elasticity and expose less surface where air oxygenates the blood. The digestive system changes very little, but its peristaltic movements and secretions tend to slow down. Kidney functions slow. The nervous system loses cells, especially after 60 years of age; this loss may help explain diminished mental activity. The immune system loses some of its ability to detect foreign substances. Menopause is the major aging event in the reproductive system. How aging affects individual organ systems will be described in later chapters.

The aging process has five fundamental features. CUPID, an acronym that reminds us more of our origins than our fates, can help you remember them. First, aging changes are **cumulative;** they accumulate gradually over time, beginning between 20 and 30 years of age. Second, these changes are **universal;** they affect all individuals. One does not age by heart disease and another person by emphysema. Aging affects everyone in more or less the same ways. Third, the changes are **progressive;** they do not reverse themselves. Exercise delays muscle and skeletal atrophy, but it does not reverse them in the long run. Skin creams, vitamins, or hormones do not prevent the inevitable. Fourth, we presume aging to be **intrinsic** to cells and tissues rather than imposed from the outside by disease or chronic environmental sources such as tobacco or air pollution. It is difficult to be entirely certain, however, that aging changes have no environmental origins. Finally, aging is **deleterious** (harmful) and ultimately leads to death.

It is not surprising that aging has no single explanation, with so many different systems and effects. Many theories of aging are **stochastic,** that is, they depend on random events like the throw of dice to explain the accumulation of deleterious changes. Stochastic theories view cell structure and metabolism as fundamentally unstable, always needing repair and replacement mechanisms to keep going. According to these theories, you age because repair mechanisms themselves tend to deteriorate. Such failure has been ascribed to cross-links that accumulate among protein molecules or to somatic mutations and other forms of DNA damage. The most powerful stochastic hypothesis is the free radical theory, in which oxygen-free radicals, highly reactive molecules that bind to other molecules with one free electron rather than the usual two, damage the organelles and cytoskeleton by attacking enzymes that enable these structures to function.

Other theories of aging regard cells and organs as **programmed,** that is, they have some way of counting metabolic events. When enough have occurred, cells, tissues, and the entire body die. The "rate of living" theory, an early version of these explanations, proposed that individuals had only limited metabolism available and died when the supply ran out. Faster use brought early death, and slower use prolonged life, according to this view.

But what particular aspect of metabolism is counted? Some investigators propose that mitochondrial damage is responsible, but it seems most likely that programmed changes will be found in the genes that regulate cell division and development. As evidence of this view, cells taken from various tissues and cultured in the laboratory live for a particular number of cell cycles before they stop dividing and then die. Cells from younger tissues proliferate longer than those from older tissues. Fibroblasts from adults, for example, proliferate for about 20 cell cycle generations, while those from fetuses continue dividing for about 48 generations. Certain varieties of cancer cells are essentially immortal. Some types have divided in culture for more than 30 years.

Not all cells continue to divide, however. Skeletal muscle fibers and neurons of the nervous system are famous for their inability to divide, but they persist into old age. In contrast, erythrocytes are especially good examples of rapidly aging cells. Erythrocytes develop from rapidly dividing cells in the bone marrow, but they lose their nucleus and die in about 120 days without dividing again.

Thus far we have assumed that the process of aging occurs within cells, but a group of **intercellular** hypotheses emphasizes that interactions between cells are instrumental in aging. Cessation of the menstrual cycle at menopause is an example of interaction hypotheses involving interactions between the nervous and endocrine systems.

- Name organ systems that age.
- Name five properties of the aging process.
- What is the main difference between stochastic and programmatic theories of aging?

SUMMARY

Figure 2.31 summarizes the major relationships between the cells and organs that are the basis of anatomy. Cell division produces clones of stem cells derived from the fertilized egg. Tissues and organs develop from these clones. Organs and tissues hold certain stem cells in reserve to replace dead cells. Mutations occasionally transform cells into cancer cells, and aging cells accumulate other changes that lead to death. Organelles enable cells to function. Hepatocytes and acinar cells are equipped with standard organelles to assemble proteins, but the pancreas has adapted these organelles to secrete zymogen granules into the gut, and the liver uses them to take up materials from the blood and to secrete others into it.

Cells perform four fundamental activities. Metabolic reactions governed ultimately by genes provide cellular energy and structure. Cells reproduce through growth and cell division. The surrounding environment influences cell activities. Cells are self-regulating. *(Page 27.)*

Organ systems employ specialized cells that have adapted the standard activities to particular functions. Skeletal muscle fibers use their energy and structure to contract. Pancreatic acinar cells devote their genes and metabolism to synthesizing digestive enzymes. Liver hepatocytes synthesize and degrade a great variety of molecules from the blood. Fibroblasts specialize in the synthesis of collagen fibers. *(Page 27.)*

Cell organelles perform standard functions. In specialized cells the functions are modified from the basic plan. A plasma membrane surrounds the entire cell. The nucleus directs metabolism with information from genes and chromosomes. Ribosomes synthesize proteins, and mitochondria provide ATP for this and other cellular processes. The endomembrane system provides surfaces for synthesis and regulation. The endoplasmic reticulum and Golgi complex assemble and package cellular products, and numerous transport vesicles distribute these materials throughout the cell. Lysosomes and peroxisomes degrade materials for metabolic recycling. *(Page 28.)*

The fibers of the cytoskeleton enable cells to

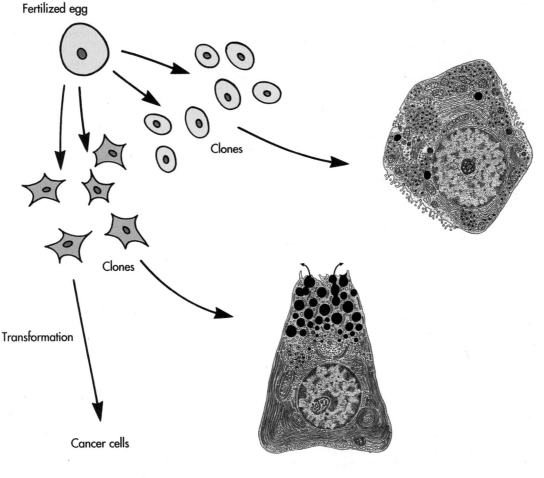

**Figure 2.31**

Cell summary. The various cells and organs derive from a single progenitor cell.

move about, change shape, and distribute organelles and stretching forces within cells. Microtubules form a cytoplasmic network that adheres to desmosomes, moves transport vesicles, and delivers chromosomes to dividing cells. Intermediate filaments form internal fiber networks that toughen skin or assemble the nuclear envelope. Actin microfilaments are contractile fibers that cleave dividing cells, move microvilli, and enable cells to move. (Page 40.)

Junctional complexes connect cells through tight junctions, desmosomes, and gap junctions. Hemidesmosomes and adhesive plaques attach cells to extracellular structures. (Page 42.)

Cilia and flagella move motile cells through fluids, and they also move fluids past stationary cells. Their basic structure is a standard 9 + 2 axoneme of microtubules. (Page 43.)

Molecules and particles that enter and leave the cell must cross the plasma membrane. The plasma membrane is selective, and this selectivity derives from the phospholipid bilayer and transport proteins. Water and other small molecules diffuse through the bilayer. Ions and large molecules cross the bilayer with the help of transport proteins that bind molecules selectively. Ions diffuse through channel proteins, and carrier proteins transport molecules passively or actively. Endocytosis engulfs particles into vesicles that enter the endomembrane system. Exocytosis releases the contents of cytoplasmic vesicles to the outside. (Page 45.)

Signal transduction pathways activate cellular responses to external signals. Two common signaling pathways are based on membrane proteins that use cyclic AMP or calcium ions to relay stimuli to protein kinases in the cytoplasm. (Page 47.)

The cell cycle enables cells to proliferate. Cells grow larger and duplicate their chromosomes during interphase; they divide by mitosis and cytokinesis into genetically identical daughter cells. This process continually replaces worn-out cells and heals wounds, and it also is the source of stem cells and clones that develop into organs. (Page 49.)

Mitosis separates identical sets of chromosomes from each other. After the chromosomes have duplicated in interphase and each one has formed two identical sister chromatids, mitosis separates the chromatids of each chromosome and delivers them to opposite poles of the mitotic spindle, where nuclei reform and cytokinesis divides the cytoplasm. Chromosomes attach to the spindle in prophase, then move to the metaphase plate. Anaphase separates and delivers sister chromatids to the poles of the spindle, where telophase restores the nuclei and both daughter cells resume interphase. (Page 51.)

Meiosis is a special form of mitosis that halves the number of chromosomes in reproductive cells from diploid to haploid. Necessary for fertilization, this reduction is accomplished by separating homolo-gous pairs of chromosomes on the mitotic spindle during meiosis I. Homologous chromosomes synapse and cross over during a prolonged prophase I. Anaphase I separates the pairs, and telophase I yields two haploid nuclei. Meiosis II separates the sister chromatids of the chromosomes and yields four haploid nuclei from the original nucleus of prophase I. (Page 52.)

Cancer is a disease of the cell cycle and signal transduction pathways. Mutations in certain genes transform normal cells into cancer cells that metastasize and invade tissues and organs. Oncogenes and tumor suppressor genes control steps in the cell cycle and signal transduction. Other genes are involved in metastasis and susceptibility to mutations. (Page 53.)

Aging is the gradual deterioration of structure and function in the body's organ systems. The process is cumulative, universal, progressive, intrinsic, and deleterious—CUPID. Aging mechanisms may be stochastic or programmed, and they may operate within cells or among cells. (Page 56.)

STUDY QUESTIONS   OBJECTIVES

1. Describe two examples of the role of genes and metabolism in cell functions. Describe how cell division leads to growth of organs. Describe an example of external stimuli that influence cells.   **1**

2. Describe the structure and functions of skeletal muscle fibers, liver hepatocytes, pancreatic acinar cells, and fibroblasts.   **2**

3. Make a table that lists cell organelles by name and function.   **3**

4. Describe the structures of the phospholipid bilayer and the location of integral proteins in the fluid mosaic cell membrane. In what ways is the membrane fluid? Mosaic?   **4**

5. Many organelles belong to the endomembrane system. Describe the relationships among these organelles.   **5**

6. What are the major components of the cytoskeleton, and what services do they provide to cells? Describe examples.   **6**

7. Draw diagrams of tight junctions, desmosomes, and gap junctions, and describe what these organelles do.   **7**

8. Draw a diagram of an axoneme as found in cilia, and label the components. Describe how the axoneme bends.   **8**

9. Inspect the specialized cells in Figures 2.1C, 2.2C, 2.3C, and 2.4B for nuclei, mitochondria, endoplasmic reticulum, ribosomes, Golgi complex, lysosomes, peroxisomes, and cytoskeleton. Which organelles occur in more than one cell?

A

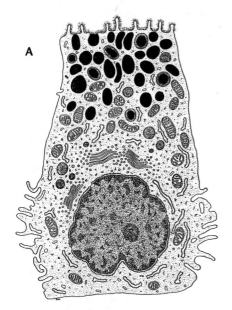

B

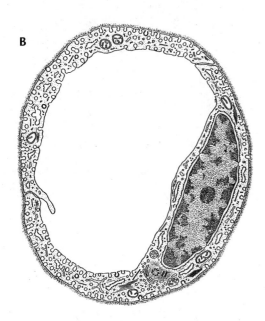

C

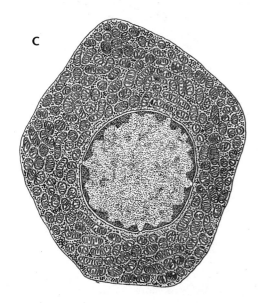

STUDY QUESTIONS continued OBJECTIVES

Are any organelles missing from these cells? Which organelles are most abundant in each cell? Are any organelles found in only one kind of cell? What do you conclude about the role of organelles in cell specialization?   **9**

10. Identify the organelles in the cells of Figure 2.32. What do your propose is the function of mucous cells, capillary endotheliall cells, and oxyphil cells?   **9**

11. List the membrane structures involved in membrane transport, endocytosis, and exocytosis. Describe what molecules or materials each structure transports.   **10**

12. Describe the steps by which skeletal muscles respond to epinephrine and to Calcium ions.   **11**

13. What contributions does the cell cycle make to growth, development, and maintenance of organs? Give examples.   **12**

14. Beginning with interphase, take a cell through the steps of mitosis and return the daughter cells to interphase.   **12**

15. Beginning with 23 pairs of duplicated chromosomes, name and describe the steps of mitosis that deliver 23 pairs of single chromosomes to each daughter cell.   **12**

16. Describe the steps in meiosis that halve the number of chromosomes.   **12**

17. What are the differences between mitosis and meiosis?   **13**

18. How do cancer and normal cells differ? What are the effects of oncogenes, tumor suppressor genes, and cancer susceptibility genes?   **14**

19. Describe examples of the effects of aging in different organ systems.   **15**

20. Describe examples of the stochastic and programmed hypotheses of aging.   **15**

 Figure 2.32

What are these cells' primary functions? A, A mucous cell from the lining of the stomach. **B,** An endothelial cell that forms the walls of the capillaries. **C,** An oxyphil (oxygen-loving) cell from the parathyroid glands of the neck. Although these cells contain similar organelles, there are some important differences that suggest each cell's function. Do you find nucleus, ribosomes, mitochondria, endoplasmic reticulum, Golgi complex, lysosomes, and peroxisomes in each cell? Are any organelles missing? From which cells? Are some organelles larger or more abundant?

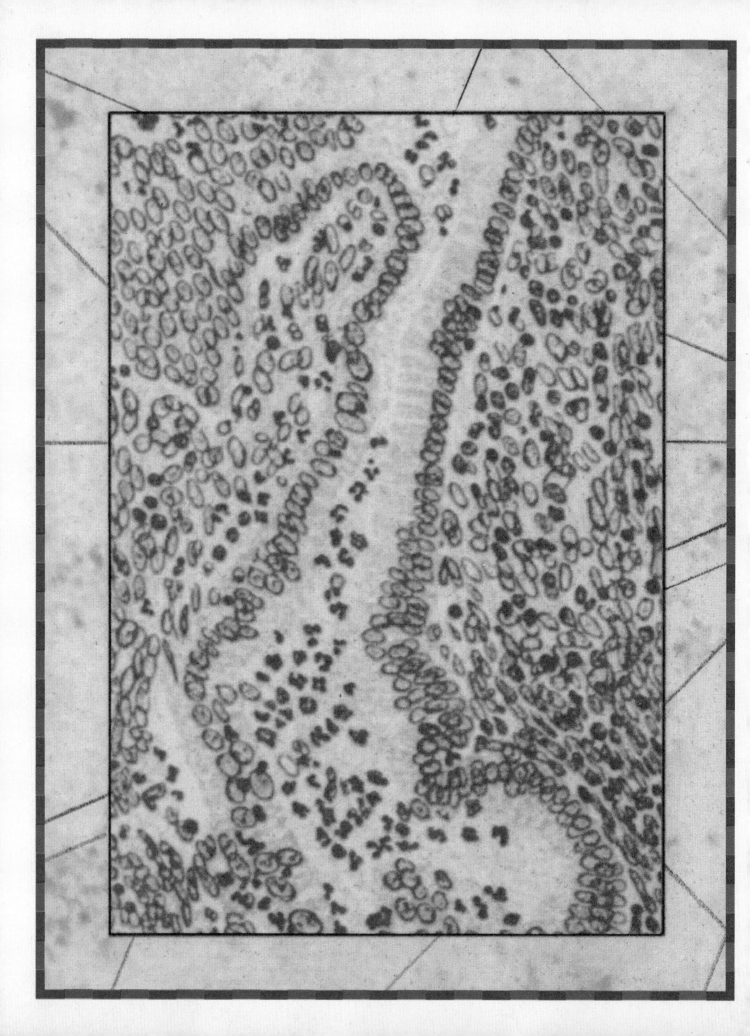

Tissues

OBJECTIVES

After studying this chapter, you should be able to do the following:

1. Define the word *tissue* and describe four primary classes of tissues.
2. Describe the anatomic relationships among the four major tissues of the intestine.
3. Identify the major types of cells found in nervous tissues and describe their functions.
4. Name and describe the three major classes of muscle cells.
5. List the seven properties of epithelial cells and give examples of each.
6. List and describe nine different forms of epithelia.
7. Classify exocrine glands according to the anatomy of their ducts and secretory portions, and by their products and secretion processes.
8. Describe the three major kinds of connective tissues.
9. Describe the structures and functions of the individual cellular and molecular components of connective tissue.
10. Describe the eight different types of connective tissues and give examples of each.
11. Cite examples of the effects of diseases on the anatomy and functions of tissues.

VOCABULARY TOOLBOX

Prefixes indicate the position of a tissue:

endo-	thelium, *G* inner, within. Endothelium is the inner lining of blood vessels.	**meso-**	thelium, *G* middle. Mesothelium (mez-zo-THEEL-ee-um) lines the peritoneal cavity.
epi-	thelium, -dermis, *G* upon, over. Epidermis (EP-ih-DER-mis) is the epithelium (EP-ih-THEEL-ee-um) that covers the skin.		

Various terms describe shapes:

acinus	*L* a berry. Acinus (AS-ih-nus) is another term for alveolus.	**cuboidal**	*G* shaped like a cube. All sides of cuboidal (kue-BOYD-al) cells have the same shape.
alveolus	*L* a cavity or pit. An alveolus (al-VEE-ol-us) is a cup-shaped cluster of cells that secrete fluids from glands.	**reticular**	*L* net, network. Reticular (reh-TIK-yu-lar) fibers form supporting networks.
amorphous	morph-, *L* form. Amorphous (ah-MORE-fuss) materials have no form of their own. Like water, they take the shape of their surroundings.	**squamous**	*L* a scale. Squamous (SKWAY-mus) cells are flat, like fishes' scales.
		stellate	*L* a star. Stellate (STELL-ate) cells have many radiating projections.
columnar	*L* a pillar. Columnar (kol-UM-nar) cells are tall, narrow cells that stand on one end.	**tubular**	*L* a pipe. Tubular (TOOB-yu-lar) glands are hollow rods.

PERSPECTIVE

Tissues are groups of similar cells that perform particular functions. Because most organs contain four universal tissues—nervous, muscle, epithelial, and connective—that together enable organs to function, tissues are keys to understanding anatomy. Histologists (those who study the anatomy of tissues; histo-, *G,* a web or tissue) and pathologists (those who study diseases; patho-, *G,* disease) think of organs and diseases in terms of these four tissues. Name an organ, and these people can tell you what tissues it contains and what role each tissue serves. Name a disease, and they will tell you the tissues it attacks. If organs had no common types of tissues, rather only unrelated cells, the task of learning anatomy would be considerably harder. Anatomy will make more sense to you when you understand the fundamental tissues and their functions.

Organs are built from four different tissues—nervous, muscular, epithelial, and connective—that are more or less standard structures with standard functions in all organs. In **nervous tissues,** long neurons that are connected to each other form pathways that carry impulses between organs and the brain, which coordinates the action of all organs and makes the body function as a whole. **Muscular tissue** contains many contractile muscle fibers that, together, move the skeleton, pump blood, propel food through the intestines, and squeeze secretions from glands. **Epithelial tissues** cover the inner and outer surfaces of organs and generally control what enters and leaves each organ; they absorb nutrients in the gut and secrete sweat from the skin, for example. **Connective tissues** join the other tissues together structurally and functionally. Connective tissue shapes the stomach and skeletal muscles with fibers and gels that adhere to other tissues, and connective tissue provides space for blood vessels to reach these tissues. Table 3.1 briefly outlines all four tissues and their functions.

Table 3.1   TISSUES OF THE BODY

Tissue	Variety	Anatomy	Function	Location
Nervous	Neurons	Cell body extends long, slender, cellular processes	Conducts and transmits impulses	Brain, spinal cord, spinal and cranial nerves
	Neuroglia	Specialized cellular processes	Supports neurons metabolically and physically	Brain, spinal cord, spinal and cranial nerves
Muscle	Skeletal muscle	Long, cylindrical, striated cells	Moves skeleton, voluntary	Skeleton
	Cardiac muscle	Short, branched, striated cells	Pumps blood, involuntary	Atria and ventricles of heart
	Smooth muscle	Spindle-shaped, nonstriated cells	Regulates blood pressure, peristalsis, glandular secretion, involuntary	Arterial walls, gut, exocrine glands, iris, uterus
Epithelia	Simple	Single cell layer on basement membrane	Secretes, absorbs, exchanges	Capillaries, gastrointestinal lining, lung alveoli, glands
	Stratified	Many cell layers on basement membrane	Protects	Epidermis, oral mucosa, esophagus, vagina
	Pseudostratified	Single cell layer resembles several layers	Secretes, senses	Nasal/tracheal mucosa, vas deferens
	Transitional	Several layers of squamous, cuboidal and columnar cells	Absorbs, protects	Bladder
	Germinal	Several layers of specialized cells	Produces gametes	Testis, ovary
Connective	Connective tissue proper	Soft or tough, dense matrix	Cushions, connects muscle and bone, stores body fluids, provides access for immune system, stores fat	Skin, fascia, tendons, ligaments, mesenteries
	Specialized connective tissue	Bone, cartilage, blood	Supports, moves, shapes, provides vascular communication	Skeleton, trachea, external ears, cardiovascular system
	Mesenchyme	Soft embryonic connective tissue	Supports, shapes, provides vascular communication, source of adult connective tissue	Head, trunk, limbs, mesenteries, gut wall of embryo and fetus

TISSUES AND ORGANS

The intestine illustrates how tissues work together within an organ. Figure 3.1 shows a transverse section through the small intestine (similar to what you saw in Figure 1.15D, Body Cavities and Membranes). An epithelium lines the lumen, the internal cavity, of the intestine. Thrown into many folds that increase its absorptive surface, this epithelium takes up nutrients from food. This epithelium also forms glands that secrete digestive enzymes and a protective film of mucus. Another epithelium, the visceral peritoneum, covers the exterior of the intestine with flat cells that lubricate this surface with watery fluid. Beneath this outer epithelium are two layers of smooth muscle tissue (called muscularis) that propel food through the gut. Between the muscularis and the inner epithelial lining lies an extensive layer of connective tissue (actually composed of several layers, described in Chapter 23, Digestive System). This soft connective tissue binds the inner and outer epithelia to the muscularis, and distributes small vessels and capillaries to all tissues. Nerves regulate secretion and motion by spreading into the connective tissue and the muscularis.

Learning tissue anatomy will help you understand how things can go wrong and appreciate what medical treatment can do to restore their functions. Many diseases attack specific tissues. For example, **multiple sclerosis** interferes with conduction in the nervous system by damaging the sheaths around neurons. **Duchenne muscular dystrophy** impairs movement by wasting skeletal muscle tissue, but this disease does not affect cardiac muscle or the smooth muscle of the digestive tract (see Chapter 9, Skeletal Muscle). **Cystic fibrosis** is an epithelial disease that causes the lungs and the digestive tract to secrete especially sticky mucus, and sweat glands to release very salty perspiration. Various connective tissue diseases, including **Marfan's syndrome,** interfere with

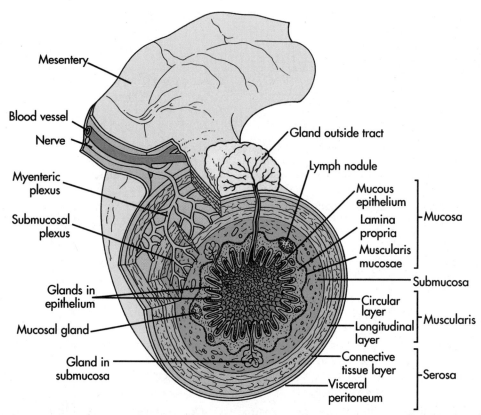

Labels in figure:
Mesentery
Blood vessel
Nerve
Myenteric plexus
Submucosal plexus
Glands in epithelium
Mucosal gland
Gland in submucosa
Gland outside tract
Lymph nodule
Mucous epithelium
Lamina propria — Mucosa
Muscularis mucosae
Submucosa
Circular layer
Longitudinal layer — Muscularis
Connective tissue layer
Visceral peritoneum — Serosa

✸ Figure 3.1

This figure illustrates a transverse section through the intestine. Epithelium (blue) covers the exterior of the intestine and lines the interior. Epithelium also extends outward and inward to form glands. The muscularis (red) lies beneath the external epithelium. Connective tissue (yellow) joins the muscularis with the inner and outer epithelium. Nervous tissue (green) enters the intestine and spreads through the muscle and connective tissue.

production of collagen fibers; skin and joints therefore become hyperextensible (stretch and extend more than normal) and wounds heal slowly without their usual supply of collagen.

- What are the names of the four major tissues of the body?
- Where are these tissues located in the intestine?
- Name a disease associated with each of these tissues.

NERVOUS TISSUE

Nervous tissue forms pathways that conduct information throughout the nervous system. The **ulnar nerve** illustrates the nature of nervous tissue, considered fully in Chapter 17, Neurons and Neuroglia. Figure 3.2A shows how this "funny bone" nerve passes from the shoulder, behind the elbow, and into the forearm where it controls movement and communicates touch and motion from the forearm and hand. Like most other nerves, the ulnar nerve contains thousands of long axons bundled together

and covered with protective sheaths, shown in Figure 3.2B. Each axon is part of an individual **neuron** (NOOR-on; Figure 3.2C) that conducts either afferent (a-, *L*, toward) impulses from sensory receptors in the skin or efferent (ef-, *L*, out, from, away) impulses from the spinal cord to muscles, somewhat like the opposite lanes of traffic on an expressway.

Neuroglia (Noor-OH-glee-ah) **is the nervous system's own connective tissue.** Whereas neurons conduct impulses, glial cells support the neurons. Figures 3.2B and C illustrate some of these interactions. **Neuroglial cells** may wrap individual neurons with myelin sheaths that speed conduction, or they may cover the entire nerve with a protective fibrous sheath. Other types of neuroglia deliver nutrients from capillaries to neurons and help to regulate transmission of impulses that pass from one neuron to the next. Nervous tissue derives from the embryonic ectoderm (see Chapter 4, Ectoderm).

- Name the two fundamental types of nervous tissues.
- Which type forms myelin sheaths?
- Which type conducts impulses?

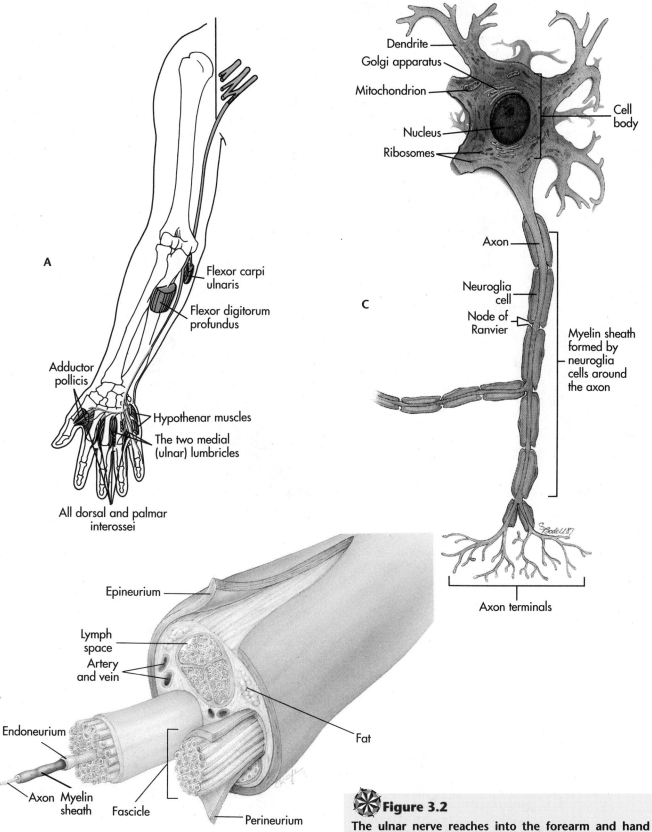

A

Flexor carpi
ulnaris

Flexor digitorum
profundus

Adductor
pollicis

Hypothenar muscles

The two medial
(ulnar) lumbricles

All dorsal and palmar
interossei

Dendrite
Golgi apparatus
Mitochondrion
Cell
body
Nucleus
Ribosomes

C

Axon

Neuroglia
cell

Node of
Ranvier

Myelin sheath
formed by
neuroglia
cells around
the axon

Axon terminals

Epineurium

Lymph
space

Artery
and vein

B

Endoneurium

Axon Myelin
sheath

Fascicle

Perineurium

Fat

Figure 3.2

The ulnar nerve reaches into the forearm and hand
(A) with thousands of axons bundled in protective
sheaths **(B)**. Individual axons **(C)** conduct afferent impulses from the skin to the spinal cord or efferent impulses from the spinal cord to muscles of the forearm
and hand. Neuroglia cells wrap axons **(C)**.

MUSCLE TISSUE

Three types of muscle tissue move various parts of the body. Skeletal muscle moves the skeleton; this is the muscle that enables us to walk and to make all the other voluntary motions living requires. Life also requires the automatic motions of cardiac and smooth muscle. In the heart, **cardiac muscle** pumps blood; **smooth muscle** squeezes glands, accommodates the pupils of our eyes to light, controls blood pressure in arteries, erects hairs, and propels food through the di- gestive tract. These muscle tissues develop from the embryonic mesoderm described in Chapter 4. Details about each type of muscle tissue are given in Chapter 9, Chapter 13, and Chapter 23.

SKELETAL MUSCLE

Skeletal muscle tissues account for approximately 40% of the body's weight. The contractile cells are called **skeletal muscle fibers.** A skeletal muscle such as the biceps muscle of the arm contains thousands of them. Skeletal muscle fibers are long, narrow, cylin-

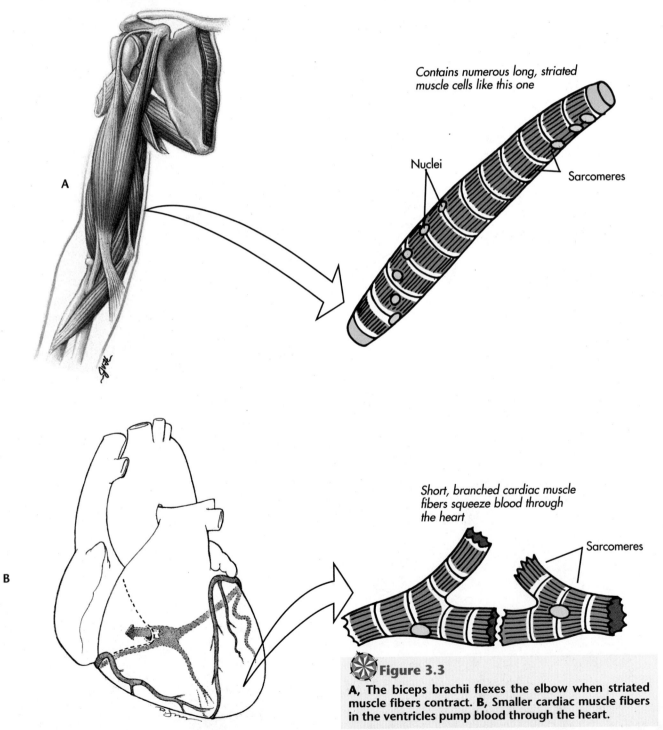

Contains numerous long, striated muscle cells like this one

Nuclei

Sarcomeres

A

B

Short, branched cardiac muscle fibers squeeze blood through the heart

Sarcomeres

Figure 3.3

A, The biceps brachii flexes the elbow when striated muscle fibers contract. **B,** Smaller cardiac muscle fibers in the ventricles pump blood through the heart.

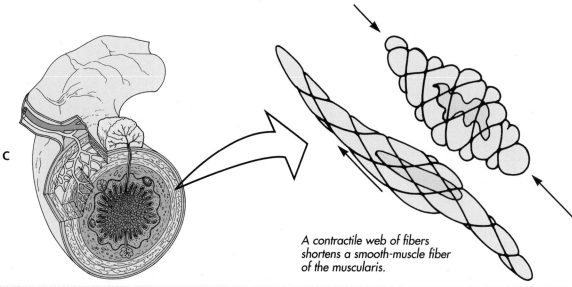

C

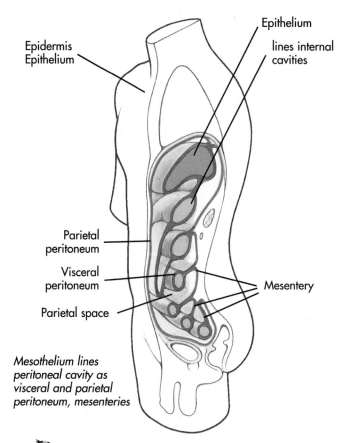

A contractile web of fibers
shortens a smooth-muscle fiber
of the muscularis.

Figure 3.3, cont'd
C, Smooth muscle fibers of the intestinal muscularis enable it to contract.

drical cells, shown in Figure 3.3A, that contain several hundred nuclei and display the characteristic cross **striations** of these fibers. As Chapter 9 explains, the striations are due to contractile structures called **myofibrils** that cause the muscle fibers to shorten and muscles to contract.

CARDIAC MUSCLE

Cardiac muscle fibers populate the walls of the heart. Cross striations and myofibrils also characterize these fibers, as you can see in Figure 3.3B, but the cells are much shorter than skeletal muscle fibers, they branch, their ends are intercalated disks, and they usually contain only one nucleus. Cardiac muscle fibers are controlled by the pacemaker of the heart without voluntary regulation.

SMOOTH MUSCLE

Smooth muscle fibers (Figure 3.3C) have no cross striations, hence their name. Smooth muscle is found in the walls of vessels, the gut, the uterus, the dermis of the skin, and the walls of glands. Smooth muscles are very different from cardiac and skeletal muscles, as Figure 3.3C shows. A web of contractile fibers instead of myofibrils enables smooth muscle cells to shorten. Smooth muscles contract involuntarily. They are capable of long, sustained contraction, as any woman who has delivered a baby can tell you.
- Name the three types of muscle tissue.
- Which one contracts voluntarily?

EPITHELIAL TISSUE

Epithelial tissues are sheets of cells that cover and line organ surfaces. The different terms anatomists use for epithelial tissues reflect the location of the tissue. **Epithelium** is the name given to epithelial tissue that covers the skin and lines internal cavities that open to the outside, such as the lumen of the gut (Figure 3.4) and the respiratory and reproductive sys-

Epithelium
lines internal
cavities

Epidermis
Epithelium

Parietal
peritoneum

Visceral
peritoneum

Parietal space

Mesentery

Mesothelium lines
peritoneal cavity as
visceral and parietal
peritoneum, mesenteries

Figure 3.4
Epithelium covers the skin and lines the visceral organs; mesothelium lines the peritoneal cavity and covers visceral organs.

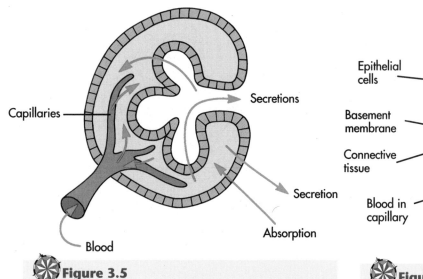

Figure 3.5

Substances cross an epithelium to enter or leave an organ, whether by secretion, absorption, or by way of capillaries.

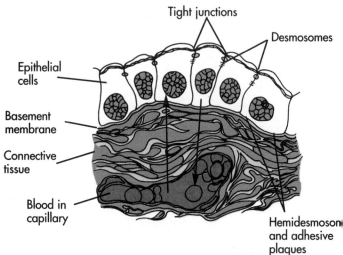

Figure 3.6

Basement membranes attach epithelia to connective tissue.

tems. The term **endothelium** is reserved for the internal lining of the heart and blood vessels. The peritoneum, pleura, and pericardium are considered **mesothelia** in recognition of their mesodermal origin (described in Chapter 4, Mesoderm). Other epithelia originate from embryonic ectoderm or endoderm (see Chapter 4).

Virtually every molecule that enters and leaves an organ must cross an epithelium (Figure 3.5) that covers its internal and external surfaces. Ultimately, all substances that enter and leave the cardiovascular system must also cross the endothelium that lines these vessels. Epithelium does not cover bones, muscles, and nerves. Instead, connective tissue binds these structures to their surroundings. Even so, molecules must cross the capillary endothelium to reach muscle and bone cells.

PROPERTIES OF EPITHELIA

Epithelial cells share several basic characteristics shown in Figures 3.6 through 3.8.

1. **Epithelial cells adhere to each other and to an underlying basement membrane.** The **basement membrane** attaches the epithelium (Figure 3.6) to the connective tissue beneath and adapts it to the shape of the organ. Tight junctions connect adjacent cells just below their external surfaces, and desmosomes and gap junctions add their connections below the tight junctions. Hemidesmosomes and adhesive plaques attach the base of the cells to the basement membrane.

2. **Epithelial cells are polar; that is, their ends differ.** The **apical, or luminal, ends** of the cells face

the exterior of the organ or the lumen, and the **basal end** attaches to the basement membrane (Figure 3.7). Tight junctions separate the cell membrane into different **apical and basal domains** that contain different integral proteins involved with transport, adhesion, and the cytoskeleton. **Microvilli** or **cilia** may decorate the apical surface, and the basal surface may extend folds that interdigitate with adjacent cells. All these structures increase the membrane surface area available for transport. **Exocytosis and endocytosis** may occur on both surfaces. In **transcytosis,** streams of transport vesicles take material in by endocytosis from one surface and release it by exocytosis at the other.

3. **Epithelia are avascular; they do not have their own blood vessels.** A further aspect of polarity is that nutrition must be transported in, and wastes transported out, through the basement membrane from capillaries in the connective tissue below, as you can see in Figure 3.6. Only rarely do capillaries cross the basement membrane into an epithelium.

4. **The cytoskeleton of epithelial cells contains cytokeratin intermediate filaments,** microtubules, and actin microfilaments, as seen in Figure 3.7. **Microtubules** extend from centrosomes and centrioles to junctional complexes in the cell membrane, and **actin microfilaments** form a **terminal web** of filaments beneath the apical surface. **Stress fibers** are located throughout the cell at points of tension, especially at desmosomes and tight junctions. To this network, **cytokeratins** add a web of

EXTERIOR

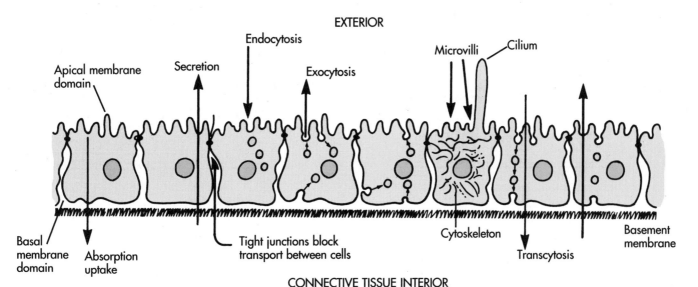

Figure 3.7

Different ends of epithelial cells face the exterior and the basement membrane. Tight junctions form boundaries between apical and basal domains of the cell membrane. Apical domains may have microvilli and cilia, and basal domains attach to the basement membrane. Materials may cross the epithelial cells in either direction, but tight junctions block transport between cells. Both domains carry out endocytosis and exocytosis. Transcytosis transports the contents of vesicles between the apical and basal ends of the cells.

fibers that links the junctional complexes with the nucleus and the apical surface of the cell. This network distributes bending and stretching forces that act on the tissue through the entire sheet of cells and the basement membrane as a whole.

5. **An epithelium may be a single layer of cells or more than one layer. Simple epithelia** contain only one layer of cells; all cells are anchored directly to the basement membrane, as Figure 3.8A illustrates. **Stratified epithelia** may have two or more cell layers, but only the innermost layer attaches to the basement membrane.

6. **Epithelia may be moist or dry and scaly.** Epithelia lining the mouth, nasal passages, and digestive tract are known as **mucosa** or **mucous membranes** because they secrete mucus that prevents them from drying out. Figure 3.8B shows how this viscous substance covers the cells with a moist, lubricating film. Mesothelia of the peritoneum are **serous membranes** that secrete thinner, watery fluids, also for lubrication. Epidermis of the skin, shown in Figure 3.8B, is dry by comparison. Thick layers of waterproof cytokeratin accumulate in this **cornified epithelia** (corn-, *L*, horny), and the dry, scaly cells continually flake away.

A

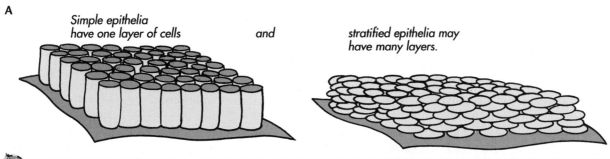

Figure 3.8

A, Epithelia may have one or many layers of cells.

Continued.

B

A mucous film may cover the surface   or   dry, scaly cells may peel
from the surface.

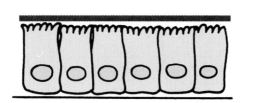

C

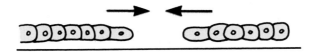

Injury removes cells.

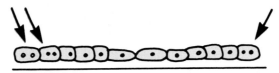

Cells spread from edges to replace
lost cells.

Cell division replaces lost cells.

Figure 3.8, cont'd
B, Although most epithelia secrete a film of fluid, some are dry. **C,** Cell spreading and cell division restore damaged epithelium.

7. **Epithelium replaces discarded cells.** Wound healing and peeling of sunburned skin are familiar examples of replacement. It takes about 4 weeks for epidermis to renew itself and even less time for the intestinal epithelium to be renewed. Next month's epidermis and gut will not contain the same cells as today's. Some cells of simple epithelia retain the ability to divide; injury and cell death stimulate them to replace the lost cells. Figure 3.8C shows cell replacement in the lining of the stomach. Stratified epithelia rely on stem cells in the basal layer for replacement because the upper layers do not divide.
- On which surfaces are epithelia, mesothelia, and endothelia located?
- What features of epithelial cells make them polar?

TYPES OF EPITHELIA

Anatomists classify epithelia by cell shape and the number of cell layers. They recognize the major types outlined below and shown in Figure 3.9.

A. Simple epithelium—single layer of cells.
  1. Squamous—thin, flat cells.
  2. Cuboidal—cube-shaped or pyramidal.
  3. Columnar—tall, narrow cells.
B. Stratified epithelium—two or more cell layers.
  1. Squamous—numerous layers of flat cells.
  2. Cuboidal—two layers of cuboidal cells.
  3. Columnar—two layers of columnar cells.
C. Pseudostratified epithelium—really a single layer of cells, but nuclei give a multilayered appearance.
D. Transitional epithelium—several layers of cuboidal and squamous cells.
E. Germinal epithelium—contains stem cells that produce sperm in the testis.

In general, thinner epithelia (simple squamous) can exchange materials by diffusion, and thicker epithelia usually secrete or absorb materials (simple and stratified columnar, pseudostratified, and stratified cuboidal). Stratified squamous epithelia are protective. Columnar and cuboidal epithelia frequently are ciliated.

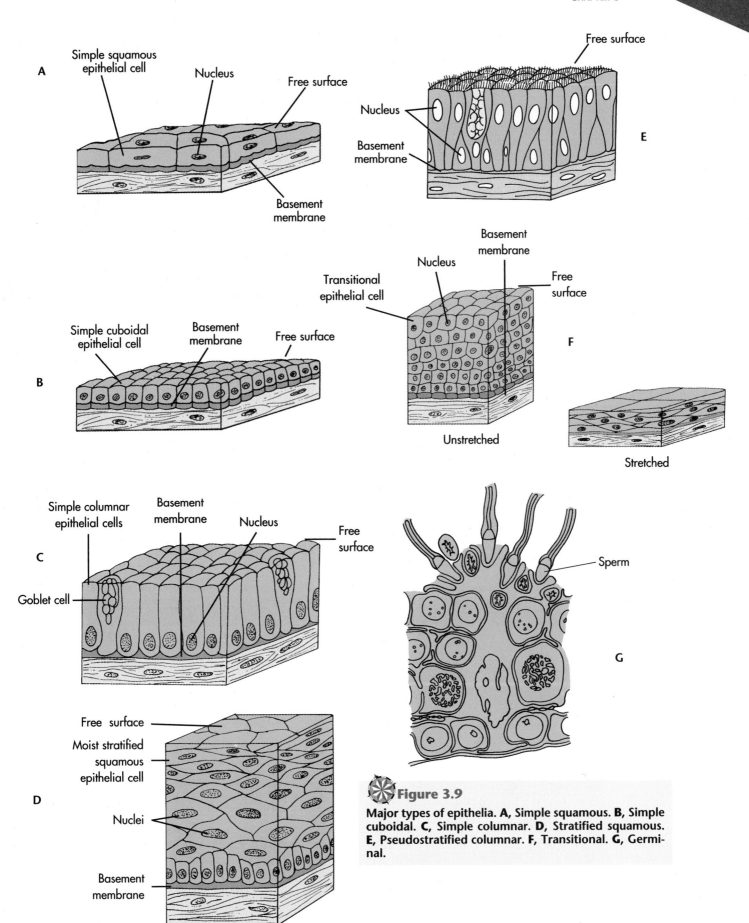

A, Simple squamous epithelial cell — Nucleus — Free surface — Basement membrane

B, Simple cuboidal epithelial cell — Basement membrane — Free surface

C, Simple columnar epithelial cells — Basement membrane — Nucleus — Free surface — Goblet cell

D, Free surface — Moist stratified squamous epithelial cell — Nuclei — Basement membrane

E, Free surface — Nucleus — Basement membrane

F, Transitional epithelial cell — Nucleus — Basement membrane — Free surface — Unstretched — Stretched

G, Sperm

✳ **Figure 3.9**

Major types of epithelia. A, Simple squamous. B, Simple cuboidal. C, Simple columnar. D, Stratified squamous. E, Pseudostratified columnar. F, Transitional. G, Germinal.

SIMPLE
SQUAMOUS

Simple squamous epithelium is a single layer of very thin cells that bind to each other at their edges and to a basement membrane underneath, as shown in Figure 3.9A. Simple squamous epithelium lines capillaries, (Figure 3.10), pulmonary alveoli, and portions of nephron loops in the kidneys. These cells are usually only 0.3 μm thick, a short enough distance (0.5 μm is the limit) for diffusion of molecules such as oxygen across the entire cell. Although squamous cells contain all of the usual organelles, their nuclei and transport vesicles are especially interesting; because these cells are so thin, (1) nuclei bulge the narrow cytoplasm and (2) coated pits and vesicles shuttle between the apical and basal surfaces as **transient pores,** which are a form of transcytosis. Tight junctions and desmosomes connect squamous cells together.

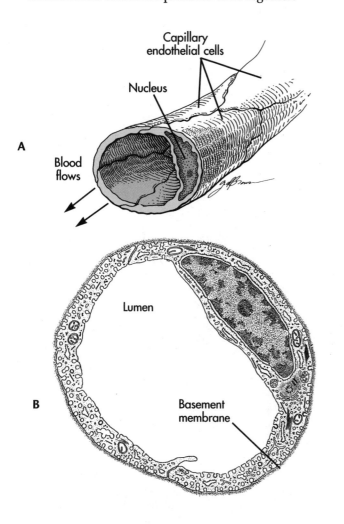

**Figure 3.10**

A, The walls of capillaries are composed of simple squamous epithelium. **B,** Numerous transient pores transport materials across the shallow cytoplasm by endocytosis and exocytosis. A basement membrane binds this capillary to surrounding connective tissue.

CUBOIDAL

Simple cuboidal epithelium lines the ducts and secretory portions of glands with a single layer of cube-shaped cells attached to each other by tight junctions and belt desmosomes (Figure 3.9B). The bronchioles of the lungs and the tubules and ducts of the kidney also are linked with cuboidal cells. Pancreatic acinar cells (Chapter 2, Cell Sketches) and cells of the parotid gland are considered simple cuboidal epithelium, even though acini and alveoli mold these cells into the pyramidal shapes seen in Figure 3.11.

COLUMNAR

Simple columnar epithelium lines the stomach, the small intestine, and the colon, where it transports nutrients and secretes digestive enzymes, hormones, and mucus (Figure 3.9C). Columnar cells also line the larger pulmonary passages, the uterine tubes, and the uterus. Nuclei and the endoplasmic reticulum of these cells, illustrated in Figure 3.12, usually occupy the basal cytoplasm, providing room for secretion granules to collect in the apical cytoplasm before being released into the lumen. **Microvilli** supported by extensive terminal webs usually populate the apical surface of simple columnar epithelia, which expands the area available for absorption. These microvilli are called the **brush border,** because they resemble the bristles of a brush. The presence of brush borders indicates that absorption and secretion are underway. Endocytotic and exocytotic vesicles enter and leave the cells between the bases of the microvilli.

STRATIFIED
SQUAMOUS

Stratified squamous epithelium contains many layers of squamous cells (Figure 3.9D). It is the thickest epithelium and may measure up to 2 mm (1/16 in) in the epidermis. The lining of the vagina, the surface of the tongue, the oral mucosa, the lining of the esophagus, and the epidermis all are stratified squamous epithelia. These epithelia are primarily protective. Their outermost cells are constantly being rubbed off through a process called **desquamation,** diagrammed in Figure 3.13, and replaced by cells from the lower layers. In dry epithelia such as the epidermis the cells dehydrate, lose their nuclei, die, and become a tough, horny, outer, cornified keratinous layer from which the cells desquamate as scaly husks. **Noncornifying epithelia** line moist surfaces such as the oral mucosa and tongue. These cells desquamate but do not dry out and cornify. Such epithelia frequently are called **keratinizing** and **nonkeratinizing epithelia,** but this terminology is confusing because both types produce cytokeratins. The difference concerns the processes that prevent moist epithelia from cornifying. Vitamin A interferes with cornification in the epidermis, for example, and this has led to its use as an antiwrinkle treatment.

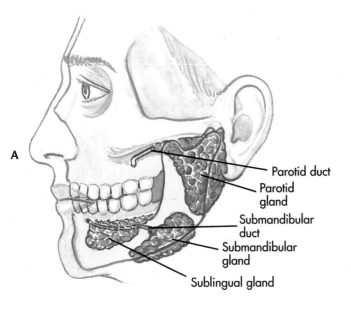

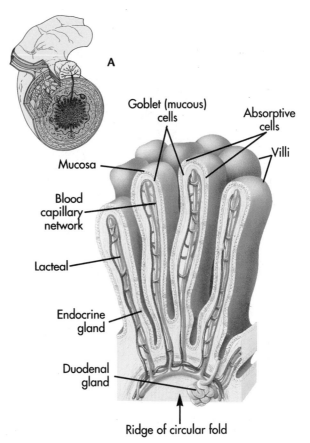

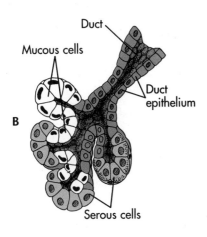

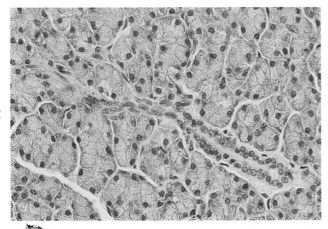

Figure 3.11

Simple cuboidal epithelium secretes fluid from the parotid gland. **A,** The parotid gland is located in the cheek. **B,** Cuboidal cells line the pouches (alveoli) of this gland. **C,** Photomicrograph of alveoli.

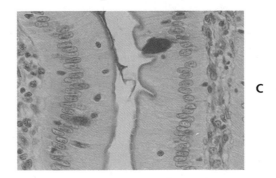

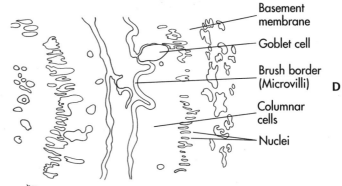

Figure 3.12

Simple columnar epithelium lines the intestine. **A,** Transverse section through the intestine locates the epithelium. **B,** Numerous projections (villi) are covered with simple columnar epithelium. **C,** Photomicrograph. **D,** Line drawing.

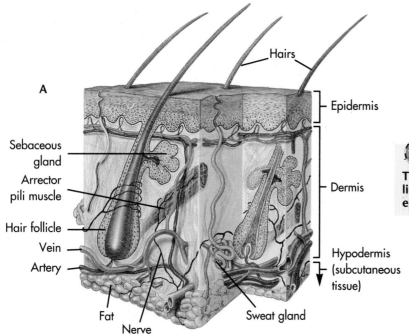

A

Hairs

Epidermis

Dermis

Hypodermis (subcutaneous tissue)

Sebaceous gland

Arrector pili muscle

Hair follicle

Vein

Artery

Fat

Nerve

Sweat gland

 **Figure 3.13**

The epidermis of the skin is stratified squamous epithelium. A, A section of skin reveals stratified squamous epithelium. **B**, Photomicrograph. **C**, Line drawing.

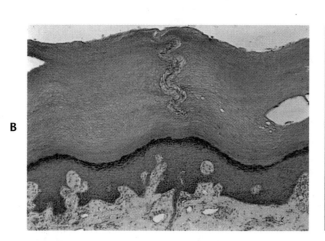

B

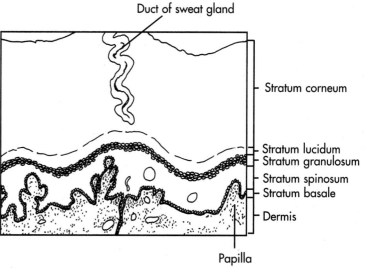

Duct of sweat gland

Stratum corneum

Stratum lucidum
Stratum granulosum
Stratum spinosum
Stratum basale

Dermis

Papilla

CUBOIDAL AND COLUMNAR

Stratified cuboidal and stratified columnar epithelia are uncommon. Stratified columnar epithelium occurs where the simple cuboidal epithelium of the salivary glands' ducts joins the stratified squamous epithelium of the oral cavity. There are usually two layers of cells.

PSEUDOSTRATIFIED

Pseudostratified columnar epithelium is really a simple columnar epithelium with all cells attached to the basement membrane, but the nuclei occupy various positions in the very tall, narrow cells, as Figure 3.9E shows. Figure 3.14 also shows how the nuclei of

nasal epithelium give the impression of several cell layers when there is really only one (SUE-doe; pseudo-, *G*, false). The lining of the trachea, the olfactory epithelium of the nasal cavity, and the lining of the vas deferens of the testis all are pseudostratified columnar epithelia.

TRANSITIONAL

Transitional epithelium lines the renal pelvis, ureters, bladder, and part of the urethra (see Chapter 25, Urinary System). It is named "transitional" because its organization changes as the epithelium stretches. When it is relaxed, up to eight lay-

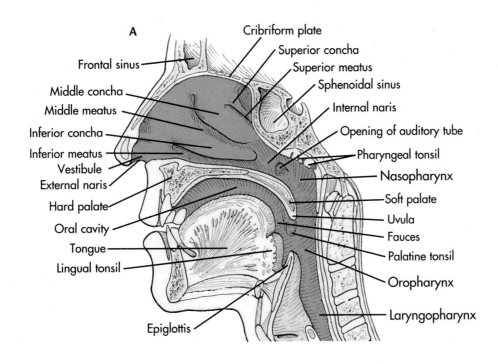

A

Frontal sinus
Middle concha
Middle meatus
Inferior concha
Inferior meatus
Vestibule
External naris
Hard palate
Oral cavity
Tongue
Lingual tonsil
Epiglottis

Cribriform plate
Superior concha
Superior meatus
Sphenoidal sinus
Internal naris
Opening of auditory tube
Pharyngeal tonsil
Nasopharynx
Soft palate
Uvula
Fauces
Palatine tonsil
Oropharynx
Laryngopharynx

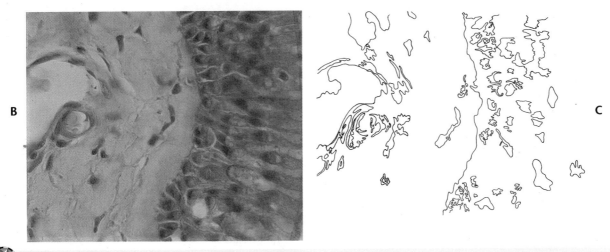

B

C

Figure 3.14

The nuclei of pseudostratified columnar epithelium give the false impression that epithelium of the nasal cavity contains several layers of cells, when in reality there is only one layer. **A**, Nasal cavity. **B**, Photomicrograph of nasal epithelium. **C**, Line drawing.

ers of dome-like squamous, cuboidal, and columnar cells line the bladder, as you can see in Figures 3.9F and 3.15. However, when this epithelium is stretched, only two or three layers of squamous and cuboidal cells appear. Stretching pulls the epithelial cells flatter as the bladder fills, and they bunch up when it empties and the epithelium relaxes. All the

layers of cells may attach to the basement membrane, although the anatomical evidence is not entirely clear on this point.

GERMINAL

Germinal epithelium produces sperm in the seminiferous tubules of the testis. This epithelium,

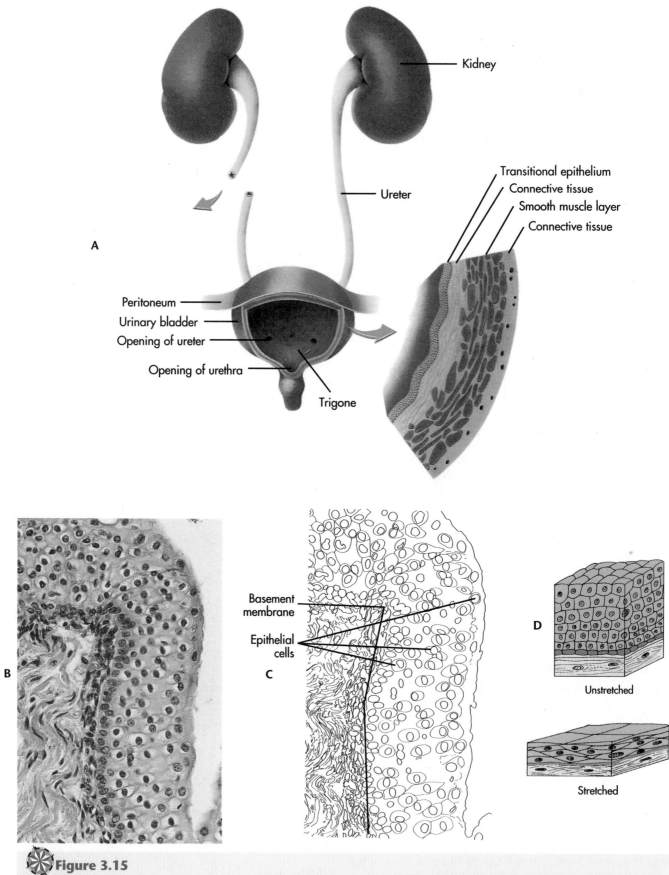

Labels in figure:
Kidney
Ureter
Transitional epithelium
Connective tissue
Smooth muscle layer
Connective tissue
Peritoneum
Urinary bladder
Opening of ureter
Opening of urethra
Trigone
A
Basement membrane
Epithelial cells
B
C
D
Unstretched
Stretched

Figure 3.15

A, Transitional epithelium lines the bladder. **B and C,** Transitional epithelium from the wall of the bladder is shown as photomicrograph and line drawing. **D,** Illustrates the changes when transitional epithelium is stretched and relaxed.

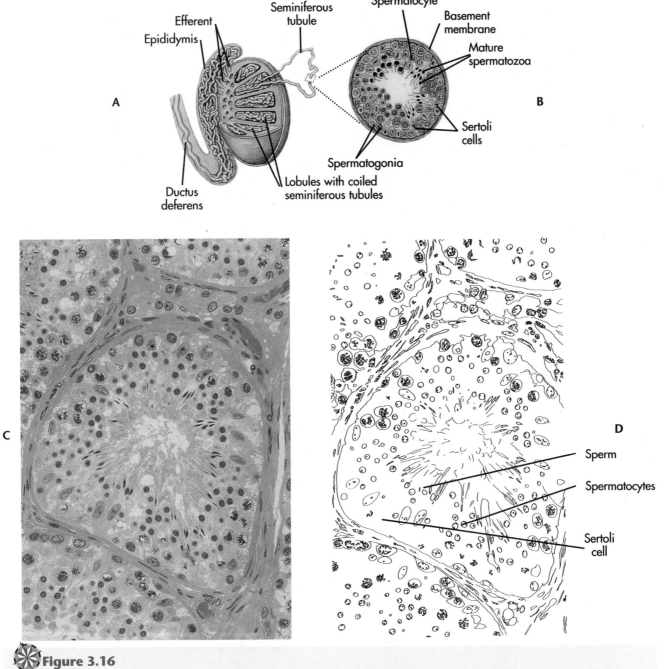

Figure 3.16

Sperm develop from the germinal epithelium that lines the seminiferous tubules of the testis. The testis (**A**) is packed with seminiferous tubules (**B**) that are lined with germinal epithelium, shown (**C**) in photomicrograph and (**D**) line drawing.

shown in Figures 3.9G and 3.16, is perhaps the most highly specialized of all epithelia. It appears stratified, but unlike stratified squamous epithelium, the layers represent stages in sperm development and meiosis. Beginning at the basal layer, dividing cells progress through meiosis until mature sperm are released from the luminal surface (see Chapter 26, Reproductive System). This epithelium is considered germinal because it is the continual source of reproductive cells.

CILIATED

Many epithelia are ciliated and move material from one region to another. Figure 3.17 shows ciliated epithelial cells of the trachea and bronchi that expel mucus from the lungs with waves of beating cilia. Ciliated columnar epithelium also drives small currents of fluid that carry embryos from the ovaries to the uterus through the uterine tubes. Some cuboidal cells such as the respiratory bronchioles of the lungs are ciliated, but squamous cells are never ciliated.

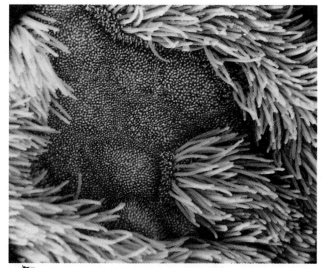

Figure 3.17

Cilia decorate the surface of some epithelial cells. (Scanning electronmicrograph by permission of R. G. Kessel and R. H. Kardon, *Tissues and organs: A text-atlas of scanning electron microscopy*, 1979, W. H. Freeman and Co.)

- Does stratified epithelium cover the skin?
- Which type of epithelium lines the intestine?

GLANDS

Glands are epithelial structures that secrete fluid products. A gland may be a single cell **(unicellular gland),** such as in the epithelial lining of the intestine (Figure 3.18A), or a large **multicellular gland,** such as the liver. The intestinal epithelium contains numerous unicellular glands, called goblet cells, that coat the epithelium with a film of mucus. Among multicellular glands are **exocrine glands** that deliver their products by way of ducts and **endocrine glands** (described in Chapter 22, Endocrine System) that secrete directly into capillaries and have no ducts. Exocrine glands differ according to their structure, products, and mode of secretion.

STRUCTURE

The secretory portion of the gland may be tubular, or it may be a small cup called an **alveolus** (or an acinus). Be careful with these terms, because the air-filled chambers of the lungs are traditionally considered alveoli and the secretory portions of the pancreatic exocrine glands are customarily called acini. Branching is less confusing. If the duct does not branch, the gland is simple. If the duct branches, the gland is compound. Follow the outline below and see Figure 3.18.

A. Simple glands—a single, unbranched duct drains the secretory portion of the gland.
 1. Tubular—secretory portion is a straight tube.
  a. Coiled—secretory portion is coiled.
 2. Alveolar—secretory portion is cup-shaped.
 3. Branched tubular—the tubules branch, and the duct drains several branches.
 4. Branched alveolar—the alveoli branch, and the duct drains several branches.
B. Compound glands—the duct branches.
 1. Tubular—branches of the duct drain the tubules.
 2. Alveolar—branches of the duct drain the alveoli.
 3. Tubuloalveolar—branches of the duct drain tubules and alveoli.

SIMPLE GLANDS

In **simple glands** a single duct passes directly to the outside of the gland. **Simple tubular glands** are just that—a blind tubular portion secretes through an un-

Figure 3.18

Exocrine glands are classified by shape. A, Unicellular gland. B, Simple tubular. C, Coiled tubular. D, Simple alveolar. E, Simple branched alveolar. F, Simple branched tubular. G, Compound tubular. H, Compound alveolar.

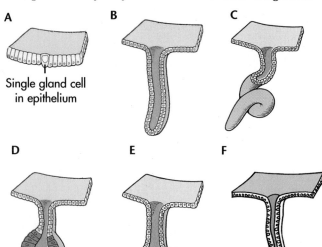

A   B   C

Single gland cell in epithelium

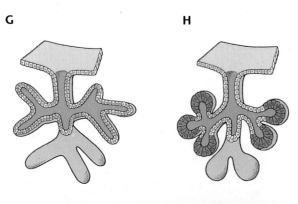

D   E   F   G   H

branched duct to the outside (Figure 3.18B). The secretory tubule is relatively straight in gastric glands, but in sweat glands it coils tightly (Figure 3.18C), forming a **simple coiled tubular gland.** Alternatively a grapelike **alveolus,** or acinus (Figure 3.18D), secretes the products, as occurs in the seminal vesicles of males.

Several alveoli or tubules may branch from the same duct, in which case the glands are (Figure 3.18E) **branched alveolar glands** (the sebaceous glands of the skin) or (Figure 3.18F) **branched tubular glands** (gastric glands of the gastric mucosa).

COMPOUND GLANDS

The ducts of **compound glands** branch many times, and each tributary conducts products downstream to exit the main duct. **Compound tubular glands** (Figure 3.18G), occur in the bulbourethral gland. Salivary glands (submandibular and sublingual) and the pancreas are **compound alveolar glands** (Figure 3.18H), although traditionally the pancreatic alveoli are called acini. The parotid gland is considered a **compound tubuloalveolar gland** because it secretes from both alveoli and tubules.

PRODUCTS

Glands may produce serous or mucous fluids. Goblet cells and **mucous glands** lubricate the lining of the gastrointestinal tract with a film of viscous mucus that also helps protect the epithelium itself from being digested. By contrast, **serous glands** such as the submandibular and sublingual secrete watery serous fluids. The parotid gland is a **seromucous gland** because it secretes both serous fluid and mucus.

SECRETION

Merocrine glands, seen in Figure 3.19A, secrete products by exocytosis and active transport. Salivary, sweat, and gastric glands are a few of the many merocrine glands in the body. **Holocrine glands** (Figure 3.19C) are less common; these glands release whole cells that break open and disperse their contents. Sebaceous glands release oil-laden cells that develop from the epithelium. Mammary glands are said to be **apocrine glands** (Figure 3.19B) because they appear to release apical cytoplasm, but experts disagree on the differences between apocrine and merocrine secretion. Apocrine secretion probably is a variation of merocrine secretion.

- Classify a sweat gland according to its form, product, and mode of secretion.
- Name an example of a compound alveolar gland.

BASEMENT MEMBRANES

Basement membranes attach epithelia to underlying tissues with a meshwork of proteins and proteoglycans. Figure 3.20 shows how the basement membrane is thought to be organized. Basement mem-

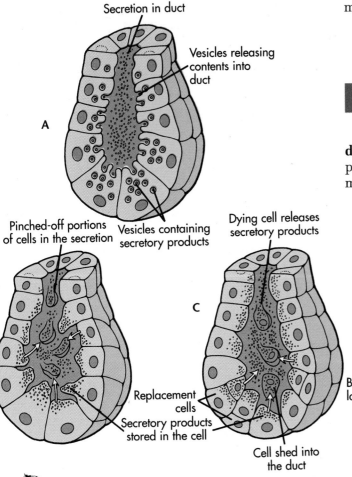

Figure 3.19

Exocrine glands are classified by mode of secretion: **(A)** merocrine, **(B)** apocrine, **(C)** holocrine.

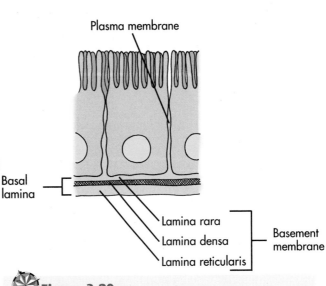

Figure 3.20

The basement membrane consists of the lamina rara, lamina densa, and the lamina reticularis.

branes usually consist of a thin **lamina rara** (light layer; lamina-, *L*, layer; rar-, *L*, rare) that binds to the epithelium and the thicker **lamina densa** (dense layer) below it. Connective tissue binds beneath the lamina densa in a zone called the **lamina recticularis** (network layer). Hemidesmosomes and adhesive plaques attach the epithelial cells to the lamina rara, and networks of collagen fibers secure connective tissue to the lamina reticularis. Some histologists refer to the first two layers as the basal lamina, and your instructor may prefer this term.

The basement membrane is a strategic but little understood boundary between epithelium and connective tissue. It contains at least 40 different proteins produced in concert by both tissues. In addition to helping regulate what materials enter the epithelia, the basement membrane appears to influence epithelial polarity and whether the epithelium is flat or in folds.

CONNECTIVE TISSUE

Connective tissue is the glue that shapes epithelium, nerve, and muscle tissues into organs. Connective tissue is characterized by an extensive **extracellular matrix** (ECM) that contains much fluid and relatively few scattered cells, unlike the closely connected cells of epithelia (Figure 3.21). Connective tissues include bone, cartilage, fascia, dermis of the skin, adipose tissue, and blood. Extracellular matrix makes up more than 90% of connective tissue volume, and this great space is responsible for the various functions and properties of connective tissue. Blood vessels and capillaries can branch into nearly all regions of the body because the matrix adheres so intimately to surrounding tissues. The interstitial fluid (between the cells) in the matrix bathes capillaries and surrounding tissues, transferring nutrients and wastes between them. Tissue inflammation attracts leucocytes and lymphocytes to migrate from the blood through the interstitial fluid to the sites of infection. The matrix is hard and rigid in bone; firm in cartilage; fibrous in fascia; flexible, tough, and soft in skin and adipose tissue; and fluid in blood. Connective tissue derives from the mesoderm (see Chapter 4, Mesoderm), and it is the most extensive and diverse tissue of the body, occurring in virtually all organs.

EXTRACELLULAR MATRIX

The properties of the extracellular matrix rely on gels and fibers. The gel (Figure 3.21) is called **amorphous ground substance**, it is shaped by surrounding tissues, has no form of its own, and cannot be seen easily with the light microscope. The properties of ground substance arise from huge, branched,

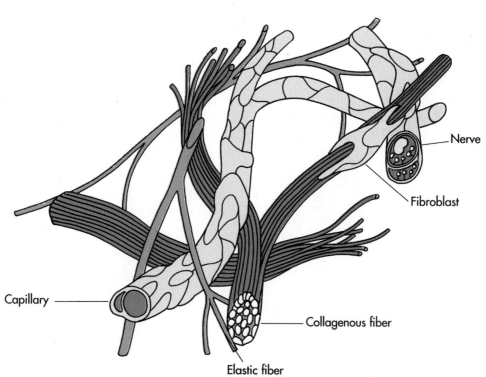

Nerve

Fibroblast

Capillary

Collagenous fiber

Elastic fiber

Figure 3.21

Connective tissue cells occupy an extracellular matrix composed of amorphous ground substance and fibers.

hydrophilic, proteoglycan and glycoprotein molecules that expand when water molecules and electrolytes bind to them. Fully 90% of connective tissue volume is amorphous ground substance with fibers and cells embedded in it. The consistency of amorphous ground substance varies from soft to hard, depending on its composition.

Three types of fibers add further variety to connective tissue. Collagen fibers are flexible but not stretchy; they especially strengthen the ground substance in tendons, ligaments, cartilage, and bone. Stretchy **elastic fibers** give the matrix elasticity in loose connective tissue. **Reticular fibers** form a flexible network that accompanies the other fibers.

COLLAGEN FIBERS

Collagen fibers are the principal fibers of connective tissue. These fibers form a tough, flexible, but inelastic network embedded in the amorphous ground substance. A family of genes governs the structure of collagen molecules, and depending on the type of collagen and quantity of fibers, this results in tough tendons and ligaments or much softer dermis and mesenchyme. Collagen molecules link with covalent and hydrogen bonds to form collagen fibers (Figure 3.22). Leather belts and shoes are composed of tanned collagen from the dermis of animals.

Some diseases interfere with production of collagen fibers. Patients with **Ehlers-Danlos syndrome** have joints that move beyond their normal limits and skin that is unusually flexible and stretchy; their injuries heal slowly, and the abdominal wall frequently herniates (HER-nee-ates; ruptures; herni-, *L*, a rupture). Improper cross-linking between collagen fibers appears to be the source of these inherited symptoms. In **rheumatoid arthritis** (ROOM-a-toyd arth-RYE-tis), enzymes called **collagenases** degrade the collagen fibers in affected joints, and in **atherosclerosis,** (ATH-er-oh-SKLAIR-oh-sis) smooth muscle fibers of arteries secrete too much collagen, which accumulates with fat in the walls of these vessels and may block them.

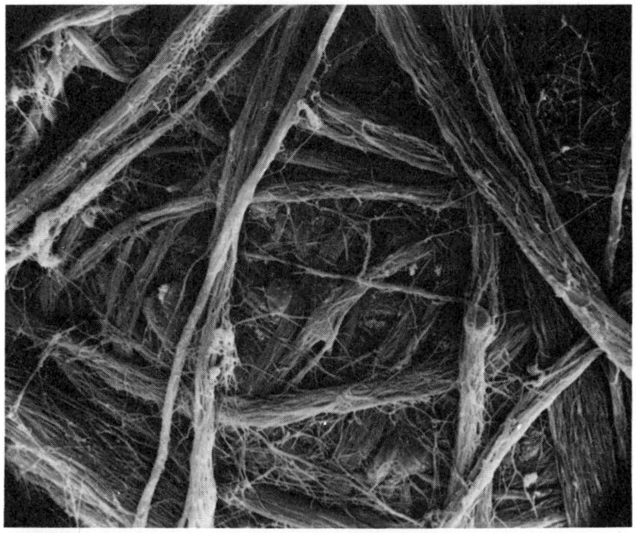

ELASTIC FIBERS

Networks of elastic fibers make the skin, vocal cords, arteries, and lungs stretchy. Compared to collagen fibers elastic fibers are poorly understood, but they appear to owe their elasticity to a network of **elastin** molecules that stretch and recoil when pulled, as shown in Figure 3.23.

RETICULAR FIBERS

Networks of delicate, branching, reticular fibers contain collagen and support the basement membranes of epithelia and vascular endothelia. These networks also support delicate liver tissue and the cells of the spleen, lymph nodes, and bone marrow. Reticular fibers lend some elasticity to tissues where they occur because the fibers can slide across one another.

CELLS

Most connective tissues contain four fundamental kinds of cells. Fibroblasts (described in Chapter 2, Cell Sketches) assemble the extracellular matrix. Other cells include macrophages, mast cells, and adipose cells that are concerned with immunological surveillance and fat storage rather than assembling the matrix. Various leukocytes and lymphocytes enter and leave via the blood and lymph vessels.

FIBROBLASTS

Fibroblasts are stellate cells that reside permanently in the matrix they synthesize. (The cellular machinery for this function is described in Chapter 2, Fibro-blasts and shown in Figure 3.24A.) Fibroblasts control the variety of connective tissues by synthesizing various mixes of fibers and ground substance. Fibroblasts develop from embryonic mesenchyme cells. Distributed throughout the body, fibroblasts continually replace old matrix with new.

All other types of cells found in connective tissue are transients that move in and out from the immune and cardiovascular systems.

Figure 3.22

Collagen fibers.(Scanning electronmicrograph by permission of R. G. Kessel and R. H. Kardon, *Tissues and organs: A text-atlas of scanning electron microscopy,* 1979, W. H. Freeman and Co.)

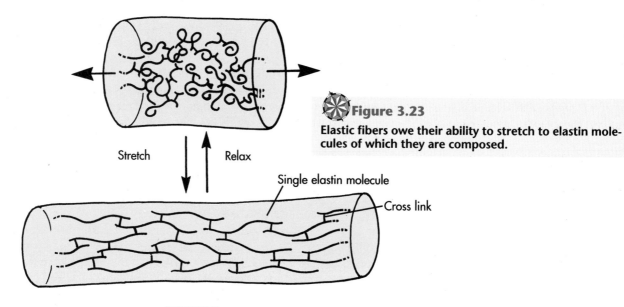

RELAXED

Stretch   Relax

Single elastin molecule

Cross link

STRETCHED

✳ **Figure 3.23**
Elastic fibers owe their ability to stretch to elastin molecules of which they are composed.

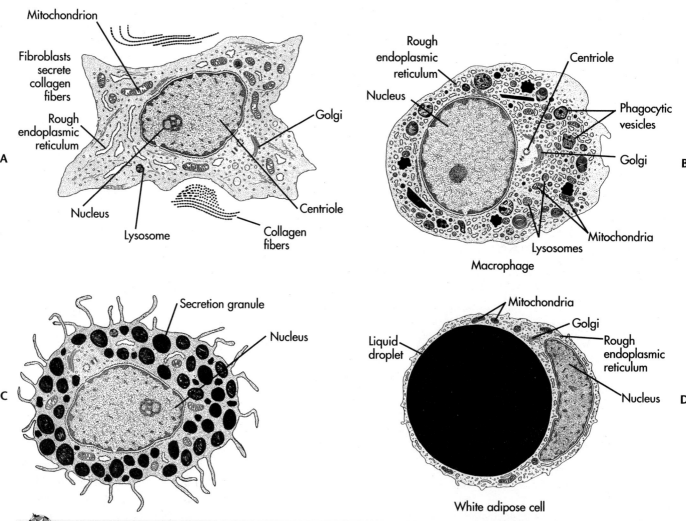

Mitochondrion

Fibroblasts secrete collagen fibers

Rough endoplasmic reticulum

Golgi

Nucleus

Centriole

Lysosome

Collagen fibers

A

Rough endoplasmic reticulum

Nucleus

Centriole

Phagocytic vesicles

Golgi

Lysosomes

Mitochondria

Macrophage

B

Secretion granule

Nucleus

C

Mitochondria

Golgi

Liquid droplet

Rough endoplasmic reticulum

Nucleus

White adipose cell

D

✳ **Figure 3.24**
Four major types of cells found in connective tissues: **(A)** fibroblast, **(B)** macrophage, **(C)** mast cell, **(D)** white adipose cell.

MACROPHAGES

Macrophages (MAK-row-faye, large eaters; macro, *G*, large, phago-, *G*, eat) are large, irregular cells (Figure 3.24B) that develop in the bone marrow and move throughout the body's connective tissues, patrolling for microorganisms, damaged cells, and foreign particles that they dispose of through phagocytosis. Macrophages in the liver and spleen consume dying erythrocytes and degrade hemoglobin into bile pigments.

MAST CELLS

Mast cells (Figure 3.24C; mast-, *G*, a breast, food) also develop in bone marrow. Mast cells are round cells that participate in inflammatory reactions, allergic responses, and pain by releasing secretion granules that contain heparin and histamine.

ADIPOSE CELLS

Adipose cells, shown in Figure 3.24D, store white or brown fat in the dermis, the mesenteries of the gut, and the colon for insulation and energy. When hundreds of adipose cells cluster together, they are considered adipose tissue (described in the next section).

- Name the components of the extracellular matrix.
- Name the main types of cells in connective tissue.

TYPES OF CONNECTIVE TISSUE

There are three fundamental categories of connective tissue—**connective tissue proper, specialized connective tissue,** and **mesenchyme.** All three are related to the matrix and cells within them, as outlined below.

A. Connective tissue proper—contains fibroblasts, macrophages, mast cells, adipose cells.
  1. Loose (areolar) connective tissue—soft and watery; superficial dermis, mesenteries, gastrointestinal tract.
  2. Dense connective tissue—firm; deep dermis, fascia, tendons, ligaments.
    a. Regular—collagen fibers oriented in direction of stress; tendons, ligaments.
    b. Irregular—fibers extend in all directions; dermis, fascia.
  3. Adipose tissue—soft, stores fat; mesenteries, dermis.
  4. Reticular connective tissue—soft; spleen, lymph nodes, bone marrow.
B. Specialized connective tissue—specialized cells deposit matrix.
  1. Bone—hard, rigid; skeleton.
  2. Cartilage—hard, flexible; skeleton, joints.
    a. Hyaline—tough, homogeneous matrix; joints, ribs, trachea, nasal septum.
    b. Elastic—prominent elastic fibers; epiglottis, external ear.
    c. Fibrocartilage—prominent collagen fiber bundles; intervertebral disks.
  3. Blood—fluid form of connective tissue based on cells derived from bone marrow.
C. Mesenchyme—soft; embryonic connective tissue.

CONNECTIVE TISSUE PROPER

Connective tissue proper is widely distributed. Dermis of the skin is a good example of connective tissue proper (Figure 3.13A); it binds the epidermis to the underlying fascia and muscle with a soft matrix and fibers. Other types of connective tissue attach intestinal epithelium to the wall of the digestive tract (Figure 3.1), join muscle fibers together into muscles, and bundle neurons into nerves. Adipose tissue stores fat beneath the skin and in the mesenteries. Tendons and ligaments are considerably denser and tougher; their bundles of collagen fibers attach muscles to bones and bones to other bones. Vessels and nerves of the cardiovascular, immune, and nervous systems are supported by connective tissue.

LOOSE

Loose connective tissue consists of a soft, watery gel. Also called **areolar connective tissue,** it supports virtually every organ of the body. The delicate mesenteries of the peritoneal cavity and the soft connective tissue walls of the gastrointestinal tract (Figure 3.25) are good examples of loose connective tissue. Amorphous ground substance takes up considerable space in loose connective tissue. Accordingly this tissue both cushions organs and stores interstitial fluid, while providing access for blood vessels, nerves, and macrophages and mast cells of the immune system. In loose connective tissue, fibroblasts produce considerable quantities of proteoglycans but relatively few elastic and collagen fibers. Fibroblasts, macrophages, mast cells, and adipose cells are scattered through the matrix and occupy a very minor volume of the tissue.

DENSE

Dense connective tissues have many bundles of collagen fibers. When the collagen bundles are oriented primarily in one direction, as with tendons, ligaments, and cornea, the tissue is considered to be **regular dense connective tissue,** as seen in Figure 3.26. Where the bundles are pulled and oriented in many directions, as with dermis and fascia, the tissue is said to be **irregular.** Dense connective tissue forms both the tough dermis of the skin and the fascia that covers muscle and bone. The protective fibrous capsules of

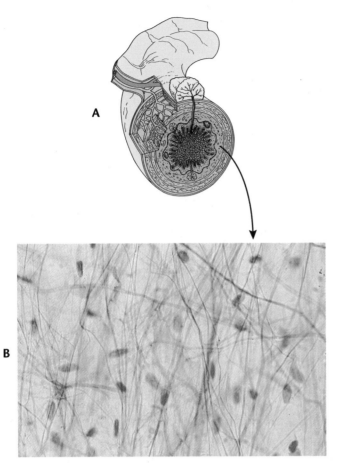

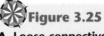

A, Loose connective tissue of the intestinal wall. **B and C**, Photomicrograph and line drawing.

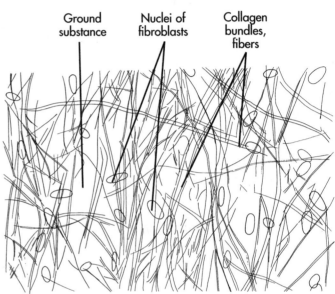

organs such as the kidneys are dense connective tissue. Less amorphous ground substance is present than in loose connective tissue, and the fibroblasts, macrophages, and mast cells occupy more space between the collagen fibers. The fibroblasts produce various proportions of elastic and collagen fibers; dermis emphasizes elastic fibers, and tendons and ligaments emphasize collagen fibers.

ADIPOSE

Adipose tissue stores fat. Although **adipose cells** (ADD-ih-pos; *L,* full of fat), also called **adipocytes,** are scattered throughout loose connective tissue, only when many of them accumulate in one place do they form adipose tissue (Figure 3.27). Humans store both **white** and **brown adipose tissues,** gradually replacing brown with white by adulthood. Newborns store major deposits of brown fat in the posterior cervical, axillary, and renal regions. Adults store white fat subcutaneously, in the mesenteries, and in retro-. peritoneal regions. In general, women store more in the dermal and abdominal regions, buttocks, and thighs than do men. Overall, fat contributes approximately 15% of body weight in men and 22% in women.

Although all adipose tissue insulates and cushions the body, white adipose tissue also is an energy source. Brown fat is primarily a heat source in infants. Spherical white adipose cells usually contain one large **fat droplet** that inflates the cell and presses its nucleus and cytoplasm to one side. Brown adipocytes, also spherical in shape, store numerous smaller fat droplets. No membrane surrounds either kind of fat droplet, since the hydrophobic nature of the contents is a sufficient barrier to the cytoplasm. Brown adipose tissue gets its appearance from a richer supply of mitochondria and capillaries, not from the color of the fat droplets, which have the same composition in both tissues.

Adipose tissue tends to increase, en-suring the body's energy reserves. A restricted diet causes fat droplets to diminish in size, but it does not reduce the number of white adipocytes, and only low temperatures reduce the size of fat droplets in brown adipose cells. Authorities agree that adipocytes are long-lived cells that, once formed, persist. These cells begin accumulating in fetal development and childhood. Overeating can add more adipocytes anytime, but restricted diets can only shrink them, not remove them.

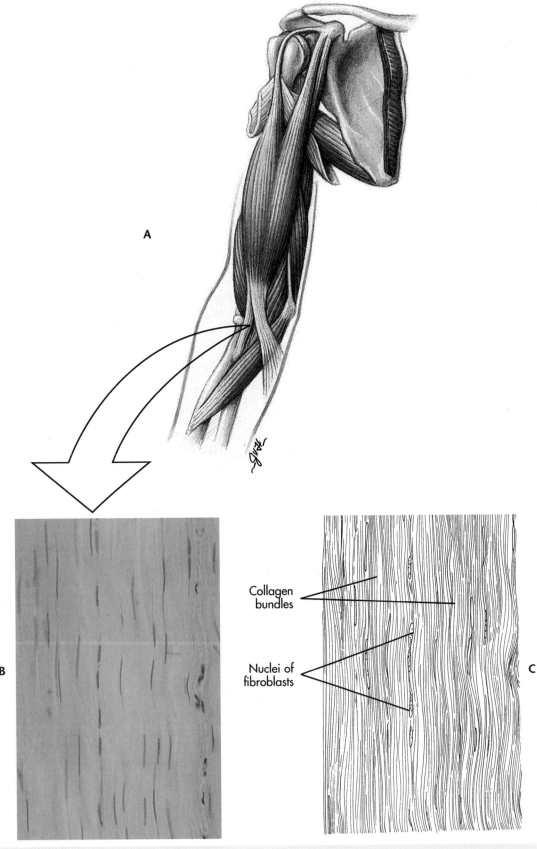

Figure 3.26
Regular dense connective tissue of the biceps tendon withstands the forces of muscular contraction. **(A)** Biceps tendon showing **(B)** microscopic structure in photomicrograph and **(C)** line drawing.

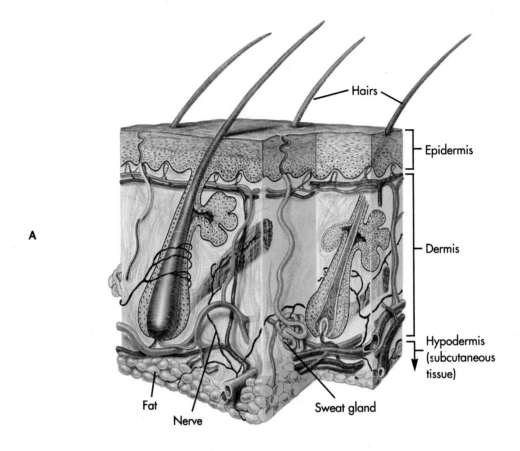

A

Hairs

Epidermis

Dermis

Hypodermis
(subcutaneous
tissue)

Fat

Nerve

Sweat gland

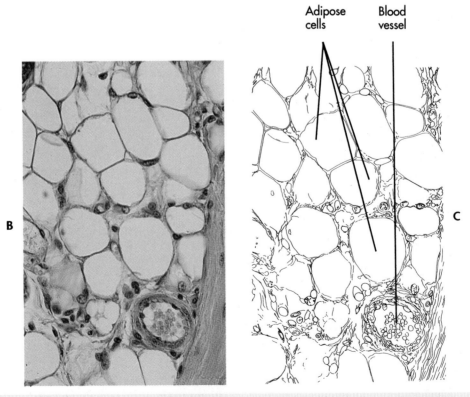

Adipose
cells

Blood
vessel

B

C

Figure 3.27

A, White adipose tissue of the dermis stores large fat droplets. **B and C,** In photomicrograph and line drawing.

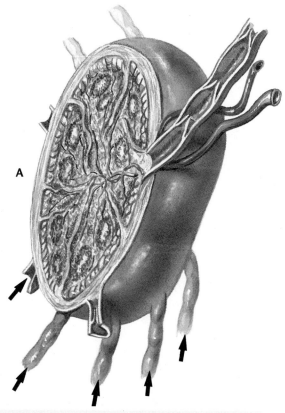

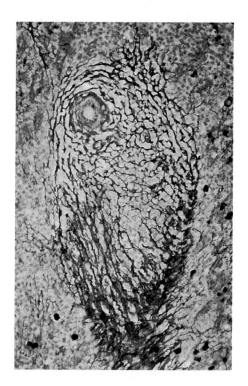

B

Figure 3.28

A, Reticular tissue from a lymph node. **B and C**, Photomicrograph and line drawing.

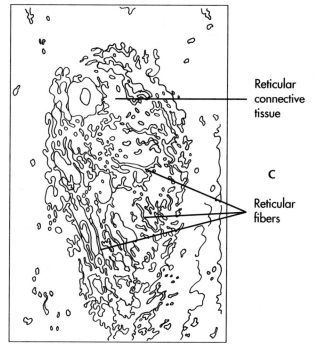

Reticular
connective
tissue

C

Reticular
fibers

RETICULAR

Reticular connective tissue contains prominent reticular fibers. This soft tissue (Figure 3.28) is found in liver and lymphoid organs such as the spleen and lymph nodes, although reticular fibers elsewhere coat the connective tissue sides of basement membranes. Reticular fibers are narrow bundles of collagen coated with proteoglycans; they are 0.5 to 2.0 μm in diameter. In the liver and lymph nodes they branch into delicate networks that support the soft epithelia of these organs and allow blood and lymph to circulate past these tissues.

- Name the varieties of connective tissue proper.
- How are they fundamentally similar?

SPECIALIZED

Bone, cartilage, and blood are considered **specialized connective tissues** because their own specialized cells rather than fibroblasts add particular components to the matrix. Chapters 5 and 15 describe all of these tissues in detail.

BONE

The extracellular matrix of bone consists of collagen fibers and amorphous ground substance, shown in Figure 3.29, that has been solidified with calcium phosphate crystals called hydroxyapatite. This matrix is deposited in layers, called lamellae, that trap osteocytes between them in small spaces called lacunae that closely fit the shape of the osteocytes. From these locations the osteocytes maintain the matrix with nutrients received from central blood vessels through slender cytoplasmic processes that connect the osteocytes together in narrow passages called canaliculi. Bone-forming cells called osteoblasts and bone-removing cells called osteoclasts sculpt this basic structure into the various bones of the skeleton. See Chapter 5 for details.

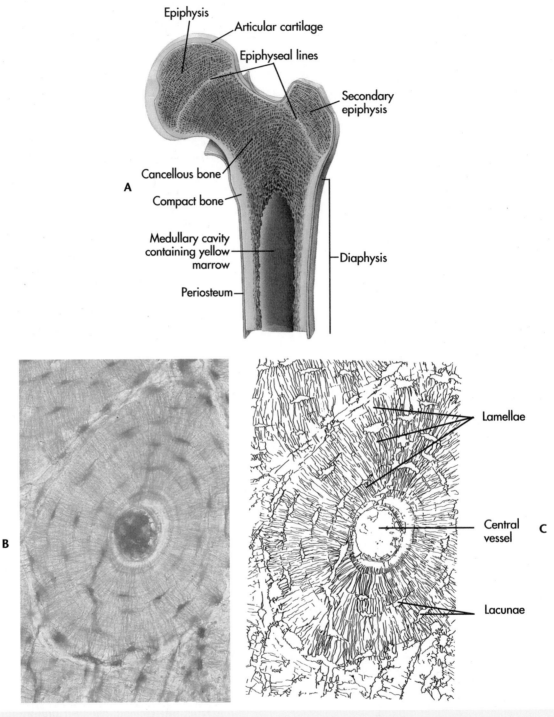

Figure 3.29

A, Bone. **B and C,** In photomicrograph and line drawing.

HYALINE CARTILAGE

Hyaline cartilage (cartilago, *L,* gristle; hyalo, *G,* glass, transparent) **supports the larynx, trachea, and bronchi,** keeping them open as air rushes into the lungs. This bluish translucent material (Figure 3.30) appears in the costal cartilages of the rib cage and also caps the articulating surfaces of bones with slippery surfaces that enable smooth movement (see Chapter 5).

The structure of hyaline cartilage resembles bone in several regards. The extracellular matrix is firm and filled with collagen fibers and proteoglycans, but it does not have hydroxyapatite to harden it. Chondrocytes maintain this matrix from within lacunae, but

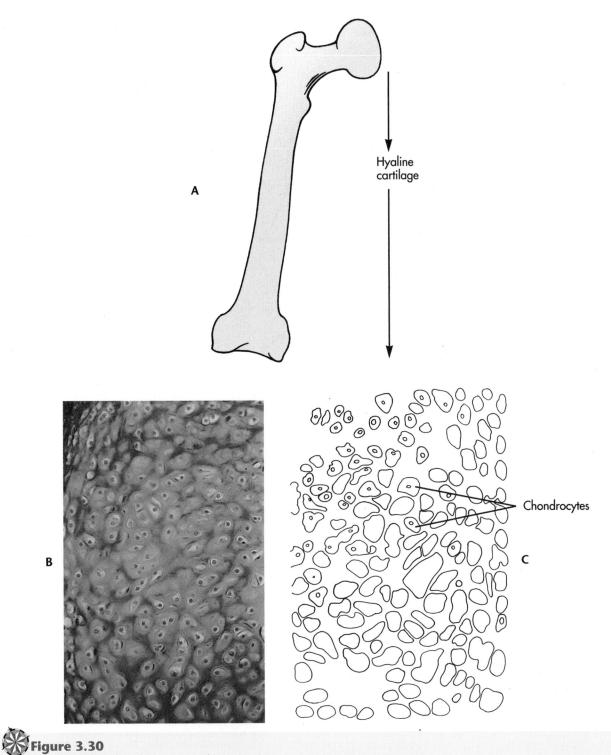

Hyaline
cartilage

A

B

C

Chondrocytes

Figure 3.30
A, Hyaline cartilage. **B and C,** In photomicrograph and line drawing.

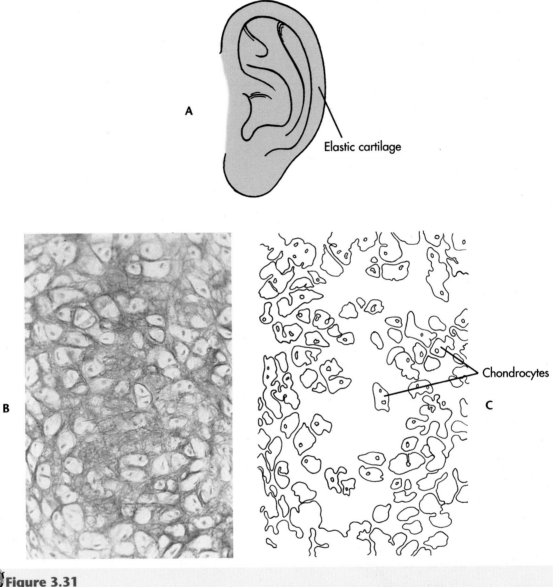

A, Elastic cartilage. Arrow points to Elastic cartilage.

B, photomicrograph. C, line drawing with label Chondrocytes.

✳ **Figure 3.31**
A, Elastic cartilage. **B and C,** In photomicrograph and line drawing.

they have no vascular supply, nor do the cells connect with each other through the matrix. Instead, nutrients diffuse through the fluid within the matrix. Unlike osteocytes, chondrocytes divide; thus lacunae commonly contain four chondrocytes in cell nests that arose from a single cell. Most of the skeleton begins as hyaline cartilage, but almost all is replaced by bone. Only the costal cartilages and articular surfaces remain from the cartilages that were present in the fetus. Tracheal cartilages are not related to bone.

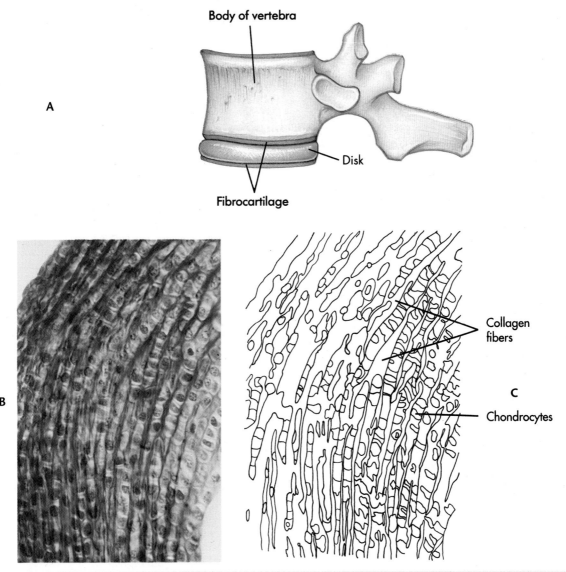

Body of vertebra

Disk

Fibrocartilage

A

B

C

Collagen fibers

Chondrocytes

Figure 3.32

A, Fibrocartilage. **B and C,** In photomicrograph and line drawing.

ELASTIC CARTILAGE

Elastic cartilage adds extra elastic fibers to the matrix of external parts of the ears and the epiglottis, two structures that snap back when bent. In these structures, elastic fibers form a dense network that is easily seen in Figure 3.31. Aside from the yellowish color of these fibers, elastic cartilage closely resembles hyaline cartilage.

FIBROCARTILAGE

Fibrocartilage contains large bundles of collagen fibers that support intervertebral disks, pubic symphyses, and the connections of tendons and ligaments to bones (Figure 3.32). Fibrocartilage is always associated with hyaline cartilage in these bony connections, because the collagen fibers extend through the hyaline cartilage into the bone. Unlike hyaline and elastic cartilage, fibrocartilage has so many collagen fibers that the ground substance fails to penetrate them. Instead, chondrocytes occupy lacunae and thin zones of ground substance squeezed between the collagen bundles. This organization resembles that of dense connective tissue, where fibroblasts occupy similar locations. In tendons and ligaments, bundles of collagen fibers tend to parallel each other. The forces they sustain all run in the same direction, in the strings of lacunae and chondrocytes you see in the figure. By contrast, in intervertebral disks and the pu-

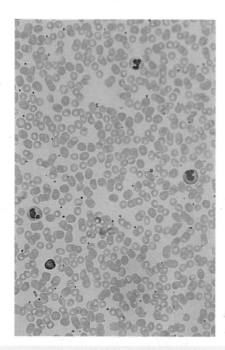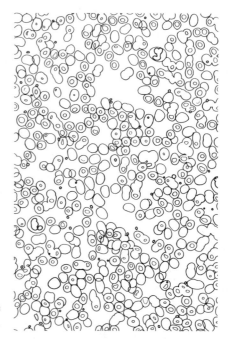

Figure 3.33
Blood smear. In photomicrograph and line drawing.

BLOOD

Plasma is the extracellular matrix of blood (Figure 3.33). Plasma contains many varieties of proteins and lipoproteins but very few proteoglycans and no collagen or elastin. The liver, rather than resident fibroblasts, is the principal source of plasma proteins. Erythrocytes, leukocytes, and thrombocytes are the cellular components of this tissue. All are transient cells that originate in the bone marrow and lymph nodes. Erythrocytes deliver oxygen from the lungs to body tissues, and various leukocytes are concerned with immune surveillance and inflammation. Thrombocytes, or platelets, activate clotting by converting fibrinogen, a prominent plasma protein, into fibrin. Chapter 15 describes all of these components of blood.

MESENCHYME

Mesenchyme is the loose connective tissue of embryos and fetuses. It is the precursor of adult connective tissues in several respects. First, this connective tissue is found where adult connective tissues will later reside. It makes up embryonic and fetal mesenteries, dermis, muscle, bone, gut wall, and virtually all connective tissue spaces until adult connective tissues begin to replace it during fetal life. **Mesenchyme cells** resemble fibroblasts (Figure 3.34), but their most important function during development is as stem cells for adult connective tissues. Adult organs may retain mesenchyme cells, which replenish cells lost from injury and disease.

- Name the specialized connective tissues.
- Which fundamental feature distinguishes them from connective tissue proper?

CANCEROUS TISSUES

Malignant tumors disrupt the basement membranes and connective tissue matrices that keep normal cells in place. Figure 3.35 shows what happens when a **squamous cell carcinoma** of the epidermis invades the dermis (see also Cancer Cells, Chapter 2). The epidermis (Figure 3.35A and B) is a stratified squamous epithelium built of many layers of squamous cells that rest on a basement membrane and the dermis beneath. In Figure 3.35C squamous cells have transformed into dark nucleated carcinoma cells, and these cells have penetrated the basement membrane, spread into the dermis, and disorganized it. Some of these cells may migrate further into the capillaries and spread through the cardiovascular system to seed more carcinomas in other dermal regions.

Figure 3.36 shows how metastasizing cells (meh-

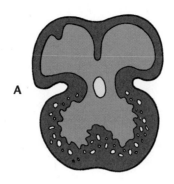

A

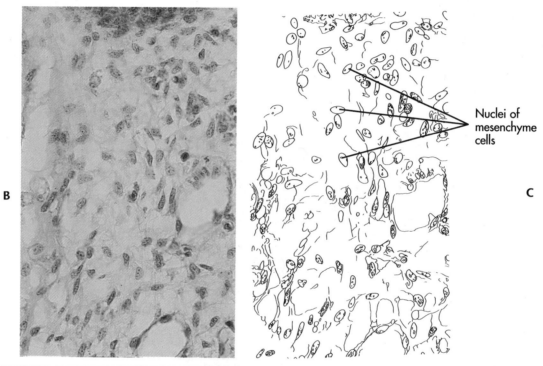

B

C

Nuclei of
mesenchyme
cells

✸ **Figure 3.34**

A, Mesenchyme. **B and C,** In photomicrograph and line drawing.

TAS-ta-size; change position; meta-, *G,* change; -stasis, *G,* standing) currently are thought to invade other tissues after they have entered the blood. They first bind to the capillary endothelium. These endothelial cells then retract, breaking the tight junctions and desmosomes that have joined them together and exposing the basement membrane. The invading cells now bind to the basement membrane and spread between it and the endothelial cells, which reseal their connections behind the invaders, thereby minimizing loss of blood from the capillaries. After the invaders have digested a passageway in the basement membrane, they can advance into the connective tissue beyond.

• Describe the main steps in metastasis.

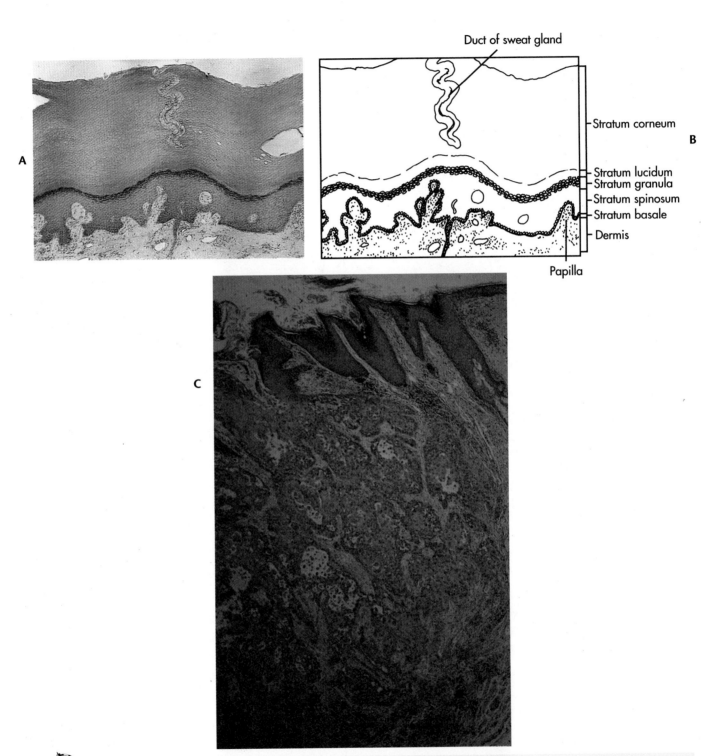

Duct of sweat gland

A

B

Stratum corneum

Stratum lucidum
Stratum granula
Stratum spinosum
Stratum basale

Dermis

Papilla

C

Figure 3.35

Invading squamous carcinoma cells have disrupted the basement membrane and the connective tissue beneath it. **A and B,** Normal epidermis in photomicrograph and line drawing. **C** Squamous cell carcinoma in photomicrograph.

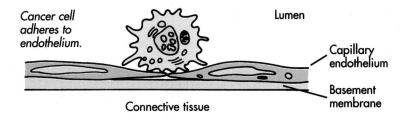

Cancer cell adheres to endothelium.

Lumen

Capillary endothelium

Basement membrane

Connective tissue

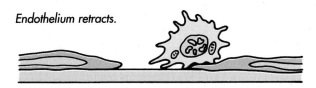

Endothelium retracts.

Cancer cell migrates into opening and slips under endothelium.

Endothelium reseals the opening and cancer cell attacks basement membrane.

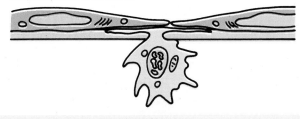

Cancer cell penetrates basement membrane and invades connective tissue.

Figure 3.36

A cancer cell penetrates the wall of a capillary and invades the surrounding connective tissue.

SUMMARY

Figure 3.37 summarizes morphological relationships among the cells of body tissues. Epithelial cells form sheets, and connective tissue cells secrete extracellular matrices that hold cells together. Muscle tissues resemble connective tissue cells that have elongated and become contractile. In nervous tissue, neurons are also elongated cells that derive from epithelia.

A tissue is a population of similar cells that performs a specific function. Nervous tissue, muscle tissue, epithelial tissue, and connective tissue are the four basic body tissues. Most organs contain all four tissues. Nervous tissues conduct impulses in the brain, spinal cord, and nerves. Muscle tissues contract; their pulling action moves the skeleton, pumps blood, and propels food through the gut. Epithelial tissues are cellular sheets that cover and line most organs. Connective tissues secrete extracellular matrices that support the body and connect its parts. *(Page 62.)*

The intestine is an example of an organ that contains all four tissues. Epithelium lines the interior and covers the exterior. Smooth muscle tissue underlies the external covering. Connective tissues join all of these layers together and provide access for nerves and vessels to spread through the connective tissue and muscle. *(Page 63.)*

Nervous tissues conduct impulses in the brain, spinal cord, and nerves. Neurons are long, slim cells that conduct impulses along pathways that connect with other neurons. Neuroglial cells support neurons, wrapping them with protective sheaths, supplying nutrition, and regulating the flow of impulses between them. Together, both types of cells form clusters and bundles of nerve fibers that extend like wires to all organs of the body *(Page 64.)*

Muscle tissues contain bundles and sheets of muscle fibers that contract and pull on whatever the fibers connect. Skeletal muscle fibers are large, cross-striated cells connected to the skeleton, and are under voluntary control of the nervous system. Cardiac muscle fibers are smaller, striated, involuntary fibers that enable the heart to pump blood. Smooth muscle fibers are neither striated nor voluntary; these cells help regulate blood flow through the cardiovascular system, propel food through the gut, and squeeze secretions from glands. *(Page 66.)*

Epithelia are sheets of cells that cover external surfaces and line internal passages. Virtually all materials that enter and leave an organ cross an ep-

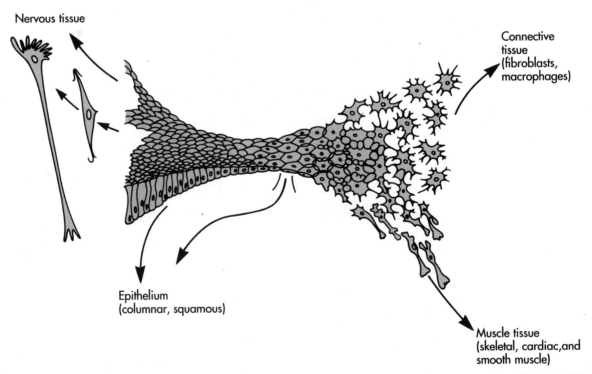

✺ Figure 3.37

Morphological relationships among the cells of nervous, muscular, epithelial, and connective tissues. Epithelial cells form sheets by attaching side-by-side; neurons connect end-to-end; extracellular matrix fills the space between connective tissue cells; and muscle fibers attach to each other side-by-side and end-to-end.

ithelium. Epithelia line the gut, cover the skin, and as pouches of tissue, form glands. Endothelia line the vessels of the cardiovascular system. Mesothelia line the peritoneal, pleural, and pericardial cavities. Epithelial cells usually are polarized to take up, secrete, and transport various materials. *(Page 67.)*

Epithelia are structurally diverse. Epithelia may be a single layer of cells or more than one layer (simple or stratified); the cells may be squamous, cuboidal, or columnar; the cells may be ciliated or without cilia; and the epithelium may be moist or dry (mucous, serous, or cornified). *(Page 70.)*

Exocrine glands secrete through ducts. Their secretory proteins may be tubular or alveolar, and their ducts may be simple or branched. The products may be mucus or serous, and the secretion process may be merocrine, apocrine, or holocrine. *(Page 78.)*

Basement membranes join epithelia to connective tissues. These mats of proteins and proteoglycans insure the integrity of the epithelium and attach the epithelium to underlying connective tissue. *(Page 79.)*

Connective tissues secrete an extracellular matrix that binds tissues together as organs. The matrix may be soft or hard, and this difference makes connective tissues as diverse as blood and bone. Fibroblasts contribute to these differences by secreting various mixes of amorphous ground substance, collagen, elastic, and reticular fibers, the principal matrix components. *(Page 80.)*

Connective tissue proper contains fibroblasts and includes loose and dense connective tissues, adipose tissue, and reticular tissue. Loose connective tissue contains soft, watery, ground substance and relatively few fibers, whereas tendons and ligaments are dense connective tissue made of especially profuse tough collagen networks. Adipose tissue stores fat, and reticular connective tissue contains delicate fiber networks that support soft organs, such as the liver and spleen. *(Page 83.)*

Specialized cells deposit the matrix of specialized connective tissues. Osteocytes maintain the hard matrix of bone, and chondrocytes maintain the matrix of hyaline, elastic, and fibrocartilage. Cells from the bone marrow and plasma proteins from the liver are the major components of blood. Embryonic fibroblasts deposit the matrix of mesenchyme, the connective tissue of embryos. *(Page 87.)*

Cancer cells invade tissues by penetrating basement membranes. Cancer cells disrupt basement membranes to which they normally have no access. *(Page 92.)*

STUDY QUESTIONS

OBJECTIVES

1. Describe the characteristic structures and functions of the four classes of tissues. **1**
2. Define the term *tissue,* and state how tissues differ from organs. **1**
3. Describe where the four different types of tissue are located in the wall of the intestine, and explain their anatomic relationships. **2**
4. List the major properties of nervous tissue, and describe the principal cells of this tissue. **3**
5. Muscle tissues contain three different types of contractile cells. Name these cells, and compare their structure and functions. In which organs is each type found? **4**
6. Describe seven properties characteristic of epithelial cells. **5**
7. Describe the different types of epithelia, and cite an example of each one. Name an organ in which your example occurs, and describe the function that the epithelium serves in that organ. **6**
8. Compare and contrast simple and stratified epithelia. **6**
9. Draw diagrams of a simple tubular gland, a compound alveolar gland, and a compound tubuloalveolar gland. Name examples of each type. **7**
10. Name a simple branched alveolar holocrine gland. **7**
11. Describe how connective tissues differ from nervous, muscular, and epithelial tissues. **8**
12. Draw diagrams of the major types of cells found in connective tissue, and describe the functions of each cell type. **9**
13. List the compounds in the extracellular matrix in connective tissue and describe their functions. **9**
14. Explain how the differences between loose and dense connective tissues are related to the composition of the matrix. **10**
15. Given your understanding of connective tissue anatomy, what changes does Ehlers-Danlos syndrome make in the composition of the extracellular matrix of connective tissue? How are cancer cells thought to penetrate basement membranes? **11**

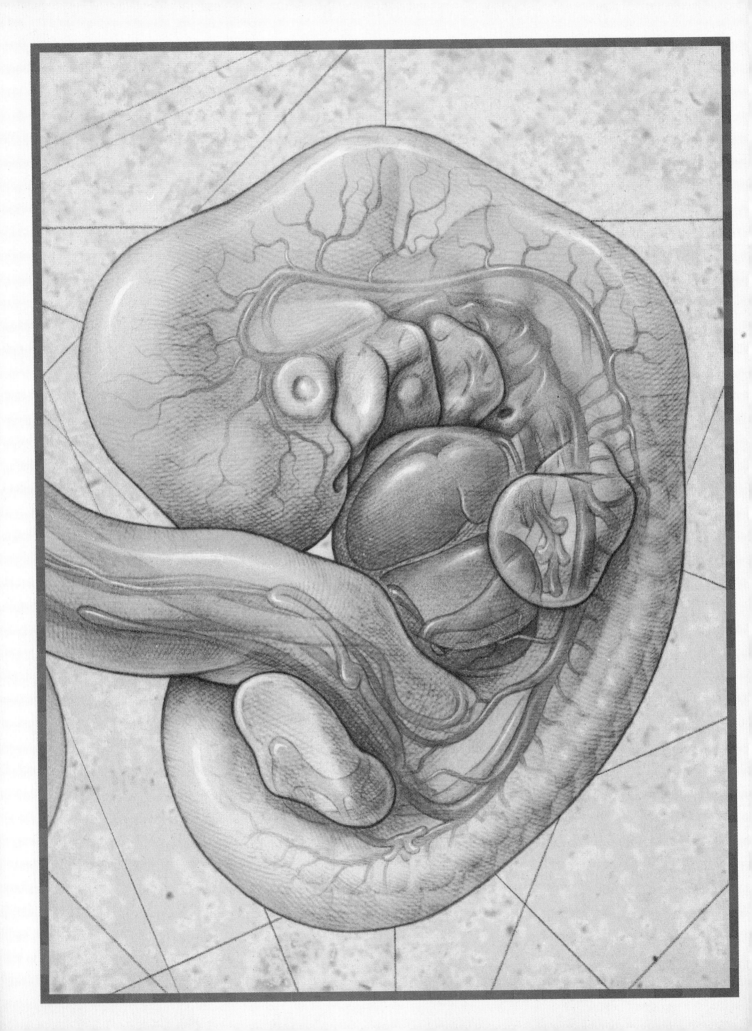

CHAPTER 4

Developmental Basics

OBJECTIVES

After studying this chapter, you should be able to do the following:

1. Name the three periods of development that contribute to the structures of the human body plan, and describe what each period accomplishes.
2. Describe the changes that transform the two-layered embryonic disk into three germ layers.
3. List the derivatives of each of the three germ layers.
4. Explain why some ectodermal structures are superficial, endodermal structures are deep, and mesodermal features are intermediate.
5. Describe the changes that transform the three-layered embryonic disk into the trunk.
6. Follow the developmental origins of the human body plan.

VOCABULARY TOOLBOX

morphogenesis   morph, *G* form; genesis, *G* origin.
morula   morul, *L* a little mulberry.

Prefixes that indicate position:

ecto-   derm, *G* outer skin. The skin develops from the external covering of the embryo.

endo-   derm, *G* inner skin. The stomach lining develops from the innermost layer.

meso-   derm, *G* middle skin. Muscle and bone develop from middle layers of the embryo.

RELATED TOPICS

- **Body cavities and membranes, Chapter 1**
- **Dermatomes, Chapters 12 and 18**
- **Fetal development, Chapter 26**
- **Human body plan, Chapter 1**
- **Myotomes, Chapters 10 and 11**
- **Pharynx, Chapter 23**
- **Stem cells, Chapter 2**

PERSPECTIVE

This chapter traces the embryonic origins of the human body plan. The first 5 weeks of development establish the fundamental structures from which the organ systems and tissues described in this text develop.

The first 5 weeks of human development establish the human body plan and lay down the embryonic tissues that will become adult organs. These tissues are called **organ rudiments** or **organ primordia** (prima, *L*, first), because each structure develops into a specific organ. The pancreas, for example, develops from the pancreatic rudiment of the gut, where it will reside in the adult. Although specific names of these precursor tissues may not literally contain the word rudiment or primordium, the principle is the same for all. All organ primordia contain stem cells (see Stem Cells, Chapter 2), the progenitor cells from which the individual organs develop.

The first 5 weeks are an especially sensitive time because genetic or environmental effects in the earliest formative stages can interfere with the basic steps by which all organ systems develop. Most spontaneous abortions occur during this period. Later effects tend to be less disastrous because they act within particular systems that already have been established.

The events that establish the embryo and its organ rudiments take place in three major periods. The first period begins with fertilization. A hollow ball of cells called the **blastocyst** is produced and implanted into the uterine lining, thus beginning embryonic development. Two different groups of cells are involved in these tasks. One group establishes the tissues that become the placenta, the embryo's connection with its mother, and the other group is a cluster of cells that will become the embryo and later the fetus.

The second period organizes the **embryonic disk** from which three germ layers develop—the ectoderm, mesoderm, and endoderm. The ectoderm gives rise to the nervous system and epidermis of the skin. From the mesoderm such structures as the heart, the vessels, kidneys and gonads, connective tissues, and the lining of the peritoneal cavity develop. The linings of the digestive system and lungs develop from endoderm.

Morphogenesis (MORF-oh-JEN-eh-sis; form formation) is the third period, in which the germ layers establish the body plan. As the tube-within-a-tube organization takes form, the organ primordia appear and set the stage for fetal development, when the organs begin to function and a new individual emerges from the organizational hierarchy (see Levels of Organization, Chapter 1) initiated by fertilization. Individual chapters describe the subsequent development of the organ systems, and Chapter 26, Reproductive System, follows a fetus from conception to birth.

- Define an organ rudiment.
- Name three germ layers.
- What major advances in embryonic development occur during the morphogenesis period?

FERTILIZATION TO BLASTOCYST

When a sperm fertilizes an egg, the new cell—the zygote—formed by this union quickly begins dividing into cells from which the blastocyst will develop 4 days later. This initial period of rapid cell division is called **cleavage** because the relatively large zygote divides into many smaller cells. First two, next three, and then four and six cells appear. The embryo becomes a **morula** (Figure 4.1A) when enough cells to form a solid berry-like ball have accumulated. Embryos of cattle and sheep can be cloned at this stage by splitting the morula in halves, but doing so with human embryos, as done in 1993, is highly controversial.

Two different populations of cells appear within the morula. Superficial cells become the **trophoblast,** the embryonic tissue that contributes to the placenta. Deeper cells become the embryo itself. On day four, when the embryo becomes a blastocyst (Figure 4.1B), a blastocyst cavity forms between the trophoblast and the **inner cell mass** of embryonic cells that remains attached to the trophoblast at one side. Since fertilization the embryo has traveled through the uterine tube to the uterus; on the fifth or sixth day the blastocyst begins to implant into the uterine mucosa (see Body Cavities and Membranes, Chapter 1). The trophoblast cells invade the mucosa by the twelfth day, when the mucosa entirely surrounds the blastocyst. The trophoblast and the surrounding uterine mucosa cells now form the placenta and the vascular connections that will sustain the fetus.

- Which developmental events do fertilization, cleavage, and the blastocyst accomplish?

EMBRYONIC DISK

The inner cell mass spreads deeper into the blastocyst during implantation, becoming a flat plate of tissue called the **embryonic disk** (Figure 4.1C). The embryonic disk consists of an upper layer of cells, the **epiblast,** from which the embryo itself develops, and a lower layer, the **hypoblast,** that produces extraembryonic tissues that are soon discarded. The **amniotic cavity,** lined by the **amnion,** fills the space above the hypoblast.

GASTRULATION

Beginning on the thirteenth day, a series of cell movements called gastrulation reorganizes the epiblast into three germ layers—the ectoderm, mesoderm and endoderm–from which the various organ rudiments develop. Gastrulation begins in the fu-

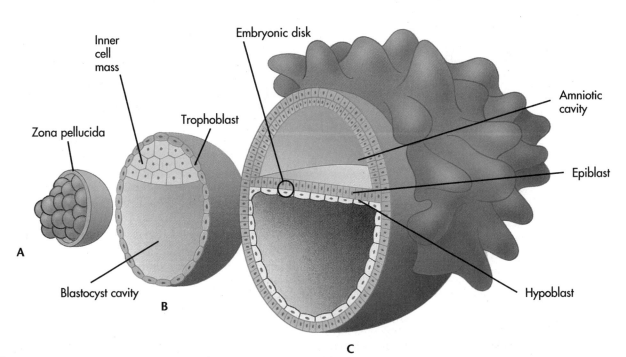

Figure 4.1

A, Morula. B, Blastocyst. C, Embryonic disk after implantation.

ture caudal end of the embryo (Figure 4.2A) when epiblast cells converge upon the midline and form the **primitive streak,** from which the cells spread outward beneath the epiblast as endoderm and mesoderm. The primitive streak is the earliest sign of bilateral symmetry in the embryonic disk (see Table 4.1 for a list of body plan features and corresponding embryonic structures). Converging cells enter the **primitive groove** and **primitive ridges** from the lateral surface of the epiblast. Cells from the cranial portion pass through the **primitive node** at the cranial end of the streak, spreading toward the future head of the embryo beneath the epiblast. Figure 4.2B shows that the endodermal cells displace the hypoblast, and mesodermal cells slip between the endoderm and the epiblast. Cells that remain in the epiblast become ectoderm. The embryonic disk is now trilaminar (three-

Table 4.1	EMBRYONIC STRUCTURES OF THE HUMAN BODY PLAN
Body Plan	**Structures**
Bilateral symmetry	Primitive streak
Dorsal hollow nerve tube	Neural tube
Vertebral column and notochord	Somites, sclerotomes, notochord
Axial segmentation	Somites
Dorsal and ventral cavities	Vertebral column, coelome
Tube-within-a-tube	Foregut, midgut, hindgut

layered); the uppermost layer is the ectoderm, the mesoderm is the intermediate layer, and the lowermost layer is the endoderm.

The positions of the germ layers preview the location of the structures that derive from them. The **ectoderm** covers the exterior of the embryo; it gives rise to the epidermis of the skin, the brain, the spinal cord, and the peripheral nerves (see Table 4.2 for a list of germ layer derivatives). From its intermediate position, **mesoderm** is the source of bone, muscle, and blood vessels, as well as the kidneys, the heart, the lining of the peritoneal cavity, and the wall of the digestive tube. **Endoderm,** the lowest layer, gives rise to deep internal structures, such as the linings of the digestive system, the liver, and the lungs. Gastrulation ends during the third week when the ectoderm and mesoderm begin to lay out additional features that place the ectoderm externally and the endoderm deep internally, with the mesoderm between.

- Name the layers that make up the embryonic disk before and after gastrulation.

ECTODERM

Two new regions appear within the ectoderm on day 17. The **neural plate** is a horseshoe-shaped region of ectoderm that is cranial to the primitive streak (Figure 4.3A) and surrounded by epidermal ectoderm. The neural plate will form the dorsal hollow nerve tube of the body plan, the tube from which the nervous system arises (see Table 4.1). **Epidermal ectoderm** forms the epidermis of the skin.

Beginning on day 18, as shown in Figure 4.4A-D, the neural plate folds into a hollow **neural tube** that encloses the **neural canal.** The anterior half of the neural tube (Figure 4.4E) develops into the brain, the posterior half becomes the spinal cord, and the neural canal develops into the ventricles of the brain and the spinal canal. **Neural crest** cells (Figure 4.4C and D) that border the neural plate form many different structures, such as the dentine of the teeth, facial bones, pigment cells of the hair and skin, and sensory neurons of the nervous system. Epidermal ectoderm

A

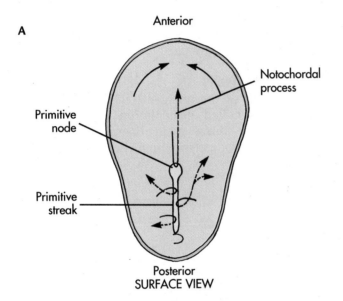

Anterior

Notochordal process

Primitive node

Primitive streak

Posterior
SURFACE VIEW

B

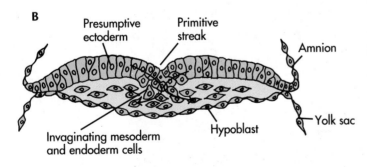

Presumptive ectoderm

Primitive streak

Amnion

Invaginating mesoderm and endoderm cells

Hypoblast

Yolk sac

TRANSVERSE SECTION

 **Figure 4.2**

Gastrulation. A, The primitive streak forms on day 13 in the midline of the embryonic disk. B, Epiblast cells converge upon the primitive streak, pass through it, and spread outward below. The endoderm displaces the cells of the hypoblast, and the mesoderm spreads between the endoderm and epiblast.

Table 4.2 SUMMARY OF GERM LAYER DERIVATIVES

Germ Layer	Structure	Derivatives
ECTODERM	Neural plate	Brain, spinal cord, motor neurons of spinal and cranial nerves
	Neural crest	Sensory neurons and ganglia of spinal and cranial nerves, neuroglia, pigment cells, adrenal medulla, bones of face, pharynx, and middle ear
	Epidermal ectoderm	Epidermis, hair, nails, sweat and mammary glands; lining of oral cavity, tongue, anus, vagina; lens, tooth enamel, semicircular canals
MESODERM	Notochord process	Notochord
	Somite mesoderm	Dermis, muscles of trunk and limbs, vertebral column
	Intermediate mesoderm	Mesonephric kidney, adrenal cortex, kidney nephrons
	Lateral plate mesoderm	Peritoneum, mesenteries; heart, vessels, blood cells; muscles of face, pharynx, larynx and neck; smooth muscle of gut; testis, ovary, oviducts, uterus
ENDODERM	Foregut	Lining of pharynx, esophagus, trachea and lung; thyroid gland; lining of Eustachian tube and tympanic membrane
	Midgut	Yolk sac; germ cells; lining of stomach, liver, gall bladder, pancreas, small and large intestine, appendix
	Hindgut	Allantois, large intestine, lining of rectum, urethra, prostate gland, urinary bladder, and vagina

From Lehman: Chordate Development, *1978, Hunter Publishing.*

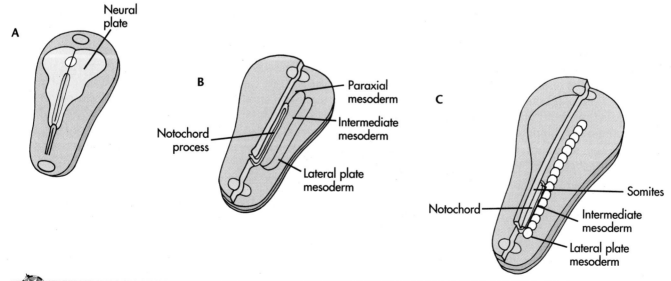

Figure 4.3

Embryonic disk. A, Neural plate. B, The notochord process and surrounding mesoderm. C, Somites.

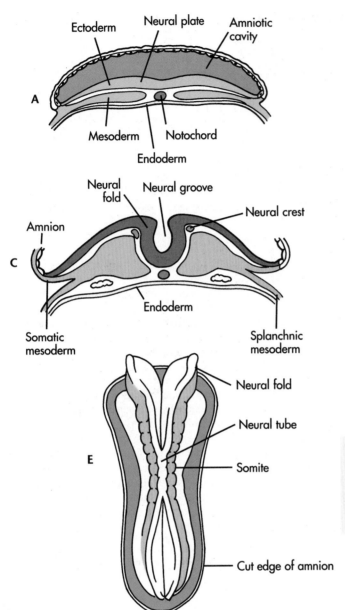

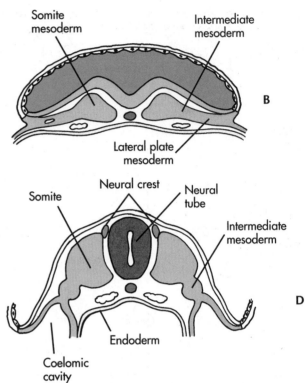

Figure 4.4

The neural tube begins to form on day 18. Seen in transverse section through the embryonic disk, the neural plate **(A)** begins to fold **(B)**. The folds approach each other **(C)** and fuse **(D)** as the neural tube. Dorsal view **(E)** shows the neural tube partially formed at day 22.

makes up the rest of the embryonic ectoderm cranially and laterally in Figures 4.3 and 4.4, as well as giving rise to the epidermis of the skin and the lenses of the eyes, among many other derivatives (see Table 4.2). While the neural tube is forming in the ectoderm, the precursors of the notochord and vertebral column appear in the mesoderm below.

- Which embryonic structure forms the neural tube?
- What structures derive from the neural tube?

MESODERM

As the mesoderm spreads away from the primitive node, it begins to segregate into several regions illustrated in Figure 4.3B. A strip of mesoderm cells that has moved cranially in the midline of the embryonic disk organizes as the **notochord process,** with the **paraxial mesoderm, intermediate mesoderm,** and **lateral plate mesoderm** arranged laterally. The noto-

chord process marks the location where the vertebral column will develop. Initially the notochord process is a hollow tube (Figure 4.4A) that opens at the primitive node and then condenses into a solid rod of tissue, which is the **notochord** itself. From notochordal process to notochord, this tissue marks the axis of the embryo, where the vertebral column later develops.

On each side of the notochord the paraxial mesoderm segregates into pairs of **somites,** shown in Figures 4.3C and 4.4C, D, and E, that contribute to the skeletal muscles, skeleton, dermis of the skin, and vertebral column. The notochord and somites are responsible for forming the vertebrae and are the principal features of axial segmentation in the human body plan (Table 4.1). Somites form sequentially, beginning cranially on day 20. A total of 44 pairs develops, and they are associated with different regions of the trunk. Each somite will form three derivatives—a **der-**

matome, a **myotome**, and a **sclerotome**—that contribute local sections of dermis to the skin, to skeletal muscles, and to the vertebral column, respectively (see Dermatomes, Chapters 12 and 18; and Myotomes, Chapters 10 and 11).

Lateral to the paraxial mesoderm, the strip of intermediate mesoderm seen in Figure 4.4B, C, and D forms parts of the urinary and reproductive systems (see Chapters 25 and 26). Lateral plate mesoderm is a plate of mesoderm on the lateral sides of the embryo. This tissue forms the visceral and parietal walls of the peritoneal cavity, the skeleton of the limbs and girdles, and, together with cells from the dermatomes, contributes to the dermis of the skin. Lateral plate mesoderm separates into two layers of mesoderm that line the **coelome,** an embryonic cavity that becomes the pleural and pericardial cavities, and the peritoneal cavity (see Table 4.2). Figures 4.4C and D show that the coelome forms between the **somatic mesoderm** and the **splanchnic mesoderm.** Together, somatic mesoderm and ectoderm are known as **somatopleura;** they become the body wall. In much the same manner, splanchnic mesoderm and endoderm are **splanchnopleura,** in which visceral organs, such as the lungs and digestive system, develop.

- Name the mesodermal structure that lies below the neural tube.
- Name the axially segmented mesodermal structures beside the neural tube.
- Name derivatives of the lateral plate mesoderm.

ENDODERM

The endoderm forms the **yolk sac** and differentiates into the digestive tube. The **foregut** extends forward beneath the neural tube and notochord to form the oral cavity, the pharynx, the esophagus, and the stomach (Figure 4.5). Simultaneously the **hindgut** extends posteriorly to become the cloaca, rectum, and large intestine. The **midgut** forms the small intestine and remains connected to the yolk sac until the sac is resorbed at about 6 weeks.

MORPHOGENESIS

BODY PLAN ATTAINED

Morphogenesis now completes the body plan, as organ rudiments appear within the germ layers. Figure 4.6 shows how the flat embryonic disk transforms into the tube-within-a-tube organization of the

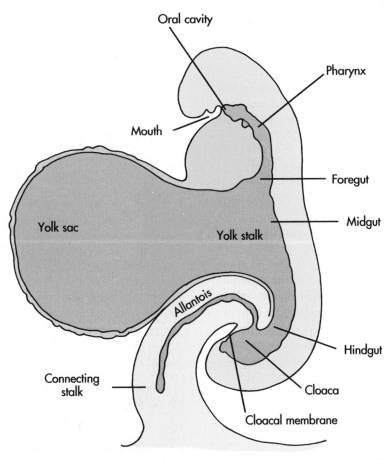

✳ **Figure 4.5**

Development of the endoderm. The foregut and hindgut extend cranially and caudally from the yolk sac.

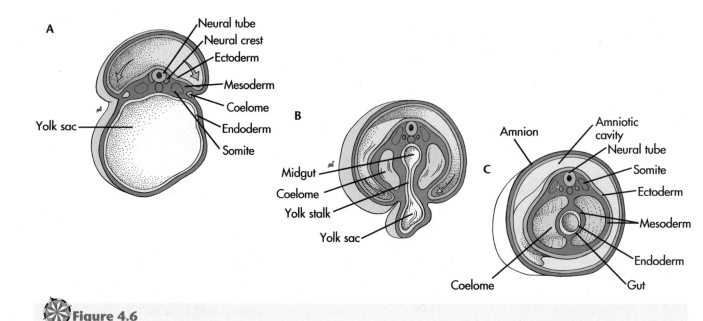

Figure 4.6

The tube-within-a-tube develops. **A,** The lateral plate mesoderm forms the coelome as the mesoderm separates into somatic and splanchnic layers. **B and C,** The body folds grow downward and envelop the digestive tube and coelome.

trunk (see Table 4.1). Figure 4.6A shows a transverse section through the embryonic disk, which is situated between the amniotic cavity and the yolk sac. The neural tube and notochord lie in the midline, with somites and lateral plate mesoderm at each side. The trunk begins to form when the somatopleura grows downward, enveloping the yolk sac between the **body folds** in Figure 4.6B. These folds meet and fuse (similar to fusion of the neural folds) in the ventral midline, retaining the digestive tube and coelome within the body wall, as Figure 4.6C illustrates. The tube-within-a-tube organization is now evident. The surrounding coelome will become the peritoneal cavity, and the sheets of splanchnopleura will become mesenteries.

Careful inspection of Figure 4.6C shows that epidermal ectoderm covers the exterior of the embryo, and the neural tube lies just beneath this ectoderm. True to its name, the mesoderm occupies intermediate positions, with somatopleura as the body wall and splanchnopleura as the dorsal mesentery and the wall of the gut. Finally, the endoderm, previously the lowermost layer of the embryonic disk, now lines the gut tube deep in the interior of the trunk. The amniotic cavity and amnion surround the entire embryo.

- Describe the motions of the somatopleura that form the trunk and body wall.

ORGAN RUDIMENTS

Organ rudiments now begin to appear within the germ layers. The precursors of familiar structures are evident on the exterior of a 5-week embryo illustrated in Figure 4.7. The most obvious features are the en-

larged head, the limb buds, and the row of myotomes that derives from the somites. The optic cup (eye) is visible cranial to the pharyngeal arches and pharyngeal slits. These arches and slits represent segmental structures of the pharynx that are described in Chapter 23, Digestive System. The first pharyngeal arch has divided into maxillary and mandibular processes,

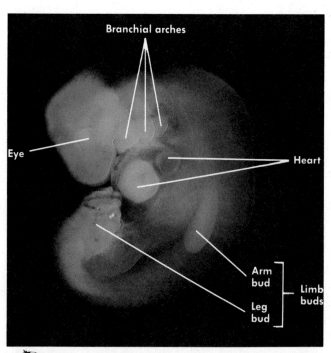

Figure 4.7

Lateral view, 37-day embryo.

from which the jaws develop. The cardiac bulge, which covers the rapidly beating heart, represents the future wall of the thorax. The developing liver rudiment forms the rounded hepatic bulge inferior to the cardiac bulge. The umbilical cord joins the embryo to the placenta, and the mesonephric kidney has appeared beneath the surface of the mesonephric bulge. The trunk and myotomes terminate in the tail, which represents the coccygeal region of the embryo (see Body Regions, Chapter 1).

ECTODERMAL DERIVATIVES

The brain enlarges and folds by the fifth week, as shown in Figure 4.8. The cranial and spinal nerves have taken positions along the brain and spinal cord that reflect the segmental arrangement of somites. Some spinal nerves already are entering the limb buds. The optic cup connects with the brain, and the otic vesicle gives rise to the inner ear. Cranial nerves communicate with the head and pharyngeal regions.

Spinal nerves derived from the neural crest and neural tube (see Table 4.2) mark segments of the spinal cord. The cervical nerves and the first few thoracic nerves form cervical and brachial plexuses that will innervate the upper extremities and pharyngeal region. Similar plexuses in the lumbar and sacral regions will innervate the lower extremities and pelvis.

- Name three ectodermal structures that appear during morphogenesis.

MESODERMAL DERIVATIVES

The somites have formed myotomes (Figure 4.8) that distribute themselves segmentally along the spinal cord. Muscles of the trunk and vertebral column develop from these myotomes, but the muscles and bones of the limbs and girdles derive from the somatic mesoderm of the lateral plate. The primordia of the facial and cervical muscles can be seen above the optic cup and in the pharyngeal arches.

The heart begins to beat on day 22. Figure 4.9 shows how the cardiovascular system has progressed by 28 days. The heart pumps blood through the aortic arches toward the head and pharyngeal regions. Two dorsal aortae (only the left one is shown) direct this

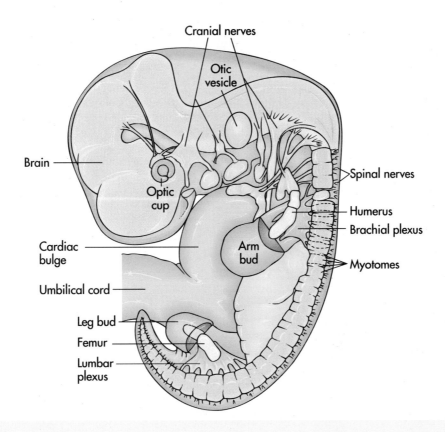

Figure 4.8
Ectodermal and mesodermal derivatives at 5 weeks.

flow to the trunk, where vitelline arteries deliver some blood to the yolk sac. Most blood, however, flows to the placenta by way of two common iliac arteries that lead to umbilical arteries in the umbilical cord. Blood returns from the placenta via the umbilical vein and from the yolk sac via vitelline veins, both of which lead through the liver to the heart.

- Name three mesodermal structures that appear during morphogenesis.

ENDODERMAL DERIVATIVES

Figure 4.9 also illustrates the digestive tube, which resembles the letter *T*. The yolk stalk is the stem of the *T* that leads to the umbilical cord. One crossbar of the *T* extends cranially as the foregut, and the hindgut is the other bar, reaching caudally toward the tail. The midgut lies at the junction of all three portions. The foregut already has differentiated into the oral cavity and a series of pharyngeal pouches between the pharyngeal arches. The eustachian tubes, palatine tonsils, parathyroid glands of the neck, and thymus develop from these pouches. Next, the lung buds project from the foregut, and the stomach, liver, gall bladder, and pancreas rudiments follow. The midgut becomes the intestine and the cephalic half of the colon. The hindgut terminates at the cloaca, which gives off the urogenital sinus, and the allantois, which extends forward into the body stalk. The hindgut forms the lining of the rectum, urethra, and urinary bladder. Beginning at the lung buds, the gut tube is derived from splanchnopleura, the composite tissue arising from endoderm and splanchnic mesoderm.

- Name the three divisions of the digestive tube, and a derivative of each one.

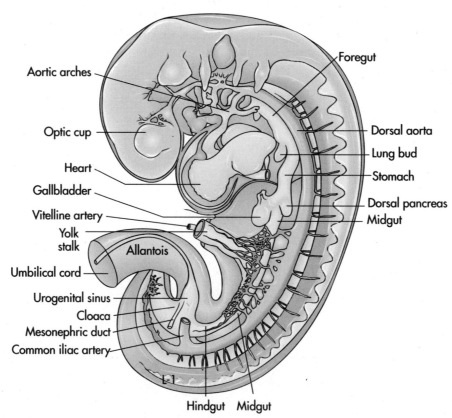

Figure 4.9
Mesodermal and endodermal derivatives at 5 weeks.

SUMMARY

The first 5 weeks of development establish the body plan and organ rudiments from which organ systems develop. Development begins at fertilization and quickly produces a large number of cells during the first week. In the second and third weeks, the embryonic disk elaborates the ectoderm, mesoderm, and endoderm—the three germ layers from which the body plan and organ rudiments arise. A period of morphogenesis transforms the flat embryonic disk into the tube-within-a-tube trunk during the fourth and fifth weeks as organs begin to develop. *(Page 100.)*

Cleavage divides the fertilized egg into cells from which the blastocyst develops. When 16 cells have formed, the embryo is a solid ball called the morula, which in turn becomes the hollow blastocyst 4 days after fertilization. Superficial cells of the blastocyst develop as the trophoblast, while the embryo itself develops from the inner cell mass. The blastocyst implants into the uterus on day 6, and the trophoblast begins to establish vascular connections with the mother by invading the uterine mucosa. The inner cell mass spreads into the blastocyst and becomes the embryonic disk, which consists of epiblast and hypoblast. *(Page 101.)*

Gastrulation reorganizes the embryonic disk into three definitive germ layers. Epiblast cells move through the primitive streak and then spread out beneath it as the endoderm and mesoderm. Endoderm displaces the hypoblast, and becomes the lowermost layer of the embryonic disk. Mesoderm spreads between the endoderm and the cells that remain in the epiblast, the ectoderm. These events establish bilateral symmetry, the first feature of the human body plan to appear. *(Page 101.)*

Germ layers develop into organs and tissues whose adult positions reflect the location of these layers in the embryo. Ectoderm, located on the uppermost surface of the embryonic disk, supplies cells that will become the brain, the spinal cord, and the epidermis. Endoderm develops into the inner linings of the gut and lungs, as well as portions of the reproductive and urinary systems. Mesoderm produces such intermediate structures as bone, muscle, vessels, and the peritoneal cavity. *(Page 102.)*

Morphogenesis reveals the structures of the human body plan as the flat embryonic disk transforms into the trunk. The neural plate forms the neural tube, the dorsal hollow nerve tube of the body plan from which the brain and spinal cord develop. Within the mesoderm the notochord and paired somites are responsible for the vertebral column, the dorsal body cavity, and axial segmentation. Intermediate mesoderm is the source of portions of the reproductive and urinary systems. The ventral body cavity arises from lateral plate mesoderm that separates into somatic and splanchnic layers that line the coelome. The endoderm forms the digestive tube-within-a-tube as the body folds meet and fuse around the gut. *(Page 105.)*

Major organ rudiments form during the fourth and fifth weeks. The brain enlarges and cranial and spinal nerves develop. Myotomes and limb buds appear lateral to the spinal cord. Pharyngeal arches appear in the head and cervical region, and the heart begins to pump blood through vessels that coalesce within the mesoderm. The endoderm forms rudimentary pharyngeal and digestive organs. All the major organs have been established and fetal development is about to begin. *(Page 106.)*

STUDY QUESTIONS   OBJECTIVES

1. What developmental advances do cleavage and the blastocyst accomplish?   **1**
2. Trace the origin of the germ layers from the epiblast and primitive streak.   **2**
3. From which germ layers do brain, epidermis, stomach, bone, heart, coelome, notochord, and skeletal muscle develop?   **3**
4. Explain why endodermal structures are deep internally and ectodermal derivatives are superficial.   **4**
5. Describe how the lateral plate mesoderm and its derivatives help to transform the embryonic disk into the tube-within-a-tube organization of the trunk.   **5**
6. Which embryonic structures are indicative of bilateral symmetry, axial segmentation, ventral body cavity, and dorsal hollow nerve tube?   **6**

PART II

SUPPORT, MOTION, AND SHAPE

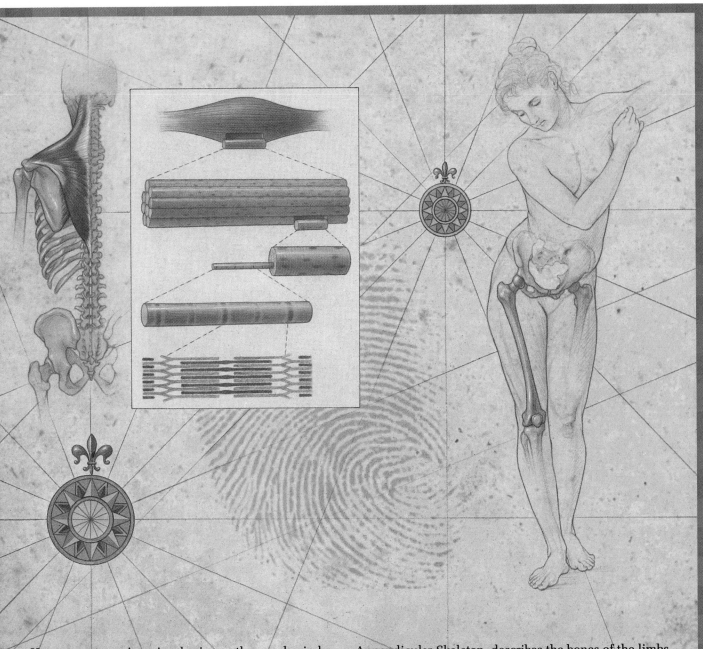

Human anatomy is a visual science; thus we begin by concentrating on the skeleton, the muscles, and the skin, which shape the human body. The rigid bones of the skeleton support and shape the body, as anyone who has broken an arm or leg clearly knows. Muscles move the body by moving the skeleton, and they also shape the body. Few of us will ever see the living muscles and bones themselves; we are content to accept them indirectly in their familiar locations beneath the skin.

Part II takes you beyond passing familiarity with the skeleton and muscles. **Chapter 5,** Bone and the Skeleton, introduces the cellular and molecular structure of bone and tells you that bones are shaped by the forces they sustain. Chapters 6 and 7 introduce the skeleton, whose bones reflect the forces that pull and twist them. **Chapter 6** is primarily concerned with the skull and vertebral column and how its architecture protects the head and supports the trunk. **Chapter 7,** Appendicular Skeleton, describes the bones of the limbs that move the body and perform its daily tasks. A set of joints is necessary if the skeleton is to flex and its vertebral column is to bend; **Chapter 8,** Articulations, covers these joints.

The next three chapters cover skeletal muscles much as the previous chapters covered bones. **Chapter 9,** Skeletal Muscle, describes the anatomy and function of muscles and muscle fibers, showing you how the contraction of individual muscle cells causes whole muscles to move the skeleton. **Chapter 10,** Muscles of the Head, Neck, and Trunk, discusses the muscles that move the head and trunk, and **Chapter 11** discusses the muscles that move the limbs. Finally, **Chapter 12** discusses the integument, which covers and conceals the bones and muscles and interacts with the organ systems that contribute energy, nutrition, raw materials, and nervous coordination to the body.

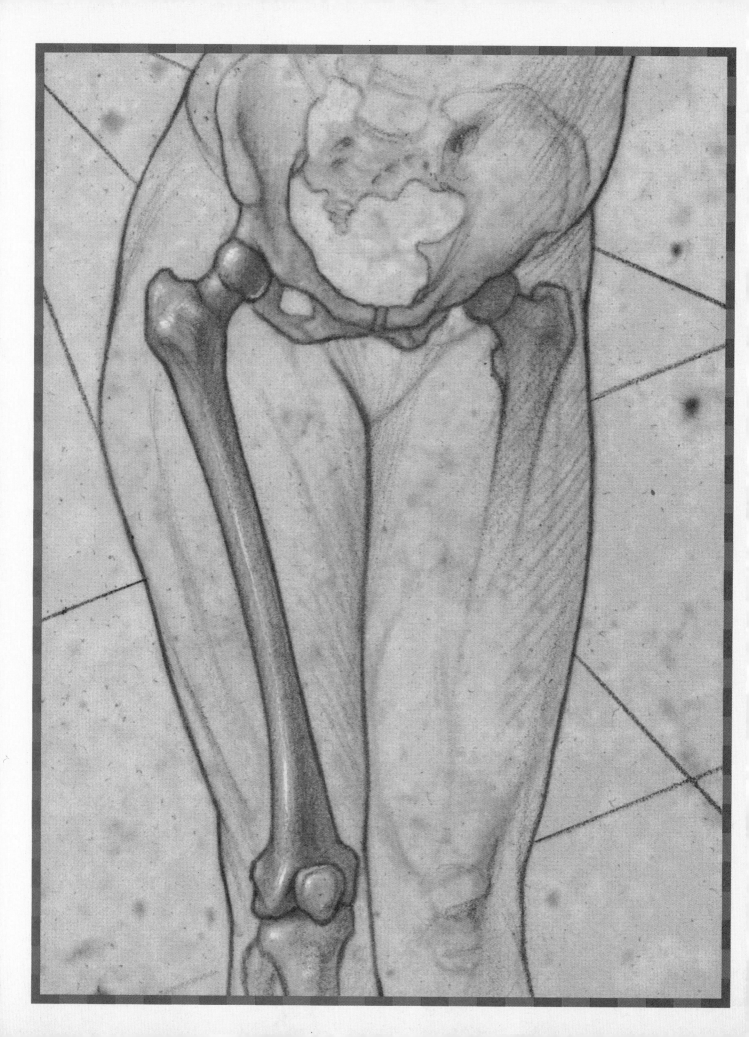

CHAPTER 5

Bone and the Skeleton

OBJECTIVES

Do the following things to understand the structure and function of bone in the human body:

1. Describe the functions of the skeleton.
2. Describe similarities and differences in the structure and function of the axial and appendicular skeletons.
3. Describe four generic features of bone.
4. Describe the five major categories of bones by shape.
5. Compare and contrast the structure and function of long, short, flat, sesamoid, and irregular bones.
6. Describe the structure and function of the molecular constituents of bone matrix.
7. Describe the cells of bone and their functions.
8. Compare and contrast the structures and functions of compact and cancellous bone.
9. Name the three types of cartilage, and describe the structure of each type.
10. Compare and contrast the steps in intramembranous and endochondral ossification.
11. Describe the properties and functions of primary and secondary ossification centers.
12. Describe calcification and decalcification.
13. Describe the action of hormones and vitamins on calcium in blood and bone.
14. Describe the steps of the remodeling cycle.
15. Describe effects of disease, aging, and inheritance on bone and what aspects of bone each affects.

VOCABULARY TOOLBOX

The following descriptive prefixes and suffixes refer to bone and cartilage:

chondro- *G* cartilage; as in **chondro**cytes (KON-droe-sites), cartilage cells.

osteo- bone; as in **osteo**cytes (OSS-tee-oh-sites), the cells of bone; and **osteo**genic, cells that form bone.

-physis *G* growth. The enlarged end of a long bone is an epi**physis** (eh-PIF-ih-sis).

The following prefixes indicate the position of bones or their components:

endo- *G* inner. **Endo**steum (end-OSS-tee-um) lines the interior surfaces of bones.

epi- *G* upon, over. **Epi**physes (eh-PIF-ih-sees) are the ends of long bones, such as the femur.

peri- *G* around. **Peri**osteum (PAIR-ee-OSS-tee-um) covers bone.

trans- *L* across. **Trans**verse processes extend from the sides of vertebrae.

RELATED TOPICS

- Articulations and ligaments, Chapter 8
- Bone and cartilage as connective tissue, Chapter 3
- Endocrine system and ossification, Chapter 22
- Homeostasis, Chapter 1
- Myeloid tissue, Chapter 15
- Skeletal muscles and tendons, Chapter 9

PERSPECTIVE

Bone is the hard, rigid material of the skeleton, and bones are the organs of this system, which supports, protects, and gives the body form and mobility. Bones also store calcium and house marrow. This chapter shows you that all bones are made of the same fundamental cells and matrix, covered with the same sheets of tissue, and nourished and stimulated by the same variety of vessels and nerves. Nevertheless, bones develop into various shapes and perform different functions. You will see how the protective bones of the skull differ dramatically from the supportive bones of the limbs. Disease, injury, and aging affect the structure and function of bones by interfering with these fundamentals. Learn these basic features of structure and function, and you will be able to describe the location of any bone in the skeleton and to estimate its function by recognizing its exterior shape and internal structure.

Steven expects to be 3 inches taller next year, not because he is growing rapidly but because he has not grown rapidly enough. At 18 years, he is 5 feet 2 inches tall, considerably below the the average height of men this age. To increase his height, Steven will invest a year of immobility and pain. His doctors plan to break the femur of one leg and the tibia and fibula of the other, then force the bones to grow by using metal frames to gradually draw the broken ends apart while they heal. If all is satisfactory in 6 months, his doctors will lengthen the remaining bones of Steven's legs. He expects to walk away a taller, perhaps happier, man.

Anatomy tells us that the shapes of bones, their location in the skeleton, and their functions are related to their internal structure and how they develop. Orthopedists, for example, can diagnose age and disease from X-ray photographs of the skeleton. By diminishing the calcium in bone, osteoporosis can dramatically weaken the skeleton and change its shape, as often occurs in older patients. It is well known that exercise enhances bone mass; ballerinas and weight lifters have heavier bones than sedentary persons and weightless astronauts.

Steven's operations are an attempt to take advantage of the ability of bone to grow and heal where it experiences mechanical stress. Steven may be so short because of insufficient growth hormone to stimulate his growth in the first place. This chapter introduces the major features of bone that Steven and his doctors are trying to use to his advantage.

Case study begins on p. 118.

BONES AND THE SKELETON

Bones are the hard, rigid organs of the skeleton; 206 of them help support, move, and shape the body as well as store calcium and house bone marrow. What each bone in the skeleton does depends on its own shape and its connections with the other bones, as well as on the muscles and tendons that move it and on vessels and nerves that supply nutrients and stimulation. Learn your way around the skeletal system by associating the functions of bones with their locations (Figure 5.1). Realize that, although all bones support the body and store calcium, the skull and vertebral column in particular protect the brain and spinal cord, and bones of the limbs enable motion.

AXIAL AND APPENDICULAR SKELETONS

The axial skeleton consists of the skull, vertebral column, ribs, and sternum. These bones support and protect the head and trunk (Figure 5.1, *yellow*). The skull mounts atop the vertebral column. It protects the brain and houses the eyes, ears, and or-

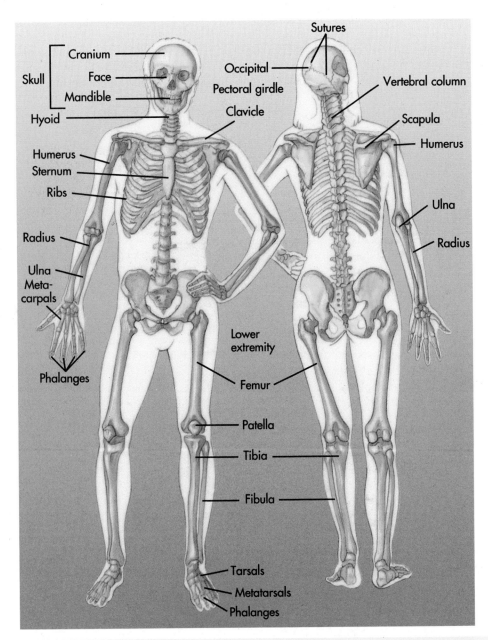

Figure 5.1

Axial and appendicular skeletons seen from anterior and posterior. The axial skeleton consists of skull, vertebral column, and ribs. Appendicular skeleton consists of the upper extremities and pectoral girdle and the lower extremities and pelvic girdle. Find the bones described in the text.

gans of balance. It also allows access to the lungs and the digestive system through the nose and mouth. Except for the mandible (the lower jaw) and hyoid bone (beneath the mandible), the bones of the skull are rigidly attached to each other, like the sides of a box, to protect the contents. (Teeth are not considered part of the skeleton because of structural and developmental differences from bone; see Chapter 23, Digestive System.)

The vertebral column supports the trunk; it behaves as if it were a rod that twists and bends with its owner's movements. It is a stack of 26 vertebrae held together by sheets of ligaments and straps of muscle. The vertebral column is the framework to which the muscles of the trunk attach. Vertebrae enclose the spinal cord and protect it from injury. Ribs articulate with the thoracic vertebrae and sternum and protect the heart and lungs.

The appendicular skeleton consists of the limbs and girdles. In contrast to the axial skeleton, the limbs of the **appendicular skeleton** (Figure 5.1, *orange*) move the body from place to place and give it dexterity. To accomplish this, the humerus, radius, and ulna of the upper limbs and the femur, tibia, and fibula of the lower limbs give the limbs length, whereas the carpal (wrist) and tarsal (ankle) bones provide flexibility for the hands and feet. Find these bones in Figure 5.1. The pelvic girdle anchors the lower limbs directly to the vertebral column and transfers the weight of the trunk to the lower limbs during walking. The pectoral girdle provides the flexibility to throw baseballs and play violins, because muscles and movable articulations (joints) join the scapulae and clavicles to the trunk. You can remember that girdles belong to the appendicular skeleton because *appendicular* has two *P*s, one for *p*elvic and one for *p*ectoral.

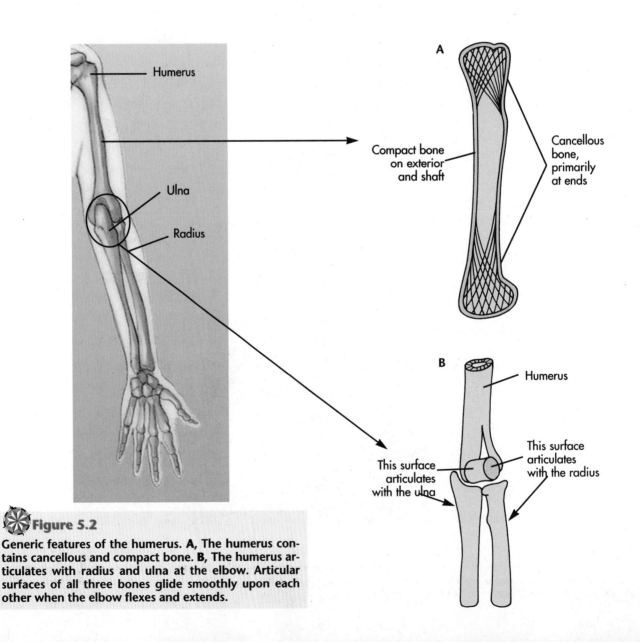

Figure 5.2

Generic features of the humerus. **A,** The humerus contains cancellous and compact bone. **B,** The humerus articulates with radius and ulna at the elbow. Articular surfaces of all three bones glide smoothly upon each other when the elbow flexes and extends.

- What division of the skeleton includes the skull and rib cage?
- Extremities and hands are located in which division?
- Name three functions of the skeleton.

GENERIC FEATURES OF BONE

Despite their differences, bones share four generic features that allow them to work together as parts of the skeleton. Because all bones have these properties, you can think of any individual bone as a variation on the basic theme. The mandible and carpals, accordingly, share these features but differ in form, location, and function.

1. **Bone consists of hard, bony extracellular matrix and cells** called osteocytes that are embedded within the matrix to maintain the bone metabolically. This arrangement takes on two principal forms (Figure 5.2A). All bones have a strong, dense outer layer of **compact bone** and an internal network of spongy **cancellous bone.** Both types distribute forces from neighboring bones to the rest of the skeleton.

2. **Most bones articulate with other bones.** Smooth, slippery **articular surfaces** allow bones to move relative to each other at joints (Figure 5.2B), as the humerus, radius, and ulna do at the elbow (see Figure 5.1). Other articulations (joints), such as the **sutures** between parietal and occipital bones of the skull (also Figure 5.1), allow little or no motion. (Chapter 8 discusses articulations in detail.)

3. **The periosteum covers all bones;** this tough sleeve of connective tissue has several functions (Figure 5.3A). Muscles, tendons, and ligaments attach to the outer **fibrous layer** of the periosteum and transfer forces into the bone by way of bundles of collagen called **perforating (Sharpey's) fibers.**

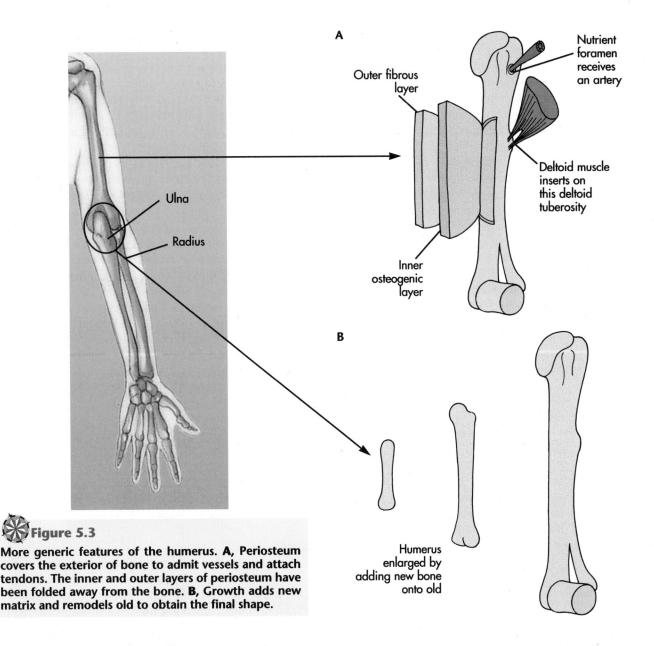

Figure 5.3

More generic features of the humerus. A, Periosteum covers the exterior of bone to admit vessels and attach tendons. The inner and outer layers of periosteum have been folded away from the bone. B, Growth adds new matrix and remodels old to obtain the final shape.

Ulna

Radius

A

Outer fibrous layer

Nutrient foramen receives an artery

Deltoid muscle inserts on this deltoid tuberosity

Inner osteogenic layer

B

Humerus enlarged by adding new bone onto old

Table 5.1	BONES CLASSIFIED		
Shape	**Name**	**Skeleton**	**Ossification**
Long	Femur, tibia, fibula, humerus, radius, ulna, metatarsals, metacarpals, phalanges	Appendicular	Endochondral
Short	Carpals, tarsals	Appendicular	Endochondral
Flat	Scapula	Appendicular	Endochondral
	Clavicle	Appendicular	Intramembranous
	Ribs, sternum	Axial	Endochondral
	Frontal, parietal, occipital, mandible	Axial	Intramembranous
Sesamoid	Patella	Appendicular	Endochondral
Irregular	Facial bones of skull, vertebrae	Axial	Endochondral
	Pelvis	Appendicular	Endochondral

Various bony knobs or **tuberosities** and **tubercles** receive these attachments. The periosteum also receives blood vessels and nerves that pass through holes, **foramina** (for-AM-in-ah, *L*, openings; singular, foramen), in the bone to communicate internally with marrow spaces and with osteocytes by way of branching canals in the matrix. The inner, **osteogenic** (bone forming) **layer**, of the periosteum contains osteoprogenitor cells, which are the source of osteocytes and bone matrix for growth. All internal canals and marrow cavities are lined with **endosteum**, which is also osteogenic.

4. **All bones grow by adding new matrix onto old** through a process called **appositional growth,** while they simultaneously remodel their shape by **resorbing** bone that has already been formed (Figure 5.3B). These two antagonistic processes are part of the **remodeling cycle** that shapes bones in response to metabolic conditions and mechanical forces. Steven's physicians are trying to stimulate appositional growth by fracturing his long bones.
   - What is the name of the cells that maintain bone metabolically?
   - What is the name of the tissue that covers all bones?
   - Name two types of joints between bones.
   - What is the name of the process that adds new bone matrix onto old?

BONES BY SHAPE

Anatomists recognize several different shapes of bones. The shape of a bone identifies what it does and whether it belongs to the axial or appendicular skeleton. Table 5.1 classifies **long, short, flat, sesamoid, and irregular bones,** according to their function and location in the skeleton. Check out the locations of these bones in Figure 5.1 and their shapes in Figure 5.4.

LONG

Long bones are noted for support and motion. The femur (FEE-mur; fem- *L*, the thigh) of the lower extremities and phalanges (FAL-anj-ees; *G*, bones of the fingers and toes) of the fingers are considered **long bones** because their length greatly exceeds their diameter. Locate them in the figures and find them in the table. A long, cylindrical **diaphysis** (DYE-a-fih-sis; shaft) provides the requisite shape. Even though phalanges are much shorter than femurs, they are nevertheless considered long bones.

The femur is a typical long bone. This bone of the thigh articulates with the pelvis and leg (see Figure 5.4A). The smooth, round, proximal **head** articulates in the acetabulum of the pelvis, and **condyles** articulate with the tibia and fibula at the knee. The **intercondylar fossa** is a depression that separates the distal condyles and receives ligaments from the tibia. A layer of hyaline cartilage covers all these heads with a slippery surface for smooth motion. **Greater** and **lesser trochanters** (troe-KANT-ers; *G*, a runner) receive tendons from muscles of the thigh. The periosteum covers the exterior of the femur's long **diaphysis,** and vessels and nerves pass in and out of the bone through **nutrient foramina.** The shaft consists of compact bone and contains a large **medullary cavity** filled with **yellow marrow.** Cancellous bone underlies the ends, or **epiphyses** of the bone and houses **red marrow.** Red and yellow marrow are different forms of the same tissue that forms erythrocytes, leukocytes, and adipose cells. Red marrow emphasizes erythrocyte formation, and yellow marrow primarily forms adipose cells. Marrow spaces are lined with **endosteum,** which is continuous with the periosteum externally by way of the nutrient foramina. A thickened disk of spongy bone between epiphyses and diaphysis marks the **epiphyseal line** where this bone stopped lengthening at maturity.

SHORT

Carpals and tarsals are short bones because they have essentially no diaphyses. The **scaphoid** bone of the wrist is a typical short bone (Figure 5.4B and Table 5.1), articulating with several other carpals for

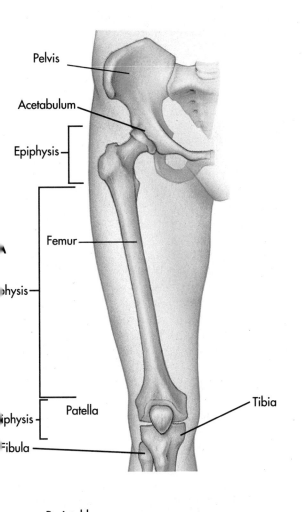

Pelvis
Acetabulum
Epiphysis
Femur
...physis
Patella
...physis
Fibula
A
Tibia

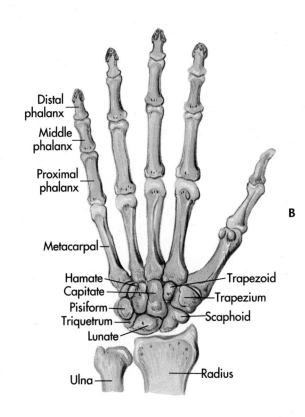

Distal phalanx
Middle phalanx
Proximal phalanx
Metacarpal
Hamate
Capitate
Pisiform
Triquetrum
Lunate
Ulna
Trapezoid
Trapezium
Scaphoid
Radius
B

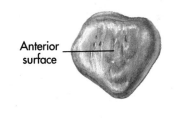

Anterior surface
D

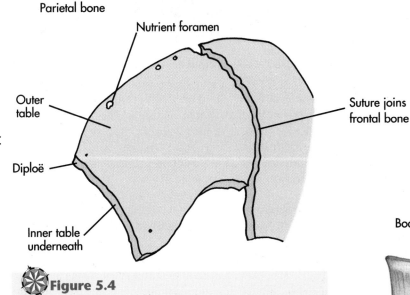

Parietal bone
Nutrient foramen
Outer table
Diploë
Inner table underneath
Suture joins frontal bone
C

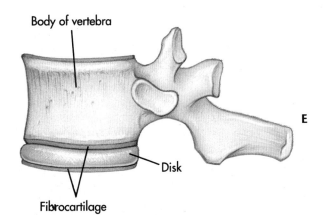

Body of vertebra
Disk
Fibrocartilage
E

✳ **Figure 5.4**

Bones according to shape. **A,** Long bones, such as the right femur (anterior view), have a long diaphysis or shaft. **B,** The scaphoid bone and other carpals are short bones that have no diaphysis. **C,** The parietal bone of the skull is a flat bone with outer table in view and inner table underneath. **D,** Sesamoid bones are oval, as is this patella of the knee. **E,** Vertebrae are irregular bones. Muscles and ligaments attach to transverse and spinous processes.

maximum flexibility in a short distance. Built with a core of cancellous bone and a compact bony exterior, the scaphoid articulates with the **radius** and through its neighbors to the thumb and **metacarpals** of the palm. Hyaline cartilage covers the smooth articular surfaces, and periosteum covers the surfaces between. Red marrow occupies the interior, and small foramina receive vessels and nerves from outside. Short bones have no epiphyses.

FLAT

Flat bones include the ribs, the sternum, clavicles, scapulae, and the bones of the cranium that surround the brain (see Table 5.1 and Figure 5.1). It is obvious that scapula and cranium are flat plates of bone, but the others might be considered long bones if it were not for their flat cross sections and cancellous interiors. Flat bones are found in both axial and appendicular skeletons.

The dome-shaped parietal bone of the cranium is a quintessentially flat bone (see Figure 5.4C). Nearly a flat plate, its surface conforms with the shape of the brain that it protects. The inner and outer surfaces of compact bone are called **tables,** and they are covered with periosteum. Cancellous bone and red marrow, which is called **diploe** in flat bones, occupies the narrow space between these tables. Nutrient foramina in both tables supply vessels and nerves. Sutures lock the edges of the parietal bone firmly into neighboring bones for a firm, protective covering for the brain. Ribs, sternum, scapulae, and clavicles have the same fundamental structure as the parietal bone, but flexible articulations allow them considerable motion.

SESAMOID

Sesamoid bones are oval, like sesame seeds. The patella, at the anterior surface of the knee in Figure 5.1, is the largest **sesamoid bone,** but others occur beneath the first metatarsal of the foot and in the metacarpals of the hand. Sesamoid bones develop inside or beside tendons. The **patella** (Figure 5.4D) is buried in the patellar tendon of the knee where it eases this strong tendon across the surface of the femur, and gives the tendon leverage when the knee flexes. Two small sesamoids beneath the ball of the first toe form a groove between them for the flexor hallucis longus tendon (hallux, *L*, the great toe) that flexes the first toe. The ball of this toe rests accordingly on the sesamoids instead of the tendon.

IRREGULAR

A bone is considered irregular when you cannot place it in other groups. Irregular bones usually indicate that muscles and bones attach to them from various directions. The vertebrae and facial bones are irregular bones of the axial skeleton, and in the appendicular skeleton the pelvic girdle consists of irregular bones (see Table 5.1 and Figure 5.1).

Vertebrae are good examples of irregular bones (see Figure 5.4E) because various **processes,** or spines, project from them in different directions to receive articulations or tendons and ligaments. In a vertebra, these processes extend from the massive drumlike **body** that supports the vertebral column. A **vertebral arch** extends around the spinal cord, and inferiorly, two notches for the intervertebral foramen conduct spinal nerves. The vertebral arch also gives off processes of its own. Some of these articulate smoothly (**articular processes**) with neighboring vertebrae and, as you might expect, have smooth surfaces of hyaline cartilage. Others, the **transverse** and **spinous processes,** receive perforating fibers from tendons and ligaments. The flat ends of the body articulate with adjacent vertebrae through **invertebral disks** of fibrocartilage that send perforating fibers into the bone. Cancellous bone and red marrow occupy the interior of each vertebra, and compact bone strengthens the exterior. As usual, periosteum and endosteum cover the exterior and interior, and nutrient foramina penetrate the body and its processes.

- What characteristic structure identifies a long bone?
- Name a flat bone.
- What shape is the patella?
- The vertebral column contains what variety of bone?

CELLS AND MATRIX OF BONE

Bones develop into the mature forms shown in Figure 5.4 through a series of processes that continue throughout the life of an individual. These processes are the subjects of later sections. The following section discusses the structure of mature bone tissue; others describe how bone ossifies and how various circumstances affect ossification and the final forms of bones.

MATRIX

Adult bone is a connective tissue (see Chapter 3, Connective Tissue) that consists of osteocytes embedded in a hard, rigid matrix of proteoglycan gel, collagen fibers, osteonectins, and the mineral hydroxyapatite, as Figure 5.5 shows. Together, the first three organic components of bone matrix are called **osteoid** (OSS-tee-oyd), to distinguish them from the inorganic constituent hydroxyapatite. **Proteoglycans** are huge macromolecular complexes of polysaccharide side chains linked by proteins to long central backbone polysaccharides. Because proteoglycans strongly bind water molecules, this fundamental amorphous ground substance of bone forms a firm gel. **Collagen**

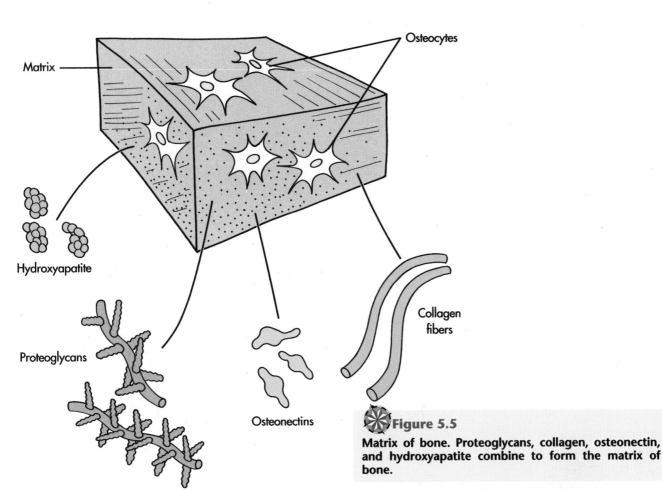

Figure 5.5

Matrix of bone. Proteoglycans, collagen, osteonectin, and hydroxyapatite combine to form the matrix of bone.

fibers occupy this gel and provide tensile strength. **Osteonectins** are proteins thought to bind collagen, proteoglycans, and hydroxyapatite together. **Hydroxyapatite** hardens the osteoid into an immensely strong material similar to fiberglass. Hydroxyapatite consists of calcium phosphate crystals, $Ca_5(PO_4)_3OH$, that deposit among the collagen fibers and proteoglycans.

Collagen fibers are arranged in two different patterns. In **lamellar bone** (la-MELL-ar; lamell-, *L,* a small plate; Figure 5.6A) these fibers are deposited in alternating layers; the fibers of one layer run perpendicular to the fibers of the next. Adult bones are lamellar, but in fetal bones the collagen fibers are deposited randomly at first. This is **woven bone** (Figure 5.6B), and lamellar bone soon replaces nearly all of it. Adults retain only traces of woven bone near sutures and tendon attachments and in the inner ear canals. How the collagen fibers become oriented in lamellar bone is unknown. More will be said of lamellar and woven bone in the Ossification section of this chapter.

CELLS

Osteocytes occupy spaces in the matrix called lacunae. Figure 5.7 shows **osteocytes** within **lacunae** (la-KUNE-ee; lacun-, *L,* a basin or space), cavities within the matrix. These irregularly shaped cells

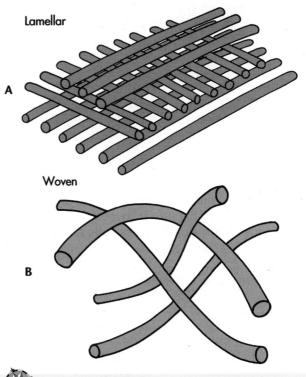

Figure 5.6

Lamellar and woven bone. A, Alternating layers of collagen fibers characterize lamellar bone. **B,** These fibers are arranged randomly in woven bone.

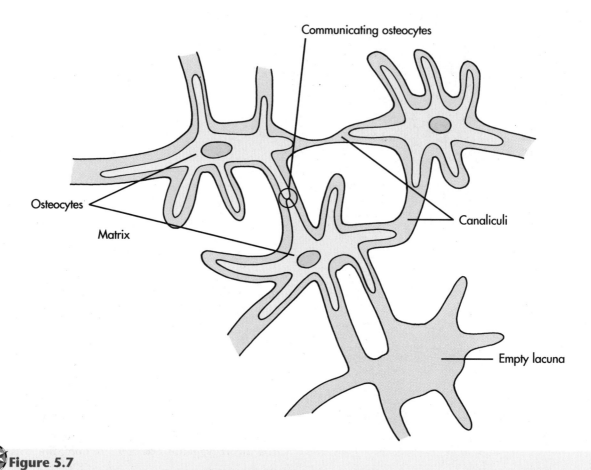

Communicating osteocytes

Osteocytes

Matrix

Canaliculi

Empty lacuna

Figure 5.7

Osteocytes occupy lacunae in the matrix. Neighboring osteocytes communicate by cellular processes that connect through canaliculi. Occasional lacunae have lost their osteocytes.

communicate along slender cellular processes that connect to each other through narrow tubules called **canaliculi** (KAN-al-IH-kue-lee; *L,* small canals). Osteocytes help maintain the matrix by depositing into it hydroxyapatite obtained from calcium and phosphate ions circulating in the blood. These substances and other metabolites reach the osteocytes through a network (described in the next section) of vessels and capillaries that enter through nutrient foramina and tunnel throughout the matrix to reach every lacuna.

Osteocytes are not the only cells of bone. Three other types of cells——the osteoblasts, osteoprogenitor cells, and osteoclasts, shown in Figure 5.8—are necessary for bone growth. **Osteoblasts** are responsible for appositional growth of bone. They reside in the periosteum and endosteum; from there they deposit new osteoid and calcify it with hydroxyapatite. Osteoblasts derive from osteoprogenitor cells (Figure 5.8A), and they become osteocytes when the calcified matrix encloses them in lacunae. **Osteoprogenitor cells** are stem cells, proliferating cells that produce osteoblasts and more osteoprogenitors by cell division. Continual proliferation of osteoprogenitors equips the perios-

teum and endosteum with osteoblasts throughout the life of an individual and thereby permits bones to remodel under mechanical stress and to heal when fractured. Needless to say, Steven will owe his additional height to his osteoprogenitor cells.

Osteoclasts help remodel bone by eroding hydroxyapatite from the old matrix. **Osteoclasts** are a different group of very large cells derived from myeloid tissue (MY-el-oyd; Figure 5.8B; see also Chapter 15, Blood) that prepare new surfaces for osteoblasts to deposit more bone. Osteoclasts develop like macrophages and muscle fibers—through fusion of precursor cells—and contain numerous nuclei. The interplay of osteoblast and osteoclast in deposition and resorption ultimately establishes the shapes of bones and is discussed later (see Remodeling Cycle). For now, however, please realize that the organization of cells and matrix described here differs in **compact** and **cancellous bone.**

- What cells are the source of all bone cells?
- Which bone cells deposit new bony matrix?
- What are the names of the organic components of bone matrix?

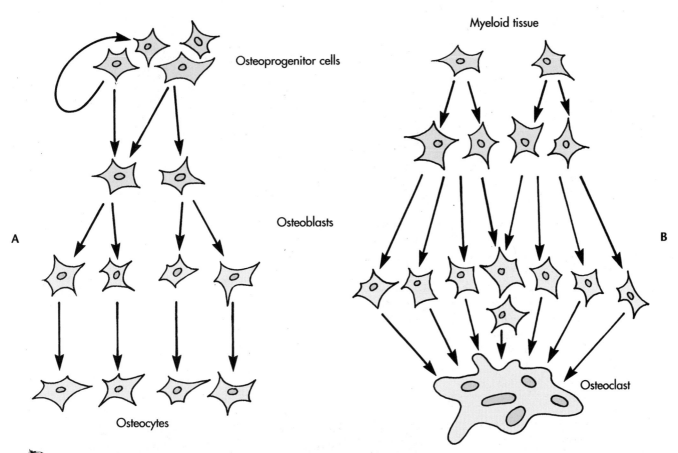

Myeloid tissue

Osteoprogenitor cells

Osteoblasts

A

B

Osteocytes

Osteoclast

✳ **Figure 5.8**

Bone cell lineage. **A,** Osteoprogenitor cells reproduce themselves and osteoblasts, which deposit matrix and become osteocytes. **B,** Proliferating myeloid cells of the bone marrow fuse together as large osteoclasts.

COMPACT BONE

Compact bone consists of osteons (OSS-tee-ahnz), cylindrical structures with several **concentric lamellae** of osteocytes and matrix surrounding a **central (Haversian) canal** (Figure 5.9A). Compact bone is lamellar bone; its collagen fibers alternate direction with each concentric lamella. Each central canal is lined with endosteum and contains an arteriole, a venule, and usually a nerve fiber. The vessels communicate with osteocytes deep within the layers of matrix. Osteocytes occupy lacunae between lamellae, connecting with each other and the central canal through canaliculi. The lamellar arrangement of osteons combines strength and rigidity with a network of microscopic passages to nourish and maintain living cells within the bone. Osteons communicate with each other and with the exterior of the bone by way of **horizontal (Volkmann's) canals** that connect the central canals and vessels of neighboring osteons

together. Some horizontal canals reach the exterior surface of the bone, emerging as nutrient foramina.

Osteons develop through appositional growth around blood vessels, as you see in Figure 5.9B. The result is compact, dense bone, with most osteons arranged in the direction of forces that the bone sustains. **Osteoblasts,** derived from the osteoprogenitor cells of the periosteum and endosteum, deposit concentric lamellae of bone around a vessel, enclosing it in a new central canal. As the osteoblasts imprison themselves in their own matrix, they become osteocytes and more osteoblasts from the central canal now deposit additional lamellae until no space for more lamellae remains.

Osteons develop inward toward the central vessel because the osteogenic cells always reside on the inside of the osteon; the outside lamella is always first to ossify and the innermost is the last. Neighboring os-

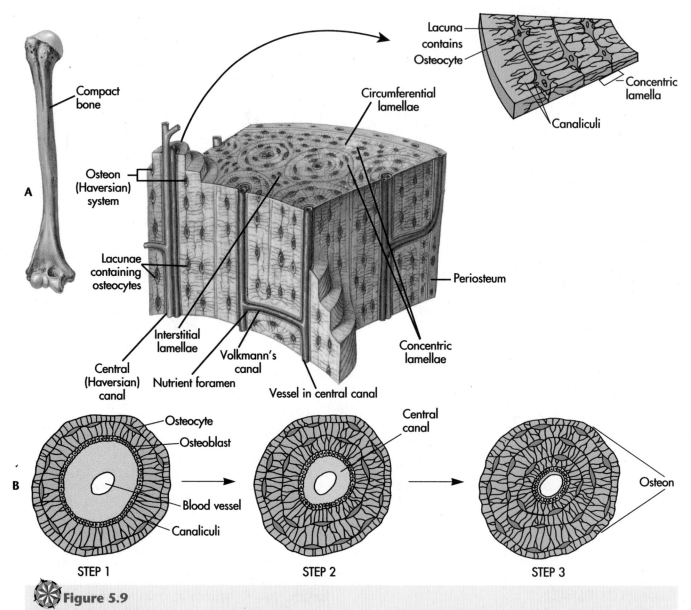

Figure 5.9

Compact bone. A, This block of compact bone from the shaft of a humerus displays several osteons. **B,** Osteons develop around blood vessels; the outermost lamella develops first. Osteoblasts deposit successive inner layers, yielding an osteon with a central canal lined by endosteum.

teons are eroded at the beginning of this process to provide space for the new osteon.

Development of new osteons is what Steven's physicians want to stimulate by fracturing the shafts of his long bones. Cancellous bone is of less concern to them because it does not add as much length to long bones.

CANCELLOUS BONE

Cancellous (spongy) bone is a network of bony trabeculae similar to the fibers of a bath sponge (Figure 5.10A). Adipose cells, red marrow, and vessels fill the spaces between the trabeculae (tra-BEK-you-lee; trab-, *L,* a beam or timber). The matrix is also lamellar

bone, but the layers are irregular and osteocytes are scattered throughout (Figure 5.10B). Endosteum covers the surface of trabeculae because there are no central canals inside them. Despite this lack, the endosteum communicates with osteocytes through canaliculi in the usual way. Osteoprogenitor cells supply the endosteum with osteoblasts for appositional growth. Again, osteoclasts erode the old matrix and osteoblasts deposit new matrix onto that site.

- What are the structural units of compact and cancellous bone called?
- Where is cancellous bone usually located?
- Why are Steven's physicians interested in promoting the formation of new osteons in his legs?

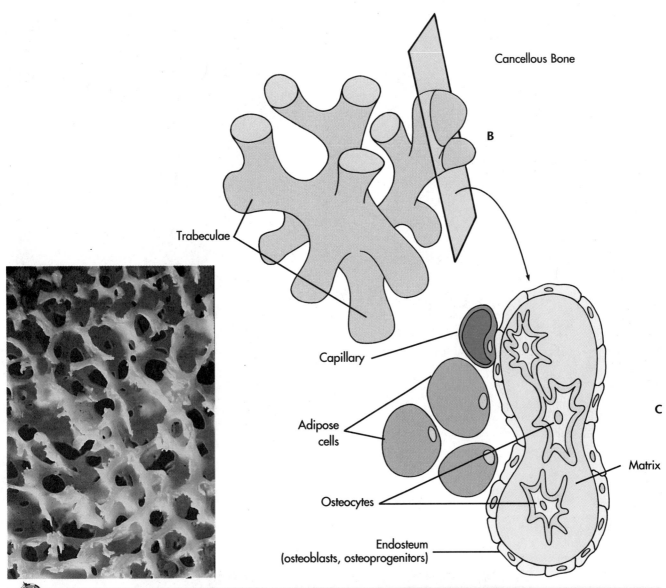

Figure 5.10

Cancellous bone is a network of fine trabeculae, seen in **(A)** photomicrograph and **(B)** illustration. A section through a trabecula **(C)** shows that it consists of lamellar bone with lacunae and osteocytes. Endosteum covers trabeculae; marrow tissue, capillaries, and adipose cells fill the spaces between trabeculae.

CARTILAGE

Cartilage is the stiff, firm connective tissue associated with joints in the skeleton and with other semirigid body parts, such as the nasal septum, the larynx, and the rings that support the trachea. Cartilage is the gristle (cartilag-, *L*, gristle) you prefer not to chew. The structure of cartilage closely resembles bone in several respects, but there are important differences that are shown in Figure 5.11A. Like the osteoblasts of bone, **chondroblasts** deposit cartilage matrix and imprison themselves in lacunae, where, like osteocytes, they become maintenance cells called **chondrocytes.** Unlike osteocytes, however, chondro-

cytes can divide, and lacunae frequently contain **cell nests** of two, four, or more daughter cells buried together in the extracellular matrix. A sleeve of **perichondrium,** analogous with the periosteum on bone, covers most varieties of cartilage. But unlike bone, cartilage lacks blood vessels and relies on diffusion from the perichondrium to supply nutrients and remove waste through the matrix.

Cartilage matrix contains proteoglycans and collagen fibers, but it lacks the hydroxyapatite crystals of bone. The usual **proteoglycan** constituents of bone are present in cartilage, and they form a firm gel that withstands compression and bending. Embedded in the proteoglycan gel are collagen fibers that give

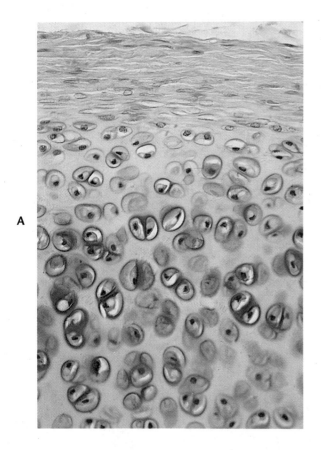

A

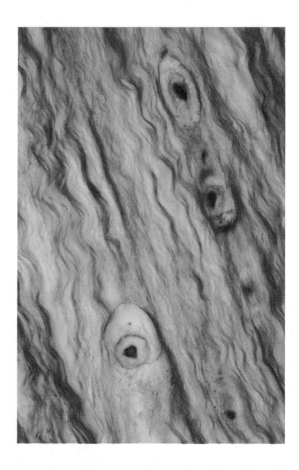

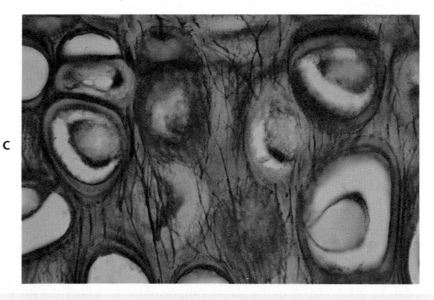

C

✦ **Figure 5.11**

Different forms of cartilage. **A,** Perichondrium covers hyaline cartilage, and nests of chondrocytes occupy many round lacunae in the matrix. **B,** Chondrocytes and lacunae are squeezed among the massive collagen fibers of fibrocartilage from an intervertebral disk. **C,** Elastic cartilage from the ear displays a network of elastic fibers in the matrix and nests of osteocytes within round lacunae.

cartilage tensile strength to resist forces that bend joints. **Chondronectin** binds the collagen and proteoglycans together, and **chondrocalcin** binds hydroxyapatite in zones of calcified cartilage adjacent to bone. Three types of cartilage are found throughout the body–**hyaline cartilage, fibrocartilage,** and **elastic cartilage.**

HYALINE CARTILAGE

This version of cartilage is the stiff, smooth, resilient, blue-white material that is widely associated with the skeleton. Fine collagen fibers penetrate the matrix, which may owe up to 70% of its weight to water. Hyaline cartilage may be covered by perichondrium, depending on the location and function of the cartilage. Hyaline cartilage stiffens the nasal septum, the larynx, and the walls of the trachea. It joins ribs to the sternum, is precursor to most of the skeleton, and remains temporarily until adulthood as epiphyseal plates and synchondroses in long bones (see Endochondral Ossification, below). In all these situations the entire surface of the cartilage binds to connective tissue, usually the bone or perichondrium. However, the hyaline cartilage that pads the articulating surfaces of bones in synovial joints (Chapter 8, Articulations) is exposed directly to the joint. The slick surface of this cartilage enables free movement, because perichondrium is absent from its open surface.

FIBROCARTILAGE

Thick, impressive bundles of collagen fibers characterize fibrocartilage, the material of intervertebral disks, pubic symphyses, and menisci of synovial joints, such as the knee (Figure 5.11B; see Chapter 8, Articulations). Fibroblasts deposit these fibers and then transform into chondroblasts that deposit small quantities of matrix between the collagen fiber bundles. This process traps the resulting chondrocytes into long, spindle-shaped lacunae and cell nests that usually follow the orientation of the fibers. There is no perichondrial covering.

ELASTIC CARTILAGE

As the name implies, elastic fibers and elastin characterize the matrix of this cartilage, which is made yellow by its fibers. Elastic fibers allow this cartilage to bend readily and return to its original shape. Flick your ears to see that this is so. In addition to your ears, elastic cartilage appears in the walls of the eustachian tubes, the external auditory canals, and the epiglottis of the larynx, where it snaps open and shut while you swallow. Elastic cartilage closely resembles hyaline cartilage. A perichondrium surrounds elastic cartilage, and round lacunae contain cell nests, as Figure 5.11C shows. Elastic cartilage develops from fibroblasts that produce fibers and then transform into chondroblasts that deposit matrix while the fibers are becoming elastic.

- In what regard does the matrix of cartilage and bone differ?
- Where is hyaline cartilage found?
- What is the major difference in composition between fibrocartilage and hyaline cartilage?
- What type of fiber characterizes elastic cartilage?

OSSIFICATION

The deposition of bone by osteoblasts and osteocytes is called **ossification.** Osteoblasts perform two principal steps in the process. First, they secrete osteoid around themselves, then they quickly **calcify** this material by causing hydroxyapatite to crystallize within it. Thereafter osteoblasts are considered to be osteocytes, which maintain the calcified state of the matrix.

Ossification occurs where osteoprogenitor cells and blood vessels encounter appropriate conditions. Bones ossify because the periosteum and endosteum provide osteoprogenitors and vessels, but what conditions start the process is uncertain. One possible stimulus may be **bone morphogenetic protein** (MORF-oh-JEN-et-ik; morph-, *G*, form; gen-, *G*, produce). This minor protein of bone matrix can stimulate osteoblasts to deposit osteoid and may be responsible for initiating ossification. Ossification also occurs pathologically in unusual locations, such as the walls of major arteries and bone spurs, which implies that osteoprogenitor cells, blood vessels, and bone morphogenetic protein can find these locations also. This information leads us to believe that the skeleton develops when and where the requisite cells and stimuli are present. Such circumstances begin in connective tissue membranes and in hyaline cartilage during the first month of fetal life and give rise to **intramembranous** and **endochondral ossification.**

INTRAMEMBRANOUS

This pattern of ossification begins in membranes (sheets) of connective tissue and produces the mandible, the clavicles, and the flat bones of the skull (Figure 5.12A). Such bones are called **membrane bones.** The parietal bone, for example, begins as a plaque of capillaries and mesenchyme tissue (see Chapter 3, Mesenchyme) within the fibrous covering of the fetal skull at the center of the bone-to-be. Some of these mesenchyme cells are osteoprogenitors and osteoblasts.

Clusters of osteoblasts now deposit matrix around themselves (Figure 5.12B), and the matrix quickly calcifies as woven bone. This structure is now a trabecula within the membrane, and the osteoblasts trapped within it are now osteocytes (Figure 5.12C). The ends of trabeculae grow outward by appositional growth toward the margin of the prospective bone, recruiting from the osteoprogenitors new osteoblasts to deposit more matrix onto what is al-

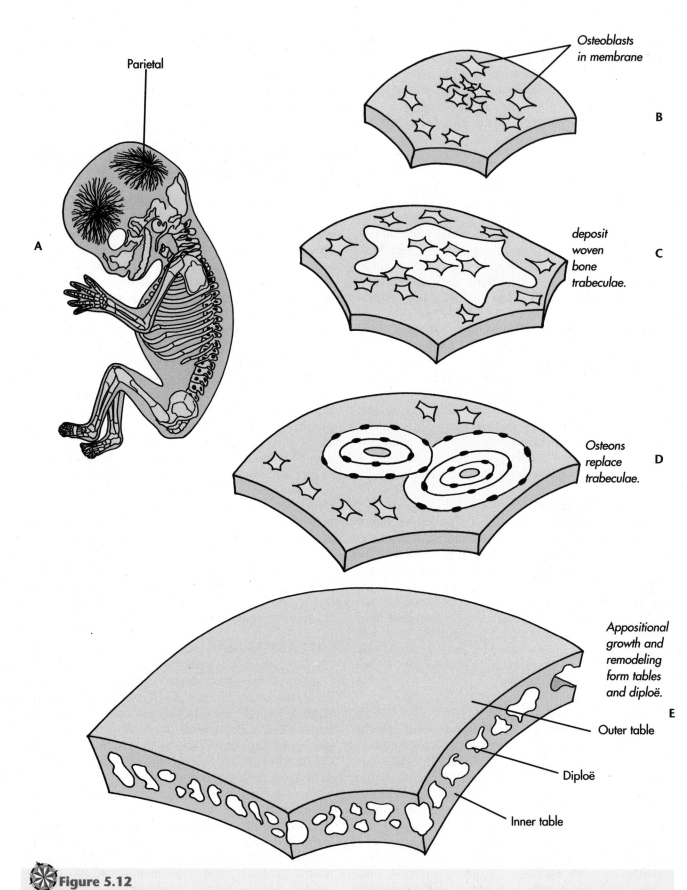

Parietal

Osteoblasts
in membrane

B

deposit
woven
bone
trabeculae.

C

Osteons
replace
trabeculae.

D

Appositional
growth and
remodeling
form tables
and diploë.

E

Outer table

Diploë

Inner table

A

Figure 5.12

Intramembranous ossification. A, The parietal bone begins to ossify in the connective tissue of the skull. B, Osteoblasts accumulate from mesenchyme cells in the membrane. C, Osteoblasts deposit woven bone. D, Osteoblasts replace it with osteons of compact bone. E, Appositional growth and remodeling form the inner and outer tables and diploe.

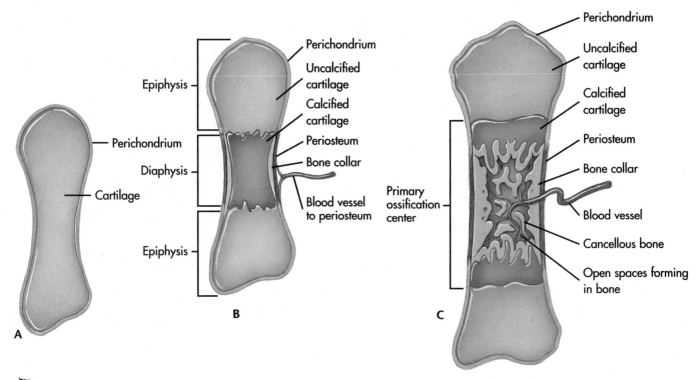

Figure 5.13
Endochondral ossification. **A,** Chondrocytes within the model die. **B,** This leaves vacant lacunae into which blood vessels enter as the primary ossification center. **C,** They bring osteoblasts that begin to deposit bone.

Continued.

ready present. Meanwhile, appositional growth also produces osteons at the center of this expanding structure (Figure 5.12D). Osteoclasts erode the old woven bone there, and osteoblasts deposit new lamellar bone around local blood vessels. These new osteons begin forming the outer and then the inner tables of compact bone (Figure 5.12E). Cancellous bone and diploe remain between these tables where osteoclasts remove old bone. After birth, the growing flat bones encounter their neighbors and form sutures where they join, but until then flexible connective tissue **fontanelles** cover the intervening space.

ENDOCHONDRAL

This process takes place within a hyaline cartilage model that resembles the mature bone. Endochondral ossification forms the long, short, and irregular bones of the skeleton that are accordingly called **chondrogenic bones.** Models of chondrogenic bones form in the second month of fetal life; **endochondral ossification** replaces them with bone and continues to enlarge the bone until adulthood. Although endochondral ossification begins in cartilage, ossification then proceeds to form trabeculae and osteons as it does within a membrane.

A long bone, such as the humerus, begins as a cylindrical cartilage model (Figure 5.13) covered with a sleeve of **perichondrium,** the cartilaginous equivalent of periosteum. **Chondrocytes** occupy lacunae

throughout the hyaline matrix of this cartilage, and the model grows longer as chondrocytes near its ends divide within their lacunae and deposit more cartilage. Eventually, descendants of these chondrocytes will produce the articular surfaces at the ends of these bones.

Preparations for ossification begin deep in the interior of the model as chondrocytes enlarge and die (Figure 5.13A), leaving huge vacant lacunae surrounded by eroded cartilage that is covered with a thin layer of calcified matrix (Figure 5.13B). Only when blood vessels invade these lacunae do osteoblasts begin to deposit bone. The first such invasion brings vessels to the center of the model and forms the **primary center of ossification.** After birth, later invasions set up **secondary ossification centers** at the ends of long bones that form the epiphyses in these locations.

Primary ossification centers form the shafts of long bones, as well as entire short and irregular bones. This process begins when capillaries envelop the middle of the perichondrium and invade the cartilage beneath, bringing osteoprogenitor cells with them (Figure 5.13B). Osteoblasts, developing from osteoprogenitor cells, deposit cancellous bone on the eroding cartilage within (Figure 5.13C), and they also deposit a collar of compact bone beneath the old perichondrium, now called the periosteum. Capillaries and ossification advance toward the growing cartilagi-

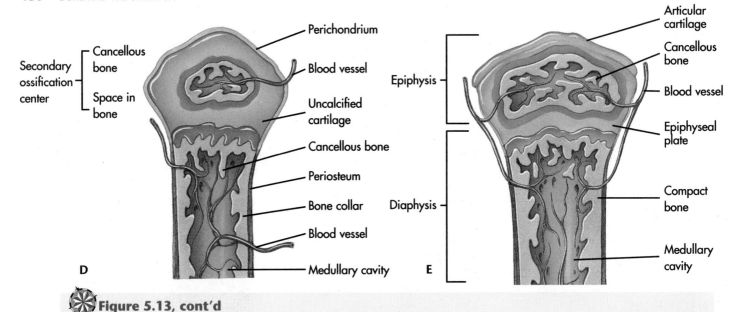

Figure 5.13, cont'd
D, Remodeling forms the medullary cavity, and secondary ossification begins in the epiphyses. **E,** Epiphyseal plates have formed between diaphysis and epiphyses.

nous ends of the model (Figure 5.13C, D), and a central medullary cavity forms behind these expanding zones as osteoclasts resorb most of the recently deposited cancellous bone (Figure 5.13D). Myeloid tissue now populates this new space and begins to form marrow (Figure 5.13E). The capillaries that originally formed the primary ossification center have now given way to nutrient arteries and veins, and their original entrances through the perichondrium are now the nutrient foramina of the shaft. At this point shortly before birth, the shaft of the humerus has ossified, but the cartilaginous ends have not.

Secondary ossification centers produce the epiphyses. Short and irregular bones usually lack epiphyses because no secondary ossification centers develop in these bones. Once again, osteoblasts replace the cartilage with cancellous bone (Figure 5.13D). Endosteum covers the interior of this cancellous bone, and a thin, external layer of chondrocytes establishes the articular cartilages on the ends of the bone. A thicker disk of growing cartilage, called the **epiphyseal plate,** still separates the epiphyses from the shaft (Figure 5.13E). Chondrocytes on the shaft side of this plate continue to proliferate and elongate the bone until the epiphyseal plate ossifies entirely in adulthood. The **epiphyseal line,** thickened cancellous bone, marks this location in mature long bones.

Ossification, or rather not enough of it, brought
▲ Steven to seek help. Since primary and secondary ossification either ceased sooner or progressed more

slowly than usual, his treatment attempts to lengthen his long bones by renewing the conditions and cells needed for ossification.

* Name the two forms of ossification.
* From which of these forms do parietal bones develop?
* Primary and secondary ossification centers occur in which shape of bone?

CALCIFICATION AND DECALCIFICATION

Calcification, you will recall, is the phase of ossification that crystallizes hydroxyapatite within the bone matrix. **Decalcification** is the opposite, the removal of hydroxyapatite. The quantity of hydroxyapatite in bone is related to the level of calcium in the blood and to the amount of phosphate present.

Parathormone and calcitonin regulate the concentration of calcium ions in the blood (Figure 5.14). These protein hormones influence whether bone cells deposit or resorb hydroxyapatite. The parathyroid glands secrete parathormone, and the thyroid gland produces calcitonin. **Parathormone (PTH)** is easy to remember; the name goes with the gland. PTH raises blood calcium by stimulating osteoclasts and probably osteocytes to release calcium from bone. When blood calcium drops below normal, the parathyroid gland releases more PTH to return calcium from bone to the blood. **Calcitonin (CT)** has the opposite effect; it lowers blood calcium by stimulat-

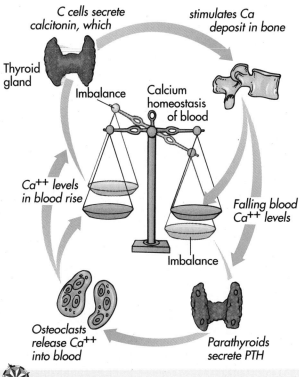

Figure 5.14

Calcium and calcification of bone. Low blood calcium levels stimulate parathormone to release calcium from bone into blood. High blood calcium stimulates calcitonin to restore calcium to bone.

ing osteocytes to deposit calcium into bone. When blood calcium is higher than normal, the thyroid gland releases more CT and the parathyroids stop releasing PTH, thereby restoring calcium from the blood to bone. The canaliculi and cellular processes that connect osteocytes to capillaries help transfer calcium between blood and bone. To summarize, the level of calcium circulating in blood reflects the balance between the antagonistic effects of calcitonin and parathormone; this in turn influences calcification and decalcification of bone. (Read more about these hormones in Chapter 22, Endocrines and Chapter 1, Homeostasis.)

Vitamin D also influences blood calcium and calcification. This vitamin stimulates the intestine to absorb calcium from the diet. Indeed, parathormone activates vitamin D in this role. Even with adequate calcium in one's food, lack of vitamin D leads to calcium deficiency because the intestine cannot transport enough into the blood. With little dietary calcium to draw from, the body decalcifies its bones to restore the deficit. See Rickets and Osteomalacia, this chapter, for details.

Finally, phosphate, the other component of hydroxyapatite, deserves mention. The enzyme **alka-**

line phosphatase (FOSS-fah-taze) plays an important role in supplying enough phosphate ions for calcium and phosphate to crystallize as hydroxyapatite. Alkaline phosphatase occurs in calcifying bone matrix, where it removes phosphate ions from various metabolic molecules. Intramembranous and endochondral ossification appear to depend on this enzyme to supply phosphate at the site of calcification.

- Which substances stimulate deposition and resorption of bone?

REMODELING CYCLE

Most anatomists agree that bones accommodate to mechanical forces. Local tension increases deposition, and pressure increases resorption. In femurs, trabeculae spread along lines of force in the epiphyses (Figure 5.15A) and transfer these forces to the thick, compact bone of the shaft. The process that focuses bone into such locations is known as **remodeling** or as the **remodeling cycle.**

In remodeling, ossification and resorption of bone cooperate with each other. Osteoblasts need a surface to guide them where to deposit new lamellae of bone, and osteoclasts provide this surface by eroding old bone from trabeculae and osteons. Resorption and ossification repeat continuously throughout life, bringing changes as aging and disease affect them. Figure 5.16 describes this cycle.

An osteon (Figure 5.16A) is eroded as osteoclasts resorb matrix from it (Figure 5.16B). Osteoblasts now accumulate on the new surface (Figure 5.16C) carved from the old osteon, and the space fills with a new osteon (Figure 5.16D). This new structure redistributes some of the mechanical force on the bone, and the new cycle continues as these new conditions constantly readjust to the forces that bone sustains.

But what establishes where bone will be removed and added? One attractive hypothesis is that minute **piezoelectric** (PIE-eh-zoh-electric; piezo-, *G*, squeeze) currents distribute trabeculae and osteons along the lines of stress they experience. Piezoelectric materials deform slightly when electric current passes through them, and vice versa, when stretched, twisted, or compressed they produce minute electric currents whose strength and direction change with the force and direction of the stress. The quartz crystal in your digital watch is piezoelectric, vibrating 32,768 times per second with current from its battery.

Considerable evidence that the piezoelectric effect does remodel bone comes from the facts that bones are piezoelectric and cancellous bone follows lines of force. Bend a long bone in Figure 5.15B slightly, and the inside of the bend, being compressed, accumulates more negative ions while positive ions collect on

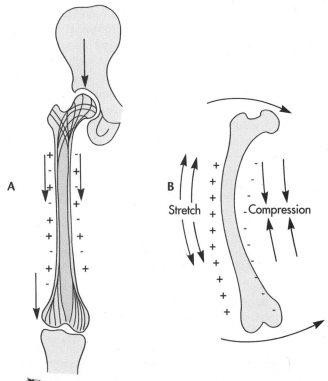

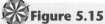**Figure 5.15**

Lines of stress and piezoelectricity in bone. **A,** Lines of stress transfer the body weight from the pelvis, through the femur, to the leg. **B,** Forces tend to bend and twist bone, and the piezoelectric effect accumulates positive ions where bone stretches and negative ions where it is compressed.

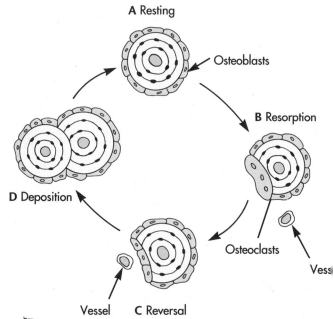

Figure 5.16

The remodeling cycle. Osteoclasts invade resting bone (**A**) and erode some of it (**B**). Osteoblasts replace the osteoclasts (**C**) reversing the erosion process. Deposition adds a new osteon and reaches a new resting phase (**D**).

the outside where it stretches. New patterns of stress can redistribute trabeculae. Although the wearer does not realize it, a new pair of high-heeled shoes can rearrange trabeculae even after a few wearings. Minute electric currents can accelerate healing of broken bones. But the central question—how do mechanical stresses direct osteoclasts and osteoblasts where to remodel bone—has no clear answer.

- What function do osteoclasts perform in the remodeling cycle of bone?
- What role might the piezoelectric effect play in the deposition of new bone on lines of force?

CONDITIONS AFFECT-ING BONE

Various circumstances affect bone, from the time the fetus first forms cartilage models to old age. Some conditions are genetic, others can be traced to dietary deficiencies, and still others are accidental. All attack some aspect of bone structure, as Figure 5.17 shows.

POLYDACTYLY AND SYNDACTYLY

Polydactyly and syndactyly are genetic conditions that change the number of fingers and toes (Figure 5.17) by controlling how many cartilage models of these bones form in early fetal life. Polydactyly (po-lee-DAK-till-ee; dactyl-, *G,* finger or toe) is caused by a recessive gene that introduces extra fingers and toes by adding extra cartilaginous models of the phalanges. Syndactyly (sin-DAK-till-ee; also recessive) fuses such models together and reduces the number of digits. Though rare in humans, these genes are common in animals. Surgery can sometimes remove the extra digits and separate fused ones.

OSTEOPOROSIS

Bone resorption outpaces ossification in osteoporosis (OSS-tee-oh-PORE-oh-sis), a disease of older people that is especially common among women after menopause. Trabeculae and osteons are thinner, and there are fewer of them (Figure 5.18A). Thus weakened, vertebrae suffer compression fractures, femurs break near the head and greater trochanter, and healing slows. Estrogen therapy, weight-bearing exercise,

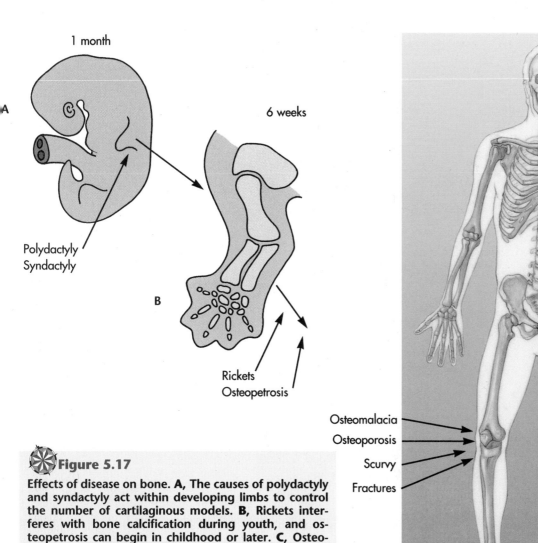

1 month

6 weeks

A

Polydactyly
Syndactyly

B

Rickets
Osteopetrosis

C

Osteomalacia
Osteoporosis
Scurvy
Fractures

Figure 5.17

Effects of disease on bone. A, The causes of polydactyly and syndactyly act within developing limbs to control the number of cartilaginous models. **B,** Rickets interferes with bone calcification during youth, and osteopetrosis can begin in childhood or later. **C,** Osteomalacia, osteoporosis, scurvy, and fractures can intervene after the remodeling cycle has established the mature forms of bone.

increased dietary calcium, and vitamin D improve but do not cure this condition. Disorders directly at the resorption site may explain how the remodeling cycle becomes unbalanced. Osteoporosis illustrates what can happen when deposition of bone fails to keep up with resorption.

RICKETS AND OSTEOMALACIA

Patients with rickets or osteomalacia do not have enough calcium available to supply the normal demands of ossification, so bones are weakened (Figure 5.18B). Vitamin D deficiency usually causes these diseases of the deposition phase of the remodeling cycle. Vitamin D stimulates the small intestine to absorb calcium into the bloodstream. Without adequate quantities of this vitamin to bring dietary calcium into the blood, osteoclasts resorb calcium from bone instead.

Vitamin D deficiency has rather different effects on children and adults. It produces rickets in children when insufficient calcium weakens the rapidly growing bone matrix and long bones bend with the weight of the body. Permanent deformation can result. In osteomalacia (OSS-tee-oh-MAL-ay-she-ah), the adult counterpart of rickets, remodeled bone calcifies poorly. Although vitamin D deficiency is frequently the cause, other disorders that interfere with the intestine's ability to absorb calcium or that accelerate its excretion by the kidneys can also lead to osteomalacia.

SCURVY

Scurvy weakens bones for lack of collagen fibers, not from insufficient calcium. Osteoblasts require dietary vitamin C to secrete collagen into the osteoid. Vitamin C deficiency causes this disease, rarely

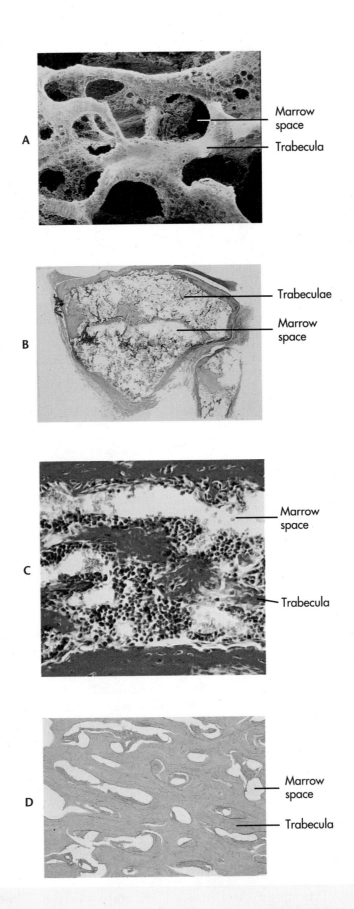

Figure 5.18

Different conditions affect the structure of bone. **A,** Osteoporosis reduces the thickness of trabeculae in cancellous bone. **B,** Rickets and osteomalacia also reduce trabeculae. **C,** Normal cancellous bone. **D,** Osteopetrosis, however, nearly obliterates the marrow spaces with very thick trabeculae.

seen in modern medicine. Calcification is excellent in scurvy; the problem is that osteoblasts deposit insufficient matrix to calcify. Vitamin C deficiency prevents osteoblasts from producing the collagen fibers of the osteoid. Hence teeth fall out, long bones break, and vertebrae suffer compression fractures as the body presses on them. Treatment supplies the missing vitamin C.

OSTEOPETROSIS

Osteopetrosis fails to remodel bone because there are no osteoclasts to do so. No weakened bones occur in osteopetrosis (OSS-tee-oh-peh-TROE-sis; petr-, *G,* stone, rock). Excessive ossification (Figure 5.18D) is the difficulty in this rare, human genetic disease of the resorption side of the remodeling cycle. The medullary cavities and epiphyses of long bones fill with thick, heavy trabeculae and nearly obliterate all marrow spaces. Judging from animal forms of this disease, a recessive gene prevents resorption for lack of osteoclasts. Treatment tries to supply the missing osteoclasts from the bone marrow of individuals who can resorb bone normally. Grafts of myeloid tissues that form osteoclasts have cured this disease in animals. Human bone marrow transplants have been only partially successful.

FRACTURES

Fractures temporarily accelerate resorption and deposition while the bone repairs itself. A simple fracture certainly and painfully illustrates the skeleton's role in support and mobility. Figure 5.19 classifies the different types of fractures a long bone can sustain. In a simple fracture the diaphysis breaks cleanly, tearing muscles and periosteum, and breaking osteons and trabeculae. A physician can align the fragments and reset them. Healing progresses through five steps illustrated by Figure 5.20. Local cells begin to die when broken vessels and clotted blood curtail their nutrition (Figure 5.20A). Acute inflammation begins within a day, as polymorphonuclear leucocytes and macrophages remove the dead and dying cells by phagocytosis (for descriptions of these cells see Chapter 15, Blood). Osteoclasts attack dead bone matrix, clearing the damage and preparing sites for new bone. Osteoprogenitor cells now proliferate (Figure 5.20B) and form a callus (a pad of fibrous connective tissue) that bridges the break with cartilage and bony trabeculae. Endochondral ossification (Figure 5.20C) replaces this cartilage with bone, and the bone heals (Figure 5.20D). When physicians use bone screws to reconnect the fragments, osteogenic cells repair the break directly and less callus forms, but osteons take longer to develop. In just the opposite strategy, Steven's operation seeks to stimulate more callus by separating the fragments while callus develops between them.

AGING

Beginning in your forties, your osteoclasts will resorb more bone than your osteoblasts will replace, and over the following 30 years, you can expect to lose about 12% of your bone mass if you are male and as much as 25% if you are female. Most of this loss comes from cancellous bone because cancellous bone is much more abundant than compact bone. There is simply more cancellous bony surface for osteoblasts to attack. Trabeculae become thinner and fewer in

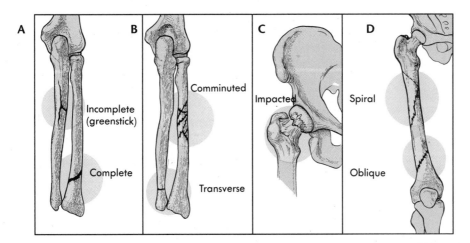

Figure 5.19

Different types of fractures. **A,** Bones may split like green sticks (incomplete) or break cleanly without puncturing the skin (complete). Compound fractures (not shown) open the skin and may introduce infection. **B,** Forces may crush bones into fragments (comminuted) or snap directly across the bone (transverse). **C,** Forces may push the ends of bones together (impacted fractures). **D,** Twisting forces produce spiral and oblique fractures

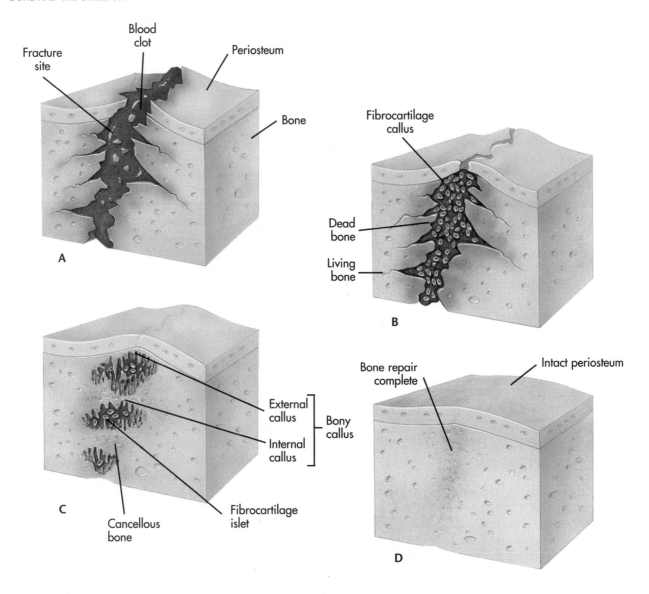

Fracture site

Blood clot

Periosteum

Bone

A

Fibrocartilage callus

Dead bone

Living bone

B

External callus

Internal callus

} Bony callus

Cancellous bone

Fibrocartilage islet

C

Bone repair complete

Intact periosteum

D

⚜ **Figure 5.20**

Healing of a simple fracture. **A,** Damaged cells are cleared from the break. **B,** Chondrocytes bridge the break with a fibrocartilage callus. **C,** Osteoblasts consolidate the callus with cancellous bone. **D,** Remodeling restores compact bone.

number, and bones can fracture under loads that would have posed no risk a few years earlier. Loss of bone is called **osteopenia** and is due in part to the "use it or lose it" nature of bone remodeling. Lack of exercise contributes to degenerative bone loss because the normal mechanical forces that stimulate bone deposition are not as intense as they were. Osteoporosis is a more extreme form of osteopenia.

These progressive changes accompany other intrinsic, cumulative, degenerative changes in muscles, tendons, and joints that are discussed in Chapters 6 through 9.

- Which conditions deposit less than the usual amount of bone matrix?
- Which conditions deposit more?

SUMMARY

Bones are the organs of the skeletal system, and their shapes are related to their function. In the axial skeleton the flat and irregular bones of the skull, the vertebral column, and the rib cage support and protect the head and trunk. The appendicular skeleton emphasizes motion and support. The long and short bones of the limbs provide mobility, and the irregular and flat bones of the girdles attach the limbs to the axial skeleton. All bones store calcium and house marrow. *(Page 115.)*

Although bones have different shapes, all share four generic properties. (1) A hard rigid matrix and cells give bones strength and shape to sustain weight and movement. (2) Bones usually articulate with other bones, thereby transferring forces and movement through the skeleton. (3) A sleeve of periosteum covers every bone and provides vessels for nutrition, bone cells for growth, and attachments for tendons and ligaments. (4) Appositional growth of new matrix and remodeling of old matrix are responsible for shaping bones. *(Page 117.)*

All bones are formed by the same cells and made of the same material—collagen, proteoglycans, osteonectins, and hydroxyapatite. Osteoblasts and osteocytes, which deposit and maintain bone, develop from osteoprogenitor cells in the periosteum and endosteum. Osteoclasts resorb bone; these cells develop from myeloid stem cells in the bone marrow. The structure of cartilage resembles bone. *(Page 120.)*

Bone assumes two major forms. Osteons are the units of compact bone; they are composed of concentric lamellae and osteocytes around a central canal. Cancellous (spongy) bone features trabeculae that consist of osteocytes embedded more or less randomly in the bony matrix. Compact bone forms the dense exterior walls of bones, and internal networks of cancellous bone support these walls. *(Page 123.)*

Two different processes ossify bones. Intramembranous ossification in mesenchyme tissue forms trabeculae and osteons of membrane bones. Endochondral ossification begins in hyaline cartilage models of long, short, and irregular bones and replaces this cartilage with bone. Primary ossification centers build the shafts, and secondary ossification centers produce the epiphyses of the larger long bones. Primary ossification is responsible for development of all other chondrogenic bones. *(Page 127.)*

The remodeling cycle regulates the balance between ossification and resorption of bone. After the initial ossification in mesenchyme or cartilage, bone is remodeled along lines of mechanical stress. This process is cyclic; it begins with resorption of previous bone and continues when new bone deposits at the site of resorption. Because this new material shifts stresses ever so slightly, it will be remodeled again as the cycle continually adjusts bone to the mechanical stresses of support and motion. The piezoelectric effect is thought to direct where remodeling occurs. Remodeling is an important part of the processes that shape bones. *(Page 131.)*

Ossification also depends on the level of calcium in the blood, to which parathormone, calcitonin, and vitamin D contribute. Parathormone raises blood calcium by resorbing this mineral from bone. Calcitonin lowers blood calcium by promoting deposition of calcium into bone. Vitamin D stimulates the gut to absorb calcium from the diet into circulating blood. *(Page 130.)*

Heredity, disease, aging, and accidents affect the shape and structure of bones. Genetics can wipe out the population of osteoclasts and prevent resorption of bone, and heredity can also change the number of digits in hands and feet. Vitamin D and vitamin C deficiencies can weaken bone for lack of calcium and collagen in the matrix. Osteoporosis weakens bone by unbalancing the remodeling cycle. Various fractures dramatically change the support and mobility a bone can provide. Aging leads to progressive loss of bone, especially cancellous bone, as osteoblasts gradually deposit less bone than osteoclasts resorb. *(Page 132.)*

STUDY QUESTIONS   OBJECTIVES

1. Describe the functions of the skeleton.   **1**
2. What is the axial skeleton? What is the appendicular skeleton? What are their functions? Which bones belong to these divisions of the skeleton?   **2**
3. What four properties do bones have in common? What function does each of these characteristics serve?   **3**
4. Name the different shapes of bones. Is each type of bone found in the axial or appendicular skeletons or in both? What functions are associated with each type?   **4**
5. How does a long bone differ from a short bone? From a flat bone? From a sesamoid bone? From an irregular bone?   **5**
6. Name and describe the molecular components of bone matrix.   **6**
7. How do lamellar and woven bone differ? Which bones are lamellar? Which are woven?   **6**
8. What are the functions of osteoblasts, osteocytes, osteoclasts, and osteoprogenitor cells?   **7**
9. Name the structural unit of compact bone. Describe the structure of compact bone. Where does this type of bone occur in a long bone? In a short bone? In a scapula? In a vertebra?   **4, 8**
10. Name and describe the structural elements of cancellous bone. Where does this type of bone occur in a long bone? In a tarsal bone? In an occipital bone? In the pelvis?   **4, 8**
11. Compare and contrast the structures of compact and cancellous bone.   **8**

12. Describe the three types of cartilage and give an example to illustrate where each type is found.   **9**
13. Describe intramembranous ossification. Which bones ossify in this manner? Does appositional growth occur in this process?   **10**
14. Describe endochondral ossification. What is a cartilaginous model? How does the model participate in endochondral ossification? Which bones ossify in this manner?   **10**
15. What is the role of primary ossification centers? Of secondary ossification centers? Do short bones have secondary ossification?   **10, 11**
16. What effect does parathormone have on calcification? How do calcitonin and vitamin D regulate calcification? What effects do high levels of blood calcium have on calcification? Low levels of blood calcium?   **12, 13**
17. Describe the steps in the remodeling cycle.   **14**
18. What aspects of bone and bone growth are Steven's physicians employing to increase his height?   **14**
19. Name and describe genetic effects on bone.   **15**
20. Describe and compare the effects of diseases and aging on bone.   **15**
21. Describe and compare different fractures. What characteristics of each would you use to identify each type?   **15**
22. Describe the steps in healing of a simple fracture.   **15**

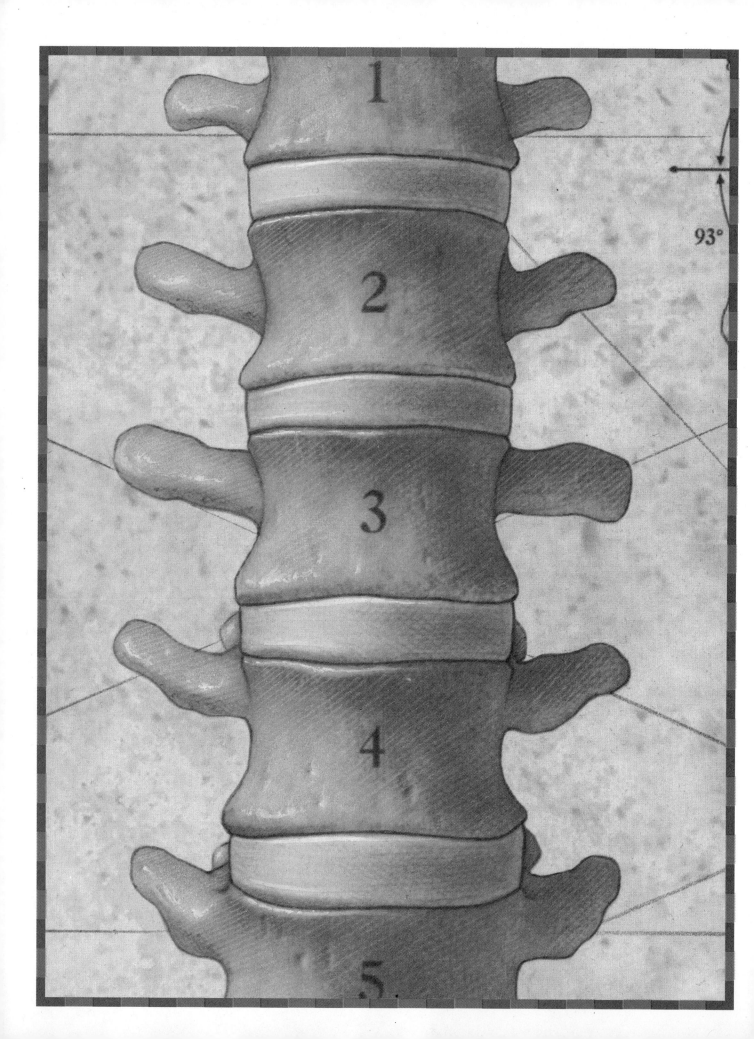

CHAPTER 6

Axial Skeleton

OBJECTIVES

After studying this chapter you should be able to do the following:

1. Identify the bones of the axial skeleton, their functions, and their locations in the axial skeleton.
2. Identify which bones are considered flat and which are irregular.
3. Identify the internal spaces of the skull, the functions of these spaces, and the entrances to them.
4. Identify and locate the various cranial and facial bones of the skull.
5. Locate foramina in the base of the cranium and identify the structures that penetrate them.
6. Describe the cranial fossae of the skull.
7. Identify the bones and sutures of the calvarium.
8. Identify the bodies and processes of the maxilla, mandible, sphenoid, temporal, and occipital bones in the skull.
9. Describe the anatomy of a typical vertebra and the processes and articulations that link vertebrae together in the vertebral column.
10. Compare and contrast the structures and functions of vertebrae from different regions of the vertebral column.
11. Identify the components of the rib cage and how they articulate with each other.
12. Describe the effects of Crouzon's syndrome and other pathological conditions of the skull.
13. Describe scoliosis and other pathological conditions of the vertebral column.
14. Describe the developmental relationships between somites, vertebrae, and intervertebral disks.
15. Describe effects of aging in the skull and vertebral column.

 VOCABULARY TOOLBOX

Many terms suggest shape or position:

falx cerebri	falx, *L* a sickle. The falx cerebri is a sickle-shaped curtain of tissue, curving between the cerebral hemispheres.	foramen **rotundum**	rotund, *L* round. This skull opening is round.
		prognathia	pro-, *G* forward, gnath- *G* jaw. Protruding jaw.
foramen lacerum	lacer-, *L* mangled, torn. This opening in the skull is rough and irregular.	**spinous** process	spin-, *L* a spine, thorn. Spinous processes are sharp.
foramen ovale	ov-, *L* egg. The foramen ovale is oval like an egg.	**trigeminal**	gemin-, *L* twin. The trigeminal nerve has three primary branches.

RELATED TOPICS

- **Articulations, Chapter 8**
- **Bones classified by shape and ossification, Chapter 5**
- **Brain and spinal cord, Chapter 18**
- **Mesoderm and somites, Chapter 4**
- **Muscles of the axial skeleton, Chapter 10**

PERSPECTIVE

The skull and vertebral column wrap the brain and spinal cord in a rigid yet flexible protective envelope, and the rib cage encloses the heart and lungs. Anyone who suffers chronic low back pain knows the axial skeleton also is involved in motion. The muscles of the back and trunk constantly balance the body in virtually all of its movements. Indeed, throwing high, inside fastballs and shooting winning, 3-point baskets are impossible without vigorously flexing and twisting the vertebral column. This chapter describes the bones of the axial skeleton and how they are involved in protection and motion.

People stand erect and walk by balancing their bodies against the forces of gravity that otherwise would topple the vertebral column (Figure 6.1). The muscles of the back and trunk adjust the vertebral column to nearly every motion of the head and limbs. When you raise your arm or move your head, muscles on the opposite side of the vertebral column compensate for the change. The additional power required to serve an ace in tennis comes from flexing and twisting the vertebral column just as the upper limbs drive the racquet forward and the lower limbs charge. Football linemen cultivate other forces; their heads and vertebral columns sustain large compressive forces when the ball is snapped. The risk to the delicate brain and spinal cord increases considerably when auto collisions at 30 mph whip the vertebral column dangerously forward, stretching its ligaments and possibly breaking vertebrae.

Table 6.1 summarizes the bones of the axial skeleton. Of all 82 bones, only the ribs, sternum, and certain skull bones are regarded as flat; the entire vertebral column and the rest of the skull are irregular bones. How these bones ossify depends on their origins. Those derived from somites are endochondral, while others may be intramembranous or both. (See Table 6.1 for further details and Chapter 5 for shapes and ossification of bone.) Consult Table 6.2 for definitions and examples of bony terms and Table 6.3 for a summary of the openings in the base of the skull.

Case studies begin on p. 148 and p. 167.

A

B

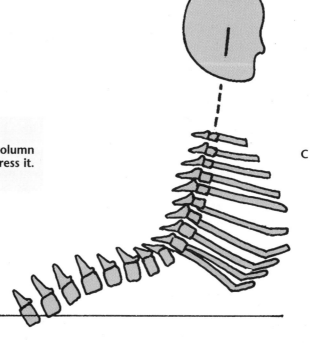

C

30 mph crash
simulation

Figure 6.1

Motions of the axial skeleton. **A,** The vertebral column
flexes and twists. **B,** It absorbs forces that compress it.
C, Severe stretching can damage it.

Table 6.1 BONES OF THE AXIAL SKELETON BY SHAPE AND OSSIFICATION

Bones	Shape	Ossification
SKULL (23)		
Cranial (8)		
Parietal (2)	Flat	Intramembranous
Frontal (1)	Flat	Intramembranous
Ethmoid (1)	Irregular	Endochondral
Sphenoid (1)	Irregular	Endochondral—body, lesser wings
		Intramembranous—greater wings, and pterygoid processes
Temporal (2)	Irregular	Endochondral—petromastoid part
		Intramembranous—squama and tympanic part
Occipital (1)	Flat	Endochondral—lateral and basal parts
		Intramembranous—squama
Facial (14)		
Nasal (2)	Irregular	Intramembranous
Lacrimal (2)	Irregular	Intramembranous
Zygomatic (2)	Irregular	Intramembranous
Maxilla (2)	Irregular	Intramembranous
Inferior nasal concha (2)	Irregular	Endochondral
Vomer (1)	Flat	Intramembranous
Palatine (2)	Irregular	Intramembranous
Mandible (1)	Flat	Intramembranous
Hyoid (1)	Irregular	Endochondral
EAR OSSICLES (6)	Irregular	Endochondral
VERTEBRAL COLUMN (26)		
Cervical (7)	Irregular	Endochondral
Thoracic (12)	Irregular	Endochondral
Lumbar (5)	Irregular	Endochondral
Sacral (1)	Irregular	Endochondral
Coccygeal (1)	Irregular	Endochondral
RIBS (24)	Flat	Endochondral
STERNUM (3)	Flat	Endochondral

TOTAL 82

Table 6.2   BONY STRUCTURES

Structure	Description	Example
Arch	A curved extension of bone, usually enclosing another structure	Vertebral arches enclose the spinal cord
Body	The central portion of a bone, usually massive, from which other parts extend	The body of the maxilla contains the maxillary sinus; drum-like part of a vertebra
Canal	A tunnel within a bone	The optic canal carries vessels and nerves to the orbit
Concha	A delicate coiled portion of bone	The inferior nasal concha
Condyle	A large, smooth, rounded articular surface	Occipital condyles receive the atlas
Crest	A sharp, curved ridge or process, receiving membranes or ligaments	The frontal crest receives the falx cerebri
Eminence	A rounded, dome-like surface	The frontal and parietal eminences of the calvarium
Facet	A small, smooth surface where bones articulate	Articular facets of vertebrae
Fissure	A large, long opening in a bone	The superior orbital fissure admits nerves and vessels to the orbits
Foramen	An opening in a bone for nerves and/or vessels, usually round	The occipital bone admits the spinal cord to the brain through the foramen magnum
Fossa	A depression on a bone; usually accommodates bones or other structures	The temporal fossa houses the muscles of the jaw
Groove	A long, rounded slot in a bone	The groove for the sagittal sinus faces the internal surface of the frontal bone
Head	Rounded projection that articulates with another bone	The heads of ribs articulate with thoracic vertebrae
Line	A long, low, narrow ridge that may attach muscles, ligaments, or fascia	The superior temporal line attaches fascia to the parietal and temporal bones
Meatus	A long, tubular opening or compartment	The external acoustic meatus opens the ear chambers in the temporal bone
Notch	A foramen that has migrated past the edge of a bone	The supraorbital foramen is frequently incomplete, a notch on the supraorbital ridge
Plate	Broad, flat portion of a bone, squama	The cribriform plate of the ethmoid bone and the plates of the palatine bones
Process	Narrow extension of a bone; sometimes named for the bone to which it connects	The zygomatic process of the temporal bone attaches to the zygomatic bone The styloid process of the temporal bone has no bony connection
Ramus	A large branch or process of a bone	The ramus of the mandible
Septum	Flat partition between skull cavities	The nasal septum
Squama	Broad, flat portion of a bone, plate	The squama of the temporal bone covers the brain
Tubercle, tuberosity	Sharp or rounded processes that receive muscles or tendons	Greater and lesser tubercles of humerus; tibial tuberosity of tibia

Table 6.3 MAJOR PASSAGES OF THE SKULL

Foramen	Bone	Conducts	Between
ANTERIOR CRANIAL FOSSA			
Cribriform plate	Ethmoid	Nerve I, olfactory	Cerebrum and nasal cavity
Optic canal	Sphenoid	Nerve II, optic, and ophthalmic artery	Cerebrum and retina
MIDDLE CRANIAL FOSSA			
Foramen rotundum	Sphenoid	Nerve V, trigeminal, maxillary ramus	Brainstem and maxilla
Foramen ovale	Sphenoid	Nerve V, mandibular ramus	Brainstem and mandible
Foramen spinosum	Sphenoid	Middle meningeal artery	From maxillary to meningeal artery
Foramen lacerum	Sphenoid, temporal, occipital	Internal carotid artery	From carotid canal to brain
Carotid canal	Temporal	Internal carotid artery	Within foramen lacerum
POSTERIOR CRANIAL FOSSA			
Jugular foramen	Temporal, occipital	Nerve IX, glossopharyngeal Nerve X, vagus, Nerve XI, accessory, Internal jugular vein	Brainstem and tongue Brainstem and visceral organs Brainstem and pharynx, larynx Sigmoid sinus and neck
Hypoglossal canal	Occipital	Nerve XII, hypoglossal	Brainstem and tongue
Foramen magnum	Occipital	Spinal cord, nerve XI, meninges, vertebral vessels, and spinal artery	Cranial cavity and vertebral canal

SKULL

BRIEF TOUR

This tour will help you relate the anatomic features of the skull to the bones that form them. Follow Figures 6.2 through 6.4 and also a laboratory specimen, if one is available. After studying this section, go on the long tour to learn the cavities of the skull and which bones form them (page 150), or go directly to the cranial and facial bones to learn their shapes and which partitions they form (page 155).

The skull encloses the brain and forms the face, with a total of 23 bones; all but two of these bones are joined rigidly together by interlocking joints called **sutures** (described in Chapter 8). This interlocking arrangement protects the brain. Only the mandible (lower jaw) and the hyoid bone (above the Adam's apple) move freely. (The bones [ossicles] of the middle ear are discussed with the ear in Chapter 20, even though they technically belong to the skull.) The brain rests inside the **cranial cavity,** protected by eight bones of the **cranium,** or braincase. The **calvarium** is the roof of the cranium. It is the skullcap or cranial vault, shown in Figure 6.3, that curves over the brain down to the **base of the cranium** beneath the brain. The calvarium includes the rounded frontal and parietal bones above, parts of the temporal bones at the sides, and parts of the occipital bone at the rear. The sphenoid bone and the ethmoid bone join with both temporals and the occipital bone to form the base of the cranium, best seen in Figure 6.4 (the ethmoid bone is too deep to be seen in these figures).

The base of the cranium also supports the **face.** Fourteen facial bones contribute to external facial features and provide spaces for the eyes, nose, mouth, and the paranasal sinuses. Turn to Figure 6.2 and find the paired **orbits** that house the eyes. These spaces are formed in part by the frontal, maxillary, and zygomatic bones. You probably know the zygomatic bones as the cheek bones. The **nasal cavity** opens at the external nares and leads internally to the **choanae,** or internal nares, posterior to the **hard palate** (Figure 6.4). The choanae in turn lead air to the pharynx, trachea, and ultimately the lungs. The **paranasal sinuses** open to the nasal cavity from spaces in the frontal and maxillary bones, as well as other bones (Figures 6.8C and D).

The floor of the nasal cavity and roof of the **oral**

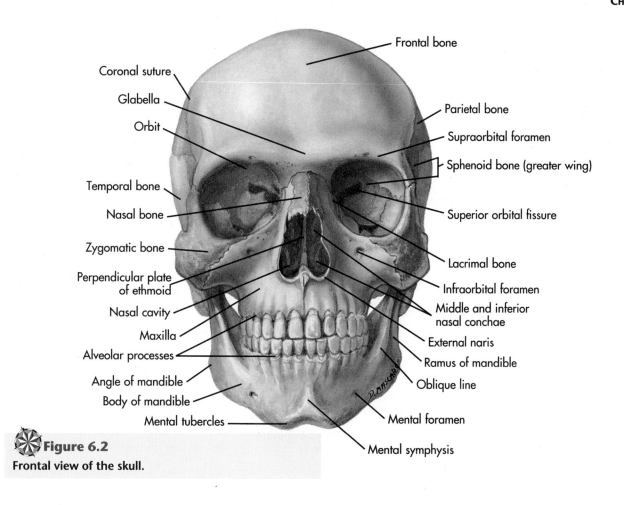

Figure 6.2
Frontal view of the skull.

Frontal bone
Coronal suture
Glabella
Orbit
Temporal bone
Nasal bone
Zygomatic bone
Perpendicular plate
of ethmoid
Nasal cavity
Maxilla
Alveolar processes
Angle of mandible
Body of mandible
Mental tubercles
Parietal bone
Supraorbital foramen
Sphenoid bone (greater wing)
Superior orbital fissure
Lacrimal bone
Infraorbital foramen
Middle and inferior
nasal conchae
External naris
Ramus of mandible
Oblique line
Mental foramen
Mental symphysis

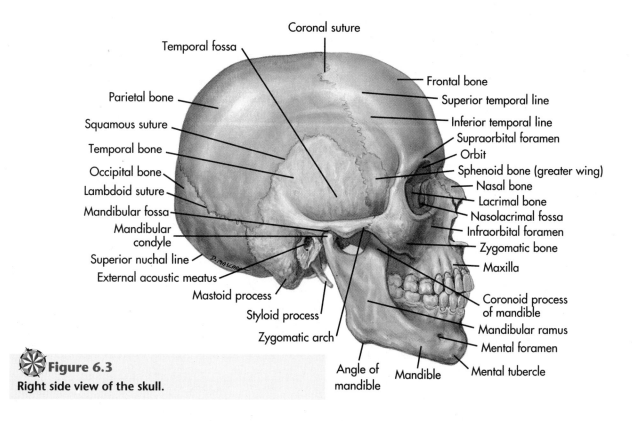

Figure 6.3
Right side view of the skull.

Coronal suture
Temporal fossa
Parietal bone
Squamous suture
Temporal bone
Occipital bone
Lambdoid suture
Mandibular fossa
Mandibular
condyle
Superior nuchal line
External acoustic meatus
Mastoid process
Styloid process
Zygomatic arch
Angle of
mandible
Mandible
Frontal bone
Superior temporal line
Inferior temporal line
Supraorbital foramen
Orbit
Sphenoid bone (greater wing)
Nasal bone
Lacrimal bone
Nasolacrimal fossa
Infraorbital foramen
Zygomatic bone
Maxilla
Coronoid process
of mandible
Mandibular ramus
Mental foramen
Mental tubercle

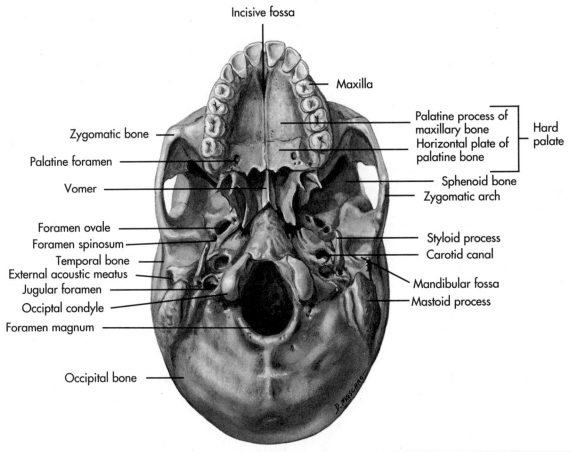

Incisive fossa

Zygomatic bone

Palatine foramen

Vomer

Foramen ovale

Foramen spinosum

Temporal bone

External acoustic meatus

Jugular foramen

Occiptal condyle

Foramen magnum

Occipital bone

Maxilla

Palatine process of maxillary bone

Horizontal plate of palatine bone

Hard palate

Sphenoid bone

Zygomatic arch

Styloid process

Carotid canal

Mandibular fossa

Mastoid process

✳ **Figure 6.4**

The base of the skull with mandible removed.

cavity are formed by the palatine bones and both maxillae, seen in Figure 6.4. Anchored in the maxillae, the upper teeth mesh with those of the **mandible,** shown in Figure 6.3. The mandible articulates with the skull in the **temporomandibular joint,** which lies forward of the **external acoustic meatus,** the entrance to the **auditory canal** of the ear. The muscles of the jaws originate from the **temporal fossa** above the **zygomatic arch** and from the zygomatic arch itself. Muscles of the back and neck attach to the **superior nuchal line** and to the **mastoid processes** on the occipital and temporal bones, respectively. The spinal cord enters the vertebral column by way of the **foramen magnum** seen in Figure 6.4. The vertebral column articulates with the skull at the **occipital condyles.**

Cavities of the skull are like rooms of a house, and the bones are its walls. Learning the skull is very much like exploring a new house; to find your way around, you need to know where doorways and rooms lead. Even though owners are more interested in the spaces and rooms, anatomy students are like builders who need to know where the partitions, wiring, and plumbing go.

CROUZON'S SYNDROME AND CRANIOSYNOSTOSES

Growth and ossification of bones and sutures determine the shape of the skull and its cavities. **Crouzon's syndrome** occurs when certain facial bones are too small (Figure 6.5A). The face is deformed because bones of the orbits, nasal cavity, and upper jaw have not grown proportionately with the mandible and cranium. There is not enough room in the orbits for the eyes, a condition called **exophthalmos.** The mandible protrudes (**prognathia;** prog-NATH-ee-ah), and the cranium may be too small for the young brain to enlarge. Crouzon's is one of many skull conditions called **craniosynostoses** (KRANE-ee-oh-SIN-os-toe-sees) that occur when bones fail to grow adequately and sutures ossify prematurely. (A synostosis is an ossified suture that solidly unites two bones as one. See Chapter 8, Articulations.) Some cases of Crouzon's syndrome are inherited through dominant genes, which may help explain how genes control skull growth. Surgery can treat some of these disfiguring conditions by enlarging the cranial cavity and moving the bones of the face forward, but this cannot be done without risk to nerves and vessels that

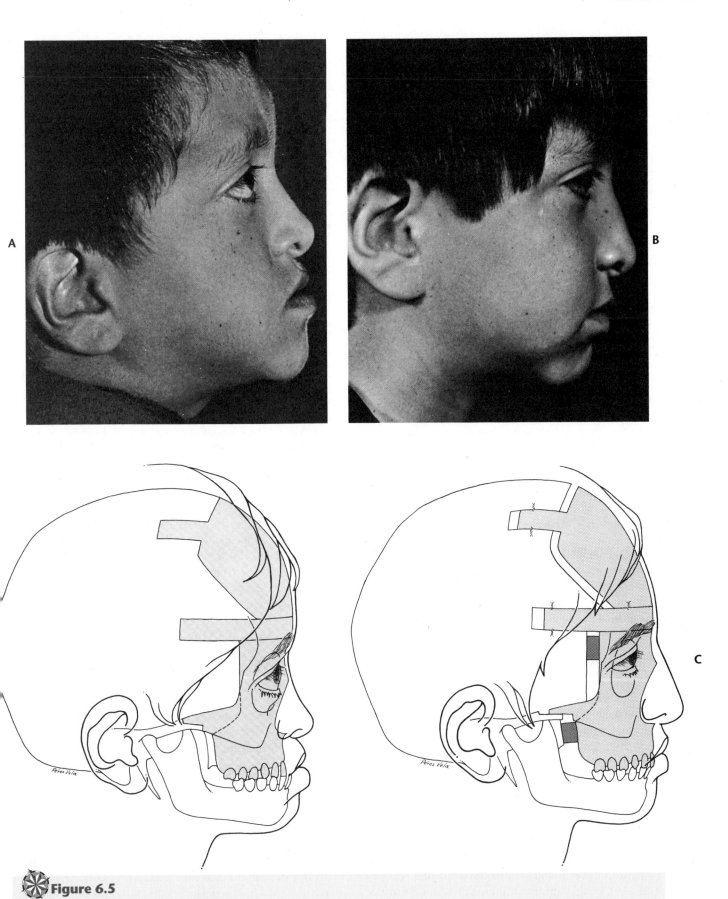

Figure 6.5

Surgical treatment of Crouzon's syndrome. **A,** Before surgery. **B,** After surgery advanced the facial bones. **C,** Cranial cavity is enlarged by cutting the coronal suture and advancing the frontal bone. Pieces of bone taken from a rib retain the facial bones in their new positions.

Table 6.4   BONES OF SKULL CAVITIES

Bones	Cavities				
	Cranial	**Orbits**	**Nasal**	**Paranasal**	**Oral**
Occipital	Posterior wall and floor	—	—	—	—
Parietal	Roof and lateral wall	—	—	—	—
Temporal	Lateral wall and floor	—	—	—	—
Frontal	Anterior wall and floor	Roof	Roof	Frontal sinus	—
Sphenoid	Floor and wall	Roof and lateral wall	Roof	Sphenoid sinus	—
Zygomatic	—	Lateral wall	—	—	—
Ethmoid	Floor	Medial wall	Roof, septum, lateral wall	Ethmoid air cells	—
Maxilla	—	Floor	Floor	Maxillary sinus	Roof
Palatine	—	—	Floor and lateral wall	Floor and lateral wall of maxillary sinus	Roof
Vomer	—	—	Roof and septum	—	—
Inf. nasal concha	—	—	Lateral wall	Medial wall of maxillary sinus	—
Nasal	—	—	Anterior roof	—	—
Lacrimal	Medial wall	Lateral wall	—	—	—
Mandible	—	—	—	—	Lateral wall

This table identifies the bones that line the cavities of the skull. Read across each row to see the contribution of each bone to the cavities. Read down the columns to find the bones of each cavity. For example, the maxilla contributes to the floor of the orbits, the floor of the nasal cavity, the entire maxillary sinus, and the roof of the oral cavity, while the orbit column shows that the maxilla and five other bones form this cavity.

supply these structures. Figure 6.5B shows what surgery can achieve.

- Which bones are associated with the cranial cavity, the orbits, the nasal cavity, and the oral cavity?

LONG TOUR

This tour thoroughly inspects the skull and its internal spaces; read it after you have completed the brief tour. Table 6.4 summarizes the bones that line the cavities of the skull. Refer to Figure 6.6, and if you have a specimen, turn to the calvarium.

CALVARIUM

The calvarium includes the frontal bone, two parietal bones, and the occipital bone at the rear (Table 6.4). The flat portions of the temporal bones complete each side of the calvarium. The outer surface of the calvarium (Figure 6.6A) is smooth where the scalp covers it and rough where muscles attach, especially · within the temporal fossa and superior nuchal line. The brain and its vessels have molded ridges and grooves (see Table 6.2, Bony Structures) on the inner surface of the calvarium. The largest of these grooves accommodates the **superior sagittal sinus** that drains blood from the brain; this sinus runs from the frontal bone posteriorly to the occipital. Ridges on each side of this groove attach the **falx cerebri,** a fibrous curtain of dura mater (the protective covering of the brain) that curves between the cerebral hemispheres of the brain.

Sutures lock the bones of the calvarium together, as Figure 6.6A illustrates. The **sagittal suture** (sagitta, *L,* arrow) between the parietal bones resembles an arrow, with the **frontal fontanelle** (fontan-, *L,* fountain) as its head and the **occipital fontanelle** as its tail. **Fontanelles** are tough, fibrous sheets of connective tissue that link the growing bones (Figure 6.6B) until they meet at the sutures. The **lambdoidal suture** between the occipital and parietal bones resembles the Greek letter *lambda,* λ. The frontal bone and parietals join at the **coronal suture,** whose name reminds us that it cuts a coronal (frontal) section across the calvarium. Surgery for Crouzon's syndrome cuts this suture to enlarge the cranial cavity (Figure 6.5C). The frontal bone ossifies from two centers that meet at the **metopic suture** (metop-, *G,* the forehead), which usually ossifies before birth. Accessory **sutural (or Wormian) bones** usually appear along the lambdoidal and sagittal sutures. Such bones spread widely over the calvarium of patients with **hydrocephaly,** whose brains and cranial cavities have enlarged with excessive cerebrospinal fluid.

BASE OF THE CRANIUM—INTERIOR

The base of the cranial cavity joins the cranium to the face and roofs the orbits and nasal cavity (Figure 6.7 and Table 6.4). The frontal, ethmoid, and the sphenoid bones, two temporal bones, and the occipital bone make up this portion of the skull. The brain rests in three **cranial fossae,** or depressions in the

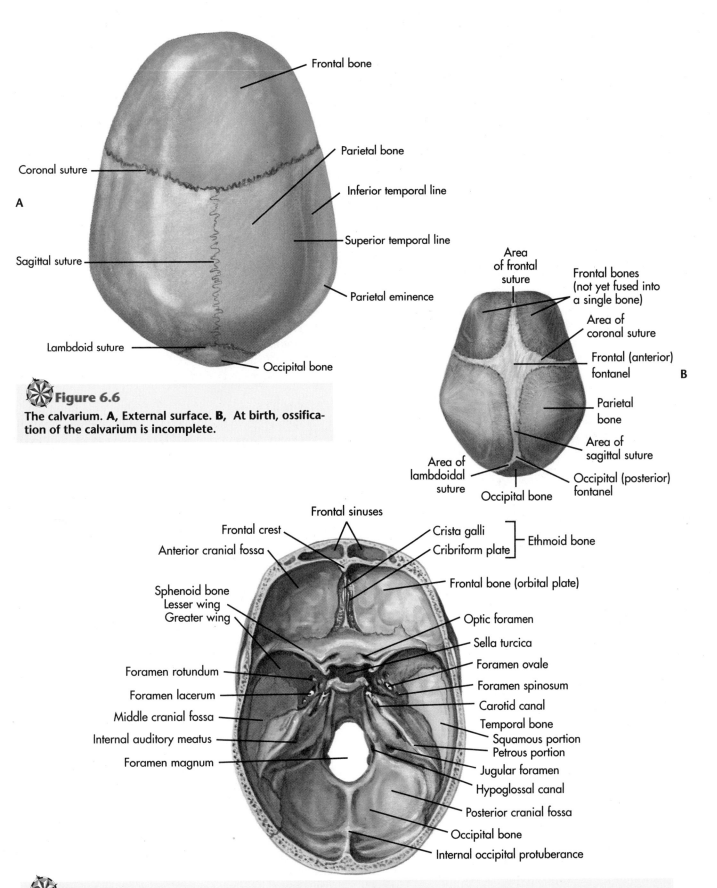

Figure 6.6

The calvarium. **A,** External surface. **B,** At birth, ossification of the calvarium is incomplete.

Figure 6.7

Floor of cranial cavity with anterior, middle, and posterior cranial fossae.

base of the cranium (the anterior, middle, and posterior cranial fossae), each of which is perforated with foramina for cranial nerves and vessels, which are summarized in Table 6.3. Chapter 19, Brain and Cranial Nerves, discusses cranial nerves thoroughly. Only the largest vessels will be mentioned here (Chapter 14 discusses vessels).

ANTERIOR CRANIAL FOSSA

This fossa is formed by the frontal, ethmoid, and sphenoid bones. It receives the frontal lobes of the cerebral hemispheres and transmits cranial nerves that emerge from the anterior region of the brain. Cranial nerve I (olfactory) is the first of these; its endings penetrate many openings in the **cribriform plate** of the ethmoid bone (crib-, *L*, sieve; see Table 6.2) onto the roof of the nasal cavity, where the endings enable olfactory receptors to detect odors. Cranial nerve II (optic) is next, communicating from the eyeballs to the brain through the **optic canals** and **chiasmatic groove** (see Table 6.2). The **crista galli** (*L*, cock's comb) reaches upward from the cribriform plate to anchor the anterior end of the falx cerebri to the base of the cranial cavity.

MIDDLE CRANIAL FOSSA.

This fossa resembles a butterfly's wings; it is formed by the sphenoid and temporal bones and receives the temporal lobes of the cerebral hemispheres. The pituitary gland rides in the **sella turcica** (*L*, Turk's saddle) between the wings of the butterfly. Five major openings penetrate the wings. Beginning with the most anterior, the **superior orbital fissure** carries considerable traffic between the brain and orbits. Cranial nerves III, IV, and VI control eyeball muscles through this opening. Cranial nerve V, the **trigeminal,** is the largest cranial nerve, and it sends three branches to the face. The ophthalmic branch reaches the optic region through the superior orbital fissure, and the maxillary and mandibular branches exit to the upper and lower jaws respectively through the small, round **foramen rotundum** and the larger, oval **foramen ovale.** The **foramen spinosum** admits the middle meningeal artery, whose branches you can see molded into the wall of the fossa. Finally, the internal carotid artery delivers most of the blood to the brain through the large **foramen lacerum** and **carotid canal.** In life, fibrocartilage plugs this foramen except for the canal itself.

POSTERIOR CRANIAL FOSSA

This fossa is the largest of the cranial fossae. It is formed by the temporal bones and the occipital bone; the cerebellum rests in the lobes of the fossa lateral to the foramen magnum; the **clivus** (clivus, *L*, a slope) of the sphenoid bone accommodates the medulla oblongata of the brain anterior to the foramen magnum. The

bony ridge that separates this fossa from the middle cranial fossa houses the ear chambers. Cranial nerves VII and VIII enter this ridge through the **internal acoustic meatus** from the posterior cranial fossa. Cranial nerve VII, the facial, passes to the muscles of the face and scalp through the **stylomastoid foramen** externally (Figure 6.4), but cranial nerve VIII, the vestibulocochlear (ves-TIB-you-low-KOKE-lee-ar), remains in the inner ear. The **jugular foramen** carries major vascular and neural traffic between the cranial cavity and the neck. Located between the foramen magnum and internal acoustic meatus, this opening both conducts blood out of the superior sagittal sinus to the jugular vein and allows cranial nerves IX, X, and XI to exit to the tongue, pharynx, neck, and peritoneal cavity. Cranial nerve XII, the hypoglossal, extends below the tongue through the **hypoglossal foramen.**

- Which cranial fossae contain the following landmarks: crista galli, foramen rotundum, internal acoustic meatus, superior orbital fissure, sella turcica, and foramen magnum?

THE ORBITS

Now turn to the orbits (Figure 6.8A). These conical cavities are surrounded by the frontal, nasal, lacrimal, ethmoid, sphenoid, maxilla, and zygomatic bones (Table 6.4). In Crouzon's syndrome the maxilla narrows the orbits and causes the eyeballs to protrude. The orbits receive their innervation and vasculature through the **superior** and **inferior orbital fissures** at the apex of each chamber. The optic nerve (II) and cranial nerves III, IV, V (ophthalmic branch), and VI all communicate through the superior orbital fissure. A branch of the maxillary ramus (RAY-mus, ram-, *L*, a branch; see Table 6.2) of cranial nerve V and several orbital vessels enter through the inferior orbital fissure. The **supraorbital foramen** (sometimes reduced to a notch; see Table 6.2) conducts nerves and vessels of the same name to the scalp. Eyelids attach to the periosteum at the superior and inferior margins of the orbits, and the **lacrimal fossa** collects tears for the nasal cavity. Certain types of surgery for Crouzon's syndrome cut through the walls of the orbits (Figure 6.5C) to advance the face, dissecting between the bone and periosteum to avoid damaging the eye itself.

NASAL CAVITY

The nasal cavity is formed by 14 bones—the frontal nasals, maxillae, lacrimals, ethmoid, sphenoid, vomer, palatines, and the inferior nasal conchae (Figure 6.8B and Table 6.4). The **external nares** and **choanae** are the external and internal openings, respectively. The vomer and ethmoid bones partition the nasal cavity into left and right portions. Three **nasal conchae (superior and middle from the ethmoid bone, and a separate inferior set)** and their underlying **meati** (me-AH-tie) curve from the lateral walls into the nasal cavity with additional surfaces to humidify passing air. The meati communicate with the **paranasal sinuses,** and the nasolacrimal duct empties into the inferior meatus from the orbit. The palatine bones and the maxillae also form the floor of the nasal cavity and the hard palate of the mouth.

Crouzon's syndrome interferes with breathing by diminishing the volume of the nasal cavity. Cleft palate occurs in this syndrome and other conditions, when the maxilla and palatine bones fail to separate entirely the oral and nasal cavities. As a result, speech, breathing, and eating are impaired.

PARANASAL SINUSES

The paranasal sinuses are paired chambers that open beside the nasal cavity (Figures 6.8C and D and Table 6.4). The **maxillary sinuses** in the maxillae beneath the orbits are the largest of these spaces (see Table 6.2). Each space communicates with the

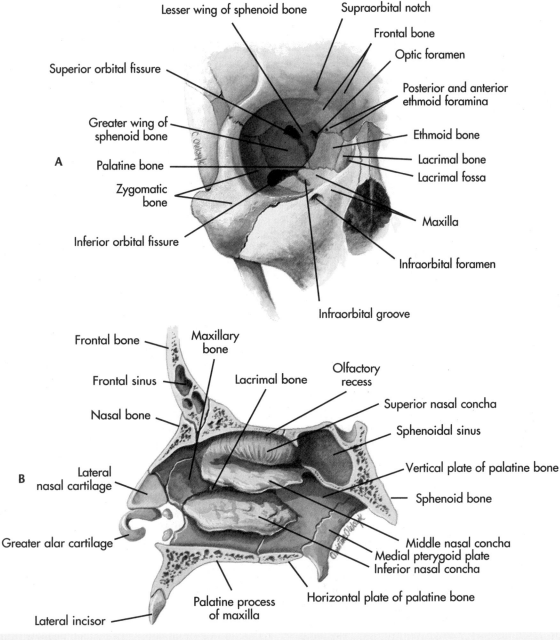

Figure 6.8
Facial cavities in the skull. **A,** The right orbit. **B,** Right nasal cavity viewed medially from the nasal septum.

Continued.

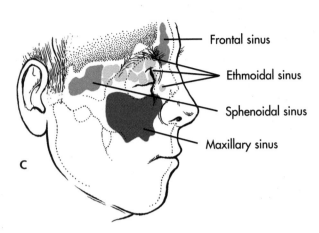

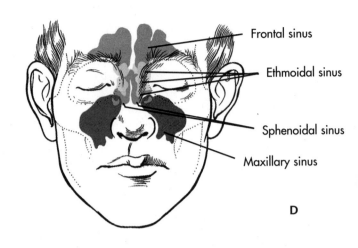

Figure 6.8, cont'd

C and **D,** Lateral and frontal views of the paranasal sinuses. Frontal sinuses are located above the orbits and the maxillary sinuses below. The sphenoid and ethmoid sinuses roof the nasal cavity. All paranasal sinuses communicate with the nasal cavity.

middle nasal meatus through two openings in the lateral wall of the nasal cavity. The **frontal sinuses** are located in the frontal bone, behind the superior orbital margin. A frontonasal duct empties each of these sinuses into the middle nasal meatus. The **ethmoid air cells** cluster in the ethmoid bone located in the lateral walls of the nasal cavity. Most air cells also open into the middle meatus, but posterior ethmoid air cells communicate with the **superior nasal meatus.** Finally, the **sphenoidal sinuses** drain into the **sphenoethmoidal recesses** (Figure 6.8B) above the superior nasal concha at the rear of the nasal cavity.

No one is entirely sure what functions the paranasal sinuses serve. The mucosal lining is continuous with the nasal cavity, and the epithelium secretes mucus, as does the nasal cavity, but air has poorer access to these spaces than to the nasal cavity itself. If the sinuses are conditioning air, why does it not flow more directly? When cleared of mucus and inflammation caused by a cold or sinusitis, the sinuses certainly improve the resonance of one's voice. These spaces may simply reflect the tendency of bone to form cavities (recall the medullary cavities of the femur, for instance), but what advantage does a 150-pound individual gain from losing a few ounces of bone?

ORAL CAVITY

Two arches of teeth in the maxillae and the mandible are the hallmarks of the oral cavity (Figure 6.4 and Table 6.4). Although both the maxillae and palatine bones roof the oral cavity, only the maxillae support the upper teeth. Both bones deliver branches of cranial nerve V to the palate through the **incisive** and **greater palatine foramina.** The mandible (Figure 6.3) articulates in the **mandibular fossa** and supports the tongue and floor of the oral cavity with muscles that insert across the arch of the mandible.

• Which internal spaces of the skull contain the following structures: mandible, vomer, lacrimal fossa, ethmoid air cells, and inferior nasal concha?

BASE OF CRANIUM—EXTERIOR

Just four bones—the sphenoid, two temporal bones, and the occipital—join the skull and the neck (Figure 6.4). The base of the skull is a complex region despite so few bones. The skull articulates with the vertebral column at the occipital condyles, and muscles of the neck insert upon the mastoid processes and nuchal lines to support the skull. The spinal cord leaves the brain through the foramen magnum and descends the vertebral column. The pharynx also descends the neck beneath the base of the skull from the nasal and oral cavities to the trachea and esophagus. In addition, an immensely complex network of arteries and veins distributes blood to the head through the base of the skull.

The choanae and foramen magnum identify the base of the skull. Most of the foramina exit through the floor of the cranial cavity between these landmarks. The **pterygoid processes** (TEAR-ih-goyd) on each side of the choanae and the **styloid processes** anchor the muscles and ligaments that suspend the pharynx and the mandible. The **external acoustic meati** and **mastoid processes** identify the lateral boundaries of the base of the skull.

On each side of the midline, the styloid process,

the occipital condyle, and the pterygoid processes form a triangular area in which important foramina open. The **foramen ovale,** the most anterior of these openings, is a large, oval exit for the mandibular branch of cranial nerve V toward the mandible. Vessels to the cranial cavity enter the next three passages. The **foramen spinosum** is located lateral and slightly posterior to the foramen ovale, where it admits the middle meningeal artery to the cranial cavity. Then comes the opening of the **carotid canal** for the internal carotid artery, which tunnels forward and medially to the **foramen lacerum.** Venous blood leaves the cranial cavity by way of the jugular vein through the **jugular foramen** that is just posterior to the carotid canal.

Cranial nerves IX, X, and XI also exit the jugular foramen to the tongue, pharynx, neck, and visceral organs. Cranial nerve VII exits toward the face from the **stylomastoid foramen** located immediately posterior to the styloid process. The **hypoglossal foramen** tunnels beneath the occipital condyles, carrying cranial nerve XII (hypoglossal) to the tongue and muscles of the hyoid bone. Finally, the cervical branch of cranial nerve XI (accessory) passes through the **foramen magnum,** along with the vertebral arteries and veins and the spinal artery.

CRANIAL BONES

Skull bones extend outward from central bodies toward other bones. In flat bones these extensions take the form of broad **parts** or **squama** (SKWAY-ma; squam-, *L,* a scale; see Table 6.2) that cover the brain, or they appear as narrow **processes** in irregular bones, such as the zygomatic arch, that attach to other bones or to ligaments and tendons. Most parts and processes of skull bones meet at interlocking sutures between different bones of the skull, but some processes, such as the styloid process, end without touching another bone. Processes are often named for the bone they meet; the zygomatic process of the maxilla meets the zygomatic bone, for example. Virtually all skull bones contribute to internal cavities or to exterior surfaces of the skull. Periosteum covers all these surfaces, attaching the scalp, muscles, or mucosal linings to the bones. Table 6.5 summarizes skull articulations.

Table 6.5   ARTICULATIONS OF SKULL BONES

These Bones	Occipital	Parietal	Temporal	Sphenoid	Frontal	Ethmoid	Zygomatic	Maxilla	Palatine	Vomer	Inf. Nasal Concha	Nasal	Lacrimal	Mandible
Occipital		*	*	*										
Parietal	*	*	*	*	*									
Temporal	*	*		*			*							*
Sphenoid	*	*	*		*	*	*		*	*				
Frontal		*		*		*	*	*				*	*	
Ethmoid			*	*				*	*	*	*	*	*	
Zygomatic			*	*	*		*							
Maxilla					*	*	*	*	*	*	*	*	*	
Palatine			*			*		*	*	*				
Vomer			*			*		*	*					
Inf. nasal concha				*				*	*				*	
Nasal				*	*		*					*		
Lacrimal				*	*		*				*			
Mandible	*													

This table traces the connections between bones of the skull. Each bone at the left articulates with the starred (*) bones at its right. For instance, the zygomatic articulates with the maxilla and the temporal bone. The same is true for the bones listed in each column; each bone at the top of this table articulates with starred bones in the column below it.

Cranial bones cluster in the upper left box, and the lower right box contains facial bones. Bones in the overlapping area, such as the sphenoid, are both cranial and facial.

Find the star where the parietal row crosses the parietal column. This star means that the left and right parietal bone articulate with each other. Can you find three other bones that self-articulate?

PARIETAL

The paired parietal bones form the roof and sides of the calvarium. The parietal bones protect the parietal (pahr-EYE-et-al; parie-, *L,* a wall) and adjacent lobes of the cerebrum. Each dome-shaped bone (Figure 6.9) is one squamous part, extending from a central **parietal eminence** (see Table 6.2) where ossification began. The parietals meet each other at the

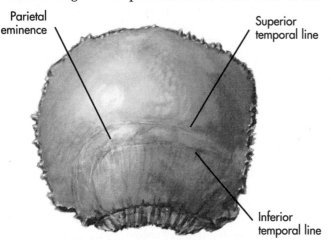

Figure 6.9

The parietal bones are flat bones. The external surface is smooth and rounded.

sagittal suture in the midline of the calvarium. The coronal suture links them with the frontal bone anteriorly, and the lambdoidal suture joins the parietals with the occipital bone in the rear. The parietals descend the wall of the temporal fossa to junctions with the sphenoid and temporal bones.

The **superior and inferior temporal lines** mark the margins of the temporal fossa on the external surface of each parietal bone (Figures 6.3 and 6.9). The temporalis muscle originates from the surface within the inferior temporal line, and the superior line is the attachment point for the fascia that covers this muscle. Near the sagittal suture, the parietal foramen conducts a vein (and sometimes infections) from the scalp to the superior sagittal sinus. Impressions of the middle meningeal artery are especially prominent on the interior surfaces of the parietal bones.

FRONTAL

The frontal bone forms the front of the cranium. It covers the frontal (front, *L,* forehead) lobes of the cerebrum (Figure 6.6A and 6.10) and curves underneath them, forming the floor of the anterior cranial fossa and the roof of the orbits. Two **frontal eminences** on the exterior make the forehead prominent and mark the centers, where ossification began. Different parts of the frontal bone extend from these landmarks.

The **squama** arches over the cranial cavity and joins the parietal bones at the coronal suture (Figures 6.6A and 6.10). The zygomatic processes descend to the zygomatic bone and form the **anterior margin of the temporal fossa** lateral to the orbits. The **nasal part** of the frontal bone reaches downward to sutures with the nasal and maxillary bones above the nose. The **frontal sinuses** lie behind the **glabella,** the slight depression between the orbits. The **orbital plate** of the frontal bone tucks sharply under the supraorbital borders, comprising the roof of the orbits and the floor of the anterior cranial fossa in Figure 6.7. This thin, bony plate articulates with the ethmoid and sphenoid bones. Internally the **frontal crest** (see Figure 6.7 and Table 6.2) receives the falx cerebri. The **groove for**

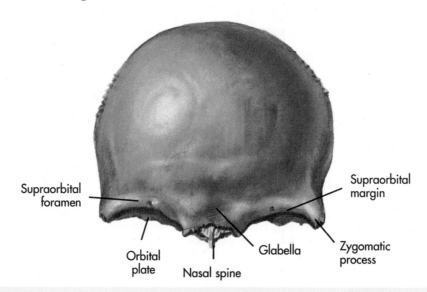

Figure 6.10

External view of frontal bone.

the **superior sagittal sinus** (Figure 6.7) begins in the midline, where the falx attaches to the frontal bone.

ETHMOID

The single ethmoid bone contributes to the base of the cranial cavity between the orbits as well as to the roof and walls of the nasal cavity (Figure 6.11A). The ethmoid bone (ETH-moyd; ethm-, *G*, sieve) also surfaces on the medial walls of the orbits. Viewed from the cranial cavity in Figure 6.11B, it presents a more or less flat body, the **cribriform plate,** perforated with openings into the nasal cavity through which pass endings of the olfactory nerve (I). The crista galli rises into the cranial cavity from the center of the cribriform plate and anchors the falx cerebri.

Ethmoid anatomy is more complicated in the nasal cavity. The **perpendicular plate** descends, like the stem of the letter *T*, from the cribriform plate into the nasal septum (Figure 6.11C), where it articulates with nasal, vomer, and sphenoid bones and with the septal cartilage. The **labyrinths** extend downward from the lateral edges of the cribriform plate as the superior and middle nasal conchae of the nasal cavity. Air cells within the labyrinths are part of the paranasal sinuses; they open into the superior and middle meati of the nasal cavity. The lateral wall of each labyrinth surfaces on the medial wall of the orbit in Figure 6.11D, where it articulates with the frontal, lacrimal, maxilla, and palatine bones. The posterior edges of the labyrinths attach to the sphenoid bone.

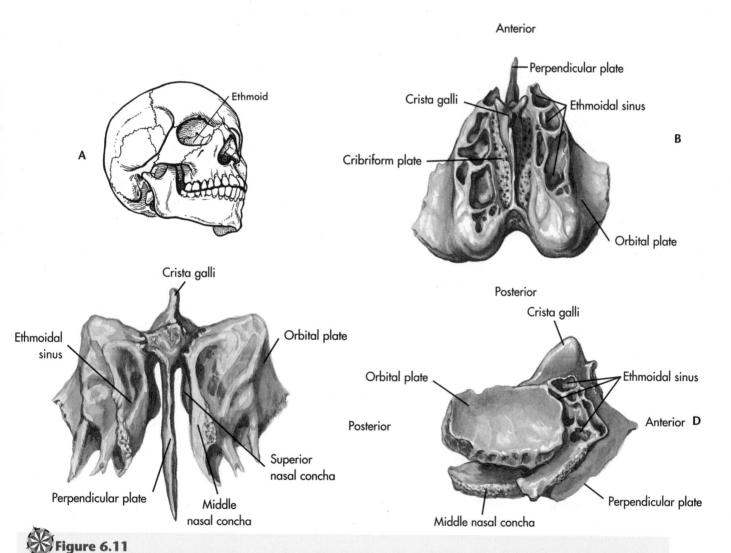

Figure 6.11

A, The ethmoid bone is located between the orbits. **B,** The cribriform plate can be seen from the cranial cavity. **C,** The T shape of the ethmoid is evident when viewed from the anterior. The perpendicular plate is the stem of the T and the air cells and orbital plates extend downward from the horizontal portion. **D,** In right lateral view, the crista galli and orbital surface of the labyrinth can be seen.

SPHENOID

The sphenoid bone is located in the base of the cranium between the frontal and occipital bones (Figure 6.12A). The sphenoid (SFEE-noyd; sphen-, *G,* a wedge) articulates with all other bones of the cranial cavity, as well as the palatine bones, the zygomatics, and the vomer. To make all these connections, the central body sends forth three paired processes that give the sphenoid its famous bat-wing appearance.

Seen from the cranial cavity in Figure 6.12A, the sella turcica identifies the **body of the sphenoid** in the middle cranial fossa. The body itself encloses the **sphenoid sinuses,** and the vomer descends from beneath them into the nasal septum.

The **lesser wings** of the sphenoid (Figure 6.12B and C) separate the anterior from the middle cranial fossae. The leading edge of each wing articulates with the orbital part of the frontal bone, and the trailing edge ends at the superior orbital fissure.

The **greater wings** spread into the middle cranial fossa, where they stretch forward onto the lateral walls of the orbits. The wing tips reach out to the temporal and infratemporal fossae, where they contribute surfaces for the origins of the temporalis and lateral pterygoid muscles of the mandible. The leading and trailing edges of the greater wings articulate with the frontal, zygomatic, parietal, and temporal bones.

The **pterygoid processes** (ptery-, *G,* a wing) extend downward, like a bat's legs, (Figure 6.12C) from the body and greater wings of the sphenoid onto the lateral walls of the choanae. More mandibular and pharyngeal muscles attach to these processes. The pterygoid processes also articulate with the palatine bones of the upper jaw and form the posterior wall of the pterygomaxillary fissure.

Sphenoid foramina are best seen on the cranial surface shown in Figure 6.12B. Beginning anteriorly, they are the optic canal, the foramen rotundum, the foramen ovale, and the foramen spinosum. See Table 6.3 for vessels and nerves that pass through these foramina.

TEMPORAL

The two temporal bones protect the temporal lobes of the cerebrum at the sides of the cranial cavity (Figure 6.13). Each temporal bone (TEM-por-al; tempor-, *L,* the temples) has three distinctive parts—the squama, the tympanic, and the petromastoid portions—but no central body.

Viewed from the exterior (Figure 6.13), each temporal spreads its smooth squama upward to meet the parietal and sphenoid bones. This portion of the temporal bone forms the wall of the temporal fossa and part of the origin of the temporalis muscle. The zygomatic process reaches forward from the squama to the zygomatic bone and completes the zygomatic arch around the temporal fossa. The mandibular fossa articulates with the mandible in the base of the zygomatic process.

The **tympanic portion** of the temporal bone surrounds the external acoustic meatus just posterior to the mandibular fossa; the slender styloid process extends downward and forward from the tympanic portion to anchor muscles and ligaments for the pharynx, mandible, and hyoid bone.

Posterior to the external acoustic meatus, the **mastoid process** identifies the **petromastoid part** of the temporal bone. This part really has two regions—the **petrous** and **mastoid portions.** The mastoid process attaches the sternocleidomastoid muscle and other muscles that rotate the head. **Mastoid air cells** fill the spongy bone beneath these attachments; they occasionally become infected through their connections with the middle ear canal. The petrous portion of the temporal bone extends deeper into the base of the skull, where it contains the inner ear chambers and the middle ear and receives the external acoustic meatus from the tympanic portion.

Only the squama and petromastoid portions of the temporal bone can be seen from inside the cranial cavity, shown in Figure 6.12A. As with other cranial bones the cranial surface of the squama conforms to the vessels and folds of the temporal lobes of the brain. The petromastoid part of the temporal bone forms the ridge between the middle and posterior cranial fossae. The internal acoustic meatus can be seen on the posterior wall of the ridge, where it admits the facial (VIII) and vestibulocochlear (VIII) nerves to the cranial cavity. The facial nerve exits the cranium from the stylomastoid foramen behind the styloid process.

Three large openings on the inferior surface of the temporal bone, seen in Figure 6.4, allow major arteries, veins, and three cranial nerves to reach the cranial cavity. They are the foramen lacerum, shared with the sphenoid and occipital bones; the carotid canal; and the jugular fossa. See Table 6.3 to recall the nerves and vessels that pass through these openings.

OCCIPITAL

The occipital bone is the bowl-shaped rear of the cranium; the occipital forms the posterior cranial fossa and articulates with the vertebral column (Figure 6.14). Virtually all features of this bone, but especially the foramen magnum, facilitate this cervical connection. Four parts of the occipital bone (ok-SIP-ih-tal; occipit-, *L,* the back of the skull) extend outward from the foramen magnum to meet the sphenoid, temporal, and parietal bones, but there is no body as such. The foramen magnum occupies the place where a body might have been.

The **squama** ascends the rear of the calvarium to the lambdoidal suture with the parietal bones. The squama is the largest portion of the occipital bone. It

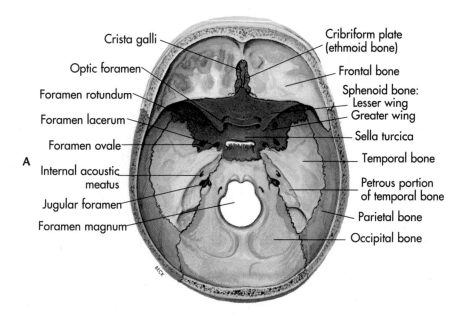

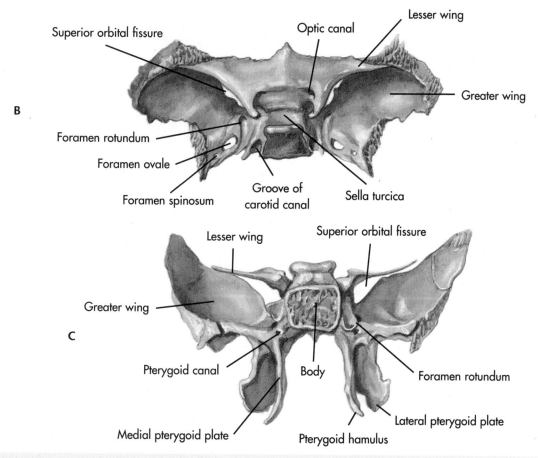

Figure 6.12

A, The sphenoid bone occupies the floor of the middle cranial fossa. **B,** The sella turcica identifies the body of the sphenoid when viewed from the cranial cavity. **C,** The greater and lesser wings and the pterygoid processes viewed from the posterior.

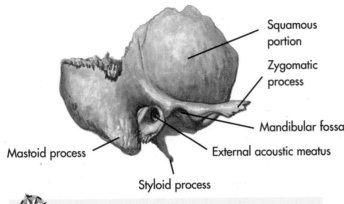

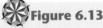

Figure 6.13

Temporal bone. Lateral view of the right temporal bone.

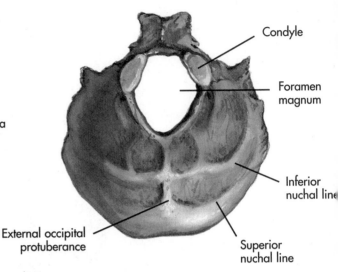

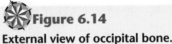

Figure 6.14

External view of occipital bone.

protects the cerebellum and occipital lobes of the brain. Ridges and a groove on the cranial surface of the occipital (Figure 6.12A) receive the falx cerebri and superior sagittal sinus. Grooves for the transverse and sigmoid sinuses reveal the course of blood to the jugular fossae. The superior and inferior nuchal lines on the external surface of the occipital bone identify the insertions of neck muscles, such as the trapezius and semispinalis capitis that raise the head.

The **basilar part** of the occipital, in Figure 6.12A, extends forward from the foramen magnum and resembles the body of a vertebra. It articulates with the sphenoid, and the pharynx attaches to its inferior surface.

The **lateral portions** of the occipital extend from each side of the foramen magnum toward the mastoid processes of the temporal bones. The foramen magnum itself transmits the spinal cord and its meninges. Close beside the foramen magnum, the occipital condyles articulate with the atlas (the first vertebra of the vertebral column) and the hypoglossal canal burrows beneath each condyle to admit the hypoglossal nerve (XII).

- Which bones of the cranium do the following structures identify: greater wings, cribriform plate, petromastoid part, superior temporal line, and superior nuchal line?

FACIAL BONES

NASAL

The paired nasal bones form the upper portion of the bridge of the nose. They are the bones most frequently broken or dislocated in nose injuries. The nasals are the second smallest bones in the skull after the lacrimals. The nasal bones (nas-, *L*, the nose) articulate with each other (Figure 6.15A) in the midline. They send a narrow crest posteriorly to the ethmoid that supports the cartilage of the nasal septum be-

neath. Each nasal bone articulates laterally with the maxilla and with the frontal bone above. Small foramina on the external surface admit nasal veins from the skin into the skull.

LACRIMAL

The lacrimals are the smallest bones of the skull. This pair of bones occupies the anterior medial wall of each orbit (Figure 6.15B); they anchor part of the orbicularis oculi muscle and form the lacrimal fossa with the maxilla. The orbicularis oculi closes the lower eyelid that spreads tears over the eye, while the lacrimal sac collects this fluid in the lacrimal fossa and sends it through the nasolacrimal duct to the nasal cavity.

The lacrimal bones (LAY-krim-al; lachrim-, *L*, tears) attach to the frontal bone above, suspend the inferior nasal concha below, and articulate with the maxillary bone laterally. The lacrimals also surface on the middle meatus of the nasal cavity and join the ethmoid bone posteriorly.

ZYGOMATIC

The two zygomatic bones are the cheek bones; they shape the cheeks and the orbits, and they serve as the origin of the masseter muscle of the mandible (Figure 6.15C). The zygomatic bone (ZYE-go-mat-ik; zygo-, *G*, a yoke) forms the lateral third of the orbital border and sends a **frontal process** above to the frontal bone and a **maxillary part** below to the maxilla. The **orbital surface** of the zygomatic bone also meets the maxilla in the floor of the orbit and extends deeper into this cavity to articulate with the sphenoid. The **temporal process** of the zygomatic meets the temporal bone posteriorly, completing the zygomatic arch.

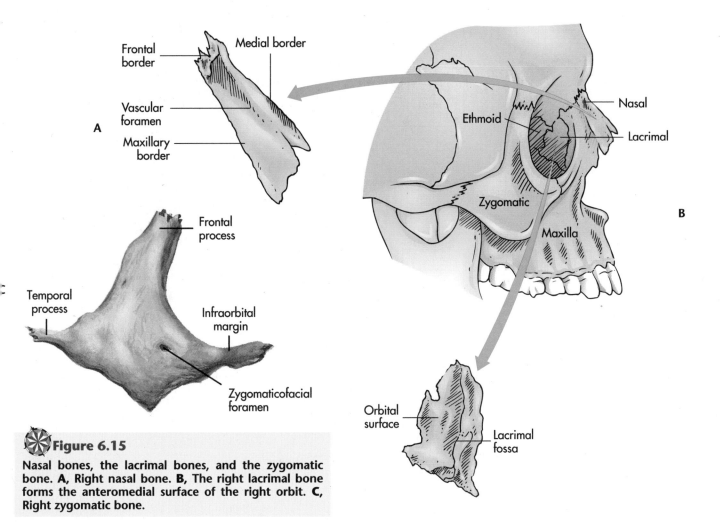

Figure 6.15

Nasal bones, the lacrimal bones, and the zygomatic bone. **A,** Right nasal bone. **B,** The right lacrimal bone forms the anteromedial surface of the right orbit. **C,** Right zygomatic bone.

MAXILLA

The maxillae are the largest bones of the face; because of this dominance they play many roles (Figure 6.16A). The maxillae form the upper jaw and the floor and walls of the nasal cavity. Both bones surround the maxillary sinuses and contribute to the floor of the orbits.

Each maxilla (maks-ILL-ah; maxill-, *L,* the jawbone) sends four processes from the body of the bone just below the orbit (Figure 6.15B and C). The **body** houses the maxillary sinus and articulates with the ethmoid bone in the roof of the nasal cavity and with the inferior nasal concha. The body of the maxilla also meets the palatine bone posteriorly and ends at the pterygomaxillary fissure.

The **frontal** and **zygomatic processes** of the maxilla reach upward to their corresponding bones. The frontal process gives a sharp border to the external nares; the process articulates with the nasal bone, and also forms the lacrimal fossa with the lacrimal bone. The zygomatic process receives the zygomatic bone. The **infraorbital foramen** conducts vessels and a branch of nerve V to the skin of the nose and lips.

The **alveolar** and **palatine processes** form the roof of the oral cavity. Each alveolar process has alveoli (sockets) for eight upper teeth, as shown in Figure 6.16B. The palatine processes form the hard palate (Figure 6.16C); each spreads medially and meets its opposite partner and the vomer, as the vomer descends in the nasal septum. The **incisive canal** behind the incisor teeth conducts the nasopalatine nerve and arteries between the oral and nasal cavities (see the vomer for details). The palatine processes articulate posteriorly with the palatine bone.

- Name an anatomical term that identifies the nasal bones, the lacrimals, the zygomatic bones, and the maxillae.

INFERIOR NASAL CONCHAE

These paired bones send delicate curved folds into the nasal cavity. The inferior nasal concha (KON-kah; conch-, *G,* a shell; see Figure 6.17A) separates the middle nasal meatus above from the inferior nasal meatus below and causes air to spiral over the richly vascularized and innervated mucosa that lines the nasal cavity. Many small perforations in the inferior conchae receive these vessels and nerves. The inferior nasal concha makes sutures with the maxilla,

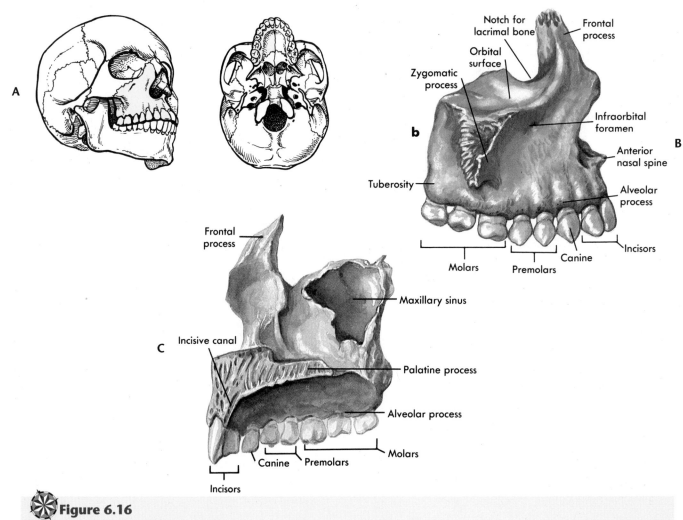

b

B

Notch for
lacrimal bone

Frontal
process

Orbital
surface

Zygomatic
process

Infraorbital
foramen

Anterior
nasal spine

Alveolar
process

Tuberosity

Incisors

Canine

Molars

Premolars

Frontal
process

Maxillary sinus

Incisive canal

Palatine process

C

Alveolar process

Molars

Canine   Premolars

Incisors

⭐ **Figure 6.16**

Maxilla. A, The right maxilla seen in skull. **B,** Lateral view. **C,** Medial view.

lacrimal, ethmoid, and palatine bones in the lateral wall of the nasal cavity.

VOMER

The vomer resembles a plowshare, which gives this bone its name (VOE-mer; *L,* plowshare). This thin plate (Figure 6.17B) forms the lower, posterior portion of the nasal septum. Inhaled air flows past either side of the vomer toward the choanae and the pharynx. The vomer descends from the sphenoid bone and angles forward like a plowshare beneath the ethmoid bone and the cartilaginous septum, where it joins the maxillae and palatine bones on the floor of the nasal cavity. **Nasopalatine grooves** on each side of the vomer distribute nerves and vessels of the same name to the incisive foramen and palate behind the incisors.

PALATINE

These *L*-shaped bones face each other, and they contribute to the lateral wall and floor of the nasal cavity at the choanae (Figure 6.17C). The palatines

(PAL-a-tines; palat-, *L,* the roof of the mouth) also reach forward into the orbit and extend behind to the sphenoid bone, delivering innervation and vessels to most of the palate and lateral walls of the nasal cavity.

The **horizontal portion** of the *L* completes the hard palate behind the maxilla. The greater palatine foramen near the third molar conducts vessels and nerves of this name forward to the mouth.

The **perpendicular portion** of the palatine bone, the stem of the *L,* ascends the lateral wall of the nasal cavity between the maxilla and the sphenoid and reaches deep into the orbit. The perpendicular portion is sculpted into three depressions that are part of the superior, middle, and inferior meati of the nasal cavity. Most of the innervation and vasculature of these walls enters through the **sphenopalatine foramen** in the middle meatus of the palatine bone.

MANDIBLE

The mandible is the bone of the lower jaw, the largest and only freely articulating bone of the

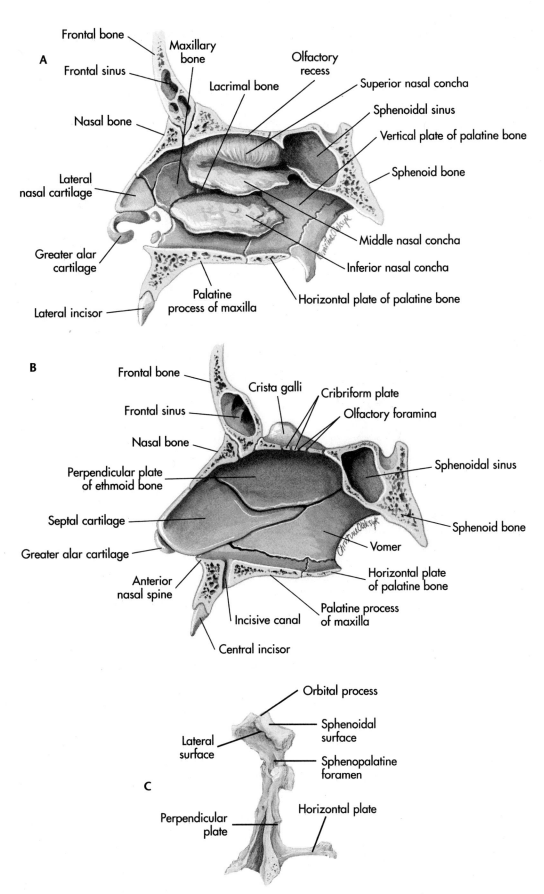

Figure 6.17
Facial bones. A, Left inferior nasal concha. B, Vomer. C, Left palatine bone.

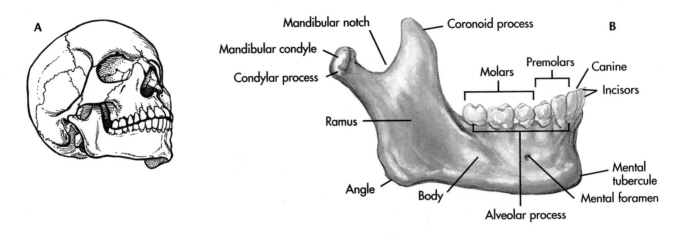

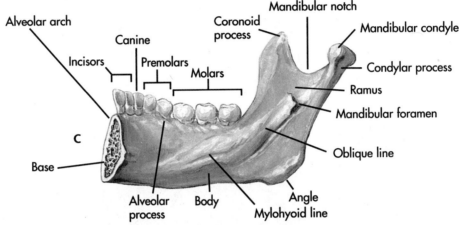

Figure 6.18

A, The mandible articulates with the mandibular fossa of the skull. **B,** Mandible, right half in lateral view. **C,** Medial view.

skull (Figure 6.18A). The mandible (MAN-dih-bl; mandib-, L, jaw) anchors the lower teeth, supports the tongue and floor of the oral cavity, and accommodates two of the three pairs of salivary glands. The **body of the mandible** (Figure 6.18B) meshes the lower teeth with their partners in the maxilla. A vertical **ramus** at each end of this arch articulates with the temporomandibular joint and receives the muscles of the lower jaw. The **angle** of the mandible marks the junction of ramus and body, and the thick **oblique line** distributes the forces of mastication between the body and rami.

The body of the mandible resolves into the thin upper **alveolar arch** and a much thicker lower **base,** seen in Figure 6.18C. The alveolar arch anchors the teeth in 16 alveoli, or bony sockets, and the base extends forward at the chin as 2 **mental tubercles.** The **mental symphysis** marks where the left and right halves of the mandible have fused. The **mylohyoid line,** seen on the medial surface of the mandible, marks the juncture of arch and base. This line anchors the mylohyoid muscle, stretching across the floor of the mouth, that supports the tongue. In the mandible

a small depression above the anterior end of the mylohyoid line receives the **sublingual glands,** while the **submandibular gland** occupies a larger fossa below the line posteriorly.

Each ramus ascends the infratemporal fossa between the zygomatic arch and the skull (see Figure 6.3). The **condylar process** articulates with the mandibular fossa, and from the **coronoid process** to the angle the entire ramus receives muscles of mastication from the skull. The mandibular nerve (from cranial nerve V) and vessels enter the **mandibular foramen,** and the **mandibular canal** carries them forward to the teeth and the tongue. Subsequent branches of the mandibular nerve reach the skin from the **mental foramen** on the external surface of the mandible beneath the premolars.

HYOID

The hyoid is a *U*-shaped bone located in the neck between the mandible and larynx. Like the mandible, it is a second bony arch beneath the skull (Figure 6.19A and B), attached by ligaments to the styloid processes. The hyoid bone (HYE-oyd; hyo-, *G*, U-

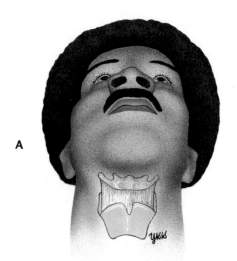

A

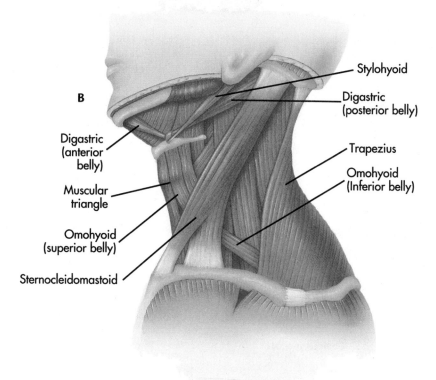

B

Stylohyoid

Digastric
(posterior belly)

Trapezius

Omohyoid
(Inferior belly)

Digastric
(anterior
belly)

Muscular
triangle

Omohyoid
(superior belly)

Sternocleidomastoid

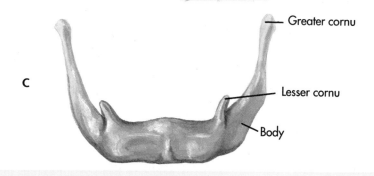

C

Greater cornu

Lesser cornu

Body

Figure 6.19

A, The hyoid bone is located beneath the mandible. **B,** There it is actuated by muscles from above and below. **C,** Anterior view of hyoid bone.

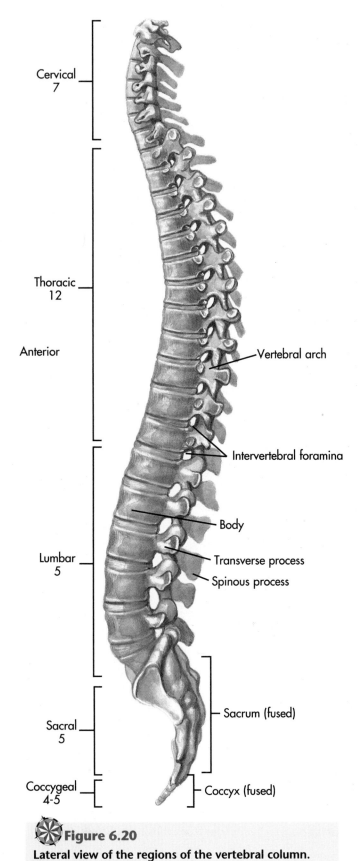

Cervical
7

Thoracic
12

Anterior

— Vertebral arch

— Intervertebral foramina

Lumbar
5

— Body

— Transverse process

— Spinous process

Sacral
5

— Sacrum (fused)

Coccygeal
4-5

— Coccyx (fused)

Figure 6.20

Lateral view of the regions of the vertebral column.

shaped) distributes swallowing muscles between the tongue, the pharynx, and the neck (Figure 6.19B). Because it suspends the larynx from the thyrohyoid ligament, swallowing raises and lowers both organs together. Crushing the hyoid, as may occur in automobile collisions and strangulation, can prevent air from reaching the lungs and result in asphyxiation. The **body of the hyoid** sends two **greater cornua** (KOR-new-ah; singular, cornu; corn, *L,* horn) toward the styloid processes on the base of the cranium. Both **lesser cornua** rise from the junction of the greater cornua and the body. Muscles of tongue, pharynx, and neck insert on the body and all cornua.

- Which of these bones—inferior nasal conchae, vomer, palatines, mandible, hyoid—contribute to the nasal septum?
- Which contribute to the hard palate?
- Which has a condylar process?
- Which extends the greater and lesser cornua?

VERTEBRAL COLUMN

BRIEF TOUR

The vertebral column is a flexible, hollow rod that bends and twists with body movements (Figure 6.20). Back muscles flex it at every stride, as the arms move and each leg leaves the ground. Twenty-six vertebrae, cushioned with intervertebral disks and sheathed with ligaments, participate in all this action. The upper regions of the vertebral column are more flexible, and the lower regions support more weight. Each vertebra supports the weight from those above it with a drum-like body. **Vertebral arches** protect the spinal cord within the **vertebral canal,** a space that is continuous with the cranial cavity. Pairs of spinal nerves communicate from the vertebral canal to the body by way of **intervertebral foramina,** the spinal equivalents of the foramina for cranial nerves in the skull. Muscles and ligaments attached to various **spinous** and **transverse processes** on each vertebra bend and twist the vertebral column.

SPINAL CURVATURES

Spinal curvatures distribute weight and motion more flexibly than if the vertebral column were absolutely straight. Five regions of the vertebral column support the head and trunk directly above the lower limbs (Figures 6.20 and 6.21). The **cervical region** (*L,* neck) is the most flexible and extends backward to take the thrust of the head directly down upon it. **Thoracic vertebrae** curve forward to support the thoracic cavity; and the **lumbar region** curves backward a second time to realign the vertebral column beneath the head. Continued too far, this curvature would also unbalance the vertebral column, but the **sacral** and **coccygeal vertebrae** (KOK-sij-ee-al; coccy-, *G,* a

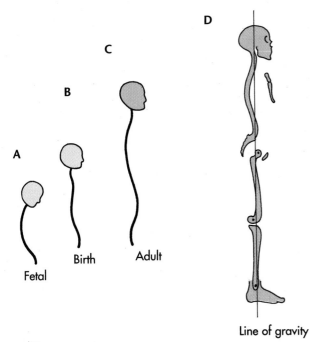

Figure 6.21

Normal spinal curvatures. The primary curvature of the fetal vertebral column (A), is modified by secondary curvatures, first in the cervical region (B), near birth, and later in the lumbar region (C). Spinal curvatures (D),support the head directly above the legs and pelvis.

cuckoo) curve forward again and bring the pelvis and legs directly below the head.

The intervertebral disks accommodate these curvatures. They, rather than the vertebrae, fit like wedges into the curves. Spinal curvatures begin in fe-

tal life as one large, forward curvature that accommodates the fetus in the uterus (Figure 6.21). The cervical curvature begins to form when babies hold up their heads and begin crawling, and the lumbar curvature appears when they learn to stand and walk. These curvatures are considered **secondary** because they begin after the **primary** thoracic and sacral curvatures have developed.

CONGENITAL SCOLIOSIS

The Martins' 2-year-old daughter had more difficulty than usual in learning to walk. She could not stand fully erect; her vertebral column curved to the right (Figure 6.22A). Joan learned to compensate for her imbalance by lowering her left shoulder and placing more weight on her left leg. Their family physician referred the Martins to a spinal surgeon. The surgeon made X-rays of Joan's vertebral column and found a 32-degree curvature in the lumbar region of the vertebral column and a milder thoracic curvature that was compensating for her lumbar curve. The right half of one lumbar vertebra was missing (Figure 6.22B).

The surgeon recommended that the hemivertebra (half vertebra) be removed and the adjacent bones fused together to straighten Joan's vertebral column. The thoracic curvature would diminish when she learned to rebalance herself. Today, Joan's condition is hard to detect externally, although an X-ray taken 10 years after surgery shows that mild curvatures remain in both regions.

There are several types of abnormal spinal curvatures (Figure 6.22A, C, and D). **Scoliosis** (SKO-lee-OH-sis; scoli-, *G*, crooked) is lateral curvature of the

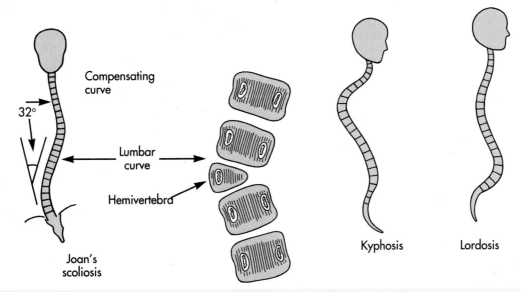

Figure 6.22

Abnormal curvatures of the vertebral column. Joan's congenital scoliosis **(A)** was caused when half of a lumbar vertebra failed to form **(B)**. Kyphosis **(C)** is an exaggerated forward curvature. Lordosis **(D)** is an exaggerated lumbar curvature.

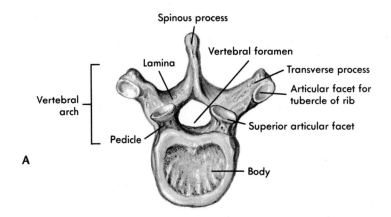

**Figure 6.23**

A typical vertebra. A, Viewed superiorly. B, Viewed laterally.

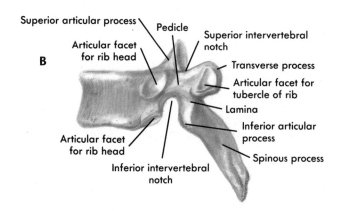

vertebral column; it is not always congenital. Disease- or accident-induced paralysis also can lead to scoliosis simply because the back muscles can no longer support the vertebral column. **Kyphosis** (kye-FOE-sis; kypho-, *G*, bent) is a forward curvature that frequently causes a humpback to form in the thoracic region. Kyphosis is especially common among osteoporosis patients who have suffered compression fractures of vertebrae. In other forms of congenital kyphosis, one or more vertebrae lack the anterior portion of the body. **Lordosis** (lor-DOE-sis; lord-, *G*, bent backwards) is an exaggerated curvature of the lumbar spine that is sometimes related to loss of vertebral arches. Spinal curvatures may become more severe with age, which is one reason why early surgery was recommended for Joan. Extreme cases of scoliosis add pulmonary and cardiovascular risks to a patient's condition because abdominal organs intrude upon the thoracic cavity and diminish lung capacity and cardiac output.

- How many vertebrae make up the vertebral column?
- Name the four spinal curvatures; in which direction does each one curve?
- What are the names given to lateral curvature of the vertebral column and to abnormally large forward curvature?

TYPICAL VERTEBRAE

Except for the first two vertebrae, all vertebrae have a central, drum-shaped body (Figure 6.23) that supports the weight of the vertebral column above. **Intervertebral disks** are cushions of fibrocartilage between the ends or **faces** of the body that articulate with the vertebrae above and below. Ligaments be-

tween the faces keep the intervertebral disks in position. The body of each vertebra sends a **vertebral arch** posteriorly to enclose the spinal cord (Figure 6.23A and B), and spines and processes extend from the arch to receive ligaments and muscles from the back. The vertebral arch begins with a pair of **pedicles** that connect with the body of the vertebra. A pair of **laminae** complete the arch beneath the **spinous process** that reaches backward and overlaps the vertebra below with connections for ligaments and muscles. **Transverse processes** make similar connections at each side of the arch where the laminae and pedicles join. The spinal cord passes through the **vertebral foramen** thus formed, and spinal nerves exit between vertebrae through **intervertebral foramina.** Each foramen is a matched pair of **vertebral notches,** one in the inferior margin of the pedicle above the foramen and the other in the superior margin of the pedicle below.

Articular processes and **articular facets** allow the arches of adjacent vertebrae to articulate (see Table 6.2). Articular processes extend superiorly and inferiorly from the junctions between laminae and pedicles in Figure 6.20. The processes end with smooth articular facets that face downward, inward,

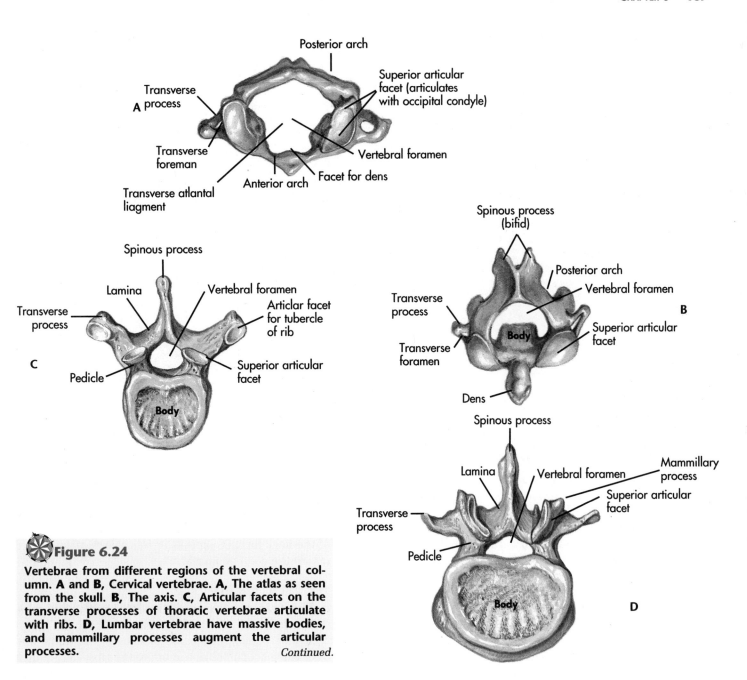

Figure 6.24

Vertebrae from different regions of the vertebral column. A and B, Cervical vertebrae. A, The atlas as seen from the skull. B, The axis. C, Articular facets on the transverse processes of thoracic vertebrae articulate with ribs. D, Lumbar vertebrae have massive bodies, and mammillary processes augment the articular processes. *Continued.*

and forward from the vertebra above, while neighboring facets of the vertebra below face upward, outward, and backward to receive them. Coated with hyaline cartilage, these facets allow the vertebral column to twist and bend. The direction of the facets limits motion. The neck can twist and bend because cervical facets articulate on an almost transverse plane. In contrast, facets in the lumbar region meet on a nearly sagittal plane that favors bending and reduces twisting. Whiplash fractures of articular processes are particularly dangerous because the broken facets may traumatize the spinal cord.

- Name the various processes that extend from a vertebra.
- What is the difference between the vertebral arch and the intervertebral foramina?

LONG TOUR

This section considers the special features of each region of the vertebral column. Refer to Figures 6.20 and 6.24.

CERVICAL REGION

Seven light, flexible cervical vertebrae are designed for mobility. These bones (Figures 6.20 and 6.24A) are easy to identify by their **transverse foramina** that pierce the transverse processes with passages that conduct the vertebral arteries to the foramen magnum and brain. With little weight to support, their oval bodies are relatively small. The vertebral foramina are large and triangular, and the articular processes have large facets that lie in a nearly transverse plane, allowing considerable bending and twist-

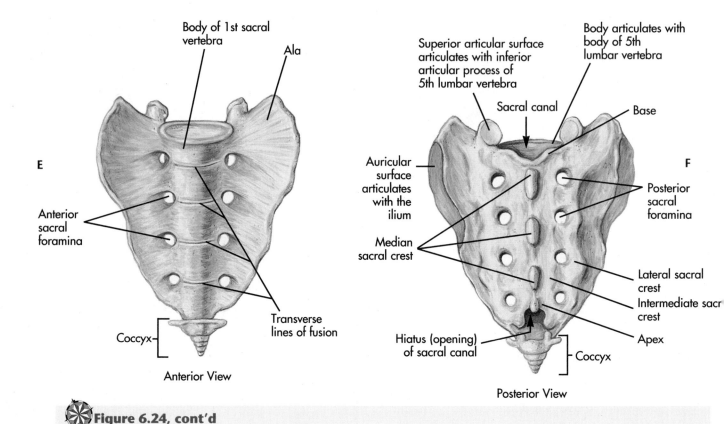

Body of 1st sacral vertebra

Ala

Anterior sacral foramina

Transverse lines of fusion

Coccyx

Anterior View

Superior articular surface articulates with inferior articular process of 5th lumbar vertebra

Body articulates with body of 5th lumbar vertebra

Sacral canal

Base

Auricular surface articulates with the ilium

Posterior sacral foramina

Median sacral crest

Lateral sacral crest

Intermediate sacr crest

Apex

Hiatus (opening) of sacral canal

Coccyx

Posterior View

✳ **Figure 6.24, cont'd**
E, Anterior view of the sacrum. **F,** Posterior view.

ing. The spinous processes are rather narrow; they project posteriorly to support the cervical curvature. The ends of the spinous processes of the third through the sixth vertebrae usually are doubled.

The atlas and the axis are greatly modified to receive the skull. The atlas (Figure 6.24A), named after the Greek god who supported the earth on his shoulders, articulates with the occipital condyles of the skull. It is the first cervical vertebra. The atlas has no body because the body has fused with the body of the axis, the second cervical vertebra, shown in Figure 6.24B. In its place the vertebral arch has formed a ring of bone with large articular facets at each side. The superior facets receive the occipital condyles and permit the head to nod forward and back. Flatter inferior facets beneath allow the skull to rotate on the axis.

The axis is aptly named; its characteristic feature, the **dens** (*L,* tooth), is the pivot around which the atlas turns. The dens develops from the body of the atlas and penetrates upward within the anterior arch of the atlas where the **transverse atlantal ligament** secures the dens in place. There is a major disadvantage to all this mobility—the dens can penetrate the spinal cord with disastrous paralytic consequences in whiplash injuries. This is one reason to wear your seat belt and use a firm head support in any car.

The seventh cervical vertebra begins the transition to the thoracic region; it resembles the thoracic vertebrae and has a large body and a longer spinous

process than other cervical vertebrae. The seventh cervical vertebra also bears the first spinous process that you can feel with your hand (palpate). Ligaments of the neck bury the spines of other cervical vertebrae inside the cervical curvature.

THORACIC REGION

Thoracic vertebrae articulate with the ribs, which form the walls of the thoracic cavity (Figures 6.20 and 6.24C). Articular facets on the body and transverse processes accomodate the ribs. Larger than cervical vertebrae but smaller than lumbar vertebrae, 12 thoracic vertebrae sustain more weight than the cervicals. The bodies are broader and heart-shaped, and the vertebral foramen is narrower and circular. The spinous processes of thoracic vertebrae are easy to palpate because they are on the convex side of the thoracic curvature. The broad laminae overlap, like shingles of a roof. The intervertebral foramina are carried entirely by vertebral notches on the posterior margin of the pedicles. The facets of the articular processes face a nearly frontal plane that permits some motion in all directions, but far less than that allowed in the neck.

LUMBAR REGION

The five lumbar vertebrae are the largest vertebrae, and they support the most weight. The bodies are longer and wider than any others, and their transverse processes are longer (Figures 6.20 and 6.24D). The spinous processes are rectangular, gaining in width what they lack in length compared with thoracic and cervical vertebrae. Despite this robustness, the laminae do not overlap as extensively as they do in the thorax; it is therefore standard procedure to tap cerebrospinal fluid from within the vertebral canal by inserting a syringe through the spaces between the lumbar laminae. The facets of the articular processes line up on a nearly sagittal plane, which permits bending in sagittal and frontal planes but little twisting. Lumbar vertebrae have special structures called **mammillary processes** that strengthen the lateral walls of the articular processes with surfaces for attachment of the deep multifidus muscles of the back.

SACRAL REGION

The five sacral (SAY-kral; sacr-, *L,* the sacrum) **vertebrae have fused with each other to become the sacrum,** which transfers weight to the hip bones and lower limbs (Figures 6.20 and 6.24E and F). Wedged between the hip bones, this triangular bone retains features of the vertebrae from which it developed. The body of the first sacral vertebra articulates with the fifth lumbar vertebra at the broad **base** of the sacrum, and that of the fifth sacral vertebra articulates with the coccyx at the **apex.** The anterior surface of the sacrum, shown in Figure 6.24E, is concave and shapes the posterior wall of the pelvic cavity. Four **transverse ridges** mark where the remaining bodies of the vertebrae would have articulated had they not fused. Lateral and superior to each of these ridges are pairs of anterior **pelvic sacral foramina,** eight altogether, that communicate nerves from the **sacral (vertebral) canal** to the anterior side of the lower extremities. These foramina represent the vertebral notches and spaces between the transverse processes. The ear-shaped **auricular surfaces** of the sacrum articulate with the hip bones as part of the pelvis.

The convex posterior side of the sacrum, shown in Figure 6.24F, displays remnants of the spinous and transverse processes. Fusion has transformed these structures into **median** and **lateral sacral crests,** respectively. The low **intermediate crest** represents the articular processes. The posterior pelvic sacral foramina deliver nerves from the vertebral canal to the back. Articular facets for the fifth lumbar vertebra reach forward from the base of the sacrum, and facets for the coccyx extend from the apex. The vertebral arch fails to ossify in the fifth sacral vertebra and leaves an opening to the sacral canal, called the **sacral hiatus.**

COCCYX

Even with the body's weight already transferred to the pelvis, there is much left for the coccyx to do. This final portion of the vertebral column (Figures 6.20 and 6.24E and F) shapes the concave floor of the pelvic cavity and receives a minor portion of the gluteus maximus muscle. The coccyx consists of four small **coccygeal vertebrae** that retain only the bodies and articular and transverse processes; there is no vertebral canal. The first coccygeal vertebra articulates with the sacrum, but it has fused with the second and third coccygeal vertebrae. The fourth is a small tag of bone to which the external sphincter muscle of the anus attaches.

The coccyx' utility lies in its connections to the levator ani muscle in the floor of the pelvic cavity. This muscle is involved in bladder control and defecation, and because it surrounds the entrance to the vagina in females, it also contributes to pleasure of sexual intercourse. The coccyx flexes away from the birth canal during childbirth.

- Which region of the vertebral column is most flexible?
- Which region supports the most weight?
- Which vertebrae articulate with ribs?
- Which vertebrae have the largest bodies?
- Which vertebrae lack a vertebral canal?

RIBS AND STERNUM

The rib cage supports the thoracic cavity. The rib cage (Figure 6.25A) is a conical structure with a narrow **superior aperture,** approximately 5 cm (2 inches) in diameter, and a much broader **inferior aperture,** 20 to 23 cm (8 to 9 inches) in depth, bounded by the diaphragm, which takes its origin from the **costal margin.** Twelve pairs of ribs (KOS-tie; costae, *L,* ribs) support the thoracic cavity, and thereby protect the heart and lungs and permit the lungs to inflate with air. Intercostal muscles enable the lungs to inhale and exhale by raising and lowering the ribs on articulations with the thoracic vertebrae and the sternum.

A typical rib curves forward from the vertebral column. Its **head** (see Table 6.2) articulates with the body of the vertebra, and at first the narrow **neck** extends posteriorly where it articulates with the transverse process, as shown in Figure 6.25B. Ligaments attach each rib to the transverse process. The body of the rib curves anteriorly from the **angle** in Figure 6.25C and ends on a **costal cartilage** that attaches to the sternum. Only the first seven ribs are true **vertebrosternal** ribs, because their costal cartilages connect directly to the sternum. The remainder are false ribs. The first three of these (8, 9, and 10) articulate with the sternum through the seventh costal cartilage and are considered **vertebrochondral** ribs. The eleventh and twelfth ribs are **vertebral** or floating ribs because they do not articulate with the sternum; their cartilages end in the muscles of the abdominal wall.

The sternum is shaped like a sword. Its **manubrium** (ma-NU-bree-um; manubri-, *L,* handle)

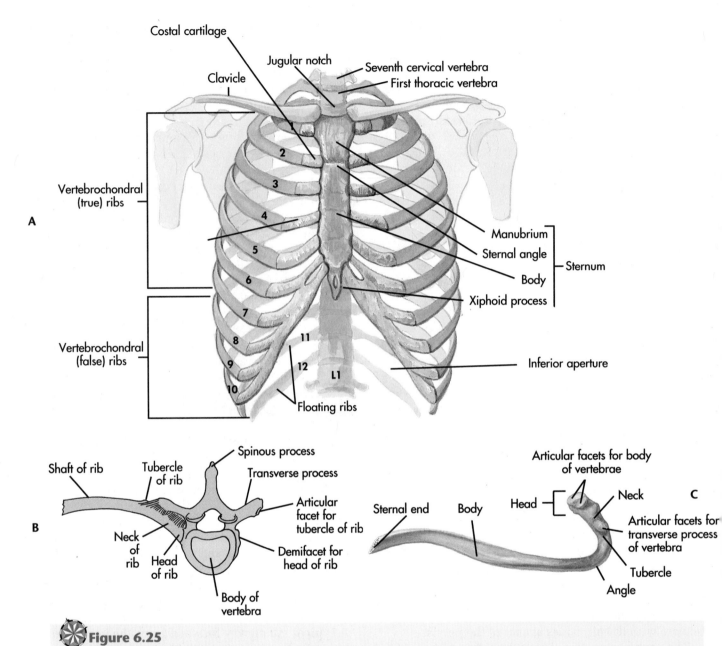

Figure 6.25

A, Rib cage and sternum, anterior view. **B,** Head and tubercle of ribs articulate with demifacets on the body and with articular facets on the transverse processes. Ligaments join the ribs to the vertebrae. **C,** Vertebrosternal rib.

receives the clavicles and first pair of ribs, as shown in Figure 6.25A. The **jugular notch** can easily be palpated between the insertions of the sternocleidomastoid muscles from the skull. At the second rib, the **sternal angle** marks the **manubrosternal joint** with the body of the sternum. This articulation is a symphysis (described in Chapter 8) that flexes with deep breathing; it marks the inferior end of the trachea. The **body of the sternum** is the blade of the sword and receives all vertebrosternal ribs except the first pair. It

develops from four cartilaginous **sternebrae** (analogous to vertebrae) that fuse together. The **xiphoid process** (ZIF-oyd; xiph-, *G*, a sword), the fibrocartilage tip of the sword, receives the vertebrochondral ribs. This process connects with the body by the **xiphisternal joint** and ossifies in early middle age.

- Name the three bones of the sternum.
- What term describes the seven ribs that articulate with the sternum?
- Which aperture of the rib cage is the larger?

DEVELOPMENT AND AGING

The vertebral column and rib cage are the most highly segmented structures of the body.

SEGMENTAL VERTEBRAE

The vertebral column develops from strips of intermediate mesoderm (described in Chapter 4) along either side of the neural tube and **notochord** (NOE-toe-kord; see Figures 6.26A, and 4.4E.) Beginning cranially, this mesoderm segments into pairs of **somites,** and three clusters of cells—the **myotome, dermotome** and **sclerotome**—develop from each somite. The sclerotomes (SKLARE-oh-tomes) produce the vertebral column and the base of the skull. (Muscles of the back develop from the myotomes, and dermis of the skin develops from the dermotomes.) Each side of a vertebra develops from the sclerotome on that side. In Joan Martin's case, lumbar sclerotomes failed to develop on her right side, leaving a gap that allowed her vertebral column to bend to the right.

At first glance, you might think that each pair of sclerotomes forms one vertebra, with the intervertebral disks between, but the opposite is true. Each disk develops from the center of a pair of sclerotomes, with the vertebrae between, as Figures 6.26A and B illustrate. Consequently the body of a vertebra represents the inferior portion of the sclerotome above and the superior portion of the sclerotome below that have fused together. The sclerotomes grow around the notochord and squeeze it into the intervertebral disk, where it becomes the nucleus pulposus, the central core of the disk.

When you study the muscles of the vertebral column in Chapter 10, you will find that many of these muscles link one vertebra directly to the next, and that these muscles overlap the vertebrae. Figure 6.26B shows that this overlapping develops because the myotomes from which the back and other trunk muscles arise overlap the developing vertebrae, an elegant way to ensure that muscles and vertebrae link properly.

REGIONAL DIFFERENCES

Differences between vertebrae can be traced to the vertebral cartilages that develop from the sclerotomes (vertebrae are chondrogenic bones; see Chapter 5). Figure 6.27 shows that three pairs of cartilage elements contribute to each vertebra. One pair becomes the vertebral arch and contributes to the transverse processes, and another pair forms the body of the vertebra (Figure 6.27A), while the third pair of **costal elements** forms the lateral wall of each pedicle. The

A   B

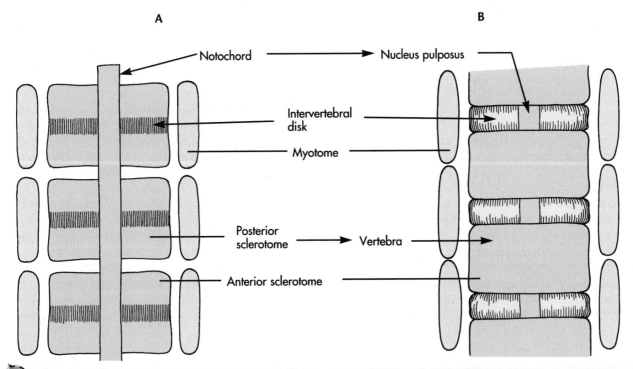

Figure 6.26

Vertebrae develop from sclerotomes. **A,** Sclerotomes cover the notochord, and intervertebral disks begin to develop in the sclerotomes. Myotomes line up beside the sclerotomes, in register with them. **B,** Vertebrae develop between intervertebral disks, from the posterior and anterior portions of adjacent sclerotomes.

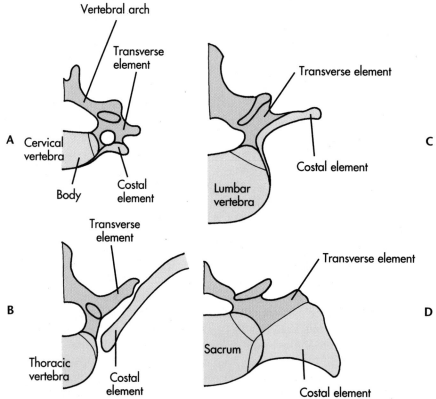

Figure 6.27

Vertebral cartilages. **A,** Transverse and costal elements form the transverse foramina of cervical vertebrae. **B,** In thoracic vertebrae, separate transverse and costal elements form the transverse processes and ribs. **C,** Transverse and costals rejoin in lumbar vertebrae to form transverse processes. **D,** The transverse and costal elements join the sacrum to the pelvis.

costal elements of cervical vertebrae contribute to transverse foramina (A), and in thoracic vertebrae they form ribs (B), but in lumbar vertebrae they form transverse processes (C) and contribute the auricular surfaces of the sacrum (D).

• Name the tissue that derives from the embryonic somites and becomes the body of the vertebrae.

AGING

The skeleton begins to lose bone after the age of 40 because more bone matrix is removed than is replaced. With less vigorous activity to promote bone deposition, bones begin to atrophy. This steady loss of bone mass has especially obvious effects that you probably will recognize in your grandparents. As their lessening heights demonstrate, you can expect to lose about 1.5 cm (0.6 inches) of height for every 20 years that elapse after 40 years of age. Most of this loss comes from the vertebral column rather than the extremities; the intervertebral disks become thinner and spinal curvatures become accentuated. As a result the extremities become disproportionately longer. The pelvis widens and the shoulders narrow into the stooped posture that many older people display.

SUMMARY

The axial skeleton protects the brain and spinal cord and supports the trunk of the body. The skull encases the brain; the face houses the eyes, ear chambers, and the nasal and oral cavities. The vertebral column and rib cage support the trunk and limbs and protect the internal organs. *(Page 142.)*

The central bodies of bones extend parts and processes outward toward other bones. Parts may be flat or irregular portions of bones, and processes are narrower, rounded projections. *(Page 155.)*

Twenty-three skull bones partition spaces for the brain, eyes, ear chambers, nasal cavity, paranasal sinuses, and oral cavity. Eight cranial bones enclose the brain, and fourteen facial bones form the orbits, nasal cavity, paranasal sinuses, and the oral cavity. Sutures lock nearly all of these bones rigidly together, except for the mandible and the hyoid bone, which move freely. *(Page 146.)*

The flat bones of the calvarium cover the roof of the cranial cavity. Three cranial fossae in the base of the cranium support the brain. Numerous foramina conduct cranial nerves and vessels through the base of the cranium. *(Page 150.)*

The face contains the orbits, nasal cavity, paranasal sinuses, and the oral cavity. The orbits are conical spaces that protect the eyeballs. The nasal cavity admits air through the external nares and choanae. The paranasal sinuses communicate with the nasal cavity through openings in the roof and walls of the nasal cavity. The hard palate, teeth, and mandible form the oral cavity, and the hyoid bone supports the larynx. *(Page 152.)*

The shape of the skull and facial features depend on growth of individual bones and sutures. Crouzon's syndrome shows us that insufficient bone growth and premature ossification of sutures can deform the skull. *(Page 148.)*

The vertebral column and rib cage support the trunk and protect internal organs. Twenty-six vertebrae compose the vertebral column. The bodies of the vertebrae support the weight of the trunk and head, and the vertebral arches protect the spinal cord. Muscles and ligaments attached to vertebral spines and transverse processes allow the vertebral column to twist and bend. Intervertebral disks cushion this motion. Normal spinal curvatures distribute these forces effectively, but these curvatures become exaggerated in kyphosis, lordosis, and scoliosis. *(Page 166.)*

The regions of the vertebral column contain different vertebrae. Cervical vertebrae support relatively little weight, are the most mobile vertebrae, and conduct vertebral arteries through transverse foramina. Thoracic vertebrae articulate with the ribs, and the lumbar vertebrae are larger and less mobile than any above them. Sacral vertebrae have fused together and transfer the weight of the trunk to the hip bones. The coccygeal vertebrae have also fused; they shape the posterior wall of the pelvic cavity and receive muscles from the floor of this cavity. *(Page 166.)*

Thoracic vertebrae, ribs, and the sternum form the rib cage. Twelve ribs articulate with the thoracic vertebrae, and the first seven ribs also articulate directly with the sternum. The next three articulate with the sternum through the seventh costal cartilage. The eleventh and twelfth ribs do not articulate with the sternum. *(Page 171.)*

Vertebrae develop from somites, which are segmental derivatives from the mesoderm. A single vertebra develops from portions of adjacent somites. Loss of bone matrix in aging people tends to reduce the height of the vertebral column and to accentuate its curvatures. *(Page 173.)*

STUDY QUESTIONS   OBJECTIVES

1. Name the three portions of the axial skeleton. How many bones does each portion have? Name the bones that protect the brain and those that form the face. What are the five regions of the vertebral column? How many bones does each region contain? In which direction do their curvatures bend?   **1**
2. Name the cavities of the skull and describe where they are located relative to each other. Which bones form the walls of these spaces? What are the names of the major openings to these cavities and where are they located?   **1**
3. Which bones of the skull are considered flat? Are they also cranial bones? Are there any flat bones in the vertebral column?   **2**
4. Name the cavities of the skull and the entrances to them, and describe the functions of these spaces.   **3**
5. Name the cranial and facial bones of the skull and find their locations in Figures 6.2 through 6.4.   **4**
6. Name the foramina in the floor of the cranial cavity and the nerves and vessels that pass through each.   **5**
7. Name the foramina in each cranial fossa of the skull.   **5**
8. Name and locate the fossae of the skull.   **6**
9. Name and locate the sutures of the calvarium. Which bones does each suture connect?   **7**
10. Select several bones of the skull and identify the body, processes, and parts of each one.   **8**
11. Name the parts of a typical vertebra. What is the function of each part?   **9**
12. Why is it advisable to tap cerebrospinal fluid from the lumbar region rather than from the thoracic region?   **10**
13. Which ribs articulate with the sternum and which do not or only indirectly meet it? How do these bones articulate with thoracic vertebrae?   **11**
14. Describe the effects of disease on the skull and the vertebral column. How can such effects be treated?   **12, 13**
15. From which portions of the sclerotomes do vertebrae and intervertebral disks develop?   **14**
16. Describe some effects of aging on the axial skeleton.   **15**

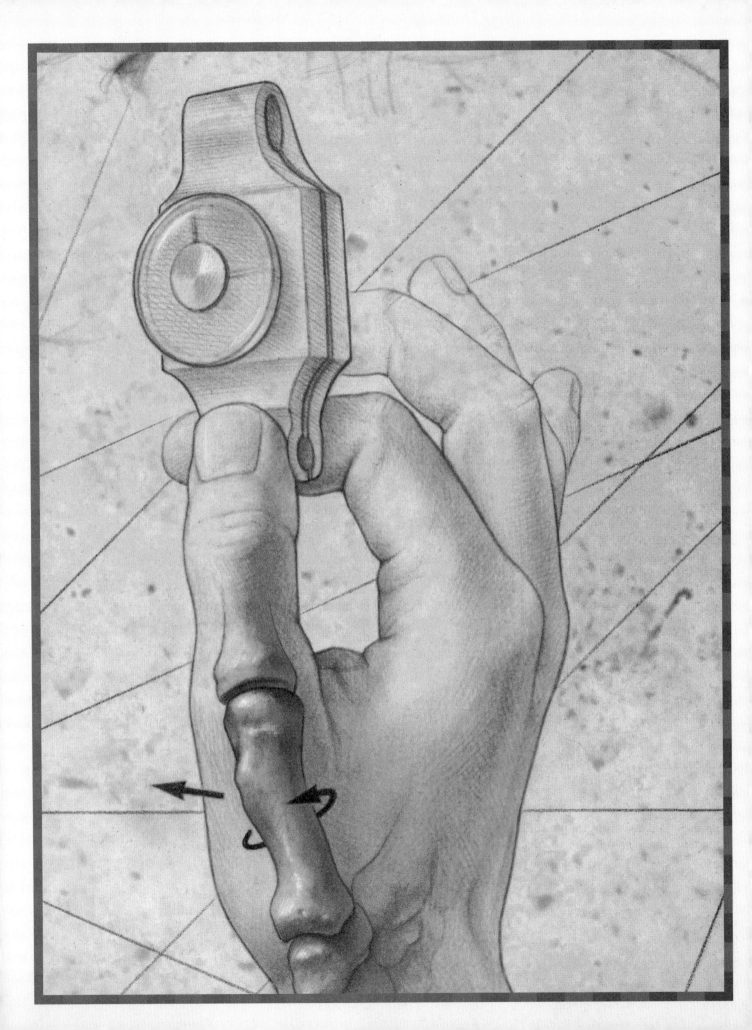

CHAPTER 7

Appendicular Skeleton

OBJECTIVES

After studying this chapter, you should be able to do the following:

1. Name the bones of the limbs and girdles.
2. Describe the motions of the limbs in walking.
3. Describe the bones of the lower limbs and their functions in walking.
4. Identify the three portions of the hip bone and their role in supporting the body.
5. Describe structures of the femur that enable walking.
6. Describe the tibia and fibula and their relationship to the knee and ankle in a walking limb.
7. Identify the groups of bones in the ankle and foot, and describe their functions in walking.
8. Name the bones of the upper limb, and describe the motions they can make.
9. Explain how the scapula and clavicle contribute to the motions of the shoulder.
10. Describe the structure of the humerus and its articulations with the shoulder and forearm.
11. Describe the anatomy of the radius and ulna, and show how they rotate the hand.
12. Name the carpal bones, and describe the motions they permit the hand to perform.
13. Describe the metatarsal bones and phalanges in the hand and the motions they permit.
14. List the similarities and differences between the skeleton of the upper limbs and the lower limbs.
15. Describe how limb buds develop into limbs.

 **VOCABULARY TOOLBOX**

Motions are described with Latin derivatives:

ab-	duction, *L* from; duc-, *L,* lead, lead from.		**ext-**	ension, *L* out, beyond; tend-, *L,* stretch.
ad-	duction, *L* toward, lead toward.		**flex-**	ion, *L* bend.
circum-	duction, *L* around, lead around.		**rota-**	tion, *L* a wheel.

RELATED TOPICS

- Articulations, Chapter 8
- Bone and ossification, Chapter 5
- Muscles of the appendicular skeleton, Chapter 11

- Skeletal muscle, Chapter 9
- Somites, mesoderm, and ectoderm, Chapter 4

PERSPECTIVE

People walk, run, feed themselves, and hug each other with their limbs. The limbs are supraorgan systems; several systems—skeleton, joints, muscle, skin, and cardiovascular and nervous systems—combine into structures that can run a mile in less than 4 minutes or find delicate melodies on a guitar. The appendicular skeleton supports the upper and lower limbs and joins them to the axial skeleton. Joints allow the muscles and the nervous system to flex and extend the limbs. The cardiovascular system provides nutrients, and the skin helps them grip and touch. This chapter introduces the anatomy of the bones that underlie these activities. Later chapters cover the other systems.

The appendicular skeleton is more mobile than the axial skeleton. The bones support the upper and lower extremities, and the girdles join the extremities to the axial skeleton (Figure 7.1). The bones of the hips and lower extremities support the body and help move it from place to place, and the bones of the shoulders and upper limbs participate in personal care and many other motions that make life interesting. In mechanical terms, our limbs resemble the arm of a "cherry picker," which can place a worker wherever required. Because most limb bones decrease in size and increase in number the farther they are from the trunk, the hands and feet are more mobile than the arms and thighs. Our upper limbs would not be very useful if their construction were reversed, with the humerus very short and the digits much longer.

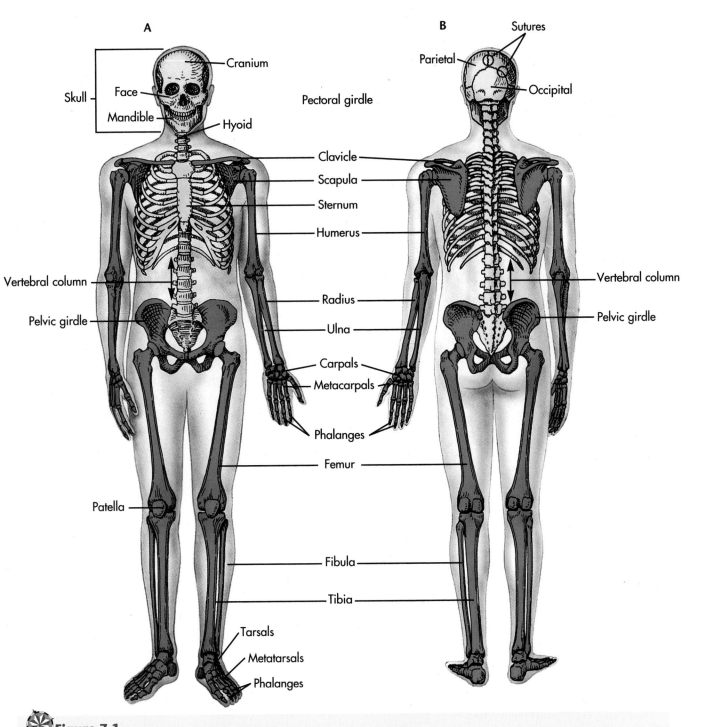

Figure 7.1
The appendicular skeleton *(blue)* supports the limbs. **A,** Anterior view. **B,** Posterior view.

LIMBS AND THE APPENDICULAR SKELETON

Table 7.1 shows that altogether the limbs contain 126 bones, almost two thirds of the 206 bones in the body. Although the upper extremities contain 64 bones and the lower 62, there are actually only half as many varieties to learn because left and right limbs

Table 7.1	BONES OF THE APPENDICULAR SKELETON	
Bones	**No.**	**Shape**
SHOULDER AND UPPER LIMBS		
Clavicle	2	Flat
Scapula	2	Flat
Humerus	2	Long
Radius	2	Long
Ulna	2	Long
Carpals	16	Short
Metacarpals	10	Long
Phalanges	28	Long
TOTAL	64	
PELVIS AND LOWER LIMBS		
Coxal bone	2	Irregular
Femur	2	Long
Patella	2	Sesamoid
Tibia	2	Long
Fibula	2	Long
Tarsals	14	Short
Metatarsals	10	Long
Phalanges	28	Long
TOTAL	62	
GRAND TOTAL	126	

are mirror images of each other. Furthermore, bones of upper and lower extremities are comparable, as you can see in the figure. The hip bone corresponds to the shoulder bones, as does the femur to the humerus. The tibia and fibula of the leg are comparable to the radius and ulna of the forearm, and the resemblance of bones of the foot to those of the hand is the most evident of all.

All five shapes of bones are present in the appendicular skeleton. You will find short bones in the ankles and wrists, flat bones in the shoulder girdle, and irregular bones at the hip. Long bones form the remaining portions of the extremities. Sesamoids appear at the knees, as well as in hands and feet. Most of the appendicular skeleton develops by endochondral ossification. See Chapter 5, Bone and the Skeleton, to refresh your memory of these shapes and how ossification proceeds.

WALKING

Because you need to understand how bones enable you and others to move, this chapter describes the anatomy of the appendicular skeleton in terms of its motion. Figure 7.2 shows how the bones of the extremities move when a person walks. Walk across the room and see for yourself what these bones do. When you are in pace with the figure, notice how your lower limbs are moving. When your right limb swings forward, the right heel touches the floor and the right side of the foot rolls forward to absorb your body's weight, while your left thigh and leg are pushing forward. The right knee locks the limb straight, and your full weight comes onto the right limb midway through the stride. The foot spreads under the weight, as the left foot rises from the floor. Now your left limb begins to swing forward. The pelvis lifts the left side upward to extend that extremity and to help the right limb push forward near the end of the stride. Your left foot returns to the floor, taking the weight from the right, as the ball and toes of your right foot finally push off. Temporarily relieved of its load, your right limb now swings forward for the next step.

Your upper limbs will swing naturally, if you put this book down while you are walking. The left, upper limb swings forward with the right, lower limb, and vice versa the right, upper limb swings forward with the left. Although you can easily walk without swinging your upper limbs, their motion helps the rhythm of the lower limbs.

MOTIONS NAMED

We throw baseballs, reach across tables for coffee, climb stairs, and scratch our heads. These motions and most others boil down to four basic movements of individual bones and joints that help you understand how the appendicular skeleton enables the body to move. This section outlines these basics, and Chapter 8, Articulations, illustrates them fully. The motions are as follows:

1. Motion forward and back, usually in the sagittal plane. Your forearm **flexes** at the elbow toward the shoulder and **extends** away from the shoulder. Your knee extends and flexes the leg in walking.
2. Side-to-side motion, usually in the frontal plane. You **abduct** your upper limb laterally away from the body, and **adduction** returns it to your side. Reaching across the table involves extension and abduction.
3. **Rotation** about the longitudinal axis. If you rotate the humerus medially and flex the forearm, your hand will cross your chest.

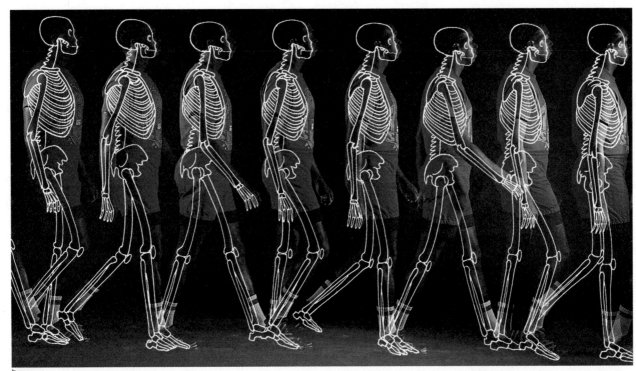

Figure 7.2
Walking movements of the appendicular skeleton. Identify the phases of your gait by matching them with this sequence.

4. **Circumduction** combines all three motions, as the upper limb of an outfielder does in returning a fly ball.

- Which part of the foot touches the floor first in walking?
- At what point does the body's full weight rest on only one limb?
- What position does the forearm take when the elbow flexes?

PELVIS AND LOWER LIMB

BRIEF TOUR

Walking motions reflect the anatomy of the limb bones. This tour follows the skeleton of the lower extremities distally from the pelvis to the feet. Refer to Figure 7.2 for walking movements and to Figures 7.3 and 7.4 for details of the bones; also examine a skeleton, if one is available. The regions of the lower extremities are hip (coxa), thigh (femoral), knee (genu), leg (crural), ankle (tarsus), foot (pes), and toes (phalanges).

The **pelvis** is the bony framework that transfers forces from the lower limbs to the vertebral column and supports the wall of the lower abdomen and pelvic cavity. Specifically the **coxal bones** (hip bones) transfer these forces to the sacrum of the axial skeleton. Each limb can swing in all directions and ro-

tate on its axis because the round **head** of the **femur** inserts into the concave **acetabulum** (ah-seh-TAB-yoo-lum; *L,* a vinegar cup) of the coxal bone, forming a highly mobile **ball-and-socket joint.** Muscles attached between the shaft of the femur and the coxal bone swing the thigh forward and recover it during each stride. Swing your lower limb and rotate it to see the effect of this all-important joint.

With every stride, body weight transfers down the shaft of the femur to the **knee** and on through the leg to the **ankle.** The articulation of the knuckle-like **condyles** (KON-diles; condyl-, *G,* knob, knuckle) of the femur with the **tibia,** the weight-supporting bone of the leg, allows flexion and extension, like a hinge. These broad connections spread body weight across the joint and supply leverage for ligaments between the **epicondyles** of each bone, which prevent side-to-side movement. The **patella** guards the anterior side of the knee, while the slender **fibula** (FIB-yoo-la; fibul-, *L,* a clasp or buckle) receives muscles of the leg and foot and guards the ankle joint. Neither patella nor fibula directly sustains body weight.

The bones and tendons of the foot form a dynamic spring called the **medial longitudinal metatarsal arch;** this spring spreads open during the stride (Figure 7.4A) and springs closed when the heel rises from the floor. The arch begins anteriorly where the **phalanges** (FAL-an-jes; phalan-, *G,* bones of fingers and toes) join the **metatarsal** bones and continues

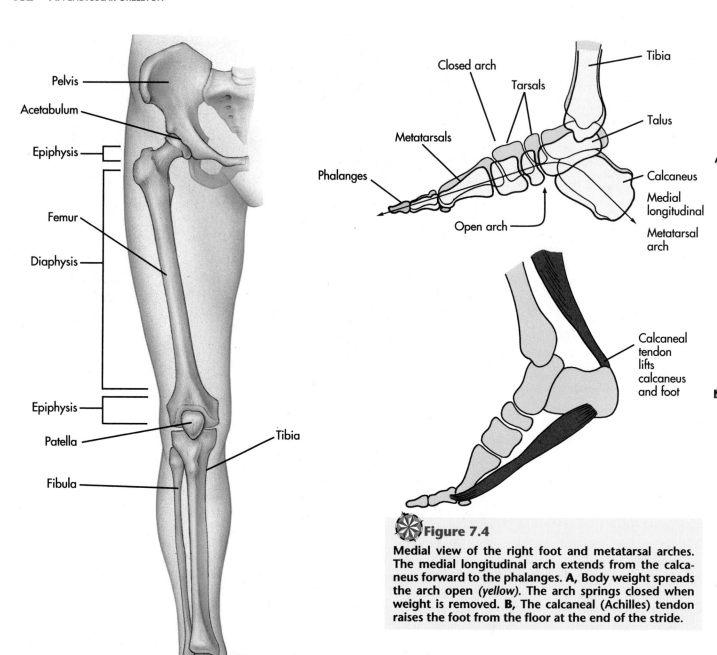

Figure 7.3

Identify the major bones of the lower limbs and pelvis in the anterior view of the right lower extremity.

Figure 7.4

Medial view of the right foot and metatarsal arches. The medial longitudinal arch extends from the calcaneus forward to the phalanges. A, Body weight spreads the arch open *(yellow)*. The arch springs closed when weight is removed. **B,** The calcaneal (Achilles) tendon raises the foot from the floor at the end of the stride.

through five **tarsal bones** (tars-, *G,* the ankle) beneath the ankle, ending at the heel where the **calcaneus** (KAL-kain-ee-us; calcan-, *L,* heel) touches ground. The calcaneus receives the calcaneal (Achilles) tendon from the gastrocnemius muscle of the calf; this

powerful muscle raises the heel at the end of the stride, as shown in Figure 7.4B. Ligaments and tendons beneath the foot close the metatarsal arch.

The ankle joint between the tibia and **talus** (TAL-us; *L,* the ankle bone) permits the foot to lift and extend but not to rotate from side to side. Although the metatarsals and tarsal bones allow the foot to rotate, the ankle is limited to hinge-like motion. This unexpectedly limited motion benefits the calcaneal tendon and other tendons of the ankle, but sprains and fractures are the occasional cost.

- Which bones support the thigh and leg?
- On what pelvic structure does the head of the femur articulate?
- What is the longitudinal metatarsal arch, and what are its component bones?

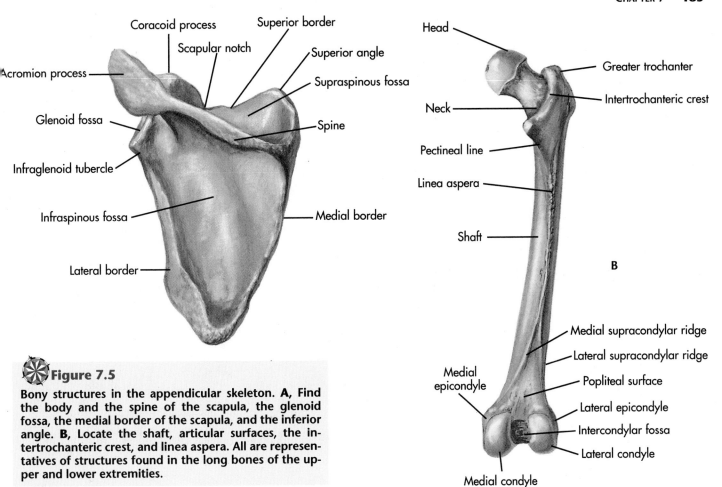

Figure 7.5
Bony structures in the appendicular skeleton. **A,** Find the body and the spine of the scapula, the glenoid fossa, the medial border of the scapula, and the inferior angle. **B,** Locate the shaft, articular surfaces, the intertrochanteric crest, and linea aspera. All are representatives of structures found in the long bones of the upper and lower extremities.

LONG TOUR

Various surfaces and processes extend from the **bodies of appendicular bones** (Figures 7.5 and 7.7) as similar structures do in the skull and vertebral column. Broad, flat portions of the scapula (Figure 7.5A) and the coxal bone (Figure 7.7A) form bony **spines** and smooth **fossae** that receive muscles and tendons. These surfaces usually end at sharp **borders** and **angles,** but the narrow **shafts** of long bones (Figure 7.5B) end in smooth **articular surfaces** that join to other bones. Rough **tuberosities, tubercles,** and long **crests** and **lines** mark the locations of muscle and tendon attachments on all of the appendicular bones. **Nutrient foramina** admit vessels and nerves to their interiors. Refer to Table 6.2 for descriptions of these bony structures.

PELVIS

The pelvis surrounds the pelvic cavity and fastens the **lower extremities to the vertebral column,** as you can see in Figures 7.1 and 7.6. The pelvis is well named; the word derives from the Latin stem pelv-, meaning basin. Two **hip bones,** or **coxal bones,** together with the sacrum and coccyx of the vertebral column enclose the lower abdominal and pelvic cavities and receive the femurs from the thighs. The **pelvic brim** marks the **superior aperture,** or entrance, **of the true pelvis,** and the sacrum and coccyx guard the **inferior aperture of the true pelvis** below. This oval space contains the colon, rectum, and bladder and, in females, the ovaries and uterus as well. The **false pelvis** above the pelvic brim supports the lower abdominal cavity and attaches muscles and ligaments to the body wall. Each coxal bone receives forces from the thigh in a prominent socket, called the **acetabulum,** on the lateral surface of this bone. Forces spread posteriorly from the acetabulum to the sacrum at the **sacroiliac joint** and anteriorly to the **pubic symphysis,** the joint that completes the anterior rim of the pelvis.

ILIUM

Each coxal bone derives from three separate irregular bones—the **ilium, ischium,** and **pubis**—that fuse together in adulthood. They resemble a figure 8 when viewed medially or laterally in Figure 7.7. The functions of the interior and the exterior of the pelvis are most obvious in the ilium, which receives the thigh laterally and faces the abdominopelvic cavity medially (Figure 7.7A). Beginning from the acetabulum on the lateral surface, where all three bones converge in the middle of the "8," the body of the ilium reaches superiorly and laterally as the **wing,** or **ala** (AIL-a; *L,* a wing) of the false pelvis. You can palpate

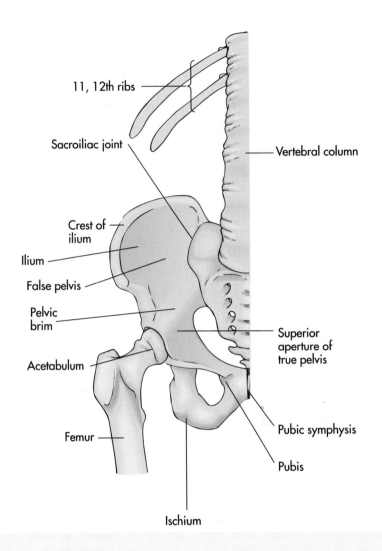

11, 12th ribs

Sacroiliac joint

Vertebral column

Crest of ilium

Ilium

False pelvis

Pelvic brim

Acetabulum

Superior aperture of true pelvis

Femur

Pubic symphysis

Pubis

Ischium

Figure 7.6

The pelvis transfers weight from the vertebral column to the hip joint and to the femur of the thigh.

the **iliac crest,** the margin of the ala, in the lower anterolateral abdomen. From above, the iliac crest (iliaco, *L,* the flank) receives the latissimus dorsi and oblique muscles of the abdominal wall; from below, it receives the **flexor** muscles of the thigh that swing the lower extremity forward in walking. The iliac crest ends anteriorly at the **anterior superior iliac spine,** which receives the sartorius muscle of the thigh and the inguinal ligament from the pubis. Just below, the **anterior inferior iliac spine** receives a tendon of the rectus femoris muscle that swings the thigh forward in walking.

Viewed from their medial surfaces in Figure 7.7B the wings of both ilia enclose the lower abdominal cavity within the **iliac fossae.** Posterior to the iliac fossa, the **body of the ilium** articulates with the sacrum, and the rough **iliac tuberosity** receives supporting sacroiliac ligaments from the lower back. The **arcuate line** below the iliac fossa marks the junction of the abdominal cavity with the true pelvic cavity.

ISCHIUM

The ischium is the "sit" bone of the true pelvis, the most inferior of the coxal bones, upon which the pelvis rests when you sit in a chair. The ischium (ISS-kee-um; *L,* the hip bone) reaches inferiorly from the acetabulum in Figure 7.7A to the **ischial tuberosity** and **ramus of the ischium,** the portions of this bone that support the pelvis in the sitting position. The **greater sciatic notch** (sciatic, *L,* of the hip) and the **lesser sciatic notch** mark the posterior margin of the ischium. Hamstring muscles of the posterior thigh originate on the ischial tuberosity, and the ramus of the ischium curves forward from the tuberosity to form the **obturator foramen** (obtur-, *L,* close, or stop up) with the pubis.

PUBIS

The two pubic bones of the true pelvis unite the left and right coxal bones anteriorly at the pubic symphysis. The **superior ramus** of the pubis curves forward

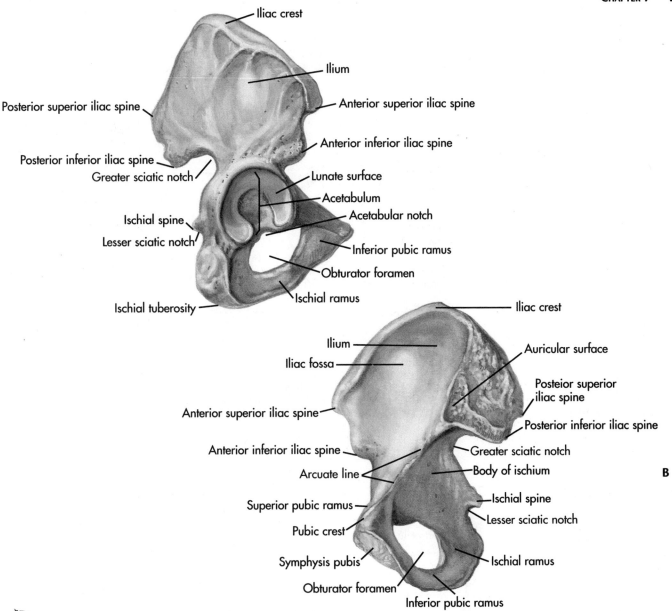

Figure 7.7
Right coxal bone. A, Lateral view. B, Medial view.

from the acetabulum (Figure 7.7A) to reach its opposite partner at the **pubic symphysis.** Near this junction the **crest of the pubis** receives the inguinal ligament that descends from the anterior superior spine of the iliac (Figure 7.8). This ligament supports the lower wall of the abdominal cavity. The **inferior ramus** extends downward below the crest of the pubis, around the obturator foramen, to meet the ramus of the ischium posteriorly.

PELVIC OPENINGS

Several openings in the pelvis accommodate muscles of the pelvis and communicate vessels and nerves to the lower extremities (Figure 7.8). In life the floor of the pelvic cavity and its levator ani muscle stretch across the inferior aperture of the pelvis, which the vagina in females and the urethra and anus

in both sexes penetrate. The **obturator membrane** closes the obturator foramen to all but the obturator nerve, artery, and vein to the thigh. The **sacrospinous ligament** completes the sciatic notch, and the sciatic nerve passes through this notch into the posterior thigh and leg. The femoral artery and vein enter the thigh above the superior ramus of the pubis through the **femoral canal.** The femoral canal is formed by the **inguinal ligament** (in-GWIN-al; inguin-, *L*, the groin) that stretches between the anterior superior spine of the ilium and the crest of the pubis.

• Name the three bony components of the coxal bone.
• Which forms the ala that supports the false pelvis?
• Which ones form the true pelvic cavity?
• Which one articulates with the sacrum?
• Name the bones associated with the obturator foramen, arcuate line, sciatic notch, and acetabulum.

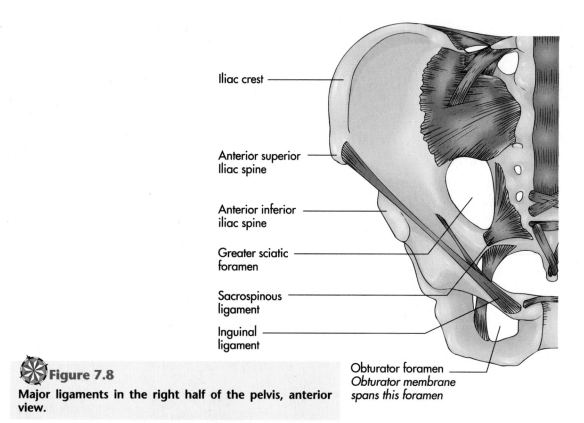

Iliac crest

Anterior superior
Iliac spine

Anterior inferior
iliac spine

Greater sciatic
foramen

Sacrospinous
ligament

Inguinal
ligament

Obturator foramen
*Obturator membrane
spans this foramen*

Figure 7.8

Major ligaments in the right half of the pelvis, anterior view.

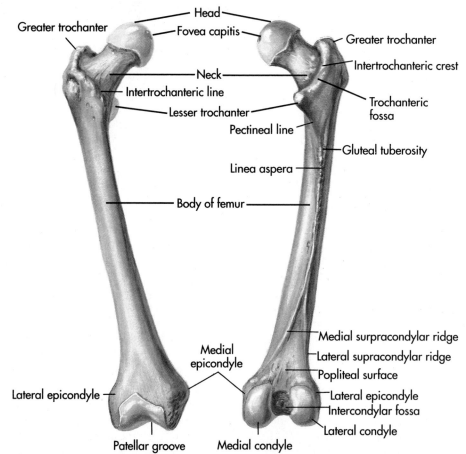

Greater trochanter

Head

Fovea capitis

Greater trochanter

Intertrochanteric crest

Neck

Intertrochanteric line

Trochanteric
fossa

Lesser trochanter

Pectineal line

Gluteal tuberosity

Linea aspera

Body of femur

Medial
epicondyle

Medial surpracondylar ridge

Lateral supracondylar ridge

Popliteal surface

Lateral epicondyle

Lateral epicondyle

Intercondylar fossa

Lateral condyle

Patellar groove

Medial condyle

Figure 7.9

The right femur. A, Anterior view. B, Posterior view.

THIGH AND LEG

The lower limb itself is supported by the femur, the patella, the tibia, and the fibula (Figure 7.3). All except the patella, which is a sesamoid bone, are long bones, and all develop by endochondral ossification. **Achondroplasia,** a condition caused by dominant mutations that particularly interfere with secondary ossification centers in long bones, illustrates how important bone length is in relation to the rest of the body. Achondroplastic limbs are about half their normal length relative to the trunk.

FEMUR

The femur is the bone of the thigh. This long bone in Figure 7.9 transfers body weight from the pelvis to the knee and leg. The **femur** (FEE-mer) is the largest bone in the body, measuring about one quarter of body height. (Divide your height by 4 to find the approximate length of your own femur.) The body's largest muscles surround this bone and move the limb through its paces. Beginning proximally at the spherical **head** that inserts into the acetabulum of the pelvis, the **neck of the femur** extends in Figure 7.9A to the **greater** and **lesser trochanters** (troe-KANT-er; *G,* a runner). You can palpate the greater trochanter through the gluteus maximus muscle as the hard lateral bulge several centimeters beneath the crest of the ilium. The gluteus medius and minimus muscles rotate the thigh medially and abduct it by pulling on the greater trochanter. Other muscles insert upon the **intertrochanteric line** on the anterior surface of the bone (Figure 7.9B) between the trochanters, and also in the **trochanteric fossa** posteriorly.

The long, cylindrical **shaft** of the femur bends medially at the trochanters (Figure 7.9A) and curves gently forward toward the knee, directing forces from the hips downward beneath the body. The anterior side of the shaft is smooth and rounded, but on the posterior side (Figure 7.9B) the **intertrochanteric crest** marks the beginning of the **linea aspera** (asper, *L,* rough), a prominent ridge extending toward the knee. Several muscles and connective tissue septa attach to the linea aspera. Distally the linea aspera divides into the **medial** and **lateral supracondylar lines** that end above the medial and lateral condyles of the knee.

The knee resembles a hinge; it permits the leg to swing forward and back, while prohibiting lateral motion. Two smooth, rounded **condyles** of the femur fit like a pair of knuckles (Figure 7.9A and B) onto the medial and lateral condyles of the tibia. Each rounded surface alone would allow forward and lateral motions, but together they allow only flexion and extension. To see this clearly, visualize the motions of a knee with only one set of condyles. The **intercondylar fossa** and the **medial** and **lateral epicondyles** also limit knee motion through ligaments between the

femur and tibia. The deep intercondylar fossa between the condyles receives the anterior and posterior cruciate ligaments from the tibia. The epicondyles, narrow prominences proximal to the margins of each condyle (Figure 7.9A), receive the collateral ligaments of the knee from the tibia. The **adductor tubercle** receives the tendon of the adductor magnus muscle less than a finger's width above the medial epicondyle. The patella fits into the **patellar surface** of the femur anteriorly between the condyles. When the knee is partially extended, you can palpate both epicondyles and the adductor tubercle 3 finger widths from the margin of the patella.

PATELLA

The patella is the knee cap (patell-, *L,* a little dish). The largest of the sesamoid bones (Figure 7.10), about 5 cm (2 inches) wide, it is enveloped by the quadriceps tendon and fits into the patellar surface of the femur, where it improves the leverage of the tendon upon the tibia. The patella is entirely cancellous bone except for a thin veneer of compact bone at its surface.

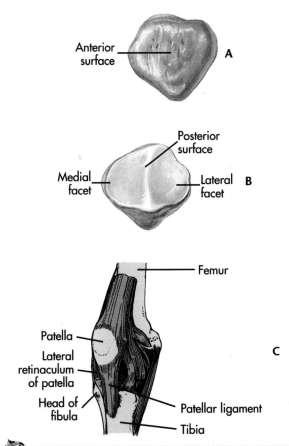

Figure 7.10

Right patella. A, The rough anterior face receives the quadriceps tendon. B, The posterior articular surface is smooth. C, The patella in position on the right knee.

TIBIA

The tibia is the shin bone, which transfers striding forces between the knee and the ankle (Figure 7.11). It is larger than its slender partner, the **fibula,** and is medial to this bone, as seen in Figure 7.3. The shaft of the **tibia** is second in length only to the femur, but unlike the femur the shaft is triangular in cross section, with three sharp **edges** (the shin itself being the anterior edge) and three **surfaces** to which muscles attach. Broad **medial** and **lateral condyles** articulate with the femur, enlarging the proximal end of

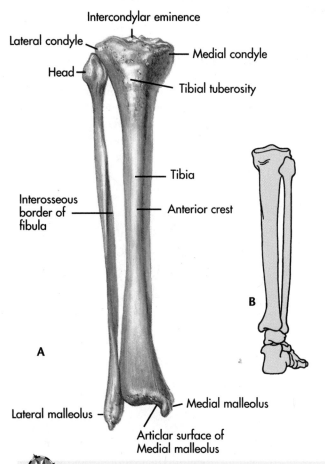

Figure 7.11

Right tibia and fibula. A, Anterior view. B, Posterior view showing the relationships of the tibia and fibula with the femur and ankle.

the tibia. The distal extremity articulates with the talus bone of the ankle. The tibia ends distally at the **medial malleolus** (malle-, *L,* a hammer).

Both condyles of the tibia in Figure 7.12 are smooth, shallow depressions on which the femoral condyles glide. The condyles of the tibia and the femur do not fit together as snugly as the head of the femur and the acetabulum, but cartilage pads called menisci achieve the same effect by conforming to both

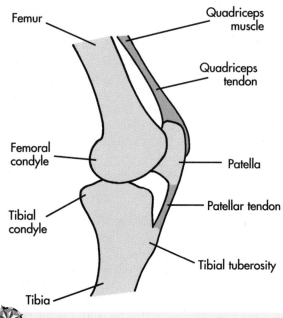

✳**Figure 7.12**

The condyles of the tibia articulate with the femoral condyles. The tibial tuberosity receives the patellar tendon from the patella.

surfaces (see Knee, Chapter 8). The quadriceps muscle of the femur and the patellar tendon extend the leg during the stride by pulling on the prominent **tibial tuberosity** on the anterior border of the tibia. Palpate this all-important attachment that is inferior to the patella with your knee partially flexed. You can feel the patellar tendon move when you extend the tibia.

The **intercondylar eminence** (Figure 7.11) protrudes between the medial and lateral condyles of the tibia into the intercondylar fossa of the femur, where the anterior and posterior cruciate ligaments attach to the tibia in front and behind. The tendons of the medial hamstring muscles of the thigh insert onto the tibia below the posterior border of the medial condyle, where they are easily palpated when the knee is flexed and the muscles are contracted.

The distal end of the tibia articulates with the talus bone of the foot. The articular surface of the talus is shaped like a spool; it fits into the concave surface of the tibia between the medial malleolus of the tibia and the **lateral malleolus** of the fibula (Figure 7.13). You can palpate both malleoli directly beneath the skin of your ankle.

FIBULA

The fibula is the slender bone lateral to the tibia seen in Figure 7.11. As its shape suggests, the fibula does not support body weight, but it attaches muscles that flex and extend the foot. The proximal head of the fibula articulates with the posterolateral surface of the lateral condyle of the tibia, and this head also receives the tendon of the lateral hamstring muscle

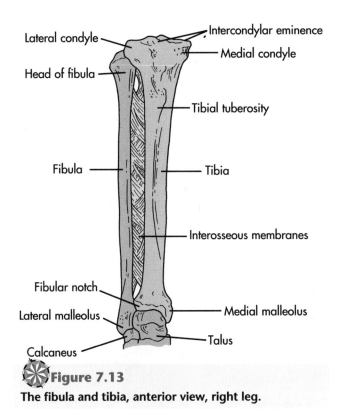

Lateral condyle
Head of fibula
Fibula
Fibular notch
Lateral malleolus
Calcaneus

Intercondylar eminence
Medial condyle
Tibial tuberosity
Tibia
Interosseous membranes
Medial malleolus
Talus

Figure 7.13
The fibula and tibia, anterior view, right leg.

sential for walking and running. Although the medial longitudinal metatarsal arch raises the medial face of the foot, the foot remains flat at its lateral margin, maintaining full contact with the ground through the **lateral longitudinal metatarsal arch** shown in Figure 7.14B. Yet a third arch, the **transverse metatarsal arch** (Figure 7.14C), also lifts the middle tarsal bones above their partners on either side, but this arch is less important in walking. Your wet footprint clearly shows the arrangement of all three arches. Muscles of the leg raise and lower the foot through tendons, shown in Figure 7.14D, that pass through the ankle and insert onto the dorsal (upper) and plantar (lower) surfaces of the phalanges and metatarsals.

TARSALS

The tarsal bones are the seven short bones of the ankle. They distribute weight from the tibia to the heel and to the ball of the foot. Figure 7.15A shows a convenient way to remember the tarsals by following the *C*s around a circle. Three **cuneiform bones** (KUE-nEE-ih-form; wedge-shaped; cune-, *L*, a wedge) begin the loop, and the **cuboid** bone (KUE-boyd) and **calcaneus** continue it to the heel. The **talus** and **navicular** (nah-VIK-yoo-lar; navicul-, *L*, a small boat) complete the circle behind the cuneiforms. The cuneiform, talus, navicular, and calcaneus bones are important elements of the medial longitudinal metatarsal arch, while the calcaneus and cuboid distribute force to the flatter, lateral metatarsal arch of the foot.

The talus is constrained by the medial and lateral malleoli beneath the tibia and fibula; there it transfers weight from the leg to the navicular and calcaneus bones. The **tarsal sinus,** a prominent groove on the inferior surface of the talus, shares an interosseous ligament with the calcaneus below.

The calcaneus is the heel bone, the largest tarsal bone. The rough, posterior end of this oblong bone (Figure 7.14B) receives the **calcaneal (Achilles) tendon** that lifts the heel in the final stages of the stride. Smooth articular surfaces meet the talus dorsally and the cuboid anteriorly where the calcaneus transmits its forces forward into the metatarsal arches of the foot. To extend the foot, the large groove on the medial surface of the calcaneus directs the tendon of the flexor hallucis muscle (hallux, *L*, the great toe) forward to the great toe, as shown in Figure 7.14D.

In Figure 7.14A it is easy to see that the navicular distributes forces from the talus and calcaneus to the cuneiform bones and thence to the three medial metatarsal bones and the toes. The cuneiform bones are wedged between the navicular and the first three metatarsals. Each cuneiform articulates with its corresponding metatarsal. The cuboid distributes forces from the calcaneus to the two lateral metatarsals.

from the thigh. Palpate this end of the fibula and its tendon insertion.

An **interosseous membrane** joins the distal extremity of the fibula to the **fibular notch** on the lateral surface of the tibia, shown in Figure 7.13. Extensor and flexor muscles of the foot occupy the space between these bones, and the interosseous ligament separates them into an anterior compartment for flexors and a posterior compartment for extensors. Beyond the fibular notch, the lateral malleolus articulates with the talus bone of the ankle.

- How do the proximal heads of the femur and tibia differ?
- How do two femoral condyles prevent sideways movement of the knee?
- What functions do the medial and lateral malleoli of the tibia and fibula perform?
- What structures attach to the tibial tuberosity and to the greater trochanter of the femur?

FOOT

The foot articulates with the leg at the ankle, transmitting forces from the tibia through the tarsal bones, the metatarsals, and the phalanges (Figure 7.14). **Tarsal bones** make up the heel and ankle; the **metatarsal bones** arch forward to the ball of the foot; and finally the **phalanges** support the toes. The tarsals and metatarsals form the springy **medial longitudinal metatarsal arch** (Figure 7.14B) that is es-

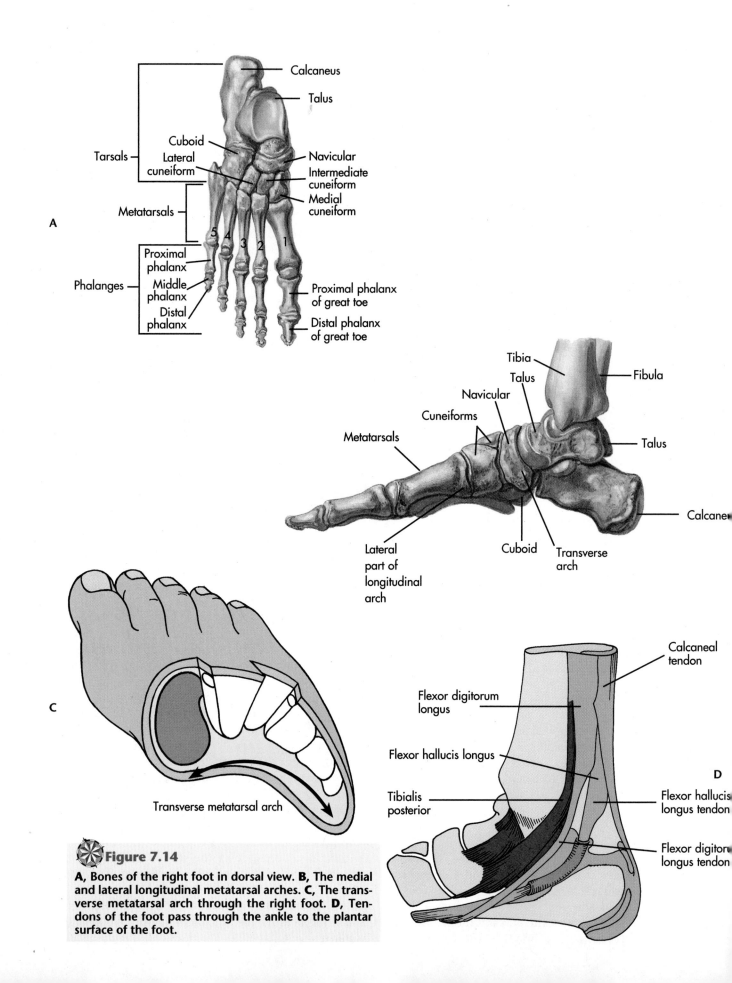

A, Bones of the right foot in dorsal view. **B,** The medial and lateral longitudinal metatarsal arches. **C,** The transverse metatarsal arch through the right foot. **D,** Tendons of the foot pass through the ankle to the plantar surface of the foot.

✴ **Figure 7.14**

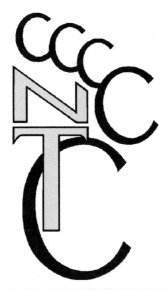

Figure 7.15

Mnemonic diagram for the tarsals. Five Cs embrace the N and T.

- How many bones does the foot contain?
- Which tarsal bone receives the calcaneal tendon?
- Which tarsal bones articulate with the cuneiform bones?
- Which metatarsal bone articulates with the great toe?

SHOULDER AND UPPER LIMB

BRIEF TOUR

Without the weight of the body to support, the upper limbs are lighter than the lower limbs, and they move more freely, with far more dexterity. The regions of the upper extremities are the pectoral girdle, arm (brachium), elbow (cubitus), forearm (antebrachium), wrist (carpus), hand (manus), and fingers (phalanges).

Beginning proximally in Figure 7.16, the **pectoral,** or **shoulder, girdle** connects the upper limb to the trunk. Two bones, the **scapula** (shoulder blade) and **clavicle** (collarbone), articulate with each other, and

METATARSALS

Five metatarsal bones bridge the metatarsal arches from the tarsal bones to the phalanges, shown in Figure 7.14A. Each metatarsal is a small, long bone whose shaft extends forward from a proximal base on a tarsal to a distal head at the base of each toe. The metatarsals are numbered, beginning with the most medial, which happens also to be the largest and connects to the great toe. Two sesamoid bones beneath the head of this metatarsal bridge the tendon of the flexor hallucis longus muscle (long flexor muscle of the great toe) as the tendon enters the toe. Each of the first three metatarsals articulates with its own cuneiform bone. The last two metatarsals share the cuboid. A large tuberosity on the fifth tarsal (which you can palpate easily on the lateral margin of your foot) extends beyond the cuboid (Figure 7.14A). A tough, fibrous plantar aponeurosis (AP-o-noor-OH-sis; broad flat tendon) radiates out from the lower surface of the calcaneus to this tuberosity and to the distal heads of the metatarsals.

PHALANGES

Fourteen phalanges shape the toes; two phalanges are found in the great toe (the hallux), and the remaining toes have three phalanges each (Figure 7.14A). The proximal base of each phalanx is concave, articulating with a convex head on the distal end of a metatarsal or on the distal head of a phalanx. Toe bones are much shorter than the phalanges of the fingers, but the toes nevertheless grasp the ground and push off from it as the stride ends.

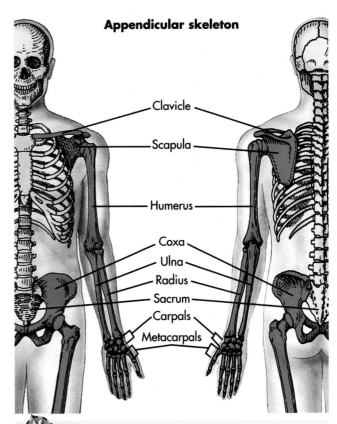

Appendicular skeleton

Clavicle
Scapula
Humerus
Coxa
Ulna
Radius
Sacrum
Carpals
Metacarpals

Figure 7.16

Brief tour of upper extremity (*blue*).

the clavicle (clavicul-, *L*, a key) articulates with the axial skeleton at the sternum. The scapula (scapul-, *L*, the shoulder blade) glides and rotates on muscles and fascia on the posterior wall of the thorax, but it does not articulate directly with the axial skeleton. The **humerus** (humer-, *L*, the shoulder) articulates with the scapula in the **glenoid fossa** (glen-, *L*, a pit or socket), a very mobile ball-and-socket joint that is much shallower than the hip joint and therefore more readily dislocated. Proceeding distally, the elbow and forearm add even more range of motion to that of the humerus. The **radius** and ulna **bones** of the forearm flex at the elbow, allowing the hands to reach the face. Furthermore the forearm can rotate along its own axis to turn the palms anteriorly (supination) or posteriorly (pronation) because the radius also rotates around the ulna, a mobility denied to the tibia and fibula of the leg. Finally, the **carpal bones** of the wrist and the **metacarpals** and **phalanges** of the hands and fingers are more mobile than the corresponding bones of the foot. Dexterity also increases because the phalanges of the hands are longer than those of the toes.

LONG TOUR—SHOULDER
SCAPULA

The scapula is a flat, triangular bone that articulates with the humerus and the clavicle at the shoulder joint. The scapula covers the posterior surface of the

thorax between the second and seventh ribs, as shown in Figure 7.17A, B, and C. The **head of the scapula** forms the shallow **glenoid fossa** of the shoulder joint with the humerus. The prominent **spine of the scapula** rises from the posterior surface of the scapula and ends laterally at the **acromion process** (ah-KROM-ee-ahn; acromi-, *G*, the point of the shoulder blade) that articulates with the clavicle (Figure 7.17A and B) and protects the shoulder joint. You can easily palpate the hard crest of the spine and the acromion process on the posterior surface of your shoulder. The **coracoid process** (KOR-a-koyd; coraco-, *G*, a crow or raven) reaches superiorly from the superior margin of the scapula, with attachments for the biceps brachii muscle of the arm; this process is anterior to the acromion but difficult to palpate. The coracoid has another important role in the shoulder joint as part of the **coracoacromial arch,** formed by the coracoacromial ligament and the coracoid and acromion processes. This arch prevents upward forces on the humerus from pushing it past the shallow glenoid fossa.

Muscles rotate the humerus from the **supraspinous fossa** above the spine of the scapula and the **infraspinous fossa** below, as well as from the **subscapular fossa** on the anterior surface of the scapula. Other muscles that insert on the borders and angles of the scapula raise the scapula, rotate it, and extend it laterally. These muscles and movements enable the

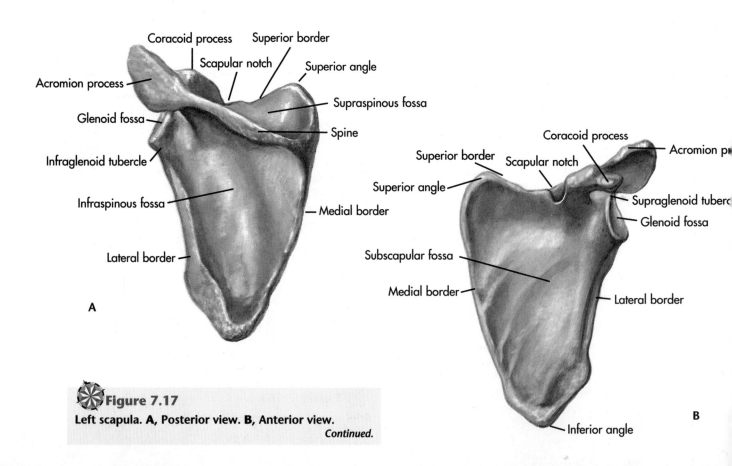

✳ **Figure 7.17**
Left scapula. A, Posterior view. B, Anterior view.
Continued.

upper limbs to reach forward or up and down; the muscles will be described in Chapter 11, Muscles of the Appendicular Skeleton.

CLAVICLE

The clavicle (collar bone) is the anterior bone of the shoulder girdle. The extremities of this mildly *S*-shaped flat bone articulate with the sternum and with the acromion of the scapula, as Figure 7.17C and D illustrates. The body of the clavicle receives muscles that raise and depress the shoulder, as well as other muscles that raise the arm. The body of the clavicle reaches horizontally above the first rib from its articulation with the **manubrium** of the sternum. At first the bone follows the thorax posteriorly. It then passes anterior to the axillary artery and vein and to the major nerves of the upper limbs; finally, the distal third bends forward to arrive at the acromion.

The clavicle is the only bone of the shoulder girdle that articulates directly with the axial skeleton. It is a strut that extends the shoulder joint from the body and prevents the shoulder from collapsing onto the chest. Unfortunately the clavicle is also the most frequently broken bone. Its middle third often is fractured in a fall. The shoulder then falls forward in precisely the motion the intact clavicle had prevented.

• The head of the humerus articulates with the scapula in what fossa?
• What are the names of the two scapular processes that help form the shoulder joint?
• Which of these processes articulates with the clavicle?
• With what bone does the proximal head of the clavicle articulate?
• Where are the infraspinous and supraspinous fossae located?

ARM AND FOREARM
HUMERUS

The humerus is the bone of the arm; this bone extends from the shoulder to the elbow, where it joins the forearm. Comparable to the femur in the thigh, this largest bone of the upper limb (Figure 7.18) accomplishes the arm's major movements through muscles attached to the shoulder and thorax. The humerus swings forward, back, and laterally, and it rotates approximately 120 degrees on its longitudinal axis. The smooth articular surface of the proximal **head** allows these movements through a ball-and-socket joint in the glenoid fossa of the scapula. The short **anatomical neck** of the humerus, shown in Figure 7.18A, connects the head with the **surgical neck,**

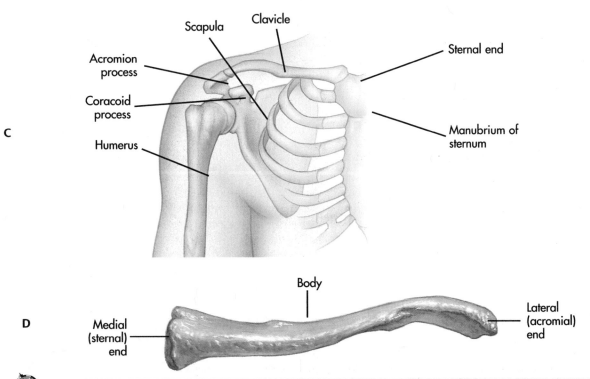

C

D

Scapula
Clavicle
Acromion process
Coracoid process
Humerus
Sternal end
Manubrium of sternum
Body
Medial (sternal) end
Lateral (acromial) end

✳ **Figure 7.17, cont'd**

C, The scapula spans the posterior wall of the thorax between the second and seventh ribs, and it articulates with the clavicle. **D,** Left clavicle, anterior view.

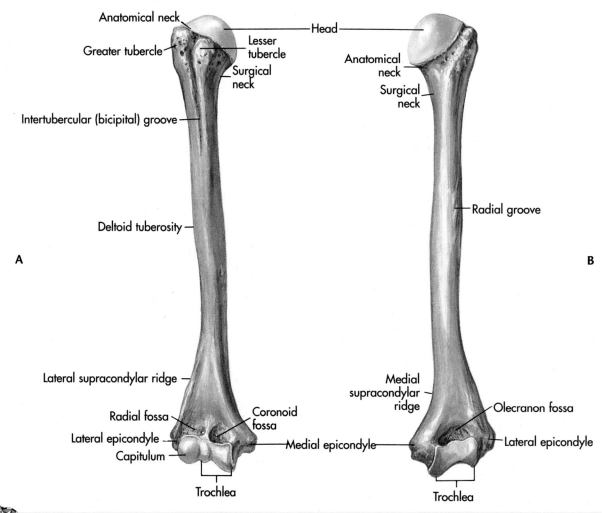

Anatomical neck
Greater tubercle
Lesser tubercle
Head
Surgical neck
Anatomical neck
Surgical neck
Intertubercular (bicipital) groove
Radial groove
Deltoid tuberosity
A
B
Lateral supracondylar ridge
Medial supracondylar ridge
Radial fossa
Coronoid fossa
Olecranon fossa
Lateral epicondyle
Capitulum
Medial epicondyle
Lateral epicondyle
Trochlea
Trochlea

Figure 7.18

Right humerus. A, Anterior view. B, Posterior view.

the more readily fractured neck, of this bone. The **greater tubercle** is located laterally between these necks and receives muscles that rotate the humerus medially and laterally. A muscle attached to the **lesser tubercle** on the anterior surface of the humerus, also rotates the humerus medially (Figure 7.18B). Both tubercles are homologous to the trochanters of the femur.

The **shaft of the humerus** is cylindrical proximally; it receives muscles from the shoulder that abduct and swing the arm. The shaft flattens distally into anterior and posterior surfaces that accommodate muscles that move the forearm; it then enlarges into the condyles and epicondyles of the elbow. The **trochlea** and **capitulum** articulate with the ulna and radius of the forearm, respectively. The trochlea (trochle, G, a pulley) is shaped like a pulley and permits the ulna and forearm only to extend and flex, limitations like those of the knee. Immediately proximal to the trochlea the **coronoid fossa** receives the

ulna on the anterior surface of the humerus when the forearm flexes (Figure 7.18A). The deeper **olecranon fossa** (olecran, G, the elbow) on the posterior surface of the humerus (Figure 7.18B) does the same when the forearm extends. Lateral to the trochlea the convex shape of the capitulum (capit, L, the head) indicates that the radius rotates, as well as swings, upon it. The shallow **radial fossa** of the humerus accommodates the radius when the elbow flexes, but there is no need for a posterior radial fossa, for reasons explained below. The **medial epicondyle** and smaller **lateral epicondyle** attach muscles of the forearm to the humerus. Palpate these rough surfaces with your elbow partially flexed and the palm of your hand facing you. The ulnar ("funny bone") nerve passes over a shallow groove on the posterior side of the medial epicondyle.

RADIUS AND ULNA

The **radius and ulna** are the bones of the forearm, homologous to the tibia and fibula, respectively,

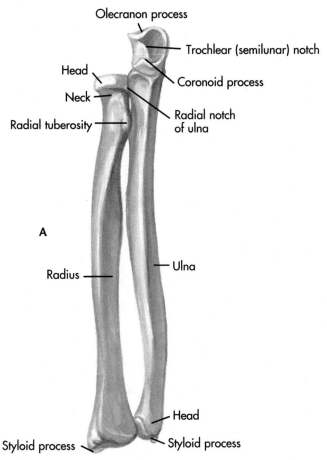

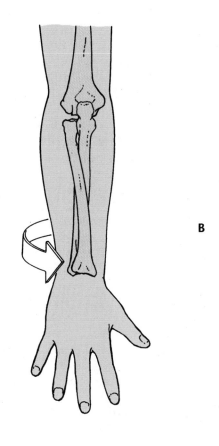

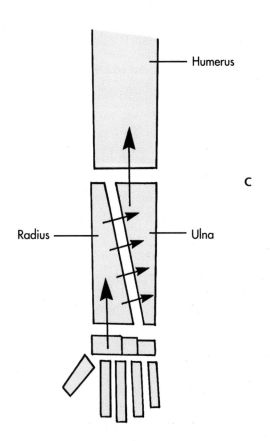

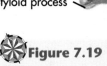**Figure 7.19**

Right radius and ulna. A, Supinated in anatomical position. B, Pronated. C, The relationship between the radius and ulna.

of the leg. They are especially mobile bones whose actions can easily be palpated. Unlike the tibia and fibula, the **radius** and **ulna** allow the forearm to rotate the hand, to supinate or pronate it so you can cup your hands to hold water or cut a steak. In the anatomical position shown in Figure 7.19A with the palm forward (supinated), the ulna is medial and the radius is lateral to it. Rotation brings the radius across the ulna and inverts the palm of the hand (pronated, Figure 7.19B). Demonstrate this for yourself by placing your upper limb in the anatomical position and flexing the elbow so you can view your hand end on. Place the fingers of your other hand on the ulna at the medial side of the wrist to verify that the ulna moves only slightly, and watch your hand as you rotate it. The radius rotates medially around the ulna from its lateral position.

Figure 7.19C demonstrates that radius and ulna can be thought of as triangles whose broad ends articulate at opposite ends of the forearm. The radius receives most of the forces from the wrist, and as the radius narrows it transfers part of them to the widening ulna and finally to the humerus. The proximal head of the ulna articulates with the trochlea of the humerus. The **trochlear notch** of the ulna fits precisely around the trochlea, and the **coronoid** and **olecranon processes** of the ulna enter the corresponding fossae of the humerus when the elbow flexes and extends. The olecranon process receives the tendon of the triceps brachii, the muscle that extends the forearm. You can feel the olecranon process on the posterior angle of your elbow. The shaft of the ulna narrows distally into a semicircular head that articulates with the medial (little finger) side of the wrist. The ulna ends at the **styloid process,** which connects to the wrist through the ulnar collateral ligament.

The radius can pronate and supinate the hand by rotating on its flat proximal head, which articulates with the capitulum of the humerus and with the **radial notch** of the ulna. The **radial tuberosity** receives the biceps brachii tendon (you can palpate the tendon) immediately anterior to the humerus. The shaft of the radius widens distally and articulates with the lateral side of the wrist through a broad, concave articular surface. This surface terminates laterally at the **styloid process of the radius,** which receives the radial collateral ligament of the wrist. The **ulnar notch** in the distal end of the radius allows the radius to rotate on the ulna. An **interosseous membrane** prevents either bone from separating.

- Name the bones of the arm and forearm.
- Which bone articulates with the shoulder?
- Which bone has a broad distal articular surface at the wrist?
- Which bone articulates through the prominent trochlear notch at the elbow?
- Are the radius and ulna crossed or parallel when the forearm is pronated?

HAND

The bones of the hand are more mobile than those of the foot, which allows the opposable thumb to help grasp an amazing variety of objects. The 29 bones of each hand (Figure 7.20A) are grouped, like the bones of the foot, into **carpals** (bones of the wrist), **metacarpals** (bones of the palm), and **phalanges** (bones of the fingers).

CARPUS

The wrist is known as the carpus, and the individual bones are carpals. Its eight short bones, arranged in two rows, are shown in Figure 7.20A, B. From the radial (thumb) side to the ulnar side of the wrist, the **scaphoid, lunate, triquetrum,** and **pisiform** bones make up the proximal row. The first two articulate with the radius; they are the primary elements of the wrist to articulate with the forearm. The next row includes the **trapezium,** the **trapezoid,** the **capitate,** and the **hamate** bones that articulate with the metacarpals and the first row of carpals.

The capitate is the central bone at the base of the third metacarpal (middle finger). The capitate articulates with many other carpals, as you can see in Figure 7.20A. The trapezium forms a saddle-shaped articular surface with the thumb ("T & T"). You can easily palpate the pisiform ("P & P"), which articulates only

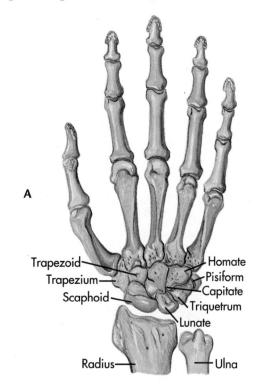

A

Trapezoid
Trapezium
Scaphoid
Radius

Homate
Pisiform
Capitate
Triquetrum
Lunate
Ulna

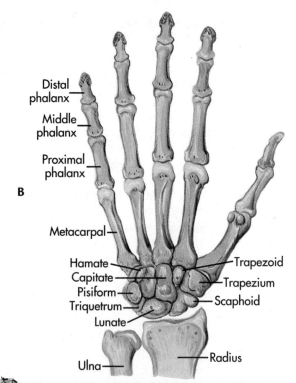

Distal phalanx
Middle phalanx
Proximal phalanx

B

Metacarpal
Hamate
Capitate
Pisiform
Triquetrum
Lunate
Ulna

Trapezoid
Trapezium
Scaphoid
Radius

Figure 7.20

Right hand, with carpals, metacarpals, and phalanges. A, Dorsal view. B, Palmar view.

with the triquetrum at the ulnar side of the palm. The **hook of the hamate bone** ("H & H") extends from the palmar surface of the hamate, shown in Figure 7.20B.

Together, the proximal surfaces of the lunate and scaphoid bones form a condyloid joint with the radius and ulna, thus permitting the wrist to bend laterally, flex, and extend the hand. The carpals form an arch for the **carpal tunnel** through which tendons, vessels, and the median nerve pass to the palm in Figure 7.21. The **flexor retinaculum ligament** spans this arch from the pisiform and the hook of the hamate across to the scaphoid and trapezium. **Carpal tunnel syndrome** occurs when repetitive motions of these tendons inflame their sheaths in the carpal tunnel (see Chapter 8, Articulations).

METACARPALS

The metacarpals are five small, long bones of the palm and thumb that fan out from the carpals to the fingers in Figure 7.20. They are numbered, beginning at the thumb; you can readily palpate the dorsal surface of each one, proximal to the knuckles. The shafts of the **metacarpals** enlarge at their bases with concave surfaces that articulate with the carpals; convex distal heads articulate with the phalanges (see the knuckles of your fist).

The metacarpals transfer forces between the fingers and the wrist. You can see which carpals receive these forces by examining the metacarpals shown in Figure 7.20. The first metacarpal (thumb) articulates with the trapezium through a very mobile joint that allows the thumb to oppose all the fingers. The remaining metacarpals move only dorsally or in the opposite direction. The second metacarpal articulates primarily with the trapezoid and capitate but also with the trapezium. At the middle of the palm the third metacarpal articulates only with the capitate, while the fourth and fifth metacarpals articulate primarily with the hamate. In the proximal row of carpals the triquetrum and lunate receive forces from the fourth and fifth metacarpals; the second and third metacarpals deliver thrust to the lunate and scaphoid, and the first (thumb) communicates to the scaphoid. Contrast this distribution with the foot, where the first three and last two metacarpals distribute separately to the talus and calcaneus.

PHALANGES

The fingers contain 14 small, long bones, called the phalanges, shown in Figure 7.20. Each finger has three rows of phalanges (proximal, medial, and distal), except the thumb, which lacks the medial row. The phalanges in each row are numbered from the thumb. The proximal phalanges are the longest, and their bases have a concave articular surface at the knuckles that receives the distal heads of the metacarpals. This joint allows the fingers to flex or extend and to spread, as you can easily demonstrate with your own hand. The medial and distal rows of phalanges only flex or extend; pairs of condyles, resembling those of the knee, almost completely prevent lateral movement at their heads and bases. Intercondylar grooves that resemble the intercondylar fossa of the knee accommodate flexor tendons on the palmar surfaces of the phalanges. Fingernails rest on the flat dorsal surfaces of the distal phalanges.

- Name the bones of the proximal row of carpals.
- Which carpal articulates with the thumb?
- Which carpals articulate with the radius?
- Which articular surfaces of the metacarpals are convex?
- Which phalanx is missing from the thumb?

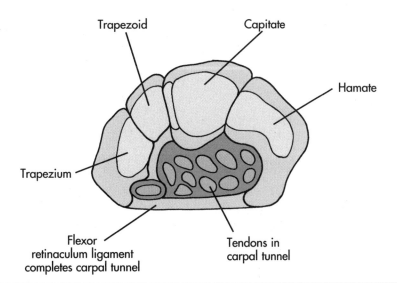

**Figure 7.21**
Carpal tunnel beneath the distal row of metacarpals, right hand. You are looking distally through this transverse section.

LIMB DEVELOPMENT

The appendicular skeleton begins developing in patches of tissue called limb placodes. Limb placodes are clusters of mesenchyme cells and ectoderm derived from the lateral plate mesoderm and ectoderm of the body wall (see Chapter 4, Organ Rudiments). Interactions between the ectoderm and mesoderm establish the proper shape and location of limbs that are so essential for daily activities.

Each limb placode becomes a **limb bud** that grows outward, forms bones and muscles, and develops into a limb (Figure 7.22). The tip of the limb bud determines the organization of the limb. The mesoderm determines whether the limb bud will develop as an upper or lower limb, and the ectoderm stimulates the limb bud to grow outward. Together, the ectoderm and mesoderm determine the order of parts in the limb.

Limb anatomy is established in proximodistal order, beginning with the girdles and ending with the

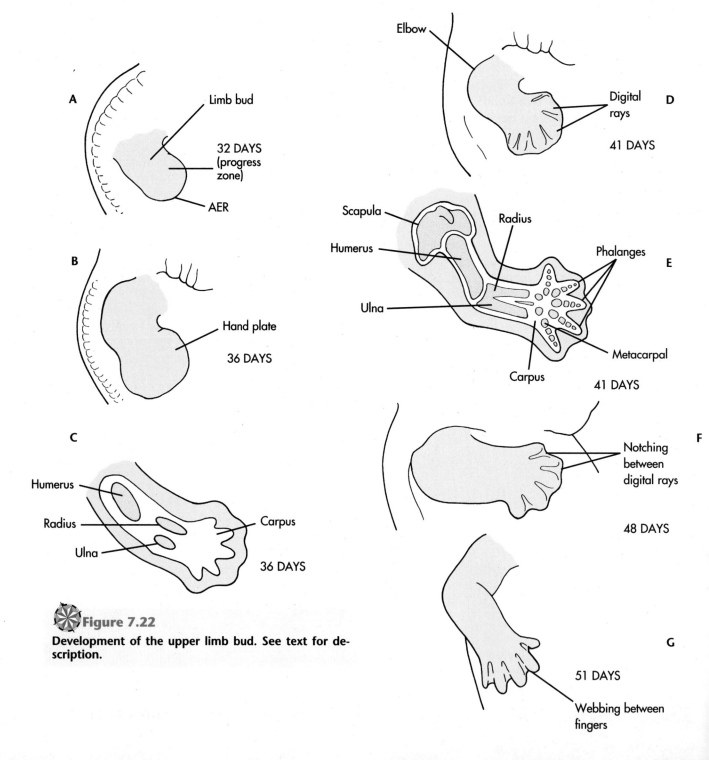

Figure 7.22
Development of the upper limb bud. See text for description.

phalanges. The upper limb placode appears on day 25, and the lower limb placode follows 2 days later. Within 24 hours of their appearance, the limb placodes have become limb buds (Figure 7.22A), and the ectodermal covering has formed a narrow **apical ectodermal ridge** along the tip of the bud. The apical ectodermal ridge (AER) is necessary for continuous limb outgrowth; development stops and distal structures fail to form if the AER is removed. Together with the apical ectodermal ridge, the mesenchyme beneath the ridge constitutes a **progress zone** that establishes the sequence of limb structures formed over the next few days. The progress zone is thought to spin off cells for each region of the limb. The first cell populations to leave the progress zone apparently are responsible for the shoulder girdle. The next populations form the humerus and arm, and later groups assemble more distal structures, until the final cells produce the distal phalanges. Having done its work in 2 or 3 days, the progress zone then disappears.

Once established, the limb regions differentiate and develop into recognizable extremities. During week 5, upper limb buds flatten and paddle-shaped **hand plates** appear at the tips of the buds (Figure 7.22B). Foot plates follow about 4 days later in the lower limbs. Cartilage models of the scapula and humerus appear in the upper limb during week 5 (Figure 7.22C), followed rapidly by the radius and ulna, and then by the carpals. **Finger rays,** the cartilaginous precursors of metacarpals and phalanges, appear in the hand plates by the end of week 6 (Figure 7.22E). Fingers begin to be sculptured during week 7 (Figure 7.22D and F), as cells die between the rays and are removed. By the eighth week fingers are well formed (Figure 7.22G).

The developing limbs bend and rotate in different directions that ultimately allow the lower limbs to walk and the upper limbs to move in many directions. In week 5 (Figure 7.23A) the palmar and plantar surfaces of the limbs face anteriorly. From this position, both limbs bend upward at the elbow and knee (Figure 7.23B) and rotate so that palms and soles face the trunk. By week 7 (Figure 7.23C) the humerus swings inferiorly, directing the forearm and elbow forward, but the femur and thigh swing anteriorly, pointing the knee and leg backward. Week 8 com-

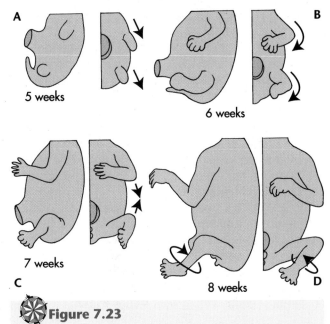

Figure 7.23

Limbs rotate in opposite directions during development. See text.

pletes the changes when the lower limbs rotate medially. This final change distributes the nerves, myotomes, and dermatomes of the lower limbs into their characteristic "barber pole" arrangement studied in Chapter 11.

Limb muscles derive from the **somites** (see Chapter 4, Somites), and their development is discussed thoroughly in Chapters 9, Skeletal Muscle, and 11, Muscles of the Appendicular Skeleton. Nevertheless, you should know that precursor muscle masses are assembling during weeks 4 and 5, while the cartilage models of the bones also are developing. These muscle masses aggregate on the anterior and posterior surfaces of the bones as the precursors of the flexor and extensor muscles respectively, which bend the joints toward the body and extend them outward. Individual muscles differentiate from these masses and follow the limbs to their attachments on bones and tendons.

- Name the structures from which limb buds develop.
- Which limb regions form first, and which last?
- Name the zone that directs this sequence of events.

SUMMARY

The body is mobile because of 126 bones of the appendicular skeleton. The pelvis and lower limbs support body weight and reach long distances in walking and running. Shoulder and upper limb bones are smaller and capable of more flexible motion than corresponding regions of the lower limbs. The bones of the feet distribute body weight to the floor and push off from it at each stride. Without such weight to support, the bones of the hands instead are specialized to grasp objects. *(Page 180.)*

Bones of the lower limbs sustain body weight when one stands and walks. The pelvis supports the pelvic cavity and lower abdomen, and the coxal bones transfer this weight to the femur at the acetabulum. The femur rotates and swings the thigh and transfers weight to the tibia at the knee. The knee allows the leg to swing forward and to lock while walking and prevents lateral motion. The tibia transfers forces to the ankle joint, which it forms with the fibula. Tarsal bones form the ankle and heel, parts of the metatarsal arches. The metatarsal bones complete the arch and articulate with the phalanges. *(Page 181.)*

The upper limbs enable a wide variety of motions. From the opposable thumb to the glenoid fossa, the organization of the upper limbs enables the hands to reach nearly every body surface. The scapula and clavicle make up the shoulder girdle. The scapula is quite mobile; it allows the shoulder to move in all directions; only the clavicle anchors the shoulder to the axial skeleton. The head of the humerus articulates freely in the glenoid fossa of the scapula, which allows the arm to circumduct and to rotate, but the elbow permits mostly hinge-like motion. Nevertheless the elbow permits the radius and ulna to pronate and supinate the forearm. The carpals add further mobility at the wrist, and the metacarpals and phalanges extend from the carpals. *(Page 191.)*

The extremities develop from limb buds. Limb bones appear in proximodistal order under control of the progress zone. Limbs begin as placodes that grow outward as limb buds. Cartilage models of the bones appear shortly thereafter, and individual fingers and toes begin to separate during the seventh week. The differences in action between the elbow and knee are established when the arms rotate posteriorly and the thighs anteriorly. *(Page 198.)*

STUDY QUESTIONS   OBJECTIVES

1. Name the bones of the upper and lower limbs and the shoulder and pelvic girdles.   **1**
2. Which bones form the hip, the knee, the ankle, and the medial longitudinal metatarsal arch? What is the action of these bones during walking?   **2, 3**
3. Where are the ilium, ischium, and pubis located in the coxal bone?   **4**
4. Where are the head, trochanters, condyles, and epicondyles located on the femur? What is the function of each of these parts?   **5**
5. What are the differences between the tibia and fibula? Where is each bone located? Which one supports body weight? With which ankle bone do the tibia and fibula articulate?   **6**
6. What are the groups of bones in the ankle and foot? How many bones are in each group? Which groups participate in the medial longitudinal metatarsal arch?   **7**
7. What are the names of the bones of the upper limb and shoulder?   **8**
8. What motions are the scapula and clavicle capable of making?   **9**
9. With which bones do the head, trochlea, and capitulum of the humerus articulate? What are the functions of the greater and lesser tubercles of the humerus?   **10**
10. Describe the motions of the radius during pronation and supination.   **11**
11. Name and locate the carpal bones of the wrist, and describe the motions of the carpals as a whole.   **12**
12. Compare the bones of the thumb with those of the other metacarpals and phalanges.   **13**
13. What similarities and differences do you find between the following pairs of bones: (1) the femur and humerus, (2) radius and tibia, (3) ulna and fibula, (4) tarsals and carpals, (5) metatarsals and metacarpals, and (6) the phalanges of fingers and toes?   **14**
14. Describe the sequence of events in the development of the upper limbs. What are the roles of the apical ectodermal ridge and the progress zone?   **15**

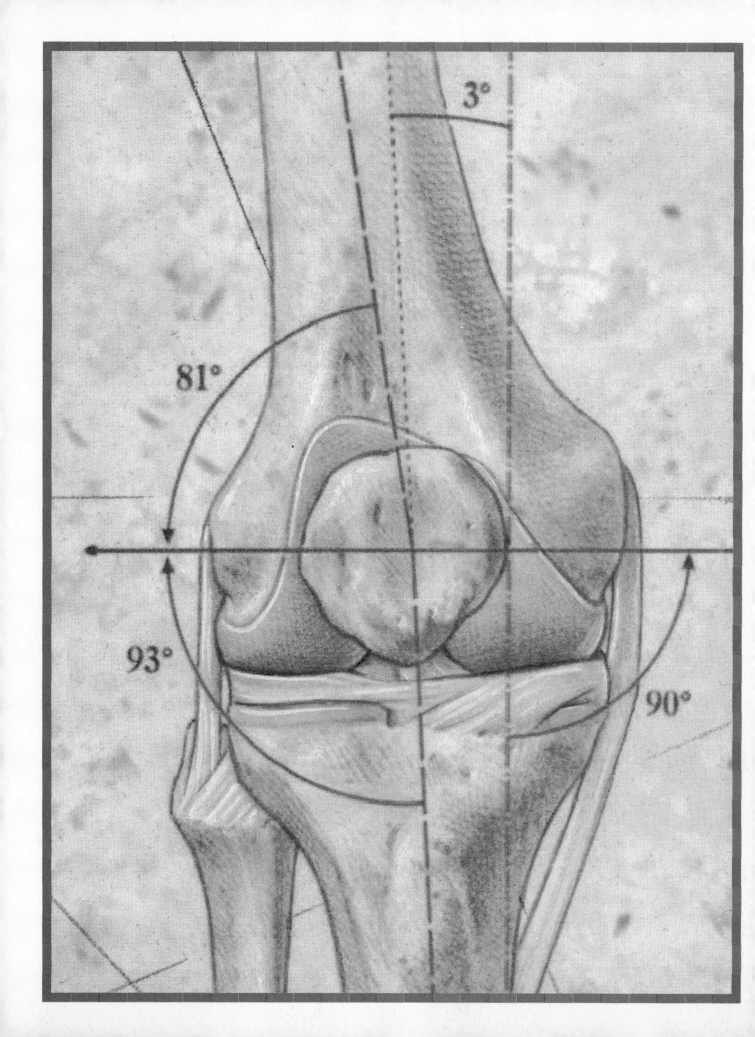

CHAPTER 8

Articulations

OBJECTIVES

After studying this chapter, you should be able to do the following:

1. Define the term *joint.*
2. Name joints according to their participating bones and locate them in the skeleton.
3. Name the three structural types of joints and the three functional categories of joints.
4. Classify joints according to their structure and their mobility.
5. Describe the different kinds of sutures.
6. State the differences between cartilaginous and fibrous synarthroses.
7. Contrast the structures of fibrous and cartilaginous amphiarthroses.
8. Describe the structure of a typical synovial joint.
9. Describe how osteoarthritis progresses in synovial joints.
10. Name and describe the four basic movements that can occur at synovial joints.
11. Give examples of synovial joints that illustrate the movements in Objective 10.
12. Distinguish between (1) linear and angular motion and (2) uniaxial, biaxial, and triaxial motion of joints.
13. Name the different structural types of synovial joints and the motions associated with each one.
14. Describe the functions of bursae and synovial tendon sheaths.
15. Briefly describe clinical conditions that affect the structure and function of joints, and describe surgical procedures that treat these conditions.

 VOCABULARY TOOLBOX

Well-connected thoughts make articulate speech; the same can be said of bones and motion:

arthr-   *G* a joint. As in arthroses, joints; arthritis, a joint disease; and arthroscopy, the technique that examines the interior of joints with flexible, fiber optic tubes.

artic-   *L* a joint, speech. As in articulation and articulate.

Several suffixes express conditions.

-itis   *G* inflammation. As in arthritis and bursitis.

-osis   *L* condition. As in arthrosis, a joint, or a structure that is a joint.

The following prefixes express mobility:

amphi-   *G* on both sides. Amphiarthroses have limited movement and are intermediate between immobile and fully mobile joints.

syn-   *G* with, together. Synarthroses are immobile joints because the bones remain tightly together.

di-   *G* separate, apart. Diarthroses are freely movable joints because a synovial cavity separates the bones of the joint.

RELATED TOPICS

- **Axial and appendicular skeletons, Chapters 6 and 7**
- **Bone and cartilage, Chapter 5**
- **Epithelial and connective tissue, Chapter 3**
- **Muscle of trunk and limbs, Chapters 10 and 11**

PERSPECTIVE

Joints connect the bones of the skeleton. The body displays a wide variety of joints; some move and others do not. You will learn what motions each variety allows. From that knowledge will come your understanding of the body's movements. You will also learn how certain diseases and other conditions interfere with the structure and function of joints.

Joints and articulations are the structures that connect bones and cartilages of the skeleton. The motion allowed by your knee or any other joint is as important as the motion it prohibits. The knee bends like a hinge, for example, but it does not swing laterally at any time. Fully extended, your knee is as rigid as a post, but if someone has ever hit you unexpectedly behind the knee (a standard middle school trick), you know how readily the knee can collapse. The body and skeleton benefit from the ability of joints to permit certain motions and to limit others.

Case study begins on p. 218.

JOINTS AND MOTION

Working together, the joints, the bones, and the skeletal muscles enable the body to move and to support itself. As you learned in Chapters 6 and 7, different bones have different functions, and the same is true of joints and muscles (see Chapters 10 and 11). Your knee is an example of the more than 200 body joints that permit the motions shown in Figure 8.1. Figure 8.2 identifies the knee and other joints in the skeleton, and Table 8.1 lists the joints discussed in this chapter. Body movement greatly depends on the structure of the joints. The wrist cannot do what the knee accomplishes because the structure and location of the wrist differ considerably from the knee, but the adaptations of the wrist neatly suit the upper limb. (Imagine the problems and benefits, if by some anatomic circumstance, wrist and knee were to exchange positions.) It is a formidable job, virtually endless in detail, to describe all motions of the body and the action of each joint, but it is possible to understand the basics.

Figure 8.1

Joints of the skeleton enable all these movements: a back flip, a figure-skater's maneuver, and the launching of a paper airplane.

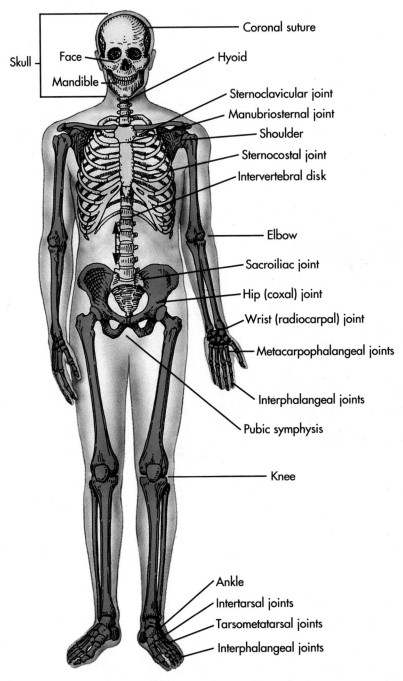

Figure 8.2

Joints of the skeleton listed in Table 8.1 and described in this chapter.

STRUCTURE AND MOTION

Anatomists classify joints according to their structure or their function, presented in Table 8.2. Structurally, joints are either **fibrous, cartilaginous,** or, if they contain a synovial cavity, **synovial** joints, as the rows of the table indicate. Sutures between the bones of the skull are fibrous joints, and intervertebral disks are cartilaginous. The knee and wrist are synovial joints. Grouped across the top of the table, according to how much movement joints permit, we have

Table 8.1   JOINTS OF THE BODY

Joint	Bones	Structure	Motion
Coronal suture	Frontal, parietal	Fibrous, serrated suture	Synarthrosis, immovable
Intermaxillary suture	Maxillae	Fibrous, plane suture	Synarthrosis, immovable
Lambdoidal suture	Occipitoparietal	Fibrous, serrated suture	Synarthrosis, immovable
Sagittal suture	Parietals	Fibrous, serrated suture	Synarthrosis, immovable
Squamosal suture	Parietal, temporal	Fibrous, squamous suture	Synarthrosis, immovable
Vomerosphenoidal suture	Vomer, sphenoid	Fibrous, schindylesis suture	Synarthrosis, immovable
Temporomandibular	Temporal, mandible	Synovial, gliding, hinge	Diarthrosis, biaxial, protraction, elevation, lateral eversion
Alveolar sockets	Teeth, mandible, maxillae	Fibrous, gomphosis	Synarthrosis, very slight movement
Atlanto-occipital	Atlas, occipital	Synovial, ellipsoid	Diarthrosis, biaxial, flexion
Medial atlantoaxial	Atlas, dens of axis	Synovial, pivot	Diarthrosis, uniaxial, rotation of head
Lateral atlantoaxial	Atlas, axis lateral condyles	Synovial, gliding	Diarthrosis, uniaxial, rotation of head
Inferior tibiofibular	Distal extremities, tibia, fibula	Fibrous, syndesmosis	Amphiarthrosis, movable
Sternocostal, first	Manubrium, first rib	Cartilaginous, synchondrosis	Synarthrosis, immovable
Sternocostal	Manubrium, sternum, second-seventh costal cartilage	Synovial, gliding	Diarthrosis, movable
Manubriosternal joint	Manubrium, sternum	Cartilaginous, symphysis	Amphiarthrosis, slight movement
Costovertebral	Heads of ribs, bodies of vertebrae	Synovial, gliding	Diarthrosis, slight rotation
Vertebral facets	Articular processes of vertebral arches	Synovial, gliding	Diarthrosis, movable
Intervertebral disks	Vertebrae	Cartilaginous, symphysis	Amphiarthrosis, limited flexion, rotation
Sternoclavicular joint	Sternum, clavicle	Synovial, gliding	Diarthrosis, biaxial, elevation, protraction, circumduction
Acromioclavicular	Clavicle, acromion process of scapula	Synovial, gliding	Diarthrosis, biaxial, elevation, rotation
Shoulder	Humerus, glenoid fossa of scapula	Synovial, ball and socket	Diarthrosis, triaxial, flexion, abduction, circumduction, rotation
Elbow (humeroulnar)	Humerus, ulna	Synovial, hinge	Diarthrosis, uniaxial, flexion
Elbow (humeroradial)	Humerus, radius	Synovial, gliding	Diarthrosis, biaxial, flexion, rotation
Elbow, proximal radioulnar	Radius, ulna	Synovial, pivot	Diarthrosis, uniaxial, rotation in pronation, supination
Distal radioulnar	Radius, ulna	Synovial, pivot	Diarthrosis, uniaxial, rotation in pronation, supination
Middle radioulnar	Shafts of radius, ulna	Fibrous, syndesmosis	Amphiarthrosis, pronation, supination
Wrist (radiocarpal)	Radius, scaphoid, lunate	Synovial, ellipsoid	Diarthrosis, biaxial, flexion, abduction, circumduction
Intercarpal	Carpals	Synovial, gliding	Diarthrosis, linear
First carpometacarpal	Trapezium, first metacarpal	Synovial, saddle	Diarthrosis, biaxial, flexion, abduction, circumduction
Carpometacarpal	Distal row of carpals, metacarpals 2-5	Synovial, gliding	Diarthrosis, uniaxial, flexion
Metacarpophalangeal	Metacarpals, proximal phalanges	Synovial, ellipsoid	Diarthrosis, biaxial, flexion, abduction, circumduction

Continued.

Table 8.1	JOINTS OF THE BODY—cont'd		
Joint	**Bones**	**Structure**	**Motion**
Interphalangeal	Phalanges	Synovial, hinge	Diarthrosis, uniaxial, flexion
Sacroiliac joint	Sacrum, ilium	Synovial/fibrous, plane	Amphiarthrosis, slight movement
Pubic symphysis	Pubic bones	Cartilaginous, symphysis	Amphiarthrosis, slight movement
Hip (coxal)	Pelvis, femur	Synovial	Diarthrosis, triaxial, flexion, abduction, circumduction, rotation
Knee (patellofemoral)	Femur, patella	Synovial, gliding	Diarthrosis, linear
Knee (tibiofemoral)	Femur, tibia	Synovial, hinge	Diarthrosis, uniaxial, flexion, some rotation
Intermediate tibiofibular	Tibia, fibula	Fibrous, syndesmosis	Amphiarthrosis, slightly movable
Ankle	Talus, tibia	Synovial, hinge	Diarthrosis, uniaxial, dorsiflexion, plantar flexion
Ankle	Talus, fibula	Synovial, hinge	Diarthrosis, uniaxial, dorsiflexion, plantar flexion
Intertarsal joints	Tarsals	Synovial, gliding	Diarthrosis, linear
Tarsometatarsal joints	Cuneiforms, cuboid, metatarsals	Synovial, hinge	Diarthrosis, uniaxial, flexion
Metatarsophalangeal	Metatarsals, proximal phalanges	Synovial, ellipsoid	Diarthrosis, biaxial, flexion, abduction, circumduction
Interphalangeal	Proximal, medial, distal phalanges	Synovial, hinge	Diarthrosis, uniaxial, flexion, extension

synarthroses, **amphiarthroses**, or **diarthroses**. Synarthroses allow little or no motion. As sutures of the skull, they immobilize bones and protect internal organs. Amphiarthroses (middle column) are slightly movable, and they include the intervertebral disks (cartilaginous, middle row), as well as syndesmoses (fibrous, top row). Diarthroses are freely movable because they are synovial joints. You should be able to identify the joints listed in the table where rows and columns coincide. Most joints of the appendicular skeleton are synovial, diarthrotic joints. The axial skeleton has numerous cartilaginous amphiarthrotic joints called symphyses.

- Name a fibrous amphiarthrotic joint.
- According to Table 8.2, do immovable synovial joints exist?

TERMS AND NAMES

The words *joint* and *articulation* refer to the structures that join bones together, and *arthrosis* is a more technical term with the same meaning. Although these words are synonymous, they are sometimes used differently. The knee joint, for example, involves three bones—the femur, the patella, and the tibia—and two articulations within the same joint. One articulation occurs between the femur and tibia and another occurs between the femur and the patella.

Although joints such as the knee and elbow have common names, all can be named according to the bones involved. The mandible articulates with the temporal bone at the temporomandibular joint that is formed by the mandibular fossa of the temporal bone (Chapter 6) and the head of the mandible. The occipitoparietal suture is part of the lambdoidal suture (Chapter 6) that joins the occipital and the parietal bones of the skull. The articulations of the knee are the tibiofemoral and patellofemoral. Convention governs which bone is named first. Table 8.1 gives the names of the joints discussed in this chapter.

- Which bones does the radiocarpal joint connect?
- What is the common name of this joint?

SUTURES

Sutures lock the margins of cranial and facial bones firmly together with fibrous connective tissue, as you can see in Figure 8.3 and Table 8.2. **Sutures** (SOO-tshur; sutur-, *L*, a seam) protect the brain and other internal structures of the skull because these joints prevent the bones from moving and pressing on the softer tissues beneath. In the **sagittal** and **lambdoidal sutures** of the cranium, you can see how the structure makes these joints immovable. As the parietal and occipital bones grew toward each other, the periosteum covering the one bone fused with the periosteum of the other, forming a continuous **sutural ligament** of fibrous connective tissue that joins the edges of the bones together. The serrated edges inter-

Table 8.2   STRUCTURE AND FUNCTION OF ARTICULATIONS

Structure	Function		
	Immovable, Synarthroses	**Slightly movable Amphiarthroses**	**Freely movable, Diarthroses**
Fibrous	SUTURES Fibrous connective tissues join margins of cranial and facial bones GOMPHOSES Anchor teeth in mandible and maxilla	SYNDESMOSES Interosseous ligaments, sheets of fibrous connective tissue, join shafts of long bones, such as radius and ulna or tibia and fibula	
Cartilaginous	SYNCHONDROSES Hyaline cartilage connects epiphyses and shafts of long bones until the joints ossify. Also join ilium, ischium, and pubis of juvenile hip	SYMPHYSES Intervertebral disks and pubic symphysis; fibrocartilage annuli surround gelatinous nucleus pulposus	
Synovial			SYNOVIAL JOINTS A synovial capsule envelops the articular surfaces and lubricates them with synovial fluid. Examples include knee, shoulder, wrist, and elbow

lock tightly, allowing the bones little opportunity to separate. Serrations not only lock the bones together, but also greatly improve the strength of the joint by increasing the surface area to which the sutural ligament adheres.

Figure 8.3 also shows that sutures interlock in different ways. Their edges may be **serrated,** as in the lambdoidal suture, or they may be slanted. The squamosal suture between the temporal and parietal bones (temporoparietal joint) of the skull is an example of this **squamous** type of suture. A **schindylesis,** (skin-dye-LEE-sis) wedges one margin between two surfaces of another bone. Perhaps you recall how the vomer fits onto the sphenoid bone in the nasal septum (Chapter 6); this joint is a schindylesis. **Plane sutures** are rough surfaces that fit together without overlapping. The junction between the maxillae in the midline of the hard palate is an example of a plane suture.

Most sutures ossify in later life, and the obliterated joints are **synostoses** (SIN-oss-TOE-sees). Bone replaces the suture and unites the originally separate bones as one. The occipital and parietals fuse in this fashion during late middle age. Originally the frontal bone was two separate bones joined in the midline by a premature synostosis. Orthopedists and forensic physicians use synostoses to verify the ages of pa-

tients and crime victims. Synostoses that occur before bone is fully grown may deform the skull, as the discussion of **craniosynostosis** and **Crouzon's syndrome** illustrates (see Chapter 6, Figure 6.5). Sutures can isolate small **Wormian** or **sutural bones** like islands in a lazy river. When the calvarium enlarges excessively, as it does in **hydrocephaly,** numerous Wormian bones fill the additional surface of the skull.

GOMPHOSES

Your teeth are anchored to the mandible and maxilla by gomphoses, which are fibrous joints shown in Figure 8.4 and Table 8.2, that suspend the teeth in alveolar sockets. **Gomphoses** derive their name from the fact that teeth appear to have been driven into the jaw like nails into wood (GOM-foe-sees; gompho-, G, nail, -osis, L, condition). The **periodontal ligament** joins the cancellous bone of the socket to the smooth surface of the root of the tooth with a web of collagen fibers (perforating fibers) that penetrate both tooth and bone. Dentists sever the periodontal ligament before extracting a tooth. Gomphoses are essentially immovable, but they are equipped with receptors that detect the forces of chewing that align

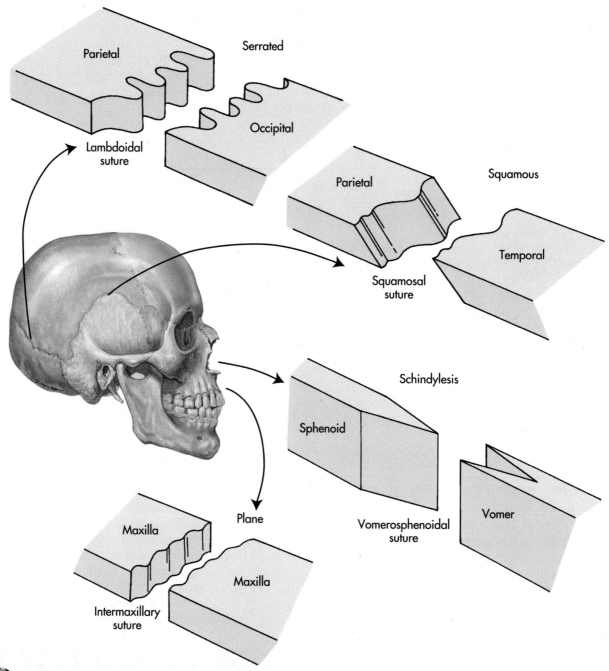

⁜ **Figure 8.3**

Sutures lock the bones of the skull together. The lambdoidal suture is a serrated suture, and the squamosal suture is squamous. In a schindylesis, the sphenoid bone fits between the alae of the vomer in the nasal septum. A plane suture joins the maxillae in the roof of the oral cavity.

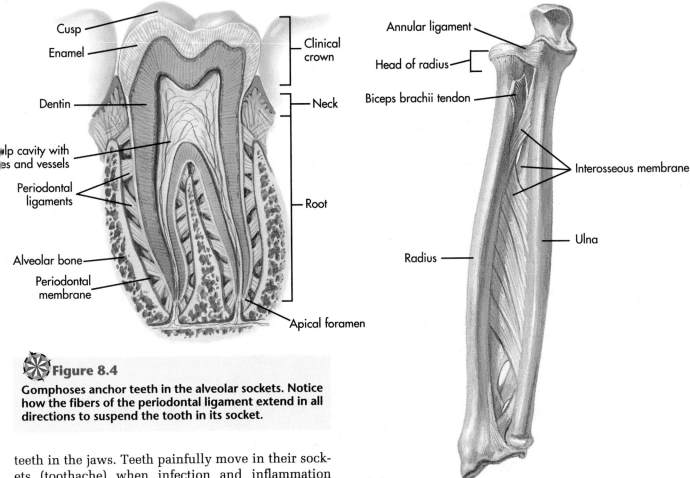

Figure 8.4

Gomphoses anchor teeth in the alveolar sockets. Notice how the fibers of the periodontal ligament extend in all directions to suspend the tooth in its socket.

Figure 8.5

The radioulnar interosseous membrane connects the shafts of the radius and ulna.

teeth in the jaws. Teeth painfully move in their sockets (toothache) when infection and inflammation swell the periodontal ligament.

- Name the suture that joins the nasal with the frontal bone.
- Why are sutures not found in the appendicular skeleton?
- What is the name of the ligament in gomphoses?

SYNDESMOSES

Syndesmoses are fibrous membranes that join moderately movable borders of certain bones together. Buried deep beneath the muscles of the forearm, a tough sheet of fibrous connective tissue links the shafts of the radius and ulna together (Figure 8.5 and Table 8.2). This particular **interosseous membrane** holds these two bones together, most notably when the forearm supinates and pronates. This membrane also separates the flexor muscles on the anterior surface of the forearm from the extensors on the posterior side. A similar syndesmosis (SIN-dez-mo-sis; -desmo-, *G*, ligament) joins the tibia and fibula of the leg in a tibiofibular joint, but there is little if any movement. Ligaments, which are also fibrous, also would be considered syndesmoses except that they are associated with the capsules of synovial joints (see Synovial Joints, this chapter).

SYNCHONDROSES

Synchondroses are the rigid cartilages that unite epiphyses to the shafts of long bones in young people (Table 8.2). Figure 8.6A shows an X-ray view of **synchondroses** in the metacarpals and phalanges of the hand. Synchondroses (SIN-kond-ROE-sees) are stiff, hyaline cartilage plates (Figure 8.6B) that represent the original endochondral models of ossification. The cartilage and its surrounding perichondrium are continuous with bone and the periosteum. Ossification usually replaces these cartilages with **epiphyseal plates** of bone by the time an individual is 25 years old.

COSTAL CARTILAGES

You might expect **costal cartilages** (Figure 8.7) to be synchondroses because they connect the distal ends of ribs to the sternum. You would be right about

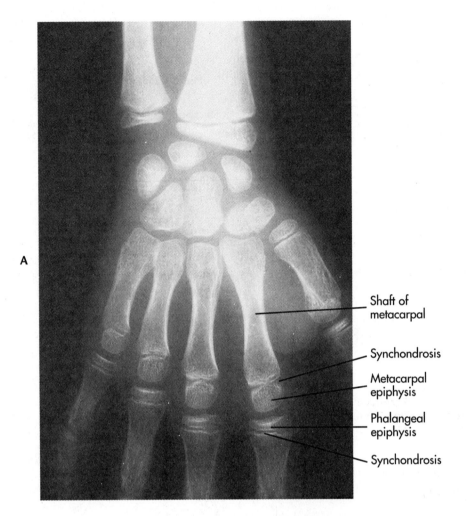

A

Shaft of
metacarpal

Synchondrosis

Metacarpal
epiphysis

Phalangeal
epiphysis

Synchondrosis

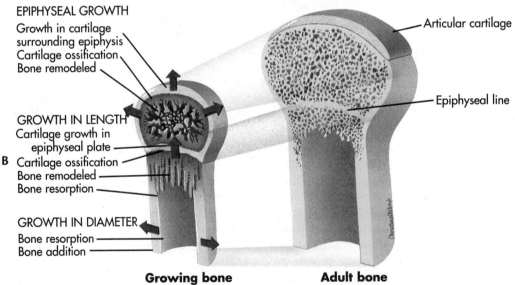

EPIPHYSEAL GROWTH
Growth in cartilage
surrounding epiphysis
Cartilage ossification
Bone remodeled

GROWTH IN LENGTH
Cartilage growth in
epiphyseal plate
B   Cartilage ossification
Bone remodeled
Bone resorption

GROWTH IN DIAMETER
Bone resorption
Bone addition

Articular cartilage

Epiphyseal line

Growing bone   **Adult bone**

✳ Figure 8.6

A, X-ray photograph of the wrist and hand of a 7-year-old boy. Synchondroses are transparent spaces in the X-ray between the proximal epiphyses and shafts of the phalanges. **B,** Synchondroses connect epiphyses to the shaft of the femur.

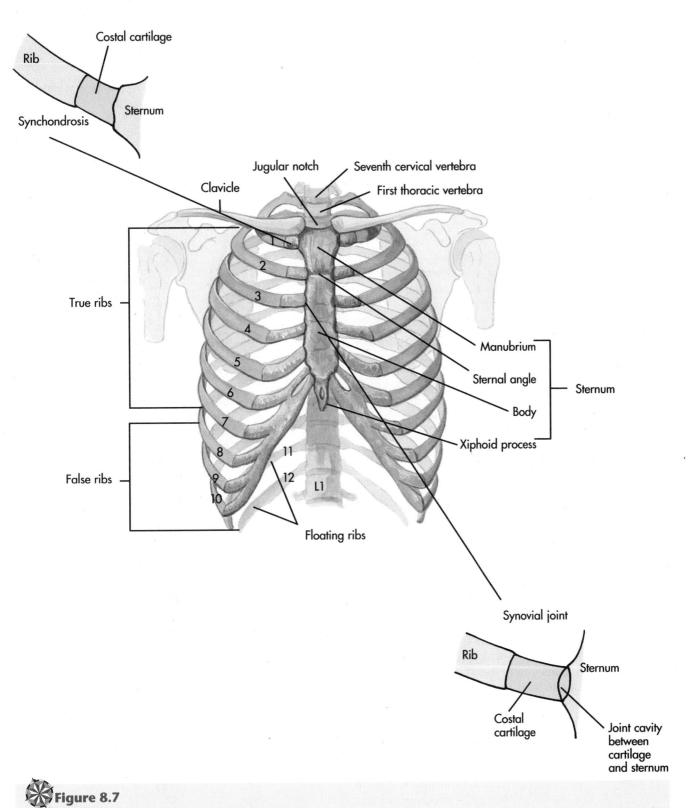

Costal cartilage

Rib

Synchondrosis

Sternum

Jugular notch

Seventh cervical vertebra

First thoracic vertebra

Clavicle

True ribs

False ribs

1
2
3
4
5
6
7
8
9
10
11
12
L1

Manubrium

Sternal angle

Sternum

Body

Xiphoid process

Floating ribs

Synovial joint

Rib

Sternum

Costal cartilage

Joint cavity between cartilage and sternum

✴ **Figure 8.7**

Costal cartilages and synchondroses. The first costal cartilage is a synchondrosis because the cartilage articulates directly with rib and sternum, but synovial joints connect other costal cartilages to the sternum.

the first cartilage but not the rest. The first seven ribs end distally in permanent hyaline cartilages that join the ribs to the sternum, but only the first cartilage is a true synchrondrosis continuous with the sternum and the rib. Although the remaining cartilages are continuous with the ribs, they end in synovial joints (see Synovial Joints, this chapter) with the sternum, which disqualifies them as synchrondroses.

The rest of the costal cartilages do not articulate with the sternum. Cartilages eight, nine, and ten articulate with the costal cartilages of the seventh and eighth ribs. Ribs eleven and twelve are "floating" ribs, that is, ribs with cartilages that do not articulate with any other ribs.

- Name the location of two syndesmoses.
- What type of cartilage makes up synchrondroses?
- How many costal cartilages are synchrondroses?

SYMPHYSES

Symphyses are slightly movable, cartilaginous joints. Intervertebral disks are the best example of symphyses (SIM-fih-sees; symphy-, *G,* growing together). These moderately flexible joints between the bodies of vertebrae in the vertebral column (Figure 8.8A and Table 8.2) are composed of pads of fibrocartilage with a soft, gelatinous core. The core is called the **nucleus pulposus,** shown in Figure 8.8B, and it derives from the notochord. Annular layers of fibrocartilage, called the **anulus fibrosus** (fibrous ring; anula, *L,* a ring) surround the nucleus pulposus, and thin films of hyaline cartilage join the margins of the disk to the vertebrae. Intervertebral disks allow the vertebral column to flex and extend, but they are sufficiently firm to shape the spinal curvatures, especially the lumbar curvature where the anterior margins of the wedge-like disks are thicker than the posterior margins (Figure 8.8A). Disks **herniate** when the fibrocartilage ruptures, as shown in Figure 8.8C, and the nucleus pulposus enters the vertebral canal or intervertebral foramen and encroaches on the spinal cord or spinal nerves. Surgical removal of the offending disk and fusion of the affected vertebrae are common treatments for this condition.

Two other joints in Figure 8.9 illustrate how symphyses vary. The **pubic symphysis** at the junction of the pubic bones in the pelvis often contains a fluid cavity, and the **manubriosternal joint,** connecting the manubrium to the sternum, begins life as a synchondrosis that is replaced subsequently by fibrocartilage.

- What are the two diagnostic structures of intervertebral disks?
- What happens to these structures when a disk herniates?

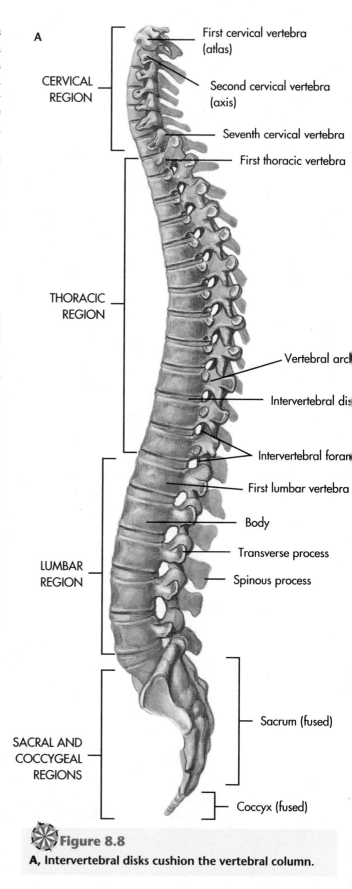

A

CERVICAL REGION

THORACIC REGION

LUMBAR REGION

SACRAL AND COCCYGEAL REGIONS

First cervical vertebra (atlas)

Second cervical vertebra (axis)

Seventh cervical vertebra

First thoracic vertebra

Vertebral arch

Intervertebral disk

Intervertebral foramen

First lumbar vertebra

Body

Transverse process

Spinous process

Sacrum (fused)

Coccyx (fused)

Figure 8.8

A, Intervertebral disks cushion the vertebral column.

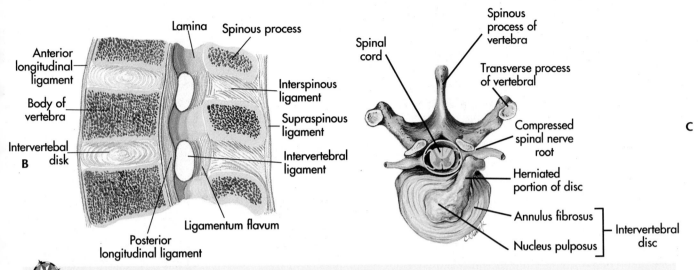

Figure 8.8 cont'd.

B, Annular layers of fibrocartilage surround the central nucleus pulposus, seen in sagittal section. **C,** The annulus fibrosus has herniated and now presses against a root of the spinal nerve.

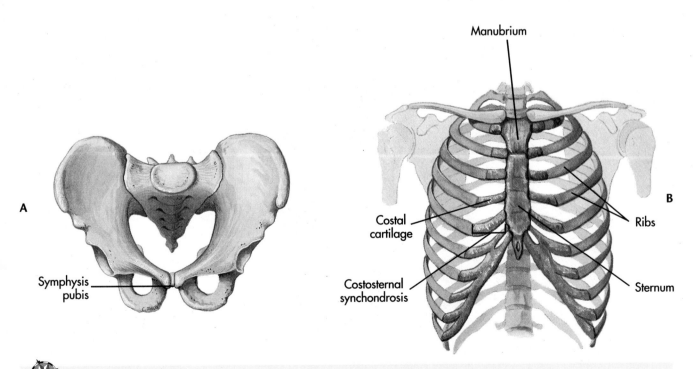

Figure 8.9

Other symphyses join the pubic bones of the pelvis (**A**), and the manubrium with the body of the sternum (**B**).

SYNOVIAL JOINTS

Most articulations are highly mobile synovial joints containing a fluid-filled cavity that reduces friction between the articulating surfaces of bones (Table 8.2). A sleeve of connective tissue, called the **synovial** or **articular capsule,** joins the bones together in Figure 8.10. The synovial capsule is composed of two layers of tissue. The outer layer, called the **fibrous capsule,** is a sheath of fibrous, connective tissue that is continuous across the joint with the periosteum of the bones. It contains ligaments that anchor the bones together and limit their movement. Internally a **synovial membrane** or **synovium** lines the **synovial cavity** between the articulating ends of the bones and secretes viscous, yellowish **synovial fluid** into the cavity. Unlike the linings of the abdomen and gut, this membrane is not an epithelium nor is there a basement membrane to bind it to underlying tissue (see Chapter 3, Epithelial Tissue). The synovial cavity actually is space within the extracellular matrix of connective tissue, and the synovial fluid equilibrates with blood supplied by capillaries to the synovium.

Thin, tough, slippery **articular cartilages** made of hyaline cartilage cover the articulating surfaces of the bones and merge with the synovial capsule at the peripheries of these surfaces. **Menisci** (men-ISS-key; menisc-, *G*, a crescent) made of fibrocartilage may protrude from the capsular wall into the synovial cavity between the articulating cartilages and cushion the joint. The fibrous capsule and its synovial membrane are highly vascularized and supplied with numerous nerve endings that regulate skeletal muscle contraction by detecting tension in the ligaments. These endings also provide "joint sense" about the position of the extremities. **Fat pads** can accumulate in the capsule and also cushion the joint as it moves. In certain joints the synovial cavity reaches beyond the joint itself as **bursae,** pockets of synovial fluid around tendons that relieve friction as the tendons slide across the surface of bones.

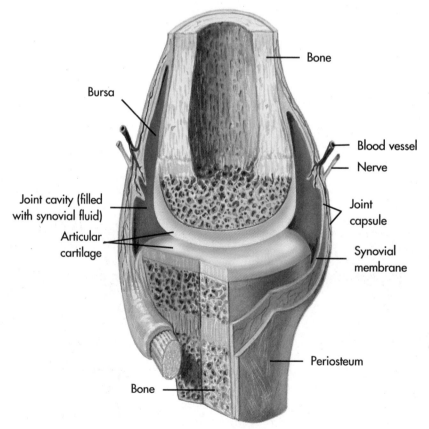

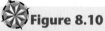Figure 8.10

Synovial joint. The fibrous capsule surrounds the articulating cartilages and synovial cavity with a tough fibrous cover and internal synovial membrane that supplies synovial fluid.

ARTICULAR CARTILAGES

Articular cartilages contain several zones of hyaline cartilage (Figure 8.11A and B) that provide a tough, slippery surface. The chondrocytes are small and sparse in the superficial layer, but are larger and more numerous in deeper zones. Although the chondrocytes are metabolically vigorous (they replace half of the proteoglycans every 8 to 17 days), they divide only rarely. Calcified cartilage connects with the underlying cortical bone. Collagen fibers anchored in the underlying bone arch through the extracellular matrix of the cartilage, transferring forces from superficial to

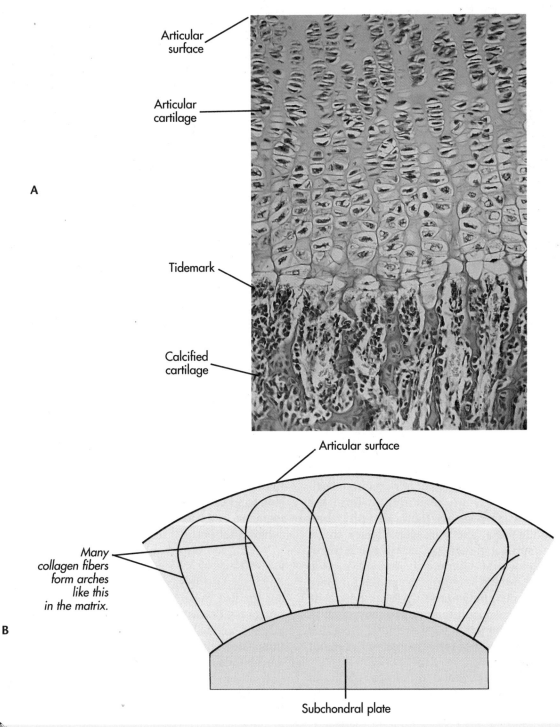

Figure 8.11

Microanatomy of articular cartilages. A, A section through the hyaline cartilage. Chondrocytes are smaller near the surface of the cartilage above the figure. Calcified cartilage joins the hyaline cartilage to the bone below. The tidemark is the boundary between hyaline cartilage and calcified cartilage. **B,** Collagen fibers arch through the hyaline cartilage.

deeper zones. A film of hyaluronic acid from the synovial fluid is thought to reduce friction at the surface of the cartilage, and proteoglycans of the cartilage matrix cushion movement to some degree, but the synovial fluid and ligaments of the articular capsule absorb most of the shock of motion.

SYNOVIAL FLUID

Synovial fluid plays several roles in synovial joints. This fluid is derived from blood plasma, thus the yellowish color. The synovial membrane increases the viscosity of synovial fluid by secreting the polysaccharide, hyaluronic acid. Confined within the capsule, synovial fluid absorbs shock by distributing forces to the ligaments and tendons in the walls of the capsule. Synovial fluid nourishes the articular surfaces when moving bones squeeze fluid in and out of the articular cartilage. Possibly most important of all (see Case History, this chapter), this fluid prevents blood vessels from invading the cartilage and forming bone in it. Recall that bone is highly vascularized and cartilage is not (see Chapter 5, Bone).

- Name the structures of the synovial capsule.
- Which one produces synovial fluid?
- Which contains ligaments?
- What part is composed of hyaline cartilage?

ARTHRITIS CASE HISTORY

Synovial fluid and articulating surfaces make synovial joints mobile, but these materials also can contribute to painful joint disease. Consider the following patient's history.

Conservative treatment can no longer manage **osteoarthritis** in Mary Jo's knees. Five years of mild exercise, weight reduction, and analgesic (pain killing) drugs have delayed deterioration of her articular cartilages, but bone deposits are beginning to limit her movement and to increase her pain. Her physician has recommended that she consider surgery to remove the bone spurs and damaged cartilages. **Arthroscopic surgery** would temporarily restore motion and reduce pain, but his long-term prognosis for Mary Jo's condition is not optimistic. The pain and deposits are likely to return, and ultimately she may need to consider replacing the joints with artificial ones that offer up to 10 or 15 years of relief before replacement becomes necessary.

Mary Jo's condition is related to the poor ability of articular cartilages to regenerate, as well as the tendency of chondrogenic tissue to form bone when it becomes vascularized. Instead of adequately regenerating, the cartilages of weight-bearing joints abrade and split with continual use. Careful diet and exercise have reduced the weight that Mary Jo's knees must bear (85% of her weight presses on each knee for about ½ second at each stride). Exercising has also promoted regeneration by continually flushing the surface of the cartilages with fresh fluid. Without a direct blood supply of their own, her articular cartilages depend on synovial fluid for nutrition and prevention of vascular invasion.

By middle age (Mary Jo is 57) the larger cracks in her cartilages probably have introduced fibrocartilage onto the slick articular surfaces. Blood vessels from the bones beneath have caused additional bone to form at the periphery of the cartilages, where they join the synovial capsule. Such **osteophytes** (OSS-tee-oh-fites; bone growths; osteo-, *G,* bone; phyt-, *G,* plant) can be very painful because they press against the nerves of the capsular wall.

Osteoarthritis is a degenerative disease of synovial joints that has no known primary cause. In addition to damaging the articular cartilages, the disease also inflames the synovial capsule and modifies the synovial fluid. Once damage has occurred, the condition can return as secondary osteoarthritis attributed to trauma, infection, inflammation of the synovial membrane, or aging. Mary Jo's case is related to her age.

- What is the immediate cause of osteophytes?
- What is the maximum weight each of your knees sustains while you are walking?
- What treatment may Mary Jo eventually need for her condition?
- What are the functions of synovial fluid?

BASIC BODY MOVEMENTS

Did you see the Boston Celtics' Larry Bird in the 1987 playoffs against the L.A. Lakers? Now, there were some moves. Figures 8.12 and 8.13 illustrate the fundamental motions Bird was using. You should be able to identify several of them in his photograph. Remember, it is joint flexibility, not the bones themselves, that permits motion.

FLEXION AND EXTENSION

Flex your forearm and draw it toward your shoulder, or flex your leg toward your buttocks. **Flexion** brings the bones toward each other in one plane, usually the sagittal plane. **Extension** is the opposite motion. The bones move away from each other as you extend your forearm and leg. Figure 8.12A shows these and other varieties of flexion and extension. Clinicians use the special terms **dorsiflexion** and **plantar flexion** to describe how the foot flexes and extends at the ankle, respectively. Dorsiflexion draws the foot toward the leg, and plantar flexion extends it away. Similar movements dorsiflex and **palmar flex** the hand at the wrist.

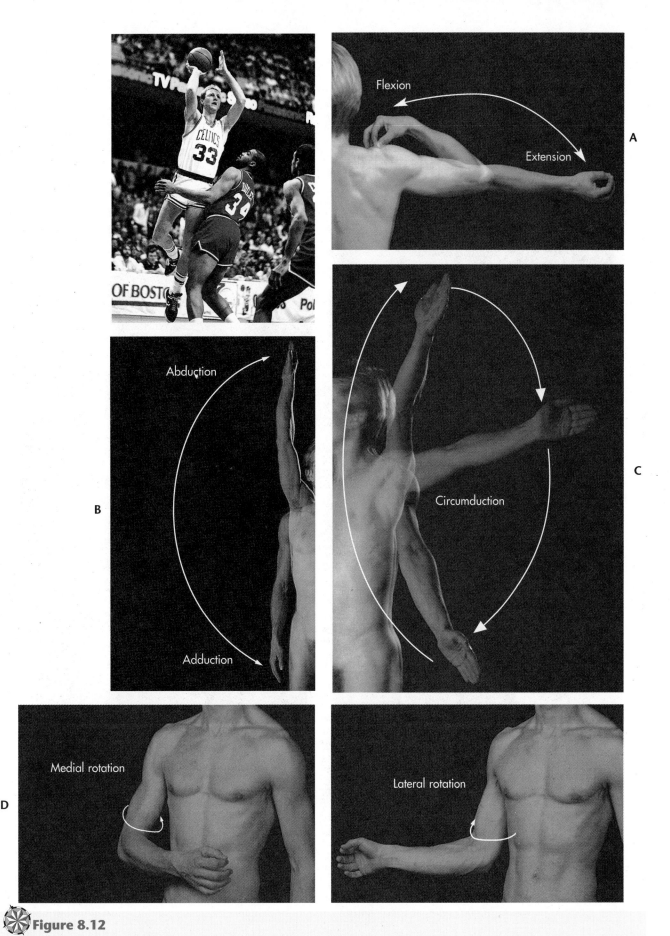

Figure 8.12

Fundamental motions. **A,** Flexion/extension of forearm. **B,** Abduction/adduction of upper or lower limbs. **C,** Circum-duction of the arm at the shoulder. **D,** Rotation of the humerus.

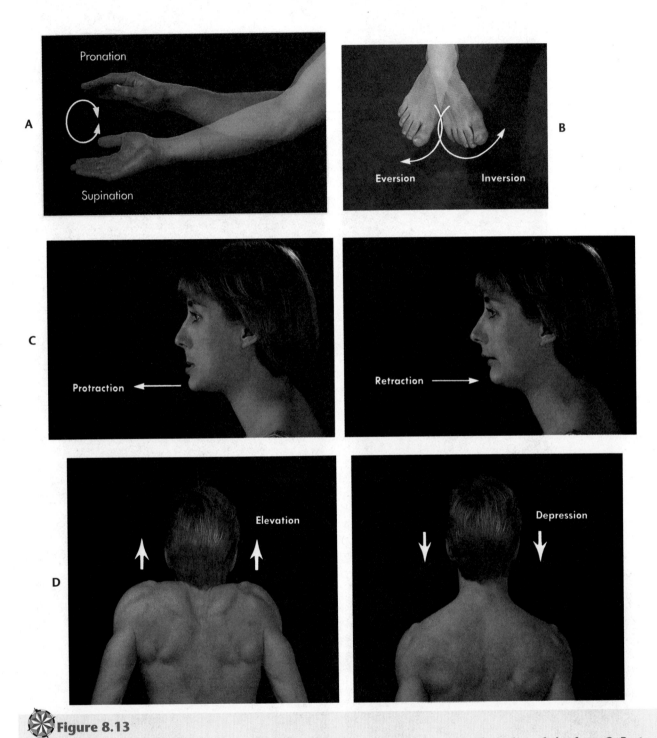

Figure 8.13

More fundamental motions. **A,** Pronation/supination of the forearm. **B,** Inversion/eversion of the foot. **C,** Protraction/retraction of the mandible. **D,** Elevation/depression of the shoulders.

ABDUCTION AND ADDUCTION

These motions resemble flexion and extension, but they move limbs away from and toward the body axis, respectively, as you can see when you raise your upper limb from your side, as in Figure 8.12B. **Abduction** (ab-, L, from, away; duct-, L, lead), increases the angle between the limb and the midline. Abduction usually moves in the frontal rather than the sagittal plane, and the trunk is the usual reference. Abduction also spreads the fingers apart; the fingers move away from the central axis of the hand rather than from the body axis.

Adduction reverses all of these movements (ad-, L, to, toward); it pulls the limbs toward the midline of the body and the fingers toward the midline of the hand.

CIRCUMDUCTION

Circumduction combines all four motions, when the arm circles around the shoulder joint in Figure 8.12C. Thighs circumduct (circum-, L, around) at the hip, and the fingers and thumbs also circumduct at the metacarpophalangeal and carpometacarpal joints. In all of these examples the moving part describes a cone in space.

ROTATION

This motion turns bones on their longitudinal axes. Turn your head from side to side or **rotate** your thighs and arms to see the effect. Figure 8.12D shows the arm rotating at the shoulder. The atlas and axis (first and second cervical vertebrae) rotate about each other, and the heads of the humerus and femur rotate in the glenoid fossa and acetabulum, respectively.

PRONATION AND SUPINATION

These terms describe the ability of the forearm to rotate the hand. When the palm faces anteriorly, as it does in the anatomical position (Figure 8.13A), and the radius and ulna are parallel, the hand is **supinated** (SOO-pin-ated; supin-, L, lying on the back). The hand also faces superiorly in this position when the forearm flexes. **Pronation** (pro-NAY-shun; pron-, L, bent forward) turns the palm in the opposite direction. The radius crosses the ulna so that palms face posteriorly from the anatomical position. The hands are partly pronated when the upper limbs are at rest by your side, but the humerus also contributes to their position by rotating medially. Can you find a position where the pronated palm faces anteriorly? Larry Bird would recognize it.

EVERSION AND INVERSION

Eversion and inversion are foot movements that superficially resemble pronation and supination but are actually very different from them. **Inversion** (invert, L, turn inward) bends the plantar surface of the foot medially (Figure 8.13B), and **eversion** (evert, L, turn outward) tilts it laterally. You can best demonstrate these motions with your leg extended. If you try it with the knee flexed to 90 degrees, you probably also will rotate the tibia and add to the motions of your foot.

Physical therapy and sports medicine frequently use the terms *pronation* and *supination* instead of *inversion* and *eversion* when describing foot movements, but these terms refer to very different motions in anatomy. Pronation and supination concern the radius and ulna, but eversion and inversion are movements of the tarsals and metatarsals, not of the tibia and fibula, which cannot be crossed.

PROTRACTION AND RETRACTION

Jut your lower jaw forward as the person in Figure 8.13C is doing. That motion is **protraction** (pro-TRAK-shun; pro-, L, forward; tract-, L, drawn). Withdrawing the mandible posteriorly is **retraction** (ree-TRAK-shun; re-, L, back). To permit these movements, the condylar process of the mandible slides in the mandibular fossa of the temporal bone of the skull. Drawing back your shoulders is also retraction. All these motions slide in a more or less straight line anteriorly or posteriorly, rather than swinging through an angle as in flexing the knee and rotating the upper limb.

ELEVATION AND DEPRESSION

These movements follow the body axis upward and downward. To raise the shoulders and lower them (Figure 8.13D) is to **elevate** and **depress** them. The mandible elevates and depresses during chewing. Motion is mostly parallel to the central body axis. Raising the upper limbs above the head is, technically speaking, not elevation because the limbs describe an arc, not a straight line.

- Give an example of (1) flexion and extension, (2) circumduction, (3) pronation and supination, (4) eversion and inversion, and (5) protraction and retraction.
- Which of these motions is Larry Bird using in Figure 8.12?

SYNOVIAL JOINT MOVEMENTS

When you think about the kinds of motion described above, rather than which body parts perform them, you can see that these motions are **linear** when they occur primarily in straight lines or **angular** when a part rotates around its longitudinal axis or swings from one end. Flexing and extending the knee is angular motion, and when the upper limb rotates at the shoulder, it is also making angular movements. Protraction and retraction of the mandible are primarily linear motions because the mandible moves forward and back.

Angular motion has further categories in the form of **uniaxial, biaxial,** or **triaxial** movement. Flexion, abduction, and eversion are uniaxial motions. Flexing the knee moves the lower limb in one plane, the sagittal plane, and swings the leg around a hinge-like axis. Abduction and eversion are uniaxial for the same reasons. Each motion occurs primarily in one plane (can you name the planes?) and swings the body part about one primary axis. Circumduction, however, is biaxial motion because the arm or thigh is moving in two planes at once (sagittal and frontal) and swinging on two axes at the shoulder or hip. Quarterbacks and pitchers earn excellent salaries from triaxial motion, by rotating their arms during circumduction. With the extra impetus from rotation, the ball can travel amazing distances and speeds. Try throwing a ball without rotating your arm; it does not work nearly so well as triaxial motion. No coach would ever consider a biaxial quarterback.

TYPES OF SYNOVIAL JOINTS

The articulating surfaces of synovial joints govern their movement, as Figure 8.14 shows. When you complete this section, you should be able to match each type of joint with the motions it allows. The shoulder is a ball-and-socket triaxial joint, for example, that enables the upper limb to move in as many as three axes simultaneously when it extends, abducts, and rotates. The actual anatomy of each category is described later.

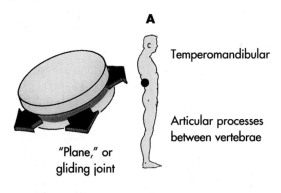

A

Temperomandibular

Articular processes between vertebrae

"Plane," or gliding joint

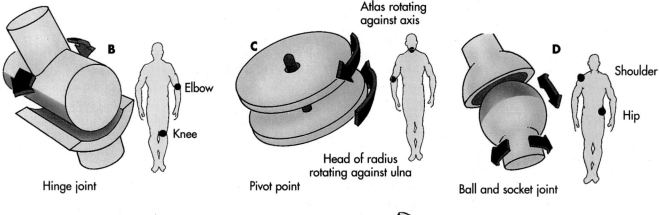

B

Elbow

Knee

Hinge joint

Atlas rotating against axis

C

Head of radius rotating against ulna

Pivot point

D

Shoulder

Hip

Ball and socket joint

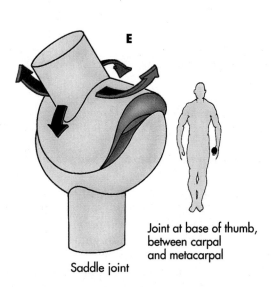

E

Joint at base of thumb, between carpal and metacarpal

Saddle joint

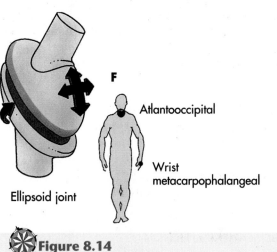

F

Atlantooccipital

Wrist metacarpophalangeal

Ellipsoid joint

✳ **Figure 8.14**

Synovial joints. **A,** Gliding. **B,** Hinge. **C,** Pivot. **D,** Ball-and-socket. **E,** Saddle. **F,** Ellipsoid.

GLIDING

In gliding joints (Figure 8.14A), two smooth, articular surfaces slide on each other, usually (but not always) in a straight line. **Gliding joints** also are called **plane joints** and **arthrodial joints** because the surfaces are very nearly flat. Articular facets on the articular processes of vertebrae that allow the vertebral column to flex and extend and also to rotate are excellent examples of gliding joints. The temporomandibular joint allows the mandible to protract and retract because it is a gliding joint, and the clavicle glides on the manubrium when the shoulders elevate and retract. Intertarsal and intercarpal joints of the feet and hands are also gliding joints.

HINGE

The elbow is a **hinge joint,** or **ginglymus** (JIN-glih-mus; ginglym-, *G,* a hinge), that allows the forearm to flex and extend at the trochlea of the humerus and the trochlear notch of the ulna, as shown in Figure 8.14B. Hinge joints enable uniaxial motion. Hinge joints connect the phalanges of the fingers and toes, and the knee is the largest ginglymus in the body.

PIVOT

In **pivot joints** (Figure 8.14C) the longitudinal axis of one bone rotates on its partner. These uniaxial joints are also called **trochoid joints** (TROE-koyd; troch-, *G,* wheel). The atlas pivots around the dens of the axis when the head rotates, and the proximal head of the radius pivots on the capitulum of the humerus when the forearm pronates and supinates. The dens actually inserts, like an axle in a bearing, into a ring formed by the atlas and its transverse atlantal ligament. In the humeroradial joint, however, the round concave head of the radius is molded around the convex capitulum.

BALL-AND-SOCKET

Ball-and-socket joints are triaxial joints that perform all four basic motions—flexion and extension, abduction and adduction, circumduction, and rotation. The hip and shoulder joints are prime examples of these articulations. All of these motions are possible because the spherical head (the ball) of a bone fits into a round socket (Figure 8.14D) in another bone.

SADDLE

The only **saddle joint** in the body lies in the thumb between the trapezium and the first metacarpal. Saddles (Figure 8.14E) are biaxial joints that allow all motions except rotation. Ball-and-socket joints rotate because the ball curves in the same direction on all sides, but saddle joints curve down on two sides and upward at the ends, which prevents rotation.

ELLIPSOID

Ellipsoid joints accommodate flexion and extension, abduction and adduction, and circumduction (Figure 8.14F). They allow primarily biaxial motion. Ellipsoid joints, as well as the carpals and radius, connect your fingers to the metacarpals at your knuckles. Because the articulating surfaces are longer than they are wide, ellipsoid (or **condyloid**) joints more readily move in one direction at a time and are not as effective as ball-and-sockets in circumduction.

- Name examples of gliding joints, a saddle joint, a joint that pivots, a ball-and-socket joint, and a hinge joint.
- What types of motion does a ball-and-socket joint permit?
- What type of motion can a hinge joint support?

ANATOMY OF SYNOVIAL JOINTS

VERTEBRAL FACETS

The simplest synovial joints are the gliding joints of vertebral facets. When the vertebral column flexes and rotates, **articular facets** on the articular processes of the vertebral arches slide forward, laterally, or superiorly, as shown in Figure 8.15. Most of the facets are oval and nearly flat. Narrow, loosely fitting articular capsules span the short distance between the articulating processes of the vertebrae, and a synovial membrane supplies synovial fluid.

Although these joints are simple, an elaborate system of accessory ligaments helps govern their movement (Figure 8.15C). The most interesting of these ligaments are the **ligamenta flava,** or yellow ligaments, that close the intervertebral foramina by connecting the lamina of the vertebral arch above each foramen with the lamina of the arch below. Elastic fibers give the ligamenta flava their yellow color and make them stretchy. Rarely found in ligaments, these elastic fibers help the vertebral column recover after flexing has stretched its ligaments. **Interspinous ligaments** connect neighboring spinous processes, and the **supraspinous ligament** runs like a cord along the tips of these spines.

TEMPOROMANDIBULAR JOINT

The lower jaw articulates with the skull at the temporomandibular joint. The head of the mandible in Figure 8.16 fits into the mandibular fossa of the temporal bone, just in front of the external acoustic meatus and the styloid process. This articulation is primarily a hinge, but it also glides. The head of the mandible slides forward onto the articular tubercle when the jaw protracts during eating and speech; in

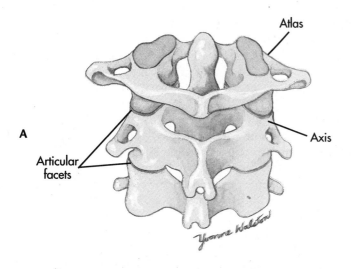

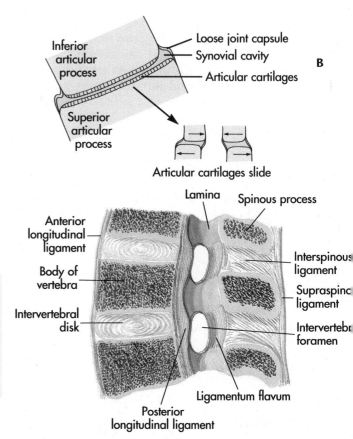

Figure 8.15

Articular facets of vertebrae. A, The facets of superior and inferior articular processes articulate between adjacent cervical vertebrae. **B,** Articular cartilages slide back and forth on each other, and the loose articular capsule allows this motion. **C,** Ligaments limit motion between vertebrae, shown in sagittal section through three lumbar vertebrae.

vigorous movement, however, the spindle-shaped head of the mandible rotates like a hinge to depress the mandible further. There is also enough freedom in the articular capsule for the mandible to grind sideways when one head and then the other glides forward and back.

Two joint cavities in the articular capsule enable these motions. The loose wall of the **upper cavity** permits the mandible to glide, and the **lower cavity** has tighter walls that permit rotation. The articular capsule attaches to the temporal bone at the margin of the mandibular fossa and the articular tubercle, reach-

ing downward around the head to the articular process of the mandible. The **articular disk** partitions the synovial cavity into an upper cavity beneath the mandibular fossa and a lower cavity over the head of the mandible. Concentric fibers support the articular disk, and synovial membranes line the upper and lower cavities.

Three ligaments and two jaw muscles in Figure 8.17 suspend the mandible as it moves. The broad **lateral ligament** connects the articular process of the mandible to the zygomatic arch, and a slender **stylo-mandibular ligament** joins the angle of the mandible

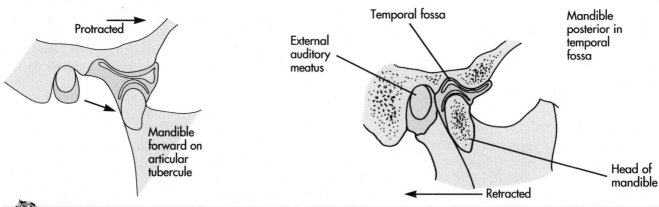

Figure 8.16

The temporomandibular joint allows the mandible to protract and retract. The mandible slides backward on the upper joint cavity into the temporal fossa and forward onto the articular tubercle. The lower cavity acts as a hinge for the head of the mandible.

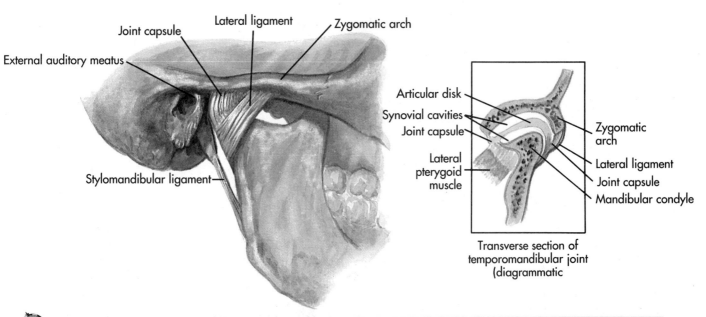

Articular disk
Synovial cavities
Joint capsule
Lateral pterygoid muscle
Zygomatic arch
Lateral ligament
Joint capsule
Mandibular condyle

Transverse section of temporomandibular joint (diagrammatic

Joint capsule
Lateral ligament
Zygomatic arch
External auditory meatus
Stylomandibular ligament

Figure 8.17
A ligamentous sling suspends the mandible from the temporomandibular joint. Right mandible, lateral view of lateral and stylomandibular ligaments.

to the styloid process, as seen in Figure 8.17A. Medially, the **sphenomandibular ligament** descends from the sphenoid bone to a point above the mandibular foramen. The masseter and medial pterygoid muscles complete this sling (described in Chapter 11, Axial Muscles).

- What anatomical type of synovial joint is the temporomandibular?
- Which two bones articulate in this joint?
- What kinds of motion does this joint permit?

ELBOW

The elbow is a complex hinge joint with three articulations—the **humeroulnar, humeroradial,** and **radioulnar.** The humeroulnar articulation is a hinge in which the trochlear notch of the ulna slides around the pulley-shaped trochlea of the humerus. In contrast the articulation between the radius and humerus is a gliding joint that allows the radius to slide upon the humerus while the joint flexes. The third articulation, the pivot joint between the proximal heads of the radius and ulna **(radioulnar joint),** aids rotation of the radius in pronation but does not add to the hinge-like action of the elbow. The ring-shaped **annular ligament** retains the circular end of the radius in a sleeve that allows the radius to rotate upon the ulna in pronation and supination.

All three articulations are considered part of the el-

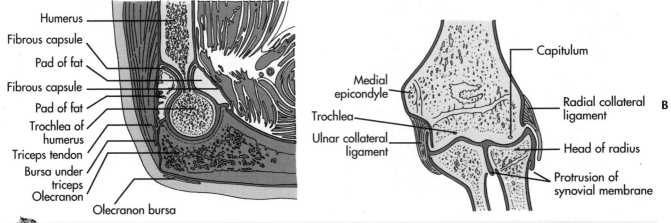

Humerus
Fibrous capsule
Pad of fat
Fibrous capsule
Pad of fat
Trochlea of humerus
Triceps tendon
Bursa under triceps
Olecranon
Olecranon bursa

Medial epicondyle
Trochlea
Ulnar collateral ligament
Capitulum
Radial collateral ligament B
Head of radius
Protrusion of synovial membrane

Figure 8.18
Elbow joint. A, Frontal section through the humeroulnar joint of the elbow shows the hinge-like articulation. B, The distal head of the humerus articulates with the radius and ulna.

bow joint because one articular capsule envelops them. Figure 8.18 shows that the capsule extends from above the radial and coronoid fossas of the humerus to the margin of the coronoid process of the ulna and to the annular ligament that surrounds the proximal head of the radius. The capsule reaches from the olecranon fossa of the humerus to the margins of the olecranon of the ulna. **Radial** and **ulnar collateral ligaments** strengthen the sides of the capsule (Figure 8.18B) between the epicondyles of the humerus and the annular ligament of the radius (permitting the radius to rotate) and the ulna. The fibers radiate from the epicondyles like spokes of a bicycle wheel, limiting extension and flexion. The anterior fibers are especially taut in extension; the posterior fibers tighten in flexion. The synovial membrane lines the articular capsule internally, following the fibrous capsule into pouches around the head of the radius and the trochlea of the humerus. Fat pads fill the olecranon, coronoid and radial fossas when they are not occupied by the ulna and radius.

KNEE

The knee is the body's largest joint. It is considered a hinge joint because of its motion, but its operation profoundly differs from that of a simple hinge (Figure 8.19). Two articulations, one between the patella and femur **(patellofemoral)** and the other between the femur and tibia **(tibiofemoral),** allow the

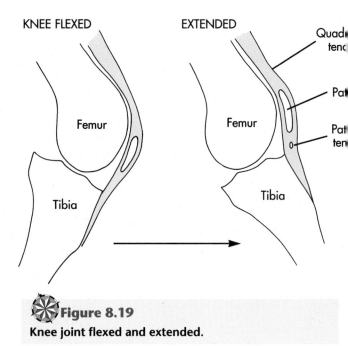

Figure 8.19

Knee joint flexed and extended.

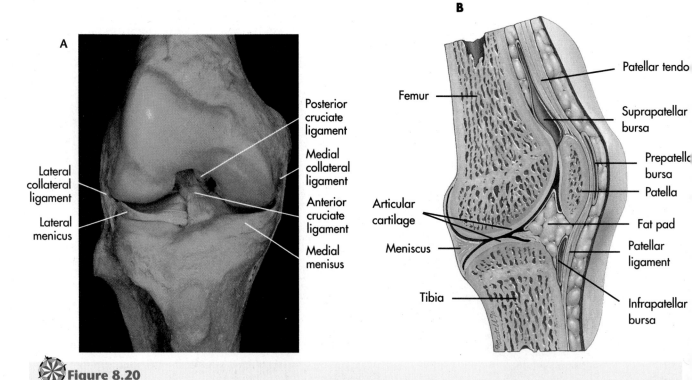

Figure 8.20

Knee. A, Two femoral condyles articulate on the tibial condyles. **B,** Sagittal section through the lateral femoral condyle of the left knee.

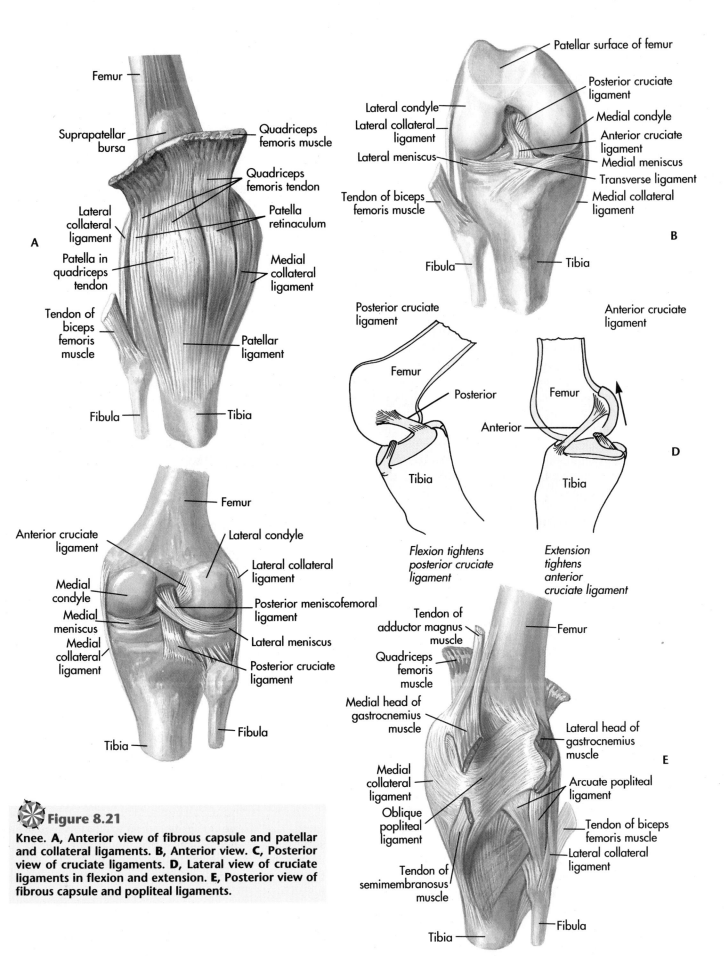

Figure 8.21

Knee. A, Anterior view of fibrous capsule and patellar and collateral ligaments. **B,** Anterior view. **C,** Posterior view of cruciate ligaments. **D,** Lateral view of cruciate ligaments in flexion and extension. **E,** Posterior view of fibrous capsule and popliteal ligaments.

leg to flex and extend. When the knee or **genu** (JEN-oo; L, the knee) is partially flexed, the tibia can rotate slightly, but when the knee extends, ligaments lock the joint rigidly in place. Its anatomy explains how the knee can be both flexible and rigid.

MOTION

Two large femoral condyles articulate with a pair of condyles on the tibia in Figure 8.20A. The medial condyles of both the femur and tibia are larger than the lateral ones. Articular cartilages cover all these articulating surfaces, but the condyles are never fully in contact with each other. The femoral condyles roll upon the tibial condyles. The femoral condyles curve gently posteriorly from the shaft, but the curves become progressively sharper until the condyles end in prominent knobs behind the axis of the femur. When the knee is flexed (Figure 8.19), the knobs contact the posterior portions of the tibial condyles, but in extension the broader part of each condyle rolls into contact with the anterior half of the tibial condyle. At the last moment of extension the medial condyle slides backward on the tibia as ligaments, tightening near the end of travel, pull the medial condyle of the tibia forward beneath the femur and slightly rotate the tibia laterally. Throughout these motions, the patella, embedded in the quadriceps tendon, slides on the broad patellar surface of the femoral condyles.

ARTICULAR CAPSULE

An articular capsule spans the knee joint. The fibrous layer of the articular capsule in Figure 8.20B is heavily modified by accessory ligaments and tendons from the muscles that activate the knee. Nevertheless, a synovial membrane envelops the joint on all sides, from the borders of the femoral condyles above to the tibial margins below. The **medial** and **lateral menisci** intrude from the capsular wall onto the tibial condyles, where they cushion the articulations by conforming to the femoral condyles while they roll and slide. Ligaments attach the menisci primarily to the tibia, but also to the femur. **Bursae** connect with the synovial cavity, and **fat pads** cushion the fibrous capsule.

LIGAMENTS

The patellar ligament extends the knee in Figure 8.19 and 8.21A. This ligament actually is a continuation of the quadriceps tendon, which extends from the patella to the tibial tuberosity. The quadriceps extends the tibia by pulling on this tendon, and the patella guides the tendon over the front of the knee.

A pair of **collateral ligaments** at the sides of the knee in Figure 8.21A acts like a hinge, allowing the knee to flex and extend, but also preventing lateral motion. Extension of the knee tightens the collateral ligaments and especially helps to lock the knee and to rotate the tibia. The **tibial (or medial) collateral ligament** connects the medial epicondyle of the femur with the medial epicondyle and shaft of the tibia. The **fibular (or lateral) collateral ligament** occupies a comparable position between the lateral epicondyle of the femur and the head of the fibula. In football, injuries to the tibial collateral ligament are more frequent than those to the fibular collateral ligament because players usually tackle and block their opponents from the side, which tends to tear the tibial collateral ligament.

The **cruciate ligaments,** in Figures 8.21B and C, limit how far the knee can extend and flex. They are so named because they cross each other. The **anterior cruciate ligament** arises from the anterior surface of the condylar eminence of the tibia and reaches in the intercondylar fossa to the posterior margin of the lateral femoral condyle. The **posterior cruciate ligament** arises posterior to the condylar eminence of the tibia and crosses forward on the medial side of the anterior cruciate ligament to the lateral margin of the medial femoral condyle. The synovial membrane follows the margins of the intercondylar fossa, so that the cruciate ligaments remain outside the synovial cavity. Neither ligament is entirely slack at any time, but, as Figure 8.21D shows, the posterior cruciate ligament tightens in flexion and the anterior cruciate ligament is especially stressed during extension. This tension is also responsible for rotating the tibia when the medial tibial condyle slides forward.

Two ligaments, the **oblique** and **arcuate popliteal ligaments** on the posterior side of the knee in Figure 8.21E, also help prevent the knee from hyperextending. Each is a strong, tough band that connects the posterior margin of the lateral femoral condyle to the tibia. Extension tightens them to their limits, and flexion releases them. On the anterior surface of the knee the **medial** and **lateral patellar retinacula** are ligamentous sheets of the fibrous capsule that complete the ring of ligaments around the articular capsule.

- What is another name for the knee?
- Which ligaments keep the femoral condyles from rolling forward during the knee's extension?
- Which ligaments prevent lateral motion of the knee?
- Which tendon extends the tibia?

MEDIAN ATLANTOAXIAL JOINT

The atlas and axis rotate upon each other at the median atlantoaxial joint, when the head turns from side to side. This joint is a pivot or trochoid joint in which the atlas pivots around the dens of the axis in Figure 8.22. Locate this joint on the skeleton in Figure 8.2. The **transverse atlantal ligament** secures the dens in a slight groove in its base in the forward portion of the cervical canal. Two small synovial capsules that slip between the dens and the atlas are the

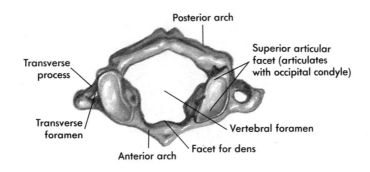

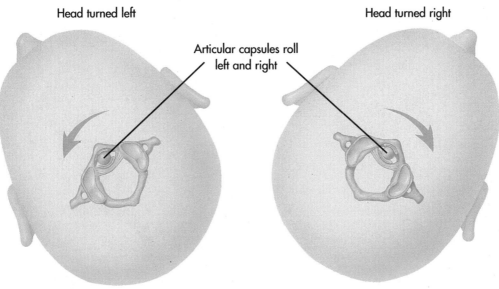

**Figure 8.22**

Median atlantoaxial joint. The dens of the axis turns in the atlas between two articular capsules that roll left and right when the head turns.

most elegant features of this joint. Both have a small synovial cavity that allows the atlas to pivot about the dens with very little friction. Figure 8.22 shows the position of these capsules when the head rotates left and right. A pair of lateral atlantoaxial articulations assists rotation on either side of the dens, but their curved facets are gliding joints.

HIP

The femur articulates with the hip in a ball-and-socket joint, illustrated in Figure 8.2. The articular surface of the femur shown in Figure 8.23A and B is shaped into two thirds of a sphere that fits snugly into the acetabulum and articulates there on the **lunate surface,** the articular surface of the acetabulum. The **acetabular notch** allows nutrient vessels to enter the joint and also admits the **ligament of the head of the**

femur (ligamentum teres), which attaches to a pit in the head. This ligament carries a small artery to the head of the femur, but the hip joint can function normally without it. A fibrocartilage lip of the acetabulum, called the **labrum,** retains the head of the femur within the acetabulum.

A fibrous capsule encircles the hip joint (Figure 8.24) with ligaments that spiral from the hip to the femur. The capsule arises from the hip bone, outside the labrum and acetabulum, and reaches the neck of the femur at the intertrochanteric line where the neck joins the shaft of the femur. The synovial membrane follows the inner capsular wall, folding around the ligament of the head of the femur and excluding this ligament from the synovial cavity in the same way that the cruciate ligaments remain outside the synovial capsule of the knee. Three ligaments make the fi-

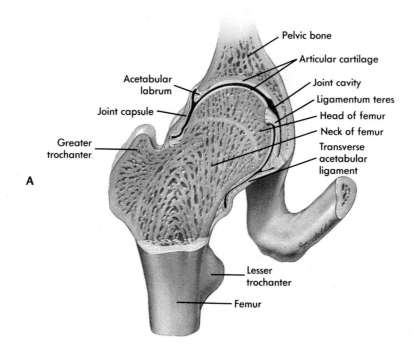

A

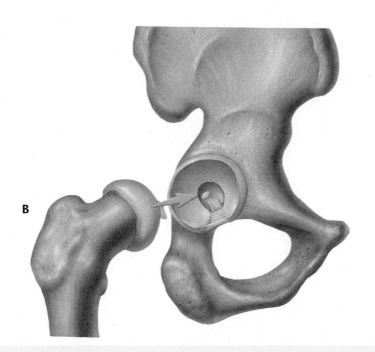

B

⬢ **Figure 8.23**

Hip joint. **A,** Frontal section through the hip joint shows the head of the femur inserted into the acetabulum of the pelvis. **B,** The articular surfaces of the acetabulum receive the head of the femur.

brous capsule of the hip especially thick and strong and limit how far the limb extends posteriorly. Named by the bones they connect, these three structures are the **iliofemoral, pubofemoral,** and **ischiofemoral ligaments.** These ligaments prevent the thigh from extending much more than 15 degrees beyond the body axis by winding tightly around the joint and locking the head and lunate surfaces to-

gether (Figure 8.24). When the thigh flexes, the ligaments unwind and the capsule becomes slack. Consequently the thigh abducts more readily when it is partially flexed.

SHOULDER

The shoulder joint is a ball-and-socket joint capable, like the hip, of extension, abduction, circumduction, and rotation. The shoulder joint is more mobile than the hip because it relies on muscles and tendons more than on ligaments between the humerus and scapula for stability. Figure 8.25A shows how the head of the humerus articulates with the glenoid fossa, which is not nearly deep enough to capture the humerus as firmly as the acetabulum receives the head of the femur. The **rotator cuff** makes up this deficit with tendons that accomplish the same function. Four broad tendons of the **subscapularis, supraspinatus, infraspinatus,** and **teres major muscles** (described in Chapter 11) form a flexible cuff (Figure 8.25B) that augments the glenoid fossa and nearly envelops the entire head of the femur. Contractions of these muscles retain the humerus within the glenoid cavity.

Inside the rotator cuff, the fibrous capsule attaches to the margin of the glenoid fossa of the scapula (Figure 8.25C) and extends to the anatomical neck of the humerus. The fibers of the capsule do not spiral about the joint and limit motion as they do in the hip. The synovial membrane lines the interior of the fibrous capsule. The tendon of the biceps muscle enters the capsule and rides in the intertubercular groove over the superior surface of the head of the humerus to an attachment on the superior rim of the glenoid fossa, further stabilizing the shoulder joint. A fold of syn-

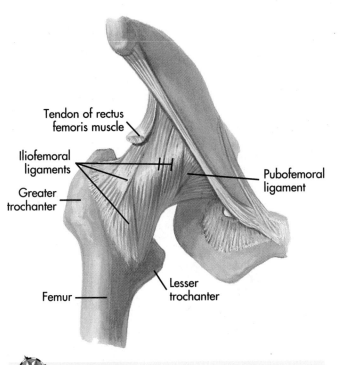

Figure 8.24

Fibrous capsule and accessory ligaments of the hip joint, anterior view. Extension of the thigh winds the ligaments tightly, and flexion slackens them.

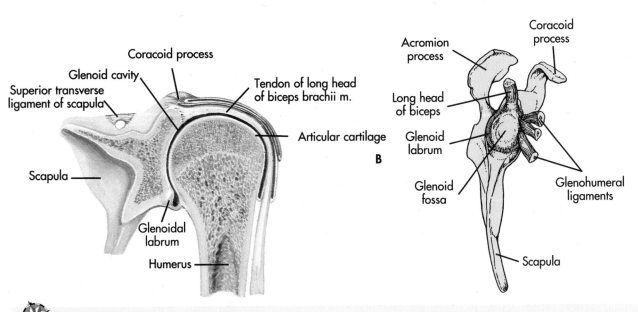

Figure 8.25

Shoulder joint. A, Frontal section through right shoulder joint, looking forward. The head of the humerus fits into the shallow glenoid fossa of the scapula. B, The articular capsule has been opened, and the humerus has been removed to show the glenoid fossa and the rotator cuff ligaments. *Continued.*

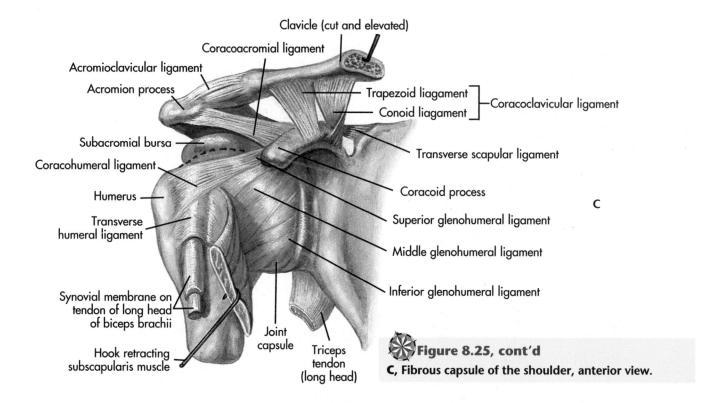

Clavicle (cut and elevated)
Coracoacromial ligament
Acromioclavicular ligament
Acromion process
Trapezoid liagament
Conoid liagament
Coracoclavicular ligament
Subacromial bursa
Transverse scapular ligament
Coracohumeral ligament
Humerus
Coracoid process
Transverse humeral ligament
Superior glenohumeral ligament
Middle glenohumeral ligament
Synovial membrane on tendon of long head of biceps brachii
Inferior glenohumeral ligament
Hook retracting subscapularis muscle
Joint capsule
Triceps tendon (long head)

C

Figure 8.25, cont'd
C, Fibrous capsule of the shoulder, anterior view.

ovial membrane forms a sheath around this tendon, and the membrane also opens to the **subscapular bursa** between the scapula and thorax. **Glenohumeral, coracohumeral,** and **coracoacromial** ligaments supplement the fibrous capsule and rotator cuff, further supporting the shoulder joint.

- What type of synovial joints are those of the hip and shoulder?
- What motions do they promote?
- What characteristic structure of these joints allows this motion?

CARPOMETACARPAL JOINT OF THE THUMB

The only saddle joint in the body connects the first metacarpal with the trapezium of the wrist (see Figure 8.2). The carpometacarpal joint enables the famous opposing motion of the thumb. Its saddle-shaped articular surfaces are complementary to each other and face medially and dorsally, with respect to the hand. As a result, the thumb flexes in a plane that parallels the palm, and it abducts dorsally, relative to the hand. Try these movements yourself. If you flex the thumb slowly, you will notice that it also rotates toward the palm.

WRIST

The wrist or radiocarpal joint is an ellipsoid joint that is capable of extension and flexion, abduction and adduction, and circumduction. Find it on the skeleton shown in Figure 8.2 and in Table 8.1. The scaphoid and lunate bones of the wrist articulate in an oval fossa formed by the distal end of the radius and the articular disk, which intervenes between the ulna and the carpals (Figure 8.26A). A fibrous capsule envelops this joint and connects with the respective bones at the margins of their articulating surfaces. The articular disk resembles that of the temporomandibular joint but merely extends the articular surface of the radius toward the triquetrum, rather than separating the radius entirely from the carpals. The ulna does not participate in the wrist joint. Figure 8.26 shows the anatomy of this situation in frontal section (A), as well as the movements of the bones during adduction and abduction (B). The carpals slide readily on the radius, with the scaphoid and lunate bones in contact during abduction and all three carpals in contact at adduction.

Five accessory ligaments strengthen the articular capsule. They are the **palmar and dorsal radiocarpal ligaments,** the **palmar ulnocarpal ligament,** and the **radial and ulnar collateral ligaments.** Figure 8.26 illustrates the collateral ligaments, and you should be able to deduce where the others are located from their names.

MOTION AT THE WRIST

Although moving the wrist seems simple enough, naming these motions is confusing because the forearm can rotate the hand away from its anatomical position of reference. In the anatomical position, abduction draws the hand away from the body, and adduction draws the hand toward the body. This is simple enough, but when the hand pronates, that former ab-

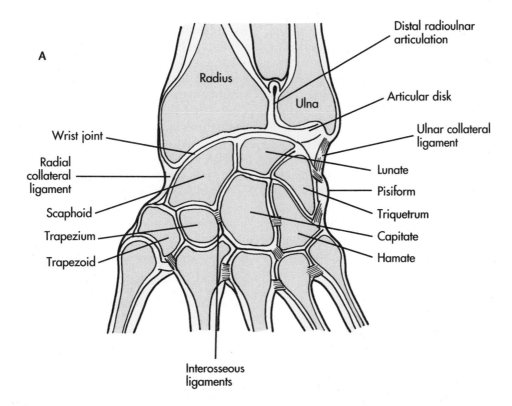

A

Radius

Ulna

Distal radioulnar articulation

Articular disk

Ulnar collateral ligament

Wrist joint

Radial collateral ligament

Scaphoid

Trapezium

Trapezoid

Lunate

Pisiform

Triquetrum

Capitate

Hamate

Interosseous ligaments

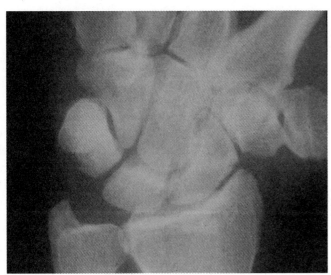

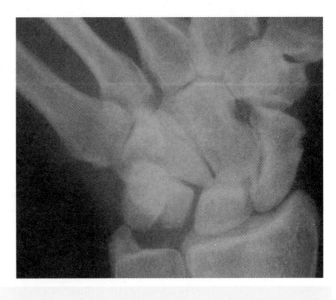

B

Figure 8.26

Wrist. **A**, Frontal section through the wrist joint and carpals. **B**, X-ray views of the palmar surface of the hand and wrist in radial and ulnar deviation.

duction movement will draw the hand toward the body when the hand faces posteriorly.

Anatomists and clinicians name these motions from two perspectives. In the first, any motion of the wrist away from the body is abduction regardless of which direction the palm is facing, and similarly, adduction draws the hand toward the body. However, *ulnar deviation* and *radial deviation,* terms based on the second point of view, are preferred when, for example, a physical therapist treating a sprained wrist is more interested in the relation of the hand to the forearm. In radial deviation, the wrist bends toward the radius at the base of the thumb. The opposite motion is ulnar deviation, and the wrist bends toward the ulnar side. These terms apply whether the hand is pronated or supinated; all you need to remember is on which side of the hand the radius and ulna articulate. Ask your instructor which terms to use.

METACARPOPHALANGEAL JOINTS

The joints between the proximal phalanges and the metacarpals of the thumb and palm are ellipsoid joints capable of circumduction and of flexion

and abduction (Figure 8.27). You can demonstrate these motions best with your fingers extended (flexing them limits abduction and adduction). Clench your fist to show that the four metacarpophalangeal joints of the palm form especially prominent knuckles. Within the joints (Figure 8.27A), shallow, concave proximal bases of the proximal phalanges fit around oval, convex distal heads of the metacarpals. Each of these latter surfaces forms a pair of small condyles on the palmar surface of the metacarpals. A loose articular capsule, collateral and palmar ligaments (Figure 8.27B), and the flexor and extensor tendons of each digit keep the metacarpophalangeal joints in position.

- Which bones articulate at the wrist and the thumb?
- What ligament forces the thumb to rotate when it flexes?

WHAT HOLDS SYNOVIAL JOINTS TOGETHER?

Several forces and structures allow joints to move without coming apart (see Sprains and Dislo-

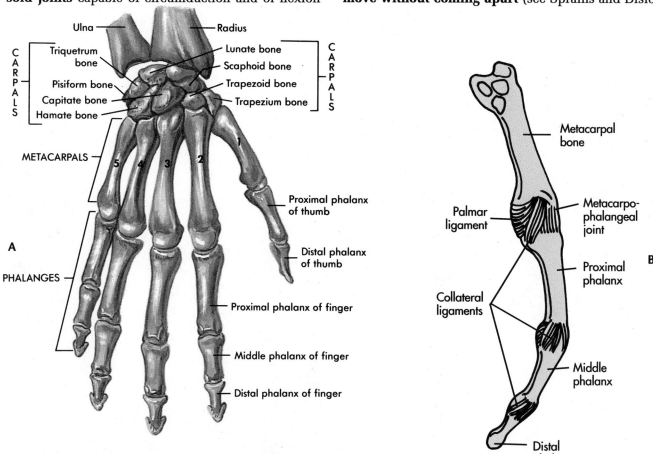

🏵 Figure 8.27

A, Five metacarpophalangeal joints connect the proximal phalanges with the metacarpals. **B,** A medial view of the middle phalanx shows the collateral and palmar ligaments.

cations). Obviously, gravity, joint capsules, ligaments, tendons of contracting muscles, and the complementary shapes of articulating surfaces help maintain contact between the bones at rest and those in motion. There is another force to consider, however; atmospheric pressure presses the bones and walls of the capsule from all directions upon the synovial fluid in the joint cavity. Arthritis sufferers know this force brings pain to sensitive joints when the weather is about to change. Atmospheric pressure is also responsible for snapping knuckles and knees when sudden motion stretches the joint capsules. Carbon dioxide bubbles form in the synovial fluid because pressure has momentarily dropped; these bubbles collapse with a snap and the gas redissolves when normal pressure returns.

BURSAE AND SYNOVIAL TENDON SHEATHS

Friction always accompanies movement, and bursae and synovial tendon sheaths relieve this source of wear and tear. These structures may be ex-

tensions of synovial cavities from nearby joints or entirely separate cavities filled with a small amount of synovial-like fluid, shown in Figure 8.28 and also Figure 8.25C. The effect resembles a water-filled balloon, whose flexible walls can move freely around the watery cushion. Figure 8.29 locates bursae and synovial tendon sheaths on the body, and Table 8.3 gives examples of them.

Bursae (burs, *L,* a purse) are located at points of wear between the skin and ligaments or bony prominences **(subcutaneous bursae),** as well as deep beneath muscles **(submuscular bursae)** and tendons **(subtendinous bursae). The prepatellar bursa** of the knee is a subcutaneous bursa that relieves friction between the skin and the patella. Beneath the gluteus maximus muscle at the hip, the **deep trochanteric bursa,** a submuscular bursa, protects the gluteus maximus where it passes over the greater trochanter. Subtendinous bursae occur at major joints, such as the shoulder, where the bursae of the teres major, latissimus dorsi, and pectoralis major muscles of the shoulder are associated with the insertions of the tendons from these muscles on the humerus.

Synovial tendon sheaths are tubular bursae that envelop tendons in a tunnel filled with synovial fluid

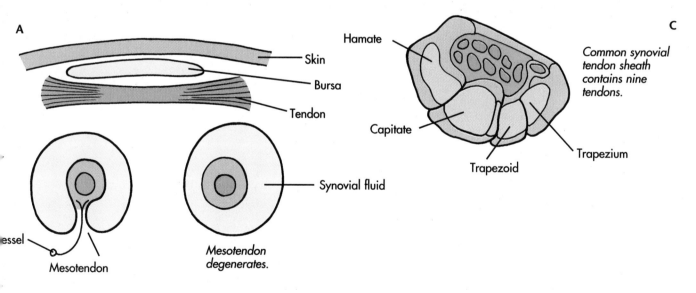

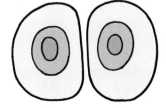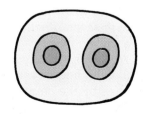

A

Skin
Bursa
Tendon
Synovial fluid

Mesotendon
Mesotendon degenerates.

essel

B

Parietal walls of adjacent tendon sheaths fuse,

and common sheath encloses two tendons.

Hamate
Common synovial tendon sheath contains nine tendons.
Capitate
Trapezoid
Trapezium

C

 Figure 8.28

Bursae and synovial tendon sheaths. A, Synovial tendon sheaths are fluid-filled tubes that wrap tendons. Bursae are synovial sacs that reduce friction where muscles and ligaments would otherwise rub skin and bones. **B,** Some tendon sheaths lose their mesotendon. **C,** Common synovial tendon sheaths, such as in the carpal tunnel, contain several tendons whose sheaths have fused together.

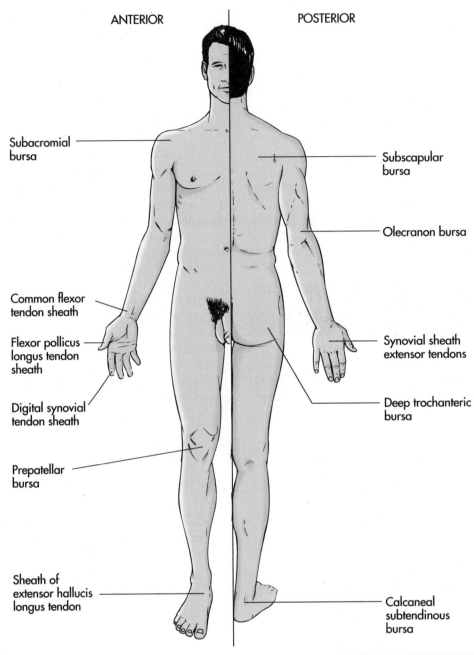

ANTERIOR   POSTERIOR

Subacromial bursa

Subscapular bursa

Olecranon bursa

Common flexor tendon sheath

Flexor pollicus longus tendon sheath

Synovial sheath extensor tendons

Digital synovial tendon sheath

Deep trochanteric bursa

Prepatellar bursa

Sheath of extensor hallucis longus tendon

Calcaneal subtendinous bursa

Figure 8.29

Locations of bursae and synovial tendon sheaths in the body. Find these structures in Table 8.3.

(Figure 8.28A). When the tendon moves, the inner sleeve of synovial membrane covering the tendon glides readily back and forth within the outer sleeve. A small mesentery, or **mesotendon,** connects the inner (visceral) and outer (parietal) sheaths and supplies nerves and vessels to the tendon (Figure 8.28B). Some mesotendons degenerate; when this occurs, fluid entirely surrounds the inner tendon sheath, now a tube-within-a-tube. Neighboring tendon sheaths also may fuse, creating a large synovial tendon sheath

with several tendons inside. All of these events have occurred in the carpal tunnel of the wrist (see Chapter 7), which contains the tendons of the muscles that flex the fingers and thumb. These tendons are enveloped by the **common flexor sheath** in Figure 8.28C, in one fluid-filled cavity, so that when you flex and extend your own fingers, these tendons slip easily beside each other.

Constant use or infection can inflame the synovial lining of bursae and tendon sheaths, conditions

Table 8.3   BURSAE AND TENDON SHEATHS

Structure	Location	Function	Type
Subscapular bursa	Shoulder, beneath tendon of the subscapularis muscle	Protects the subscapularis tendon from shoulder joint capsule	Subtendinous
Subacromial bursa	Shoulder, between acromion and tendon of supraspinatus muscle of scapula	Facilitates motion of suprapinatus muscle beneath the acromion	Subcutaneous
Olecranon bursa	Elbow, between olecranon and skin	Allows skin to slide over ulna forward of the olecranon	Subcutaneous
Common flexor tendon sheath	Palmar surface, wrist and hand	Facilitates flexion of wrist	—
Digital synovial tendon	Palmar surface, digits	Facilitates flexion of digits	—
Synovial sheath of flexor pollicus longus	Palmar surface, thumb	Lubricates flexor tendon of thumb	—
Synovial sheath of extensor tendons	Dorsal surface, hand	Lubricates extensor tendons of hand and digits	—
Deep trochanteric bursa	Hip, between gluteus maximus muscle and greater trochanter	Facilitates movement of gluteus maximus on greater trochanter	Submuscular
Prepatellar bursa	Knee, anterior surface of patella	Protects anterior surface of patella	Subcutaneous

known as **bursitis** and **tenosynovitis. Tennis elbow** and **housemaid's knee** are inflammations of the humeroradial bursa of the elbow and the prepatellar bursa of the knee, respectively. The synovial capsule swells with excess synovial fluid, and it may thicken with fibrous deposits from cells proliferating in the synovial membrane. Both conditions usually subside when the joint is immobilized and rested. Anti-inflammatory steroids may be recommended.

Carpal tunnel syndrome is a consequence of tenosynovitis and other conditions in the common flexor sheath of the wrist. Overuse may inflame the sheath and cause it to press on the median nerve, which also passes through the carpal tunnel (but outside the flexor sheath) to the hand. Numbness in the medial three fingers and burning pain are the usual symptoms of carpal tunnel syndrome. Treatment attempts to remove the source of irritation, and immobilizing the wrist or corticosteroid injections frequently restore normal use.

CLINICAL MATTERS

INFLAMMATION

Joints are subject to many forms of inflammation because they are composed of connective tissue. Synovial joints are not the only targets. In **ankylosing spondylitis** (ankyl-, *G*, crooked, bent; spondyl-, *G*, vertebra), inflammation of the cartilaginous joints of the vertebral column and pelvis causes pain and stiffness in the back, chest, and around the shoulders and pelvis. Because it fuses adjacent vertebrae together,

this condition severely reduces motion of the trunk. **Syndesmophytes,** the cartilaginous form of osteophytes, accumulate in the joint cartilages and can lead to ossification of ligaments and fusion of the vertebrae in a crooked vertebral column. **Ankylosis** (ANK-ee-LOW-sis) is the immobilization of joints by fibrous or bone tissue. These tissues usually deform the joint as they pathologically replace its cartilages and ligaments.

Rheumatoid arthritis is considered to be an autoimmune disease of synovial joints. Characterized by **rheumatoid nodules** (ROO-ma-toyd; rheum-, *G*, watery flow) that surround the affected joints (Figure 8.30), this disease usually begins with an acute arthritic attack in early adulthood and subsequently lingers in the smaller joints, especially those of the hands and feet. It appears three times more frequently in women than men. The synovial membrane becomes inflamed and the joint swells. Lymphocytes and macrophages collect in the lining of the synovial cavity, an apparent immune reaction to changes in the membrane. Granular tissue intrudes from the margin of the articular cartilages, spreads over the surfaces of the cartilages as a mass called **pannus** (pann-, *L*, rags), and eventually replaces the cartilages and immobilizes the joints. Without movement, local muscles atrophy and bone is reabsorbed, leaving the skin stretched smooth and tight over the nodular joints. Treatment attempts to retain joint mobility with exercise and to reduce inflammation and pain with analgesics and corticosteroids, if necessary. Joint replacement (see Surgical Procedures) may be recommended if these lesser treatments fail.

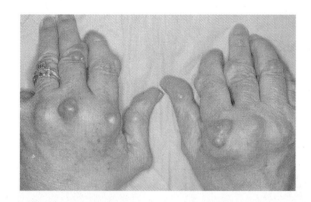

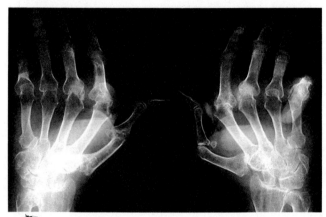

Figure 8.30
Hands of a rheumatoid arthritis patient.

Gouty arthritis is associated with deposition of uric acid crystals in the synovial cavities, especially of the fingers and toes. Uric acid also crystallizes in the kidney and cartilages. Uric acid is an end product of purine and nucleic acid metabolism; it is normally present in lower quantities and excreted by the kidney. Why uric acid accumulates is unknown, but it surely reaches the synovial fluid by way of the blood and synovial membrane, where inflammatory responses encapsulate the crystals in fibrous nodules called **tophi** (TOE-fee; toph-, *L*, a porous stone). Fibroblasts and lymphocytes accumulate around crystal deposits, forming nodules that press painfully on the nerve endings of the articular capsule and obstruct blood flow. Uric acid crystals also erode the articular cartilages, which leads to severe osteoarthritis and may immobilize the joints. Treatment attempts to reduce pain and inflammation and to lower blood uric acid levels.

SPRAINS AND DISLOCATIONS

An injury that sprains an ankle, wrist, neck, or any other joint stretches the articular capsule and accessory ligaments enough to tear them slightly. Sprains are frequent athletic injuries that usually result from rapid, unexpected twisting of the joint. A sprain may even rupture the joint cavity, but the bones do not separate. Local tissues may swell and vessels may hemorrhage, but the joint continues to function despite pain and stiffness. Elevating the joint, bandaging it, or packing it with ice minimizes swelling and tenderness, and immobilizing it eases the pain.

Severe sprains may tear a ligament or a meniscus entirely, as frequently happens to the knee. This type of injury compromises the synovial capsule and damages the articulation itself. Surgery may be needed to repair the ligament or to remove the meniscus. Special bandaging and bracing usually support the joint during recovery.

Dislocations disconnect the articulating surfaces from each other. Also known as **luxations** (luxa-, *L*, dislocate), these injuries usually disrupt the joint capsule. They may tear its accessory ligaments and also damage local nerves and circulation. The joint does not move normally and may appear deformed. Treatment rearticulates the bones surgically or manually, depending on the circumstances. The dislocated finger in Figure 8.31 was restored by pulling the tendons and ligaments until the joint snapped back into place. **Subluxations** only partially dislocate articulations, but treatment is similar to that for a full dislocation.

VALGUS AND VARUS

People with genu valgus are knock-kneed. Normally the lower limb supports body weight along a straight line (the mechanical axis) from the hip joint, through the knee, directly down the tibia to the foot, as illustrated in Figure 8.32. The anatomical axis of the femur forms an angle with this line that is usually 5 to 12 degrees in men and slightly more in women. Genu valgus increases this angle. The knees may touch each other during the stride and the legs may bend outward. Childhood cases are often associated with diet, and symptoms usually disappear by age 6. In older patients the angle can be improved with bracing or surgery to remove a wedge of bone proximal to the medial femoral condyle (see Osteotomy). **Hallux valgus,** bunion of the great toe, is comparable to genu valgus and responds to similar surgery.

Genu varus is the opposite deformity of the knee, bowlegs, in which the femur angles laterally and the leg medially. The angle between the mechanical and anatomical axes reverses. As in genu valgus, the condition is frequently dietary in young children and may require surgical treatment if severe.

SURGICAL PROCEDURES

Several kinds of surgery can treat pathological conditions of joints. Procedures range from relatively

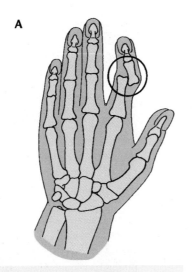

A

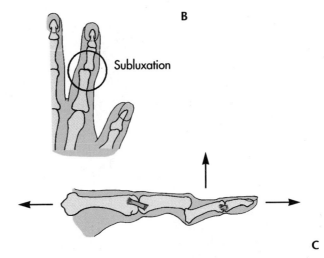

B

Subluxation

C

Joint realigned

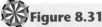

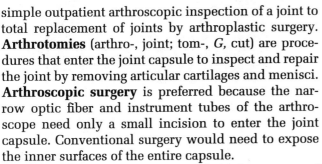

✴ Figure 8.31

Dislocations and subluxations of the first phalanx are corrected by stretching the muscles, tendons, and ligaments.

simple outpatient arthroscopic inspection of a joint to total replacement of joints by arthroplastic surgery. **Arthrotomies** (arthro-, joint; tom-, *G*, cut) are procedures that enter the joint capsule to inspect and repair the joint by removing articular cartilages and menisci. **Arthroscopic surgery** is preferred because the narrow optic fiber and instrument tubes of the arthroscope need only a small incision to enter the joint capsule. Conventional surgery would need to expose the inner surfaces of the entire capsule.

Arthroplasty (-plasty, *G*, molding) is a group of procedures that remodel and even replace a joint damaged by arthritis or trauma. Total hip and knee replacement arthroplasty removes the affected joint and its capsule and substitutes an implant that can restore painless and free motion for 10 to 15 years before replacement is needed again. Prosthetic joints attempt to reproduce the shape and motion of the original joint (Figure 8.32). Titanium or stainless steel femoral heads fit into polyethylene acetabular sockets in artificial hips, and plastic tibial condyles accommodate metal femoral condyles at the knee. Total hip replacement removes the affected parts and prepares the remaining bone to receive the new artificial acetabulum and femoral head. These new parts are cemented into spaces reamed into the bone, but more recently, porous surfaces on the metal pieces allow the bone to grow into the prosthesis for a truer biological union. The principal complications in total hip replacement are loosening of the implant with age and use, as well as deep infection of the operation site, a feared condition sometimes difficult to treat. Some circumstances of joint disease lead surgeons to recommend **arthrodesis,** or fusion of a joint (desis, *G*, binding), rather than replacement, especially in younger patients who

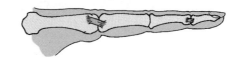

5°   Anatomical axis   12°   Mechanical axis

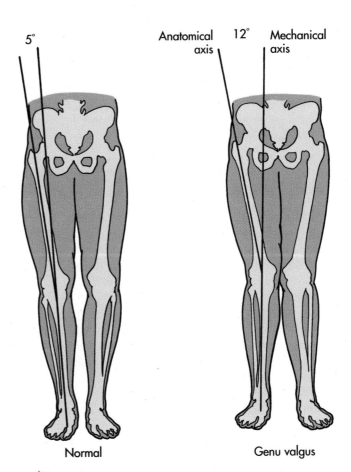

Normal

Genu valgus

✴ Figure 8.32

Genu valgus, or knock knees.

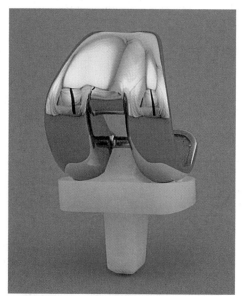

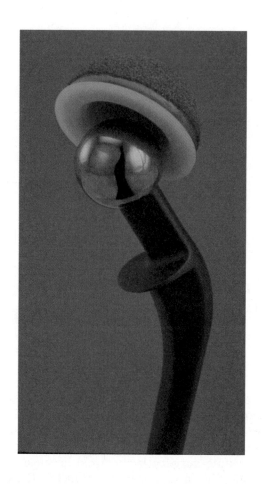

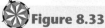

Figure 8.33

Knee and hip prostheses attempt to duplicate the natural joints.

might otherwise require several renewals of an artificial implant. **Osteotomy,** cutting the bone outside the joint capsule, is useful in cases of knock knees, bowlegs, and other conditions.

- What type of inflammation is responsible for rheumatoid arthritis?
- What damage occurs to a mildly sprained joint?
- Which surgical procedure implants artificial replacements into hips and knees?

SUMMARY

Joints are the junctions between bones. Classified by anatomy, joints fall into fibrous, cartilaginous, or synovial categories. By motion, they are grouped as synarthroses, amphiarthroses, and diarthroses. Joints may be named for the bones they connect. *(Page 205.)*

Sutures are fibrous, synarthrotic joints of the skull. Collagen fibers rigidly join the bones of the cranial cavity and face. Some sutures have rough edges (serrated), others are relatively flat (plane), and still others are beveled in several ways (squamous and schindylesis). *(Page 208.)*

Gomphoses are fibrous, synarthrotic joints between the roots of teeth and the jaw bones. A fibrous periodontal ligament locks each tooth into its alveolar socket in the mandible and maxilla. *(Page 209.)*

Syndesmoses are moderately movable fibrous joints. These joints are represented by the interosseous membranes that join the long bones of the forearm and leg. *(Page 211.)*

Synchondroses are cartilaginous synarthroses made of stiff hyaline cartilage that join epiphyses to the shafts of growing long bones. *(Page 211.)*

Symphyses, such as the intervertebral disks, are also cartilaginous, but are considered amphiarthroses because they allow some movement. Symphyses are composed of fibrocartilage with a gelatinous core. *(Page 214.)*

Synovial joints contain a fluid-filled synovial cavity, which classifies them as highly movable diarthrotic joints. A synovial capsule envelops these joints with an outer, fibrous layer of ligaments and an internal membrane that secretes synovial fluid. Bones themselves meet at articular cartilages made of hyaline cartilage. Menisci, bursae, and fat pads are accessory structures. *(Page 216.)*

Osteoarthritis is a degenerative disease of synovial joints related to the tendency of articular surfaces to regenerate poorly and to form bony deposits when blood vessels invade them. Weight reduction, mild exercise, and analgesic drugs treat the first stages of this disease, but prognosis is usually not optimistic. Surgery to remove the cartilages or replace the joint may be indicated eventually. *(Page 218.)*

Synovial joints permit a wide range of movements based on flexion and extension, abduction and adduction, circumduction, and rotation. Special motions of particular joints are pronation and supination of the forearm, eversion and inversion of the foot and hand, protraction and retraction of the mandible, and elevation and depression of the shoulders. *(Page 218.)*

Different types of synovial joints allow the motions listed above. They are (1) gliding joints, usually allowing linear motion, (2) hinge and pivot joints in uniaxial movement, (3) saddle and ellipsoid joints in biaxial movement, and (4) ball-and-socket joints in triaxial motion. The temporomandibular joint and vertebral facets are gliding joints. The elbow and knee are hinge joints, the atlantoaxial joint is a pivot, and the shoulder and hip are ball-and-socket joints. Finally, the carpometacarpal joint of the thumb is a saddle joint and the wrist is an ellipsoid joint with the radius. *(Page 221.)*

Bursae are connective tissue pockets of synovial fluid that reduce friction between tendons and ligaments and between ligaments and bone. Synovial tendon sheaths are tubular bursae that permit tendons to slide over neighboring surfaces. *(Page 235.)*

Inflammation and trauma are the sources of many joint disorders. In rheumatoid arthritis the patient's own immune system attacks the synovial membrane. In gouty arthritis, uric acid crystals inflame the joint cavity. Fibrous deposits accumulating around the irritation can immobilize joints and lead to their fusion. Trauma may sprain or dislocate joints. Arthroplasty, arthrodesis, and arthrotomy can correct these conditions. *(Page 237.)*

STUDY QUESTIONS

OBJECTIVES

1. What is the definition of the term *joint?* 1
2. Name five joints and find their locations in the skeleton. 2
3. What joints connect the bodies of vertebrae together in the vertebral column? To which structural and functional groups do they belong? 2, 3
4. To what categories of structure and function does the temporoparietal joint belong? 2, 3, 4
5. Where are sutures found in the skeleton? Where is the frontoparietal suture located? This suture has another name; find it in Chapter 6. How do serrated and squamous sutures differ? Give an example of each type. 5
6. Describe the different types of fibrous joints. What types of motion does each support? 3, 4, 6, 7

7. What is an amphiarthrosis? Give an example of a syndesmosis and a symphysis, and describe the differences between these two types of joints. 4, 7
8. What are the fundamental features of synovial joints? Name five synovial joints. 8
9. What types of damage does osteoarthritis cause in synovial joints? 9
10. Describe examples of the four basic motions of synovial joints. Perform these motions yourself. When you reach over to the TV set and turn it on, what motions have you performed? Name five of the joints involved. 2, 10, 11
11. Pronate your hand and then supinate it. What articulations have moved? Refer to Figures 8.5, 8.13A, 8.18, Table 8.1, and the discussion of this motion in the text. 11, 12
12. Evert and invert your foot. Protract and retract your mandible, and elevate and depress your shoulders. What joints are involved in these movements? Consult the text descriptions, Figure 8.13B, and Table 8.1. 10
13. What is the name of the motion performed by the elbow? Is it angular or linear, uniaxial or biaxial? 10, 12
14. Flex your knee, and then extend it until it is rigid. What are the following ligaments doing during extension: tibial collateral, patellar, anterior cruciate, and posterior cruciate? 10, 13
15. Abduct your lower limb and then return it to the anatomical position, then extend it as far as you can. What is happening to the ligaments of the hip joint in these two separate movements? 10, 13,
16. What is bursitis and tenosynovitis, and how do these conditions interfere with the bursae and synovial tendon sheaths? 14, 15
17. What damage do gouty and rheumatoid arthritis do to synovial joints? What damage occurs in moderate and severe sprains of a joint, such as the wrist? What is the difference between a dislocated and a subluxated joint? 15
18. What are the differences between arthrotomy, arthroplasty, and arthrodesis? 15

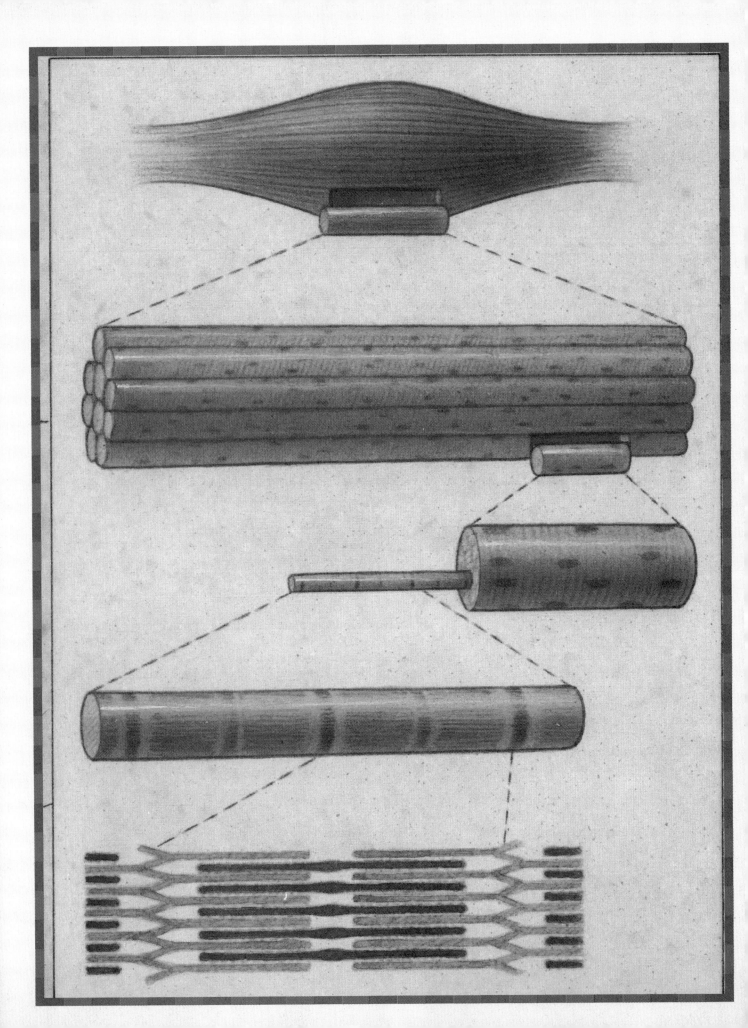

CHAPTER 9

Skeletal Muscle

OBJECTIVES

To understand the structure of skeletal muscle and how it works, do the following:

1. Describe how a whole muscle contracts gradually and smoothly despite the fact that its individual muscle fibers contract in all-or-none fashion.
2. Describe the actions of main mover, antagonist, synergist, and fixator skeletal muscles.
3. Select a muscle discussed in this chapter and describe its structure and movement.
4. Describe the hierarchical relationships between the entire muscle, its fascicles, and its muscle fibers.
5. Describe how neurons and blood vessels communicate with muscle fibers.
6. Describe the structure of a skeletal muscle fiber and its myofibrils and sarcomeres.
7. Describe the sequence of events in a muscle fiber that make it contract.
8. Explain what is meant by all-or-none contraction, motor endplates, motor units, and muscle tone.
9. Compare and contrast the properties of white, red, and intermediate muscle fibers.
10. Explain how muscle spindles measure contraction and how the stretch reflex uses this information.
11. Describe the structure of Golgi tendon organs and their role in the tendon reflex.
12. Describe how muscles produce and dissipate heat.
13. Describe the steps in development of muscle fibers.
14. Describe some effects of exercise and drugs on muscle.
15. Describe the effects of aging and disease on the structure and function of muscle.

VOCABULARY TOOLBOX

Muscle terms frequently use these stems:

myo- *G* a muscle (or a mouse). Myoblasts are cells that become muscle fibers, and myosin is one of the major contractile muscle proteins.

mys- *G* a muscle. Epimysium and endomysium are muscle connective tissues.

sarco- *G* flesh. Sarcomeres are the contractile organelles of muscle fibers, and a plasma membrane, the sarcolemma, covers these cells.

You will encounter the following positional terms again:

endo- *G* within, inner. Endomysium is the inner-most layer of muscle connective tissue.

epi- *G* upon, over. Epimysium is the exterior covering of muscles.

peri- *G* around. Perimysium surrounds bundles of muscle fibers.

RELATED TOPICS

- Active transport and exocytosis, Chapter 2
- Bones, tendons, Chapter 5
- Brain and motor coordination, Chapter 19
- Cells and Aging, Chapter 2
- Joints, Chapter 8
- Neurons, neural impulses, and action potentials, Chapter 17
- Skeletal muscles described, Chapters 10 and 11
- Spinal reflexes, Chapter 18

PERSPECTIVE

The skeletal muscle system moves the skeleton and holds it in position. Each muscle is an organ of this system, pulling but never pushing the tendons and bones attached to it. Coordinated by the brain and reflexes, muscles make the body walk, run, stand, and sit. Although motion may be slow or fast and strong or weak, the individual muscle fibers within each muscle contract fully and rapidly or not at all. The nervous system obtains smooth overall motion from these jerky sources by coordinating which fibers and muscles contract and when they do so. Various circumstances affect this performance. Exercise and drugs can enhance muscle strength, and disease or accident can interfere with contraction by depriving muscles of neural stimulation; other diseases directly attack muscle fibers and their contractile machinery. This chapter shows how these individual parts make the muscular system work and how changes in them can affect its motion.

In May, 1985, Anne strapped on a computer, attached a set of electrodes to her legs, and walked forward at commencement to receive her Master's degree from Wright State University. Her gait was awkward, but she was walking, which her paralyzed limbs could not have done alone. In the computer, a program was sending stimuli to her lower extremities to replace the communication between muscles and brain that an automobile collision had interrupted several years before.

Walking, like the action of almost all skeletal muscles, is voluntary in theme but involuntary in detail. You think, "Walk," and your brain automatically coordinates the smooth, orderly motion shown in Figure 9.1. You do not have to command the individual muscles to act. It is as if the brain specifies a plan of action, and monitors progress, adjusting as necessary to obtain the goal. Walking is a complex interplay of plan and progress that computer programs can partially replace.

The anatomy of muscles helps us understand what the computer program has simulated and how each part of the system contributes to its overall motion. Muscles pull the skeleton through its movements by orchestrating the small pulls of individual muscle fibers into smooth, purposeful motion. The action most of us take for granted is extraordinarily complex in detail but simple in overall principle.

Case study begins on p. 244 and p. 247.

Figure 9.1

A strobe light lets you see the movements of walking. The nervous system coordinates when and where muscles and muscle fibers contract in the limbs to obtain this smooth rhythm. Computer analyses of the motion you see here can help athletes improve their performance by disclosing inefficient movements.

MUSCLES—ORGANS OF MOTION

ANATOMICAL BASICS

Muscles are organs of the skeletal muscle system; they move the body, hold it in position, give it shape, and heat it. Altogether, nearly 600 skeletal muscles contribute 40% of the body's weight, slightly more in men and less in women. Figure 9.2 shows some of the superficial muscles that move the skeleton. To understand the impact of the skeletal muscle system on body form, consider how weight training dramatically enhances the body's musculature or how muscular dystrophy wastes it away. Run a few laps to see how working muscles heat the body. The nervous system coordinates muscular movement, the respiratory system supplies oxygen and removes carbon dioxide, and the cardiovascular system communicates these and other substances to working muscles and dissipates heat from them through the skin.

A massive muscle like the quadriceps femoris (four-headed muscle of the femur; *yellow* in Figure 9.2) on the anterior surface of the thigh has four large, central **bellies** that contract and pull **tendons** at the ends. These forces extend the leg because they operate on the anterior side of the thigh and leg. Think of the pull as originating at the more or less stationary ends of this muscle, called **origins,** where three bellies attach to the femur and a tendon attaches the fourth to the pelvis. **Insertions** attach the moving ends of muscles to bones or to other muscles. The quadriceps tendon collects the pulling forces from all four portions of the muscle, directs them around the patella, and inserts them, as the patellar tendon, on the anterior surface, rather than on the posterior surface of the tibia. (What motion would the quadriceps produce if, for some reason, the patellar tendon inserted on the posterior surface of the tibia?)

Muscles work in groups. Main movers such as the quadriceps are called **agonists** or **prime movers.** Stand up, raise your knee, and extend the leg to see this action. The contracting quadriceps extends the leg forward (Figure 9.3). The biceps femoris (two-headed muscle of the femur, *blue* in Figures 9.2 and 9.3) is an **antagonist** to the quadriceps because it produces the opposite motion when it flexes the leg by pulling from the posterior side of the thigh (Figure

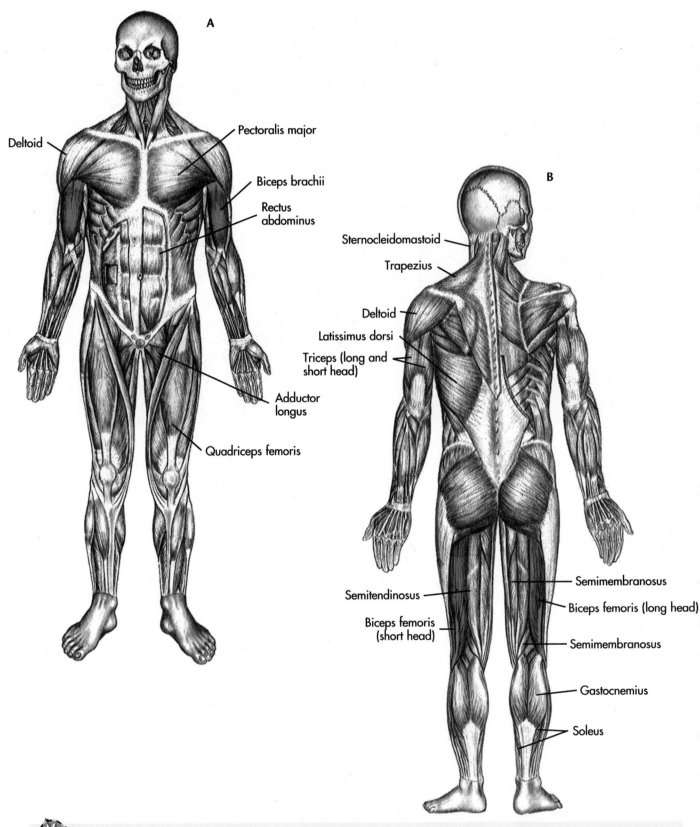

Figure 9.2
Superficial skeletal muscles seen from anterior and posterior. In the thigh, find the quadriceps femoris, adductor longus, and the biceps femoris *(yellow, green,* and *blue,* respectively). This chapter also discusses triceps *(yellow)* and biceps brachii *(blue)* muscles of the arm.

9.3). Flex your leg fully and feel the tension in the biceps. Together, quadriceps and biceps are an **antagonistic pair** because they pull the tibia in opposite directions, from their positions on opposite sides of the thigh. The quadriceps extends the leg when it contracts and the biceps femoris relaxes; then while the quadriceps relaxes, the biceps flexes the leg. (Chapter 11 describes these muscles in detail.) In walking, neither muscle relaxes fully or the limb would collapse under the weight of the body, nor do these muscles fully contract. Indeed, working skeletal muscles never reach full contraction or relaxation during normal movement.

Not all muscles are antagonistic pairs. Synergists produce similar motions, and their combined pull is stronger than that of either muscle alone. The vastus medialis (large medial muscle) and rectus femoris (upright muscle of the femur) are synergists (Figure 9.2). These muscles are actually parts of the quadriceps; they originate on the femur (vastus) and the ilium (rectus), and both insert upon the tibia. **Fixators** hold parts of the body in position while others move. The adductor longus muscle (long adductor; *green* in Figure 9.3) helps to flex your thigh while the quadriceps and biceps work; it also helps flex your thigh as you walk. It originates on the pubis and inserts onto the shaft of the femur. (Chapters 10 and 11

describe these and other kinds of muscular action in detail.)

The muscular system obtains a wide variety of motions from relatively few muscles by allotting several roles to a single muscle. It is misleading to call individual muscles only antagonists, synergists, or fixators because each muscle can assume many roles. Do not think that the rectus femoris is only an antagonist. It also pulls as a synergist and a fixator in a walking limb.

Still more economy and versatility arise from the fact that comparable muscles yield similar movements in other regions of the body. The triceps brachii and biceps brachii are antagonistic muscles that extend and flex the forearm and correspond to the quadriceps and biceps femoris, respectively. Figure 9.2 shows these muscles in *yellow* and *blue.*

A STRAINED MUSCLE

Something popped in the back of Dave's thigh as he followed the ball toward the fence. The sharp, warm pain nearly stopped him before he returned the ball to the infield where the hitter had already scored.

Pulled muscles **(strains),** like Dave's hamstring, are common athletic injuries that can lead to myositis (inflammation). Rapid treatment reduces the chance of pulling the muscle again, and it restores performance almost entirely. The most dangerous complication in this kind of injury occurs if osteoblasts from torn periosteum deposit bone in the damaged tissue, a condition called **myositis ossificans** (MY-ose-itis OSS-ih-fih-KANS; ossifying muscle irritation). The **RICE program** is the proper treatment for muscle strains. **Rest,** use and stretch the muscle as little as possible. **Ice** the injury immedi-

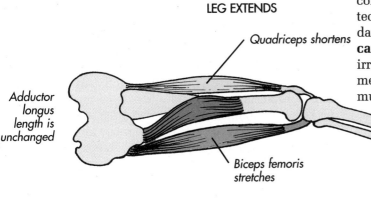

LEG EXTENDS

Quadriceps shortens

Adductor longus length is unchanged

Biceps femoris stretches

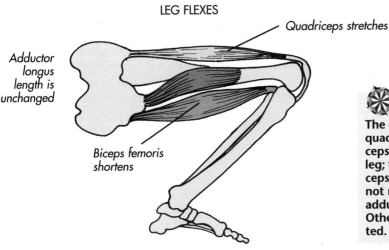

LEG FLEXES

Quadriceps stretches

Adductor longus length is unchanged

Biceps femoris shortens

⚹ **Figure 9.3**

The quadriceps and biceps femoris are antagonists. The quadriceps *(yellow)* has shortened and stretched the biceps to extend the leg. The opposite motion flexes the leg; the biceps *(blue)* contracts and stretches the quadriceps. The adductor longus *(green)* is a fixator and has not moved in the meantime. Origins of quadriceps and adductor have been moved to illustrate these muscles. Other muscles that assist this motion have been omitted.

ately to minimize hemorrhage and inflammation. **Compression** bandages reduce these problems over the long term. **Elevate** the injured limb to minimize stress on the healing tissue.

RICE had Dave playing again in a month. Muscle fibers had separated in his left biceps femoris, rupturing capillaries and damaging nerve endings in the local connective tissue. In **acute myositis** (avulsion) the tendons may tear from their insertions with excruciating pain, whereas simply overstretching the muscle without much tearing is a much milder form. RICE works by promoting conditions that repair muscle fibers and connective tissue.

- Skeletal muscle makes up approximately what percent of the body's weight?
- Name an example of a main mover muscle.
- What muscle is its antagonist?
- What is the function of fixator muscles?
- What is the condition called myositis ossificans?

STRUCTURE OF SKELETAL MUSCLES

An individual muscle may contain as many as **several million muscle fibers,** the contractile cells of skeletal muscle. Each fiber contracts in an **all-or-none** manner, that is, it contracts fully and rapidly or is fully relaxed. There is no state of partial contraction in an individual skeletal muscle fiber. If an entire muscle contracted in this manner, skeletons would move jerkily or not at all. The smooth rhythmic action of muscles in walking is accordingly a matter of signaling muscle fibers to contract at different times, adding each fiber's contribution to the others.

HIERARCHICAL ORGANIZATION

The organization of skeletal muscle is a hierarchy whose various ranks show how individual fibers contribute to overall motion. At the top of the organization are entire muscles, the rectus femoris in Figure 9.4A, for example; each is covered with a sleeve of connective tissue called the epimysium (Figure 9.4C) that attaches this muscle to the deep fascia of neighboring muscles or to the superficial fascia of the skin. **Deep fascia** (FAH-shya; fasci-, *L,* bundle) forms partitions, called **intermuscular septa,** between muscle groups. These partitions separate prime movers from antagonists so that they can move individually. The epimysium (ep-ih-MISS-ee-um) is continuous with the periosteum of the bones that receive the muscle. Vessels and nerves pass through the epimysium to nourish the muscle fibers within and to coordinate movement, as you will see in successive levels of the hierarchy.

Beneath the epimysium, muscle fibers are em-

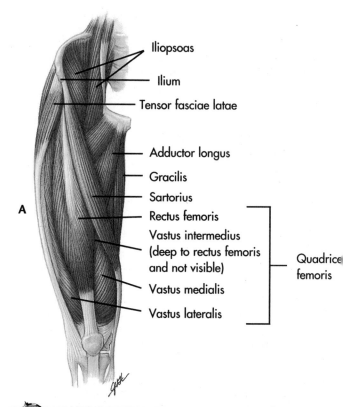

Labels: Iliopsoas, Ilium, Tensor fasciae latae, Adductor longus, Gracilis, Sartorius, Rectus femoris, Vastus intermedius (deep to rectus femoris and not visible), Vastus medialis, Vastus lateralis, Quadriceps femoris

A

Figure 9.4

Hierarchical organization of skeletal muscles. **A,** The rectus femoris, again.

bedded in a network of connective tissue. The next level of the hierarchy is represented by **fascicles** (Figures 9.4C), which are bundles of **muscle fibers,** the all-or-none cellular units of contraction. Large muscles may contain several thousand fascicles, and each fascicle may contain several hundred muscle fibers. Each fascicle is connected to its neighbors by more connective tissue, the **perimysium.** Perimysium forms **intramuscular septa** between fascicles. These septa distribute the pull of each fascicle to the tendons, while nerves and vessels form the epimysium branch through the septa toward individual muscle fibers within each fascicle. Inside the fascicles, the **endomysium** connects muscle fibers to each other, while it insulates them from the impulses that cause neighboring fibers to contract. This tissue is continuous with the perimysium, communicating capillary networks and nerves to each fiber and transferring contraction to the perimysium and thence to the tendons. Collagen fibers in all three "mysia" distribute the forces of contraction throughout the muscle, and elastic fibers help smooth its action by stretching somewhat when muscle fibers pull. The perimysium and endomysium tore somewhere in Dave's hamstring, causing capillaries to hemorrhage and producing local inflammation. He also tore muscle fibers loose from their connections with the endomysium.

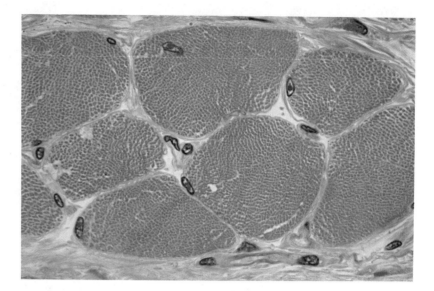

B

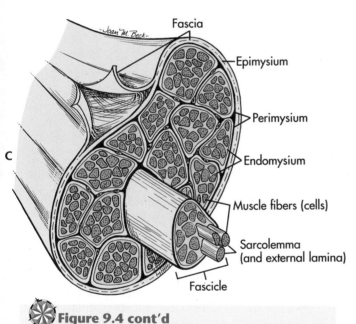

Figure 9.4 cont'd

B, Cross-sections through muscles reveal numerous muscle fibers embedded in connective tissue. **C,** Fiber-within-fascicle-within-muscle organization.

SKELETAL MUSCLE FIBERS

Skeletal muscle fibers are long, cylindrical cells that taper at their ends (see Figure 9.5A). Representing the third rank in the hierarchy, they are between 20 and 100 μm in diameter and range from a few millimeters to many centimeters in length. Some reach from one end of a muscle to the other. All display characteristic cross-striations (Figure 9.5B) caused by the contractile structures inside. Numerous nuclei lie beneath the **sarcolemma** (SAR-ko-LEM-ma; -lemma, *G,* sheath; flesh sheath), the plasma

membrane that covers these cells. A basement membrane binds the sarcolemma and the entire fiber to the surrounding endomysium. Collagen fibers from the basement membrane anchor the cell into the connective tissue or directly onto tendons and bone, as the case may be. The sarcolemma also distributes impulses triggered by nerve endings to the contractile machinery within the cell.

Inside the muscle fiber, the workings of the next hierarchical level consist of as many as several hundred long, contractile **myofibrils** (MY-oh-FYE-brils; Figure 9.5C) packed side-by-side within the **sarcoplasm,** the cytoplasm of this cell. **Mitochondria,** the **sarcoplasmic reticulum** (SAR-ko-PLAZ-mik reh-TIK-yoo-lum), and the **T-tubule system** occupy the space between myofibrils. Mitochondria are respiration centers that produce ATP for contraction. The sarcoplasmic reticulum is a network of membranes and internal spaces called **cisternae** (singular, cisterna; SIS-ter-ni; cist-, *G,* a box, chest) that cloaks virtually every myofibril in the cell and releases calcium ions, which are considered responsible for triggering the myofibrils to shorten. The T-tubule (transverse tubule) system delivers impulses for contraction from the sarcolemma to the sarcoplasmic reticulum. RICE treatment works at this level to help repair this highly organized structure when muscle fibers are damaged.

Myofibrils are about 1 to 2 μm diameter and may be as long as muscle fibers (Figure 9.5C). They exhibit repeating cross-striations (Figure 9.5B). The dark bands are called **A-bands,** and the light regions between are **I-bands,** names based on the underlying molecular organization when the fibers are viewed with polarized light in the microscope (isotropic, mostly random organization in I-bands; anisotropic, nonrandom organization in A-bands). Careful exami-

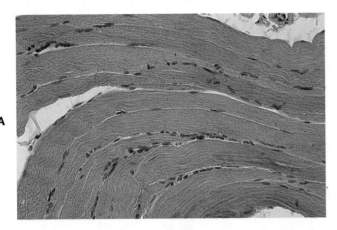

A   B

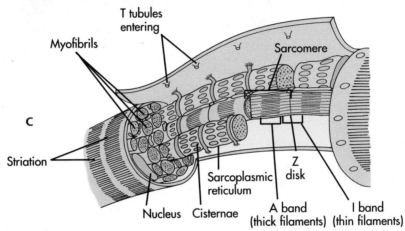

C

T tubules entering

Myofibrils

Sarcomere

Striation

Sarcoplasmic reticulum

Z disk

Nucleus   Cisternae

A band (thick filaments)   I band (thin filaments)

Figure 9.5

Skeletal muscle fibers. A, Muscle fibers. **B,** Characteristic cross-striations. **C,** A muscle fiber has been opened to show the fine structure of the sarcoplasm within. The sarcolemma covers the cell, and cross-striations of sarcomeres can be seen beneath it. T-tubules enter the sarcoplasm from the sarcolemma. The cut surface at left reveals numerous myofibrils and mitochondria, and just beneath the sarcolemma is a nucleus. All myofibrils are wrapped with sarcoplasmic reticulum that communicates to the sarcolemma by way of T-tubules. The sarcoplasmic reticulum has been removed from one myofibril to reveal its sarcomeres, and sections through several sarcomeres show thick and thin myofilaments and Z-disks.

nation of the figure also shows that **Z-disks** reside in the middle of each I-band.

Each myofibril is a chain of sarcomeres. Sarcomeres are the structural units of contraction in skeletal muscle fibers, and as the fifth level of structure, they are responsible for each muscle fiber's all-or-none contraction. An individual A-band marks the center of each sarcomere in Figure 9.6A, and the Z-disks identify the boundaries at each end of the sarcomere that join adjacent sarcomeres together. This arrangement means that neighboring sarcomeres share individual I-bands that span the distance between the A-bands of one sarcomere and the next.

The A-band contains **thick myofilaments** packed hexagonally, as close to as many others as possible (Figure 9.6A and B). The midpoints of these filaments are held in register at the **M-band.** The I-bands contain **thin myofilaments** that extend between the thick filaments of neighboring A-bands (Figure 9.6A and B), where numerous cross bridges hold them in position. You can see where thick and thin myofilaments overlap at each end of the A-band. The **H-zone** is the mid-

dle portion of the A-band between the ends of the thin filaments. The thin filaments from each sarcomere are not actually continuous, but end instead at alternating locations on opposite sides of the Z-disks. This situation gives Z-disks their characteristic zig-zag appearance. When muscle fibers contract, the I-bands shorten and the Z-disks and A-bands come closer to each other, but the A-bands themselves remain the same length.

The sarcoplasmic reticulum and T-tubules communicate to every sarcomere. The sarcoplasmic reticulum is a network of internal membranes that appears to trigger contraction of sarcomeres. Figure 9.6D shows that tubules of the sarcoplasmic reticulum drape every A- and I-band. These tubules join larger **terminal cisternae** at the junctions between these bands. The space within tubules and cisternae stores calcium ions that are considered to trigger contraction, as you will see later in this chapter. The sarcoplasmic reticulum also wraps mitochondria on the surface of the I-bands.

Terminal cisternae do not actually come into con-

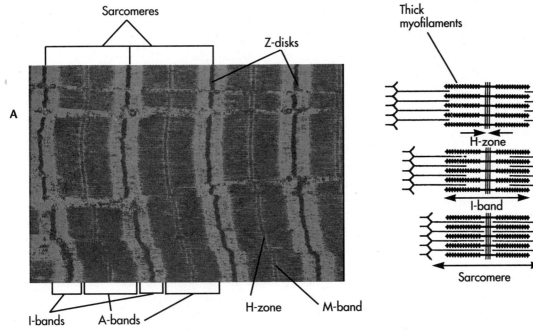

Sarcomeres

Z-disks

A

I-bands A-bands H-zone M-band

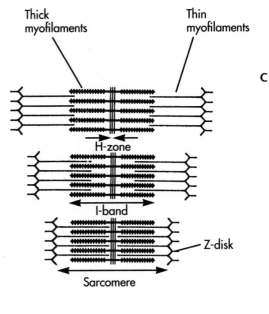

Thick myofilaments Thin myofilaments

C

H-zone

I-band

Z-disk

Sarcomere

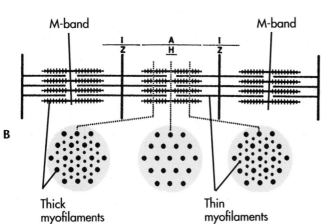

M-band M-band

B

Thick myofilaments Thin myofilaments

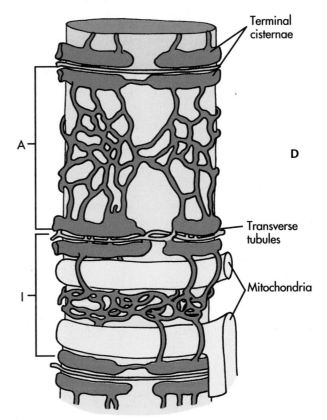

Terminal cisternae

A

D

I

Transverse tubules

Mitochondria

Figure 9.6

Sarcomeres. **A,** The electron micrograph shows thick and thin filaments of the A- and I-bands in a contracted muscle fiber. **B,** Thick and thin myofilaments. **C,** Thin filaments slide along the thick filaments during contraction and shorten the H-zone, the I-band, and the sarcomere, but the dimensions of A-bands do not change. The H-zone and I-band elongate when the sarcomere relaxes. **D,** The sarcoplasmic reticulum and the T-tubules wrap sarcomeres.

tact with each other at the junctions of the A- and I-bands but are interconnected by the T-tubules (**transverse tubule system,** Figure 9.6D). These tubules are extensions of the sarcolemma that reach from the cell surface throughout the sarcoplasm and end between the terminal cisternae where numerous "feet" connect them to the cisternae walls. The nervous system, and in Anne's case her computer program, triggers impulses for contraction to flow from the surface of the

muscle fibers, through the T-tubules, to the sarcoplasmic reticulum.

- What muscular structures does the perimysium connect?
- What are the dimensions of skeletal muscle fibers?
- What structural units cross-striate skeletal muscle fibers?
- Which type of myofilaments are in A- and I-bands?
- What ions does the sarcoplasmic reticulum store?

SLIDING FILAMENT HYPOTHESIS

Myofibrils and sarcomeres shorten when skeletal muscle fibers contract. In 1954, H. E. Huxley and his coworkers were the first to publish electron micrographs of skeletal muscle like that shown in Figure 9.6A. According to Huxley's **sliding filament hypothesis** of contraction, contractile forces slide thin myofilaments along the thick ones (Figure 9.6C) toward the center of the A-bands in every sarcomere and myofibril, thereby shortening the whole muscle fiber. The forces that slide the filaments make entire muscles contract and the skeleton move. Myofilaments return to their original positions when muscle fibers relax. Walking, whether coordinated normally or artificially by computer programs such as Anne's, ultimately controls the forces that make filaments slide and sarcomeres shorten.

Contraction arises from cross-links between overlapping thick and thin myofilaments (Figure 9.7). Thick filaments are bundles of **myosin** (MY-oh-sin) molecules that can cross-link to **actin** molecules of the thin filaments. Myosin molecules are shaped like golf clubs (Figure 9.7A), with long, narrow shafts and club-shaped heads. Actually, there are two heads located side by side. Packed together, the myosin molecules become thick myofilaments with the myosin heads pointing outward. Actin thin filaments (Figure 9.7B) assemble into helices of globular actin molecules, and elongated **tropomyosin** (TROE-poe-MY-o-sin; tropo-, *G*, change) molecules reside near the grooves of each helix, blocking sites to which myosin heads can bind. **Troponin** (TRO-poe-nin) molecules bind to tropomyosin at regular intervals along the helix. This organization represents the final, or molecular, level of muscle hierarchy in resting muscle fibers.

Contraction can begin when each myosin head actually does form a cross-link to an actin-binding site. The myosin head bends forward, thrusting the actin filament onward like a ratchet. Then the cross-link breaks, and the myosin head cocks backward toward another binding site. By repeatedly making and breaking these cross-links, thin filaments slide along the thick, shortening an entire muscle fiber in less than ¹⁄₂₀ of a second. To form cross-links, however, tropo-

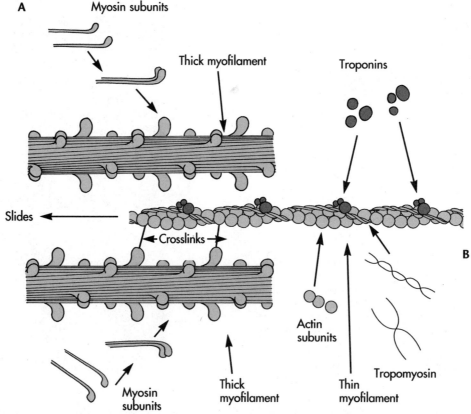

✳ Figure 9.7

Structure of thick and thin filaments. A, Pairs of myosin molecules pack together as thick myofilaments, with the myosin heads arranged spirally around the surface. **B,** Actin thin filaments are assembled from globular actin molecules, elongated tropomyosin, and small troponin units. The structure is a helix, and tropomyosin lies in (or near) the helical grooves associated with six or seven actin molecules. Troponin associates with tropomyosin at regular intervals. Cross-links form between myosin heads and actin-binding sites near the tropomyosins so that each filament binds with several others.

myosin must be displaced from the sites on actin where myosin heads can bind, and that event initiates contraction, as explained below.

ALL-OR-NONE CONTRACTION

Calcium ions released by the sarcoplasmic reticulum trigger contraction. In resting muscle fibers the sarcoplasmic reticulum actively transports Ca++ ions continuously from the myofibrils (see Transport Proteins, Chapter 2), and accumulates these ions in its cisternae. When a muscle fiber is stimulated, the sarcoplasmic reticulum releases this Ca++ through calcium ion channels (see Chapter 2) back into the myofibrils, where it binds with troponin and displaces tropomyosin from the binding sites. Active transport continues nevertheless, but with the ion channels open, calcium ions flood the myofibrils faster than the leaky cisternae. Now, with binding sites available, cross-linking proceeds, and the muscle fiber remains shortened as long as the muscle fiber is stimulated. When these stimuli cease, calcium ion channels in the sarcoplasmic reticulum close, Ca++ ions accumulate in the cisternae once again, tropomyosin blocks the binding sites, and the muscle fiber begins to relax. Antagonistic muscles will stretch the fibers to their original length.

In resting muscle fibers, cross-links from the previous contraction bind actin thin filaments to some myosin heads. To prepare for contraction, ATP molecules from local mitochondria bind to the myosin heads and break cross-links (Figure 9.8A). ATP then splits into ADP and phosphate, and the liberated energy cocks each myosin head backwards in its high-energy configuration toward another binding site. Cross-links form now, (Figure 9.8B) when Ca++ ions have unblocked the binding sites on the actin filaments. In this condition the myosin heads now return to their low-energy configuration (Figure 9.8C). The heads snap forward for the power stroke, releasing their stored energy and drawing the actin filament some 7 nm further along the thick filament. ADP and phosphate now dissociate from the myosin heads, allowing molecules of ATP to bind there again (Figure 9.8D) and a new cross-linking cycle to begin. These ATP molecules now split, cocking the myosin heads backward once again to catch the next actin binding sites (Figure 9.8E). As long as enough ATP energy and Ca++ ions are present to displace tropomyosin from the binding sites, this cross-linking cycle continues to slide actin filaments forward. A single myosin head can repeat this cycle up to 50 times a second.

If all cross-links were to break simultaneously while filaments slide, what would prevent muscle tension from pulling these filaments backward? One can argue that some filaments prevent this motion by forming cross-links, while other filaments break theirs, but it is difficult to see how entire filaments

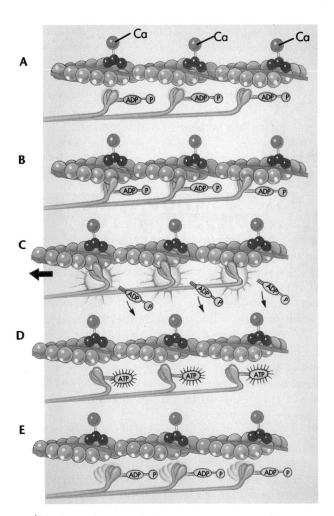

Figure 9.8

Cross-linking cycle. Calcium ions have already caused troponin and tropomyosin to unblock the binding sites on actin. **A,** ATP molecules bind to myosin heads and split into ADP and phosphate. The energy released cocks the heads into position for new cross-links to form. **B,** Cross-links now form between myosin heads and actin binding sites that have been opened by Ca++ ions. Removing Ca++ blocks this step, and contraction stops. **C,** Power stroke. Myosin heads now bend forward, advancing the actin-thin filament as the energy in the heads is converted into motion. ADP and phosphate dissociate from the heads. **D** and **E,** This cycle repeats as long as calcium ions are present and ATP molecules bind to the myosin heads.

could behave so differently. On the other hand, cross-linking may not occur simultaneously along a filament. But if some cross-links break while others form, how does a filament advance? Although we can see how Ca++ ions and ATP affect individual cross-links, we do not yet fully understand how cross-links work together in muscle contraction.

- Where in a sarcomere do thick and thin myofilaments form cross-links?
- What roles do troponin and tropomyosin have in muscular contraction?
- What events convert the heads of myosin molecules to the high-energy configuration?
- Do cross-links form between this form of myosin and actin?
- How fast can cross-links form and break?

MOTOR ENDPLATES AND MOTOR UNITS

Motor neurons stimulate skeletal muscle fibers to contract. The axons of motor neurons, extending from the spinal cord, associate with skeletal muscle fibers at endings, called **motor endplates,** located near the center of each muscle fiber. (See Chapter 17 for details of motor neurons and axons.) Each muscle fiber usually receives only one motor endplate (Figure 9.9A), but very long fibers may have two or three. When the motor endplate stimulates a muscle fiber, the fiber contracts.

The axon conducts an impulse along its plasma membrane in the form of an action potential (see Chapter 17 for details), and the motor endplate transmits the effects of this impulse to the muscle fiber. A new impulse then propagates like a wave (Figure 9.9B) away from the motor endplate along the sarcolemma of the muscle fiber, entering any T-tubules it encounters. Traveling rapidly down the T-tubules, the impulse reaches the sarcoplasmic reticulum within milliseconds and causes the sarcoplasmic reticulum to release Ca^{++} ions, which in turn trigger myofibrils to contract. Approximately 0.01 second has elapsed since the stimulus left the spinal cord.

Motor endplates chemically cause skeletal muscle fibers to contract by releasing **acetylcholine** (Figure 9.9C). Motor endplates are a variety of **chemical synapse** (discussed in Chapter 17, which you should read for background on how synapses operate). Acetylcholine (AS-eh-til-KOE-leen) is one of many neurotransmitter substances that transfer neural impulses from neuron to neuron, and it is also characteristic of motor endplates. The axon endings of motor endplates continually synthesize acetylcholine and package it into **synaptic vesicles** (SIN-ap-tik; synap-, *G,* a union) for storage until action potentials cause the endings to release the contents of the vesicles. At an individual motor endplate, the arrival of an action potential causes a pulse of calcium ions to enter the axon ending. In response, the ending releases acetylcholine by exocytosis (Chapter 2) from some of the vesicles, into the narrow **synaptic cleft** (50 nm across) between the axon ending and sarcolemma of the muscle fiber. Acetylcholine diffuses across this short distance to the sarcolemma, where it binds briefly to acetylcholine receptors before being degraded there. When enough acetylcholine has bound to the receptors, the sarcolemma initiates its own impulse that propagates away from the motor endplate and triggers the muscle fiber to contract.

Neural impulses do not jump across motor endplates from axon to muscle fiber, as many students commonly think. The neural impulse stops at the axon ending. Motor endplates would not function, were it not for the ability of the axon ending to release acetylcholine in response to this impulse and the ability of the muscle fiber to propagate its own impulse in response to acetylcholine. Indeed, certain drugs interfere with or enhance this chemical transmission process.

A single motor neuron may innervate many muscle fibers, delivering stimuli to each one, and making them all contract as a group. Such a group is a **motor unit** (see Figure 9.10), in contrast to a muscle fiber, which is the structural unit of muscle contraction. Whether a muscle contracts massively or delicately depends partially on the size of its motor units. Large motor units with 700 fibers are found in the quadriceps femoris and other large, strong muscles that participate in running and walking. At the other extreme, motor units of 20 to 30 fibers are found in muscles of fingers that control delicate, finely tuned actions such as typewriting or playing a violin. Two or three fibers per motor unit provide the muscles of the eyeball very delicate eye motion. Most motor units contain about 500 muscle fibers. How growing neurons find the muscle fibers with which to connect is discussed later.

MUSCLE TONE

When muscles relax, motor endplates cease firing and muscle fibers stop contracting, as if the nerves to the muscle had been severed. Unlike a denervated muscle, however, motor endplates of normal muscles at rest remain ready to fire when the muscle is palpated or recruited voluntarily. This state of preparedness or responsiveness is called **tone,** and it enables muscles to act quickly. Physicians and physical therapists know this state as local stiffness and resistance their fingertips feel when they palpate a patient's relaxed limb muscles. In contrast, paralyzed muscles remain entirely flaccid.

Most working muscles never relax completely. At any one time, some fibers in a working muscle are resting while others are contracting. This state of partial contraction is an attribute of muscle tone that relieves fatigue by shifting contraction and relaxation among the motor units. Resting fibers will contract

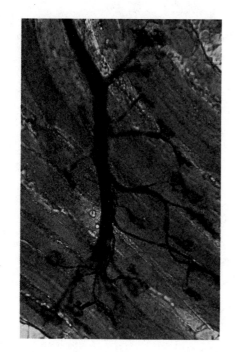

A

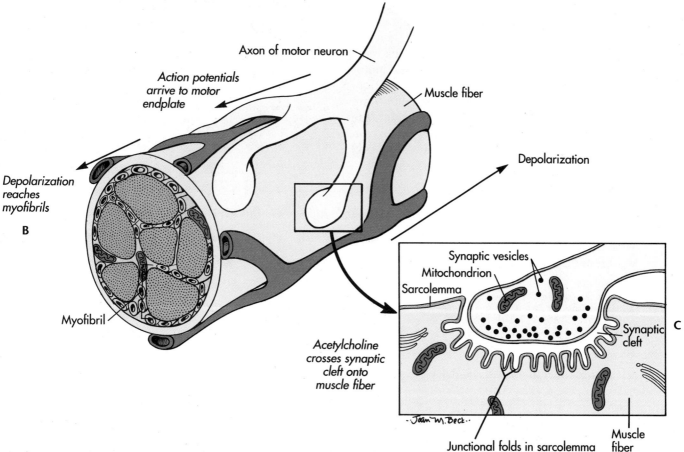

Axon of motor neuron

Action potentials
arrive to motor
endplate

Muscle fiber

Depolarization

Depolarization
reaches
myofibrils

B

Myofibril

Acetylcholine
crosses synaptic
cleft onto
muscle fiber

Synaptic vesicles

Mitochondrion

Sarcolemma

Synaptic
cleft

C

- Joan M. Beck -

Junctional folds in sarcolemma

Muscle
fiber

Figure 9.9

Motor endplate. **A,** A single motor neuron delivers to a motor endplate on the surface of a skeletal muscle fiber. **B,** Impulse advances over the sarcolemma, penetrates T-tubules, and reaches the sarcoplasmic reticulum, where calcium ions stimulate contraction. **C,** A cross-section through the motor endplate reveals an intimate association between neuron and sarcolemma. Synaptic vesicles release acetycholine into the synaptic cleft, where these molecules diffuse onto acetylcholine receptors in the sarcolemma and depolarize it. By increasing its surface area, junctional folds provide the sarcolemma with more receptors to bind acetylcholine.

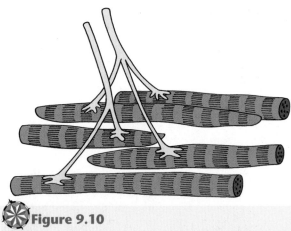

Figure 9.10

A motor unit is a group of skeletal muscle fibers stimulated by a single motor neuron. Two small motor units are shown here as they would be found intermingled within a muscle fascicle.

Figure 9.11

White and red fibers. This cross-section through a few fibers from a muscle fascicle shows that white *(light)* and red fibers *(dark)* are scattered among each other. Endomysium fills the narrow space between them. These fibers have been treated with a fluorescent antibody that specifically detects white fiber myosin and leaves red fibers unstained.

later to relieve those contracting now, as the nervous system stimulates first one motor unit and then the next. The postural muscles that hold your back erect are partially contracted, stimulating enough motor units to hold position but not enough for motion. Soon, other units will contract while fatigued ones recover. If contracting muscle fibers emitted light, the muscle would flicker as different motor units take on the load. Muscles lose tone and atrophy (waste away) when loss of motor innervation paralyses them.

- What substance transmits stimuli across motor endplates to muscle fibers?
- How does the effect of this stimulus reach the sarcoplasmic reticulum?
- What is a motor unit?
- What is muscle tone?

WHITE, RED, AND INTERMEDIATE FIBERS

Good marathon runners are usually poor sprinters, and a good sprinter has little chance of winning a marathon. A similar distinction is true of muscle fibers. Muscles contain **white, intermediate,** and **red muscle fibers** (Figure 9.11) that contract differently and thereby allow muscles a wide range of action. General-purpose muscles use all three types, but specialized muscles favor certain fibers. Some of the differences between sprinters and long-distance runners concern the proportion of red and white fibers in the muscles of the upper and lower extremities.

Different fibers trade-off advantages of speed, power, and fatigue resistance. Table 9.1 summarizes

Table 9.1 PROPERTIES OF SKELETAL MUSCLE FIBERS

	White	Intermediate	Red
Twitch	Fast	Fast	Slow
Maximum tension	High	Medium	Low
Fiber size	Medium	Large	Small
Fibers in motor unit	750	440	540
Myoglobin content	Low	Medium	High
Fatigue resistance	Very low	Medium	High
Oxidative metabolism	Anaerobic	Aerobic	Aerobic

From Shephard RJ : Exercise physiology, Toronto, 1987, BC Decker Inc.

these properties. White fibers contract more rapidly and forcefully, are larger than red fibers, and belong to larger motor units that fire when the nervous system demands rapid, powerful motion. They also are called fast twitch fibers. By comparison, red fibers (sometimes called slow twitch fibers) are smaller, contract more slowly and weakly, and belong to smaller motor units that are fired during slower, delicate movements. Red fibers contain much larger quantities of myoglobin, the red, oxygen-storing protein of muscles, than do white fibers.

An intermediate group of fast, red fibers combines the advantages of white and red fibers to contract rapidly with moderate force and fatigue resistance. Intermediate fibers are larger than red and white fibers, have the smallest motor units, and store moderate amounts of myoglobin. On average, 50% of the fibers in a muscle are red, 35% intermediate, and 15% white, but this composition varies from muscle to muscle. The postural muscles of the trunk and neck have up to 90% red fibers that maintain tone and position, while limb muscles, such as the quadriceps, tend to have more white and intermediate fibers that allow rapid action.

Each type of fiber offers advantages, according to its source of energy. White fibers derive ATP by converting glucose to lactic acid in the absence of oxygen. They store large quantities of glycogen to fuel this anaerobic process, rapidly producing and splitting ATP molecules. In contrast, red fibers receive more capillaries than white fibers and store large amounts of oxygen in their myoglobin. This richer supply of oxygen enables red fibers to produce ATP aerobically, by oxidizing glucose more completely to carbon dioxide and water instead of lactic acid. Red fibers contract more slowly because they split ATP slowly. White fibers fatigue easily because accumulating lactic acid interferes with contraction. Red fibers resist fatigue because they do not produce this product.

Sprinters and marathoners are born, not made. As much as 80% of the fibers in limb muscles of long-distance runners may be red, compared to 60% white fibers among sprinters. Training does not change this muscle composition, but it develops what is already there. Power exercising can increase the size of white fibers but not their numbers, and endurance work can convert some white fibers into intermediate, but no way is known for normal, healthy humans to increase the number of red fibers.

Whether a fiber is red or white appears to depend on the motor neurons that innervate it. White fibers behave as if they were red when they are experimentally reinnervated by neurons that have stimulated red fibers. The reverse is also true; red fibers convert to white when innervated with neurons that were previously connected to white motor units. Reinnervation will not make you a sprinter. Besides the totally impractical surgery required, there is another impressive reason why switching red fibers for white would not work. Do you know why?

- Which muscle fibers are more resistant to fatigue?
- Which ones are larger and anaerobic?
- What is the proportion of red, intermediate, and white fibers in a typical muscle?
- Why is it an advantage for large prime movers to have white muscle fibers?

COORDINATION

MUSCLE SPINDLES

If the brain is to coordinate motion, it must be aware of muscular movement. **Muscle spindles** and **Golgi tendon organs** provide this information by detecting contraction.

Muscle spindles are scattered within fascicles of skeletal muscles (Figure 9.12A); there they inform the brain about contraction. Delicate muscles have relatively more spindles than large prime movers, such as the quadriceps. Spindles, about 4 mm long and 0.2 mm in diameter, are covered by connective tissue capsules (Figure 9.12B) that anchor them to the endomysium between muscle fibers. Each spindle contains 2 to 20 small, modified skeletal muscle fibers to which sensory and motor nerve endings attach. Only the ends of the fibers contract because sarcomeres are missing from the centers. These fibers are called **intrafusal fibers** (intra-FUZE-al; intr-, *L,* inside; fus-, *L,* a spindle) to distinguish them from the regular muscle fibers outside, the **extrafusal fibers.**

Spindles contain two types of intrafusal fibers that measure how much contraction has occurred and how fast it is taking place, rather as your odometer and speedometer tell the distance traveled and the speed of your car. **Nuclear chain fibers** disclose contraction (termed static sensitivity), and **nuclear bag fibers** also detect its rate (dynamic sensitivity). Every spindle contains one or two large nuclear bag fibers in which the nuclei cluster at the middle of the cell. Nuclear chain fibers, on the other hand, are smaller and more numerous and display a central string of nuclei.

Both types of fibers use spring-like **annulospiral endings** (ANN-yoo-loe-SPY-ral; annul-, *L,* a ring; also called primary sensory endings) to detect the rate of contraction. Wrapped around the nuclear region, these endings fire rapidly when they are stretched rapidly and more slowly when stretched slowly, telling the brain how fast the muscle is moving through the intensity of their signals. **Flower spray endings** (secondary sensory endings) usually are attached to nuclear chain fibers at both ends of the annulospiral endings; they emit a slow, steady stream of stimuli that informs the brain that contraction is occurring. As a result the brain receives continually changing information from each muscle regarding how much motion has been accomplished (flower sprays) and how fast it is happening (annulospiral); it then decides how much more contraction is needed. (Cerebral and cerebellar coordination of muscle contraction is discussed in Chapter 19.)

None of this would work if muscle spindles relaxed while the extrafusal fibers are contracting. The spindles would not be very sensitive to contraction, and the brain would not receive up-to-date information. Separate nerve pathways help spindle fibers adjust their length to the extrafusal fibers. Large alpha motor neurons innervate the extrafusal fibers, and smaller gamma motor neurons stimulate intrafusal fibers. Gamma motor neurons end on the striated portions of nuclear bag and nuclear chain fibers. Being smaller in diameter, gamma neurons conduct impulses more slowly to intrafusals so that extrafusal fibers tend to lead the intrafusals through contraction. The spasms in a pulled muscle are caused in part by damaged muscle spindles that are no longer able to control contraction.

STRETCH REFLEX

When a muscle is stretched, it pulls harder, resisting the extra force. You see this reflex work when you pick up a heavy load. Your muscles lift harder as they take up the weight of the package. This response, called the **stretch reflex,** keeps the length of muscles constant despite increasing loads. Your physician knows this reflex and tests for it by tapping the patellar tendon with a hammer, as illustrated in Figure 9.13. Muscle spindles detect this slight stretching, and the quadriceps muscle responds with the familiar knee jerk.

Reflex arcs work involuntarily, relaying stimuli from sensory endings to the spinal cord and back to muscle fibers near the source of the stimulus, integrating muscular action automatically, before the brain is aware of what has happened. Muscle spindles detect the stretch and trigger impulses through sensory (afferent) neurons into the spinal cord. There, synapses link the sensory neurons to motor (efferent) neurons that relay stimuli to additional motor units in the same muscle to bring about more contraction and to inhibit antagonistic muscles, such as the biceps femoris, from contracting. (More detail about these reflexes is given in Chapter 18.)

GOLGI TENDON ORGANS AND THE TENDON REFLEX

Golgi tendon organs help prevent damage by inhibiting contraction when tension is strong enough to tear tendons and muscles. Located where muscle and tendon meet, **Golgi tendon organs** are capsules of connective tissue (Figure 9.14A) that are structurally similar to muscle spindles but are filled with a few collagen fibers and innervated by sensory neurons that branch onto these fibers. Because of their location, Golgi tendon organs can summarize the tension of entire muscles, in contrast to the local information sent by individual muscle spindles. Golgi (GOAL-jee) tendon organs communicate changes in muscular tension to the spinal cord, activating a response that causes the muscle to relax and its antagonist to contract (Figure 9.14B). This response is called the **tendon reflex.**

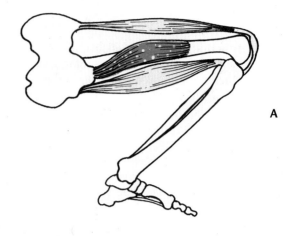

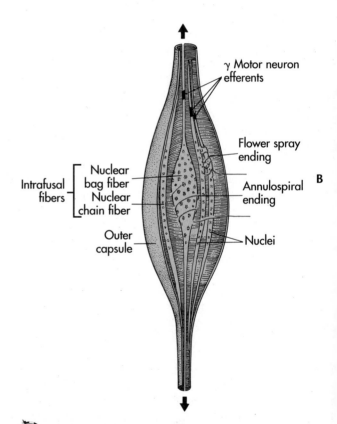

✳ Figure 9.12

A, Muscle spindles *(yellow)* are scattered in the endomysium throughout muscle fascicles, where they detect contraction and inform the brain by sending impulses along afferent neurons to the spinal cord. **B,** Each spindle contains intrafusal fibers inside a connective tissue envelope, and extrafusal fibers surround the spindle. Annulospiral nerve endings about nuclear chain and nuclear bag fibers detect the rate of contraction, and flower spray endings signal the degree of contraction from nuclear chain fibers. Alpha and gamma efferent neurons stimulate extrafusal and intrafusal fibers, respectively. The diameter of these fibers has been exaggerated to show detail.

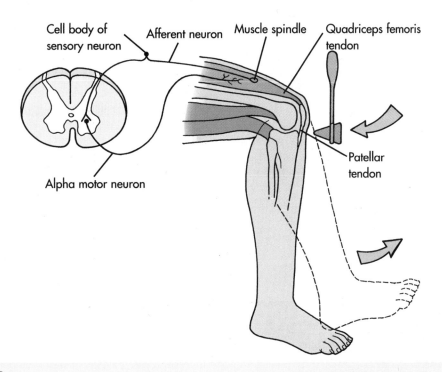

Figure 9.13

Stretch reflex. Muscle spindles deep in the quadriceps femoris detect stretching and fire impulses through afferent neurons (only one is shown) to the spinal cord. This neuron stimulates certain other neurons in the spinal cord that cause the quadriceps to contract further and inhibit the biceps femoris from contracting.

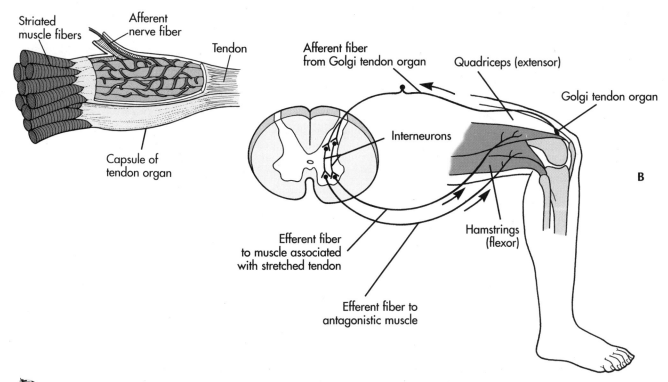

Figure 9.14

Tendon reflex. **A,** Golgi tendon organs detect stretching within the quadriceps tendon. **B,** They inhibit the quadriceps muscle from further contraction, while stimulating the biceps femoris to contract. Tendon organs stimulate afferent neurons that relay the impulse within the spinal cord to efferent neurons that inhibit contraction in the quadriceps and stimulate the biceps.

STEPPING REFLEX

Limbs can retain surprising reflex control even when spinal injury prevents communication to the brain. When such spinal animals are held in the walking position, their limbs extend in support of the body and flex in the walking motion, first one limb and then the opposite. These **stepping reflex** actions indicate that the brain normally coordinates movement through extensive reflex reactions that operate within the skeletal muscles and spinal cord.

- What aspects of muscle contraction do nuclear bag and nuclear chain muscle fibers measure?
- What is the effect of this information on muscle contraction in the stretch reflex?
- What is the function of Golgi tendon organs?

MUSCLE AS HEAT SOURCE AND INJECTION SITE

Skeletal muscles are better heaters than movers. Working muscles convert 75% of their energy into heat and only 25% into motion. If movement were not so important, we would consider skeletal muscles primarily heat machines. Vigorous exercise can increase the temperature of your muscles as much as 2 to 3 degrees Celsius within 10 minutes. Derived from production of ATP and its splitting into ADP and phosphate during contraction, this heat is dissipated by your cardiovascular system through the lungs and skin to prevent your body's core temperature from rising. Although these matters limit vigorous exercise, they also can benefit performance. Warming up beforehand usually improves your performance by heating the muscles and preparing your cardiovascular system for the heat that it will have to dissipate later.

The same capillaries that dissipate heat also can deliver medication. Large muscles, such as the deltoid and gluteus maximus (see Figure 9.2), can hold larger doses of drugs for longer times than subcutaneous injections can, simply because these muscles have more extensive capillary beds. Because the clinician injecting the drug can avoid major vessels in these muscles, the medication will disperse relatively slowly through capillaries in the perimysium and endomysium.

DEVELOPMENT AND REPAIR OF MUSCLE

Skeletal muscle fibers develop from myoblasts. Myoblast cells proliferate rapidly and then fuse to make giant cells, the muscle fibers, that contain numerous nuclei but do not divide again. Developing muscle fibers assemble sarcomeres from actin and myosin molecules that they begin to synthesize during fusion.

Somites are a principal source of myoblasts. Somites (Figure 9.15) are segmented clusters of myoblasts (MY-oh-blasts; blast, *G*, a sprout or bud; muscle bud) and other cells that lie beside the spinal cord during the first weeks of embryonic development. Proliferating myoblasts migrate from the somites along pathways beneath the epidermis into the limbs and body wall where they fuse and develop into muscle fibers. Other myoblasts from the somites surround the vertebral column and become the muscles of the trunk and neck. In all of these places, muscle fibers organize into muscles while motor neurons, growing outward from the spinal cord, establish permanent motor endplates on them.

How growing neurons recognize muscle fibers is largely unknown, but there are some clues to the precision required from what you now know of muscle structure. An individual muscle fiber usually receives only one motor endplate, which is located in the middle of that fiber. This arrangement implies that myoblasts continue fusing onto the ends of the fiber after the endplate forms. Motor units contain the same kind of fibers, never a mixture. That is, motor units consist of only white, exclusively red, or only intermediate fibers, precisely what would be expected if neurons were specifying what type of fiber develops. The consequences of this recognition process are a normally functioning system of muscles that we take for granted until something goes wrong. Clearly, neurons send more than simply contractile stimuli to muscle fibers.

Because skeletal muscle fibers do not proliferate, muscles grow by enlarging the fibers that development has already produced. As the skeleton grows, muscle fibers enlarge by adding more myofibrils and sarcomeres. Myofibrils apparently grow by separating lengthwise and adding new material to themselves. In Dave's torn hamstring, this process presumably restored damaged muscle fibers. Exercise also stimulates these processes, as you will see below.

Satellite cells can replace dead and damaged muscle fibers. It is especially important to promote this process when treating muscle injury. While the number of muscles and muscle fibers already has been established by the time of birth, **satellite cells** can replace muscle fibers that die from aging or injury. Satellite cells lie beside muscle fibers within the same basement membrane. These myoblast-like stem cells proliferate in the endomysium and become new muscle fibers by fusing together and reestablishing neural connections as the original fibers did. RICE treatment attempts to stimulate this replacement after macrophages have removed traumatized and dead muscle fibers during the inflammatory reaction.

Duchenne muscular dystrophy, which wastes skeletal muscles, prevents satellite cells from replacing lost muscle fibers.

- From what cells do skeletal muscle fibers develop?
- Why are muscle fibers multinucleated?
- What is the role of the satellite cells?
- How do developing muscle fibers enlarge?

FACTORS AFFECTING MUSCLE

EXERCISE AND DRUGS

Exercise induces anatomic and metabolic changes in skeletal muscles, which can improve athletic performance. Figure 9.16 summarizes these

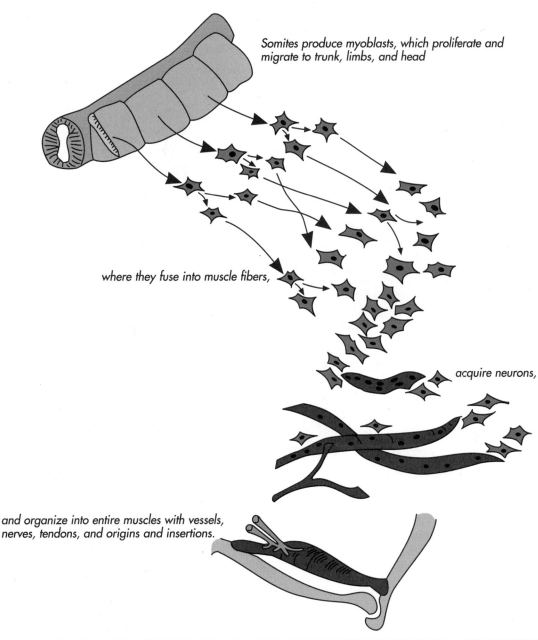

Somites produce myoblasts, which proliferate and migrate to trunk, limbs, and head

where they fuse into muscle fibers,

acquire neurons,

and organize into entire muscles with vessels, nerves, tendons, and origins and insertions.

Figure 9.15

Skeletal muscle development. Proliferating myoblasts originate in the somites and migrate to sites of muscle differentiation in the limbs and elsewhere, where they fuse together as muscle fibers; find neurons, vessels, origins, insertions; and organize into skeletal muscles.

effects and others discussed in this section. Increasing the frequency, duration, and intensity of muscular action can improve power and endurance, especially of targeted muscle groups. Exercise does not increase the number of muscle fibers but strengthens them by enlarging the diameter of these cells with more myofibrils. Endurance exercises tend to develop the aerobic aspects of muscular action; red fibers hypertrophy (enlarge) and store more myoglobin, and their mitochondria produce more ATP. On the other hand, power exercising enhances the anaerobic functions of muscle; white fibers enlarge, the capacity of muscle and blood to accommodate lactic acid increases, and so does your tolerance of fatigue.

Certain treatments and drugs offer athletes a competitive edge by extending their physiological and anatomical limitations. At what point their use be-

comes abuse is a matter of intense debate. Few deny that protecting a damaged knee with tape or a brace legitimately sustains needed performance, but many deny that anabolic steroids are ethically and medically appropriate ways to build muscle mass, particularly in light of the severe side effects that develop from continual use. Consider the problems of monitoring ethics with chemical tests in the following situations. Decisions concerning ethical use have to come from competitors and the public rather than from chemistry and anatomy.

Middle-distance runners can improve their speed through careful administration of **sodium bicarbonate.** Because this material is the body's principal buffer of blood and tissue fluids, its use is virtually undetectable. Additional sodium bicarbonate reduces fatigue by more rapidly clearing lactic acid from mus-

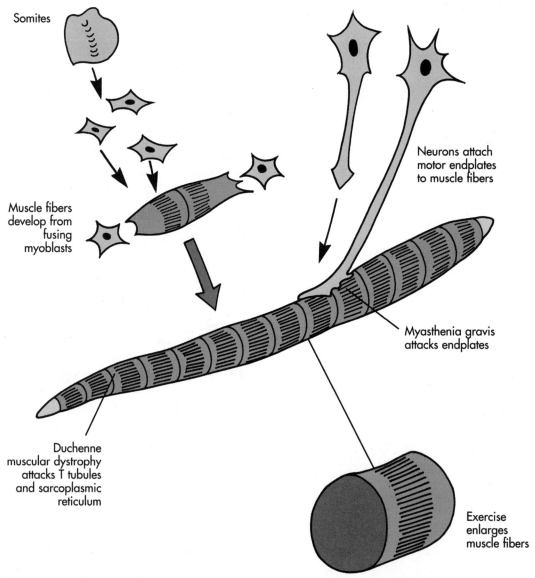

Somites

Muscle fibers develop from fusing myoblasts

Neurons attach motor endplates to muscle fibers

Myasthenia gravis attacks endplates

Duchenne muscular dystrophy attacks T tubules and sarcoplasmic reticulum

Exercise enlarges muscle fibers

Figure 9.16

Diseases and drugs act on different sites and stages of muscle development.

cle fibers. The breakdown products of **anabolic steroids,** however, are readily detectable in urine tests. Sustained massive doses (10 or more times the physiological dose) enlarge muscle fibers and may improve an athlete's confidence at the expense of testosterone production, which leads to temporary infertility in men and masculinization in women. Other side effects, such as liver damage, prostate tumors, liver cancer, and hypertension, have been reported. **Blood doping** can improve performance by enlarging the cardiovascular system's supply of erythrocytes to transport oxygen. This extra margin is obtained by withdrawing and storing up to ½ liter of a person's blood and then transfusing the erythrocytes back into the body after the marrow has replaced the erythrocytes. With more erythrocytes from which to draw oxygen, a runner's muscle fibers presumably gain an extra margin of ATP for contraction.

DISEASES

Numerous diseases and conditions attack muscles and their innervation (Figure 9.16). **Duchenne muscular dystrophy** (DO-shenn; DIS-troe-fee; dys-, *G,* bad, troph-, *G,* nourish) appears to interfere with the sarcolemma of skeletal muscle fibers, resulting in debilitating loss of muscle tissue and bringing death to patients in their early teens. This form of muscular dystrophy usually affects boys who have inherited the gene from their mothers on the X chromosome. Parents usually notice symptoms when their child is weak and slow to walk. By the time the disease is apparent, muscles have failed to regenerate fibers that die in the normal course of fetal life and childhood. A gene for Duchenne's has been isolated from the X chromosome; it codes for a protein named dystrophin. Located where T-tubules link with the sarcoplasmic reticulum, dystrophin apparently is required to return Ca++ ions to the cisternae of the reticulum. Somehow, its lack leads more cells to die than satellite cells can regenerate, and the patient steadily loses muscular control. Nerves, muscle spindles, and blood vessels apparently are not affected.

Myasthenia gravis (MY-as-THEE-nee-ah GRA-vis; asthen, *G,* weak; grav-, *L,* grave, heavy) is a motor endplate disease in which the immune system attacks acetylcholine receptors on motor endplates. With fewer receptors to receive stimuli, muscle fibers fail to contract when they should, and patients suffer muscular weakness, especially while exercising. Drugs, such as neostigmine, can improve muscular activity by allowing endplates to build up enough acetylcholine to depolarize muscle fibers and make them contract. Neostigmine inhibits the destruction of acetylcholine on these receptors so that acetylcholine remains bound to more receptors longer. Myasthenia gravis usually affects adults and is apparently not inherited.

AGING

Muscle mass and muscle strength decline after 30 to 40 years of age. The tendency of older people to be less active is responsible for some of this loss, as muscle fibers in once-robust muscles diminish in numbers, size, and strength. Lack of exercise is not an intrinsic cause of physical decline, as the CUPID definition of changes caused by aging would require (see Cells and Aging, Chapter 2). Connective tissue and eventually adipose tissue replace lost muscle fibers that satellite cells themselves can no longer replace. Red and white muscle fibers are thought to age differently. White fibers are lost and diminish in size because they lose the innervation that sustains them in active muscles. Red fibers, however, appear simply to be lost; their size apparently does not diminish. Energy metabolism of the remaining muscle fibers diminishes, and the metabolic differences between red and white fibers also decrease. Nevertheless, exercise can temporarily restore strength to disused muscles in the elderly, and it has been suggested that some of this improvement may be due to reinnervation of muscle fibers.

- Does exercise increase the size or number of muscle fibers in muscles?
- What other effects does exercise have on muscle fibers?
- What are the effects of the use of anabolic steroids?
- What muscle structures do Duchenne muscular dystrophy and myasthenia gravis attack?
- What is the major effect of aging on skeletal muscles?

SUMMARY

The skeletal muscle system moves the body, holds it in position, gives it form, and generates heat. Skeletal muscle motion is voluntary in theme but involuntary in detail, as if one thinks, "Walk," and the nervous system coordinates progress until the goal is obtained. Muscles are the organs of this system that pull the skeleton through its motions, and the nervous system coordinates their contraction. Groups of muscles act together as antagonists, synergists, or fixators. Although individual muscle fibers contract all-or-none, the nervous system coordinates entire muscles and muscle groups to obtain smooth motion. *(Page 245.)*

Muscles are hierarchies of structure in which muscle fascicles, muscle fibers, myofibrils, sarcomeres and myofilaments represent successively finer levels where contraction and coordination take place. A pulled muscle disrupts this organization. Sarcomeres are the structural units of contraction. They consist of thick myosin filaments and thin actin filaments that slide along each other during contraction and relaxation. *(Page 248.)*

Forces of contraction derive from cross-links between these filaments. Cross-links pull thin filaments along the thick filaments by breaking and reforming repeatedly as the filaments move. According to the sliding filament theory of contraction, ATP provides the energy for contraction, and calcium ions trigger this process. *(Page 252.)*

Muscle fibers receive impulses from motor neurons that transmit stimuli across motor endplates to the surface of muscle fibers. Each fiber responds, either with a rapid wave of contraction or not at all. In contraction the sarcolemma and T-tubule system conduct the action potential to the sarcoplasmic reticulum, which releases calcium ions into the sarcomeres, where these ions are thought to trigger contraction. *(Page 253.)*

Muscle fibers are innervated by single motor neurons, but one such neuron can deliver stimuli to endplates on as many as 700 fibers, causing them to contract together as a motor unit. White, red, and intermediate types of muscle fibers contract at different speeds and resist fatigue differently, providing a versatile range of capabilities to entire muscles. All the muscle fibers in an individual motor unit are of the same type. Muscle tone allows entire muscles to contract and resist fatigue at the same time by alternating contraction and relaxation among motor units. *(Page 254.)*

Muscle spindles and tendon organs measure contraction and inform the brain of progress. Stretch reflexes tend to keep the length of a muscle constant when it encounters additional tension, and tendon reflexes prevent overcontraction. These and other reflexes coordinate movements, such as walking, even without participation of the brain. But the brain does intervene at higher levels. Somatomotor areas in the cerebrum initiate voluntary movements, and the cerebellum coordinates them, recruiting many motor units swiftly for rapid motion or recruiting them gradually for slower movement. *(Page 257.)*

Muscles produce heat when they contract, warming the body, and the excess is dissipated by breathing and perspiration. Muscles are excellent sites for drug administration because they are richly vascularized with capillary beds that distribute sustained doses from the site of injection. *(Page 260.)*

Muscles develop embryonically from myoblasts that fuse together and form muscle fibers. Fibers receive their innervation and organize into muscles with remarkable precision, and muscles acquire their origins and insertions with similar accuracy, but how they do so is poorly understood. Although muscle fibers do not divide, satellite cells can replace dead fibers with new ones. *(Page 260.)*

Exercise and drugs can enhance muscular performance, but numerous diseases and other circumstances interfere with muscle development and function. Weight training improves the strength of muscles by increasing the number of myofibrils in muscle fibers, but it does not raise the number of fibers, nor can it convert white fibers into red. Continued innervation and stimulation is necessary to maintain muscle tone. Loss of innnervation through spinal cord injury or other causes diminishes tone, and muscles then atrophy. Muscular dystrophy leads to atrophy through lesions in the T-tubule and sarcoplasmic reticulum, but myasthenia gravis affects motor endplates rather than the contractile machinery itself. Because aging muscles gradually lose their ability to replace muscle fibers, muscle mass declines as muscle fibers diminish in size, and some are lost entirely. *(Page 261.)*

STUDY QUESTIONS

1. Identify an antagonistic and a synergistic pair of muscles. What does each do? How are they different? How are they similar?   **1, 2**

2. Does a skeletal muscle contain more fascicles, muscle fibers, sarcomeres, or motor endplates? Discuss the relative numbers of these structures.   **3, 4**

3. What layer of connective tissue covers the biceps femoris? What does this layer do? Which layer surrounds muscle fibers with capillaries?   **3, 4, 5**

4. Draw the structure of myosin thick filaments and actin thin filaments from memory. Show how cross-links are thought to move one along the other.   **6**

5. Which dimensions change in a contracting sarcomere, and which do not? Draw diagrams of relaxed and contracted sarcomeres.   **6**

6. Describe the structure and role in contraction of T-tubule system, sarcoplasmic reticulum, calcium ions, sarcolemma, motor endplates, myosin, troponin, tropomyosin, actin, and ATP.   **6, 7**

7. Describe the sequence of events that breaks and forms cross-links during muscle contraction.   **7**

8. Describe how an impulse reaches a muscle fiber from the axon of a motor neuron. Draw a diagram of a section through a motor endplate.   **8**

9. Explain what is meant by all-or-none contraction of muscle fibers.   **8**

10. Make a diagram of a simple motor unit that contains three muscle fibers.   **8**

11. What are the properties of white muscle fibers, and how do they differ from the red variety? From the intermediate form?   **8**

12. Which type of fiber would you want to predominate in your muscles as a sprinter? As a pianist?   **9**

13. Describe the structure of a muscle spindle. What are the similarities and differences between nuclear bag fibers and nuclear chain fibers? What do flower spray endings and annulospiral endings detect?   **10**

14. Describe the actions of stretch and tendon reflexes.   **11**

15. What are the differences between Golgi tendon organs and muscle spindles? What are their similarities?   **10, 11**

16. How do muscles produce and dissipate heat? Why are this ability and the structures that provide it useful?   **12**

17. How do muscle fibers develop from myoblasts?   **13**

18. Describe the effects of exercise and certain drugs on muscle performance and structure. What happens to muscle structure when a muscle, such as the hamstring, is injured?   **14**

19. Does exercise increase the number of muscle fibers or myofibrils in a muscle?   **14**

20. What changes does aging bring to skeletal muscles?   **15**

21. Describe the effects of muscular dystrophy and myasthenia gravis on the action of muscles. Trace each disease to its apparent anatomical defect.   **15**

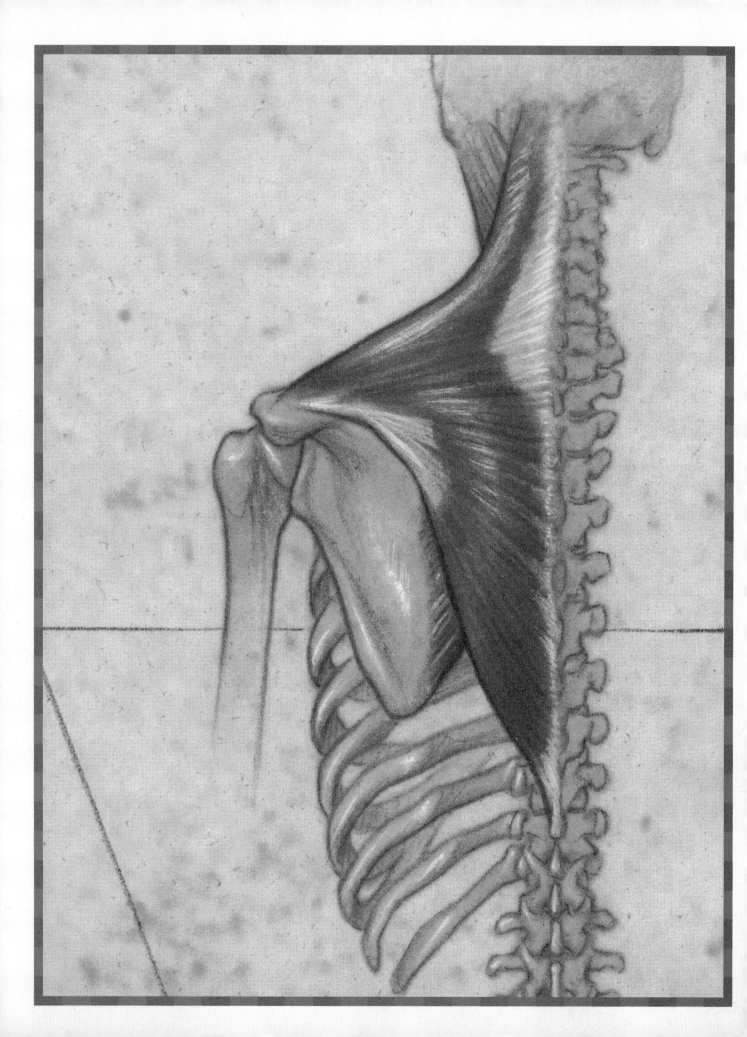

Muscles of the Head, Neck, and Trunk

OBJECTIVES

After studying this chapter, you should be able to do the following:

1. Describe the action and location of muscles when given their names, and vice versa, from information about their actions and locations, name the muscles.
2. Use the following terms to describe the action of head and trunk muscles: origin, insertion, prime mover, muscle group, antagonistic pair, synergist, and fixator.
3. Use the following terms to describe the action of head and trunk muscles: flexor, extensor, lateral flexor, and rotator.
4. Describe the three types of skeletal levers moved by muscles.
5. Describe and give examples of the different shapes of muscles.
6. Describe the actions of the major muscle groups of the head and anterior neck.
7. Describe the locations and actions of the five major groups of muscles in the trunk.
8. Describe the locations and actions of three muscles from each of the major groups in the trunk.
9. Explain the role of myotomes in development of skeletal muscles.

 **VOCABULARY TOOLBOX**

Use the Toolbox to decipher the names of muscles according to shape, location, direction, attachment, and action.

SHAPE

biceps	(BY-seps) -ceps, *G* head; two-headed muscle.
digastric	(dy-GAS-trik) -gast, *L* belly; two bellies.
pennate	(PEN-ate) penn-, *L* a feather.
quadrate	(KWAD-rate) quadr-, *L* four; quadriceps, four-headed muscle.
scalene	(SKAY-leen) scalen-, *G* lumpy.
serratus	(sair-RATE-us) *L* a saw, serrated.
trapezius	(tra-PEEZ-ee-us) trapez-, *G* a table; a four-sided muscle.

LOCATION

capitis	(KAP-it-is) *L* of the head.
cervicis	(SER-vih-sis) *L* of the neck.
spinalis	(spy-NAL-is) *L* of the spine.

DIRECTION

longitudinal	(LON-jih-TOO-din-al) long-, *L* lengthwise.
rectus	(REK-tus) *L* straight, upright.
transverse	(TRANS-vers) *L* across, transverse.

ATTACHMENTS

iliocostalis	(ILL-ee-oh-kos-TAL-is) origin on the ilium, insertion on ribs. Origin is named first.

ACTION

erector	(ee-REK-tor) erect, *L* upright, raise upright, straighten.
levator	(le-VAY-tor) leva-, *L* raise, lift.
tensor	(TEN-sor) tens-, *L* stretched.

RELATED TOPICS

- **Anatomy of vertebral column and skull, Chapter 6**
- **Human body plan, Chapter 1**
- **Movements of joints, Chapter 8**
- **Myotomes and somites, Chapter 4**
- **Skeletal muscle tissue, Chapter 9**

PERSPECTIVE

This chapter introduces the muscles that move the head, neck, and trunk. You will learn to identify these muscles by their shapes and location in the skeleton and by their actions. When you are able to associate the actions of muscles with their shapes and locations, you will begin to see how the body moves.

Your instructor wants you to recognize the positions and actions of muscles in the head, neck, and trunk in order to understand how these parts move. The first three sections of this chapter—Muscles and Motion, Levers and Mechanical Advantage, and Shapes of Muscles—introduce principles of action that apply to all muscles. Read these sections first to understand what features of muscles to look for in the rest of the chapter. Then learn the origins and insertions of individual muscles, their shapes, and their actions from the descriptions, figures, and tables in each section. The sections are arranged from the head down to the pelvis, considering the superficial muscles of each region first, then the deeper muscles. Find pronunciations of names in the tables. Check your knowledge by answering the study questions at the end of this chapter.

MUSCLES AND MOTION

Muscles are the organs of motion. Approximately 650 skeletal muscles, representing about 40% of the body's weight, move the entire skeleton. Many of the muscles you see in Figure 10.1 are superficial muscles that move the extremities, but many other less obvious, deeper muscles move the trunk itself, as well as the head and neck. Both groups move the axial skeleton.

Your vertebral column and the tower shown in Figure 10.2 stand erect for basically the same reasons. Like the guy wires that support the tower, muscles on all four sides erect the vertebral column, flex it forward and laterally, and extend it posteriorly. These muscles are described in the sections on cervical muscles (page 281), vertebral muscles of the neck (page 284), and deep muscles of the back (page 292). The vertebral column itself, together with the rib cage and pelvis, supports the trunk with muscular body walls that extend from the vertebral column in Figure 10.1A and B. Description of these muscles begins on page 294. Discussion of muscles of facial expression, chewing, swallowing, and speaking begins on page 275.

MUSCLE GROUPS

Muscles work in groups whose actions are determined by their respective positions on the skeleton. Muscles in the side of the neck, for example, flex the neck laterally, and, as you might expect, posterior muscles extend the neck dorsally. Coordinated, purposeful actions of many different **muscle groups** move the body. When you associate the action of each group with its location, you will see how muscle groups cooperate. Bending to get a candy bar from a vending machine uses the same groups of back and abdominal muscles that came into play when you walked to the machine, for example, but bending requires more work from the back muscles than walking does.

Muscle groups are classified according to their primary action. They may be **flexors, extensors, lateral flexors,** and **rotators,** to name a few examples (see Chapter 8, Articulations, for descriptions of these motions). Muscles that are most responsible for a particular motion are called **prime movers.** The scalene muscles of the lateral neck, shown in Figure 10.3, are lateral flexors that support the neck and bend it from side to side. You can anticipate this motion from the locations of the three scalenes—**scalenus anterior, s. medius,** and **s. posterior**—on the ribs and cervical vertebrae. The scalenus muscles (SKAY-leen-us) have their **origins** on the cervical vertebrae, from which tendons **insert** onto the first two ribs. The scalenes are **antagonistic pairs** of muscles; that is, the members of one side of the neck counteract those on the other side. When the tension on both sides is equal, the scalenes behave as **fixators,** holding the neck erect (Figure 10.3A), but when one set shortens more than the other, the neck flexes to that side (Figure 10.3B). What action would return the neck vertically? Members of the same set are also **synergists** because, in pulling from the same side of the neck, each member adds to the force of the others. Although lateral action is their main motion, the scalenes also provide anterior support because they insert forward of the vertebral column.

- Define the term *muscle group,* and describe how the scalenes act as a muscle group.
- How do antagonistic and synergistic muscles act?

LEVERS AND MECHANICAL ADVANTAGE

Physical therapists and bioengineers think of the skeleton and joints as a system of levers moved by the muscles. This analogy helps them to analyze the forces that move the skeleton and to design exercises that improve athletic performance or hasten rehabilitation after injuries. Moving bones resemble the seesaw in Figure 10.4A. Each person can raise the other, provided both players balance about a central support, or **fulcrum,** that permits the seesaw to tilt. The player who then moves farther away from the center gains an advantage because it is easier to raise the partner. This player has the greater **mechanical advantage** in lifting the other player.

Bones are the levers in the skeleton, joints the fulcrums, and muscles provide the forces that move the loads. In physics there are three classes of levers, and interestingly enough, the body uses all of them. Knowing these classes is important because some classes produce much motion and others a great deal of power, which gives the body versatile power and movement. As you will see, no single class of lever can maximize both power and motion. Almost all muscles operate third-class levers that maximize motion, especially in the extremities.

FIRST CLASS

First-class levers place the fulcrum between the effort and the load, as do seesaws and scissors. The muscles of the neck provide the only first-class lever system in the body (Figure 10.4A), by balancing the weight of the skull on the occipital condyles. The condyles are the fulcrum between the weight of the skull and muscles of the neck. This arrangement produces considerable motion but is not very powerful because the weight of the skull has the mechanical advantage; its weight concentrates farther from the occipital condyles than the effort of the muscles.

Text continued on p 272.

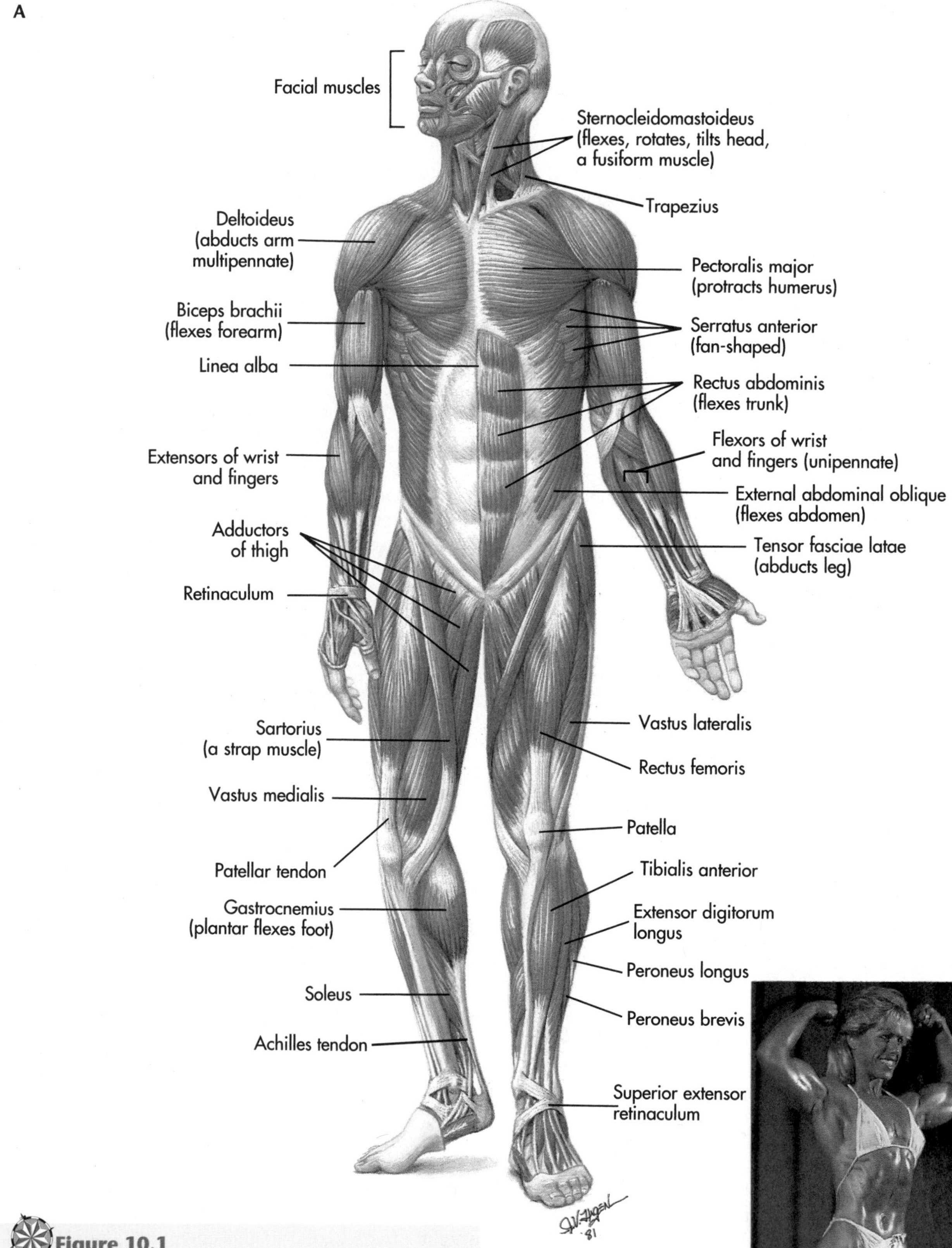

A

Facial muscles

Sternocleidomastoideus
(flexes, rotates, tilts head,
a fusiform muscle)

Trapezius

Deltoideus
(abducts arm
multipennate)

Pectoralis major
(protracts humerus)

Biceps brachii
(flexes forearm)

Serratus anterior
(fan-shaped)

Linea alba

Rectus abdominis
(flexes trunk)

Flexors of wrist
and fingers (unipennate)

Extensors of wrist
and fingers

External abdominal oblique
(flexes abdomen)

Adductors
of thigh

Tensor fasciae latae
(abducts leg)

Retinaculum

Vastus lateralis

Rectus femoris

Sartorius
(a strap muscle)

Vastus medialis

Patella

Patellar tendon

Tibialis anterior

Gastrocnemius
(plantar flexes foot)

Extensor digitorum
longus

Peroneus longus

Soleus

Peroneus brevis

Achilles tendon

Superior extensor
retinaculum

✸ **Figure 10.1**

Superficial muscles of the body. A, Anterior. B, Posterior. Find the sternocleidomastoid, trapezius, rectus abdominis, and external oblique muscles of the trunk discussed in the text. Most of the muscles in view move the limbs and girdles, and they obscure the deeper muscles that move the trunk and neck.

B

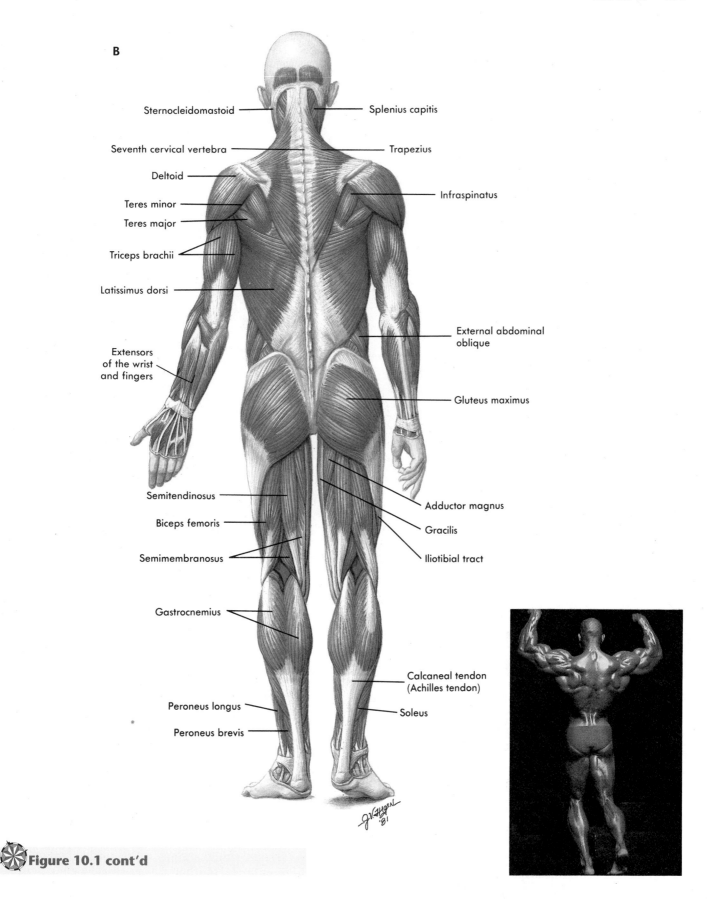

Sternocleidomastoid

Splenius capitis

Seventh cervical vertebra

Trapezius

Deltoid

Infraspinatus

Teres minor

Teres major

Triceps brachii

Latissimus dorsi

External abdominal oblique

Extensors of the wrist and fingers

Gluteus maximus

Semitendinosus

Adductor magnus

Biceps femoris

Gracilis

Semimembranosus

Iliotibial tract

Gastrocnemius

Calcaneal tendon (Achilles tendon)

Peroneus longus

Soleus

Peroneus brevis

⊛ **Figure 10.1 cont'd**

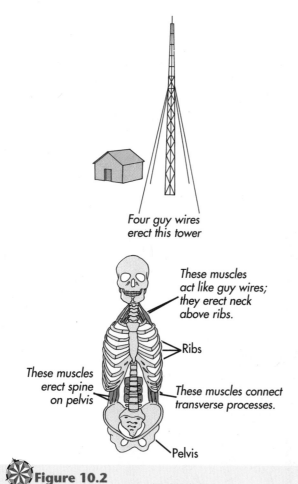

Four guy wires erect this tower

These muscles act like guy wires; they erect neck above ribs.

Ribs

These muscles erect spine on pelvis

These muscles connect transverse processes.

Pelvis

Figure 10.2

The muscles of the trunk are like guy wires that hold towers erect. Muscles connect transverse processes of vertebrae together, and they support the vertebral column above the ribs and pelvis.

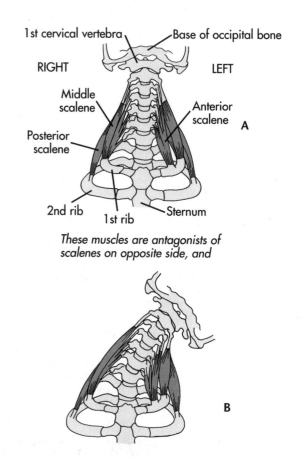

1st cervical vertebra — Base of occipital bone

RIGHT — LEFT

Middle scalene

Anterior scalene **A**

Posterior scalene

2nd rib — 1st rib — Sternum

These muscles are antagonists of scalenes on opposite side, and

B

when they shorten and their opposing partners stretch, the neck flexes left.

Figure 10.3

A, The scalene muscles support the neck from left and right sides. **B,** The neck flexes left when the muscles on that side shorten more than those on the right.

SECOND CLASS

This class places the load between the fulcrum and the effort. Wheelbarrows and doors are second-class levers whose action resembles the calcaneal (Achilles) tendon and the ankle, illustrated in Figure 10.4B. The gastrocnemius (GAS-troe-NEE-mee-us) muscle and its tendon raise the ankle above the fulcrum at the ball of the foot, just as if you lifted the handles of the wheelbarrow. This arrangement is very powerful because the tendon is farther from the ball of the foot than the ankle and has the greater mechanical advantage.

THIRD CLASS

Third-class levers produce considerable motion, but they have poor mechanical advantage because the muscle is between the load and the fulcrum, as when you lift a shovel or squeeze a pair of forceps. Muscular force must be very large to move even small loads. The biceps brachii muscle of the arm, which inserts between the elbow and the hand, operates an

excellent third-class lever, the forearm (Figure 10.4C). Third-class levers are useful because short, powerful muscles can move loads a long distance. Flex your elbow to see that the forearm moves much farther than the biceps. Despite their inherent mechanical disadvantage, most muscles operate third-class levers and thereby gain much motion.

- Give an example for each of the three classes of lever.
- Where are the fulcrum, effort, and load located in each case?
- Describe the motions of each class of lever.

SHAPES OF MUSCLES

The various shapes of muscles reveal their power and motion. Anatomists recognize four shapes based on the direction of muscle fibers—**parallel, pennate, fan-shaped,** and **sphincter**—shown in Figure 10.5. Find examples of each type in Figure 10.1A and B as you follow the discussion.

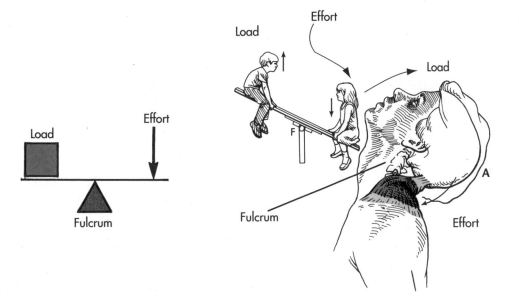

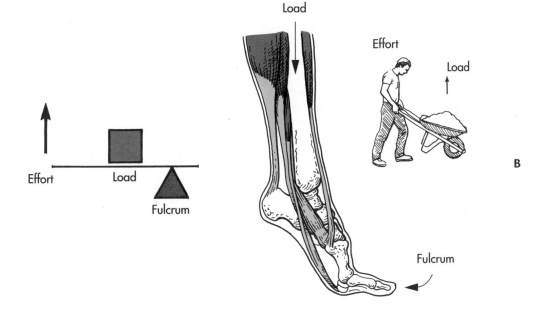

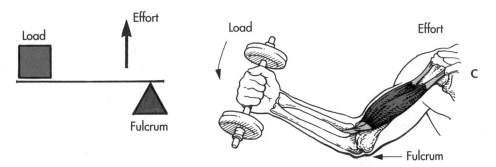

Figure 10.4

Three classes of levers and their actions in the skeleton. **A,** First class. **B,** Second class. **C,** Third class.

PARALLEL

Most muscles are parallel because their fibers run longitudinally, attaching to tendons at either end. Within this group there are three varieties—**fusiform, strap,** and **flat muscles.** Fusiform (FEW-sih-form) muscles are spindle-shaped, with a substantial central **belly** that tapers to narrow **tendons** at either end. The sternocleidomastoid muscles (Figure 10.1A) and scalenes (Figure 10.3) are fusiform muscles. The diameter of the belly, or more precisely, the cross-sectional area of the belly perpendicular to the fibers, reflects the number of muscle fibers within and hence the power of the muscle. The larger the belly, the stronger the muscle.

Bellies of strap muscles are more nearly uniform in diameter. The **sartorius muscle** of the thigh (sar-TOR-ee-us) is a strap muscle (Figure 10.1A). Flat muscles are thin and broad, as are the rectus abdominis and the external oblique muscles of the abdomen in Figure 10.1A. They take their origins and insertions from broad, fibrous, sheet-like tendons called **aponeuroses** (APO-noor-oh-sees). As their shape indicates, flat muscles spread their forces over broad areas, whereas fusiform and strap muscles focus their power onto small, bony targets. A few muscles have two bellies, such as the **omohyoid muscle** of the neck in Figure 10.9A (Table 10.4). All muscles can shorten from one third to one half of their resting length.

PENNATE

Pennate muscles are very powerful because their fibers run obliquely like the rays of a feather, attaching to tendons at the center and sides of the muscle. This construction improves power by increasing the cross-sectional area, with little additional bulk. The fibers of **unipennate muscles** (YOU-nih-PEN-ate), such as the flexor muscles of the wrist and fingers in Figure 10.1A, run obliquely from the tendon of origin on one side of the muscle to the tendon of insertion on the opposite side, producing a muscle with more strength than its width indicates. In **bipennate muscles,** such as the triceps brachii (Figure 10.1B), fibers spread in two directions from a central tendon, collecting forces from both margins of the muscle. **Multipennate muscles,** such as the deltoid (Figure 10.1A), simply add more tendons and chevron-like ranks of fibers.

FAN-SHAPED

Members of this group combine the features of flat and fusiform muscles because they originate on broad aponeuroses and converge on a tendon (or vice versa)

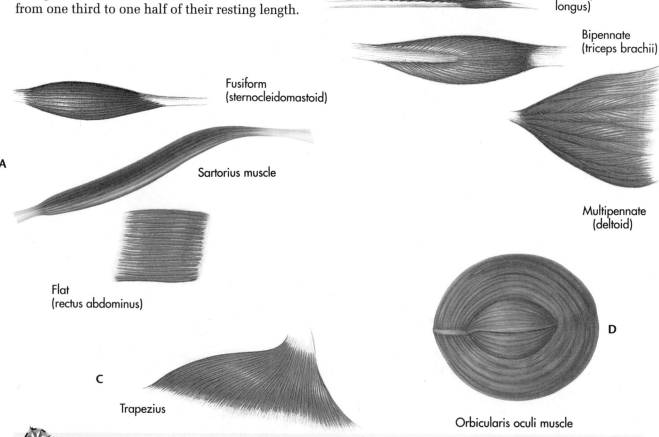

Fusiform
(sternocleidomastoid)

Sartorius muscle

A

Flat
(rectus abdominus)

C

Trapezius

Unipennate
(flexor pollicis longus)

Bipennate
(triceps brachii)

B

Multipennate
(deltoid)

D

Orbicularis oculi muscle

Figure 10.5

Muscles are classified as (A) parallel, (B) pennate, (C) fan-shaped, or (D) sphincters.

at the opposite end of the muscle. The trapezius and pectoralis major muscles in Figure 10.1A and B are examples; they focus forces collected from the thorax onto the scapula and the humerus.

SPHINCTERS

Circular sphincter muscles surround the eye and mouth openings and the exits of the urethra and anus. Sphincters (SFINK-ters), such as the orbicularis oculi muscles in Figure 10.1A and 10.6, are really endless strap muscles. The fibers of these muscles insert and originate on themselves, closing the eyelids when the fibers shorten.

- Name examples of parallel and pennate muscles.
- Why are pennate muscles powerful?
- Name examples of fan-shaped and sphincter muscles.

MUSCLES OF HEAD AND NECK

The muscles of the head and neck are versatile and complex. These muscles mold our facial expressions and move our eyes; they enable us to chew, swallow, and talk; and they hold the head erect and move it delicately or vigorously in all the ways we need. We will follow these muscles and their actions from the head downward, superficial muscles first, then the deeper muscles.

MUSCLES OF FACIAL EXPRESSION

Most muscles of the face and scalp originate from the skull and insert into the superficial fascia beneath the skin, where their motions create smiles, frowns, laughing, and all the other facial communications of emotions and speech illustrated in Figure 10.6 and listed in Table 10.1. Circular muscles of the eyes and mouth dominate the face because many other facial muscles insert upon them and pull the skin into various expressions. Branches of the facial nerve, cranial nerve VII, supply all the facial muscles.

EXTRINSIC MUSCLES OF THE EYEBALLS

Three pairs of muscles rotate the eyeballs in all directions—roll them in disbelief, look up to see a traffic light, look down at this book, or look forward to find your friend in a crowd. Six delicate fusiform muscles shown in Figure 10.7 and listed in Table 10.2 perform all these motions. The **superior** and **inferior rectus muscles** rotate the eyeballs superiorly and inferiorly and adduct them slightly. **Medial** and **lateral rectus** muscles rotate each eyeball left or right, and the **superior** and **inferior obliques** rotate the eyeballs on their anterior-posterior axes and abduct them. All are considered extrinsic muscles because their origins lie outside the eyeball. Cranial nerves III, IV, and VI serve them (Table 10.2).

All four rectus muscles rotate the eyeball vertically or horizontally because they insert on their respective

Table 10.1   FACIAL MUSCLES

Muscle	Origin	Insertion	Action	Innervation
Epicranius frontalis (ep-ih-KRANE-ee-us fron-TAL-iss)	Galea aponeurotica	Muscles of skin and eyebrow	Wrinkles forehead; raises scalp and eyebrows	Facial nerve; all facial muscles are innervated by this nerve (cranial nerve VII)
Epicranius occipitalis (ep-ih-KRANE-ee-us ok-SIP-ih-TAL-iss)	Occipital bone	Galea aponeurotica	Draws scalp posteriorly	—
Orbicularis oculi (or-BIK-you-LAR-iss OK-you-lee))	Self	Self	Closes the eye in sleep, blinking, squinting, and frowning	—
Orbicularis oris (OR-iss)	Muscles surrounding mouth	Skin at lip and corners of mouth	Closes, opens, purses, everts lips; speech movements	—
Nasalis (nay-SAL-iss)	Maxilla below ala of nose	Nasal septum and ala	Dilates nasal aperture	—
Levator labii superioris alaeque nasi (le-VAY-tor LAY-bee-eye su-PEER-ee-OR-iss AL-ek-wee NAY-see)	Frontal process of maxilla	Nasal septum, skin of nose, corners of mouth	Raises upper lip; dilates nostrils	—

Continued.

Table 10.1   FACIAL MUSCLES—cont'd

Muscle	Origin	Insertion	Action	Innervation
Levator labii superioris	Lower margin of orbit; maxilla; zygomatic bone	Orbicularis orbis of upper lip	Raises upper lip	—
Levator anguli oris (ANG-you-lee OR-iss)	Canine fossa	Angle of the mouth	Raises angle of the mouth	—
Zygomaticus minor (ZYE-go-MAH-tih-kus)	Zygomatic bone	Orbicularis oris	Draws lips up and outward	—
Zygomaticus major	Zygomatic bone	Angle of mouth	Draws angle of mouth upward and backward, as in laughing	—
Mentalis (men-TAL-iss)	Incisive fossa of mandible	Skin of chin	Protrudes lower lip; wrinkles skin of chin, as in pouting	—
Depressus labii inferioris (de-PRESS-us LAY-bee-eye in-FEAR-ee-or-iss)	Lower border of mandible	Skin of lower lip, orbicularis oris	Depresses lower lip	—
Depressor anguli oris (de-PRESS-or ANG-you-lee OR-iss)	Lower border of mandible	Muscle at angle of mouth	Depresses angle of mouth	—
Risorius (rih-ZOR-ee-us)	Fascia of masseter muscle (Table 10.3)	Muscles and skin at angle of mouth	Retracts angle of mouth, as in grin and dimples	—
Buccinator (BUK-sin-ate-or)	Alveolar process of maxilla and mandible	Orbicularis oris at angle of mouth	Retracts angle of mouth; compresses cheek; mastication	—
Platysma (PLAH-tiz-ma)	Fascia of chest and shoulder	Fascia of face, lower jaw, and corners of mouth	Wrinkles skin of neck and upper chest	—
Auricularis (superior, anterior, and posterior divisions) (or-IK-you-lar-iss)	Fascia of temporal region	Auricle (pinna)	Raises, retracts and depresses ear	—

From Moore PJ: Muscular and associated neurological systems. In Hilt NE and Cogburn SB, editors: Manual of orthopedics, St. Louis, 1980, CV Mosby Co.

Table 10.2   MUSCLES THAT MOVE THE EYEBALLS

Muscle	Origin	Insertion	Action	Innervation
Superior rectus (REK-tus)	Fibrous ring encircling the optic nerve from margin of optic foramen	Sclera, superior margin of eyeball	Rotates eyeball upward	Oculomotor (cranial nerve III)
Inferior rectus	Same	Sclera, inferior margin, eyeball	Rotates eyeball downward	Oculomotor (cranial nerve III)
Lateral rectus	Same	Sclera, lateral margin, eyeball	Rotates eyeball laterally	Abducens (cranial nerve VI)
Medial rectus	Same	Sclera, medial margin, eyeball	Rotates eyeball medially	Oculomotor (cranial nerve III)
Superior oblique (oh-BLEEK)	Margin of optic foramen by way of trochlea	Sclera, behind and between insertions of superior and lateral recti	Rotates and abducts eyeball, directs view downward and laterally	Trochlear (cranial nerve IV)
Inferior oblique	Orbital surface of maxilla	Sclera, behind and between insertions of inferior and lateral recti	Rotates and abducts eyeball, directs view upward and laterally	Oculomotor (cranial nerve III)

From Gray H: Anatomy of the human body, ed 30, Clemente CD, editor, Philadelphia, 1985, Lea & Febiger.

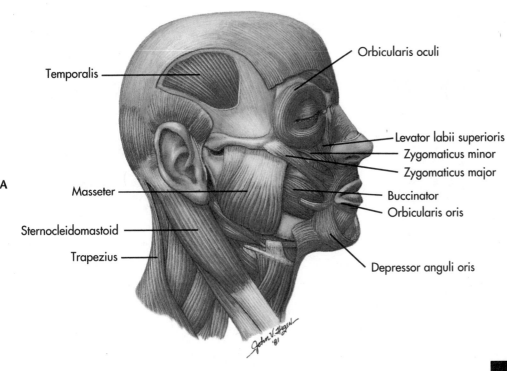

Temporalis

Orbicularis oculi

Levator labii superioris
Zygomaticus minor
Zygomaticus major

Masseter

Buccinator
Orbicularis oris

Sternocleidomastoid

Trapezius

Depressor anguli oris

A

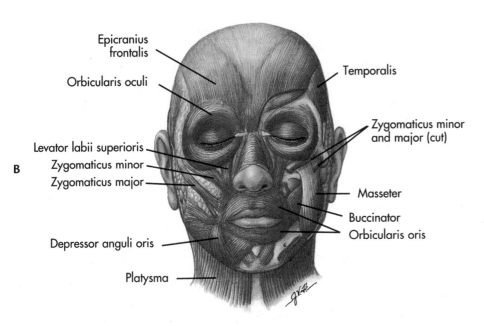

Epicranius
frontalis

Orbicularis oculi

Temporalis

Zygomaticus minor
and major (cut)

Levator labii superioris
Zygomaticus minor
Zygomaticus major

Masseter

Buccinator
Orbicularis oris

Depressor anguli oris

Platysma

B

**Figure 10.6**

Muscles of facial expression. A, Right lateral view. B, Anterior view.

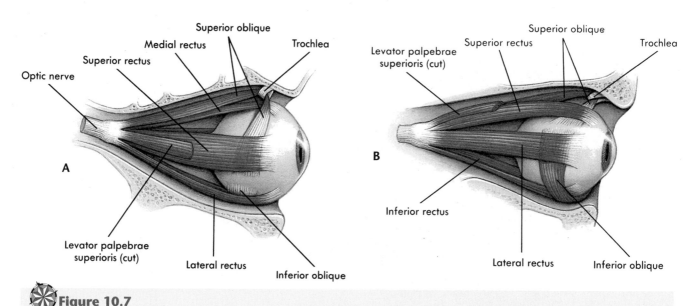

Figure 10.7

Muscles that move the eyeballs. **A**, Superior view. **B**, Lateral view of right eyeball and orbit.

sides of the eyeball and pull directly toward the back of the orbit, where their origins arise from a fibrous ring around the optic nerve. The oblique muscles, however, pull from very different directions than the recti. The superior oblique muscle takes its origin at the back of the orbit near the origin of the rectus muscles and inserts behind the insertion of the superior rectus on the upper surface of the eyeball. From this location the superior oblique would rotate the eyeball upward, if it were not for a fibrous pulley known as the **trochlea** (TROKE-lee-ah; troch-, *G,* a wheel) that diverts the muscle medially and causes the muscle to rotate the upper surface of the eyeball medially. No trochlea assists the inferior oblique muscle, however. It reaches under the eyeball from the orbital surface of the maxilla and inserts behind the insertion of the lateral rectus, where it rotates the eyeball laterally in the opposite direction from the superior oblique. Chapter 20, Eye, discusses how the extrinsic muscles of the eyeball coordinate eye movements.

- What motions do the superior and inferior rectus muscles produce?
- Which muscles swing your view from left to right?
- How does the position of the superior oblique enable it to rotate the eyeball?

MUSCLES OF MASTICATION

The muscles of mastication (chewing) move the mandible when you eat and talk. Four muscles, together with ligaments, form a sling that enables the jaw to open and close by sliding and rotating on the temporomandibular joint. Find these muscles in Figure 10.8 and Table 10.3. All are innervated by the

mandibular branch of the trigeminal nerve (cranial nerve V).

The **temporalis,** the **masseter,** and the **medial** and **lateral pterygoid muscles** are prime movers of the mandible. Other muscles in the floor of the oral cavity help depress the mandible (Figure 10.9A and Table 10.4). The temporalis is a thin, fan-shaped muscle with its origin in the temporal fossa of the skull and its insertion on the coronoid process of the mandible, which enables it to retract and elevate the mandible in chewing. The masseter, a wide fusiform muscle, originates forward of the zygomatic arch and inserts on the lateral surfaces of the coronoid process and ramus of the mandible, where it protracts and elevates the mandible. The temporalis and masseter can easily be seen at work; look in a mirror when you clench your jaw, and watch them contract. The medial pterygoid muscle (Figure 10.8B) is located on the medial surface of the mandible, where it assists the masseter. Acting together, all three muscles close the mouth and mesh the lower teeth with the upper teeth. The masseter and medial pterygoid also bring the jaw forward on the temporomandibular joint to bite with the incisors, and the temporalis retracts it.

The lateral pterygoid muscle, shown in Figure 10.8B, completes the muscular sling of the mandible and is the only muscle of the four that can depress the mandible. The lateral pterygoid pulls forward and medially upon the temporomandibular joint and the neck of the mandibular condyle from the lateral pterygoid process of the skull. This motion tilts the ramus of the mandible forward and depresses the jaw as it rotates about the angle of the mandible. The lateral

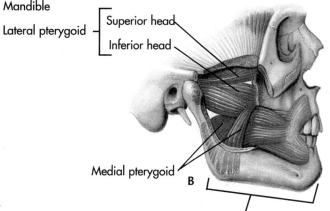

Temporalis

Zygomatic arch (cut)

Zygomatic arch
cut to show
tendon of temporalis

Coronoid process

Buccinator

Orbicularis oris

Masseter (cut)

Mandible

Lateral pterygoid — Superior head

Inferior head

Medial pterygoid

A

Lateral ligament
of temporomandibular
joint

Ramus

B

Angle of mandible;
masseter inserts
here

Figure 10.8

Muscles of mastication. **A,** Right lateral view; masseter
muscle and zygomatic arch have been removed to show
the insertion of the temporalis muscle onto the
mandible. **B,** Temporalis has been removed to expose
the lateral and medial pterygoid muscles.

Table 10.3	MUSCLES OF MASTICATION			
Muscle	**Origin**	**Insertion**	**Action**	**Innervation**
Temporalis (TEMP-or-AL-iss)	Temporal fossa and temporal fascia	Ramus and coronoid process of mandible	Elevates and retracts the mandible; keeps mouth closed, teeth apposed	All innervated by mandibular division of trigeminal nerve (cranial nerve V)
Masseter (mah-SEAT-er)	Zygomatic process of maxilla and zygomatic arch	Ramus and coronoid process of mandible	Elevates and protracts mandible	—
Medial pterygoid (TEAR-ih-goyd)	Medial face of lateral pterygoid process, palatine bone, tuberosity of maxilla	Ramus and angle of mandible	Elevates and protracts mandible	—
Lateral pterygoid	Greater wing of sphenoid; lateral face of lateral pterygoid process	Neck of mandibular condyle and articular disk of temporomandibular joint	Depresses, protracts mandible; moves mandible sideways	—

From Moore PJ: Muscular and associated neurological systems. In Hilt NE and Cogburn SB, editors: Manual of orthopedics, St. Louis, 1980, CV Mosby Co.

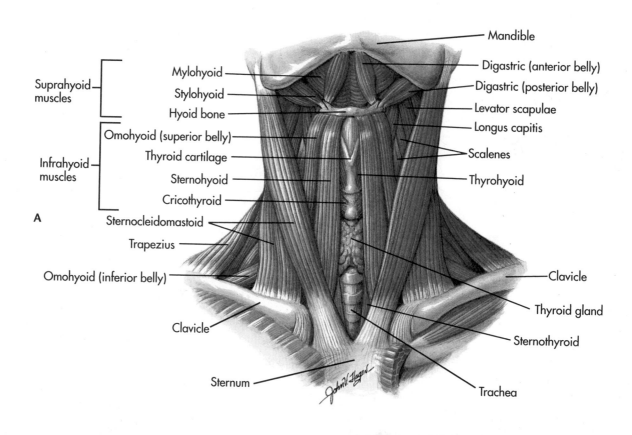

A

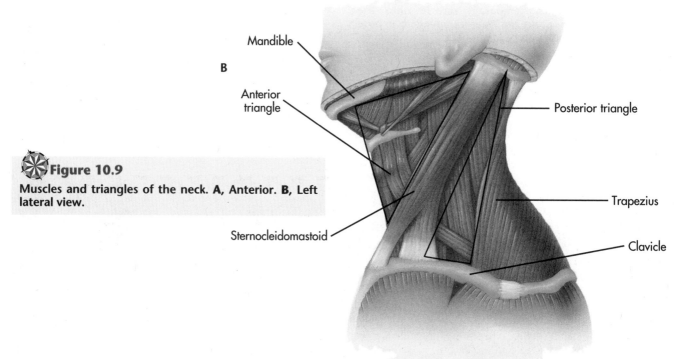

B

Figure 10.9
Muscles and triangles of the neck. A, Anterior. B, Left lateral view.

pterygoids also move the mandible sideways and grind the upper and lower molars together when alternate sides contract.

- Which muscles close the mouth?
- Which two muscles protract the mandible?
- What muscle would you use to move the mandible from side to side?

CERVICAL TRIANGLES
The sternocleidomastoid muscles divide the neck into anterior and posterior cervical triangles, the landmarks of the cervical muscles shown in Figure 10.9B. The anterior triangle is bordered by the sternocleidomastoid muscle, the inferior margin of the mandible, and the anterior midline of the neck.

Superficial muscles of the neck, together with muscles of the hyoid bone and larynx and the deeper anterior cervical muscles, reside in the anterior triangle. The lateral vertebral muscles lie within the posterior triangle that follows the posterior border of the sternocleidomastoid, the anterior border of the trapezius muscle, and the medial third of the clavicle.

Turn your head. Tilt it. Raise it up and look at the ceiling. Lower it to your chest. Rotate it through all these positions; you are using the muscles of the head and neck described in this section, just as, right now, they are holding your head stationary. Like the guy wires of a broadcasting tower, these muscles are stationed around all sides of the neck. By working together in various combinations, they move the head and neck. You will find the anterior and lateral cervical muscles in Table 10.4 and the posterior muscles in Table 10.10.

- Which cervical muscle forms the boundary between the triangles of the neck?
- Describe the boundaries of each triangle.

LATERAL CERVICAL MUSCLES

The sternocleidomastoid and trapezius muscles assist your head and neck motions. The sternocleidomastoid muscles (STER-no-KLY-doe-MAS-toyds) form a *V* on the front of the neck, shown in Figure 10.9 (Table 10.4), as they ascend from the manubrium and clavicle to the mastoid processes behind the ears. Together with the muscles of the back (Table 10.10), sternocleidomastoids flex the head when both muscles contract equally; they rotate or tilt the head when one pulls more strongly than the other (Figure 10.10).

The trapezius (tra-PEEZ-ee-us; Figure 10.9B) flexes the neck laterally and extends it when the shoulders are held stationary. The trapezius is a flat, fan-shaped muscle with origins that extend from the skull to the thoracic vertebrae and insertions on the lateral third of the clavicle, the acromion, and the scapular spine. Both sternocleidomastoid and trapezius are innervated by the accessory nerve and the cervical nerves. Chapter 11 describes how the trapezius also moves the shoulders.

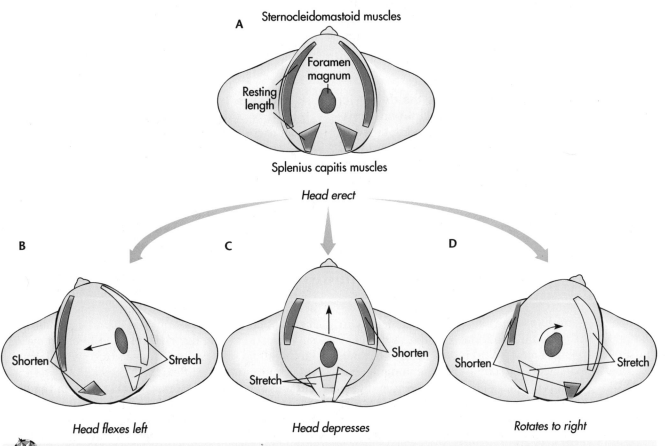

Figure 10.10

The action of sternocleidomastoids and splenius capitis in head movement, as seen from above through a transparent skull. **A,** When head is erect, muscles are at resting length. To move the head from this position, different sets of muscles shorten and stretch. **B,** Shortening the muscles on the left side and stretching them laterally on the right flexes the head to the left. **C,** Shortening the anterior muscles and stretching the posterior depresses the head. **D,** In rotation to the right, the left sternocleidomastoid and right splenius capitis shorten and the right sternocleidomastoid and left splenius capitis stretch. Contracted muscles are shown in *red* and stretched in *yellow*. What actions will turn the head to its left, and what additional motions are needed to depress it while turning?

ANTERIOR CERVICAL MUSCLES

Nearly everything moves nearly everything else in the anterior muscles of the neck. The **hyoid bone** is the center of the action, as Figure 10.9A illustrates. How these muscles of the anterior cervical triangle act is easier to understand when you realize that the mandible and the hyoid form a pair of arches below the skull in front of the vertebral column (Figure 10.11). The **mandible,** of course, frames the oral cavity. The hyoid bone forms a smaller arch below the mandible, where it supports the pharynx and larynx. In turn, styloid ligaments and the suprahyoid and infrahyoid muscles tether the hyoid to the skull and thorax. Your hyoid and larynx move up and down from this floating position when you swallow.

The **suprahyoid muscles** (Figure 10.9A and Table 10.4) attach the hyoid to the skull, so that when these muscles contract, the hyoid elevates. Suprahyoid muscles raise your larynx while you are swallowing, and they help to open your mouth. Below the hyoid, the **infrahyoid group** of muscles moors the hyoid bone to the sternum and clavicle and depresses the bone when the suprahyoids relax. When you lift your head from lying on your back, however, both groups pull upon the hyoid and the head rises.

SUPRAHYOID MUSCLES

The four suprahyoid muscles are the **digastric, stylohyoid, geniohyoid,** and the **mylohyoid** muscles, illustrated in Figure 10.9A and B and detailed in Table 10.4. All are strap muscles that raise the hyoid bone and draw it forward and back. They are innervated by branches of the trigeminal, facial, and first cervical nerves. The **digastric** is the most unusual because it has two bellies. Beginning near the mental symphysis of the mandible and at the mastoid notch of the temporal bone respectively, the anterior and posterior bellies insert on a tendon that joins the muscle to the body of the hyoid bone. The **stylohyoid** muscles suspend the hyoid from the styloid processes of the skull. With its origin also near the mental symphysis, the **geniohyoid** muscle extends posteriorly to the body of the hyoid bone, deep to the mylohyoid muscle described below. Working together, all three muscles raise the hyoid bone and the tongue, but when the geniohyoid and anterior belly of the digastric contract more firmly than the others, they protract the hyoid. The stylohyoid retracts the hyoid bone. All help depress the mandible when the infrahyoid muscles stabilize the hyoid.

The **mylohyoid** muscles form the floor of the oral cavity. These two broad, flat sheets of muscle converge from the interior surface of the mandible toward each other at the midline and insert upon the hyoid bone at the rear. Contraction of the mylohyoids raises the hyoid and the tongue. When the hyoid is held in place by the infrahyoid muscles below it, the mylohyoids help depress the mandible.

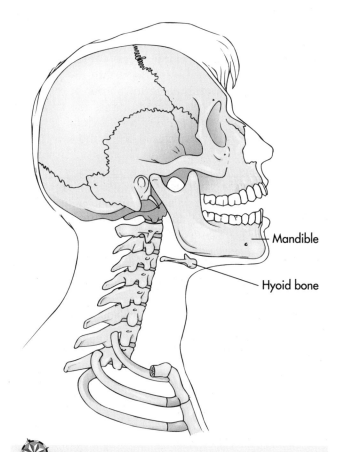

Mandible

Hyoid bone

✳ **Figure 10.11**

The mandible and hyoid bone form two bony arches in front of the vertebral column.

INFRAHYOID MUSCLES

Four infrahyoid muscles occupy the anterior triangle of the neck below the hyoid bone. As their names indicate and Figure 10.9A and B illustrates, the **sternohyoid, sternothyroid, thyrohyoid,** and **omohyoid** muscles (Table 10.4) tether the hyoid bone or thyroid cartilage of the larynx to the sternum, larynx, or scapula (omo-, *G,* shoulder). All are strap muscles. As a group, the infrahyoid muscles are antagonists of the suprahyoids when they depress the hyoid bone and larynx after swallowing. They also are synergists with the same muscles in flexing the neck. The sternohyoid and sternothyroid muscles originate on the manubrium of the sternum and insert, respectively, on the hyoid bone and on the thyroid cartilage of the larynx. Thyrohyoid muscles connect the larynx and hyoid bone. When the hyoid bone is stationary, the thyrohyoid muscle raises the larynx, and just the reverse occurs when the larynx is stabilized. The omohyoid muscle has two bellies like those of the digastric muscle. One inserts upon the hyoid bone and leads beneath the sternocleidomastoid muscle by way of an intermediate tendon to the second belly, which takes its origin from the scapula. Fibers from the first three cervical nerves serve the infrahyoid muscles.

Table 10.4 CERVICAL MUSCLES

Muscle	Origin	Insertion	Action	Innervation
LATERAL CERVICAL MUSCLES				
Sternocleidomastoid (STER-no-CLY-do-mast-oyd)	Anterior surface of manubrium and medial third of clavicle	Lateral surface of mastoid process, temporal bone	Together, flex head toward chest; separately, flexes neck and head laterally and turns face to opposite side	Spinal accessory (cranial nerve XI) and anterior primary rami of cervical nerves 2 and 3
Trapezius, superior fibers (tra-PEEZ-ee-us; See Table 11.1 for entire muscle)	External occipital protuberance, superior nuchal line, ligamentum nuchae	Lateral third of clavicle and acromion	Together, retract head; separately, draws head to same side and turns face to opposite side	Spinal accessory and cervical nerves 3 and 4
ANTERIOR CERVICAL MUSCLES				
Suprahyoid Group				
Digastric, anterior belly (dye-GAS-trik)	Inner surface, inferior border of mandible, near mental symphysis	Intermediate tendon, by way of suprahyoid aponeurosis to body and greater cornu of hyoid bone	With posterior belly, elevates hyoid bone, and by depressing the mandible, helps open mouth	Branch of mandibular division of trigeminal nerve (cranial nerve V)
Digastric, posterior belly	Mastoid notch of temporal bone	Intermediate tendon, by way of suprahyoid aponeurosis to body and greater cornu of hyoid bone	With anterior belly, elevates hyoid bone, and by depressing the mandible, helps open mouth	Branch of facial (cranial nerve VII)
Stylohyoid (STY-low-HY-oyd)	Styloid process	Body of hyoid bone at junction with greater cornu	Elevates and retracts hyoid bone, elongates floor of oral cavity	Branch of facial (cranial nerve VII)
Mylohyoid (MY-lo-HY-oyd)	Mylohyoid line, medial surface of mandibular arch	Posterior third on body of hyoid, anterior two thirds on median raphe with opposite partner	Elevates hyoid bone and tongue, helps depress mandible	Branch of mandibular division of trigeminal nerve (cranial nerve V)
Geniohyoid (JEEN-ee-oh-HY-oyd)	Inferior mental spine near mental symphysis deep to mylohyoid	Anterior body of hyoid bone	Elevates and protracts hyoid bone, shortens oral cavity; helps depress mandible	First cervical nerve via hypoglossal (cranial nerve XII)
Infrahyoid Group				
Sternohyoid	Medial end of clavicle and superior manubrium	Lower border body of hyoid bone	Depresses hyoid bone after swallowing; antagonist to elevators of hyoid and larynx	Fibers from anterior rami of cervical nerves 1-3
Sternothyroid	Posterior surface of manubrium and cartilage of first rib	Lamina of thyroid cartilage of larynx	Depresses larynx after swallowing	Fibers from anterior rami of cervical nerves 1-3
Thyrohyoid	Lamina of thyroid cartilage of larynx	Inferior border of greater cornu of hyoid bone	Depresses hyoid bone and elevates thyroid cartilage	Anterior rami of cervical nerve 1 via hypoglossal nerve (XII)
Omohyoid (OH-mo-HY-oyd)	Inferior belly, from upper border of scapula to intermediate tendon	Superior belly, from intermediate tendon to lower border of body of hyoid	Depresses hyoid bone after swallowing	Fibers from anterior rami of cervical nerves 1-3

From Moore PJ: Muscular and associated neurological systems. In Hilt NE and Cogburn SB, editors: Manual of orthopedics, St. Louis, 1980, CV Mosby Co.

Table 10.5 VERTEBRAL MUSCLES OF THE NECK

Muscle	Origin	Insertion	Action	Innervation
POSTERIOR VERTEBRAL MUSCLES	See Table 10.10 for the cervicus and capitis members of the semispinalis, spinalis and longissimus groups.			
ANTERIOR VERTEBRAL MUSCLES	These muscles flex the head and neck from origins on anterior surfaces of lower cervical and upper thoracic vertebrae and insertions on the occipital bone in front of the foramen magnum and on the upper cervical vertebrae.			
Longus colli (LON-gus KOAL-ee)	The longus colli muscle is a weak flexor of the neck; it slightly rotates and laterally flexes the neck. All three portions of the longus coli are innervated by branches from cervical nerves 2 through 6.			
Superior oblique portion	Transverse processes of cervical vertebrae 3-5	Anterior arch of atlas	—	—
Inferior oblique portion	Bodies of first three thoracic vertebrae	Transverse processes of 5th-6th cervical vertebrae	—	—
Vertical portion	Bodies of last three cervical and first three thoracic vertebrae	Anterior surface of bodies of cervical vertebrae 2-4	—	—
Longus capitis (LON-gus KAP-ih-tiss)	Transverse processes of cervical vertebrae 3-6	Basilar part of occipital bone	Flexes head and neck	Cervical nerves 1-3
Rectus capitis anterior (REK-tus)	Anterior surface of lateral mass of atlas	Basilar part of occipital bone anterior to foramen magnum	Flexes head and helps stabilize atlanto-occipital joint	First cervical nerve
Rectus capitis lateralis	Superior surface of transverse processes of atlas	Jugular process of occipital bone	Flexes head laterally and helps stabilize atlanto-occipital joint	Cervical nerves 1 and 2
LATERAL VERTEBRAL MUSCLES				
Scalenus anterior (SKAY-leen-us)	Transverse processes of cervical vertebrae 3-6	Inner border and upper surface of first rib	Raises first rib; flexes neck forward and laterally, rotates it toward opposite side	Cervical nerves 5 and 6 (sometimes 4)
Scalenus medius	Transverse processes of cervical vertebrae 3-8 and atlas	Superior surface of first rib	Similar; weaker rotation	Cervical nerves 3-8
Scalenus posterior	Transverse process of cervical vertebrae 4-6	Outer surface of second rib	Raises second rib; flexes neck forward and laterally, rotates it slightly toward opposite side	Cervical nerves 6-8

From Moore PJ: Muscular and associated neurological systems. In Hilt NE and Cogburn SB, editors: Manual of orthopedics, St. Louis, 1980, CV Mosby Co.

- Name the four suprahyoid muscles.
- What is unusual about the anatomy of the digastric muscle?
- Is there a double-bellied infrahyoid muscle?

VERTEBRAL MUSCLES OF THE NECK

These cervical muscles flex the head and neck and hold it in position by inserting on the vertebral column and skull. They lie deep to the muscles of the anterior and lateral cervical triangles. All are innervated by the cervical nerves (see Table 10.5).

ANTERIOR VERTEBRAL MUSCLES

These muscles can flex the neck and head because they are located anterior to the cervical vertebrae. Four fusiform muscles make up this group—the **longus colli,** the **longus capitis,** and the **recti capitis anterior** and **lateralis.** The longus colli is the largest and deepest muscle of the group; its three divisions flex the neck forward from origins on the bodies and insertions on the transverse processes of cervical and thoracic vertebrae. The longus capitis continues the action of the longus colli onto the skull.

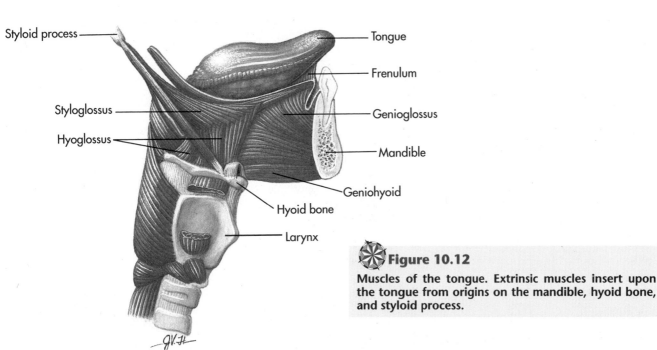

Styloid process —
Tongue —
Frenulum —
Styloglossus —
Genioglossus —
Hyoglossus —
Mandible —
Geniohyoid —
Hyoid bone —
Larynx —

Figure 10.12

Muscles of the tongue. Extrinsic muscles insert upon the tongue from origins on the mandible, hyoid bone, and styloid process.

In contrast to the longus muscles, the rectus capitis muscles are relatively broad, short muscles of the occipital bone that flex the head from origins on the atlas. These recti muscles cooperate with comparable rectus capitis posterior muscles (not shown) on the posterior surface of the skull to erect the head and stabilize the atlas.

LATERAL VERTEBRAL MUSCLES

All three muscles—**scalenus anterior, medius, and posterior**—arise from transverse processes of cervical vertebrae and descend to insertions upon the first or second rib at locations described by their names (see Figure 10.3A). From these locations the scalenes flex the neck forward and laterally. They also can rotate the neck by virtue of their origins on the tips of the transverse processes. Scalenes also raise the ribs, a movement that comes into play when you inhale deeply.

- What are the names of the anterior vertebral muscles?
- What role do the scalenus muscles play?

MUSCLES OF THE TONGUE, PHARYNX, AND LARYNX

TONGUE

The extrinsic and intrinsic muscles of the tongue maneuver food in the mouth and shape speech. The extrinsic muscles originate outside the tongue, and the intrinsic muscles reside entirely within the tongue. Being outside, the extrinsic muscles raise and lower the tongue and draw it forward and back in the mouth. Intrinsic muscles change the shape of the tongue (Figure 10.12 and Table 10.6).

EXTRINSIC MUSCLES

Four muscles belong to this group—the **genioglossus, hyoglossus, styloglossus,** and **palatoglossus.** Their names indicate that they insert onto the tongue (gloss-, *G,* the tongue; Figure 10.12) from various origins. All are innervated by the hypoglossal nerve (cranial nerve XII), except the palatoglossus muscle, which is served by the vagus nerve (cranial nerve X).

It is easy to visualize the action of each muscle when you are familiar with its origin. The fan-shaped genioglossus muscles (geni, *G,* the chin) reach from the mental symphysis of the mandible into the base of the tongue, where they can protrude the tongue and retract it into the oral cavity. When all fibers of the genioglossus shorten, the tongue becomes a trough that conveys liquid to the rear of the oral cavity where it is swallowed. The hyoglossus (hyo, *G,* hyoid) depresses the tongue, especially its sides, because this muscle extends from the hyoid bone onto the lateral base of the tongue. The styloglossus is a thin muscle arising from the styloid process above and behind the oral cavity; it elevates and retracts the tongue toward the back of the oral cavity. Finally, the palatoglossus muscle (Figure 10.14A), which enters the lateral base of the tongue from the soft palate, raises the tongue and depresses the soft palate—actions that lead food into the pharynx and the esophagus. The interplay of these structures in swallowing is described in Chapter 23, Digestive System.

INTRINSIC MUSCLES

Four intrinsic muscles change the shape of the tongue. The **superior** and **inferior longitudinal muscles** run longitudinally from the back of the tongue to its apex. The superior longitudinals curl the tongue

Table 10.6 MUSCLES OF THE TONGUE

Muscle	Origin	Insertion	Action	Innervation
EXTRINSIC GROUP	These muscles move the tongue within the oral cavity by inserting into the tongue from external origins.			
Genioglossus (JEEN-ee-oh-GLOSS-us)	Superior mental spine at mental symphysis	Body of hyoid bone and base of tongue, blends with intrinsic muscles	Protracts and retracts tongue, depresses trough in tongue when sucking	Hypoglossal nerve (cranial nerve XII)
Hyoglossus (HY-oh-GLOSS-us)	Lateral body and greater cornu of hyoid bone	Lateral base of tongue, blends with intrinsic muscles	Depresses tongue and its sides, flattens tongue	Same
Styloglossus (STY-low-GLOSS-us)	Styloid process and styloid ligament	Lateral base of tongue, blends with intrinsic muscles	Elevates and retracts tongue toward fauces	Same
Palatoglossus (PAL-a-toe-GLOSS-us)	Anterior surface of soft palate	Lateral base of tongue	Elevates, retracts base of tongue toward fauces in swallowing; depresses soft palate	Spinal accessory nerve (cranial nerve XI)
INTRINSIC GROUP	These muscles originate and insert entirely within the tongue; they change its shape.			
Superior longitudinal	Median septum and fibrous tissue near epiglottis	Anterior margins of tongue	Shortens, curls tongue upward	Hypoglossal nerve (cranial nerve XII)
Inferior longitudinal	Base of tongue	Apex of tongue	Shortens, curls tongue downward	Same
Transverse	Median septum	Lateral margins	Narrows, helps extend tongue	Same
Vertical	Anterior undersurface	Anterior upper surface	Flattens and broadens tongue	Same

From Moore PJ: Muscular and associated neurological systems. In Hilt NE and Cogburn SB, editors: Manual of orthopedics, *St. Louis, 1980, CV Mosby Co.*

upward, the inferior bend it downward, and the tongue shortens when both contract. **Transverse fibers** that are perpendicular to the longitudinals narrow the tongue and help extend it when longitudinal fibers relax. **Vertical fibers** that are perpendicular to the others flatten and broaden the tongue.

- What is the fundamental difference between the intrinsic and extrinsic muscles of the tongue?
- To say the letter *L*, the tongue extends forward to the roof of the mouth behind the incisor teeth. Which muscles would you expect to participate in this motion?

SOFT PALATE

The soft palate is a valve that separates the oral cavity from the nasal cavity (Figure 10.13). In the resting position, shown in the figure, the soft palate admits air through the nasal cavity but prevents air from flowing into the oral cavity. With the soft palate raised, air and food flow between the mouth and pharynx. You cough, blow, and say words such as "Pow" and "Bang" with the soft palate raised. When physicians ask you to say, "Ah," they see your soft palate in Figure 10.14A as a curtain, or **palatine velum** (VEEL-um; vel-, *L*, a veil), that separates the oral cavity from the pharynx behind. The conical **uvula** (YOU-view-la; uvul-, *L*, the palate) hangs from the midline of the velum. The **fauces** is the passage beneath the velum into the pharynx. Figure 10.14A and C and Table 10.7 illustrate the muscles of the soft palate, and Chapters 23 and 24, Digestive and Respiratory Systems, detail their operation.

Three muscles of the soft palate itself—the **levator veli palatini, tensor veli palatini** (elevator and tensor of the palatine velum), and the **uvulae** (shown in Figure 10.14A)—open and close

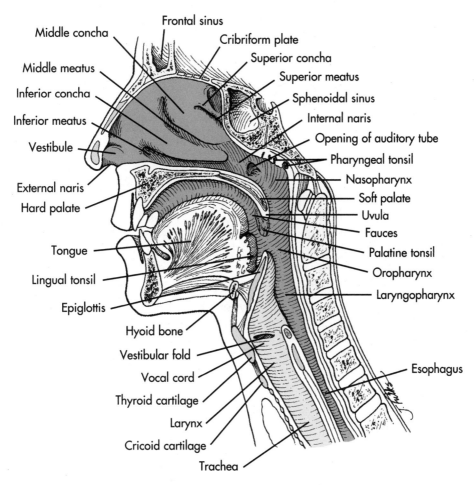

Middle concha
Frontal sinus
Cribriform plate
Middle meatus
Superior concha
Superior meatus
Inferior concha
Sphenoidal sinus
Inferior meatus
Internal naris
Vestibule
Opening of auditory tube
Pharyngeal tonsil
External naris
Nasopharynx
Hard palate
Soft palate
Uvula
Fauces
Tongue
Palatine tonsil
Lingual tonsil
Oropharynx
Epiglottis
Laryngopharynx
Hyoid bone
Vestibular fold
Vocal cord
Thyroid cartilage
Esophagus
Larynx
Cricoid cartilage
Trachea

Figure 10.13
Regions of the pharynx, midsagittal section viewed from right side.

the soft palate. The **palatoglossus** muscle of the tongue and the **palatopharyngeus** muscle of the pharynx help depress the soft palate. The accessory nerve (cranial nerve XI) innervates all of these muscles, except the tensor veli palatini, which receives from the trigeminal nerve (cranial nerve V).

- Which muscle elevates the soft palate and which ones depress it?
- What position does the palatine velum take when you cough?
- Which muscles accomplish this?
- Which muscles allow you to breathe through the nasal cavity?

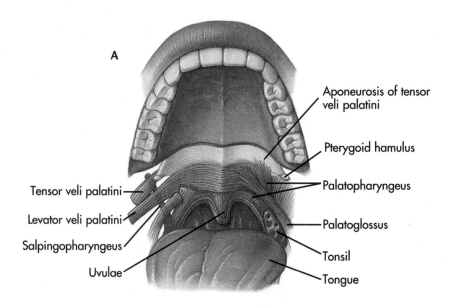

A

Aponeurosis of tensor veli palatini

Pterygoid hamulus

Palatopharyngeus

Palatoglossus

Tonsil

Tongue

Tensor veli palatini

Levator veli palatini

Salpingopharyngeus

Uvulae

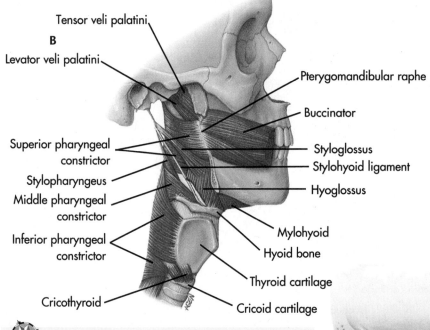

B

Tensor veli palatini

Levator veli palatini

Pterygomandibular raphe

Buccinator

Superior pharyngeal constrictor

Stylopharyngeus

Middle pharyngeal constrictor

Inferior pharyngeal constrictor

Cricothyroid

Styloglossus

Stylohyoid ligament

Hyoglossus

Mylohyoid

Hyoid bone

Thyroid cartilage

Cricoid cartilage

Figure 10.14

Muscles of the soft palate, pharynx, and larynx. A, Soft palate, viewed from the oral cavity. **B,** Muscles of the pharynx and larynx. The stylopharyngeus muscle raises the hyoid bone and larynx during swallowing, and the infrahyoid muscles (not shown) return it to its position afterwards. The pharyngeal constrictor muscle propels food down the pharynx. **C,** The pharynx has been opened along its posterior midline to show the flow of food and air. Air flows between the choanae and the larynx, and food flows between the oral cavity and the esophagus.

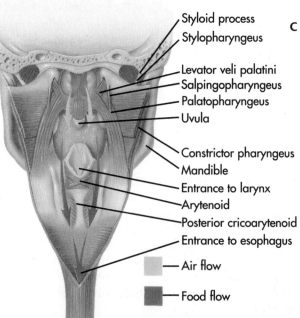

C

Styloid process

Stylopharyngeus

Levator veli palatini

Salpingopharyngeus

Palatopharyngeus

Uvula

Constrictor pharyngeus

Mandible

Entrance to larynx

Arytenoid

Posterior cricoarytenoid

Entrance to esophagus

Air flow

Food flow

Table 10.7   MUSCLES OF THE SOFT PALATE

Muscle	Origin	Insertion	Action	Innervation
Levator veli palatini (le-VAY-tor VELL-ee PAL-ah-tee-nee)	Apex of petrous part of temporal bone and medial wall of auditory tube	Palatine velum	Elevates soft palate	Accessory nerve (cranial nerve XI) via vagus nerve (cranial nerve X)
Tensor veli palatini (TEN-sore, etc.)	Medial pterygoid plate, sphenoid spine, and lateral wall of auditory tube	Aponeurosis of soft palate, posterior hard palate (horizontal part of palatine bone)	Tenses soft palate	Branch of mandibular ramus, trigeminal nerve (cranial nerve V)
Uvulae (YOU-view-lee)	Posterior nasal spine of palatine bone	Uvula	Elevates, shortens uvula	Accessory nerve via vagus
Palatoglossus	Anterior surface of soft palate	Lateral base of tongue	Elevates, retracts base of tongue toward fauces in swallowing; depresses soft palate	Same
Palatopharyngeus (PAL-ah-toe-fahr-IN-jee-us)	Soft palate	Posterior border of thyroid cartilage	Elevates pharynx to capture food boli in swallowing, and depresses soft palate	Same

From Moore PJ: Muscular and associated neurological systems. In Hilt NE and Cogburn SB, editors: Manual of orthopedics, St. Louis, 1980, CV Mosby Co.

PHARYNX

The pharynx is the tube that connects the oral and nasal cavities to the esophagus and trachea, as illustrated in Figure 10.13. The muscles of the pharynx act mostly when you swallow; they are relatively inactive in breathing (see Figure 10.14B and C and Table 10.8). Pharyngeal muscles raise and dilate the pharynx to receive food from the oral cavity, simultaneously closing off the nasal cavity and larynx. The pharynx then constricts around the bolus (pl. boli), or mass, of food and drives it downward into the esophagus, where other muscles conduct it to the stomach. Constrictor muscles propel liquids and solids downward, and longitudinal muscles raise the pharynx and larynx into position. Branches of the vagus nerve, the glossopharyngeal, and the accessory nerve innervate the pharyngeal muscles.

The **constrictor pharyngeus** muscle is a circular layer of three thin, muscular sheets that originate in front of the pharynx and spread around its walls to meet on the posterior midline (Figure 10.14B and C). These muscles relax below descending food boli and constrict above them; the second action presses the food into the space prepared by the first.

Three longitudinal muscles suspend the pharynx from the skull and raise the pharynx when you swallow. The **stylopharyngeus** (Figure 10.14B and C) descends from the styloid process, down the entire pharynx, deep to the constrictors, and inserts onto the larynx. The **salpingopharyngeus** (Figure 10.14A) takes a similar path to the base of the pharynx. Both muscles blend with the **palatopharyngeus** (Figure 10.14A and C) that descends from the soft palate to the larynx and the posterior wall of the pharynx, where it meets the fibers of the salpingopharyngeus and stylopharyngeus muscles.

- Name the circular muscle of the pharynx.
- Which muscles raise the pharynx?
- How does their longitudinal position assist this motion?

MUSCLES OF THE LARYNX

The larynx guards against the entry of food into the trachea during swallowing, and it houses the vocal cords that supply the sounds of speech (Figure 10.13). As in the tongue, extrinsic and intrinsic groups of muscles attend these duties. **Extrinsic muscles of the larynx** (LAHR-inks) include the suprahyoid and infrahyoid muscles (Figure 10.9A and Table 10.4) that originate outside the larynx. They raise and lower the larynx in swallowing. Other muscles, including the stylopharyngeus and palatoglossus listed in Table 10.9, also are involved in swallowing. Five **intrinsic muscles** originate and insert in the larynx, where they vary the pitch and texture of sound by controlling the length, tension, and distance between the vocal cords

Table 10.8   MUSCLES OF THE PHARYNX

Muscle	Origin	Insertion	Action	Innervation
CIRCULAR LAYER				
Constrictor pharyngeus (fahr-IN-jee-us)	Three overlapping, fan-shaped parts of this muscle originate on larynx, hyoid bone, and medial pterygoid plate and insert upon the dorsal midline of the pharynx.			
Superior constrictor	Medial pterygoid plate, pterygomandibular raphe, alveolar process of mandible	Dorsal midline of pharynx	Constricts pharynx around food boli in swallowing, drives boli down pharynx	Branches of accessory nerve (cranial nerve XI) via pharyngeal plexus and vagus nerve (cranial nerve X)
Middle constrictor	Greater and lesser cornua of hyoid bone, and stylohyoid ligament	Posterior portion, dorsal midline of pharynx	Same	Same
Inferior constrictor	Sides of cricoid and thyroid cartilages of larynx	Posterior portion, dorsal midline of pharynx	Same	Same
LONGITUDINAL LAYER				
Stylopharyngeus (STY-low-fahr-IN-jee-us)	Medial side, base of styloid process	Posterior border of thyroid cartilage, larynx	Draws sides of pharynx upward and laterally; dilates pharynx to receive food boli in swallowing	Branch of glossopharyngeal (cranial nerve IX)
Salpingopharyngeus (sal-PIN-joe-FAHR-in-jee-us)	Inferior surface, auditory tube	Palatopharyngeus muscle	Draws lateral walls of pharynx upward and inward; dilates pharynx	Branches of pharyngeal plexus
Palatopharyngeus (PAL-ah-toe-fahr-IN-jee-us)	Soft palate	Posterior border of thyroid cartilage	Elevates pharynx to capture food boli, and depresses soft palate in swallowing	Accessory (cranial nerve XI) via vagus nerve (cranial nerve X)

From Moore PJ: Muscular and associated neurological systems. In Hilt NE and Cogburn SB, editors: Manual of orthopedics, *St. Louis, 1980, CV Mosby Co.*

as air passes them (Figure 10.15, Table 10.9). All are innervated by branches of the vagus nerve (cranial nerve X). The **cricothyroid muscle** connects the thyroid and cricoid cartilages and levers them open, stretching the vocal cords for higher pitch. The **thyroarytenoid** muscles antagonize this action, shortening and relaxing the cords for lower pitch. The **lateral** and **posterior cricoarytenoid** muscles, seen in Figures 10.15C and 10.14C, are also antagonists of each other; they respectively narrow and widen the opening between the vocal cords, which makes the voice louder or quieter. The **arytenoid muscle** is the only unpaired muscle in the neck. It crosses between opposite arytenoid cartilages, and draws them closer together, constricting the entrance to the glottis during swallowing. See Chapter 24, Respiration, for the structure of the larynx.

- In what directions do the extrinsic muscles move the larynx?

- What is the function of the intrinsic laryngeal muscles?

MUSCLES OF THE TRUNK

Five major groups of muscles move the trunk. (1) The deep muscles of the back extend the vertebral column, flex it laterally, and also rotate it by pulling upon the ribs and vertebrae. (2) The abdominal muscles retain the abdominal and pelvic organs, and also flex and rotate the vertebral column by drawing the rib cage toward the pelvis. (3) Muscles of the wall of the thorax and of the diaphragm assist breathing. Muscles in the floor of the pelvis (4) and the perineum (5) also help to retain abdominal and pelvic organs. The perineal muscles also control the anal and urethral orifices, and participate in intercourse as part of the external genitalia. Tables 10.10 through 10.13

Table 10.9   MUSCLES OF THE LARYNX

Muscle	Origin	Insertion	Action	Innervation
EXTRINSIC GROUP	Members of the extrinsic group insert upon the hyoid bone or larynx from origins elsewhere, and these muscles therefore elevate and depress both structures. The suprahyoid muscles belong to this group (see Table 10.4). Others are pharyngeal muscles, also listed in Table 10.8.			
Stylopharyngeus (STY-low-fahr-IN-jee-us)	Medial side, base of styloid process	Posterior border of thyroid cartilage, larynx	Draws sides of pharynx upward and laterally. Dilates pharynx to receive food boli in swallowing	Branch of glossopharyngeal nerve (cranial nerve IX)
Palatopharyngeus (PAL-ah-toe-fahr-IN-jee-us)	Soft palate	Posterior border of thyroid cartilage	Elevates pharynx to capture food boli, and closes nasopharynx in swallowing	Accessory nerve (cranial nerve XI)
Pharyngeus, inferior constrictor (fahr-IN-jee-us)	Sides of cricoid and thyroid cartilages of larynx	Posterior portion, dorsal midline of pharynx	Constricts pharynx around food boli in swallowing, drives boli down pharynx	Branches of pharyngeal plexus, from vagus (cranial nerve X) and glossopharyngeal nerves
Pharyngeus, medial constrictor	Greater, lesser cornua of hyoid, and stylohyoid ligament	Same	Same	Same
INTRINSIC GROUP	The origins and insertions of muscles in this group lie entirely within the larynx. These muscles open and close the glottis during swallowing, or they regulate tension in the vocal cords for speech.			
Cricothyroid (KRY-ko-THIGH-royd)	Anterior and lateral part of cricoid cartilage	Inferior cornu and lamina of thyroid cartilage	Stretches vocal cords	External laryngeal branch of vagus (cranial nerve X)
Thyroarytenoid (THIGH-roe-ar-ih-TEE-noyd)	Inferior half of the angle of the thyroid cartilage	Base and anterior surface of arytenoid cartilage	Shortens and relaxes vocal cords; antagonizes cricothyroid	Recurrent laryngeal nerve, a branch of vagus
Posterior cricoarytenoid (KRY-ko-ar-ih-TEE-noyd)	Posterior surface of lamina of cricoid cartilage	Muscular process of arytenoid cartilage	Opens glottis, draws vocal cords apart	Recurrent laryngeal nerve
Lateral cricoarytenoid	Superior border of arch of cricoid cartilage	Muscular process of arytenoid cartilage	Closes glottis, draws vocal cords together; antagonizes posterior cricoarytenoid	Recurrent laryngeal nerve
Arytenoid (ar-ih-TEE-noyd)	Originates and inserts between arytenoid cartilages		Constricts entrance to glottis, bends epiglottis over glottis	Recurrent laryngeal nerve

From Moore PJ: Muscular and associated neurological systems. In Hilt NE and Cogburn SB, editors: Manual of orthopedics, *St. Louis, 1980, CV Mosby Co.*

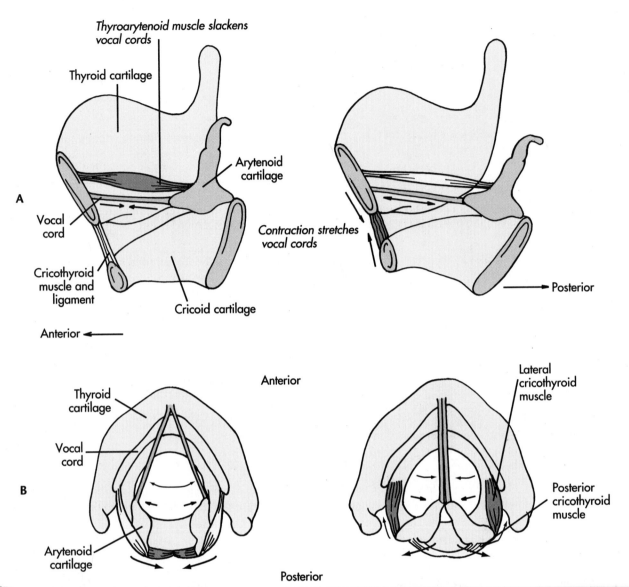

Thyroarytenoid muscle slackens vocal cords

Thyroid cartilage

Arytenoid cartilage

Vocal cord

Contraction stretches vocal cords

Cricothyroid muscle and ligament

Cricoid cartilage

Anterior ◄—

Posterior —►

A

Anterior

Thyroid cartilage

Vocal cord

Lateral cricothyroid muscle

Posterior cricothyroid muscle

Arytenoid cartilage

Posterior

B

✵ Figure 10.15

Intrinsic muscles of the larynx adjust the tension of the vocal cords and the opening between them. The thyroid cartilage of the larynx fits over the cricoid cartilage and the vocal cords stretch between both cartilages. **A,** Sagittal plane, right side. When the cricothyroid muscle contracts, the vocal cords tighten. Contraction of the thyroarytenoid muscle releases the tension of the vocal cords. **B,** View from above. The posterior and lateral cricoarytenoid muscles attach to each arytenoid cartilage. When the posterior cricothyroids contract, the vocal cords swing apart, and when the lateral cricothyroids shorten, the cords approach closely. Contracted muscles are shown in *red,* stretched in *yellow.*

give their origins, insertions, actions, and innervations.

DEEP MUSCLES OF THE BACK

The deep muscles of the back move and support the vertebral column. These muscles are covered by the more familiar superficial muscles that move the shoulders and upper limbs (Figure 10.1B). Four groups of strap-shaped muscles—the **erector spinae,** the **transversospinalis,** the **interspinal-intertransverse,** and the **splenius**—lace the ribs and vertebral

column with muscular support for the trunk. All are supplied by the spinal nerves (Table 10.10).

The erector spinae muscle spreads up the back, as shown in Figure 10.16, passing on either side of the vertebral column from its origins on the sacrum and ilium, and reaches all the way to the base of the skull. As its name indicates, the erector spinae holds the vertebral column erect. It also flexes the column laterally and extends it posteriorly (arches the back) from insertions among the ribs and vertebrae. Three overlapping subdivisions of the erector spinae—the **ilio-**

Table 10.10   DEEP MUSCLES OF THE BACK

Muscle	Origin	Insertion	Action	Innervation
Erector spinae (ee-REK-tor SPINE-eye)	This large, diverse muscle supports the vertebral column. It originates from the sacrum and ascends the trunk, attaching to vertebrae, ribs, and cranium and dividing into three parallel columns—the iliocostalis (lateral), longissimus (intermediate), and spinalis (medial). Each column itself subdivides into three regions of the trunk.			All innervated by posterior primary ramus of spinal nerves
Iliocostalis (ILL-ee-oh-kos-TAL-iss)	Sacrospinalis tendon of sacrum, the lumbar vertebrae, and the posterior iliac crest	Ribs and vertebrae	All divisions extend and flex vertebral column laterally	—
I. lumborum (lum-BOR-um)	Sacrum, lumbar, and T11-T12 vertebrae, posterior iliac crest	Lower six or seven ribs	Also depresses ribs	—
I. thoracis (thor-AH-kiss)	Lower six ribs	Upper six ribs and transverse process of C7 vertebra	—	—
I. cervicis (SIR-vih-kiss)	Ribs 3-6	Transverse processes of C4 to C6 vertebrae	—	—
Longissimus (lon-GISS-ih-mus)	Sacrospinalis tendon, thoracic and cervical vertebrae	Lumbar, thoracic, and cervical vertebrae and skull	Extends and flexes vertebral column and head laterally	All divisions, posterior primary ramus of spinal nerves
L. thoracis	Sacrospinalis tendon	Transverse processes of lumbar and thoracic vertebrae and lower 9 or 10 ribs	Extends and flexes vertebral column laterally; depresses ribs	—
L. cervicis	Transverse processes, T4, T5 vertebrae	Transverse processes, C2 to C6 vertebrae	Same	—
L. capitis (CAP-ih-tiss)	Upper 4 or 5 thoracic and lower 3 or 4 cervical vertebrae	Mastoid process	Extends head, flexes laterally, and rotates head	—
Spinalis (spy-NAL-iss)	Between L2 and C7 vertebrae	T4 to T8 vertebrae, axis and occipital bone	Extends vertebral column and rotates head	All divisions, posterior primary ramus of spinal nerves
S. thoracis	Spinous processes of T11 to L2 vertebrae; merges with longissimus	Spinous processes of T4 to T8 vertebrae	Extends vertebral column	—
S. cervicis	Ligamentum nuchae and spinous processes of C7 vertebra	Spinous process of axis	Extends vertebral column	—
S. capitis	Transverse processes, C7 to T7 vertebrae	Occipital bone, between superior and inferior nuchal lines	Extends head and rotates it to opposite side	—
Semispinalis	Between T10 and C7 vertebrae	Cervical and thoracic vertebrae and occipital bone	Extends and rotates vertebral column and head	All divisions, posterior primary ramus of spinal nerves
S. thoracis	Transverse processes, T6 to T10 vertebrae	Spinous processes of T1 to T4 and C7 to C8 vertebrae	Extends and rotates vertebral column	—
S. cervicis	Transverse processes, T1 to T5 or T6 vertebrae	Spinous processes from axis to C5 vertebrae	Extends and rotates vertebral column to opposite side	—
S. capitis (CAP-ih-tiss)	Transverse processes, C7 to T7 vertebrae	Occipital bone, between superior and inferior nuchal lines	Same; Extends and rotates head to opposite side	—

From Moore PJ: Muscular and associated neurological systems. In Hilt NE and Cogburn SB, editors: Manual of orthopedics, St. Louis, 1980, CV Mosby Co.

Continued.

Table 10.10 DEEP MUSCLES OF THE BACK—cont'd

Muscle	Origin	Insertion	Action	Innervation
Multifidus (mul-tih-FYE-dus)	Sacrum, iliac spine, transverse processes of lumbar, thoracic, and lower 4 cervical vertebrae	Spinous processes of second, third, or fourth vertebrae above origin	Extends vertebral column and rotates it to opposite side	Posterior primary ramus of spinal nerves
Rotatores (ROTE-ah-tore-es)	From transverse process of each vertebra	To base of spinous process of next vertebra above	Extends vertebral column and rotates it to opposite side	Posterior primary ramus of spinal nerves
Interspinales (inter-spy-NAL-eez)	From spinous process of each vertebra	To spinous process of next vertebra above	Extends vertebral column	Posterior primary ramus of spinal nerves
Intertransversarii (inter-TRANS-ver-SAIR-ee-ee)	From tubercles of transverse process of each vertebra	To tubercles of transverse processes of next vertebra above	Flexes vertebral column laterally	Anterior primary ramus of spinal nerves
Splenius group (SPLEEN-ee-us)	Thoracic and cervical vertebrae	Cranium and cervical vertebrae	Draws head and neck dorsally, laterally, and rotates them	All divisions, posterior primary ramus of spinal nerves
S. capitis	Spinous processes, C7 to T4 vertebrae	Superior nuchal line and mastoid processes	Same	—
S. cervicis	Spinous processes, T3 to T6 vertebrae	Spinous processes, C1 to C4 vertebrae	Same	—

From Moore PJ: Muscular and associated neurological systems. In Hilt NE and Cogburn SB, editors: Manual of orthopedics, St. Louis, 1980, CV Mosby Co.

costalis, the **longissimus** and the **spinalis**—ascend the lumbar, thoracic, and cervical regions of the vertebral column, respectively. The iliocostalis is the most lateral and has the strongest leverage to bend the column laterally; it reaches from the ilium to the ribs and cervical vertebrae. The longissimus takes an intermediate position over the ribs and transverse processes of vertebrae, where it assists extension and lateral flexing. Finally, nearest the midline the spinalis connects the spinous processes of vertebrae; it extends the vertebral column and helps to rotate the head.

Deep to the erector spinae, the transversospinalis group of muscles (**semispinalis, multifidus,** and **rotatores**) rotates the vertebral column and also helps the erector spinae extend and laterally flex it. These muscles link the transverse processes of a vertebra to the spinous processes of the next two vertebrae above (Figure 10.16 and 10.17), so that when muscles of both sides contract equally, the vertebral column extends, but when one side pulls harder than the other, the column rotates toward that side.

The interspinal-intertransverse group of muscles also extends and flexes the vertebral column but does not rotate it. These muscles lie deep to the rotatores (Figure 10.17). The interspinals are extensors that connect spinous processes of adjacent vertebrae, and the intertransverse muscles help flex the vertebral column laterally by connecting adjacent transverse processes of vertebrae.

The splenius muscles are cervical and upper thoracic muscles that extend the neck and rotate the head by connecting the spinous processes of cervical vertebrae to each other and to the occipital region of the skull.

- What is the name of the largest deep muscle of the back?
- What are the actions of this muscle?
- For what motions are the rotatores, the intertransverse group, and the splenius muscles responsible?
- Which muscles help extend the back?

ABDOMINAL MUSCLES

The abdominal wall contains broad, flat muscles whose fibers run obliquely, transversely, or longitudinally. Their contraction exerts pressure on the abdomen from all these directions (see Figure 10.18 and Table 10.11) and can flex and twist the abdomen when one muscle pulls more than another. The abdominal muscles also help expel contents of the abdominal cavity during vomiting, defecation, micturition (urination), and parturition (childbirth). All are innervated by branches of the thoracic spinal nerves.

The **external oblique** is the most superficial of three flat abdominal muscles. Its fibers originate from the lower ribs on right and left sides and course obliquely downward and medially across the ab-

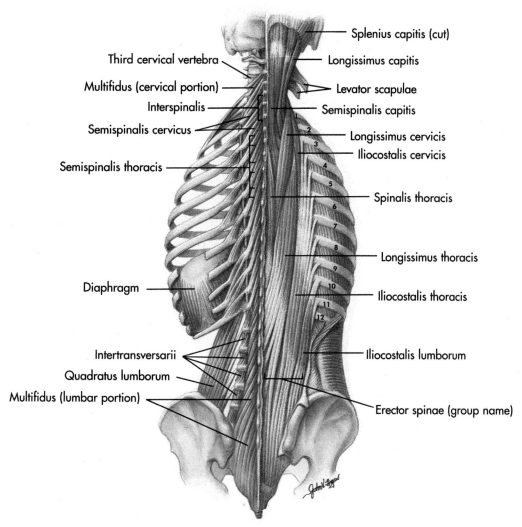

Splenius capitis (cut)

Third cervical vertebra

Longissimus capitis

Multifidus (cervical portion)

Levator scapulae

Interspinalis

Semispinalis capitis

Semispinalis cervicus

Longissimus cervicis

Iliocostalis cervicis

Semispinalis thoracis

Spinalis thoracis

Longissimus thoracis

Diaphragm

Iliocostalis thoracis

Iliocostalis lumborum

Intertransversarii

Quadratus lumborum

Iliocostalis lumborum

Multifidus (lumbar portion)

Erector spinae (group name)

✳ Figure 10.16

Deep muscles of the back. Erector spinae and its iliocostalis, longissimus, and spinalis divisions are shown on the right, but they have been removed to reveal elements of the multifidus, semispinalis, and intertransverse on the left side.

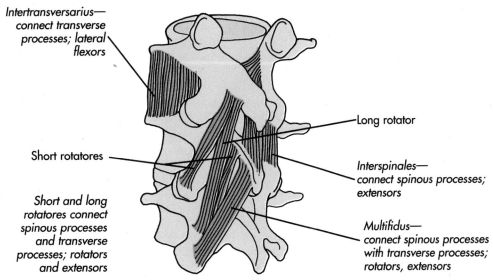

Intertransversarius— connect transverse processes; lateral flexors

Long rotator

Short rotatores

Interspinales— connect spinous processes; extensors

Short and long rotatores connect spinous processes and transverse processes; rotators and extensors

Multifidus— connect spinous processes with transverse processes; rotators, extensors

✳ Figure 10.17

Three cervical vertebrae reveal the multifidus, rotatores, and interspinal-intertransverse muscles.

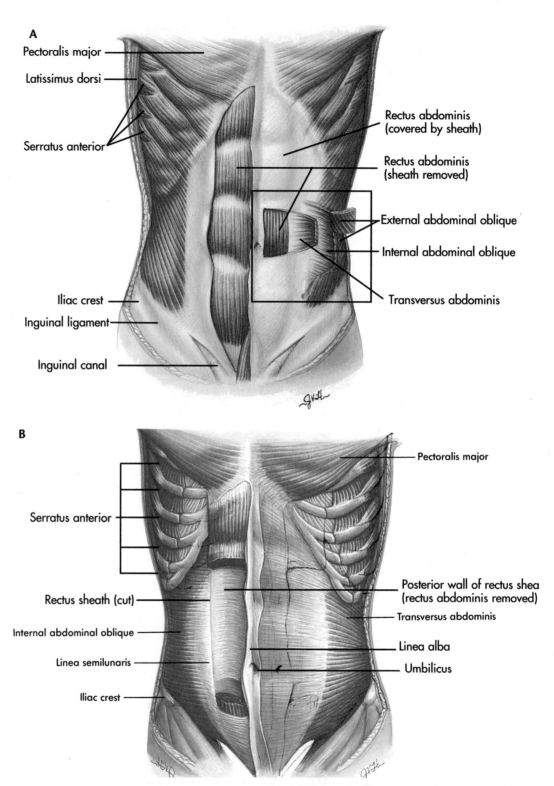

A

Pectoralis major

Latissimus dorsi

Serratus anterior

Rectus abdominis (covered by sheath)

Rectus abdominis (sheath removed)

External abdominal oblique

Internal abdominal oblique

Transversus abdominis

Iliac crest

Inguinal ligament

Inguinal canal

B

Serratus anterior

Pectoralis major

Posterior wall of rectus shea (rectus abdominis removed)

Transversus abdominis

Rectus sheath (cut)

Internal abdominal oblique

Linea semilunaris

Linea alba

Umbilicus

Iliac crest

⊛ **Figure 10.18**

Muscles of the abdomen. A, External oblique and rectus abdominis. The fibrous sheath around the rectus has been re-moved on the right side to show the muscle within. B, Transverse abdominis. External and internal obliques and the rectus abdominis have been removed on both sides to show the transverse abdominis beneath.

Table 10.11   MUSCLES OF THE ABDOMINAL WALL

Muscle	Origin	Insertion	Action	Innervation
Rectus abdominis (REK-tus ab-DAHM-in-is)	By tendon from the crest of the pubis and symphysis pubis	Fifth, sixth, and seventh costal cartilages	Flexes vertebral column, depresses thorax, elevates pelvis; tenses anterior abdominal wall, compresses abdominal contents in parturition, vomiting, urination, defecation	Seventh to twelfth intercostal nerves
External oblique	Inferior borders of lower eight ribs	Anterior iliac crest, and by aponeurosis to the linea alba and inguinal ligament	Flexes, bends, rotates vertebral column; tenses abdominal wall and assists forced exhalation, parturition, vomiting, urination, and defecation	Seventh to twelfth intercostal nerves
Internal oblique	Inguinal ligament, iliac fascia, iliac crest, and lumbar aponeurosis	Lower 3 or 4 costal cartilages, and by aponeurosis to the linea alba	Same as external oblique	Seventh to twelfth intercostal nerves and first lumbar nerve
Transverse abdominis	Inguinal ligament, iliac crest, thoracolumbar fascia, and cartilages of lower six ribs	By aponeurosis to the linea alba	Constricts abdominal wall and assists forced exhalation, parturition, vomiting, urination, and defecation	Seventh to twelfth intercostal nerves and first lumbar nerve
Quadratus lumborum (KWAD-rate-us lum-BOR-um)	Iliolumbar ligament and iliac crest	Inferior border 12th rib and transverse processes of first four lumbar vertebrae	Laterally flexes or extends the lumbar spine	Twelfth thoracic and first three lumbar nerves.
Cremaster (KREE-mast-er)	Inguinal ligament	Crest of pubis and sheath of rectus abdominis	Elevates testis toward inguinal ring	First and second lumbar nerves
Pyramidalis (peer-AM-id-al-iss)	Anterior surface of pubis	Linea alba midway between pubis and the umbilicus	Tenses linea alba; sometimes absent	Twelfth thoracic nerve

From Moore PJ: Muscular and associated neurological systems. In Hilt NE and Cogburn SB, editors: Manual of orthopedics, *St. Louis, 1980, CV Mosby Co.*

domen to broad aponeuroses that insert on the **linea alba** (LIN-ee-ah AL-ba; *L,* white line), a narrow, white, median band of fascia that extends from the sternum to the pubis. This fascia is a **raphe** (raph, *G,* a seam; RAY-fee) that formed where developing body folds fused and closed the abdomen. Other fibers of the external oblique muscle insert upon the inguinal ligament and the iliac crest of the pelvis. The **internal oblique** muscle is deep to the external oblique; its fibers reach medially upward from the iliac crest and lumbar region to insert upon the lower costal cartilages and upon an aponeurosis that also leads to the linea alba. The fibers of both oblique muscles can flex and twist the trunk because they run nearly perpendicular to each other. The **transverse abdominis,** deepest of these three, constricts the abdomen with transverse fibers that originate from the lower ribs, the lumbar region, and the pelvis and insert medially on the linea alba by way of an aponeurosis.

Two longitudinal muscles—the **rectus abdominis** and **quadratus lumborum**—bend the trunk. The recti abdominis are a pair of long, flat muscles that arise from the pubis and reach upward to the costal cartilages at the inferior end of the sternum (Figure 10.18A). These muscles lie on each side of the linea alba in the **rectus abdominis sheath,** which is formed by the aponeuroses of the obliques and transversus. The recti gain their impressive appearance from three **transverse tendinous bands** that interrupt their bellies. The quadratus lumborum occupies the posterior abdominal wall adjacent to the lumbar region, as shown in Figure 10.16. The quadratus lumborum muscles originate on the iliac crest and insert on the lowest rib and first four lumbar vertebrae, where they flex the vertebral column laterally and extend it.

The **cremaster** muscle is a continuation of the internal oblique muscle. In males, fibers of the cremaster loop downward, like a sling, into the scrotum and draw it upward when they contract. The cremaster is missing or much reduced in females. Because the first and second lumbar spinal nerves innervate the cremaster, rather than thoracic spinal nerves, physicians use this reflex action to help locate spinal injuries in accident victims.

- Name three abdominal muscles that bend the trunk and vertebral column forward.
- Which of these abdominal muscles also twist the trunk?
- Which muscles must especially contract to twist the trunk to the right?

THORACIC WALL

The intercostal muscles and the diaphragm are responsible for ventilating the lungs. As their name indicates, the intercostal muscles are located between the ribs. The muscles of the diaphragm help partition the thoracic cavity from the abdominal cavity (Figure 10.19 and Table 10.12). Both sets of muscles, but primarily the diaphragm, force air in and out of the lungs when you inhale and exhale. Branches of the thoracic spinal nerves innervate all intercostal muscles, but the diaphragm is innervated by branches of cervical spinal nerves 3 through 5 that are carried in the phrenic nerve.

Three sets of intercostal muscles are located between the ribs, as you can see in Figure 10.19A and B. The first two sets—the **external intercostal muscles** and the **internal intercostals**—can be thought of as continuations of the abdominal obliques, interrupted by the ribs. The external intercostal muscles are the most robust, each sending oblique fibers inferiorly and medially to the adjacent rib. Deep to the external intercostals, the **internal intercostals** send fibers inferiorly and laterally. Although it seems obvious that the external intercostals raise the ribs when you inhale and the internal intercostals depress them when you exhale, their actual roles are more complicated and uncertain because the external intercostals also contract early in exhalation.

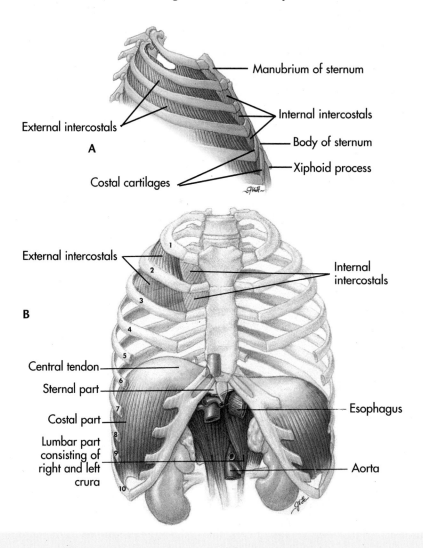

Figure 10.19

Muscles of the thorax. A, A portion of the upper rib cage shows external and internal intercostal muscles that occupy the intercostal spaces between all ribs. **B,** The diaphragm takes its origins from the margins of the rib cage and from the vertebral column. Locate the costal, sternal, and lumbar parts of this muscle.

A third series of smaller strap muscles inserts on the ribs from origins on the sternum or the vertebrae. The **transverse thoracis** muscles (Figure 10.19B) help depress the ribs, and the **levatores costorum** and **serratus posterior superior** elevate the ribs from origins on the vertebral processes. Lower in the thorax the **serratus posterior inferior** helps stabilize the diaphragm by drawing the last four ribs outward and downward when the diaphragm contracts.

The **diaphragm muscle** lies in the dome-shaped fibromuscular **diaphragm,** illustrated in Figure 10.19C, that partitions the thoracic cavity from the abdominal cavity. This broad, flat muscle originates from the inferior margin of the rib cage and from lumbar vertebrae and inserts upon the broad **central tendon** of the diaphragm. (A better name for this tendon would have been aponeurosis. Why?) Costal, sternal, and lumbar parts of this muscle are associated with the ribs, sternum, and the lumbar vertebrae, respec-

tively. Contraction of the diaphragm muscle depresses the tendon during inhalation and thereby causes air to rush into the lungs; relaxation of the muscle helps exhalation.

- Which two sets of muscles are responsible for inhalation of air into the lungs?
- Where are their origins and insertions located?

FLOOR OF THE PELVIS

The floor of the pelvis spans the pelvic outlet with muscle and fascia that retain the pelvic organs within the pelvic cavity. These organs rest on the **pelvic diaphragm,** a kind of sling or hammock within the floor of the pelvis that is suspended from the true pelvis. The **urogenital diaphragm,** external to the pelvic diaphragm, reinforces it against the considerable pressure of the abdominal and pelvic organs. The muscles of these diaphragms are concerned with defecation, micturition, intercourse, and childbirth because the anus, urethra, and the vagina in fe-

Table 10.12   MUSCLES OF THE THORACIC WALL

Muscle	Origin	Insertion	Action	Innervation
External intercostals (inter-KOS-tals)	Inferior border of rib	Superior border of next rib below	Elevates ribs	Intercostal branches of thoracic nerves T1 to T11
Internal intercostals	Longitudinal ridge on inner surface of ribs and costal cartilages	Superior border of next rib below	Elevates costal cartilages of first four ribs during inhalation, depresses all ribs in exhalation	Intercostal branches of thoracic nerves T1 to T11
Transversus thoracis (trans-VER-sus thor-AH-Kiss)	Inner surface of sternum and xiphoid process;sternal ends of costal cartilages of ribs 3-6	Inner surfaces and inferior borders of costal cartilages 3-6	Depresses ribs	Intercostal branches of thoracic nerves T3 to T6
Levatores costorum (le-VATE-ors KOST-or-um)	Ends of transverse processes of vertebrae C7, T2 to T12	Outer surface of angle of next rib below origin	Elevates ribs, bends vertebral column laterally	Intercostal branches of thoracic nerves
Serratus posterior superior (ser-RATE-us)	Ligamentum nuchae, and spinous processes of C7 and T1 to T2 or T3 vertebrae	Superior borders lateral to angles of ribs 2-5	Elevates upper ribs	Branches from anterior primary rami of nerves T1 to T4
Serratus posterior inferior	Spinous processes of T10 to T12 and L1 to L3 vertebrae	Inferior borders lateral to angles of last four ribs	Counteracts inward pull of diaphragm by drawing last four ribs outward and downward	Branches from anterior primary rami of nerves T9 to T12
Diaphragm (DYE-ah-fram)	Circumference of thoracic inlet from xiphoid process, costal cartilages 6-12, and lumbar vertebrae	Central tendon of diaphragm	Depresses and draws central tendon forward in inhalation; reduces pressure in thoracic cavity and raises pressure in abdominal cavity	Phrenic nerve, from cervical nerves 3 to 5

From Moore PJ: Muscular and associated neurological systems. In Hilt NE and Cogburn SB, editors: Manual of orthopedics, St. Louis, 1980, CV Mosby Co.

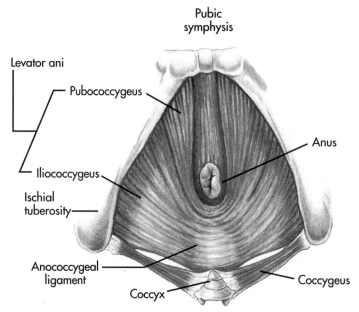

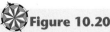

Figure 10.20

Floor of the female pelvis viewed from below, consists of the pelvic diaphragm, which contains the levator ani and coccygeus muscles that support the pelvic organs.

males penetrate one or both. All are innervated by branches of the sacral spinal nerves.

The pelvic diaphragm consists of two muscles—the levator ani and the coccygeus muscles (Table 10.13 and Figure 10.20)—that raise and lower the floor of the pelvis and the coccyx. These muscles are attached to the pubis, the ischium, and the coccyx, which mark the pelvic outlet. Viewed from below in the figure, the **levator ani** is a broad, fan-like sheet of muscle whose fibers converge from the pubis and ischium upon the coccyx and the **anococcygeal ligament** (AY-no-KOK-

sid-jee-al) that links the anus to the coccyx. Like the linea alba of the abdomen, the anococcygeal ligament is a raphe that closed the developing pelvis. The **iliococcygeus** portion of the levator ani originates from the ilium posteriorly, and the **pubococcygeus** originates anteriorly from the pubis. The **coccygeus** muscle is also fan-shaped, but its fibers diverge in the opposite direction from the spine of the ischium to the coccyx and the sacrum.

Surprisingly, for so basic a function the actions of the levator ani and coccygeus muscles in defecation

Table 10.13	MUSCLES OF THE FLOOR OF THE PELVIS			
Muscle	**Origin**	**Insertion**	**Action**	**Innervation**
Levator ani (le-VAY-tor AIN-eye)	Inner surface of the ischial spine and tendinous arch below brim of true pelvis	Lower two coccygeal vertebrae, external anal sphincter, and anococcygeal ligament	Hammock-like pelvic diaphragm; retains and elevates pelvic organs	Fibers of pudendal plexus from sacral nerve 4 and sometimes from 3 to 5
Iliococcygeus portion	Spine of the ischium	Lower two coccygeal vertebrae and anococcygeal ligament	Same	Same
Pubococcygeus portion	Tendinous arch	Coccyx, anococcygeal ligament, and external anal sphincter	Same	Same
Coccygeus (kok-SID-jee-us)	Inner surface of ischial spine and sacrospinous ligament	Margin of coccyx and lower sacrum	Draws coccyx anteriorly after defecation or parturition; retains pelvic organs	Fibers of pudendal plexus from sacral nerves 4 and 5

From Moore PJ: Muscular and associated neurological systems. In Hilt NE and Cogburn SB, editors: Manual of orthopedics, *St. Louis, 1980, CV Mosby Co.*

are not well understood because it is difficult to see them in action. The levator ani relaxes, allowing the rectum to fill and to descend more deeply into the pelvis. As the rectum fills, abdominal muscles pressurize the contents. Muscle fibers in the wall of the rectum contract involuntarily and the anal sphincter opens, while the levator ani, now contracting around the extended rectum, aims the feces away from the body.

- What is the pelvic diaphragm?
- How does it resemble a sling?
- What are the names of the two major muscles of this diaphragm?
- When does the levator ani relax and contract during defecation?

PERINEUM

The perineum is the diamond-shaped region external to the floor of the pelvis that is bounded by the pubis, the coccyx, and the ischial tuberosities (Figure 10.21), the same landmarks that identify the floor of the pelvis internally. The muscles of the perineum (PAIR-in-ee-um) extend like spokes of a wheel from its borders toward the central hub, the **perineal body,** which is a fibrous disk that is held in tension by all seven perineal muscles. The perineal body also unites with the urogenital diaphragm and with the pelvic diaphragm beneath it. Through their many links to the floor of the pelvis, the perineal body and the perineal muscles affect the orifices of the anus, urethra, and vagina and help erect the penis and clitoris. Rupturing the perineal body can release the abdominal and pelvic organs, a potentially disastrous situation called **prolapse.**

Find the **external anal sphincter** muscle in both sexes (Figure 10.21A and B and Table 10.14, bottom). This muscle originates from the anococcygeal ligament, passes on each side of the anus, and inserts on the perineal body, to close the anal opening. This sphincter relaxes voluntarily in defecation, together with the **internal anal sphincter** that lies deep to it (not shown). The **superficial** and **deep transverse perineus** muscles hold the perineal body in position with fibers from the ischial spine (review the ischial spine in Figure 7.7).

The **bulbospongiosus** and **ischiocavernosus** muscles differ considerably between the sexes. The bulbospongiosus muscle originates from the perineal body and inserts either on the penis or the clitoris. In males the bulbospongiosus helps erect the penis by encircling the corpus spongiosum (erectile shaft of the penis, Chapter 26), as well as squeezing the urethra empty after urination and ejaculation. This muscle also constricts the vaginal opening in females. Originating from the ischial tuberosity and ramus of the ischium, the ischiocavernosus muscle inserts at the base of the penis or clitoris, where it promotes erection of either structure by preventing blood from leaving the corpora cavernosae (see Chapter 26). The **urethral sphincter** muscle (not shown in Figure 10.21) also helps eject urine from the urethra after micturition. In both sexes it envelops the membranous portion of the urethra that passes through the urogenital diaphragm. In males it also ejects semen remaining in the urethra after ejaculation.

Older people may become incontinent when the perineal muscles weaken, but exercise can help

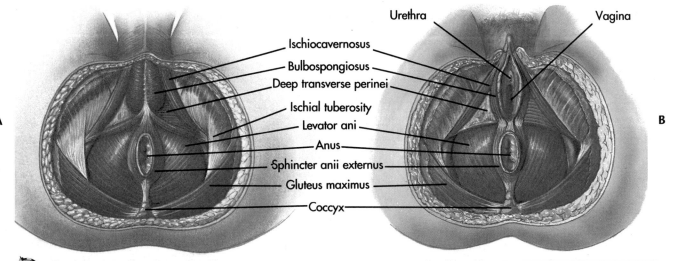

A   B

Urethra   Vagina

Ischiocavernosus
Bulbospongiosus
Deep transverse perinei
Ischial tuberosity
Levator ani
Anus
Sphincter anii externus
Gluteus maximus
Coccyx

Figure 10.21
Muscles of the perineum seen from below. A, Male. B, Female.

Table 10.14   MUSCLES OF THE PERINEUM

Muscle	Origin	Insertion	Action	Innervation
MALE, SUPERFICIAL GROUP				
Superficial transverse perineus (PAIR-in-ee-us)	Inner anterior ischial tuberosity	Perineal body	Stabilizes perineal body	All muscles by perineal branch of pudendal nerve from sacral nerves 2-4
Bulbospongiosus (BUL-bo-SPON-jee-oh-sus)	Perineal body	Encircles corpus spongiosum, meets corpora cavernosae of penis	Empties urethra after micturition, helps erect penis	
Ischiocavernosus (ISS-she-oh-KAV-ern-oh-sus)	Inner surface of ischial tuberosity and rami of ischium and pubis	Sides and undersurface of crus of corpora cavernosae of penis	Compresses crus penis and maintains erection by retarding emptying of corpora cavernosae	
DEEP GROUP				
Deep transverse perineus	Ramus of the ischium inferior surface	Forms median raphe (RAH-fay) with opposite partner in urogenital diaphragm	Supports urogenital diaphragm and compresses membranous portion of urethra	All muscles by perineal branch of pudendal nerve from sacral nerves 2-4
Urethral sphincter	Junction of rami of ischium and pubis, and local fascia	Envelops membranous portion of urethra and meets opposite partner	Sphincter of the membranous urethra, ejecting urine after micturition and seminal fluid in ejaculation	
FEMALE, SUPERFICIAL GROUP				
Superficial transverse perineus	Inner anterior ischial tuberosity	Perineal body	Stabilizes perineal body	All muscles by perineal branch of pudendal nerve from sacral nerves 2-4
Bulbospongiosus	Perineal body	Corpora cavernosae of clitoris	Constricts vaginal orifice	
Ischiocavernosus	Inner surface of ischial tuberosity and ramus of ischium	Sides and undersurface of crus of clitoris	Compresses crus of clitoris, and maintains erection by retarding emptying of corpora cavernosae	
DEEP GROUP				
Deep transverse perineus	Ramus of the ischium, inferior surface	Walls of vagina	Supports urogenital diaphragm and stabilizes perineal body	All muscles by perineal branch of pudendal nerve from sacral nerves 2-4
Urethral sphincter	Margin of inferior ramus of pubis	Envelops membranous portion of urethra and meets opposite partner	Sphincter of the membranous urethra, ejects urine after micturition	

From Moore PJ: Muscular and associated neurological systems. In Hilt NE and Cogburn SB, editors: Manual of orthopedics, *St. Louis, 1980, CV Mosby Co.*

them regain control. Women know they are controlling the bulbospongiosus and levator ani muscles when they constrict the vaginal opening. When a man moves the levator ani muscle while standing, his penis rises and falls slightly. Verify these statements by examining Figure 10.21A and B again.

- What are the names of the perineal muscles?
- Where is the perineal body located among these muscles?
- What are the actions of the bulbospongiosus and ischiocavernosus muscles?

DEVELOPMENT

The muscles of the trunk and vertebral column develop from somites, clusters of cells that form the skeleton, dermis, and skeletal muscles. Some 44 pairs of somites (SO-mites) appear beside the spinal cord and prospective vertebral column by the fourth week of fetal development, where they represent axial segmentation in the human body plan (see Chapter 1 and Chapter 4, Somites, Figure 4.7). Each somite produces a strip of muscle-forming tissue called a myotome (MY-oh-tome) that, together with other myotomes, envelops the trunk with muscles (see Chapter 9, Devel-opment of Muscle). As a result, different myotomes provide the muscles of the cervical, thoracic, lumbar, sacral, and coccygeal regions (Figure 10.22). Four occipital pairs of myotomes furnish certain muscles of the head, although some skeletal muscles, especially in the neck, do not arise from myotomes.

Each pair of myotomes becomes innervated by a corresponding pair of nerves from the spinal cord, so that cervical nerves, for example, innervate cervical myotomes and thoracic nerves supply the thoracic myotomes. This correspondence develops because one pair of spinal nerves innervates each pair of myotomes. Even though a myotome may migrate from its site of origin, it tows its spinal nerve along with it, so you can deduce the source of a muscle if you know which spinal nerve innervates it. This relationship is most obvious in the intercostal muscles. Each set of intercostal muscles (Table 10.12) receives from its own thoracic spinal nerve. The first intercostals are innervated by the first thoracic nerve and so on. Other muscles, such as the diaphragm, which is innervated by cervical nerves 3 through 5 (same table), have migrated from their original cervical myotomes.

- How many pairs of spinal nerves innervate each pair of myotomes?

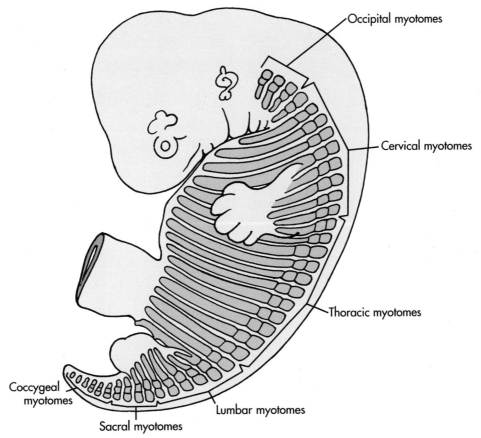

✳ Figure 10.22

Myotomes of a 6-week-old embryo, left side. Each myotome is innervated by a corresponding spinal nerve.

SUMMARY

Muscles are the organs of motion. They act in groups that flex and extend the joints of the skeleton. The positions of muscle groups on the skeleton determine their actions. Prime movers on anterior surfaces flex the trunk and neck, but muscle groups on the posterior surfaces extend the vertebral column and trunk. Antagonistic pairs of muscles usually work from opposite sides of the member they move, whereas synergists operate from the same side. Fixators hold parts stationary while other parts are moving. *(Page 269.)*

Muscles, bones, and joints act as levers. In first-class levers the muscle and load are on opposite sides of the joint. The load is between the joint and muscle in second-class levers, and in third-class levers the muscle lies between the load and joint. Most skeletal muscles operate third-class levers that enable short muscles to move bones long distances. *(Page 269.)*

A muscle's action depends on its shape. Muscles composed of parallel fibers concentrate their forces directly on narrow tendons or spread them across broad aponeuroses. The oblique fibers of pennate muscles are very powerful. Fan-shaped muscles focus forces onto narrow insertions or, conversely, spread them out from a small origin. Sphincters control orifices through circular bands of fibers. *(Page 272.)*

Muscles of the head include the facial muscles, extrinsic muscles of the eyeballs, and the muscles of mastication. Facial muscles govern facial expressions, extrinsic muscles rotate the eyeballs, and the muscles of mastication move the mandible. *(Page 275.)*

The anterior and lateral cervical triangles are landmarks of the cervical muscles. The anterior triangle contains the suprahyoid and infrahyoid muscles that move the hyoid bone and larynx. The anterior and lateral vertebral muscles of the neck in both cervical triangles flex the neck. Muscles of the tongue, soft palate, pharynx, and larynx govern swallowing and speech. *(Page 280.)*

Five major groups of muscles move the trunk. (1) Deep muscles of the back extend and rotate the vertebral column. (2) Abdominal muscles flex and rotate the trunk and retain the abdominal organs. (3) Thoracic muscles participate in breathing. (4) Muscles in the floor of the pelvis retain the pelvic organs, and (5) the perineal muscles control the urethral, vaginal, and anal orifices and assist intercourse. *(Page 290.)*

Muscles of the head, neck, and trunk develop from somites and myotomes, which represent segments in the human body plan. The muscles that develop from a particular myotome are innervated by branches of the corresponding spinal nerve. *(Page 303.)*

STUDY QUESTIONS
OBJECTIVES

1. Where are the erector spinae, sternocleidomastoid, rectus abdominis, levatores costorum, levator ani, male and female bulbospongiosus, scalenus anterior, orbicularis oris, superior rectus, temporalis, styloglossus, and stylopharyngeus muscles located?   **1**
2. Describe the action(s) of each muscle listed in Question 1.   **1**
3. Use the following terms: origin, insertion, prime mover, muscle group, antagonistic pair, synergist, and fixator to describe the action of the eyeball muscles, illustrated in Figure 10.7 and listed in Table 10.2.   **2**
4. List examples of flexor, extensor, lateral flexor, and rotator muscles in the trunk and neck.   **3**
5. What are the differences between the three classes of levers that muscles move in the skeleton? Describe the action of a muscle in each class.   **4**
6. Name the different shapes of muscles and describe an example of each type.   **5**
7. What is the shape of each of these muscles: the external oblique muscle of the abdomen, the temporalis, the scalenus anterior, orbicularis oculi, and the levator ani muscle?   **5**
8. Which muscles of the orbits rotate the eyeballs upward?   **2, 6**
9. Which muscles close the mandible?   **6**
10. Which intrinsic and extrinsic muscles protract the tongue?   **6**
11. Describe several prime movers that are responsible for the smile in Figure 10.6.   **6**
12. Which muscles elevate the hyoid bone?   **6**
13. Which muscles of the larynx and pharynx raise the larynx in swallowing?   **6**
14. Which muscles depress the larynx after swallowing?   **6**
15. Suppose that an injury severs the first three cervical nerves. With which muscles and actions would this injury interfere?   **6**
16. Name the five major muscle groups of the trunk and their actions.   **7**
17. What are the actions of the iliocostalis lumborum, the external oblique, the external intercostals, the levator ani, and the ischiocavernosus (of both sexes) muscles?   **8**
18. Which deep muscles of the back rotate the vertebral column? How do the origins and insertions of these muscles enable rotation?   **8**
19. Which muscles operate in deep breathing and to what motions do they contribute?   **2, 8**
20. Which muscles of the back and abdomen do you use when you bend over to reach an object on the floor?   **2, 8**
21. Which muscles of the thorax, abdomen, and perineum operate during defecation?   **2, 8**
22. What are the roles of somites, myotomes, and spinal nerves in muscle development?   **9**

CHAPTER 11

Muscles of the Extremities

OBJECTIVES

After studying this chapter, you should be able to do the following:

1. Name muscles of the upper and lower limbs and describe their action; vice versa, given an action, such as adducting the lower limb or flexing the elbow, name the muscles responsible for the action.
2. Know the differences between positioning and manipulative motions in the limbs.
3. Name the four fundamental muscle compartments of the limbs, and describe where they are located and what movements they accomplish.
4. Describe the proximal to distal organization of muscles in the limbs.
5. Describe the actions of the muscles that attach the upper extremity to the axial skeleton.
6. Describe the actions of the muscles of the scapula that move the humerus.
7. Name the muscles of the arm that flex and extend the forearm. Give their origins and insertions, as well as examples of motion in which these muscles are involved.
8. Describe the location of flexors and extensors of the hand and digits.
9. Name and give the actions of the thenar and hypothenar muscles of the hand.
10. Describe how electromyography reveals the action of muscles.
11. Describe the major events in walking and the phases of the stride.
12. Describe the action of the iliopsoas and gluteal muscles in walking.
13. Describe the action of thigh muscles in walking.
14. Describe the action of leg and foot muscles in a walking limb.
15. Describe the actions of muscles that develop on different sides of the limbs.

 VOCABULARY TOOLBOX

ACTION
It is not surprising that limb muscles are named for their actions.

abductor hallucis	ab-, *L* from, away; duc-, *L* lead. Abducts, spreads the great toe.	**extensor digitorum**	ex-, *L* out, from; tens, *L* stretched. Extends, straightens the fingers.
adductor hallucis	ad-, *L* to, toward. Adducts, pulls the great toe toward the others.	**flexor digitorum**	flex-, *L* bend. Flexes, curls the fingers.

SIZE AND SHAPE
When several muscles have the same actions, size and shape distinguish among the members of the group.

brevis	brev-, *L* short. The flexor pollicis brevis is the short flexor of the thumb.	**magnus**	magn-, *L* great, large. The adductor magnus is the largest adductor muscle of the thigh.
longus	long-, *L* long. The extensor pollicis longus is the long extensor of the thumb.	**minimus**	minim-, *L* the least. The extensor digiti minimi extends the little finger.
lumbrical	lumbric-, *L* earthworm. The lumbrical muscles of the phalanges resemble earthworms.	**teres**	tere-, *L* round, smooth. The teres major is the large, rounded muscle of the scapula.

LOCATION
Different muscles cannot occupy the same place.

profundus   profund-, *L* deep. The flexor digitorum profundus is the deep flexor of the fingers.

superficialis   super-, *L* on the surface. The flexor digitorum superficialis is the superficial flexor muscle of the fingers.

STRUCTURES
Muscles move body parts and joints.

digitorum   digit, *L* of the digits, fingers or toes.

hallucis   hallux, *L* the great toe.

palmaris   palm-, *L* the palm of the hand.

plantaris   plant-, *L* sole of the foot.

pollicis   pollex, *L* the thumb.

RELATED TOPICS
- **Limb buds, Chapter 7**
- **Motions of the joints, Chapter 8**
- **Muscle tissue, Chapter 9**
- **Muscles of the axial skeleton, Chapter 10**
- **Somites and myotomes, Chapters 4 and 10**
- **Spinal nerves and plexuses, Chapter 18**

PERSPECTIVE

Most of us take the muscles of our limbs for granted until they are injured or we need to exercise them. Dancers learn exercises that extend the limbs, and body builders develop pectoral muscles that ripple over the chest. Perhaps a groin pull has sidelined your favorite player, or maybe you have just been told what you already knew, that your hamstrings aren't supple enough for you to touch your toes without bending your knees. Understanding these situations begins with anatomy of the muscles that move the limbs. In this chapter you will learn to identify these muscles and their actions. Practitioners of sports medicine and physical therapy use this knowledge to design and administer exercises that improve performance and rehabilitate injured muscles.

Muscles are the currency of motion. Some 106 muscles of the lower limbs are involved in walking and running, and at least 30 more swing the upper limbs. To understand how so many muscles work together, you must know the names of the muscles, where they are located, and what motions each produces. With this information you can easily understand, for example, that the deltoid muscle on your shoulder abducts the arm. As you learn more muscles, you will begin to see how entire limbs, such as those shown in Figure 11.1, move.

This chapter describes the muscles of the upper limbs, then those of the lower limbs. The text follows the anatomy and action of muscles distally through the limbs, region by region, giving examples of familiar motions that each region enables. Look for systematic details of the origin, insertion, action, and innervation of each muscle in the tables that accompany each section. You will find pronunciations in these tables, and the Vocabulary Toolbox helps with frequently used terms. After each section, you are asked to identify specific muscles and to name the muscles involved in various motions. See Chapter 8, Joints, to review motions of limbs.

First, however, you need to know some general principles about limb muscles.

B

C

✳ **Figure 11.1**

People in motion. A, Michael Jordan. B, Larry Bird. C, Rock climber.

LIMB MUSCLES AND MOTION

POSITIONING AND MANIPULATIVE MOVEMENTS

Many muscles collaborate to place a hand or foot where you want it. Begin by moving your upper limbs. Go ahead, reach for the telephone or pick up a pencil and write with it. Your arm and shoulder muscles, such as those shown in Figure 11.2, place your forearm and hand where the fingers and thumb can grasp the target. These proximal motions are also called **positioning movements** because they move the forearm so that the hand can grasp the object with **manipulative motions.** Proximal positioning allows distal dexterity. The lower extremities also perform these movements, but the toes and feet are much less dextrous than the hands, although the rock climber in Figure 11.1 might disagree.

MUSCLE GROUPS AND COMPARTMENTS

Muscles are grouped as flexors, extensors, adductors, or abductors, according to their locations in the limbs. In general these four groups surround the bones as they do in Figure 11.3A, but one or more groups may be missing in particular regions. An **adductor** muscle, such as the pectoralis major of the chest, usually inserts along with other adductors on the medial surface of the limb, where it draws the limb toward the body. **Abductors** occupy the opposite side and draw the limb away from the body when a muscle, such as the deltoid muscle of the shoulder (Figure 11.2), raises the limb. **Flexor** muscles, such as the biceps brachii, frequently work on the anterior surfaces of a limb, and **extensors,** the triceps brachii, for example, attach to the posterior surfaces. Find the muscles in Figure 11.2 that helped you pick up the phone, and assign them to these groups.

Connective tissue partitions called intermuscular septa separate these groups into different **muscle compartments.** The adductor, extensor, and flexor muscles of the femur occupy separate compartments, shown in Figure 11.3C. Adductor muscles fill the adductor (or medial) compartment on the medial surface of the thigh, and extensors reside on the anterior surface in the extensor (or anterior) compartment. Similarly the posterior compartment (the flexor compartment) contains the thigh muscles that flex the leg. The abductor (lateral) compartment does not contain muscles in the thigh. Flexor and extensor compartments develop on opposite sides of the limb buds, described later in Muscle Development.

PROXIMAL TO DISTAL ORGANIZATION

Muscles span joints from proximal origins to distal insertions. When you trace a muscle from its origin to its insertion, you usually proceed distally across one or more joints and two or more bones. The deltoid muscle, for example, crosses the glenohumeral joint from its origin on the scapula and clavicle to its insertion on the humerus (Figure 11.3B).

Muscles overlap. As you move along the upper extremity, you find that insertions of proximal muscles overlap the origins of distal muscles in Figure 11.3B. Muscles of the arm originate on the humerus or scapula and insert on the forearm, while forearm muscles originate on the humerus and insert on the hand and digits. This organization integrates the motions of each region of the upper and lower extremities with those of adjacent regions, so that positioning and manipulative movements work together.

Proximal muscles usually are larger and stronger than distal muscles because they sustain more weight and stress. Like the bones of the limbs, the proximal muscles support the weight distal to them, as seen in Figure 11.2. Distal muscles tend to be smaller. They also are more numerous because hands and feet have more bones than the pelvis and shoulder.

- What kind of motion places the hand on a door handle, and what kind of motion grasps the handle to open the door?
- On which sides of a limb are you most likely to find adductor and extensor muscles?
- Where would you look for proximal muscles of the upper extremity?
- Why are many proximal limb muscles larger than distal muscles?

SHOULDER AND UPPER LIMB

BRIEF TOUR

The muscles of the shoulder are gross positioners of the upper limb; they anchor the clavicle and scapula to the trunk and to the arm (Figure 11.4). The scapula and clavicle form an arch with which the arm articulates at the glenohumeral joint in Figure 11.5. Shoulder muscles elevate this arch to raise the arm, as shown in the figure. They protract the arch forward when the arm flexes (in quest of telephones and pencils). The shoulder muscles also retract the scapula and clavicle when the arm extends backward, as well as depress the arch.

Superficial and deep muscles maneuver the shoulder and upper limb, and spinal nerves innervate all of them. Tables 11.1 through 11.7 give details.

SUPERFICIAL MUSCLES OF THE SHOULDER

The superficial muscles are powerful flexors, extensors, and adductors of the arms; all are fan-shaped.

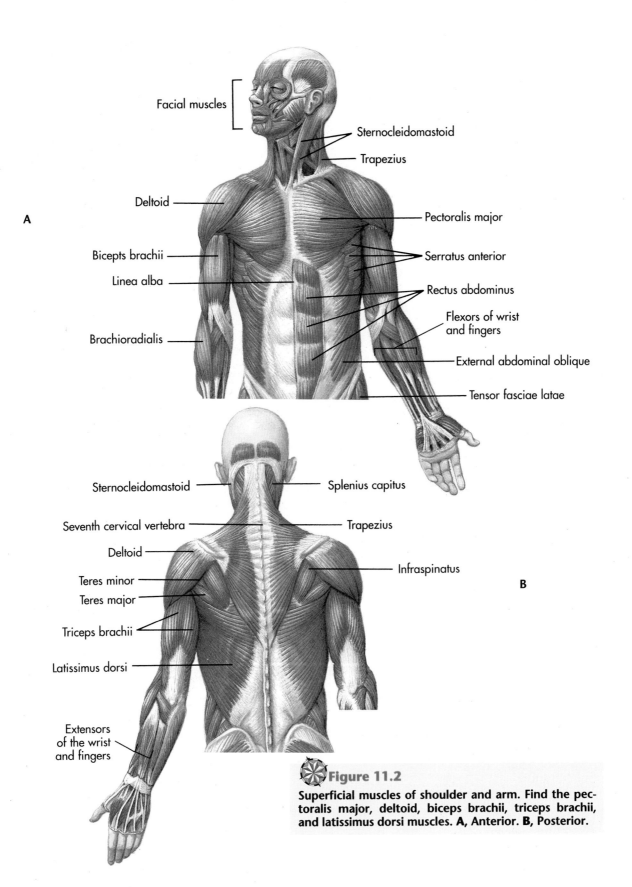

A

B

Facial muscles

Sternocleidomastoid

Trapezius

Deltoid

Pectoralis major

Bicepts brachii

Serratus anterior

Linea alba

Rectus abdominus

Brachioradialis

Flexors of wrist
and fingers

External abdominal oblique

Tensor fasciae latae

Sternocleidomastoid

Splenius capitus

Seventh cervical vertebra

Trapezius

Deltoid

Infraspinatus

Teres minor

Teres major

Triceps brachii

Latissimus dorsi

Extensors
of the wrist
and fingers

Figure 11.2

Superficial muscles of shoulder and arm. Find the pectoralis major, deltoid, biceps brachii, triceps brachii, and latissimus dorsi muscles. A, Anterior. B, Posterior.

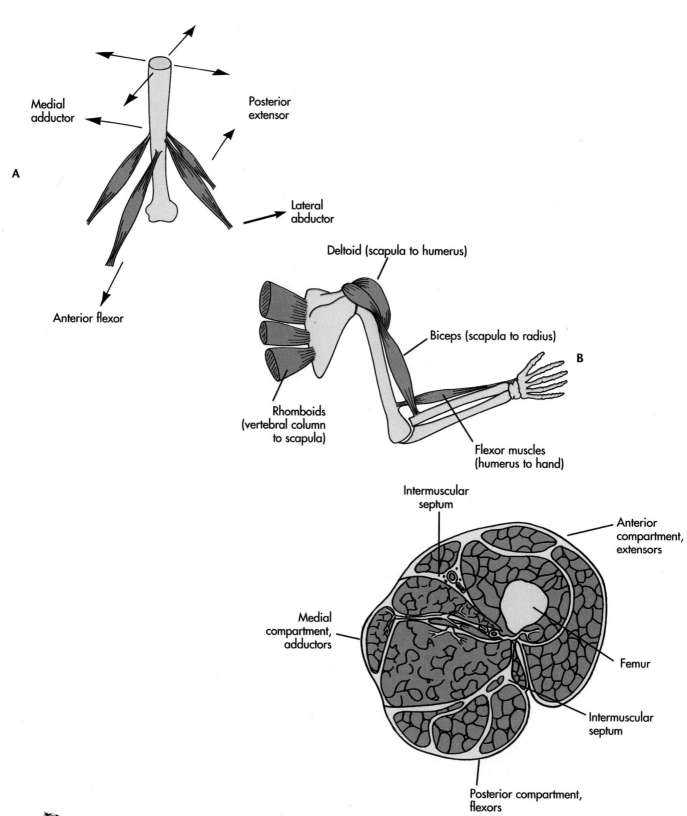

A

Medial adductor

Posterior extensor

Lateral abductor

Anterior flexor

Deltoid (scapula to humerus)

Biceps (scapula to radius)

B

Rhomboids (vertebral column to scapula)

Flexor muscles (humerus to hand)

Intermuscular septum

Anterior compartment, extensors

Medial compartment, adductors

Femur

Intermuscular septum

Posterior compartment, flexors

✳ Figure 11.3

A, Muscle groups surround the bones of limbs like guy wires of towers. Adductors are located medially, abductors laterally; flexors are anterior, extensors posterior. Some groups may be absent from certain regions of the limbs. The elbow has no adductor or abductor muscles, for example. **B,** Muscles overlap, so that the insertions of proximal muscles frequently are distal to the origins of more distal muscles. For instance, the deltoid inserts on the humerus, and the biceps brachii originates on the scapula. Find other muscles in this figure discussed in the text. **C,** Muscle compartments in the thigh, transverse section.

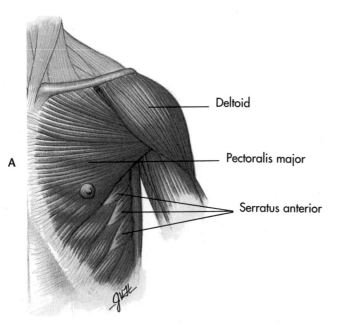

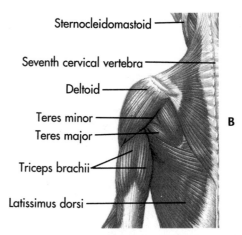

Figure 11.4

Superficial muscles of the shoulder. A, Anterior. B, Posterior.

Beginning with the **pectoralis major** (Figure 11.4A), they sweep up the thorax, over the shoulder to the **trapezius** (Figure 11.4B), and down the back to the **latissimus dorsi.** The first and last of these muscles form the anterior and posterior walls of the axilla, as you can demonstrate by palpating these muscles when you raise your arm from the side.

PECTORALIS MAJOR

The pectoralis major (large muscle of the breast) adducts the humerus, as shown in Figure 11.4A and Table 11.2. Its fibers converge on the greater tubercle of the humerus from the clavicle, sternum, costal cartilages, and the external oblique muscle of the abdomen. With so broad an origin, the pectoralis major is a powerful adductor of the arm that you use when you clamp a book under your arm or push yourself upward from a chair. The pectoralis major also rotates the humerus medially because its insertion curls over the anterior surface of the bone. Impressive "pecs" are body builders' stock-in-trade.

TRAPEZIUS

The fibers of the trapezius muscle (shaped like a trapezoid; see Table 11.1) converge on the bony arch formed by the clavicle, acromion, and spine of the scapula. The broad origin of the trapezius, extending from the superior nuchal line of the skull down to the spinous processes of all thoracic vertebrae, enables the trapezius to pull the shoulder in several directions. The upper fibers elevate the shoulder, and the lower fibers rotate the scapula upward at the medial end of the spine of the scapula, as you can see in Fig-

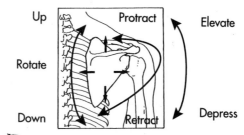

Figure 11.5

The clavicle and scapula form an arch that reaches from the sternum to the back and supports the upper limb. The arrows show how the shoulder moves.

ure 11.4B. You comb your hair or hug your friends with these motions. When all fibers shorten, the "trap" retracts your shoulders.

LATISSIMUS DORSI

The latissimus dorsi (broadest muscle of the back) is a powerful adductor and medial rotator that also extends the arm. The latissimus (see Table 11.2) reaches from the lumbar and sacral regions of the back in Figure 11.4B and curls across the medial surface of the humerus to insert on the intertubercular groove. Sprint swimmers and rock climbers depend on powerful "lats."

- Which bones do the trapezius and latissimus dorsi muscles move?
- Where do these muscles originate and insert?
- How do the pectoralis major, trapezius, and latissimus dorsi act when you cross your hands behind your head?

Table 11.1   MUSCLES OF THE SHOULDER

Muscle	Origin	Insertion	Action	Innervation
ANTERIOR MUSCLES	These muscles draw the clavicle and scapula forward, depress them, and rotate the scapula upward. They work from origins on the ribs and insertions upon the clavicle, the coracoid process, and the vertebral border of the scapula. All are innervated by way of the brachial plexus.			
Subclavius (sub-KLAVE-ee-us)	First rib and costal cartilage	Inferior surface of clavicle	Draws shoulder forward and caudally	Cervical nerves 5 and 6 via brachial plexus (anterior rami)
Pectoralis minor (PEK-tor-al-iss)	Third through fifth ribs, near costal cartilages	Coracoid process of scapula	Draws scapula forward and caudally; rotates scapula inferiorly	Spinal nerves C8 and T1 via median pectoral nerve
Serratus anterior (ser-RATE-us)	First 8 or 9 ribs and aponeuroses of intercostal muscles	Vertebral border of scapula, anterior surface, from superior to inferior angle	Rotates scapula upward, draws vertebral border medially	Spinal nerves C5 to C7 via long thoracic nerve
POSTERIOR MUSCLES	These muscles attach the shoulder to the vertebral column. They insert on the clavicle, acromion, and scapular spine, or on the vertebral border of the scapula, where they rotate, elevate, depress, and adduct the scapula. All are innervated by cervical nerves from the cervical or brachial plexuses, with a contribution from the spinal accessory nerve. The trapezius is superficial to others of this group.			
Trapezius (TRAP-eez-ee-us)	Occipital bone, nuchal line, nuchal ligament, spinous processes of C7, and all thoracic vertebrae	Posterior border, lateral third of clavicle; medial acromion and spine of scapula	Rotates, adducts, elevates and depresses scapula; elevates glenoid cavity in abduction and flexion of arms; elevates clavicle; rotates, extends head	Motor via spinal accessory (cranial nerve XI), and sensory via anterior rami of cervical nerves 3 and 4
Levator scapulae (le-VAY-tor SKAP-you-lie)	Transverse processes of first four cervical vertebrae	Vertebral border of scapula, posterior surface, between superior angle and scapular spine	Raises scapula, draws it medially, rotates it	Anterior rami of C3 and C4 from cervical plexus, and C5 via dorsal scapular nerve
Rhomboideus major (rom-BOYD-ee-us)	Spinous processes and supraspinal ligament of thoracic vertebrae 2-5	Vertebral border of scapula, posterior surface, at scapular spine	With rhomboideus minor adducts scapula, rotates vertebral border upward, and depresses glenoid cavity	Brachial plexus via dorsal scapular nerve (C5)
Rhomboideus minor	Inferior surface of nuchal ligament and spinous processes of C7 and T1 vertebrae	Vertebral border of scapula, posterior surface, between scapular spine and inferior angle	With rhomboideus major adducts scapula, rotates vertebral border upward, and depresses glenoid cavity	Same

From Moore PJ: Muscular and associated neurological systems. In Hilt NE and Cogburn SB, editors, Manual of orthopedics, *St. Louis, 1980, CV Mosby.*

Table 11.2   MUSCLES THAT MOVE THE HUMERUS

Muscle	Origin	Insertion	Action	Innervation
AXIAL MUSCLES	These muscles take origins from the axial skeleton and insert upon the humerus. Innervated through the brachial plexus, they flex, extend, adduct, and rotate the arm.			
Pectoralis major (PEK-tor-al-iss)	Medial half of clavicle, anterior surface of sternum, costal cartilages 2-6, and aponeurosis with external oblique	Humerus, crest of greater tubercle	Flexes, adducts, and rotates arm medially	Nerves C5 through T1, via median and lateral pectoral nerves
Latissimus dorsi (la-TISS-ih-mus DOR-see)	By aponeurosis from spinous processes of T7 to L5 vertebrae, crests of sacrum and ilium, lower four ribs	Humerus, bottom of intertubercular groove	Extends, adducts, and rotates arm medially. Draws shoulder backward and caudally	Cervical nerves 6-8, via thoracodorsal nerve
SCAPULAR MUSCLES	These muscles flex, extend, abduct, and rotate the arm by acting from origins on the scapula and insertions on the humerus, mostly the greater and lesser tubercles. Cervical nerves 5 and 6 innervate these muscles through the brachial plexus.			
Deltoid	Lateral third of clavicle, acromion, scapular spine	Deltoid tuberosity of humerus	Acromial portion abducts arm; clavicular portion flexes arm; spinous portion extends arm	Nerves C5 and C6 via axillary nerve
ROTATOR CUFF				
Supraspinatus (SOO-prah-spy-NAH-tus)	Supraspinous fossa	Greater tubercle of humerus	Abducts arm	Nerve C5 via suprascapular nerve
Infraspinatus (IN-frah-spy-NAH-tus)	Infraspinous fossa	Greater tubercle of humerus	Rotates arm laterally	Nerves C5 and C6 via suprascapular nerve
Teres minor (TEAR-eze)	Upper two thirds of axillary border of scapula	Greater tubercle of humerus	Rotates arm laterally and weakly adducts it	Nerve C5 via axillary nerve
Subscapularis (sub-SKAP-you-LAR-iss)	Axillary border of scapula and subscapular fossa	Lesser tubercle of humerus and joint capsule	Rotates arm medially, stabilizes joint	Nerves C5 and C6 via upper and lower subscapular nerves
Teres major (TEAR-eze)	Inferior angle of scapula, posterior surface	Lesser tubercle of humerus	Adducts, extends, and rotates arm medially	Nerves C5 and C6 via lower subscapular nerve

From Moore PJ: Muscular and associated neurological systems. In Hilt NE and Cogburn SB, editors, Manual of orthopedics, St. Louis, 1980, CV Mosby.

DEEP MUSCLES OF THE SHOULDER
ANTERIOR MUSCLES

Three deep muscles insert upon the clavicle and scapula from origins on the anterior thorax. Figure 11.6 has removed the pectoralis major to reveal the **subclavius, pectoralis minor,** and **serratus anterior** muscles. As a group, these muscles (Table 11.1) draw the clavicle and scapula forward and depress them, and they also rotate the scapula upward. The subclavius (beneath the clavicle) presses the clavicle into the sternoclavicular joint, reaching laterally beneath the

clavicle from the first costal cartilage. The pectoralis minor (small muscle of the breast) reaches upward from ribs 3 through 5 to the coracoid process of the scapula, which it draws downward and forward when the arm reaches forward. Neither of these muscles is visible beneath the pectoralis major, but the sawtoothed origins of the serratus anterior (anterior serrated muscle) can be seen on the rib cages of swimmers and body builders. This muscle inserts on the vertebral border of the scapula, where its primary action pulls the body of the scapula laterally and down-

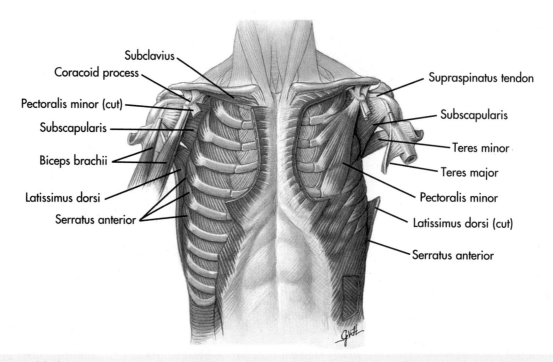

Subclavius

Coracoid process

Pectoralis minor (cut)

Subscapularis

Biceps brachii

Latissimus dorsi

Serratus anterior

Supraspinatus tendon

Subscapularis

Teres minor

Teres major

Pectoralis minor

Latissimus dorsi (cut)

Serratus anterior

Figure 11.6

Anterior deep muscles of the left shoulder. Find the subclavius, pectoralis minor, and serratus anterior.

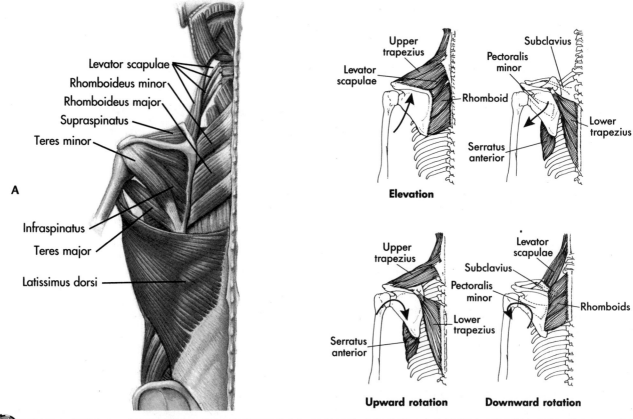

Levator scapulae

Rhomboideus minor

Rhomboideus major

Supraspinatus

Teres minor

Infraspinatus

Teres major

Latissimus dorsi

A

Upper trapezius

Levator scapulae

Rhomboid

Serratus anterior

Subclavius

Pectoralis minor

Lower trapezius

Elevation

Upper trapezius

Serratus anterior

Subclavius

Pectoralis minor

Lower trapezius

Levator scapulae

Rhomboids

Upward rotation   **Downward rotation**

Figure 11.7

A, The levator scapulae and rhomboideus muscles join the vertebral border of the scapula to the vertebral column. B, Coordinated muscular action elevates the scapula and rotates it upward and downward.

ward when you reach forward or elevates the scapula when you raise your arms. The serratus anterior resembles the trapezius in its ability to pull from several directions.

POSTERIOR MUSCLES

Three muscles on the posterior thorax also move the scapula. The **levator scapulae, rhomboideus minor,** and **rhomboideus major** adduct and elevate the scapula. These flat muscles (Figure 11.7A and Table 11.1) reach downward and laterally from the cervical and thoracic vertebrae to the vertebral border of the scapula. The levator scapulae (elevator of the scapula), the most superior, inserts above the **scapular spine,** followed by the rhomboideus muscles below. The narrow rhomboideus minor (smaller, four-sided) inserts at the spine of the scapula, and the broader rhomboideus major inserts below the spine. All three muscles elevate the vertebral border of the scapula, and with help from the serratus anterior, they can adduct the entire bone. Figure 11.7B shows how these posterior muscles plus the serratus anterior and pectoralis minor elevate and rotate the scapula in positioning the upper limbs.

- Which muscles rotate the scapula upward when the arm is raised?
- Which muscles protract the scapula when you reach forward?

MUSCLES THAT MOVE THE ARM

Seven short scapular muscles flex, extend, abduct, and rotate the arm by inserting on the humerus. They are the **deltoid, subscapularis, supraspinatus, infraspinatus, teres major and minor,** and the **coracobrachialis** muscles. Together with the pectoralis major and latissimus dorsi, these muscles maneuver the arm in all directions (Table 11.2).

The deltoid (triangular) muscle gives the shoulder its rounded appearance, shown in Figure 11.4. This muscle originates on the clavicle, the acromion, and the spine of the scapula, and it inserts onto the deltoid tuberosity on the lateral surface of the humerus. The deltoid's broad origin makes it a versatile muscle. Short, powerful, multipennate fibers form the **acromial portion** at the middle of the muscle that abducts your arms because it is directly above the glenohumeral joint. Anteriorly the **clavicular por-**

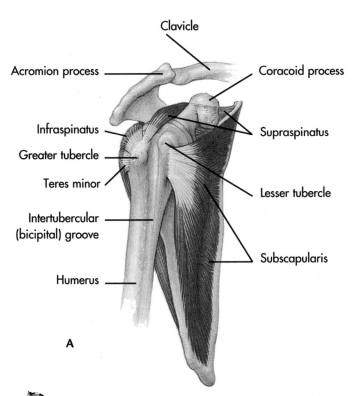

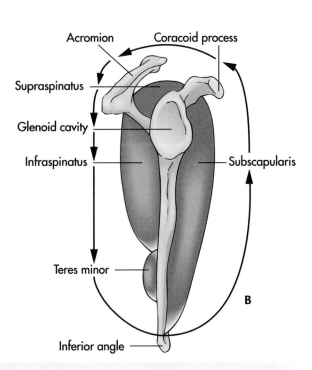

✷ **Figure 11.8**
A, Deep posterior scapular muscles of left shoulder. **B,** Rotator cuff muscles encircle the body of the scapula.

tion helps the pectoralis major flex the arm, and posteriorly the **spinous portion** assists the latissimus dorsi to extend the arm. The latter portions of the deltoid play little role in lifting the arm initially, because they are below the glenohumeral joint. Only when the acromial portion has raised the arm part way and the clavicular and spinous portions rise above the joint, do they begin to lift the arm.

Four rotator cuff muscles stabilize the glenohumeral joint (Figure 11.8A) and help rotate the humerus, as their name suggests. These muscles—the **supraspinatus, *i*nfraspinatus, *t*eres minor,** and **sub-scapularis** (SITS)—form a cuff around the articular capsule of the shoulder joint that retains the humerus within the glenoid fossa of the scapula. The rotator cuff muscles rotate around the scapula, as you can see in Figure 11.8B. Beginning counterclockwise with the supraspinatus (above the spine), this muscle reaches from the supraspinous fossa, over the shoulder joint, to the greater tubercle of the humerus, pressing the head of the humerus into the glenoid cavity (Figures 11.8A and 11.7A). Next, the infraspinatus (below the spine) also inserts onto the greater tubercle from the infraspinous fossa, followed by the teres minor (smaller rounded muscle) from the lateral border of the scapula. Finally, on the anterior surface of the scapula the triangular subscapularis (beneath the scapula) muscle converges onto the lesser tubercle in front of the humerus. It is easy to see from their positions that the infraspinatus and teres minor rotate the humerus laterally, and that the subscapularis rotates it medially.

The teres major (larger rounded muscle; Figure 11.7A) extends from the inferior angle of the scapula to the intertubercular groove of the humerus. The teres major helps the latissimus dorsi adduct and extend the arm, and it also helps rotate the arm medially.

The coracobrachialis flexes and adducts the arm by reaching from the coracoid process of the scapula midway down the humerus (Figure 11.9B and Table 11.3). The coracobrachialis is a weak adductor and flexor, compared with the pectoralis major.

- Which muscles belong to the rotator cuff?
- What muscle raises the arm?
- Which muscles rotate the arm laterally and medially?

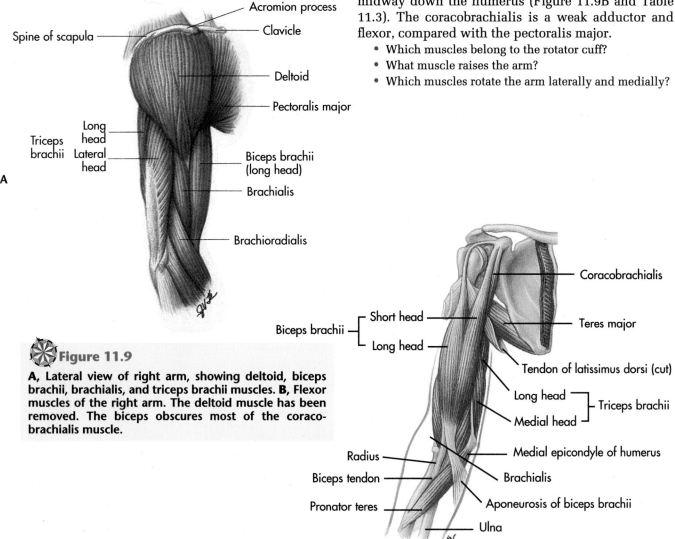

Figure 11.9

A, Lateral view of right arm, showing deltoid, biceps brachii, brachialis, and triceps brachii muscles. **B,** Flexor muscles of the right arm. The deltoid muscle has been removed. The biceps obscures most of the coracobrachialis muscle.

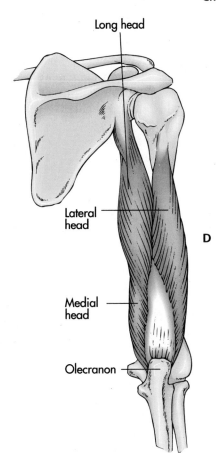

Long head

Lateral head

D

Medial head

Olecranon

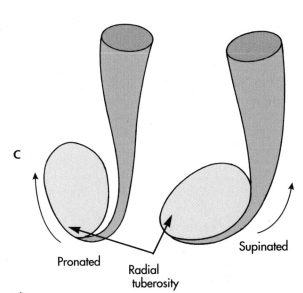

C

Pronated

Radial tuberosity

Supinated

Figure 11.9, cont'd

C, Pronation of the hand winds the biceps tendon around the radial tuberosity, and supination unwinds the tendon. **D,** Triceps brachii of right arm. Find all three heads.

Table 11.3   MUSCLES OF THE ARM

Muscle	Origin	Insertion	Action	Innervation
ANTERIOR COMPARTMENT				
Biceps brachii (BI-seps BRAY-kee-ee)		Posterior portion of radial tuberosity	Flexes arm and forearm, supinates hand	Musculocutaneous nerve (C6, C7) from brachial plexus
Short head	Coracoid process			
Long head	Supraglenoid tuberosity			
Brachialis (BRAY-kee-AL-iss)	Distal half of anterior humerus	Tuberosity of ulna in front of coronoid process	Flexes forearm	Musculocutaneous nerve (C6, C7) from brachial plexus
Coracobrachialis (KORE-a-koe-BRAY-kee-AL-iss)	Apex of coronoid process	Middle of medial surface of humerus	Flexes and adducts arm	Musculocutaneous nerve (C6, C7) from brachial plexus
POSTERIOR COMPARTMENT				
Triceps brachii (TRY-seps BRAY-kee-ee)		Posterior portion of olecranon process	Extends forearm; long head extends and adducts arm	Radial nerve (C7, C8) from brachial plexus
Long head	Infraglenoid tuberosity of scapula			
Lateral head	Shaft of humerus, posterior surface			
Medial head	Shaft of humerus, posterior surface			

From Moore, PJ: Muscular and associated neurological systems. In Hilt NE and Cogburn SB, editors, Manual of orthopedics, *St. Louis, 1980, CV Mosby.*

MUSCLES OF THE ARM

Three muscles of the arm move the forearm. Two flexors—the **biceps brachii** and **brachialis**—operate in the anterior compartment of the humerus, and the only extensor—the **triceps brachii**—fills the posterior compartment (Figure 11.9A and Table 11.3). Axons from cervical nerves 5 through 8 innervate these muscles via the musculocutaneous and radial nerves.

ANTERIOR MUSCLES

The two-headed biceps brachii muscle of the arm flexes the elbow and supinates the forearm. Both heads originate on the scapula (Figure 11.9B); the **short head** descends the humerus directly and joins the **long head,** whose tendon detours over the glenohumeral joint and the intertubercular groove. The biceps supinates the forearm because its insertion on the radial tuberosity of the radius bone rotates the radius laterally. When the forearm pronates, the radial tuberosity winds the biceps tendon part way around the radius, like a ribbon on a spool in Figure 11.9C. The biceps then unwinds the tendon and rotates the radius into the supine position. As a few chin-ups prove, the biceps also is a powerful flexor of the forearm when the forearm is supinated but not when it is pronated. Palpate the movements of the biceps tendon when you supinate your forearm, and then as you flex it.

The brachialis (of the arm) muscle takes up the load when the forearm is pronated. The brachialis originates on the distal humerus, deep to the biceps, and inserts onto the ulnar tuberosity, where pronation and supination do not affect its action.

POSTERIOR MUSCLES

Three heads of the powerful triceps brachii muscle extend the forearm, as shown in Figure 11.9A and D. The **long head** arises from the scapula, and the **medial** and **lateral heads** take their origins from the posterior surface of the humerus. All heads converge on the broad triceps tendon that inserts upon the olecranon process of the ulna.

- Which muscle is the major supinator of the forearm?
- Which flexor inserts upon the ulna?
- Which heads of the triceps and biceps brachii are two-joint muscles, that is, cross two joints between origin and insertion?
- Why is it more difficult to do chin-up exercises with forearms pronated than with them supinated?

ANTERIOR MUSCLES OF THE FOREARM— FLEXORS

The forearm is packed with two groups of flexor and extensor muscles that maneuver the wrist and the hand and its digits. Pronators and supinators also rotate the hand, and several more muscles help flex the elbow. **Flexors** grasp objects and palmar flex the wrist; **extensors** do the opposite by opening the hand

and dorsiflexing the wrist. Altogether, 19 forearm muscles cooperate when you grasp a coffee cup, a pencil, or a telephone.

Flexors and extensors work from opposite sides of the forearm, flexors on the anterior and extensors on the posterior (Figure 11.10). The flexors pass from the medial epicondyle of the humerus near the ulna into the anterior compartment of the radius and ulna, onto

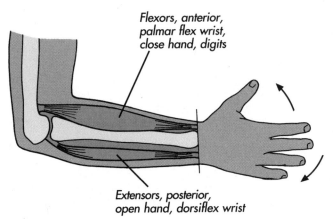

Flexors, anterior, palmar flex wrist, close hand, digits

Extensors, posterior, open hand, dorsiflex wrist

✺ Figure 11.10

Flexors on the anterior aspect of the left upper limb close the hand and fingers, whereas extensors on the posterior side open them.

the palmar surfaces of the carpals, metacarpals, and phalanges in Figure 11.11 and Table 11.4. Similarly, extensors begin on the lateral epicondyle of the humerus near the radius bone and pass into the posterior compartment to the dorsal surfaces of metacarpals and phalanges. **Pronators** work from the anterior side of the forearm and **supinators** from the posterior. As a group, the forearm muscles receive innervation from several spinal nerves—the fifth cervical nerve to the first thoracic.

SUPERFICIAL LAYER

There are three layers of anterior muscles in the forearm (Figure 11.11 and Table 11.4). The superficial layer originates and inserts proximally to the deep layers, so that the superficial layer flexes the wrist and the deep layer flexes the phalanges. Muscle names indicate size, position, and which digits they control. The flexor digitorum profundus is the deep flexor of the fingers, for example.

Beginning from the ulnar side, the superficial muscles consist of the flexor carpi ulnaris, palmaris longus, flexor carpi radialis, pronator teres, and the brachioradialis. Find these muscles in Figure 11.11A and B. Except for the brachioradialis, all fan out onto the forearm from the medial epicondyle of the humerus. Trace the tendons of the **flexor carpi radialis,** the **palmaris longus,** and **flexor carpi ulnaris**

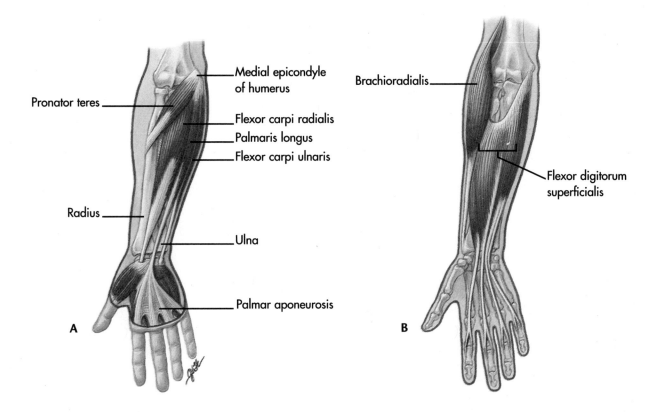

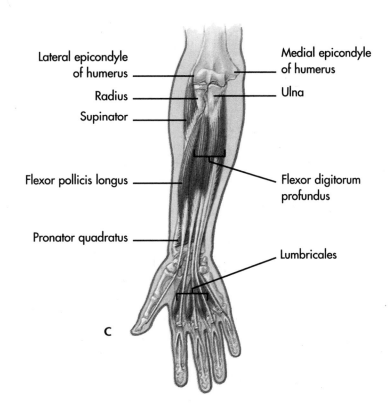

A, Superficial layer, labels: Pronator teres, Medial epicondyle of humerus, Flexor carpi radialis, Palmaris longus, Flexor carpi ulnaris, Radius, Ulna, Palmar aponeurosis

B, Intermediate layer, labels: Brachioradialis, Flexor digitorum superficialis

C, Deep layer, labels: Lateral epicondyle of humerus, Medial epicondyle of humerus, Radius, Ulna, Supinator, Flexor pollicis longus, Flexor digitorum profundus, Pronator quadratus, Lumbricales

Figure 11.11

Flexors and pronators on the anterior surface of the right forearm. A, Superficial layer. B, Intermediate layer. C, Deep layer.

Table 11.4   ANTERIOR MUSCLES OF THE FOREARM

Muscle	Origin	Insertion	Action	Innervation
SUPERFICIAL MUSCLES	The superficial muscles flex the hand and digits and pronate the hand. Served by the ulnar and median nerves, these muscles take origins from the distal humerus, proximal ulna, and the radius. They insert upon the distal radius, certain carpals and metacarpals, and the distal two rows of phalanges.			
Flexor carpi radialis (flex-or KAR-pee RAY-dee-al-iss)	Medial epicondyle of humerus	Proximal base of second metacarpal	Flexes and abducts hand	Median nerve (C6, C7)
Palmaris longus (pal-MAR-iss LONG-us)	Medial epicondyle of humerus	Flexor retinaculum and palmar aponeuro-sis	Flexes hand	Median nerve (C6, C7)
Flexor carpi ulnaris (flex-or KAR-pee ul-NAR-iss)		Pisiform, hamate, and fifth metacarpal bones	Flexes and adducts hand	Ulnar nerve (C8, T1)
Humeral head	Medial epicondyle of humerus			
Ulnar head	Medial surface, coro-noid process of ulna			
Pronator teres (PRO-nay-tor TEAR-eez)		Mid-shaft, lateral sur-face of radius	Pronates hand	Median nerve (C6, C7)
Humeral head	Above medial epicon-dyle			
Ulnar head	Medial surface of coro-noid process of ulna			
Brachioradialis (BRAY-kee-oh-RAY-dee AL-iss)	Supracondylar ridge of humerus	Lateral base of radial styloid process	Flexes forearm	Radial nerve (C5, C6)
INTERMEDIATE MUSCLES				
Flexor digitorum superficialis (DIH-jih-TOR-um)		Lateral sides of middle phalanx of all four digits	Flexes middle phalanx of each finger and en-tire hand	Median nerve (C7, C8, T1)
Humeral head	Medial epicondyle of humerus			
Ulnar head	Medial coronoid pro-cess			
Radial head	Oblique line and middle third, lateral border of radius			
DEEP MUSCLES	Beneath the superficial muscles, deep muscles take their origins from the proximal radius and ulna, and insert more distally upon the radius and phalanges. Innervated via the median and ulnar nerves, these muscles flex the fingers and thumb or pronate the hand.			
Flexor digitorum profundus (FLEKS-or DIH-jih-TOR-um pro-FUND-us)	Coronoid process and proximal body of ulna	Base of distal phalanges	Flexes distal phalanx of each finger and fingers together and slightly flexes the hand	Median and ulnar nerves (C8, T1)
Flexor pollicis longus (POLL-ik-iss)	Body of radius, interos-seous membrane, coro-noid process	Base of distal phalanx of thumb	Flexes proximal and dis-tal phalanx of thumb; flexes, adducts metacar-pals	Median nerve (C8, T1)
Pronator quadratus (PRO-nay-tor KWAD-rate-us)	Pronator ridge and dis-tal fourth of ulna	Distal fourth of lateral border of radius	Pronates hand	Median nerve (C8, T1)

From Moore, PJ: Muscular and associated neurological systems. In Hilt NE and Cogburn SB, editors, Manual of orthopedics, *St. Louis, 1980, CV Mosby.*

into the hand in Figure 11.11A and Table 11.4, to see which structures they control. The flexor carpi radialis (flexor of the radial carpals) abducts the hand, and the palmaris longus (long muscle of the palm) flexes three central metacarpals. The flexor carpi ulnaris (flexor of the ulnar carpals) adducts the hand. Table 11.6 summarizes the locations of these insertions on the hand.

Next is the **pronator teres** (rounded pronator), which pronates the hand by inserting onto the lateral body of the radius. Finally, on the radial side of the forearm the **brachioradialis** (of the arm and radius) muscle is the only flexor of the forearm located on the forearm. The brachioradialis links the humerus with the distal radius and, together with the biceps and brachialis of the arm, flexes the forearm.

DEEP LAYER

In the intermediate layer of anterior muscles the **flexor digitorum superficialis** (superficial flexor of the digits) reaches the **middle phalanges** of the fingers. The deep layer originates from the ulna and radius (Figure 11.11C). In this group are the **flexor digitorum profundus** (deep flexor of the digits) and the **flexor pollicis longus** (long flexor of the thumb), which control all five digits through insertions on the distal phalanges. Table 11.6 locates the insertions of these muscles on the hand.

Pronator quadratus (four-sided pronator) is the deepest, most distal muscle on the anterior surface of the forearm. Find it in Figure 11.11C and in Table 11.4. Usually included with the deep layer, the quadratus is the primary pronator of the forearm despite its small size.

- Which muscles send several tendons or branches to the central palm and digits?
- Which muscles actuate the ulnar side of the hand; the radial side?
- Name the superficial muscles of the forearm, beginning from the radial side.
- Which muscles insert upon the carpals; the metacarpals; the phalanges?

POSTERIOR MUSCLES OF THE FOREARM— EXTENSORS

Muscles of the posterior compartment of the forearm extend the wrist and digits. They also supinate the forearm and adduct and abduct the wrist (Figure 11.12 and Table 11.5). Overall, these muscles correspond with their antagonists in the anterior compartment. Superficial muscles fan out from the lateral epicondyle of the humerus and extend the metacarpals and phalanges. Deep muscles supinate the forearm or actuate the thumb and index finger.

Five superficial muscles extend the hand and adduct or abduct the wrist. Beginning from the radial

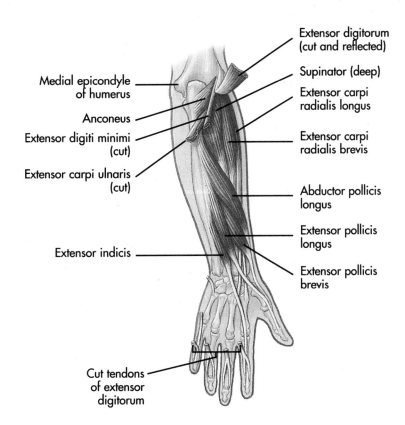

Medial epicondyle of humerus

Anconeus

Extensor digiti minimi (cut)

Extensor carpi ulnaris (cut)

Extensor indicis

Cut tendons of extensor digitorum

Extensor digitorum (cut and reflected)

Supinator (deep)

Extensor carpi radialis longus

Extensor carpi radialis brevis

Abductor pollicis longus

Extensor pollicis longus

Extensor pollicis brevis

Figure 11.12

Extensors and supinator of the posterior surface of the right forearm.

Table 11.5   POSTERIOR MUSCLES OF THE FOREARM

Muscle	Origin	Insertion	Action	Innervation
SUPERFICIAL MUSCLES	Except for the brachioradialis and anconeus, all the superficial muscles on the posterior surface of the forearm extend the hand, fingers, and forearm. All arise at or near the lateral epicondyle of the humerus; their tendons insert upon the wrist, metacarpals, phalanges, and the ulna. The radial nerve supplies all innervation.			
Extensor carpi radialis longus (KAR-pee)	Distal third, lateral supracondylar ridge of humerus	Radial side of base of second metacarpal	Extends and abducts (radial flexes) hand	Radial nerve (C6, C7)
Extensor carpi radialis brevis (BREV-iss)	Lateral epicondyle of humerus via common extensor tendon	Radial side of base of third metacarpal	Extends and may abduct (radial flex) hand	Radial nerve (C6, C7)
Extensor digitorum	Lateral epicondyle of humerus via common extensor tendon	Middle and distal phalanges of fingers	Extends phalanges and wrist	Deep radial nerve (C6 to C8)
Extensor digiti minimi (DIH-jih-tee MIN-ih-mee)	Extensor tendon and ulnar side of extensor digitorum	Proximal phalanx of little finger	Extends little finger	Deep radial nerve (C6 to C8)
Extensor carpi ulnaris	Lateral epicondyle of humerus	Ulnar side of base of fifth metacarpal	Extends and adducts (ulnar flexes) hand	Deep radial nerve (C6 to C8)
DEEP MUSCLES	The deep muscles extend and abduct the thumb, extend and adduct the index finger, and supinate the hand. All are innervated by the deep branch of the radial nerve. They insert by way of tendons onto the thumb, index finger, and radius from origins on the posterior surfaces of the radius and ulna and from the lateral epicondyle of the humerus.			
Anconeus (an-koe-NEE-us)	Lateral epicondyle of humerus	Lateral side of olecranon and body of ulna	Extends forearm	Radial nerve (C7, C8)
Supinator (SOOP-in-ATE-or)	Lateral epicondyle of humerus; radial collateral ligament and ulna	Proximal radius; radial tuberosity, oblique line, dorsal and lateral surfaces of body	Supinates hand	Deep radial nerve (C6)
Abductor pollicis longus (POLL-ik-iss)	Posterior surfaces of ulna, interosseous membrane, and middle third of radius	Radial side of first metacarpal	Abducts (radial flexes) thumb and wrist	Deep radial nerve (C6, C7)
Extensor pollicis brevis	Posterior surfaces of interosseous membrane and radius distal to abductor pollicis longus	Base of proximal phalanx of thumb	Extends thumb; abducts (radial flexes) hand	Deep radial nerve (C6, C7)
Extensor pollicis longus	Posterior surfaces of interosseous membrane and ulna, distal to abductor pollicis longus	Base of distal phalanx of thumb	Extends distal phalanx of thumb; abducts (radial flexes) hand	Deep radial nerve (C6 to C8)
Extensor indicis (IN-dih-kiss)	Posterior surfaces of interosseous membrane and ulna, distal to extensor pollicis longus	Extensor digitorum tendon to index finger	Extends and adducts (ulnar flexes) index finger	Deep radial nerve (C6 to C8)

From Moore, PJ: Muscular and associated neurological systems. In Hilt NE and Cogburn SB, editors, Manual of orthopedics, *St. Louis, 1980, CV Mosby.*

Table 11.6   FOREARM MUSCLES THAT MOVE THE HAND

	DIGITS				
	Radial side I (Thumb)	**II (Index)**	**III (Middle)**	**IV (Ring)**	**Ulnar Side V (Little)**
EXTENSORS	←————————————————— E. digitorum —————————————————→				
Superficial layer		E. carpi radialis longus	E. carpi radialis brevis		E. digiti minimi E. carpi ulnaris
Deep layer	E. pollicis longus E. pollicis brevis	E. indicis			
ABDUCTOR	A. pollicis longus				
FLEXORS	F. pollicis longus	←————————— F. digitorum profundus —————————→			
Deep layer			F. digitorum superficialis		
Intermediate layer		F. carpi radialis			
Superficial layer		←————————— Palmaris longus —————————→			F. carpi ulnaris

This table summarizes the forearm muscles that move the hand. The table represents your right hand with its palmar side downward, phalanges and metacarpals across the top, and the flexors and extensors at the left, going from the posterior to the anterior compartments of the forearm. Arrows stretching right and left indicate muscles that serve several digits such as the extensor digitorum. Conversely, each column lists the muscles that move an individual digit.

Abbreviations: E.-extensor, A.-abductor, and F.-flexor.

side in Figure 11.12, they are the **extensor carpi radialis longus** (long extensor of the radial carpals), the **extensor carpi radialis brevis** (short extensor of the radial carpals), the **extensor digitorum** (extensor of the digits), the **extensor digiti minimi** (extensor of the little finger), and the **extensor carpi ulnaris** (extensor of the ulnar carpals). All take their origins from the lateral epicondyle of the humerus but diverge over the hand. The extensor carpi radialis longus extends and abducts the hand; its partner, the extensor carpi radialis brevis, extends the hand. The extensor digitorum divides into four tendons, one for each individual finger. Approaching the ulnar side, the extensor digiti minimi extends the little finger. Finally, the extensor carpi ulnaris extends and adducts the hand. Notice that the lateral and medial muscles abduct and adduct the wrist, respectively, while the central muscles extend the hand, as summarized in Table 11.6.

DEEP MUSCLES

Six deep muscles extend and abduct the thumb and index finger or supinate the forearm. Beginning proximally from the radial side in Figure 11.12 and Table 11.5, the muscles are the anconeus, supinator, abductor pollicis longus, extensor pollicis brevis, extensor pollicis longus, and extensor indicis. Unlike the superficial muscles that diverge from a common

origin, the deep posterior muscles converge on the thumb and index finger from origins between the lateral epicondyle of the humerus and the distal ulna. The **anconeus** is an enigma because it has very little leverage; it may stabilize the elbow joint. The **supinator** inserts on the radius and supinates the forearm with help from the biceps brachii. Next come three muscles of the thumb. The **abductor pollicis longus** (long abductor of the thumb) abducts and extends the thumb and also abducts the wrist. The **extensor pollicis brevis** (short extensor of the thumb) also extends the thumb. The **extensor pollicis longus** (long extensor of the thumb) assists the brevis by extending the distal phalanx of the thumb. Finally, the **extensor indicis** (extensor of the index finger) extends the second digit. Verify the locations of these muscles in Table 11.6.

- Which muscles extend the little finger?
- Which muscles extend the thumb?
- Which muscles extend the hand and the central fingers?
- Which muscles insert onto metacarpal bones?
- Which muscles actuate the proximal phalanges?
- Which muscles abduct or adduct the wrist?

INTRINSIC MUSCLES OF THE HAND

The intrinsic muscles act entirely within the hand and are responsible for most of its dexterity, es-

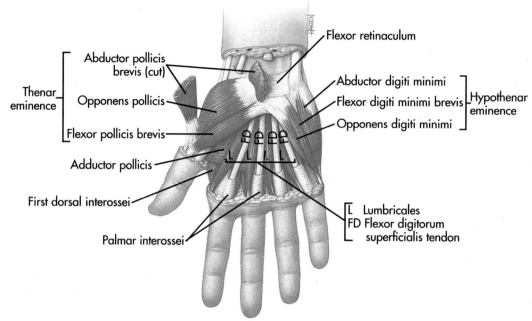

Thenar eminence
- Abductor pollicis brevis (cut)
- Opponens pollicis
- Flexor pollicis brevis

Adductor pollicis

First dorsal interossei

Palmar interossei

Flexor retinaculum

Hypothenar eminence
- Abductor digiti minimi
- Flexor digiti minimi brevis
- Opponens digiti minimi

L Lumbricales
FD Flexor digitorum superficialis tendon

✳ **Figure 11.13**
Superficial intrinsic muscles of the right hand, palmar surface.

pecially that of the thumb. Find these muscles in Figure 11.13 and Table 11.7. **Intermediate muscles** control the three central fingers, and the **thenar** and **hypothenar muscles** (thenar, *G,* palm of the hand) maneuver the thumb and little finger, respectively. See these movements in Figure 11.14.

THENAR MUSCLES

Four thenar (palm of the hand) muscles form the **thenar eminence,** the muscular pad on the palmar surface of the first metacarpal in Figure 11.13. The **abductor pollicis brevis, opponens pollicis, flexor pollicis brevis,** and **adductor pollicis** oppose the thumb to the little finger. All insert on the **proximal phalanx or metacarpal** of the thumb, and all work against the deep extensors on the posterior surface of the forearm. The abductor pollicis brevis (short abductor of the thumb) is the most superficial thenar muscle; it draws the thumb perpendicular to the palm when you grasp a coffee cup. Immediately distal to it, the flexor pollicis brevis (short flexor of the thumb) flexes the thumb toward the palm. Deep to these muscles, the two heads of the adductor pollicis reach from the capitate bone and the middle metacarpal, also drawing the thumb toward the palm. The opponens pollicis (opposing muscle of the thumb) draws the thumb away from the palm and rotates it by inserting on the radial side of the first metacarpal.

HYPOTHENAR MUSCLES

The hypothenar muscles resemble the thenar muscles in reverse. These four muscles in Figure 11.13 form the **hypothenar eminence** that pads the ulnar border of the palm. The **palmaris brevis** (short palmar) muscle (not shown in the figure) is the most superficial, taking origin from the flexor retinaculum; it tightens the skin of the ulnar surface of the hand. The **abductor digiti minimi** (abductor of the little finger) and **flexor digiti minimi brevis** (short flexor of the little finger) insert onto the proximal phalanx of the little finger, which they abduct and flex. Like the opponens pollicis, the **opponens digiti minimi** (opposer of the little finger) inserts on the metacarpal of the little finger and also abducts the little finger.

INTERMEDIATE MUSCLES

Intermediate muscles adduct and abduct the fingers shown in Figures 11.13 and 11.15. Four **lumbrical muscles** (resembling earthworms) originate from the flexor digitorum profundus tendons; they flex the proximal phalanges and extend the middle and distal phalanges. The **interossei muscles** (between the bones) are located between the metacarpals. The **interossei palmares** (palmar) muscles draw the second, fourth, and fifth digits toward the middle finger, and the **interossei dorsales** (dorsal) abduct the three central fingers.

- Name the thenar muscles and their actions.

Text continued on p.330.

Table 11.7   INTRINSIC MUSCLES OF THE HAND

Muscle	Origin	Insertion	Action	Innervation
THENAR MUSCLES (THEE-nar)	These muscles move the thumb.			
Abductor pollicis brevis (ab-DUK-tor POLL-ik-iss BREV-iss)	Scaphoid bone, transverse carpal ligament, and trapezium	Base of proximal phalanx of thumb, radial surface	Abducts (radial flexes) thumb	Median nerve (C6, C7)
Opponens pollicis (oh-PONE-ens POLL-ik-iss)	Trapezium and flexor retinaculum	First metacarpal, radial surface	Abducts (radial flexes), flexes, rotates first metacarpal	Same
Flexor pollicis brevis				
Superficial portion	Trapezium	Base of proximal phalanx of thumb, radial surface	Flexes and adducts (ulnar flexes) thumb	Median nerve (C6, C7)
Deep portion	First metacarpal, ulnar side	Base of proximal phalanx of thumb, ulnar surface	Flexes and adducts (ulnar flexes) thumb	Ulnar nerve (C8, T1)
Adductor pollicis				
Oblique head	Capitate; bases of second and third metacarpals	Base of proximal phalanx of thumb, ulnar surface	Adducts (ulnar flexes) thumb; draws thumb toward palm	Ulnar nerve (C8, T1)
Transverse head	Third metacarpal, distal third of palmar surface	Same	Same	Same
HYPOTHENAR MUSCLES	These are intrinsic muscles that abduct and flex the little finger.			
Palmaris brevis (pal-MAR-iss BREV-iss)	Transverse carpal ligament and palmar aponeurosis	Skin on ulnar border of hand	Draws skin on ulnar surface of hand toward palm	Ulnar nerve (C8)
Abductor digiti minimi (ab-DUK-tor DIH-jih-tee MIN-ih-mee)	Pisiform bone	Base of proximal phalanx of little finger, ulnar surface	Abducts (radial flexes) little finger; flexes proximal phalanx	Ulnar nerve (C8, T1)
Flexor digiti minimi brevis	Hamulus of hamate bone	Same	Flexes little finger	Same
Opponens digiti minimi (oh-PONE-ens)	Same	Ulnar margin of fifth metacarpal	Abducts (radial flexes), flexes, and rotates fifth metacarpal	Same
INTERMEDIATE MUSCLES	These muscles move the index, middle, and ring fingers.			
Lumbricals (LUM-brih-kals)	Four small muscles are associated with the tendons of the flexor digitorum profundus.			
First and second	Flexor tendons of index and middle fingers	Ulnar surface of proximal phalanx	Flex metacarpophalangeal joint and extend middle and distal phalanges	Median nerve (C6, C7)
Third	Flexor tendons of middle and ring fingers	Same	Same	Ulnar nerve (C8)
Fourth	Flexor tendons of ring and little fingers	Same	Same	Same
Interossei palmares (INT-er-OSS-ee-eye pal-MAR-es)	These are three small muscles of the internal surfaces of the index, ring, and little fingers.			
First	Ulnar surface of second metacarpal	Ulnar surface of index finger	Adducts finger and flexes proximal phalanx	Ulnar nerve, (C8, T1)
Second and third	Radial surfaces of fourth and fifth metacarpals	Radial surface of ring and little fingers	Same	Same

From Moore, PJ: Muscular and associated neurological systems. In Hilt NE and Cogburn SB, editors, Manual of orthopedics, *St. Louis, 1980, CV Mosby.*

Table 11.7   INTRINSIC MUSCLES OF THE HAND—cont'd

Muscle	Origin	Insertion	Action	Innervation
Interossei dorsales (INT-er-OSS-ee-eye dor-SAL-es)	These four small, bipennate muscles arise between the metacarpal bones, and each inserts upon the proximal phalanx of its respective finger. All abduct the fingers; they flex the metacarpophalangeal joints, and they extend the distal phalanges. The ulnar nerve (C8, T1) serves the interossei dorsales.			
First	First and second metacarpals	Radial surface, index finger	—	—
Second	Second and third metacarpals	Radial surface, middle finger	—	—
Third	Third and fourth metacarpals	Ulnar surface, middle finger	—	—
Fourth	Fourth and fifth metacarpals	Ulnar surface, ring finger	—	—

From Moore, PJ: Muscular and associated neurological systems. In Hilt NE and Cogburn SB, editors, Manual of orthopedics, *St. Louis, 1980, CV Mosby*

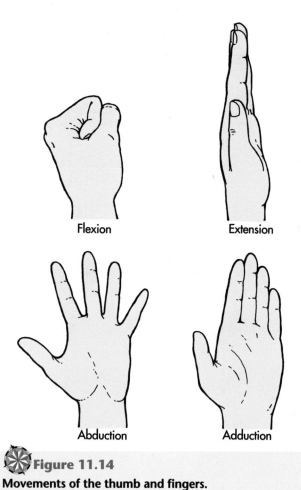

Flexion   Extension

Abduction   Adduction

Figure 11.14

Movements of the thumb and fingers.

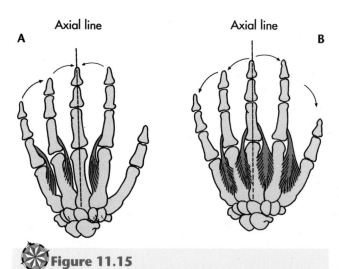

Axial line   Axial line

A   B

Figure 11.15

Interossei muscles. A, Palmar interossei muscles adduct the index, ring, and little fingers toward the middle finger. **B,** Dorsal interossei abduct the three central fingers.

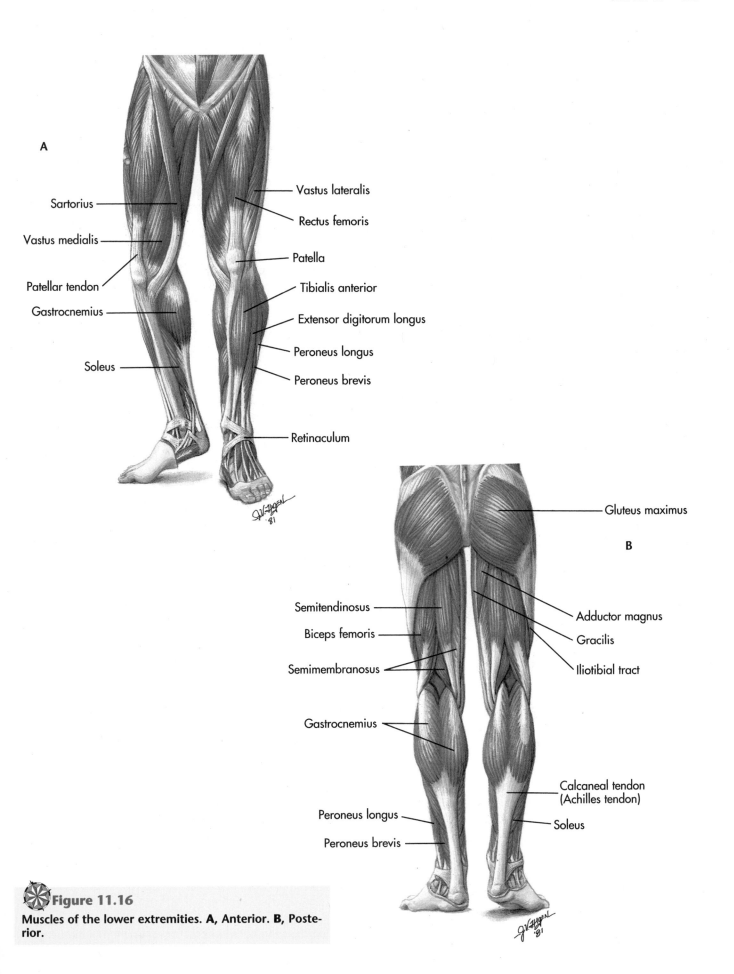

A

Sartorius

Vastus medialis

Patellar tendon

Gastrocnemius

Soleus

Vastus lateralis

Rectus femoris

Patella

Tibialis anterior

Extensor digitorum longus

Peroneus longus

Peroneus brevis

Retinaculum

Gluteus maximus

B

Semitendinosus

Biceps femoris

Semimembranosus

Gastrocnemius

Peroneus longus

Peroneus brevis

Adductor magnus

Gracilis

Iliotibial tract

Calcaneal tendon
(Achilles tendon)

Soleus

✳ **Figure 11.16**
Muscles of the lower extremities. A, Anterior. B, Posterior.

- What are the actions of the interossei and lumbrical muscles?
- Which are the hypothenar muscles?
- Which muscles originate on the flexor retinaculum?
- Which of these muscles is especially useful when you clench your fist?

PELVIS AND LOWER LIMB

BRIEF TOUR—WALKING LIMBS

Walking, running, and standing are the principal actions of the muscles in the lower limbs, shown in Figure 11.16A and B. Walk across the room to demonstrate how these limbs alternately swing forward and push backward with each stride (see Figure 11.17). An entire **stride** begins when one foot (the right foot in the figure) touches the floor and ends when that foot returns to the floor. Your right limb swings forward to take your weight from the left limb. The stride begins when your right heel strikes the floor **(heel strike)** and begins to take up your weight from the left limb. Your foot rolls forward on its right side, absorbing your weight, while your left limb pushes you forward, and the left foot rises from the floor **(toe-off).** Your right limb straightens, and your foot spreads under your full weight **(stance phase),** as your left limb begins to swing forward **(swing phase).** The left side of your pelvis lifts and rotates forward to extend this limb and to help the right limb push backward near the end of its stance phase. Now your left foot returns to the floor (heel strike) and takes up the weight from the right, as the ball and toes of your right foot finally push off (toe-off). Temporarily relieved of its load, your right limb now swings forward (swing phase) for the next step, as the pelvis lifts this limb. The stride ends when your right heel returns to the floor.

Figure 11.17 points out some of the important muscles that control walking. For example, the gastrocnemius muscle of the calf and calcaneal (Achilles) tendon raise the heel before toe-off, as you can see in the figure and feel by palpating this muscle as you walk. Unfortunately, palpation gives only a crude impression of activity and cannot reveal the motions of deep muscles at all.

ELECTROMYOGRAPHY

Electromyography is a more effective means of characterizing muscle action. This technique records contractile activity through electrodes placed in a patient's muscles. The recording pen traces a graph that represents electrical depolarizations of the muscle fibers and motor units at the tip of the electrode. Volleys of sharp peaks on the chart register activity, and flat lines indicate muscular silence. Clinicians, especially in physical therapy and sports medicine, use electromyography (EMG) to diagnose muscle damage and the success of treatment. As discussed in Chapter 9, electromyography was the source of the information that programmed Anne's computer to enable her to walk.

Except that the pelvis is firmly attached to the axial skeleton, the principles of muscular action in the lower limbs are the same as those operating in the upper limbs. Flexors, extensors, abductors, and adductors occupy specific compartments in the lower limbs. The muscles of adjoining regions overlap, allowing integrated control of motion. Although the feet are far less dextrous than the hands, the motions of feet and

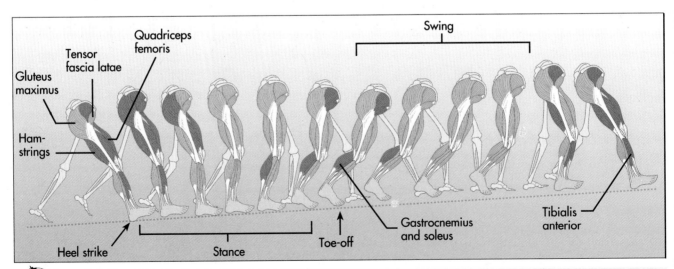

Figure 11.17

Phases in the stride follow the description in the text. Match the walking figure with the swing and stance phases and heel-strike and toe-off. Walk a few paces to identify these phases for yourself and to feel the muscles moving.

toes still are considered manipulative. Muscles of the hip, thigh, and leg are concerned with positioning movements.

- Name the major events that occur during one stride of walking.
- Describe two methods of detecting muscular action.

ILIAC AND GLUTEAL REGIONS OF THE PELVIS

The iliopsoas muscle and the gluteus group of muscles move the thigh. The **iliopsoas** is a major flexor of the femur (Figure 11.18 and Table 11.8), notably at the swing phase of the stride. The iliopsoas is really two muscles. The **psoas** muscle (of the loin) arises on each side of the vertebral column, and the **iliacus** muscle (of the ilium) originates in the iliac fossa. Both descend beneath the inguinal ligament of the pelvis to a tendon that inserts on the lesser trochanter.

The gluteal region of the pelvis (the rump), illustrated in Figure 11.19, contains 10 muscles that help to extend and rotate the femur during walking and running. The **gluteus maximus** muscle is the largest (Part A of the figure) and most superficial. Originating on the posterior of the ilium and sacrum, the gluteus maximus extends the thigh by inserting on the gluteal tuberosity of the femur, and also helps abduct the thigh (see Thigh, tensor fasciae latae, this chapter). The gluteus maximus opposes the iliopsoas and helps to extend the thigh early in the stance phase. The fibers of the **gluteus medius** (Figure 11.19A) and **glu-**

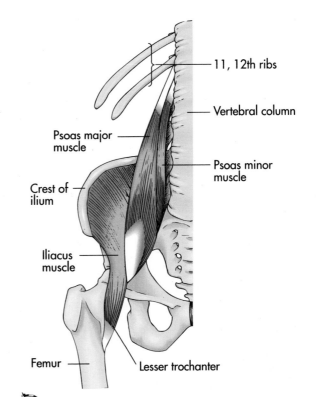

Figure 11.18

The iliopsoas is actually two muscles—the iliacus and psoas—that insert on the lesser trochanter of the femur and help flex and rotate the thigh. The psoas minor muscle is absent in many individuals.

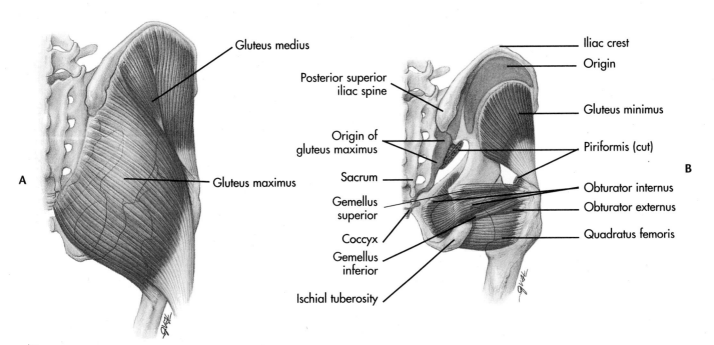

Figure 11.19

Muscles of the gluteal region, right lower limb. **A,** Glutei maximus and medius muscles are superficial. **B,** Removal of these exposes the deep gluteal muscles. The piriformis has been cut.

teus minimus (Figure 11.19B) converge on the greater trochanter from the ilium to support the pelvis during the stance phase, while the opposite limb is off the ground. Beneath these muscles, the **piriformis** (pear-shaped), **obturators** (closing muscles), **gemellus** (twin muscles), and **quadratus femoris** (four-sided muscle of the thigh) form a series that reaches behind the ischium from the rami of the pubis and ischium to insert on the inner surface of the greater trochanter (Figure 11.19B). Their action is obvious; they rotate the femur laterally, which keeps the limb pointing forward as it and the pelvis swing

forward. These muscles give another, extra push by rotating the limb laterally at toe-off. The **tensor fasciae latae** (tensor of the wide fascia) flexes the thigh; it contracts during the stance and early swing phases, when the weight of the body is passing over the limb, and also at toe-off (see Thigh, Abductors, next section)

- Name the muscles of the pelvis that flex and extend the thigh.
- Which muscles are responsible for raising the opposite side of the pelvis, and which ones rotate the femur laterally during the stride?

Table 11.8   PELVIC MUSCLES THAT MOVE THE FEMUR

Muscle	Origin	Insertion	Action	Innervation
ILIAC REGION				
Iliopsoas (ILL-ee-oh-SO-as)				
Psoas major	Bodies, intervertebral discs of T12 to L5, and transverse processes of L1 to L5 vertebrae	Lesser trochanter	Flexes thigh; flexes lumbar region	Lumbar plexus L2, L3
Iliacus (ILL-ee-AK-us)	Upper iliac fossa, iliac crest	Lateral side of psoas tendon	Flexes thigh	Femoral nerve (L2, L3)
GLUTEAL REGION				
Gluteus maximus (GLUE-tee-us)	Ilium and iliac crest; sacrum and coccyx	Iliotibial band and gluteal tuberosity of femur	Extends and laterally rotates thigh	Inferior gluteal nerve (L5, S1, S2)
Gluteus medius	Between iliac crest and gluteal lines	Greater trochanter	Abducts and medially rotates thigh	Superior gluteal nerve (L4, L5, S1)
Gluteus minimus	Same	Same	Same	Same
Piriformis (PEER-ih-FOR-miss)	Anterior sacrum	Greater trochanter	Abducts and laterally rotates thigh; helps extend thigh	Sacral nerve S1 or S1 and S2
Obturator externus (OB-tur-ATE-or)	Pubic rami, ischial ramus, and orburator membrane	Trochanteric fossa	Rotates thigh laterally	Obturator nerve (L3, L4)
Obturator internus	Superior and inferior rami of pubis, ramus of ischium, obturator membrane	Same	Rotates thigh laterally; extends, abducts thigh when it is flexed	Sacral plexus, (L5 to S2)
Gemellus superior (JEM-ell-us)	Ischial spine	Same	Same	Sacral plexus, (L3, L4)
Gemellus inferior	Ischial tuberosity	Same	Same	Sacral plexus, (L4 to S1)
Quadratus femoris (kwad-RATE-us)	Same	Intertrochanteric ridge	Same	Same
Tensor fasciae latae (TEN-sore FAH-she-eye LAT-eye)	Iliac crest, anterior superior iliac spine	Iliotibial band	Flexes thigh and helps rotate it medially	Superior gluteal nerve (L4, L5, S1)

From Moore, PJ: Muscular and associated neurological systems. In Hilt NE and Cogburn SB, editors, Manual of orthopedics, *St. Louis, 1980, CV Mosby.*

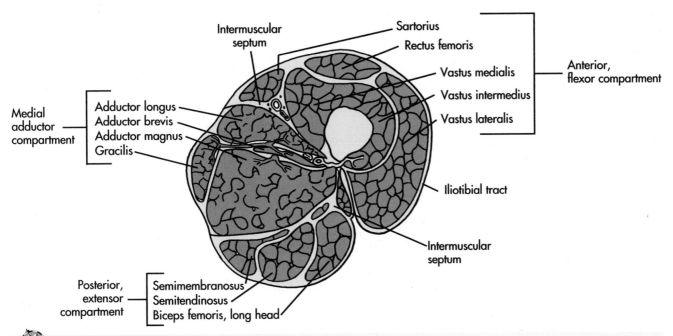

Figure 11.20

Midway down the thigh, this transverse section reveals anterior, posterior, and medial muscle compartments separated from each other by intermuscular septa. In Table 11.9, you will find that different nerves supply each group.

THIGH

The thigh contains flexor, extensor, and adductor muscles in three compartments (Figure 11.20). The anterior compartment contains the quadriceps femoris, the major extensor of the leg; the posterior compartment contains the hamstring muscles, flexors that correspond to the biceps brachii and brachialis of the arm. Adductor muscles, whose action resembles the coracobrachialis of the arm, reside in the medial compartment. The iliotibial tract of fascia is linked to the tensor fasciae latae and gluteus maximus, abductor muscles higher on the hip. **Intermuscular septa** separate these compartments from each other and deliver vessels and nerves to the muscles in each compartment.

ANTERIOR COMPARTMENT—EXTENSORS

In the anterior compartment the massive four-headed **quadriceps femoris** extends the tibia (Figure 11.21); but one head, the **rectus femoris**, originates on the iliac spine and also helps flex the thigh. You use this portion of the quadriceps to shift your foot between the accelerator and brake pedals when you drive a car. The remaining heads of the quadriceps—the **vastus medialis, vastus intermedius,** and **vastus lateralis**—only extend the tibia because they originate on the femur (Table 11.9).

Although most flexor muscles occupy the posterior

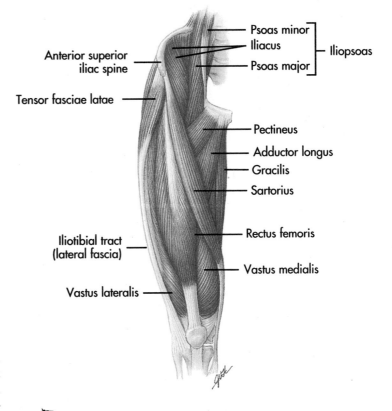

Figure 11.21

Flexor muscles of the right thigh. Find these muscles in the transverse section.

Table 11.9 MUSCLES OF THE THIGH

Muscle	Origin	Insertion	Action	Innervation
ANTERIOR COMPARTMENT				
Sartorius (sar-TOR-ee-us)	Anterior superior iliac spine	Proximal medial body of tibia	Flexes thigh, rotates it laterally; flexes leg, rotates it medially	Femoral nerve (L2, L3)
Quadriceps femoris (KWAD-rih-seps FEM-or-iss)		Tibial tuberosity via quadriceps tendon	Extends leg; rectus femoris also flexes thigh	Femoral nerve (L2 to L4)
Rectus femoris	Anterior inferior iliac spine, posterior brim of acetabulum	Base of patella		
Vastus lateralis	Intertrochanteric line and linea aspera	Lateral border, patella		
Vastus medialis	Intertrochanteric line and linea aspera	Medial border, patella		
Vastus intermedius	Anterior and lateral body of femur	Deep border of patella		
POSTERIOR COMPARTMENT				
Biceps femoris				
Long head	Ischial tuberosity	Lateral side of head of fibula and tibia	Flexes, rotates leg laterally; also extends and rotates thigh laterally	Sciatic nerve (S1 to S3)
Short head	Linea aspera	Same	Flexes leg and rotates it laterally	Sciatic nerve (L5 to S2)
Semitendinosus (SEM-ee-TEN-din-oh-sus)	Ischial tuberosity	Medial body of tibia	Flexes leg and rotates it medially after flexion; extends thigh	Same
Semimembranosus (SEM-ee-MEM-bran-oh-sis)	Same	Medial tibial condyle, and lateral femoral condyle	Same	Same
MEDIAL COMPARTMENT				
Gracilis (GRASS-il-iss)	Pubic symphysis and pubic arch	Body of tibia distal to medial condyle	Adducts thigh, flexes leg, and helps rotate leg medially	Obturator nerve (L3, L4)
Pectineus (PEK-tih-NEE-us)	Pectineal line of pubis	Between lesser trochanter and linea aspera	Adducts, flexes, and medially rotates thigh	Femoral nerve (L2 to L4)
Adductor longus	Anterior pubis, symphysis pubis	Linea aspera	Same	Obturator nerve (L3, L4)
Adductor brevis	Inferior ramus of pubis	Between lesser trochanter and linea aspera	Same	Same
Adductor magnus	Inferior rami of pubis and ischium, ischial tuberosity	Linea aspera and adductor tubercle	Adducts thigh; upper portion flexes, rotates thigh medially; lower portion extends, rotates laterally	Obturator nerve (L3, L4), sciatic nerve

From Moore, PJ: Muscular and associated neurological systems. In Hilt NE and Cogburn SB, editors, Manual of orthopedics, St. Louis, 1980, CV Mosby.

compartment, one resides in the anterior compartment. The narrow **sartorius** (tailor) muscle, shown in Figure 11.21, spans the forward surface of the thigh from the iliac crest, across the hip and knee joints, to the medial side of the tibia. Its motions are complex and not well understood. It helps flex both the thigh and the knee. Most active after toe-off in walking, the sartorius is electromyographically silent in the swing phase and when you stand still.

POSTERIOR COMPARTMENT—FLEXORS

The hamstring muscles on the posterior thigh enable several motions because they, too, cross both hip and knee joints. The **biceps femoris** is the **lateral hamstring.** It originates on the ischial tuberosity and the femur (Figure 11.22 and Table 11.9) and inserts on

stance phase begins. All three muscles flex the knee; they also rotate the leg when the knee is flexed, as you can demonstrate by swinging your foot medially and laterally while sitting down. You can touch your toes without bending your knees if your hamstrings are supple, but if they are tight, your knees will bend.

MEDIAL COMPARTMENT—ADDUCTORS

The **gracilis, pectineus,** and the **adductors longus, brevis,** and **magnus** all originate near the inferior ramus of the pubis, from which they fan out to insertions on the linea aspera of the femur in Figure 11.23 or, in the case of the gracilis, below the medial tibial condyle. See Table 11.9 for details. These muscles of the medial compartment draw the thigh toward the central axis of the body and also can rotate

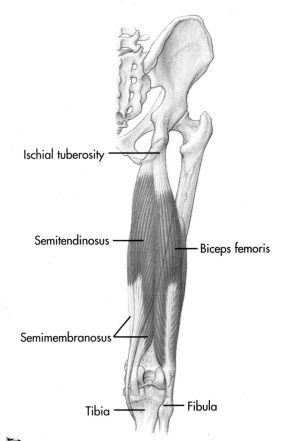

 **Figure 11.22**

Hamstrings, the extensor muscles of the right thigh. Locate these muscles in the transverse section in Figure 11.20.

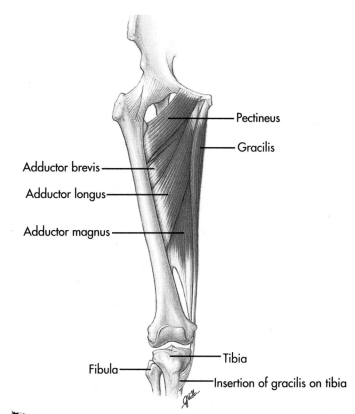

Figure 11.23

Adductor muscles of the right thigh. Find these muscles in the transverse section in Figure 11.20.

the fibula. The **semimembranosus** and **semitendinosus** muscles are the **medial hamstrings.** They also originate from the ischial tuberosity, but they insert below the medial condyle of the tibia. Hamstrings slow the forward swing of the leg when the heel touches the floor, and they extend the thigh as the

the thigh medially. Electromyography shows that they are silent midway through each swing phase, but the adductor longus and gracilis are especially active at toe-off.

ABDUCTORS

The **tensor fasciae latae** and **gluteus maximus** muscles of the gluteal region abduct the thigh by pulling together on the **iliotibial tract,** a band of fascia that extends down the lateral side of the thigh and inserts upon the tibia, as shown in Figure 11.24. Resembling a gun belt and holster, the tensor fasciae latae (tensor of the broad fascia) and gluteus maximus

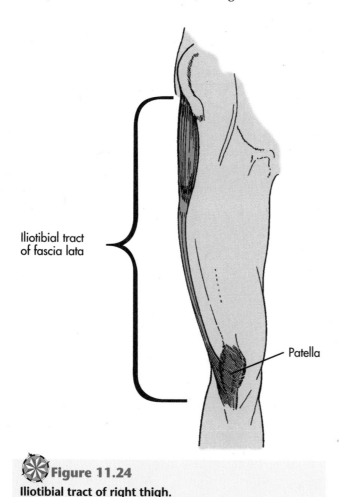

Iliotibial tract of fascia lata

Patella

Figure 11.24
Iliotibial tract of right thigh.

suspend the iliotibial tract beside the thigh and tighten it when abducting the thigh.

- Name several two-joint muscles of the thigh.
- Give the names and action of the four portions of the quadriceps femoris muscle.
- How many hamstring muscles are there?
- Which are the lateral and medial hamstrings?
- What is the action of hamstrings?
- When are these muscles most active?

LEG

The anatomy of leg muscles resembles that of the forearm. Extensors and **flexors** of the foot and toes occupy the anterior and posterior compartments, respectively, in front of and behind the tibia and fibula. Their relative locations are reversed, however, because the lower limbs have rotated medially during development, and the flexors now face posteriorly (see Limb buds, Chapter 7). A third **lateral compart-**

ment also contains flexor muscles. The positions of the major layers of muscles resemble those of the forearm; superficial proximal muscles take origins at the knee and insert on the tarsals, while deeper distal muscles originate on the shafts of the tibia and fibula and insert on the phalanges. Flexor tendons like those passing through the carpal tunnel in the wrist pass under the medial and lateral malleoli around the ankle, traveling in grooves and bursae on the calcaneus and talus bones. The fourth and fifth lumbar nerves and the first two sacral spinal nerves innervate the muscles of the leg.

ANTERIOR COMPARTMENT—EXTENSORS

The anterior compartment contains four versatile extensor, or dorsiflexor, muscles that also evert the foot (Figure 11.25A and B). These muscles—tibialis anterior, extensor digitorum longus, peroneus tertius, and extensor hallucis longus—raise the foot and toes from the floor each time the leg swings forward. They also ease the foot back onto the floor and evert it when the heel touches, and shortly afterward, they help pull the leg over the foot as the stride continues. Fibers from lumbar nerves 4 and 5 and from the first sacral nerve supply all four muscles through the deep peroneal nerve.

The **tibialis anterior** can be felt lateral to the anterior border of the tibia. This large muscle dorsiflexes and inverts the foot at toe-off by way of insertions on the medial cuneiform bone and first metatarsal. It is electromyographically most active at heel-strike, which suggests to some researchers that this muscle absorbs landing shock by lowering the foot onto the floor. The tibialis anterior is the landmark for the other three anterior muscles.

The **extensor digitorum longus** (long extensor of the digits) descends the leg lateral to the tibialis anterior, its tendon dividing into four tendons that extend the lesser toes. The **peroneus tertius** (third peroneal muscle) inserts upon the fifth metatarsal bone from its origin lateral to the extensor digitorum longus. The **extensor hallucis longus** (extensor of the great toe) is deep to the tibialis anterior and the extensor digitorum; its tendon inserts on the distal phalanx and extends the great toe. The tendons all pass beneath the **extensor retinaculum** onto the dorsal surface of the foot, where the tendons of the extensor digitorum longus radiate to the toes (Figure 11.25A).

POSTERIOR COMPARTMENT—FLEXORS

Seven flexor muscles fill the posterior compartment of the leg. From superficial to deep, they are the gastrocnemius, soleus, plantaris, popliteus, tibialis posterior, flexor hallucis longus, and flexor digitorum longus (Table 11.10 and Figure 11.26). The last two lumbar and first two sacral spinal nerves innervate this group by way of the tibial nerve.

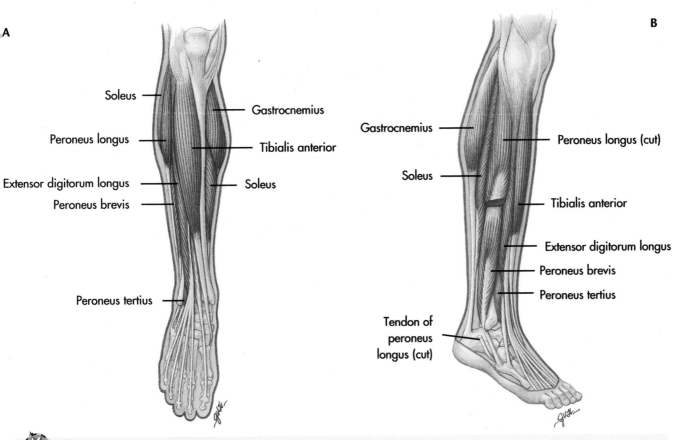

Figure 11.25
Muscles of the anterior **(A)** and lateral compartments **(B)**, right leg.

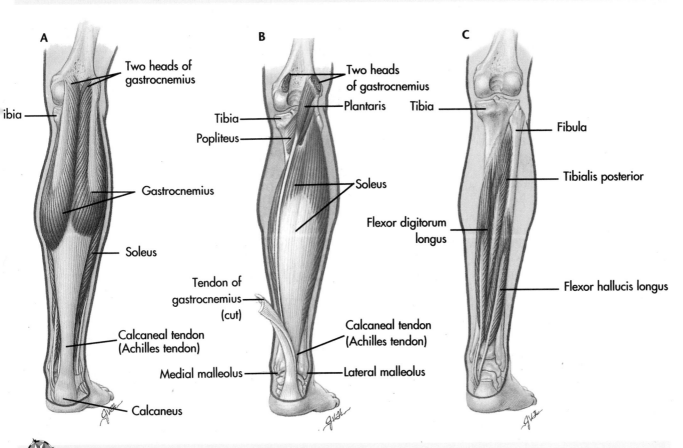

Figure 11.26
Posterior muscles of the right leg. **A,** The gastrocnemius is a superficial muscle. **B,** When removed, it reveals the soleus, plantaris, and popliteus muscles. **C,** Deep muscles lie beneath the soleus.

Table 11.10   MUSCLES THAT MOVE THE FOOT

Muscle	Origin	Insertion	Action	Innervation
ANTERIOR COMPARTMENT				
Tibialis anterior (TIB-ee-AL-iss)	Lateral tibial condyle and proximal tibia	First cuneiform and first metatarsal	Dorsiflexes, adducts, and inverts foot	Deep peroneal nerve, (L4 to S1)
Extensor digitorum longus (dih-jih-TOR-um)	Lateral tibial condyle and anterior body of fibula	Middle and distal phalanges of lesser four toes	Extends proximal phalanges; dorsiflexes and abducts foot	Same
Peroneus tertius (per-oh-NEE-us TER-she-us)	Distal third of fibula	Base of fifth metatarsal	Dorsiflexes and abducts foot	Same
Extensor hallucis longus (HAL-uh-kiss)	Anterior surface of fibula and interosseous membrane	Base of distal phalanx of great toe	Extends proximal phalanx; dorsiflexes and adducts foot	Same
POSTERIOR COMPARTMENT				
Gastrocnemius (GAS-troe-NEEM-ee-us)	Medial and lateral femoral condyles	Calcaneal tendon with soleus	Plantar-flexes foot	Tibial nerve (S1, S2)
Soleus (SOL-ee-us)	Head and proximal third of fibula; medial border of tibia	Calcaneal tendon with gastrocnemius	Same	Same
Plantaris (plan-TAR-iss)	Linea aspera	Posterior calcaneus	Plantar-flexes foot; flexes leg	Tibial nerve, (L4 to S1)
Popliteus (POP-lit-ee-us)	Lateral femoral condyle	Popliteal line of tibia	Flexes and rotates leg medially	Same
Tibialis posterior (TIB-ee-AL-iss)	Posterior body of fibula and interosseous membrane	Navicular, calcaneus, cuneiforms, cuboid bones; 2nd to 4th metatarsal	Inverts, adducts, plantar-flexes foot	Same
Flexor hallucis longus (HAL-uh-kiss)	Distal fibula	Distal phalanx, great toe	Flexes middle phalanx, great toe; everts and plantar-flexes foot	Tibial nerve, (L5 to S2)
Flexor digitorum longus (dih-jih-TOR-um)	Body of tibia below popliteal line	Distal phalanges, lesser four toes	Flexes distal phalanges; adducts and plantar-flexes foot	Tibial nerve, (L5 to S1)
LATERAL COMPARTMENT				
Peroneus longus (per-oh-NEE-us)	Lateral tibial condyle, head and lateral surface of fibula	Base of first metatarsal and medial cuneiform bone	Everts, adducts, plantar-flexes foot	Superficial peroneal nerve (L4 to S1)
Peroneus brevis	Distal fibula, lateral surface	Base of fifth metatarsal	Same	Same

From Moore, PJ: Muscular and associated neurological systems. In Hilt NE and Cogburn SB, editors, Manual of orthopedics, *St. Louis, 1980, CV Mosby.*

SUPERFICIAL

The **gastrocnemius** and **soleus** muscles are the major flexors, or plantar flexors, of the foot (Figure 11.26A and B). These powerful muscles lift the heel from the ground near the end of the stride. The gastrocnemius (muscle of the calf) reaches from the femoral condyles and the soleus reaches from the proximal tibia and fibula to the calcaneal (Achilles) tendon and calcaneus bone at the heel. Between the gastrocnemius and soleus, the very small **plantaris** flexes the foot weakly. How so small a muscle benefits its strong partners is a mystery.

The position of the **popliteus muscle** (back of the knee, Figure 11.26B) corresponds to the pronator teres of the forearm, but the popliteus does not pronate the leg. Instead, the popliteus unlocks the extended knee when you are standing, by rotating the femur laterally upon the tibia. The popliteus reverses the motion that, you will recall from Chapter 8, rotates the tibia laterally and locks the knee when the leg extends.

DEEP

The tibialis posterior, flexor hallucis longus, and flexor digitorum longus muscles lie deep to the soleus muscle (Figure 11.26C). The **tibialis posterior** (posterior muscle of the tibia), which resembles the palmaris longus muscle of the forearm, originates on the superior half of the interosseous membrane and the adjacent tibia and fibula. The tibialis posterior descends on the medial side of the leg, behind the medial malleolus of the tibia. The tendon radiates onto the undersides of the tarsals and the last four metatarsals, where it plantar flexes and inverts the foot. During the stance phase, the tibialis posterior helps prevent the foot from everting, or turning outward. The **flexor hallucis longus** (long flexor of the great toe) and **flexor digitorum longus** (long flexor of the digits) lie on each side of the tibialis posterior, and their tendons descend the medial side of the ankle, respectively, to the distal phalanges of the great and lesser toes. These muscles correspond to the flexors pollicis longus and digitorum profundus of the forearm. The flexor hallucis longus is most active during stance, when the foot bears the body's full weight, but the flexor hallucis apparently does not help the gastrocnemius and soleus lift the heel.

LATERAL COMPARTMENT—MORE FLEXORS

The **peroneus longus** and **peroneus brevis** (long and short peroneus) muscles occupy the lateral compartment on the fibular side of the leg, shown in Figure 11.25B, where they evert and plantar-flex the foot when the heel rises from the floor. With the more proximal origin (Table 11.10), the peroneus longus also inserts more proximally on the foot. Its tendon passes the lateral malleolus of the ankle, across the sole of the foot, to the medial cuneiform bone and the first metatarsal. The peroneus longus everts the foot more effectively from this angle than does the peroneus brevis, which inserts on the fibular side at the fifth metatarsal. The superficial peroneal nerve supplies both muscles with fibers from the last two lumbar and first sacral spinal nerves.

- To which compartments do the soleus, tibialis anterior, and peroneus longus belong?
- Which muscles are responsible for raising the heel just before toe-off?
- Which muscle is required to lift the foot from the floor at toe-off?
- On which bones does the tibialis anterior insert?
- Which muscles of the anterior and posterior compartments insert upon the distal phalanx of the great toe?

INTRINSIC MUSCLES OF THE FOOT

Muscles of the foot resemble those of the hand, but they bear the body's weight and are far less dextrous (Figure 11.27 and Table 11.11). All but the **extensor digitorum brevis** occupy the plantar surface. Muscles on the tibial and fibular sides move the great and little toes, respectively, and medial muscles control the middle digits. Four layers of plantar muscles support the longitudinal and transverse metatarsal arches of the foot. The fourth and fifth lumbar nerves, along with the first two sacral nerves, supply innervation.

FIRST LAYER

The **abductor hallucis, flexor digitorum brevis,** and **abductor digit minimi** arise from the calcaneus and reach forward to the great toe, central digits, and the little toe, respectively, as shown in Figure 11.27A. The flexor digitorum brevis flexes the four lesser toes; its partners abduct the great and little toes.

SECOND LAYER

The **quadratus plantae** (four-sided muscle of the sole of the foot) and **lumbrical muscles** (worm-like) comprise the second layer of muscles, framed by the abductors hallucis longus and digiti minimi. The quadratus plantae redirects the pull of the flexor digitorum longus tendon, so that when both muscles contract, they pull directly posterior toward the heel. The lumbricals resemble those of the hand; they flex the four lesser toes.

THIRD LAYER

Control of the middle digits gives way to the great and little toes in the third layer of muscles, which resemble the thenar and hypothenar muscles of the hand. The **flexor hallucis brevis** (short flexor of the great toe) and **adductor hallucis** (adductor of the

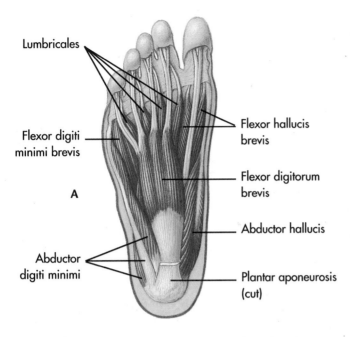

Lumbricales

Flexor digiti
minimi brevis

A

Abductor
digiti minimi

Flexor hallucis
brevis

Flexor digitorum
brevis

Abductor hallucis

Plantar aponeurosis
(cut)

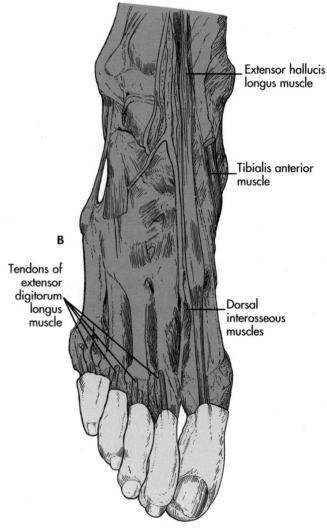

Extensor hallucis
longus muscle

Tibialis anterior
muscle

B

Tendons of
extensor
digitorum
longus
muscle

Dorsal
interosseous
muscles

Figure 11.27
Intrinsic muscles of the foot. A, Plantar surface. B, Dorsal interossei.

great toe) maneuver the great toe, and the **flexor digiti minimi brevis** (short flexor of the little toe) flexes the proximal phalanx of the little toe.

FOURTH LAYER

This layer contains the dorsal and plantar **interossei muscles** (between the bones). The bipennate **dorsal interossei** muscles (Figure 11.27B) abduct the three central toes, and the **plantar interossei** adduct the last three toes.

• Which muscles originate from the calcaneus bone?

• Name the muscles of the third layer and their actions.
• How do the interossei muscles of the hand and foot differ?

MUSCLE DEVELOPMENT

Limb muscles develop from myotomes, which are strips of muscle-forming tissue derived from somites, as you will recall from Chapter 10. Muscle-forming cells called myoblasts migrate from these my-

Table 11.11   INTRINSIC MUSCLES OF THE FOOT

Muscle	Origin	Insertion	Action	Innervation
DORSAL COMPARTMENT				
Extensor digitorum brevis (DIH-jih-TOR-um)	Calcaneus	Proximal phalanx of great toe; tendons of extensor digitorum longus	Extends proximal phalanges of first four toes	Deep peroneal nerve, (L5 to S1)
PLANTAR COMPARTMENT				
First layer				
Abductor hallucis (HAL-uh-kiss)	Tuberosity of calcaneus	Proximal phalanx, tibial side, great toe	Abducts great toe	Medial plantar nerve (L4, L5)
Flexor digitorum brevis	Same	Middle phalanges of four lesser toes	Flexes middle phalanges of four lesser toes	Same
Abductor digit minimi (DIH-jih-tee MIN-ih-mee)	Same	Proximal phalanx, fibular side, little toe	Abducts little toe	Lateral plantar nerve (S1, S2)
Second layer				
Quadratus plantae (PLAN-tie)				
Medial head	Medial surface of calcaneus	Flexor digitorum longus tendons	Flexes distal phalanges of four lesser toes	Lateral plantar nerve (S1, S2)
Lateral head	Lateral surface of calcaneus	Same	Same	Same
Lumbricals (4) (LUM-brih-cals)	Flexor digitorum longus tendon	Flexor digitorum longus tendons at proximal phalanx	Flex proximal and extend both distal phalanges of four lesser toes	Medial plantar nerve (L4, L5) lateral plantar nerve (S1, S2)
Third layer				
Flexor hallucis brevis	Cuboid, lateral cuneiform bones	Proximal phalanx, great toe	Flexes proximal phalanx of great toe	Medial plantar nerve (L4 to S1)
Adductor hallucis				
Oblique head	Second to fourth metatarsals	Lateral side of proximal phalanx, great toe	Adducts great toe	Lateral plantar nerve (S1, S2)
Transverse head	Plantar metarsophalangeal ligaments of toes 3-5	Same	Same	Same
Flexor digiti minimi brevis	Base of fifth metatarsal	Proximal phalanx, fifth toe	Flexes proximal phalanx of little toe	Lateral plantar nerve (S2, S3)
Fourth layer				
Dorsal interossei (4) (IN-ter-OSS-ee-eye)	Between metatarsals	Proximal phalanges; 1, medial side of second toe; 2-4, lateral sides of second to fourth toes	Abduct toes; flex proximal, extend distal phalanges	Lateral plantar nerve (S1, S2)
Plantar interossei (3)	Medial side, metatarsals 3-5	Medial side, proximal phalanges of toes 3-5	Adduct toes; flex proximal, extend distal phalanges	Same

From Moore, PJ: Muscular and associated neurological systems. In Hilt NE and Cogburn SB, editors, Manual of orthopedics, *St. Louis, 1980, CV Mosby.*

otomes into the developing limb buds, where they form muscles. Limb bud development was described in Chapter 7, Appendicular Skeleton. Muscles that develop on the ventral sides of the upper limb buds become the flexors and supinators, and dorsal myoblasts form the extensors and pronators. As a result, the biceps brachii muscle, a flexor and supinator, develops from ventral myoblasts, and the triceps brachii develops from dorsal myoblasts. The lower extremities follow similar rules. The quadriceps and other extensors, together with adductor muscles, develop on the dorsal side of the lower limb buds, while the flexors, including the hamstrings, and abductors develop on the ventral side. Certain muscles violate these rules because they migrate to the opposite side of the limb bud. The short head of the biceps femoris that migrates from the dorsal side of the limb bud is one example of this contradiction.

- On which side of the upper limb bud do the flexors of the wrist and hand develop?
- Does the soleus muscle develop from dorsal or ventral myoblasts?

SUMMARY

Positioning movements of the extremities place the hands and feet where they can manipulate objects. Large proximal muscles are responsible for positioning the limbs, while smaller distal muscles extend and flex the digits with manipulative motions. These muscles are grouped into anterior, posterior, medial, and lateral compartments around the limbs. *(Page 310.)*

The shoulder, arm, forearm, wrist, carpals, and digits are the major regions of the upper extremity. Superficial and deep muscles of the thorax join the trunk to the shoulder and the arm. These muscles elevate and depress, as well as protract and retract, the scapula when the limb reaches away from the body. Muscles, such as the deltoid, teres major, and those in the rotator cuff, connect the scapula to the arm and raise, extend, flex, adduct, and rotate this limb. *(Page 310.)*

Biceps and triceps brachii muscles and the brachialis and coracobrachialis muscles of the arm

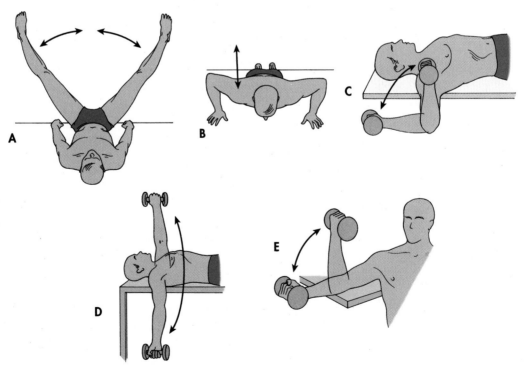

⬱ Figure 11.28

Muscles and exercises. Which muscles do these exercises stimulate? **A,** Adduct and abduct the lower limbs. **B,** Push-ups. **C,** Rotate arm. **D,** Extend and flex arm. **E,** Forearm curls. (From Alter MJ: *Science of stretching,* Champaign, Ill, 1988, Human Kinetics Books.)

connect to the scapula and flex and extend the forearm. The biceps also supinates the forearm. Anterior forearm muscles flex the hand and digits from a common origin on the medial epicondyle of the humerus. Posterior muscles that originate from the lateral epicondyle extend the hand and digits. Intrinsic muscles of the hand maneuver the digits, especially the thumb. *(Page 320.)*

A stride begins when the heel touches down and ends the next time it touches the floor. Electromyography reveals when limb muscles contract during walking and other exercises. The iliopsoas and gluteus maximus muscles help flex and extend the femur. The iliopsoas works at the beginning of the swing phase, and the gluteus maximus works during the stance phase. *(Page 330.)*

The quadriceps femoris and hamstring muscles extend and flex the leg. They also help flex and extend the thigh during the swing and stance phases, respectively. The gastrocnemius and soleus muscles raise the heel, and the tibialis anterior lifts the foot from the floor at toe-off, when stance ends and swing begins. The tibialis posterior helps keep the foot from everting during the stance phase. The flexor hallucis longus and flexor digitorum longus help the toes grasp the floor when the leg begins to push off from the floor. Intrinsic muscles of the foot help to distribute pressure from it to the floor. *(Page 333.)*

Generally speaking, flexors develop on the ventral sides of limbs and extensors develop on the dorsal sides. *(Page 342.)*

STUDY QUESTIONS
OBJECTIVES

1. Give the names of four muscles that move the upper extremity, and describe their motions. Do the same for the lower extremity.   **1**
2. What muscles elevate the scapula, abduct the arm, flex the elbow, and flex the little finger? What muscles flex the thigh, extend the tibia, plantar-flex the foot, and abduct the lesser toes?   **1**
3. Describe a positioning motion of the upper limb, and give an example of manipulative motions. What uses do a ballerina and rock climber have for positioning and manipulative motions?   **2**
4. Where is the adductor compartment of the thigh located? What motion do the muscles in this location impart to the limb? Where is the extensor compartment of the arm, and what motion do its muscles produce?   **3**

5. Why are proximal limb muscles frequently larger than distal ones? What benefits come from the fact that limb muscle origins and insertions overlap?   **4**
6. Which muscles move the scapula and humerus? Where are the origins and insertions of these muscles located?   **5, 6**
7. What muscle extends the forearm, and where are its origins and insertions located? Which muscles flex the forearm?   **7**
8. Which muscles of the forearm flex and extend the digits? Which forearm muscles flex and extend the thumb? How do these muscles act when you grasp a glass of water?   **8**
9. Which thenar muscles abduct, adduct, flex, and extend the thumb? On which bone do these muscles insert? How do these muscles act when you oppose your thumb to the little finger of the same hand?   **9**
10. Describe the action of shoulder and arm muscles when you reach forward to turn on the faucet at a sink. Which muscles are required to protract the shoulder, to flex the arm, to extend the forearm, to grasp the faucet, and twist the handle? How much pronation of the hand and rotation of the arm is necessary to turn the faucet? Describe which arm muscles are moving in Figure 11.28.   **5-9**
11. What does electromyography disclose about muscle contraction that mere palpation does not show?   **10**
12. Beginning with the right heel striking the floor, describe the movements of walking that take place during one stride.   **11**
13. When are the gluteal muscles and the iliopsoas active in walking? What motions do they accomplish?   **12**
14. Name the muscles of the adductor, flexor, and extensor compartments of the thigh, and describe their actions. In which phases of the stride are these groups of muscles active?   **13**
15. Heel-strike and toe-off use different muscles. Describe which ones contract at each event.   **14**
16. Which lower limb muscles are being exercised in Figure 11.28?   **12-14**
17. Which classes of muscles develop on the dorsal and ventral surfaces of limbs?   **15**

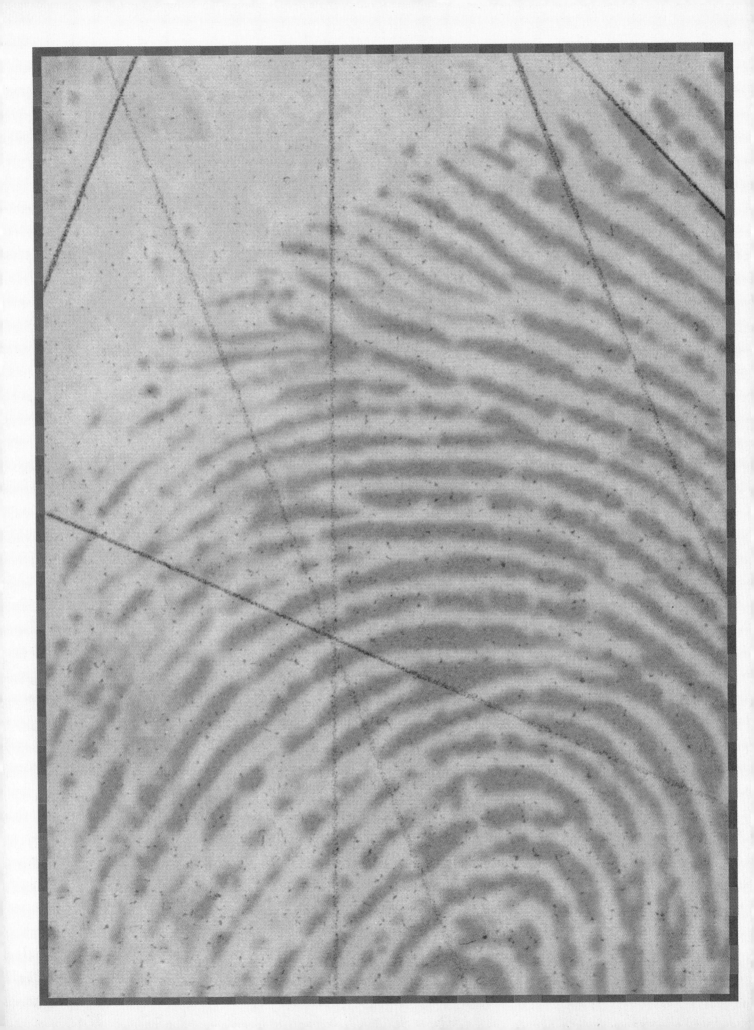

CHAPTER 12

The Integument

OBJECTIVES

After studying this chapter, you should be able to do the following:

1. Describe the major functions of the skin.
2. Describe the major features of the epidermis and dermis, and compare these tissues with each other. Show how each tissue contributes to the functions of the skin.
3. Describe the anatomy of the four tissue layers of the epidermis and how they are related to the processes of keratinization and desquamation.
4. Describe how the balance between keratinization and desquamation in the epidermis is affected by disease.
5. Describe the detailed anatomy of dermis.
6. Describe the epidermal appendages, their functions, and their anatomical similarities.
7. Describe the anatomy of dermal ridges.
8. Name and describe the sensory nerve endings of the skin, and list the stimuli each type is thought to detect.
9. Trace the development of skin and epidermal appendages.
10. Describe the damage that first, second, and third degree burns inflict upon the skin.
11. Describe various diseases and conditions of the skin.
12. Explain how transdermal drug delivery systems work, and give examples.

 **VOCABULARY TOOLBOX**

SKIN TERMS

cutaneous   of the skin; cut-, (say cute) *L* skin. Also **cuti**cle and sub**cut**aneous.

epidermis   the outer layer of skin; derm-, *G* skin; as in **derm**al ridges, **derm**atoglyphics, and **derm**atology.

keratin   a major protein of skin; kera-, *G* horn; as in **kera**tinocyte and **kera**tinization.

skin   Old Norse, skin.

POSITIONAL TERMS

apo-   *G* from, away, off. **Apo**crine secretion releases products from cells.

epi-   *G* upon, over. **Epi**dermis—the outer layer of skin—lies upon the dermis.

hypo-   *G* under, beneath. The **hypo**dermis lies beneath the dermis.

per-   *L* through. **Per**spiration moves through the skin.

peri-   *G* around. **Peri**derm surrounds the fetal skin.

RELATED TOPICS

- Dermatomes, Chapter 18
- Ectoderm and mesoderm, Chapter 4
- Epithelia and connective tissue, Chapter 3
- Keratins, Chapter 2

- Neurons, nerve endings, and myelin sheath, Chapter 17
- T-lymphocytes, Chapter 16

PERSPECTIVE

As the largest organ of the body, the skin accomplishes a great variety of tasks. Its tough, waterproof covering protects against fluid loss, infection, and injury. Its sweat glands and capillaries help to regulate body temperature, and its receptors communicate temperature, touch, pressure, and pain to its owner's brain. The skin also displays to others information about its owner's age, sex, physical condition, and emotional state.

These functions depend on the epidermis and dermis, the two fundamental layers of the skin. Interactions between the epidermis and dermis produce sweat glands, hairs, nails, and the skin's own protective surface, but these relationships are still poorly understood. No artificial skin, for example, is yet able to reconstitute hairs and sweat glands. In learning the anatomy of the skin you will be investigating a dynamic organ that may someday also communicate its inner workings.

The skin covers the body. Composed of approximately 20 square feet of tough, resilient tissue, the skin weighs at least 20 lbs and averages about 2 mm thick; it is the body's largest organ by size and weight. Because it is so visible, skin displays a wealth of features that no internal organ discloses (Figure 12.1). Babies' skin has soft hair and waxy texture (Figure 12.1A). Most of adult skin is hairy (Figure 12.1B), especially the scalp, but epidermal ridges instead of hair cover the palms of the hands and soles of the feet (Figure 12.1C). The skin usually is smooth and pleasing to touch, but disease can turn it scaly and hard. Texture is in the eye of the observer, however, for even normal skin is anything but smooth when seen at high magnification (Figure 12.1D). Although skin is waterproof and often referred to as "dry," sweat glands cool it with perspiration that keeps its surface moist. In addition to its usual colors the skin may flush with extra blood, be pallid white in albinos, turn ashen grey in death, or appear bluish when bruised or in physiological shock. At its thinnest point, on the eyelids (Figure 12.1E), skin is only 0.6 mm deep, but calluses may make it more than 10 times thicker. Although the skin appears to change little from day to day, it continually sheds old cells and replaces them with new ones, as fading suntans and healing cuts attest. Most features of the skin relate to its four fundamental functions—protection, regulation, communication, and synthesis and secretion.

Case study begins on p. 349.

A

B

C

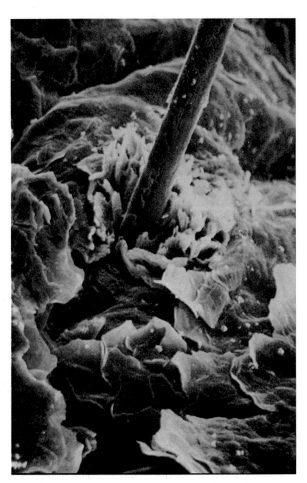

D

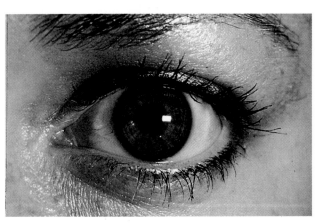

E

 Figure 12.1

A, Newborns arrive with waxy vernix caseosa and lanugo hairs covering their skin. **B,** Hair is found on the arms and elsewhere. **C,** There is no hair on the palms of the hands. **D,** Magnified 700 times, the skin appears rough and uneven. **E,** The skin of the eyelids is quite thin and features specialized hairs.

ANATOMY—BRIEF TOUR

TWO LAYERS

The skin consists of two fundamental layers of tissue—a thin, outer epithelium called the **epidermis,** and the **dermis,** a much thicker connective tissue layer beneath the epidermis—as you can see in Figure 12.2. The epidermis (EP-ih-DER-miss) is only 75 to 600 μm thick, whereas the dermis may be 10 times thicker. A **basement membrane** attaches the epidermis to the dermis and sharply defines the boundary between the two layers. The dermis blends with the **hypodermis,** or **subcutaneous fascia,** which is more connective tissue beneath the skin that stores fat and binds the skin to underlying bones and muscles.

Barrier is the key word for epidermis. The epidermis toughens the skin with a cytoplasmic network of keratin filaments (KARE-ah-tin) that **keratinocytes** (KARE-ah-tin-oh-sites), the principal epidermal cells, assemble by a process called **keratinization** (KARE-ah-tin-ih-ZAY-shun). Hair follicles, nail grooves, and sebaceous glands adapt this process to form solid hairs, tough nails, and oily liquids. Melanocytes pigment the skin and protect it from ultraviolet light, and Langerhans cells respond to irritations and infections. Because the epidermis contains no capillaries, it depends on dermal vessels for nutrients that pass through the basement membrane. Nerve endings do reach the epidermis, however, through the basement membrane.

Quite a different term, *access,* describes the dermis. As a connective tissue (see Chapter 3, Tissues), the dermis consists almost entirely of **extracellular matrix** that carries capillaries, nerve endings, and tissue fluid between the epidermis and the interior body. The epidermis receives nutrients and discharges products such as vitamin D through the interstitial fluid and blood vessels of the dermis. Macrophages migrate from the capillaries into the extracellular matrix and converge around cuts and scrapes to combat infection. Collagen fibers make the skin tough and resilient, and elastic fibers allow it to fold and stretch. Leather is made from the dermis fibers of cattle and other animals.

FOUR FUNCTIONS

PROTECTION

The skin is a barrier against injury, dehydration, infection, and sunlight. Keratin fibers in the epidermis and collagen fibers in the dermis resist

Figure 12.2

The epidermis covers the skin, and dermis supports the epidermis. The illustration shows a block of hairy skin from the arm.

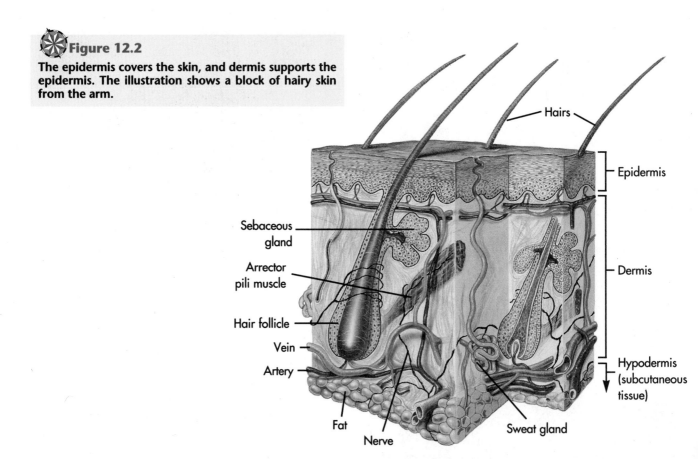

Hairs

Epidermis

Sebaceous gland

Arrector pili muscle

Hair follicle

Vein

Artery

Fat

Nerve

Sweat gland

Dermis

Hypodermis (subcutaneous tissue)

physical injury. This toughness also helps prevent microorganisms from entering the body, but when they inevitably do penetrate, lymphocytes of the skin attack them. Epidermal waterproofing prevents water and electrolytes from escaping. Two of physicians' first concerns when treating burn patients are to administer intravenous saline solution to restore lost fluids and ions, and to prevent infections with topical antibiotics. Melanin pigment shields skin cells from the damaging effects of sunlight and ultraviolet radiation, and subdermal adipose tissue cushions the body against injury.

REGULATION

The skin regulates heat loss. When warm surroundings or exertion cause more blood to flow through the dermal capillaries, more heat radiates from the body. Perspiration from sweat glands cools the skin further by evaporating from its surface. Skin also conserves body heat. Its adipose tissue helps to insulate against heat loss, and its capillaries can reduce heat loss by restricting the flow of blood.

COMMUNICATION

The skin communicates external and internal conditions. The skin is laced with sensory receptors that detect external changes interpreted by the brain as heat or cold, touch or pain. The texture and color of skin also help disclose the owner's age, physical condition, sex, race, and emotions. In the United States we spend some $18.5 billion annually to cleanse, deodorize, and decorate our skins.

SYNTHESIS AND SECRETION

Skin cells synthesize a variety of products that are released inside or outside the body. In its most fundamental activity, epidermis engages in massive

growth and synthesis of keratin proteins that renew the surface of the skin, secrete hairs and nails, and heal wounds. Ultraviolet light stimulates the skin to convert 7-dehydrocholesterol into the active form of vitamin D, which stimulates the small intestine to absorb calcium ions. Sebaceous glands secrete oils and waxes that lubricate the skin and may inhibit microbial growth. Sweat glands secrete small quantities of urea wastes along with perspiration. Pigment cells synthesize melanin, the pigment that colors the hair and epidermis, and subdermal adipose tissue synthesizes and stores fat.

- Name the four major functions of the skin.
- Which functions do the epidermis and dermis serve?

SKIN ANATOMY AND BLISTER DISEASE

About 1 in 50,000 people in the United States suffers from **epidermolysis bullosa simplex** (**EBS**; BULL-osa; bulla, *L,* bubble), a condition that causes the skin to blister from even gentle touch. As you can imagine, the skin remains tender and painful because of unavoidable daily contact. Blistering may be especially severe on the palms of the hands and soles of the feet. Although most of the blisters heal readily, more quickly appear. These victims have lost the benefits of normal skin's toughness and resilience.

EBS helps illustrate epidermal anatomy because it reveals the underlying processes that form the protective barrier. Unlike normal epidermal cells, EBS cells lack the strength to resist physical contact because their network of imperfect keratin protein filaments fails to support the cells against daily wear and tear. The disease has been traced to a defective gene located on chromosome 17 that produces defective keratin that weakens the intermediate filament network. The skin blisters when epidermal cells break open (cytolysis—cell splitting). Patients usually inherit EBS on a dominant allele, but occasionally the disease arises spontaneously as a mutation. Cortisone treatment reduces inflammation.

EPIDERMIS

KERATINIZATION AND DESQUAMATION

Epidermis is a keratinizing epithelium, which means that unlike other epithelia, it consists of **keratinocytes** (KARE-ah-tin-oh-SITES) that strengthen and waterproof the skin with a tough, horny outer covering (see Epithelia, Chapter 3). Some anatomists conceptualize the epidermis as a wall of bricks and mortar, as shown in Figure 12.3. For strength, keratinocyte "bricks" assemble a web of cytoplasmic ker-

SURFACE OF SKIN

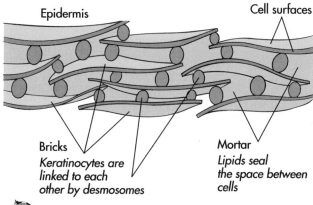

Epidermis

Cell surfaces

Bricks
Keratinocytes are linked to each other by desmosomes

Mortar
Lipids seal the space between cells

Figure 12.3
Brick and mortar model of the epidermal barrier.

atin filaments and embed these proteins in a proteo-glycan matrix within their cytoplasm. Then the cells seal themselves together with waterproofing lipids—the "mortar." Eventually the filaments and lipids break down. Dry squamous husks then become un-stuck and flake away (desquamation) from the surface of the skin into your clothing, wash water, and the air, coating the air conditioning and heating systems of the world with trillions of tiny fragments of people.

Keratinization and desquamation occur in stages that can be seen in four or five cell layers within the epidermis, illustrated in Figure 12.4. Keratinocytes originate in the lowest layer and slough away from the outermost layer, protecting the skin in the process. It takes about 27 days for a keratinocyte to pass through all the epidermal layers.

BASAL LAYER

Keratinocytes originate in the basal layer, or stratum basale (bay-SAL-eh) of the epidermis, shown in Figures 12.4 and 12.5. This layer of proliferating cuboidal stem cells, one or two cells thick, is attached to the basement membrane. When a typical stem cell divides, one daughter cell becomes a keratinocyte that stops dividing, while the other daughter remains a stem cell and divides in this manner again, thus as-suring a continuous supply of both keratinocytes and stem cells. Besides producing keratinocytes, these stem cells are also responsible for healing wounds by providing new cells that reestablish the epidermis.

SPINOUS LAYER

Young keratinocytes acquire desmosomes that give them a spiny appearance. Such cells compose the spinous layer, also called the stratum spinosum or prickle cell layer (Figures 12.4 and 12.5). **Desmosomes** (see Junctional Complexes, Chapter 2) con-nect the keratinocytes together, and interstitial fluid fills the space between them. Young keratinocytes begin to assemble a cytoplasmic network of inter-mediate filaments while they are in the **spinous layer.** Composed of **alpha keratin** proteins, this net-work appears to distribute physical stresses among the epidermal cells through desmosomes that connect cells.

Alpha keratin proteins belong to a family of more than 20 **intermediate filament (IF) proteins** (de-scribed in Chapter 2). Related to the **hard keratins** of hairs and nails, the softer alpha keratins are either acidic or basic. Keratinocytes synthesize acidic and basic molecules that bind together in helical pairs to form intermediate filaments.

GRANULAR LAYER

Keratinocytes mature in the granular layer. When keratinocytes reach this layer, also known as the stratum granulosum, **keratohyalin granules** and **membrane coating granules** appear in the cytoplasm (Figure 12.5). Keratohyalin (KARE-ah-toe-HY-ah-lin) granules have no membranes to isolate their con-tents from the cytoplasm. Instead, bunches of

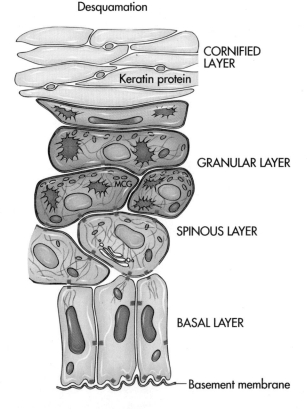

![Figure 12.4]

Figure 12.4

Four layers of epidermis. The diagram identifies the four layers of epidermis shown by the photomicro-graph.

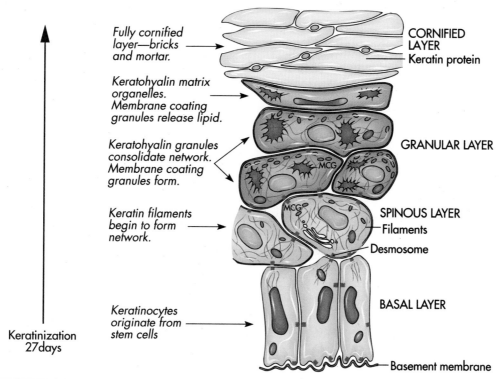

Figure 12.5

Cellular changes in keratinization. (From Montagna W: *The structure and function of skin,* ed 3, New York, 1974, Academic Press.)

keratin filaments spread throughout the cytoplasm to form the protective epidermal covering. Membrane coating granules contain lipid molecules that become the "mortar" that waterproofs the epithelium.

LUCID LAYER

The lucid layer, or stratum lucidum, is found in the thick skin of palms and soles. This thin, homogeneous, transparent layer of cells represents a brief step in the formation of the keratohyalin matrix. Precisely how it forms is unclear because the lucid layer is missing from most epidermis.

CORNIFIED LAYER

This layer contains the tough, squamous husks of dead keratinocytes that waterproof the epidermis and protect it against injury (Figures 12.4 and 12.5). Keratinocytes in this layer, also known as the stratum corneum, abruptly flatten and dehydrate. Cell nuclei and all other organelles disintegrate, and a **keratohyalin matrix** of tightly packed filaments derived from the keratin filaments and keratohyalin granules fills the cytoplasm, as illustrated in Figure 12.5. The plasma membrane also undergoes changes that preserve and toughen it. Overlapping like shingles, flat squamous keratinocytes are held together by the desmosomes that formed in the spinous layer and by intercellular lipids released from the membrane-

coating granules. Cells remain in this layer for about 14 days before they are lost when desquamation finally peels them from the outer surface of the cornified layer. The molecular basis of both adhesion and desquamation is poorly understood.

Normally, keratinization produces enough cells to balance those that desquamation removes, but various conditions can tip the balance one way or another. Abrasion and friction stimulate keratinization, resulting in **calluses** that thicken the cornified layer with extra cells (Figure 12.5). **Ichthyosis** (IK-thee-oh-sis; scaly; ichthy-, *G,* fish) is the name of a variety of conditions in which the squamous cells peel away as large, thick, scaly plates (Figure 12.5). Conversely, **psoriasis** (sor-EYE-ah-sis; psor- *G,* itch) tilts the balance toward desquamation (Figure 12.5). This condition inflames the skin, making it itchy and scaly as the result of cells that whiz through the epidermis in only 4 days. Psoriasis speeds up cell proliferation and leaves few keratinized cells to protect the skin. Dividing cells in the basal layer of the epidermis are thought to push the keratinocytes above them through the epidermis, but calluses and psoriasis illustrate that the process is not nearly that simple. In psoriasis, faster cell division alone would thicken the skin with more cells, not fewer. The key to understanding this condition lies in faster shedding of cells.

- What is a keratinocyte?
- Name the four cell layers of the epidermis.

- Which steps in differentiation of keratinocytes does each layer represent?
- How does EBS interfere with the cytoplasmic network of keratin filaments?

PIGMENT AND MELANOCYTES

Three sources of pigment influence the colors of skin and hair. Melanin, the pigment of melanocytes, is the primary pigment in all black, brown, and white skin colors. **Blood** circulating in the dermis adds red tones to the skin, especially in the lips where the epidermis is thin and capillaries are profuse. Yellow **carotenes** (CAR-oh-teenz) are especially abundant in Asian populations. Various influences modify the color of one's skin. Suntanning darkens all colors of skin, as comparison of exposed and unexposed areas readily shows. Bruises turn the skin purplish with blood that leaks into the dermis. Infection and inflammation also can redden the skin (**erythema,** air-ih-THEE-mah) as tiny capillaries proliferate. Patches of intensely red or purple skin (**Port wine stain** or **capillary nevus**) occasionally found in newborns are also due to profuse dermal capillaries. When blood drains from the skin, as in physiological shock, all skin colors take on an ashen gray appearance.

Melanocytes synthesize melanin pigment. These cells reside in the basal layer of the epidermis and extend long cytoplasmic processes among the neighboring cells, as Figure 12.6 illustrates. Inside the melanocyte's (MEL-ah-no-sites; melan-, *G,* black) cytoplasm, pigment-forming organelles called **melanosomes** (or melanin granules) synthesize melanin (MEL-ah-nin) pigment through reactions catalyzed by the enzyme tyrosinase. Melanocytes color the epidermis by secreting their melanosomes directly into neighboring cells, and pigment the hair in a similar process (see Hair Follicles, this chapter). Continual production assures that the epidermis remains pigmented. In contrast, tattooing places pigment permanently in superficial regions of the dermis. Melanin granules shield the cell nuclei from ultraviolet radiation by clustering around them until the cells leave the epidermis. Each melanocyte is responsible for pigmenting a particular cluster of keratinocytes known

Two epidermal melanin units. Each receives melanin pigment from the melanocyte at its base.

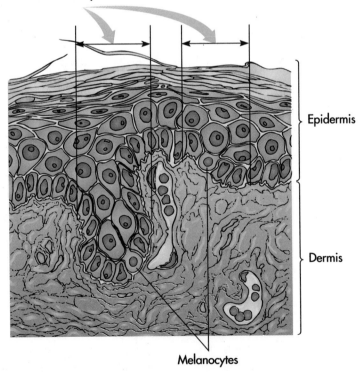

Epidermis

Dermis

Melanocytes

Figure 12.6

Melanocytes and epidermal melanin units (EMU) in the epidermis. Melanocytes in the basal layer supply melanosomes to developing keratinocytes; transverse section through epidermis.

as an **epidermal melanin unit** (EMU), whose cells extend upward from the basal layer in a column above the melanocyte. Figure 12.6B shows that Melanocytes and EMUs are packed closely and uniformly in the epidermis.

Melanocytes produce two types of melanin. **Eumelanin** (you-MEL-ah-nin; eu-, *G,* true, real) is brown or black, and **phaeomelanin** (FEE-oh-MEL-ah-nin; phae-, *G,* dusky) is yellow. Blonde and red-haired individuals produce phaeomelanin, and darker skin colors synthesize brown and black melanin. Adults of all races have between 1000 and 2000 melanocytes per square mm of skin, most densely packed in the face, other exposed areas, the nipples, labia, and scrotum. Darker skin pigment is related to melanin pro-

duction rather than the density of melanocytes, which is rather constant. Consequently, black skin contains more melanosomes than white skin, not more melanocytes.

Melanocytes originate in the neural crest and migrate into the epidermis, where they pigment either skin or hair, according to their destinations. Both genetics and environment influence pigmentation. The skin of **albinos** (al-BINE-ohs; alb-, *L,* white) is light pink, and their hair is white because such individuals have inherited recessive alleles that prevent pigment production. Without the usual browns and yellows that obscure dermal blood, the skin appears pink and is very sensitive to sunlight and ultraviolet radiation (Figure 12.7). Even though albino melansomes cannot

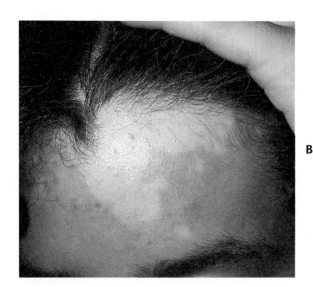

B

C

**Figure 12.7**
Pigment conditions of the skin. **A,** Albinism prevents melanin production. **B,** Vitiligo depigments patches of skin. **C,** Freckles have extra melanosomes.

make pigment, melanocytes populate the epidermis and export their transparent melanosomes into the keratinocytes. In other conditions, patches of skin may lack pigment. In **vitiligo** (VIT-ih-LYE-go; *L,* a type of skin irritation), patches of skin lose their pigment and become lighter. The skin apparently reacts to its own melanocytes and replaces them with Langerhans cells (see below). The opposite happens in **freckles;** clusters of melanocytes produce more melanosomes than usual. Besides producing excess pigment, melanocytes can become malignant and migrate in the dermis and epidermis, resulting in melanoma, a cancer (see Diseases and Disorders, this chapter).

OTHER EPIDERMAL CELLS

Langerhans cells resemble melanocytes. Langerhans (LANG-er-hans) cells, or nonpigmented granular dendrocytes, are dendritic (branched) cells of the basal and spinous layers, shown in Figure 12.8. They also are found in hair follicles, sebaceous and apocrine glands, and the oral mucosa. They lack melanosomes, desmosomes, and intermediate filaments, but they contain tennis racquet-shaped organelles called **Birbeck granules,** whose function is unknown. Current opinion views Langerhans cells as **antigen-presenting cells** that stimulate the immune system to attack foreign antigens (see Chapter 16, Immunity) in the epidermis, where blood vessels cannot reach. Langerhans cells are thought to present allergens to lymphocytes in allergic contact dermatitis.

Merkel cells resemble keratinocytes, but their associations with the nervous system suggest that they are sensory. Their actual role is unknown. Residents of the basal epidermal layer throughout the skin (Figure 12.8), Merkel cells contain desmosomes, intermediate filaments, and dense cytoplasmic granules. These cells are associated with axons (see Merkel Endings, this chapter), and they are especially abundant in the fingers, where they are thought to respond to touch.

- How do keratinocytes become pigmented?
- Where are melanocytes, Langerhans cells, and Merkel cells located in the epidermis?
- What is the function of Langerhans cells?

DERMIS

Dermis connects the epidermis to the underlying fascia, muscles, and bones. Dermis is composed almost entirely of extracellular matrix, a network of **collagen fibers** and **elastic fibers** embedded in a soft **ground-substance** gel made of proteoglycan molecules (Figure 12.9), that gives the skin its suppleness

Cornified layer

Granular layer

Spinous layer

Basal layer

Dermis

Langerhans cells are found in basal and spinous layers, but Merkel cells are basal.

Sensory nerve ending

✳ **Figure 12.8**
Langerhans and Merkel cells in the epidermis.

✳ **Figure 12.9**
Collagen fibers in the reticular layer of the dermis, seen with the scanning electron microscope. (Scanning electronmicrograph by permission of R.G. Kessel and R.H. Kardon, *Tissues and organs: A text-atlas of scanning electron microscopy,* 1979, W.H. Freeman and Co.)

and toughness. Most of the dermis' volume consists of water bound by the ground substance. All of these materials are synthesized and maintained by **fibroblasts** that are derived from the embryonic mesoderm. Macrophages, lymphocytes, and mast cells migrate through the matrix but, unlike the fibroblasts, are not permanent residents.

Slight changes in the characteristics of the dermis are responsible for many specialized regional features of the skin. Dermis is especially thick on the back, hands, and soles and especially thin at the eyelids, penis, scrotum, and labia minora. It tends to be thicker on the lateral surfaces of the extremities than on their medial surfaces. Dermis controls the differences between the hairy and glabrous (GLA-brus; hairless; glab-, *L,* smooth) regions of the skin (see Epidermal-Dermal Interactions, this chapter). Dermis also promotes the development of hairs, nails, and glands in the epidermis.

The papillary and reticular layers of the dermis reflect these special features. The principal structures of the papillary layer are millions of **dermal papillae,** (pah-PILL-ee; papill-, *L,* a nipple) that increase contact with the epidermis, as shown in Figure 12.2. Additional surface area provided by these dermal projections promotes the dermis' adhesion to the basement membrane and exchange of nutrients and wastes with the epidermis. Each papilla contains fine capillaries of the **papillary plexus** that nourish the epidermis and radiate heat through it. Papillary dermis also is involved in forming the epidermal ridges and dermatoglyphic patterns of the glabrous skin on hands and feet (see Dermal Ridges, this chapter).

Deeper in the dermis the reticular layer contains an extensive network of large collagen fibers (Figure 12.9) that strengthens the skin. These fibers are extraordinarily strong—one fiber only 1 mm in diameter will support 80 lbs—but they do not stretch. Stretchiness comes from **elastic fibers** interspersed among the collagen fibers. The effect of the **elastin** proteins in these fibers is evident in conditions such as **cutis laxa** (KUE-tiss LAX-ah), a hereditary disease of elastic fibers in which skin does not snap back into place

after it is stretched. Near the subcutaneous layer, small vessels of the **subcutaneous plexus** communicate with the papillary plexus above and the internal body below. At this point the skin becomes continuous with the underlying subcutaneous fascia and the deep fascia that envelops the bones and muscles; no internal epithelium separates the skin from the rest of the body.

- Which cells of the dermis produce the extracellular matrix?
- Name the various layers of the dermis.
- What structures characterize each layer?
- What is the function of the papillary layer of dermis?

EPIDERMAL APPENDAGES

Hair follicles, nail beds, and sweat glands are specialized extensions of the epidermis. Their cells have penetrated into the dermis, where they secrete solids and liquids by modifying the basic keratinization and desquamation of the epidermis. Slender **hair follicles** secrete solid hairs from deep within the dermis, and tubular **eccrine sweat glands** secrete perspiration from the same layer (Figure 12.2). The oily secretions from **sebaceous glands,** milk from **mammary glands,** and the milky liquids of **apocrine sweat glands** arise from epidermal cells that also have penetrated the dermis. Table 12.1 summarizes the properties of these structures, except the mammary glands, which are discussed in Chapter 26, Reproduction.

HAIRS, HAIR FOLLICLES, AND PILOSEBACEOUS APPARATUS

Hairs cover the entire skin, except the palms and soles, penis, clitoris, nipples, dorsal surfaces of distal phalanges, and body orifices. Fine **vellus hairs** cover most of the body, but the scalp, upper lips, beard, eyebrows, eyelashes, and the axillary and pubic regions produce coarser, **terminal hairs.** Some 2 million hair follicles populate the skin, with 100,000 located on the scalp alone. All hairs grow from hair follicles (Fig-

Table 12.1 EPIDERMAL APPENDAGES

	Epithelium	Hair	Nail	Sebaceous gland	Apocrine sweat gland	Eccrine sweat gland
Product	Soft keratin	Hard keratin	Hard keratin	Sebum	Milky liquid	Watery liquid
Process	Keratinization/ desquamation	Keratinization	Keratinization	Holocrine secretion	Apocrine secretion	Merocrine secretion
Source	Basal layer	Hair matrix	Matrix cells	Germinal cell layer	Columnar cells	Clear, dark cells
Regulation	—	—	—	Testosterones	Sympathetic nervous system	Sympathetic nervous system
Growth	Continuous	Periodic	Continuous	Continuous	Continuous	Periodic

ure 12.10), which are long, slender, epidermal tubes that reach into the dermis. A single hair projects up through the neck of each follicle. One or more sebaceous glands bulge from the sides of the follicle, and an **arrector pili muscle** connects the follicle with the papillary layer of the dermis. These smooth muscle fibers raise "goose bumps" on the flesh when the sympathetic nervous system causes them to contract. An apocrine sweat gland also accompanies each hair follicle in the genital and axillary regions. Taken together, the hair follicle, its glands, and its arrector pili muscle form a **pilosebaceous apparatus.**

FOLLICLE

Hair growth is intimately related to the anatomy of the hair follicle, just as keratinization is to that of the epidermis. Dermatologists divide the follicle into three regions marked by sebaceous glands and arrector pili in Figure 12.10. The **infundibulum** extends from the epidermis to the openings of sebaceous glands, and the **isthmus** continues to the insertion of the arrector pili muscle. Beyond this point is the **inferior portion** of the follicle, where hair growth and keratinization take place. The hair grows from the matrix cells of the enlarged **hair bulb** that keratinize in the **keratogenous zone** (KARE-ah-TODJ-en-us) of the follicle. A **dermal papilla,** enveloped by the hair bulb, stimulates this growth and supports it with nutrients from a sleeve of papillary dermis and capillaries that followed the follicle into the dermis. Consequently the oldest portion of any hair is its tip and the newest, as much as several years younger, is forming in the hair bulb. Melanocytes in the hair bulb pigment terminal hairs, but vellus hairs are not pigmented. Whether hairs are straight or curly depends on genes that influence the shape of hair follicles. Straightnecked follicles produce straight hairs, and twisted follicles produce curly or kinky hair.

Three major tissue layers make up the neck of the follicle, as you can see in Figure 12.11. Not sur-

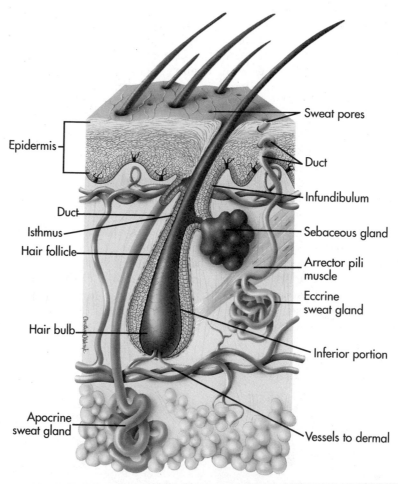

Epidermis

Duct
Isthmus
Hair follicle

Hair bulb

Apocrine
sweat gland

Sweat pores
Duct
Infundibulum
Sebaceous gland
Arrector pili
muscle
Eccrine
sweat gland
Inferior portion
Vessels to dermal

✶ **Figure 12.10**

Pilosebaceous apparatus with hair follicle, apocrine sweat gland, sebaceous gland, and arrector pili muscle. Infundibulum, isthmus, and inferior portion divide the follicle into three regions. The eccrine sweat gland is not part of the pilosebaceous apparatus.

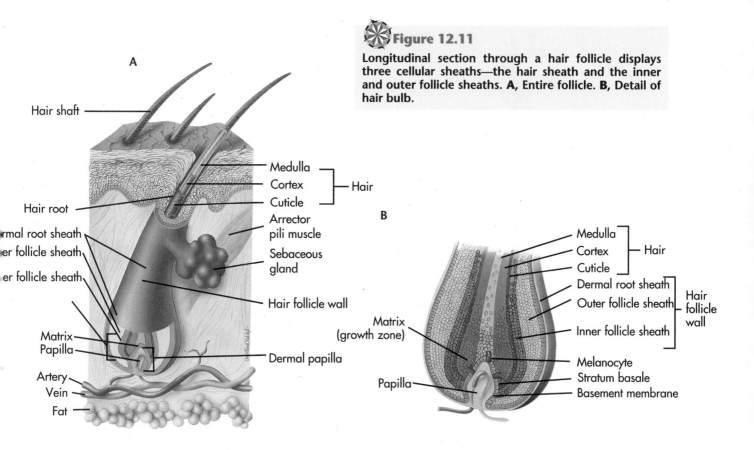

Figure 12.11

Longitudinal section through a hair follicle displays three cellular sheaths—the hair sheath and the inner and outer follicle sheaths. A, Entire follicle. **B,** Detail of hair bulb.

A

Hair shaft

Hair root

rmal root sheath

er follicle sheath

er follicle sheath

Matrix
Papilla

Artery
Vein

Fat

Medulla
Cortex   — Hair
Cuticle

Arrector
pili muscle

Sebaceous
gland

Hair follicle wall

Dermal papilla

B

Matrix
(growth zone)

Papilla

Medulla
Cortex   — Hair
Cuticle

Dermal root sheath
Outer follicle sheath   Hair
follicle
wall
Inner follicle sheath

Melanocyte
Stratum basale
Basement membrane

prisingly, these tissues resemble the layers of the epidermis, because they derive from the epidermis. The **hair shaft** itself is the innermost layer. Next is the **inner follicle sheath** that disintegrates before reaching the surface of the epidermis. Finally, the **outer follicle sheath** is continuous with the lower epidermal layers. All derive from proliferating matrix cells in the hair bulb.

The hair shaft itself contains three layers—the **medulla, cortex,** and **cuticle**—named from the inside out. The medulla, missing in vellus hairs, contains connective tissue and may not keratinize, but the cells of the cortex fill with hard keratin and melanosomes. Hard keratin does not desquamate, so the hair elongates. The external cuticle is one cell thick, with squamous cells overlapping like shingles pointing up toward the tip of the hair. This arrangement makes hairs feel smooth when stroked toward the tip and rough in the opposite direction, which is why bowstrings (made of horse hairs pointing in opposite directions) can draw music from violins and cellos. The inner follicle sheath (also called the inner root sheath) degenerates before it reaches the upper follicle because it contains soft keratin like that of the epidermis.

HAIR GROWTH CYCLE

Hair follicles pass through three cyclic phases of growth, shown in Figure 12.12. A period of sustained hair growth, called **anagen** (AN-ah-gen; ana-, *G,* up), begins the cycle and may last anywhere from 2 to 6 years. Individual scalp hairs may grow for 2 or 3 years before the follicle ceases production and enters a period of **catagen** (CAT-ah-gen; cata-, *G,* down, breakdown), during which the follicle regresses. The inferior portion of the follicle atrophies and shortens to the insertion of the arrector pili muscle, but the hair shaft remains. After 2 or 3 weeks, the follicle enters its resting, or **telogen** (TELL-o-gen; telo-, *G,* end), phase of 3 or 4 months, without further atrophy or growth. When anagen resumes, the inferior end of the follicle lengthens to its old depth, matrix cells resume growth, and a new hair dislodges the old one.

If your hairbrush has been gathering loose hairs lately, be encouraged; new hairs will soon replace them. They may not be long, terminal hairs, however, if you are suffering **male pattern alopecia** (AL-o-PEE-sha), or hair loss. Influenced by high levels of testosterones and inheritance of a dominant gene that predisposes for baldness, too many new hairs return as fine vellus. The scalp normally loses approxi-

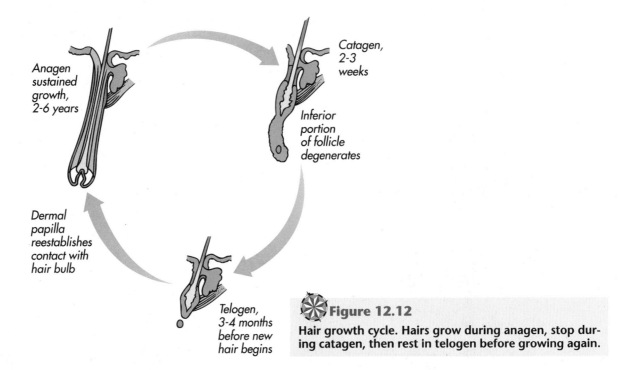

Anagen sustained growth, 2-6 years

Catagen, 2-3 weeks

Inferior portion of follicle degenerates

Dermal papilla reestablishes contact with hair bulb

Telogen, 3-4 months before new hair begins

Figure 12.12

Hair growth cycle. Hairs grow during anagen, stop during catagen, then rest in telogen before growing again.

mately 100 hairs a day through the hair growth cycle.

All the follicles you will ever have are present at birth, but they are capable of forming different types of hairs, depending on time and place. Newborns are covered with a fine pelt of delicate **lanugo** (la-NEW-go; lanu-, *L*, wool, down) hairs, soon to be replaced with vellus and terminal hairs from the same follicles. At puberty, dormant pilosebaceous apparatuses become active and add terminal hairs in the pubic and axillary regions.

- What are the components of the pilosebaceous apparatus?
- What cell layers make up the inferior portion of a hair follicle?
- What portion of the follicle assembles the hair shaft?
- Name the three phases of hair growth.
- What does the follicle do during catagen?

NAILS

Nails are essentially broad hairs that help us grip with our hands and feet. Instead of a slender hair a wide **nail plate,** about 0.5 mm thick (Figure 12.13), grows from a **nail groove** approximately 5 mm deep in the dermis of the distal phalanx of each finger and toe. Dense collagen fibers connect the nail to the underlying **nail bed. Matrix cells** in the nail groove produce the nail that pushes out over the nail bed. The **lunula** (LOON-you-la), a white crescent at the base of all nails, but not usually visible on the little finger, marks the location of the matrix. Lack of vessels or incomplete keratinization of the nail may account for the paleness of the lunula. The distal end of the nail protrudes over a shallow groove of skin called the **hyponychium** (HY-po-NEEK-ee-um). At the proximal end, the **eponychium** (EH-po-NEEK-ee-um), or cuti-

Figure 12.13

A nail grows outward from the nail groove.

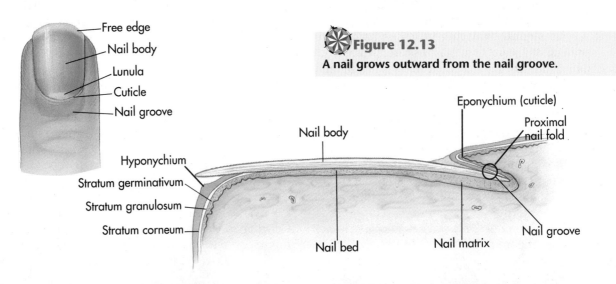

Free edge

Nail body

Lunula

Cuticle

Nail groove

Hyponychium

Stratum germinativum

Stratum granulosum

Stratum corneum

Nail body

Nail bed

Nail matrix

Eponychium (cuticle)

Proximal nail fold

Nail groove

cle, seals the **proximal nail fold** to the dorsal surface of the nail plate. **Lateral nail folds** join the nail plate to the skin on either side.

Like a hair, the nail itself consists of cornified cells and hard keratin fibers. Matrix cells deep in the nail groove form the dorsal surface of the nail, and cells at the distal margin of the lunula form the ventral surface. This arrangement recalls the pattern of cornification in the hair matrix, where medial cells form the medulla of the hair and more lateral cells, the follicle sheaths. The nail bed is analogous to the inner follicle sheath of the hair. Permanent desmosomes and tight junctions join the cells of the nail together so the nail grows continually without desquamating, about 0.1 mm per day. Nail growth proceeds faster in the fingers than the toes and faster on the dominant hand. Nails need trimming more frequently in summer, when their growth accelerates.

- What is the function of finger and toe nails?
- What part of a nail bed is analogous to the matrix of a hair follicle?

ECCRINE SWEAT GLANDS

Eccrine sweat glands cool the skin through perspiration. Independent of hair follicles, approximately 3 million of these simple, coiled, tubular glands cover the skin. At some 3000 per square mm, they are especially profuse in the palms, soles, axillae, and forehead, but they are not present on the external genitalia or the lips. A basement membrane and a bed of fine dermal capillaries support the tightly coiled **secretory portion** of the gland, shown in Figure 12.14. A single, narrow **duct** coils upward at first and then leads secretions directly to the epidermis, where cornified cells line the duct and open onto the surface at a **sweat pore.**

Two layers of cells, each one cell thick, form the secretory portion of the gland. The innermost layer, which secretes perspiration into the lumen of the gland, consists of **clear serous cells** (SEER-us; ser-, *L*, watery) and **dark mucous cells** that secrete different components of perspiration. Clear cells contain abundant mitochondria and glycogen for energy production, but their numerous secretory granules and rich endoplasmic reticulum suggest that dark cells also vigorously synthesize and secrete sweat products. **Microvilli** stud the luminal surfaces of both cells, and **canaliculi** (KAN-ah-LIK-you-lee, small canals) conduct products between the cells into the lumen. A layer of scattered **myoepithelial cells** (MY-o-epih-THEEL-ee-al) is thought to compress products from the gland. The lumen of the duct is narrower than the lumen of the secretory portion and consists of a stratified cuboidal epithelium two cells thick.

An individual working in severe heat can produce as much as 2 to 3 l of perspiration per hour. This remarkable capacity requires that the sweat glands ac-

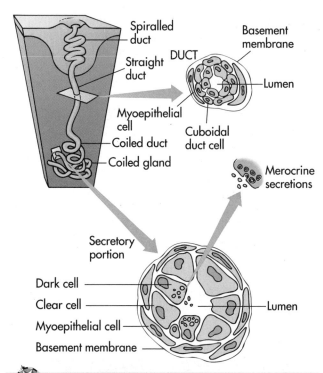

Figure 12.14

Eccrine sweat gland. Clear and dark cells secrete salts and water into the lumen of the secretory portion of the gland, and cuboidal duct cells reabsorb some of the salts as perspiration travels up the duct and empties through a sweat pore onto the epidermis.

tively transport products from the blood of the dermis to the lumen of the gland. This process requires energy in the form of ATP to excrete products; they do not filter across the membrane passively, as occurs in the kidney. **Perspiration** is a clear, hypotonic solution whose principal solutes are sodium, potassium, and chloride ions, plus lactate and urea. Smaller amounts of amino acids, proteins, and heavy metals accompany these constituents; certain drugs may also be present. **Sodium-potassium pumps** in the membranes of clear cells transport sodium ions into the canaliculi, and water and chloride ions follow the sodium ions isotonically into the lumen of the gland. The duct, however, resorbs a greater proportion of these ions than of water, so that perspiration flows onto the surface of the skin as a hypotonic solution that evaporates and cools the skin. Sodium chloride crystals can accumulate on the skin because the duct does not remove all of these ions. The perspiration of **cystic fibrosis** (SIS-tik FI-bro-sis) patients frequently is saltier than normal because their sweat glands do not reabsorb sodium ions. A defective gene encodes a defective sodium membrane transport protein that prevents reabsorption.

Temperature and emotions control sweating through the sympathetic nervous system. When body temperature increases, the hypothalamus stimulates sweat glands over most of the skin to release more perspiration. Sweat glands on the palms, soles, and axillae also are mediated by the hypothalamus, but they respond more directly to emotions than to temperature.

- What steps in sweat secretion take place in the secretory portion and duct of an eccrine sweat gland?

APOCRINE SWEAT GLANDS

Apocrine sweat glands resemble eccrine sweat glands but are associated with the pilosebaceous apparatus. Like sweat glands, apocrine glands (Figure 12.15) are simple, coiled, tubular glands, but the duct empties into the infundibulum of the follicle rather than onto the exterior surface of the epidermis. **Apocrine sweat glands** (AP-oh-krin) populate the axillae, genital, and anal regions of the skin, as well as the areolae, umbilicus, eyelids, and external ear canals. The

coiled, secretory portion of the gland lies deep in the reticular layer of the dermis, where a single layer of columnar cells secretes a milky-white mixture of proteins, carbohydrates, ammonia, lipids, and fatty acids that have uncertain functions in humans. An additional layer of myoepithelial cells periodically squeezes products from the gland when stimulated by sympathetic neurons.

Apocrine sweat glands begin to develop in all hair follicles, but most regress by the fifth month of fetal life, leaving survivors that begin secreting at puberty. The products of apocrine sweat glands are normally sterile and odorless until modified by microorganisms, the agents responsible for body odor. Microorganisms make it difficult to characterize these secretions because skin bacteria quickly metabolize the constituents before they can be analyzed. The term *apocrine* is controversial because it is uncertain that the columnar cells actually release apical cytoplasm,

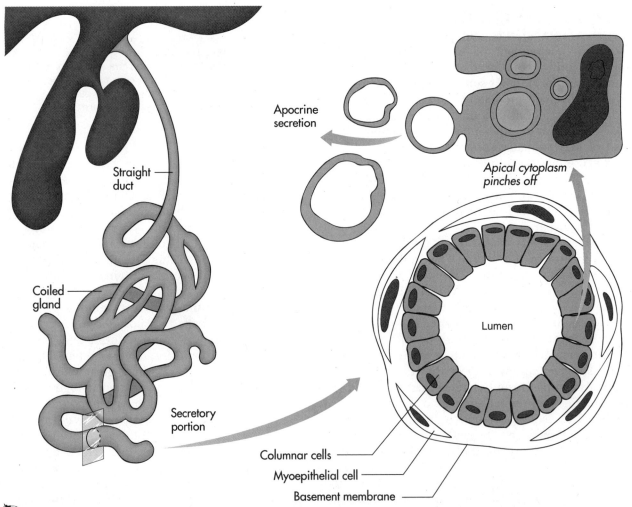

Apocrine
secretion

Apical cytoplasm
pinches off

Straight
duct

Coiled
gland

Lumen

Secretory
portion

Columnar cells

Myoepithelial cell

Basement membrane

✳ **Figure 12.15**

Apocrine sweat gland. Columnar cells secrete products into the lumen of the gland, and a straight duct releases these materials into the neck of a hair follicle. Apocrine secretion pinches off portions of cytoplasm.

as the strict definition of these glands requires (see Glands, Chapter 3).

SEBACEOUS GLANDS

Sebaceous glands secrete an oily film of sebum from the walls of hair follicles. Sebaceous glands are absent from the palms and feet, where no hair follicles exist, but the glands do occur without hairs in the areolae, lips, eyelids, labia minora, and prepuce. The densest and most active populations occupy the face and scalp, and the largest glands secrete from the upper back, forehead, and nose.

Secretion of **sebum** (SEE-bum; seb-, *L,* grease) resembles keratinization and desquamation more closely than eccrine or apocrine secretion. Like the epidermis, secretion is continuous, but it follows the **holocrine** (HOL-oh-krin; hol-, *G,* whole) pattern of release, as Figure 12.16 shows. Sebaceous cells grow

from the periphery of lobular sebaceous glands. Dividing cells push into the centers of the lobules and fill with lipid droplets. The cells finally break down, and the gland disgorges these oily contents through a sebaceous duct into the infundibulum of the hair follicle and thence to the external surface of the skin. The papillary layer of the dermis supports sebaceous glands. Sebaceous glands develop by the fifth month of fetal life, but they remain dormant until puberty.

Sebum is an enigmatic substance. Although its general composition is known, its functions are not. Various lipids, notably triglycerides, wax esters, squalene, cholesterol esters, and cholesterol, are its primary components, but it is uncertain whether they act as barriers to permeability, lubricants, bacteriostatic agents, or have other functions. Teenagers know the zits and blackheads of **acne vulgaris** (AK-nee vul-GAR-iss; vulga, *L,* common) very well, however. Stimulated by increasing levels of testosterones at puberty, sebum frequently accumulates and blocks hair follicle openings, especially of the very active glands on the face and upper back. These obstructions are called **comedoes** (KOM-eh-does) or blackheads. With the usual exits blocked, microorganisms infect the glands; the ensuing inflammations may permanently damage the skin, although the condition usually improves with time. Treatment with antibiotics and keratolyzing agents, such as Retin-A, improves such conditions by controlling infection and reducing blockage.

- What products do apocrine sweat glands and sebaceous glands secrete?
- Where do these glands connect to the pilosebaceous apparatus?
- In what regions of skin are these glands located?

DERMAL RIDGES

Dermal ridges are the source of fingerprints. When you look at the palmar surfaces of your hands and fingers, you can see the loops and whorls raised on the surface by epidermal ridges that follow dermal ridges beneath. These distinctive **dermatoglyphic patterns** (DER-mat-oh-GLIF-ik; glyph, *G,* carved) owe their individuality to complex inheritance and environmental conditions. In Figure 12.17, each dermal ridge is separated from its neighbor by a trough of epidermis that has grown down into the dermis, reminiscent of much deeper penetrations by hair follicles. A smaller downgrowth, the **interpapillary peg,** runs along the apex of each dermal ridge and divides the ridge into **secondary dermal ridges** on each side. Individual dermal papillae populate each secondary ridge, and ducts of eccrine sweat glands open on the apices of the epidermal ridges.

- How are dermal ridges different from dermal papillae?

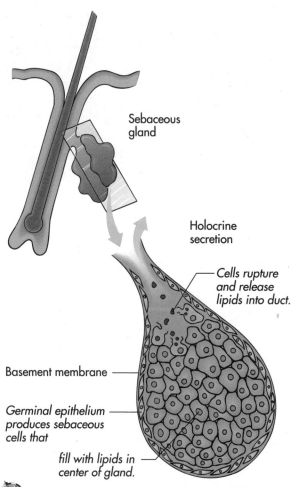

Sebaceous
gland

Holocrine
secretion

*Cells rupture
and release
lipids into duct.*

Basement membrane

*Germinal epithelium
produces sebaceous
cells that*

*fill with lipids in
center of gland.*

Figure 12.16

Sebaceous gland. Sebaceous cells develop from a germinal epithelium and accumulate in the center of the gland, where they break open and release their contents through a short duct into the hair follicle.

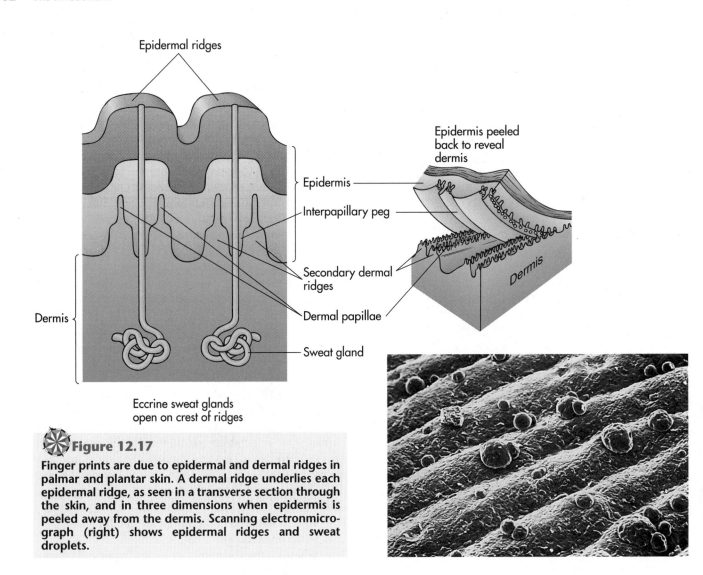

Figure 12.17

Finger prints are due to epidermal and dermal ridges in palmar and plantar skin. A dermal ridge underlies each epidermal ridge, as seen in a transverse section through the skin, and in three dimensions when epidermis is peeled away from the dermis. Scanning electronmicrograph (right) shows epidermal ridges and sweat droplets.

SENSORY ENDINGS

The sensations of touch, pain, and temperature begin in nerve endings that detect pressure and thermal changes in the skin. A wealth of **nerve endings** in the skin stimulate sensory neurons (described in Chapter 17, Neurons) that conduct impulses the brain interprets as touch, pain, heat, or cold, depending on the nature of the ending. Particular nerve endings transduce particular forms of energy (mechanical, chemical, or thermal) into neural impulses that in turn stimulate sensory neurons. For example, many receptors that respond to pressure changes are involved in the sense of touch because the impulses from these receptors are conducted to centers in the brain where perceptions of touch are formulated.

Nerve endings fall into two major anatomical classes—**free** and **encapsulated endings.** Encapsulated endings cover the ends of sensory neurons inside fibrous capsules. These endings contribute to the sense of touch because most of them respond to pressure. Free endings have no capsules; they probably respond to all types of stimuli. Nerve endings that register heat and cold have not been identified satisfactorily. It is especially difficult to trace the sensation of pain to a particular type of ending because all endings can register pain when overstimulated. Table 12.2 itemizes nerve endings and their actions, and Figure 12.18 locates them in the skin. Although this table may give you the impression that a particular receptor responds to a particular stimulus, the overall sensations that we perceive through our skins are interpreted and synthesized collectively in the brain from impulses received simultaneously from many millions of sensory nerve endings.

FREE NERVE ENDINGS

Free nerve endings are by far the most plentiful sensory endings in the skin. They are found throughout the dermis and the lower layers of the epidermis in hairy and glabrous skin. Free nerve endings wrap the isthmus of many hair follicles. (Tweak the hairs of

Table 12.2 SENSORY RECEPTORS OF SKIN

Receptor	Location	Sensation
Pacinian (lamellated corpuscles)	Dermis and subcutaneous layer; especially abundant in digits, breasts, genitalia, mesenteries, and periosteum	Vibration
Meissner (touch corpuscles)	Papillary layer of dermis; digits, genitalia, nipples, and eyelids	Light pressure?
Ruffini	Deep dermis and subcutaneous layer; plantar surfaces and joint capsules	Tension?
Mucocutaneous	Subpapillary layer of dermis; lips, eyelid, genitalia, perianal region, and oral mucosa	Unknown
Merkel (tactile disks)	Basal layer of epidermis; palmar and plantar surfaces	Pressure?
Pinkus	Basal layer of epidermis; hairy skin	Pressure?
Free	Epidermis and dermis; all regions of body	Mechanical, thermal, and pain?

From Sinclair D: Mechanisms of Cutaneous Sensation, *Oxford University Press, New York 1981.*

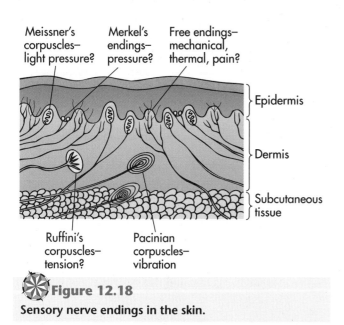

Meissner's corpuscles—light pressure?
Merkel's endings—pressure?
Free endings—mechanical, thermal, pain?
Epidermis
Dermis
Subcutaneous tissue
Ruffini's corpuscles—tension?
Pacinian corpuscles—vibration

Figure 12.18
Sensory nerve endings in the skin.

your arm to find the innervated ones.) Unlike encapsulated endings, the ends of these neurons have no cellular covering, as you can see in Figure 12.19A. Only a basement membrane intervenes between the neuron and its surroundings. As a group, free nerve endings may transmit all thermal, pain, and some mechanical sensations. The locations of free nerve endings sometimes suggest their functions. For example, endings close to the surface are more likely than deeper endings to detect an insect crawling on your skin.

ENCAPSULATED NERVE ENDINGS
PACINIAN CORPUSCLES
These oval nerve endings are the largest in the body. They may be up to 1 cm in diameter, though 1 to

4 mm is more common. As their anatomy and distribution suggest, **Pacinian corpuscles** (pa-SIN-ee-an KOR-pus-els; also called lamellated corpuscles; Figure 12.19B), respond to vibration and to a wide range of pressures. A fibrous capsule and concentric layers of flattened cells surround the single neuron ending in the center of the corpuscle. Larger corpuscles contain up to 70 cellular lamellae, but small ones have only a few. Size also is related to depth in the skin; larger corpuscles are located deep in the dermis and subcutaneous layers, and smaller ones are more superficial. Small Pacinian corpuscles are especially abundant in the fingers (more than an estimated 100 per digit), external genitalia, and breasts, but they also are plentiful in the mesenteries of the abdominal cavity and the periosteum of bones. The concentric lamellae may filter or focus mechanical pressure on the nerve ending, but they are not essential since the nerve ending will respond when the lamellae are missing.

MEISSNER'S CORPUSCLES
Meissner's corpuscles evidently respond to light mechanical pressure. The fingers and toes contain many **Meissner's corpuscles** or touch corpuscles (MICE-nerz; Figure 12.19C), atop dermal ridges. These endings also are found in other sensitive areas of the body, including the eyelids, external genitalia, and nipples. Several layers of flattened cells that resemble the lamellae of Pacinian corpuscles extend along the 80 μm length of each corpuscle. These layers taper toward the apex of the papillae, apparently accommodating the shape of the dermal papillae. A connective tissue capsule encloses the exterior of each corpuscle, and elastic fibers connect the entire corpuscle with the epidermis, further suggesting delicate mechanical reception.

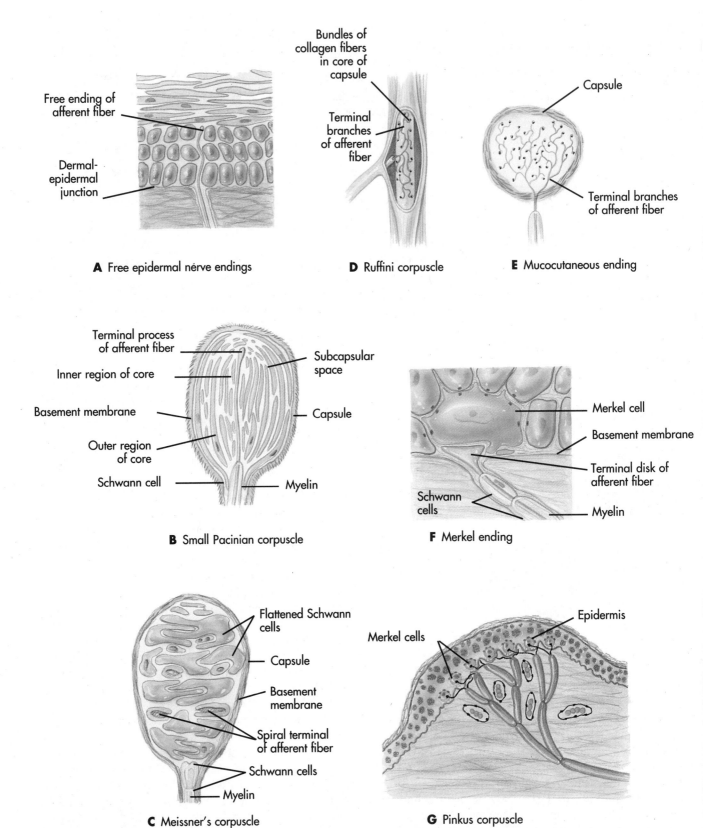

Free ending of afferent fiber

Dermal-epidermal junction

A Free epidermal nerve endings

Bundles of collagen fibers in core of capsule

Terminal branches of afferent fiber

D Ruffini corpuscle

Capsule

Terminal branches of afferent fiber

E Mucocutaneous ending

Terminal process of afferent fiber

Inner region of core

Basement membrane

Outer region of core

Schwann cell

Subcapsular space

Capsule

Myelin

B Small Pacinian corpuscle

Merkel cell

Basement membrane

Terminal disk of afferent fiber

Schwann cells

Myelin

F Merkel ending

Flattened Schwann cells

Capsule

Basement membrane

Spiral terminal of afferent fiber

Schwann cells

Myelin

C Meissner's corpuscle

Merkel cells

Epidermis

G Pinkus corpuscle

✳ **Figure 12.19**

Structure of sensory nerve endings. A, Free nerve ending. **B,** Pacinian corpuscle. **C,** Meissner's corpuscle. **D,** Ruffini corpuscle. **E,** Mucocutaneous ending. **F,** Merkel ending. **G,** Pinkus corpuscle.

RUFFINI CORPUSCLES

Ruffini corpuscles are narrow, spindle-shaped structures about 1 mm long that fire volleys of impulses when slow mechanical forces deform them. These endings are abundant deep in the dermis and subcutaneous layer of the plantar surfaces of the feet. They occur frequently in joint capsules but are relatively scarce elsewhere in the skin. An external connective tissue capsule receives profuse terminal branches from a single neuron, illustrated in Figure 12.19D. These endings attach to collagen fibers running the length of the capsule and appear to respond to tension in the fibers. A few supporting cells interweave among the collagen fibers.

MUCOCUTANEOUS CORPUSCLES

Formerly known as Krause's endbulbs, mucocutaneous corpuscles resemble Meissner's corpuscles, but usually are smaller (100 μm maximum) and lack lamellae. They are not covered by a definite external capsule; instead, a ball of connective tissue (shown in Figure 12.19E) condenses around the ends of several sensory neurons. Most **mucocutaneous endings** are located beneath the papillary layer of the dermis at body orifices. Typical locations are the lips, eyelids, external genitalia, and perianal region, the conjunctiva of the eyes, and in the mucous membranes of the mouth and pharynx. Mucocutaneous corpuscles are not found in hairy skin or in palms and soles. Their function is unknown.

MERKEL ENDINGS

These endings differ from all other encapsulated endings because a neuron ends on a specific type of cell, the Merkel cell, as seen in Figure 12.19F. Despite this distinctive anatomy, the function of **Merkel endings** (or tactile disks) is ambiguous; most investigators consider them mechanical receptors. Merkel endings are located in the basal layer of epidermis in the hands and feet.

PINKUS CORPUSCLES

Pinkus corpuscles are disks that contain approximately 50 Merkel endings. These composite structures (Figure 12.19G) are also known as **Merkel's touch corpuscles,** or tactile disks, from their location in the basal layer of the epidermis, but they respond more strongly to mechanical deformation than to gentle touch. They are present in hairy skin, but not on the face, the external genitalia, or the palms and soles. The upper abdomen displays the richest supply, with one or two per square cm.

- Describe the functions and locations of each of the sensory endings discussed above.

DEVELOPMENT

EPIDERMIS AND DERMIS

The skin begins developing from two layers of tissue in the second week of embryogenesis. The

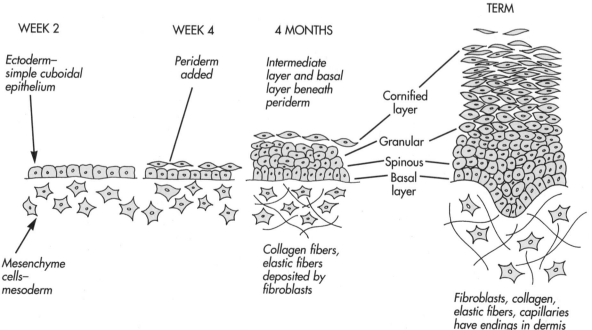

Figure 12.20

Development of skin. Skin develops from ectoderm and mesoderm, which become epidermis and dermis, respectively.

epidermis and its appendages derive from the **ecto-derm** that covers the blastoderm after gastrulation. The dermis arises from **mesenchyme** of the **lateral plate mesoderm** and **somite** mesoderm (Figure 12.20). Langerhans cells, fibroblasts, vessels, and adipose cells derive from the mesoderm, but melanocytes, neurons, and sensory endings derive from neural ectoderm and neural crest (see Chapter 4, Development).

The four layers of mature epidermis begin as a simple cuboidal epithelium that quickly forms a thin, squamous layer called the **periderm** (PAIR-ih-derm; peri-, *G,* around) by the fourth week. The periderm protects the fetus until the epidermis cornifies at the end of the second trimester. Keratinization begins in the **basal layer** of cells beneath the periderm, and three additional layers appear successively as young keratinocytes progress through their development. The spinous, granular, and cornified layers of cells appear after the fifth month. Melanocytes enter the epidermis during the eighth week, and by the sixth month they have begun pigmenting the epidermis. Epidermal appendages begin appearing after the sec-

ond month (see below), and Langerhans cells appear in the fourteenth week.

Dermis begins as diffuse mesenchyme cells and extracellular matrix, which develop into fibroblasts and begin to assemble a mat of collagen fibers. Papillary and reticular layers organize in this meshwork, and neuron processes begin to penetrate the dermis in the fifth week. Capillary beds form during the third month, and by the fourth month, definitive dermis has appeared.

Sensory endings of skin are innervated by spinal nerves that enter zebra-like stripes of skin called dermatomes (Figure 12.21). As Chapter 18 describes, a single spinal nerve communicates with each dermatome, segmenting the skin into invisible strips of dermis. **Dermatomes** (DERM-ah-tomes; -tome, *G,* cut; a skin slice) originate from somites (see Somites, Chapter 4). Cells migrate from the somites laterally beneath the epidermis into the flanks and onto the limbs of the embryo, where they establish the dermatomes.

- From what embryonic layers do epidermis and dermis develop?
- When does the epidermis begin to keratinize?

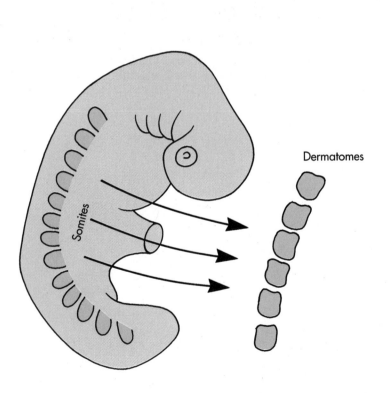

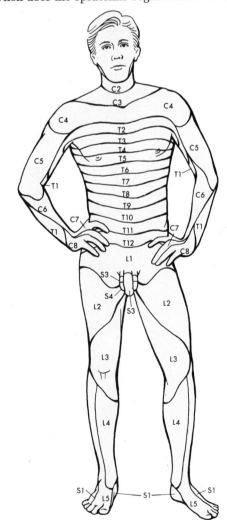

Dermatomes

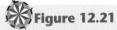

Figure 12.21

Dermatomes. One spinal nerve receives stimuli from sensory nerve endings in each dermatome shown in the skin *(right)*. Each dermatome develops from one somite.

EPIDERMAL-DERMAL INTERACTIONS

Hair follicles, glands, and nail grooves all begin as placodes (PLA-kode; plac, *G,* plate), thickenings in the epidermis and dermis, as shown in Figure 12.22. Beginning in the third month, **hair placodes** appear where epidermal and dermal cells aggregate. The germinal layer of the follicle grows deeper into the dermis, spinning off cells that form the neck of the follicle. The growing tip of the young hair expands as a hair bulb and engulfs a tuft of dermal capillaries, and the young hair shaft begins to develop. Sebaceous glands and apocrine sweat glands bud off from the neck of the follicle, and an arrector pili muscle attaches to its insertion. Melanocytes migrate into the hair bulb and pigment the hair. By the twentieth week, a coat of fine hairs, the lanugo, covers the body. Nails, eccrine sweat glands, and dermal ridges form in the same manner.

Why are some regions of the skin hairy while others are not, and why do different kinds of hairs and sensory endings populate different regions of the skin? The answers to these questions have to do with interactions between the epidermis and dermis, illustrated in Figure 12.23. Hair follicles develop where the dermis induces them to do so. For example, when embryonic dermis from hairy regions of a mouse embryo, such as the back, is experimentally coupled with epidermis from the pads of the feet, where hairs do not normally develop, the epidermis nevertheless produces hair follicles. Obviously this epidermis has the ability to produce hairs even though it does not normally do so. Dermis from glabrous skin, however, lacks the ability to induce the epidermis to form hair follicles. When dermis from the footpads is coupled with epidermis from the back, the epidermis does not produce hair follicles, even though it does in its normal setting.

Such epidermal and dermal influences regulate hair and gland development. The epidermis is responsible for the different varieties of hair and controls the difference between vellus hairs and the terminal hairs of scalp and beard. Injury to the skin interferes with interactions between the epidermis and dermis.

- What structure would you expect to develop if fingernail dermis were transplanted beneath epidermis of the forearm?
- When do hair follicles and eccrine sweat glands first appear?

BURNS

Each year approximately 10,000 people in the United States die of burns. Scalds and flammable liquids are the most frequent killers; young children and men 18 to 25 years old are the most frequent victims. Burns are traumatic injuries to the skin (and sometimes to the respiratory system) caused by contact with heat, electricity, caustic chemicals, or even sunlight. Whether a burn is life-threatening depends on how deeply the skin has been attacked and how large the injured area is. The victim's survival also depends on skillful treatment. Since 1964, hospital burn cen-

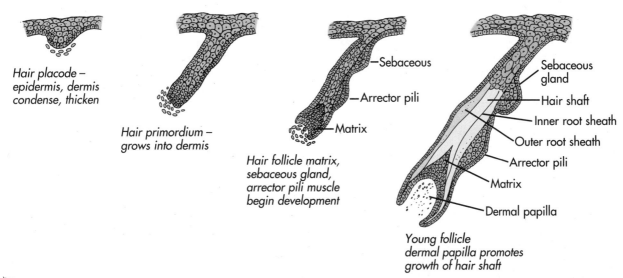

Hair placode –
epidermis, dermis
condense, thicken

Hair primordium –
grows into dermis

Hair follicle matrix,
sebaceous gland,
arrector pili muscle
begin development

—Sebaceous

—Arrector pili

—Matrix

Sebaceous gland

Hair shaft

Inner root sheath

Outer root sheath

Arrector pili

Matrix

Dermal papilla

Young follicle
dermal papilla promotes
growth of hair shaft

✳ **Figure 12.22**
Hair follicles and other epidermal appendages begin as placodes that extend into the dermis.

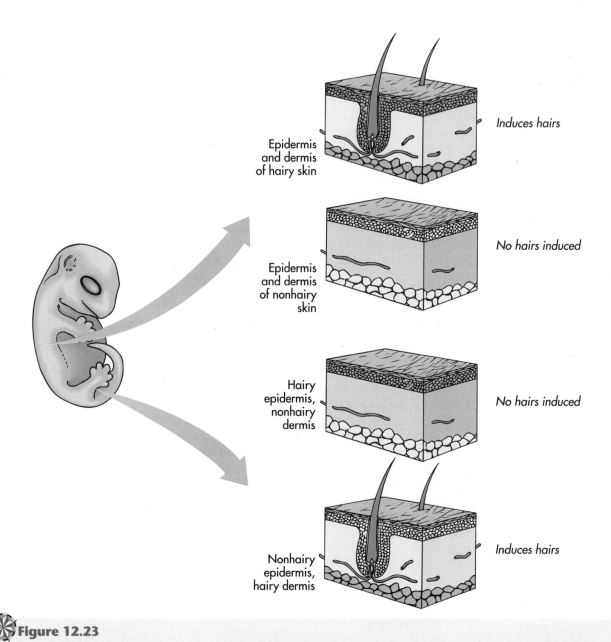

Epidermis and dermis of hairy skin

Induces hairs

Epidermis and dermis of nonhairy skin

No hairs induced

Hairy epidermis, nonhairy dermis

No hairs induced

Nonhairy epidermis, hairy dermis

Induces hairs

Figure 12.23

Dermis induces the epidermis to form hair follicles. Dermis of hairy skin *(blue)* induces, but dermis from hairless skin *(yellow)* does not. It makes no difference whether epidermis is from hairy or hairless skin; both form hair follicles when exposed to dermis from hairy skin.

ters have reduced the mortality from 30% to 10% among patients suffering second or third degree burns on 50% of the body. Treatment attempts to restore the protective barrier that the body has lost. Replacing body fluids and electrolytes, protecting against infection and heat loss, and of course replacing the damaged skin are its primary goals.

Burns are rated as first, second, or third degree according to the depth of skin they damage. First degree burns (Figure 12.24) affect the epidermis and do not compromise its protective barrier. These injuries, frequently caused by scalds or sunburn, are painful for several days. The epidermis peels but is quickly replaced from the basal layer without scarring. **Second degree** burns remove the epidermis and part of the dermis beneath. When these burns damage just the papillary layer they are considered **superficial,** but **deep dermal** burns penetrate the reticular layer. Damage to nerve endings makes second degree burns the most painful variety, and epidermal regeneration becomes more difficult because only hair follicles and glands remain to reestablish the basal layer of epidermis. Skin grafts usually are needed to close these wounds and to start recovery. Scarring will occur, especially in deep dermal burns. **Third degree,** or full-thickness burns, destroy the entire epidermis

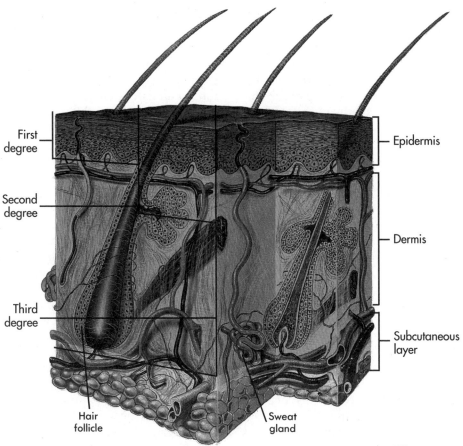

**Figure 12.24**

Burns and skin damage. First degree burns affect the epidermis, second degree burns penetrate the dermis, and third degree burns destroy both epidermis and dermis.

and dermis, leaving an open wound with no adequate source of regenerating epithelial cells. These burns require skin grafts to close them, but they are relatively painless because nerve endings have been destroyed. Both second and third degree burns require careful control of infection, fluid loss, and body temperature. Artificial skin, containing the patient's own cultured epithelial cells spread on an adherent membrane, provides a promising source of new epithelium to close such burns and to restore the protective barrier.

It is important to know how much of the skin has been injured because risk increases with a burn's body coverage. Clinicians quickly estimate coverage by the **"rule of nines"** that divides the body into 11 convenient sections, each covering approximately 9% of the surface, as Figure 12.25 shows. The entire head is 9%, and the anterior and posterior surfaces of the trunk are each 18%. The anterior and posterior surfaces of the lower extremities also rate 9% each, and a burn involving the entire upper limb is an additional 9%. Body proportions differ in children, and other charts take these differences into account.

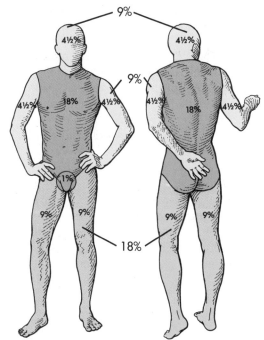

Figure 12.25

Rule of nines. Anterior and posterior surfaces of the head rate 4.5% each, with 9% for the entire head. Upper limbs are rated in the same way, but an entire lower limb rates 18% because it is larger, with 9% for each anterior and posterior surface. A burn of the back and posterior lower limbs would cover 36% of the body.

Burn scars disfigure the skin and immobilize joints. Hairs and glands do not regenerate in this tissue because the epidermal interactions that promote their formation are missing. Minimizing scarring is a major cosmetic and psychological concern. **Scar tissue** derives from fibroblasts and myofibroblasts that populate the field of injury. **Myofibroblasts** spread over the wound and draw its edges together because these cells are contractile. The myofibroblasts also produce extracellular matrix that shrinks with the wound, and the scar tissue becomes more rigid than its surroundings. Pressure garments minimize scarring by reducing the flow of blood that brings new myofibroblasts into the wound, but restoring original texture to deeply burned skin is not yet possible.

- What tissue layers do first, second, and third degree burns damage?
- How much body surface would a burn of the face, entire left arm, and anterior trunk involve?

DISEASES AND DISORDERS

CARCINOMAS
Squamous and basal cell carcinomas are the most common cancers of nonpigmented epithelial cells (see Figure 12.26). **Squamous cell carcinomas** derive from keratinocytes, but the source of basal cell carcinoma is uncertain. Squamous cell carcinoma is a serious tumor that can spread aggressively to local lymph nodes, if it is not surgically removed. In contrast, **basal cell carcinoma** rarely metastasizes. Dividing cells disrupt the epidermal layers in both conditions, and cells cross the basement membrane into the dermis. Both carcinomas frequently occur on skin that is exposed to sunlight, and basal cell lesions appear mostly on the face.

MELANOMA
Epidermal melanocytes are the source of **cutaneous malignant melanomas** (MEL-ah-NO-ma), in which epidermal melanocytes resume dividing and penetrate the dermis. The prognosis for curing this cancer depends on how deeply and far these cells migrate from their source. Malignant melanoma also is related to exposure to sunlight, which is thought to be the reason for the rapid increase of these tumors over the last 40 years. Persons with light skin, blue eyes, and blond or red hair frequently are affected, but this condition seldom affects African Americans.

KERATOSES
Actinic keratoses are scaly, darker lesions on the skin where the normal pattern of keratinization has been disrupted by **parakeratosis** (PARA-kare-ah-TOE-sis; para-, G, beside). Nucleated keratinocytes move into the cornified layer of the epidermis where they complete the final steps in keratinization. Once

again, sunlight promotes the development of actinic keratoses (ak-TIN-ik KARE-ah-TOE-sees; sun induced keratinization; actin-, G, a ray; -osis, L, condition), and fair-skinned individuals are at greater risk. Over several years the lesions may transform into squamous cell carcinomas. Cryotherapy to remove the lesions or application of fluorouracil creams (inhibitors of DNA replication) are the usual treatments for actinic keratoses. Regular use of powerful sunblocking agents may help prevent actinic keratoses.

KELOIDS
Keloids are benign growths of collagen scars on the skin. Frequently induced by abrasions, insect bites, burns, or surgical incisions and more common in African Americans than whites, keloids (KEE-loyds) have uncertain origins and imperfect treatment. Most often found on the chest, back, and shoulders and more common on the abdomen in women than in men, keloids range from small, circular irritations to large, cosmetically unpleasant fields of doughy, shiny, raised skin that lacks hairs and sweat glands. Hormones may have a role in their development because keloids rarely appear before puberty. Fibroblasts of the dermis synthesize excessive quantities of collagen that accumulate beneath the epidermis as large collagen fibers. Prevention is the most effective treatment. Therapy first attempts to reduce collagen production by injection of cortisone into the keloidal tissue. If necessary, keloids then are removed surgically.

AGING
Aging changes the appearance and structure of the skin. Although your own skin changes little from day to day, it nevertheless will alter over time. The wrinkles and changes in your parents' skin will eventually be yours also, brought about by intrinsic aging and skin disorders, such as those described above. Most changes are associated with slower metabolism. Epidermal growth declines, and cuts and bruises take longer to heal. Small vessels in the dermis regress, and the skin becomes drier and may itch, as sweat glands and sebaceous glands secrete less lubrication. Dermal papillae tend to flatten, thus the epidermis and dermis have less surface in which to adhere to each other. Skin sags because the elastic network in the dermis is less firm. Certain sensory endings, such as Pacinian corpuscles, diminish, but free nerve endings remain abundant. Skin and hair alike lose melanocytes, as graying hairs and increased sensitivity to sunlight attest. Hairs and nails grow more slowly. Scalp hairs spend more time resting in the telogen phase. Not all hairs are so affected though; terminal hairs begin to appear in men's ears and on women's lips and chins.

- Which skin cancers derive from basal cells, and which from melanocytes?

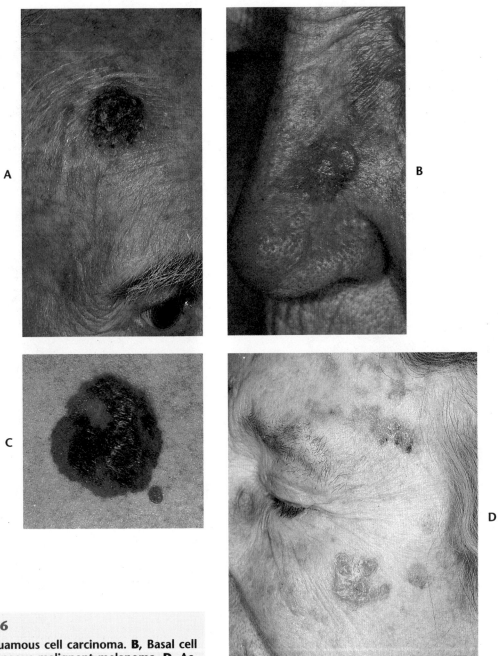

Figure 12.26

Skin lesions. **A,** Squamous cell carcinoma. **B,** Basal cell carcinoma. **C,** Cutaneous malignant melanoma. **D,** Actinic keratosis.

TRANSDERMAL DRUG DELIVERY

Perhaps you've traveled with someone who wore a scopolamine patch behind an ear to prevent motion sickness. The patch painlessly and conveniently delivers a continual, low dose of this tranquilizing drug through the skin. The traveler removes the patch when it is no longer needed or applies a new one when the drug is exhausted. What could be simpler?

Transdermal drug delivery combines several features of other methods of drug delivery. Injecting a drug beneath the skin or into a muscle establishes a reservoir that diffuses slowly into the vascular system. The patch places the reservoir outside the body and delivers the drug gradually through the skin. Figure 12.27 illustrates the process.

Transdermal drug delivery passes drugs through the keratinized barrier of squamous cells that waterproof the skin. That certain substances pass, albeit slowly, is as much a tribute to the effectiveness

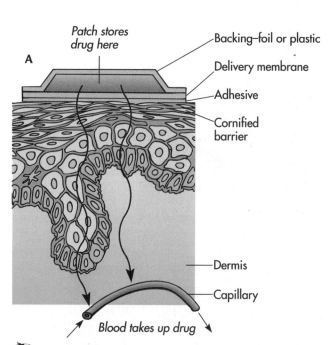

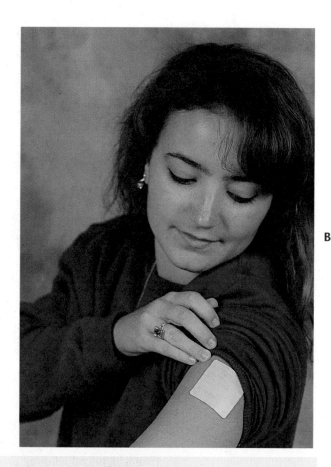

✳ Figure 12.27

Transdermal drug delivery. A, Drug stored in the patch diffuses through the delivery membrane and cornified layer of the epidermis into dermal capillaries. **B,** Woman wearing a transdermal patch.

of the barrier as it is to the nature of the chinks in the skin's armor. The barrier limits the speed of entry and the drugs that can be administered. Transdermal drugs must be hydrophobic enough to diffuse through the intercellular lipids between the squamous cells, but they also must be soluble enough in water to traverse the interstitial fluids of the dermis. In practice this requires small doses of drugs that are effective at very low concentrations. You would need to sit all day in a jacuzzi full of aspirin, for example, to receive the dose of one tablet taken orally. The most frequently administered transdermal drugs in the United States are scopolamine, nitroglycerin (for angina), clonidine (antihypertensive), estradiol (to suppress ovulation), and nicotine (to ease withdrawal from tobacco use).

- Why can transdermal delivery supply only low doses of drugs?
- What characteristics must a drug possess to be delivered transdermally?

SUMMARY

Figure 12.28 summarizes many of the processes that affect the anatomy of the skin.

The skin is a protective barrier composed of epidermis and dermis. The epidermis protects against physical injury, infection, sunlight, and fluid loss. Dermal capillaries and epidermal sweat glands regulate the flow of heat from the skin. Numerous sensory nerve endings detect stimuli that are interpreted as pain, touch, and temperature. Various epidermal glands and appendages synthesize and secrete solid hairs, nails, and liquid oils onto the surface of the skin. Epidermis and dermis contribute to the texture of skin, which helps to indicate an individual's age, sex, and physical condition. *(Page 348.)*

Epidermal keratinization and desquamation form the protective barrier of the skin. Four cell layers in the epidermis represent stages in the development of keratinocytes that form a tough, cornified, waterproof layer of cells. Keratinocytes develop in the basal layer and begin to lay down keratin intermediate filaments in the spinous layer. Keratohyalin granules appear in the granular layer. In the cornified layer, keratinocytes release intercellular lipids that waterproof the cells as the keratinocytes dehydrate and die. Cornified cells desquamate from the surface of the epidermis. *(Page 349.)*

Skin diseases affect the balance between keratinization and desquamation. Psoriasis diminishes the cornified layer by accelerating both these processes, but ichthyosis and calluses delay desquamation. Epidermolysis bullosa simplex weakens the epidermis by impairing the network of keratin filaments. *(Page 351.)*

Epidermal cells protect against sunlight and infection. Melanocytes reduce damage from ultraviolet light through a shield of pigmented melanosomes. Langerhans cells present antigens to T-lymphocytes, but the function of Merkel cells is unknown. *(Page 352.)*

The extracellular matrix of the dermis makes the skin resilient. Fibroblasts in the dermis synthesize and deposit the extracellular matrix of collagen and elastic fibers embedded in proteoglycan ground-substance gel. Macrophages and mast cells patrol the dermis for foreign antigens, and capillaries in the reticular and papillary layers deliver blood. Sensory nerve endings lace the dermis with receptors for heat, pressure, and pain. *(Page 354.)*

Epidermal appendages produce hairs, nails, sebum, and sweat. Hair follicles, nail grooves, and glands are downgrowths of the epidermis into the dermis that produce hard keratins, oils, and perspiration. Dermal ridges resemble epidermal appendages, but they remain in the papillary layer of the dermis. *(Page 355.)*

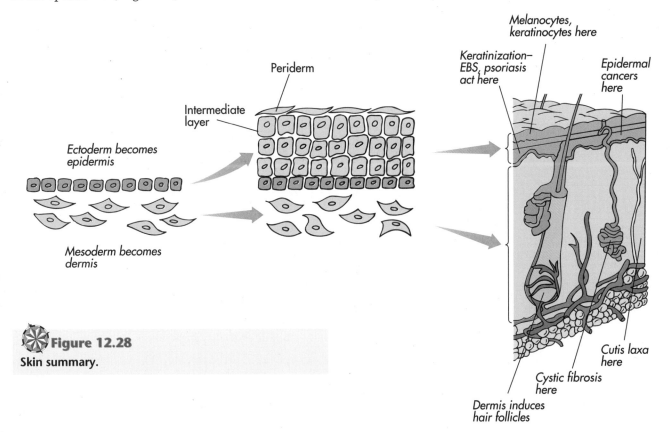

Figure 12.28
Skin summary.

Encapsulated and free nerve endings in the skin communicate temperature, pain, and touch. Specific thermal and pain receptors are difficult to identify, but most of the encapsulated endings probably are concerned with touch because they respond to mechanical forces. Examples of these endings are Pacinian, Meissner, Ruffini, mucocutaneous, and Pinkus corpuscles and Merkel endings. *(Page 362.)*

Skin develops from ectoderm and mesoderm. The epidermis begins as a simple cuboidal epithelium of ectoderm that differentiates into four cell layers and begins to keratinize the skin during the fifth month of fetal development. Dermis derives from dermatome and lateral plate mesoderm. The dermis controls the development of epidermal appendages. *(Page 365.)*

Burns damage the skin and force it to regenerate. First degree burns damage only the epidermis, but second degree burns reach the dermis, and third degree burns remove both layers. The deeper the burn, the more difficult it is for the epithelium to regenerate because fewer hair follicles and glands remain to reestablish the basal layer. Dermis regenerates scar tissue because the original processes that promoted skin development in the fetus are destroyed. *(Page 367.)*

Carcinomas and melanomas occur when keratinocytes and melanosomes transform into malignant cells. Actinic keratoses are precancerous epidermal tissue. Metabolism slows in aging skin. *(Page 370.)*

Transdermal drugs cross the epidermal barrier. Adhesive transdermal skin patches can deliver low doses of drugs across this barrier. *(Page 371.)*

STUDY QUESTIONS OBJECTIVES

1. Describe the four major functions of the skin, and name the layer or layers of tissue responsible for each.   **1**
2. How are the epidermis and dermis similar? How are they different?   **2**
3. Describe the four cellular layers of the epidermis. Follow the differentiation of keratinocytes through these layers. Which steps are considered keratinization, and which desquamation?   **3**
4. What properties of the cornified layer of epidermis protect the body against injury, infection, and fluid loss?   **3**
5. Which phase of keratinization and desquamation do epidermolysis bullosa simplex, psoriasis, calluses, and ichthyosis affect? How do these conditions modify the balance between the processes of desquamation and keratinization?   **4**
6. What are the roles of fibroblasts, elastic and collagen fibers, and the papillary and reticular layers of the dermis?   **5**
7. Describe the structure of a pilosebaceous apparatus.   **6**
8. Compare and contrast the anatomy of eccrine sweat glands with that of apocrine sweat glands and sebaceous glands.   **6**
9. What are the similarities and differences between hair follicles and nail beds?   **6**
10. Explain why the epidermis can be thought of as a flat, broad hair.   **6**
11. How do dermal ridges produce finger prints, or dermatoglyphics? What are some of the differences between hairy skin and glabrous skin from the palm of the hand?   **7**
12. What external stimuli do sensory endings in the skin detect? Which endings detect which stimuli?   **8**
13. Describe how the anatomy and location of sensory nerve endings in the skin adapt them to detect different forms of mechanical pressure.   **8**
14. What are the differences between free and encapsulated nerve endings?   **8**
15. Describe the changes that occur in developing epidermis and dermis.   **9**
16. Describe the development of a hair follicle and hair. Which parts of the hair are oldest? Which are youngest?   **9**
17. What levels of epidermis and dermis do first, second, and third degree burns of the skin damage? What functions of the skin does each type of burn impair?   **10**
18. How do epidermis and dermis contribute to healing burns?   **10**
19. What epidermal processes and structures do carcinomas, melanomas, keloids, and keratoses affect?   **11**
20. Cite changes that indicate that metabolism slows in aging skin.   **11**
21. How do transdermal delivery systems deliver drugs through the skin? What major barrier must be passed? Why can these systems deliver only low doses of a drug?   **12**

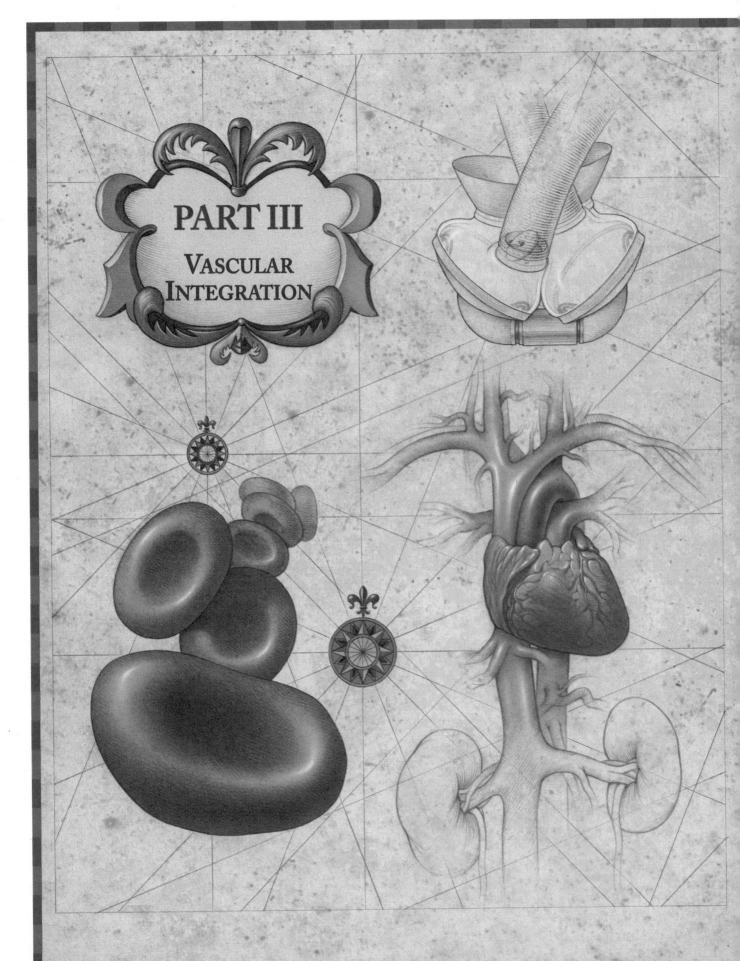

PART III

VASCULAR INTEGRATION

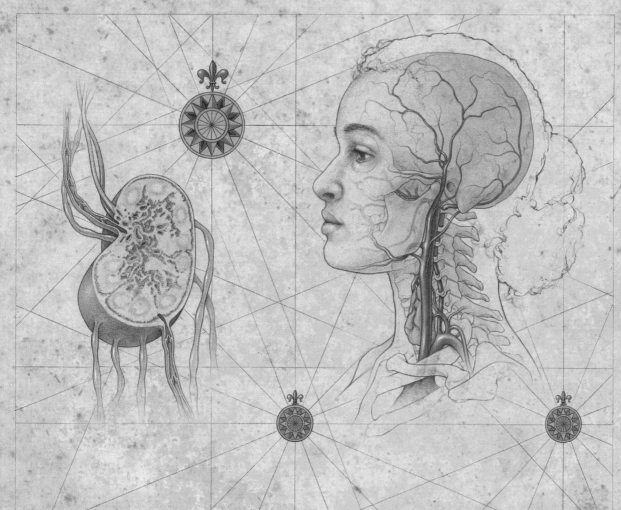

Part III describes the cardiovascular system, which conducts blood to nearly every cell of the body. **Chapter 13,** Heart, describes the organ that pumps blood through the vessels of the system. **Chapter 14** concentrates on the vessels themselves, and **Chapter 15** describes the constituents of blood. **Chapter 16,** Lymphatic System and Immunity, describes the body's second circulatory system and the tissues and organs that combat infections and other challenges from foreign cells.

Part III is about integration. To integrate things is to make them work together and to see those workings as a whole. Obviously the body parts work together, and the cardiovascular system enables them to do so. The first person who bled from an injury was not aware that blood transports numerous materials among the tissues. Today, we take this idea for granted and wonder what the mystery was. Pumped by the heart through an extensive network of vessels, blood communicates with nearly all body tissues, taking up and discharging molecules as it passes. A drop of blood circulates oxygen from the lungs to respiring tissues. As blood returns to the heart, the small intestine charges that same drop with glucose, now freshly loaded with nutrients for the tissues the next time around. Urea wastes from tissues are discharged in the kidneys. The kidneys release into this traffic their own hormones that regulate blood pressure and production of erythrocytes in the bone marrow. Lymph nodes remove infecting bacteria and viruses.

The cardiovascular system resolves the surface volume dilemma presented by our large bodies—too little external surface to support the volume inside—by communicating ultimately to every cell. Long ago, bodies became too large to exist without a cardiovascular system and they do so again every time an embryo develops. Blood begins to form during the third week of life, and the heart begins to pump a few days later when the embryo is only 2 mm long, because the surface-volume limits have already been approached. Part III describes how the cardiovascular system links the cells together and integrates them into a body.

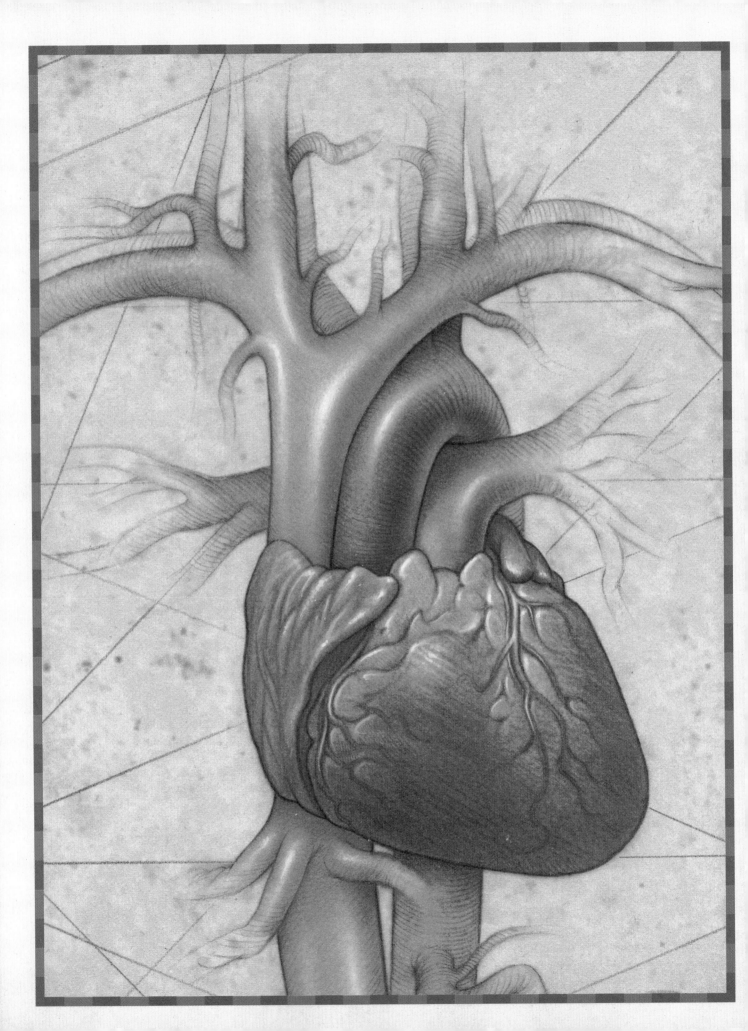

CHAPTER 13

Heart

OBJECTIVES

After studying this chapter, you should be able to do the following:

1. Follow the flow of blood through the chambers and valves of the heart, from the pulmonary to the systemic side and back again.
2. Describe various conditions of heart failure.
3. Describe the external landmarks of the heart and its location in the thorax.
4. Describe the four chambers and valves of the heart, how the atria differ from the ventricles, and how the pulmonary and systemic sides of the heart differ.
5. Describe the functions of the three layers of tissue that make up the wall of the heart.
6. Describe the anatomy of cardiac muscle fibers; relate it to the ability of the heart muscle to contract.
7. Trace the events of the heart cycle by following the action of the chambers and valves through systole and diastole.
8. Take your pulse and estimate your heart rate.
9. Describe the actions of the pacemaker and specialized conducting fibers in the heart cycle.
10. Relate the electrocardiogram trace to the normal heart cycle and to certain dysrhythmias.
11. Describe how blood circulates in the heart wall and some consequences of blocking this flow.
12. Describe how the heart develops from cardiac mesoderm.
13. Describe the changes at birth that finally separate the pulmonary and systemic sides of the heart.
14. Describe how congenital defects in the heart affect its blood flow and the delivery of oxygen to the tissues.
15. Describe changes that occur in hearts of aging people.

 VOCABULARY TOOLBOX

The following terms refer to the heart and associated structures:

aorta — aort-, *G* the great artery. The aorta carries blood from the heart.

atrium, atrial — atri-, *L* a vestibule. The atria receive blood for the ventricles.

cardiac, cardial, cardium — card-, *G* the heart. The great cardiac vein drains the heart wall.

carnae — carn-, *L* flesh, meat. Trabeculae carnae are muscular ridges on the inner surface of the ventricles.

coronary — coron-, *L* a crown. The coronary sulcus and arteries resemble a crown.

cusp — cuspi-, *L* a point. Cusps are the surfaces of cardiac valves.

diastole — dia-, *G* separate, apart. Diastole is the period between contractions.

lunar — lun-, *L* the moon. Semilunar valves guard the exit from the ventricles.

lunule — *L* a crescent. Crescent-shaped lunules seal semilunar valves shut.

pulmonary — pulmo-, *L* the lung. Pulmonary trunk and arteries deliver blood to the lungs.

septum — sept-, *L* a fence. The interatrial septum divides the right atrium from the left.

sinus — sinu-, *L* a fold, a hollow. The coronary sinus receives blood from the heart wall.

sulcus — sulc-, *L* a groove. The coronary sulcus is a groove encircling the heart.

systemic	system-, *G* a system. Systemic circulation delivers blood to body tissues.
systole	systol-, *G* a contraction. The period when the heart muscle contracts.
vein	ven, *L* a vein; cava, *L* hollow, cave. The vena cava is the major vein of the trunk.
ventricle	ventricul-, *L* the belly, ventricle. The ventricles swell with blood when full.

RELATED TOPICS

- **Autonomic nervous system and control of heart rate, Chapter 21**
- **Blood, Chapter 15**
- **Body cavities, pericardial cavity and mediastinum, Chapter 1**
- **Mesoderm and germ layers, Chapter 4**
- **Skeletal muscle, Chapter 9**
- **Vessels and cardiovascular disease, Chapter 14**

PERSPECTIVE

Only in the last 20 years has ultrasound technology made it possible to record the heart's sustained motions on videotape. Early observers had only a few minutes to observe heart movements before their experimental animals expired. William Harvey, who is credited with the first descriptions of human circulation, wrote:*

> There are as it were at one time two motions, one of the ears [the atria and auricles], and another of the ventricles themselves, for they are not just at one instant, but the motion of the ears goes before, and the motion of the heart follows; and the motions seem to begin at the ears and to pass forward to the ventricles.

O nce a minute, on the average, a drop of blood makes a complete trip around your cardiovascular system. Approximately 6 l of blood (1½ gal), the entire amount of blood in the body, is pumped each minute—about 85 ml at each beat, 70 beats a minute, and 3 billion beats in a lifetime. As do all mammalian hearts, ours has four chambers—two **atria** that receive blood and two **ventricles** that pump blood into the arteries to the tissues of the body. Blood flows through all four chambers and through the tissues on each circuit. Figure 13.1 shows the sequence—right atrium, right ventricle, lungs, left atrium, left ventricle, body tissues, and back to the right atrium.

Case study begins on p. 382.

*The Anatomical Exercises of Dr. William Harvey, 1653, p. 13.

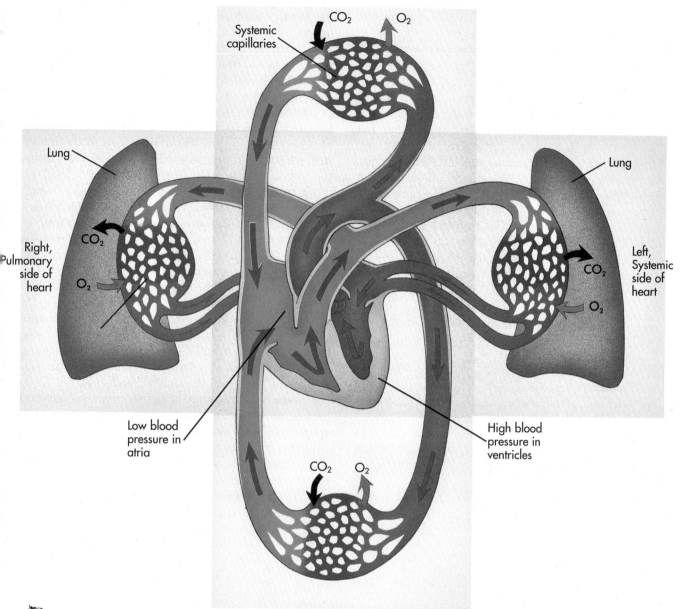

Figure 13.1

Basic circuitry of the heart and cardiovascular system. This anterior view places the anatomic right side on the left and the anatomic left side on the right. Follow the circuits, beginning on the pulmonary side in the right atrium. Oxygenated blood is shown in *red,* deoxygenated blood in *blue.*

HEART AND CARDIOVASCULAR SYSTEM

The heart actually is a double pump. Two pumps located side-by-side force a torrent of blood first through the lungs and then to the body. Blood passes first through the right, or **pulmonary,** side of the heart, to be oxygenated in the lungs. This flow begins in the right atrium and enters the right ventricle, which pumps the blood to the lungs where it takes on

oxygen and discharges carbon dioxide. Freshly charged with oxygen, blood now returns to the heart through the left, or **systemic,** side, which boosts pressure still higher for the circuit around the body. The left atrium receives **oxygenated blood** from the lungs, and the left ventricle pumps it into the arteries of the head, trunk, and extremities. Tissues take oxygen from the blood and release carbon dioxide into it as it passes. Deoxygenated blood (only 40% of the oxygen remains) returns to the pulmonary side of the heart to recharge with oxygen. The pulmonary and systemic circuits are efficient because they separate the processes that oxygenate and deoxygenate the blood.

Blood flows from higher to lower pressure. High-pressure blood flows from the ventricles out through the tissues and returns to the atria at much lower pressure, as if it were water flowing downstream. (Blood will not flow any more readily toward a region of higher pressure than water will flow uphill). Normal blood pressure in the left ventricle registers 120 mm of mercury on the sphygmomanometer (SFIG-mo-ma-NOM-eh-ter; sphygm-, *G,* the pulse; an instrument used with a pressure cuff to measure blood pressure), but almost 0 mm at the atria. Blood would flow directly back from the ventricles into the atria, taking the path of least resistance, if valves did not block its return. Consequently, blood flows out through the arteries. Blood flows because of the pressure the heart places upon it, but the valves control its direction of flow.

- What are the names of the chambers of the heart?
- Which chambers receive blood from the lungs and body?
- Which chambers pump blood to the lungs; which chambers pump to the body?
- Which side of the heart pumps deoxygenated blood; which side pumps oxygenated blood?

HEART FAILURE

▲ **Failing hearts do not pump enough blood to sustain the body with oxygen.** Heart failure alerts us to the crucial anatomical features that sustain normal hearts. For example, when the systemic side pumps less blood than the pulmonary side, the blood that would have entered the left atrium remains in the lungs. Fluid leaks into the air passages, and the patient complains of shortness of breath. This situation is known as **left heart failure** (Figure 13.2), and it illustrates how important it is for both sides of the heart to pump the same amount of blood. **Right heart failure,** when the pulmonary side pumps less blood than the systemic side, also disturbs the normal balance between left and right. No fluid accumulates in the lungs because the left side of the heart quickly drains what blood the right ventricle sends to it, but fluid accumulates nevertheless in the tissues and veins that lead into the right side of the heart. Patients with this condition suffer **edema,** in which the excess fluid usually swells the lower limbs, but they do not have difficulty breathing.

Heart failure takes many other forms. In **congestive heart failure,** fluid that would normally pass through the heart collects instead in the lungs and main vessels that return blood to the heart because the weakened heart cannot pump enough blood. A different cause of failure may occur rapidly in a **heart attack** when a clot blocks blood flow to a large area of a ventricle's wall and the muscle cannot contract. In contrast, **coronary artery disease** may gradually reduce blood flow through the ventricle walls until the remaining healthy tissue can no longer pump adequately. Heart valves that fail to open or close fully also lead to heart failure when the ventricles attempt

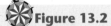**Figure 13.2**

Left and right heart failure. When the left ventricle fails, the fluid that it cannot pump accumulates in the lungs and pulmonary vessels. The converse occurs when the right ventricle fails to pump as much blood as the left ventricles; the extra fluid accumulates in the veins and tissues of the body.

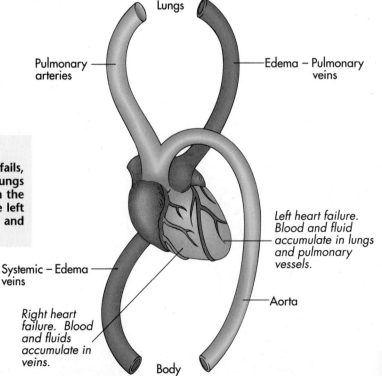

to compensate for the narrowed openings or leaky valves through additional effort (see Stenosis and Regurgitation, this chapter). Chapters 14, Vessels, and 15, Blood, discuss how some of these circumstances develop.

- Edema in the lungs and pulmonary vessels is a symptom of failure on which side of the heart?
- When the opposite side fails, where do fluids collect?

HEART AND PERICARDIAL CAVITY

MEDIASTINUM AND PERICARDIAL CAVITY

The heart is located behind the sternum in the mediastinum, the connective tissue partition that separates the left and right lungs from each other in the thoracic cavity. The heart occupies the lower half of the mediastinum (MEE-dee-ah-STY-num), between the second and sixth ribs, as Figure 13.3A shows. It is larger than your fist, weighs less than a pound, and packs impressive power. Unlike the great vessels, esophagus and trachea, that are embedded in the connective tissue, the heart is suspended within the **pericardial cavity** (PEAR-ih-KAR-dee-al) by the great vessels that enter and leave the heart. This arrangement allows the beating heart to expand and contract within the smooth, slippery linings of the pericardial cavity. The **visceral pericardium,** also known as the epicardium, covers the external surface of the heart with a thin, fibrous mesothelium that secretes a lubricating serous fluid into the pericardial cavity. Externally the **parietal pericardium** isolates the pericardial cavity from the mediastinum itself and from the pleural cavities. This structure is a fibrous, connective tissue overlain with thin, serous mesothelium.

EXTERNAL LANDMARKS

The walls of the right atrium and right ventricle dominate the anterior surface of the heart, shown in Figure 13.3B. Embryonic development displaces the heart to the left and rotates it slightly so that it intrudes upon the left lung and the right ventricle partially obscures the left ventricle. Both the left and right ventricles rest on the superior surface of the diaphragm that forms the floor of the pericardial cavity. The atria lie above the ventricles. The lobular anterior margins of the atria are known as auricles because of their resemblance to dogs' floppy ears. If you were able to look directly into the chest cavity from the anterior, as thoracic surgeons do, you would see several major anatomical landmarks of the heart.

1. The **apex** of the heart is the blunt conical end that points downward, forward, and toward the left. It is formed by the ventricles; the left ventricle lies behind and to the left of the right ventricle.
2. The **base** of the heart is the broad, posterior, superior surface, formed by the walls of the atria and the great vessels of the heart, that suspends the heart from the mediastinum into the pericardial cavity. The base is easier to see in Figure 13.3C, which shows the posterior surface of both atria.
3. The **right margin** of the heart is the long, curved outline that follows the thin, fleshy wall of the right atrium downward along the reddish, muscular wall of the right ventricle to the apex, visible in parts B and C of Figure 13.3. The **left margin** is shorter than the right, but it follows the corresponding outline on the left side of the heart, descending the walls of the atrium and the ventricle to meet the right margin at the apex. Find this margin in both parts of the Figure.
4. The **coronary sulcus** (SUL-cus) is the prominent groove that encircles the heart between the atria above and the ventricles below. This groove marks the internal location of the atrioventricular valves that admit blood from atria into the ventricles. Major coronary arteries and veins in this sulcus distribute blood to and collect blood from the heart wall (see Coronary Circulation, this chapter). The right coronary artery marks this sulcus on the anterior surface (Figure 13.3B), and the great cardiac vein and coronary sinus identify it on the posterior surface (Figure 13.3C). Adipose tissue accompanies these vessels in the coronary sulcus and other locations on the surfaces of the heart.
5. A shallow **interventricular groove** on the powerful muscular walls of the ventricles marks the location of the **interventricular septum,** the internal partition between the left and right ventricles and the pulmonary and systemic sides of the heart. The anterior interventricular artery marks the groove on the anterior surface (Figure 13.3B), and the posterior interventricular artery does the same on the posterior (Figure 13.3C).

GREAT VESSELS

The heart sends and receives blood through the great vessels, so named because they are the largest in the body, approximately 2.5 cm (1 inch) in diameter. Blood returns to the right atrium from the head and upper extremities by way of the **superior vena cava** and from the trunk and lower extremities through the **inferior vena cava** (VEE-na KAY-va). The right atrium hides the entry of these vessels in Figure 13.3B, but part C of the figure illustrates how these vessels join at their entrance to the posterior surface of the atrium. Blood flows from the right ven-

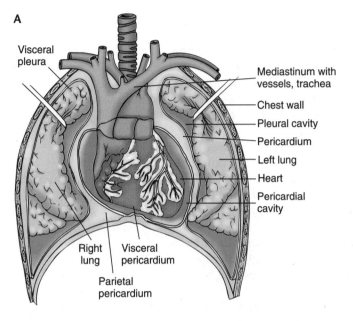

A

Visceral pleura

Mediastinum with vessels, trachea

Chest wall

Pleural cavity

Pericardium

Left lung

Heart

Pericardial cavity

Right lung

Visceral pericardium

Parietal pericardium

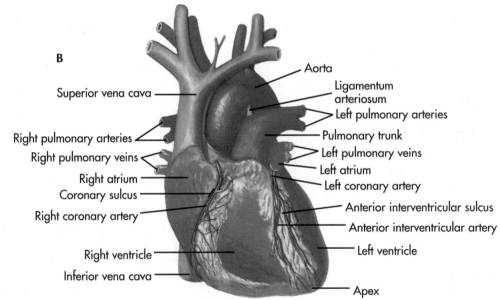

B

Superior vena cava

Right pulmonary arteries

Right pulmonary veins

Right atrium

Coronary sulcus

Right coronary artery

Right ventricle

Inferior vena cava

Aorta

Ligamentum arteriosum

Left pulmonary arteries

Pulmonary trunk

Left pulmonary veins

Left atrium

Left coronary artery

Anterior interventricular sulcus

Anterior interventricular artery

Left ventricle

Apex

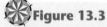

Figure 13.3

A, The heart fills the pericardial cavity within the mediastinum. **B** and **C**, Anterior and posterior views of the heart removed from the pericardial cavity.

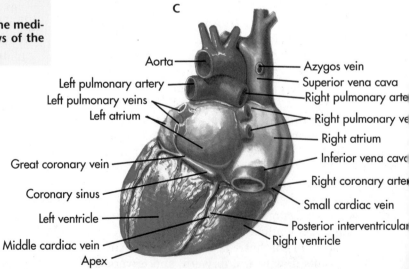

C

Aorta

Left pulmonary artery

Left pulmonary veins

Left atrium

Great coronary vein

Coronary sinus

Left ventricle

Middle cardiac vein

Apex

Azygos vein

Superior vena cava

Right pulmonary arte

Right pulmonary ve

Right atrium

Inferior vena cava

Right coronary arte

Small cardiac vein

Posterior interventricular

Right ventricle

tricle through the **pulmonary trunk,** seen in part B, that divides into right and left **pulmonary arteries** to the lungs. The right pulmonary artery branches under the aorta on its way to the right lung. Part C of the figure shows this branching more clearly. Four smaller **pulmonary veins** deliver blood from the lungs to the left atrium. Each vein enters separately, as you can see in part C. Two veins from the left lung enter the left side of the atrium and two from the right lung enter the right side. The **aorta** (ay-OR-tah) is the dominant vessel. It ascends from the left ventricle (ascending aorta) and bends posteriorly and to the left (arch of the aorta), giving off branches to the neck, head, and arms, and then descends to the trunk and lower extremities. Figure 13.3B shows the ascending aorta and the arch. Part way along the arch, a narrow **ligamentum arteriosum** (LIG-ah-MEN-tum ar-TEER-ee-OH-sum; *L,* ligament of the artery), whose significance is described in Development of the Heart, this Chapter, joins the aorta and pulmonary trunk.

- What internal structures do the coronary sulcus and interventricular groove mark?
- Where are the base and apex of the heart located?
- What are the names of the great vessels of the heart?

FOUR CHAMBERS

The four chambers of the heart are separated by partitions and valves that ensure that blood does not leak backward through the heart or mix between pulmonary and systemic sides. Figure 13.4 shows a frontal section through all these structures. Identify the walls of the ventricles by their thick, muscular tissue. The **interventricular septum** separates the left and right ventricles. This partition continues between the left and right atria as the thinner, much less muscular, **interatrial septum.**

Valves direct blood through the heart. Two **atrioventricular (AV) valves** between the atrium and ventricle on each side of the heart ensure that blood

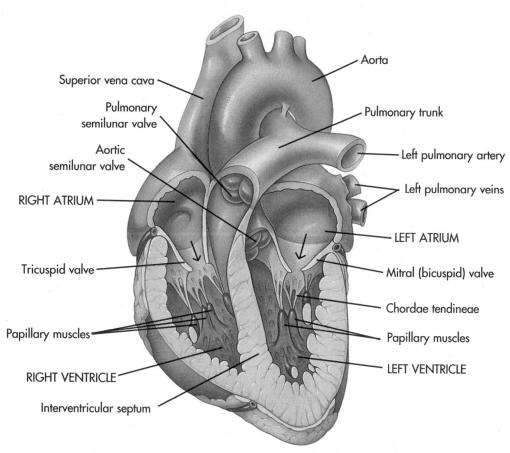

Figure 13.4
Chambers and valves of the heart.

flows from the atrium into the ventricle and not in the opposite direction. The right AV valve is also called the **tricuspid valve** (try-KUS-pid) because three flaps, or **cusps,** open and close as blood passes. (Earlier names, such as the tricuspid valve, that recognize valve shapes have been supplanted by the functional name, atrioventricular.) Two cusps serve the left AV valve, also called the **mitral valve** (MY-tral) or less commonly the bicuspid valve, because the cusps resemble a bishop's mitre, or hat. The **aortic** and **pulmonary semilunar valves** (SEM-ee-LOON-ar; half moon) guard the exits of the left and right ventricles against backflow from the aorta and pulmonary trunk.

RIGHT ATRIUM

The right atrium has two parts—the main portion and the auricle, which is a conical extension of the main portion. A slick, smooth surface lines both regions, but narrow slips of muscle, called **musculi pectinati,** form small ridges beneath the lining of the auricle. Both sections fill and expand with blood. Blood from the head and upper extremities enters the main portion of the atrium through the **superior vena cava,** and the **inferior vena cava** delivers blood from the trunk and lower extremities. The much smaller **coronary sinus** returns blood from the heart wall itself (see Coronary Circulation, this chapter). A slight, oval depression in the interatrial septum, called the **fossa ovalis** (FOSS-ah oh-VAL-iss; *L,* oval depression), marks the foramen ovale, an opening that allowed blood to pass from the right to the left atrium before birth, bypassing the undeveloped lungs. Contraction of the atrium helps blood flow into the right ventricle through the right atrioventricular (tricuspid) valve.

RIGHT VENTRICLE

This chamber inflates with blood from the right atrium and pumps the blood into the **pulmonary trunk** and the lungs. The wall of the right ventricle is thinner and less powerful than that of the left ventricle, developing lower blood pressure than the left ventricle. The right ventricle also is crescent-shaped in cross section, wrapped partially around the circular left ventricle. Although it is as slippery as the atrial wall, the internal surface of the right ventricle is laced by many muscular ridges and cords called **trabeculae carnae** (tra-BEK-you-lee KAR-nee). The largest are **papillary muscles** (PAP-ill-air-ee; papill-, *L,* a nipple) that help support the AV valves during closure (see Valves, below). When the right ventricle contracts and the pressure on the blood begins to exceed that in the atrium, the AV valve slams shut and blocks any backward flow. Shortly afterward, rising pressure pushes the **pulmonary semilunar** valve open and blood flows into the pulmonary trunk. From there it follows the left and right **pulmonary arteries** into the lungs where it is oxygenated.

LEFT ATRIUM

The left atrium receives freshly oxygenated blood from the lungs by way of the **pulmonary veins** and discharges it through the left AV valve into the left ventricle. Four pulmonary veins, two from each lung, enter the main portion of this chamber from the posterior wall of the left atrium. The left auricle extends forward like that of the right atrium, and the fossa ovalis also can be seen on the interatrial septum. Musculi pectinati also corrugate the left auricle. The entire atrium fills with returning blood that then flows into the left ventricle past the left AV (mitral) valve.

LEFT VENTRICLE

The left ventricle sends blood into the **ascending aorta** when the muscular wall of this ventricle contracts. Pressure closes the left AV (mitral) valve, preventing backflow, and opens the **aortic semilunar valve.** This latter valve snaps closed when the ventricle begins to relax and fill with blood for the next beat. As in the right ventricle, trabeculae carnae weave a network of muscle fibers on the inner surface of the left ventricle and two papillary muscles control the mitral valve.

- Which openings deliver blood to the right atrium?
- What valve guards the exit from the right atrium?
- Which valve guards the exit from the left atrium?
- Which valve prevents leakage of blood back into the left ventricle?

FOUR VALVES

VALVES AND FIBROUS SKELETON OF THE HEART

The skeleton of the heart is a set of four fibrous rings that support the valves. Figure 13.5A shows a pair of large atrioventricular valves and two smaller semilunar valves as seen from the atria and great vessels. Each valve displays **cusps,** that is, flaps or edges that point in the direction of flow. Blood flows through the valve by pushing the cusps aside, but flow in the opposite direction is blocked when the blood forces the cusps together and closes the valve. Cusps of the semilunar valves point upward in the direction of blood flow into the aorta and pulmonary trunk, but the cusps of the atrioventricular valves point downward into the ventricles. These valves are set into **fibrous rings** in the upper wall of the ventricles from which the cusps extend. Made of tough, rigid, fibrous connective tissue and fibrocartilage, these rings connect with each other, and together they form the **skeleton of the heart** to which the atria and ventricles attach. The muscles of the atria and ventricles attach to the rings in somewhat the same way as the origins and insertions of skeletal muscles connect to bones.

A ALL VALVES CLOSED

Cusps of pulmonary
semilunar valve

Cusps of aortic
semilunar valve

Cusps of the
left atrioventricular
valve (bicuspid)

Cusps of the
right atrioventricular
valve (tricuspid)

C SEMILUNAR VALVE

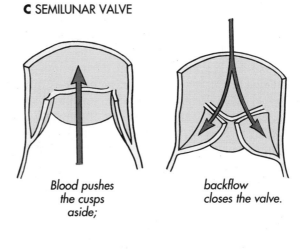

Blood pushes
the cusps
aside;

backflow
closes the valve.

B ATRIOVENTRICULAR VALVE

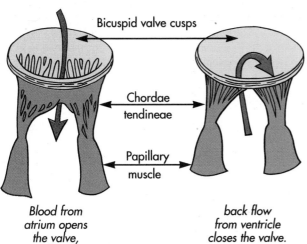

Bicuspid valve cusps

Chordae
tendineae

Papillary
muscle

Blood from
atrium opens
the valve,
but

back flow
from ventricle
closes the valve.

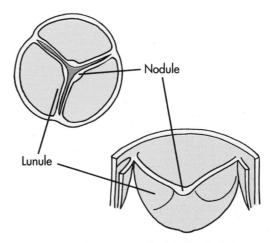

Nodule

Lunule

Flexible lunule seals
against lunules of other
cusps.

**Figure 13.5**

A, Atrioventricular and semilunar valves. Mounted on fibrous rings in the skeleton of the heart, the cusps of all
valves are closed. Anterior surface of heart at top of figure. **B,** Operation of atrioventricular valves. Blood flows
into the ventricles by pushing the cusps aside. Papillary
muscles and chordae tendineae prevent the cusps from
bursting open when blood pushes the cusps closed.
C, Lunules seal the cusps of semilunar valves when blood
flow closes the valves.

ATRIOVENTRICULAR VALVES

The orifice, or opening, of each atrioventricular valve is circular, the right being about 4 cm (1.6 inches) in diameter, somewhat larger than the left. Three cusps identify the right AV valve (tricuspid), and two cusps identify the mitral or left AV (bicuspid) valve. These cusps occupy anterior, posterior, and septal positions (from their relationship to the interventricular septum) around the orifice of the tricuspid valve. When closed, their edges form a Y, shown in Figure 13.5A, in which the stem of the Y points laterally between the anterior and posterior cusps. The cusps of the mitral valve align on the anterior and posterior sides of its ring, and their junction runs laterally across the orifice.

The edges of the cusps are attached by numerous fine, strong **chordae tendineae** (KORD-ee TEN-din-ee; *L*, tendinous cords), shown in Figure 13.5B, to the **papillary muscles** in the wall of the ventricles. Despite what you might think, the papillary muscles do not pull the atrioventricular valves open. Instead, they support the valves against the force of the blood when the ventricles contract. The arrangement works somewhat like a parachute in which air inflates the chute and the jumper pulls on the shrouds to adjust the edges. Papillary muscles vary in size and number, but each ventricle has two principal papillary muscles. In the right ventricle the anterior muscle controls chordae to the anterior and posterior cusp, whereas the posterior muscle serves the septal and the posterior cusp. The arrangement is similar in the left ventricle, where anterior and posterior muscles receive chordae from both cusps.

SEMILUNAR VALVES

The orifices of the semilunar valves are about 2.5 cm (1 inch) in diameter, smaller than those of the atrioventricular valves. The cusps of semilunar valves are fibrous pockets that jam closed when blood fills them; no chordae tendineae support them. Both valves have three cusps, and the free margins of the cusps meet in Y-shaped patterns similar to that of the right AV valve. The cusps of the pulmonary valve occupy right, left, and anterior positions around the fibrous ring, and those of the aortic valve occupy right, left, and posterior locations. The right and left coronary arteries exit the aorta toward the wall of the heart through openings behind the right and left cusps of the aortic valve, respectively.

Figure 13.5C illustrates why semilunar valves are so named. This view shows the cusps of the aortic valve spread out around the orifice of the valve. Each cusp has a central, fibrous **nodule** that seals the center of the Y where all three cusps meet. Many supporting collagenous fibers fan out from each nodule toward the wall of the valve, but these fibers are missing from the **lunules** (LOON-yules), delicate crescent moon-shaped edges that seal the cusps together.

STENOSIS AND REGURGITATION

Disease may damage the cusps of valves, causing scar tissue that interferes with their opening and closing. In **stenosis** (STEN-oh-sis; sten-, *G*, narrow) the valves do not open fully, as illustrated in Figure 13.6. Mitral stenosis indicates, for example, that the mitral valve admits less blood than normal because scar tissue has partially closed the valve. Surgeons can treat

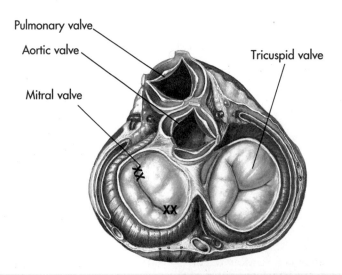

Pulmonary valve
Aortic valve
Mitral valve
Tricuspid valve

✴ **Figure 13.6**

Stenosis and regurgitation in the atrioventricular valves. Same view as in Figure 13.5A. Stenosis blocks the mitral valve; the valve cannot open fully because scar tissue has shortened its orifice. Cross hatching shows scar tissue. Scar tissue prevents the tricuspid valve from closing; the valve regurgitates. Scar tissue is less flexible and prevents the cusps from sealing shut.

stenosis by cutting the scar tissue and opening the valve. Severe stenosis may require that the valve be replaced with an artificial one. **Regurgitation** (ree-GUR-jih-TAY-shun; gurg-, *L,* a whirlpool) is the name given to backflow of blood through a poorly sealed valve. In **rheumatic fever** (ru-MAT-ik) the valves (and other tissues) become inflamed following strep throat infections, leaving minute scars that can cause stenosis or regurgitation if they are large enough; a valve may show both conditions.

- The size of the orifice is related to the pressure of the blood passing through the valve. Compare the size of AV and semilunar valves, and the pressure of blood (see Heart and Cardiovascular System) to figure out this relationship.

VALVE SOUNDS AT FOUR CORNERS OF THE HEART

Valves of healthy hearts slam shut with characteristic sounds. "Lubb-dupp, lubb-dupp," the physi-cian hears, as first the AV and then the semilunar valves close. Knowing where to place the **stethoscope** to detect these and other sounds most effectively is one of the first arts clinicians learn. Figure 13.7A shows the landmarks on the chest wall that help locate the valves and the best sites for **auscultation** (aws-KUL-tay-shun; auscult-, *L,* listening). Although all four valves lie in the skeleton of the heart that crosses the sternum from your upper left to lower right, these locations are not the most informative places to listen. Sounds are heard better at the four **corners of the heart,** where the blood-filled chambers and vessels direct these vibrations closer to the chest wall. The "lubb" sound of the right AV valve is best heard with the stethoscope placed in the fifth intercostal space on the right margin of the sternum. Find this location on your own chest, and then move across the sternum to the left side to hear the "lubb" sound of the left AV valve at the apex of the heart, also in the fifth intercostal space, 10 cm (2.5 inches)

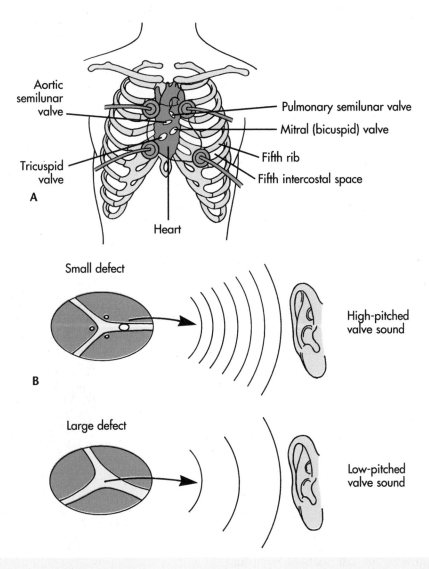

Aortic semilunar valve

Pulmonary semilunar valve

Mitral (bicuspid) valve

Tricuspid valve

Fifth rib

Fifth intercostal space

A

Heart

Small defect

High-pitched valve sound

B

Large defect

Low-pitched valve sound

Figure 13.7

A, Four points of the heart. **B,** Heart murmurs.

from the left margin of the sternum. The "dupp" sound of the pulmonary valve is heard in the second intercostal space at the left margin of the sternum, just superficial to the pulmonary trunk. Moving the stethoscope to the right of the sternum at the same level reveals the "dupp" sound of the aortic valve as blood enters the ascending aorta.

Valves squeal and rumble when their cusps do not fit together properly, making noises called **heart murmurs**. Irregularities from scar tissue on the cusps allow the cusps to vibrate when blood regurgitates through them, making sounds that reflect the size of the defect (Figure 13.7B). High-pitched squeals generally indicate small leaks, and low frequency rumbles are signs of larger damage.

- Which valves have three cusps?
- Which valves are controlled by chordae tendineae and papillary muscles?
- Which pair of valves issues the "lubb" sound when they close?
- Where can the aortic valve best be heard with the stethoscope?

CARDIAC MUSCLE AND WALL OF THE HEART

Ventricles and atria contain three layers of tissue—the endocardium, epicardium, and myocardium—illustrated in Figure 13.8.

ENDOCARDIUM

The **endocardium** lines the interior of the atria and ventricles and covers the cusps of the valves. This tissue is a web of thin, seemingly insubstantial, simple squamous endothelium that is directly exposed to the blood. The endocardium is continuous with the endothelium of the great vessels. Endothelia of the pulmonary veins and venae cavae blend with the endocardium at the entrance of the atria. Endocardium covers both surfaces of the AV valves but only the ventricular surfaces of the pulmonary and aortic semilunar valves. Endothelium covers the arterial sides of these valves. Cardiac endothelium derives from embryonic mesoderm.

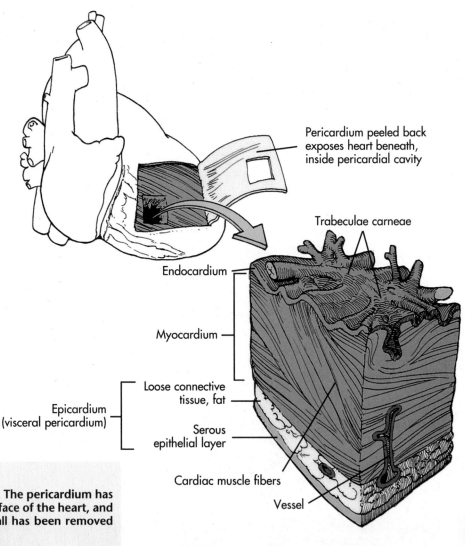

Pericardium peeled back exposes heart beneath, inside pericardial cavity

Trabeculae carneae

Endocardium

Myocardium

Loose connective tissue, fat

Serous epithelial layer

Epicardium (visceral pericardium)

Cardiac muscle fibers

Vessel

✳ **Figure 13.8**

Myocardium and wall of the heart. The pericardium has been peeled back to reveal the surface of the heart, and a section of the right ventricle wall has been removed to show the internal structure.

EPICARDIUM

As its name indicates, the **epicardium** covers the external surface of the heart. This tissue is a thin sheet of serous mesothelium, made slippery with serous fluid, that promotes motion inside the pericardium. Adipose tissue and vessels lie between the epicardium and the myocardium. Epicardium is continuous with the pericardium at the base of the heart. When infection enters the pericardial cavity, the epicardium may become inflamed and irritable, a condition called **pericarditis.**

MYOCARDIUM AND CARDIAC MUSCLE FIBERS

Myocardium is the muscular tissue of the heart. Its anatomy reflects both the pressure this muscle develops in contraction and its constant demand for oxygen to sustain repeated contractions without fatigue (Figure 13.9). The **myocardium** is thickest in the left ventricle, which develops the highest pressure in the heart, but it is only one third as thick in the right ventricle, where pressure is much lower. Myocardium is composed almost entirely of cardiac mus-

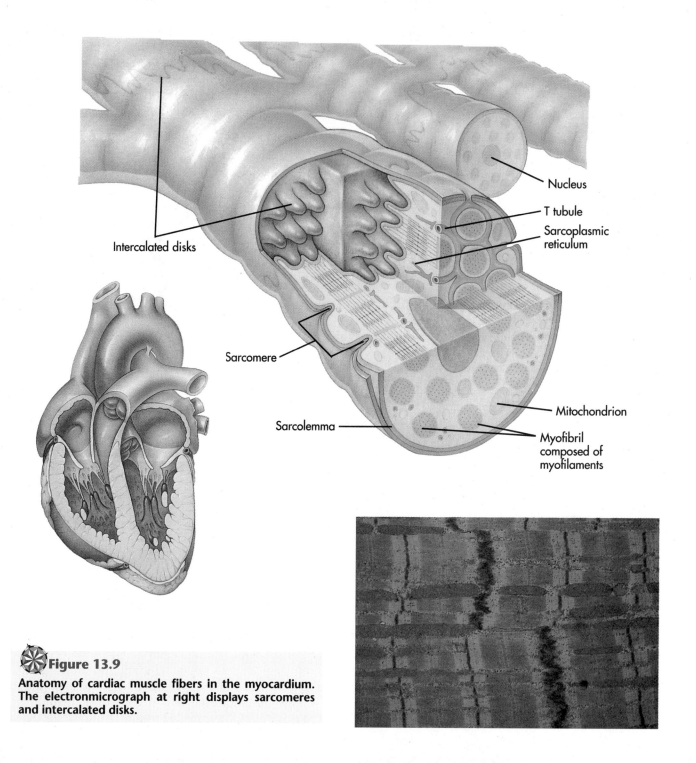

Figure 13.9

Anatomy of cardiac muscle fibers in the myocardium. The electronmicrograph at right displays sarcomeres and intercalated disks.

cle fibers interspersed with ample blood vessels, lymphatics, and innervation.

Cardiac muscle fibers are short, branched, cylindrical structures with numerous **myofilaments** and **sarcomeres** that give them a striated appearance. These fibers are distinguished by their ability to contract repeatedly without voluntary control, for which they demand continuous high quantities of oxygen. The presence of large numbers of **mitochondria** (10 times more than in skeletal fibers) indicates these needs. Most cardiac muscle cells contain a single, large nucleus or occasionally two, with numerous cytoplasmic lipid droplets and glycogen granules that store carbohydrates for cellular respiration. A profuse **sarcoplasmic reticulum** triggers contraction, with one T-tubule entering at the Z-disks between sarcomeres (see Skeletal Muscle Fibers, Chapter 9, to review these structures).

Intercalated disks, shown in Figure 13.9, join the ends of fibers together into extensive webs of contractile tissue. These disks are unique features of the heart; they pass stimuli from cell to cell and simultaneously attach the cells together. The disks occur where Z-disks would be expected between sarcomeres. The disks employ **desmosomes** to connect the cells and **gap junctions** for intercellular communication. The desmosomes connect thin myofilaments of the I-bands between adjacent cells.

A **basement membrane** envelops each cardiac fiber and joins it with **endomysium,** connective tissue that delivers capillaries, lymphatics, and neurons throughout the myocardium (Figure 13.10). The conduction system of the heart stimulates cardiac fibers to contract by promoting an inflow of calcium ions

across the **sarcolemma.** This flow continues from cell to cell through the gap junctions in the intercalated disks and causes the sarcoplasmic reticulum to release its own internal calcium as the immediate trigger of contraction. Although cardiac muscle fibers develop strength by increasing their diameter much as skeletal fibers do **(hypertrophy),** cardiac fibers do not divide. Without an obvious supply of **satellite cells** that skeletal muscles use to replace losses, the adult heart contains all the muscle fibers it will ever have. This apparent deficiency is one reason why heart muscle does not repair itself after heart attacks.

The network of branching cardiac fibers weaves itself into superficial and deep bundles of fibers that originate and insert upon the fibrous skeleton of the heart. Atrial bundles reach out from the atrioventricular rings around the venae cavae and pulmonary veins and then loop back. Figure 13.11 shows how superficial bundles spiral along the walls of the ventricles from the atrioventricular rings to the apex and back.

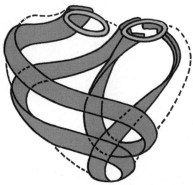

Figure 13.11

Bundles of cardiac muscle fibers originate on the fibrous rings, loop outward in the walls of the ventricles, and insert on the rings again.

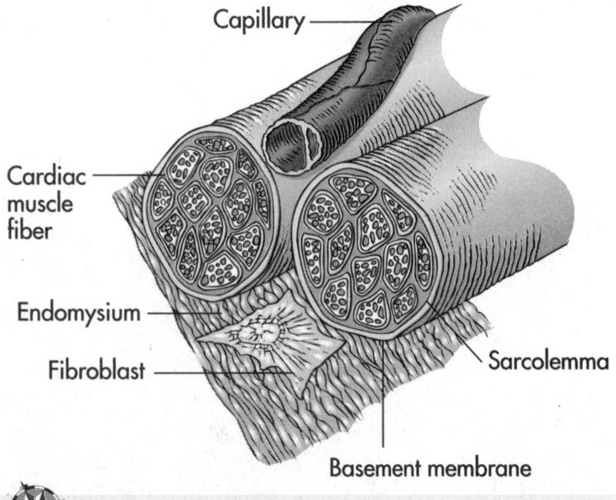

MYOCARDIUM AND INFARCTION

Damage to the myocardium further illustrates how myocardial structure is related to its function. When a coronary artery is obstructed, nearby myocardial tissue sometimes dies for lack of oxygen. Such patches of dead tissue, illustrated in Figure 13.12, are called **infarcts** or infarctions (IN-farkt; farc, *L,* stuffing). It is puzzling that a tissue that depends so heavily on rapid metabolism cannot replace the dead cells with new muscle fibers. The affected area immediately stops contracting, which can threaten life if enough tissue is involved. A zone of **ischemic** (iss-KEM-ik; isch-, *G,* suppress) tissue, living but damaged

Capillary

Cardiac muscle fiber

Endomysium

Fibroblast

Sarcolemma

Basement membrane

Figure 13.10

Endomysium and basement membranes envelop cardiac muscle fibers.

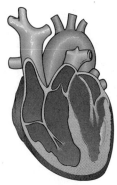
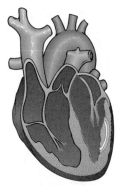
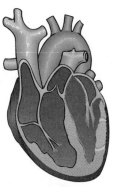
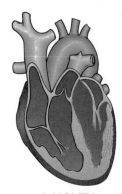

ONSET
*Injury and
ischemia
beneath
endocardium.*

DAY 1
*Injury and ischemia
reach epicardium and
infarct develops in inner
myocardium.*

DAY 2, 3
*Infarct reaches
epicardium.*

1 MONTH
*Fibrous scar
replaces infarct.*

 **Figure 13.12**
Progress of a transmural infarction.

from lack of blood, usually surrounds the infarct but can repair itself without permanent damage. Within the infarct, however, cells die during the next few days following the blockage. Fibrous scar tissue replaces the dead tissue during the following 3 or 4 weeks, but because cardiac muscle fibers do not invade this scar tissue, the damage is permanent. Scars from **transmural infarcts** (trans-MYUR-al) penetrate the entire wall and can be seen as white disks on the exterior of the heart. The scars of **subendocardial infarcts** may not be seen, however, because they do not obliterate all the muscle tissue. Infarcts may affect the heart cycle, and their effects can be detected by electrocardiograms.

- What are the names of the inner and outer tissue layers of the heart?
- Why is the myocardium of the left ventricle thicker than that of the right ventricle?
- Name three identifying features of cardiac muscle fibers.
- What type of tissue replaces infarctions in the myocardium?

HEART CYCLE

SYSTOLE AND DIASTOLE

The heart beats about 70 times per minute when you are resting. The sequence of events in each beat is called the **heart cycle.** The cycle begins as both atria are filling with blood. This blood flows into the ventricles, which expand as the blood enters. The atria then contract, adding more blood to the expanding ventricles. The ventricles now contract, pumping blood into the great vessels. The atria immediately begin to fill again while the ventricles are contracting.

Finally, the ventricles relax and take on blood from the atria. As William Harvey found long ago, the actions of the atria precede those of the ventricles; the atria contract before the ventricles, and the atria begin to fill while the ventricles are contracting.

Your pulse reflects these events, particularly the contraction of the left ventricle, where pressure is highest. Feel your pulse in your radial artery proximal to your thumb. Use your index and middle fingers as shown in Figure 13.13 but not your thumb; its own pulse will confuse you. Count the number of beats for 10 seconds and multiply by 6 to estimate your heart rate, the number of beats per minute. Do this several times until you get consistent numbers.

Each pulse marks the contraction of the ventricles and is commonly called systole (SIS-tow-lee). **Dias-**

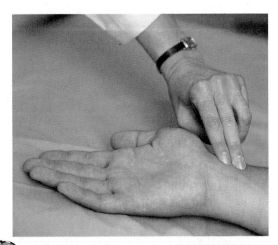

 **Figure 13.13**
Taking one's pulse. Follow directions in the text.

tole (di-AH-sto-lee) is the period between pulses, when the ventricles and atria are refilling with blood. Because the ventricles dominate the action, it is easy to forget that the atria also enter their own periods of systole and diastole before the ventricles. More precisely, the pulse in your wrist represents **ventricular systole.** The intervening period finds the atria and ventricles relaxing and filling in **atrial diastole** and **ventricular diastole.** Diastole ends when **atrial sys-**tole, the atrial contraction, begins. Overall, systole begins when the atria contract (atrial systole) and culminates in ventricular systole, when the ventricles contract. Diastole resumes when the ventricles begin to relax.

When you follow the course of blood through the heart, you can see how the movements and sounds of the valves are related to systole and diastole. Follow the details in Figure 13.14. The AV valves open as the

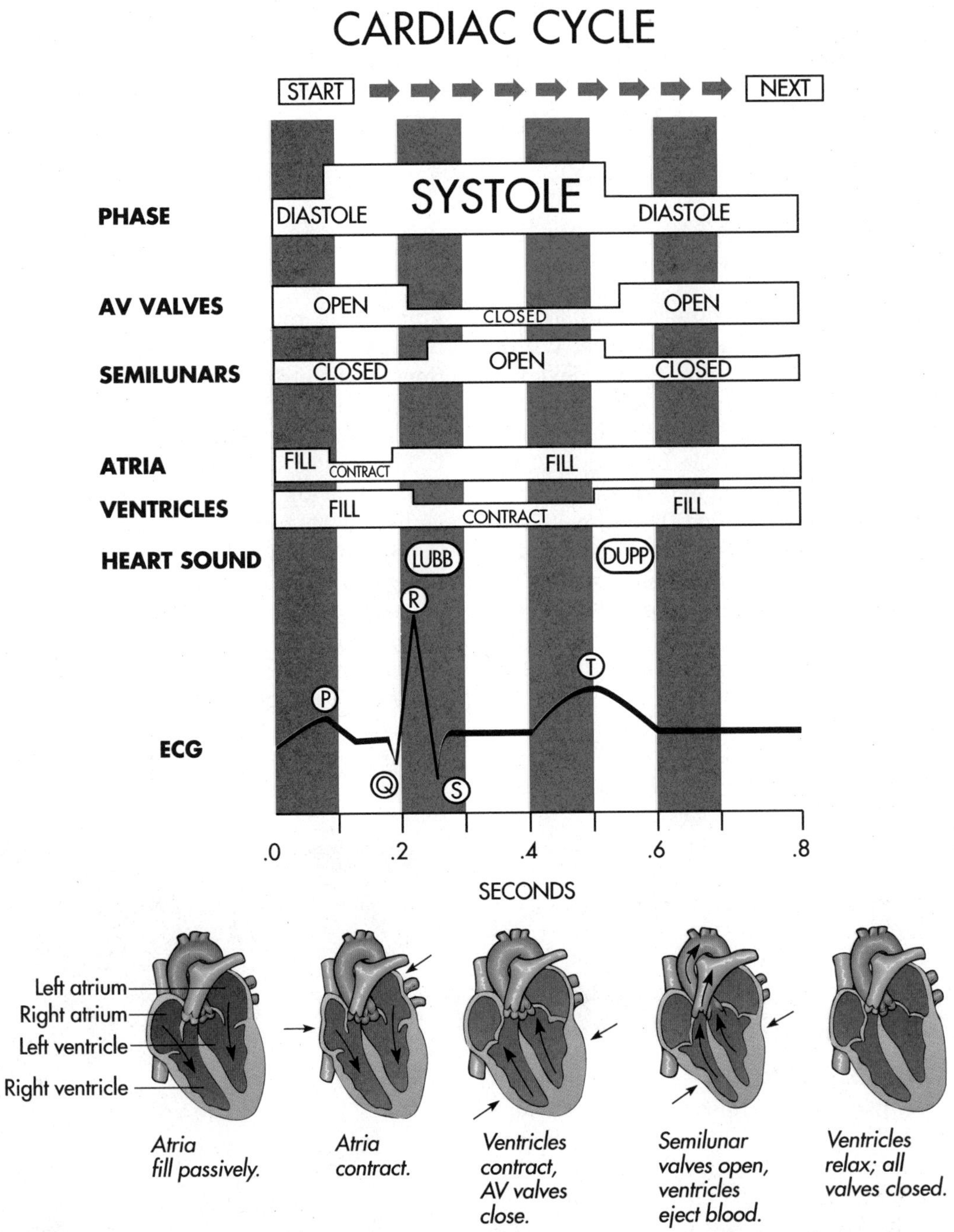

CARDIAC CYCLE

PHASE	DIASTOLE	SYSTOLE	DIASTOLE
AV VALVES	OPEN	CLOSED	OPEN
SEMILUNARS	CLOSED	OPEN	CLOSED
ATRIA	FILL / CONTRACT	FILL	
VENTRICLES	FILL	CONTRACT	FILL
HEART SOUND		LUBB	DUPP

ECG: P, Q, R, S, T

SECONDS .0 .2 .4 .6 .8

Left atrium
Right atrium
Left ventricle
Right ventricle

Atria fill passively.

Atria contract.

Ventricles contract, AV valves close.

Semilunar valves open, ventricles eject blood.

Ventricles relax; all valves closed.

Figure 13.14

Phases of the heart cycle. Be able to relate the diagrams of blood flow and heart motion (below) to the events in systole and diastole (above).

atria are filling, allowing blood into the ventricles while the semilunar valves are closed. Ventricular systole closes the AV valves with a sharp "lubb" sound, and the semilunars open, allowing blood to exit the heart. The "dupp" sound of the semilunars closing signals the beginning of diastole, when the ventricles relax. If you follow the valve movements carefully on the chart, you will see that both sets of valves never open at the same time. What would happen to blood flow if all were to open simultaneously?

PACEMAKER AND SPECIALIZED CONDUCTING FIBERS

Cardiac muscle fibers have an intrinsic ability to contract, a tendency coordinated overall by the pacemaker, or **sinoatrial node** (SIGN-oh-AYE-tree-al), and the conduction system of the heart. The sinoatrial (SA) node is a cluster of modified cardiac muscle fibers located beneath the endocardium in the right atrium at the entrance of the superior vena cava (Figure 13.15). Nodal cells have fewer sarcomeres and myofilaments than cardiac muscle and lack intercalated disks. The cell membranes self-depolarize about 70 times a minute, and each depolarization radiates over the walls of the atria, causing them to contract and the heart to beat.

The depolarizations also cause the ventricles to contract by regulating their own control center, the **atrioventricular (AV) node.** The AV node is located in the right atrium between the right AV valve and the coronary sinus, and its cells resemble those of the SA node. The AV node can set its own rhythm when the SA node fails but is slower; normally it is governed by the faster SA node. The SA node "catches" the AV node part way through the cycle and causes it to fire sooner than it otherwise would.

Larger cells called **specialized conducting fibers** conduct depolarizations from the atrioventricular node to the ventricles. These cells are especially large, modified cardiac fibers whose greater diameters and fibrous, connective tissue insulation enable faster conduction. The largest bundle of fibers, the **atrioventricular bundle,** or bundle of His (pronounced "hiss"; named for Wilhelm His, a nineteenth century German anatomist), leads from the AV node into the interventricular septum where it branches into two **left bundle branches** to the left ventricle and one **right bundle branch** to the right ventricle. **Purkinje fibers** in all the bundle branches descend the interventricular septum to the apex of the heart, conducting impulses that cause contraction, and then radiate into the lateral walls, where they finally approach the coronary sulcus. The apices of the ventricles thus contract before the side walls, projecting blood quickly through the semilunar valves and leaving the walls to wring remaining blood from the ventricle at the end of the stroke. In the Purkinje fibers, disorganized sarcomeres and myofilaments surround a central core of cytoplasm that stores numerous glycogen granules as an energy source. Each cell has a central

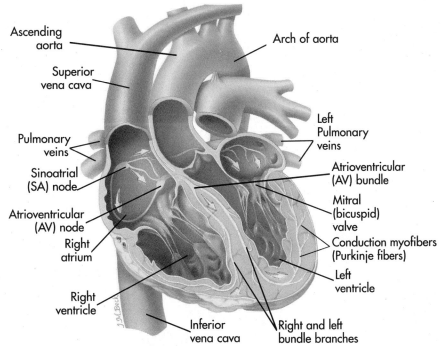

Figure 13.15
Conduction system of the heart.

nucleus, a few intercalated disks, and numerous gap junctions, but no T-tubules.

INNERVATION

Innervation of the heart increases or slows the heart rate by regulating the sinoatrial node. The **autonomic nervous system** (Chapter 21) supplies the heart with separate nerve fibers that accelerate or decelerate its beat, as Figure 13.16 illustrates. The vagus nerve supplies **parasympathetic neurons** from the cardioinhibitory center in the medulla of the brain to the SA and AV nodes, but some of its nerve fibers also go to the atria and a few to the ventricles. These parasympathetic fibers inhibit the heart rate by causing potassium ions to exit the muscle fibers. Conversely, the cardioaccelerator center in the medulla sends **sympathetic neurons** primarily to the ventricles and atria, where they accelerate heart rate by causing the muscle fibers to become more permeable to sodium and calcium ions. The cardioaccelerator and cardioinhibitory centers themselves respond to receptors in the aorta and carotid artery of the neck that monitor blood pressure. The overall effect is a negative feedback loop that adjusts the heart rate so that when blood pressure rises, the heart rate falls; and vice versa, when blood pressure falls, the heart rate rises. These feedback loops tend to maintain a constant heart rate, but the heart rate may increase to 150 beats per minute with exertion or decline to 50 beats or fewer per minute when you sleep, depending on demands from the body's tissues for oxygen. When oxygen levels in the blood decline, another group of oxygen receptors in the carotid artery accelerates the heart rate by way of the cardioaccelerator center.

ELECTROCARDIOGRAM

The electrocardiogram (ECG) traces events in the cardiac cycle. The peaks and valleys shown in Figure 13.14 represent depolarizations of the SA node and the contraction of the myocardium as recorded by electrodes placed on the patient's chest. As the chart paper rolls beneath the pen, the pen traces a record of these events. From the timing of the spikes, their heights, and their shapes, cardiologists can diagnose many heart conditions.

The **P wave** is the first feature in a normal electrocardiogram. It records the wave of depolarization passing from the SA node across the atria as they contract. The trace remains flat while the AV node sends depolarizations into the bundle branches of the ventricles, but contraction of these muscles records a sharp **QRS complex** of spikes on the paper. The trace remains flat once again until the **T wave** signals that the ventricular muscles are recovering for the next contraction. The atria actually produce a similar recovery wave, but the QRS complex masks it.

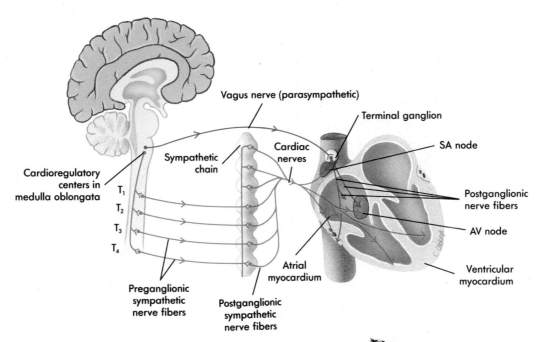

Figure 13.16

Autonomic innervation regulates the heart rate. Sympathetic fibers are shown in *blue,* parasympathetic in *red.*

The P wave signals the beginning of atrial systole, and the QRS signals the beginning of ventricular systole. The heart then enters diastole, its chambers beginning to fill before the T wave, until the next P wave signals that systole has resumed.

DYSRHYTHMIAS

There are many varieties of dysrhythmias, or irregular heart cycles. The ventricles or atria may contract too rapidly or irregularly, or one ventricle may lag behind the other. Some dysrhythmias (dis-RITH-mee-ahs) are life threatening, and others are harmless. The examples below in Figure 13.17 give you an impression of how the relationships between the pacemaker, heart cycle, and ECG may be modified. In **bradycardia** (BRAY-dee-KAR-dee-a; brady, *G,* slow), for example, the pacemaker slows the heart rate and the intervals between P waves on the ECG chart in-

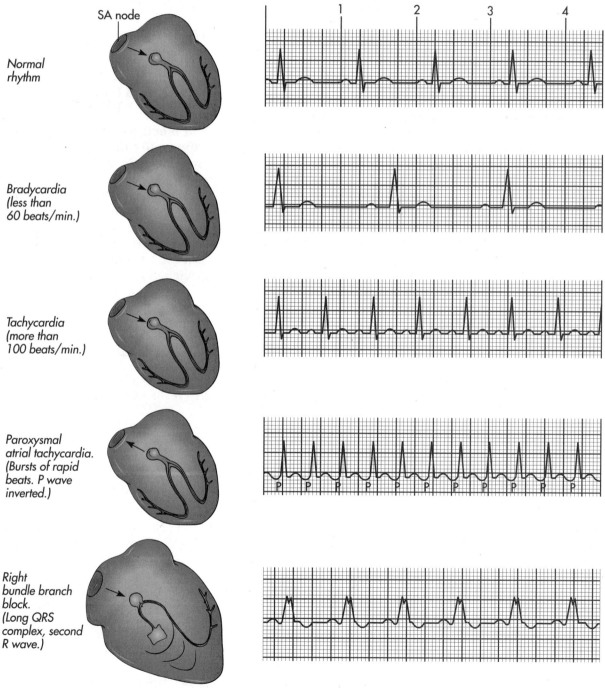

Figure 13.17
Dysrhythmias.

crease accordingly. The heart rate increases in **tachycardia** (TACK-ih-KAR-dee-a; tach-, *G*, fast). In **paroxysmal atrial tachycardia** (PAT; PAR-ox-izz-mal) the heart experiences episodes (paroxysms) of very rapid atrial contraction (tachycardia), sometimes 200 per minute, passing in a few seconds or lasting for hours. P waves occur more frequently and some may be inverted, but the rest of the pattern is normal. PAT is the most common benign dysrhythmia. All three of these situations involve the atria, but the next case is potentially serious because it affects the ventricles. The atrioventricular bundle and the right and left bundle branches may become blocked, delaying depolarization of the ventricles. In **right bundle branch block,** the right hand bundle branch cannot conduct impulses from the AV node, and the ventricle contracts only when depolarization from the left bundle branch reaches the right side, a fraction of a second later. This delay is evident in a longer-lasting QRS complex that frequently displays a second R spike, representing right ventricle depolarization. Right bundle branch block is less dangerous than **left bundle branch block** because the right side of the heart pumps only to the lungs. Left block delays the stronger ventricle that then relies on the weaker right ventricle to stimulate its own contraction. Such blocks can interrupt blood flow through the heart when the ventricles contract independently of the

atria, but this situation can be controlled by implanting an artificial pacemaker to provide normal timing of depolarization.

- What occurs in the atria and ventricles during systole?
- What event in the pulse signals that systole is occurring?
- What part of the ECG trace indicates that ventricular diastole is beginning?
- What structure initiates the heart cycle?
- What types of cells conduct impulses to the ventricles?
- What effect does a right bundle branch block have on contraction of the ventricles?

CORONARY CIRCULATION

The torrent of blood passing through the heart does not directly supply the myocardium because blood cannot cross the endocardium to reach it. Even if direct supply were possible, the right ventricle would still be disadvantaged because it receives deoxygenated blood. Instead, the heart's own coronary circulation, illustrated in Figure 13.18, supplies freshly oxygenated blood to the entire heart wall. Blood enters from two **coronary arteries** that exit the ascending aorta and returns to the right atrium by way

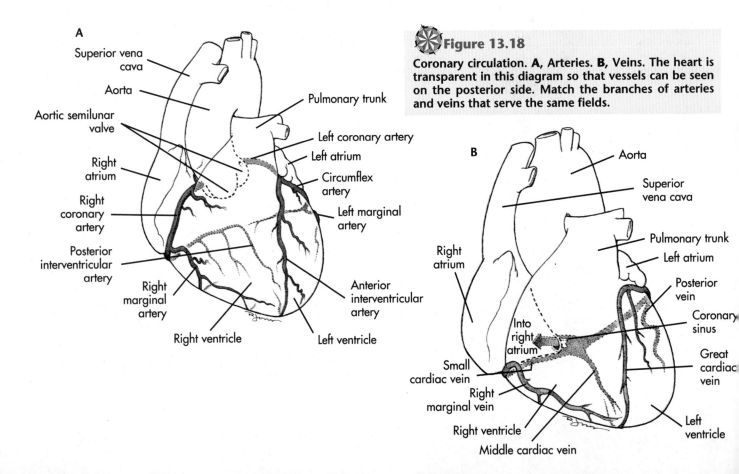

Figure 13.18

Coronary circulation. A, Arteries. B, Veins. The heart is transparent in this diagram so that vessels can be seen on the posterior side. Match the branches of arteries and veins that serve the same fields.

of the **aortic sinus** (SIGN-us). To visualize this anatomy, picture arterial and venous trees wrapped around the heart like a crown, with their trunks in the coronary sulcus and branches reaching into the ventricles and atria.

CORONARY ARTERIES

The **left coronary artery** leaves from the left aortic sinus behind the aortic valve and reaches behind the pulmonary trunk to the coronary sulcus, shown in Figure 13.18A, where it gives off branches to the left atrium and also divides into the **anterior interventricular** and **circumflex branches.** As its name indicates, the first of these branches descends the anterior interventricular sulcus to the apex of the heart, carrying blood for the interventricular septum and adjacent portions of the walls of the left and right ventricles. The circumflex branch also functions as its name implies. It passes posteriorly in the coronary sulcus past the left margin of the heart and flexes downward over the surface of the left ventricle, providing blood to this wall and to the left atrium.

The **right coronary artery** gives rise to three main branches on the right side of the heart. Exiting the ascending aorta from the right coronary sinus, the right coronary artery courses forward into the coronary sulcus and around the right margin of the heart, onto the diaphragmatic surface of the right ventricle. Along the way, it delivers several small branches to the right atrium. It supplies blood to the wall of the right ventricle by way of the **marginal branch** that reaches along the lateral wall of this ventricle to the apex of the heart. The **posterior interventricular branch** proceeds toward the apex in the posterior interventricular groove, providing blood for the interventricular septum and the ventricular walls on each side. The interventricular septum thus receives blood from both coronary arteries by way of their interventricular branches. Similarly, both arteries feed the lateral walls of both ventricles through circumflex and marginal branches.

CORONARY SINUS

In the heart's own circulation, most venous blood returns to the right atrium by way of the coronary sinus, located in the coronary sulcus on the posterior surface of the heart in Figure 13.18B. This large vessel receives blood from tributaries that for the most part follow the branches of the coronary arteries. The **great** and **middle cardiac veins** follow the anterior and posterior interventricular sulci, respectively, from the apex of the heart, where they drain fields supplied by interventricular branches of the coronary arteries. The **small cardiac vein** in the coronary sulcus receives blood from the posterior wall of the right atrium and has no large arterial counterpart. Two other veins on the lateral walls of the heart drain the

fields of the circumflex and marginal branches of the coronary arteries. The **posterior vein of the left ventricle** follows the circumflex branch and drains into the coronary sinus, and the **right marginal vein** receives from the field of the marginal branch of the right coronary artery on the lateral wall of the right ventricle. This vessel drains to the small cardiac vein and thence to the coronary sinus.

- Which coronary artery delivers to the left side of the heart?
- What branches supply blood to the lateral wall of the right ventricle?
- Which vein drains this latter territory?
- What vein collects blood and returns it to the right atrium?

DEVELOPMENT OF THE HEART

MESODERMAL ORIGIN

The heart develops from sheets of mesoderm that transform into a four-chambered heart. (You should review the origin of mesoderm and coelome in Chapter 4 before reading this section.) After gastrulation has inserted the mesoderm between the ectoderm and endoderm of the embryo, two strips of **cardiac mesoderm** develop within the mesoderm, beneath the margins of the neural plate. Figure 13.19 shows that each strip forms a delicate **endocardial tube** and that both tubes then move toward each other and fuse together as one large **cardiac tube.** Two layers of cardiac mesoderm are already present in the cardiac tube; internally, the delicate endocardium will become the inner lining of the heart, and the epicardium and myocardium will develop from the thicker **epimyocardium** on the outside of the tube. Figure 13.20 views the cardiac tube from below and shows the vessels at either end that become the great vessels of the heart. The cardiac tube occupies the **pericardial coelome** (SEE-lome; coel-, *G,* hollow), a space that will become the pericardial cavity. Many changes will take place, however, before definitive chambers and valves appear.

The cardiac tube elongates and bends toward the right side of the embryo (Figure 13.20) and begins to pump blood. Two vitelline veins discharge blood from the yolk sac and placenta into the **sinus venosus.** Immediately anterior to the sinus venosus lie the **atrium** and the **ventricle,** the latter identified by its rightward bend. The tapering **conus arteriosus** (KONE-us; con-, *G,* a cone) discharges blood from the ventricle into the narrower **truncus arteriosus** (TRUNK-us) and thence to the **ventral aorta** that leads to the head and pharyngeal regions. Some of this blood circulates through embryonic vessels back

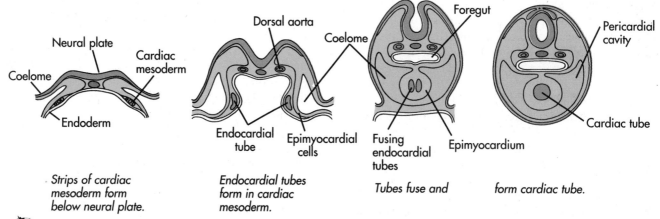

Neural plate

Coelome

Endoderm

Cardiac mesoderm

Dorsal aorta

Coelome

Endocardial tube   Epimyocardial cells

Foregut

Fusing endocardial tubes   Epimyocardium

Pericardial cavity

Cardiac tube

Strips of cardiac mesoderm form below neural plate.

Endocardial tubes form in cardiac mesoderm.

Tubes fuse and

form cardiac tube.

Figure 13.19

Formation of the cardiac tube. Transverse sections, cranial view. Strips of cardiac mesoderm form endocardial tubes that converge, fuse, and become the cardiac tube.

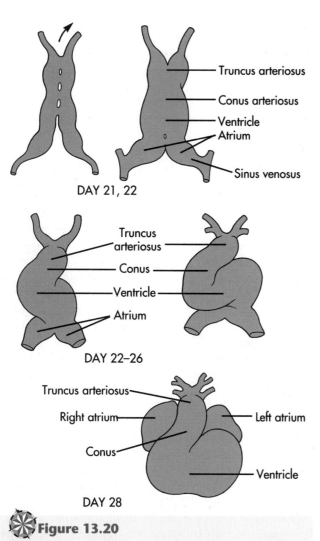

DAY 21, 22

Truncus arteriosus

Conus arteriosus

Ventricle

Atrium

Sinus venosus

DAY 22–26

Truncus arteriosus

Conus

Ventricle

Atrium

Truncus arteriosus

Right atrium

Conus

Left atrium

Ventricle

DAY 28

Figure 13.20

Fusion of the endocardial tubes and development of atria and ventricles. View corresponds to anterior view of the heart in Figure 13.3A.

to the sinus venosus, but more continues to the placenta and yolk sac before returning to the sinus venosus. At this point in development, the heart is a single tube that pumps blood to both left and right sides of the body. True pulmonary circulation awaits the development of the lungs and the changes that will partition the heart into systemic and pulmonary sides.

These changes progress rapidly. During the fourth week, the heart assumes an S-shape, with two atria at one bend and ventricles at the other. Figure 13.21 shows these changes internally from the fifth week of development. The **coronary sulcus** encircles the young heart as the chambers enlarge above and below. The epimyocardium of the ventricle has already thickened, and **trabeculae carnae** are forming. Septa and valves begin to partition the heart as the atria and ventricles expand. The **interatrial septum** appears between the left and right atria. The **endocardial cushion** and **interventricular septum** also begin to separate the ventricle into left and right portions, but both septa leave temporary openings that allow blood to pass between both sides of the heart. Because blood primarily enters the right atrium, the **foramen ovale** (oh-VAL-ee) permits the left atrium to fill until it receives its own flow from the lungs and pulmonary veins at birth. Blood flows from both atria into the ventricles, past the endocardial cushion and the atrioventricular valves that are forming at left and right. Once in the ventricles, blood mixes freely between left and right through the **interventricular foramen.** Although anatomically four-chambered, the heart still is functionally one-sided, a systemic heart, until the foramina close (see Changes at Birth, in this chapter).

Separation into functional pulmonary and systemic circuits nevertheless is underway, as shown in

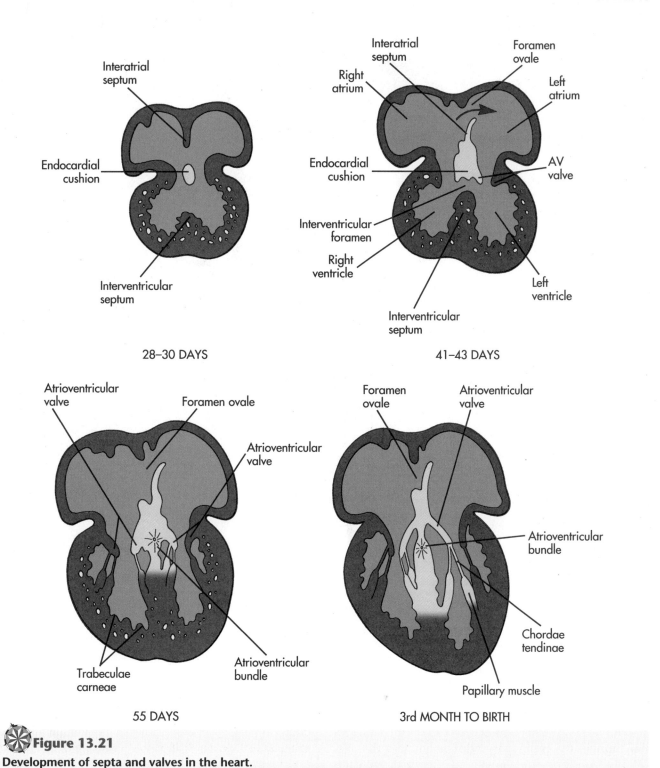

Figure 13.21
Development of septa and valves in the heart.

Figure 13.22. The interventricular foramen closes during the third month of development, forcing blood from each ventricle separately into the conus arteriosus. In the conus and truncus arteriosus, two parallel ridges called the **bulbar septum** (BUL-bar) further separate these torrents into pulmonary and systemic flows that spiral around each other in the pulmonary trunk and the ascending aorta. Later, when the **pulmonary trunk** and **ascending aorta** become separate vessels, they remain twisted around each other (Figure 13.3B) because of the spiral path taken by the bulbar septum .

During the third month, aortic and pulmonary valves develop on opposite sides of the bulbar septum

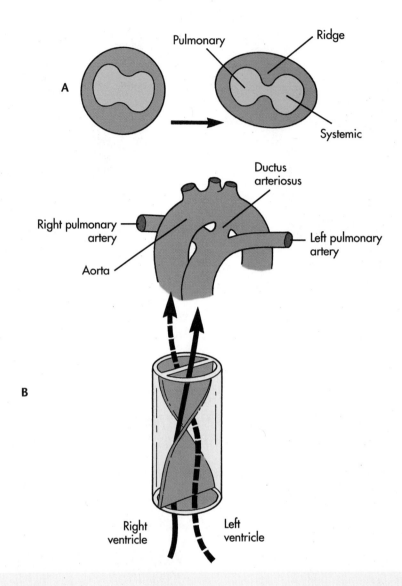

Figure 13.22

The bulbar septum partitions the conus arteriosus and truncus arteriosus into the pulmonary trunk and ascending aorta. **A,** Ridges form the bulbar septum in the truncus and conus arteriosus. **B,** The bulbar septum turns 180 degrees, twisting the pulmonary trunk to the left and the aorta to the right.

at the exits from the ventricles. Because the lungs are still only rudimentary, most pulmonary blood flows not through the pulmonary arteries but into the systemic aorta through a connection known as the **ductus arteriosus.** Until the ductus closes at birth, it allows the right side of the heart to mature without sending blood through the lungs. Only after the ductus and the foramen ovale close does the heart become a fully functional, two-sided pump.

CHANGES AT BIRTH

The pulmonary and systemic sides of the heart become separate at birth. Figure 13.23 shows the changes that completely partition the heart into pulmonary and systemic sides. (Other circulatory changes that disconnect the fetus from maternal circulation are discussed in Chapter 14, Vessels.) Within minutes following birth, smooth muscle in the ductus arteriosus constricts the opening to the aorta, now directing all blood into the pulmonary arteries where the lungs oxygenate it. Connective tissue seals off the ductus altogether during the next 6 to 8 weeks, leaving a rudimentary connection, the **ligamentum arteriosum.** The foramen ovale also closes, leaving only a slight depression formed by the **fossa ovalis** to indicate that a passage once existed. Since blood can no longer mix between left and right sides, deoxygenated blood now enters the right atrium and leaves from the right ventricle. It returns, oxygenated, to the left atrium and exits the left ventricle, passing out the as-

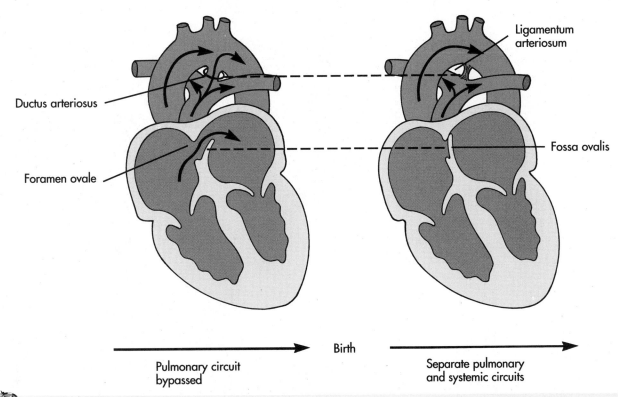

Ligamentum
arteriosum

Ductus arteriosus

Foramen ovale

Fossa ovalis

Birth

Pulmonary circuit
bypassed

Separate pulmonary
and systemic circuits

Figure 13.23

The ductus arteriosus and foramen ovale close at birth, finally separating the pulmonary from the systemic circulation.

cending aorta to the system. The pulmonary and systemic sides of the heart are now separate.

- From what germ layer does the heart develop?
- When do the two endocardial tubes fuse together as one?
- What chamber delivers blood to the atria, and which section of the heart receives blood from the ventricle?
- What openings allow blood from the right side of the heart to mix with that of the left side?
- What changes finally separate the pulmonary and systemic sides from each other?

CONGENITAL DEFECTS

Several common heart defects may lead to heart failure in early childhood or later life. Each can have many complications, and only the simplest examples are described here.

PATENT DUCTUS ARTERIOSUS

In this condition, the ductus arteriosus does not close. It remains open (PAY-tent; paten-, *L*, open), and some of the oxygenated blood that should flow out the aorta crosses into the pulmonary circulation for an unnecessary trip through the lungs (Figure 13.24B),

compared with the normal circuit shown in Figure 13.24A. This diversion can cause the left side of the heart to fail as it attempts to keep up with the demand for oxygen, and fluid may accumulate in the lungs. The circumstances of failure vary considerably, and a patent ductus may not be detected until adulthood. Surgical closure of the ductus is a safe and relatively simple cardiac operation.

ATRIAL SEPTAL DEFECTS

In the most common atrial septal defect, the foramen ovale remains open, and blood passes from the left atrium, where pressure is higher, into the right atrium, as shown in Figure 13.24C. Blood and extra fluid therefore accumulate in the lungs, and the right ventricle enlarges, but the heart seldom fails. Repair of this relatively simple defect is quite safe, although the prognosis in larger defects is less optimistic.

VENTRICULAR SEPTAL DEFECTS

When the interventricular septum fails to close, as in Figure 13.24D, holes remain in the septum, and blood flows into the right ventricle from the higher-pressure left ventricle. The effect is similar to that of atrial defects; the pulmonary circulation engorges with blood, and the left side of the heart may fail.

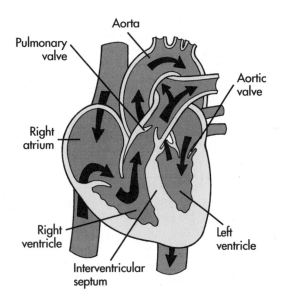

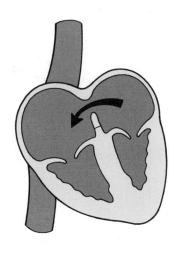

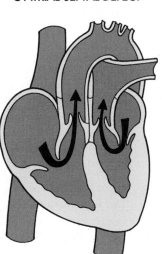

A NORMAL HEART

Aorta

Pulmonary valve

Aortic valve

Right atrium

Right ventricle

Left ventricle

Interventricular septum

B PATENT DUCTUS ARTERIOSUS

Ductus arteriosus

C ATRIAL SEPTAL DEFECT

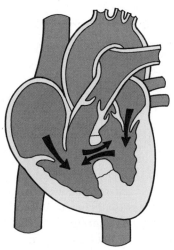

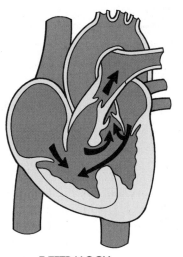

D VENTRICULAR SEPTAL DEFECT

E TETRALOGY OF FALLOT

F TRANSPOSITION OF GREAT VESSELS

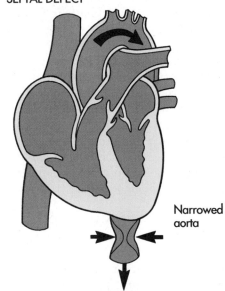

G COARCTATION OF AORTA

Narrowed aorta

Figure 13.24

Congenital defects of the heart (see discussion in text). **A**, The normal heart. **B**, Patent ductus arteriosus. **C**, Atrial septal defect. **D**, Ventricular septal defect. **E**, Tetralogy of Fallot. **F**, Transposition of the great vessels. **G**, Coarctation of the aorta.

Small holes usually have little effect, but larger holes impose extra loads on the pulmonary circulation and the left ventricle as it attempts to satisfy the demand for blood.

TETRALOGY OF FALLOT

In this most common anomaly among "blue babies," deoxygenated blood flows into the aorta and newborns are cyanotic because their tissues receive too little oxygen (Figure 13.24E). Four anomalies (tetra, *G,* four) drive deoxygenated blood from the right ventricle into the aorta. An interventricular septal defect (1) shifts the aorta toward the right ventricle (2), and stenosis narrows the pulmonary valve (3), diverting blood into the aorta instead. The right ventricle enlarges (4) as it attempts to pump against the narrow exit and high pressure from the left ventricle. Tetralogy is difficult to repair surgically, but surgery can relieve cyanosis by redirecting blood from the aorta back to the pulmonary artery through an implanted bypass.

TRANSPOSITION OF THE GREAT VESSELS

Only rarely do the pulmonary trunk and ascending aorta switch positions during development, but when they do (Figure 13.24F), the left ventricle pumps blood to the pulmonary trunk and the right ventricle sends deoxygenated blood to the aorta. A disaster results; the body receives deoxygenated blood, and the left side of the heart cycles oxygenated blood back to the lungs. Septal defects are a benefit in these babies because oxygenated blood can reach the tissues by leaking into the right side of the heart. Transposition is caused when the bulbar septum fails to twist. Lacking any twist, the right ventricle connects to the aorta and the left ventricle joins to the pulmonary trunk.

COARCTATION OF THE AORTA

In this condition (Figure 13.24G), a section of the aorta narrows and the left ventricle sustains extra pressure in pumping against this resistance. The left side of the heart will fail if the aorta has narrowed considerably. Surgical correction removes the obstructed portion and rejoins the cut ends or inserts an artificial section, if the ends are too short.

- Which congenital anomaly allows blood to flow from the aorta into the pulmonary trunk?
- Which congenital defect is responsible for many "blue babies"?
- What events have failed to occur in ventricular and atrial septal defects?
- In which conditions does the left ventricle pump oxygenated blood to the lungs?

AGING

Surgeons cannot easily deduce ages from the anatomy of individuals' hearts because remarkably few changes occur. The wall of the left ventricle thickens, but the heart's resting capacity does not change. Reserve capacity and maximum heart rate decline, which is probably related to cell losses in the conduction system. The sinoatrial node begins to lose cells at about 60 years, and only 20% may remain by 75 years of age. Cells of the atrioventricular bundle and Purkinje cells also diminish, while fat and collagen fibers accumulate. These changes are associated with dysrhythmias and bundle blockages. Collagen fibers and calcium deposits appear in the heart valves, which can lead to stenosis or regurgitation and heart failure. Coronary artery disease, heart attack, congestive heart failure, and disorders of the valves are the principal cardiac disorders of elderly people.

SUMMARY

Figure 13.25 summarizes the development and action of the heart. The two-sided heart develops from a single tube that folds and forms four chambers that alternately fill and empty during the heart cycle. The sinoatrial node initiates systole, which is fol-lowed by diastole, as the elements of the ECG trace and the arrows indicate. Dysrhythmias in the atria disturb the heart cycle, and bundle branch blocks may delay ventricular contraction. Hearts fail when one or both ventricles pump inadequate quantities of blood, when stenosis blocks valves, or when valves leak.

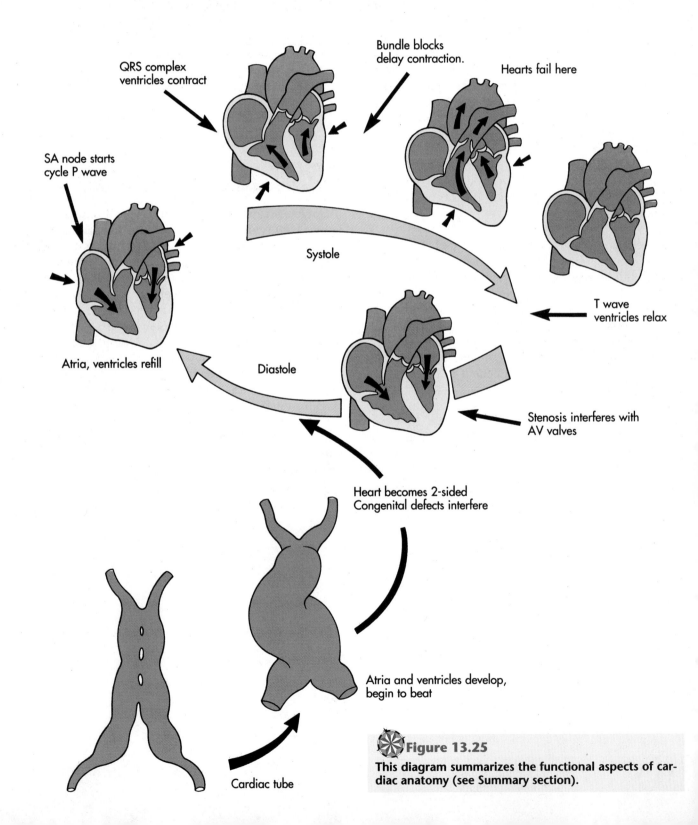

QRS complex
ventricles contract

Bundle blocks
delay contraction.

Hearts fail here

SA node starts
cycle P wave

Systole

T wave
ventricles relax

Atria, ventricles refill

Diastole

Stenosis interferes with
AV valves

Heart becomes 2-sided
Congenital defects interfere

Atria and ventricles develop,
begin to beat

Cardiac tube

Figure 13.25

This diagram summarizes the functional aspects of car-diac anatomy (see Summary section).

The heart is a double pump, two pumps side-by-side. Blood flows through four chambers—from the right atrium, to the right ventricle, out to the lungs, back to the left atrium, to the left ventricle, out to the body tissues, and back to the right atrium. The right, or pulmonary, side of the heart receives deoxygenated blood from the tissues and sends it to the lungs for oxygenation. From there it returns to the left, or systemic, side and is pumped to the systems of the body. *(Page 381.)*

Blood flows from high pressure in the ventricles, through the vessels, to low pressure in the atria. The heart valves oblige blood to circulate through the vessels by preventing backflow from ventricles into the atria. *(Page 382.)*

Failing hearts are unable to supply adequate blood. In congestive heart failure, fluids accumulate in the lungs and tissues because the weakened heart muscle cannot pump enough blood to carry these fluids from the tissues. In left heart failure, fluid accumulates in the lungs; when the right side fails, fluid collects in the body tissues. *(Page 382.)*

The heart is suspended in the pericardial cavity behind the sternum, between the second and sixth ribs. The pericardial cavity allows the heart free movement. The base of the heart, where the great vessels enter and leave, suspends the heart by the posterior walls of the atria. The ventricles form the apex of the heart, and two grooves—the coronary sulcus and interventricular groove—identify the location of atrioventricular valves and interventricular septum. *(Page 383.)*

Thin-walled, low pressure atria receive blood from the veins, and thick-walled, high pressure ventricles expel blood from the heart into the arteries. The right atrium receives deoxygenated blood from the inferior and superior venae cavae and from the coronary sinus, and it sends this blood through the right AV valve into the right ventricle. This ventricle pumps blood through the pulmonary semilunar valve into the pulmonary trunk and to the lungs. Oxygenated blood returns from the lungs into the left atrium and passes the left AV (mitral) valve into the left ventricle. This most muscular of chambers sends the blood through the aortic semilunar valve into the ascending aorta. The systemic side of the heart is more powerful than the pulmonary side. *(Page 385.)*

The skeleton of the heart supports the valves with four fibrous rings. Atrioventricular valves have two or three cusps, and both semilunar valves have three cusps. Papillary muscles and chordae tendineae prevent blood from regurgitating past the cusps of the AV valves. Semilunar valves have no chordae tendineae; instead, blood pressure jams the cusps of the valve shut. Scar tissue can narrow valve openings (stenosis) or cause regurgitation when the scarred cusps leak. Valve sounds are heard best at the four corners of the heart. *(Page 386.)*

Endocardium, myocardium, and epicardium compose the heart wall. Endocardium lines the interior, epicardium covers the exterior, and the myocardium is the muscular layer responsible for contraction. Bundles of muscle fibers curve and loop across both ventricles from the fibrous skeleton of the heart. These bundles are networks of branching cardiac muscle fibers held together by endomysium and basement membranes. Cardiac muscle fibers can enlarge, but they cannot divide or regenerate. *(Page 390.)*

The anatomy of cardiac muscle fibers reflects their ability to contract repeatedly and their demand for oxygen. Cardiac fibers are cross-striated, involuntary, and rich with mitochondria. In further distinction from skeletal muscle fibers, cardiac fibers branch and intercalated disks connect the cells end to end. Stimulated to contract by extracellular calcium, the sarcoplasmic reticulum triggers contraction by releasing calcium ions internally. *(Page 391.)*

The atria and ventricles contract during systole and fill during diastole. The heart cycle begins in diastole, while the atria are filling with blood and the ventricles are relaxing. Systole begins when the atria contract and fill the ventricles. Then the ventricles contract, closing the AV valves and opening the semilunar valves. As blood rushes from the ventricles and they begin to relax, the semilunar valves snap closed and systole comes to an end. Diastole resumes, the AV valves open, and all chambers fill. *(Page 393.)*

The sinoatrial node initiates the heart cycle, and the autonomic nervous system accelerates and decelerates it. The sinoatrial node depolarizes the walls of the atria and causes them to contract. The atrioventricular node, regulated by the sinoatrial node, sends stimuli via specialized conducting fibers and Purkinje fibers to the myocardium of the ventricles, which contracts. *(Page 395.)*

Electrocardiograms record systole and diastole. The P wave records the depolarization of the atria, the QRS complex records the depolarization of the ventricles, and the T wave signals repolarization of ventricular myocardium for the next cycle. Each depolarization triggers contraction of the myocardium. Tachycardia is an increase in the heart rate, and bradycardia is a decrease. Paroxysmal atrial tachycardia involves premature contraction of the atria, and bundle blocks delay ventricular contraction by blocking the Purkinje fibers. *(Page 396.)*

Coronary arteries and veins deliver blood to the heart itself. Two coronary arteries conduct oxygenated blood from the aorta to branches that serve the left and right sides of the heart. Tributaries from these regions of the heart wall drain into the coronary sinus that returns venous blood to the right atrium. When an obstruction blocks the coronary arteries or their branches, the myocardium may form infarcts, patches of dead tissue downstream from the blockage. *(Page 398.)*

The heart develops from cardiac mesoderm. Atria and ventricles develop from a pair of endocardial tubes that fuse as one. The interatrial and interventricular septa divide the atria and ventricles left and right, but these partitions remain open, allowing blood entering the right atrium to reach both sides of the heart. AV valves develop between the atria and ventricles, and aortic and pulmonary valves develop at the exit from the ventricles into the truncus arteriosus. *(Page 399.)*

Separate pulmonary and systemic sides of the heart begin to function at birth. Even though the interventricular foramen has been closed since the third month of development, the pulmonary circulation remains rudimentary until the baby begins to breathe. At this time the foramen ovale and ductus arteriosus close, directing blood into the lungs for oxygenation. *(Page 400.)*

Congenital defects retain some embryonic features of the heart. Patent ductus arteriosus and defects of the atrial and ventricular septa retain the fetal openings that allow oxygenated blood to mix with deoxygenated blood in the pulmonary circuit. Tetralogy of Fallot causes deoxygenated blood to enter the systemic circuit; such babies suffer cyanosis ("blue baby" syndrome). When the pulmonary trunk and aorta are transposed, oxygenated blood cannot reach the tissues because it never leaves the pulmonary circuit, and deoxygenated blood never reaches the lungs. *(Page 403.)*

Aging introduces few overt anatomical changes in the heart, but its conduction system deteriorates and collagen fibers and calcium deposits may accumulate in the valves. The incidence of heart attack, coronary artery disease, dysrhythmias, and heart failure increases with age. *(Page 405.)*

STUDY QUESTIONS
OBJECTIVES

1. Beginning with the right atrium, list in sequence the chambers through which blood flows in the heart. **1**
2. Heart failure can cause fluids to accumulate in the tissues. How can you tell from the location of this fluid whether the left or right side of the heart is failing? **2**
3. If you were asked to draw the image of the heart on a diagram of the thorax, which landmarks would you use to locate it? Where are the base and apex of the heart located? Where are the coronary sulcus and interventricular groove located on the heart?**3**
4. What anatomic and functional features distinguish the atria from the ventricles? Describe anatomic and functional differences between the pulmonary and systemic sides of the heart. What is the skeleton of the heart? In which respects are the four valves similar to and different from each other? How would you distinguish a pulmonary valve from a mitral valve? Where, specifically, on the chest would you place a stethoscope to hear the sounds of each valve? **4**
5. Identify the cardiac structures shown in a transverse section through the heart at the level of the atria in Figure 13.26. **4**

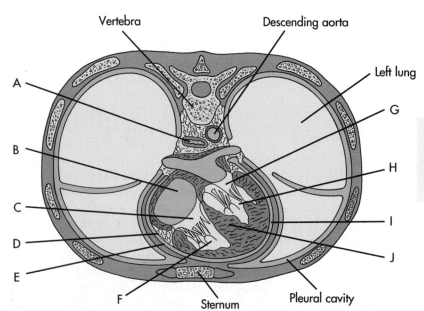

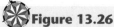**Figure 13.26**

Transverse section through the atria and ventricles of the heart. Identify the lettered structures from descriptions in the text. Answers (upside down) on page 409.

STUDY QUESTIONS cont'd   OBJECTIVES

6. Describe the three tissue layers of the heart. **5**
7. Cardiac muscle fibers are unique. What anatomic features of these cells set them apart from other muscle fibers? How are mitochondria and intercalated disks related to the ability of cardiac muscle tissue to beat repeatedly? **6**
8. What events occur during atrial systole, ventricular systole, and diastole? Decide which chambers fill and empty, what valves open and close, and what valve sounds are made. **7**
9. Make three readings of your pulse while sitting down, then exercise vigorously for 5 minutes and make three readings again. How much has your pulse changed? **8**
10. Follow the wave of depolarization from the sinoatrial node over the walls of the atria to the AV node, bundle branches, and the ventricles. Describe the events that cause the myocardium of the atria and ventricles to contract. **9**
11. What fundamental events of the heart cycle do the P wave, QRS complex, and T wave signify to someone reading an electrocardiogram? What effects on the ECG and pulse would tachycardia and bradycardia have? **10**

12. Follow a drop of blood from the aorta, into the heart wall, and back to the right atrium. What arteries and veins might it traverse to reach the left and right ventricles and the interventricular septum? What region of the heart would blockage of the circumflex and right marginal arteries affect? Match the veins with the fields of the arteries they drain. **11**
13. What developmental changes convert the cardiac mesoderm into a single cardiac tube? Name the embryonic chambers of the heart in order of blood flow. What adult structures do they form? What structures separate the right side of the heart from the left? From what structures do papillary muscles derive? **12**
14. What events finally separate the pulmonary and systemic circuits at birth? **13**
15. Describe the abnormality and effect on circulation of the following: patent ductus arteriosus, atrial septal defect, ventricular septal defect, Tetralogy of Fallot, transposition of the great vessels, and coarctation of the aorta. **14**
16. Describe conditions that affect the hearts of aging people. **15**

Figure 13.26, cont'd

Answers:
A. Esophagus
B. Right atrium
C. Right atrioventricular (tricuspid) valve
D. Epicardium, visceral pericardium
E. Parietal pericardium
F. Right ventricle
G. Left atrioventricular (mitral) valve
H. Left ventricle
I. Pericardial cavity
J. Interventricular septum

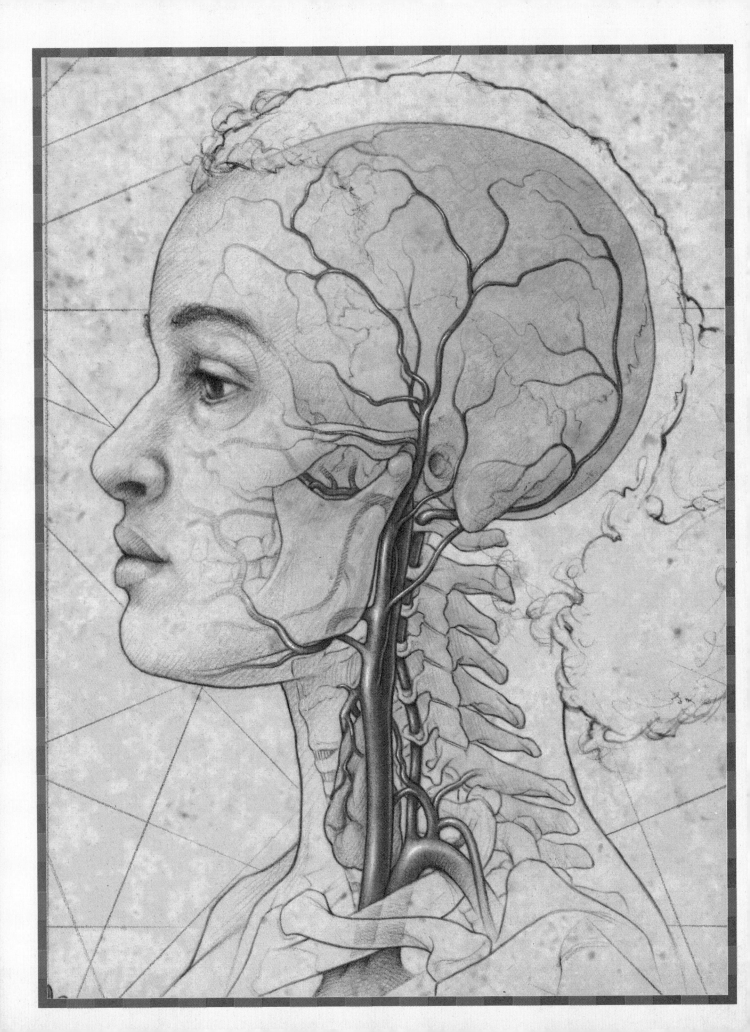

CHAPTER 14

Vessels

OBJECTIVES

After studying this chapter, you should be able to do the following:

1. Recognize and describe the various types of vessels that make up the vascular tree.
2. Describe the characteristics and functions of the tunics of the arteries, veins, and capillaries.
3. Relate differences between arteries and veins to the pressure of blood.
4. Describe the effects of atherosclerosis on the tunics of arteries and on circulation.
5. Describe the anatomy by which oxygen, glucose, and carbon dioxide molecules cross capillary walls.
6. Trace the flow of blood through the vascular tree. You should be able to start in the heart and name in proper order the vessels that supply blood to any organ and those that return it to the heart.
7. Starting in any organ, trace a pathway through which blood can reach any other organ.

 **VOCABULARY TOOLBOX**

Directional terms.

circumflex	circum-, *L* around; -flex, *L* bend; fold or bend around. The humeral circumflex artery bends around the head of the humerus.	**sub**	*L* below. The subclavian artery is beneath the clavicle.
retro	*L* back, behind. The retromandibular artery is behind the mandible.	**supra**	*L* above. The suprarenal artery is superior to the kidneys.

Frequently encountered vessels.

aorta	aort-, *G* the greatest artery.	**jugular**	jugul-, *L* the throat, collarbone
artery	arteri-, *G* an artery.	**portal**	port-, *L* carry. The hepatic portal vein carries blood from gut to liver.
carotid	carot-, *G* stupor, deep sleep. It was thought that the carotid arteries induced sleep.	**saphenous**	saph-, *G* clear, apparent. The small saphenous vein is easily seen on the calf of the leg.
celiac or coeliac	coelia, *G* the abdominal cavity. The celiac trunk delivers to digestive organs.	**tunica**	tunic-, *L* a covering, cloak. The tunica adventitia covers blood vessels.
ductus	arteriosis or venosus; duct-, *L* to lead.	**vein, venous**	ven-, *L* a vein.

RELATED TOPICS

- Blood, Chapter 15
- Development, Chapter 4
- Heart and coronary circulation, Chapter 13
- Surface-volume problem, Chapter 1
- Systemic and pulmonary circulation, Chapter 13

PERSPECTIVE

It is important to know how blood flows between the heart and the organs in order to understand how the cardiovascular system links the body together. Vascular pathways knit the body together, transporting nutrients and delivering stimuli that coordinate the various organs into one active organism.

Vessels are medically important because they allow physicians to trace the causes of disease and to treat them. Blood samples are withdrawn in order to diagnose disease in organs that release their products into the blood, and injections are given to deliver drugs to the site of disease. When vessels are impaired, there can be severe consequences. Strokes can occur when obstructions block the flow of blood to the brain, and heart attacks can occur when coronary arteries are blocked.

The heart pumps blood through the branches of a great, vascular tree to nearly every tissue in the body. Blood rushes away from the heart through an arterial (ar-TEER-ee-al) tree, and returns through the venous (VEEN-ous) tree, illustrated in Figures 14.1A and B. In this way, the cardiovascular system connects the organ systems of the body. A drop of blood flowing from the heart, will follow first the trunk of the arterial tree, then branches, and finally small twigs, until it reaches an organ, perhaps the brain, where its erythrocytes discharge oxygen acquired from the lungs and take up carbon dioxide. The blood may take up other essential substances, such as gonadotropin hormones from the pituitary gland. Returning from the head, this blood mixes with blood from other parts of the body and enters the heart. On the next cycle, chance may find the same drop of blood in another branch of the tree, delivering oxygen to another organ, perhaps the ovary or testis, where the gonadotropins from the brain stimulate egg or sperm development.

Blood vessels connect virtually every organ with every other organ through links like these. It takes about 1 minute for an average drop of blood to make a complete circuit out from and back to the heart. As you may guess, its trip can be longer or shorter, depending on the actual pathway taken.

Each year in the United States some 700,000 people die when their blood flow is blocked and essential tissues downstream die from lack of oxygen. Cardiovascular disease is this country's number 1 killer, accounting for 48% of all fatalities. Most deaths occur when a plugged coronary artery precipitates a heart attack; strokes kill many more when a major artery of the brain is blocked. This chapter introduces the pathways that circulate blood through different regions of the body. When you have finished it, you will be able to follow the course of blood from any organ in the body to any other. You also will begin to understand the structure of arteries and veins and how cardiovascular disease interferes with circulation.

Case study begins on p. 419.

VASCULAR TREES

Vascular trees consist of arteries, capillaries, and veins. Arteries of course distribute blood from the heart to the tissues (Figure 14.2A), where thin-walled capillaries (KAP-ill-air-eez) communicate more directly with the cells. Veins collect blood from the capillaries and return it to the heart. The aorta, shown in Figure 14.1A, is the trunk of the tree, conducting blood to different regions of the body, where smaller arteries branch into the local organs and finally branch into small arterioles (ar-TEER-ee-oles), the twigs of the tree, that deliver blood to the capillaries. To describe this progression, we say that arteries branch, distribute, send, or deliver blood to the capillaries, as if the blood were flowing up the trunk and out the branches of a tree. Arteries are usually located deep within the body and appear near its surface only near joints.

Capillaries are the business end of the cardiovascular system. They are fine networks (beds) of minute vessels that take up and deliver materials across truly immense surface areas. In the lungs alone, approximately 140 m² (1500 ft²) of capillary surface, about two thirds the area of a tennis court, take up oxygen and release carbon dioxide. Think of these areas as fields supplied by arteries and drained by veins.

Blood pressure is very much lower in the veins than in the arteries and capillaries, a difference reflected in the thin, readily collapsible walls of these vessels. Small venules (VEN-yools) drain the capillaries and become tributaries that, like streams and rivers, converge with others to form larger veins. These veins in turn receive other tributaries, and ultimately the largest veins deliver to the heart. We say that veins join, coalesce, or unite with others to drain or receive blood, as if you were climbing back down a tree to the trunk. Major veins and arteries frequently accompany each other, side by side, deep in the body, as you can see by matching the patterns of arteries and veins in Figure 14.1. Look at the femoral vein of the thigh (Figure 14.1B), for example, and you will see that it occupies the same location as the femoral artery (Figure 14.1A). Match the course of the aorta with that of the superior and inferior venae cavae (VEEN-ee KAVE-ee) in the same figure. Many other veins are superficial and can easily be seen beneath the skin. Their accessibility and low blood pressure allow veins to be injected with medication or sampled for blood.

BLOOD PRESSURE AND FLOW

Blood flows from high to low pressure. Blood pressure declines as it flows through the tree in Figure 14.2B, and the pulse disappears as blood approaches the capillaries. These changes are related to the sur-face area of vessels exposed to the blood. As branches become smaller and more numerous, their total surface area increases, something you can judge by comparing the surface of all capillaries in Figure 14.2 with the surface of the aorta that supplies them. Friction from the linings of these vessels, especially in the arterioles, dissipates this pressure and the pulse. The blood returns to the heart at zero pressure relative to the surrounding tissues. At any time, fully 64% of the blood is flowing in the veins but only 15% is moving through the arteries.

Blood flows most slowly in the capillaries because the capillaries have the largest total cross-sectional area. Imagine that all the capillaries supplied by the aorta were packed together side by side. Their entire cross-sectional area would be several thousand times larger than the aorta itself. This situation is like water flowing through a narrow tube into a wide pipe, illustrated Figure 14.2D. The same amount of fluid that zips through the tube slows dramatically when it spreads out in the pipe. Blood flow speeds up somewhat in the veins because the cross-sectional area decreases.

- Why are vascular trees so named?
- What are the names of the different sizes of arteries and veins in the vascular tree, and where in the tree are they found?
- Which vessels sustain the highest blood pressure? The lowest?
- Where is the greatest surface area found in the tree?

VASCULAR TUNICS

Three tissue layers called tunics are responsible for the special features of arteries, capillaries, and veins (Figure 14.3). The **tunica intima** (TUNE-ik-ah IN-tim-ah), or tunica interna, lines the lumen of these vessels with a thin endothelium that intimately contacts the blood. The **tunica media,** the middle layer, is composed of collagen fibers and various quantities of elastic and smooth muscle fibers. This layer accounts for many of the differences between larger and smaller arteries and between the structure of arteries and veins. Externally, vessels are covered with connective tissue called the **tunica adventitia** (tunica externa) that connects the vessels to the surrounding tissues. Table 14.1 summarizes the tunics throughout the vascular tree.

TUNICA INTIMA

The tunica intima consists of a thin endothelium underlain with a basement membrane and supported by **subendothelial connective tissue** and a layer of elastic fibers known as the **internal elastic lamina** (Figure 14.3A). The **endothelium** is a single layer of

Text continued on p. 418.

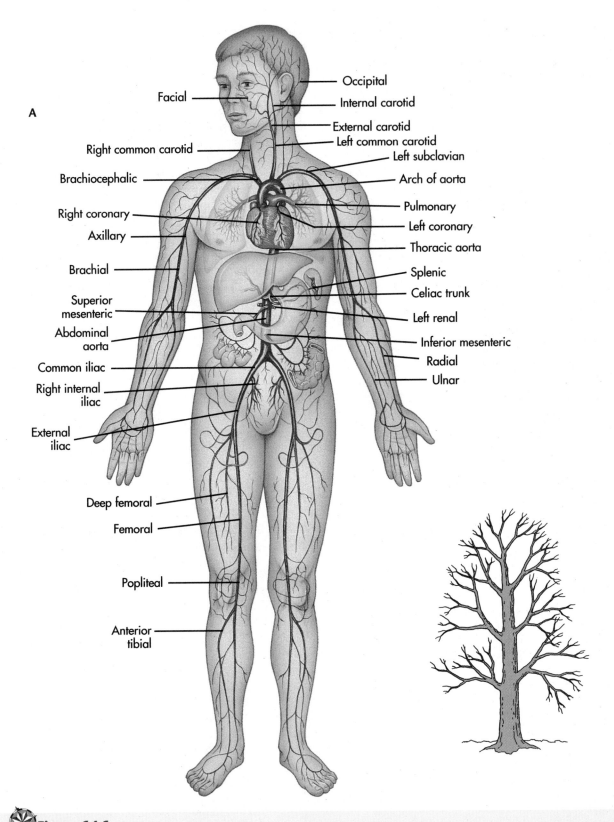

A

Occipital

Facial

Internal carotid

External carotid

Left common carotid

Right common carotid

Left subclavian

Brachiocephalic

Arch of aorta

Right coronary

Pulmonary

Axillary

Left coronary

Thoracic aorta

Brachial

Splenic

Superior mesenteric

Celiac trunk

Abdominal aorta

Left renal

Common iliac

Inferior mesenteric

Right internal iliac

Radial

Ulnar

External iliac

Deep femoral

Femoral

Popliteal

Anterior tibial

Figure 14.1

A, Arterial tree. **B,** Venous tree. The major arteries reach from the heart to all regions of the body, and veins return to the heart. Many veins and arteries accompany each other; compare the location of the femoral artery with that of the femoral vein, for example.

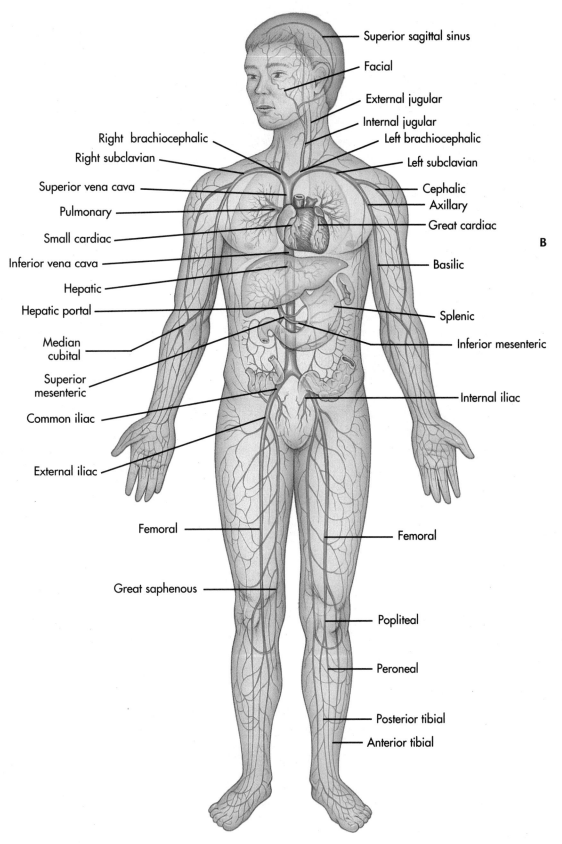

Superior sagittal sinus

Facial

External jugular

Internal jugular

Right brachiocephalic

Left brachiocephalic

Right subclavian

Left subclavian

Superior vena cava

Cephalic

Pulmonary

Axillary

Small cardiac

Great cardiac

Inferior vena cava

B

Hepatic

Basilic

Hepatic portal

Splenic

Median cubital

Inferior mesenteric

Superior mesenteric

Internal iliac

Common iliac

External iliac

Femoral

Femoral

Great saphenous

Popliteal

Peroneal

Posterior tibial

Anterior tibial

Figure 14.1, cont'd

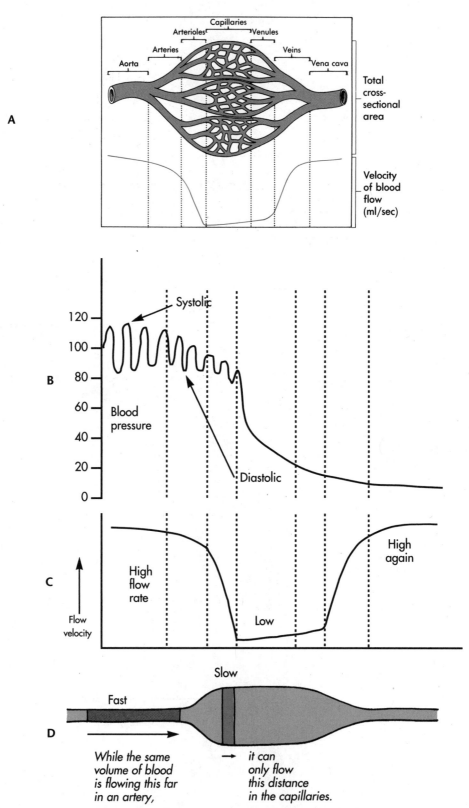

Aorta

Arteries

Arterioles

Capillaries

Venules

Veins

Vena cava

Total cross-sectional area

Velocity of blood flow (ml/sec)

A

Systolic

Blood pressure

Diastolic

B

High flow rate

Low

High again

Flow velocity

C

Fast

Slow

While the same volume of blood is flowing this far in an artery,

it can only flow this distance in the capillaries.

D

Figure 14.2

Structure and blood flow in the vascular tree. A, Divergent branches lead arteries from the heart to the capillaries, and convergent veins return to the heart. **B,** Blood flows from high to low pressure. **C,** Blood flows most slowly through the capillaries because they have the largest cross-sectional area. **D,** This situation resembles flow from a narrow tube into a wide pipe.

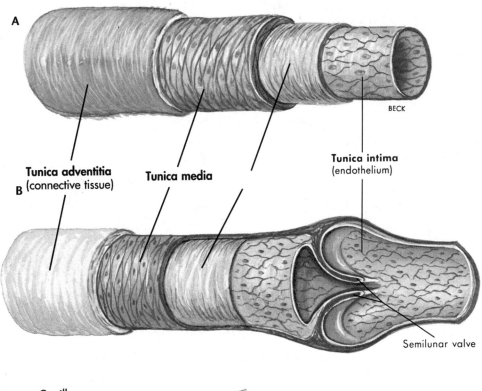

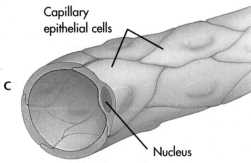

A

Tunica adventitia
B (connective tissue)

Tunica media

Tunica intima
(endothelium)

BECK

Semilunar valve

Capillary
epithelial cells

C

Nucleus

Figure 14.3

Tunics of the vessels. **A,** An artery with tunica intima, tunica media, and tunica adventitia. **B,** The tunica media of veins is much thinner than in arteries of the same size. **C,** Capillaries consist of tunica intima.

Table 14.1 TUNICS OF THE VESSELS

Tunic	Arteries, Arterioles	Capillaries	Venules, Veins
T. INTIMA			
Endothelium	Continuous	Continuous, fenestrated, or discontinuous	Continuous
Basement membrane	Continuous	Continuous, discontinuous, or absent	Continuous, very thin in veins
Subendothelial layer	Thick connective tissue (thin in arterioles); longitudinal muscle (absent in arterioles)	Connective tissue; pericytes present	Thin connective tissue
Internal elastic lamina	Present; fenestrated in arterioles	Absent	Absent in venules, present in veins
T. MEDIA			
Elastic fibers	Abundant in elastic arteries; reduced in muscular arteries, absent in arterioles	Absent	Absent
Smooth muscle fibers	Abundant; 30-40 layers in muscular arteries, 1-2 layers in arterioles	Absent	Absent or few thin layers
External elastic lamina	Present	Absent	Absent
T. ADVENTITIA			
Adventitia	Thick in muscular arteries, thin in elastic arteries, arterioles; longitudinal muscle; nerves; vasa vasorum except arterioles	Thin connective tissue	Connective tissue thin in venules to very thick in large veins; nerves; vasa vasorum in veins.

From Simionescu N and Simionescu M: The cardiovascular system. In Wiess L, editor, Cell and tissue biology: a textbook of histology, Baltimore, 1988, Urban and Schwarzenberg.

squamous cells, 0.2 to 0.4 µm thick, that adhere tightly to each other and to the basement membrane, lest blood peel them away like shingles from a roof in a storm. Flowing blood also shapes these endothelial cells like spindles, with their longitudinal axes parallel to the flow. When they are experimentally placed crosswise, the cells reorient parallel to the flow. A slick **glycocalyx** (gly-ko-KAIL-iks) made by the endothelium prevents blood from clotting on the endothelium. It is important that the endothelium remain intact because thrombi (THROM-by; thromb-, *G*, blood clots; thrombus, singular) form on the underlying tissue when the epithelium is damaged. Endothelial cells rarely divide in mature vessels, but they will do so in damaged or growing vessels.

Endothelial cells secrete a number of products that influence circulation. Among them are enzymes that regulate blood pressure by activating angiotensin and inactivating the neurotransmitter substances—norepinephrine and serotonin. The endothelium also synthesizes prostaglandins, collagen, and blood group antigens. Various binding sites on its surface attach the low-density lipoprotein (LDL) molecules that carry cholesterol; other receptors bind the pancreatic hormone, insulin; and still others recognize circulating lymphocytes. The endothelium also releases blood-clotting factors and enzymes that digest blood clots. Most importantly the endothelium is both a barrier and transport tissue for substances that exchange between the blood and the surrounding tissues.

The collagen fibers of the **basement membrane** and the **subendothelial connective tissue** form mats that resist blood pressure, even as elastic fibers in the connective tissue allow the intima to stretch and recoil with each pulse. Smooth muscle fibers also weave throughout the extracellular matrix of the connective tissue, where, in addition to contracting, they synthesize the collagens, elastin, and proteoglycans of the extracellular matrix. The **internal elastic lamina** is a substantial layer of elastic fibers that also allows the tunica intima to stretch. High-pressure arteries have robust internal elastic laminae, as you might expect, but this layer is missing in capillaries and all but the largest veins.

TUNICA MEDIA

The tunica media also accounts for many differences related to blood pressure. The tunica media is thickest in arteries (Figure 14.3A), thinner in large veins (Figure 14.3C), much diminished in arterioles, and absent from capillaries (Figure 14.3B) and smaller veins. An **external elastic lamina** similar to the internal elastic lamina of the tunica intima is the outermost surface of the tunica media. Although collagen fibers are always present in the tunica media, differing proportions of elastic lamellae (layers) and smooth muscle fibers adapt it to different situations. Many concentric elastic lamellae allow elastic arteries such as the aorta to stretch and recoil with pulsing blood, and smooth muscle fibers enable muscular arteries to constrict and dilate. Relatively few smooth muscle fibers are present among the elastic and collagen fibers of the tunica media in comparably sized veins. Several layers of smooth muscle fibers support the walls of superficial veins in the extremities, however.

TUNICA ADVENTITIA

The tunica adventitia is fibrous connective tissue continuous with the fascia of surrounding tissues. Relatively thin in arteries and minute or missing from capillaries and small veins, the adventitia becomes thicker in larger veins (Figure 14.3B). The adventitia holds veins in position and frequently pairs them with arteries; it also admits nerves and vessels to the tunica media. These small arteries and veins called **vasa vasorum** (VAH-sa vah-ZOR-um; vasa, *L,* vessels of the vessels) are necessary to supply blood to the thick walls of larger vessels because the intima is relatively impermeable. Nerves from the autonomic nervous system control blood flow by causing vessels to constrict or dilate and to stimulate the endothelium to release its products.

- Which tunic of an artery contains the endothelium?
- Which tunic supports the vasa vasorum, and in which vessels is this tunic especially thick?
- Which tunic is responsible for the differences between muscular and elastic arteries?
- Which tunics are absent from capillaries?

TUNICS AND CARDIOVASCULAR DISEASE

Physicians know that vessels deteriorate with age and that "You are as old as your arteries." The tunica intima of smooth, flexible, youthful arteries thickens and stiffens with calcium deposits and smooth muscle cells that deposit collagen in the subendothelial connective tissue. These changes are referred to as **arteriosclerosis** or hardening of the arteries and as **phlebosclerosis** in veins (scler-, *G,* hard). Phlebosclerosis has received far less medical attention than arteriosclerosis. In contrast, **atherosclerosis** is a pathological condition in which cholesterol and collagen fibers thicken patches of tunica intima and tunica media in the walls of arteries; by middle age, important arteries such as the carotid (to the head) and coronary (heart) arteries may become blocked, as Figure 14.4 illustrates.

How these patches, plaques, or **atheromas** develop provokes considerable controversy about the origin of the cells that store cholesterol. One hypothesis, the **lipid hypothesis,** emphasizes that elevated cholesterol levels in the blood contribute to formation and enlargement of **foam cells,** the cholesterol-storing cells. Other aspects of plaque formation seem better explained by the **endothelial injury hypothesis.** According to this view, injury (physical, viral, or immunological) to the endothelium causes plaques to thicken and foam cells to form. Both hypotheses undoubtedly contain some truth because they describe parts of a complex series of causes that involve cholesterol and endothelial damage, described below. (Cholesterol will be discussed in Chapter 15, Blood.)

Atheromas pass through three stages that can ultimately block arteries. Their development may begin at birth as thin **fatty streaks** (Figure 14.4A) that are found in the large arteries of teens and even newborns. Fatty streaks already contain foam cells filled with cholesterol, but the streaks are too thin to impede blood flow. Foam cells form in the subendothelial connective tissue from smooth muscle fibers of the tunica media and from macrophages from the blood. In the next step, thicker **sclerotic plaques** (sklair-OT-ik) form and begin to narrow the arteries (Figure 14.4B). More smooth muscle cells from the tunica media now begin to surround the foam cells inside large collagen capsules, as if to isolate the foam cells from circulating blood. Sclerotic plaques progress gradually; they can occur in teenagers' large arteries and spread to smaller arteries during middle age. High blood pressure (hypertension), smoking, and elevated blood cholesterol are major risk factors in sclerotic plaque formation. As cholesterol and collagen accumulate, the foam cells die and the endothelium deteriorates faster than it can be repaired. With sticky undersurfaces of the tunica intima exposed to the blood, **complicated plaques** (Figure 14.4C) now form, and the chances that platelets will form clots on the naked tunica intima are greater. Two damaging consequences may now occur. Thrombi may plug a vessel, but more ominously, fragments may break loose and become **emboli** (EM-bo-lee; embol-, *G,* a wedge). Carried downstream, such emboli may lodge in a narrow branch and cut off local flow (Figure 14.4D). In a cerebral artery, emboli or thrombi can cause strokes, and in a coronary artery, heart attacks can occur.

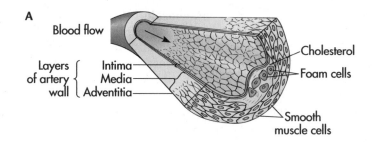

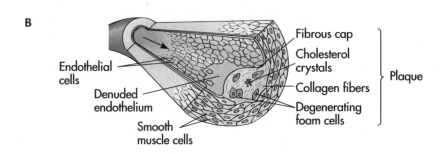

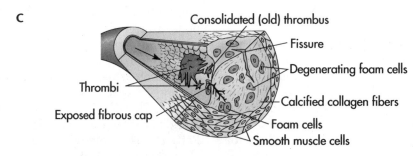

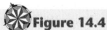**Figure 14.4**
Development of an atheroma. A, Fatty streak. B, Sclerotic plaque. C, Complicated plaque.

- What is the difference between arteriosclerosis and atherosclerosis?
- Which tunic is the source of smooth muscle fibers in atherosclerotic plaques?
- What are the names of the three stages of atheroma formation?
- What features distinguish each stage?

VESSELS

This section relates the structure of vessels to their roles and locations in the vascular tree. Figure 14.1 locates most of the vessels for you, and Figure 14.3 illustrates their structure.

LARGE ARTERIES

The pulmonary trunk and the aorta are the two largest arteries of the body, about 2.5 cm (1 inch) in diameter. These arteries are the trunks of the pulmonary and systemic arterial trees. These and the main branch arteries, such as the brachiocephalic, common carotid, subclavian, and common iliac (ILL-ee-ak) arteries, shown in Figure 14.1A, are known as **conducting arteries** because they are main lines from the heart to different regions of the body. They are also **elastic arteries;** their walls are thick with elastic fibers that stretch and recoil with each pulse. The tunica media of the aorta contains some 50 to 70 concentric, elastic lamellae.

MEDIUM ARTERIES

Smaller distributing arteries branch into the organs and tissues of the head, abdomen, and elsewhere in the body. These vessels range from 1 cm down to 0.3 mm in diameter. Smooth muscle fibers replace most of the elastic fibers in the tunica media of these **muscular arteries.** In some arteries, this transition is abrupt, as in the celiac trunk and superior and inferior mesenteric arteries, but other arteries gradually become muscular. One example is the axillary (AX-ill-ary) artery of the shoulder that enters the axilla as an elastic artery and reaches the arm as the fully muscular brachial (BRAY-key-al) artery (find these arteries in Figure 14.1A). Muscular arteries are distributive in another sense because they control blood flow into the tissues. The autonomic nervous system reduces flow by causing the tunica media to constrict (**vasoconstriction; VAH-zo-**) or increases flow by allowing it to dilate **(vasodilation)**. At any time the amount of blood flowing through these arteries depends on the balance between vasoconstriction and vasodilation.

ARTERIOLES

Arterioles are the smallest, most numerous arteries, less than 0.3mm in diameter. Too fine to show in Figure 14.1 but easy to see in Figure 14.5, these ves-

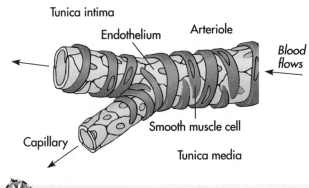

Figure 14.5
Arterioles. A sparse tunica media surrounds the endothelium and tunica intima.

sels connect muscular arteries to the capillary networks of tissues. Arterioles are very important local flow regulators because there are so many of them. A thin layer of smooth muscle allows the tunica media to vasoconstrict and to vasodilate.

CAPILLARIES

With their immense surface area, capillaries are the business end of the cardiovascular system. Networks of these very fine vessels (5 to 10 μm diameter) lace nearly every tissue and wrap almost every cell with intimate streams of essential molecules. A very thin endothelium, only 0.2 to 0.3 μm thick, and a basement membrane (Figure 14.3B) permit rapid diffusion and transport of molecules to the tissues. There is no tunica media or tunica adventitia. See Capillary Exchange, this chapter, for full details.

VENULES

Venules drain the capillary networks and begin to return blood down the venous tree to the heart (Figure 14.2A). **Venules** are the venous counterparts of arterioles in diameter and frequently accompany them.

A relatively thick tunica adventitia covers most venules. Larger venules have a few layers of smooth muscle fibers, but most venules have none.

SMALL AND MEDIUM-SIZED VEINS

Veins become larger as they receive more tributaries, but their walls remain thin compared to arteries of the same size, a difference reflecting the greatly diminished blood pressure inside them. Arbitrarily placed into categories of 0.2 to 1 mm and 1 to 10 mm in diameter, respectively, **small veins** and **medium-sized veins** have few other differences. The tunica media is thin and sparse, but the tunica adventitia is thick. Medium-sized veins, such as the femoral (FEMor-al), renal (REE-nal), and brachial veins, drain the visceral organs and distal extremities in Figure 14.1B, where they accompany the corresponding arteries.

Because the walls of small veins are much thinner than in arteries of the same diameter and blood pressure in them is very low, blood would accumulate in the veins of the hands and feet when you stand, were it not for venous valves in the extremities. The skeletal muscles in Figure 14.6 act as a **muscular venous pump** that squeezes the blood upward past each valve to the trunk. The soleus and gastrocnemius muscles of the calf pump blood as you walk. When your stride pushes forward, these muscles compress the veins that fill with blood when the muscles relax as the leg swings forward. The soles of the feet also contain a plantar pump. Figure out when this one pumps and fills. The veins involved in these pumps are described in the Lower Extremities section of this chapter. As you might expect, most veins located above the heart rely on gravity to drain them and have no valves. Surprisingly, there are no valves in the inferior vena cava and other large veins of the trunk, which rely directly on the heart to drain them.

LARGE VEINS AND VENOUS SINUSES

Large veins are more than 1 centimeter in diameter. They accompany the conducting arteries and drain the regions of the body supplied by these arteries. The brachiocephalic and common iliac veins and the superior and inferior venae cavae all are large veins. Venous sinuses of the brain (find the superior sagittal and transverse sinuses in Figure 14.13D) collect blood in vessels covered by the dura mater, the outer layer of meninges. See Head and Neck, this chapter, for details of venous sinuses and also Chapter 18, Spinal Cord and Spinal Nerves, for description of the dura mater.

END-ARTERIES AND ANASTOMOSES

If arteries branched as in an actual tree, we would call them end-arteries, illustrated in Figure 14.7A, because each branch would be the sole access to the

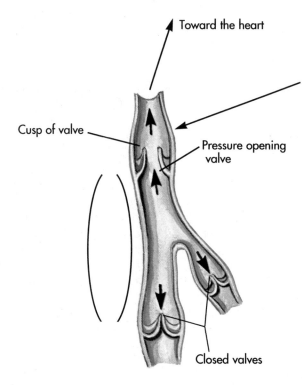

Toward the heart

Cusp of valve

Pressure opening valve

Blood flows through this valve when skeletal muscles squeeze vein but the same pressure closes these valves and prevents backward flow. These valves open, like the one above, when muscles below them contract.

Closed valves

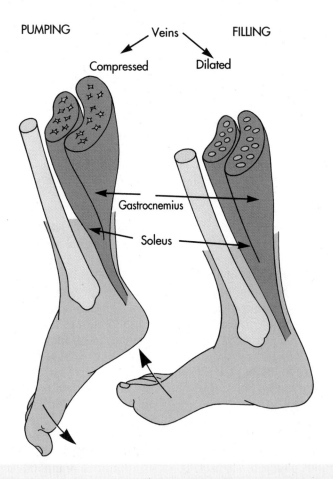

PUMPING   FILLING

Veins

Compressed   Dilated

Gastrocnemius

Soleus

✸ **Figure 14.6**
Muscular venous pump. Venous valves allow blood to flow forward but not backward, when skeletal muscles squeeze veins. The crural pump of the calf pumps when the foot plantar-flexes and fills when the foot dorsi-flexes.

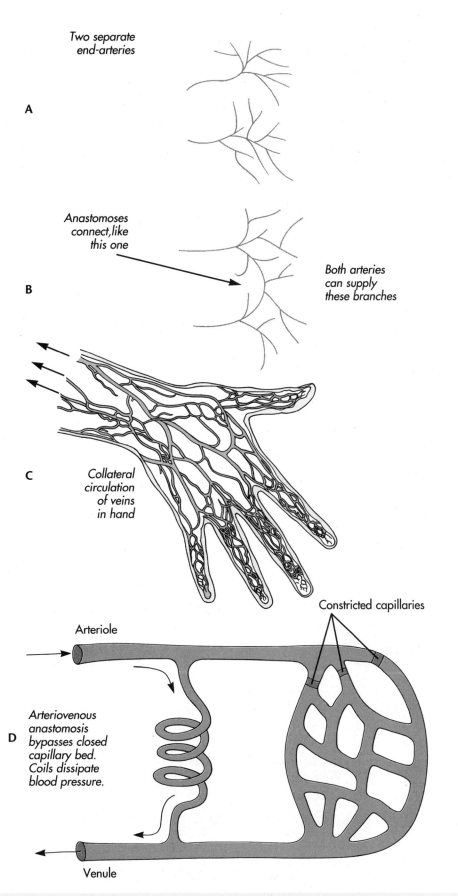

Two separate
end-arteries

A

Anastomoses
connect, like
this one

Both arteries
can supply
these branches

B

C

Collateral
circulation
of veins
in hand

Constricted capillaries

Arteriole

Arteriovenous
anastomosis
bypasses closed
capillary bed.
Coils dissipate
blood pressure.

D

Venule

Figure 14.7

A, End-arteries. **B,** Anastomoses. **C,** Collateral circulation in the hand. Anastomoses provide more than one pathway to the forearm from each finger. **D,** Arteriovenous anastomoses bypass capillary beds.

branches derived from it. The capillary fields supplied by a true end-artery have no other access, so a block in the main branch deprives that field of blood. The body has few true end-arteries, however. Tissues usually have more than one path that blood can follow in and out, providing alternatives in case one passage is blocked. Coronary arteries of the ventricles are said to be end-arteries, but even these have numerous alternative entries. Such alternatives are known as **anastomoses** (ah-NAS-toe-mo-seez; anastomosos, *G*, coming together) in which different parts of the tree connect with each other, as shown in Figure 14.7B.

Veins anastomose profusely. Figure 14.7C shows anastomoses in the veins of the hands and forearms. This situation is called **collateral circulation** because it provides alternate pathways for blood to flow easily at low pressure and to detour around obstructions. One especially important version of collateral circulation bypasses capillary beds (Figure 14.7D). Direct **arteriovenous anastomoses** between small arteries and veins detour the blood around the capillary beds when the arterioles constrict, and this allows blood to return to the veins without blocking flow in other parts of the tree. Recall the last time you were in a traffic jam and needed to detour around a wreck, to understand the usefulness of anastomoses. Arteriovenous anastomoses are especially common in the dermis of the skin, where they allow blood to bypass the delicate capillaries beneath the epidermis and to conserve body heat. Their coils dissipate blood pressure that the capillaries themselves would have dissipated.

- Why are large arteries known as conducting arteries?
- Which arteries correspond in size to venules?
- What type of vein usually is directly downstream of venules? Upstream?
- What is an anastomosis?

CAPILLARY EXCHANGE

Blood exchanges some of its components for products from the surrounding tissues when it arrives in a capillary bed. The tissues take up materials from the blood, and the blood takes up products from the tissues. This process is called **capillary exchange;** it is the fundamental function of the cardiovascular system. You have studied the familiar example of oxygen and carbon dioxide exchange. Tissue respiration takes up oxygen that must cross the capillary walls from the blood to reach the tissues, and the tissues give off carbon dioxide that crosses the capillary endothelium in the opposite direction.

The anatomy of capillaries reflects the imperatives of exchange. Capillary walls are thin, minimizing the distance between blood and tissue cells. As you have seen, the total surface area of capillaries is quite large, which resolves the surface-volume problem confronting all tissues (see discussion in Chapter 1). Surface and thinness are illustrated in Figure 14.8. Branches and anastomoses weave the capillaries into a network between the cells in Figure 14.8A. There is usually a main **thoroughfare channel** through the maze, led by a **metarteriole** that branches into the

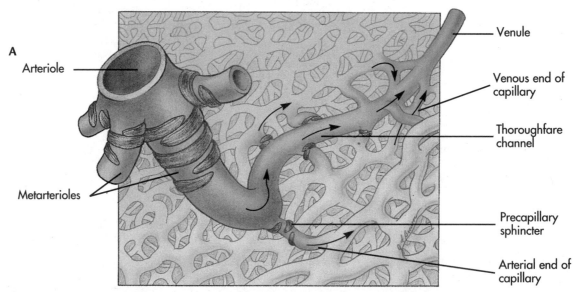

Figure 14.8
Capillary exchange. **A,** Metarterioles lead to capillary beds.

surrounding capillaries. **Precapillary sphincters** (smooth muscle fibers) coiled around the metarteriole and its capillaries control blood flow into the bed. The lumen of capillaries is typically 5 to 10 µm in diameter, and each capillary is itself 20 to 100 mm long from arteriole to venule. Capillary walls themselves are composed of simple squamous endothelium, some 0.2 to 0.4 µm thick, as illustrated in Figure 14.8B. The capillary wall is considerably thinner than the nuclei of the endothelial cells that bulge into the lumen. A basement membrane connects the endothelium to the surrounding tissue, and tight junctions and gap junctions join the squamous cells to each other at their margins. There are no tunica media or tunica externa; capillaries appear to have shed these layers to improve exchange.

Indeed, the capillary endothelium features a variety of structures for exchange, illustrated in Figure 14.9. Capillaries may be continuous, fenestrated, or discontinuous. In **continuous capillaries** (Figure 14.9A) the endothelium and basement membrane have no gaps, which obliges materials to cross the plasma membrane and cytoplasm. This is the most common kind of capillary, found in all types of muscle tissue, the nervous system, and connective tissue. **Fenestrated capillaries** (Figure 14.9B) allow substances to pass through fenestrae (FEN-es-tree; pores, *L,* windows) in the endothelium. A thin diaphragm usually stretches across each opening and apparently regulates what molecules pass. Endocrine glands, the kidneys, stomach and intestines, and the choroid plexus of the brain have fenestrated capillaries. **Discontinuous capillaries** are unusual in having large tears or gaps in the endothelium and the basement membrane, as in the sinusoids of the liver and bone marrow (see Liver, Chapter 23, and Bone Marrow, Chapter 13).

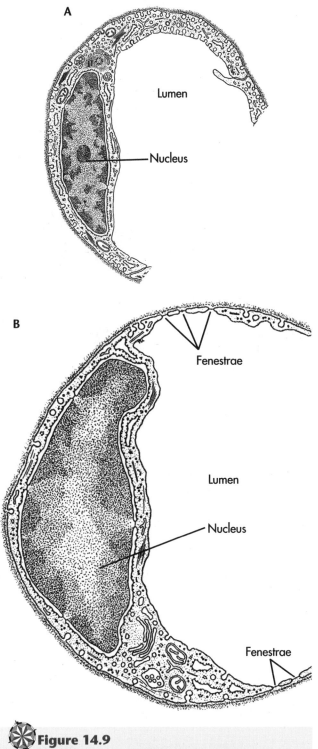

Figure 14.9

Capillary endothelium. Cross section through (A), continuous endothelium and (B), fenestrated endothelium.
Continued.

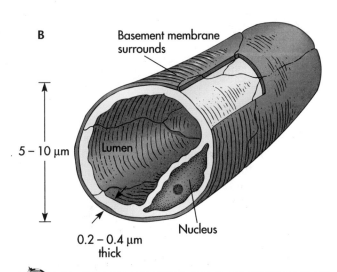

Figure 14.8, cont'd

B, Capillary endothelium is thin.

Different substances cross the endothelium by different mechanisms, as illustrated in Figure 14.9C. Water and dissolved metabolites, drugs, and hormones cross by **diffusion.** Blood pressure enhances this movement by pushing these substances out of the

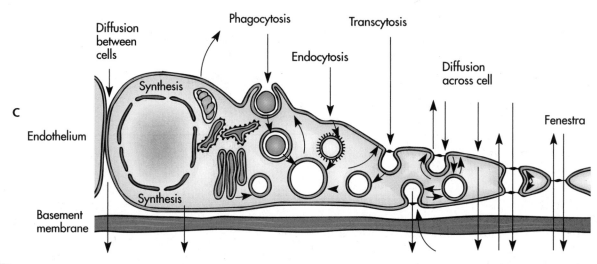

Lumen—blood plasma

Figure 14.9, cont'd
Mechanisms of capillary exchange (C).

arterial ends of the capillaries, and osmotic pressure pulls water and dissolved molecules back in at the venous ends. As a result, solutes flow through the capillary itself and through the interstitial fluid of the tissue, giving cells ample exposure to these substances. Large and small molecules cross the epithelium by **transcytosis,** in which membrane vesicles shuttle back and forth between the luminal and external plasma membranes as **transient pores,** shown in Figure 14.9 (see Endocytosis and Exocytosis, Chapter 2). **Receptor mediated endocytosis** also takes up certain products from the blood, such as low density lipoproteins (LDL) that transport cholesterol. The capillary endothelium of bone marrow and liver also **phagocytoses** microorganisms and spent cells.

• What features of capillary walls promote transport of materials between blood and tissues?

VASCULAR PATHWAYS

PRINCIPLES

Learning vascular routes is like following a road map. You simply need to trace blood flow from the heart to an organ and back. There are differences, however, between a map and the cardiovascular system. Most vessels have only local names, even though they may be parts of a direct line, such as Interstate 5

running between California and Washington. Such local names reflect anatomists' and surgeons' interest in the immediate surroundings of a vessel, even though they know the vessel leads elsewhere. Only a few vessels have throughway names; the axis artery of the upper limb is one of these exceptions. Even so, as this main vessel leaves the thorax and descends the arm, it is known first as the subclavian artery beneath the clavicle, then the axillary artery as it enters the axilla, and finally as the brachial artery after it passes the tendon of the teres major muscle in the upper limb.

Veins and arteries usually accompany each other side by side and bear similar names because of their geography. This simplifies matters considerably because one name tells, for example, that the renal artery and renal veins both serve the kidneys. You will encounter important exceptions, however. Many of these variants arise because veins tend to anastomose more freely than arteries and large veins run both superficially and deeply, whereas large arteries seldom venture near the surface of the body. Deep veins tend to run in pairs known as **venae comitantes** (VEEN-ee KOM-ih-TAN-tes), one vein on each side of the corresponding artery, as Figure 14.10 illustrates. Anastomoses communicate back and forth between the partners, nearly burying the artery beneath them. Like muscular venous pumps, pulsing arteries

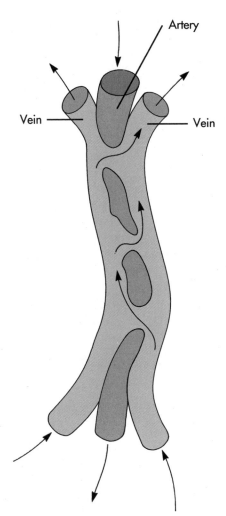

Figure 14.10

Venae comitantes. Two veins follow an artery and nearly bury it with anastomoses.

help force blood past venous valves in the venae comitantes.

Finally, anatomists traditionally describe all the arteries of the entire body first and then all of the veins. In this text, however, you will find the arteries and veins of each region described together, so that you can follow an arterial tree from the heart to the tissue and return through the venous tree with untraditional ease.

Our discussion begins with systemic circulation, then follows the pulmonary circuit, and ends with fetal circulation. Visualize the main routes described in the Overview sections and shown in the Figures, then follow the flow of blood through the regions, making lists of names, like directions to your house, of the vessels named along the way. Study Questions 8 through 23 provide plenty of practice. The Vocabulary Toolbox gives derivations of frequently used names.

AORTA AND THE VENAE CAVAE

OVERVIEW

The aorta and the venae cavae are the main conducting vessels between the heart and all regions of the body, except the lungs, which are on the pulmonary circuit (see Heart, Chapter 13). All parts of the systemic circuit begin and end with these vessels, thus you will need to follow every pathway into the body regions by beginning with the aorta and ending with the vena cava. The heart itself, the head, neck, thorax, abdomen, and extremities all depend on systemic arteries and veins. Together, these vessels hold about 95% of the blood; the remaining 5%, about 200 to 300 milliliters, circulates through the pulmonary tree in the lungs.

Three major systemic trees communicate with the heart. The **aorta** (ay-ORT-ah) illustrated in Figure 14.11A, is the trunk of the arterial tree that sends blood to the body. Two venous trees return to the heart by way of the superior vena cava and the inferior vena cava (Figure 14.11B). The **superior vena cava** (VEE-nah KAVE-ah) drains blood from the upper half of the body above the diaphragm, and blood returns from below the diaphragm through the **inferior vena cava.** Coronary arteries and veins have already been described in Chapter 13, Heart.

AORTA

The aorta is the large, elastic artery (2.5 cm, 1 inch, diameter) that exits the left ventricle of the heart. This conducting artery consists of three portions. The **ascending aorta** proceeds superiorly behind the sternum from the aortic semilunar valve and immediately gives off the left and right coronary arteries to the heart wall. The **arch of the aorta** curves posteriorly to the left and first gives off the brachiocephalic artery to the right arm, thorax, and right side of the head and neck, then the left common carotid artery (kar-AH-tid) branches to the left side of the head and neck, and finally, the left subclavian artery leads toward the left arm and thorax. The **descending aorta** conducts blood down the trunk, sending branches to the organs and walls of the thorax and abdomen, where it is known locally as the thoracic or abdominal aorta, before branching into the common iliac arteries to the pelvis and lower extremities.

The **thoracic aorta** descends the left side of the vertebral column and penetrates the aortic opening of the diaphragm at the twelfth thoracic vertebra. It sends **bronchial arteries** to the lungs, **esophageal arteries** (ee-SOF-ah-jeel) to the esophagus, **posterior intercostal arteries** to the posterior wall of the tho-

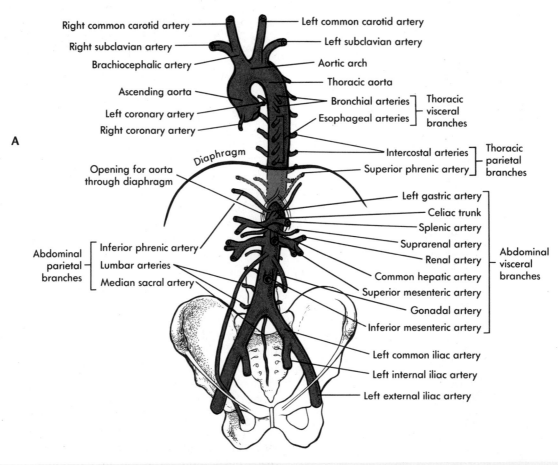

Right common carotid artery — Left common carotid artery
Right subclavian artery — Left subclavian artery
Brachiocephalic artery — Aortic arch
— Thoracic aorta
Ascending aorta — Bronchial arteries ⎤ Thoracic
Left coronary artery — Esophageal arteries ⎦ visceral branches
Right coronary artery —

A

— Intercostal arteries ⎤ Thoracic
Diaphragm — Superior phrenic artery ⎦ parietal branches
Opening for aorta
through diaphragm

— Left gastric artery ⎤
— Celiac trunk
— Splenic artery
— Suprarenal artery
Abdominal ⎡ Inferior phrenic artery — Renal artery ⎥ Abdominal
parietal ⎢ Lumbar arteries — Common hepatic artery ⎥ visceral
branches ⎣ Median sacral artery — Superior mesenteric artery ⎦ branches
— Gonadal artery
Inferior mesenteric artery ⎦

Left common iliac artery
Left internal iliac artery
Left external iliac artery

✦ **Figure 14.11**
Systemic trees. A, Aorta. Parietal branches to body wall, visceral branches to organs.

rax, and a pair of **superior phrenic arteries** (FREN-ik) to the upper surface of the diaphragm. Below the diaphragm the **abdominal aorta** descends in front of the vertebral column to the fourth lumbar vertebra, where it divides into the common iliac arteries. Its principal branches are a pair of **inferior phrenic arteries** to the lower surface of the diaphragm; **middle suprarenal arteries** to the adrenal glands; the **celiac trunk** (SEAL-ee-ak) to the liver, stomach, and spleen; the **superior mesenteric artery** to the small intestine and colon; two **renal arteries** to the kidneys; **gonadal arteries** to the testes or ovaries; and the **inferior mesenteric artery** to the colon and rectum. The **common iliac arteries** lead to the pelvis and lower extremities. Most of these branches are muscular, distributive arteries that exit the aorta at the level of their target organs.

SUPERIOR VENA CAVA

The superior vena cava returns blood from the upper half of the body. This vessel begins at the level of the first thoracic vertebra in the confluence of the **right** and **left brachiocephalic veins**, illustrated in Figure 14.11B. The superior vena cava lies in the right side of the mediastinum in front of the trachea and esophagus. The right brachiocephalic vein receives blood that entered the head and shoulders through the branches of the brachiocephalic artery, and the left brachiocephalic vein returns blood from the left common carotid and left subclavian arteries. The superior vena cava descends to the right atrium of the heart, receiving blood by way of the azygos vein (AZ-ih-goss) from the thorax wall and from the connective tissue of the lungs, but not from pulmonary surfaces exposed to air.

INFERIOR VENA CAVA

The inferior vena cava begins at the confluence of the common iliac veins. It ascends from the fifth lumbar vertebra along the right side of the abdominal aorta, passes through the caval opening of the diaphragm into the thorax, and enters the right atrium. The inferior vena cava receives seven major tributaries along its abdominal course. Except in the gastrointestinal tract, these vessels generally follow the branches of the abdominal aorta. Two **common iliac**

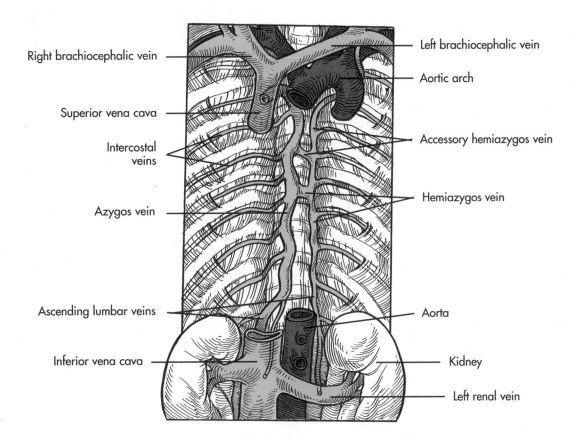

Right brachiocephalic vein

Superior vena cava

Intercostal veins

Azygos vein

Ascending lumbar veins

Inferior vena cava

Left brachiocephalic vein

Aortic arch

Accessory hemiazygos vein

Hemiazygos vein

Aorta

Kidney

Left renal vein

B

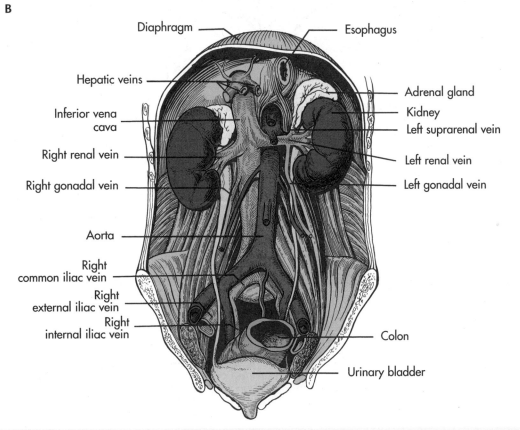

Diaphragm

Hepatic veins

Inferior vena cava

Right renal vein

Right gonadal vein

Aorta

Right common iliac vein

Right external iliac vein

Right internal iliac vein

Esophagus

Adrenal gland

Kidney

Left suprarenal vein

Left renal vein

Left gonadal vein

Colon

Urinary bladder

Figure 14.11, cont'd
B, Superior and inferior venae cavae.

veins drain the lower extremities and the pelvis. The **right gonadal vein** drains the testis or ovary on that side, but the **left gonadal vein** discharges into the left renal vein. Both **renal veins** return blood from the kidneys, and the left renal vein also receives from the left adrenal gland. The **right suprarenal vein** drains from the right adrenal gland. The **hepatic vein** returns blood from the liver itself and from gastrointestinal tract. Blood from the gastrointestinal tract reaches the liver by way of the **hepatic portal vein,** described later, rather than returning directly to the inferior vena cava. Finally, the **inferior phrenic veins** receive blood from the lower surface of the diaphragm.

- What are the names of the three major systemic vessels, and what regions of the body do they serve?
- Name the main branches of the aorta and the territories they serve.
- Name the venous tributaries that drain blood from each of these fields.

HEAD AND NECK

OVERVIEW

The arteries and veins of the head and neck are quite complex, but in general the face and brain have separate blood supplies that are nevertheless united by numerous anastomoses. Blood enters the head through the common carotid arteries and vertebral arteries. The **common carotid arteries** in Figure 14.12A ascend the neck carrying blood from the aorta, the right by way of the brachiocephalic artery and the left directly from the aorta. Near the angle of the mandible, both common carotids branch into the **internal and external carotid arteries.** The internal carotid arteries primarily serve the brain, and the external carotids supply the face, cranium, and the dura mater covering the brain. A pair of **vertebral arteries** ascends from the subclavian arteries by way of the transverse foramina of the cervical vertebrae, enters the foramen magnum, and

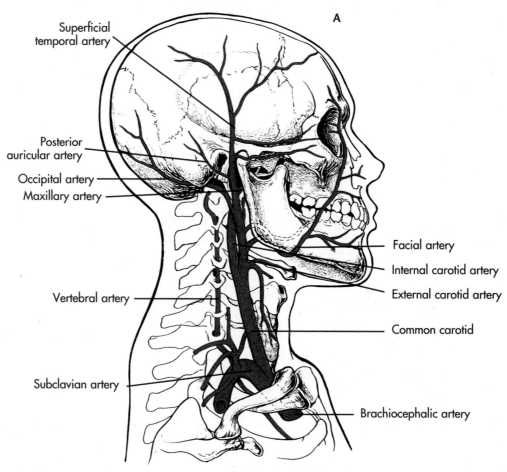

✳️**Figure 14.12**
Head and neck. A, Arteries.

fuses together as the single **basilar artery** (BAH-zil-ar; see Figure 14.13A).

Venous blood returns from capillary plexuses throughout the head and neck by way of the **internal and external jugular veins** (JUG-you-lar; Figure 14.12B) that lead to the brachiocephalic veins and superior vena cava. The internal jugular vein receives from the brain, but it also has stolen some circulation from the face, which leaves the external jugular to drain the sides of the head and the deeper regions of the mouth and face. A pair of **vertebral veins** follows the vertebral arteries down the cervical vertebrae but drains relatively little blood from the brain. The most notable venous specializations in the head are the **venous sinuses** of the meninges that collect blood from the brain and cranial cavity before releasing it to the internal jugular vein.

FACE AND SKULL
COMMON CAROTID ARTERIES

The right and left common carotid arteries arise from the right brachiocephalic artery and the arch of the aorta, respectively. Both enter the neck behind the sternoclavicular joints and ascend beside the trachea, under cover of the sternocleidomastoid muscle, to the level of the larynx (Figure 14.12A). Here, they bifurcate into **internal** and **external carotid arteries** near the angle of the mandible. The carotid body, a receptor that monitors oxygen levels and blood pressure, occupies the narrow V in the wall between these vessels.

The external carotid artery delivers blood to the face and cranium. It sends branches to the face and occipital (OK-sip-it-al) region before dividing into two terminal branches behind the ramus of the mandible. The **superficial temporal artery** is the posterior terminal branch. It ascends in front of the ear canal, where you can feel its pulse, to supply the frontal, parietal, and temporal regions of the scalp and cranium. The **maxillary artery,** the anterior terminal branch, tucks beneath the temporomandibular joint and branches to the anterior and middle meninges, to the orbits, and to the muscles of mastication, the maxilla, zygoma, and the floor of the oral cavity.

Additional branches leave from the external carotid artery itself. Beginning inferiorly, the **superior**

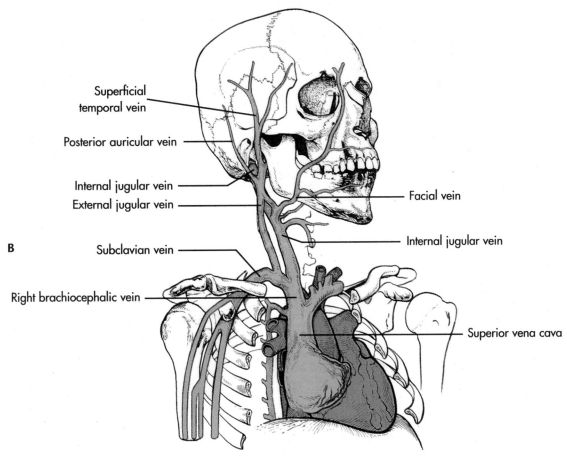

Superficial temporal vein

Posterior auricular vein

Internal jugular vein

External jugular vein

Subclavian vein

Right brachiocephalic vein

Facial vein

Internal jugular vein

Superior vena cava

B

Figure 14.12, cont'd
B, Veins.

thyroid artery delivers to the larynx, thyroid gland, and nearby muscles. The **lingual artery** is the main supply to the tongue. Next, the **facial artery** crosses the inferior border of the mandible to the nose, the orbits, and the mouth. The **ascending pharyngeal artery** serves the pharynx, palatine tonsil, auditory tube, and meninges. Posteriorly the **occipital artery** reaches to the muscles and meninges of the occipital region, thereby completing the supply begun by branches of the maxillary artery. Finally, the **posterior auricular artery** supplies the external ear, tympanic membrane, semicircular canals, and mastoid air cells of the temporal bone.

VEINS

The internal and external jugular veins serve the face and skull. The external jugular vein descends the side of the neck, where it may be seen be-

neath the skin, to the medial portion of the clavicle, where it joins the subclavian vein (Figure 14.12B). Its tributaries drain the fields of the corresponding arteries. The external jugular vein arises posterior to the ramus of the mandible by the confluence of the **posterior auricular** and **retromandibular veins.** The retromandibular vein in turn receives from the **superficial temporal** and **maxillary veins** (not shown). The **occipital vein** usually discharges into the posterior auricular vein (or-IK-u-lar). The external jugular vein also has an important communication with the facial vein via the **anterior retromandibular vein.**

The internal jugular vein also drains the face, but its main source is the brain and cranial cavity. This vein emerges from the jugular foramen (see Cerebral Circulation, below) and descends the neck in company with the internal carotid and common carotid arteries to the medial end of the clavicle, where it

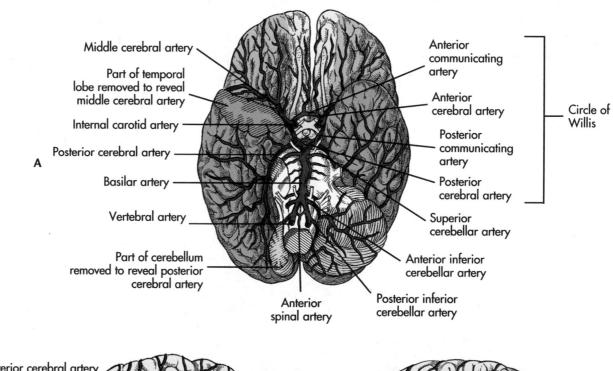

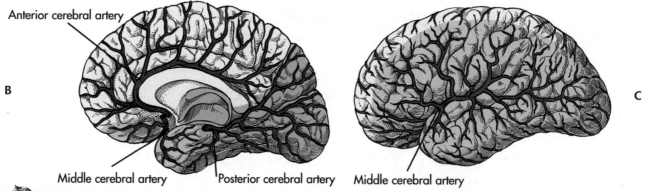

Figure 14.13

Cerebral circulation. **A,** Circle of Willis and cerebral arteries. **B,** Cerebral arteries, midsagittal view. **C,** Cerebral arteries, left lateral view.

joins the brachiocephalic vein. You will recognize many tributaries of the internal jugular vein that correspond to branches of the external carotid arteries; they are the large **facial vein** (face, nose, and orbits), the **lingual veins** (tongue), and the **superior thyroid veins** (larynx and thyroid gland).

- Which arteries deliver to the face? To the brain?
- What are the names of the vessels that drain blood from the face?
- Into which veins do these latter vessels drain?

CEREBRAL CIRCULATION
ARTERIES

The brain receives most of its blood from the internal carotid arteries and the basilar artery. The **internal carotid arteries** enter the cranial cavity through the carotid canal of the cranium and then divide into anterior and middle cerebral arteries. The **basilar artery** divides into the left and right posterior cerebral arteries. The basilar artery is formed by the union of the **vertebral arteries** that ascend the transverse foramina of the first six cervical vertebrae, enter the foramen magnum, and unite. The basilar and internal carotid arteries converge on the **circle of Willis** beneath the hypothalamus of the brain, shown in Figure 14.13A. **Anterior** and **posterior communicating arteries** link all three cerebral arteries, so blood from any entry can reach any part of the brain.

The cerebral arteries continue over the surfaces of the cerebral hemispheres from the circle of Willis. The **anterior cerebral artery** and its branches (Figure 14.13A and B) supply the orbits, the eyeballs, and eyelids, as well as the frontal sinus and ethmoid air cells. The thalamus and hypothalamus of the brainstem, and the cerebral nuclei, corpus callosum, frontal and parietal lobes of the cerebral hemispheres all receive from the anterior cerebral artery. **Middle cerebral arteries** reach the lateral surfaces of the frontal, parietal, and occipital lobes (Figure 14.13), while the **posterior cerebral arteries,** shown in Figure 14.13A and B, supply the inferior and medial surfaces of the temporal and occipital lobes.

VEINS AND VENOUS SINUSES

The internal jugular vein drains the majority of blood from the brain. Venous sinuses collect this flow from capillary plexuses and cerebral veins and conduct it to the internal jugular vein.

The **venous sinuses** of the brain (Figure 14.13D) are unique tubes and irregular spaces within the outer covering of the brain, the dura mater (see Meninges, Chapter 18). These chambers receive from cerebral veins located on the surface and interior of the brain. The endothelial lining of these sinuses is continuous with the endothelium of the veins, but dura rather than tunica media and tunica adventitia forms the walls. Some sinuses are paired and others lie in the midline unpaired, as you can see in Figure 14.13D.

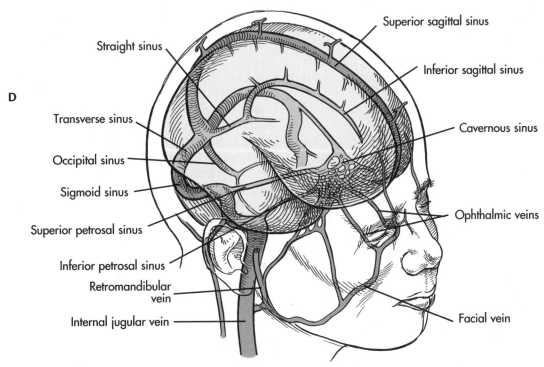

D

Straight sinus
Transverse sinus
Occipital sinus
Sigmoid sinus
Superior petrosal sinus
Inferior petrosal sinus
Retromandibular vein
Internal jugular vein

Superior sagittal sinus
Inferior sagittal sinus
Cavernous sinus
Ophthalmic veins
Facial vein

✳ **Figure 14.13, cont'd**
D, Major venous sinuses of the brain.

The **superior** and **inferior sagittal sinuses** and the **straight sinus,** unpaired sinuses between the cerebral hemispheres, drain the fields of the anterior and posterior cerebral arteries. These sinuses also drain the cranium and scalp by way of emissary veins through the skull. The **occipital sinus** drains from the cerebellum. All unite in the **transverse sinuses** that flow to the **sigmoid sinuses** and thence to the internal jugular vein. As this flow leaves the cranium, it joins blood from the **cavernous sinuses** beneath the brainstem. These paired spaces lie in the middle cerebral fossa on each side of the body of the sphenoid bone, where they collect from the face and cerebrum and also receive hormones from the pituitary gland.

- Which arteries deliver to the circle of Willis, and which arteries receive blood from the circle?
- Which venous sinuses drain the fields served by the anterior and middle cerebral arteries?

THORAX

OVERVIEW

All the vessels that send blood to the body pass through the thorax. The thoracic wall and diaphragm and the heart, lungs, and esophagus all receive from the descending aorta (Figure 14.14A) and from the internal thoracic artery. Blood returns to the heart by way of the azygos vein and the superior vena cava (Figure 14.14B). Axial segmentation strongly influences the pattern of branching in these vascular trees because the veins and arteries follow the ribs.

ARTERIES

The aorta is of course the dominant thoracic artery in Figure 14.14A. The arch of the aorta supplies the **brachiocephalic artery** with blood for the head and right upper extremity, and the **left common carotid** and **subclavian arteries** supply the left side of the head and the left upper extremity. Some of this blood immediately returns to the thorax through the internal thoracic arteries, described below. The thoracic aorta gives off branches to the posterior thorax wall and mediastinum before penetrating the diaphragm into the abdominal cavity. The **posterior intercostal arteries** dominate this supply. Nine pairs serve the intercostal spaces of the third to the eleventh ribs, sending blood to the vertebral column, spinal cord, and local muscles and bones. The first two ribs are served by branches from the subclavian arteries, and the twelfth rib receives separately from the **subcostal artery.** The **superior phrenic arteries** serve the upper surface of the diaphragm. Several visceral branches of the thoracic aorta supply the esophagus, surrounding mediastinum (esophageal arteries), and the bronchi of the lungs (bronchial arteries). The mediastinum, trachea, and anterior thoracic wall all re-

ceive from the **internal thoracic arteries,** which are branches of the subclavian arteries that return to the anterior wall of the thorax on left and right sides. **Anterior intercostal arteries,** counterparts to the posterior intercostals, branch from the internal thoracic arteries.

VEINS

Thoracic circulation returns to the heart through the superior vena cava. Formed by the union of the brachiocephalic veins (Figure 14.14B), the superior vena cava receives thoracic blood directly from the **azygos vein** and indirectly from the **internal thoracic veins.** The single azygos vein and its adjunct, the **hemiazygos vein,** drain the vertebral column, the posterior thorax, and the mediastinum. The azygos vein ascends the right side of the vertebral column and receives from the hemiazygos on the left side. The internal thoracic veins receive from their corresponding arterial fields in the intercostal spaces of the anterior and lateral chest. Each internal thoracic vein drains into the subclavian vein of its side and thence to the brachiocephalic veins and superior vena cava.

- Which arteries deliver blood to the wall of the thorax?
- What vessel does the azygos vein drain into, and what vessels of the thorax drain into the azygos?

UPPER EXTREMITIES

OVERVIEW

A single artery, the axis artery, delivers blood to the shoulder and upper limb. This artery leaves the thorax in Figure 14.15A as the subclavian artery, continues across the axilla as the axillary artery, and enters the arm as the brachial artery. Branches of the axillary and subclavian arteries supply the shoulder with blood. At the elbow the brachial artery branches into the radial and ulnar arteries located on their respective sides of the forearm. Venous plexuses collect this blood and return it from the hands both through deep veins that follow the arteries (Figure 14.15B) and through superficial veins that drain into the subclavian vein, and finally, into the brachiocephalic vein and superior vena cava to the heart.

ARTERIES

The subclavian arteries begin differently on the left and right sides. The right subclavian artery branches from the brachiocephalic artery in Figure 14.14A, and the left subclavian branches from the arch of the aorta. Both arteries leave the thorax above the first rib and course beneath the clavicle (hence the name) toward the axilla. In this short distance, five branches enter the neck, head, shoulder, and thorax. The **vertebral artery,** the most proximal of these five,

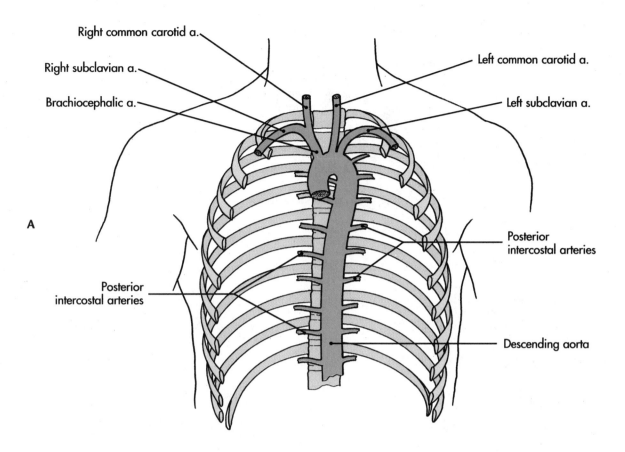

Right common carotid a.

Right subclavian a.

Brachiocephalic a.

Left common carotid a.

Left subclavian a.

A

Posterior intercostal arteries

Posterior intercostal arteries

Descending aorta

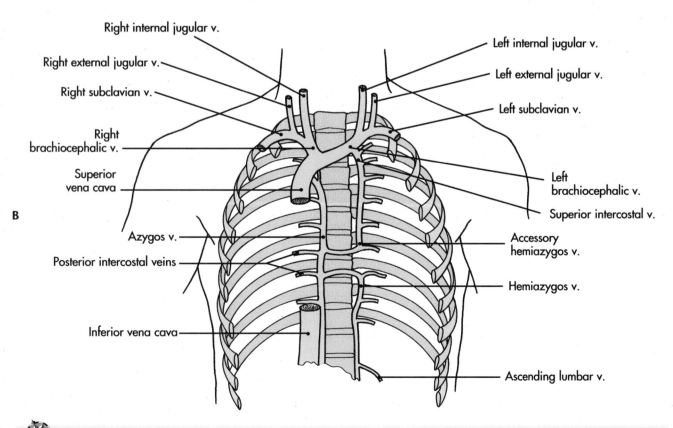

Right internal jugular v.

Right external jugular v.

Right subclavian v.

Right brachiocephalic v.

Superior vena cava

B

Azygos v.

Posterior intercostal veins

Inferior vena cava

Left internal jugular v.

Left external jugular v.

Left subclavian v.

Left brachiocephalic v.

Superior intercostal v.

Accessory hemiazygos v.

Hemiazygos v.

Ascending lumbar v.

Figure 14.14
Major arteries (A), and veins (B), of the thoracic wall.

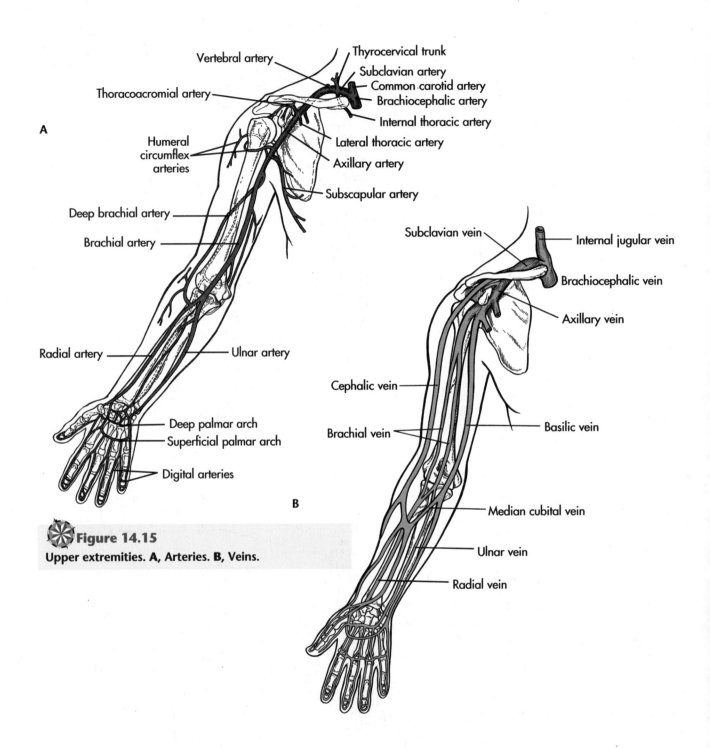

Figure 14.15
Upper extremities. A, Arteries. B, Veins.

ascends the neck as described in Head and Neck, this Chapter. Next, the **thyrocervical trunk** immediately gives off branches to the neck, the scapula, and neighboring muscles of the shoulder in Figure 14.15A. The **internal thoracic artery** sends branches to the pectoralis major muscle and breast. Next, the **costocervical trunk** serves the first two ribs and intercostal spaces and muscles of the back of the neck. Finally, the **descending scapular artery** leads directly from

the subclavian artery or from the thyrocervical trunk to the trapezius and other muscles of the scapula.

The subclavian artery becomes the axillary artery at the border of the first rib, where it crosses beneath the pectoralis major muscle into the axilla. Along this course the axillary artery sends branches to the pectoral muscles and thorax **(supreme thoracic artery);** to the breast, deltoid muscle, and sternoclavicular joint **(thoracoacromial artery);** and to the serratus anterior, pectoral muscles, subscapularis muscle, breast, and axillary lymph nodes (lateral thoracic artery). The axillary artery sends other branches to teres major and minor muscles, latissimus dorsi, and the long head of the triceps brachii muscles (subscapular artery); to the deltoid muscle and shoulder joint; and to the head of the humerus (humeral circumflex arteries).

The axillary artery becomes the brachial artery at the tendon of the teres major muscle and reaches the elbow, where it branches into the radial and ulnar arteries. Along its course the brachial artery serves the humerus and local muscles. Its largest branch, the deep brachial artery, serves the deltoid muscle, triceps brachii, and the elbow.

The ulnar and radial arteries supply the forearm, the wrist, and the hand. The ulnar artery is the larger of the two. It reaches along the ulnar side of the forearm from the elbow to the flexor retinaculum and arches across the palm of the hand to the radial side as the superficial palmar arch. The radial artery extends from the cubital (KUE-bit-al) fossa of the elbow joint, along the radial side of the forearm, to become the deep palmar arch across the hand to the ulnar side. The deep palmar arch gives off palmar digital arteries that lead between the metacarpal bones onto the lateral and medial sides of the digits. Perforating branches of the deep palmar arch reach the dorsal side of the hand and fingers.

VEINS

Some of the blood from the palmar arches drains through deep veins on the opposite side of the forearm from which it entered (Figure 14.15B). Superficial veins receive the rest of this blood and ultimately return it to the subclavian vein. One superficial vein, the median cubital vein of the forearm, is the standard vessel for withdrawing blood. Valves in both types of veins enable blood to flow up the limb.

DEEP

The radial veins drain the deep palmar venous arch (Figure 14.15B), and the ulnar veins receive from the superficial palmar venous arch. Radial and ulnar veins form venae comitantes around the radial and ulnar arteries of the forearm. The radial and ulnar veins join at the elbow and become the brachial veins, which also receive from the median cubital vein and ascend the humerus beside the brachial artery, where at the confluence with the basilic vein all three become the axillary vein. The axillary vein flows into the subclavian vein; both collect tributaries from the corresponding arteries of the shoulder and thorax before entering the thorax wall and the brachiocephalic vein.

SUPERFICIAL

Dorsal and palmar venous networks are the principal venous centers of the hand. Blood from the fingers (digital veins) drains into these plexuses, and two or three venous channels conduct this flow up the forearm, anastomosing with each other and with the deep vessels as they go. The basilic vein drains the dorsal venous network of the hand and ascends the posterior ulnar side of the forearm and the median surface of the biceps muscles, where it joins the brachial vein and becomes the axillary vein. The cephalic vein (seh-FAL-ik), bearing blood from the thumb and also from the dorsal venous network, ascends the radial side of the forearm and joins the axillary vein beneath the clavicle. The median cubital vein receives from variable veins of the forearm and joins the brachial vein at the cubital fossa.

- Name the three sections of the axis artery.
- Which arteries serve the superficial and deep palmar arches?
- Do the veins that drain the superficial and deep palmar venous arches correspond to the arteries that deliver to the corresponding arterial arches?
- Which vein can be seen on the anterior surface of the deltoid and biceps brachii muscles?

ABDOMINAL VISCERA AND HEPATIC PORTAL SYSTEM

OVERVIEW

The abdominal aorta delivers blood to the abdomen and the inferior vena cava returns this blood to the heart. Three unpaired arteries—the celiac, the superior mesenteric, and the inferior mesenteric—supply the gastrointestinal tract and spleen. A single vessel—the hepatic portal vein—drains this circulation into the liver before the blood returns to the vena cava. Paired arteries and veins serve the paired abdominal organs and the abdominal wall and diaphragm.

UNPAIRED VISCERAL VESSELS
ARTERIES

The celiac trunk is the most superior of the unpaired arteries. This short, muscular artery quickly branches into the left gastric, splenic, and common hepatic arteries, carrying blood for the stomach, spleen, pancreas, liver, duodenum, and gall bladder, as you can see in Figure 14.16. One cm (0.4 in) below the celiac trunk, the superior mesenteric artery enters the mesentery opposite the duodenum. This artery sends numerous branches, named in

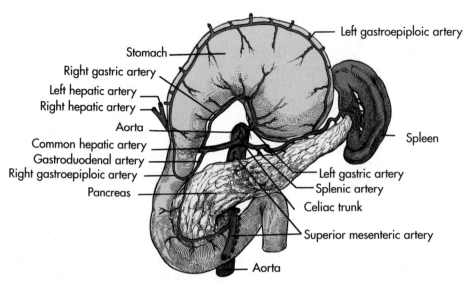

**Figure 14.16**
Branches of the celiac trunk.

Figure 14.17A and B, to the duodenum, pancreas, jejunum, ileum, and large intestine. The **inferior mesenteric artery,** in keeping with its name, supplies the remainder of the large intestine and the rectum.

VEINS

The hepatic portal vein returns all of this circulation to the liver, as illustrated in Figure 14.18. Portal veins are unique in draining one set of capillaries (of the digestive tube, in this instance) into a

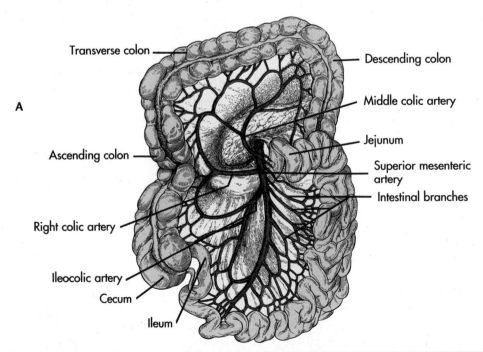

Figure 14.17
Superior and inferior mesenteric arteries send branches and numerous anastomoses into the abdominal mesentery. A, The middle colic artery sends blood to the transverse colon; the right colic artery, to the ascending colon, the iliocolic artery, to the ileum and ascending colon; and intestinal branches, to the small intestine.

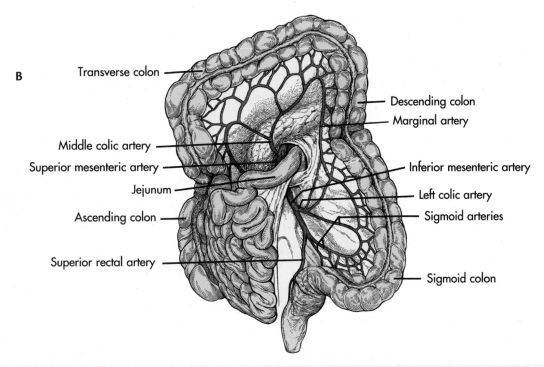

B

Transverse colon

Descending colon

Marginal artery

Middle colic artery

Superior mesenteric artery

Jejunum

Ascending colon

Superior rectal artery

Inferior mesenteric artery

Left colic artery

Sigmoid arteries

Sigmoid colon

Figure 14.17, cont'd

B, The left colic artery delivers to the descending colon; the sigmoid artery, to the sigmoid colon; and the rectal artery, to the rectum.

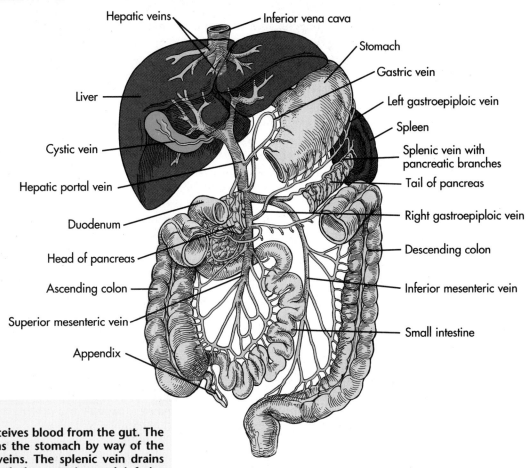

Hepatic veins

Inferior vena cava

Stomach

Gastric vein

Left gastroepiploic vein

Spleen

Splenic vein with pancreatic branches

Tail of pancreas

Right gastroepiploic vein

Descending colon

Inferior mesenteric vein

Small intestine

Liver

Cystic vein

Hepatic portal vein

Duodenum

Head of pancreas

Ascending colon

Superior mesenteric vein

Appendix

Figure 14.18

Hepatic portal system receives blood from the gut. The hepatic portal vein drains the stomach by way of the gastric and left gastric veins. The splenic vein drains spleen and pancreas, and the superior and inferior mesenteric veins receive from the small intestine and colon.

second set (in the liver), without returning to the heart. Another portal system supplies blood to the pituitary gland (see Chapter 22, Endocrines). As you can see in the figure, the **hepatic portal vein** receives the **superior** and **inferior mesenteric veins,** counterparts of the inferior and superior mesenteric arteries. The **left gastric** and **splenic veins** also deliver to the hepatic portal vein. Blood from all these tributaries enters the tissue of the liver, which takes up glucose and other nutrients absorbed by the intestines and releases serum proteins and other products into the blood (see Liver, Chapter 23). There are no venous valves in the hepatic portal system. The **hepatic veins** direct this conditioned blood to the **inferior vena cava.**

PAIRED VESSELS

Beginning immediately below the diaphragm in Figure 14.19, a pair of **inferior phrenic arteries** supplies blood to the diaphragm and the adrenal glands. Venous circulation returns to the inferior vena cava by way of the **inferior phrenic veins** and through the suprarenal vein on the left side (see below).

Five pairs of **lumbar arteries** branch on each side into the lumbar vertebrae and the surrounding muscles of the lower back. These vessels are the abdominal continuation of the posterior intercostal arteries of the thorax. Blood returns to the **azygos vein** from the upper two intercostal arteries, the next two return to the **inferior vena cava,** and the fifth pair returns to the **iliolumbar vein,** described in Gluteal and Pelvic Regions, this chapter. **Ascending lumbar veins** anastomose with all members on both sides and communicate to the azygos vein and thorax. The ascending lumbar veins supplement the inferior vena cava, and each provides an alternative return should the vena cava become blocked. Finally, the large **common iliac arteries** branch to the pelvis and the lower limbs, leaving the small unpaired **median sacral artery** to supply the coccygeal region. The **common iliac veins** and the **median sacral vein** return to the inferior vena cava.

Find the **suprarenal arteries** in the superior portion of the abdomen in Figure 14.19. This pair of arteries serves the suprarenal glands; immediately below these vessels, the **renal arteries** branch to the kidneys at the level of the superior mesenteric artery. **Renal veins** drain the kidneys. The left renal vein is longer than the right and passes in front of the abdominal aorta, where it also receives from the left suprarenal gland and left gonadal vein.

The **gonadal arteries** branch from the abdominal aorta, approximately halfway between the superior and inferior mesenteric arteries. Each slender vessel descends the abdomen to the ovaries or testes, whence circulation returns by way of gonadal veins (testicular or ovarian veins). The right gonadal vein empties into the vena cava at the same level as the corresponding artery, but the left gonadal vein discharges into the renal vein above the point of departure from the aorta.

- Name the three unpaired branches of the abdominal aorta that serve the gastrointestinal tract and the vein that receives from them.
- Which abdominal regions and organs are served by paired arteries and veins?

PELVIC AND GLUTEAL REGIONS

OVERVIEW

The common iliac arteries and veins serve the pelvic cavity and the gluteal regions. After giving off an internal iliac artery to each side of the pelvis (Figure 14.20), the common iliac artery becomes the external iliac artery and passes into the thigh as the femoral artery. Venous blood returns to the inferior vena cava by way of corresponding iliac veins. Both arterial and venous vessels are quite variable.

ARTERIES

As its name implies, the external iliac artery leads outside the pelvis and descends over the ilium into the thigh (Figure 14.20), but the **internal iliac artery** remains inside the pelvis, sending anterior and posterior divisions onto its wall and to the visceral organs. In the most common variation the **posterior division** sends the **inferior gluteal artery** to the gluteus muscles and proximal thigh. The **superior gluteal artery** branches separately to the gluteal muscles. **Iliolumbar arteries** deliver to the lumbar vertebrae, and **lateral sacral arteries** supply the sacrum.

The **anterior division** of the internal iliac artery sends five branches, three to the visceral organs and two into the pelvic wall. Among the visceral arteries are the **superior** and **inferior vesical arteries** (VEH-sih-kal; bladder; not to be confused with vesicle [VES-ih-kl; a small bladder]) that supply the bladder in both sexes. In males (Figure 14.20A) the inferior vesical artery reaches the prostate gland, and in females (Figure 14.20B) the **uterine** and **vaginal arteries** (VAH-jin-al) supply blood to the vagina, cervix, uterus, and ovaries. The **middle rectal artery** serves the rectum. The remaining two arteries leave the pelvic cavity. The **obturator artery** (OB-tyur-AY-tor) passes through the obturator foramen to the head of the femur and hip joint. Finally, the external anal sphincter and the pelvic diaphragm receive blood from the **internal pudendal artery** (pew-DEN-dal).

The external iliac artery is larger than the internal iliac artery and reaches from the lumbosacral

A

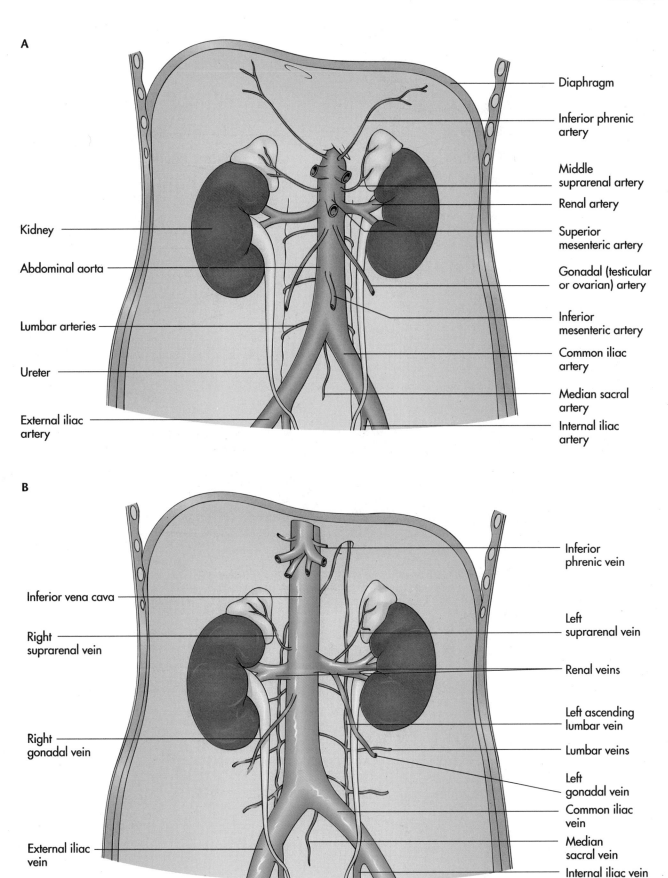

Kidney

Abdominal aorta

Lumbar arteries

Ureter

External iliac
artery

Diaphragm

Inferior phrenic
artery

Middle
suprarenal artery

Renal artery

Superior
mesenteric artery

Gonadal (testicular
or ovarian) artery

Inferior
mesenteric artery

Common iliac
artery

Median sacral
artery

Internal iliac
artery

B

Inferior vena cava

Right
suprarenal vein

Right
gonadal vein

External iliac
vein

Inferior
phrenic vein

Left
suprarenal vein

Renal veins

Left ascending
lumbar vein

Lumbar veins

Left
gonadal vein

Common iliac
vein

Median
sacral vein

Internal iliac vein

Figure 14.19
Paired arteries (A), and veins (B), of the abdomen.

A

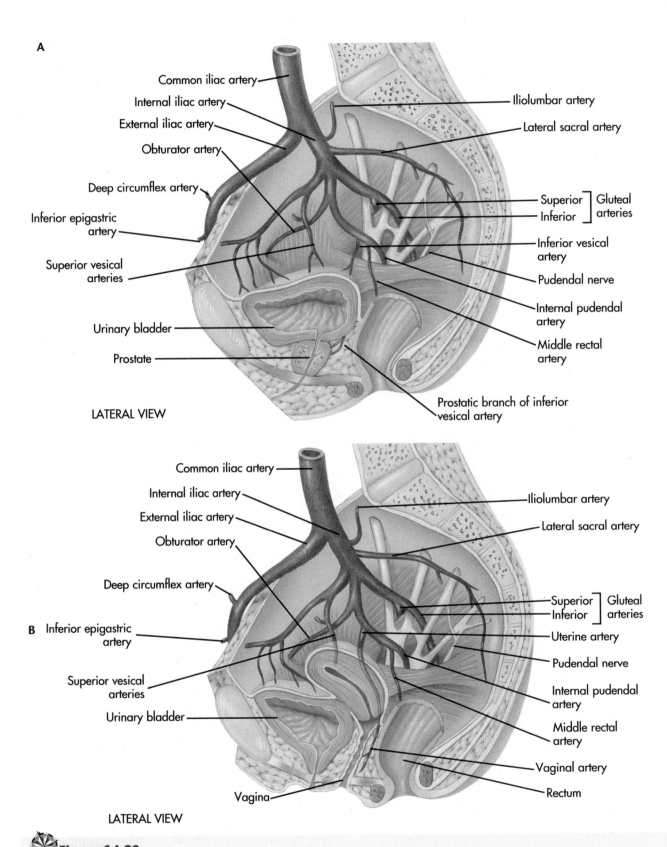

Common iliac artery

Internal iliac artery

External iliac artery

Obturator artery

Deep circumflex artery

Inferior epigastric artery

Superior vesical arteries

Urinary bladder

Prostate

Iliolumbar artery

Lateral sacral artery

Superior ⎤ Gluteal
Inferior ⎦ arteries

Inferior vesical artery

Pudendal nerve

Internal pudendal artery

Middle rectal artery

Prostatic branch of inferior vesical artery

LATERAL VIEW

Common iliac artery

Internal iliac artery

External iliac artery

Obturator artery

Deep circumflex artery

B Inferior epigastric artery

Superior vesical arteries

Urinary bladder

Vagina

Iliolumbar artery

Lateral sacral artery

Superior ⎤ Gluteal
Inferior ⎦ arteries

Uterine artery

Pudendal nerve

Internal pudendal artery

Middle rectal artery

Vaginal artery

Rectum

LATERAL VIEW

Figure 14.20
Arteries of the pelvic and gluteal regions. A, Male. B, Female.

disk to the inguinal ligament of the pelvis, giving off two branches to the anterior abdominal wall. The rectus abdominis and external oblique muscles receive from the **inferior epigastric artery** that anastomoses with the superior epigastric and posterior intercostal arteries from the thorax. The **deep circumflex iliac artery** serves the transversus abdominis and the internal oblique muscles.

VEINS

Blood returns from the pelvis through the internal iliac vein and from the lower limb by way of the external iliac vein. Both join to form the common iliac vein (Figure 14.21A and B). Numerous venous plexuses in the pelvic organs collect blood and direct it to veins that correspond with the arteries that deliver the flow. Most of the veins surround these arteries as venae comitantes. Thus the **internal iliac vein** receives the **anterior** and **posterior divisions** that serve their arterial counterparts. Tributaries of the posterior division drain blood from the gluteal muscles **(superior and inferior gluteal veins).** Sacral venous plexuses and the fifth lumbar vein drain to the internal iliac vein by way of the **sacral** and **iliolumbar veins,** respectively.

The anterior division of the internal iliac vein receives from the rectal plexus **(middle rectal vein),** from the perineum and penis or clitoris **(internal pudendal vein)** and from the venous plexuses of the bladder and prostate **(superior** and **inferior vesical veins).** In females (Figure 14.21B) the uterine and vaginal plexuses drain to the corresponding veins and thence to the anterior division. The anterior division also receives from the **obturator vein** and the hip joint.

The external iliac vein, continuous with the femoral vein, receives from the **deep circumflex iliac** and **inferior epigastric veins** that accompany the arteries described above.

- What are the two terminal branches of the abdominal aorta?
- Which divisions of the internal iliac artery serve primarily visceral organs and which primarily serve parietal structures?
- If the vesical artery supplies the prostate gland in a male, what are the corresponding arteries in a female?

LOWER EXTREMITIES

OVERVIEW

Blood enters the lower extremities through the femoral artery and returns to the pelvis through the femoral vein. One long artery (Figure 14.22A), known

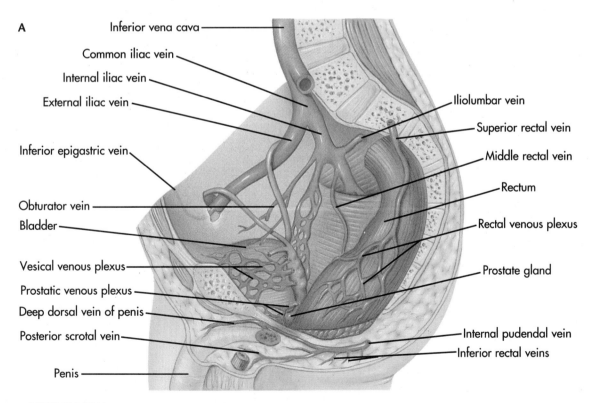

A

Inferior vena cava
Common iliac vein
Internal iliac vein
External iliac vein
Inferior epigastric vein
Obturator vein
Bladder
Vesical venous plexus
Prostatic venous plexus
Deep dorsal vein of penis
Posterior scrotal vein
Penis

Iliolumbar vein
Superior rectal vein
Middle rectal vein
Rectum
Rectal venous plexus
Prostate gland
Internal pudendal vein
Inferior rectal veins

LATERAL VIEW

Figure 14.21
Pelvic and gluteal veins. A, Male.

Continued.

as the femoral or popliteal (pop-LIT-ee-al) for regions it passes, delivers blood to the thigh and knee before branching into the anterior and posterior tibial arteries of the leg and ankle. Both vessels become broad plantar arches in the foot that spin off branches to the metatarsals and digits. Blood returns from all regions of the leg through deep or superficial tributaries of the femoral vein (Figure 14.22B). Deep veins follow the arteries in name and location, and two long superficial veins—the great and small saphenous (sah-FEE-nus) veins—also return blood to the femoral vein. In these veins, numerous muscular venous pumps sustain the upward flow of blood, which otherwise would collect in the feet and ankles.

ARTERIES

After piercing the pelvic wall, the external iliac artery becomes the femoral artery. Together with the femoral vein, this vessel follows the medial side of the femur in Figure 14.22A and then penetrates the adductor magnus muscle to emerge posteriorly between the condyles of the femur and tibia as the **popliteal artery.** The femoral artery also sends the **deep femoral branch** to the femur and to the adductors and hamstring muscles. The **lateral circumflex femoral artery** delivers blood to the vastus lateralis and quadriceps femoris muscles. After crossing the

knee, the popliteal artery bifurcates into the **anterior** and **posterior tibial arteries.** The anterior tibial artery descends between the tibia and fibula and continues as the **dorsal pedis artery** (PEE-diss) onto the dorsal surface of the foot. The posterior tibial artery gives off the **peroneal artery** to the medial side of the leg, deep to the peroneus muscle. The posterior tibial artery descends the posterior muscle compartment to the medial malleolus of the tibia, and enters the plantar surface of the foot as the lateral and medial plantar arteries. The **medial plantar artery** serves the great toe, and the **lateral plantar artery** forms the **plantar arch** that curves across the foot to the dorsalis pedis artery on the medial side. The plantar arch sends branches between the metatarsal bones and the neighboring digits, where **digital plexuses** distribute blood to the toes themselves.

VEINS

Blood returns to the femoral vein through deep and superficial veins. Blood that has reached the foot collects in the **deep plantar venous arch,** shown in Figure 14.22B, that accompanies the arterial plantar arch. Anastomoses carry this blood deeply to the **anterior** and **posterior tibial veins** and superficially to the **saphenous veins** at either side of the foot. The tibial veins are tributaries to the popliteal vein, whose

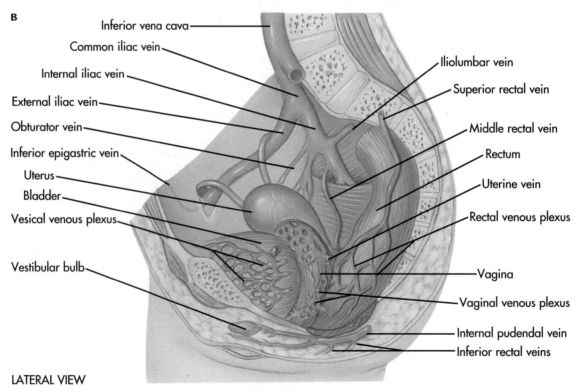

B

Inferior vena cava
Common iliac vein
Internal iliac vein
External iliac vein
Obturator vein
Inferior epigastric vein
Uterus
Bladder
Vesical venous plexus
Vestibular bulb

Iliolumbar vein
Superior rectal vein
Middle rectal vein
Rectum
Uterine vein
Rectal venous plexus
Vagina
Vaginal venous plexus
Internal pudendal vein
Inferior rectal veins

LATERAL VIEW

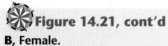

Figure 14.21, cont'd
B, Female.

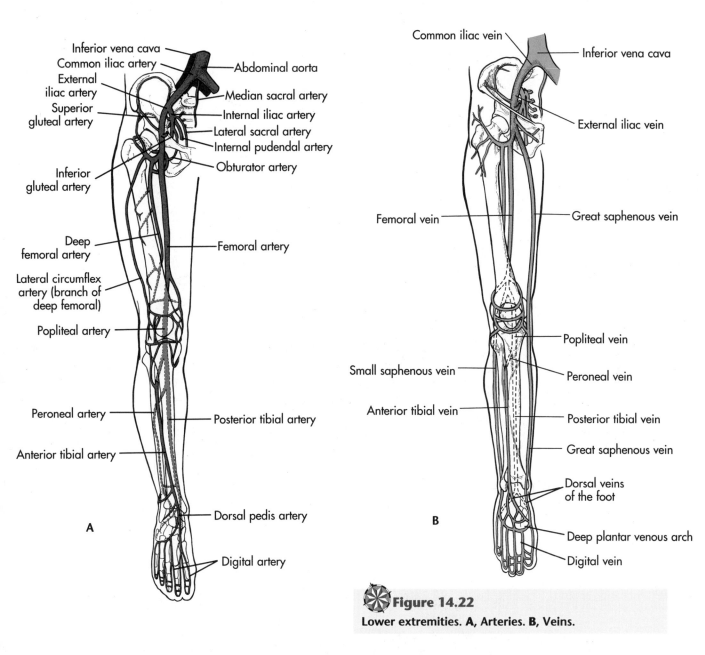

Common iliac vein

Figure 14.22
Lower extremities. A, Arteries. B, Veins.

course follows the popliteal artery to the femoral vein. The **small saphenous vein** ascends directly up the gastrocnemius muscle from the lateral malleolus of the fibula to the popliteal fossa and popliteal vein. The small saphenous vein is easy to find on the calf above the calcaneal (Achilles) tendon. The **great saphenous vein** is the longest vein in the body. Beginning among the superficial veins of the dorsal surface of the foot, it ascends the anteromedial surface of the leg, passes behind the knee, and reaches up the medial thigh to a few centimeters below the inguinal ligament before joining the femoral vein. The femoral vein becomes the external iliac vein, and then the common iliac vein after passing the inguinal ligament.

- Give the regional names of the artery that delivers from the abdominal aorta to the popliteal artery.
- What is the name of the longest vein in the body?

- The femoral vein is continuous with what vein after passing the inguinal ligament of the pelvis?

PULMONARY CIRCULATION

OVERVIEW

The lungs receive both pulmonary and systemic blood. Air rushes into each lung through bronchi that branch into each of 8 to 10 segments of the lung and to a few million thin-walled chambers called alveoli in each segment (Figure 14.23). Pulmonary blood arrives at the alveoli through segmental branches of the **pulmonary arteries** that end in capillary plexuses, which cover the alveoli of each lung segment. As

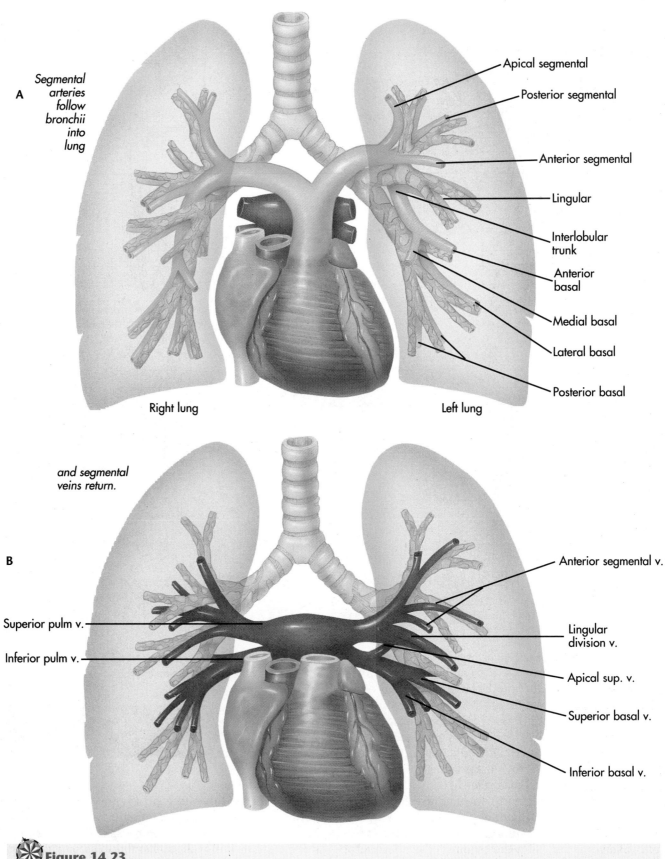

A

Segmental arteries follow bronchii into lung

Apical segmental

Posterior segmental

Anterior segmental

Lingular

Interlobular trunk

Anterior basal

Medial basal

Lateral basal

Posterior basal

Right lung

Left lung

B

and segmental veins return.

Anterior segmental v.

Lingular division v.

Apical sup. v.

Superior basal v.

Inferior basal v.

Superior pulm v.

Inferior pulm v.

Figure 14.23
Pulmonary circulation. A, Arteries. B, Veins.

blood and air flow past each other on opposite sides of delicate capillary walls, they exchange gases. The blood emerges rich with oxygen, and air takes on carbon dioxide. Venules quickly convey the oxygenated blood to **pulmonary veins** that return to the left atrium of the heart. The pulmonary circuit is exceptional in the adult vascular system because the arteries conduct deoxygenated blood and veins oxygenated blood, precisely reversing the usual condition. (Remember, arteries carry blood away from and veins conduct blood toward the heart.) The bronchi themselves receive freshly oxygenated systemic blood from the heart by way of the **bronchial arteries** described in Thorax, this chapter.

PULMONARY TRUNK

The pulmonary trunk arises from the right ventricle (Figure 14.23A) and ascends in front of and to the left of the ascending aorta. The pulmonary trunk (3 cm) is wider than the ascending aorta (2.5 cm), and its wall is one third as thick. Beneath the arch of the aorta, the trunk divides into the left and right **pulmonary arteries** that travel horizontally through the mediastinum to the root of the lung on each side. Upon entering the lungs, both arteries divide into **segmental branches** that follow the bronchi into each segment of the lung. Pulmonary arterioles conduct blood to the **pulmonary capillaries,** where gasses exchange.

PULMONARY VEINS

Segmental veins drain oxygenated blood from the pulmonary capillaries, as Figure 14.23B shows. In turn these vessels deliver to two **pulmonary veins** for each lung. The superior veins collect from the superior and middle segments, and the inferior veins collect from the inferior segments of the lungs. Both left and right pairs of pulmonary veins exit the lungs into the mediastinum and the left atrium of the heart.

- Which structures of the lung receive oxygenated blood from the systemic circuit?
- Which arteries deliver this blood?
- By contrast, what surfaces receive deoxygenated blood via the pulmonary trunk?
- In what respect does pulmonary circulation reverse the usual pattern of vessels carrying oxygenated and deoxygenated blood?

FETAL CIRCULATION AND CHANGES AT BIRTH

Newborn infants begin to use their own circulation at birth when the placenta and umbilical (um-BIL-ih-kal) cord disconnect from mother and fetus (Figure 14.24). Until then (Figure 14.24A), the **placenta** re-

lays oxygen from the mother's blood to the fetus through the **umbilical vein** and **umbilical cord.** (Remember, veins conduct toward the heart, the fetal heart in this case.) This oxygenated blood enters the liver by way of the **ductus venosus** (DUK-tus veh-NO-sus) and joins deoxygenated blood from the hepatic portal vein and inferior vena cava, before entering to the right atrium. The **foramen ovale** and **ductus arteriosus** (ar-TEER-ee-oh-sus) allow this now partially oxygenated blood to bypass the lungs and enter the **aorta** (See Chapter 13, Heart). The aorta delivers to the body of the fetus, but the **umbilical arteries** divert a major part of the flow from the **common iliac arteries** back to the placenta, where the mother oxygenates the blood once again.

The vascular changes that occur when the baby begins to breathe and its umbilical cord is cut (Figure 14.24B) leave remnants of the original vessels. The ductus venosus and the umbilical vessels close, leaving the **ligamentum venosum** (lig-ah-MEN-tum; venous ligament), the **ligamentum teres** (TEAR-eez; round ligament), and **medial umbilical ligaments** to mark the position of the old ductus venosus, umbilical vein, and umbilical arteries, respectively. Fully oxygenated blood now flows through the body because the foramen ovale and ductus arteriosus also close, which obliges blood in the right atrium to enter the lungs before returning to the left side of the heart. The ductus arteriosus constricts within 15 hours of birth, and the foramen ovale fuses by 3 months.

- Which fetal vessels connect the placenta with the fetus?
- What structures do these arteries and veins become after birth?
- What organs receive pulmonary blood after the foramen ovale and ductus arteriosus close?

VESSEL DEVELOPMENT

How do vessels, such as the aorta and capillaries, develop? Do they sprout like branches from a tree, or do they simply coalesce from cells that happen to be at the right place at the right time (see Figure 14.25A and B)? How do vessels attach to the appropriate organs, and how does the lumen form inside the new vessels? Embryologists have some broad answers to these questions, but no one knows the details of how vascular connections develop.

If you and I were going to build a plumbing system for a house, we would buy the pipes and install them where the plans indicate. Embryos seem to have no plans, but nevertheless the appropriate cells arrive at the appropriate locations at the right time for vessels to form. Furthermore, vessels are not permanent installations; flowing blood continually remodels arter-

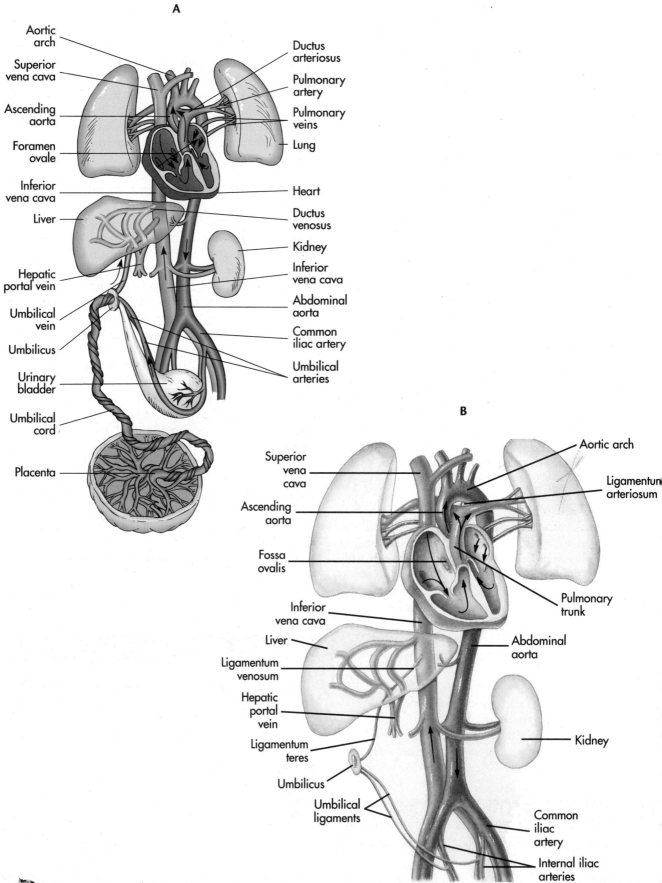

A

Aortic arch

Superior vena cava

Ascending aorta

Foramen ovale

Inferior vena cava

Liver

Hepatic portal vein

Umbilical vein

Umbilicus

Urinary bladder

Umbilical cord

Placenta

Ductus arteriosus

Pulmonary artery

Pulmonary veins

Lung

Heart

Ductus venosus

Kidney

Inferior vena cava

Abdominal aorta

Common iliac artery

Umbilical arteries

B

Superior vena cava

Ascending aorta

Fossa ovalis

Inferior vena cava

Liver

Ligamentum venosum

Hepatic portal vein

Ligamentum teres

Umbilicus

Umbilical ligaments

Aortic arch

Ligamentum arteriosum

Pulmonary trunk

Abdominal aorta

Kidney

Common iliac artery

Internal iliac arteries

✸ **Figure 14.24**
Fetal circulation and changes at birth. A, Before birth. B, After birth.

ies and veins so that they form or disappear, according to cellular demands. An excellent example of this plasticity is the ability of tumors to attract vessels to them. **Angiogenin,** a protein produced by certain tumors, causes capillaries to grow toward the tumor in Figure 14.25C and thereby supply it with blood. Nor-

mal tissues are thought to release factors like angiogenin that guide vessels to them.

The vascular system develops both by sprouting branches and by coalescence. The vertebral artery, for example, sprouts from a trunk that has already formed, as illustrated in Figure 14.25A. Sprouting

A   SPROUTING–ANGIOGENESIS

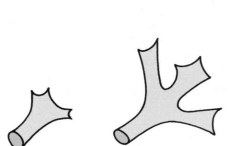

Endothelial cells at the tip proliferate and spread into the tissue, where they hollow out and become vessels.

COALESCENCE–VASCULOGENESIS

Angioblasts coalesce and form new vessel

C   ANGIOGENIN

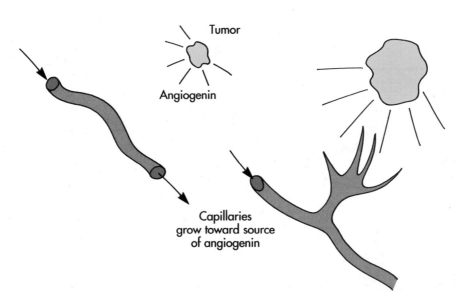
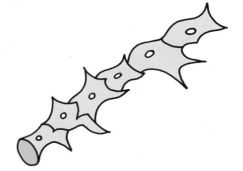

Tumor

Angiogenin

Capillaries grow toward source of angiogenin

Figure 14.25
Vascular development. A, Angiogenesis. B, Vasculogenesis. C, Capillaries grow toward sources of angiogenin.

from an existing vessel is called **angiogenesis.** Other vessels such as the vena cava coalesce from portions of previous veins. These processes tell little about how the first vessels formed. In vessels such as the aorta we come closer to understanding how development begins. The aorta coalesces from smaller clusters of cells by the process of **vasculogenesis,** shown in Figure 14.25B.

The cardiovascular system develops from mesoderm (see Mesodermal Derivatives, Chapter 4). Its precursor cells are called **angioblasts** for their ability to proliferate and for their commitment to develop into the endothelial lining of all vessels. Angioblasts first appear in the yolk sac at 3 weeks of development, when the trophoblast can no longer sustain the enlarging embryo. Clusters of angioblasts form **blood islands** surrounded by endothelium (Figure 14.26), and these islands merge with their neighbors, forming a network of vessels and primitive blood cells in the wall of the yolk sac. Angioblasts that form the aorta differentiate from local mesoderm within the embryo and form solid cords of cells that hollow out and become the endothelium of the aorta. The tunica media and tunica adventitia derive from other mesenchyme cells that accrete onto the endothelium.

- Do vessels sprout or coalesce from angioblasts in vasculogenesis?
- What is the name of the cells from which vessels develop?
- Why is angiogenin an appropriate name for the tumor protein that attracts capillaries?
- Where do blood vessels first develop?

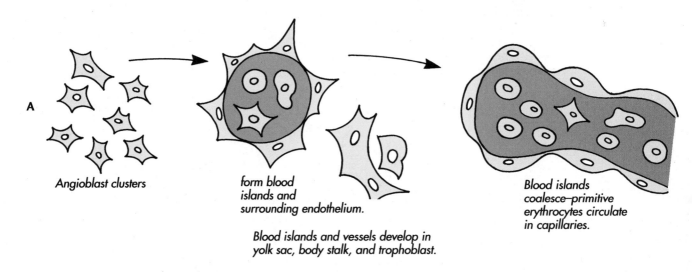

Angioblast clusters

form blood islands and surrounding endothelium.

Blood islands and vessels develop in yolk sac, body stalk, and trophoblast.

Blood islands coalesce—primitive erythrocytes circulate in capillaries.

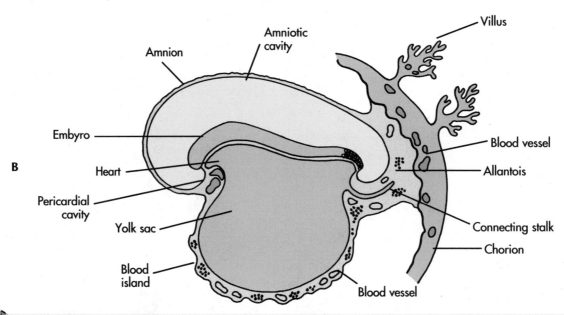

Figure 14.26

A, Vessels begin as clusters of angioblasts that coalesce into primitive vessels. **B,** Blood islands appear first in the yolk sac and trophoblast at 3 weeks.

SUMMARY

Arteries, capillaries, and veins are trunks and branches of the vascular tree. Branching arteries carry blood from the heart to capillaries in the tissues, and veins drain blood from the capillaries to the heart. Capillaries are the business end of the cardiovascular system because they expose immense surface areas to tissues that exchange oxygen and other essential molecules. *(Page 413.)*

Vessels diminish in diameter the farther from the heart they are in the vascular tree. The largest arteries conduct blood to smaller muscular arteries that branch and diminish in size but increase in numbers until they become arterioles, the smallest arteries. Capillaries, the smallest, most abundant vessels, receive blood from the arterioles. Venules drain the capillaries into progressively larger veins that return blood to the heart. *(Page 413.)*

Three tunics are responsible for the different properties of vessels. Tunica intima (interna) lines all vessels. Tunica media is thickest in arteries and may be elastic, muscular, or a combination of both, but the tunica media of veins is relatively thin and only weakly muscular. The tunica externa (adventitia) retains vessels in position and supplies the vasa vasorum to the larger veins and arteries. *(Page 413.)*

Atherosclerosis can plug arteries with cholesterol-laden plaques. Macrophages and smooth muscle cells can gradually thicken the tunica intima and tunica media of arteries with deposits of collagen fibers and foam cells laden with cholesterol. The risk that a plaque will block an important artery and cause stroke or heart attack increases during life as plaques accumulate and enlarge. *(Page 419.)*

The structure of vessels reflects blood pressure. Large elastic and muscular arteries sustain the high, pulsing pressure of blood. Much of this pressure dissipates before blood reaches the thin-walled capillaries, which have no media or externa. Tunica media and externa reappear in the veins, but pressure continues to decline and the thin walls readily collapse when the veins empty. Venous valves promote blood flow in the larger veins. *(Page 420.)*

The capillary endothelium transports molecules between the blood and tissues. Tissues and capillaries communicate across an immense internal exchange surface formed by the walls of capillaries that weave among the cells of the tissues. Small molecules diffuse across this surface, and transcytosis transports larger molecules in membranous vesicles. Capillary endothelium also takes up materials by phagocytosis and receptor-mediated endocytosis. *(Page 424.)*

Blood circulates out the arterial tree from the heart, to the organs, and back. Arteries and veins usually have local names that identify their immediate surroundings rather than the ultimate destination of the blood they carry. Veins and arteries frequently have the same names because these vessels tend to accompany each other through the same territories. Arteries and especially veins anastomose, providing alternate pathways for blood to and from the tissues. *(Page 426.)*

Systemic circulation delivers blood from the left side of the heart to virtually all body systems except the respiratory surfaces of the lungs, which the pulmonary circulation supplies from the right side of the heart. The aorta is the largest systemic artery; it conducts blood to branches that distribute to the head, neck, upper extremities, thorax, abdomen, pelvis, and lower extremities. The superior and inferior venae cavae return blood from regions above and below the diaphragm, respectively, to the heart. *(Page 427.)*

Blood reaches the head and neck by way of the common carotid and vertebral arteries that serve the face, cranium, and brain. Venous sinuses, the internal and external jugular veins, and the vertebral veins return blood from the head to the brachiocephalic veins and superior vena cava. *(Page 430.)*

Branches from the arch of the aorta and from the thoracic aorta distribute blood to the upper limbs and thorax. The subclavian arteries and veins serve the shoulders and upper limbs, while the internal thoracic arteries and branches of the thoracic aorta supply the thoracic wall and visceral organs. Blood returns from these fields by way of the azygos vein and internal thoracic veins that drain into the superior vena cava and the brachiocephalic veins, respectively. *(Page 434.)*

Below the diaphragm, the abdominal aorta and inferior vena cava serve the gastrointestinal tract and paired organs. The celiac trunk and superior and inferior mesenteric arteries serve the gastrointestinal tract and spleen, which are drained into the liver by the hepatic portal vein and thence to the inferior vena cava. Paired arteries supply the diaphragm, adrenal glands, kidneys, gonads, and vertebral column. Paired veins return this blood to the inferior vena cava. *(Page 437.)*

The common iliac arteries and veins serve the pelvic and gluteal regions and the lower extremities. Branching pairwise from the abdominal aorta, the common iliac arteries supply the pelvis through the internal iliac artery and its anterior and posterior divisions. Venous return is similar to this arterial outflow; blood from the rectum, bladder, genitalia, and pelvic walls returns to the inferior vena cava by way of anterior and posterior divisions of the internal iliac veins, which are tributaries to the common iliac veins and the inferior vena cava. *(Page 440.)*

Blood enters the lower extremities through the femoral artery and returns to the pelvis through the femoral vein. On piercing the wall of the pelvis, the external iliac artery becomes the femoral artery and then the popliteal artery that subsequently branches into the leg and foot. Deep veins that correspond with these arteries return this flow to the external iliac vein. Superficial veins, notably the great and small saphenous veins, also contribute to this flow. *(Page 443.)*

The lungs oxygenate blood in the pulmonary circuit. The pulmonary trunk and pulmonary arteries deliver blood to the alveoli of the lungs, where air and blood exchange oxygen and carbon dioxide. This blood returns through the pulmonary veins to the heart. *(Page 445.)*

A fetus receives nutrition and oxygen from the placenta. A single umbilical vein delivers blood from the placenta to the fetal heart; the blood bypasses the lungs and flows back to the placenta by way of two umbilical arteries. These vessels close when birth disconnects the baby from its mother, and the pulmonary and systemic circulations become separate. *(Page 447.)*

STUDY QUESTIONS

1. Follow in sequence the types of vessels an erythrocyte encounters on the way from the heart to the capillaries and back to the heart.   **1**

2. What are the differences between the tunics of the vessels? In what ways does the intima differ from the media, and the media from the externa?   **2**

3. How do the tunica media of elastic and muscular arteries differ? What differences in the tunica externa does one encounter in going from the capillaries to large veins?   **2**

4. What differences in the tunics of large arteries and veins are related to blood pressure?   **3**

5. Describe the changes that culminate in blockage of an artery by atherosclerosis.   **4**

6. How do substances cross the capillary endothelium?   **5**

7. What anatomical features show that capillaries are the business end of the cardiovascular system?   **5**

8. To which organs do the following arteries deliver blood: internal thoracic, pudendal, facial, subclavian, celiac, hepatic, femoral, renal, and ovarian?   **6**

9. Which organs or tissues do the following veins drain: vesical, inferior phrenic, hepatic portal, brachiocephalic, internal jugular, superior sagittal sinus, azygos, and small saphenous?   **6**

10. Beginning in the ascending aorta, trace blood flow into the middle cerebral artery of the brain and its return via the jugular vein to the superior vena cava.   **6**

11. Beginning with the left common carotid artery, trace blood to the face and its return via the external jugular vein to the brachiocephalic vein.   **6**

12. Deliver blood from the left ventricle of the heart to the superior surface of the diaphragm and return the blood to the right atrium.   **6**

13. Follow blood from the ascending aorta into the right upper limb to the digits and back by way of the cephalic vein to the right brachiocephalic vein.   **6**

14. Deliver blood from the left ventricle to the small intestine and return it through the hepatic portal vein to the right atrium of the heart.   **6**

15. Follow blood from the abdominal aorta to the right ovary or testis and return to the right atrium of the heart.   **6**

16. Beginning again with the abdominal aorta, follow the vascular pathway to the bladder and return to the inferior vena cava.   **6**

17. Deliver blood to the left foot from the heart and return to the inferior vena cava.   **6**

18. Follow deoxygenated blood from the right ventricle, through the left lung, and return oxygenated blood to the left atrium.   **6**

19. What pathway conducts oxygenated blood from the right lung to the muscles of the right thigh?   **7**

20. How does urea produced by the liver reach the kidneys? Outline the vascular pathway between these two organs.   **7**

21. The cavernous sinus receives gonadotropin hormones from the pituitary gland. Trace the flow of these hormones to the ovaries or testes by way of the gonadal arteries.   **7**

22. The brain consumes large quantities of glucose, which is absorbed from the small intestine. Follow the path taken by these molecules from the duodenum to the anterior cerebral artery.   **7**

23. Muscles of the left upper extremity release carbon dioxide into the veins that return to the heart and lungs. Follow this deoxygenated, carbon dioxide-enriched blood from the left upper extremity to the lungs. What path would carbon dioxide follow from the gastrocnemius muscle of the calf to the lungs?   **7**

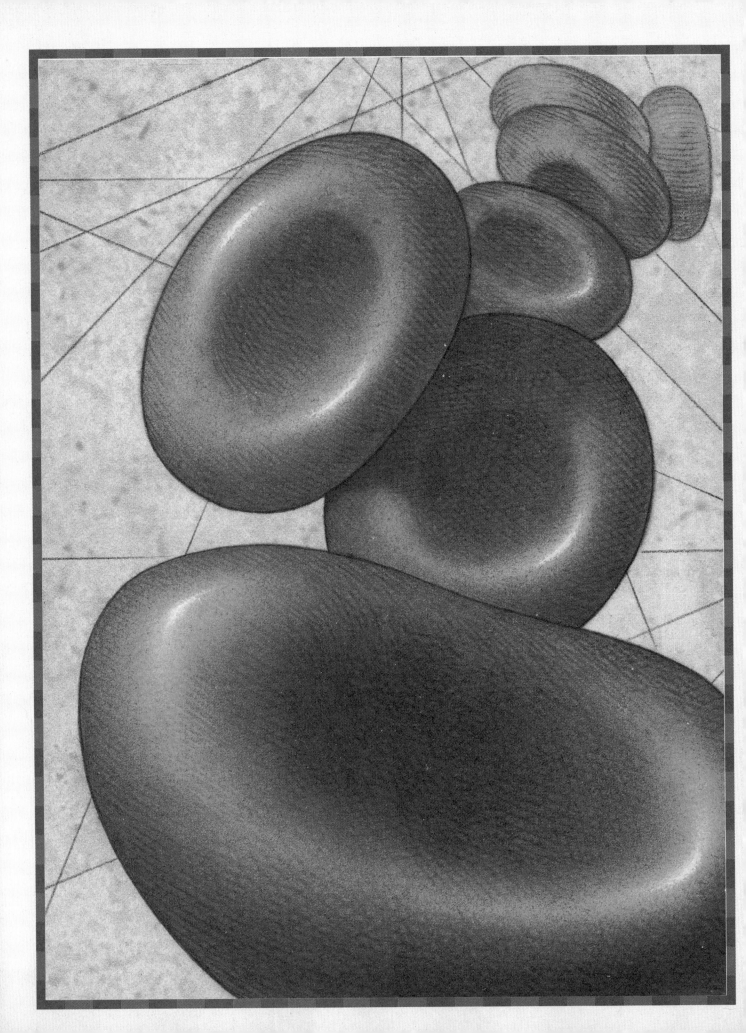

CHAPTER 15

Blood

OBJECTIVES

After studying this chapter, you should be able to do the following:

1. Name the major components of blood plasma and their functions.
2. Name the cellular components of blood and their functions.
3. Describe what information a hematocrit, a differential cell count, and a plasma profile disclose about a sample of blood.
4. Outline the lineages from which blood cells develop.
5. Describe the events in the life of an erythrocyte.
6. Describe the cellular features and functions of the various granulocytes.
7. Describe the appearance and function of monocytes and the life cycle of these cells.
8. Describe the appearance and function of lymphocytes, and trace the events in their life cycle.
9. Describe the cytoplasmic anatomy and cellular function of platelets.
10. Describe the link between elevated blood cholesterol and atherosclerosis.

VOCABULARY TOOLBOX

Blood cell terms.

erythro-	*G* red. **Erythro**cytes are red cells.
hemo-, hemato-	*G* blood. **Hemo**globin is the red blood pigment. **Hemato**poiesis is the process that produces blood cells.
karyo-	*G* nucleus. Mega**karyo**cytes have large nuclei.
leuko-	*G* white. **Leuko**cytes are white blood cells.
lympho-	*L* water. **Lymph** is the watery fluid of the lymphatic system, and **lympho**cytes are its cells.
macro-	*G* large. **Macro**phages are large cells that engulf other cells.
mega-	*G* large. **Mega**karyocytes have large nuclei.
myelo-	*G* marrow. **Myelo**id cells develop in the bone marrow.
phago-	*G* eat. **Phago**cytes and macro**phages** engulf other cells.
-phil	*G* love, loving. Neutro**phils** are blood cells that bind neutral dyes.
pluri-	*L* several. **Pluri**potential stem cells have several different fates.
-poiesis	*G* make, produce. Erythro**poiesis** is the process that produces red cells.
thrombo-	*G* a blood clot. **Thrombo**cytes cause blood clots to form.

RELATED TOPICS

- **Lymphatic and immune system, Chapter 16**
- **Tissues, Chapter 3**
- **Vessels, Chapter 14**

PERSPECTIVE

Blood is an excellent coordinator of the body's tissues since, like traffic on an interstate, everything in blood is going somewhere. No other tissue displays more information about an individual's metabolism than does the blood, because its components are readily measured in health and disease. To understand the function of blood, you need to trace the origin and fate of blood cells and molecules and how their production and consumption is balanced, much as you would follow trucks and cars to learn their destinations. Without the usual tissue connections to hold them together, cells and molecules of blood enter and leave the traffic at various locations, according to their functions. All of the cells develop outside the blood itself and then circulate with it for various times, carrying out specific functions—transporting oxygen, defending against infection, or controlling bleeding—before finally entering the tissues.

B lood is the red fluid that circulates through your arteries and veins to all tissues of your body. On average, about 4 to 5 liters of blood circulate in women and 5 to 6 liters in men. Blood is a connective tissue; the body functions as a whole partly because blood communicates between its parts.

Blood connects tissues metabolically by circulating the products of different organs to other tissues that need these substances. These connections are as diverse as the cells and molecules in blood. For example, plasma delivers glucose from the gut for cellular respiration in virtually every tissue, and erythrocytes carry oxygen from the lungs throughout the body, as Figure 15.1 shows. The lungs also remove carbon dioxide from the blood, and the kidneys excrete urea wastes that arise in all tissues. Erythropoietin is an example of a molecule that regulates a process downstream from its source. Produced in the kidneys, erythropoietin stimulates erythrocyte development when it reaches the bone marrow. Certain transport molecules carry other molecules that do not readily dissolve in blood. One such group is made up of the lipoproteins that carry cholesterol molecules, essential constituents of all cell membranes. The liver is the source of fibrinogen, the principal clotting protein that prevents bleeding; and the lymph nodes and bone marrow circulate leukocytes that patrol the body for foreign cells and substances.

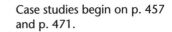

Case studies begin on p. 457 and p. 471.

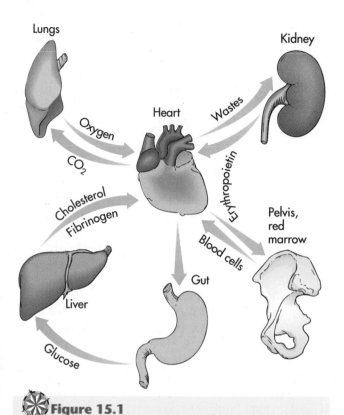

Figure 15.1

Blood from all organs mixes in the heart. Oxygen reaches all tissues from the lungs. The kidneys collect wastes from the tissues and produce erythropoietin that is sent to all tissues, but stimulates only the bone marrow to produce erythrocytes. The intestines absorb glucose, which circulates to the liver. The liver produces cholesterol and fibrinogen.

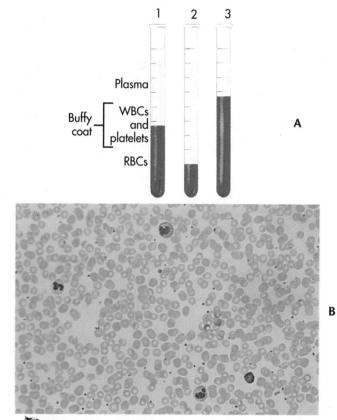

Figure 15.2

A, A patient's blood provides samples for a hematocrit(1), a differential cell count and blood smear(2), and a profile of plasma components(3). **B,** Photomicrograph of blood smear.

CELLS AND PLASMA

With hundreds of circulating components, it is not surprising that a sample of blood reveals more about your health than any other tissue. When a clinician draws a sample of blood from a patient (Figure 15.2), the laboratory divides it into three portions. A small portion drawn into a narrow glass tube and then spun in a centrifuge reveals the two major components of blood—**plasma,** the fluid portion, and **formed elements,** the cells that are suspended in the plasma. Most of the remaining blood goes into a **plasma profile,** or analysis of the plasma components, but a third portion is used to make a **differential cell count** that registers the different types of circulating cells.

The centrifuge sediments the cells, mostly erythrocytes (red blood cells), to the bottom of the centrifuge tube, and the clear, straw-colored plasma remains above them. The height of the red column of cells may reveal that cells occupy, for example, 45% of the patient's blood volume, and the plasma occupies 55%. Called **hematocrit** (HE-mat-oh-crit), this measurement quickly discloses the proportion of cells and plasma in the blood as a whole by their height in the tube. Normal hematocrit values are 36 to 45 in women and 42 to 50 in men. A lower hematocrit can indicate that too few erythrocytes are circulating, and that the patient may suffer from anemia.

PLASMA

As the extracellular matrix of blood, plasma is 91.5% water. Electrolytes, gases, numerous metabolites, and various proteins dissolved in the plasma make up the rest of the volume. Table 15.1 lists the components revealed in a routine plasma profile of a blood sample drawn from an actual patient. Sodium, potassium, and chloride ions are major **electrolytes** that maintain membrane potentials for membrane signaling and membrane transport in all cells. The levels of calcium, magnesium, and phosphate indicate the quantities of these ions available to bone and other tissues. Iron atoms bind oxygen in hemoglobin molecules, and the level found in this profile indicates that the patient has adequate quantities circulating.

Plasma levels indicate the balance between production and consumption of metabolite molecules. The glucose reading for example, indicates that the

Table 15.1 PLASMA PROFILE

Name	Quantity*	Normal Range
ELECTROLYTES		
Sodium	139 mmol/l	134-143
Potassium	4.70 mmol/l	3.60-5.30
Chloride	103 mmol/l	95.0-107
Calcium	9.50 mg/dl	8.90-10.3
Magnesium	2.00 mg/dl	1.60-2.10
Phosphate	3.30 mg/dl	2.20-4.20
Iron	131 µg/dl	50.0-160
METABOLITES		
Glucose	92.0 mg/dl	65.0-130
Triglycerides	64.0 mg/dl	50.0-200
Cholesterol	242 mg/dl	150-246
HDL Cholesterol	35.0 mg/dl	39.0-73.0
Urea	17.0 mg/dl	9.00-24.0
Bilirubin	0.71 mg/dl	0.30-1.40
PROTEINS		
Albumin	4.30 gm/dl	3.70-4.80
Globulin	2.90 gm/dl	2.20-3.50
ENZYMES		
Alkaline phosphatase	79.0 units/l	54-158
Glutamyl transpeptidase	8.00 units/l	1.00-70.0
Lactate dehydrogenase	146 units/l	115-250
Transaminase	23.0 units/l	1.00-45.0

*Abbreviations: mmol, millimoles; l, liter; mg, milligrams; dl, deciliter (1/10 liter); µg, micrograms; gm, grams.

patient's tissues are receiving sufficient supplies of this fundamental carbon source for cellular respiration. Because lipid metabolism assembles fatty acid molecules and glycerol into triglycerides, the triglyceride value indicates that lipid metabolism is operating normally. Our patient has too much cholesterol (ko-LES-ter-ole) in his blood, however, and the quantity indicates that the liver is not absorbing as much as it should. Physicians consider patients with this level to have a borderline high risk of atherosclerosis and coronary heart disease. Furthermore, HDL cholesterol, the so-called good cholesterol, is below normal levels. This means that high density lipoprotein (HDL) carrier molecules in the plasma are transporting less cholesterol than normal from the tissues (see Cholesterol and Atherosclerosis, this chapter).

Metabolism of proteins and nucleic acids in the tissues produces ammonia. The liver takes up the ammonia and converts it into urea that returns to the plasma. The patient's urea reading indicates that production is balanced by excretion from the kidneys. When red

blood cells are destroyed in the spleen and liver, a portion of each hemoglobin molecule is converted to bilirubin (BIL-ee-roo-bin), which is circulated through the blood and then stored in the gall bladder in the form of bile salts that assist digestion. The plasma profile indicates that the gall bladder is consuming about as much bilirubin as the liver and spleen are producing. Excessive bilirubin gives the skin a bronze color recognized as **jaundice** (JAWN-diss; jaune, *F*, yellow).

Blood proteins fall into three major groups— albumins, globulins, and clotting proteins. **Albumins** (al-BYOO-mins), together with bicarbonate ions, buffer the blood plasma in the narrow range between pH 7.35 and 7.45. Synthesized primarily in the liver, albumins are important proteins because they are more abundant in plasma than in tissue fluid and thereby tend to draw water from the tissues into the blood osmotically. The **globulins** (GLO-byoo-lins) serve many roles. Perhaps the most important is as circulating antibody molecules that remove foreign materials. Our patient has normal levels of albumin and globulin. The globulins also contain numerous enzymes from the liver and other tissues. Alkaline phosphatase (FOS-fate-ace), transaminase (trans-AM-in-ace) and lactate dehydrogenase (de-HY-dro-JEN-ace) are examples of these enzymes; each is circulating at normal levels. These three substances are indicator enzymes that help identify different disorders, according to which levels rise or fall. **Fibrinogen** (fye-BRIN-oh-jen) is the principal clotting protein that platelets convert into fibrin clots. The components of plasma come from several locations; the liver is the principal source of globulins and the clotting proteins, but lymph nodes are responsible for circulating antibodies, and numerous tissues release enzymes.

- Name the two major components of blood revealed by a hematocrit.
- Name the classes of substances dissolved in plasma.
- What metabolic processes balance the quantities of these constituents circulating in the plasma?

CELLS

Blood contains three major classes of cells— erythrocytes (eh-RITH-row-sites; red blood cells), **leukocytes** (LOO-ko-sites; white cells with no pigment of their own), and platelets, the cells responsible for hemostasis (HEE-mo-stay-sis; prevention of bleeding). Later sections describe each class in detail, but this brief overview helps you see their relationships. When a drop of blood is spread on a microscope slide for a differential cell count, allowed to dry, and then stained, the dyes color the various cells as shown in Figure 15.3. Most numerous in this **blood smear** are the **erythrocytes,** which the dyes stain pale pink. Erythrocytes sustain cellular respiration by ferrying oxygen from the lungs and removing carbon dioxide.

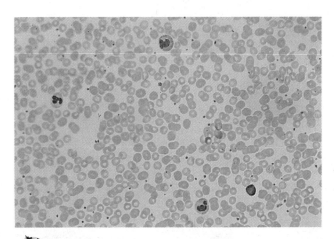

Figure 15.3

Erythrocytes, leukocytes, and platelets in a blood smear. Photomicrograph.

Leukocytes and platelets take up little blood volume compared with the erythrocytes, as you can see in Figure 15.2 by finding the narrow "buffy coat" on top of the erythrocytes in the hematocrit tube.

Larger than erythrocytes and more vigorous metabolically, **leukocytes** are nucleated cells that defend against foreign cells (1) by ingesting them, (2) by producing antibodies against them, and (3) by promoting inflammation reactions in the tissues. With such different functions, it is not surprising that there are many types of leukocytes. **Granulocytes** (GRAN-you-low-sites) contain cytoplasmic granules that stain brilliantly, as well as lobular nuclei, for which these cells are also known as **polymorphonuclear leukocytes** (POLY-mor-fo-NUK-lee-ar; many-formed nuclei). Granulocytes, or polymorphs as they are frequently called, kill cells and promote inflammation by releasing the contents of their granules. **Lymphocytes** (LIMF-oh-sites) and **monocytes** are leukocytes that have finer cytoplasmic granules and nuclei that take on fewer forms than do those of polymorphs. Lymphocytes produce antibodies against foreign cells and molecules. Monocytes, with their typical U-shaped nucleus, are precursors to the macrophages that patrol body tissues, ingesting spent and foreign cells.

Figure 15.3 also displays a few small, bluish **platelets,** the cells that prevent bleeding by sealing ruptured vessels. The smallest cellular elements, platelets are fragments of cytoplasm that have pinched off of large leukocytes called **megakaryocytes** (MEG-a-KAR-ee-oh-sites; large nucleated cells).

CELL LINEAGE

The cells in the blood smear represent a balance between the development of new cells and loss of the old

from circulation. The situation somewhat resembles that in plasma, but cell division and differentiation in the bone marrow, rather than metabolic synthesis and degradation, continually produce enough mature cells to replace those lost on patrol and through routine cell death. Anemia is an example of imbalances; in this disorder the marrow does not supply enough red cells and hemoglobin to meet the needs of the tissues. Later sections follow the development of blood cells in more detail, but a short overview now will help you to see how balance is achieved.

All blood cells develop from pluripotential stem cells, progenitor cells whose progeny can become erythrocytes, granulocytes, monocytes, platelets, or lymphocytes. This versatility is the reason why bone marrow transplantation can replace a patient's diseased marrow with healthy cells. Figure 15.4 shows the lineage by which blood cells develop from their common progenitors. Colonies of **pluripotential stem cells** (PLOOR-ih-PO-ten-shal; capable of several fates) proliferate in the bone marrow. The progeny commit to either the **lymphoid line,** from which lymphocytes develop, or the **myeloid line** (MY-eh-loyd), which produces all other types of blood cells. Once committed to each line, the cells continue to divide but their progeny can only mature into cells of that particular line. They are said to be **unipotent** (one fate). Thus a myeloid progenitor cell might differentiate as an erythrocyte or granulocyte but not as a lymphocyte. As a result, each type of cell that enters the blood matures from a specific lineage of cell division and differentiation. Most important, each lineage is regulated differently, which is why anemic patients can be deficient in erythrocytes but not in granulocytes.

Different names identify the development of different lineages. **Hematopoiesis** (he-MAT-oh-POY-ee-sis) includes the development of all blood cells. Hematopoiesis begins in the blood islands of the yolk sac during the third week of development (see Chapter 14, Vessels) and moves with these cells to the liver during the third month. Hematopoiesis also begins in the young thymus, spleen, and lymph nodes during this time, and in the fourth month it begins in the bone marrow. By birth, hematopoiesis has spread through the red marrow, where it resides in adults (more details in Bone Marrow, this chapter). **Erythropoiesis** concerns only the erythrocyte lineage; **granulopoiesis,** the granulocytes and monocytes; and **thrombopoiesis,** the platelets **Lymphopoiesis** is the name given to lymphocyte development.

- What are the names of the five major classes of blood cells?
- What types of cells develop from the myeloid lineage?
- How many different cell lines can a pluripotent and a unipotent precursor cell produce?

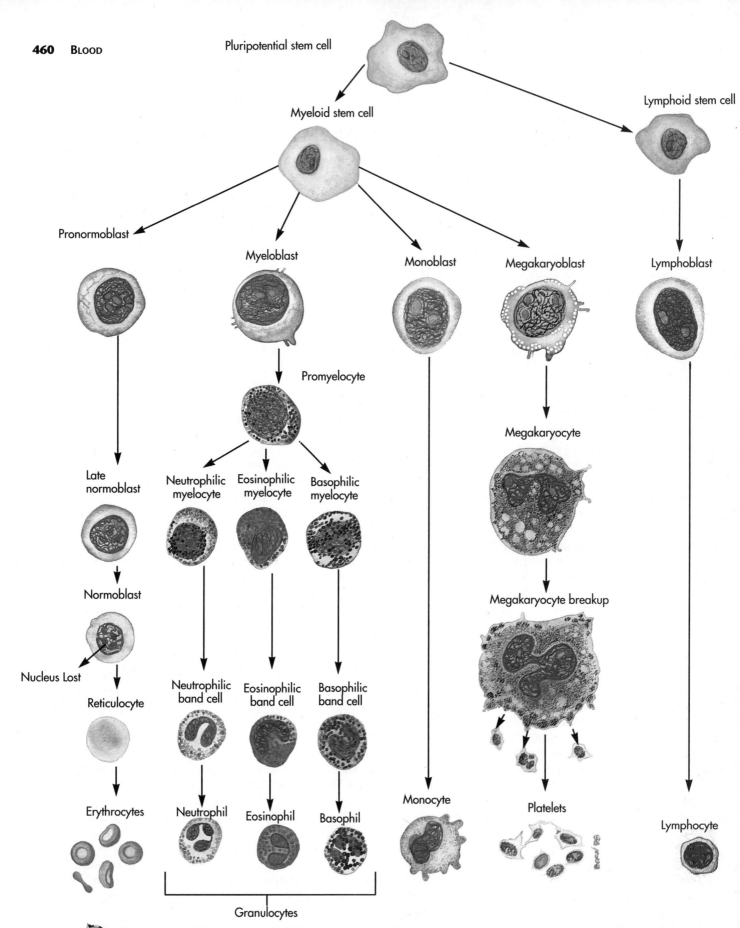

Pluripotential stem cell

Myeloid stem cell

Lymphoid stem cell

Pronormoblast

Myeloblast

Monoblast

Megakaryoblast

Lymphoblast

Promyelocyte

Late
normoblast

Neutrophilic
myelocyte

Eosinophilic
myelocyte

Basophilic
myelocyte

Megakaryocyte

Normoblast

Nucleus Lost

Reticulocyte

Neutrophilic
band cell

Eosinophilic
band cell

Basophilic
band cell

Megakaryocyte breakup

Erythrocytes

Neutrophil

Eosinophil

Basophil

Monocyte

Platelets

Lymphocyte

Granulocytes

Figure 15.4

Blood cell lineage. Pluripotential stem cells establish the different lineages. Once committed to its lineage, precursor cells are unipotent and mature accordingly. Myeloid stem cells, myeloblasts, and promyelocytes are pluripotent cells. Pronormoblasts, the various granular myelocytes, monoblasts, megakaryoblasts, lymphoid stem cells, and lymphoblasts are unipotent cells.

ERYTHROCYTES—FUNCTION AND LIFE CYCLE

ANATOMY

Approximately 25 to 40 trillion erythrocytes, about 5 trillion/l of blood, ferry oxygen and carbon dioxide throughout the circulatory system. The hematocrit and blood smear from our patient indicate how important this role is; nearly all of the cells and 45% (about 2 l) of the blood volume are erythrocytes. These small (7.5 to 8.5 μm diameter), flexible, cellular disks (Figure 15.5A) contain hemoglobin, the molecule that carries oxygen and carbon dioxide and colors blood red. A network of **spectrin** fibers in the cytoplasm (Figure 15.5B) molds the cell in the characteristic double concave shape that improves the efficiency of exchange across the cell membrane. Figure 15.5C shows that circulating erythrocytes tend to follow each other single file through capillaries and other narrow vessels. Blood bends them into cone-like forms because it flows faster in the center of vessels than at the walls.

HEMOGLOBIN AND OXYGEN TRANSPORT

About 97% of the protein molecules in an erythrocyte are hemoglobin. A single hemoglobin molecule consists of four subunits linked together (Figure 15.5D). Two different subunits—two alpha chains and two beta chains—make up the hemoglobins of most adults. This arrangement enables hemoglobin to transport oxygen molecules readily because the subunits interact with each other to take on and discharge oxygen molecules. Each chain, whether it be an alpha or beta, consists of a protein molecule, the globin, bound to a porphyrin ring, the heme portion of the molecule that binds oxygen and colors hemoglobin red. Each heme firmly binds an iron atom into its central core, and the iron atom in turn readily binds with a single molecule of oxygen. As a result, every fully charged hemoglobin molecule (oxyhemoglobin) that circulates out of the lungs carries four oxygen molecules, one in each heme ring and globin chain. Hemoglobin molecules release oxygen molecules when hemoglobin encounters large quantities of carbon dioxide in the tissues. Carbon dioxide binds to hemoglobin (carbaminohemoglobin), thereby releasing one or more of the oxygen molecules. Fetal hemoglobin contains alpha and gamma chains that together bind oxygen more firmly so that it can be transferred from mother to fetus. Mutant chains that bind oxygen poorly are found in certain anemias (see Disorders, p. 462).

ERYTHROPOIESIS

Erythrocytes differentiate in about 9 days. During the process known as **erythropoiesis,** erythrocyte precursor cells or **pronormoblasts,** derived from pluripo-

tential stem cells, proliferate in the bone marrow (Figure 15.4). The progeny of the pronormoblasts mature into normoblasts and eject their nuclei. Without their nuclei, normoblasts become **reticulocytes** (reh-TICK-you-low-sites), the cells that actually produce hemoglobin. Reticulocytes require 1 or 2 days to synthesize hemoglobin, and they slip into the circulation during this time. Hemoglobin production shuts down when enough has accumulated in the cell's cytoplasm, and the membranous organelles that have driven synthesis in the reticulocytes disappear. The reticulocyte then becomes an erythrocyte, a remnant piece of cytoplasm with a rudimentary metabolism sufficient only to maintain its cell membrane. Mature erythrocytes circulate for about 120 days, until changes in the plasma membrane signal macrophages in the liver and spleen to remove them by phagocytosis. Phagocytosis conserves the iron and heme portions of the hemoglobin molecule, but it degrades the globin portion to amino acids for assembly into other proteins. The liver stores the iron atoms in **ferritin,** one of the globulin proteins, until new reticulocytes use the iron once again. The liver and spleen convert the heme portion into **bilirubin** that is stored in the gall bladder as bile pigments and released into the intestine to facilitate fat digestion.

FEEDBACK REGULATION

Erythropoiesis adjusts to the body's demand for oxygen. Insufficient oxygen stimulates production of red cells, as attested by the extra quantities of erythrocytes in the blood of persons who live at high altitudes. **Erythropoietin** (ee-RITH-ro-POY-eh-tin) is a protein, coded by a gene located on chromosome 7, that stimulates erythropoiesis through a negative feedback loop illustrated by Figure 15.6. This protein stimulates pronormoblasts to become normoblasts and thence erythrocytes, by binding to receptors on the plasma membranes of the cells. This boost would be very short-lived, if the depleted pool of pronormoblasts did not also stimulate stem cells to replenish the pool. Erythropoietin is made in the kidney by the endothelium of peritubular capillaries and interstitial cells of the cortex.

DISORDERS

Disorders of erythropoiesis cause some but not all forms of anemia. Remember that anemia is a broad spectrum of conditions in which the blood has insufficient hemoglobin or erythrocytes to supply the tissues with enough oxygen. Two major forms of this disorder are **aplastic anemia,** in which the marrow fails to produce enough erythrocytes, and **hemolytic anemia,** wherein the red cells die prematurely. As you might expect, the causes of aplastic anemia interfere with erythropoiesis by attacking erythrocyte stem cells and the maturation of normoblasts and reticulocytes. The source of trouble may be exposure to drugs

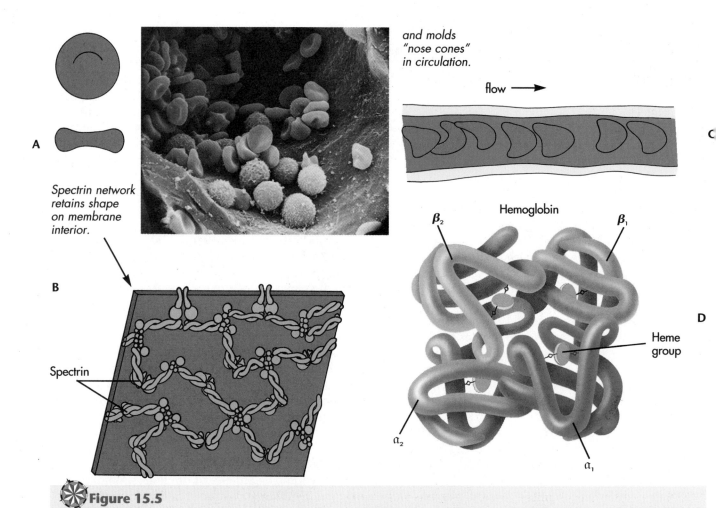

and molds "nose cones" in circulation.

flow ⟶

Spectrin network retains shape on membrane interior.

Spectrin

Hemoglobin

β₂ β₁

Heme group

α₂ α₁

A

B

C

D

★ Figure 15.5

Erythrocytes. A, Cross section through an erythrocyte similar to those seen in the scanning electron micrograph. Irregular cells are leukocytes. (Scanning electronmicrograph by permission of R.G. Kessel and R.H. Kardon, *Tissues and organs: A text-atlas of scanning electron microscopy,* 1979, W.H. Freeman and Co.) **B,** A network of spectrin protein fibers on the inner surface of the plasma membrane shapes erythrocytes into their discoid form. **C,** Blood flow bends erythrocytes into "nose cone" shapes. Blood flows to the right. **D,** Hemoglobin molecules consist of alpha and beta polypeptide chains that each bind an oxygen molecule to an iron atom housed in a heme ring.

---

or other chemicals to which the patient is sensitive. Radiation, infections, and inherited conditions may also interfere with stem cell development. Erythropoietin levels frequently are high, which suggests that the feedback loop that stimulates erythropoiesis has been broken. In hemolytic anemia, however, fewer red cells circulate because they lyse (break open) prematurely. The cell membrane may be especially fragile, or immunological processes may destroy it.

Other forms of anemia, mostly hereditary, affect the hemoglobin molecule itself. In **sickle cell anemia,** the best known of these **hemoglobinopathies** (HE-mo-GLOW-bin-oh-PATH-eez), genetically altered hemoglobin binds oxygen poorly, forcing the erythrocytes into unusual sickle shapes. The **thalassemias** (THAL-a-SEEM-ee-ahs; thalasso, *G,* sea; this disease was first found in Greece and Italy around the Mediterranean Sea) interfere with the synthesis of hemoglobin, so that erythrocytes have insufficient quan-

tities or abnormal combinations of hemoglobin subunit molecules.

- What is the shape and size of an erythrocyte?
- Which type of precursor cell loses its nucleus, and which type synthesizes hemoglobin?
- What is the fate of an erythrocyte?
- What is the name of the protein that regulates erythropoiesis?
- What is the difference between aplastic and hemolytic anemia?

GRANULOCYTES— PHAGOCYTOSIS AND INFLAMMATION

OVERVIEW

Granulocytes exist in three major forms—neutrophils, eosinophils, and basophils. Figure 15.7

Pluripotential
stem cell

Myeloid stem cell

*Erythropoietin
stimulates
division and
differentiation*

*Depletion
stimulates
replenishment*

Normoblast

Reticulocyte

Erythrocytes

*Oxygen,
hemoglobin
tissue demand*

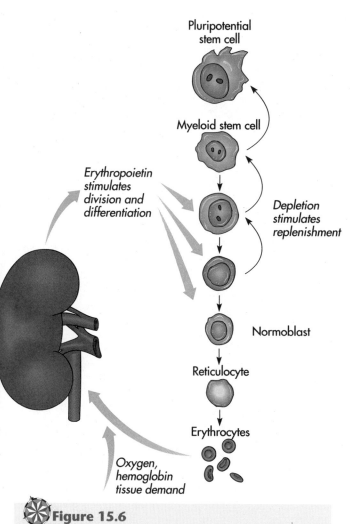

Figure 15.6

Erythropoietin regulates erythropoiesis through negative feedback. Insufficient oxygen stimulates the kidney to increase production of erythropoietin. Additional erythropoietin stimulates pronormoblasts and normoblasts to form additional erythrocytes. Myeloid stem cells replenish the pool of pronormoblasts depleted by this burst of erythropoiesis.

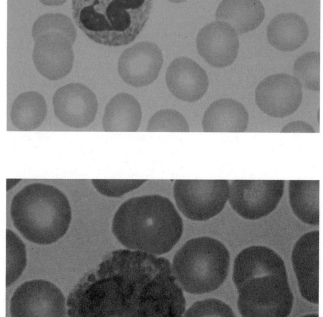

A

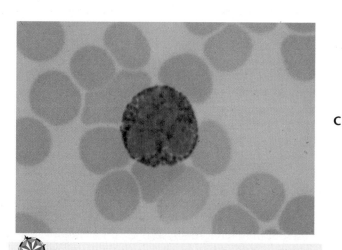

B

C

shows examples of each type taken from our patient's blood smear. These names indicate that the cells have affinity for different dyes in the stain. The large cytoplasmic granules of granulocytes take on different colors that are clues to their contents and functions, as you will see below. **Neutrophil** (NEW-tro-fil) granules usually are blue; those of **eosinophils** (EE-oh-SIN-oh-fils) are brilliantly orange or red; and **basophil** (BAYS-oh-fil) granules are intensely violet. The shape of the nucleus also helps identify polymorphonuclear granulocytes. Always dark blue in color, neutrophil nuclei may have as many as five lobes, but eosinophils and basophils usually display only two or three. In females one small, nuclear lobe contains inactive sex chromatin from one of the two X-chromosomes. When you learn the various features of nucleus and cytoplasm, you will see how the cytoplas-

Figure 15.7

Granulocytes. **A,** Neutrophils. **B,** Eosinophils. **C,** Basophils.

Table 15.2 CELLULAR ELEMENTS OF BLOOD

Cell	Characteristics	Function	Count	Life Span
ERYTHROCYTES	Biconcave disks, no nucleus, 7.5-8.5 μm, 97% of protein is hemoglobin	Ferry oxygen and carbon dioxide between lungs and tissues	$4.60 \times 10^6/mm^3$ women $5.21 \times 10^6/mm^3$ men	9 days development, 120 days circulation
LEUKOCYTES				
Granulocytes				
Neutrophils	Round, polymorpho-nuclear, 2-5 lobes, blue granules, pink cytoplasm, 12-15 μm diameter	Phagocytose bacteria	$4300/mm^3$, 55% of leukocytes	2 weeks development, several hours circula-tion, 1-2 days in tissues
Eosinophils	Round polymorpho-nuclear, 3 lobes, red-orange granules, 12-15 μm diameter	Destroy parasite larvae, antigen-antibody complexes, suppress tissue inflammation	$230/mm^3$, 3% of leukocytes	2 weeks development, 8-12 hours in circulation
Basophils	Round, polymorpho-nuclear, 2-3 lobes, violet granules, 12-15 μm diameter	Induce tissue inflam-mation, release heparin, histamine, serotonin	$40/mm^3$, 0.5% of leukocytes	2 weeks develop-ment, 8-12 hours in circulation
Nongranulocytes				
Monocytes	Round, mononuclear, U-shaped nucleus, 18 μm diameter	Precursors of macrophages	$500/mm^3$, 6% of leukocytes	Several days devel-opment, 2 days in circulation, months in tissues as macrophages
Lymphocytes	Round, mononuclear, 6-15 μm diameter	Humoral and cellular immunity	$2700/mm^3$, 33% of leukocytes	2 weeks development, weeks to years survival
PLATELETS	Cytoplasmic fragments, 2-4 μm diameter	Hemostasis	$150,000-350,000/mm^3$	Several days develop-ment, 8-10 days circulation

From Beck WS, editor: Hematology, *ed 5, Cambridge, 1991, The MIT Press; Weiss L, editor:* Cell and tissue biology: A textbook of histology, *ed 6, Baltimore, 1988, Urban and Schwarzenberg.*

mic granules function in each type of granulocyte. Table 15.2 summarizes granulocytic form and function.

NEUTROPHILS

Neutrophils phagocytose bacteria, killing them with a variety of molecules from their cytoplasmic granules. At 4300 cells/mm³ of blood, neutrophils are the most numerous granulocytes, making up 55% of the total leukocytes. Neutrophils are 12 to 15 μm in diameter; they have small, blue granules in a pink cytoplasm when stained, and between two and five lobes in the nucleus (Figure 15.7A). Bacteria and inflammation attract neutrophils to sites of infection.

There the neutrophils lose their spherical shapes, become mobile, and slip through the endothelium of the vessels into the inflamed tissue. Neutrophils bind to the bacterial cells, engulfing them into their cytoplasm as **phagosomes** (FAG-oh-soams), membranous vesicles that pinch off from the cell membrane. This arrangement separates the digestive process that now occurs inside the phagosomes from the rest of the cytoplasm, which would otherwise be degraded along with the bacteria. Lysosomes now release an arsenal of enzymes that attack the invader. **Lysozyme** (LIE-so-ZIME) digests the bacterial cell wall and lyses its cell membrane. **Myeloperoxidase** (MY-eh-lo-per-OX-ih-days) activates oxygen molecules that kill the bacteria, and **lactoferrin** (LAK-toe-FAIR-in) kills by sequestering iron needed by bacterial energy metabolism. Lactoferrin also adjusts neutrophil production to the quantity of bacteria. When too few bacteria are present to bind all the lactoferrin molecules, the excess lactoferrin inhibits neutrophil development, but when enough bacteria are present to take up most of the inhibitory lactoferrin, neutrophil development resumes.

EOSINOPHILS

Eosinophils probably have a broad role in tissue inflammation, but they are best known as defenders against parasites, even though they are not phagocytic in the latter role. With the same round shape and size as neutrophils and with abundant large, red-orange cytoplasmic granules and a three-lobed nucleus (Figure 15.7B), eosinophils suppress parasitic infections by attacking the parasite's larvae. Only about 230 eosinophils/mm³ of blood are found, accounting for up to 3% of all leukocytes. These mobile granulocytes home in on the larvae by following trails of lymphokines and other substances released by inflamed tissues or by the larvae themselves. (Lymphokines are products of lymphocytes, described later.) Eosinophils cluster around the larva, which is far bigger than any one of them, and the cells eject the contents of their granules onto the prey. Chief among these constituents is **major basic protein**, which binds to the larva and leads to its destruction. This protein is responsible for the brilliant orange color of the granules. Eosinophils also suppress inflammation by inactivating histamine, a substance that induces the inflammation process.

BASOPHILS

Like neutrophils, basophils kill bacteria by phagocytosis, but they also induce tissue inflammation by releasing the contents of their cytoplasmic granules. Basophils are the rarest granulocytes, at only 40/mm³ and 0.5% of the leukocytes. They resemble neutrophils and eosinophils in size (12 to 15 μm diameter), but their nuclei have only two or three lobes, and their cytoplasm contains prominent violet granules (Figure 15.7C). These granules release **heparin, histamine,** and **serotonin,** which promote allergic inflammation reactions, such as bronchial asthma and anaphylactic responses to bee and fire ant stings and to penicillin. Heparin (HEP-a-rin), a mucopolysaccharide, is an anticoagulant that promotes the entry of blood into inflamed tissues. Histamine and serotonin (sear-oh-TONE-in) promote swelling by increasing the permeability of vessels to plasma, so that albumins leak into the tissues and draw water with them. The serotonin-containing granules of basophils are related to those of mast cells in connective tissue (see Connective Tissues, Chapter 3). Once basophils have entered an inflamed area,

they also release agents that attract eosinophils to the site.

DEVELOPMENT AND DISEASE

Granulocytes develop from myeloid stem cells that have derived in turn from pluripotential stem cells. These myeloid cell populations, illustrated in Figure 15.4, produce myeloblasts committed to each of the neutrophil, eosinophil, and basophil lines. Each type of myeloblast acquires its cytoplasmic granules and becomes a myelocyte that then loses its ability to divide when the nucleus condenses and becomes lobular. Mature neutrophils, eosinophils, and basophils are smaller than their progenitors primarily because their nuclei have diminished in size.

These developmental stages of granulocytes represent pools of cells that are drawn upon, like a bank account, to combat infection. The most mature stages are mobilized first, and the cells surge out of the marrow to the target. Later, as these cells are consumed, demand works back through the lineage and deeper feedback loops stimulate the myeloid stem cell precursors to replenish the expended populations. The quantity of circulating leukocytes, principally neutrophils, rises in response to infection, a condition known as **neutrophilia** or **leukocytosis** (more than 10,000 leukocytes/mm³). Neutrophilia is a common signal that a patient may be combating infection. On the other hand, **neutropenia** (NEW-tro-PEEN-ee-ah) concerns low neutrophil counts (less than 1500/mm³). This situation resembles anemia and may result from reduced granulopoiesis or more rapid loss of neutrophils from circulation.

- What forms of cellular defense do neutrophils, eosinophils, and basophils perform?
- Which type of granulocyte has up to five lobes in its nuclei?
- Which one is the rarest of the three types?
- What color do eosinophil granules typically stain?
- Which group of stem cells forms the granulocytes?

MONOCYTES AND MACROPHAGES

Monocytes are large, round, phagocytic leukocytes with a horseshoe-shaped nucleus and finely granular cytoplasm (Figure 15.8 and Table 15.2). These cells circulate for about a day in the blood, before spending the rest of their long lives as **macrophages** (MAK-row-fayjes; large eaters) in the tissues. About 18 μm in diameter, monocytes are the largest leukocytes, and they are more numerous (500/mm³, 6% of leukocytes) than eosinophils and basophils. As precursors of macrophages, monocytes are the source of a family of phagocytic cells that

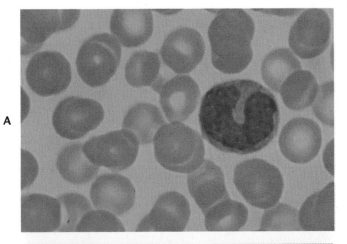

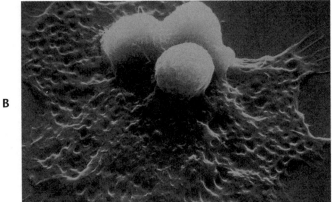

Figure 15.8

A, Monocytes. **B,** A macrophage.

LYMPHOCYTES AND IMMUNITY

LYMPHOCYTES

Lymphocytes are the cellular agents of the immune system. Your lymphocytes recognize and remove cells and molecules that differ from your own by making antibodies that bind to the interlopers and remove them from circulation. Such cells are known as **immune competent cells** because of this very precise ability to distinguish self from nonself. One of the many consequences of this ability is that your own lymphocytes differ from those of everyone else, unless you have an identical twin. Chapter 16, Lymphatic System, describes immunity. Despite their ability to distinguish whether a cell is foreign or one's own, lymphocytes are rather nondescript cells. They are abundant, nongranular, nucleated leukocytes (Figure 15.9 and Table 15.2) that resemble the monocytes and have a rather simple cytoplasm. About one in every three leukocytes is a lymphocyte; approximately 2700/mm³ circulate in adult blood. The nucleus is spherical except for a shallow crease on its surface. The cytoplasm contains the usual synthetic organelles and mitochondria. Most lymphocytes are quite small, only slightly larger than erythrocytes (6 to 9 vs. 7.5 to 8.5 μm diameter), which leaves room for only a thin skin of relatively simple cytoplasm and ribosomes around the nucleus. Large and intermediate lymphocytes are considerably bigger (9 to 15 μm diameter). Their cytoplasm contains more endoplasmic reticulum, Golgi apparatus, and mitochondria, suggesting a more vigorous metabolism. Whatever the significance of this difference in size, all lymphocytes are motile and can squeeze past the vascular endothelium into the tissues.

Lymphocytes develop from the same pluripotential bone marrow stem cells that supply all other lines of leukocytes. They belong to the lymphoid cell line, however, rather than to the myeloid, as shown in Figure 15.4. The **lymphoblasts** that descend from these pluripotential progenitors are unipotent cells committed to develop as lymphocytes. These lymphoblast stem cells proliferate, and their progeny differentiate into lymphocytes that enter the blood and tissues and later recirculate back into the blood through the lymph nodes and vessels of the lymphatic system, as you will see in Chapter 16. Because lymphocytes live from several weeks to many years and recirculate millions of times, they have plenty of opportunities to encounter foreign cells and molecules.

Two major lineages of lymphocytes circulate in the blood. B-lymphocytes produce circulating antibody molecules that bind to foreign proteins in the

clear spent red cells, bacteria, cellular debris, cancer cells, and dying cells from the tissues of the body. Monocytes (Figure 15.8) trace their lineage from the same myeloid stem cells that give rise to other leukocytes (Figure 15.4). The progeny of the monocyte lineage populate all tissues, but especially the lungs, liver, marrow, spleen, and lymph nodes where they become the **mononuclear phagocytic system** (formerly known as the reticuloendothelial system). Such macrophages become the endothelium of the sinusoids of the bone marrow and liver. They become dust cells of the alveoli in the lung, and the osteoclasts of bone develop from them. Other members of this system are microglial cells of the central nervous system and macrophages of the lamina propria lining of the digestive system. Most of these tissues serve as filtering centers where the macrophages intercept targets that enter through the blood (marrow, bone, nervous system) or from the outside (lungs and digestive system).

- Describe several differences between monocytes and macrophages.

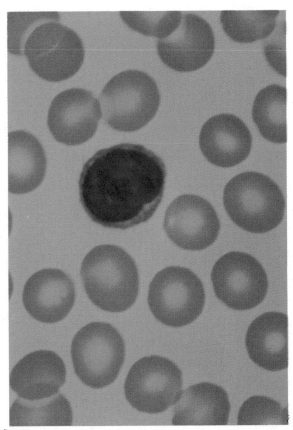

Figure 15.9

Small lymphocyte.

blood and interstitial fluid, marking them for removal by other cells. B-cells also attack bacteria. The name B-lymphocyte (or B-cell) arose from the discovery that in chickens, B-cells develop in the bursa of Fabricius in the cloaca. In humans, B-cells mature in the fetal liver and adult bone marrow, but the original name remains. **T-lymphocytes** are named for their origins in the thymus gland. Foreign cells such as bacteria, cancer cells, dead cells, parasites, and others cause T-cells to produce lymphokines, signal molecules that cause T-cells and others to destroy the interlopers. Chapter 16 describes how T- and B-lymphocytes acquire these remarkable abilities to distinguish between the body's own cells and other so-called foreign cells.

LEUKEMIA

Leukemia is a general name for a group of diseases in which the blood and marrow contain abnormally large quantities of leukocytes. Individual lines or entire lineages of leukocytes may be affected, as stem cell populations proliferate out of control and obliterate other cell lines in the marrow. About one new case is diagnosed for every 1000 persons each year. About half are **acute leukemias,** usually fatal, in which early steps of the lineages are affected. The remainder are **chronic leukemias,** often treated with considerable success, that affect the later stages of leukocyte maturation. Most leukemias affect the lymphoid cell lineage; myeloid leukemias are relatively uncommon. **Acute lymphocytic leukemia** usually attacks children, and **acute myelocytic leukemia** is a disease of adulthood; chronic leukemias are seldom seen in children. All leukemias are accompanied by chromosomal changes in the leukocytes and activation of oncogenes. Heredity may predispose individuals to the disease, but exposure to radiation, viral infection, environmental chemicals, or antitumor drugs also contributes to the incidence of leukemia. The main goal of treatment is to suppress proliferation of stem cells with chemotherapeutic drugs.

- How do lymphocytes compare in size and shape to monocytes?
- Name the two major classes of lymphocytes.
- How do acute and chronic leukemia differ?

PLATELETS AND HEMOSTASIS

Platelets prevent bleeding; they maintain hemostasis, the normal circulation of blood. Platelets, also known as thrombocytes, are the smallest formed elements of the blood, as shown in Figure 15.10 and Table 15.2. At 2 to 4 µm diameter, they are cytoplasmic fragments that split away from **megakaryocytes** in the bone marrow and circulate in the blood for 8 to 10 days before being removed by the spleen. Platelets lack nuclei. Each cubic millimeter of blood contains 150,000 to 350,000 platelets, second only to the number of erythrocytes, but their small size makes platelets difficult to find in the blood smear of Figure 15.3. Look for small, pale blue fragments among the erythrocytes.

Platelets seal off small breaks in vessels by clustering onto the damaged tissue and forming clots (Figure 15.11) that prevent blood from escaping. This process is called **platelet aggregation.** As more and more cells join the cluster, they release **clotting factors** that convert the plasma protein **prothrombin** into the protein **thrombin.** Thrombin in turn converts **fibrinogen** that circulates in the plasma into insoluble **fibrin** molecules that form a mesh of fibers, called a clot or **hemostatic plug,** around the platelets. The clot dissolves later, after the injury has begun to heal.

A

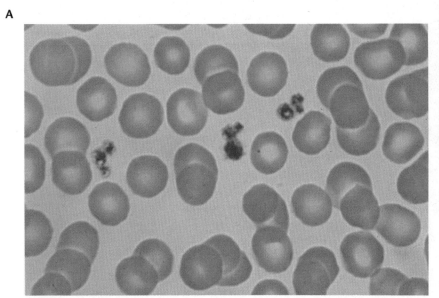

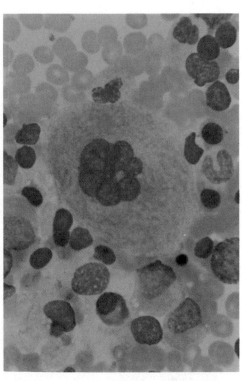

B

Figure 15.10

Platelets and megakaryocytes. A, Platelets in a blood smear. **B,** Megakaryocytes are huge cells, compared with erythrocytes.

Platelets also release factors that promote healing of vessels and cause endothelial cells to divide. They also maintain the integrity of the endothelium; when too few platelets are present, blood leaks into the tissues. Clotting must be carefully controlled; too little clotting risks disastrous bleeding, and too much can plug vessels entirely, a condition called **thrombosis.**

Part of the delicate balance between these outcomes lies in the platelet plasma membrane and how it releases the cytoplasmic contents that mobilize clotting. Figure 15.10B shows the internal structure of a platelet and its relation to the plasma membrane. A ring of **microtubules** gives circulating platelets a rigid disk shape. This ring collapses when the cells adhere to the damaged tissue, and **actin microfilaments** cause the cell to send out filopodia that feel their way over the wound. An **open canalicular sys-**tem of membrane pockets and tubes leads to the **dense tubular system** of modified endoplasmic reticulum deeper in the cytoplasm that stores calcium ions. This arrangement resembles the T-tubule system of skeletal muscle fibers, and it enables the plasma membrane to signal the release of cytoplasmic granules.

Three main types of granules are employed in clotting. Lysosomes release hydrolases and promote inflammatory reactions that clear away debris. Each platelet also has about 10 **dense granules** that contain calcium, ATP, ADP, serotonin, and vasopressin. Release of ADP recruits other platelets into the growing platelet aggregate. **Alpha granules** store a wide variety of proteins that operate in clotting and inflammation. Notable members of this important group are **fibrinogen,** a local source of fibrin in the aggre-

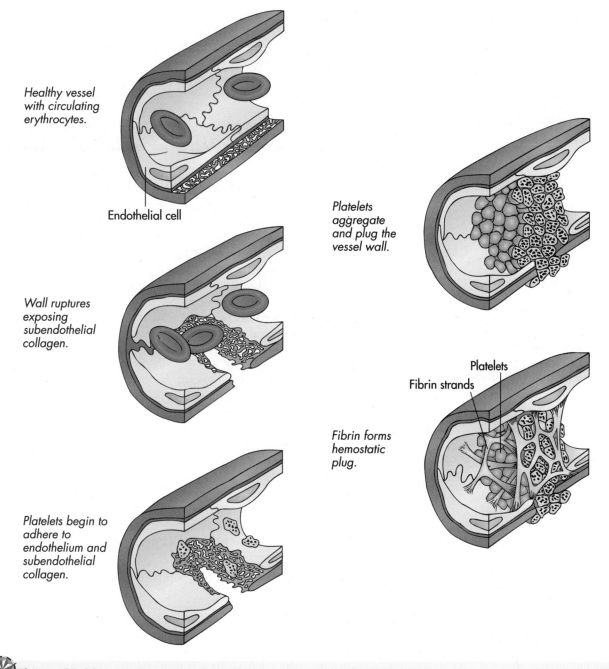

Healthy vessel with circulating erythrocytes.

Endothelial cell

Wall ruptures exposing subendothelial collagen.

Platelets begin to adhere to endothelium and subendothelial collagen.

Platelets aggregate and plug the vessel wall.

Fibrin forms hemostatic plug.

Platelets

Fibrin strands

Figure 15.11
Formation of a hemostatic plug.

gate; **thrombospondin,** which binds fibrin fibers to the surface of platelets; and **platelet-derived growth factor** (PDGF), which stimulates endothelial cells and smooth muscle cells to proliferate in many tissues, including atheromas (see Chapter 14, Vessels). Among the many effects of **aspirin** is its ability to inhibit clotting, which it does by preventing release of granules.

THROMBOPOIESIS

Platelets derive from the same pluripotent stem cells as all other blood cells. Differentiation commits progeny of these cells to **thrombopoiesis** in which the progeny proliferate further and become megakaryocytes (large nucleus), huge cells 50 μm in diameter whose nuclei have up to 64 times more DNA than the usual diploid amount (Figure 15.12). This ex-

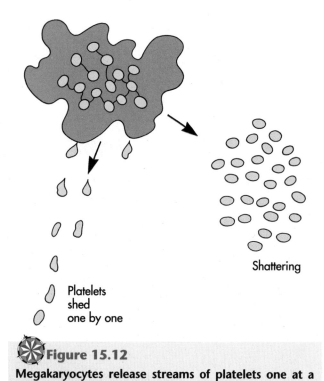

Platelets
shed
one by one

Shattering

Figure 15.12

Megakaryocytes release streams of platelets one at a time, or they may split into many fragments at once.

tra genetic power enables each megakaryocyte to form 1000 to 2000 platelets. Upon completing this feat of synthesis the megakaryocytes migrate to the sinusoids of the bone marrow and release platelets into the circulation. Just how release occurs is uncertain. Some observations suggest that streams of platelets peel away from the megakaryocyte, while others suggest that the entire cell shatters into fragments all at once.

The blood usually contains more than enough platelets for aggregation. Within 8 hours of entering the blood, all new platelets enter the spleen, where about one third are stored in a pool for emergencies. Unless an injured vessel captures them, the rest of the platelets circulate for 8 to 10 days before macrophages finally destroy them upon their return to the spleen. **Thrombopoietin** stimulates platelet development through a feedback loop, in the same manner that erythropoietin regulates erythropoiesis. When platelet levels decline, thrombopoietin first mobilizes the platelet pool from the spleen, then promotes replenishment of megakaryocytes from stem cells. In **thrombocytopenia** (THROM-bo-site-oh-PEEN-ee-ah; low platelet count) a patient's platelets fall below 150,000/mm³, allowing pinpoint hemorrhages called **petechiae** (pe-TEE-kee-ee) to form in the skin and mucous membranes. Underproduction of platelets or their accelerated loss are common causes of thrombocytopenia. **Thrombocytosis** is the opposite condition in which the platelets climb past 450,000 mm³ this extra supply may cause unwanted aggregation and clotting, even thrombosis.

- How big and how numerous are platelets?
- What is hemostasis?
- How do platelets maintain hemostasis?
- What is a megakaryocyte?
- What do petechiae indicate about the amount of platelets in circulation?

BONE MARROW

Bone marrow occupies the cancellous spaces and medullary cavities of bones. It is a soft connective tissue, containing hematopoietic and adipose cells within a network of soft reticular fibers. Red marrow takes its color from the erythrocytes it forms, and adipose cells populate yellow marrow, which has no significant hematopoiesis. All marrow is red at birth, but in adults hematopoiesis has retreated to the cancellous bone found in the proximal epiphyses of larger long bones, sternum, ribs, clavicle, flat bones of the skull, bodies of vertebrae, and pelvis. Clinicians take samples of red marrow from the sternum and crest of the ilium because these bones are conveniently near the surface of the body. Figure 15.13 shows the internal structure of red marrow from the epiphysis of a femur.

Myeloid tissue, the red marrow, occupies the spaces between trabeculae of cancellous bone. A sheet of endosteum covers the trabeculae and lines these spaces. Clusters of hematopoietic stem cells and developing erythrocytes, leukocytes, and platelets surround marrow **sinusoids,** small vessels that branch and anastomose freely into the marrow spaces. Sinusoids consist of an endothelial lining, a basement membrane, and an external adventitia (see Chapter 14, Vessels). Two features of sinusoids are important for hematopoiesis. The adventitia helps support the marrow by extending cytoplasmic processes and **reticular fibers** (see Chapter 3, Tissues) out among the clusters of marrow cells. **Fenestrated endothelium** and basement membrane of sinusoids evidently allow mature cells to enter the blood, as Figure 15.13 shows. These fenestrae, or gaps between neighboring endothelial cells, are not covered with basement membrane. Sinusoids are supplied by capillaries from the nutrient artery of the bone. Large, central sinusoids drain the smaller sinusoids into venous tributaries that exit the bone by way of the nutrient vein.

Clusters of proliferating and differentiating hematopoietic cells are packed around the sinusoids. These cells are not necessarily fixed in position, and clusters usually contain cells from several different lineages. Despite this mixing the marrow does contain particular regions, called **hematopoietic inductive microenvironments** (HIM), in which molecular conditions favor granulopoiesis, thrombopoiesis, or development of other cell lineages. Most hematopoietic

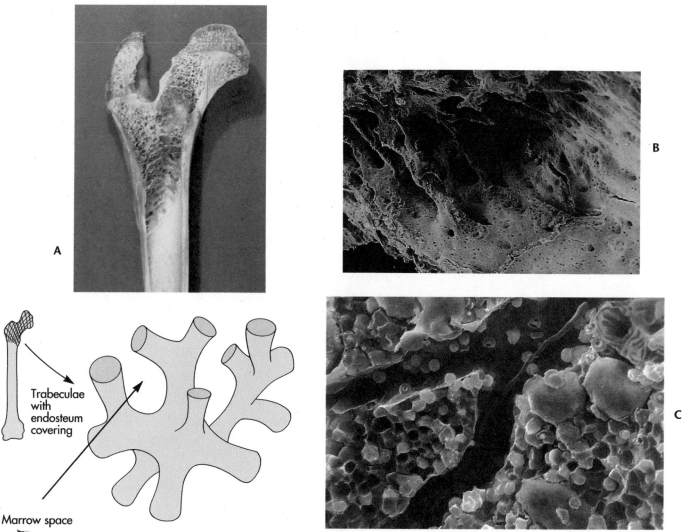

Figure 15.13

A, Hematopoiesis in red bone marrow from the epiphysis of a femur. **B,** Marrow occupies the spaces between trabeculae. **C,** Bone marrow sinusoids serve the myeloid tissue and receive newly maturing cells.

cells are round, but those close to the sinusoids develop **microvilli** that may help them pass through the endothelium into the blood. The marrow receives no lymphatic circulation; autonomic neurons control vascular tone.

- Where is red bone marrow found in adults?
- Of what cell layers are marrow sinusoids made?
- How do mature blood cells enter circulation?

CHOLESTEROL AND ATHEROSCLEROSIS

High levels of cholesterol circulating in the blood increase the risk of atherosclerosis. Elevated blood cholesterol in our patient implies that either too much cholesterol is entering the blood, not enough is leaving, or both. Cholesterol travels throughout the body on lipoprotein carrier molecules that enter cells by receptor-mediated endocytosis. Once cholesterol molecules enter an atherosclerotic plaque, evidently few ever leave and the plaque slowly grows, increasing the risk of heart attack or stroke.

Most cholesterol enters the blood through the small intestine or through synthesis in the liver. Most adults take in about 400 mg of cholesterol from their daily diet, and the liver synthesizes another 1200 mg. The intestinal epithelium (Figure 15.14) assembles cholesterol molecules into large lipoprotein particles called **chylomicrons** (KY-low-MY-krons). Carried to the liver by the lymph (see Chapter 16), chylomicrons en-

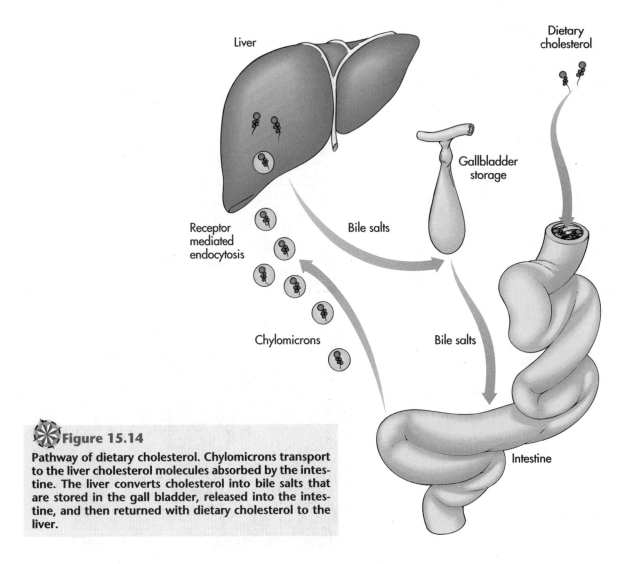

Figure 15.14

Pathway of dietary cholesterol. Chylomicrons transport to the liver cholesterol molecules absorbed by the intestine. The liver converts cholesterol into bile salts that are stored in the gall bladder, released into the intestine, and then returned with dietary cholesterol to the liver.

ter the cytoplasm of liver cells through endocytosis. Most of these cholesterol molecules are converted to bile salts, stored briefly in the gall bladder, and released again into the intestine to facilitate digestion of fats. Most of the cholesterol released is absorbed again by the intestine and cycled back to the liver; very little leaves in the feces.

Lipoprotein molecules (Figure 15.15) shuttle cholesterol synthesized in the liver between the liver and the tissues for assembly into cell membranes or conversion into steroid hormones. The vascular endothelium and atherosclerotic plaques take up cholesterol from these lipoproteins on either leg of the trip. These lipoproteins are known as **low density lipoproteins** (LDL), **very low density lipoproteins** (VLDL), or **high density lipoproteins** (HDL), depending on how many relatively light cholesterol molecules bind to the relatively dense protein. Very low density lipoproteins are the least dense because they carry the most cholesterol; high density lipoproteins have the least cholesterol. LDL is the major cho-

lesterol carrier in the blood. To oversimplify, LDL (and to lesser extent VLDL) are the "bad" forms of cholesterol that tend to raise the level of cholesterol in the blood by transporting cholesterol from the liver, and HDL is the "good" lipoprotein that tends to remove cholesterol from the blood by returning it to the liver.

What causes cholesterol levels to rise? Some of the many factors that can raise blood cholesterol involve the synthesis of cholesterol and the cellular receptors that recognize LDL as it enters cells. The amount of cholesterol free in the cytoplasm of liver cells exerts a negative feedback control over synthesis of cholesterol and LDL receptors. High levels of cholesterol in the liver cytoplasm (but not in the blood) inhibit synthesis of cholesterol and LDL receptors by the liver, and low cytoplasmic cholesterol levels enhance these syntheses. This effect tends to keep cholesterol levels in the liver constant. High-cholesterol diets raise cholesterol levels in the blood by reducing the number of LDL receptors in the liver. As more

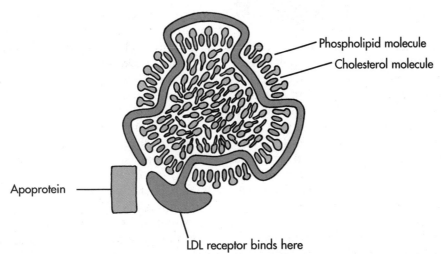

Phospholipid molecule

Cholesterol molecule

Apoprotein

LDL receptor binds here

Figure 15.15

LDL lipoprotein. LDL contains a central core of cholesterol molecules and triglycerides, surrounded with phospholipids and the LDL protein molecule itself. LDL receptor molecules on the surfaces of cells bind to the exposed portion of the protein. A smaller apoprotein (AP-oh-PRO-teen) molecule regulates binding to the receptor.

cholesterol enters the liver cytoplasm from the intestines, cholesterol and receptor synthesis decrease. With fewer receptors on liver cell membranes to remove LDL and HDL, blood cholesterol rises by more than the amount absorbed from the diet alone. This same feedback loop also lowers plasma cholesterol levels when dietary cholesterol declines.

Several drugs reduce cholesterol levels. One of them, **lovastatin** (LOW-va-STA-tin), inhibits cholesterol synthesis in the liver, thereby reducing the liver's contribution to blood cholesterol. Another drug, **colestipol** (ko-LES-tih-pole), prevents the intestine from reabsorbing bile salts and the cholesterol molecules they carry. These salts therefore are eliminated from the body with the feces. Colestipol further reduces blood cholesterol by increasing the amount of LDL receptors in the liver. LDL receptors increase be-

cause cytoplasmic cholesterol declines when less cholesterol enters from the intestine.

This discussion has concerned the levels of cholesterol, which merely make more or less cholesterol available in the blood to enter cells. New drugs will be targeted closer to the source of atherosclerotic plaques themselves by preventing cholesterol from entering the vascular endothelial cells, once the entry mechanisms are discovered.

- What tissues are the primary sources of circulating cholesterol?
- Which lipoproteins carry cholesterol from the liver? To the liver?
- Why are elevated levels of cholesterol in the blood considered to increase the risk of heart attack or stroke?
- Which drugs lower blood cholesterol?

SUMMARY

Blood coordinates the body functions by circulating among the tissues, so that the products of various tissues affect the metabolism of others. Blood is a connective tissue; its cells are suspended in an extracellular matrix called plasma. Dissolved in the plasma are electrolytes, gases, metabolites, and three classes of proteins—the albumins, the globulins, and the clotting proteins. The amount of each solute is balanced by how much enters and leaves the blood. Because of this relationship, blood plasma profiles reveal considerable information about an individual's metabolic condition. *(Page 456.)*

Erythrocytes, leukocytes, and platelets are the principal classes of blood cells. All are formed in the bone marrow, and they circulate through the vessels until they enter the tissues. Erythrocytes have no nuclei, but their cytoplasm is loaded with hemoglobin, the red protein that ferries oxygen and carbon dioxide between the lungs and the tissues. Leukocytes lack pigment but not nuclei. Most of these cells contain characteristic cytoplasmic granules. Leukocytes defend the body by phagocytosis, by promoting inflammation, and by forming antibodies against foreign cells and substances. The leukocytes themselves consist of granulocytes, monocytes, and lymphocytes. Platelets are nonnucleated fragments of cytoplasm responsible for hemostasis. *(Page 459.)*

All blood cells develop from pluripotent stem cells in the bone marrow through a process of cell division and differentiation called hematopoiesis. Each type of cell is the product of a separate lineage that produces mature erythrocytes, granulocytes, monocytes, lymphocytes, or platelets. Each line is regulated separately, so unusually large or small quantities reflect an imbalance between cell proliferation and differentiation in that lineage and the processes that remove those cells from circulation. *(Page 459.)*

Erythrocytes are the largest cellular contingent in blood, which reflects the importance of transporting oxygen. The discoid shape of erythrocytes maximizes diffusion of oxygen and carbon dioxide across the cell membrane. Erythrocytes develop by erythropoiesis from pluripotential stem cells in the bone marrow. Maturing cells eject their nuclei and synthesize hemoglobin before entering the blood. Erythrocytes live about 120 days before macrophages remove them from circulation and recycle their hemoglobin, conserving iron and forming bile pigments. *(Page 461.)*

Erythropoietin adjusts erythropoiesis, according to the tissues' demand for oxygen. The kidneys stimulate red cell production, releasing more erythropoietin when insufficient oxygen reaches the tissues. Disorders of erythropoiesis cause anemia by producing too few red cells or defective hemoglobin molecules. *(Page 461.)*

Granulocytes display polymorphic nuclei and cytoplasmic granules used in cellular defense. Many varieties of granules are lysosomes, but other granules contain special products for inflammation or cellular defense. All granulocytes develop by granulopoiesis from myeloid cell lines. *(Page 463.)*

Neutrophils are the most abundant leukocytes. They contain small, blue-staining granules and nuclei with up to five lobes. Neutrophils kill bacterial cells by phagocytosis. *(Page 464.)*

Eosinophils attack parasites and suppress inflammation of tissues. They release the contents of their orange-stained granules onto the surface of parasitic larvae. Next to basophils, eosinophils are the least abundant granulocytes. *(Page 465.)*

Basophils consume bacteria and promote tissue inflammation by releasing the contents of their prominent, violet-stained granules. Basophils resemble mast cells of connective tissue in releasing serotonin and other products from their cytoplasmic granules. Basophils are the rarest granulocytes. *(Page 465.)*

Monocytes are the precursors of macrophages, mobile cells that patrol tissues for dead, spent, or foreign cells that they remove by phagocytosis. Monocytes are the largest leukocytes; they have a U-shaped nucleus and fine cytoplasmic granules. Monocytes are also derived from myeloid stem cells. *(Page 465.)*

Lymphocytes are the cellular agents of the immune system. These cells recognize and remove foreign cells and molecules. Lymphocytes develop from the lymphoid cell lineage as round cells with an indented nucleus and the usual synthetic organelles required to synthesize antibody molecules. Leukemia occurs when leukocytes or lymphocytes are overproduced. *(Page 466.)*

Platelets maintain hemostasis by sealing off small leaks in vessel walls. Second to erythrocytes in abundance, platelets are anucleate fragments of megakaryocytes. Platelets bind to damaged vessel walls and release thrombin that converts fibrinogen in the plasma to fibrin clots. *(Page 467.)*

Elevated levels of blood cholesterol are one of the major risk factors in atherosclerosis. Cholesterol circulates between the liver and other organs in a variety of lipoprotein carrier molecules. Dietary cholesterol enters circulation from the intestine, is taken up by the liver, and converted to bile salts that recycle from the intestine back to the liver. The liver also synthesizes cholesterol that is transported to the tissues for production of cell membranes and steroid hormones. A high-cholesterol diet raises blood cholesterol levels by interfering with production of LDL receptors in the liver. Drugs lower blood cholesterol by inhibiting synthesis by the liver or by sequestering bile salts in the intestine. *(Page 471.)*

STUDY QUESTIONS

1. What are the two major components of blood? What does an hematocrit tell about the relative amount of each? **1**
2. Describe the function of various components of the blood plasma. **1**
3. How do the components of blood plasma reveal a person's metabolic condition? **1**
4. Describe the major classes of blood cells and the functions they perform. **2**
5. What information about blood do a hematocrit, a differential cell count, and a plasma profile provide? **3**
6. From what cells do blood cells develop, and why can various blood disorders raise or lower the levels of one type without affecting others? **4**
7. What are the differences between the following terms: myeloid and lymphoid cell lines; hematopoiesis, erythropoiesis, granulopoiesis, and thrombopoiesis? **4**
8. What features make erythrocytes different from other blood cells? **5**
9. Describe the differentiation of erythrocytes. **5**
10. Describe the feedback loop by which erythropoietin regulates erythropoiesis. **5**
11. Erythropoietin levels are high in many forms of anemia. What does this suggest has happened to the feedback control of erythropoiesis? **5**
12. What are the different functions of neutrophils, eosinophils, and basophils? What cytoplasmic features distinguish these cells from each other? **6**
13. Describe the life cycles of the different varieties of granulocytes. **6**
14. What are the cytoplasmic properties and functions of monocytes? How do these cells differ from granulocytes? **7**
15. Why are monocytes considered progenitors of the mononuclear phagocytic system? **7**
16. What is the role of lymphocytes? How are these cells distinguished anatomically from other leukocytes? From which cell lineage do these cells develop? **8**
17. What is leukemia? Describe the difference between chronic and acute leukemia. **8**
18. What cytoplasmic properties distinguish the platelets from other leukocytes? How do platelets participate in hemostasis? **9**
19. Describe the process of thrombopoiesis. **9**
20. What are the sources and functions of cholesterol that circulates in the blood? **10**
21. Follow a molecule of cholesterol carried on lipoproteins from the liver to the tissues and its return to the liver. **10**
22. How can diet raise the cholesterol level in blood, and how do colestipol and lovastatin lower it? **10**

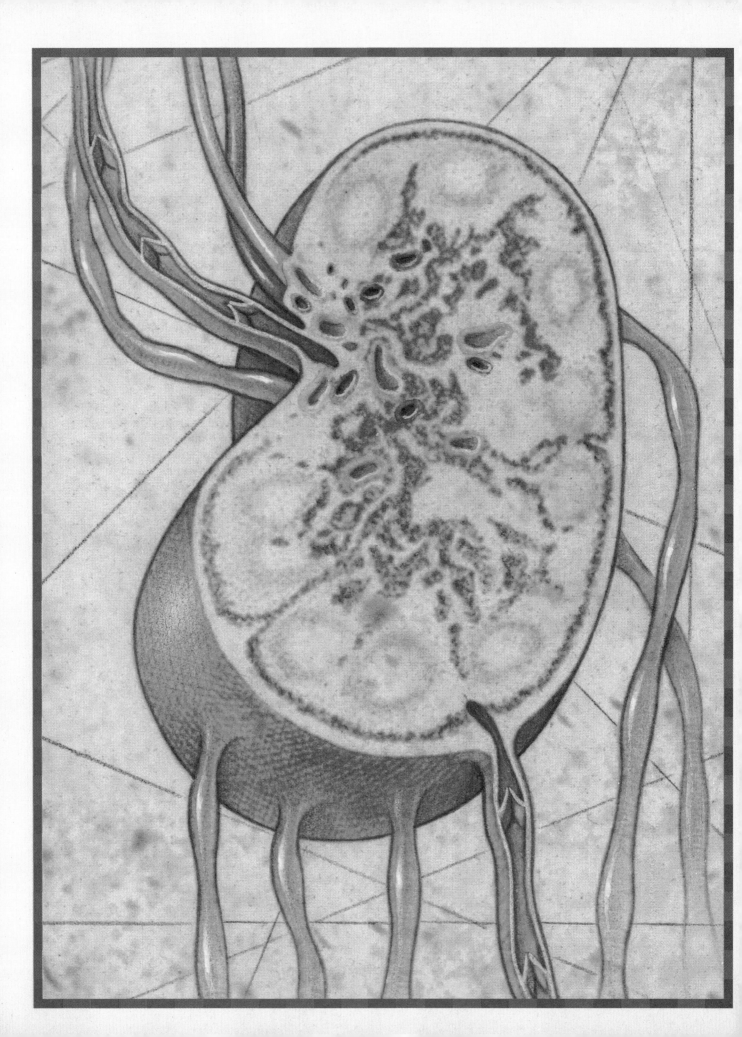

CHAPTER 16

Lymphatic System and Immunity

OBJECTIVES

After studying this chapter you should be able to do the following:

1. Describe the various roles of lymphocytes in immunity. Distinguish between B- and T-lymphocytes.
2. Describe what is meant by the terms antigen, antibody, immunoglobulin, immunological recognition, removal, memory, and self vs. nonself.
3. Describe the relationships between circulating blood and lymph.
4. Describe the four varieties of lymph vessels that return lymph to the blood, as well as the regions of the body drained by the thoracic duct and the right lymphatic duct.
5. Compare the composition of lymph with that of blood plasma.
6. Describe the anatomy of lymphatic nodules and the functions of lymphocytes within them. Be able to recognize lymph nodules in other lymphatic tissues and organs.
7. Describe the anatomy of a Peyer's patch and how it differs from lymphatic nodules and lymph nodes.
8. Describe the anatomy of a lymph node, and relate it to the flow of blood and lymph through the node and to the immune functions of the node.
9. Describe the anatomy of the thymus gland and how it relates to development of T-lymphocytes and immunity. Trace the flow of blood and lymph through this gland.
10. Describe the anatomy of the spleen; trace the flow of blood and lymph through it. Describe the features of white pulp that relate to immunity, and the anatomy of red pulp that stores platelets and reclaims red cells.
11. Outline the development of the lymphatic system.

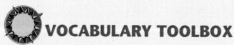 **VOCABULARY TOOLBOX**

Directional terms.

afferent	ad-, *L* toward; fer, *L,* carry. **Aff**erent lymphatics bring lymph to lymph nodes.	**postnodal**	post-, *L* after. **Post**nodal vessels are downstream from the first lymph node.
efferent	ef-, *L* away. **Eff**erent vessels conduct lymph away from nodes.	**prenodal**	pre-, *L* before. **Pre**nodal vessels are upstream from lymph nodes.
paracortical	para-, *G* beside, near. The **para**cortical region of the thymus is beneath the cortex.		

Structures.

antigen	anti-, *G* against; gen, *G* produce. **Anti**body **gen**erator.	**crypt**	crypt-, *G* hidden, concealed.
		germinal	ger-, *L* to bear, carry.
cisterna chyli	cist-, *G* box, chest, tank; chyl-, *G* juice.	**hilum**	hil-, *L* a little thing.

VOCABULARY TOOLBOX—cont'd

Structures.

hist-	*G* a web or tissue. **Hist**ocompatibility molecules distinguish self from nonself.
immune	immun-, *L* safe, free.
lymph	lymph-, *L* water.
reticular	reticul-, *L* a network.
septa	sept-, *L* a fence.
spleen	splen-, *G* the spleen.
thymus	thymo, *G* the mind, spirit, thymus gland.
tonsil	tonsilla, *L* a tonsil.

RELATED TOPICS

- Chylomicrons, Chapter 15
- Lymphocytes, macrophages, lymphoid stem cells, Chapter 15
- Vascular tunics and vessel development, Chapter 14
- Veins and capillaries, Chapter 14

PERSPECTIVE

The lymphatic system protects the body against infection and invasion by foreign organisms. Lymphocytes and macrophages patrol most of the tissues for invading viruses, bacteria, foreign proteins and toxins, dying cells, and even foreign tissue grafts. Lymph vessels communicate to all these tissues, transporting the lymph fluid that carries the immune cells to lymph nodes and to organs that remove the intruders and replenish lymphocytes lost in the process. The effectiveness of the lymphatic system can be judged by the consequences of suppressing it. Heart, kidney, and bone marrow transplants cannot survive unless it is suppressed, and conversely, it must be revived to save those infected with HIV.

The lymphatic system is the body's second circulatory system, illustrated in Figure 16.1A. It resembles the cardiovascular system, with fluid and vessels, but it lacks an active pump, such as the heart. Together, both systems provide cells and organs for the immune system that combats infections and removes foreign materials from the body. The lymphatic system's vessels, called **lymphatics,** resemble veins and capillaries. Lymphatics drain **lymph** fluid from the head, trunk, and extremities into veins at the base of the neck, where the lymph mixes with blood of the cardiovascular system that the heart pumps to the body tissues and capillaries. Instead of returning to the heart through the venous tree, however, some cells and fluid drain into lymphatic capillaries. From there, larger lymphatics conduct this lymph through **lymph nodes** and **lymphatic organs,** such as the thymus and spleen, before the lymph rejoins the blood. One to two liters of clear, straw-colored lymph collect from the body's tissues every day, flowing several thousand times slower than the blood.

Lymphatic cells and organs are the agents of immunity. Lymphocytes and macrophages, carried by blood and lymph from the bone marrow, patrol nearly all tissues of the body, detecting viruses, bacteria, tumor cells, and damaged, infected, or otherwise foreign cells and substances in the body's fluids and tissues. Lymph nodes and lymphatic organs sustain this surveillance by producing additional lymphocytes and removing the destroyed targets.

The next section outlines the cellular processes of immunity that underlie the anatomy of the lymphatic system. If you want to pursue anatomy directly, skip ahead to Lymph Vessels, page 482.

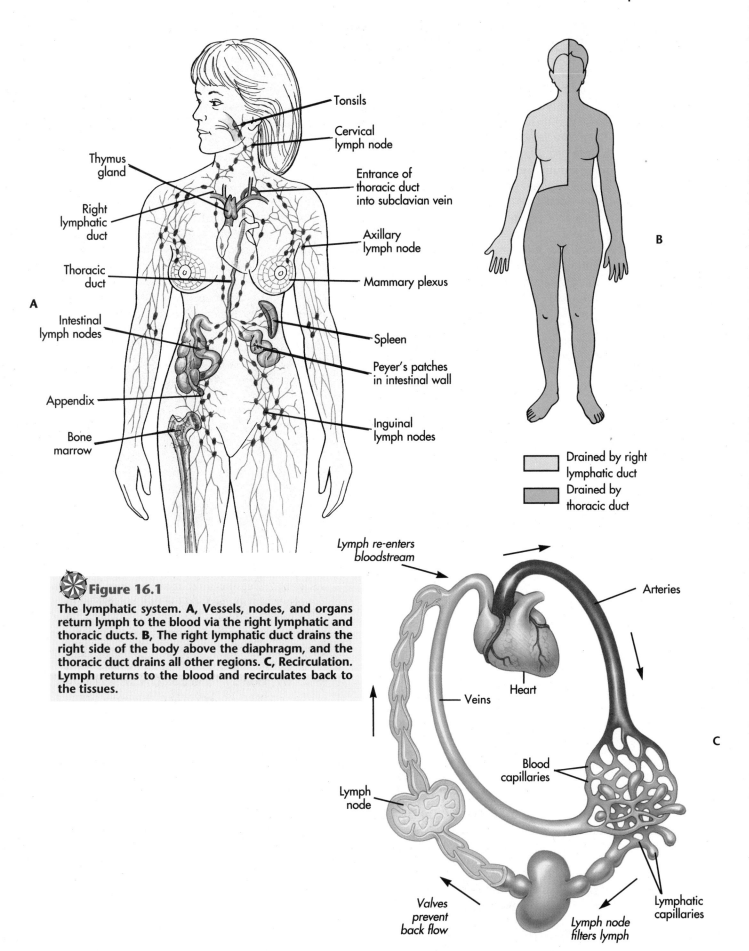

Tonsils

Cervical
lymph node

Entrance of
thoracic duct
into subclavian vein

Axillary
lymph node

Mammary plexus

Spleen

Peyer's patches
in intestinal wall

Inguinal
lymph nodes

Thymus
gland

Right
lymphatic
duct

Thoracic
duct

Intestinal
lymph nodes

Appendix

Bone
marrow

A

B

Drained by right
lymphatic duct

Drained by
thoracic duct

*Lymph re-enters
bloodstream*

Arteries

Heart

Veins

Blood
capillaries

Lymph
node

Lymphatic
capillaries

*Valves
prevent
back flow*

*Lymph node
filters lymph*

C

Figure 16.1

The lymphatic system. A, Vessels, nodes, and organs
return lymph to the blood via the right lymphatic and
thoracic ducts. **B,** The right lymphatic duct drains the
right side of the body above the diaphragm, and the
thoracic duct drains all other regions. **C,** Recirculation.
Lymph returns to the blood and recirculates back to
the tissues.

IMMUNITY

FOREIGN CELLS AND SUBSTANCES

The immune system protects the body against the effects of a great variety of foreign cells and substances. These responses range from the sneezing and runny nose of hay fever allergies and the inflammation of an infected finger, to rejecting transplanted kidneys or to autoimmune attacks against your own tissues. The macrophages and lymphocytes of the immune system respond to these assaults with three fundamental steps that (1) **recognize** cells or substances as foreign, that is, different from your own, (2) **remove** the offending material, and in many instances (3) store **immunological memory** of the exposure, which makes future encounters swifter and more vigorous (Figure 16.2). All of these immune responses

An individual's cells display the same proteins, all are self.

A

Immune cells also display antibodies that can bind with foreign antigens.

Foreign cells display different proteins — they are "nonself," foreign antigens.

Antibodies bind to and kill foreign cells.

B

Immune cells secrete antibodies that bind to cellular antigens

or to soluble antigens, but not to self-proteins.

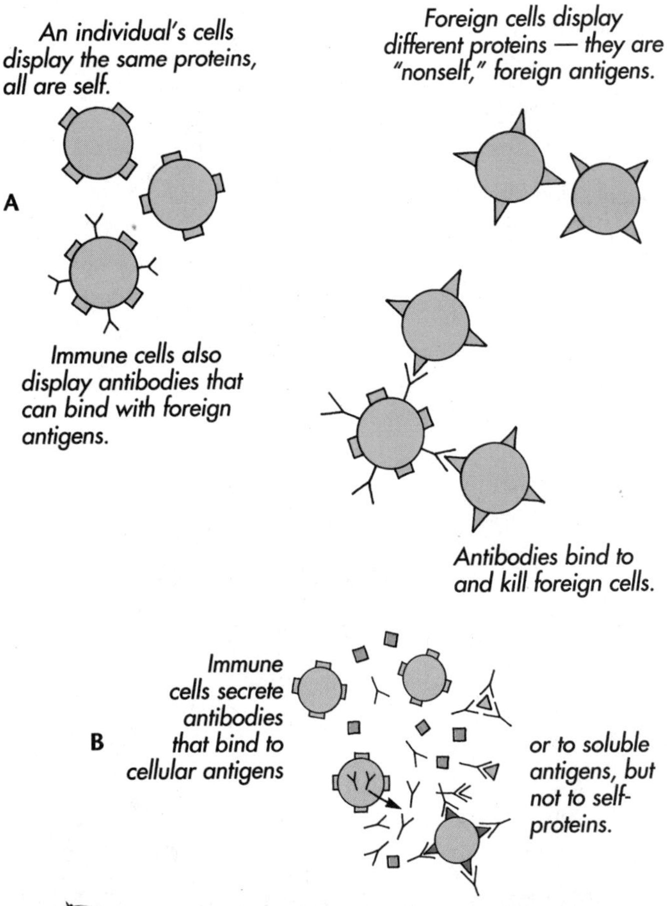

Figure 16.2

Antigens and antibodies. A, Cellular immunity. B, Humoral immunity.

distinguish between your own (self) healthy tissues and other (foreign or nonself) cells or substances, such as viruses, bacteria, parasites, your own infected cells, cancer cells, blood of a different type than your own, or spent and dying cells. These responses are

subtle enough to distinguish between brother and sister but not between identical twins. Your immune system recognizes your sister's cells as foreign, for example, and your cells are foreign to hers. All three steps involve interactions between your own and foreign cells that are mediated by foreign antigens and cellular receptors on your own cells that bind with antigens.

ANTIGENS

Antigens (ANT-ih-jens) are foreign, nonself, proteins or polysaccharides, dissolved in body fluids or located on the surfaces of cells, that stimulate immune responses. Two major classes of lymphocytes— the B- and T-lymphocytes—respond to antigens because they have receptors on their membranes that bind to antigens. Your own **B-lymphocytes** recognize bacterial cells because **immunoglobins,** membrane receptor proteins on their own cell surfaces, bind to the bacterial antigens and mark the bacteria for disposal. B-cells also become **plasma cells** that secrete **antibodies** (ANT-ih-bod-eez), soluble versions of immunoglobins, that also mark antigens dissolved in body fluids for removal. B-lymphocytes develop in the bone marrow, but they are named for their discovery in the bursa of Fabricius in chickens (see Lymphocytes, Chapter 14). **T-lymphocytes,** named for their origin in the thymus, detect intracellular antigens, such as viruses multiplying in infected cells and fragments of bacteria engulfed by macrophages, and they secrete lymphokines that mobilize B- and other T-lymphocytes to recognize and remove these invaders. Instead of antibodies, T-cells employ **membrane receptor proteins** (T-cell receptors) that bind with fragments of antigen molecules that macrophages have engulfed. Together, antibodies and T-cell receptors distinguish millions of different foreign molecules from the vast array of proteins and polysaccharides on the surfaces of your own cells.

B-LYMPHOCYTE RESPONSE TO BACTERIA

Recognition begins when pneumococcus bacteria, for example, encounter B-lymphocytes in the lymph nodes of the lungs (Figure 16.3). The immunoglobin molecules on the B-cells bind with the polysaccharide coating of the bacterial cells. This binding activates the B-cells to multiply into swarms armed with enough of the same immunoglobin receptors and antibodies to bind to the invaders. Thus marked with antibodies, these bacteria are ingested by macrophages and removed. Some B-cells are set aside as memory cells that produce more lymphocytes and memory cells upon later exposure to the same antigens.

T-LYMPHOCYTE RESPONSE TO VIRUS

T-lymphocytes attack cells infected with virus by recognizing fragments of the virus that infected cells

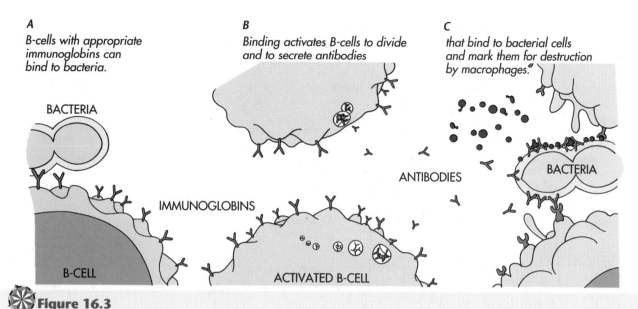

A
B-cells with appropriate immunoglobins can bind to bacteria.

B
Binding activates B-cells to divide and to secrete antibodies

C
that bind to bacterial cells and mark them for destruction by macrophages.

BACTERIA

IMMUNOGLOBINS

B-CELL

ACTIVATED B-CELL

ANTIBODIES

BACTERIA

Figure 16.3

B-lymphocytes attack pneumococcus bacteria. (A) Bacterial cells activate B-lymphocytes to (B) multiply and produce antibodies that (C) mark the bacteria for destruction by macrophages.

present on their surfaces (Figure 16.4). When virus particles multiply in your own cells, the endomembrane system (see Chapter 2) transports some fragments of the viruses to the cell surface on carrier proteins known as **major histocompatibility complex (MHC) proteins.** On their plasma membranes, infected cells present these viral fragments on their MHC proteins to any T-lymphocytes whose receptors can bind with the MHCs and their viral fragments. Removal begins with activation of such lymphocytes in the lymph nodes to produce more T-cells with the

same receptors. These T-cells produce both lymphokines that inhibit viral replication and toxins that kill infected cells. Even though your own cells are killed in the process, viral replication halts. Once again, memory cells are set aside to respond vigorously to future challenges from the same virus.

SELF VS. NONSELF

How does the immune system "learn" to recognize the vast variety of antigens among the forest of proteins sprouting on the surface of your own cells and

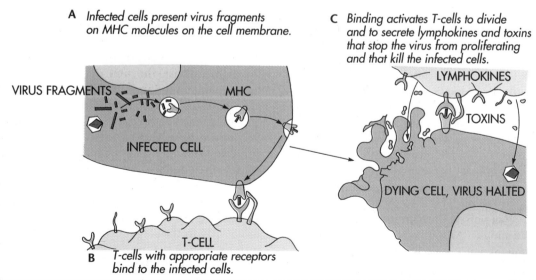

A Infected cells present virus fragments on MHC molecules on the cell membrane.

C Binding activates T-cells to divide and to secrete lymphokines and toxins that stop the virus from proliferating and that kill the infected cells.

VIRUS FRAGMENTS

MHC

INFECTED CELL

LYMPHOKINES

TOXINS

DYING CELL, VIRUS HALTED

T-CELL

B T-cells with appropriate receptors bind to the infected cells.

Figure 16.4

T-lymphocytes attack viruses multiplying in infected cells. (A) MHC proteins present fragments of the virus onto the surface of the infected cell. (B) Infected cells activate T-lymphocytes whose receptors bind to the virus fragments and MHC proteins. (C) Activated lymphocytes proliferate and release lymphokines that inhibit viral multiplication and kill the infected cells.

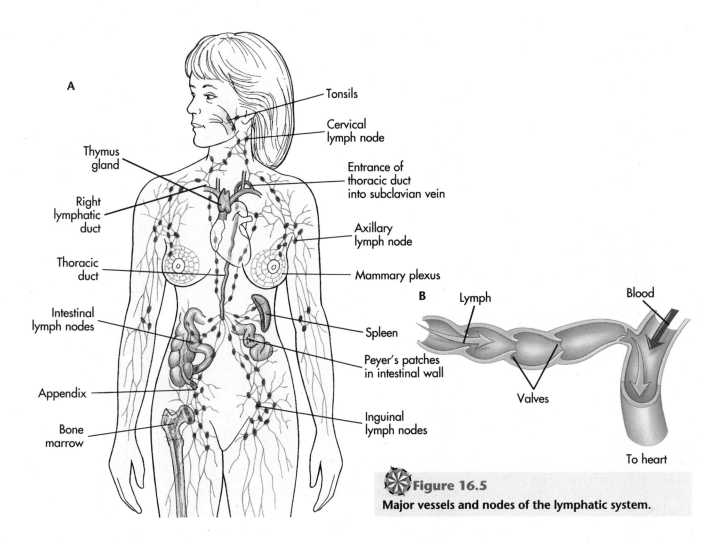

Figure 16.5

Major vessels and nodes of the lymphatic system.

in the pool of soluble antigens circulating in the blood and lymph? First of all, the immune system produces a vast array of lymphocytes, each equipped with a particular antibody or receptor different from all others. This immense variety is potentially capable of binding with any antigens that may be encountered, including your own. Early development, however, eliminates self-recognizing B-lymphocytes in the bone marrow and T-lymphocytes in the thymus. Surviving lymphocytes recognize only nonself substances, the molecules we call antigens. Each such lymphocyte is capable of recognizing a particular antigen and responding by producing a clone of identical lymphocytes. This process equips lymph nodes with lymphocytes poised to repel invaders.

- What is an antigen? An antibody? Immunoglobins?
- How do the roles of T- and B-lymphocytes differ?

LYMPH VESSELS

OVERVIEW

The vessels of the lymphatic system resemble veins. Lymph flows strictly one way, from the tissues toward the heart, as Figure 16.5A shows. Flow begins in narrow **lymphatic capillaries** (lim-FAH-tik), about 100 µm in diameter, and drains into **collecting lymphatics** where it encounters the first of many lymph nodes. Then it moves through **trunks,** and finally flows through the **thoracic duct** or the **right lymphatic duct,** each about 5 mm in diameter. From these ducts, lymph mixes with blood in the brachiocephalic veins and circulates once again to the tissues. A thin **endothelium** lines all lymphatic vessels, and an **adventitia** (AD-ven-TIH-sha) covers them externally. Larger vessels insert between these layers a **tunica media** (see Chapter 14, Vessels) with smooth muscle fibers that propel the lymph past numerous **one-way flap valves** (Figure 16.5B). These three layers are well defined only in the larger vessels. Valves and **lymphatic pumps** located throughout the lymphatic vessels maintain a steady flow of about 100 ml per hour in the thoracic duct. A major valve at the entrance to the brachiocephalic veins prevents blood from entering the lymph.

Clusters of **lymph nodes** filter out foreign cells and antigens as lymph passes through them. Find the cervical, axillary, iliac, and inguinal nodes in Figure

16.5. Lymphocytes from these nodes, the **thymus** (THIGH-muss; Figure 16.1A), and the **spleen** patrol the tissues as the blood and lymph circulate, but erythrocytes, platelets, and other leukocytes normally do not reach the lymph. Lymphatics enter all tissues except epithelia, brain and spinal cord, and bone marrow. A few connective tissues (such as cartilage and the cornea) that have no blood capillaries also lack lymphatics.

CAPILLARIES

Elaborate networks of lymph capillaries drain interstitial fluid from the tissues but do not connect with the vessels of the cardiovascular system. Instead, lymph capillaries have openings in their walls that admit the interstitial fluid and antigens that percolate among the tissue cells (Figure 16.6A). Interstitial fluid that has entered the capillaries is called lymph. There is ample opportunity to enter because capillaries reach virtually every cell.

Lymph capillaries are difficult to see in the microscope because they are so delicate. The walls, illustrated in Figure 16.6A and B, are a thin **squamous endothelium** (SKWAY-mus) bound firmly to the tissue cells by patches of **basement membrane** and numerous **anchoring filaments** that reach like cables among the neighboring cells. Tight junctions join the epithelial cells themselves, but they leave many flap-like in-

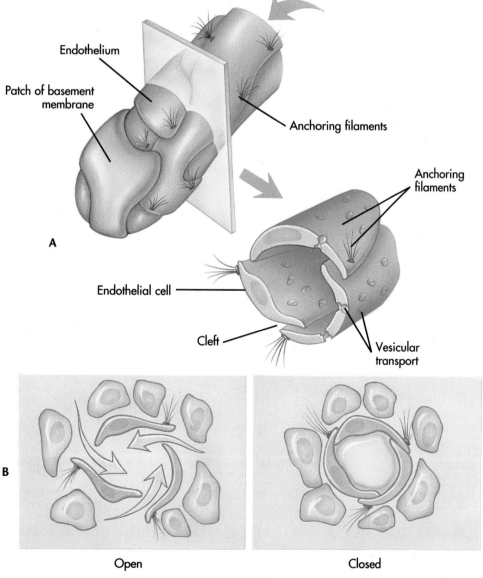

A

Endothelium

Patch of basement membrane

Anchoring filaments

Anchoring filaments

Endothelial cell

Cleft

Vesicular transport

B

Open   Closed

Figure 16.6

Lymphatic capillaries. **A,** Lymphatic capillaries are blind tubes of squamous endothelium. Intercellular clefts admit lymph and cells. **B,** Tissue movements draw lymph into the capillaries by opening and closing the clefts.

tercellular clefts through which lymph and particulates enter. The only other entry is by way of **vesicular transport** directly across the endothelium. Lymph capillaries are irregular in diameter (up to 100 μm) because they follow spaces between cells that are continually changing. Some experts think that normal tissue movement draws the intercellular clefts open and presses them closed like flap valves, as shown in Figure 16.6B. Others contend that the movement of tissues presses water but not solutes from the capillaries, which draws lymph into the capillaries osmotically. Both mechanisms probably contribute to the process.

COLLECTING VESSELS

Collecting vessels drain lymph from capillaries in the head, extremities, and visceral organs, generally following the course of veins and arteries. Collecting vessels are larger (1 to 3 mm) than capillaries because they receive from larger territories, but their walls are thinner than those of veins of similar diameter. Lymph nodes stationed along the vessels filter lymph as it passes (Figure 16.7A). Upstream from the first lymph node a vessel is considered **prenodal,** and downstream it is **postnodal,** even though it may pass several lymph nodes before it reaches a trunk. As you can see in Figures 16.1A and 16.5A, lymph nodes es-

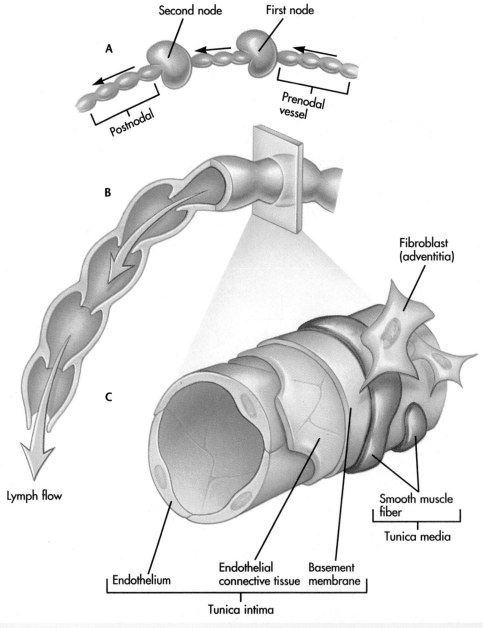

🏵️ **Figure 16.7**
Collecting vessels drain lymph nodes (A) propel lymph past valves (B) and contain three tunics (C).

pecially cluster in the popliteal and inguinal regions of the lower extremities, as well as in the abdominal, axillary, and cervical regions.

Collecting vessels introduce several structures that continue in the trunks and ducts, shown in Figure 16.7B and C. The endothelium now becomes a **tunica intima** supported by a thin, subendothelial connective tissue and a fragmentary basement membrane (Figure 16.7C). Intercellular clefts are no longer present to admit lymph. A **tunica media,** one or two circular layers of **smooth muscle fibers,** appears for the first time; these cells propel lymph through the vessels. **Fibroblasts** in the adventitia build a light coating of collagen and elastic fibers around the vessels that accommodates small blood vessels and autonomic neurons. All of these structures are quite variable, frequently interrupted, and difficult to see, even with the electron microscope. Collecting vessels are most robust and easiest to see as they enter and leave lymph nodes.

The endothelium forms numerous valves, as Figure 16.7B shows. As do venous valves, the cusps of **lymphatic valves** point in the direction of flow, ultimately toward the heart. Forward flow presses the thin cusps open and backflow closes them, forcing lymph forward. There are so many valves that collecting vessels have a beaded appearance when lymph inflates the regions between the valves. Although the smooth muscle of the tunica media is an important **lymphatic pump,** skeletal muscles and general body motion also squeeze lymph past the valves.

TRUNKS

Trunks drain collecting vessels into the right lymphatic duct or thoracic duct. As Figure 16.8 shows, there are five vessels large enough (more than 2 mm diameter) to be considered trunks. The right and left **jugular trunks** drain collecting vessels of the head and neck, and the **subclavian trunks** receive from the upper limbs. A pair of **bronchomediastinal trunks** drains the lymph nodes of the trachea and bronchi. In the abdomen an **intestinal trunk** and two **lumbar trunks** deliver to the cisterna chyli from the intestines and the lower extremities, respectively. All these vessels are quite variable in location and number. Internally, trunks have the same layers and valves found in collecting vessels.

DUCTS

The thoracic and right lymphatic ducts discharge lymph into the brachiocephalic veins at the base of the neck. The thoracic duct is the larger of the two, about 40 cm (16 inches) long and 0.5 cm (⅕ inch) in diameter. Figure 16.1B shows that below the diaphragm, it drains all lymph nodes and tissues (except portions of the liver), and above the diaphragm it drains the left side of the body. The thoracic duct be-

gins at the level of the second lumbar vertebra (Figure 16.8) in a sac called the **cisterna chyli** (sis-TER-nah KYE-lee) that receives one or more intestinal trunks and two lumbar trunks, left and right. The duct ascends beside the aorta and passes through the diaphragm into the thorax, where it receives intercostal vessels from the rib cage. At the fifth thoracic vertebra, the thoracic duct curves toward the left **internal jugular** and **subclavian veins,** where it receives the jugular, subclavian, and bronchomediastinal trunks before entering at the junction of these veins. The right lymphatic duct is rarely more than 1 cm long. It forms by the confluence of the jugular, subclavian, and bronchomediastinal trunks on the right side (although one or more may enter separately) and discharges into the blood at the union of the right internal jugular and right subclavian veins, beneath the sternal end of the right clavicle. Bicuspid valves prevent blood from flowing into both ducts.

- What are the names of the four varieties of lymph vessels?
- How does lymph enter a lymphatic capillary?
- What three layers of lymph vessels first appear in collecting vessels?
- Which vessels do lymphatic trunks drain?
- Where does the thoracic duct discharge into the blood, and what territories of the body does this vessel drain?
- What structures prevent blood from leaking back into the thoracic duct?

LYMPH FLUID

Lymph is interstitial fluid that accumulates in the lymphatic vessels, and for this reason its composition resembles that of blood plasma. There is one major difference—lymph and interstitial fluid have lower concentrations of protein than blood, about 3% vs. 7%. The difference arises because blood capillaries are relatively impermeable to proteins, compared with smaller, readily exchangeable molecules, such as glucose, electrolytes, and water. Relatively little protein leaves the blood plasma and virtually none returns by reabsorption.

Metabolism of fats and proteins in the small intestines and liver changes the local composition of lymph, and the content of the thoracic duct reflects these contributions. After meals the thoracic duct is milky with **chylomicrons** (see Cholesterol, Chapter 15, Blood) from the intestine. Chylomicrons are clusters of phospholipids and triglycerides formed by the endothelium of the intestines, that are released into special lymphatic capillaries called **lacteals** (LAK-teels; see Intestinal Villi, Chapter 23, Digestive System). Being lipids, the chylomicrons are suspended, not dissolved, in the lymph and turn the fluid milky.

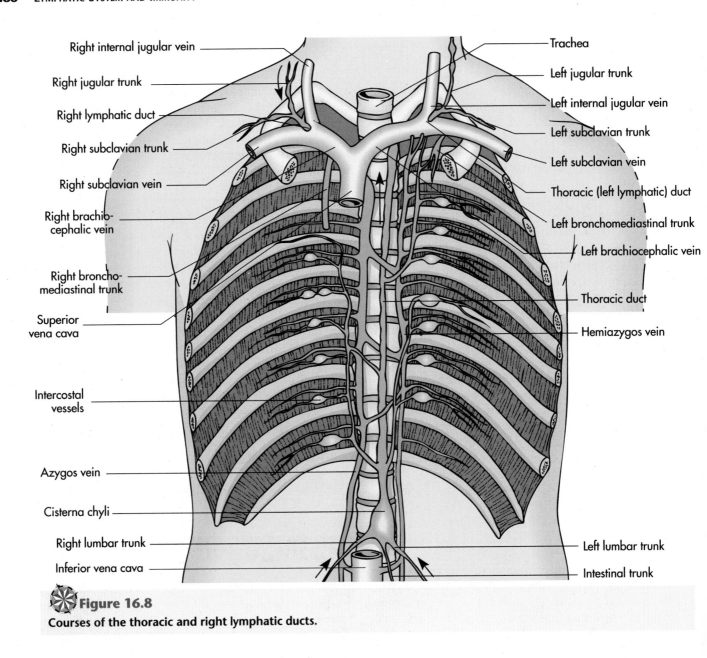

Right internal jugular vein

Right jugular trunk

Right lymphatic duct

Right subclavian trunk

Right subclavian vein

Right brachio-cephalic vein

Right broncho-mediastinal trunk

Superior vena cava

Intercostal vessels

Azygos vein

Cisterna chyli

Right lumbar trunk

Inferior vena cava

Trachea

Left jugular trunk

Left internal jugular vein

Left subclavian trunk

Left subclavian vein

Thoracic (left lymphatic) duct

Left bronchomediastinal trunk

Left brachiocephalic vein

Thoracic duct

Hemiazygos vein

Left lumbar trunk

Intestinal trunk

Figure 16.8
Courses of the thoracic and right lymphatic ducts.

The liver synthesizes albumin proteins that it discharges into the thoracic duct at a concentration of 6%, very near the blood plasma level. As a result the protein content of lymph in the thoracic duct is higher than that in most of the vessels this duct drains.

- Why is the protein content of most lymph lower than that of plasma?
- What local differences in its composition do the intestines and liver produce?

LYMPHATIC TISSUES AND ORGANS

OVERVIEW

Lymphatic tissues and organs range from simple to complex. The simplest are **lymphatic nodules,** clusters of lymphocytes (Figure 16.9) that form at sites of infection and then disappear. Next in complexity, **Peyer's patches** (PIE-ers) in the walls of the intestines

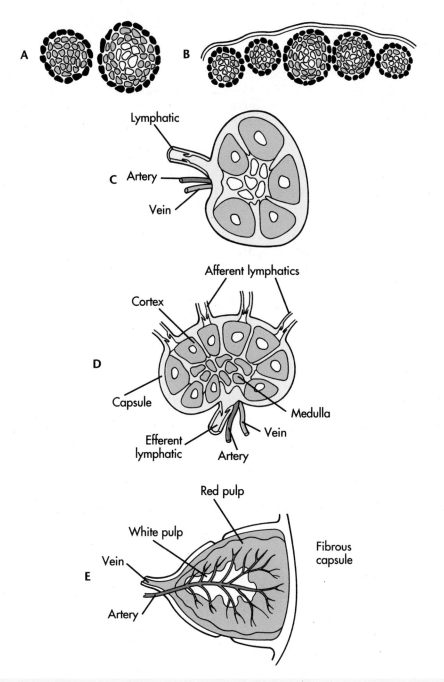

Figure 16.9

Comparison of lymphatic nodules (A), Peyers patches (B), nodes (C), thymus (D), and spleen (E).

cover larger nodular clusters with a thin epithelium, and the tonsils enfold more nodular tissue into pockets of mucosa in the pharynx. Both structures are nicely placed to intercept antigens from the digestive and respiratory tracts. **Lymph nodes** encapsulate many lymphatic nodules within a tougher capsule supplied with blood vessels and lymphatics. From sites in the limbs and trunk, these nodes filter lymph from larger territories than nodules because blood and lymph bring antigens directly to them. The **thy-**

mus and **spleen** filter blood and lymph from the entire body. The thymus resembles a cluster of lymph nodes and produces the body's first populations of T-lymphocytes that thereafter populate all lymphatic tissues. The spleen stores platelets, removes old erythrocytes, and its B-cells produce antibodies.

• Name the various lymphatic tissues and organs.
• Identify three trends toward increasing complexity in the anatomy of these tissues and organs.

LYMPHATIC NODULES

Lymphatic nodules are the simplest lymphatic tissues. They are transient clusters (about 1 mm diameter) of lymphocytes embedded directly within mucous membranes at sites of infection (Figure 16.10). No capsule or external covering separates the nodules from the surrounding cells and fluid, which percolate directly into the nodules. The figure shows that **primary nodules** consist entirely of densely stained small lymphocytes, and **secondary nodules** also contain lightly staining **germinal centers** full of large lymphocytes. Each cluster represents one or more clones of B-lymphocytes in different stages of response to antigens from the infection nearby. Antigens have not yet activated the small lymphocytes, but the response has progressed further in the germinal centers where macrophages are presenting antigens and B-cells are multiplying to form memory cells and large plasma cells that secrete antibodies.

- What is the difference between the structure of primary and secondary lymphatic nodules?

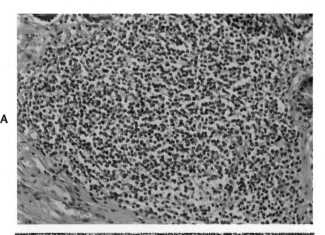

A

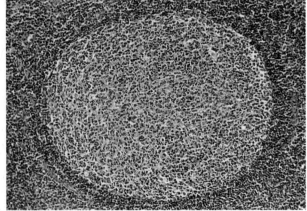

B

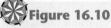

Figure 16.10

A, Primary lymphatic nodules. **B,** Secondary lymphatic nodules.

GALT AND BALT

Gut associated lymphoid tissues (GALT) are permanent clusters of lymphatic nodules that populate the intestines **(Peyer's patches),** pharynx **(tonsils),** and **appendix,** as Figure 16.11 shows. Antigens enter these tissues through special channels directly from the lining of the gut. The lungs also benefit from **bronchial associated lymphoid tissues** (BALT).

PEYER'S PATCHES

Peyer's patches are penny-sized disks scattered in the mucosal lining of the intestines. Antigens enter from the lumen by transcytosis across bell-shaped **M-cells** that enclose B-lymphocytes beneath them, as Figure 16.11B shows. Lymphocytes that recognize the antigens leave the M-cells and multiply in the germinal centers of adjacent nodules. (Other lymphocytes quickly replace the migrants beneath the M-cells.) Progeny of the lymphocytes leave the patch, recirculate through lymph and blood, and return to the patch as mature antibody-producing plasma cells.

TONSILS

Three sets of tonsils collect antigens from the nasal and oral cavities. The pharyngeal tonsils (FAHR-in-jee-al; also called adenoids) are located at the back of the nasal cavity (Figure 16.11A); the base of the tongue houses the **lingual tonsils,** and the **palatine tonsils** are recessed in folds of mucosa behind the hard palate. These folds, or **crypts** (KRIP-ts), especially prominent in the palatine tonsils, are lined with epithelium and lymphatic nodules. Bacteria and other foreign materials enter the crypts, where B-cells attack them. Large infections may become trapped in the crypts and inflame them enough to cause secondary infections, possibly prompting surgical removal of the tonsil.

APPENDIX

A Peyer's patch surrounds the lining of the appendix. The appendix is a narrow, blind tube that extends from the cecum near the junction of the ileum and the ascending colon. Lymphatic nodules fill the mucosal lining, and receive bacterial, viral, or parasite antigens through M-cells like those of Peyer's patches. When the appendix becomes inflamed and neutrophils accumulate in it, its owner may require surgery to relieve **acute appendicitis.**

- What common cellular feature is found in all GALT?

LYMPH NODES

Lymph nodes are bean-shaped structures approximately 1 cm long scattered in the connective and adipose tissues of blood vessels and visceral organs. Figure 16.12 shows the structure of an axillary lymph node, typical of all others. A **fibrous capsule** covers the exterior of every node, and blood vessels enter

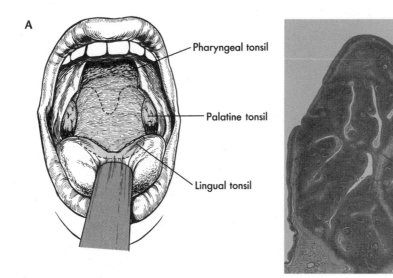

A

Pharyngeal tonsil

Palatine tonsil

Lingual tonsil

B

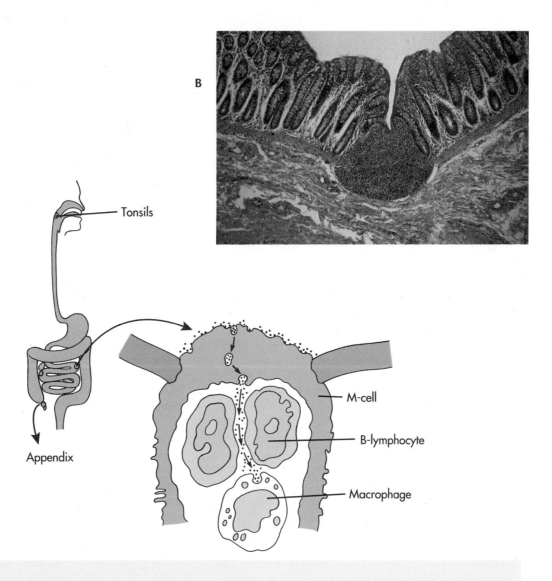

Tonsils

Appendix

M-cell

B-lymphocyte

Macrophage

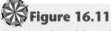Figure 16.11

Gut-associated lymphatic tissues. A, The tonsils are located at the openings of the pharynx. Lymphatic nodules line the crypts of the palatine tonsils. **B,** The lymphatic nodules of Peyer's patches are sheltered beneath the intestinal epithelium; M-cells transport antigens to the nodules.

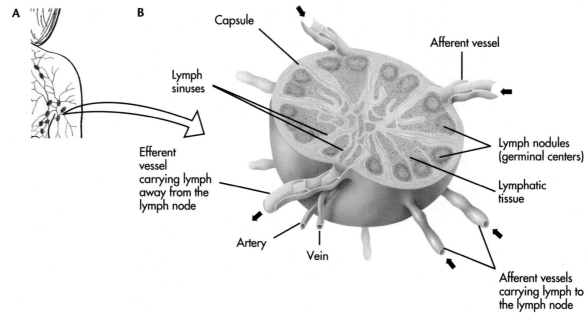

Figure 16.12

An axillary lymph node. A, Axilla. B, Lymph node.

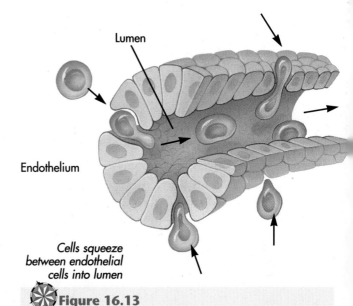

Figure 16.13

High endothelial vessels. Cuboidal endothelium admits lymphocytes to the lymph fluid.

and leave at an indentation known as the **hilum. Afferent lymphatics** (incoming collecting vessels) enter the node elsewhere, bringing cells and lymph from capillaries or nodes upstream; **efferent lymphatics** drain from the hilum downstream to other nodes or trunks. Axillary lymph nodes, for example, receive from collecting vessels of the arm and from axillary arteries; the cells passing through these vessels return to the blood by way of the subclavian trunk and thoracic duct.

Internally a **cortex** of lymphatic nodules and an underlying **paracortical layer** surround a central **medulla** in all lymph nodes. **Reticular fibers** and **reticular cells** (reh-TIK-you-lar) support and suspend all of this tissue from the capsule and from connective tissue partitions called **trabeculae** (tra-BEK-you-lee) that reach deep into the medulla from the capsular wall. Lymph and its recirculating lymphocytes reach all layers of the node through **subcapsular sinuses, cortical sinuses,** and **medullary sinuses** (MED-you-lair-ee). Cellular partitions called **medullary cords** separate the medullary sinuses. Efferent lymphatics drain the medullary sinuses, and blood capillaries penetrate the cortical and medullary tissue from arteries that branch within the trabeculae. Special venules, the **high endothelial venules** (HEV) shown in Figure 16.13, admit lymphocytes into the paracortical layer. Receptors on the HEV recognize the lymphocytes and draw them into the lymph, but erythrocytes and other cells do not enter. B-lymphocytes proceed to the cortical layer and T-lymphocytes to the paracortical layer; both types populate the medulla.

Lymph nodes filter lymph and produce lymphocytes and antibodies. Your axillary lymph nodes release an impressive arsenal of defenses against an infected cut on your hand. B-lymphocytes recognize and remove free bacteria that may reach these nodes. Your lymph nodes may swell with macrophages that have ingested bacteria and are presenting fragments of them to T-lymphocytes in the paracortical layer. Dendritic reticular cells (den-DRIH-tik) also present antigens to T-lymphocytes in the same layer. Recognition may activate cytotoxic T-cells that kill infected cells in the wound, or helper T-cells may stimulate B-cells to secrete antibodies in the medulla.

- What is the main function of a lymph node?
- What aspects of this role do the cortex, paracortical layer, and medulla perform?

THYMUS

The thymus resembles a cluster of lymph nodes. Located beneath the sternum at the base of the neck, this paired, triangular organ (Figure 16.14A) was named for leaves of the thyme plant. A thin **epithelium** covers numerous branching **lobes** of lymphatic tissue. Figure 16.14B shows that small septa subdivide each lobe into **lobules,** and a network of **epithelial reticular cells** reaches throughout, supporting the tissue of both lobes and lobules. Unlike lymph nodes and the spleen, however, no reticular fibers support this mesh of cells. Small lymphocytes make up the darkly stained **cortex** in the lobules of the thymus (Figure 16.14C), and deeper, lightly stained cells of the **medulla** unite lobules and lobes. **Hassall's corpuscles** are scattered through the medulla, but they have no known function. The thymus is largest at birth but begins **involution** at puberty. Adipose cells and fibroblasts replace the lymphocytes and reticular cells, and the lobes and lobules diminish in size and number. Even though the thymus loses its importance, T-lymphocytes continue to develop in the bone marrow.

Thymic arteries supply blood to the thymus, but no lymph enters the thymus. The arteries enter the medulla, and arterioles branch outward to the boundary between the cortex and medulla, where capillary loops supply the cortex. Venules drain these capillaries back into the medulla and finally discharge through **thymic veins.** Because no **afferent lymphatics** enter the thymus, no lymph capillaries or lymph nodes drain into this gland. Nevertheless, **efferent lymphatics** drain the medulla of lymph accumulated from the blood.

The cortex of the thymus is the body's primary source of T-lymphocytes. Derived from lymphoid stem cells in the fetal thymus, these lymphocytes proliferate continuously in the cortex, shielded from

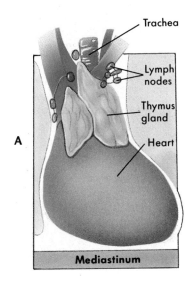

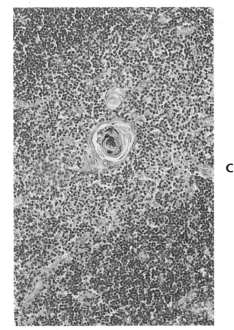

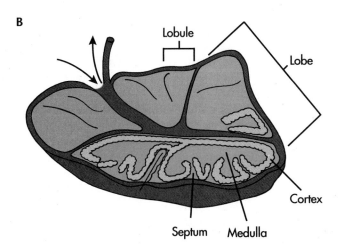

Figure 16.14

The thymus gland, **(A)** is located at the base of the neck **(B)**, contains cortex and medulla, and **(C)** shows darker cortex surrounding the lighter medulla, which contains a Hassall's corpuscle.

antigens by an epithelial **blood-thymus barrier** around the cortical vessels (Figure 16.14B). **Thymopoietin** (THIGH-mo-poy-et-in) from the epithelial reticular cells stimulates this intense proliferation. From the same source, **thymosin** induces the progeny cells to express T-lymphocyte receptors on their surfaces. Fully 95% of T-lymphocytes formed by the cortex die there, apparently because their receptors recognize self MHC molecules. New T-cells leave the thymus without entering the medulla or being exposed to antigens. The spleen completes their maturation by exposing them to antigens and sending them on to lymphatic tissues, where further exposure activates them for defense. B-cells become activated in the medulla of the thymus, but they do not enter the cortex.

- What is the major function of the thymus?
- What region of this gland is isolated antigenically from the blood?
- What events in T-cell development occur in this region?

SPLEEN

The spleen is the soft, purple, oval organ about the size of your fist tucked between the stomach, liver, kidney, and ribs 9 through 11 on the left side of the abdomen, as shown in Figure 16.15A and B. As does the thymus, it filters the entire blood supply, but its antigenic targets are quite different. The spleen traps and destroys large quantities of spent erythrocytes and other damaged cells, recycling bilirubin pigments and iron from their hemoglobin. These erythrocytes are responsible for the spleen's color. The spleen is also the center for humoral immunity because it stimulates B-lymphocytes to develop into plasma cells that produce circulating antibodies. The spleen also stores platelets and sometimes erythrocytes, as well as assisting monocytes to transform into macrophages. Removing the spleen after trauma or disease usually does not interfere with these processes because the bone marrow can act as a substitute.

The spleen resembles a lymph node on a larger scale. A **fibrous capsule** covers the spleen, and **splenic arteries, splenic veins,** and **efferent lymphatics** pass through the **hilum** on the medial surface. As in the thymus, there are no afferent lymphatics; all lymph cells enter from the blood. Internally a meshwork composed of **trabeculae, reticular fibers,** and **reticular cells** supports the lymphatic tissue in **filtering beds,** shown in Figure 16.15C, called **red pulp** and **white pulp.** Unlike the thymus, there are no branching lobes or lobules and no cortex or medulla separate white and red pulp. As good filters will do, red and white pulp cluster around the arteries and veins that supply blood to them, with white pulp usually innermost. Plasma cells develop in the white pulp through which lymph circulates, and blood circulates through

the red pulp that filters erythrocytes and platelets. This difference is responsible for the colors of red and white pulp.

Upon entering the hilum, the splenic artery branches into **trabecular arteries** that spread in the trabeculae and enter the white pulp as **central arteries.** These small, muscular arteries are so named because sheaths of white pulp form around them, as shown in Figure 16.15C and D. Central arteries send branches through the white pulp into the **marginal zone** that surrounds the white pulp. The marginal zone dispatches lymph to the white pulp and blood to the red pulp. Lymph drains from the white pulp into efferent lymphatics that return through the trabeculae and hilum (not shown in the figure). Blood filters through the red pulp and is collected by **venous sinuses** and **pulp veins** that also return to the trabeculae via **trabecular veins** and exit through **splenic veins.** Blood also enters the red pulp directly from branches of trabecular arteries. Red pulp has no lymphatic drainage.

WHITE PULP

This tissue resembles thymic and lymph nodal tissue because it contains **lymphatic nodules** replete with **germinal centers, B-lymphocytes,** and **plasma cells. T-lymphocytes** accumulate in the sheath around the central arteries. Unless they are activated by incoming antigens, these cells leave the spleen several hours after arriving and cycle back through the blood for later encounters with antigens. Activated T-lymphocytes help B-cells produce plasma cells and circulating antibodies.

RED PULP

The red pulp filters erythrocytes and platelets through a maze of filtration passages, shown in Figure 16.15D, that empty by way of **venous sinuses** into the pulp veins. Reticular cells line these passages with narrow partitions called **splenic cords.** The cords trap platelets and macrophages, and spent erythrocytes are removed here. The cords are unusual because they are passages where blood flows unrestrained by the endothelium that elsewhere lines the vascular system. This part of the circulation is considered "open" because the usual endothelium does not close it off from surrounding tissues.

Venous sinuses collect blood from the red pulp. These vessels act as sieves, keeping spent erythrocytes and platelets from leaving the spleen. The sinuses are constructed of long, narrow endothelial cells (Figure 16.15E) wrapped by straps of basement membrane, like barrel staves bound by hoops. Strips of basement membrane are not the only structures holding these cells together, however. Tight junctions connect the endothelial cells at intervals big enough for macrophages and other cells to slip through into

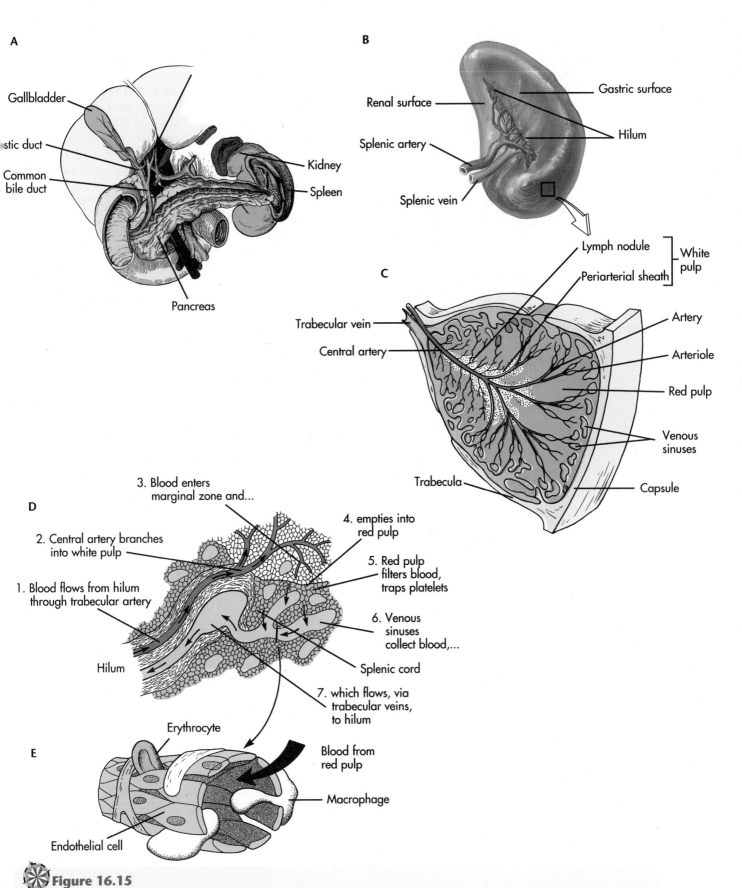

A

Gallbladder

stic duct

Common
bile duct

Kidney

Spleen

Pancreas

B

Gastric surface

Renal surface

Hilum

Splenic artery

Splenic vein

Lymph nodule ⎤
                ⎥ White
Periarterial sheath ⎦ pulp

C

Trabecular vein

Central artery

Artery

Arteriole

Red pulp

Venous
sinuses

Trabecula

Capsule

D

3. Blood enters
marginal zone and...

2. Central artery branches
into white pulp

4. empties into
red pulp

5. Red pulp
filters blood,
traps platelets

1. Blood flows from hilum
through trabecular artery

6. Venous
sinuses
collect blood,...

Hilum

Splenic cord

7. which flows, via
trabecular veins,
to hilum

Erythrocyte

Blood from
red pulp

E

Macrophage

Endothelial cell

✳ **Figure 16.15**

The spleen. **A,** Location. **B,** External anatomy. **C,** Red and white pulp in a filtering bed. **D,** Circulation in red and white
pulp. **E,** Details of a venous sinus.

the blood. As in the HEV capillaries of lymph nodes, epithelial surface receptors prohibit cells from leaving the red pulp. Once inside the venous sinuses, blood leaves the red pulp by way of trabecular veins and exits the spleen through splenic veins.

- What are the functions of the spleen?
- Which of these do the white pulp and the red pulp perform?
- How do blood cells enter venous sinuses?

DEVELOPMENT OF THE LYMPHATIC SYSTEM

Lymph nodes and the major lymphatic vessels develop from lymph sacs that accompany the veins of the cardiovascular system. Sacs grow into the surrounding mesenchyme and coalesce into thoracic and lymphatic ducts as their tributaries spread further into the tissues. Lymphocytes enter these vessels and concentrate in local clusters that become lymph nodes. Whether lymphatic sacs sprout from embryonic veins by **angiogenesis** (see Vessel Development, Chapter 14) or coalesce from clusters of mesenchyme by **vasculogenesis** is controversial.

Whatever the mechanism, in week 6 of development a series of sacs that will become the thoracic duct and the right lymphatic duct begins forming. Jugular lymph sacs (Figure 16.16) appear first. Then, working caudally, the retroperitoneal sac and cisterna chyli follow, and finally the posterior lymph sacs. At 9 weeks, all sacs link together to form a large plexus in the wall of the peritoneal cavity, and the major channels condense into the ducts and the lumbar trunks.

The vessels and nodes of the lymphatic system result from this spreading and consolidation. These vessels are lined by endothelium formed by the lymph sacs and covered with an external media and adventitia from the invaded mesenchyme and connective tissues. The capsules and trabeculae of lymph nodes derive from these coverings wherever lymphatic nodules aggregate. Lymphoid stem cells settle, proliferate, are joined by more immigrants, and fill a capsule that expands around them. Trabeculae branch into the interior, bringing blood vessels with them, but the afferent and efferent lymphatic vessels are of course already present.

- Which are the first lymph sacs to form?
- Which sacs form the thoracic duct?
- How do lymph nodes form?

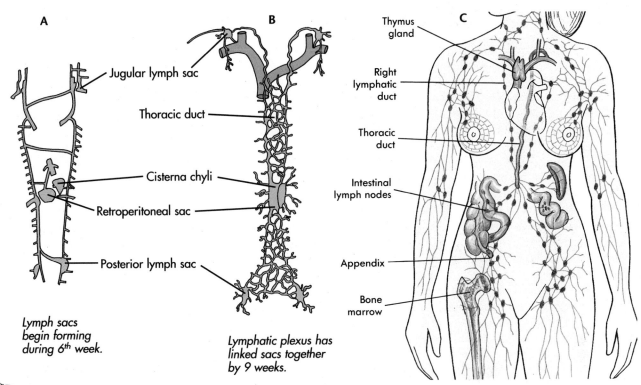

A

Jugular lymph sac

Thoracic duct

Cisterna chyli

Retroperitoneal sac

Posterior lymph sac

Lymph sacs begin forming during 6th week.

B

Lymphatic plexus has linked sacs together by 9 weeks.

C

Thymus gland

Right lymphatic duct

Thoracic duct

Intestinal lymph nodes

Appendix

Bone marrow

Figure 16.16
Development of the lymphatic system.

SUMMARY

The lymphatic system is the body's second circulatory system. Lymphatic vessels return lymph from the tissues to the blood for recirculation through the vascular system to the tissues. *(Page 478).*

The lymphocytes and tissues of the lymphatic system are agents of immunity. Two broad classes of lymphocytes—B-cells and T-cells—patrol the blood and lymph for foreign cells and substances that they recognize and remove. B-cells mark soluble antigens and bacteria for disposal, and T-cells recognize fragments of foreign materials that have been ingested by macrophages. T-cells secrete lymphokines that induce other cells to remove foreign material. *(Page 480).*

Lymph circulates from the tissues back to the veins and arteries and then returns to the tissues. Lymphatic vessels collect lymph fluid that enters from the tissues. Lymphatic nodules and nodes filter the lymphocytes and antigens from the lymph and release antibodies and new lymphocytes into it as the lymph returns to the blood. *(Page 482).*

Lymphatic vessels resemble veins. Lymphatic capillaries collect lymph from tissues in thin-walled structures consisting of endothelium and adventitia. Intercellular clefts admit lymph to the lumen of the capillaries. Collecting vessels introduce the tunica media and smooth muscle fibers and form endothelial valves that propel lymph toward the heart when the vessels constrict. Larger lymphatic trunks drain collecting vessels into the thoracic duct or the right lymphatic duct. The thoracic duct drains the abdomen and lower extremities and the left side of the body above the abdomen. The right lymphatic duct drains the right side of the head, neck, thorax, and right arm. *(Page 483).*

The lymph is a clear, straw-colored fluid that resembles blood plasma except for its lower protein content. Only about 100 ml of lymph pass through the thoracic duct each hour, a rate several thousand times slower than blood flow. Lymph carries lipid-laden chylomicrons from the small intestine, and the liver releases albumin proteins. Thoracic duct lymph appears milky, and its proteins are more concentrated for this reason. *(Page 485).*

Lymphatic nodules are the simplest lymphatic tissues. Individual nodules appear temporarily at infec-

tion sites, and others become permanent clusters in the gut-associated lymphoid tissues (GALT), lymph nodes, thymus, and spleen. Small B-cells and T-cells surround germinal centers where activated lymphocytes secrete antigens and form memory cells. *(Page 488).*

Peyer's patches cover clusters of lymphatic nodules with an epithelium and admit antigens from the lumen of the intestines directly to B-cells by transcytosis through M-cells. Other members of GALT and BALT resemble Peyer's patches. *(Page 488).*

Lymph nodes encapsulate nodules within a tough, fibrous capsule that admits blood and lymph from larger territories and greater distances than do nodules and Peyer's patches. Collecting vessels and small arteries and veins supply the circulation. B-cells and lymphatic nodules accumulate in the cortex, T-cells do so in the paracortical region, and both types populate the medulla. Reticular fibers, reticular cells, and trabeculae support these lymphatic tissues. *(Page 488).*

The thymus is the source of T-lymphocytes. These cells acquire the capacity to distinguish self from nonself as they develop from lymphoid stem cells in the cortex. B-lymphocytes are activated in the medulla, which resembles the germinal centers of lymph nodules. Blood and lymph carry new lymphocytes away from the thymus, but only blood enters this gland. Reticular cells support the internal lobes and lobules of tissue, and vessels branch within. *(Page 491).*

The spleen is the most highly modified lymphatic organ. Regions of white and red pulp, rather than cortex and medulla, characterize this organ. Networks of reticular fibers and cells support its tissue, and blood vessels branch throughout it. Red pulp stores platelets and reprocesses spent erythrocytes. White pulp, where B-cells and plasma cells produce antibodies, resembles lymphatic nodules. *(Page 492).*

Vessels and nodes of the lymphatic system develop from lymph sacs that coalesce into the thoracic duct and right lymphatic duct. Tributaries spread from these beginnings, and lymph nodes form where stem cells aggregate in the vessels. The endothelium of vessels and nodes derives from the lining of the lymph sacs, whereas the media and adventitia of the vessels and the capsules of nodes are formed by mesenchyme from surrounding tissues. *(Page 494).*

STUDY QUESTIONS

1. What roles do T- and B-cells play in immunity?   **1**
2. Describe the three steps of an immune response.   **2**
3. What is meant by the ability of lymphocytes to distinguish between self and nonself cells? What molecules provide this ability?   **2**
4. Outline the vascular pathway a lymphocyte would follow in order to return to a lymphatic nodule located at an infection (A) in the right hand and (B) in the left leg.   **3, 4, 6**
5. Describe the three layers of tissue that compose the walls of lymph vessels. What are the structural differences between lymph capillaries and collecting vessels?   **4**
6. How does the composition of lymph differ from that of blood plasma?   **5**
7. What phases of the immune response take place in germinal centers and in the outer layer of small lymphocytes in lymphatic nodules?   **6**
8. Describe the anatomical differences between Peyer's patches and lymphatic nodules.   **6, 7**

9. Outline the similarities and differences among the gut-associated lymphoid tissue.   **7**
10. Which vessels enter and leave lymph nodes? Which regions of the node contain T- and B-cells? Which lymph nodes will lymphocytes from the extremities enter?   **8**
11. Follow the circulation of blood and lymph through a cervical lymph node.   **8**
12. Trace the development of T-lymphocytes in the thymus gland. Describe how blood vessels and lymphatics reach the cortex and medulla of the thymus.   **2, 9**
13. Describe the anatomy of white pulp in the spleen. How does blood reach it and flow from it? How does the anatomy of this pulp explain the production of humoral antibodies?   **10**
14. Describe the anatomy and function of the red pulp. How does blood reach this portion of the spleen, and how does it leave?   **10**
15. High endothelial vessels and venous sinuses have similar structures and functions in the thymus and spleen. Describe the similarities and differences between them.   **8, 10**
16. From what tissues do lymphatic vessels and lymph nodes develop?   **11**

PART IV

NEURAL INTEGRATION

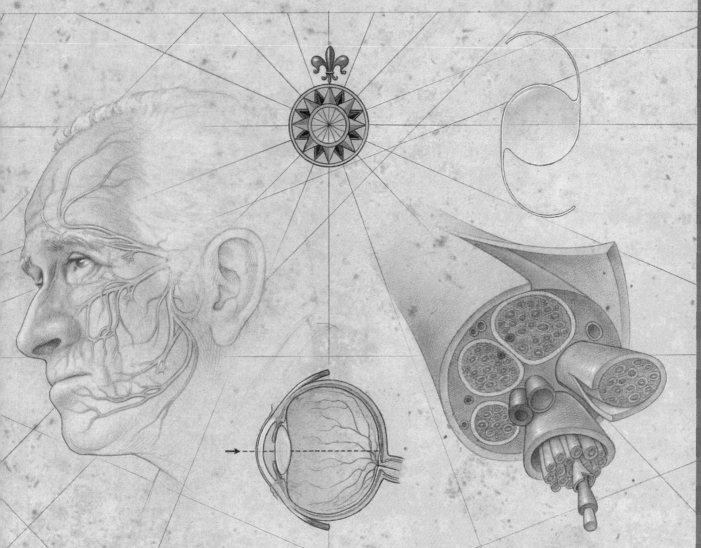

The basis of our personality and behavior lies in the cells of the nervous system. Connections between these cells and information that flows through them somehow result in the memories, ideas, and actions that make us individuals. Neuroanatomy reveals that the cells and pathways that conduct and store such information are precise, and when they are affected by disease, injury, or drugs, our behavior changes accordingly. Neuroanatomy probably will not entirely disclose the nature of mental processes any more than the anatomy of your telephone line can reveal the conversations it carries. Even though these signals may be beyond the limits of anatomy, the structure of the nervous system and its cells nevertheless helps us see how the system works.

Part IV introduces the fundamentals of the nervous system and how its various cells and tissues enable the body to work as an individual. **Chapter 17,** Neurons, Neuroglia, and Neural Pathways, describes the cells that conduct impulses throughout the nervous system and the glial cells that support them. This chapter illustrates how connections between neurons form neural path-

ways through which streams of impulses flow that coordinate the action of muscles and glands. **Chapter 18** constructs higher levels of neural organization in the pathways of the spinal cord and spinal nerves that communicate between the brain and body. The effects of spinal injury illustrate how normal function depends on different neural pathways. In **Chapter 19,** Brain and Cranial Nerves, the regions of the brain are surveyed and neural traffic is traced through some of them. This flow conveys sensory impulses from the skin and muscles to the brain and sends out motor stimuli that coordinate muscular motion and glandular secretion. **Chapter 20** examines the origin of this traffic in sensory receptors and in the organs of special sense. The next chapter, Autonomic Nervous System, focuses on the anatomy of nerves that deliver efferent motor impulses to the heart, gut, and other visceral organs. Finally, **Chapter 22** explores the endocrine system, which coordinates metabolic functions by secreting hormones in a manner that resembles how the nervous system communicates from one neuron to the next.

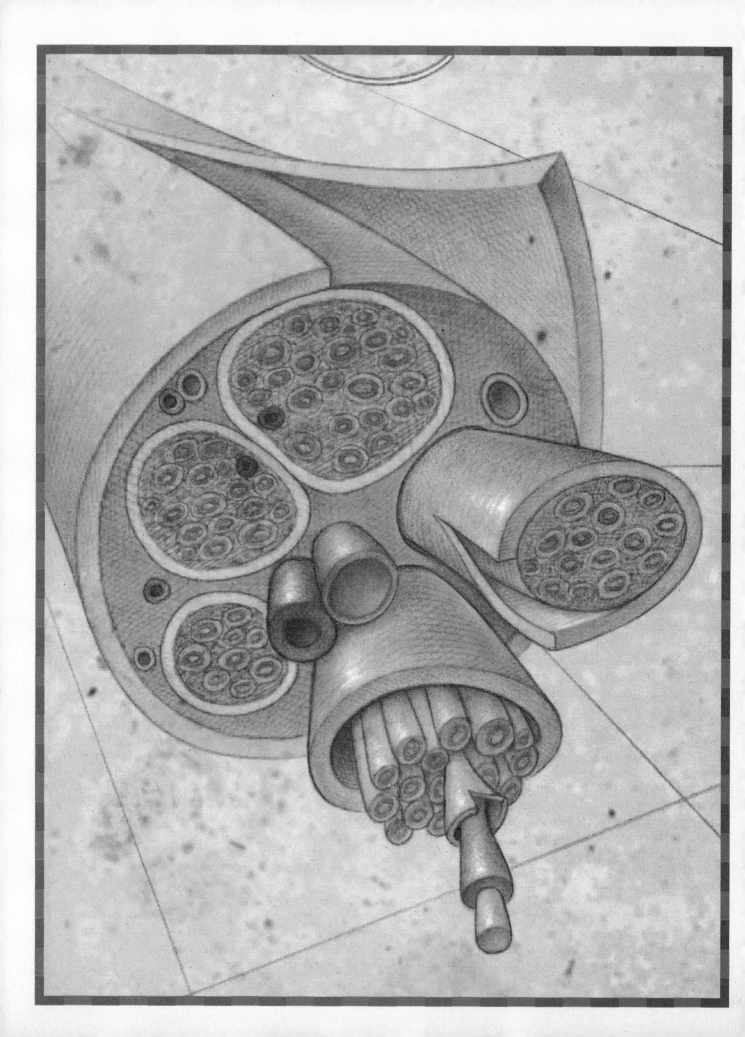

CHAPTER 17

Neurons, Neuroglia, and Neural Pathways

OBJECTIVES

To understand the anatomy and function of neural tissues, you should be able to do the following:

1. Know what functions neurons and neuroglia perform.
2. Trace the optic pathway from the retina to the brain.
3. Describe the divisions of the nervous system, and classify neurons in these categories.
4. Explain the roles and structures of the cell body, dendrites, and axon of a neuron.
5. Explain what happens in the membrane of a neuron as an action potential passes.
6. Describe the structure of a synapse and the events that transmit impulses across it.
7. Classify neurons by their shape and direction of conduction.
8. Describe the various types of neuroglia cells, their functions, and how they differ from neurons.
9. Describe the structure of the myelin sheath of peripheral nerves and how it speeds conduction of impulses.
10. Describe the structure and function of the nodes of Ranvier in neuronal conduction.
11. Describe the internal organization of peripheral nerves.
12. Tell which embryonic tissues are the sources of neurons and neuroglia cells.
13. Describe the structure of growth cones and their role in development of neurons and neural pathways.
14. Describe how neurons regenerate.
15. Describe various conditions that interfere with the functions of neurons and the nervous system.

 VOCABULARY TOOLBOX

Directional terms.

afferent neurons	ad-, *L* toward. Carry impulses toward the brain and spinal cord.	**perikaryon**	peri-, *G* around; kary-, *G* the nucleus.
efferent neurons	e-, *L* out, from. Transmit impulses away from the brain and spinal cord.	**postsynaptic**	post-, *L* after. After the synapse.
		presynaptic	pre-, *L* before. Before the synapse.
		telodendria	tel-, *G* end; dendr-, *G* a tree. End branches.

Neural terms.

arborization	arbor-, *L* a tree. Tree-like branching.	**dendrite**	dendr-, *G* a tree. Dendrites branch tree-like from the neuron cell body.
autonomic	auto-, *G* self; nomi-, *G* custom. Customarily automatic.	**ependymal**	ependyma, *G* a tunic. The ependymal layer lines the interior of the spinal cord and brain.
axon	axon, *G* an axle, an axis. The main extension of a neuron.	**myelin**	myel-, *G* the spinal cord.

PERSPECTIVE

The nervous system acts at many levels. The aspirin you took for yesterday's headache brought relief by slowing the release of substances that stimulate pain receptors. Sensing less pain, you felt better. At the lowest levels of organization, sensitive neurons were being stimulated and, several steps up, neuronal pathways conducted these impulses to the brain. At still higher levels the brain recognized these stimuli as painful and recalled experiences with other headaches, and you decided to take an aspirin this time. Together, these interactions represent progressively higher plateaus in the anatomy and action of the nervous system described piece by piece and level by level in this and following chapters. We begin with neurons and neuroglia, relatively simple pieces of the puzzle that are parts of a more complex structure.

Neurons are the fundamental units of the nervous system. They are the cells that conduct, transmit, and store its information. To do this, **neurons** (NOOR-ons; Figure 17.1) connect to other neurons, forming "pathways" that conduct information. Neurons receive signals at their **dendrites** and central **cell bodies;** impulses then flow outward along **axons** to **synapses,** where the impulses signal the dendrites of other neurons. Axons can reach long distances, up to 3 feet between the spinal cord and toes, but dendrites are much shorter. **Neuroglia** cells (NOOR-oh-glee-ah) are the nervous system's supporting cells that cover and provide nutrition to neurons and remove cellular debris, to name just a few services.

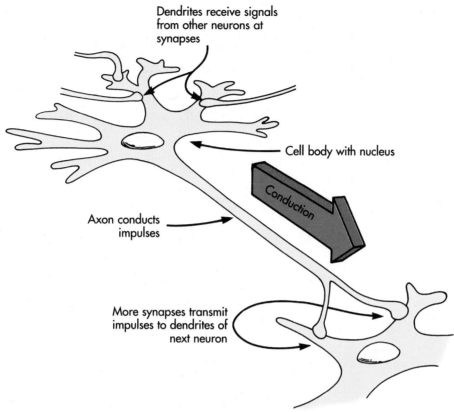

Dendrites receive signals from other neurons at synapses

Cell body with nucleus

Conduction

Axon conducts impulses

More synapses transmit impulses to dendrites of next neuron

Figure 17.1

This neuron is part of a chain that transmits stimuli from one neuron to the next. Dendrites receive impulses from other neurons at synapses. Impulses flow out the axon to synapses that transmit to the dendrites of the next neuron.

NEURONS AND NEURAL PATHWAYS

Chains and networks of neurons carry information. Most of the nervous system's neural pathways remain unknown, but the optic pathway is a good example of what is known. This pathway projects images of the car in Figure 17.2 from the retina onto the primary visual areas of the brain, so that the brain receives a direct counterpart of the car, somewhat like TV images from a video camera. In the retina, rods and cones receive incoming light and stimulate neurons that relay the signals to ganglion cells. From the retina, these ganglion cells conduct impulses through the optic nerve to the brain, where a third set of neurons finally relays into the primary visual cortex in the occipital lobes of the cerebral hemispheres. Other areas of the brain form images from this information and interpret them.

The brain interprets the image because the arrangement of neurons in the visual cortex corresponds to the arrangement of rods and cones in the retina, where the stimuli began. The image of the sports car spreads into both hemispheres. The right visual cor-

tex sees the rear end of the car because this side of the cortex receives stimuli from the right half of each retina, and the left cortex sees the front end of the car from the left halves of the retinas. Binocular vision exists because each cortex receives stimuli from both eyes. The anatomy you are about to study gives some clues about how this and many other neural pathways operate.

- Name the cells that conduct impulses in the nervous system.
- Which type of cell supports conducting cells?
- Where does the optic pathway begin and end?

DIVISIONS OF THE NERVOUS SYSTEM

Neuroanatomy distinguishes between the central and peripheral nervous systems (Figure 17.3A and B). The neurons of the brain and spinal cord belong to the **central nervous system (CNS)** and are encased in the skull and vertebral column. The central nervous system communicates with the body by way of the **peripheral nervous system (PNS)**, through

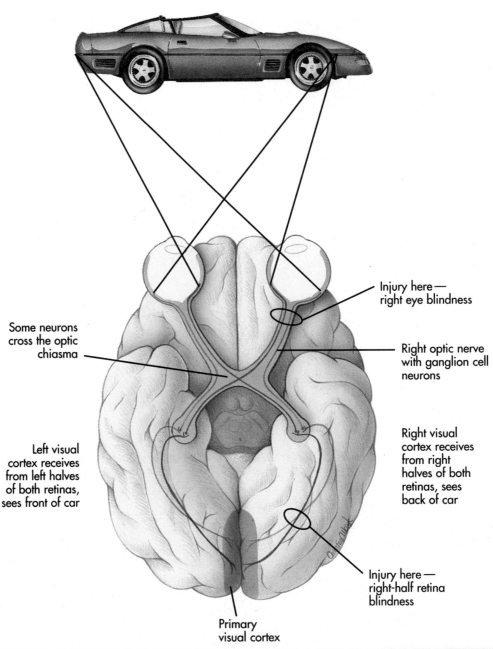

Injury here—
right eye blindness

Some neurons
cross the optic
chiasma

Right optic nerve
with ganglion cell
neurons

Left visual
cortex receives
from left halves
of both retinas,
sees front of car

Right visual
cortex receives
from right
halves of both
retinas, sees
back of car

Injury here—
right-half retina
blindness

Primary
visual cortex

Figure 17.2

The optic pathway communicates stimuli from the retina to the brain. Trace the pathway from the retina, to ganglion neurons, through the optic nerves, into the primary visual cortex. Ganglion cells cross at the optic chiasm. Each side of the visual cortex receives signals from rods and cones in the corresponding half of each retina—right half to right cortex, left half to left cortex.

cranial and spinal nerves (Figure 17.3) that connect the brain and spinal cord to skeletal and smooth muscles, to the glands, and to the eyes, ears, and other sensory receptors throughout the body.

In turn, the PNS has somatic and visceral divisions. Nerves in the **somatic division** communicate with the skin, skeletal muscles, and skeleton, whereas those of the **visceral division** convey impulses between the CNS and the **visceral organs**, vessels, smooth mus-

cles, and glands. In each division, neural traffic travels in both directions. **Afferent neurons** convey impulses from sense organs in the skin and muscles to the spinal cord and brain. For example, somatic afferent neurons convey impulses in the optic pathway to the brain, while visceral afferents from the stomach convey impulses that signal stretching of the wall of the stomach. **Efferent neurons** conduct from the CNS to the somatic and visceral organs. Somatic efferent

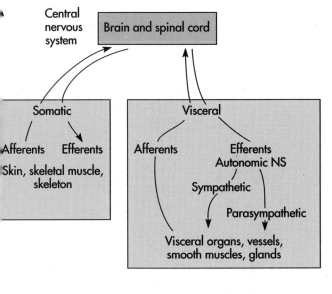

Central nervous system

Brain and spinal cord

Somatic

Afferents   Efferents

Skin, skeletal muscle, skeleton

Visceral

Afferents   Efferents
Autonomic NS

Sympathetic

Parasympathetic

Visceral organs, vessels, smooth muscles, glands

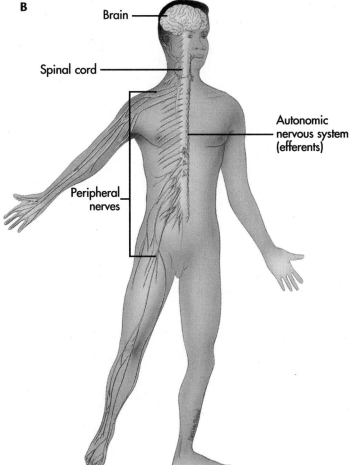

B

Brain

Spinal cord

Autonomic nervous system (efferents)

Peripheral nerves

**Figure 17.3**

A, Divisions of the nervous system. B, Find the brain and spinal cord (CNS) and the nerves of the peripheral and autonomic nervous systems.

neurons govern voluntary muscle action, but efferent neurons to the visceral organs may stimulate secretion of gastric juice or regulate heart rate. These visceral efferents have special pathways that constitute a separate autonomic (auto-NOM-ik) **nervous system (ANS).**

The autonomic nervous system (seen in Figure 17.3B) is physically and functionally separated into **sympathetic** and **parasympathetic** divisions. Visceral efferents of the sympathetic system accelerate heart rate, for example, and parasympathetic neurons retard it. These opposing functions of the ANS are discussed in Chapter 21.

• Which portion of the nervous system communicates between the spinal cord and the body?

• What elements of the peripheral nervous system conduct impulses that control walking?

• Which elements relay sensations of a stomach ache to the spinal cord and brain?

NEURONS

Cells communicate in many ways, though most often by secreting substances that cause changes in other cells. Endocrine cells secrete hormones that are broadcast through the bloodstream over long distances to their target tissues. Nerve cells solve this problem differently. They secrete signal molecules across very short distances between the ends of long, thin extensions of the nerve cells themselves. Figure 17.4 shows the structure of a typical nerve cell or **neuron.**

CELL BODY

Genes in the cell body govern synthesis of cytoplasmic constituents for the dendrites and axons. The cell body is known also as the soma, or perikaryon (PAIR-ih-KARRY-ahn). Especially large clusters of rough endoplasmic reticulum called **Nissl bodies** (rhymes with thistle) accumulate in the cell body and assemble protein molecules. **Golgi bodies** package these products into **secretion vesicles** for distribution to the cell membrane, the lysosomes, and the dendrites and axon. Neuron cell bodies contain especially large and abundant **mitochondria** that supply ATP for these syntheses. The brain and spinal cord require a continual supply of oxygen to sustain this energy demand; neurons die if oxygen is interrupted for even a few minutes.

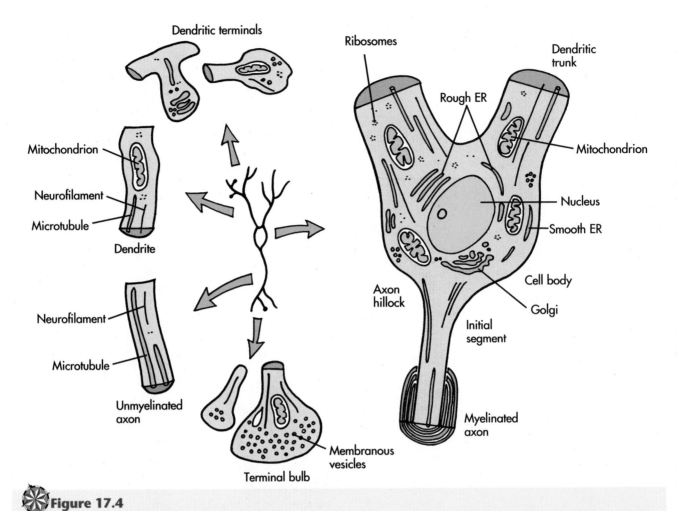

✳ **Figure 17.4**

Internal anatomy of a typical neuron. Important sections are enlarged to show the details of the cell body: dendrites and their tips; and the initial segments, shafts, and terminal bulbs of axons.

Neurons have extensive cytoskeletons that give them support and shape and also transport materials through the cytoplasm. **Microtubules** weave throughout the cytoplasm of the cell body, extending into the dendrites and axon (see Axonal Transport, below). A network of **neurofilaments** accompanies the microtubules, but its function is unclear. Neurofilaments belong to the category of **intermediate filaments,** a group of proteins (see Chapter 4, Epithelial Tissue) concerned with cytoplasmic support. Clusters of neurofilaments accumulate in the neurons of patients suffering from Alzheimer's disease. Neurons also contain **actin microfilaments,** the third major component of the cytoskeleton. These contractile proteins are involved in axonal growth (see Neuronal Outgrowth, this chapter) but apparently are not directly concerned with transport.

DENDRITES

Dendrites receive chemical signals from other neurons. Several dendrites extend from the cell body in Figure 17.4; each one branches many times and be-

comes narrower. Tiny **dendritic spines** stick out along each branch. Such branches and spines increase the surface available to receive synapses from other neurons. Nissl bodies, Golgi bodies, ribosomes, and mitochondria are present in the primary branches near the cell body, but only free ribosomes are seen in the tips. Neurofilaments and microtubules extend throughout the dendrites, in keeping with their postulated role in cytoplasmic transport.

AXONS

Axons can initiate impulses in response to stimuli from dendrites. These impulses are then conducted away from the cell body toward the axon terminals. Axons are very different from dendrites. The single axon in Figure 17.4 extends from the cell body at a tapering region of clear cytoplasm called the **axon hillock.** Ribosomes and Nissl bodies are missing from this region, but neurofilaments and microtubules are present, and they extend the entire length of the axon. The axon hillock narrows into the **initial segment** of the axon, where impulses can begin. Ion channels also are concentrated there, as well as at synapses. Axons have a nearly constant diameter, even though they may branch and be up to several feet long. Axon branches are called **collaterals.** Numerous small branches called **telodendria** (tello-DEN-dree-ah) are located at the end of the axon; the entire network of these branches is called the **terminal arborization** (AR-bor-eye-ZAY-shun). Each branch ends with a **terminal bulb** (variously called a **terminal bouton, end bulb,** or **terminal knob**) located close to its target, which is most often the dendrite or cell body of another neuron.

Axonal cytoplasm contains microtubules and neurofilaments, occasional mitochondria, and numerous membranous vesicles, but there is no endoplasmic reticulum or Golgi body to indicate that the axon itself assembles macromolecules. Axonal transport apparently provides these molecules from the cell body.

AXONAL TRANSPORT AND AXOPLASMIC FLOW

Neurons transport vesicles, proteins, and transmitter substances from the cell body to sustain the axons and dendrites. Traffic also proceeds toward the cell body, so that axons and dendrites are two-way streets in this regard. As proof, dyes injected into axon terminals reach the dendrites at the opposite end of the cell. **Axoplasmic flow** of cytoplasm is slow, about 1 mm per day. **Axonal transport** is more rapid (100 to 400 mm per day) and requires ATP, but it is never as fast as an impulse speeding along the membrane. Microtubules (and perhaps also neurofilaments) participate in axonal transport (Figure 17.5). Microtubules serve as intracellular highways for cytoplasmic vesicles to move in both directions along axons and dendrites, but it is not yet clear exactly what mechanisms transport which neuronal constituents. Synthesis of new cytoplasm by the cell body probably produces axoplasmic flow.

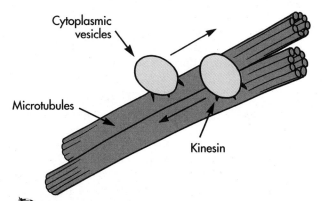

Figure 17.5

Axonal transport along microtubules. Kinesin, a protein that binds vesicles to the tubules, transports cytoplasmic vesicles in both directions along microtubules.

- In which direction do impulses flow in a neuron?
- What parts of a neuron receive signals from other neurons?
- What are the functions of neurofilaments and microtubules in an axon?

ACTION POTENTIAL

Neurons conduct impulses called action potentials, which travel along the membrane as the "wave" proceeds at a baseball game. The wave flows around the stadium because people stand up when it approaches, then sit down and wait for the next time, as shown in Figure 17.6A. The comparable elements of an action potential are sodium and potassium ion channels that open and close as the action potential flows past. Sodium ions can enter the cell through **sodium ion channels,** and potassium ions can exit through **potassium ion channels,** but when the membrane is at rest, most of these channels are closed. These channels are packed quite closely in the membrane, only 20 nm apart. In an action potential the sodium channels open briefly and sodium ions enter the cell (Figure 17.6B). The channels close automatically, but they are open long enough to cause neighboring channels to admit sodium ions also. The inflow of sodium ions spreads along the membrane toward the tip of the axon as each sodium channel opens and closes.

The membrane expends energy to set up the conditions for action potentials. The plasma membranes of neurons and most cells contain **sodium-potassium pumps** (Figure 17.7A), integral protein molecules that export sodium ions and import potassium ions, using energy obtained by hydrolyzing ATP into ADP and phosphate. These proteins also are known as the enzyme **Na$^+$K$^+$-ATPase** for this reason. Their work **polarizes,** or distributes, ions unequally onto the cytoplasmic and external sides of the membrane, resulting in a difference in charge that is called the **resting potential** (Figure 17.7A). On the average, the pump moves three sodium ions out for every two potassium ions that enter. Together with ions already on both sides of the membrane, this difference makes the internal surface of the membrane slightly more negative than its exterior. Neuronal resting potentials are commonly –70 mV, which means that the cytoplasmic side is more negative, with an electric potential of 70 mV less than the outside. Consequently, membrane potentials reflect the **concentration gradients** of sodium and potassium ions across the membrane. The pumps must work continuously or the resting potential will disappear in several hours. Sodium ions are thus poised on the exterior of the neuron to diffuse down their concentration gradient into the cytoplasm, and potassium ions are prepared

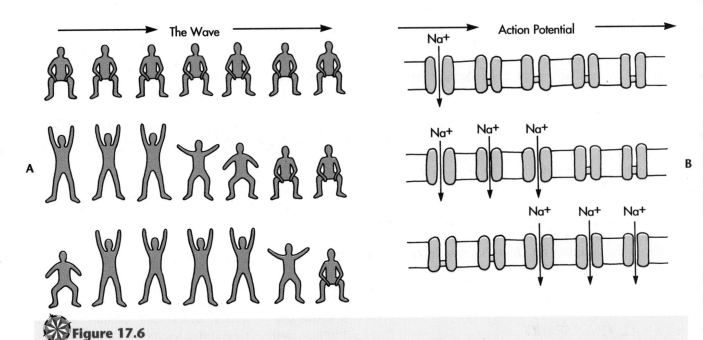

Figure 17.6
"The wave" analogy and sodium ion channels. A, "The wave" proceeds when people stand up and then sit down, causing others nearby (who haven't already done so) to do the same. **B,** Sodium ion channels are near enough to each other so that ions flowing through several channels cause neighboring channels to open and then close.

likewise to diffuse outward when a stimulus opens the ion channels.

Action potentials begin when the membrane depolarizes. The sodium and potassium ion channel proteins are sensitive to the local membrane potential. When enough sodium ions flow inward through nearby channels to **depolarize** the membrane, that is, reverse the local membrane potential, neighboring sodium channels open (Figure 17.7B). This causes the next channels along the membrane to open. Sodium channels are said to be **voltage-gated;** that is, when the membrane depolarizes, gates blocking the sodium channels open. At this point the action potential has begun, and the membrane potential has risen to about +30 mV. The membrane potential has reversed; the internal face of the membrane is now more positive than the external face.

How is the resting potential restored and the membrane readied for the next action potential? The sodium gates automatically close very quickly (Figure 17.7C), but not before enough sodium has entered to depolarize the membrane and open the voltage-gated potassium channels. These channels remain open longer than sodium channels, just long enough to allow sufficient potassium ions to exit and to repolarize the membrane, restoring the resting potential. Repolarizing the membrane makes the sodium channels able to open once again. Action potentials begin in the

initial segment of the axon where Na$^+$K$^+$-ATPase and sodium and potassium channel proteins have accumulated in the membrane. When ionic currents from the dendrites are strong enough to depolarize the initial segment and open its sodium channels again, a new action potential begins to propagate along the axon toward the terminal arborization.

- How many millivolts is the resting membrane potential?
- Which side of the membrane is more positive at rest?
- What ions and ion pump make it so?
- What change in the membrane opens the sodium ion channels at the beginning of an action potential?
- What does the term "voltage-gated" mean?
- Where does an action potential begin in a neuron?

SYNAPSES

Synapses transmit signals from one neuron to the next. At a typical synapse, a terminal bulb of a **presynaptic** neuron comes close to a dendrite or cell body (only rarely, close to the axon) of the next, or **postsynaptic**, neuron. A single neuron may make as many as 20,000 synapses with terminal bulbs of other axons converging onto its dendrites and cell body. Transmission across each synapse may excite or inhibit the postsynaptic cell. Generally, excitation de-

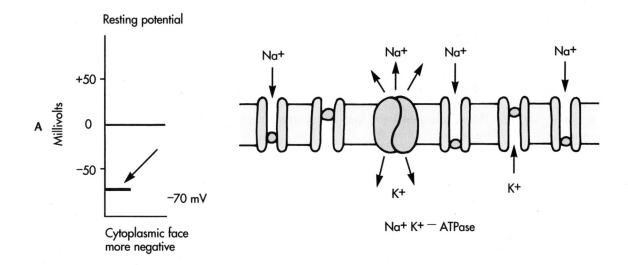

Resting potential

+50

Millivolts

0

−50

A

−70 mV

Cytoplasmic face
more negative

Na+   Na+   Na+   Na+   Na+   K+

Na+ K+ — ATPase

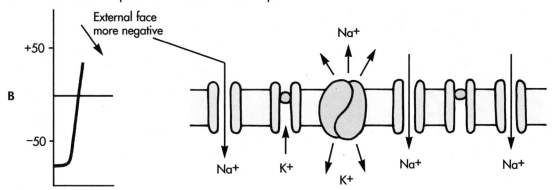

Membrane depolarizes—sodium channels open

External face
more negative

+50

B

−50

Na+   Na+   K+   K+   Na+   Na+

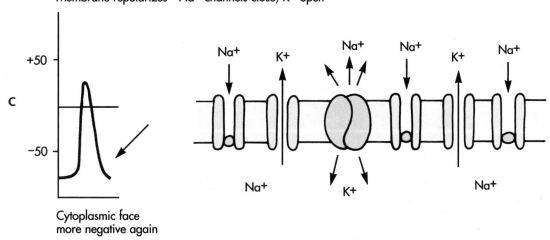

Membrane repolarizes—Na+ channels close, K+ open

+50

C

−50

Cytoplasmic face
more negative again

Na+   K+   Na+   Na+   K+   Na+   Na+   K+   Na+

Figure 17.7

Components of the action potential. **A,** Resting potential. Sodium potassium pumps in the axonal membrane establish the resting potential by pumping sodium ions out and potassium ions in. The cytoplasmic face is more negative than the external face of the membrane. **B,** Membrane depolarization. An action potential is arriving. Sodium ions enter the cell, briefly reversing the polarity of the membrane. The cytoplasmic face of the membrane becomes more positive than the exterior for a short time. The membrane potential rises rapidly from −70 mV to +30 mV. **C,** Membrane repolarizes. About 0.2 msec have passed since **B,** and the action potential is passing. The membrane potential returns to −70 mV because the sodium channels close and the potassium channels open. As the potassium ions rush out, the cytoplasmic face of the membrane becomes more negative again and the membrane potential is restored to its original value.

polarizes and inhibition hyperpolarizes the postsynaptic membrane. When excitation exceeds inhibition, the axon generates and propagates an action potential.

Most synapses are **chemical synapses** (Figure 17.8A and B) that transmit signals by way of neurotransmitter molecules. **Electrical synapses** are specialized connections that permit ions to flow directly between cells through matched ion channels in the presynaptic and postsynaptic membranes (Figure 17.8C). In chemical synapses a **synaptic cleft,** usually 20 nm across, separates the membranes of the presynaptic and postsynaptic cells. Receptor proteins for acetylcholine, the best known **neurotransmitter substance,** pack this region of the postsynaptic membrane and are retained there by a dense protein un-

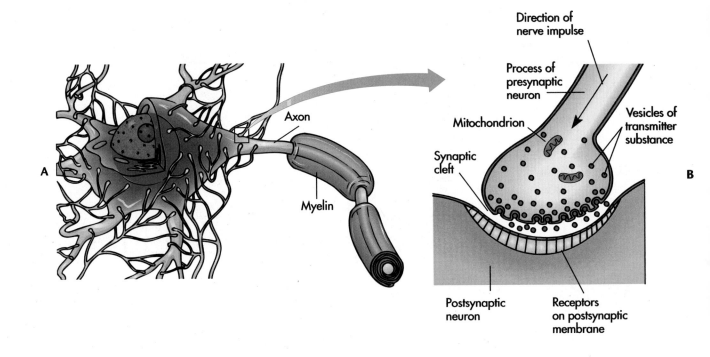

A

Axon

Myelin

B

Direction of
nerve impulse

Process of
presynaptic
neuron

Mitochondrion

Synaptic
cleft

Vesicles of
transmitter
substance

Postsynaptic
neuron

Receptors
on postsynaptic
membrane

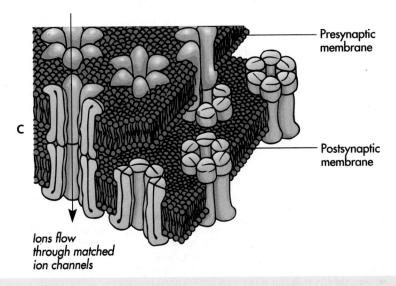

C

Presynaptic
membrane

Postsynaptic
membrane

Ions flow
through matched
ion channels

Figure 17.8

Synapse. **A,** Neurons receive many synapses. **B,** A chemical synapse. **C,** Matched ion channels permit ions to flow directly between cells at electrical synapses.

dercoat. **Synaptic vesicles** are concentrated in the terminal bulb. These vesicles, about 40 nm in diameter, store **acetylcholine** (ah-SEET-il-KOE-leen) and release it by exocytosis (at A in Figure 17.9) into the synaptic cleft when action potentials depolarize the terminal bulb (at B in Figure 17.9). Calcium ions facilitate this release.

The acetylcholine quickly diffuses across the cleft and binds to acetylcholine receptors (at C in Figure 17.9), causing sodium channels to open in the membrane of the postsynaptic cell. When acetylcholine binds to enough receptors, the postsynaptic membrane depolarizes as sodium ions enter. This depolarization flows toward the cell body, where it may induce an action potential in the initial segment of the axon. In the synapse the enzyme **acetycholinesterase** (ACHase; ah-SEET-il-KOE-lin-ES-ter-aze) rapidly splits acetylcholine on the postsynaptic membrane into choline and acetate, so that each burst of acetylcholine stimulates the postsynaptic neuron only briefly. Choline is returned across the synaptic cleft to the presynaptic neuron (at D in Figure 17.9) and recycled into new synaptic vesicles for later transmissions. The acetate diffuses away. Synaptic transmission takes about 1 msec. The fewer the synapses in a neuronal pathway, the faster the pathway.

Not all neurotransmitter substances excite their postsynaptic cells. **Gamma-aminobutyric acid** (GABA; ah-MEEN-oh-BEW-teer-ik) is an inhibitory neurotransmitter that causes postsynaptic cells to polarize further (hyperpolarize), often by allowing more potassium ions to leave the cell. This makes the inside of the cell even more negative and thus less likely to generate an action potential.

- How many synapses can typical neurons receive?
- Name the class of molecules chemical synapses use to change the polarization of a postsynaptic cell.
- What are the roles of synaptic vesicles and receptors in synaptic transmission?

CLASSES OF NEURONS

Most neurons are either unipolar, bipolar, or multipolar. These types have one, two, or more cytoplasmic processes, respectively, that extend from the cell body. All have one axon, and the remaining processes are considered dendrites. **Unipolar neurons** extend a single axon from the cell body (Figure 17.10A). True unipolar neurons are rare in our nervous system, but in peripheral afferent neurons the process extends in two directions as an axon between sensory receptors in the skin or skeletal muscles and in the spinal cord. There are no true dendrites at the tips of these axons, just receptors. (Some texts call this type a pseudounipolar neuron.) **Bipolar neurons** possess one axon and one dendrite (Figure 17.10B). Bipolar cells are common in the gray matter of the spinal cord. In the retina, these cells communicate stimuli from the rods and cones to the ganglion cells

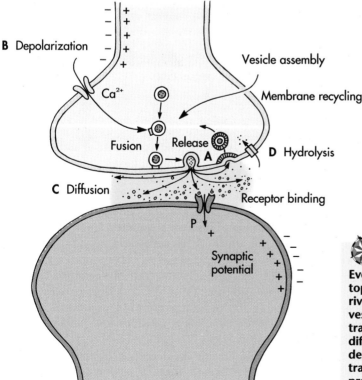

Figure 17.9

Events at chemical synapses. **A,** Axon tip assembles cytoplasmic vesicles containing neurotransmitter. **B,** Arriving action potentials admit calcium ions, causing vesicles to fuse with the membrane and release neurotransmitter into the synaptic cleft. **C,** Neurotransmitter diffuses across the cleft and binds with receptors, which depolarizes the postsynaptic membrane. **D,** Neurotransmitter is hydrolyzed or returned intact to the presynaptic cell for repackaging inside recycled vesicles.

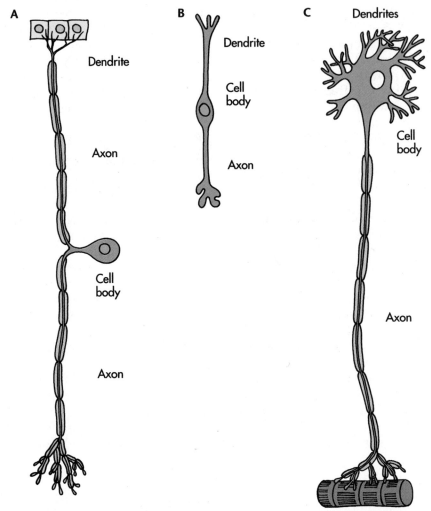

Figure 17.10

Classes of neurons. A, Unipolar cells send out one process that acts as an axon. B, Bipolar neurons are short cells with one axon and one dendrite. C, Multipolar neurons have several dendrites and a single axon.

of the optic nerve. **Multipolar neurons** have many dendrites and one long axon, allowing them to receive signals from many other cells (Figure 17.10C). Efferent neurons of the peripheral nervous system are multipolar neurons, with cell bodies and dendrites in the spinal cord and axons in the spinal nerves.

Many different neurons modify these basic forms; the differences help reveal each cell's function. **Purkinje cells** (purr-KIN-jee; Figure 17.10D) are multipolar neurons that help the cerebellum coordinate muscular movement by collecting continuous streams of stimuli with their profuse dendritic trees. The dendrites of a **small (Golgi) neuron** (Figure 17.10E) in the spinal cord branch sparsely by comparison, but the axon sprouts more collaterals in a shorter distance. These short neurons link much longer neurons together into reflex arcs in the spinal cord (see Chap-

ter 18). Finally, multipolar **pyramidal cells** (Figure 17.10F) in the cerebral hemispheres communicate between the superficial and deep layers of the cerebral cortex, as you will see in Chapter 19, Brain.

Axons of large diameter conduct more rapidly than do those of narrow diameter. Table 17.1 classifies neurons in the PNS according to the diameter of their axons and their speed of conduction. Type A-alpha neurons conduct most rapidly because they are the largest, up to 22 μm diameter. Larger-diameter axons enable the nervous system to communicate rapidly with the peripheral muscles and sensory receptors.

- How many axons and dendrites do bipolar neurons have? Multipolar neurons? Unipolar neurons? Give an example of each type.
- Which class of neuron conducts the fastest? Slowest?

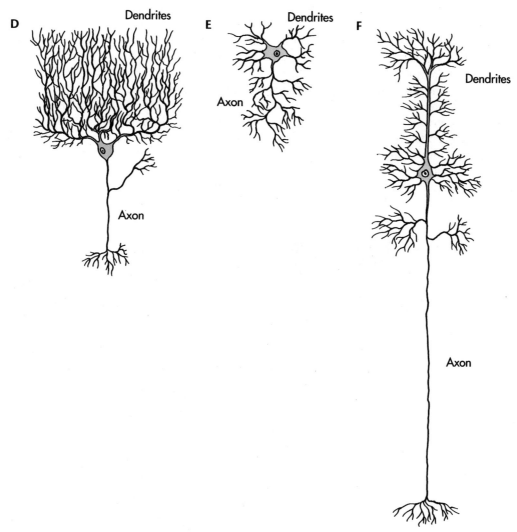

⚙ **Figure 17.10, cont'd**

D, Purkinje cell dendrites branch profusely. **E,** The axon and dendrites of this small Golgi neuron branch frequently but not as intensely as Purkinje dendrites. **F,** Pyramidal cells feature a conical cell body, a typical axon, and a primary dendrite with secondary branches.

Table 17.1	PERIPHERAL NERVE FIBERS		
Name	**Function**	**Diameter (μm)**	**Conduction (m/sec)**
A-alpha	Afferent—positional and muscle sensation (muscle spindles, Golgi tendon organs)	12-22	70-120
	Efferent—skeletal muscle motor neurons		
A-beta	Afferent—pressure, touch, vibratory sensation, stretch (muscle spindles)	5-12	30-70
A-gamma	Efferent—intrafusal fibers, muscle spindles	2-8	15-30
A-delta	Afferent—touch, pain, temperature	1-5	5-30
B	Efferent—ANS, sympathetic preganglionic neurons	<3	3-15
C*	Afferent—touch, pain, temperature	0.1-1.3	0.6-2.0
	Efferent—ANS, sympathetic postganglionic neurons (smooth muscle, glands)		

All neurons listed are myelinated, except Type C.

Abbreviations: ANS, autonomic nervous system; μm, micrometers; m, meters; sec, seconds.

From Afif, AK and Bergman RA: Basic neuroscience, ed 2, Baltimore, 1990, Urban and Schwarzenberg.

NEUROGLIA

Neuroglial cells support and nourish neurons and help regulate their function. Neuroglia are usually smaller than neurons, but there are nearly 10 times more of them and they make up about half the volume of the nervous system. Neuroglia extend cytoplasmic processes to neurons and other cells, but unlike neurons, neuroglia cells do not conduct action potentials. These "nerve glue" cells do influence axonal conduction and synaptic transmission, however. Neuroglia provide pathways for growing and regenerating axons, supply nutrients, and remove cellular debris. Unlike neurons, neuroglia cells can divide readily. Five main varieties of neuroglia cells are recognized, with considerable variation within each group.

EPENDYMAL CELLS

Ependymal cells (Figure 17.11A) line the ventricles of the brain and the central canal of the spinal cord with ciliated cuboidal epithelium, which helps produce and circulate the cerebrospinal fluid in these spaces. Differentiating neurons divide in the ependyma during development, then migrate away from it as the nervous system matures. **Ependymal**

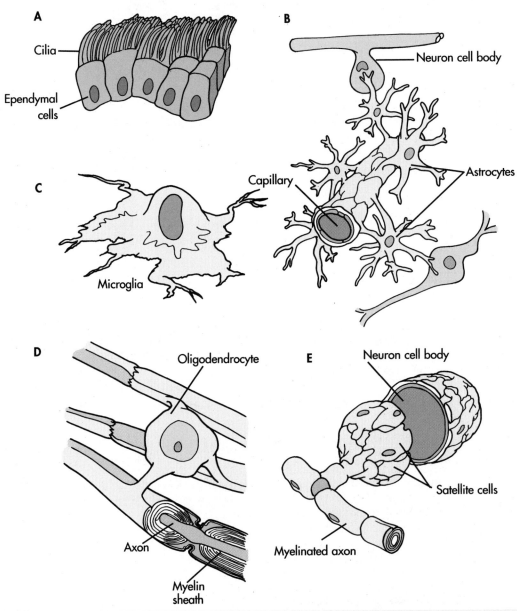

Figure 17.11

Classes of neuroglia cells. **A,** Ependymal cells line the ventricles. **B,** Protoplasmic astrocytes connect capillaries with neuron cell bodies. **C,** Microglial cells phagocytose spent cells. **D,** An oligodendrocyte has myelinated several neurons. **E,** Satellite cells cover cell bodies of neurons.

(eh-PEN-dih-mal) cells also form sheets of endothelium at the choroid plexuses of the brain, where they produce cerebrospinal fluid (see Chapter 19, Brain and Cranial Nerves).

ASTROCYTES

Slender cytoplasmic processes extend from the cell bodies of astrocytes, which accounts for their name, "star cell" (Figure 17.11B). Astrocytes assume many different forms and probably as many functions. **Fibrous** and **protoplasmic astrocytes** are two major varieties. Narrow fibrous processes are the hallmarks of the former, which are abundant in the gray matter of the brain and spinal cord, whereas protoplasmic astrocytes sprout heftier processes in the white matter. (Chapter 18, Spinal Cord, describes gray and white matter.) Although both types of cells have Golgi bodies, mitochondria, neurofilaments, and microtubules, the large quantities of rough endoplasmic reticulum indicate that astrocytes probably synthesize proteins as vigorously as do neurons. One long-standing clue to astrocyte function has been the fact that their processes end on neurons and on capillaries, as seen in Figure 17.11B. This has suggested that astrocytes are intermediaries in the transport of oxygen and nutrients to neurons (see Blood-brain Barrier, Chapter 19). More recently, investigators have shown that astrocytes resynthesize neurotransmitter substances at synapses, and this suggests that astrocytes may regulate synaptic transmission by influencing the supply of neurotransmitter. Other roles are likely.

MICROGLIA

Microglia are phagocytic and can wander throughout the nervous system, removing dead or damaged cells (Figure 17.11C). Microglia are small, nondividing cells found throughout the gray and white matter of the CNS. Trauma to the nervous system can transform them into mobile macrophages that scavenge cellular debris. It is controversial whether microglia develop as monocytes from myeloid stem cells of the blood or from the neural tube and neural crest (see Origins of Neurons and Neuroglia, this chapter).

NEURILEMMACYTES

Neurilemmacytes cover axons with a myelin sheath and neurilemma (NOOR-ih-LEM-ma) that electrically insulates and protects the axon and also speeds up axonal conduction. **Schwann cells** (sh-WAHN) supply the myelin of the peripheral axons (see below), but **oligodendrocytes** (Figure 17.11D) myelinate neurons of the brain and spinal cord. In adults, Schwann cells promote axon regeneration after injury, while oligodendrocytes in the CNS actively retard axon regeneration.

SATELLITE CELLS

These Schwann cell variants cover the cell bodies and initial segments of axons in the peripheral nerves (Figure 17.11E), but they do not produce a myelin sheath. In the autonomic nervous system, axon terminals penetrate between these cells to synapse with the cell bodies. A basement membrane covers the external surface of satellite cells and joins them with the surrounding tissue.
- What does the name "neuroglia" mean?
- Name five types of neuroglia and their functions.

MYELIN SHEATH

The myelin sheath covers axons with layers of plasma membrane that speed conduction by electrically insulating axons. The nervous system contains numerous **myelinated axons** (MY-eh-lin-ATE-ed); this covering gives these tissues a glistening white appearance because of the phospholipids in the membranes. The white matter of brain and spinal cord primarily contains myelinated axons, whereas most axons in the gray matter are **unmyelinated.** Schwann cells myelinate peripheral axons, and oligodendrocytes myelinate central ones.

In Figure 17.12A, the axon of a peripheral neuron has been wrapped by a series of Schwann cells, each one covering a length of about 1 mm. Nearly 1000 cells are needed to myelinate a single long axon in the limbs. The junctions between adjacent Schwann cells are called **nodes of Ranvier** (RON-vee-aye), and the portion of the axon covered by each Schwann cell is the **internode.** The areas of the Schwann cells adjoining the nodes of Ranvier are the **paranodal** (beside the node) regions. The myelin sheath itself is innermost, wrapping the axon; the nucleus and the bulk of the Schwann cell cytoplasm is farthest from the axon cylinder. A **basement membrane** around the whole connects with the surrounding tissue. The external sleeve of Schwann cell membrane, cytoplasm, and basement membrane is considered the **neurilemma.** Together with its sheath, an axon is often called a **nerve fiber** for convenience.

Seen in transverse section (Figure 17.12B), the plasma membrane of the Schwann cell clearly spirals around the axon, like a roll of paper towels wrapped around the tube. Follow the layers from inside out; there are as many as 10 to 20 turns, fewer for smaller axons. Wrapping begins when the Schwann cell body revolves around the axon (Figure 17.12C), spinning out a tongue of plasma membrane behind. The portion of the membrane attaching this tongue to the cell body is called the **mesaxon** because of its resemblance to mesenteries of the gut. Each layer represents one complete trip around, but the actual mechanism for rotation is unknown. **Myelin basic protein,**

A

Dendrites

Neurilemma

Schwann cell

Cell body

Axon hillock

Axon

Myelin sheath

Node of Ranvier

B

Myelin sheath

Axon

C

Nucleus of Schwann cell

Myelin sheath

Axon

D

Plasma membrane

Integral protein

Extracellular surface

Cytoplasm

Myelin basic protein

E

Axons

Schwann cell

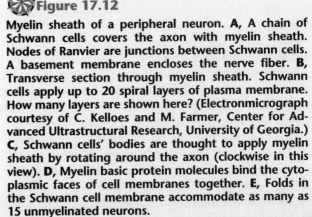

★ **Figure 17.12**

Myelin sheath of a peripheral neuron. **A,** A chain of Schwann cells covers the axon with myelin sheath. Nodes of Ranvier are junctions between Schwann cells. A basement membrane encloses the nerve fiber. **B,** Transverse section through myelin sheath. Schwann cells apply up to 20 spiral layers of plasma membrane. How many layers are shown here? (Electronmicrograph courtesy of C. Kelloes and M. Farmer, Center for Advanced Ultrastructural Research, University of Georgia.) **C,** Schwann cells' bodies are thought to apply myelin sheath by rotating around the axon (clockwise in this view). **D,** Myelin basic protein molecules bind the cytoplasmic faces of cell membranes together. **E,** Folds in the Schwann cell membrane accommodate as many as 15 unmyelinated neurons.

which attaches successive layers of membrane to each other, is a major component of the myelin sheath in the brain (Figure 17.12D). Similar molecules serve this function in peripheral nerves. The principal phospholipid of the myelin sheath is **sphingomyelin** (SFIN-joe-MY-eh-lin), a nonpolar molecule that helps insulate the neurons from each other's action potentials and minimize "cross-talk" between axons within a nerve.

Individual oligodendrocytes may myelinate portions of up to 50 different axons in the brain and spinal cord (see Figure 17.11D). Because an oligodendrocyte cannot rotate its cell body around one axon without hopelessly tangling the other axons, neurobiologists have proposed that the cell body remains in place while the tip of each "tongue" advances on the inside of the spiral. Perhaps you can think of another method; the actual mechanism is unknown. Whatever the difference in these mechanisms, the central and peripheral nervous systems have myelin sheaths that are similar in appearance.

The two systems show considerable differences in unmyelinated axons, however. In the peripheral nervous system, axons enfold themselves into the membrane of Schwann cells (Figure 17.12E), several axons per cell, without forming a myelin sheath. A basement membrane surrounds these Schwann cells; it and the cytoplasm of the Schwann cells act as the neurilemma. In contrast the central nervous system leaves its unmyelinated neurons naked, with only a basement membrane attached.

Myelination begins in the fourteenth week of gestation and wraps axons that have already developed. It continues while motor coordination develops and the child learns to crawl and walk. Axons can grow longer inside their myelin sheaths. Each Schwann cell deposits more membrane as the internode gets longer, but new Schwann cells must also be added to accommodate growing axons.

NODES OF RANVIER

Nodes of Ranvier speed conduction over long distances by allowing the action potential to be regenerated only at each node and not along the entire membrane (Figure 17.13C). Dense undercoatings of the axon membrane indicate that sodium ion channels are especially concentrated at nodes of Ranvier. As at the initial segment, a node generates action potentials and the inflowing current of sodium ions speeds toward the next node, confined to the axon by the myelin sheath. The next node sets off a new action potential that stimulates the following node, and so on. This **saltatory, or jumping, conduction** speeds the impulse down the axon. Without it, conduction would be much slower.

Nodes are formed by the folds of the plasma membrane in the myelin sheath (Figure 17.13A). Only at

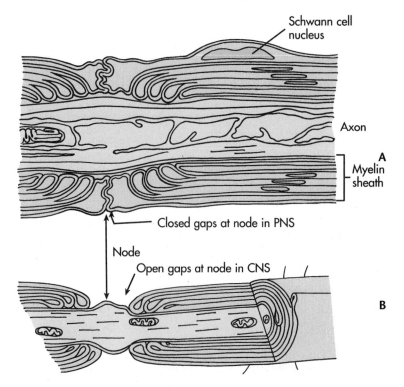

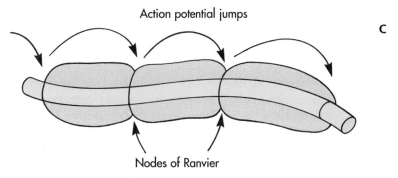

✳ **Figure 17.13**

Nodes of Ranvier. **A,** In peripheral nerves, overlapping tongues of cytoplasm from each layer of myelin sheath join Schwann cells together without large gaps between. Longitudinal section. **B,** Large gaps between oligodendrocytes characterize nodes of Ranvier in the central nervous system. **C,** The action potential skips from node to node.

nodes in the CNS is the axon actually bare (Figure 17.13B) because the margins of neighboring oligodendrocytes are more separated.

- What is the role of the myelin sheath in impulse conduction?
- Which kinds of neuroglial cells myelinate peripheral and central neurons?
- What is a nerve fiber?

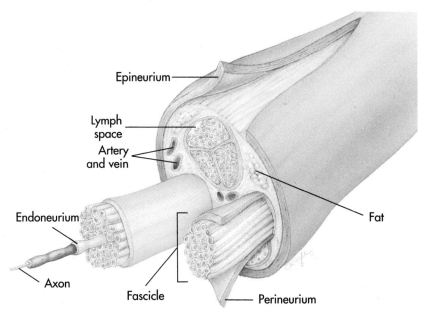

Epineurium

Lymph space

Artery and vein

Endoneurium

Axon

Fascicle

Perineurium

Fat

Figure 17.14

Structure of a peripheral nerve. Epineurium covers the exterior of this nerve, and perineurium envelops fascicles. Endoneurium surrounds individual nerve fibers.

PERIPHERAL NERVES

Peripheral nerves are bundles of nerve fibers that carry impulses between the central nervous system and the body. Three layers of connective tissues support these fibers (Figure 17.14). The tough, slick, outermost layer, the **epineurium,** is the most fibrous and covers the external surface of large nerves. Neurosurgeons suture the epineurium of larger nerves to promote regeneration of severed axons. Fascia attaches the epineurium to skeletal muscles and neighboring vessels that these nerves frequently follow. The external location of the epineurium is comparable to the epimysium of skeletal muscles (see Chapter 9, Skeletal Muscle).

The second layer of tissue, the **perineurium,** is analogous to the perimysium because it bundles individual nerve fibers together as **fascicles.** Loose connective tissue fills in the spaces between fascicles. **Endoneurium** is the delicate innermost layer of connective tissue around individual nerve fibers. What connective tissue in muscle corresponds with endoneurium? Neurons, neuroglia, and connective tissue all receive their blood supply by way of the **vasa nervorum** (VAZ-ah NER-vor-um; vessels of the nerves) that penetrate the epineurium and branch into capillary beds within the endoneurium.

- Which connective tissue layer sheaths entire nerves? Fascicles? Nerve fibers?

ORIGIN OF NEURONS AND NEUROGLIA

Glial cells and neurons derive from the neural tube and neural crest, as Figure 17.15 outlines. During the second week of development, the **neural plate** appears along the axis of the embryo (Figure 17.15A). The anterior half of the plate becomes the brain, and the remaining half becomes the spinal cord. The **neural crest** forms at the margin of the neural plate. The neural plate soon folds into the **neural tube** (Figure 17.15B), from which the spinal cord and brain develop (see Chapter 18, Spinal Cord). The neural crest cells remain dorsal to the neural tube and migrate ventrally and laterally toward destinations in the viscera and the body wall.

Stem cells in the neural crest produce afferent neurons of sensory ganglia, cells of the autonomic nervous system, and secretory cells of the adrenal medulla, as Figure 17.16 indicates. Schwann cells and perhaps microglia derive from both the neural crest and neural tube. Efferent neurons, astrocytes, oligodendrocytes, and ependymal cells develop from stem cells in the innermost layer of the neural tube.

- The neural tube develops from which structure?
- Which portion of the neural tube becomes the brain?
- What are the origins of astrocytes, afferent neurons, and Schwann cells?

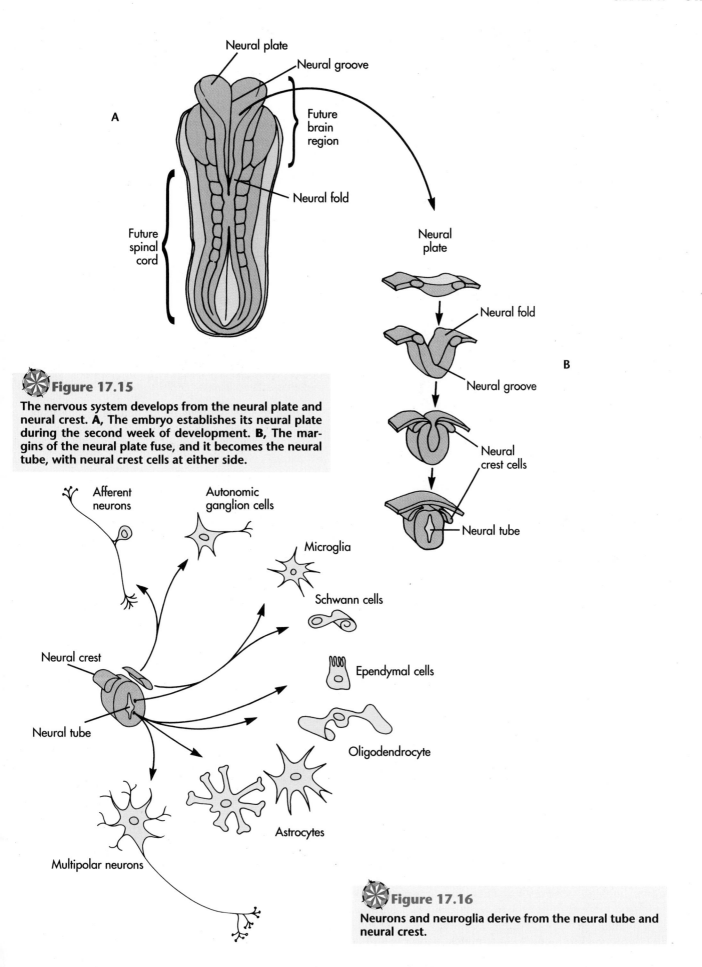

A

Neural plate
Neural groove
Future brain region
Neural fold
Future spinal cord

Neural plate
Neural fold
Neural groove
Neural crest cells
Neural tube

B

Figure 17.15

The nervous system develops from the neural plate and neural crest. **A,** The embryo establishes its neural plate during the second week of development. **B,** The margins of the neural plate fuse, and it becomes the neural tube, with neural crest cells at either side.

Afferent neurons
Autonomic ganglion cells
Microglia
Schwann cells
Ependymal cells
Neural crest
Oligodendrocyte
Neural tube
Astrocytes
Multipolar neurons

Figure 17.16

Neurons and neuroglia derive from the neural tube and neural crest.

NEURONAL OUTGROWTH

"Reach out and touch someone," says AT&T. Clearly, getting the connections correct is absolutely critical if telephone and nervous systems are to work properly. How neurons make their connections is one of the most intriguing questions of neurobiology because the answers disclose how neural pathways are assembled. Neurons are led to their targets by **growth cones** (Figure 17.17), the semiautonomous growing tips of axons. When a growth cone meets an appropriate target cell, it transforms into a terminal bulb and forms a synapse with the postsynaptic cell.

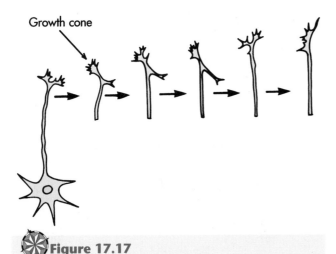

Growth cone

Figure 17.17

Neuronal outgrowth. Growth cones advance by extending and withdrawing filopodia. This cone advanced half a millimeter in 38 minutes.

Growth cones extend filopodia that pull the axon forward. Cell adhesion molecules and fibronectins attach these filopodia (narrow cytoplasmic processes) to their surroundings; when they are secure, actin and myosin microfilaments in the cytoskeleton draw cytoplasm into the filopodia and the growth cone advances. Filopodia that do not adhere are withdrawn, and new ones search for the next advance, as shown in Figure 17.17. Growth cones add new cell membrane and elongate their microtubules as they proceed. All these activities probably are subtly controlled by calcium ions that flow into the cytoplasm as the growth cone proceeds.

Growth cones reach their targets by following a series of cues that brings the axon near its destination and prepares it to recognize its prospective partner. Some cues are genetic. For example, mutations that interfere with locomotion in mice (reeler, staggerer, and weaver) scramble connections between neurons in the cerebellum, which coordinates locomotion. Lo-

cal conditions also are cues. **Pioneer neurons** grow outward from the spinal cord and establish pathways for others to follow. A neuron's local environment can determine whether its synapses use noradrenaline or acetylcholine as neurotransmitter substances.

The history of developing neurons is important. Neurons somehow record the history of their early cell divisions and contacts with other cells, and this information is sometimes called forth later. Regeneration of the optic nerve in frogs is a good example of developmental history at work. When this nerve is cut and the eye inverted, the retinal neurons reestablish essentially original connections with the brain, despite the new orientation of the eye. This indicates that neurons retain a cellular "memory" of their original connections and can sort themselves out accordingly.

- What structure leads the outgrowth of axons?
- Name one type of cue that axons use to find their targets.

NERVE REGENERATION

The nervous system is notoriously sensitive to injury because it does not replace lost neurons. Approximately 10,000 neurons die daily in the brain and spinal cord. In the epidermis or among erythrocytes this loss would never be noticed because stem cells continually replace the dying cells. But in the CNS, no stem cells fill the breach. Olfactory neurons are the only neurons known to divide. In humans, it is not yet possible to entice other neurons to divide, though recent studies with birds indicate that some neurons can be stimulated to divide after injury to the brain.

In contrast, fetal nervous tissue proliferates vigorously because neural stem cells are still multiplying. In the treatment of **Parkinson's disease,** such tissue has been used to replace cells that have lost the ability to produce DOPA, a neurotransmitter precursor the brain uses for motor coordination. These particular treatments have not always been successful and certainly raise complex ethical questions.

Even though lost cells cannot be replaced, axons and perhaps dendrites of damaged cells can regenerate. Severed axons in peripheral nerves regenerate themselves and sometimes reestablish appropriate connections, but those in the central nervous system cannot do so. In both the CNS and PNS the distal portion of the severed axon will die (Figure 17.18), but only in the PNS can the proximal portion (connected to the nucleus and cell body) regenerate a new axon. While this ability offers hope to victims whose severed fingers and hands have been reattached, little can be done to restore mobility to patients with spinal cord injuries. The reasons for these differences in regenerative ability are poorly understood.

When a peripheral nerve is severed, the distal por-

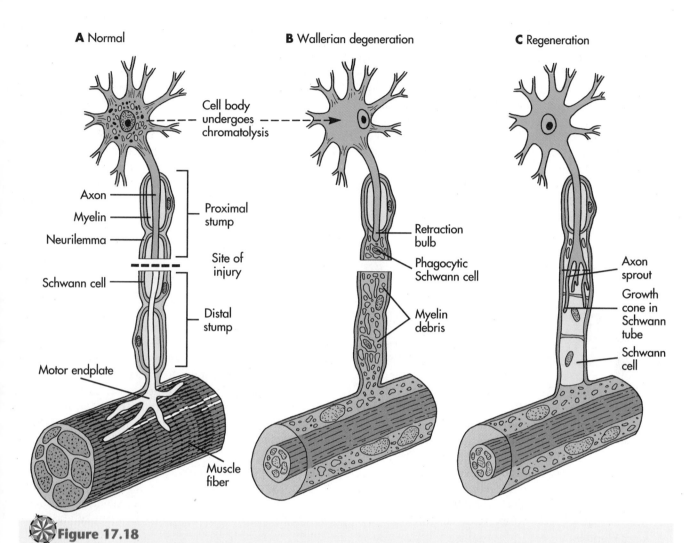

A Normal

Cell body undergoes chromatolysis

Axon

Myelin

Neurilemma

Proximal stump

Site of injury

Schwann cell

Distal stump

Motor endplate

Muscle fiber

B Wallerian degeneration

Retraction bulb

Phagocytic Schwann cell

Myelin debris

C Regeneration

Axon sprout

Growth cone in Schwann tube

Schwann cell

✦ Figure 17.18

Regeneration of a motor neuron. **A,** The axon is severed with a scalpel. **B,** Wallerian degeneration begins in the distal segment, as Schwann cells and macrophages attack the myelin sheath through phagocytosis. The muscle fibers also degenerate. Chromatolysis is underway in the cell body, and the axon has formed a retraction bulb. **C,** Regeneration begins, as a new growth cone explores reorganized Schwann cells within the Schwann tube. Muscle fibers will regenerate when the growth cone reconnects to them.

tions of neurons die by a process called **Wallerian degeneration** (wahl-EER-ee-an). Immediately after an axon is cut, its membrane seals off leaking cytoplasm and becomes a **retraction bulb** (Figure 17.18B). The retraction bulb becomes a growth cone when regeneration is successful, but when regeneration fails, the bulb and axon die back to the next proximal collateral or even to the cell body. Wallerian degeneration can even extend across synapses to other neurons, where it is known as **indirect Wallerian degeneration.** This process tends to damage entire neural pathways; even the second and third neurons downstream from the original site of damage may die. **Anterograde degeneration** extends in the direction of neural conduction across the synapse into the next cells, and **retrograde degeneration** progresses upstream to presynaptic cells.

When regeneration is successful, the retraction bulb transforms into a growth cone within a few days and sends out filopodia in search of contacts. Over the next month the Nissl bodies break down and reestablish themselves, the cell body swells, and the nucleus attaches to the cell membrane; these changes are known as **chromatolysis** (kro-MAT-oh-LY-sis; Figure 17.18B). Whether or not regeneration is successful, glial cells remove debris through phagocytosis or by encapsulating it with myelin. They also form scar tissue across the damage site, which may interfere with regeneration. Preventing scarring is one of the primary goals of treating spinal cord and other neural injuries.

Schwann cells form new myelin sheaths while chomatolysis is underway. Macrophages and Schwann cells engulf large pieces of degenerating myelin from damaged Schwann cells. These injured Schwann cells do not necessarily die but recover by enlarging and proliferating (Figure 17.18C). They appear to guide regenerating growth cones into new cellular processes called **Schwann tubes** that form inside the basement membrane of the old nerve fiber. The axons once again fold into the Schwann cells as growth cones explore these new tubes, and a new myelin sheath forms.

- Why does the central nervous system not replace dead neurons?
- When an axon is cut, which part regenerates and which part degenerates?
- How do Schwann cells assist regeneration?
- How does Wallerian degeneration damage neural pathways?

CONDITIONS AFFECTING NERVOUS TISSUE

MULTIPLE SCLEROSIS

Multiple sclerosis (MS) is a demyelinating disease of the brain and spinal cord. MS is difficult to diagnose because it has many subtle symptoms, but it is usually identified by the patient's repeated episodes of physical disability caused by multiple lesions (multiple scleroses) of the fibers in the CNS that have lost their myelin sheath. Oligodendrocytes, especially dividing ones, rather than neurons, are the apparent targets of MS. Macrophages attack these cells and phagocytose their myelin sheath. Demyelination retards and blocks axonal conduction, damage that seems to underlie such varied symptoms as double vision, numbness and coldness in limbs, loss of motor coordination in the legs or arms, and paroxysmal loss of speech and the ability to swallow. It is not clear, however, why these physical disabilities wax and wane.

Twenty percent of MS cases are familial, stemming from a genetic background that is related to the major histocompatibility gene complex (see Chapter 16, Lymphatic System and Immunity). The patient's own macrophages may attack new antigens on the oligodendrocytes that are remyelinating the axons. Nongenetic causes have been even more difficult to identify. No cure is known; treatment attempts to manage the symptoms with physical therapy, drugs for pain and loss of motor ability, and MS support groups.

DRUG EFFECTS

Some of the many substances that affect the nervous system are known to act directly on neurons by interfering with action potentials and synaptic transmission. The examples in Figure 17.19 show neuroanatomy in the light of the effects of some drugs.

LOCAL ANAESTHETICS

Local anaesthetics, such as procaine, lidocaine, and cocaine, prevent axons from starting or conducting action potentials. Applied directly, these drugs prevent the initial inrush of sodium ions through the membrane by blocking sodium ion channels. As a result, axons remain relatively silent. The euphoric effects of cocaine are thought to be caused by the failure of neurons to take up the neurotransmitter substance, dopamine, from the synaptic clefts into which it is released. Dopamine therefore accumulates in the synapse and overstimulates the postsynaptic neurons, especially in the brain.

ANTICHOLINESTERASES

Numerous agents interfere with synaptic transmission, among them the **"nerve gases"** and certain **agricultural insecticides** that attack synapses where acetylcholine is the neurotransmitter. Compounds such as parathion and DFP (diisopropylfluorophosphate) block **acetylcholinesterase,** the enzyme in the synaptic cleft that degrades acetylcholine. Acetylcholine then accumulates in the cleft and continues to stimulate the postsynaptic cell, even though the presynaptic neuron may have stopped releasing more neurotransmitter.

CURARE

First found in the famous arrow poisons from South America, this group of natural and synthetic compounds blocks motor endplates in skeletal muscles by binding to the acetylcholine receptors on the muscle fiber membrane. Consequently, muscles relax or, given high doses of the drug, become completely paralyzed.

ETHANOL

Ethyl alcohol, or ethanol, releases personal inhibitions and self-restraint, apparently by depressing neural circuits that govern behavior. **Ethanol** is a central nervous system depressant that acts by dissolving into the lipid membrane of neurons, where it evidently interferes with ion channels and membrane signal transduction.

BENZODIAZOPINES

Clinicians use the benzodiazopines (BEN-zo-dye-AZO-peens) as sedatives, to induce hypnosis, or to control anxiety. These drugs are thought to increase the effects of GABA, an inhibitory neurotransmitter substance (see Synapses, this chapter) that acts throughout the spinal cord and brain.

OPIOIDS

This group includes heroin, methadone, and codeine, all of which are chemical relatives of morphine. The body makes its own painkillers, the

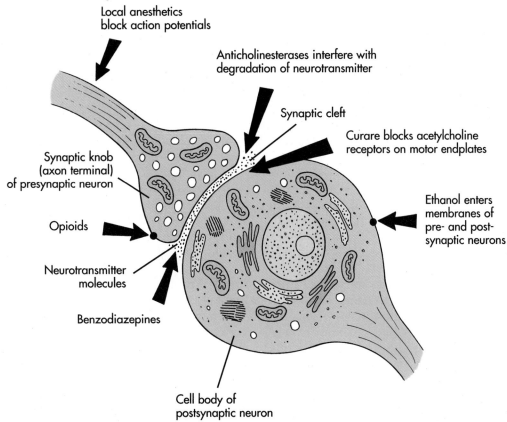

Local anesthetics
block action potentials

Anticholinesterases interfere with
degradation of neurotransmitter

Synaptic cleft

Curare blocks acetylcholine
receptors on motor endplates

Synaptic knob
(axon terminal)
of presynaptic neuron

Ethanol enters
membranes of
pre- and post-
synaptic neurons

Opioids

Neurotransmitter
molecules

Benzodiazepines

Cell body of
postsynaptic neuron

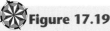

Figure 17.19
Drug action on neurons and synapses.

endorphins (en-DOR-fins); both groups act by binding to opioid (OH-pee-oyd) membrane receptor proteins distributed through the brain and spinal cord. Opioids reduce pain by inhibiting neurons in afferent pathways from releasing neurotransmitter substances at synapses in the spinal cord and brain.

- What structures and physiological processes does multiple sclerosis attack?
- Which drugs inhibit action potentials or interfere with synaptic transmission?

SUMMARY

Neurons and neuroglia are the two primary classes of cells that form the nervous system. Neurons conduct, transmit, and store information. They operate through neural pathways that communicate within the spinal cord and brain and between the CNS and the body. The optic pathway is an example of such neural pathways. Neuroglia support neurons by covering them with myelin sheaths that speed conduction, by providing them with nutrition, and by removing damaged cells. *(Page 503.)*

Neurons of the CNS reside within the brain and spinal cord; those of the PNS have all or parts of the

cell outside the brain and spinal cord. The brain and spinal cord constitute the central nervous system, and the nerves of the peripheral system conduct impulses between the body and the central nervous system. The somatic portion of the PNS communicates with skeletal muscles, the skeleton, and the skin via afferent and efferent neurons. Afferent and efferent neurons also innervate the visceral organs, smooth muscle, vessels, and glands in the visceral portion of the PNS. The autonomic nervous system is the efferent portion of the visceral division. *(Page 504.)*

Neurons communicate through long cytoplasmic processes that extend from the cell body. Dendrites carry impulses toward the cell body. In response the initial segment of the axon can generate action potentials that travel to the ends of the axon, which synapses with other neurons. Axons and dendrites transport materials that support conduction and transmission. Afferent neurons conduct impulses toward the CNS, and efferent neurons conduct them away from it. *(Page 505.)*

Axons conduct action potentials. Action potentials are waves of depolarization propagated by opening and closing voltage-gated sodium and potassium ion channel proteins in the plasma membrane. Na$^+$K$^+$-ATPase maintains the resting potential by pumping sodium ions outward and potassium ions inward across the membrane. When the resting potential at the initial segment of an axon is diminished, voltage-gated sodium and potassium ion channels propagate this rapid depolarization by allowing first sodium and then potassium ions to cross the membrane. The wave of depolarization advances along the membrane toward the axon terminals. *(Page 507.)*

Most synapses transmit chemically. Neurotransmitter substances are secreted into the synaptic cleft as a consequence of depolarization of the terminal bulb. Neurotransmitter molecules communicate with the postsynaptic cell by opening ion channel proteins in the dendrites. When channels open, the postsynaptic membrane depolarizes. *(Page 508.)*

Anatomically, neurons may be unipolar, bipolar, or multipolar, according to the number and nature of cytoplasmic processes that extend from the cell body. Every neuron has one axon. Unipolar neurons have no dendrites. Bipolar neurons have one dendrite, and multipolar neurons have two or more dendrites. *(Page 511.)*

Neuroglia are the nervous system's connective tissue. Ependymal cells line the ventricles of the brain and the spinal canal, producing and circulating cerebrospinal fluid. Astrocytes nourish neurons through connections to capillaries and also resynthesize neurotransmitter molecules. Microglia are macrophages that patrol the nervous system, destroying damaged and foreign cells. *(Page 514.)*

Oligodendrocytes and Schwann cells insulate axons with myelin sheaths and form the neurilemma. Nodes of Ranvier in the myelin sheath speed axonal conduction over long distances by allowing action potentials to jump the internodes.

Satellite cells are modified Schwann cells that enclose the cell bodies of peripheral neurons. *(Page 515.)*

Three layers of connective tissue cover the neurons of peripheral nerves. Fibrous epineurium protects the outside of all but the smallest nerves. Perineurium wraps nerve fibers into fascicles, and a delicate endoneurium envelops all myelinated and unmyelinated nerve fibers. Capillaries of the vasa nervorum supply blood to neurons and Schwann cells. *(Page 518.)*

Neurons and neuroglia originate from the neural plate and neural crest. Axons grow outward from the cell body by extending growth cones that develop into synapses upon reaching their target cells. Axons identify their targets according to genes inherited by the individual, according to the history of the developing cell, and in response to the local environment encountered by the growth cone. *(Page 518.)*

The nervous system does not replace lost neurons, but damaged axons can regenerate. Unlike the stem cells of other tissues, neuron stem cells cease dividing and do not replace lost neurons. Axons can regenerate in the peripheral nervous system but are prevented from doing so in the central nervous system. The distal segment of the severed axon or dendrite degenerates, microglia remove debris and myelin, and in the PNS, Schwann cells reestablish pathways for regenerating growth cones to follow. *(Page 520.)*

Various circumstances and conditions that affect nervous tissue help reveal how the nervous system functions. Multiple sclerosis interferes with physical and mental processes by demyelinating axons of the central nervous system. Various drugs interfere with action potentials or with the release, reception, or recycling of neurotransmitters at synapses. *(Page 522.)*

STUDY QUESTIONS

1. What do neurons do in the nervous system? What are the functions of dendrites, cell body, and axon? How does each of these components contribute to neuronal function?   **1, 4**
2. Trace the optic pathway in Figure 17.2 from retina to primary visual cortex. Which neurons receive from the rods and cones? Which neurons receive from ganglion cells?   **2**
3. Name the divisions of the nervous system, and describe their components.   **3**
4. Name the cytoplasmic organelles and structures of neurons. Are they located in dendrites, cell body, initial segment, axon, or terminal bulbs?   **4**
5. Describe the conditions of the resting potential. What membrane protein is responsible for maintaining the resting potential? What does it do to establish this potential?   **5**
6. Which membrane proteins are responsible for the action potential? What changes do these molecules experience when the membrane depolarizes? How do these changes propagate the action potential?   **5**
7. How is the action potential similar to "the wave" in a stadium?   **5**
8. Describe the structure of a chemical synapse. What organelles release neurotransmitter from the axon terminal bulb? What interactions between neurotransmitter and receptors depolarize the postsynaptic membrane? What recycling processes assure that synaptic responses are brief?   **6**
9. Classify the neurons in Figure 17.10 by their shapes (unipolar, bipolar, or multipolar). Locate them in the nervous system (PNS or CNS), according to their descriptions in the text.   **7**
10. Name the various categories of neuroglial cells, and describe their functions in the nervous system. How are neurons and neuroglia cells different?   **8**
11. Describe the structure of the myelin sheath and neurilemma in a peripheral nerve. Identify the components, where each is located, and its function. Which cells deposit the myelin sheath in the CNS and PNS?   **9**
12. What are nodes of Ranvier? Describe their structure. What are paranodal regions and internodes, and how do these regions differ? In what respect do nodes of Ranvier differ between the central and peripheral nervous systems?   **10**
13. What functions of the nodes of Ranvier and the myelin sheath increase the rate of conduction over long distances? Describe these events.   **9, 10**
14. Where are epineurium, perineurium, and endoneurium located in peripheral nerves? What function does each perform for the nerve and neurons? How do these layers resemble the connective tissue coverings of skeletal muscles?   **11**
15. From which tissues—neural tube, neural crest, or both—do the following cells develop: afferent neurons, sympathetic ganglion neurons, efferent neurons, astrocytes, ependymal cells, microglia, Schwann cells, oligodendrocytes, and satellite cells?   **12**
16. Describe a growth cone. How does it accomplish neuronal outgrowth? What information directs growth cones to their targets?   **13**
17. Outline the sequence of events in nerve regeneration. What is Wallerian degeneration? What degenerates? What regenerates? What roles do macrophages and Schwann cells play? Does regeneration occur in the PNS? In the CNS?   **14**
18. What cells appear to be the primary targets of multiple sclerosis?   **15**
19. Describe the action of several drugs on the nervous system. How are their actions similar? Different?   **15**

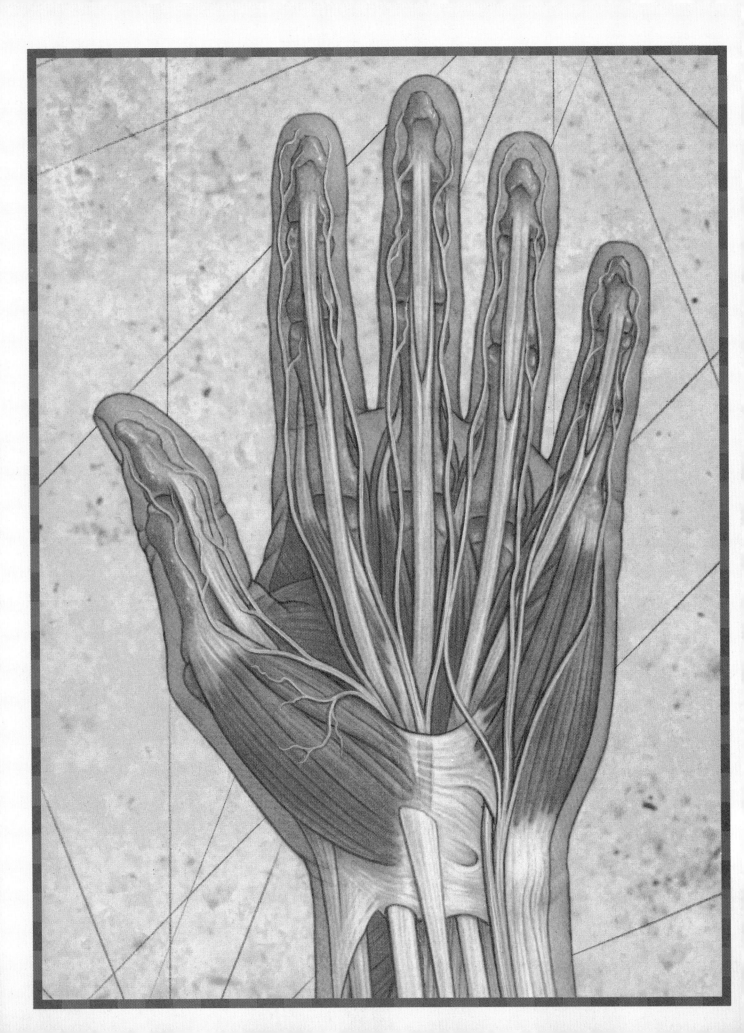

CHAPTER 18

Spinal Cord and Spinal Nerves

OBJECTIVES

Do the following exercises to be sure that you understand this chapter.

1. Draw a diagram of the neural tree, showing the spinal nerves and the five regions of the spinal cord.
2. State which spinal nerves innervate which dermatomes of the skin.
3. Predict how injuries that sever the spinal cord at high or low levels will affect an individual's ability to control bodily functions.
4. Describe the anatomy of the spinal cord.
5. Name and describe the three layers of the meninges.
6. Describe how vessels circulate blood to the anterior and posterior sides of the spinal cord.
7. Give the number of spinal nerves and vertebrae, and describe how these nerves exit the vertebral column.
8. Name and follow the roots and rami of spinal nerves between the spinal cord and peripheral organs.
9. Name the five plexuses and list the spinal nerves involved in each one. Trace the cervical nerves through the brachial plexus into the skin of the forearm.
10. Outline the paths of major peripheral nerves in the limbs.
11. Trace the paths of various tracts in the spinal cord.
12. Trace the neural pathway of a reflex arc from receptors in the skin, into the spinal cord, and back to muscles of the limbs.
13. Describe the developmental origins of the spinal cord and nerves.

VOCABULARY TOOLBOX

alar	ala, *L* wing.	meninges, meninx	mening, *L* a membrane. The membranes of the spinal cord.
arachnoid	arachn-, *G* a spider web.		
decussate, decussation	decuss-, *L* a crossing.		
dermatome	derm, *G* skin; tom-, *G* slice; skin slice.	pia mater	pi, *L* tender; mater, *L* mother; tender mother.
dura mater	dura, *L* hard; mater, *L* mother; tough mother.	plexus	plex-, *L* interwoven, a network.
fasciculus, fasciculi	fasci-, *L* a bundle; groups of axons in the spinal cord or spinal nerves.	radicular	radix, *L* a root.
		ramus, rami	ram-, *L* branch, branches.
		sulcus	sulc-, *L* groove, furrow.
funiculus, funiculi	funicul, *L* a rope or cord.	trabecula	trab-, *L* beam, timber.
isthmus	isthm-, *L* a narrow passage.		

RELATED TOPICS

- Divisions of the nervous system, Chapter 17
- Flexor and extensor muscles, Chapter 11
- Neural pathways, Chapter 17
- Neural plate and neural tube, Chapter 17
- Neurons, nerve fibers, nerves, regeneration, Chapter 17
- Vertebral column, Chapter 6

PERSPECTIVE

The spinal cord and spinal nerves resemble a telephone cable and wires that communicate between the body and brain. Because axons are packed like wires in the main cable, it is not surprising that injuries to the spinal cord can be devastating. Damage the cord in the neck, and all four limbs are paralyzed; injuries at lower levels, however, may involve only those axons that reach the legs. Neurologists can determine the sites of such injuries because they know where many neural pathways are located in the cable; the functions lost indicate where the cord is injured. This chapter introduces the anatomy of the spinal cord and spinal nerves and follows some of these pathways.

Every year in the United States some 14,000 persons, most often young men 16 to 25 years old, suffer a gunshot, automobile, or other injury to the spinal cord. The consequences can be devastating. Terry Franklin is one of these victims; he has agreed to let us examine his injury so that others may understand the immense personal challenges and triumphs of his rehabilitation. Last year Terry (not his real name) fell from a ladder while painting his house and was paralyzed from the chest down. The landing crushed his third thoracic vertebra and severed the cord at that level. He lost all skin sensations and muscular movement below the armpits. He can tell the position of his lower body only if he looks. However, he has full use of his arms and hands, which are his tickets to independence; he can use a wheel chair, drive a car, feed himself, and attend to his own hygiene. He has much less endurance because he can breathe only with his diaphragm; the thoracic and abdominal muscles that assist breathing no longer function. He is unable to cough and is continually susceptible to respiratory infections. Below the injury, Terry's skin tends to cool to room temperature because he loses heat from vessels in his skin that no longer constrict. His rehabilitation therapists showed him how to compensate for his lost bladder and bowel control and how to check his skin daily for pressure ulcers that movement once prevented. He has no cutaneous sensation during sex, although he continues to produce sperm and have erections.

Case study begins on p. 529.

The spinal cord and spinal nerves serve the entire body except the face and parts of the neck (Figure 18.1). Neural pathways within the spinal cord and spinal nerves conduct to the brain impulses that produce sensations of touch and pain from the skin and of body sense from the muscles and joints. Other pathways carry commands for muscle movement down the cord from the brain to the muscles. Terry's injury cut the cord high in the thorax, so that the brain cannot send or receive information concerning the parts of the body below the point of the break.

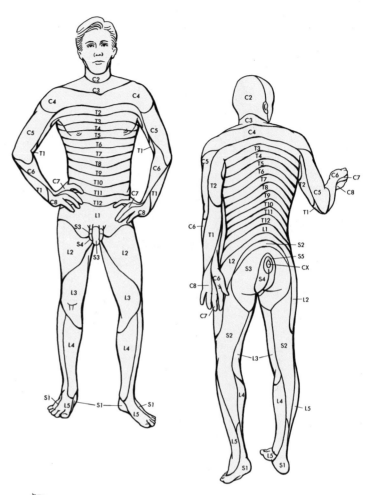

Figure 18.2

Dermatomes wrap the trunk and limbs. *C* = cervical, *T* = thoracic, *L* = lumbar, *S* = sacral, *CX* = coccygeal.

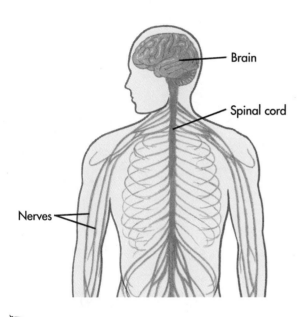

Figure 18.1

Spinal cord and spinal nerves. Spinal nerves communicate between the body and spinal cord, and the spinal cord links these branches with the brain.

Spinal nerves innervate sections of the skin called dermatomes. **Dermatomes** (DER-ma-TOMES) cover humans with invisible stripes such as those in Figure 18.2A. Each spinal nerve supplies one dermatome. Although the eye cannot see them, the prick of a pin can locate them. Neurologists usually can locate a spinal injury by testing which dermatomes retain the senses of touch and pain. Testing for fine touch with soft balls of cotton and for pain by pricking with needles, Terry's neurologist found no responses below the second thoracic dermatome (Figure 18.2A). Terry would have lost skin sensation from the medial sides of his arms if the injury had been at the first thoracic level, because the first and second thoracic dermatomes (T1, T2) innervate skin on the medial surfaces of the arms, as you can see in the figure.

- Why can spinal injuries be debilitating?
- What are dermatomes?

SPINAL CORD

GROSS ANATOMY

The spinal cord descends from the medulla oblongata of the brain through the foramen magnum of the skull, extending some 43 to 45 cm (16 to 17 inches) down the **vertebral canal** within the vertebral column and sprouting spinal nerves as it goes (Figure 18.3). These spinal nerves reach the body through foramina between the vertebrae. For most of its length the spinal cord is about 25 mm (1 inch) in diameter, but **cervical** and **lumbar enlargements,** which contain neurons for the arms and legs, increase it to about 35 mm (1.5 inches). The cord tapers to an end, the **conus medullaris** (KONE-us MED-you-LAR-iss), at the level of the first and second lumbar vertebrae, amid a fan of spinal nerves to the legs and lower trunk. These nerves are known as the **cauda equina**

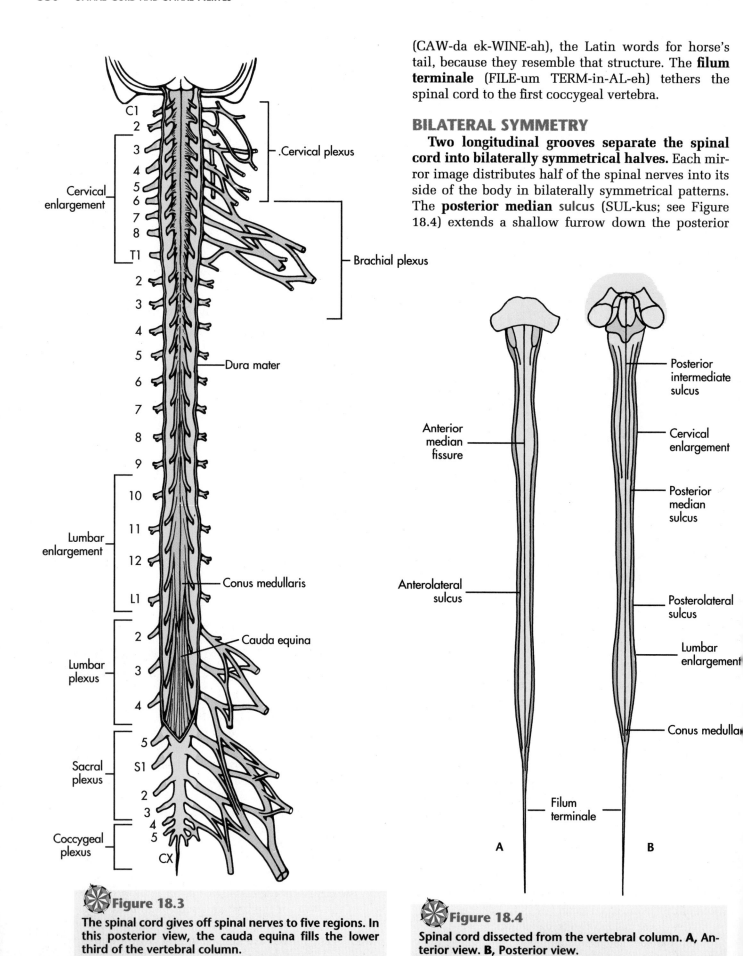

(CAW-da ek-WINE-ah), the Latin words for horse's tail, because they resemble that structure. The **filum terminale** (FILE-um TERM-in-AL-eh) tethers the spinal cord to the first coccygeal vertebra.

BILATERAL SYMMETRY

Two longitudinal grooves separate the spinal cord into bilaterally symmetrical halves. Each mirror image distributes half of the spinal nerves into its side of the body in bilaterally symmetrical patterns. The **posterior median** sulcus (SUL-kus; see Figure 18.4) extends a shallow furrow down the posterior

Figure 18.3

The spinal cord gives off spinal nerves to five regions. In this posterior view, the cauda equina fills the lower third of the vertebral column.

Figure 18.4

Spinal cord dissected from the vertebral column. **A**, Anterior view. **B**, Posterior view.

midline of the cord from the medulla to the conus. Its deeper counterpart, the **anterior median sulcus,** creases the anterior surface. **Posterior intermediate sulci** (SUL-ki) extend caudally on each side of the posterior median sulcus down to the thoracic region; they mark separate bundles of axons ascending to the brain (see Major Tracts, this chapter).

Figure 18.5 shows that the spinal cord is oval in cross section. A narrow isthmus (ISS-thmus) between the anterior and posterior median sulci connects the left and right sides. The **spinal canal** occupies the center of the isthmus and contains cerebrospinal fluid produced in the brain. Although the posterior median sulcus is not as deep as the anterior median sulcus, the **posterior median septum** continues just as deeply as the anterior median sulcus. Neurons communicate across the anterior and posterior **commissures** through axons that cross the midline in front of and behind the spinal canal, respectively. Anterior and posterior **roots** of the spinal nerves enter the cord at **anterolateral** and **posterolateral grooves** on each side, respectively.

WHITE AND GRAY MATTER

The white matter and gray matter are roughly concentric columns of neural tissue that run the length of the cord. The color refers to the appearance of the tissues after death; in life, white and gray matter are light pink and brownish pink as a result of their blood supply. Figure 18.5 shows that the **white matter** surrounds the butterfly-shaped **gray matter.** On each side the white matter contains three masses of myelinated axons—the posterior, lateral, and anterior **funiculi.** These axons carry impulses up and down the spinal cord. These are the important parts of the cord that were cut when Terry fell (see Tracts and Nuclei, this chapter). The gray matter contains the cell bodies of neurons whose axons occupy the white matter. Phospholipids in myelin sheaths give the white matter its color; neuroglia, neuron cell bodies, dendrites, and terminal branches appear gray.

The gray matter contains four major regions. The **intermediate substance,** where the wings of the butterfly join, connects each side across the isthmus; the **posterior** and **anterior gray horns** are the wings. Neurons in the posterior horn form synapses with sensory axons from the posterior roots of the spinal nerves. The anterior horn contains the cell bodies of motor neurons that exit through the anterior roots. The **lateral horns** project from the intermediate substance laterally, and they contain cell bodies of neurons in the autonomic nervous system.

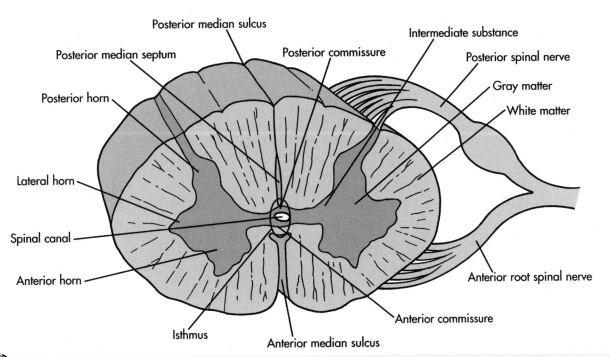

Figure 18.5

Cervical segment of the spinal cord, sectioned to show internal anatomy.

- There are 31 pairs of spinal nerves; how many dermatomes are there?
- Where are the white and gray matter located in the spinal cord?

MENINGES
THREE LAYERS

The meninges surround the delicate spinal cord and cushion it from contacting the walls of the vertebral canal as the vertebral column moves. The meninges (MEN-in-jez) consist of three sleeves of tissue, as you can see in Figure 18.6. The inner layer or **meninx** (MEN-inks) covers the spinal cord and the outer layer attaches to the vertebrae. The middle meninx traps cerebrospinal fluid between it and the inner layer, and it cushions the cord like a waterbed.

The delicate **pia mater** (PEA-ah MAY-ter; tender mother), the innermost meninx, intimately covers the spinal cord and roots of the spinal nerves with a thin epithelium that follows every groove and fold. Arteries and veins run through the pia into the cord and roots. Caudally the pia mater becomes the filum terminale, which anchors the inferior end of the spinal cord to the bottom of the vertebral canal.

The **dura mater** (tough mother) is the outermost sleeve. It is actually two layers with a space between them. The outer layer is the **periosteal dura,** the periosteum of the vertebral canal. The inner layer, the **meningeal dura,** is a tough, fibrous, connective tissue that envelops the cord and the roots of the spinal nerves, as well as serving as the epineurial covering of the nerves (see Roots, this chapter). Collars of dura mater at the foramen magnum and intervertebral foramina join the two layers. These connections seal off the **epidural space** (EP-ih-DUR-al) between the layers of dura. This space contains a fatty outer cushion that supports the cord and receives blood vessels.

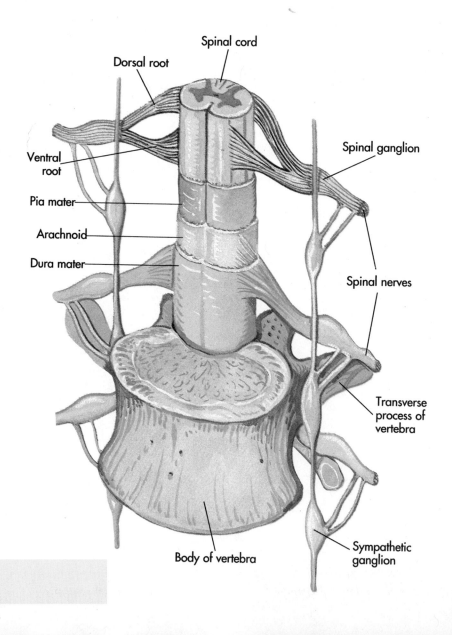

Labels: Dorsal root, Spinal cord, Spinal ganglion, Ventral root, Pia mater, Arachnoid, Dura mater, Spinal nerves, Transverse process of vertebra, Body of vertebra, Sympathetic ganglion

✳ **Figure 18.6**
Meninges envelop the spinal cord.

The arachnoid (ah-RAK-noyd) and pia mater fill the space between the dura mater and the brain with a "waterbed" of cerebrospinal fluid. The arachnoid itself is a thin, fibrous, connective tissue membrane separated from the dura mater by the **subdural space.** The arachnoid is like the top of a waterbed, and the pia mater resembles the bottom. Cerebrospinal fluid fills the large subarachnoid space between the arachnoid and the pia mater. Numerous fine threads, or filamentous trabeculae (tra-BEK-you-lee) in the subarachnoid space, join the pia mater to the arachnoid. Cerebrospinal fluid circulates around these threads and through the meshwork.

Other structures contribute to suspension of the spinal cord within the vertebral canal. The **posterior arachnoid septum** (not shown in Figure 18.6) moors the arachnoid to the dura mater at the posterior median fissure, and **denticulate ligaments** (den-TIK-you-late) do so laterally. Like very large trabeculae, these ligaments arise from the pia mater and extend through the subarachnoid space onto the meningeal dura. Twenty-one pairs of denticulate ligaments occupy the intervals between the spinal nerves in the cervical and thoracic regions. The spinal nerves themselves also help anchor the cord by penetrating between the vertebrae.

SPINAL TAPS AND SADDLE BLOCKS

The spaces between lumbar vertebrae allow clinicians to access the meninges in the vertebral canal (Figure 18.7). A syringe can be inserted into the subarachnoid space between the second and third or third and fourth lumbar vertebrae to withdraw samples of cerebrospinal fluid; this is called a **spinal tap.** This procedure involves little risk to the cord itself because the spinal cord ends at the second lumbar vertebra. Spinal taps can disclose whether the cerebrospinal fluid contains infectious microorganisms such as those that cause spinal meningitis. Spinal anesthetics can be administered into the epidural space through this same route to block uterine and vaginal pain during childbirth or other surgery; this procedure is called **saddle block.**

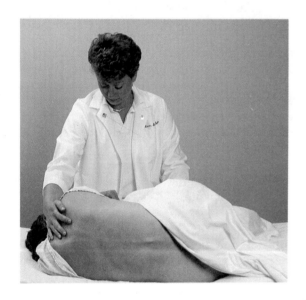

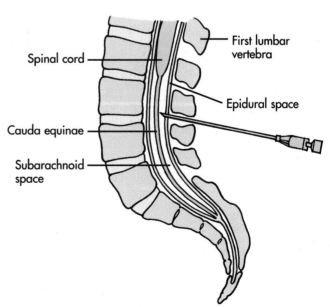

 **Figure 18.7**

Spinal taps and saddle blocks. Flexing the vertebral column opens spaces between the laminae of the lumbar vertebrae through which a syringe can penetrate the epidural and subarachnoid spaces.

CIRCULATION

The spinal cord receives circulation from its posterior and anterior sides. Posterior and **anterior spinal arteries** in Figure 18.8 supply blood to a network of smaller vessels in the pia mater on the corresponding surfaces of the cord. **Radicular arteries** (ra-DIK-you-lar) deliver blood from the intervertebral foramina to this same network through the anterior and posterior roots of the spinal nerves. From the pia mater, smaller branches enter the posterior white matter and the posterior gray horn, and others reach into the anterior white matter from the front of the spinal cord. The anterior spinal artery also sends branches through the anterior median sulcus deep into the gray matter. All of these branches deliver blood to their own territories without connecting with others; consequently the posterior cord shares little circulation with the anterior portion. Venous drainage returns through all these entry points into sinuses in the median fissures and drains into the **intersegmental veins.** See Chapter 14, Vessels, for the sources of these vessels.

- What is the role of: the dura mater, the arachnoid, the pia mater?
- Which arteries serve the spinal cord?

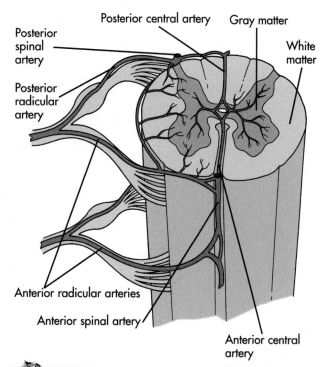

Posterior central artery   Gray matter

Posterior spinal artery

White matter

Posterior radicular artery

Anterior radicular arteries

Anterior spinal artery

Anterior central artery

✳️**Figure 18.8**

Arterial circulation to the spinal cord.

SPINAL NERVES

NERVES AND VERTEBRAE

Thirty-one pairs of **spinal nerves** pass adjacent to thirty vertebrae, as shown in Figure 18.3: eight **cervical nerves** (C1 through C8), twelve **thoracic** nerves (T1 through T12), five **lumbar** nerves (L1 through L5), five **sacral** nerves (S1 through S5), and one **coccygeal** nerve (Cx1; COK-sid-jee-al). Spinal nerves exit between the skull and vertebrae. All thoracic, lumbar, sacral, and coccygeal spinal nerves are given the name and number of the vertebra above them. Because the skull is above the first cervical nerve, it and the remaining cervical spinal nerves are given the name and number of the vertebra below their exit. The spinal nerve between the seventh cervical and the first thoracic vertebrae thus is named C8. For example, thoracic 2 exits below the second thoracic vertebra and innervates the second thoracic dermatome in Figure 18.2. This connection between spinal nerves, their dermatomes, and their spinal segments is the basis for locating the site of Terry's injury between nerves T2 and T3.

The spinal cord is shorter than the vertebral column. Figure 18.9A shows that earlier in development, column and cord were the same length. Rapid, lengthwise growth of the vertebral column made the difference. As the column began to outdistance the cord in the third month of fetal development, it towed the spinal nerves along with it. Because each spinal nerve still passes between the same vertebrae, the nerves had to grow longer. This process results in the cauda equina alone occupying the lower lumbar vertebral canal. The displacement explains how clinicians can tap the lumbar region of the vertebral column without damaging the spinal cord itself (see Meninges, this chapter).

ROOTS

Anterior and posterior roots connect each spinal nerve to the cord, as Figure 18.10 shows. **Posterior roots** contain afferent axons that conduct impulses to the posterior half of the spinal cord. All spinal afferent fibers take this route into the spinal cord, whether they come from the viscera, the skeletal muscles, or the skin. The cell bodies of these neurons occupy the **posterior root ganglion** (GANG-lee-ahn), an enlargement midway along the root. These cells are **unipolar neurons** (see Chapter 17) of somatic and visceral types. Somatic afferent neurons receive stimuli from skin, muscles, and the peritoneum, and central branches of their axons make synapses in the posterior horns of the gray matter or ascend through the white matter of the spinal cord to the brain. Visceral afferent axons send impulses, arising in the visceral organs, to synapses in the lateral horns and along collateral branches that also ascend the cord. Afferent

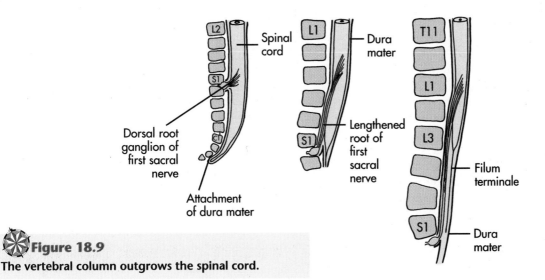

Figure 18.9

The vertebral column outgrows the spinal cord.

fibers enter each segment as rootlets that form a continuous strip along the entire cord at the posterolateral groove, much like the teeth in a comb. Posterior roots are larger than anterior roots because they contain more axons. Had his injury affected only the anterior and posterior roots of a few thoracic spinal nerves, Terry would have lost sensation and muscular control only in a few segments of the thorax, not his entire lower body. Why would this be so?

Anterior roots carry efferent axons (conducting impulses away from the cord) from the anterior side of the cord to skeletal muscles and to the viscera. Only efferent nerve fibers travel these roots. Somatic efferents are multipolar neurons (see Chapter 17, Neurons), with cell bodies in the anterior horn of the gray matter. They innervate skeletal muscles in the neck, trunk, and the limbs. Visceral efferents are multipolar neurons with cell bodies in the lateral horns of the gray matter and axons that synapse with neurons in the ganglia of the autonomic nervous system (see Chapter 21, Autonomic Nervous System). Anterior rootlets are fewer in number than posterior ones, and

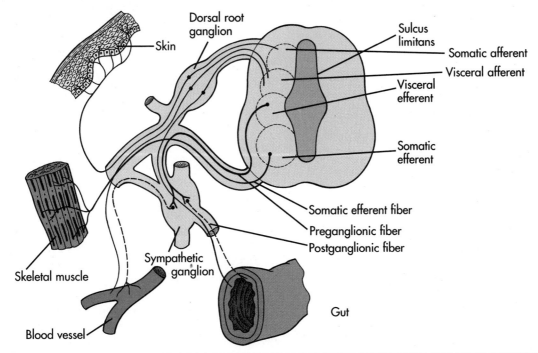

Figure 18.10

Axons of spinal nerves. Afferent neurons are shown in red; efferent are shown in blue. Somatic neurons are indicated by a solid line; visceral neurons are indicated by a dashed line. Afferents enter posterior roots, efferents leave through anterior roots, and spinal nerves carry both afferent and efferent axons.

they emerge near the middle of each spinal segment at the anterolateral groove.

The anterior and posterior roots of each segment converge in the intervertebral foramina to form the spinal nerve, and their axons intermingle as the nerve passes between the vertebrae. From this point onward, nerve fascicles carry both afferent and efferent axons, and the nerve therefore is said to be mixed. A tough, fibrous sheath, or **epineurium,** covers the exterior of all spinal nerves. The epineurium thins as the nerves branch, but it disappears only when axons reach their targets. The **perineurium** surrounds fascicles within the nerve, and a delicate third sheath, the **endoneurium,** envelops each nerve fiber.

TWO RAMI

Each spinal nerve innervates portions of three principal territories—the back, the trunk and limbs, and the viscera, as Figure 18.11 shows. On emerging from the vertebral column, every spinal nerve branches into anterior and posterior primary **rami** (RAM-eye). The axons of the **posterior primary ramus** (RAY-mus) turn posteriorly to the skin of the back and to axial muscles. The much larger **anterior primary ramus** sends branches within the body wall to the muscles, as well as superficially to the skin and deeper to the peritoneum. This ramus also innervates the limbs through the appropriate cervical, lumbar, and sacral spinal nerves. Cutaneous branches from

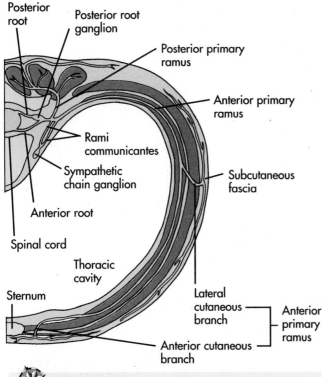

Posterior root
Posterior root ganglion
Posterior primary ramus
Anterior primary ramus
Rami communicantes
Sympathetic chain ganglion
Subcutaneous fascia
Anterior root
Spinal cord
Thoracic cavity
Sternum
Lateral cutaneous branch
Anterior cutaneous branch
Anterior primary ramus

✳ Figure 18.11

Spinal nerves send two major branches to the body. Transverse section through the thorax.

both the posterior and anterior primary rami supply the dermatome of each spinal nerve. Terry's diaphragm still functions because anterior rami from cervical nerves (C3 through C5) above his injury supply it with motor and sensory axons.

- Which vertebra is immediately above cervical spinal nerve 8? Which is below?
- How is growth of the vertebral column responsible for forming the cauda equina?
- What parts of spinal nerves carry afferent, efferent, and mixed impulse traffic?
- Which ramus of a spinal nerve delivers fibers to the limbs and anterior trunk?

PLEXUSES

A plexus is a network of peripheral nerve fibers outside the vertebral column that distributes axons toward their destinations, somewhat as interchanges on an interstate highway distribute traffic. **Plexuses** near the shoulder and pelvis, shown in Figure 18.12, distribute the anterior primary rami of spinal nerves to the shoulders, arms, and legs. The **cervical plexus** delivers the first four cervical nerves (C1 through C4) to the head, neck, and shoulders. Beneath the clavicle, the **brachial plexus** sends C5 through T1 and part of C4 into the arms, the forearms, and the hands. The lumbar and sacral plexuses do the same for nerves innervating the lower limbs. The **lumbar plexus** is located in the posterior abdominal wall; it directs four lumbar nerves (L1 through L4) to the abdomen and to the thighs. The **sacral plexus,** located in the posterior pelvis, is the main distribution center for the legs; it sends five spinal nerves (L4 through S3) to each one. The **coccygeal plexus** does not supply any limbs, but it sends branches of three spinal nerves (S4 through Cx1) to the skin and to the muscles of the coccyx.

BRACHIAL

The brachial plexus is the best example of what you should understand about plexuses. Located beneath each clavicle (Figure 18.13), this plexus distributes axons from the spinal nerves to the anterior and posterior surfaces of the arms. Nerves emerge from the plexus as anterior and posterior divisions, sending axons to the skin, muscles, and bones on their respective sides of the arms. This separation is especially important for skeletal muscles because most flexors are on the anterior surface of the limb and most extensors on its posterior. You will find similar divisions in the lumbar and sacral plexuses.

Distribution takes place stepwise rather than at once. If a plexus were a freeway, you would drive through a rapid series of interchanges, as traffic meets and diverges. In a plexus, these sections are called roots, trunks, divisions, and cords. **Roots** are the first

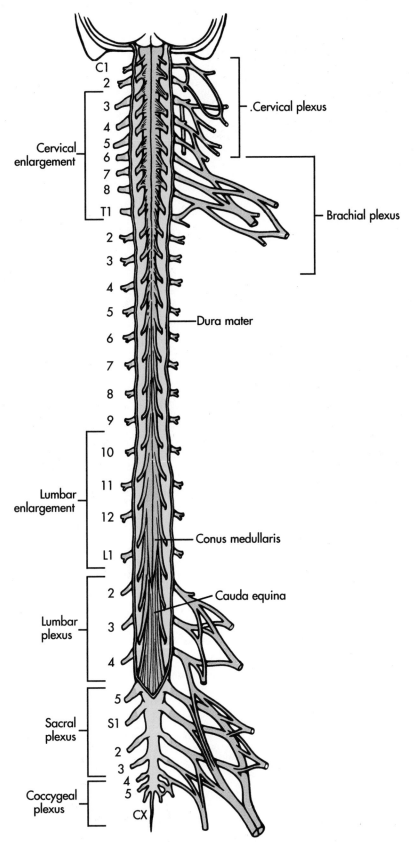

C1
2
3
4
5
6
7
8
T1
2
3
4
5
6
7
8
9
10
11
12
L1
2
3
4
5
S1
2
3
4
5
CX

Cervical enlargement

Lumbar enlargement

Lumbar plexus

Sacral plexus

Coccygeal plexus

.Cervical plexus

Brachial plexus

Dura mater

Conus medullaris

Cauda equina

Figure 18.12

Plexuses of the spinal nerves.

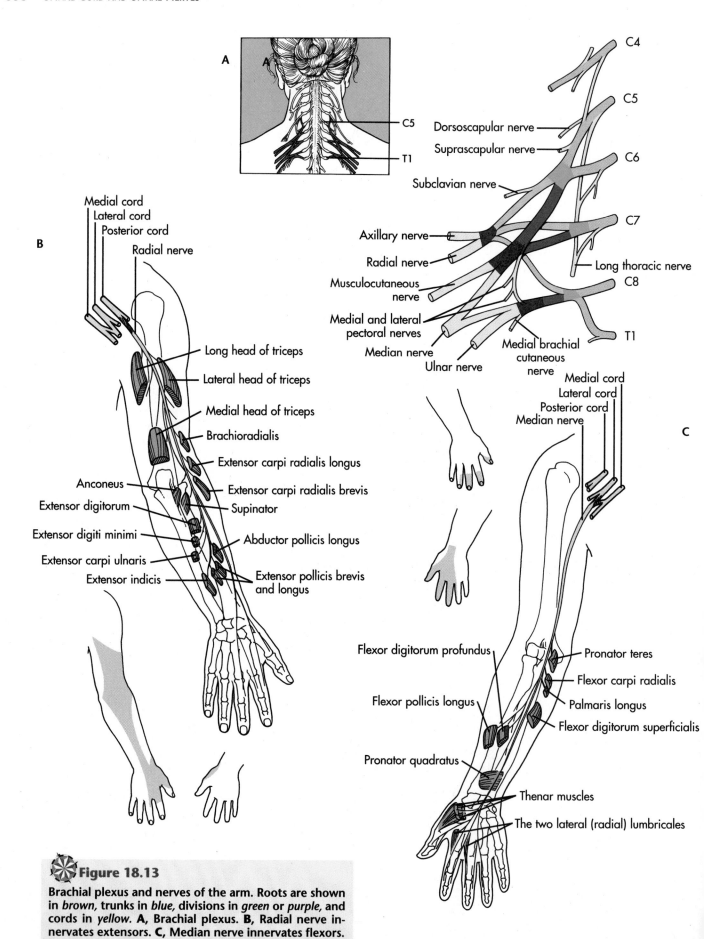

A

C5
T1

B

Medial cord
Lateral cord
Posterior cord
Radial nerve

Long head of triceps
Lateral head of triceps
Medial head of triceps
Brachioradialis
Extensor carpi radialis longus

Anconeus
Extensor digitorum
Extensor digiti minimi
Extensor carpi ulnaris
Extensor indicis

Extensor carpi radialis brevis
Supinator

Abductor pollicis longus

Extensor pollicis brevis
and longus

C4
C5

Dorsoscapular nerve
Suprascapular nerve

C6

Subclavian nerve

C7

Axillary nerve

Radial nerve

Long thoracic nerve

Musculocutaneous
nerve

C8

Medial and lateral
pectoral nerves

Median nerve

Ulnar nerve

Medial brachial
cutaneous
nerve

T1

Medial cord
Lateral cord
Posterior cord
Median nerve

C

Flexor digitorum profundus

Pronator teres

Flexor carpi radialis

Flexor pollicis longus

Palmaris longus

Flexor digitorum superficialis

Pronator quadratus

Thenar muscles

The two lateral (radial) lumbricales

Figure 18.13

Brachial plexus and nerves of the arm. Roots are shown in *brown*, trunks in *blue*, divisions in *green* or *purple*, and cords in *yellow*. A, Brachial plexus. B, Radial nerve innervates extensors. C, Median nerve innervates flexors.

sections; they are the anterior rami of the spinal nerves in Figure 18.13A. **Trunks** are next; they usually form at junctions of roots. For example, the superior trunk of the brachial plexus receives C4, C5, and C6, and the inferior trunk receives C8 and T1; but C7 travels straight through, alone, as the middle trunk. Trunks become **divisions.** Each trunk splits into anterior and posterior divisions, and the members of each division join each other as **cords.** The posterior cord collects posterior divisions from all three trunks. The lateral cord receives anterior divisions from the superior and middle trunks, and the medial cord receives only one anterior division from the inferior trunk. Having received axons from all six roots, the posterior cord serves the extensor muscles by way of the **radial** and **axillary nerves.** The lateral and medial cords also give rise to the **musculocutaneous, radial,** and **ulnar nerves** that generally speaking innervate the flexors on the anterior side of the upper limb in Figure 18.13B and C. Overall, the brachial plexus recombines and directs axons of the spinal nerves to the appropriate anterior (flexor) or posterior (extensor) muscles in the upper limb.

- Name the plexuses of the spinal nerves.
- Which spinal nerves contribute to each plexus?
- What is the proximodistal sequence of sections in the brachial plexus?

CERVICAL

Small and complicated, the cervical plexus directs the anterior rami of the first four cervical nerves (C1 through C4) to the neck, shoulders, and diaphragm. This plexus (Figure 18.14) is lateral to the first four cervical vertebrae, deep beneath the sternocleidomastoid and medial scalene muscles. Fibers of this plexus communicate with the viscera, skin of the neck, and muscles of the pharynx and tongue by way of the vagus, accessory, and hypoglossal cranial nerves described in Chapter 19. Of particular importance, axons pass through this plexus to innervate those muscles that elevate and depress the hyoid bone when swallowing. The **phrenic nerve** (FREN-ik) innervates the diaphragm muscles with fibers from C3 through C5, which is why Terry can breathe without help from his paralyzed intercostal muscles.

LUMBOSACRAL

Many anatomists and physicians consider the following subjects as one large plexus, the **lumbosacral plexus.** Combining the lumbar and sacral plexuses simplifies your job of following the nerves, because their dermatomes wrap around the leg as you follow the nerves in Figure 18.15B. Beginning with L1, the dermatomes sweep around the limb from proximal and medial positions at the groin to the posterior, distal bands of S2 on the back of the leg and foot. The pattern twists because the lower limbs rotated during development (see Limb Development, Chapter 7). It is easier to see which fibers reach these territories by following one plexus at a time. The lumbar plexus involves the first four dermatomes, from L1 to L4, and the sacral plexus continues from L4 to S3.

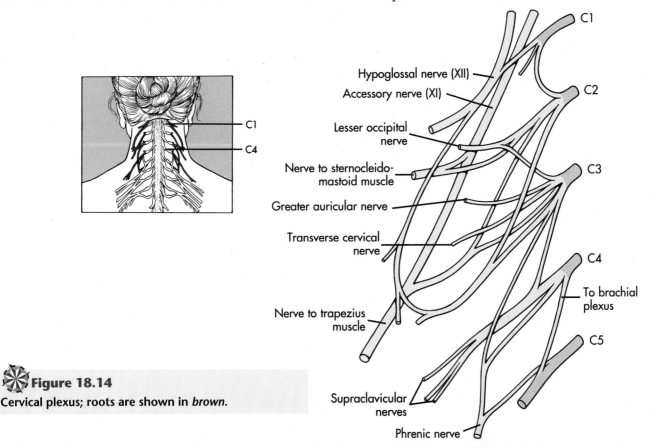

Figure 18.14
Cervical plexus; roots are shown in *brown.*

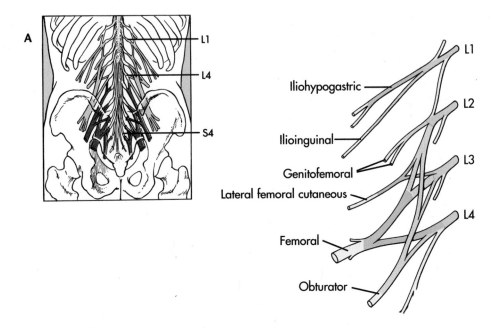

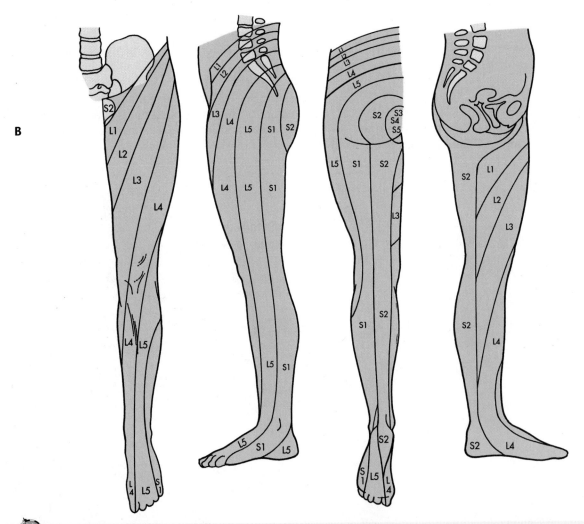

Figure 18.15

A, Lumbar plexus; roots shown in *brown,* posterior divisions in *green,* and anterior divisions in *purple.* **B,** Dermatomes of the lower limb.

LUMBAR

The lumbar plexus leads the first three lumbar nerves and most of L4 to the lower abdomen, pelvis, thigh, and genitalia. Figure 18.15A shows this plexus located lateral to the lumbar vertebrae, between the transverse processes of these bones and the fascicles of the psoas muscle. Five main nerves leave the plexus. The three largest distribute axons mainly to the muscles and skin of the thigh. As you work downward from the most superior of these three, you will find that their dermatomes shift laterally and distally over the anterior thigh as lumbar spinal nerves L2 through L4, shown in Figure 18.15B. The lateral femoral cutaneous and femoral nerves contain the posterior divisions of the plexus, and the obturator nerve carries part of the anterior division. The **lateral femoral cutaneous nerve** (L2, L3) innervates skin from the greater trochanter of the femur, over the anterior and medial surface, to the knee. Next, the **femoral nerve** (L2 through L4, Figure 18.16A), the largest of this series, supplies the major flexor muscles of the thigh, extensor muscles of the knee, and skin anteriorly and medially. The longest branch of this nerve, the **saphenous nerve** (sa-FEE-nus), continues to the skin (but not to the muscles) past the knee, down the tibial side of the leg to the ankle and the ball of the great toe. Finally, the **obturator nerve** (OB-tyur-aye-tor; L2 through L4; Figure 18.16B) penetrates the obturator foramen of the pelvis to the adductor muscles and skin of the medial thigh. Lesser

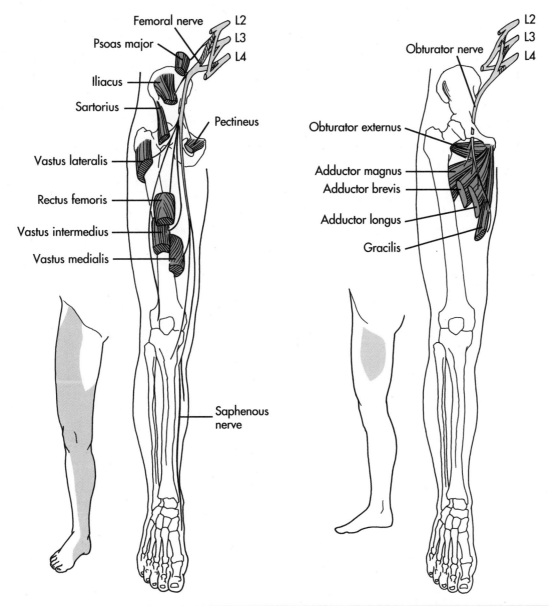

Figure 18.16

Nerves of the lumbar plexus. **A,** Femoral nerve. **B,** Obturator nerve.

nerves of the lumbar plexus supply the anterior abdominal wall (**iliohypogastric nerve;** ILL-ee-oh-HIGH-po-GAS-trik), and the scrotum in males or labia majora in females (**ilioinguinal nerve;** ILL-ee-oh-IN-gwin-al), from the first lumbar nerve. From nerves L1 and L2 the **genitofemoral nerve** (JEN-ih-toe-FEM-or-al) also serves the scrotum in males or the round ligament of the uterus in females.

- In which direction do dermatomes from the first lumbar to the third sacral nerves wrap the leg?

SACRAL

The sacral plexus furnishes nerves to the thigh, leg, and foot from the last two lumbar spinal nerves (L4 and L5) and the first three sacral spinal nerves (S1 through S3) in Figure 18.17. The largest nerve of the body, the **sciatic nerve,** some 2 cm in diameter (¾ inch), is the main supply for this massive territory. The sacral plexus sorts out the fibers for this and four other nerves on the posterior lateral wall of the pelvis, where the roots emerge from the pelvic sacral foramina of the sacrum and from between the lumbar vertebrae. Most roots separate into anterior and posterior divisions that converge as the tibial and common fibular nerves, respectively. These two nerves continue together as the sciatic nerve (sigh-AT-ik) into the thigh, through the greater sciatic notch of the pelvis. Look for dermatomes L4 through S3 in Figure 18.15B to find the general destination of axons from the sciatic nerve. Remember, as you descend the series of spinal nerves, the dermatomes go from medial, proximal locations to more posterior and distal areas.

The sciatic nerve descends the posterior thigh beneath the biceps femoris muscle in Figure 18.18 and separates into the common peroneal and tibial nerves above the knee. (In rare individuals these nerves remain separate, and there is no sciatic nerve as such.) The **tibial nerve** descends deep inside the leg in Figure 18.18A, innervating skin and muscles as it goes. It passes the medial malleolus of the tibia, traverses the ankle, and enters the plantar surface of the foot where it separates into the **medial** and **lateral plantar nerves.** The **common peroneal nerve** (Figure 18.18B) leads to the peroneus muscles and others in the anterior compartment of the leg. After passing beneath the head of the fibula, this nerve splits into the **deep peroneal nerve** and the **superficial peroneal nerve,** which give off branches to the muscles already mentioned, and then proceed through the flexor retinaculum to the dorsal surface of the foot. Cutaneous branches of these nerves serve the overlying skin (see the dermatomes in Figure 18.15B).

Other axons of the sacral plexus deliver to the gluteal muscles (**superior and inferior gluteal nerves,** L4 through S1 and L5 through S2), to the skin of the perineum and posterior thigh and leg (**posterior femoral cutaneous nerve,** S1 through S3), and to the skin of the buttocks (**perforating cutaneous nerve,** S2, S3). The **pudendal nerve** (pew-DEN-dal; S2 through S4) supplies the floor of the pelvic cavity, specifically the external anal sphincter, the anal skin, and the external genitalia.

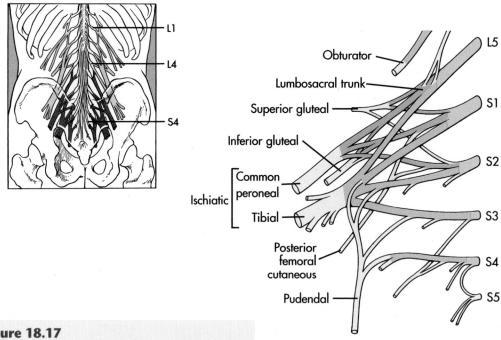

Figure 18.17

Sacral plexus; roots are shown in *brown,* posterior divisions in *green,* and anterior divisions in *purple.*

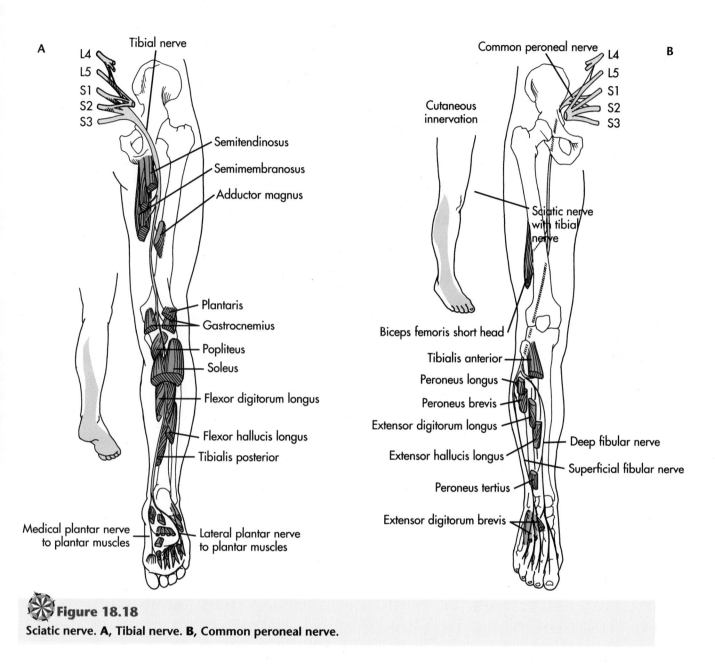

Figure 18.18
Sciatic nerve. **A,** Tibial nerve. **B,** Common peroneal nerve.

COCCYGEAL

This small plexus innervates the skin over the coccyx from the **coccygeal nerve** and portions of S4 and S5, shown in Figure 18.17.

COMMUNICATION IN THE SPINAL CORD

COMMUNICATION AND COORDINATION

The spinal cord is a communication and coordination center controlled by the brain. As you have seen, each segment of the spinal cord communicates with its dermatome by way of its pair of spinal nerves. The segments also communicate with each other through neurons that connect neighboring segments above and below, as illustrated in Figure 18.19. This higher, spinal level of control helps coordinate the muscles of neighboring segments in the limbs, enabling them to move effectively. The brain exerts still higher orders of control, because it communicates with every segment through neurons that ascend and descend to all levels of the cord. This organization allows the brain to modulate the actions of reflex arcs within each segment of the spinal cord.

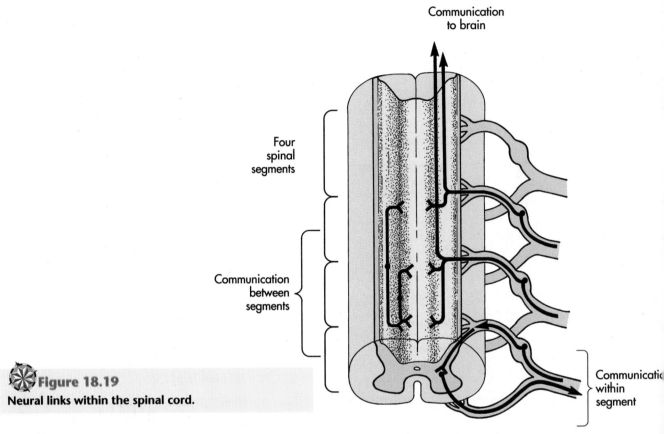

Communication
to brain

Four
spinal
segments

Communication
between
segments

Communication
within
segment

Figure 18.19
Neural links within the spinal cord.

▲ In Terry's case, communication to areas below the second thoracic segment and the brain has stopped, but some segmental reflexes and intersegmental communication remain below his injury.

REFLEX ARCS

Reflex arcs link afferent neurons with efferent neurons in each segment of the spinal cord. A pinched finger on the right hand is quickly withdrawn, as in Figure 18.20, before the brain is aware of any danger, because **interneurons** link the afferent sensory neurons from the skin to efferents that retract the upper limb. Stimuli from pain receptors in the finger return through the brachial plexus and enter the posterior horn of the gray matter; there interneurons excite impulses in motor neurons in the anterior gray horn that then travel out the anterior root, through the brachial plexus, to the muscles of the upper limb. Interneurons also extend axons across the commissures to the left side of the cord and to neighboring segments of the cord, marshaling all the flexor muscles that withdraw the right arm and inhibiting the extensors. Walking uses such contralateral (opposite side) reflexes to swing and then recover the upper and
▲ lower limbs. Terry regained some of these reflexes in a few weeks, but he has no voluntary control over them because axons of neurons no longer connect the reflex loops with the brain. Although he would never consciously feel a pinched toe, reflexes nevertheless would attempt to withdraw Terry's leg.

PLATES, FUNICULI, TRACTS, AND NUCLEI
PLATES

The spinal cord develops as two large functional divisions called plates (or laminae). The anterior halves of each side of the spinal cord (Figure 18.21A) primarily contain efferent neurons that serve motor functions and emerge through the anterior roots of the spinal nerves; these regions are the **basal plates.** In contrast, alar plates (AIL-ar) are largely sensory and contain the endings of afferent neurons whose axons enter through the posterior roots of the spinal nerves. The alar plates resemble butterfly wings, hence their name. Alar plates occupy the posterior half of the cord, beginning at the intermediate substance of the gray matter, which they share with the basal plates. Two narrow **roof** and **floor plates** form the isthmus of the spinal cord above and below the spinal canal, respectively. Remember: alar—afferent, sensory, posterior root; basal—efferent, motor, anterior root.

FUNICULI

Axons ascend or descend in the white matter of the spinal cord in groups that reflect their origins and destinations. This organization resembles that of fascicles in peripheral nerves, but there are important differences, as you will see. Funiculi (few-NIK-you-lee) are the largest groups in the spinal cord, and Figure 18.21A shows three of them surrounding the gray matter. The **posterior funiculi** are medial to the pos-

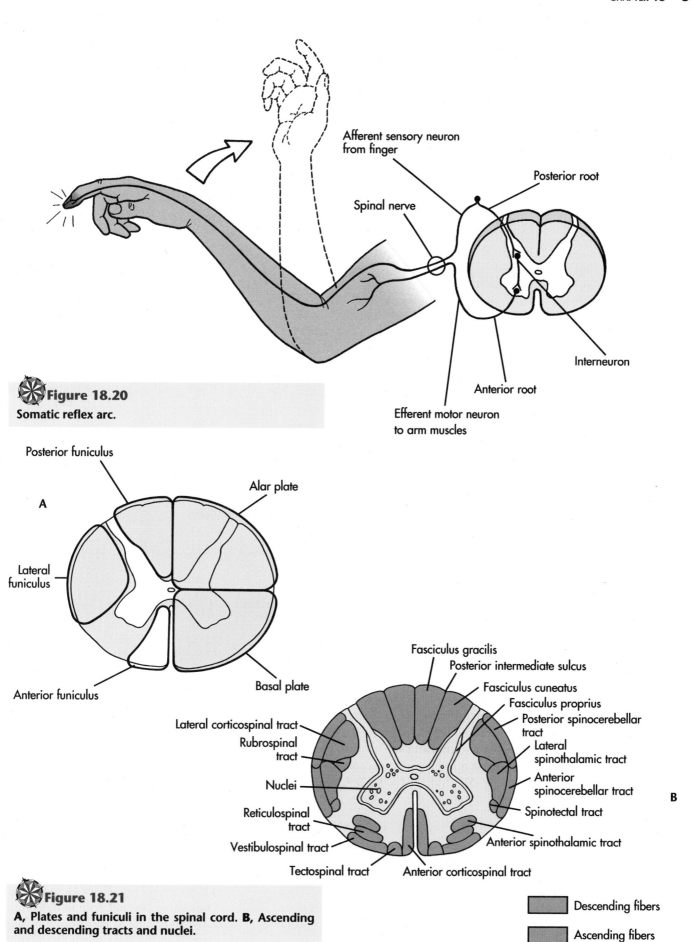

Figure 18.20
Somatic reflex arc.

Figure 18.21
A, Plates and funiculi in the spinal cord. **B,** Ascending and descending tracts and nuclei.

terior horns of the gray matter in the alar plate; the **lateral funiculi** are lateral to the anterior and posterior horns, and they occupy both the alar and basal plates. The **anterior funiculi** are medial to the anterior horn in the basal plate.

TRACTS AND NUCLEI

Tracts ascend and descend in these funiculi. Tracts are groups of nerve fibers in the spinal cord and brain that, like fascicles of peripheral nerves, have common sources and destinations (Figure 18.21B). Having no perineurium to separate them, most tracts do not have precise boundaries. **Nuclei** are clusters of functionally related neuronal cell bodies, with their associated synapses and glial cells (same figure) in the gray matter of the spinal cord and the brain. Nuclei (not to be confused with nuclei of cells) are comparable to ganglia of the peripheral nervous system. Nuclei integrate and relay signals among different tracts. Think of them as junctions and switching points where neural pathways can branch or intersect.

- Give an example of nerve fibers that are linked together in a reflex arc.
- Which plate is primarily sensory?
- What is the difference between funiculi and tracts?

MAJOR TRACTS

▲   Terry has lost communication through all spinal tracts below the second thoracic segment. Figure 18.21B shows the locations of major tracts of the spinal cord and Table 18.1 summarizes them, so that you may see some effects of his injury. Ascending tracts carry impulses up the spinal cord to the brain. They are usually located in the alar plate, in the posterior and lateral funiculi. Descending tracts carry impulses from the brain to the cord, usually in the anterior and lateral funiculi of the basal plate.

Many tracts are named by the origin and the destination of their axons. For example, the spinothalamic tract carries ascending (sensory) nerve fibers from the spinal cord to the thalamus in the brain. The prefix, spino-, indicates the origin of neurons, and the suffix, -thalamic, indicates their destination. Names also reflect special features or the location of a tract, such as the fasciculus gracilis, the slender tract that parallels the posterior median sulcus.

ASCENDING

Ascending tracts carry impulses up the spinal cord to the brain. These tracts originate either from afferent neurons that enter the posterior root of the spinal nerves and ascend the cord directly, or from secondary neurons that synapse with afferents in the gray matter of the cord. Some tracts remain on the

same side of the cord as their origin and **decussate** (de-KUSS-ate), that is, cross to the opposite side in the brain, while others cross in the spinal cord.

POSTERIOR COLUMNS

The posterior funiculus contains two ascending tracts, the fasciculus gracilis and the fasciculus cuneatus, also called the posterior columns. The **fasciculus gracilis** (GRAH-sil-iss) carries ascending axons to the brain (see Figures 18.21B and 18.22) from

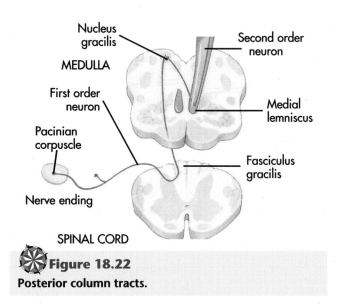

Figure 18.22
Posterior column tracts.

spinal nerves that enter the lower half of the cord, below spinal segment T6. This fasciculus gradually enlarges as it ascends until T6 where it receives no more axons from spinal nerves. The **fasciculus cuneatus** (KUE-nee-AYE-tus; wedge-shaped) is lateral to the fasciculus gracilis and separated from it by the posterior intermediate sulcus. This fasciculus carries ascending axons from the thoracic and cervical regions (T6 and above) to the brain. As its name indicates the fasciculus cuneatus widens steadily as more fibers enter it. Both fasciculi carry sensory information about discriminative touch and body position from receptors in the skin, tendons, joints, and muscles on the same side of the body as the tract. Discriminative touch is the ability to distinguish two simultaneous tactile stimuli. Most people are able, when blindfolded, to detect pin pricks 5 mm or more apart on the fingers, the most sensitive areas. Body sense from the right arm thus rises through the fasciculus cuneatus on the right side of the spinal cord. Once in the brain, axons cross to the left side. The left side of the brain therefore perceives the right side of the body.

Table 18.1   SPINAL TRACTS

Tract	Source	Decussation	Destination	Stimuli
ASCENDING				
Fasciculus gracilis	Afferent neurons with cell bodies in posterior root ganglion below T6 nerve	None, subsequent axons cross in medulla	Gracile nuclei of medulla	Body sense and discriminative touch
Fasciculus cuneatus	Afferent neurons with cell bodies in posterior root ganglion above T6 nerve	None, subsequent axons cross in medulla	Cuneate nuclei of medulla	Same
Anterior spinothalamic	Cell bodies in posterior gray horn	Anterior white commissure several segments above origin	Cerebral cortex of opposite side	Light touch
Lateral spinothalamic	Afferent neurons with cell bodies in posterior root ganglion of spinal nerves	Anterior white commissure several segments above origin	Cerebral cortex of opposite side	Pain and thermal sensations
Anterior spinocerebellar	Cell bodies in posterior gray horn and intermediate substance	Within intermediate gray substance at segments of origin	Cerebellar cortex of opposite side	Body sense from tendons
Posterior spinocerebellar	Cell bodies in intermediate gray substance between C8 and L2 nerves	None	Cerebellar cortex of vermis	Body sense; muscles, tendons, joints
Fasciculi proprii	Gray matter of spinal cord	Local segments	Gray matter of adjacent segments	Various
DESCENDING				
Lateral corticospinal	Cell bodies in cerebral cortex	In pyramids of medulla	Anterior horn of gray matter, opposite side	Voluntary motor control
Anterior corticospinal	Cell bodies in cerebral cortex	In spinal cord at level of exit	Same	Same
Rubrospinal	Cell bodies in red nucleus of midbrain	In midbrain	Posterior gray horn, opposite side	Motor coordination
Lateral vestibulospinal	Cell bodies in lateral vestibular nucleus of pons (equilibrium)	None	Intermediate substance, anterior gray horns	Facilitates extensor motor neurons
Medial vestibulospinal	Cell bodies in medial vestibular nucleus of pons (equilibrium)	None	Same	Facilitates flexor motor neurons
Reticulospinal	Reticular formation of pons and medulla	Some decussations in cord	Intermediate substance, anterior horn, gray matter	Facilitates extensor, flexor motor neurons
Tectospinal	Cell bodies in superior colliculus of midbrain (visual reflexes)	In midbrain	Intermediate substance, anterior gray horn, opposite side	Turns head toward light

From Afifi AK and Bergman RA: Basic neuroscience, ed 2, Baltimore, 1986, Urban and Schwarzenberg.

SPINOTHALAMIC TRACTS

The anterior and lateral spinothalamic tracts carry touch, pain, and thermal stimuli to the brain from the opposite sides of the body, as Figure 18.23 shows. These axons enter the posterior roots of the spinal nerves and cross the white commissure to the next higher segment of the cord; there they ascend the anterior and lateral funiculi to the medulla oblongata of the brain, ultimately reaching the thalamus. Pain stimuli from the right hand reach the left side of the brain through this tract (see Table 18.1).

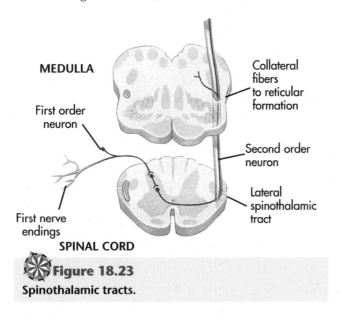

⚝**Figure 18.23**
Spinothalamic tracts.

SPINOCEREBELLAR TRACTS

The spinocerebellar tracts travel the surface of the lateral funiculus, shown in Figure 18.21B, carrying muscle and joint senses from the lower limbs, pelvis, and trunk for coordination by the cerebellum. (The fasciculus cuneatus carries this information from above nerve C8.) Muscle senses convey information about strength of contraction, as well as its duration and speed (see Muscle Spindles, Chapter 9). Most axons in the spinocerebellar tract begin at cell bodies in the thoracic nucleus (see Major Nuclei, this chapter). The axons of the **posterior spinocerebellar tract** travel on the same side of the body on which they originated, through the medulla into the cerebellum.

FASCICULI PROPRII

These thin tracts cover the horns of the gray matter with short axons that reach a few segments above and below their cell bodies in the gray matter (Figure 18.21B). The **fasciculi proprii** (PRO-pree-ee) coordinate reflexes between segments of the spinal cord.

DESCENDING

These tracts descend the cord with motor impulses from their origins in the brain. Decussation may be within the brain or in the cord. The destinations in all cases are synapses on the cell bodies of motor neurons in the intermediate substance or anterior horns of the gray matter.

CORTICOSPINAL TRACTS

The corticospinal tracts descend from the cerebral cortex of the brain, conducting motor impulses for voluntary control of the skeletal muscles (Figure 18.24). The severing of these tracts is the principal reason why Terry is unable to control his trunk and leg muscles. Axons of the **lateral corticospinal tracts** cross to the opposite side of the spinal cord as they pass through the medulla oblongata (see Chapter 19, Brain), but those of the **anterior corticospinal tract** cross in all segments of the spinal cord. These crossings, or **decussations** (DE-cuss-aye-shuns), result in each cerebral hemisphere controlling muscles on the opposite side of the body. Corticospinal tracts have little role in the reflexes that withdrew the arm, but they moderate any voluntary motion afterward, such as popping an injured finger into your mouth or shaking it vigorously.

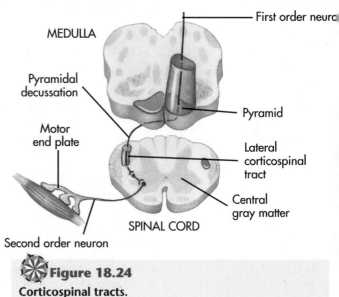

⚝**Figure 18.24**
Corticospinal tracts.

ORGANIZATIONAL RULES

The amount of white matter is roughly proportional to the number of nerve fibers at any level of the spinal cord, as Figure 18.25A shows. The white matter of the cervical cord is two and one half times larger than that in the sacral region. Ascending tracts

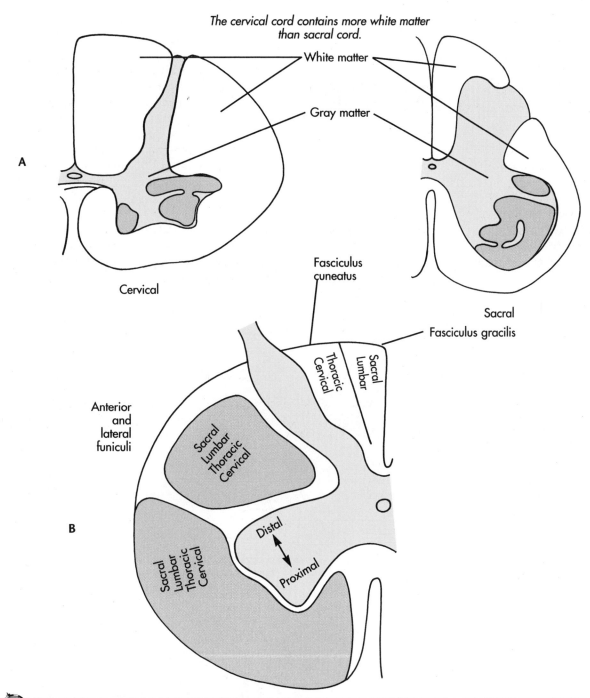

The cervical cord contains more white matter than sacral cord.

White matter

Gray matter

A

Cervical

Fasciculus cuneatus

Fasciculus gracilis

Sacral

Anterior and lateral funiculi

Thoracic
Cervical

Sacral
Lumbar

Sacral
Lumbar
Thoracic
Cervical

Distal

Proximal

Sacral
Lumbar
Thoracic
Cervical

B

✳ **Figure 18.25**

Neuronal organization of the spinal cord. A, White matter is larger in cervical than in sacral segments. B, Tracts of sacral regions are deeper than those of cervical regions.

usually enlarge as they ascend the spinal cord because they receive progressively more fibers from spinal nerves. The fasciculus cuneatus is an excellent example of this relationship; its wedge shape reflects the steady accumulation of axons from the spinal nerves. For similar reasons, descending tracts narrow as they descend the spinal cord and fibers leave the tract to find their targets.

Axons of higher segments more proximal to the brain lie closer to the gray matter than those of lower segments, as shown in Figure 18.25. Superficial damage to the cord often affects only a few dermatomes far below the injury. Deeper tracts serving segments caudal to the injury often escape.

• Name an ascending tract that carries discriminative touch and body sense impulses to the brain.

• What are the origin and destination of the spinothalamic tracts?
• What function is served by the corticospinal tracts?

SPINAL NUCLEI

Figure 18.26 illustrates several prominent spinal cord nuclei. Like tracts, nuclei are named for their locations, other special features, and the targets they innervate.

CERVICAL AND SACRAL MOTOR NUCLEI

Nuclei in the cervical and lumbar enlargements contain cell bodies of motor neurons that serve the muscles of the extremities, as seen in Figure 18.26A and B. Nuclei for different groups of muscles

occupy characteristic locations; they are often called motor pools. **Anteromedial nuclei** in the anterior horn innervate the proximal limb muscles, whereas **posterolateral nuclei** drive the distal muscles. This relationship arises because the anteromedial nuclei and the proximal muscles develop first. Muscles of the fingers and toes develop later and receive axons that originate in the most posterior and lateral reaches of the lower cervical and lumbar enlargements. Smaller nuclei in the appropriate segments innervate the muscles of the neck, the thorax, and the perineum.

THORACIC SENSORY NUCLEUS

The thoracic nucleus contains cell bodies of neurons whose axons ascend the spinal cord in the posterior spinocerebellar tract. This nucleus, seen

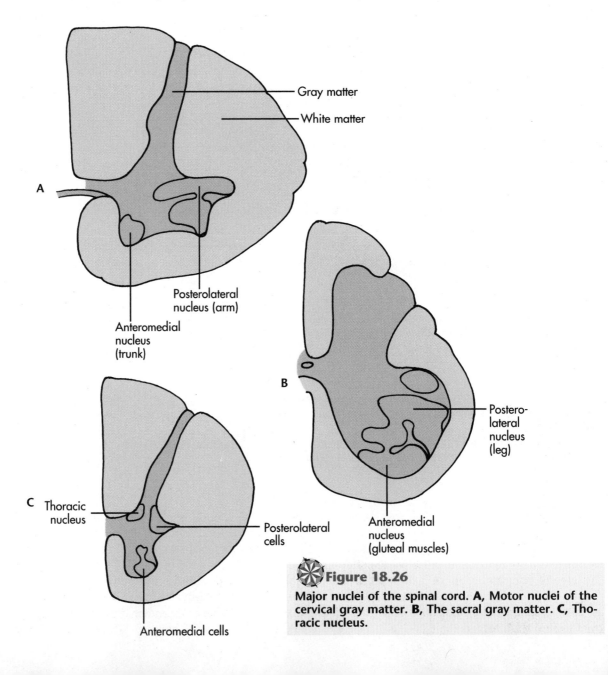

Figure 18.26

Major nuclei of the spinal cord. **A,** Motor nuclei of the cervical gray matter. **B,** The sacral gray matter. **C,** Thoracic nucleus.

in Figure 18.26C, is a prominent strip of cell bodies located in the alar plates between the T1 and L2 levels of the cord. More precisely, the thoracic nucleus is found in the posterior horn of the gray matter, at its junction with the intermediate substance of the gray matter. From muscle spindles and tendon organs in the trunk and lower extremities, the thoracic nucleus receives sensory impulses that the posterior spinocerebellar tract conducts to the cerebellum.

RETICULAR FORMATION

This prominent nucleus extends into the brainstem from the fifth cervical segment. The **reticular formation** (reh-TIK-you-lar) is a network of cell bodies and axons located in the posterior horns of the gray matter, which links ascending and descending axons from the brain to the spinal cord. Its role in activating the brain is discussed in Chapter 19, Brain and Cranial Nerves, but it is mentioned here to emphasize that the spinal cord is continuous with the brain.

- Which nucleus would you expect to contain cell bodies of motor neurons innervating the muscles of the arm?
- Which region of the lower limb does the anteromedial nucleus of the lumbar enlargement innervate?
- Where is the reticular formation located?

NEURAL PATHWAYS

This chapter began by saying that the brain communicates with the body through neural pathways. So far, you have explored the principal tracts and nuclei of these pathways. Terry Franklin's disabilities show that when pathways are interrupted, the functions they support fail but unaffected parts remain fully active. This section summarizes the flow of impulses through the pathways that connect the arms and legs with the brain. Chapter 19, Brain and Cranial Nerves, continues tracing these pathways inside the brain.

OVERVIEW

Receptors in the skin, muscles, and joints send streams of impulses to the spinal cord by way of peripheral axons that pass through the brachial, lumbar, and sacral plexuses. Reflex arcs relay these stimuli back to the flexor muscles that withdraw the limbs from danger or make them move in concert with the other limbs. At the same time, the streams of signals ascending the posterior columns inform the brain of body position and touch. In the spinothalamic tracts, similar streams alert the brain to pain, touch, and thermal changes, but the spinocerebellar tracts coordinate muscular action automatically without actually informing the conscious portion of the brain what is happening. The corticospinal tracts are under conscious control, and they send patterns of impulses that carry out movements directed by the brain. Figure 18.27 traces the pathways that react when Terry pinches his right thumb.

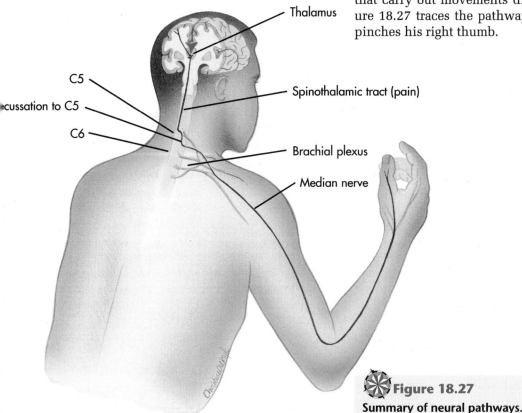

Thalamus

C5

cussation to C5

C6

Spinothalamic tract (pain)

Brachial plexus

Median nerve

Figure 18.27

Summary of neural pathways.

DETAILS

The thumb belongs to the sixth cervical dermatome; receptors there return their impulses through the median nerve and the brachial plexus through cervical nerve 6, where they enter the spinal cord by way of the posterior root. Collateral branches of these axons enter the posterior gray horn or ascend to the brain. Synapses in the gray matter connect with interneurons leading to motor neurons in the anterior horn that activate the flexor muscles. Connections also go to the fasciculi proprii, which mobilizes the flexors of neighboring segments above and below C6. These impulses activate motor neurons whose axons travel in the major nerves of the arm that withdraw the arm involuntarily by activating its flexor muscles and inhibiting the extensors.

Meanwhile, the spinothalamic tracts have been conducting pain impulses to the thalamus. After entering the spinal cord, collateral branches of the afferent axons cross into the left side of the next higher segment, beginning at C5. From there they ascend to the medulla oblongata and the thalamus on that side. Synapses in the thalamus continue the pathway to third order neurons that enter the cerebral cortex, where impulses inform the brain of injury, and establish an emergency. To enable Terry to examine his thumb, the cortex fires impulses back down the corticospinal tracts in the spinal cord, which cross back to the right side at the medulla or cervical levels, to synapse with motor neurons in the anterior horn that raise the arm for inspection.

- Information from pain receptors in the right thumb enters and ascends the spinal cord through which spinal nerves? What roots? Which primary rami? What tracts?

DEVELOPMENT

Terry's intricate and rapid responses to pinching his thumb and the delicate mechanisms his spinal injury has disrupted both originate in early embryonic development. The spinal cord and spinal nerves develop from the neural plate and neural crest. The **neural plate** produces the cord and the efferent neurons of the spinal nerves, as well as the brain and efferent portions of the cranial nerves. Neural crest cells produce the sensory neurons of the posterior root ganglion and autonomic ganglia.

The margins of the neural plate fold upward toward each other, as shown in Figure 18.28A, as the neural plate elongates and temporarily forms a **neural groove** and **neural folds** (Figure 18.28B) in the midline of the embryo. The neural groove becomes a **neural tube** (Figure 18.28C) when the margins of the neural folds meet and fuse. The anterior half of this tube becomes the brain, and the spinal cord develops from the remaining half. The spinal canal and ventricles of the brain develop from the cavity captured by the neural folds when the open ends close, seen in Figure 18.28D.

Developing neurons and neuroglial cells divide and thus expand the lateral walls of the neural tube, and the shallow **sulcus limitans** (LIM-ih-TANS) divides the wall longitudinally into **alar** and **basal plates** (Figure 18.28F) that extend the entire length of the tube. The alar plate receives afferent sensory axons through the posterior root of the spinal nerve, and the basal plate sends efferent axons out through the anterior roots. The roof and floor of the neural tube remain narrow and become the **roof** and **floor plates** of the spinal cord, passing white and gray commissures between each side. Later development dramatically expands the alar and basal plates (Figure 18.28G), and the thin roof and floor plates become the isthmus.

Afferent neurons and spinal ganglia develop from the neural crest, a strip of tissue that has accompanied each margin of the neural plate when it folds. While neurons and glia develop in the neural tube, certain neural crest cells migrate downward and laterally, as you can see in Figure 18.28E and F, to form the posterior root ganglia of the spinal nerves. Cell bodies and glia cluster into the developing ganglia and send axons toward the skin and into the marginal layer of the alar plate in the spinal cord. Other neural crest cells move deeper into the body where they become neurons in the sympathetic ganglia of the autonomic nervous system. Figure 17.16 summarizes the derivatives of the neural crest.

The white and gray matter of the spinal cord develop from different cell layers in the neural tube (Figure 18.29). Cells that will become neurons and neuroglia begin proliferating while the neural tube forms. Known as **neuroblasts** and **glioblasts,** these are bipolar cells with their ends anchored on the inner and outer surfaces of the tube. Their cell nuclei take up different positions as they develop. Nuclei descend to the **ventricular** (innermost) layer when they divide, and they shuttle back up to the **intermediate layer** during interphase. The intermediate layer becomes the gray matter when the neuroblasts stop dividing and differentiate into neurons. Few cell bodies occupy the **marginal layer** (the outermost), which becomes the white matter when myelinated axons and glia invade it. Glioblasts of the **subventricular layer** between ventricular and intermediate layers also become part of the gray matter. Cells of the ventricular layer become ependymal cells that line the spinal canal.

- From what embryonic structure does the spinal cord develop?
- What is the developmental source of posterior root ganglia?
- Which cellular layer of the developing cord becomes gray matter?

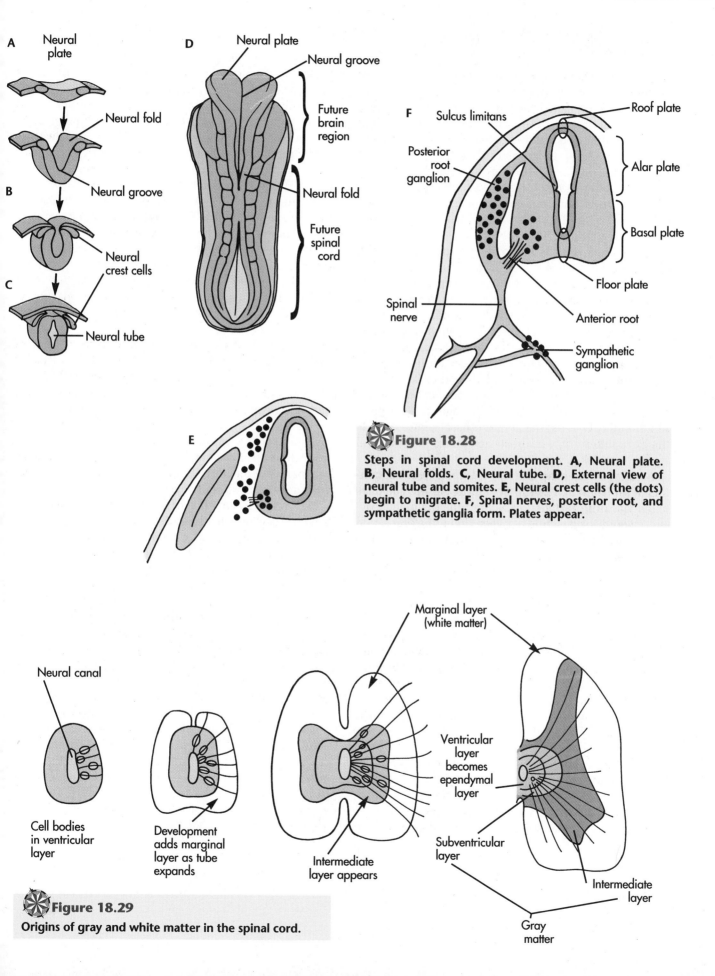

Figure 18.28

Steps in spinal cord development. **A,** Neural plate. **B,** Neural folds. **C,** Neural tube. **D,** External view of neural tube and somites. **E,** Neural crest cells (the dots) begin to migrate. **F,** Spinal nerves, posterior root, and sympathetic ganglia form. Plates appear.

Figure 18.29

Origins of gray and white matter in the spinal cord.

SUMMARY

The spinal cord and spinal nerves communicate between the brain and the body. Sensory information ascends the spinal cord to the brain, and impulses descend the cord to the spinal nerves. The spinal nerves conduct information between the body and the spinal cord. Afferent axons in these nerves conduct sensory information to the cord from dermatomes in the skin, skeletal muscles, and visceral organs. Axons of efferent neurons conduct motor impulses through spinal nerves to the skeletal muscles and to smooth muscles and glands. *(Page 529.)*

The spinal cord is an extension of the medulla oblongata of the brain, and it descends in the vertebral canal two thirds of the way down the vertebral column. The cord is bilaterally symmetrical; its left and right halves are connected by a narrow isthmus created by two longitudinal sulci. Other sulci mark the surface boundaries of tracts in the cord, and cervical and lumbar enlargements contain additional neurons for the limbs. Internally, cell bodies of neurons occupy the gray matter and myelinated axons ascend or descend funiculi of the white matter. Two great functional divisions, the alar and basal plates, divide the cord anteriorly and posteriorly into sensory and motor domains. *(Page 529.)*

Meninges cushion the spinal cord against contact with the walls of the vertebral canal. The dura mater lines the vertebral canal and the pia mater covers the cord itself, while between them, the arachnoid mater and cerebrospinal fluid cushion the cord. The spinal nerves, the denticulate ligaments, and the posterior arachnoid septum tether the cord in the vertebral canal. Arterial circulation enters from the anterior and posterior sides of the cord. *(Page 532.)*

Thirty-one pairs of spinal nerves represent the segments of the spinal cord. Afferent sensory impulses enter the posterior roots, and efferent motor impulses leave the anterior roots. Both roots merge to form spinal nerves as they pass through each intervertebral foramen. Spinal nerves conduct afferent and efferent impulses through two rami that innervate the back, trunk, limbs, and viscera. Spinal nerves emerge between vertebrae. *(Page 534.)*

Plexuses outside the spinal cord recombine and distribute axons from the anterior primary rami of spinal nerves to the flexor and extensor muscles of the limbs. Nerve fibers are redistributed along functional lines by successive roots, trunks, divisions, and cords, emerging as the peripheral nerves of the extremities. *(Page 536.)*

Axons ascend and descend the white matter of the spinal cord in funiculi and tracts that direct these fibers to particular destinations. Funiculi and tracts are comparable to fasciculi of spinal nerves. Tracts are usually named for their origin and destination. Spinothalamic tracts ascend the cord to the thalamus, and corticospinal tracts descend from the cerebral cortex to the spinal cord. *(Page 544.)*

Spinal nuclei are clusters of nerve cell bodies in the gray matter of the cord, whose axons innervate particular targets. Nuclei are integrative and relay centers in which synapses relay signals along pathways from presynaptic neurons to postsynaptic neurons. The nuclei (motor pools) for the various muscles of the limbs occupy particular locations in the cervical and lumbar enlargements of the spinal cord. *(Page 550.)*

Two levels of communication and coordination occur in the spinal cord. Reflex arcs automatically link sensory stimuli to motor responses within segments. Fasciculi proprii connect adjacent segments, and ascending and descending tracts connect these segments of the cord with the brain for higher, voluntary control of movement, responding to information from the muscles, skin, eyes, and ears. Spinal injury can paralyze an individual by depriving him of this voluntary control. *(Page 551.)*

Afferent sensory impulses flow from pain receptors in the hand to the brain via the median nerve, brachial plexus, posterior roots of cervical nerves, and the spinothalamic tract. Efferent impulses descend the corticospinal tracts to cervical motor nuclei, anterior roots, anterior primary rami, brachial plexus, and nerves of the arm to reach the muscles. *(Page 552.)*

The spinal cord and spinal nerves develop from the neural plate and neural crest. The gray matter originates from the intermediate layer of the neural tube, and the funiculi and tracts of the white matter develop from myelinated axons of the marginal layer. Axons of motor neurons grow outward from nuclei in the basal plate into the anterior root of the spinal nerve, and they reach skeletal muscle targets in the trunk and limbs. Afferent neurons arise from neural crest cells that migrate into the posterior root ganglia and sympathetic ganglia. These neurons extend axons into the ascending tracts of the spinal cord and outward to sensory receptors in the skin, muscles, and viscera. *(Page 552.)*

STUDY QUESTIONS

1. Trace the path of impulses from a stubbed right toe to the brain. **1, 12**
2. What effects would you expect from an injury that severs the spinal cord at the fifth cervical vertebra? **2, 3**
3. Diagram a transverse section through the spinal cord at the level of the cervical nerves. Show the various regions of gray matter and the funiculi of white matter, along with the grooves and sulci that mark the surface of this region of the cord. **4**
4. What do the locations of the basal plate, cell bodies of motor neurons, and the anterior funiculus of the spinal cord have in common? **4**
5. What are the individual functions of the dura mater, arachnoid, and pia mater? **5**
6. Which vessels supply the anterior and posterior halves of the spinal cord? **6**
7. How many spinal nerves and vertebrae are there? What is the anatomical relationship between them? **7**
8. Describe several important differences between spinal nerves and spinal tracts. **8, 11**
9. Trace the path of axons that will comprise the radial nerve through the brachial plexus (identifying roots, trunks, divisions, and cords). **9**
10. Outline the path of the sciatic nerve from the sacral plexus to the ends of its branches in the leg. Which segments supply axons to this nerve? Outline similar pathways for other nerves, such as the musculocutaneous nerve of the arm and the femoral nerve of the thigh. **10**
11. Which tract is affected in a patient who has lost body sense and touch in the entire left lower limb? What level is suspected in this tract, and which side of the spinal cord has sustained damage? **11**
12. Show the location of the ascending and descending tracts that pass through the cord in question 11. **11**
13. Which structures in the spinal cord and spinal nerves derive from the neural tube, and which derive from the neural crest? **13**
14. Which portions of the neural tube give rise to the brain and spinal cord? **13**

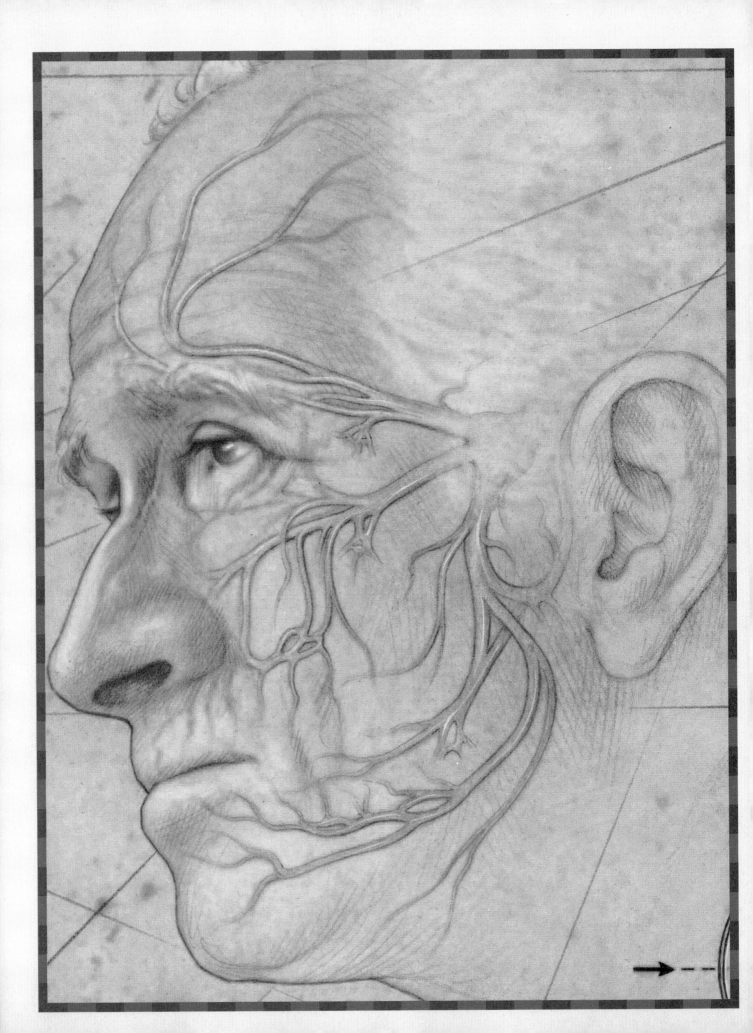

CHAPTER 19

Brain and Cranial Nerves

OBJECTIVES

After studying this chapter, you should be able to do the following:

1. Name and describe the location and functions of the divisions of the brain.
2. Relate these regions to their precursors in the neural tube.
3. Locate the ventricles of the brain in their appropriate regions.
4. Show where different portions of the brain reside in the cranial cavity.
5. Describe the principal components and functions of the medulla, and name its cranial nerves.
6. Describe the components of the pons and the cerebellum.
7. Describe the functions of the midbrain and the location of its major tracts and nuclei.
8. Describe the location and functions of the thalamus and hypothalamus.
9. Draw a map of the cortical centers of the cerebral hemispheres and describe their functions.
10. List the types of neurons that communicate among the various parts of the brain.
11. Trace the flow of stimuli through major ascending and descending tracts of the brain and spinal cord, from their origins to their destinations.
12. Relate the functions of basal ganglia, the reticular formation, and the limbic system to the anatomy of each structure.
13. Name the three layers of the cranial meninges; how do these layers support and protect the brain?
14. Name the cranial nerves and their target structures, and describe their functions.

 **VOCABULARY TOOLBOX**

Regions of the brain.

encephalon	encephal-, *G* the brain.
	tel-, *G* the end. Telencephalon is the endbrain.
	di-, *G* across. Diencephalon is the "between" brain.
	mes-, *G* middle. Mesencephalon is the middle brain.
	met-, *G* after. The metencephalon is the "after" brain.
	myel-, *G* the spinal cord. The myelencephalon resembles the spinal cord.

Structures and functions.

cerebellum	cerebell-, *L* little brain.	**falx**	falx, *L* a sickle.
cerebrum	cereb-, *L* the brain.	**lemniscus, lemnisci**	lemnisc-, *L* a ribbon. The medial lemniscus is a ribbon of neural tissue.
chiasm	chias-, *G* cross, crosswise. Optic fibers cross at the optic chiasm.		
		peduncle, peduncles	peduncul-, *L* a little foot.
choroid	chor-, *G* a place.	**pons**	pons, *L* a bridge.
ependyma	ependyma, *G* a tunic. The ependyma lines the ventricles.	**tectum**	tectum, *L* a roof.
		thalamus	thalam-, *G* an inner room or chamber.
		vermis	verm-, *L* a worm.

PERSPECTIVE

"How can I ever make sense of all those lobes and fissures and bumps? They just sit there. If they moved or pumped blood like the heart, at least I could see what is happening."

Brain anatomy is formidable but manageable, especially when you relate the neural traffic inside to the anatomical landmarks on its surface. The key to brain anatomy is visualizing the traffic that flows through its neurons, just as you would follow blood through the heart. The key to brain function is recognizing that although individual cells have similar functions, it is the connections between neurons that assemble pathways and dictate how information is processed. When you learn something, you have subtly altered the connections in your own brain. When you study this chapter, you will find your own brain making itself understandable. Among other things, you will be able to trace pathways and structures that control motion and speech, and to understand disorders of the brain.

The brain uses the language of nerve impulses to communicate with the body through the spinal cord and cranial nerves. Communication is a two-way street. It requires that the brain react to impulses it receives by sending impulses of its own. The brain gets its general information about touch, pain, temperature, and body sense from the trunk and extremities by way of the spinal cord, and from the head through the cranial nerves. Some cranial nerves also inform the brain of visceral functions, such as whether blood pressure is high or low and when it is time to eat. The cranial nerves also supply the all-important special senses of smell, vision, hearing, balance, and taste. These ever-changing stimuli cause streams of neural traffic to flow in all regions of the brain. Some traffic goes into memories, some makes decisions and plans, and some adds our emotions to the mixture we call personality. The results are our behavior: streams of impulses that return from the brain to the cranial nerves and spinal cord in control of facial expressions, speech, body language, and our daily activities.

GUIDE TO THE BRAIN

Think of the brain as a highly modified spinal cord whose segments have undertaken the various specialized functions of the head. At first glance the brain shown in Figure 19.1 resembles a mushroom rather than a spinal cord. The massive cerebral hemispheres and cerebellum obscure nearly all other structures except the brainstem. The brainstem resembles the spinal cord most closely. Ascending and descending pathways pass through the brainstem to the cerebral hemispheres and the spinal cord. Cranial nerves enter and exit the brainstem, much as spinal nerves communicate with the spinal cord. The cerebral hemispheres and cerebellum represent highly specialized outgrowths of the neural tube from which the brain and spinal cord both develop.

BRAIN AND NEURAL TUBE

The brain develops from the neural tube that expands and folds to form five major regions of the brain, as you can see in Figure 19.2. The cerebral hemispheres develop from the **telencephalon** at the anterior end of the neural tube. The **diencephalon** becomes the thalamus and hypothalamus and receives the optic nerves from the retinas. The brainstem begins behind the diencephalon. The midbrain, or **mesencephalon,** remains tubular and bends forward to accommodate the cranial cavity. The brainstem continues as the **metencephalon,** to become the pons and cerebellum. Finally, the myelencephalon becomes the medulla that connects the brainstem and the spinal cord.

All these embellishments obscure the underlying **segmentation** of the central nervous system, most evident in the spinal nerves and segments of the spinal cord. The cranial nerves show that the brainstem also is segmented, because these nerves closely resemble spinal nerves. The complexity of the cerebral hemispheres has scrambled whatever segmentation existed, however, and little if any evidence of segments remains. The cerebral hemispheres and diencephalon differ from the spinal cord in part because they lack an obvious basal plate, the spinal cord region associated with motor activities.

FOUR VENTRICLES

The brain contains four cavities, called **ventricles,** filled with **cerebrospinal fluid** that supports the brain from the inside and helps maintain the constant ionic conditions the brain requires to function. As Figure 19.3 shows, the ventricles are continuous with the central canal of the spinal cord because all these spaces develop from the neural tube and neural canal. All are lined with **ependymal cells,** a glial cell epithelium. All the ventricles also contain a **choroid plexus** (KORE-oid) that produces cerebrospinal fluid, though it is most elaborate in the ventricles. Ependymal cells are neuroglia cells, described in Chapter 17.

Moving upward from the medulla, the central canal of the spinal cord expands into the **fourth ventricle** beneath the cerebellum. Two recesses project laterally from the center of the ventricle, and a third recess leads dorsally to the base of the cerebellum. The fourth ventricle reaches across the pons, then narrows into the **cerebral aqueduct** that curves forward into the midbrain and enters the diencephalon, where it opens into a deep, narrow vault, the **third ventricle.** Suprapineal and infundibular recesses extend this chamber above and below. The

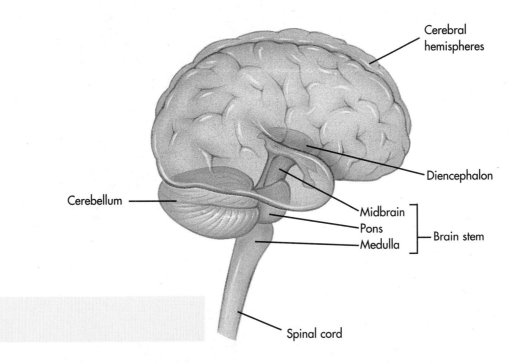

Cerebral hemispheres

Diencephalon

Cerebellum

Midbrain
Pons
Medulla

Brain stem

Spinal cord

Figure 19.1
The brain.

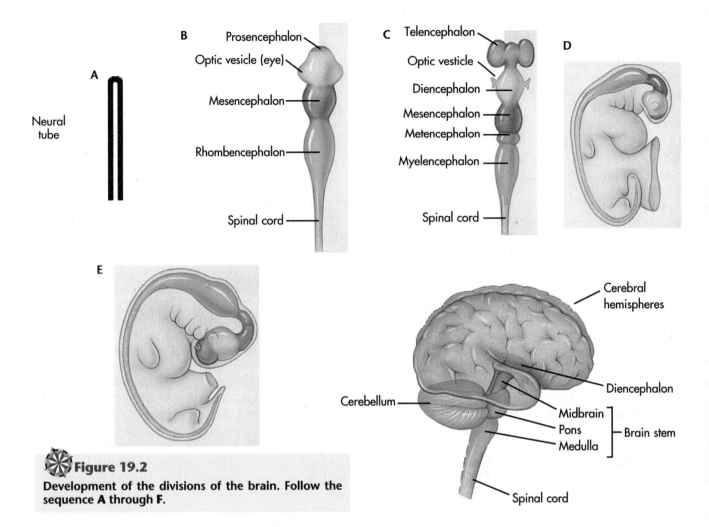

Neural tube

B
Prosencephalon
Optic vesicle (eye)
Mesencephalon
Rhombencephalon
Spinal cord

C
Telencephalon
Optic vesicle
Diencephalon
Mesencephalon
Metencephalon
Myelencephalon
Spinal cord

D

E

Cerebral hemispheres
Diencephalon
Cerebellum
Midbrain
Pons — Brain stem
Medulla
Spinal cord

✳ Figure 19.2
Development of the divisions of the brain. Follow the sequence A through F.

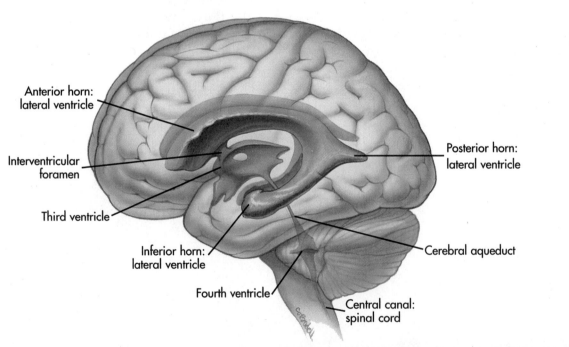

Anterior horn: lateral ventricle
Interventricular foramen
Third ventricle
Inferior horn: lateral ventricle
Fourth ventricle
Posterior horn: lateral ventricle
Cerebral aqueduct
Central canal: spinal cord

✳ Figure 19.3
Four ventricles of the brain.

third ventricle joins with the **lateral ventricles** of the cerebral hemispheres by way of the **interventricular foramina.** The lateral ventricles curl like ram's horns. The interventricular foramen enters the body of each ventricle, near the **anterior cornu** (KORnew). The body curls back, gives off a **posterior cornu,** and continues downward and forward to end in the **inferior cornu.** The lateral ventricles are not numbered. It is too easy to forget whether number one is left or right.

CENTERS AND TRACTS

The brain communicates with the world through primary sensory and motor centers in the cerebral hemispheres. Sight, hearing, balance, smell, and taste derive from sensory centers, illustrated in Figure 19.4, that receive impulses through the cranial nerves from the eyes, ears, nose, and tongue. The **somatosensory center** also receives information about touch, temperature, pain, and body position (body sense) from sense organs in the skin, muscles, and joints. **Association areas** elsewhere in the cerebral hemispheres interpret this information. When these areas call for action, the **somatomotor center** controls voluntary motion, and **speech centers** put thoughts into spoken words.

Tracts reach the cerebral hemispheres through the brainstem, carrying information from the spinal cord and cranial nerves. Other tracts conduct motor commands down the brainstem to the muscles and glands. Not all of these tracts go directly up and down. Most stop at **nuclei** in the cerebellum or diencephalon, where their signals are coordinated with others before going further. All such neurons are called **projection neurons** because, like a projectile, they are aimed at a target.

Commissural axons communicate between opposite sides of the brain, allowing each hemisphere to

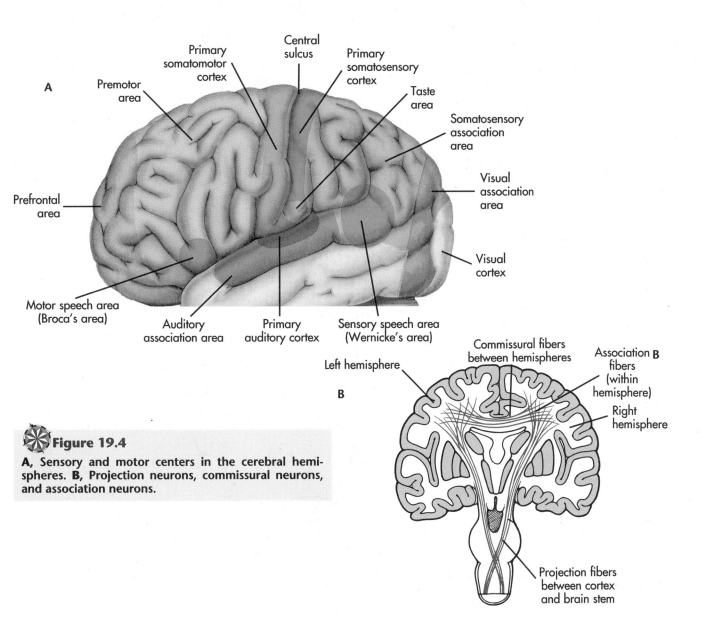

Figure 19.4

A, Sensory and motor centers in the cerebral hemispheres. **B,** Projection neurons, commissural neurons, and association neurons.

coordinate with the other. **Association neurons** connect different areas in the same hemisphere. When you speak, all three types of neurons are needed to formulate your thoughts into speech.

- Name the three divisions of the brain in sequence, beginning anteriorly.
- Which regions of the brain does each division contain?
- What do projection neurons do in the cerebral hemispheres?
- Which divisions contain the ventricles of the brain?

BRAIN IN THE SKULL

The brain is approximately 1½ kg (3.3 lbs) of soft, delicate tissue that lacks the support from collagen fibers that most other tissues have. Fresh brains deform in the autopsy pan, but a bowl of fluid supports them nicely. This is precisely why the brain must float in cerebrospinal fluid inside the skull. Cushioned by the meninges and a thin layer of cerebrospinal fluid, the brain fits snugly (Figure 19.5A)

within the cranial cavity (see Meninges, this chapter, for details). Figure 19.5B shows that the frontal lobes of the cerebrum rest in the **anterior cranial fossa** of the skull, the **middle fossa** supports the temporal lobes, and the cerebellum fills the **posterior cranial fossa.** The brainstem rests on the sphenoid and occipital bones, and the medulla fills the **foramen magnum.** Cranial nerves exit through **foramina** in the floor of the cranial cavity. To recall a few of them from Chapter 6, the optic nerve enters the diencephalon through the optic canal; the trigeminal nerve reaches the face by way of the foramen rotundum and foramen ovale, and the vagus nerve leaves the medulla and exits through the jugular foramen. Figure 19.5C allows you to match the brain's underside with the floor of the cranial cavity. The roof of the cranial cavity is comparatively simple. The cerebrum marks it only with impressions of cranial vessels and with sulci and gyri. The molding of the skull to the cerebrum gave rise to the discredited "science" of phrenology, or "reading" personalities and characteristics from the bumps on people's heads.

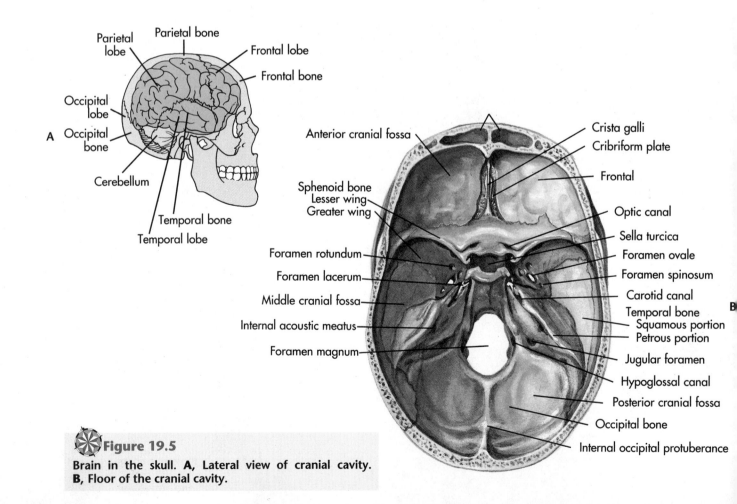

✸ Figure 19.5

Brain in the skull. A, Lateral view of cranial cavity. B, Floor of the cranial cavity.

MEDULLA

The medulla oblongata is a major reflex center and the site where great tracts cross to the opposite sides of the brain. Of all brain regions the **medulla oblongata** is the most like the spinal cord. It shares the spinal cord's roughly cylindrical shape and its segmentation (Figure 19.6A and B). **Anterior and posterior median sulci** divide the medulla left and right as they do in the cord, and **cranial nerves** appear at intervals that resemble spinal segments. There are important differences, however. The rostral end of the medulla (closest to the nose) enlarges to accommodate additional tracts and nuclei of the cranial nerves, and the H-shaped pattern of spinal gray matter disperses. Evidence of these changes are the **pyramids** on the anterior surface, and the **inferior olives** lateral to them. On the posterior side the **posterior columns** of the spinal cord give way to the thin **roof of the fourth ventricle,** and the **inferior cerebellar peduncles** at each side carry nerve fibers to the cerebellum.

The medulla communicates with its surroundings through five pairs of cranial nerves that are the basis of medullary reflexes. Beginning rostrally (Figure 19.5C), the ear communicates impulses concerning sound and balance through the **vestibulocochlear nerve** (ves-TIB-you-low-KOKE-lee-ar). Taste and salivation are controlled next through the **glossopharyngeal nerve** (GLOSS-oh-fahr-IN-jee-al). The **vagus nerve** (VAY-gus) follows, carrying impulses that control swallowing and many visceral actions in the trunk. Nerve XI, the **accessory,** mediates pitch and tone of the larynx. Last and most caudal is the **hypoglossal nerve** (HIGH-po-GLOSS-al) that reaches upward to the muscles of the tongue that control speech, mastication, and swallowing. Because they supply the pharynx, neck, and trunk, the last four cranial nerves mediate the vomiting, hiccuping, swallowing, and gag reflexes, as well as other reflexes that regulate heart rate and blood pressure.

The tracts that conduct touch, body sense, and pain impulses and control the body's movements, all pass through the medulla on their way up and down the brainstem (Figure 19.6C). Many of these fibers cross in the medulla to the opposite side, so that the left side of the brain controls and senses the right side of the

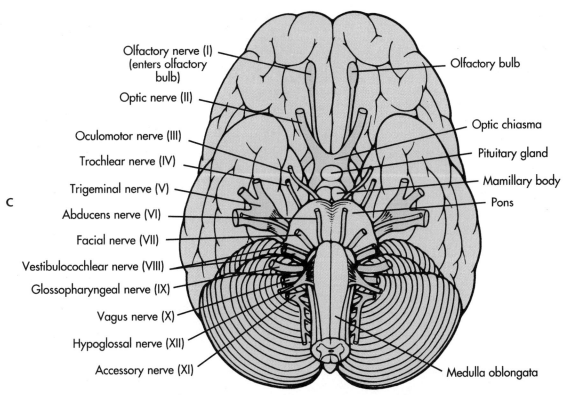

C

- Olfactory nerve (I) (enters olfactory bulb)
- Optic nerve (II)
- Oculomotor nerve (III)
- Trochlear nerve (IV)
- Trigeminal nerve (V)
- Abducens nerve (VI)
- Facial nerve (VII)
- Vestibulocochlear nerve (VIII)
- Glossopharyngeal nerve (IX)
- Vagus nerve (X)
- Hypoglossal nerve (XII)
- Accessory nerve (XI)

- Olfactory bulb
- Optic chiasma
- Pituitary gland
- Mamillary body
- Pons
- Medulla oblongata

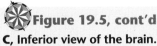

Figure 19.5, cont'd
C, Inferior view of the brain.

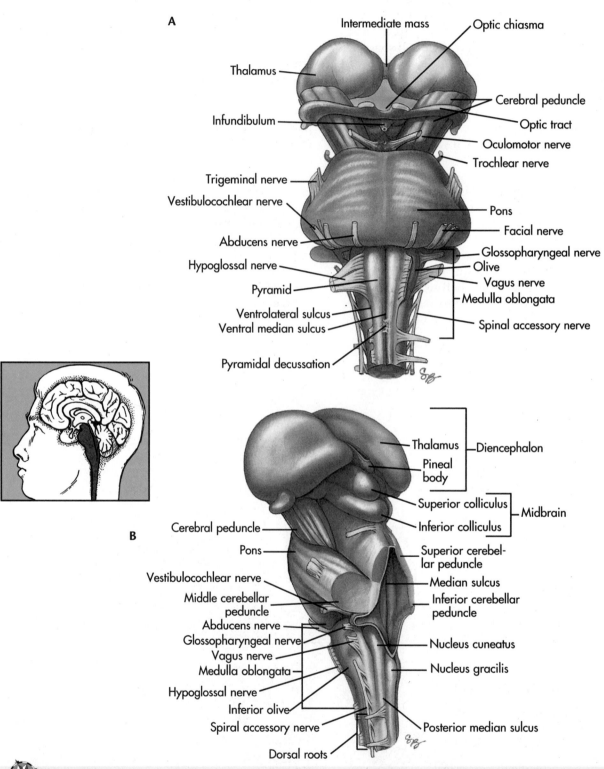

A

Intermediate mass

Optic chiasma

Thalamus

Cerebral peduncle

Infundibulum

Optic tract

Oculomotor nerve

Trochlear nerve

Trigeminal nerve

Vestibulocochlear nerve

Pons

Facial nerve

Abducens nerve

Glossopharyngeal nerve

Hypoglossal nerve

Olive

Vagus nerve

Pyramid

Medulla oblongata

Ventrolateral sulcus

Ventral median sulcus

Spinal accessory nerve

Pyramidal decussation

B

Thalamus — Diencephalon

Pineal body

Cerebral peduncle

Superior colliculus — Midbrain

Pons

Inferior colliculus

Vestibulocochlear nerve

Superior cerebellar peduncle

Middle cerebellar peduncle

Median sulcus

Abducens nerve

Inferior cerebellar peduncle

Glossopharyngeal nerve

Vagus nerve

Nucleus cuneatus

Medulla oblongata

Nucleus gracilis

Hypoglossal nerve

Inferior olive

Spiral accessory nerve

Posterior median sulcus

Dorsal roots

Figure 19.6

Medulla oblongata in the brainstem. A, Anterior. B, Posterolateral view.

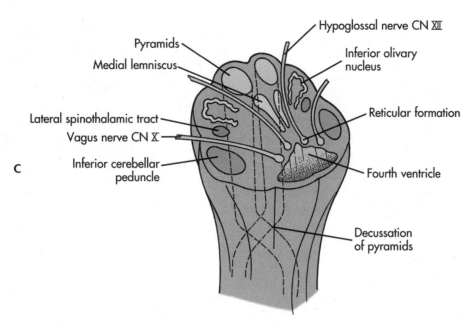

Pyramids

Medial lemniscus

Lateral spinothalamic tract

Vagus nerve CN X

Inferior cerebellar
peduncle

C

Hypoglossal nerve CN XII

Inferior olivary
nucleus

Reticular formation

Fourth ventricle

Decussation
of pyramids

**Figure 19.6, cont'd**

C, Internal structures. Find the pyramidal tracts (pyramids) and the lateral spinothalamic tract.

body, and vice versa. The pyramids carry the pyramidal tracts that conduct motor commands to the **lateral corticospinal tracts** of the spinal cord. The **pyramidal tracts** cross the anterior median sulcus to the opposite side of the medulla on their way to the lateral corticospinal tracts in the spinal cord. The **posterior columns** also cross in the medulla after synapsing in the cuneate and gracile nuclei. Axons from cells in these nuclei descend deep into the interior and cross to the opposite side, thereby forming the **medial lemnisci** (lem-NIH-skee) that continue up the brainstem to the diencephalon.

As tracts rise upward toward their destinations, the gray matter they grow through becomes dispersed into several groups of nuclei. The inferior olives (Figure 19.6B) mark the location of **olivary nuclei** that relay coordinating impulses between the spinal cord, cerebellum, and cerebral cortex. The **cuneate** and **gracile nuclei** connect the fibers of the posterior columns to those of the medial lemniscus. The cranial nerve nuclei connect to most other regions of the brain. Interspersed among these nuclei is the **reticular formation,** a network of nuclei and fibers throughout the brainstem that is important in the arousal response that summons you from sleep and directs your attention to whatever needs doing. Look for the reticular formation again in the pons and midbrain.

- Which structures carry ascending tracts through the medulla?
- Where do descending tracts reside?
- What reflex actions does the medulla control?

- Which cranial nerves are involved in these reflexes?
- Where is the reticular formation located?

PONS AND CEREBELLUM

The pons and cerebellum lie between the midbrain and medulla, as illustrated in Figure 19.7A and B. The pons bulges anteriorly, separated from the medulla by the inferior pontine sulcus. The pons represents the floor and basal plates of the spinal cord. Posteriorly the cerebellum fills the space between the cerebral hemispheres and the medulla. The cerebellum is an expansion of the roof plate and shares the fourth ventricle with the pons and medulla.

PONS

The **pons** (PAHNS) is aptly named. This Latin term for bridge means that neural traffic between the medulla and midbrain crosses the pons, flows to the cerebellum, and travels back to the brainstem. The anterior portion is known as the **base of the pons,** and the deeper tissue beneath the floor of the fourth ventricle is the **tegmentum** (teg-MEN-tum). As Figure 19.7C shows, corticospinal tracts and the medial lemniscus pass through the base of the pons, and the reticular formation continues toward the midbrain in the tegmentum. Most of the base of the pons is filled with fibers of the **pontocerebellar tract** (PON-toe-sair-eh-BELL-ar) that pass from the cerebral hemi-

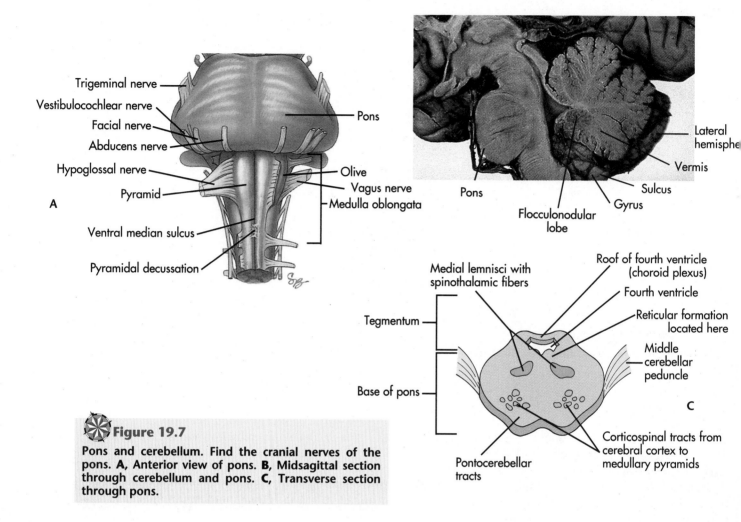

Figure 19.7

Pons and cerebellum. Find the cranial nerves of the pons. A, Anterior view of pons. **B,** Midsagittal section through cerebellum and pons. **C,** Transverse section through pons.

spheres into the base of the pons, where they cross the midline and pass laterally into the cerebellum.

Three cranial nerves enter the brainstem in the **inferior pontine sulcus** between the pons and medulla, and a fourth enters the superior lateral wall. Beginning at the caudal end (Figure 19.7A), the target nuclei of the **vestibulocochlear nerve** from the ear and semicircular canals are found in the pons, even though the nerve emerges from the medulla. Next is the **facial nerve,** carrying fibers to the facial muscles, tongue, and salivary glands. The **abducens** (ab-DOO-sens) is nerve VI, a small nerve that controls one muscle concerned with eyeball motion. The huge **trigeminal nerve** (V) relays touch, pain, and temperature from the face and mouth and also controls muscles of mastication. Burning the roof of your mouth with hot pizza is a trigeminal sensation, but the pizza's taste is facial.

CEREBELLUM

The cerebellum is responsible for fine motor control. In effect, it measures muscular progress and calculates adjustments that are needed for smooth motion. The **cerebellum** (SAIR-eh-BELL-um) lets you

press buttons on a phone and button clothing without fumbling. In contrast, someone with cerebellar damage has jerky handwriting because his neural circuitry cannot smoothly integrate the flow of impulses.

Two **cerebellar hemispheres** reach from the roof of the brainstem in Figure 19.8A, but they are joined into one continuous structure by the **vermis** (VER-miss). The **transverse sulci** that cross the vermis and hemispheres are responsible for the vermis' name; they give it the appearance of the segments of an earthworm. More important, these sulci greatly increase the area of gray cortex available for motor coordination. Viewed in sagittal section (Figure 19.8B), the cerebellum resembles a cauliflower because the cortex has branched out from the base. The medulla, pons, and cerebral hemispheres communicate with the cortex through tracts that radiate through the white matter. Three **cerebellar peduncles** (ped-UNK-lz) bring these tracts to the cerebellum, as shown in Figure 19.8C. The **superior cerebellar peduncle** connects to the midbrain; the **middle peduncle** connects to the pons (carrying the pontocerebellar tract), and the **inferior peduncle** connects to the medulla (carrying the spinocerebellar tracts).

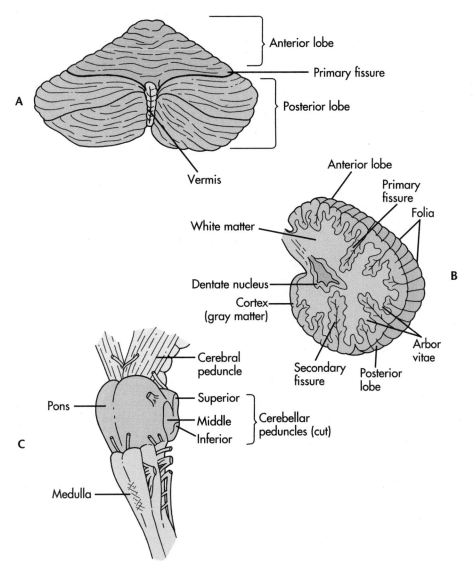

Figure 19.8

Cerebellum. **A,** Posterior view of cerebellar hemispheres and vermis. **B,** Midsagittal section through cerebellum. **C,** Cerebellar peduncles connect with the brainstem.

- Where are the pons and cerebellum located in the brainstem?
- What is the major function of the pons? Of the cerebellum?
- Which cranial nerves enter the pons?
- What structures carry ascending tracts and descending tracts through the pons?

MIDBRAIN

The midbrain is the rostral end of the brainstem; it connects the pons and cerebellum with the diencephalon, as shown in Figure 19.9. A relatively simple structure for so high a position in the brain-

stem, its thick walls (Figure 19.10) surround a narrow **cerebral aqueduct** that leads forward to the third ventricle in the diencephalon. Tracts ascend and descend these walls, and the reticular formation also continues forward through them. The midbrain sends cranial nerves to muscles of the eyes, and it governs ocular and auditory reflexes from the corpora quadrigemina, four bumps protruding from the roof of the midbrain.

The **tectum** (TEK-tum) of the midbrain consists of a roof over the cerebral aqueduct, shown in Figure 19.10, where the **corpora quadrigemina** (KORE-por-ah KWAD-rih-JEM-in-ah) are located. Below the aqueduct the **tegmentum** houses the reticular formation and other nuclei and tracts. Beneath the tegmentum

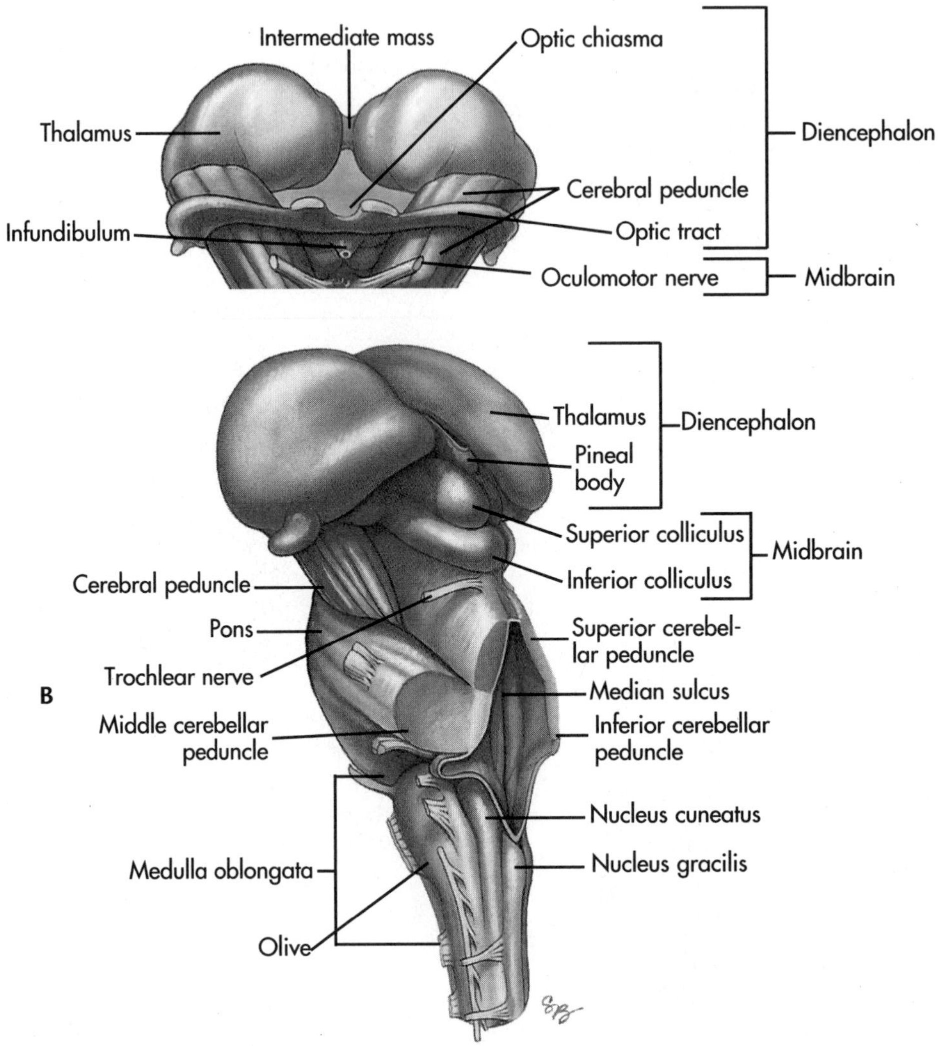

A

Intermediate mass

Optic chiasma

Thalamus

Diencephalon

Cerebral peduncle

Optic tract

Infundibulum

Oculomotor nerve — Midbrain

B

Thalamus — Diencephalon

Pineal body

Superior colliculus — Midbrain

Inferior colliculus

Cerebral peduncle

Pons

Trochlear nerve

Superior cerebellar peduncle

Median sulcus

Inferior cerebellar peduncle

Middle cerebellar peduncle

Nucleus cuneatus

Medulla oblongata

Nucleus gracilis

Olive

Figure 19.9

Midbrain. Find the cranial nerves. A, Anterior view. B, Dorsolateral view.

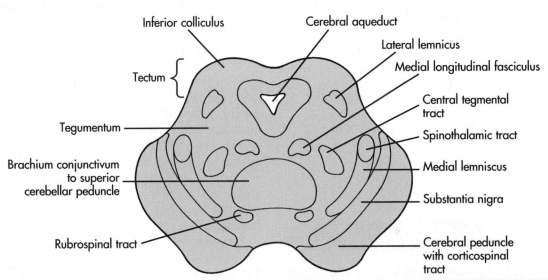

Inferior colliculus

Cerebral aqueduct

Lateral lemnicus

Medial longitudinal fasciculus

Tectum

Central tegmental tract

Spinothalamic tract

Tegumentum

Medial lemniscus

Brachium conjunctivum to superior cerebellar peduncle

Substantia nigra

Rubrospinal tract

Cerebral peduncle with corticospinal tract

Figure 19.10

Transverse section through midbrain. Find the spinothalamic and corticospinal tracts.

are two massive **cerebral peduncles,** one on each side, that carry corticospinal tracts from the cerebrum to the pons and on to the medulla and spinal cord. Two special nuclei—the **red nucleus** and the **substantia nigra** (SUB-stan-cha NY-gra)—are located between the peduncles and tegmentum. Both nuclei are concerned with motion; fibers link them to the cerebrum, diencephalon, and cerebellum, as well as to the spinal cord. The medial lemnisci continue to carry spinothalamic tracts forward to the diencephalon, but they are now located laterally, deep to the substantia nigra.

The midbrain controls eye motions through the oculomotor and trochlear nerves. Your ability to focus on an object while moving your head, or to aim your focus at different subjects, resides in the midbrain. The **oculomotor nerve** (Figure 19.9) orchestrates three of the rectus muscles and the inferior oblique muscles of the eyeball, while the **trochlear nerve** (TROKE-lee-ar) controls the superior oblique (see Extrinsic Muscles of the Eyeball, Chapter 10). These muscles move the eyeball in all directions except laterally, a direction controlled from the pons by the abducens nerve and the lateral rectus muscle. As you would expect, the nuclei of these nerves receive input from the cerebrum to transmit commands for voluntary motion, and with the auditory centers and the cerebellum, to direct your eyes toward the source of sudden sounds and to coordinate with head movements. The oculomotor and trochlear nerves also connect with the **superior colliculi** (KOL-ik-you-lee). These two mounds on the tectum (Figure 19.9) are the first two members of the corpora quadrigemina; they manage visual reflexes that focus the lens, dilate or constrict the pupils, and cause the eyes to converge on near subjects. The **inferior colliculi,** the other members of the corpora quadrigemina, have connections with the cerebellum and auditory pathways that help determine the location of voices or noises.

- Where is the midbrain located in the brainstem?
- What structures carry its ascending and descending tracts?
- What cranial nerves enter the midbrain, and what actions do they control?
- What are the roles of the corpora quadrigemina?

DIENCEPHALON

The diencephalon is the gateway to the cerebral hemispheres. Every entering tract communicates with nuclei in the diencephalon, allowing the diencephalon to control everything from emotional expressions to heart rate. In addition, all descending tracts exit this gate, although many depart without visiting a nucleus.

The diencephalon is difficult to examine because the cerebral hemispheres nearly surround it, as you can see in Figure 19.11A. The **pineal gland** (PINE-ee-al) protrudes posteriorly from the roof of the diencephalon, and the **anterior choroid plexus** extends beneath the cerebrum as the thin roof of the **third ventricle.** This chamber reaches forward to join the lateral ventricles of the cerebrum above and extends downward as a narrow cavity between thick lateral walls to the **infundibulum** (in-fun-DIB-you-lum), another projection from which the pituitary gland is suspended in the sella turcica of the sphenoid bone (see Chapter 6, Axial Skeleton). A thin **lamina terminalis** covers the anterior end of the diencephalon with commissural fibers that cross between the cerebral hemispheres. The **optic nerve** is the only cranial nerve that enters the diencephalon, and its chiasm (KI-asm) where some optic fibers cross, is easy to see in front of the infundibulum. Two small **mamillary bodies** bulge from the posterior sides of the infundibulum.

Figure 19.11B shows that the walls of the third ventricle in the diencephalon form the thalamus (THAL-a-mus), a large collection of functionally separate nuclei and the great relay center for sensory information from the body and face. Below, the **hypothalamus** coordinates visceral reflexes and neurosecretion from the pituitary gland. The cerebral peduncles of the midbrain continue around these structures as the **internal capsules,** an apparent misnomer until you realize that they lead inside the cerebral hemispheres.

THALAMUS

The thalamus relays sensory information (except smell) to primary sensory and association centers of the cerebral cortex. It also communicates with the somatomotor center and other motor areas in the basal ganglia, midbrain, and cerebellum. Approximately a dozen nuclei cluster into four groups on each side of the thalamus, as shown in Figure 19.12. The **ventral portion of the lateral nuclear group** relays signals from ascending tracts to the somatosensory center in the cerebral hemispheres and to other thalamic nuclei. Fibers of the spinothalamic tract form synapses here with projection neurons that continue to the somatosensory center. The **dorsal portion of the lateral nuclear group** appears to have a role in speech; it relays signals from other thalamic nuclei to the auditory, visual, and speech association areas. The **medial nuclear group** is involved with emotional behavior and memory through communications with the somatomotor center and prefrontal cortex. At the rear of the thalamus, the **intralaminar** and **reticular nuclei** help arouse the cerebral cortex by relaying motor and sensory impulses throughout the thalamus and cerebrum. Protruding from the lateral surfaces, the **lateral geniculate bodies** contain

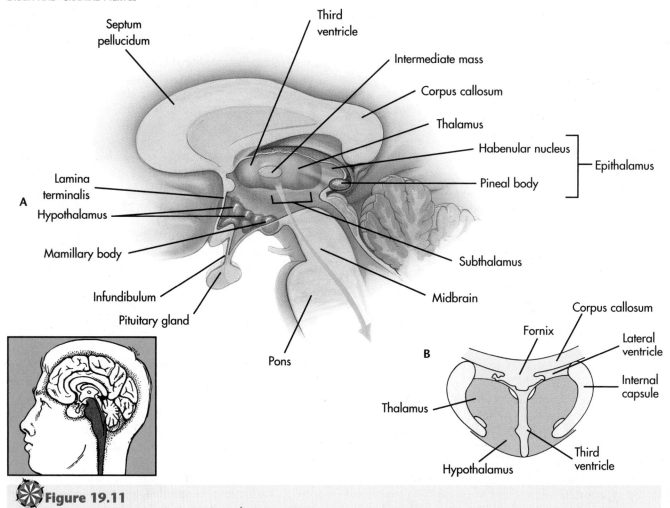

A

- Septum pellucidum
- Third ventricle
- Intermediate mass
- Corpus callosum
- Thalamus
- Habenular nucleus ⎤
- Pineal body ⎦ — Epithalamus
- Lamina terminalis
- Hypothalamus
- Mamillary body
- Infundibulum
- Pituitary gland
- Subthalamus
- Midbrain
- Pons

B

- Fornix
- Corpus callosum
- Lateral ventricle
- Internal capsule
- Thalamus
- Hypothalamus
- Third ventricle

Figure 19.11

Diencephalon. A, Midsagittal section through the third ventricle. B, Transverse section. Find the internal capsule.

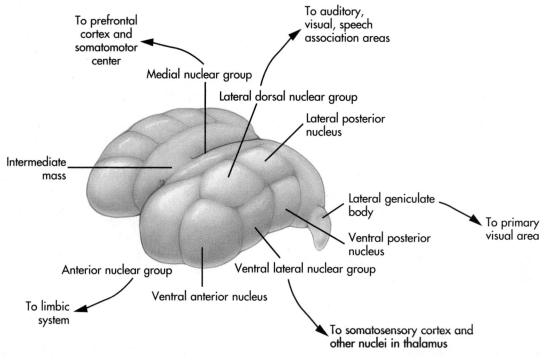

- To prefrontal cortex and somatomotor center
- To auditory, visual, speech association areas
- Medial nuclear group
- Lateral dorsal nuclear group
- Lateral posterior nucleus
- Intermediate mass
- Lateral geniculate body
- To primary visual area
- Ventral posterior nucleus
- Ventral lateral nuclear group
- Anterior nuclear group
- To limbic system
- Ventral anterior nucleus
- To somatosensory cortex and other nuclei in thalamus

Figure 19.12

Nuclei of the thalamus.

cell bodies that receive synapses from fibers of the optic nerve; these cells direct their own axons to the visual cortex in the cerebrum. The **anterior nuclear group** at the opposite end of the thalamus is part of the limbic system and is concerned with emotion and memory (see Limbic System, this chapter).

HYPOTHALAMUS

The **hypothalamus, the inferior portion of the diencephalon, coordinates neurosecretion and visceral reflexes** through its own nuclei, illustrated by Figure 19.13. The hypothalamus is an important center of the autonomic nervous system (see Chapter 21), as well as for all homeostatic functions. Named for their location above the optic chiasm, the **suprachiasmatic nuclei** (SOO-pra-KI-as-MAH-tik) are concerned with neurosecretion in the infundibulum, where their axons release oxytocin (causing uterine contraction) and vasopressin (vasoconstriction and diuresis; also known as antidiuretic hormone) into the posterior pituitary. The suprachiasmatic nuclei also mediate daily sleep cycles. When nuclei from mice adapted to long days (25 hours) are transplanted to mice adapted to shorter days (22 hours), the recipients become "long-day" mice. The hypothalamus also regulates the anterior pituitary gland through several releasing hormones that cause the gland to release gonadotropins and adrenocorticotropic hormone (ACTH).

As you will learn in Chapter 21, the hypothalamus is considered a main center of the autonomic nervous system. This system controls the viscera through antagonistic influences from its sympathetic and parasympathetic divisions. One division may accelerate functions and the other inhibit them. The hypothalamus is a microcosm of this system because its anterior and posterior portions elicit opposite responses to heat and other stimuli. The anterior portions stimulate sweating and dilate vessels when body temperature rises; when it falls, the posterior portions trigger shivering and vascular constriction and inhibit sweating. The **ventromedial nuclei** contain a satiety center; lesions to this area stimulate excessive eating, but lesions to the lateral hypothalamus abolish all desire for food. Similarly the lateral hypothalamus stimulates thirst. The ventromedial nucleus elicits rage, poignancy, and fear. Responding to life-threatening crises, the hypothalamus causes blood pressure to rise, hairs to stand on end, the pulse to race, and may even trigger defecation.

- Why is the diencephalon called the gateway to the cerebral hemispheres?
- What are the roles of the thalamus and hypothalamus, and where are they located?
- Which ventricle is found in the diencephalon?

CEREBRAL HEMISPHERES

The **cerebrum is the largest part of the brain.** Its neural traffic is responsible for our memories, our thoughts, and in short, much of our personalities.

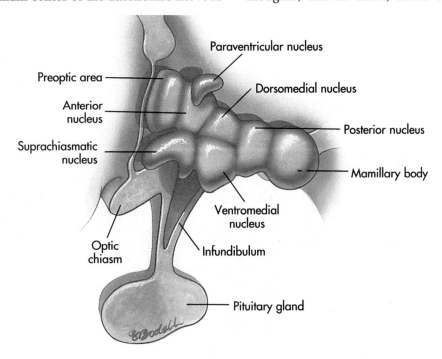

Figure 19.13
Nuclei of the hypothalamus.

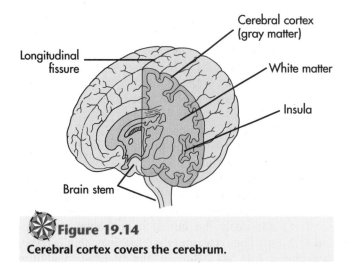

Figure 19.14
Cerebral cortex covers the cerebrum.

Most of these intricate connections reside in the **cerebral cortex,** illustrated in Figure 19.14, a thin layer of gray matter about 2 mm thick that covers the exterior of the cerebrum. The **cerebrum** (SER-ee-brum), has grown upward and posteriorly from the alar plates of the diencephalon, leaving only small areas of cortex attached to the base while the rest expands and folds within the cranial cavity. The result is a pair of cerebral hemispheres with **sulci** and **gyri,** grooves and folds that increase the surface area of the cortex to about 2400 cm², about the same area as the seat cushion of an easy chair. Most of this area is folded deep in the sulci. Large nuclei, the **basal ganglia,** lie deep in

the cerebral hemispheres, below the lateral ventricles and next to the thalamus.

FISSURES, LOBES, SULCI, AND GYRI

The longitudinal fissure divides the cerebrum into left and right hemispheres, also shown in Figure 19.14. Each hemisphere has four lobes—**frontal, parietal, occipital,** and **temporal**—that for the most part underlie the corresponding bones of the skull (Figure 19.15). Each lobe is folded in characteristic patterns of sulci and gyri that identify the locations of sensory centers and association areas in the cerebral cortex. The **lateral sulcus,** the largest and deepest, separates the temporal lobe from the frontal and parietal lobes. When this sulcus is gently spread open, the folds of the **insula** (IN-soo-lah), cortical tissue that remains deep in the base of the cerebrum, can be seen. The **central sulcus** is the boundary between the frontal and parietal lobes. On either side of this sulcus the **precentral** and **postcentral gyri** mark the positions of the primary somatomotor and somatosensory areas, respectively, of voluntary muscle control. Not all the important sulci are as large. Because no convenient sulcus divides the occipital lobe from its neighbors, anatomists and surgeons use the **parieto-occipital sulcus** (look carefully in the figure) together with the **preoccipital notch,** to mark the arbitrary boundary of the occipital lobe on the lateral surface of the cerebral hemispheres.

Some sulci and gyri extend onto the medial surface of the cerebral hemispheres, as Figure 19.16 shows. Three notable sulci are the parieto-occipital sulcus,

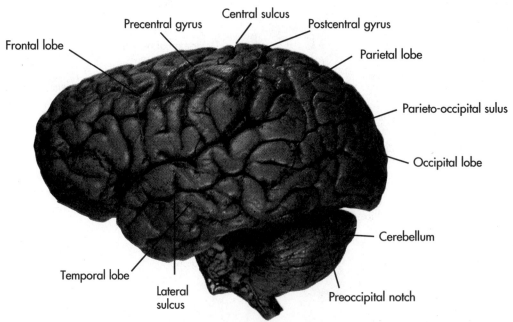

Figure 19.15
Lobes, sulci, and gyri of the left cerebral hemisphere.

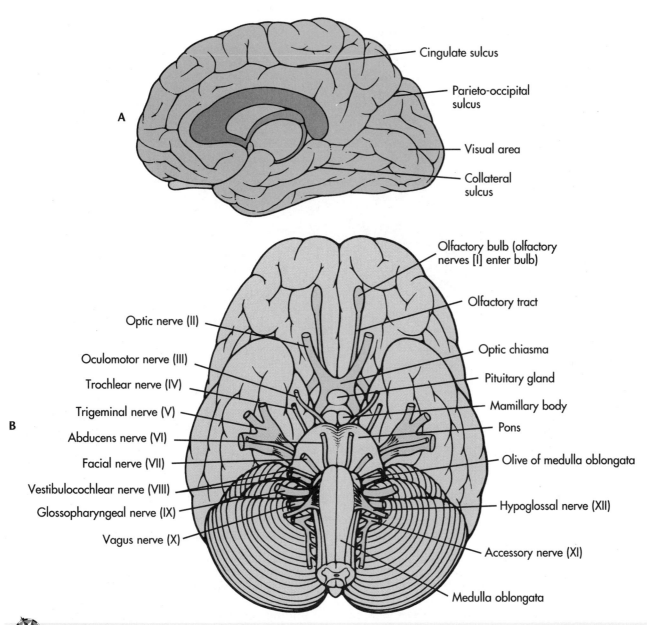

A

Cingulate sulcus

Parieto-occipital sulcus

Visual area

Collateral sulcus

Olfactory bulb (olfactory nerves [I] enter bulb)

Olfactory tract

Optic nerve (II)

Oculomotor nerve (III)

Trochlear nerve (IV)

Trigeminal nerve (V)

B

Abducens nerve (VI)

Facial nerve (VII)

Vestibulocochlear nerve (VIII)

Glossopharyngeal nerve (IX)

Vagus nerve (X)

Optic chiasma

Pituitary gland

Mamillary body

Pons

Olive of medulla oblongata

Hypoglossal nerve (XII)

Accessory nerve (XI)

Medulla oblongata

Figure 19.16

A and **B,** Medial and ventral views of cerebral hemispheres.

the cingulate sulcus, and the collateral sulcus. The **cingulate sulcus** (SIN-gyou-late) marks the cingulate gyrus in the frontal and parietal lobes. On the temporal lobe, the **collateral sulcus** identifies the parahippocampal gyrus that will be discussed as part of the limbic system.

CORTICAL AREAS

Because the cortex has so many landmarks on its surface, early neuroanatomists attempted to match them with particular personality traits. They failed because the cortex is not organized this way (Figure 19.17). In 1909 Korbinian Brodmann mapped the cortex according to its cellular organization (see Cortical Microanatomy, this chapter). His **Brodmann areas** represent different cortical structures and still are used as a reference map, even though many of them do not correspond with particular functions.

Following the lead of Wilder Penfield, surgeons map the cortex by stimulating it during neurosurgery. Penfield showed, for example, that weak electrical stimulation of the visual area causes the patient to "see" flashes of light, even though his eyes may be closed. Stimulating the motor area causes fingers to move or the nose to twitch, depending on which part is excited. Stimulation in the somatosensory area may cause the patient to report pain in the shoulder or the impression that a finger has moved.

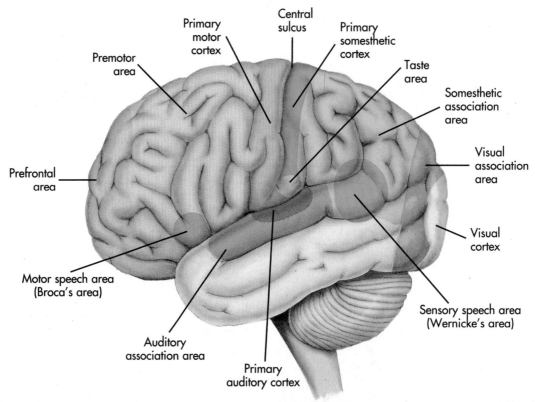

Premotor
area

Primary
motor
cortex

Central
sulcus

Primary
somesthetic
cortex

Taste
area

Somesthetic
association
area

Prefrontal
area

Visual
association
area

Visual
cortex

Motor speech area
(Broca's area)

Sensory speech area
(Wernicke's area)

Auditory
association area

Primary
auditory cortex

Figure 19.17
Primary motor and sensory areas of cerebral hemispheres.

The cortex contains several primary sensory and motor areas that are linked together by association areas. Exactly where these areas are located, with respect to sulci, normally differs among individuals without adverse effects. Using Figure 19.17, find the **primary visual cortex** at Brodmann's area 17 in the occipital lobe of each hemisphere; both receive input through the visual pathway, splitting the image between hemispheres. The right cortex receives from the right half of each retina, and the left halves of the retinas send to the left visual cortex. The **visual association area,** surrounding the primary visual centers, helps join the halves into a single image.

Sound is perceived in the **auditory cortex** of the temporal lobe (Brodmann's areas 41 and 42). Each center receives primarily from the opposite ear because most fibers from the vestibulocochlear nerve cross to the opposite side before reaching the auditory center. An **auditory association area** nearby (Brodmann 22) interprets speech. Equilibrium is sensed near the auditory center, as you might expect since the vestibulocochlear nerve carries both auditory and vestibular impulses. Stimulating the temporal lobe anterior to the auditory center induces vertigo, as do stimuli in the facial region of the somatosensory center.

Discriminating sweet from sour and sensing other flavors is accomplished in the **primary taste area,** near or in the tongue portion of the somatosensory center in the parietal lobe. The **olfactory centers** of temporal lobes perceive odors. The tips of the temporal lobes receive from the olfactory tract, and the surrounding cortex is the **olfactory association area.**

Figure 19.18 shows **homunculi** (HO-mun-KUE-lee, L, little man) that represent the proportion of cortical neurons in the somatosensory and somatomotor areas devoted to particular functions or regions of the body. Larger features indicate larger numbers of neurons. The hands and face have considerably more cortical representation than feet and trunk. Can you relate this central representation to the peripheral muscles and motor units, as well as to the tactile sensitivity of these areas? Notice also that cranial structures are represented in the lateral cortex, and caudal structures are medial. If a neurosurgeon were to stimulate the lateral surface of the postcentral gyrus, what would result?

Various association areas link sensory and motor functions together. The speech areas of the left hemisphere are excellent examples of the associations that speech and the use of language require. **Wernicke's speech area** (WER-nih-kez) in the parietal lobe en-

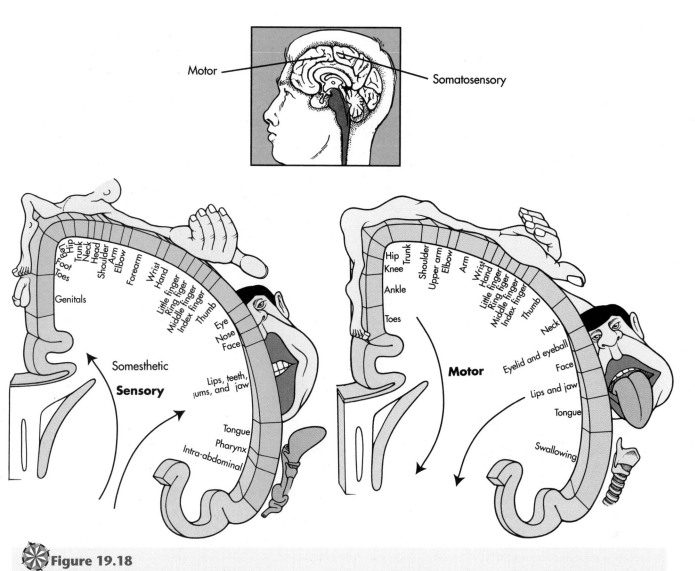

Figure 19.18

Primary somatosensory and somatomotor areas of the cerebral cortex. Somatosensory center illustrated on left, somatomotor center on right.

ables you to formulate thoughts into words, and **Broca's speech area** (BROE-kaz) in the frontal lobe enables you to say them. Conversation with your friends uses visual and auditory areas to receive the discussion; Wernicke's area assembles into words the images and thoughts from memory stored elsewhere in the cortex, and Broca's area converts them to speech. Memories themselves are not so obviously localized in the cerebral cortex.

- Name the primary sensory areas of the cerebral cortex.
- What do these areas do?
- Why is Wernicke's speech area considered an association area?

CORTICAL MICROANATOMY

The cerebral cortex is arranged in six cellular layers that represent a vast communication network. An estimated 5 trillion neurons work here. These cells are clues to the processes of memory and thought, but like your telephone, their form does not disclose what information flows through them. Figure 19.19 shows how the layers weave through the cortex, and it gives a more detailed view of the cells themselves. The four outer layers primarily receive afferent axons from other regions of the cortex and the brainstem, and layers five and six, the innermost, are primarily composed of efferents to other regions of the cortex and brain. The outermost layer, the **molecular**

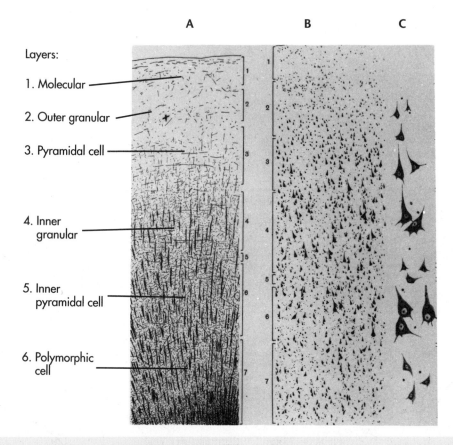

Layers:

1. Molecular

2. Outer granular

3. Pyramidal cell

4. Inner granular

5. Inner pyramidal cell

6. Polymorphic cell

Figure 19.19

Microanatomy of the cerebral cortex. Section through cortex showing cell layers. A, Emphasizes axons. B, Cell bodies. C, Large cell bodies.

layer, is a network of axons and dendrites from the cells below. Billions of synapses here process the impulse traffic of the cortex. Synapses continue in deeper layers, where cell bodies of the fibers above occur. These layers are the **outer granular layer, pyramidal layer,** and the **inner granular layer,** named for the appearance of the cell bodies, as you can see in Figure 19.19. The fifth and sixth layers are the **inner pyramidal layer,** with more pyramidal cells, and the **polymorphic cell layer,** named for the differing cell types found there. These cells communicate with others above and below them, as shown by axons on the left side of the figure. In effect, the cortex is organized into millions of cell columns standing side by side, somewhat like soft drink cans. The pyramidal cells in layer five are an example of this organization because their dendrites synapse in more superficial layers, and their axons extend to other areas of the cortex or to the brainstem.

INTERNAL ANATOMY

Figure 19.20 shows the internal anatomy of the cerebrum. This frontal section passes through Broca's speech area in the parietal lobe, just behind the somatosensory center. **Projection fibers** from (and to) the brainstem pass through the **internal capsule** deep

in the cerebral hemispheres and radiate outward through the white matter to the cortex. These fibers fan out as the **corona radiata,** spreading forward to the frontal lobe and posteriorly to the occipital lobe. The corona radiata of each side carries sensory impulses from the thalamus to the hemisphere on the same side. As you know, the left side of the brain communicates with the right, especially in control of speech, through Broca's and Wernicke's areas in the left hemisphere. The communication is accomplished through the **corpus callosum** (KAL-oh-sum), a strip of white matter between the hemispheres below the sagittal fissure. The corpus callosum carries commissural (crossing) fibers across the brain in both directions.

The lateral ventricles lie below the corpus callosum, separated from each other by the **septum pellucidum** (pel-LOOSE-ih-dum) and the **fornix,** a component of the limbic system, described later. The **choroid plexuses** of the lateral ventricles intrude upon the ventricles from the septum pellucidum and release cerebrospinal fluid into the ventricles. **Interventricular foramina** empty into the third ventricle between the thalami on each side. The **basal ganglia,** described later in this chapter, lie lateral to the thalamus, and the **insula,** a portion of the cerebral cortex deep within the **lateral fissure,** is lateral to the basal ganglia.

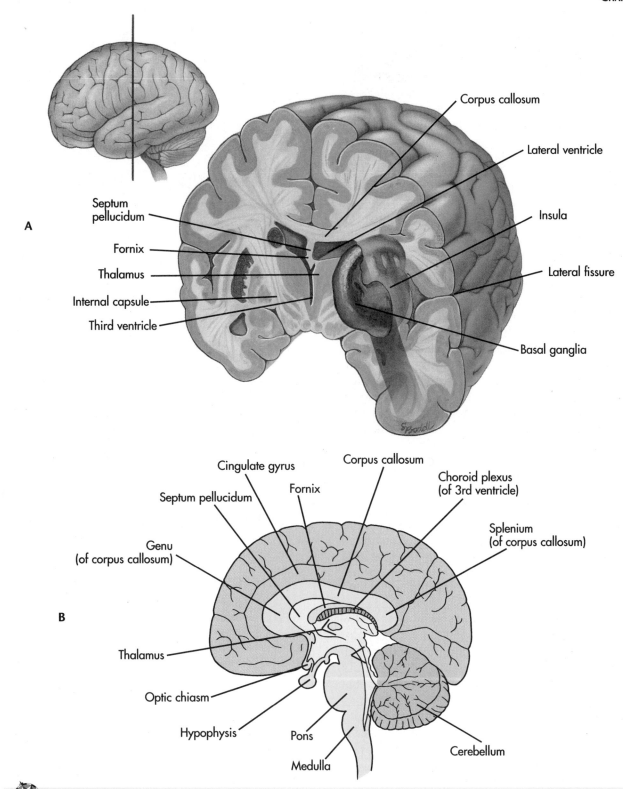

Figure 19.20

Internal anatomy of the cerebral hemispheres. A, Frontal section. B, Midsagittal section.

Figure 19.20B illustrates the parts of these cerebral structures visible in a midsagittal section through the brain. As you can see, the corpus callosum arches over the diencephalon; it bends downward at the **genu** (jeh-NU) and ends posteriorly at the **splenium** (SPLEEN-ee-um). The septum pellucidum is a curtain of tissue that separates the lateral ventricles, and the choroid plexus of the right ventricle can be seen below the fornix.

- How many layers make the cerebral cortex?
- What are the functions of the corpus callosum, corona radiata, and lateral ventricles?

MAJOR TRACTS OF THE BRAIN

Thus far you have been seeing tracts of the brain region by region. To see them whole, visualize some of the neural traffic involved in walking. To review: neural pathways resemble routes on a road map, made up of tracts connected together at nuclei by synapses. Traffic flows from the originating neuron, called the first order neuron, to the second order neuron, and so on, passing from tract to tract, until the destination is reached. The **posterior column pathway** carries conscious body sense and touch to the brainstem, thalamus, and then to the cerebral hemispheres. **Spinocerebellar tracts** carry muscle senses you are unconscious of to the cerebellum. Touch, pain, and thermal stimuli ascend the **spinothalamic tracts** from the body to the thalamus and thence to the cerebral cortex. **Corticospinal tracts** carry traffic in the opposite direction, conducting motor commands from the somatomotor cortex to the spinal nuclei and through spinal nerves to body muscles. Along with these voluntary commands go messages for involuntary coordination, traveling from the cerebellum by way of the **rubrospinal tracts.**

POSTERIOR COLUMNS

Posterior column pathways carry touch and body sense to the somatosensory centers in the cerebral cortex. In walking, as in most other activities, you are aware of your limbs and their motion because muscle spindles and stretch receptors in tendons and joints relay this information to the cerebrum, and receptors in the skin and hair follicles detect clothing rubbing and air flowing over the surface of the skin. Beginning in the right leg, for example, **first order neurons** conduct these impulses (Figure 19.21) into the sacral or lumbar segments of the spinal cord, where they ascend in the posterior funiculus of the cord to the posterior columns of the medulla. Any fibers that enter below the sixth thoracic segment, including sacral fibers, follow the fasciculus gracilis; those with higher entrances are more lateral in the fasciculus cuneatus. Whichever pathway they follow, ascending fibers make synapses in the nucleus cuneatus or gracilis with **second order neurons,** whose axons penetrate deep into the medulla and cross to the medial lemniscus of the left side. These fibers continue through the tegmentum of the pons and midbrain to arrive at the thalamus, where synapses link them with **third order neurons** in the ventral posterior portion of the lateral nucleus. This last series of neurons projects through the internal capsule onto the primary somatosensory center of the left hemisphere, particularly to medial areas of the postcentral gyrus associated with the right lower limb (see the sensory homunculus

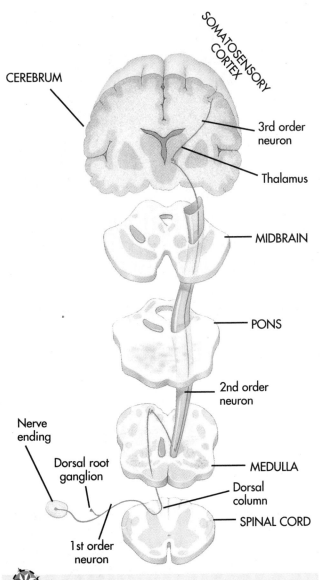

Figure 19.21

Posterior column pathway for conscious body sense and touch.

in Figure 19.18). Consequently the left side of the brain is aware of the right leg swinging through its strides. A mirror image of this pathway serves the left leg on the right side of the cerebrum.

SPINOCEREBELLAR TRACTS

The cerebellum uses information it receives from the spinocerebellar tracts to coordinate muscular motion. Posterior and anterior spinocerebellar tracts conduct impulses from the trunk and extremities through separate pathways to the vermis of the cerebellum. You are unaware of this information because it never reaches the cerebral cortex. Although both pathways conduct body sense impulses, the posterior spinocerebellar tract receives from muscle spindles and Golgi tendon organs in tendons and joints, but the anterior spinocerebellar tract receives only from Golgi tendon organs. Lesions in these pathways cause you to walk with legs wide apart and to stagger.

Beginning in the right leg in Figure 19.22, axons of **first order neurons** enter the spinal cord and make

synapses in the gray matter. The posterior spinocerebellar tracts begin in the thoracic nucleus of the spinal cord between segments C8 and L2. **Second order neurons** ascend the lateral funiculus of the cord and pass through the medulla into the cerebellum by way of the inferior cerebellar peduncle, without crossing to the opposite side.

In contrast the anterior corticospinal tract crosses twice. The fibers cross the cord and ascend in the lateral funiculus on the left side, through the medulla and pons. In the midbrain the fibers double back to the cerebellum through the superior cerebellar pe-

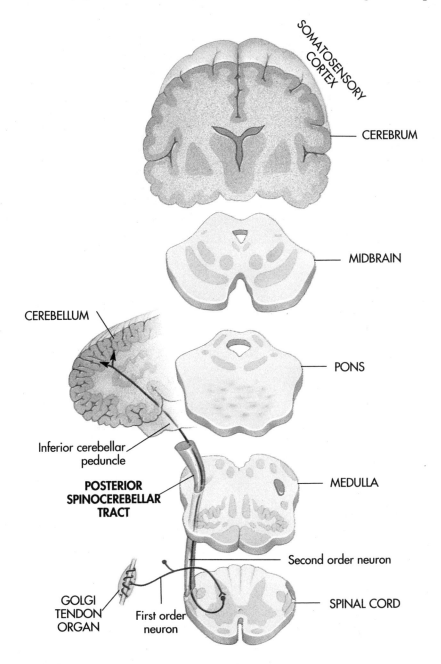

Figure 19.22
Spinocerebellar tracts for nonconscious body sense.

duncle, crossing once again in the cerebellum, back to the right side, for an unusual double decussation.

SPINOTHALAMIC TRACTS

If an arthritic knee or an old injury protests as you walk, another series of connections relays that message to the brain. The axons of **first order neurons** (Figure 19.23) traverse the right lumbar and sacral plexuses to synapses with interneurons **(second order neurons)** in the intermediate substance of the gray matter of the spinal cord. These in turn make

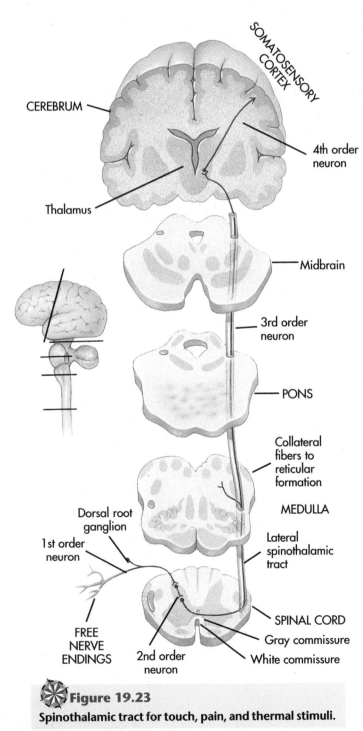

CEREBRUM

SOMATOSENSORY CORTEX

4th order neuron

Thalamus

Midbrain

3rd order neuron

PONS

Collateral fibers to reticular formation

MEDULLA

Dorsal root ganglion

1st order neuron

Lateral spinothalamic tract

FREE NERVE ENDINGS

2nd order neuron

SPINAL CORD

Gray commissure

White commissure

✳ Figure 19.23
Spinothalamic tract for touch, pain, and thermal stimuli.

synapses with **third order neurons** that send axons across the anterior white commissure in the segments immediately above, where they enter the lateral and anterior spinothalamic tracts in the left half of the spinal cord. The fibers of the spinothalamic tracts remain on this side, opposite from the right knee. The fibers ascend the spinal cord, pass through the lateral wall of the medulla, traverse the tegmentum of the pons, and arrive in the ventral posterior portion of the lateral nucleus of the thalamus. Synapses connect them to **fourth order neurons,** whose axons enter the internal capsule and project upon the left somatosensory cortex, along with impulses from the posterior column pathway. Although fibers of the spinothalamic tract and posterior columns cross at different levels, impulses nevertheless arrive in the opposite sensory cortex from the side on which they began. Association areas now process this stream of signals, and the brain recognizes that the knee is painful and should perhaps be rested. When pain is unrelenting, the lateral spinothalamic tract sometimes is severed in an attempt to interrupt the constant stream of impulses that signal pain to the brain.

CORTICOSPINAL TRACTS

Voluntary motor commands control walking and other activities through the corticospinal tracts that provide direct connections to the motor neurons that stimulate the muscles. One of the largest tracts in the brain, the corticospinal tracts contain about 10 million nerve fibers. Part of the commands to continue walking despite a painful knee come from the left precentral gyrus, just anterior to the somatosensory center that registers the pain. **First order neurons** project down the spinal cord from cell bodies in the fifth and sixth layers of the somatomotor cortex in the medial surface of the right cerebral hemisphere (Figure 19.24; see the motor homunculus, Figure 19.18). The fibers pass through the internal capsule, into the left cerebral peduncle of the midbrain, and down the base of the pons to the left pyramid of the medulla. Up to 90% of the fibers cross to the right side at the decussation of the pyramids and continue down the pyramidal tracts as the lateral corticospinal tract of the lateral funiculus of the spinal cord. At sacral levels the fibers leave the white matter and link by synapses with lower motor neurons in the anterior horn of the gray matter. These **third order neurons** then conduct impulses to the right lower limb. Another 8% to 10% of fibers reach this same destination by crossing at lower levels of the cord. These fibers descend the cord in the anterior corticospinal tract on the left side. In the sacral segments (or others, as the case may be) the fibers enter the gray matter and cross to the anterior horn on the right side, where they form synapses with motor neurons **(third order)** of the right lower limb.

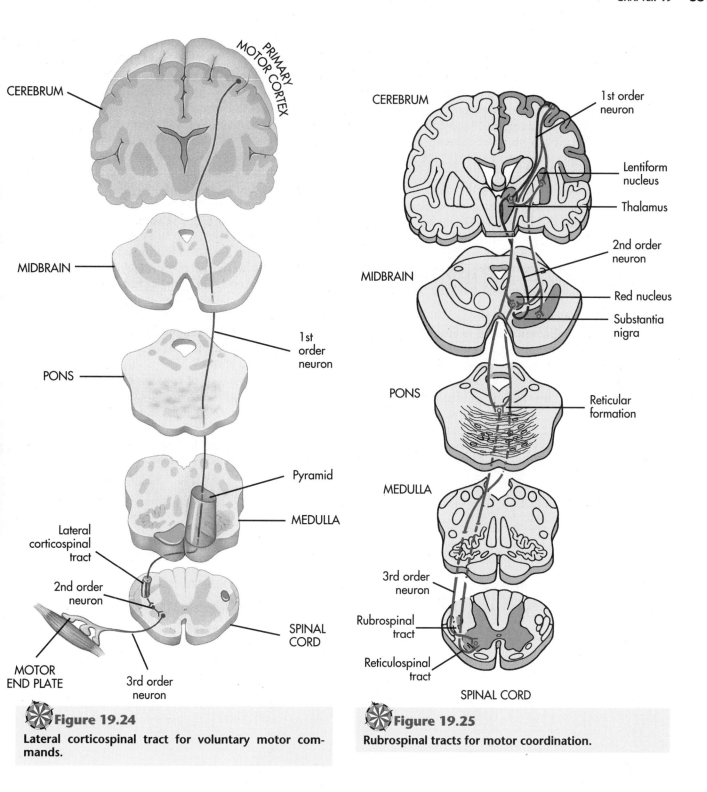

Figure 19.24

Lateral corticospinal tract for voluntary motor commands.

Figure 19.25

Rubrospinal tracts for motor coordination.

RUBROSPINAL TRACT

The rubrospinal tract supplements the corticospinal tract. Originating in the red nucleus (hence the name), this tract descends the brainstem and accompanies the lateral corticospinal tracts in the spinal cord. Fibers of the rubrospinal tract end at all levels of the spinal cord, where they make

synapses with motor neurons in the gray matter of the cord. Because the rubrospinal tract is part of a pathway from the somatomotor cortex and cerebellum, it adjusts actual muscular movements of the arms and legs to commands from the somatomotor center. Figure 19.25 shows that the pathway begins in the motor cortex as **first order neurons** project onto the thalamus, where **second order neurons** continue to the red nucleus. The rubrospinal tract itself begins as **third order neurons** in this pathway that leave the red nucleus and cross in the tegmentum of the midbrain before descending the pons and medulla on the opposite side from their origin. These fibers enter the lateral funiculus of the spinal cord and accompany corticospinal fibers to synapses with motor neurons in the spinal gray matter.

- What information does the spinothalamic tract carry?
- Where do the corticospinal tracts cross the midline?
- To what destination does the posterior column pathway lead?

OTHER SYSTEMS

BASAL GANGLIA

The basal ganglia are actually nuclei that help coordinate muscular motion. Deep in the cerebral hemispheres, lateral to the ventricles, is a hook-shaped cluster of nuclei known as the **basal ganglia.** The hook follows the lateral ventricles, as Figure 19.26 shows. Its base is the **lentiform nucleus,** from which the **head** and **tail of the caudate nucleus** (KAW-date) follow the lateral ventricle to its inferior horn, where the tail of the caudate nucleus joins the

amygdala (ah-MIG-dah-lah). The lentiform nucleus itself is composed of the **putamen** (PUTT-ah-men) laterally and **globus pallidus** (GLOW-bus PAL-ih-dus) medially. The fibers of the **internal capsule** pass between the caudate nucleus and the lentiform nucleus. The basal ganglia receive impulses from both motor and sensory areas of the cortex, and they communicate to the thalamus and substantia nigra of the midbrain and thence to the reticular formation, cerebellum, and spinal cord.

Basal ganglia appear to inhibit general muscular tone, in contrast to the somatomotor center that promotes contraction, so that coordination arises from interactions between these signals. How the basal ganglia actually coordinate motion is as yet poorly understood. One part of the puzzle is that, for particular motions, neurons in the basal ganglia fire before the somatomotor center, in response to sensory impulses from the thalamus and brainstem. If the basal ganglia initiate movement, what role does that leave the motor cortex?

Clues to the action of the basal ganglia come from animal studies and disease or injury to the ganglia and substantia nigra. Certain lesions of the globus pallidus make it difficult to place the hands (or feet) where the person wants them because the individual cannot control the proximal muscles of the limbs.

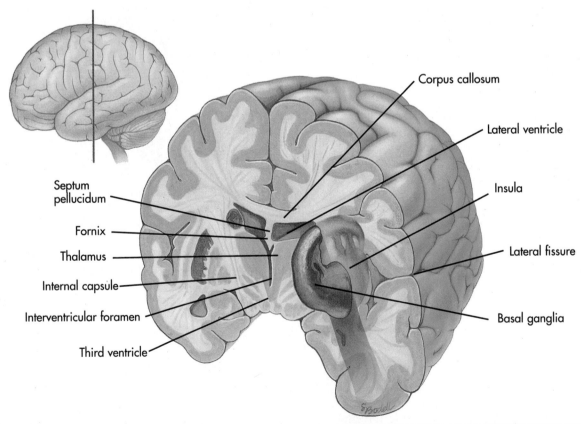

Labels: Corpus callosum · Lateral ventricle · Insula · Lateral fissure · Basal ganglia · Septum pellucidum · Fornix · Thalamus · Internal capsule · Interventricular foramen · Third ventricle

✶**Figure 19.26**
Basal ganglia of cerebral hemispheres, shown in frontal section.

Athetosis (AH-theh-TOE-sis) is a similar condition of the globus pallidus, in which the distal parts of extremities perform involuntary, writhing motions. **Parkinson's disease** is a disorder of the basal ganglia that causes general immobility and trembling when the muscles rest. The fundamental cause of these symptoms appears to lie in the degeneration of the substantia nigra that exchanges signals with the caudate nucleus of the basal ganglia.

- What are the components of the basal ganglia, and where are they located in the cerebrum?

RETICULAR ACTIVATING SYSTEM

The reticular activating system uses the reticular formation to help regulate sleep-wake cycles and to recall ideas and information (Figure 19.27). This system somehow alerts pathways that find memories in the cerebral cortex. **Coma,** the unconscious state that results from inactivity of brain systems, is the opposite condition and can be viewed as a failure or suppression of the reticular activating system. The reticular formation, you will recall, is a network of nuclei and fibers that extends from the medulla, through the midbrain, into the thalamus. These cells have widespread connections throughout the cerebral cortex and diencephalon, enabling them to rouse the brain from sleep and to keep it active. Numerous impulses, particularly from the spinothalamic tract, cause the reticular activating system to alert the brain.

LIMBIC SYSTEM

The limbic system evokes memories and emotions. It is an association area of the cortex that controls behavior and memory. This system is a more or less complete ring of gray matter, surrounding the thalamus at the lower margins of each cerebral hemisphere. The cerebral cortex of this area is evolutionarily old and contains only three of the six cortical layers. The newer cortex expanded through this ring to enlarge the cerebral hemispheres.

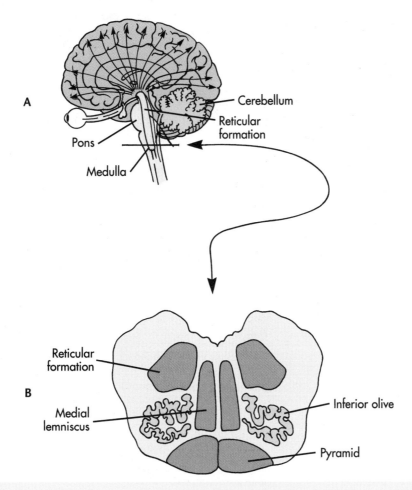

A

Cerebellum
Reticular formation
Pons
Medulla

B

Reticular formation
Medial lemniscus
Inferior olive
Pyramid

✳ **Figure 19.27**

Reticular activating system. A, Connections with the brain. B, Reticular formation of the medulla, transverse section.

Figure 19.28 shows that the main features of the limbic system are the **cingulate gyrus** on the medial surface of each hemisphere and, continuing to the medial surface of the temporal lobe, the **hippocampus, parahippocampal gyrus, amygdala,** and **uncus** (UN-kus), near the tip of the temporal lobe. These areas connect to nearly every other cortical region in the hemispheres, especially with the **anterior nuclei** of the thalamus and **mamillary bodies** of the hypothalamus. Most sensations cause neurons in the hippocampus and amygdala to fire volleys of impulses to the hypothalamus and other regions of the limbic system.

Exactly how the limbic system is involved in memory is uncertain, but it is clear that at least one hippocampus is needed to store new information. Surgical removal of both structures, occasionally done to relieve the symptoms of epileptic seizures, leaves the patient with **anterograde amnesia;** that is, it prevents new memories from being stored. Such patients quickly forget current events, even what happened only minutes ago, but their memories before surgery remain intact and vivid. Removing the hippocampus on only one side has no such effect.

The amygdala seems to do for behavior what the cerebellum does for motor control—coordinate available choices of behavior with what is appropri-ate. Removing the amygdala from both temporal lobes results in excessive sex drive, loss of fear, and aggressiveness. Like the hippocampus, the amygdala has extensive communications to other parts of the limbic system and cerebral cortex, most especially to the hypothalamus. Stimulating the amygdala excites many hypothalamic functions, such as heart rate, micturition, peristalsis, and release of pituitary hormones.

- Where is the reticular formation situated in the brain?
- The amygdala, cingulate gyrus, and hippocampus are important components of the limbic system. Where are these structures located in the cerebral hemispheres?

MENINGES AND CEREBROSPINAL FLUID

DURA, ARACHNOID, AND PIA

The meninges cushion the brain with cerebrospinal fluid. Floating in the fluid, 1500 gm of brain weigh only 50 gm, but even so cushioned, a blow to the head can make the brain collide with the skull, usually on the side opposite from the injury. Shaking a very young child in anger also can produce this po-

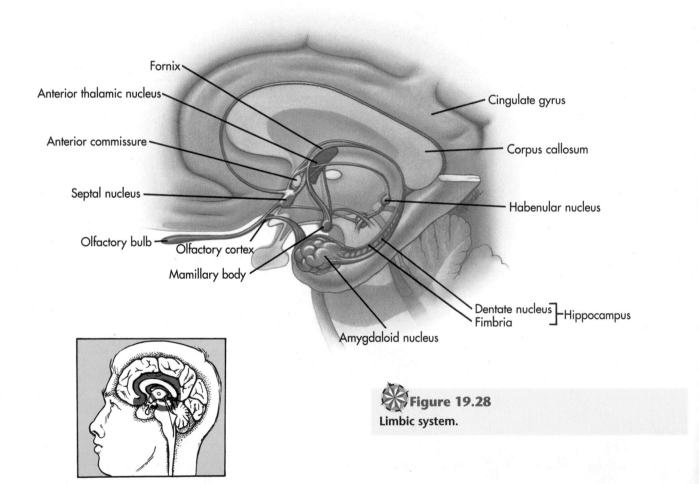

✳ **Figure 19.28**
Limbic system.

tentially tragic collision. The three layers of meninges are continuous with those of the spinal cord described in Chapter 18. The outermost layer is the tough, fibrous **dura mater,** shown in Figure 19.29, that neurosurgeons encounter first when exposing the brain during surgery. The dura is really two layers. The **endosteal dura** (end-OS-tee-al) adheres firmly to the periosteum of the bones of the skull so that, unlike the spinal cord, there is no **epidural space** between the dura and skull. The **meningeal dura** (men-IN-jee-al) is attached thoroughly to its partner, except where venous sinuses of the brain intrude between them.

The meningeal dura forms partitions in the great fissures of the brain that divide the cranial cavity into smaller supportive compartments, as shown in Figure 19.29B. These structures attach to the cranium, as described in Chapter 6, Axial Skeleton. The **falx cerebri** (FALKS SEH-ree-bree) is a fold of meningeal dura that stretches between the cerebral hemispheres, resembling a sickle about to cut the brain in half. The anterior point anchors on the crista galli in the anterior cerebral fossa, and the blade curves up and over the corpus callosum to the tentorium cerebelli. The **falx cerebelli** continues into the posterior cranial fossa between the hemispheres of the cerebellum. The **tentorium cerebelli** (TEN-tore-ee-um) stabilizes the falx

cerebri and falx cerebelli by extending laterally in the fissure between the cerebrum and cerebellum. Inside the falx cerebri, **superior** and **inferior sagittal sinuses** collect blood and cerebrospinal fluid (CSF) from the brain. The **occipital sinus** does the same in the falx cerebelli; all three sinuses drain into the **transverse sinus** that runs in the tentorium cerebelli to the jugular foramina on either side of the cranial cavity.

Between the meningeal dura and brain are the **subdural space** and the **arachnoid,** with its underlying cerebrospinal fluid (Figure 19.29A). The subdural space is a potential space between the dura and arachnoid, opened only in surgery or by injury, when the dura is withdrawn from the surface of the brain. Bleeding into this space produces a **subdural hematoma,** which can exert pressure on the underlying brain. The arachnoid itself is a smooth, delicate membrane that separates the subarachnoid space beneath it from the subdural space. Cerebrospinal fluid fills the subarachnoid space, and **arachnoid trabeculae** reach through this fluid to attach the arachnoid to the **pia mater.** The pia mater is directly on the surface of the brain. Vessels reach the brain through the subarachnoid space, penetrating deep into the sulci, where the pia mater finally gives way to astrocytes that cover the exterior of capillaries.

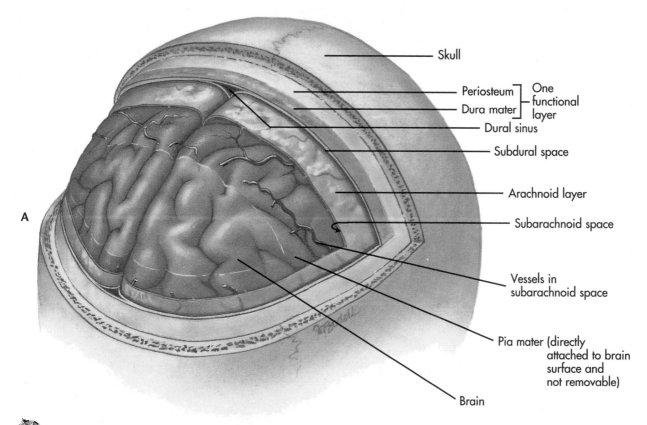

A

- Skull
- Periosteum ⎤ One
- Dura mater ⎦ functional layer
- Dural sinus
- Subdural space
- Arachnoid layer
- Subarachnoid space
- Vessels in subarachnoid space
- Pia mater (directly attached to brain surface and not removable)
- Brain

 **Figure 19.29**

Cranial meninges. A, Three meninges cover the brain.

Continued.

BLOOD-BRAIN BARRIER

The central nervous system maintains a remarkably consistent internal environment by means of a barrier that blocks vascular changes. More than 100 years ago, it was known that dyes injected into the blood do not enter the brain or spinal cord. Only since the 1960s has it become clear that the epithelium of brain capillaries is the barrier. To enable the brain to maintain adequate levels of glucose for metabolism and low levels of potassium ions for neural conduction, the capillary endothelium selectively transports glucose from the blood into brain tissue and pumps potassium ions out by means of a sodium potassium ATPase (see Resting Potential, Chapter 17). To prevent these materials and others from leaking across, **tight junctions** seal brain capillary endothelial cells securely together, a function capillaries of other tissues do not provide.

CIRCULATION OF CEREBROSPINAL FLUID

Cerebrospinal fluid renews itself three times a day. The turnover is not at any specific times, of course, but in the course of a day the brain produces about 500 ml of fluid, about three times the meninges' capacity of 160 ml. About 80% of this production comes from the **choroid plexuses** in the ventricles,

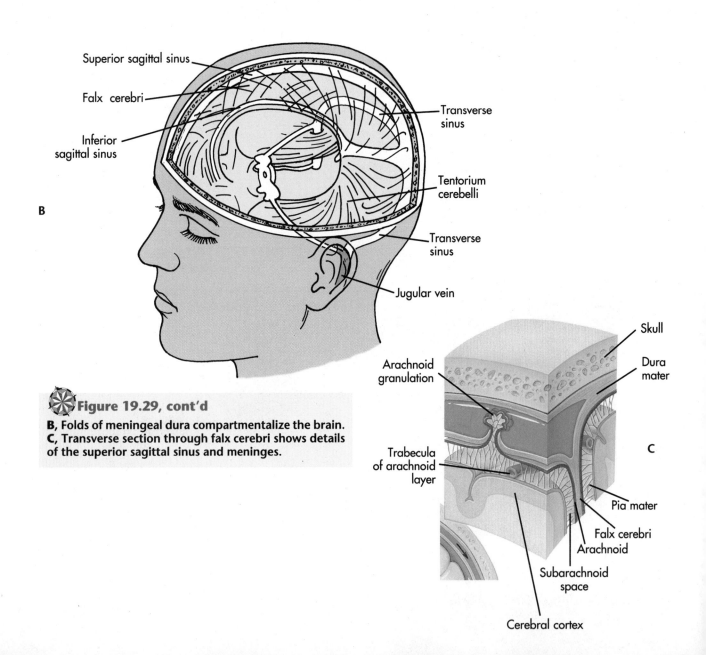

Figure 19.29, cont'd
B, Folds of meningeal dura compartmentalize the brain. **C,** Transverse section through falx cerebri shows details of the superior sagittal sinus and meninges.

and the rest leaks from the blood through the **interstitial spaces** within the brain and spinal cord. This process is not surprising because the brain has no lymphatic vessels to drain accumulating fluid. Cerebrospinal fluid and the subarachnoid space substitute for lymphatics. Figure 19.30 shows that production in the ventricles circulates within minutes through the **lateral** and **median apertures** of the fourth ventricle, out into the subarachnoid spaces around the brain and spinal cord. The superior sagittal sinus reabsorbs much of this outflow through numerous **arachnoid villi,** tufts of arachnoid that penetrate the meningeal dura directly into the blood of the sinus. Only the arachnoid epithelium itself partitions cerebrospinal fluid from the blood. How cerebrospinal fluid circulates in the spinal cord is uncertain; normal body movements may power some of its flow, and the **arachnoid cuffs** around spinal nerves at the intervertebral foramina may reclaim more fluid as the spinal nerves emerge from the vertebral column.

Blocked circulation can cause too much cerebrospinal fluid to accumulate, a condition called **hydrocephalus** (water head). When the cerebral aqueduct is blocked, as occasionally happens during development, the ventricles inflate and press the cerebrum against the skull, greatly expanding the imma-

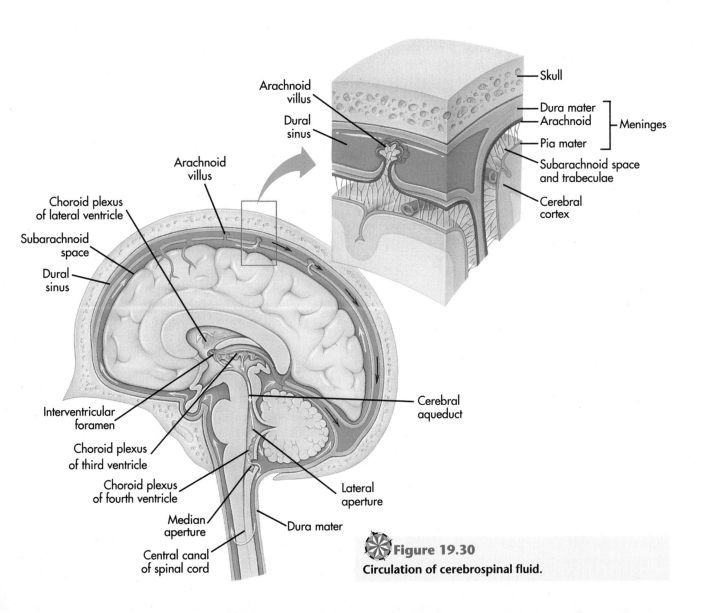

Figure 19.30
Circulation of cerebrospinal fluid.

ture bones of the calvarium. This condition is called **noncommunicating,** or **internal, hydrocephalus** because cerebrospinal fluid is trapped in the ventricles and cannot reach the subarachnoid space. In **communicating,** or **external, hydrocephalus,** blockage (often caused by meningitis) occurs in the subarachnoid spaces, swelling the spaces and pressing inward against the brain and outward against the cranium. The pressure can be relieved by draining cerebrospinal fluid through a tube surgically inserted into the CSF and passed through the neck into veins returning blood to the heart.

- Name the three layers of meninges.
- Which layer contains cerebrospinal fluid?
- What are the functions of CSF?

CRANIAL NERVES

This section discusses the cranial nerves, that special group of nerves that communicates directly between the brain and the head and neck. Table 19.1 summarizes all 12 cranial nerves, and Table 6.3 matches them with skull foramina. There are several mnemonics, some printable, that will help you remember them. Ask your instructor or invent your own. Cranial nerves are numbered (Roman numerals) in the rostrocaudal order by which they enter or leave the brain. They are named by function (olfactory, optic), by target (facial, hypoglossal), or by some other notable feature (the trigeminal nerve has three branches, for example). Their rostrocaudal order usually matches the location of their target organs, but beware of exceptions. As in the spinal nerves, most cranial nerves are mixed nerves, with both afferent (sensory) and efferent (motor) fibers, but some, such as the optic and vestibulocochlear, are exclusively sensory. We will not keep track of the precise location of nuclei.

OLFACTORY (I)

The olfactory pathway (ohl-FAK-tor-ee) is the only sensory input that enters the cerebral cortex directly without first synapsing in the thalamus. The pathway is exclusively sensory and conducts unmyelinated axons of olfactory cells in the nasal lining through the cribriform plate of the ethmoid bone to the olfactory bulb beneath the frontal lobes of the cerebral hemispheres. From synapses there, main fibers lead to the olfactory centers in the same hemisphere. In humans the olfactory neurons are the only ones known to replace themselves by cell division.

OPTIC (II)

The optic nerves carry afferent fibers from the retina to the thalamus. Both nerves enter the cranial cavity through the optic canal and join the optic chiasm in front of the infundibulum. Axons recombine

here, and some cross the midline to the opposite side. From the chiasm the tracts curl around the thalamus and enter posteriorly, forming synapses with cells in the lateral geniculate nucleus of the thalamus. From here, fibers project to the primary visual cortex in the occipital lobe of each hemisphere.

OCULOMOTOR (III)

The oculomotor nerve (OK-you-low-MO-tor) controls eyeball motions through all the eyeball muscles except the superior oblique and lateral rectus. It also innervates the superior levator palpebrae muscles that raise the upper eyelids. Although its name emphasizes these motor actions, the oculomotor nerve also carries sensory information from muscle spindles in its targets. The oculomotor nerves leave the midbrain from the medial surface of the cerebral peduncles and, without crossing, enter the orbits through the orbital fissure. Damage to this nerve causes the eyelid on the same side as the injury to droop and the eye to turn downward and laterally. Consult the anatomy of eyeball muscles in Table 10.2 to see why the eyeball should move in this manner. Parasympathetic outflow also controls the pupil and the lens of the eye.

TROCHLEAR (IV)

The trochlear nerve is also a midbrain nerve in charge of eyeball motion. This smallest cranial nerve innervates only the superior oblique muscle that abducts and depresses the eyeball. The name derives from the action of this muscle around the trochlea (trochle-, *G* a pulley) in the orbit (see Muscles that Move the Eyeballs, Chapter 10). The **trochlear nerve (TROKE-lee-ar)** is unusual because it emerges from the dorsal side of the brainstem, behind the inferior colliculi, after its fibers have already crossed from the opposite side. As a result the left side of the midbrain controls the right superior oblique muscle, and trauma to one side of the midbrain affects motion of the opposite eye. Such patients frequently complain of double vision because the eyes cannot converge on the same subject. The trochlear nerve is a mixed nerve, with afferents from muscle spindles in the superior oblique muscles.

TRIGEMINAL (V)

This largest cranial nerve has three great branches, hence its name. The **trigeminal nerve** (try-JEM-in-al) contains sensory fibers from the skin and mucous membranes of the face, as well as motor fibers to the muscles of mastication. The names of the trigeminal branches reflect their peripheral territories. The **ophthalmic nerve** serves the skin of the forehead, eyelids, and nose and also structures deep to these areas—the eyeball, conjunctiva of the eye, the lacrimal gland, and parts of the nasal and paranasal

Text continued on p. 594.

Table 19.1   CRANIAL NERVES

Name	Function	Neural Traffic	Target
CEREBRUM			
I. Olfactory	Smell	Afferent	Olfactory epithelium, ethmoid recess, nasal cavity
THALAMUS			
II. Optic	Vision	Afferent	Retina
MIDBRAIN			
III. Oculomotor	Eyeball movement; accommodation, dilation of pupil	Mixed	Superior, medial, and inferior rectus; inferior oblique muscles of eyeballs. Pupillary sphincter of iris and ciliary muscles of lens

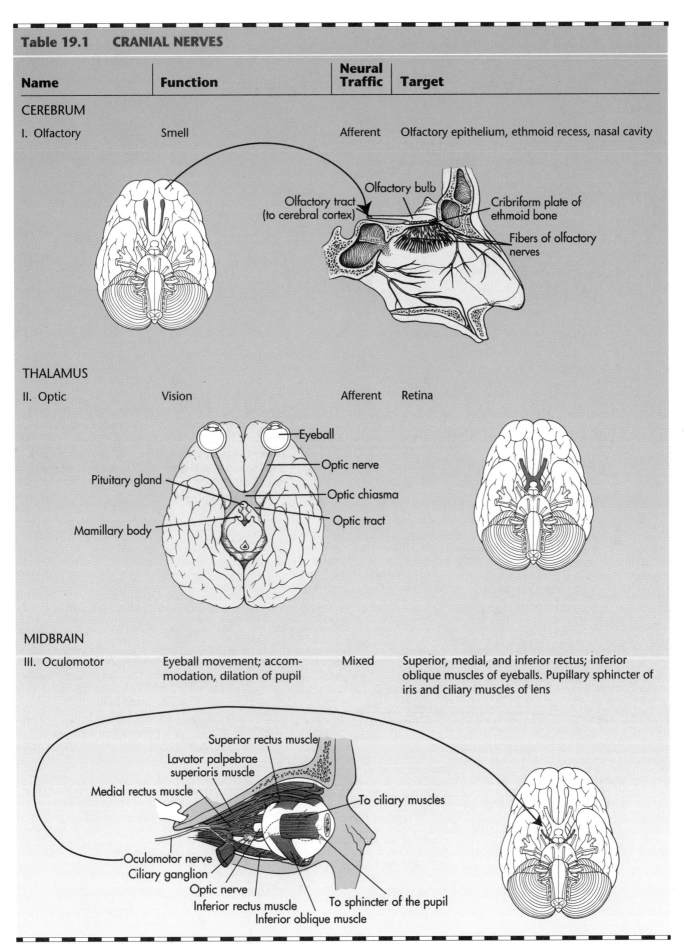

Continued.

Table 19.1   CRANIAL NERVES—cont'd

Name	Function	Neural Traffic	Target
MIDBRAIN—cont'd			
IV. Trochlear	Eyeball movement	Mixed	Superior oblique muscles of eyeballs
PONS			
V. Trigeminal	Facial sensation, mastication	Mixed	Three branches to ophthalmic, maxillary, and mandibular regions of face *Efferents*—muscles of mastication, tensors tympani and palatini, mylohyoid and anterior belly of digastric *Afferents*—pain, temperature, and touch from face and forehead, and from deep structures of face (teeth)

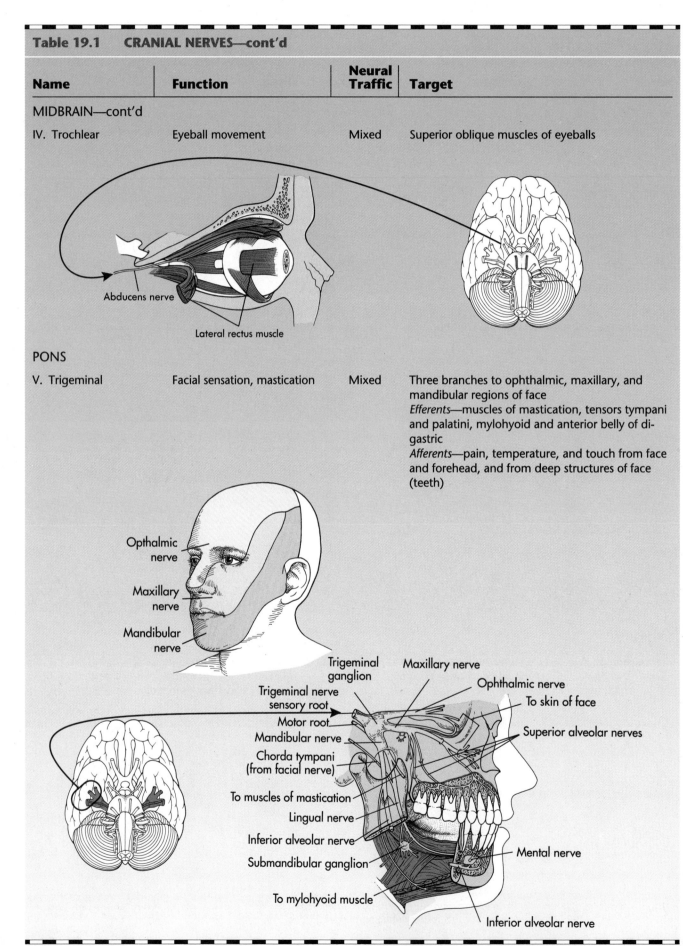

Table 19.1   CRANIAL NERVES—cont'd

Name	Function	Neural Traffic	Target
PONS—cont'd			
VI. Abducens	Eyeball movement	Mixed	Lateral rectus muscles of eyeballs, muscle spindles

Abducens nerve

Lateral rectus muscle

VII. Facial	Facial expression, taste, secretion	Mixed	*Afferents*—taste from anterior two thirds of tongue and exteroceptive from pinna of ear *Efferents*—muscles of facial expression, and stapedius, stylohyoid, and posterior belly of digastric

Geniculate ganglion

Trigeminal ganglion

Pterygopalatine ganglion

Facial nerve

To lacrimal gland and nasal mucous membranes

To occipitofrontalis

To forehead muscles

Chorda tympani (for salivary glands, sense of taste)

To orbicularis oculi

To digastric and stylohyoid muscles

To buccinator, lower lip, and chin muscles

To orbicularis oris and upper lip

To platysma

Continued.

Table 19.1   CRANIAL NERVES—cont'd

Name	Function	Neural Traffic	Target
MEDULLA—cont'd			
VIII. Vestibulocochlear	Hearing and equilibrium	Afferent	Vestibular division from semicircular canals, saccule, utricle; cochlear division from hair cells in organ of Corti, inner ear

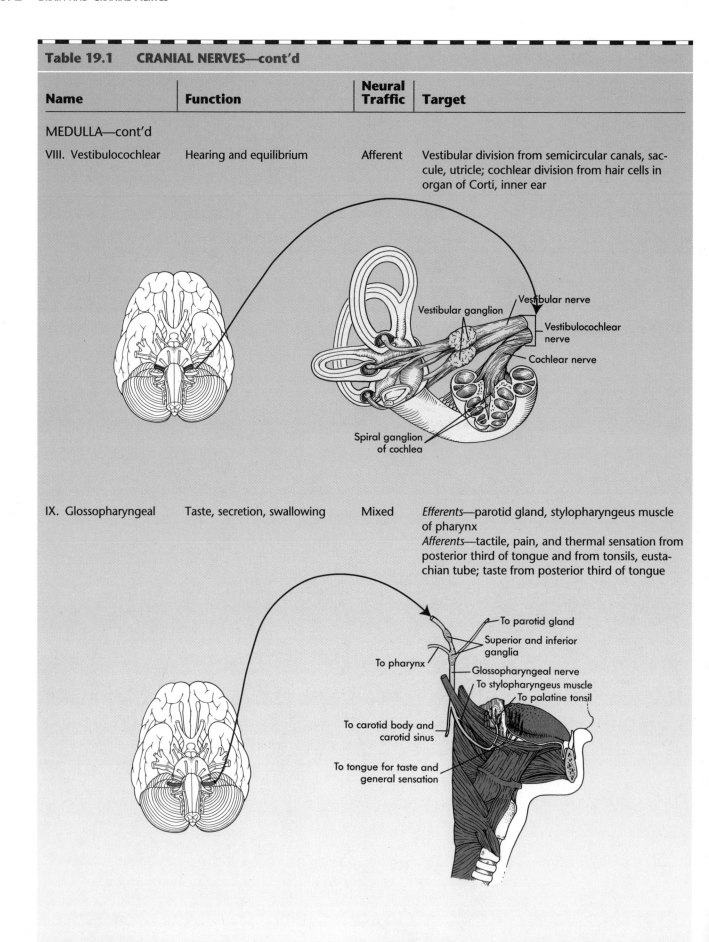

Vestibular nerve
Vestibular ganglion
Vestibulocochlear nerve
Cochlear nerve
Spiral ganglion of cochlea

Name	Function	Neural Traffic	Target
IX. Glossopharyngeal	Taste, secretion, swallowing	Mixed	*Efferents*—parotid gland, stylopharyngeus muscle of pharynx *Afferents*—tactile, pain, and thermal sensation from posterior third of tongue and from tonsils, eustachian tube; taste from posterior third of tongue

To parotid gland
Superior and inferior ganglia
To pharynx
Glossopharyngeal nerve
To stylopharyngeus muscle
To palatine tonsil
To carotid body and carotid sinus
To tongue for taste and general sensation

Continued.

Table 19.1   CRANIAL NERVES—cont'd

Name	Function	Neural Traffic	Target

MEDULLA—cont'd

X. Vagus	Visceral muscle and glands; taste, swallowing, speech	Mixed	*Afferents*—from pharynx, larynx, esophagus, trachea, thoracic and abdominal viscera *Efferents*—thoracic and abdominal viscera, pharyngeal and laryngeal muscles

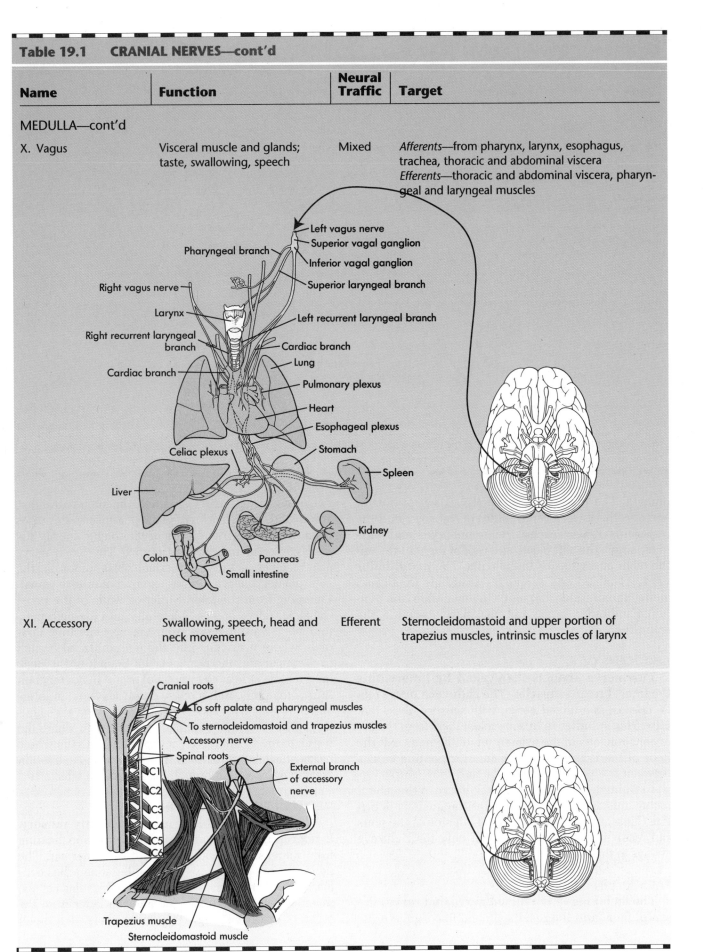

Left vagus nerve
Superior vagal ganglion
Inferior vagal ganglion
Pharyngeal branch
Superior laryngeal branch
Right vagus nerve
Left recurrent laryngeal branch
Larynx
Right recurrent laryngeal branch
Cardiac branch
Cardiac branch
Lung
Pulmonary plexus
Heart
Esophageal plexus
Celiac plexus
Stomach
Liver
Spleen
Kidney
Colon
Pancreas
Small intestine

XI. Accessory	Swallowing, speech, head and neck movement	Efferent	Sternocleidomastoid and upper portion of trapezius muscles, intrinsic muscles of larynx

Cranial roots
To soft palate and pharyngeal muscles
To sternocleidomastoid and trapezius muscles
Accessory nerve
Spinal roots
External branch of accessory nerve
C1
C2
C3
C4
C5
C6
Trapezius muscle
Sternocleidomastoid muscle

Continued.

Table 19.1   CRANIAL NERVES—cont'd

Name	Function	Neural Traffic	Target
MEDULLA—cont'd			
XII. Hypoglossal	Tongue motion in speech and mastication	Mixed	Intrinsic and extrinsic muscles of tongue

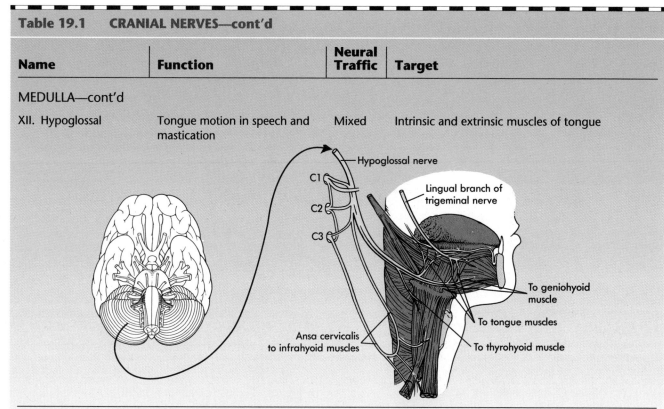

From Clemente CD, editor: Gray's anatomy of the human body, *ed 30, Philadelphia, 1985, Lea and Febiger; Afifi AK and Bergman RA: Basic neuroscience,* ed 2, Baltimore, 1986, Urban and Schwarzenberg.

mucosa. The **maxillary nerve** also is sensory; it innervates the lower eyelids, sides of the nose, and the upper lips. It also reaches the nasopharynx and maxillary sinus, the soft palate and roof of the mouth, and the teeth and gums of the maxilla. The **mandibular nerve** is mixed and serves the mandible (teeth and gums), the muscles that move the mandible, the temporomandibular joint, and the skin overlying the mandible.

ABDUCENS (VI)

This nerve abducts the eyeball by innervating the lateral rectus muscles. The abducens nerve (ab-DOO-sens) is a mixed nerve, with sensory fibers that arise from spindles in this muscle. Fibers from the abducens nucleus in the tegmentum of the pons exit the floor of the brainstem in the inferior pontine sulcus between the pons and medulla, and they course forward through the superior orbital fissure to the lateral rectus muscle on the same side as the nucleus of origin. Consequently, lesions on the left side of the pons can result in double vision when only the right eye swings to the left.

FACIAL (VII)

The facial nerve is a mixed nerve that serves the scalp, face, and tongue. Its efferent fibers govern the muscles of facial expression and the submandibular, sublingual, and lacrimal glands. Sensory impulses from the anterior two thirds of the tongue supply the taste center of the cerebral cortex. The facial nerve also receives tactile and thermal sensations from the soft palate, pharynx, and external auditory canal. Emerging from the lateral inferior walls of the pons, the facial nerve proceeds laterally into the petromastoid bone and emerges from the stylomastoid foramen, where it divides into the temporofacial branch to the head and the cervicofacial branch to the neck on the same side as the nuclei of origin in the tegmentum of the pons. **Bell's palsy** paralyses facial muscles and interferes with taste, salivation, and the ear's ability to filter out loud sounds, depending on where the lesion to the facial nerve occurs. Damage to the whole nerve affects all functions. If only taste and salivation are damaged, which branch has the lesion affected?

VESTIBULOCOCHLEAR (VIII)

The vestibulocochlear nerve is strictly sensory. It returns impulses concerning balance and hearing from the vestibule and cochlea of the inner ear. The nerve has two divisions—the vestibular and the cochlear. The former arises from the semicircular canals, saccule and utricle (balance), and the latter from the organ of Corti (hearing). The nerve fibers then travel through the internal auditory meatus, into the medulla at the level of the inferior peduncle. Most of the cochlear fibers cross to the opposite side, ascending through the pons and also to the inferior colliculus of the midbrain. Fibers lead from the pons to the thalamus and auditory cortex of the cerebral hemispheres. The vestibular division synapses in the vestibular nucleus, from which axons project to the cerebellum and to the vestibular cortex of the cerebral hemispheres. Most vestibular axons remain on the side of their origin.

GLOSSOPHARYNGEAL (IX)

The glossopharyngeal nerve innervates the tongue (glosso-) and the pharynx (-pharyngeal); it is a mixed nerve. Afferent fibers convey taste and general sensations from the posterior third of the tongue and general sensory impulses from the eustachian tube, tonsils, and behind the external ear. Efferent fibers stimulate secretion by the parotid gland and also control the stylohyoideus muscle of the pharynx, which is essential to the gag reflex. The fibers of the glossopharyngeal nerve remain on the side of their origin. A glossopharyngeal lesion on one side blocks the gag reflex; it also blocks taste from the posterior third of the tongue and displaces the uvula to the undamaged side.

VAGUS (X)

As its Latin name indicates (vagus, wandering), branches of the vagus nerve reach throughout the neck, thorax, and abdomen. The vagus is a mixed nerve, with parasympathetic visceral efferent and afferent fibers innervating the digestive tract and most other visceral organs. The only somatic fibers are afferents bearing touch, pain, and thermal impulses from the external ear and acoustic meatus. Among the sensory functions of the vagus are sensations of stretching and pain from the pharynx, larynx, esophagus, trachea, gastrointestinal tract, heart, and lungs. The motor functions of the vagus nerve belong to the **parasympathetic portion** of the autonomic nervous system (see Chapter 17, Divisions of the Nervous System). These fibers slow the heart cycle, increase peristaltic movements and secretion in the gastrointestinal tract, and contract the bladder wall, among many other actions (see Chapter 21, Autonomic Nervous System). The vagus nerve also supplies most of the voluntary muscles of the palate, pharynx, and larynx. Trauma to the vagus nerves on both sides is fatal because the pharynx and larynx become blocked, and the victim suffocates. Damage on one side makes breathing and swallowing difficult, and the voice becomes hoarse. The vagus nerve exits the medulla between the olive and inferior cerebellar peduncle and passes through the jugular foramen of the skull before branching into the neck and trunk.

ACCESSORY (XI)

Sometimes called the spinal accessory, because it has spinal and cranial roots, the accessory nerve is a motor nerve that sends spinal fibers to the sternocleidomastoid and trapezius muscles, as well as axons from cranial neurons to the pharynx and the intrinsic muscles of the larynx. This nerve assists the vagus with swallowing and breathing, but it contributes only moderately to neck motion because cervical nerves also supply the sternocleidomastoid and trapezius. The spinal portion collects roots from the first five cervical segments of the spinal cord as it ascends beside the cord into the foramen magnum, where it joins the cranial roots. The cranial portion collects four or five small roots from beside the medulla as it ascends, together with the spinal portion, to the jugular foramen. There it exits the skull with the vagus nerve. In fact the cranial fibers travel with the vagus nerve to the pharynx and larynx, but the spinal fibers descend into the neck. Since the fibers do not cross to the opposite side, damage to the spinal fibers of the accessory causes the shoulder on the same side to drop and the head to turn away from this side.

HYPOGLOSSAL (XII)

Hypoglossal means "below the tongue," and that is the key to this cranial nerve. The hypoglossal nerve controls the tongue and thereby assists you to eat, speak, and swallow. Both efferent and afferent fibers supply the intrinsic and extrinsic tongue muscles. The fibers pass from the floor of the fourth ventricle, lateral to the medial lemniscus and pyramids, without crossing, to exit from the anterior surface of the medulla between the pyramids and the olives. From there the hypoglossal passes through the hypoglossal canal, follows the vagus nerve briefly, and enters the base of the tongue above the hyoid bone. Since the nerve does not cross, the effects of lesions to the hypoglossal remain on the same side as the injury. Injury on the left side paralyses the left half of the tongue and causes it to deviate to the left side of the mouth when the patient "sticks out his tongue." The tongue thus points to the side of the hypoglossal lesion. Patients with this problem have difficulty swallowing and speaking.

SUMMARY

The brain communicates with the body by way of the spinal cord and cranial nerves. The brain receives and sends a spectrum of impulses that deal with personality, memory, and simple reflexes that regulate blood pressure, for example. The cerebral hemispheres are more advanced evolutionarily and handle memory, voluntary movement, and sensory functions, while the brainstem coordinates reflexes and channels neural pathways to the cerebral hemispheres. *(Page 558.)*

Both the brain and spinal cord develop from the neural tube. The regions of the brain are more highly specialized and modified than those of the spinal cord. The cerebral hemispheres are the most highly modified derivatives of the neural tube. Segmentation is more evident among the cranial nerves of the brainstem as one descends to the spinal cord. *(Page 559.)*

The ventricles of the brain are enlargements and specializations of the neural canal. The lateral ventricles are located in the cerebral hemispheres. The diencephalon contains the third ventricle, but the midbrain contains only the cerebral aqueduct. The fourth ventricle is shared by the pons and medulla and is continuous with the spinal canal. Ependymal cells line the ventricles, and at the choroid plexuses in each ventricle, these cells produce cerebrospinal fluid that circulates within the brain. *(Page 559.)*

The brain fits snugly into the cranial cavity, where it floats on a thin cushion of cerebrospinal fluid held in position by the meninges. The cranial fossas support the brain from below, and the calvarium protects it from above. Various foramina admit vessels and cranial nerves, and the spinal cord exits through the foramen magnum. *(Page 562.)*

The medulla is a reflex center that connects the brainstem with the spinal cord. The medulla closely resembles the spinal cord. The posterior columns ascend, and pyramidal tracts descend, the medulla. Cranial nerves enter this region from the head, neck, and viscera. Cranial nerve nuclei and the reticular formation communicate with higher portions of the brain. *(Page 563.)*

The pons bridges the distance between the medulla and midbrain, and the cerebellum coordinates muscular motion. Three cerebellar peduncles connect the cerebellum with the pons and adjacent regions. Four cranial nerves communicate between the pons and the ear, face, and mouth. The pons and cerebellum share the fourth ventricle with the medulla. *(Page 565.)*

The midbrain is the most rostral section of the brainstem. Ascending tracts continue through the midbrain, and other tracts descend from the cerebrum and diencephalon. The corpora quadrigemina are centers for ocular and auditory reflexes, and the cranial nerves of the midbrain control eye movements. The midbrain contains the cerebral aqueduct but no ventricle. *(Page 567.)*

The diencephalon is the gateway to the cerebral hemispheres and the brainstem. Situated atop the brainstem beneath the cerebral hemispheres, the nuclei of the thalamus communicate with nearly every tract that enters the cerebrum. The hypothalamus regulates neurosecretion and coordinates visceral reflexes. The optic nerve is the only cranial nerve of the diencephalon. The narrow third ventricle separates the diencephalon into the left and right walls, both of which house the thalamus and hypothalamus. *(Page 569.)*

The primary sensory and motor centers of the cerebral cortex receive conscious sensory impulses from the cranial nerves and spinal cord, and they send voluntary motor commands to the skeletal muscles. Association areas of the cortex interpret this sensory inflow into images and associations that are the basis of reasoning, memory, and personality. Although many functions, such as the special senses, motor control, and speech, can be localized to cortical areas, memory is distributed generally through the cortex. *(Page 573.)*

Typical cerebral cortex tissue contains six layers of neurons organized in vertical columns. Projection neurons communicate between the brainstem and cortex through the internal capsule and corona radiata. Opposite cerebral hemispheres communicate across the corpus callosum by way of commissural neurons, and association neurons connect different areas of the cortex within each hemisphere. The fourth ventricles are located deep in cerebral hemispheres, and the caudate nucleus of the basal ganglia curls beside them. The cerebral hemispheres receive the olfactory nerves. *(Page 576.)*

Tracts of the brain communicate between the cerebral cortex and the brainstem and spinal cord. The posterior column pathway conducts impulses concerning body sense and touch from the spinal cord to the cerebral cortex, making synapses in the medulla and thalamus before projecting onto the somatosensory cortex. Fibers of the spinothalamic tracts conduct pain, touch, and thermal stimuli received in the spinal gray matter, up the cord to the thalamic nuclei, and thence to the somatosensory cortex. Corticospinal tracts descend from the somatomotor center through the pyramids of the medulla to the spinal cord, where they make synapses with motor neurons. *(Page 578.)*

The cortex is not the only functional area of the cerebrum. The basal ganglia are known to be involved in motion because damage to the ganglia introduces abnormal movements, such as athetosis, that the ganglia normally inhibit. The limbic system is a ring of cortical gray matter plus portions of the thalamus and hypothalamus concerned with storing new information and choosing appropriate forms of behavior. *(Page 582.)*

The meninges support and protect the brain. The dura mater attaches the meninges to the surface of the cranial cavity. Beneath the dura, the arachnoid contains the cerebrospinal fluid that cushions the brain. The arachnoid is attached in turn to the pia mater, which is intimately bound to the external surfaces of the brain. Arteries and capillaries enter the brain after coursing through the subarachnoid space and pia mater. The blood-brain barrier lies in the walls of these capillaries and maintains constant internal conditions for the brain. Cerebrospinal fluid circulates from the choroid plexuses in the ventricles to the subarachnoid space and returns to the blood in the superior sagittal sinus. *(Page 584.)*

The brain communicates with the head, neck, trunk, and special sense organs by way of 12 pairs of cranial nerves. Most of these nerves contain afferent and efferent fibers, but some are strictly afferent, and a few are only efferent. Olfactory sensation reaches the cortex without going through the thalamus. The optic nerves supply vision by way of the diencephalon. The midbrain is concerned with ocular motion and reflexes. The pons is more diverse; it communicates with the face, orbits, oral cavity, tongue, and mandible. The medulla covers the largest territory; it receives hearing and balance impulses; it communicates with the tongue, pharynx, larynx, muscles of the neck, and the thoracic and abdominal viscera. *(Page 588.)*

STUDY QUESTIONS   OBJECTIVES

1. Name the three great divisions of the brain. Where are they located in the brain, and what functions does each one perform? Which divisions form the brainstem?   **1**
2. To which region of the embryonic brain does each division correspond? Which regions of the brain develop from these sections?   **2**
3. In which regions of the brain are the ventricles, aqueducts, and canals located?   **3**
4. What portion of the cerebrum is found in the anterior cranial fossa? Which fossa contains the cerebellum? Which fossa supports the temporal lobes? Where is the brainstem located in the cranial cavity?   **4**
5. Which cranial nerves enter the medulla oblongata? Which tracts descend through the pyramids? What are the posterior columns, and to what tract do they connect in the medulla?   **5**
6. Why is the medulla considered the part of the brainstem most like the spinal cord?   **5**
7. Which tracts pass through the base of the pons? Into the cerebellum? Which cranial nerves enter the pons? Do any enter the cerebellum itself?   **6**

8. What structure connects the third and fourth ventricles? What tracts do the cerebral peduncles and tegmentum of the midbrain carry? What effects would you expect a lesion of the superior colliculi to have on eye motion?   **7**
9. What parts of the diencephalon can be seen from the underside of the brain? Which cranial nerve enters here?   **8**
10. What are the functions of the various groups of thalamic nuclei? The functions of these nuclei require them to communicate with other regions of the brain. Which regions complement the functions of each group?   **8**
11. Draw a map of the primary motor and sensory areas of the brain, and name the cranial nerves that communicate to each one. What neurons connect these centers with others in the same hemisphere, in the opposite hemisphere, and with the brainstem?   **9, 10**
12. Beginning with the neurons of origin, trace the fibers of the spinothalamic tracts, posterior column pathway, spinocerebellar tracts, and corticospinal tracts through the brain. Be sure to indicate the sequence of neurons (first order, second order, etc.) in the pathway that connects each tract with its origin and destination.   **11**
13. How do we know that the basal ganglia are involved in muscular coordination? What are the major components of the limbic system? What role has the hippocampus in this system? The reticular formation is the agent of the reticular activating system of the brain. What features of the reticular formation enable this relationship?   **12**
14. Why is the brain so much "lighter" in cerebrospinal fluid than in air? How do the meninges take advantage of this circumstance to support the brain?   **13**
15. What is the blood-brain barrier? Why is it necessary? What tissue serves as the barrier, and how is this tissue specially adapted for its function?   **13**
16. Name the entire sequence of cranial nerves. Match these nerves with the regions of the brain they enter or exit.   **14**
17. Which cranial nerves are concerned with the following functions: eye movements, taste, swallowing, sensory input from the face and oral cavity, motor impulses to the muscles of facial expression and mastication, and hearing and balance?   **14**

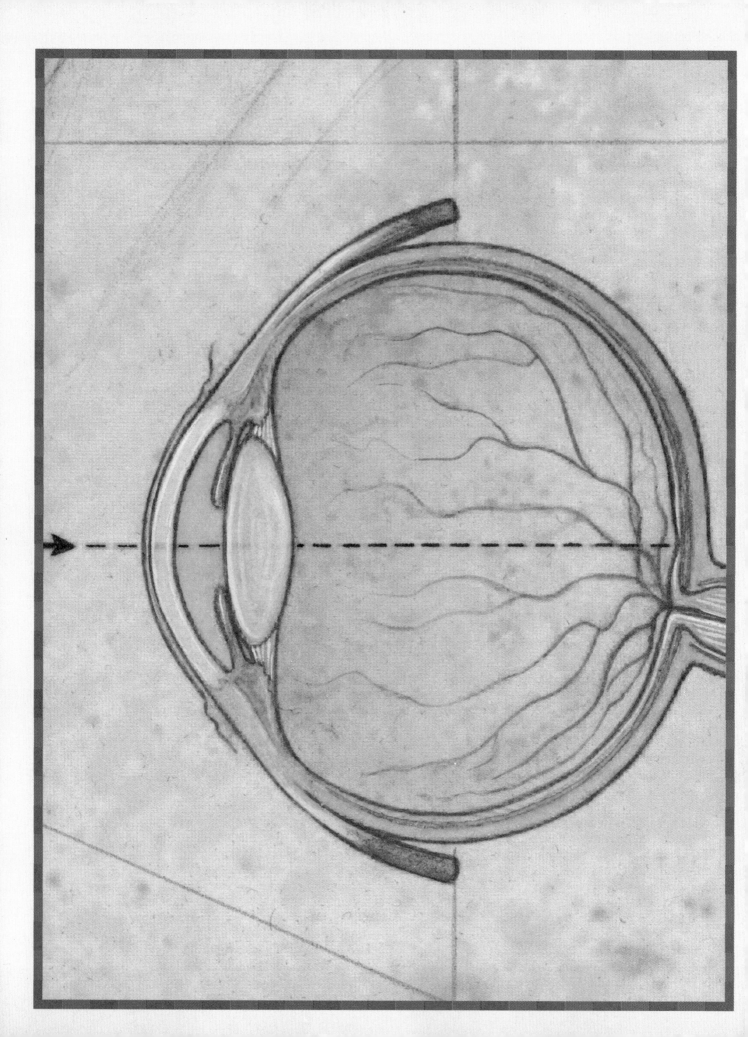

CHAPTER 20

Receptors and Special Sensory Organs

OBJECTIVES

After studying this chapter, you should be able to do the following:

1. Describe the perceptions resulting from stimulation of general and special sensory organs.
2. Decide whether a stimulus is detected by chemoreceptors, mechanoreceptors, or photoreceptors.
3. Name and describe the three tunics of the eyeball and the specialized structures they form. Know which meninges of the brain each tunic represents.
4. Describe the respective functions of the conjunctiva, lacrimal apparatus, cornea, iris, lens, retina, and aqueous humor.
5. Describe the action of the extrinsic muscles of the eyeballs.
6. Describe how the lens accommodates for near and far vision.
7. Describe how the pupillary light reflex works.
8. Trace the path of light into the eye to the retina, and trace the optical nerve pathway from the retina to the primary visual cortex. Know which portions of the visual cortex correspond to which portions of the retina.
9. Describe the individual functions of the three divisions of the ear. Know which structures conduct sound waves in each division.
10. Trace the path of sound waves from the external and middle ears to the organ of Corti in the inner ear, and outline the main features of the auditory pathway to the auditory centers of the brain.
11. Describe how the cochlea and organ of Corti distribute the sound waves that hair cells and cochlear neurons detect.
12. Describe how the semicircular canals and vestibule detect changes in motion.
13. Describe the anatomy and action of hair cells in the cochlea and in the vestibule and semicircular canals.
14. Describe the anatomy of olfactory epithelium and olfactory neurons, and trace the path of olfactory stimuli to the brain.
15. Describe the structure and distribution of taste buds, and trace the gustatory pathway to the brain.

 VOCABULARY TOOLBOX

These terms refer to sight, hearing, taste, or smell.

gustatory	gust-, *L* taste. Related to taste.	**optic**	opt-, *G* the eye, vision. Of the eye.
ocular	ocul-, *L* an eye. Of the eye.	**otic**	ot-, *G* the ear of the ear.
olfaction	olfact-, *L* smell. The process that detects odors.	**otitis**	-itis, *G* inflammation of the ear.
		otolith	-lith, *G* stone. Otoliths are ear stones.

RELATED TOPICS

- Autonomic nervous system, Chapter 21
- Cilia, Chapter 2
- Nasal mucosa, nasal cavity, Chapter 24
- Neurons, synapses, action potentials, Chapter 17
- Oral papillae, Chapter 23
- Pharyngeal arches, grooves, pouches, Chapter 4

PERSPECTIVE

The neural traffic that links the body with the brain begins flowing at sensory receptors throughout the body. Sensory receptors translate many kinds of stimuli, such as light, sound, odors, or pressure, into the neural impulses that the brain interprets as vision, hearing, smell, or touch. Your perception of these qualities depends as much on the ability of receptor cells to respond to a particular kind of stimulus as it does upon the centers in the brain that interpret these stimuli. What would be perceived if rods and cones of the retina somehow were connected to the auditory centers of the brain? Would you "hear" this chapter? Decide for yourself as you read the descriptions of sensory receptors and the pathways that convey their impulses to the brain.

Luscious popcorn from the microwave, sweet frozen yogurt at TCBY, and a frantic Springsteen concert all have their special delights. The cells that make us aware of these and all other perceptions are sensory receptor cells that detect the molecules and waves we interpret as odors, taste, sound, and light. Virtually all of our perceptions of the world inside and outside our bodies begin with sensory receptor cells. Receptor cells are not alone in responding to stimuli; all cells respond to changes around them. Sensory receptors are unique, however, because their responses produce action potentials (impulses) in the nervous system. Popcorn smells so good because certain molecules from the toasted kernels cause our olfactory receptors to stimulate pathways to the brain.

Three categories of receptors operate while you eat that frozen yogurt; each spoonful is cold, tastes sweet, and begs you to dig in and savor it. The simplest receptors are **free nerve endings,** shown in Figure 20.1A. These are specialized terminals of axons from individual neurons that end in the skin, muscles, and visceral organs, without cells to cover or support them. Many sensory receptors are **encapsulated nerve endings** in which supporting cells form small capsules around the tip of the axon (Figure 20.1B). Both free and encapsulated endings are **general sense organs** that provide the general senses of touch, pain, temperature, and body sense. These senses guide your spoon and assure you that the yogurt is indeed cold. The third group of receptors is assembled into elaborate **special sense organs** of the eyes, ears, tongue, and nose that supply the **special senses**—sight, hearing, balance, taste, and smell—that tell you how sweet and smooth the yogurt is.

Most receptors respond to light **(photoreceptors)** or to particular stimuli that are either chemical **(chemoreceptors)** or mechanical **(mechanoreceptors)** in nature. Chemoreceptors (KEEM-oh-reh-SEP-tors) detect chemical changes that underlie such perceptions as taste and olfaction. Mechanoreceptors (meh-KAN-oh-reh-SEP-tors) provide touch, body sense, hearing, and balance. Photoreceptors of course are the rod and cone cells of the eye. All three groups detect changes through the ion channels and membrane transduction pathways illustrated in Figure 20.1C. Chemoreceptors open or close ion channels when molecules bind to the cell membrane, and mechanoreceptors are arranged so that stress on cell membranes opens or closes ion channels. Light opens and closes ion channels on rod and cone cells.

We also speak of thermoreceptors, nociceptors, and others that describe the perception rather than the structure of the receptor or the kind of stimulus.

Thermoreceptors do just what their name says; they are chemoreceptors or mechanoreceptors that detect the flow of heat. **Nociceptors** (NO-see-SEP-tors) noc-, *L*, harm, harmful) provide the sense of various types of pain.

We also can distinguish receptors by their location. External sensors, such as the eyes and touch receptors in the skin, are **exteroceptors** (EX-ter-oh-SEP-tors). Sensations from within the body arise from **interoceptors,** or **visceroceptors,** that monitor internal conditions and are located in the visceral organs. Muscle spindles and stretch receptors in tendons, ligaments, and the stomach are interoceptors (IN-ter-oh-SEP-tors). The carotid body, which detects the oxygen content of the blood is another interoceptor.

- Which sensations are considered to be general senses, and which are special senses?
- What stimuli do mechanoreceptors, photoreceptors, and chemoreceptors detect?
- Where are interoceptors and exteroceptors located?

GENERAL SENSORY ENDINGS

FREE NERVE ENDINGS

Free nerve endings are the tips of axons that have no myelin sheath or Schwann cell covering. Only a basement membrane joins the axonal membrane to surrounding tissue. Figure 20.1A shows that free nerve endings slip in between epithelial cells of the skin, where they are capable of responding to heat and cold or to mechanical stimuli.

ENCAPSULATED NERVE ENDINGS

Supporting cells cover encapsulated nerve endings. Pacinian corpuscles are excellent examples of encapsulated endings. Each is an axon tip enclosed in concentric layers of supporting cells covered by a fibrous capsule. These spheroidal nerve endings are surrounded by concentric layers of modified Schwann cells, and also by supporting cells that resemble onion leaves when cut longitudinally, as in

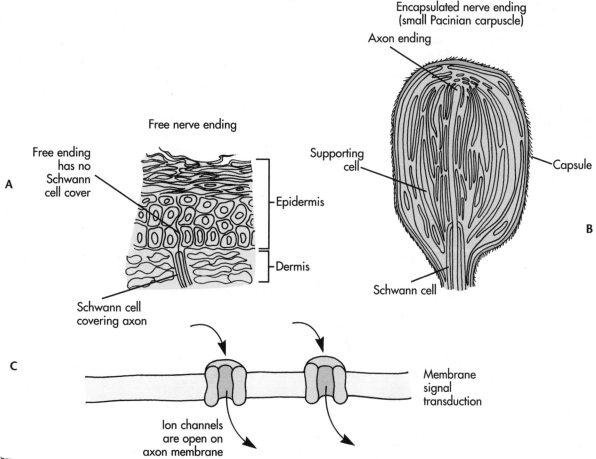

Figure 20.1

Nerve endings are either free **(A)** or encapsulated **(B)**. All endings employ ion channels **(C)** and membrane signal transduction pathways to initiate action potentials.

Figure 20.1B. Most Pacinian corpuscles (pah-SIN-ee-an) are 1 to 2 mm long, but large ones may be 1 cm in length. Located in the skin, mesenteries, external genitalia, and visceral organs, Pacinian corpuscles are mechanoreceptors that detect vibration and pressure. Chapter 12, Integument, describes other encapsulated endings of the skin.

Muscle spindles and **Golgi tendon organs** are mechanoreceptors that provide the sense of the body's position (described in Chapter 9, Skeletal Muscles). Muscle spindles illustrated in Figure 20.2A are scattered throughout skeletal muscles, where they detect the rate and extent of contraction. Special annulospiral and flower spray endings detect the stretch and contraction of small intrafusal muscle fibers inside the fibrous capsule that anchors the spindle in position. Golgi (GOAL-jee) tendon organs (Figure 20.2B) are found in tendons, ligaments, and joint capsules, where they detect tension by which the brain recognizes body position and controls muscle movement.

PAIN

Acute loud noises, stubbed toes, pin pricks, chronic headaches, and low backaches are all painful. They are unpleasant experiences associated with injury, actual or imagined. Unlike sight and hearing, pain is an elusive perception because the usual direct pathway between receptors and distinct regions of the brain is linked to many other influences. Minor injuries experienced during the heat of a game may only become painful afterward, for example. **Nociceptors** are free nerve endings distributed widely in the skin, periosteum of bones, muscles and joints, and the visceral organs. These receptors are generally silent unless they are stimulated by more heat and cold, pressure, or noxious substances than most sensory receptors regularly detect. Holding a warm coffee mug is not painful, but it becomes painful when the mug is filled with boiling water because nociceptors respond to the greater heat.

GATE THEORY

Perception of pain is modified by interactions between nociceptor neurons and other neurons in the dorsal horn of the spinal cord gray matter (Figure 20.3A). Certain interactions that sensitize our perception of pain "open the gate" to passage of impulses, whereas others inhibit or "close the gate." Pain from a cut in the skin of the arm is a good example of the gate theory at work. Nearby nociceptors that would not have responded before the injury, now sensitize the surrounding skin so that even gentle touch is painful, an example of gates opening. Who has not rubbed his or her shin after striking it against a coffee table? The rubbing (gentle stimulation) seems to help the pain (gates closing). This interaction between mechanoreceptors and nociceptors takes place in the spinal cord. Similarly, painkilling pathways from the brain can inhibit such stimuli (see the next section).

ANALGESIA

Analgesics are drugs that diminish pain (an, *G*, not; algia, *G*, pain) by acting either at the source of the painful stimuli or on the pathways that conduct these stimuli to the brain. Pain impulses travel to the brain through neurons that enter the dorsal roots of the

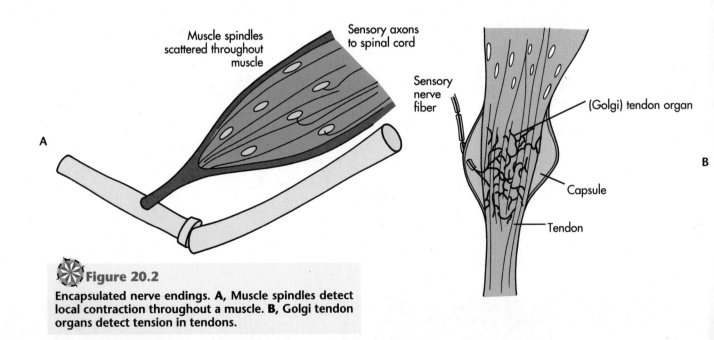

Figure 20.2

Encapsulated nerve endings. A, Muscle spindles detect local contraction throughout a muscle. **B,** Golgi tendon organs detect tension in tendons.

spinal cord and ascend the spinothalamic tracts to the thalamus and thence to the somesthetic center of the cerebral cortex (see Chapter 18, Spinal Cord). The nervous system produces its own analgesics—**enkephalin, dynorphin,** and **beta endorphin** (en-KEF-ah-lin; die-NOR-fin; en-DOR-fin)—peptides that are stored and released at many locations, especially in the dorsal horn of the spinal cord where nociceptive afferent neurons enter. Descending pathways from the cerebral cortex and the lateral corticospinal tracts release these painkillers in the dorsal horn, where they inhibit conduction of pain impulses, thereby closing the "gate." External analgesics, such as **aspirin,** relieve pain by interfering with the release and production of prostaglandins, substances that can promote inflammation and pain. Local anesthetics for tooth pain, for example, block local nerve conduction, so that fewer impulses reach the brain.

REFERRED PAIN

We perceive visceral pain in different locations from the actual source. Pain is often referred to cutaneous areas of the body that belong to the same spinal segment or dermatome as the source. Figure 20.3B shows that cardiac pain is perceived as being from the left arm and lateral thorax, both regions innervated by the same set of thoracic spinal nerves. Pain from the lungs is perceived as coming from the neck because both are innervated from the same segments of the

cervical spinal cord. The mechanisms of referred pain are controversial, but Figure 20.3C shows one reasonable explanation of how the brain can perceive two separate sources of pain in the same somesthetic area. Two different visceral and somatic afferent neurons may synapse with the same second order neurons ascending in the spinothalamic tracts. As a less likely alternative to this two-on-one convergence, single primary afferent neurons may have visceral and somatic collateral branches.

- Name examples of free and encapsulated nerve endings.
- What are analgesics?
- Name several examples of them.
- How does the gate theory explain pain?
- What is referred pain?

EYES AND VISION

BRIEF TOUR

Vision is our most precious sense. Every year in the United States the 10,000 people who lose their vision learn to face the fears we all have of living without this essential connection to the world. The images that dominate our lives begin in the retina, a thin sheet of tissue at the rear of the eye that detects light and begins to interpret patterns for the brain. It is not surprising that the retina helps process patterns, be-

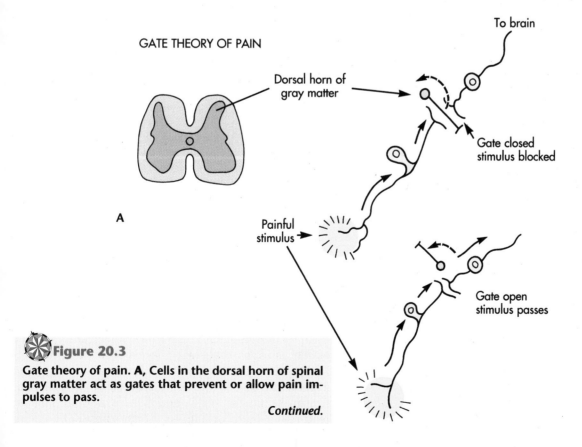

GATE THEORY OF PAIN

Dorsal horn of gray matter

To brain

Gate closed
stimulus blocked

Painful
stimulus

Gate open
stimulus passes

A

✳ **Figure 20.3**

Gate theory of pain. A, Cells in the dorsal horn of spinal gray matter act as gates that prevent or allow pain impulses to pass.

Continued.

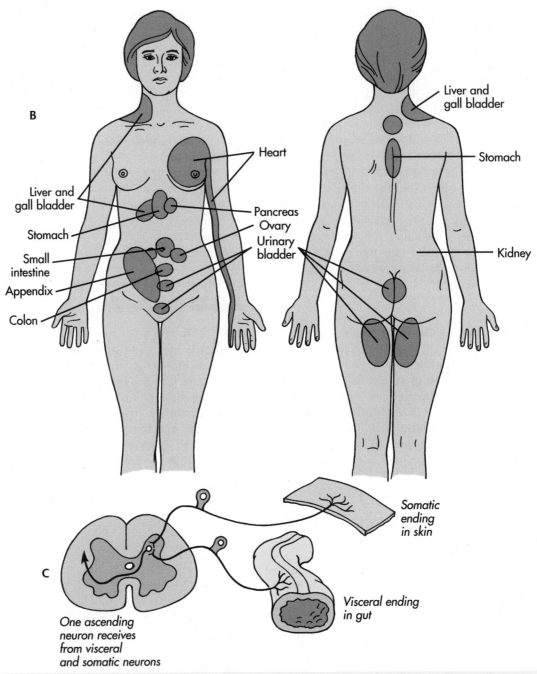

Figure 20.3, cont'd.

B, Referred pain. **C,** Two different somatic and visceral neurons that stimulate the same ascending fiber can explain referred pain.

cause it develops from the diencephalon (see Eye Development, this chapter).

The eyes direct light to the retina. The eyes, as seen in Figure 20.4A, reside in the **orbits,** protected by the skull from injury and cushioned on pads of fat. The **eyelids** guard the front of the eye, and **lacrimal glands** lubricate the front of the eyeball with tears. Light passes through the **palpebral fissure** between the eyelids before entering the eye itself. The **conjunctiva** covers the front of the cornea and the back of

the eyelids, thereby sealing off the eyeball and orbit from the outside. The eyeball itself (Figure 20.4B) consists of three layers or **tunics** of tissue. The outermost tunic is the **sclera** (SKLAIR-ah), a tough, opaque, fibrous material to which attach the **extrinsic ocular** muscles that train the eyes on their subject. Light enters the transparent **cornea** in front of the eyeball and passes through the **anterior chamber** before reaching the **pupil** and **iris,** which are special features of the second tunic, the **choroid** (KORE-oyd).

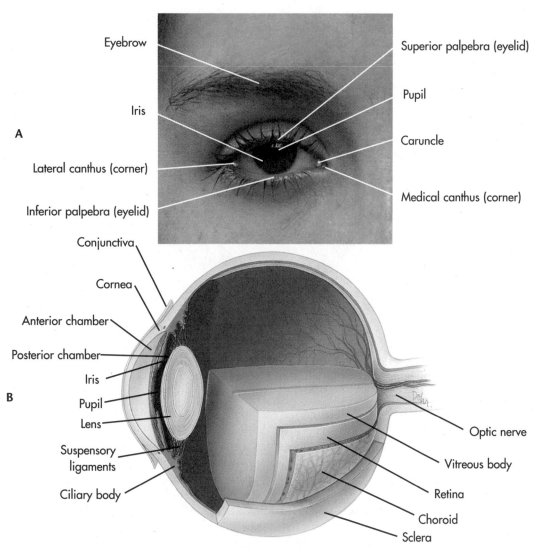

Figure 20.4

A, The eye. **B,** A sagittal section through the right eye. Light enters the eye from the left of the figure.

Highly vascular, the choroid is responsible for delivering circulation to the eye. The choroid forms the iris, which regulates how much light enters the pupil, and also the **ciliary body,** which focuses the **lens** and produces **aqueous humor,** the fluid that fills the anterior chamber. The narrow **posterior chamber** lies between the ciliary body and iris.

Emerging from the lens, light enters the transparent, gelatinous **vitreous body** (vitr-, *L,* glass, glassy) that fills most of the eyeball, and finally reaches the **retina,** the innermost tunic. Here, **rods** and **cones** transduce the light into neural impulses that, processed by the brain, result in the visual images we live by. Optic neurons collect these impulses and conduct them through the **optic nerve** and visual pathway to the **primary visual centers** of the brain, where image assembly and interpretation begins. The **fovea** (FOE-vee-ah) is an especially dense cluster of cones for acute vision at the optical axis of the eyeball; nearby,

the **optic disk,** or **blind spot,** covers the exit of the axons that form the optic nerve.

ORBITS

Each orbit is a pear-shaped opening in the skull whose walls taper toward an apex at the rear. Each eye peers out of the **orbital aperture** that is framed by the frontal, zygomatic, and maxillary bones, shown for the right eye in Figure 20.5. The lacrimal gland tucks behind the dorsolateral border of the orbit, and the lacrimal bone in the medial border directs the lacrimal ducts to the nasal cavity. The ethmoid and sphenoid bones enlarge the orbit behind these structures to accommodate the eyeball and ocular muscles. The orbit ends at the **optic foramen,** where the optic nerve leaves and the ophthalmic artery enters through the sphenoid bone. More vessels and nerves enter the **superior** and **inferior orbital fissures,** just forward of the optic foramen, on

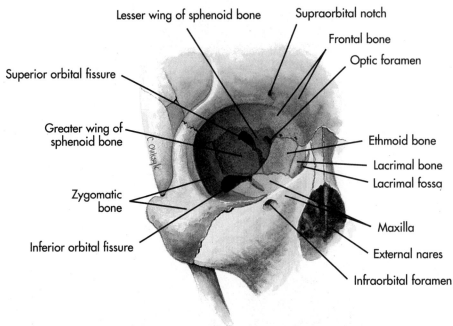

Superior orbital fissure
Lesser wing of sphenoid bone
Supraorbital notch
Frontal bone
Optic foramen
Greater wing of sphenoid bone
Ethmoid bone
Lacrimal bone
Lacrimal fossa
Zygomatic bone
Maxilla
Inferior orbital fissure
External nares
Infraorbital foramen

Figure 20.5
The right orbit. Find each of the bones that surface in the orbit, and find the orbital fissures and optic foramen.

course to the eyelids, extrinsic ocular muscles, lacrimal apparatus, and conjunctiva.

EXTRINSIC OCULAR MUSCLES

Six extrinsic ocular muscles aim each eyeball. Your brain processes images only when your eyes focus on a particular subject. Both eyes aim on the same subject in binocular vision, converging slightly for near objects, so that both images fuse together in a three-dimensional view. Double vision, or **strabismus** (stra-BIZ-muss; strab-, *G,* squint, cross-eyed), occurs when the images fail to match. As you read this, your eyes are flicking down the page in **saccadic motion** (sa-KAY-dik), forming momentary images on the retina as they go. Even when your head or the subject moves, your eyes keep the subject in constant view. This ability to search and follow is called **nystagmus** (nih-STAG-muss; nystagm-, *G,* nodding the head). Abnormal or exaggerated nystagmus often indicates neurologic disease.

Four rectus muscles and two oblique muscles control all eyeball motions, as shown in Figure 20.6. The **superior** and **inferior recti** elevate and depress the eyes (controlled by the oculomotor, cranial nerve III), and the **medial** and **lateral recti** swing the eyes horizontally (oculomotor and abducens, nerves III and VI, respectively). Move your eyes left, right, up, and down to work these muscles. All four muscles attach to the eye and to the **annular ring** that surrounds the optic nerve at the optic foramen. The **superior** and **inferior oblique** muscles rotate the eyeballs. The superior oblique (trochlear, nerve IV) rotates them medially, and the inferior oblique (oculomotor) rotates them laterally. The superior oblique is a remarkable muscle. It slides around the **trochlear ligament** (TROKE-lee-ar; troch-, *G,* a wheel) like a rope on a pulley and draws the upper surface of the eyeball forward and toward the midline. From below, the inferior oblique rotates the eyeball in the opposite direction, but there is no trochlear ligament for this muscle. A seventh muscle, the **levator palpebrae superioris** (LEV-ah-TORE PAL-peh-BREE su-PEER-ee-OR-iss), raises the upper eyelid.

The motions of the rectus muscles are easy enough to understand, but what is the function of the obliques? Figure 20.7 shows that when the eyes are looking straight ahead (you see them from above in the figure) and the superior recti elevate them, most of their pull goes into elevating the eyeballs, but the recti also adduct and rotate the eyeballs slightly. This motion would disrupt binocular vision if the obliques did not keep the eyes on target. The inferior obliques nicely cancel adduction and rotation by pulling from below. Likewise the superior obliques correct the inferior rectus muscles when they depress the eyes. The obliques also cancel abduction when the eyes swing laterally. Similar problems and solutions occur when the lateral and medial recti swing depressed or elevated eyes. The oblique muscles are indispensable for binocular vision because each rectus muscle, in whatever direction it is pulling, needs an oblique to adjust its effects. One reason why all these adjustments are necessary is that the axes of the

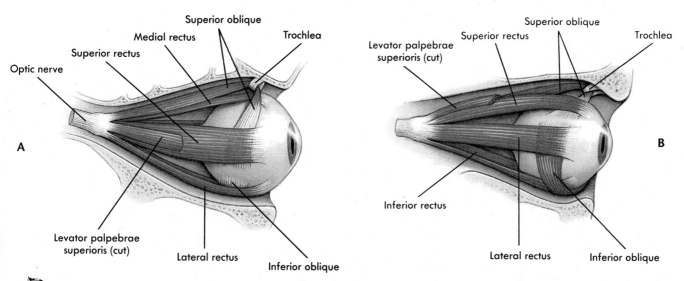

 Figure 20.6

The extrinsic ocular muscles. **A,** Superior view of the right eye. **B,** Lateral view of the right eye. Find the superior rectus muscle, the medial and lateral recti, and the superior oblique muscle. The inferior oblique appears in **B.**

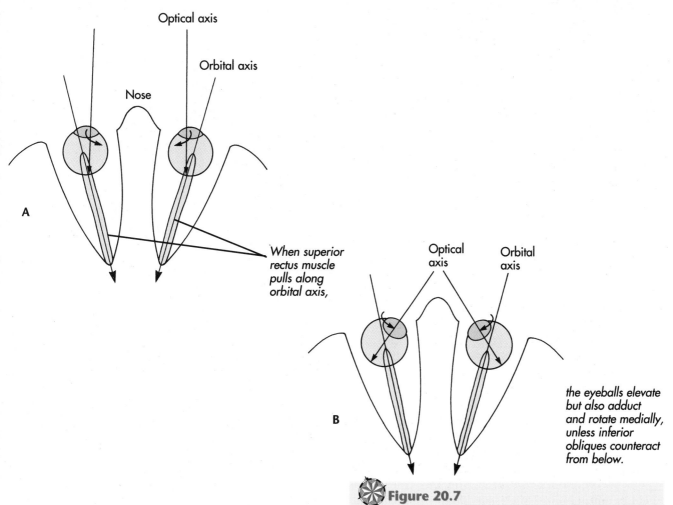

Figure 20.7

A, Superior rectus muscle in resting position. **B,** When this muscle contracts, it elevates the eyeball but also adducts and rotates it toward the nasal septum (curved arrows).

eyes and the orbits are not the same. The orbits diverge by 45 degrees, but the optical axes (through the lens and fovea) are parallel.

- Which rectus muscle elevates the eyes, and which depresses them?
- Which recti swing the eyes to the right? To the left?

EYELIDS, CONJUNCTIVA, AND LACRIMAL APPARATUS
EYELIDS

The eyes see through the opening between the eyelids known as the palpebral fissure (palpebr-, *L*, an eyelid, wink, blink). The eyelids themselves are folds of skin and muscle from the superior and inferior margins of the orbit that join at the **medial** and **lateral canthi** (KAN-thy; canth, *G*, the corner of the eye) on either end of the palpebral fissure (Figure 20.4A). The eyelids protect the eyes from infection and irritations, and they lubricate the cornea when the eyelids blink. Sebaceous glands in the **lacrimal caruncle** (KAR-un-cl), a fold of skin at the medial canthus, produce the scaly secretions that accumulate during sleep. The **semilunar fold,** a portion of conjunctiva that resembles the nictitating membrane that protects the eyes of many animals, has no known function in humans. Two or three rows of **eyelashes** decorate the margin of each lid and help prevent small foreign objects, such as dust and insects, from entering the eye.

Tarsal plates (tars-, *G*, a flat surface) support each eyelid, as you can see in Figure 20.8A. Shaped like half-moons, these fibrous connective tissue plates ride up and down each time the lids open and close. **Palpebral ligaments** join the ends of the plates to each other and to the wall of the orbit behind the canthi. Each plate contains approximately 30 **tarsal (Meibomian) glands** (meh-BOME-ee-an) that secrete a film of oil over the surface of the eye to help prevent evaporation. The levator palpebrae muscle raises the upper eyelid by pulling the upper tarsal plate, and gravity opens the lower plate. Two visceral, or **Muller's, muscles** also help withdraw the lids. (Chapter 21, Autonomic Nervous System, discusses the role of these muscles.)

CONJUNCTIVA

The conjunctiva lines the eyelids and the cornea. The conjunctiva is an epithelial sac that opens through the palpebral fissure (Figure 20.8B) and covers the inner surface of the eyelids and outer surface of the cornea (corn-, *L*, horny) with one continuous sheet of epithelium. This tissue is known as the **palpebral conjunctiva** as it lines the eyelids, but it becomes the **bulbar conjunctiva** that covers the cornea and sclera. Numerous goblet cells in the palpebral conjunctiva lubricate the surface of the eye. The bul-

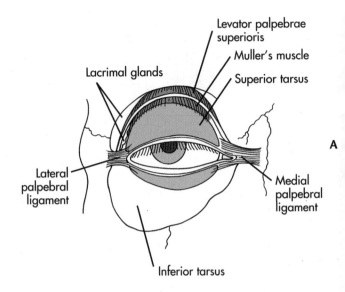

A

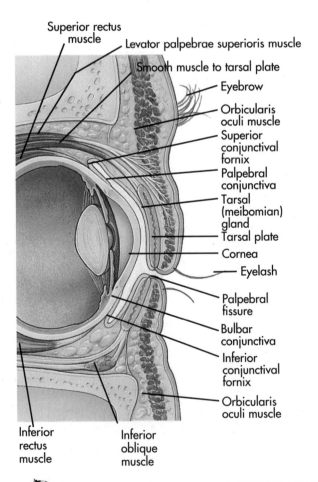

Figure 20.8

A, Forward view of tarsal plates and palpebral ligaments of the right eye. **B,** Sagittal section through the eye and eyelids.

bar conjunctiva is firmly attached to the cornea, but looser connections at the bases of the eyelids allow the conjunctiva to move with the eyeball. Eyes become bloodshot when irritation dilates the capillaries and small vessels of the bulbar conjunctiva.

LACRIMAL APPARATUS

The lacrimal gland secretes tears into the conjunctival sac. Approximately six ducts, shown in Figure 20.9, continuously release a thin film of tears from the superior lateral wall of the orbit. Every time they blink, the eyelids sweep this optically smooth film across the eye toward the medial canthus, where two **lacrimal canaliculi** drain into the lacrimal sac.

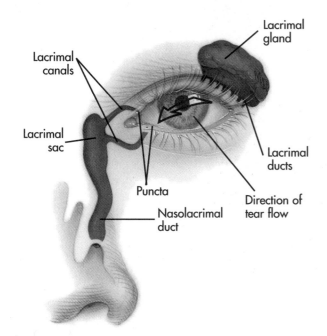

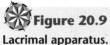
Figure 20.9
Lacrimal apparatus.

The **lacrimal punctum** drains from the eyelid at the margin of the semilunar fold, and the **lacrimal sac** collects tears behind the medial canthus. The **nasolacrimal duct** delivers to the inferior meatus of the nasal cavity (see Chapter 24, Respiratory System) by way of a shallow groove in the lacrimal bone. Usually, you are not conscious of this flow of tears because the nasolacrimal duct carries it away, but when you cry the lacrimal glands discharge more than the duct can drain. Somatic and autonomic nerve fibers serve the lacrimal gland through branches of the trigeminal nerve V.

- Where are the medial and lateral canthi located?
- What is the name of the opening of the conjunctival sac?
- Which structure drains tears to the nasal cavity?

SCLERA AND CORNEA—THE FIBROUS TUNIC

The sclera is the fibrous tunic of the eye; it is homologous with the dura mater of the brain. The **sclera** (scler-, *G*, hard) is opaque except for the **cornea** in front of the eyeball, seen in Figure 20.4B. Both structures are poorly vascularized mats of collagen fibers about 0.4 to 1.0 mm thick. The sclera itself surrounds nearly five sixths of the eyeball. Extrinsic ocular muscles insert upon the sclera, and a thin membrane called **Tenon's capsule** allows the sclera to rotate smoothly on fatty pads in the orbit. The cornea joins the sclera at the **limbus** (*L*, an edge, a head band). Because few vessels enter either tissue, the fibrous tunic relies on the choroid beneath and external membranes for respiration and nourishment. This is an advantage in surgery because the cornea and sclera do not bleed. The optic nerve pierces the sclera through a sieve-like plate called the **lamina cribrosa.** Optic nerve fibers pass through holes in the plate. In the disease **glaucoma** (glauc-, *G*, gray, bluish-gray), high pressure within the eyeball often causes blindness by squeezing the fibers against this plate until they are unable to conduct impulses and may die.

The shape of the cornea helps focus light on the lens because the cornea curves more sharply and is thicker at the limbus (1.0 mm) than the center (0.5 mm). When the shape of the cornea is imperfect, ophthalmologists may attempt to improve it with **radial keratotomy** (KAIR-ah-toe-TOE-mee; tomy, *G*, cut), a procedure that reshapes the cornea through minute slits cut by a laser.

FIVE LAYERS

The cornea owes its transparency to its ability to absorb water from the aqueous humor (aqua-, *L*, water). The aqueous humor also supplies nutrients, and oxygen diffuses through the bulbar conjunctiva from the outside to compensate for the lack of corneal vessels. Contact lenses must therefore be very permeable to oxygen, or the epithelium may ulcerate and the cornea may become opaque. Prolonged wearing of contacts can induce capillaries to grow into the cornea. This is why your ophthalmologist or optometrist urges you to keep your contact lenses clean and to remove them at night.

The outermost layer of cornea (1) shown in Figure 20.10 is the bulbar conjunctiva, a **stratified epithelium** five to six cells thick that remains transparent because it does not cornify. Microridges make the surface optically smooth by trapping a thin film of tears. Proliferating cells renew the epithelium approximately every week. A scratched cornea is painful because numerous free nerve endings make it sensitive.

The next layer (2) is **Bowman's membrane,** an extracellular matrix of randomly oriented collagen fibers that ends abruptly at the limbus and anchors the epithelium to the substantia propria.

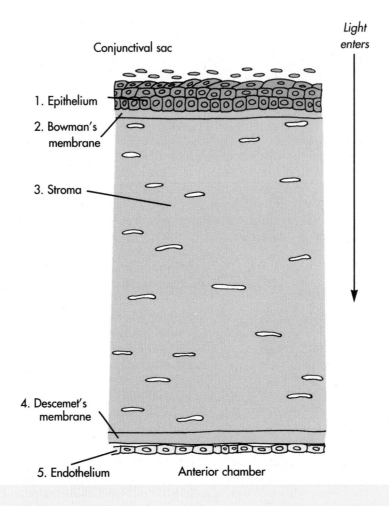

Conjunctival sac

Light enters

1. Epithelium

2. Bowman's membrane

3. Stroma

4. Descemet's membrane

5. Endothelium

Anterior chamber

Figure 20.10
Five layers of the cornea, viewed in transverse section. Light enters from above.

The **stroma** (3) is the main transparent layer of the cornea. Except for occasional keratinocytes and numerous macrophages, the substantia propria is composed of alternating strata of **collagen fibers** and elastic fibers with mucopolysaccharides interspersed among them. There are no vessels. This arrangement also ends sharply at the limbus, where the scleral fibers become more random and some vasculature returns. The mucopolysaccharides tend to take up water and thus provide transparency, but too little or too much hydration causes the substantia propria to lose its clarity.

Like Bowman's membrane, **Descemet's membrane** (des-eh-MAYS; 4) anchors the substantia propria to thin, hexagonal cells of the corneal endothelium.

The **endothelium** (5) adjusts the water content of the substantia propria by pumping water back against the flood diffusing from the anterior chamber. Numerous **transcytotic vesicles** vigorously move across the endothelium. The possibility of breaks in the en-dothelium is a substantial risk in radial keratotomy because the substantia propria quickly swells and clouds over wherever a break occurs.

- What are the functions of the sclera?
- Which portion of the cornea accounts for its transparency?
- What tissues regulate transparency?

CHOROID, IRIS, AND CILIARY BODY— THE VASCULAR TUNIC

The **vascular tunic is a modified extension of the arachnoid and pia mater** from the meninges of the brain. Also known as the **uveal tract** (YOU-vee-al), it is made up of the choroid, iris, and ciliary body.

CHOROID

The **choroid is sandwiched between the sclera and retina,** and it reaches forward as the ciliary body and iris (Figure 20.4B). Pigment cells give the choroid (*G,* like a membrane) the color of dark wine grapes (uvea; uv-, *L,* a grape). As in the pia mater, a fine

Fovea centralis　　Macula lutea

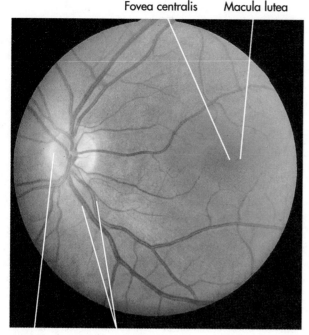

Optic disc　　Retinal vessels

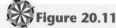

Figure 20.11

Fundus of the eyeball. This photograph reveals what ophthalmologists see of the retina and choroid at the rear of the eyeball.

network of capillaries called the **choriocapillaris** (KORE-ee-oh-KAP-ill-ar-iss) invades the inner surface of the choroid, bringing blood to the rods, cones, and pigment cells of the retina. Figure 20.11 shows that blood also enters through the posterior ciliary arteries at the optic nerve and through anterior ciliary arteries at rectus muscle insertions. Vortex veins drain from the posterior eyeball. Damage to the choroid can be dangerous because ruptured vessels can detach the retina from its source of nutrition.

IRIS

The iris regulates the amount of light entering the eyeball. It is a delicate screen in front of the lens, as seen in Figure 20.4B. Its round **pupil** dilates in dim light and constricts in bright light. Smooth muscle fibers and autonomic neurons regulate the pupillary opening. Parasympathetic neurons excite a ring of **sphincter fibers** around the pupil to constrict it. **Radial fibers** contract and thereby dilate the pupil under sympathetic stimuli (see Chapter 21, Autonomic Nervous System). Epithelium (Figure 20.12) covers the anterior and posterior surfaces of the iris and encloses a connective tissue **stroma** that supports the pupillary muscle fibers and contains pigment cells, nerves, and vessels.

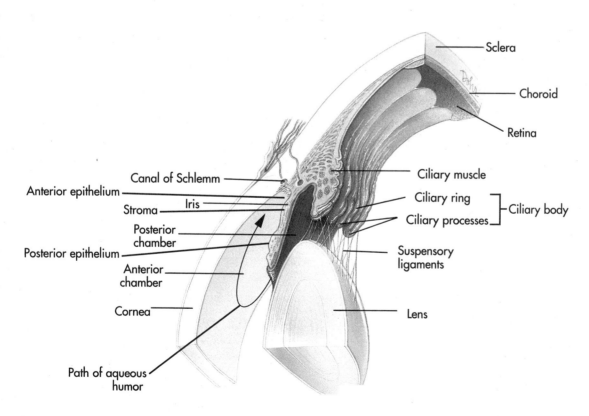

Sclera

Choroid

Retina

Canal of Schlemm

Anterior epithelium

Stroma

Iris

Ciliary muscle

Ciliary ring

Ciliary processes

Ciliary body

Posterior chamber

Posterior epithelium

Anterior chamber

Suspensory ligaments

Cornea

Lens

Path of aqueous humor

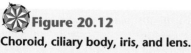

Figure 20.12

Choroid, ciliary body, iris, and lens.

Immense quantities of **melanin granules** (melanosomes) produced by melanocytes block light from passing through the posterior epithelium. Although these melanin granules usually are brown, the iris does not always appear so. Eye color depends on the distribution of melanin. Brown granules in the epithelium, viewed through the stroma of the iris, make the iris appear blue because the stroma scatters light rays and only shorter blue waves reflect out of the iris to an observer. When the stroma also contains melanin granules, these granules also absorb blue light and scattered green light bounces back from the iris. The iris appears brown when the stroma contains enough melanin granules to absorb all light. If the stroma is especially dense but lacks melanin granules, the iris looks gray.

LENS AND CILIARY BODY

The lens focuses light upon the retina. The lens is a resilient, epithelial sac about 9 mm in diameter and 5 mm thick, as shown in Figure 20.13. The anterior wall of the sac is a single, thin layer of cuboidal cells known as the **lens epithelium.** Tall, narrow,

highly refractile **lens fiber cells** form the posterior wall. Like other epithelia, the lens has no vasculature or connective tissue. It relies on the aqueous humor for nutrients, which must cross a thick basement membrane, called the **lens capsule,** that envelops the entire lens.

The lens focuses light from near or far objects by changing its shape, a process called **lens accommodation.** Far objects come into focus when the lens flattens (Figure 20.12), and when the lens is nearly spherical, it is focused on near objects. Tension from two sources determines the shape of the lens. First, the resilience of the lens itself tends to make it spherical. Second, the lens is suspended by numerous **zonular filaments** (ZONE-you-lar), like the hub of a bicycle wheel. These filaments run like the spokes of the wheel between the lens capsule and the **ciliary body,** a band containing smooth muscle around the root of the iris. Increased tension on the zonular filaments tends to flatten the lens, and reduced tension allows the lens to round up.

Circular fibers in the ciliary muscle adjust the balance between flattening and rounding. Their contraction releases tension on the zonular fibers and the lens rounds up, but their relaxation places more tension on the fibers and the lens flattens. The fatigue from continual contraction of these fibers is why close reading tires the eyes. Because parasympathetic neurons of the oculomotor nerve (III) stimulate the circular fibers to contract, ophthalmologists use cholinergic blockers, such as cyclopentolate, to prevent the lens from accommodating while they examine a patient's eyes. Lenses eventually lose their refractility and their ability to accommodate to near objects. This condition, called **presbyopia** (PREZ-bee-OPE-ee-ah; presby-, *G,* old), awaits you in middle age, when you will probably need reading glasses to compensate for your aging lenses.

Adult lenses continue to grow and enlarge by adding more lens fibers. New cells in the **germinal band** of epithelium (Figure 20.13) synthesize crystalline proteins that give the lens its refractile ability. Lens fibers persist for life, with the ones from infancy clustered in the **central nucleus** of the lens, while newer fibers accumulate around them. The newest of course are forming now at the equator of the lens.

CILIARY BODY AND AQUEOUS HUMOR

The ciliary body also produces aqueous humor that nourishes the lens and cornea, as illustrated in Figure 20.12. The ciliary epithelium secretes about 3 ml of aqueous humor every day. Approximately 70 highly vascular **ciliary processes** extend the epithelium among the zonular filaments, increasing its secretory surface. **Sodium-potassium ATPases** in the cell membranes secrete sodium ions into the **posterior chamber,** and chloride ions and water follow os-

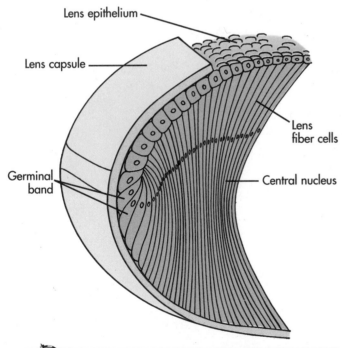

Lens epithelium

Lens capsule

Germinal band

Lens fiber cells

Central nucleus

✳ **Figure 20.13**

This section through the lens reveals lens fibers and epithelium.

motically from the ciliary vessels beneath the epithelium. Most of this fluid passes through the pupil into the anterior chamber, but some enters the vitreous body behind the lens, bathing the lens as it flows past. The aqueous (as clinicians frequently call it, for short) bathes the endothelium of the cornea and the iris, and it is drawn toward the periphery of the anterior chamber, where it finally exits through the **canal of Schlemm** into veins. Blockage in this canal is a common contributing cause of glaucoma because accumulating aqueous humor increases the intraocular pressure and squeezes optic nerve fibers into the lamina cribrosa, the openings in the sclera for the fibers of the optic nerve. One surgical treatment drains the excess aqueous through tiny tubes implanted at the base of the cornea.

- Which layer of meninges does the choroid represent?
- Why do some irises appear blue even though their pigment granules are brown?
- Which shape of the lens, round or flattened, focuses on near objects?
- How can blocking the canal of Schlemm lead to glaucoma?

RETINA—THE NEURAL TUNIC

The retina is the light-sensitive tissue that registers images for the brain. The retina actually is composed of two layers of epithelium, shown in Figure 20.4B. The innermost **neural retina** is a thin, transparent sheet of tissue that contains the rods and cones and other neurons. The outer layer is the **retinal pigment epithelium,** which backs the neural retina with melanocytes that trap extraneous light. Although both layers extend forward over the ciliary body as pouches, called the **ora serrata,** and over the posterior surface of the iris, these areas are not light-sensitive. At the opposite pole of the retina, the **fovea** (L, a pit) is a depression that is packed with cone cells for especially acute vision. Just medial to the fovea is the **optic disc,** the retina's blind spot; no image is recorded here because the axons of the optic nerve have displaced the rods and cones. To find your blind spots, look straight ahead, close one eye, and hold your finger at arm's length slightly off center. Slowly move your finger horizontally until you find the location where the tip disappears. Do the same for the other eye.

Light passes through most of the neural retina before reaching the rods and cones, as Figure 20.14 shows, but we will follow the layers of the retina in the opposite direction along the path of neural impulses to the optic nerve. The **retinal pigment epithelium** is the outermost of ten layers. These cells are so densely packed with melanin granules that it is difficult to see other organelles in the cytoplasm. These pigment cells enfold the rods and cones, absorbing extraneous light and also degrading discarded photoreceptor membranes, as described below. **Rods** and **cones** are tall, narrow, modified neurons that extend through the next four layers of the retina. Light-sensitive **outer segments** and supportive **inner segments** occupy the first of these layers. An **outer limiting membrane** separates these segments from the **outer nuclear layer,** where nuclei of rods and cones are located. Axon-like processes extend into the **outer plexiform layer** (plex-, L, interwoven, a network), where they synapse with **bipolar** and **horizontal cells.** The cell bodies of these neurons in turn are the **inner nuclear layer.** Next, the **inner plexiform layer** is the site of more synapses between the bipolar cells, **amacrine cells** (AM-ah-KRIN), and dendrites of **ganglion cells.** Axons of ganglion cells form the optic nerve. The nuclei and cell bodies of the ganglion cells are found in the **ganglion cell layer,** and their axons lie in the adjacent retinal **nerve fiber layer.** The **internal limiting membrane** of the retina abuts the vitreous body. Interspersed among all these receptors and neurons are tall, narrow **Muller cells** (MULE-er), glial cells whose opposite ends spread out to form the limiting membranes and to support the other cells.

RODS AND CONES

Rods and cones are photoreceptors. Rods, illustrated in Figure 20.15, are slightly taller and more sensitive to light (they can detect single photons), but their cylindrical outer segments record only intensity, not color. Cones distinguish color, but their conical outer segments require stronger light, which is why colors seem to disappear in very dim light. Cones use different photopigment molecules governed by different genes; some photopigments detect red, others green, still others blue light. Each retina contains approximately 120 million rods and 6 million cones.

Outer rod and cone segments are modified cilia with centrioles and the typical 9 + 2 arrangement of microtubules (see Cilia, Chapter 2). Stacks of **membranous disks** contain the light-sensitive **photopigments—rhodopsin** in rods, **iodopsin** in cones—that capture photons of light. Although individual rods and cones persist for life, they renew their disks from components assembled by the inner segments. Rods continually form new disks at the base of the stack and discard old ones at the tip of the outer rod segment. The pigment epithelium destroys the discarded disk membranes by phagocytosis. In contrast, disks of cones are permanent structures but are renewed in place with new iodopsin molecules.

Rhodopsin and iodopsin photopigments control gated ion channels in the membrane of the outer and inner segments of the rods and cones. Rods and cones maintain a constant sodium ion current that flows out from ion channels in both segments. When photopigments absorb light, they trigger membrane signaling systems and second messenger molecules (cyclic guanosine monophosphate, cGMP) in the disks to

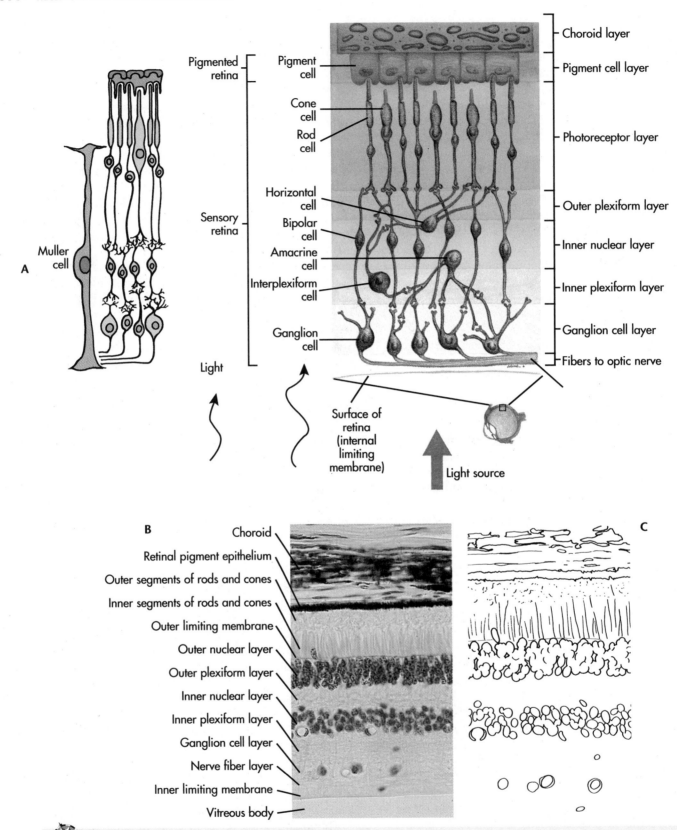

Figure 20.14

Transverse section through the retina. **A,** Rods, cones, and neurons occupy different layers of the retina. **B,** Photomicrograph of cellular layers. **C,** Interpretive drawing identifies the layers

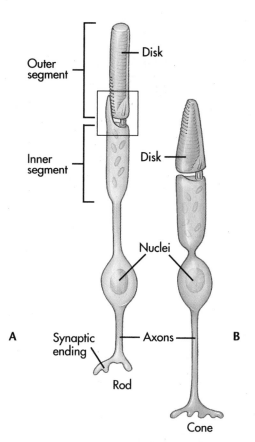

Figure 20.15

Anatomy of rods and cones. Light enters from the bottom of the figure and is stopped by the pigmented retina at the top.

close the sodium ion channels in the outer segments. These momentary decreases in current stimulate the bipolar cells, and a stream of action potentials starts toward the brain.

VISUAL PATHWAY

The retina and primary visual cortex resemble TV screens, with every portion of the retina represented in the cortex as shown in Figure 20.16. Each rod and cone "sees" a small point in the visual field, like each dot on the screen, and transmits impulses to comparable points in the visual cortex, which assembles pictures of the entire scene. The image of the car in Figure 20.16 drapes from right to left, upside down, across the **primary visual cortex.** The right halves of each retina see the left half of the visual field, that is, the rear of the car. The rest of the car is recorded by the left half of each retina. The lens reverses and inverts the image, but this does not matter because the brain learns to interpret the images in an upright world. The right halves of the retinas deliver to the

right primary visual cortex in the right occipital lobe, and the left halves of the retinas deliver to the left visual cortex. Axons cross as appropriate to the opposite side at the **optic chiasm** (chias-, *G*, cross).

A series of three neurons connects each rod or cone to cells in the visual cortex. Impulses from a rod or cone travel (1) first through a bipolar cell, (2) then to a ganglion cell that conducts through the optic nerves to the thalamus (lateral geniculate nucleus), (3) where the axon of a third order neuron delivers to the primary visual cortex. Horizontal and amacrine (ama, *G*, together) cells integrate the signals from clusters of rods and cones, shown in Figure 20.17, so that each ganglion cell receives stimuli from approximately 130 rod and cone cells in each cluster. Rods predominate in the periphery of the retina (which is why peripheral vision is more sensitive in dim light), but cones and smaller clusters predominate at the fovea, allowing sharper vision in color.

Ganglion cell axons converge toward the optic disk and become myelinated when they exit as the **optic nerve.** The optic nerve exits the rear of the eyeball, passes through the optic canal of the skull, and meets its opposite partner at the optic chiasm, just anterior to the diencephalon in the chiasmatic groove of the sphenoid bone.

The first steps toward binocular vision occur in the optic chiasm, where axons from both eyes meet (Figure 20.16). Axons from the nasal halves of each retina cross to the opposite side across the chiasm, but those of the temporal halves do not cross. Consequently, each **optic tract** exits the chiasm with axons from the corresponding half retinas—the left tract with axons from the left halves and the right tract with those from the right. The regrouped fibers run as the optic tract to the **lateral geniculate nuclei** (je-NIK-you-late) in the posterior wall of the thalamus, where the next step in binocular vision takes place.

In the lateral geniculate nuclei, fibers of the optic tracts synapse with third order neurons that project to the primary visual cortex and also to the midbrain to control the pupillary light reflex. Each nucleus matches fibers from corresponding locations in each retina, but the images do not yet fuse; that occurs in the visual cortex itself. Each nucleus contains **columns** of cell bodies of **third order neurons** that receive impulses from clusters of rods and cones from corresponding points in each retina. A single column will receive input from the cells in both retinas that "see" a point on the hood ornament in Figure 20.16, for example. The **optic radiations** connect these columns with the primary visual cortex.

The primary visual cortex fuses the images from both retinas. The optic radiations wrap the retinal image of the car across the visual cortex in Figure 20.16. The anterior end of the left cortex records the front end of the car. Impulses from the fovea, which

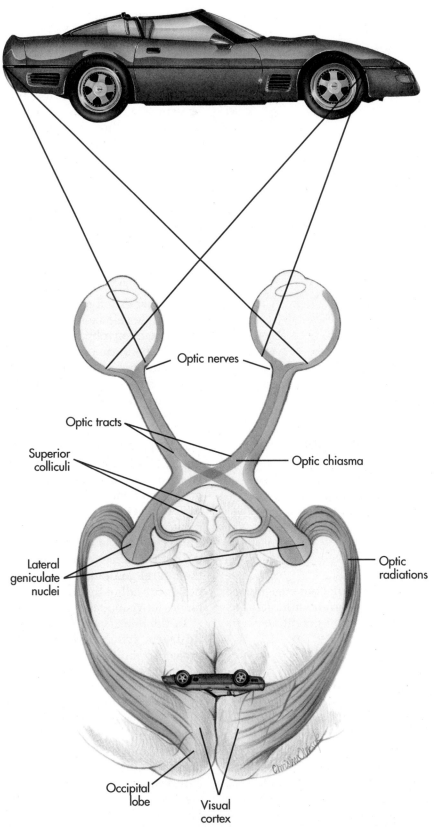

Optic nerves

Optic tracts

Optic chiasma

Superior colliculi

Lateral geniculate nuclei

Optic radiations

Occipital lobe

Visual cortex

✳ **Figure 20.16**

The optic pathway. The right half of each retina views the rear of the sports car, and optic neurons project this image onto the right primary visual cortex. Similarly the left half of each retina views the forward half of the car and projects this image onto the left visual cortex. The fovea projects onto the posterior of each cortex, and the lateral retina projects to the anterior cortex.

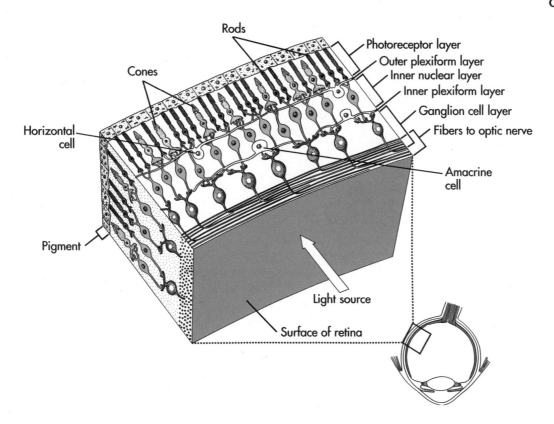

Figure 20.17

Neural circuitry within the retina. Find a ganglion cell body in the ganglion cell layer and trace back from it to the rods and cones that stimulate this cell. Each cluster like the one you have found actually contains about 130 rods and cones.

concentrates on the middle of the car, are recorded in the posterior cortex of both sides, while the rear of the car is perceived in the anterior portion of the right cortex. Trace the figure back to the retina to be sure you understand the pathway. Remember: right cortex, right half retinas, left half visual field; left cortex, left half retina, right half visual field.

The cells of the visual cortex form alternating **ocular dominance columns** that resemble zebra stripes, as shown in Figure 20.18. Each column (approximately 1 mm wide) receives axons from the half retina of one eye, such that left and right eye inputs alternate with each other across the cortex. Connections between the columns fuse the separate images into three dimensions.

- How many neurons connect a rod photoreceptor with its corresponding location in the visual cortex?
- What does the left side of the primary visual cortex "see"?
- Which axons of the optic nerve cross the optic chiasm, and which ones do not?

EYE DEVELOPMENT

The eye begins developing when optic vesicles grow out from the diencephalon and induce ecto-

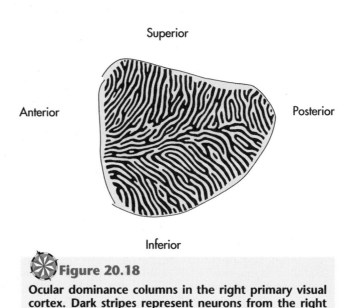

Figure 20.18

Ocular dominance columns in the right primary visual cortex. Dark stripes represent neurons from the right eye and light stripes, neurons from the left eye.

derm to form a lens, as you can see in Figure 20.19A. The **optic vesicles** appear at the end of the third week of development. In the following week, they enlarge, contact the ectoderm, and induce the lens to form. By week 5, the neural and pigmented retina begin to

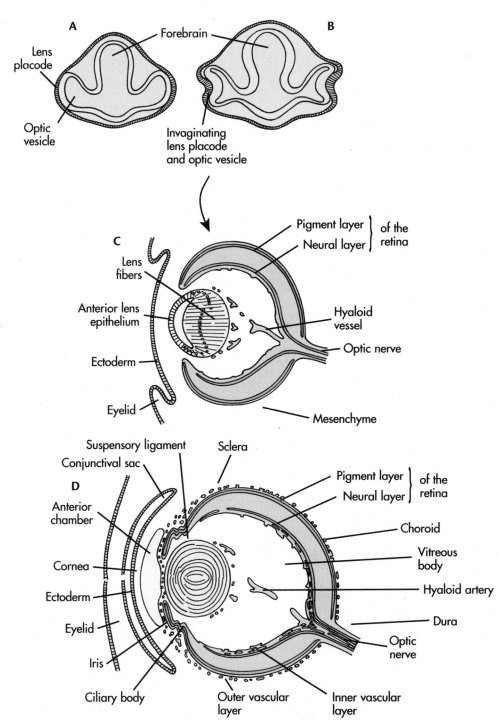

Figure 20.19

Four stages in eye development. **A,** Optic vesicles grow outward from the brain. **B,** Optic cup and lens form. **C,** Mesenchyme condenses around the optic cup as choroid and sclera. **D,** Conjunctiva, cornea, and anterior chamber make their appearances.

form, when the wall of the optic vesicle folds inward and becomes the double layered **optic cup** at the end of the **optic stalk** (Figure 20.19B). The lens forms in the opening of the optic cup and detaches from the epidermis. In weeks 6 and 7 (Figure 20.19C), the **hya-**

loid artery appears in the **choroid fissure,** a crease in the optic stalk, and begins to supply blood to the lens. The hyaloid artery degenerates later during fetal life, when the aqueous humor takes over the job of supplying nutrients to the lens. Mesenchyme condenses outside the optic cup and begins to form the choroid and the sclera. This mesenchyme forms the substantia propria of the cornea in front of the lens vesicle, and the epidermis becomes the conjunctival sac as the **eyelid folds** spread across the cornea. These folds temporarily fuse together and isolate the conjunctival sac until the sixth month, when the sac reopens and eyelashes and sebaceous glands appear along the palpebral fissure.

By 15 weeks (Figure 20.19D), the anterior chamber of the eye is forming in the mesenchyme behind the cornea, and the advancing retinal tissue forms the ciliary body and iris in front of the lens. Ganglion cells in the retina begin to grow toward the optic disk and to enter the optic nerve. Because peripheral ganglion cells are deeper in the nerve fiber layer of the retina than central ganglion cells, they assume superficial positions in the optic nerve, leaving space deeper in the nerve for axons from the center of the retina. Some axons will cross at the optic chiasm; all will reach the thalamus. Bipolar cells, rods, and cones appear after the ganglion cells, and connections begin to form among them, so that by 6 months the retina responds to light.

- What structure induces the lens to form?
- What structure forms the neural retina?
- Which tissue forms the choroid and sclera?

EAR AND HEARING

Different parts of the ear detect sound or motion. The **outer ear** (Figure 20.20) collects sound waves, and the **middle ear** conducts these vibrations to the **inner ear,** where they are converted to neural impulses that travel to the brain. There are three regions in the inner ear itself, but only the **cochlea** (KOKE-lee-ah) is concerned with hearing. The brain senses equilibrium with the other two regions—the **vestibule** and **semicircular canals.** Even though the perceptions of hearing and equilibrium are different, they both begin with similar **hair cell mechanoreceptors.** We will follow sound to the cochlea first and discuss balance in the next section.

Tweak one cochlear hair cell and a certain note registers in the brain. Tweak another and the brain "hears" a different pitch. The structures that sort sound waves into their various frequencies and deliver those waves to the appropriate hair cells are the essence of the anatomy of hearing.

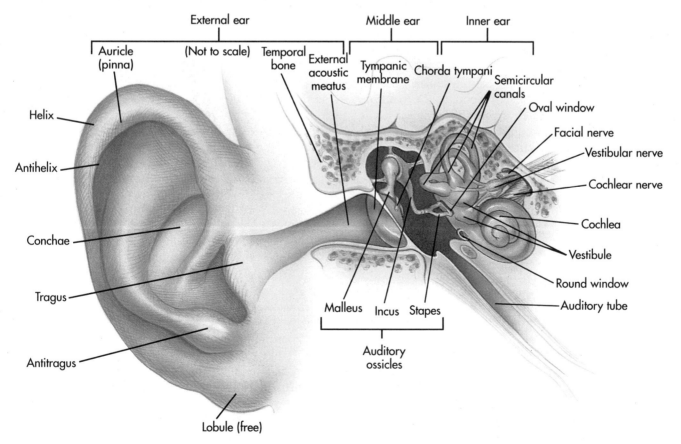

Figure 20.20
The ear.

OUTER EAR

The pinna is the part of the ear visible at the sides of the head. Inside the pinna the **auricular cartilage** (or-IK-you-lar) conducts sound to the **auditory canal,** with help from various highly individual folds (Figure 20.20). Three ligaments attach this single piece of elastic cartilage to the skull. Skin attaches firmly to the cartilage, except at the **lobule,** which hangs freely. A single gene determines whether the lobule is free (dominant), as shown in the figure, or attached to the head (recessive). Gently rub the lobule and then the superior border of the pinna with your finger, to see how much better the auricular cartilage conducts sound than skin alone.

For a relatively unimportant structure, the pinna is richly innervated by five nerves from the brainstem and spinal cord. Part of the reason for this profusion is that the forward and after portions of the pinna derive from opposite sides of the first pharyngeal pouch that forms the auditory canal (see Chapter 23, Pharyngeal Development). Blood circulates to the pinna through the external carotid artery and is drained through the external jugular vein.

The auditory canal leads inside the temporal bone to the tympanic membrane. This passage runs medially and forward into the temporal bone from the central **concha** of the external ear, shown in Figure 20.20. Fibrocartilage forms the distal portion of the canal, and the temporal bone forms its remainder. Stratified squamous epithelium lines the entire canal, while fine hair follicles and ceruminous glands line the canal only as far as the bony portion. **Ceruminous glands** (se-ROOM-in-us) are modified sweat glands that secrete earwax and fluid. **Otitis externa** (oat-EYE-tiss) is a microbial infection of the epithelium that can inflame and close the canal.

The tympanic membrane (tympan-, *G,* a drum) vibrates with sound waves deep at the proximal end of the auditory canal, some 25 to 30 mm from the external opening. This membrane is a shallow conical disk stretched obliquely across the auditory canal, so that the slightest sound causes it to vibrate. The membrane is about 0.1 mm thick and nearly circular, measuring 11 mm (½ inch) in its longer dimension. Its thin, stratified squamous epithelium is continuous with that of the ear canal, and its internal surface also is continuous with the low cuboidal epithelium of the middle ear. Radial and circular layers of collagen fibers act as a drum head between the epithelia.

As in the front and rear sides of the pinna, different nerves and vessels supply each surface of the tympanic membrane because the middle and outer ears develop separately (see Ear Development, this chapter). The trigeminal, facial, and vagus nerves supply the external surface of the membrane, and the glossopharyngeal innervates the internal surface.

• Name three components of the outer ear.

• What kind of epithelium lines the outer ear?
• What structure inside the ear canal registers sound waves?

MIDDLE EAR

The middle ear is the air-filled chamber on the medial side of the tympanic membrane. Figure 20.20 shows that the middle ear consists of the **tympanic cavity** and the **auditory (eustachian) tube** (you-STAY-kee-an) that opens to the pharynx. The auditory tube opens when you swallow, which allows air pressure to equalize on both sides of the tympanic membrane. The tympanic cavity houses three all-important bones called **ossicles** (ossic-, *L,* a little bone) that transfer and amplify sound to the inner ear. Ligaments and synovial joints anchor the ossicles to each other, and more ligaments anchor them to the walls of the tympanic cavity. The **malleus** (*L,* hammer), mounts on the inner surface of the tympanic membrane and receives vibrations directly from it. The **tensor tympani muscle** adjusts tension in the tympanic membrane according to the intensity of sound, much as the pupil compensates for light. The muscle protects the inner ear from intense sound by relaxing and loosening the membrane, whereas weaker sound tightens and sensitizes the membrane. The malleus conducts to the **incus** (incu-, *L,* anvil) and the incus to the **stapes** (STAY-peas; *L,* stirrup), which directs its vibrations onto the **oval window** in the wall opposite the tympanic membrane. The **stapedius muscle** also dampens the effects of loud sounds on the motion of the stapes. The oval window and **round window** resemble the tympanic membrane but lead to the inner ear.

Infections of the middle ear can affect nearby structures. Normally, air sweeps into the tympanic cavity when the **levator veli palatini** muscle (see Pharyngeal Muscles, Chapter 10) opens the auditory tube, and the ciliated lining of the tube clears out any fluid and bacteria that may have entered. When inflammation blocks the tube, however, the tympanic cavity can fill with fluid drawn in by the partial vacuum formed when air dissolves in the mucous membrane lining. This condition, called **otitis media** (inflammation of the middle ear), is the most common source of temporary deafness in children because the fluid dampens the ossicles' motion. If enough fluid accumulates, the pressure rises painfully and the tympanic membrane bulges into the auditory canal.

Unless antibiotics can reduce the infection, it may be necessary to cut a small opening in the tympanic membrane to release the fluid and relieve the pain. Hearing returns immediately, and the small opening heals readily in the highly vascular membrane. If infection remains in the **mastoid air cells** that communicate with the tympanic cavity, it may become necessary to implant permanent tubes in the membrane

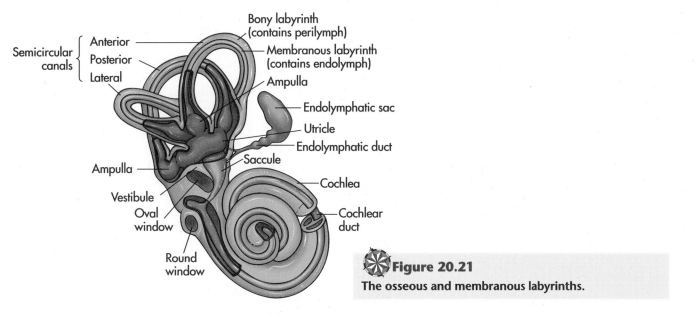

Figure 20.21
The osseous and membranous labyrinths.

to drain the middle ear. In another disorder, known as **otospongiosus** (OH-toe-SPON-jee-oh-sus), cancellous bone fuses the stapes to the bone surrounding the oval window, causing permanent deafness.

- What are the names of the ossicles of the middle ear?
- Which ossicle conducts sound to the oval window?
- What is the name of the tube that connects the tympanic cavity with the pharynx?

INNER EAR

The oval window conducts sound waves to the cochlea, which sorts out various frequencies. Together with the vestibule and semicircular canals, the cochlea (cochl-, *L,* snail shell, spiral) is part of the bony chamber of the inner ear also known as the **osseous labyrinth** (OSS-ee-us; Figure 20.21). The **membranous labyrinth** lines these passages with sleeves of tissue that fit snugly into the space eroded in the petrous portion of the temporal bone. **Perilymph** surrounds the membranous labyrinth, and **endolymph** fills the interior. Perilymph is derived from cerebrospinal fluid, and endolymph is stored in the **endolymphatic sac.** The cochlea can function because endolymph is rich in potassium ions but perilymph and cerebrospinal fluid are richer in sodium ions, as you will see.

COCHLEA

The cochlea coils upon itself two and a half times, as Figure 20.22 shows, narrowing steadily to its apex like a snail's shell. The membranous labyrinth lines the cochlea with three tubes (Figure 20.22B). The central tube is called the **cochlear duct.** It is filled with endolymph and contains the **spiral organ (of Corti)** that detects sound. The cochlear duct

also is called the **scala media** (SKAH-lah), which helps explain the names of the **scala vestibuli** and the **scala tympani,** the tubes on either side of it (scal-L, a ladder). Perilymph fills these latter tubes, which are separated from the endolymph in the cochlear duct by the **vestibular membrane** (scala vestibuli) and the **basilar membrane** (scala tympani). Each tube narrows progressively toward the apex of the cochlea. The scala vestibuli and tympani are continuous with each other at the apex, where they connect through a small passage called the **helicotrema** (HEL-ih-coe-TREE-ma).

- How do the membranous labyrinth and the osseous labyrinth differ?
- What is the sound-detecting organ of the cochlea?

COCHLEAR DUCT AND ORGAN OF CORTI

The organ of Corti rests on the basilar membrane inside the cochlear duct and spirals with the duct through the cochlea. The **spiral lamina** receives the medial wall of the cochlear duct and **cochlear nerve axons** from Corti's organ. The cell bodies of these axons reside in the **spiral ganglion,** and the axons themselves enter the **vestibulocochlear nerve** to the brain. The thick lateral wall of the cochlear duct is lined by the **stria vascularis** (STRY-ah VASS-kue-LAR-iss; vascular strip), which enriches the endolymph with potassium ions from capillaries behind the stria.

The organ of Corti transduces the motion of the basilar membrane into neural impulses. Two strips of **hair cells,** illustrated in Figure 20.22C, extend the length of the organ on either margin of the basilar membrane, converting sound into neural impulses wherever the membrane resonates. About 3500 hair

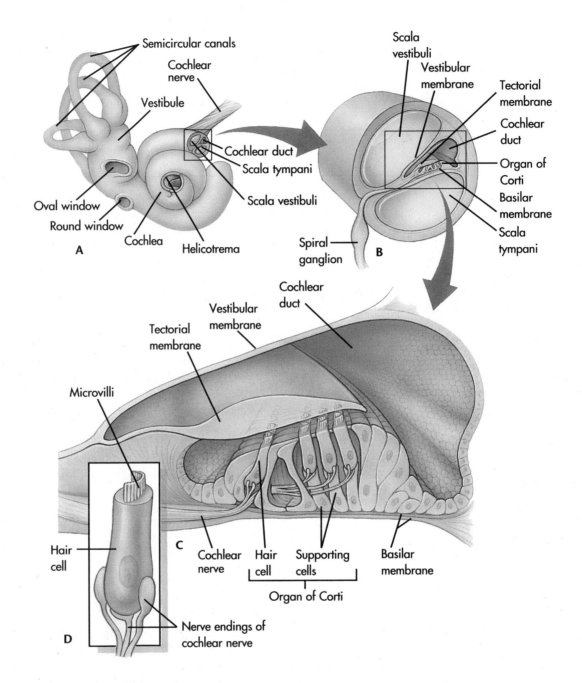

Figure 20.22

The cochlea and organ of Corti. Arrows show the movement of perilymph between the oval and round windows. A, Membranous labyrinth and cochlea. B, Transverse section through cochlea reveals the cochlear canal and the scalae media and tympani. C, Transverse section through organ of Corti. D, Hair cell.

cells form a row of **inner hair cells** nearest the spiral lamina, and three more rows of 20,000 **outer hair cells** complete the receptive machinery. (Figure 20.22D shows a hair cell.) Various **phalangeal** (FAL-an-jee-al) and **pillar cells** support the hair cells with cytoplasmic microtubules on either side of the **tunnel of Corti,** and the **tectorial membrane** (TEK-tor-ee-al; tect-, *L,* a roof, covering) covers the hair cells with a matrix of collagen fibers. A single cochlear axon

synapses with each hair cell. Wherever sound resonates a particular region of the basilar membrane, the hairs deflect against the tectorial membrane and trigger action potentials in the cochlear fibers of that section, which ultimately register as corresponding pitches in the primary auditory cortex of the brain.

You can understand how the cochlear duct sorts different sound waves along the organ of Corti when you imagine that the cochlea has been unwound, as in

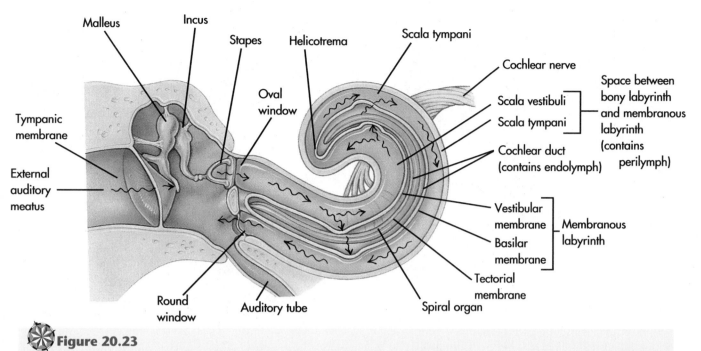

Figure 20.23

The organ of Corti absorbs waves of sound that travel into the cochlea. High frequency waves reach the basal regions of the cochlea, and low frequency sound reaches the apex before it is absorbed.

Figure 20.23. When sound waves arrive, the ossicles vibrate the oval window in and out at the base of the scala vestibuli. When they push the oval window in, the perilymph bulges the **round window** out at the base of the scala tympani because there is nowhere else for the perilymph to go; it is not compressible. The perilymph therefore vibrates back and forth along the scalae and through the helicotrema, causing the walls of the cochlear duct, especially the basilar membrane, to vibrate. Because different regions of the basilar membrane pick up different frequencies, sound waves travel into the cochlea until they encounter those corresponding regions. At these points, the waves cause the basilar membrane to resonate and to absorb their sound energy, so that the waves go no deeper into the cochlea. High frequencies resonate near the base of the membrane where it is thin and narrow, but lower frequencies travel deeper into the cochlear duct where the basilar membrane is thicker and wider.

HAIR CELLS AND STEREOCILIA

Hairs are really rows of stereocilia, and each row is taller than the next. Figure 20.24 shows both rows of hair cells in the organ of Corti and individual hairs. Stereocilia resemble microvilli, as you can see in Figure 20.24B. Each **stereocilium** is anchored in the **cuticular plate** of the hair cell, linked by microfilaments to its neighbors. All are linked to a single **kinocilium** (KI-no-sil-ee-um) in the center of the tallest row. Although the kinocilium (kino-, *G,* moving) is a true, but nonmotile, cilium with a 9 + 2 axoneme of microtubules, the stereocilia (stereo-, *G,* sol-

id) have no internal microtubules, but only a core of actin microfilaments. Although the tectorial membrane may rub in all directions, hair cells fire more rapidly when stereocilia bend toward the highest row (Figure 20.24C) and more slowly when the stereocilia bend away from that row. Bending toward the highest row opens potassium ion channels in the tips of the stereocilia that depolarize the cell membrane as potassium ions flow inward and fire action potentials in the cochlear nerve fiber of each hair cell. Now you can see why it is important that endolymph be potassium-rich—to provide a concentration gradient, allowing potassium to diffuse rapidly into the hair cells.

- Do high frequency or low frequency sound waves travel deeper into the cochlear duct?
- What are the names of the three tubes within the cochlea?
- Which way must stereocilia bend to stimulate cochlear nerve fibers?

AUDITORY PATHWAY

Four orders of neurons conduct impulses from the cochlea to the auditory cortex in the temporal lobes of the cerebrum. As you might expect from the presence of so many levels, the pathway is complex and redundant (Figure 20.25). You can damage up to 75% of axons and retain normal hearing, in part because each cortex receives from both ears. Axons cross at three different locations in the brainstem. Most axons are tuned to particular sound frequencies and map to particular areas of the auditory cortex in much the same way that optic neurons map onto the visual cortex.

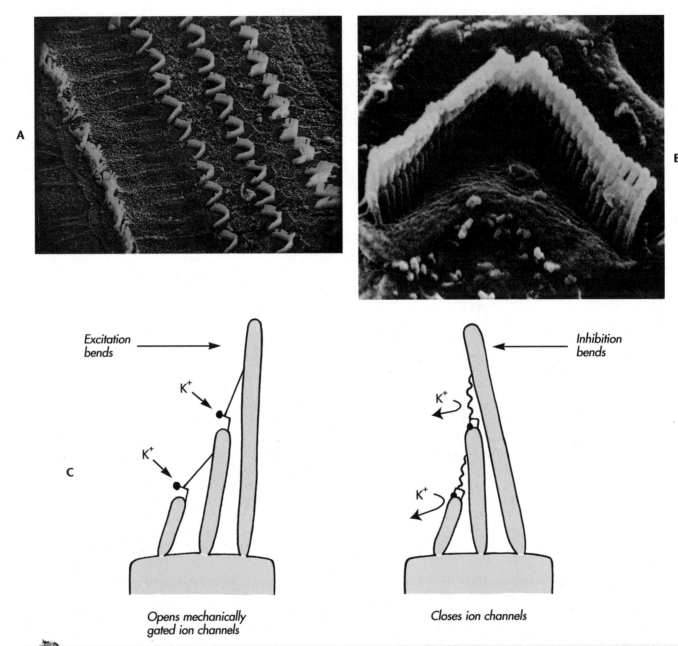

Excitation
bends

K⁺

K⁺

C

Opens mechanically
gated ion channels

Inhibition
bends

K⁺

K⁺

Closes ion channels

Figure 20.24

Hair cells and the organ of Corti. A, Scanning electronmicrograph of the inner and outer rows of hair cells. **B,** Scanning electronmicrograph of an individual hair cell, showing kinocilium and stereocilia. (Scanning electronmicrographs in A & B by permission of R.G. Kessel and R.H. Kardon: *Tissues and organs: A text-atlas of scanning electron microscopy,* 1979, W.H. Freeman and Co.) **C,** Bending toward the kinocilium opens ion channels, but bending in the opposite direction closes them.

The auditory pathway begins with hair cell neurons that exit the organ of Corti through the **spiral (cochlear) ganglion** and **cochlear nerve** to synapses in the cochlear nuclei of the medulla oblongata. (Axons from the vestibule and cochlea join as the vestibulocochlear, cranial nerve VIII.) Second order neurons conduct to the opposite side of the medulla (superior olivary nucleus), and third order neurons rise through the lateral lemnisci to the **inferior colliculus** (co-LIK-yule-us) of the midbrain. From here, axons of fourth order neurons project into the thalamus, where they synapse in the medial geniculate nucleus with dendrites of fifth order neurons whose axons project to the temporal lobes and auditory cortex. This pathway communicates to the **reticular activating system,** ensuring that unusual sounds will rouse the brain to attention. Axons from the inferior colliculus communicate with the visual pathway to locate the sources of sudden noise, and other axons warn the cerebellum to move the body away from danger.

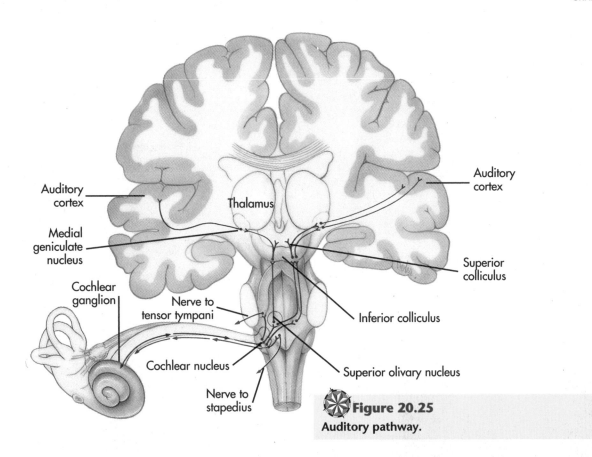

Figure 20.25
Auditory pathway.

DISORDERS

Hearing disorders can affect most of the structures of the inner ear and auditory pathways. Hereditary failure of the cochlea to develop (**Michel's aplasia;** aye-PLAYS-ee-ah) brings total deafness. But in the most common developmental failure **(Scheibe's aplasia),** some low frequency hearing remains even though the tectorial membrane is smaller, the scala media may collapse, or the organ of Corti may be entirely lost.

Hereditary disorders usually affect both ears, but traumatic injury may attack one or both. **Noise-induced hearing loss** (NIHL) affects outer hair cells first, swelling them or damaging their stereocilia. The effects of chronic exposure to severe noise progress to the inner hair cells next and will attack auditory neurons after the hair cells have been destroyed. High frequency (basal) regions of the cochlea are most sensitive. Old age also interferes with hearing (**presbycusis;** PREZ-bee-KUE-sis). Over time, demyelination reduces the ability of nerve fibers to conduct auditory impulses. Multiple sclerosis, one form of demyelinating disease, also may interfere with the auditory pathways in the brainstem. Efforts to replace damaged organs of Corti with **cochlear prostheses** have had some success but not yet enough to allow their recipients to hear speech without also reading lips.

- With what regions of the brain do the auditory pathways communicate?
- How does noise affect the organ of Corti?

ORGANS OF EQUILIBRIUM

The vestibule and semicircular canals adapt hair cells to detect motion. Hair cells virtually identical to those in the cochlea send continual streams of impulses to equilibrium centers of the cerebrum. The brain combines this information with vision and muscle sense to signal the skeletal muscles that maintain your balance. The semicircular canals detect changes in rotation when you turn your head in any direction, and the vestibule senses linear motion (forward and back, sideways, and vertical). Stereocilia initiate the signals to the brain in both cases. **Vertigo** is one of the results of impaired equilibrium, when the cerebellum receives inappropriate signals from the brain or ear.

SACCULE AND UTRICLE

The **saccule** (SAK-yule) **and utricle** (YOU-trih-cl) **contain patches of hair cells** called the **macula sacculi** (SAK-you-lee) and **macula utriculi** (you-TRIK-you-lee). Endolymph fills the saccule and utricle in Figure 20.26, and perilymph bathes their exteriors, as in the cochlea. The macula utriculi is a 2 × 2 mm patch of epithelium (macula; *L,* a spot) that lies horizontally in the floor of the utricle (Figure 20.26B; utricul, *L,* a leather bag). The macula sacculi is slightly larger (3 × 2 mm), and lines the medial wall of the sacculus (sacc-, *L,* a sack). Both maculae resemble the organ of Corti in several important respects.

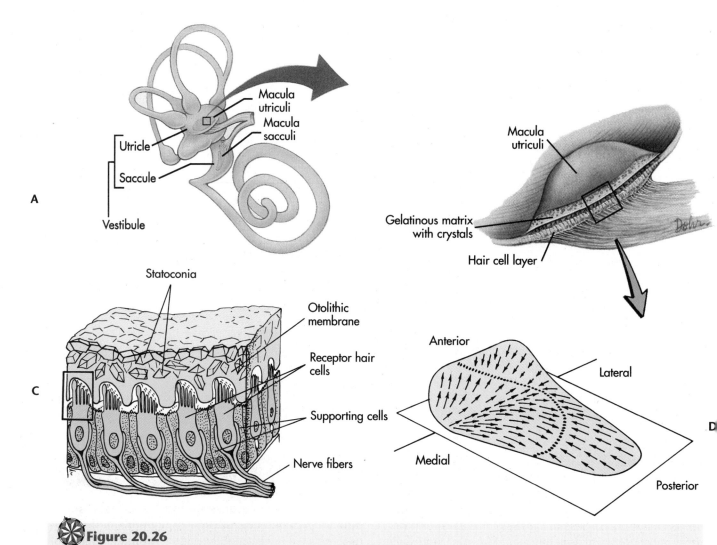

※ **Figure 20.26**

Vestibule. A, Utricle and saccule. **B,** Macula utriculi on the floor of the utricle. **C,** Otolithic membrane covers the hair cells with a gelatinous matrix and statoconia. **D,** Most hair cells in the macula utriculi point anteriorly or posteriorly toward the highest row of stereocilia.

Supporting cells immobilize the hair cells, and vestibular neurons innervate each hair cell. As shown in Figure 20.26C, a gelatinous **otolithic membrane** (OH-toe-LITH-ik), softer than the tectorial membrane of Corti's organ, covers the surface of each macula, embedding tufts of stereocilia within it. Calcium carbonate crystals called **statoconia** or **otoliths** (lith-, *G,* a stone; ear stones) cover the outer surface of the otolithic membrane, giving the membrane enough inertia so that it momentarily lags behind when your head changes motion. (Statoconia are three times denser than the membrane itself.) However brief, this lag causes certain hair cells to fire more rapidly and others more slowly, so that the pattern of signals flowing from the vestibular nerve informs the brain about the direction of movement and whether you are accelerating or decelerating.

Maculae detect motions preferentially. The macula utriculi is most effective in detecting anterior and posterior motion because the macula is horizontal, and most of its hair cells point in these directions (Figure 20.26D). Hair cells point toward the highest rank of stereocilia, the direction that excites them to fire action potentials. The function of the macula sacculi is uncertain. This macula lies on the sagittal plane, its hair cells also pointing forward and back, which suggests that it responds like the macula of the utricle. One would expect both maculae to inform the brain about the position of the head, as well as its motion, because gravity pulls the otolithic membrane in different directions when one lies down or stands erect.

- What is an otolith?
- How does the macula utriculi detect anterior and posterior motion?

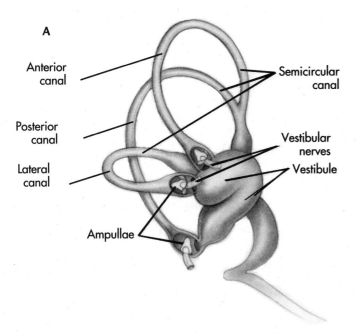

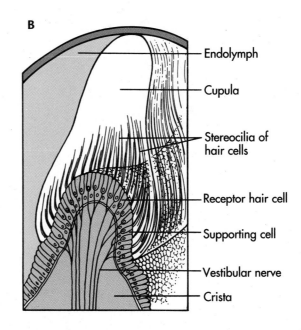

A

Anterior canal

Posterior canal

Lateral canal

Ampullae

Semicircular canal

Vestibular nerves

Vestibule

B

Endolymph

Cupula

Stereocilia of hair cells

Receptor hair cell

Supporting cell

Vestibular nerve

Crista

Figure 20.27
Semicircular canals. **A,** Anterior, lateral, and posterior canals with ampullae. **B,** Section through crista ampullaris.

SEMICIRCULAR CANALS

The semicircular canals use hair cells to detect rotational motion. Three semicircular canals arch away from the utricle along three separate perpendicular planes (Figure 20.27). The lateral canal returns to the utricle directly, but the anterior and posterior canals join together before returning. The anterior canal detects vertical rotation because it lies near the sagittal plane, like the macula utriculi. Nodding your head will stimulate its hair cells. Posterior canals register frontal rotation (tilt your head), and the lateral canals alert the brain of any turns to left or right. All of this permits the brain to formulate three-dimensional images of its own movement.

A small swelling, the **ampulla** (am-PUL-la), marks each canal as it leaves the utricle. Inside, a flap of tissue called a **crista ampullaris** (AM-pew-LAR-iss; crist-, *L,* a crest) sweeps forward and back under the momentary difference between the inertia of the endolymph and the rotation of the head. Ice in your cold drink does the same thing when you rotate the glass; the pieces remain in place until the liquid transfers the glass's motion to them; vice versa, when you stop turning the glass, the ice continues to swirl until the liquid dissipates its momentum. The crista is neatly lined with hair cells and covered with a form-fitting gelatinous **cupula** that bends its stereocilia. All the hair cells in the lateral canal point toward the utricle, so that in the right ear they fire more rapidly when your head turns right. This situation is reversed in the

left ear; the hair cells fire more slowly because the crista bends in the opposite direction.

Visualize the signals that equilibrium centers receive in the brain. Both centers receive the same impulses when the head is at rest (Figure 20.27C), not changing its motion. But when the head accelerates (starts to turn, or turns faster), certain cristae fire more rapidly and others on the opposite side fire more slowly. The right side fires faster when the head is turning right; the left side fires faster when it is turning left. Deceleration reverses these effects. The details quickly become complicated, but the same principle works for anterior and posterior canals, except that the hair cells point away from the utricle. Nodding your head downward will fire hair cells of the anterior canal more rapidly, and tilting to the right will fire them more rapidly in the right posterior canal. Figure 20.28 illustrates the vestibular pathways of the brain.

VERTIGO

Vertigo is the false report of motion that the semicircular canals and vestibule can deliver to the brain. When your head is at rest, the brain receives symmetrical signals from both organs that are compared to the signals that result when head motion changes. One side now transmits faster, the other slower. Certain circumstances, such as infection or injury on one side, may cause this difference in signals to happen when the head is actually at rest, but the

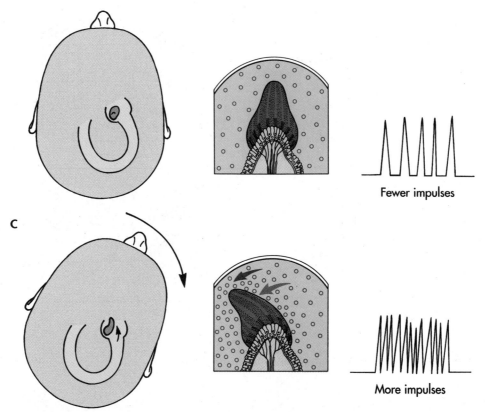

Figure 20.27, cont'd.
C, At rest the crista does not move, but it nevertheless sends slow streams of impulses. When the head turns right, however, the endolymph bends the crista forward and impulses are more frequent.

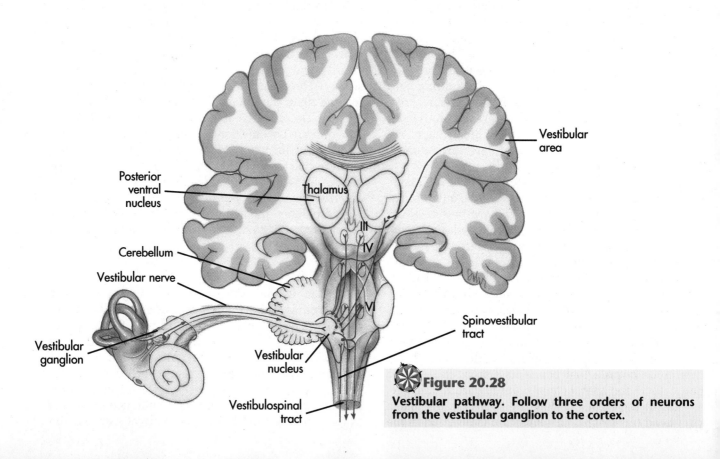

Figure 20.28
Vestibular pathway. Follow three orders of neurons from the vestibular ganglion to the cortex.

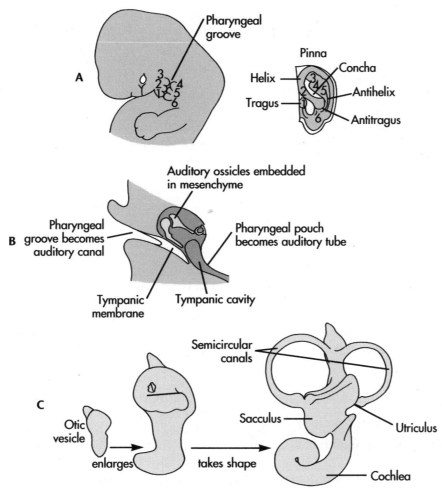

Figure 20.29

Ear development. **A,** The pinna derives from each wall of the first pharyngeal groove. **B,** The pharyngeal pouch grows outward toward the pharyngeal groove. **C,** The membranous labyrinth develops from the otic vesicle.

brain reacts as if the head were in motion. The person loses balance and stumbles, while reflexes try to compensate for motion that is not occurring. The eyes search for a stable image (nystagmus), and the individual becomes nauseated and may vomit. Motion sickness is a mild form of vertigo. Chronic sufferers attempt to move as little as possible.

- What is the name of the structure, similar to the maculae of the vestibule, that detects motion in the semicircular canals?
- What pattern of stimuli do the semicircular canals send at rest, and when the head turns to the right?

EAR DEVELOPMENT

The ear develops from three sources. The external ear develops inward from the first **pharyngeal groove,**

and the middle ear derives from the first **pharyngeal pouch,** which pushes out from the pharynx to form the tympanic membrane where pouch and groove meet. The chambers of the inner ear develop from the **otic vesicle** (OH-tik). The external ear develops from the first and second pharyngeal arches, which are the anterior and posterior walls, respectively, of the first pharyngeal groove, shown in Figure 20.29A (see also Figure 4.7, Chapter 4, Development). The arches become the pinna, and the groove develops into the auditory canal. Figure 20.29B shows that the first pharyngeal pouch grows outward from the pharynx into the mesenchyme tissue of the head toward the deepening pharyngeal groove. The ossicles condense in this mesenchyme between the groove and pouch, so that the pouch hollows out around the ossicles and becomes the tympanic cavity and auditory tube. The tympanic membrane, the round window, and the oval window form in the thin partitions that separate the middle from the inner and outer ears.

While these changes are underway, the otic vesicle appears beside the midbrain and molds itself into the chambers of the membranous labyrinth (Figure 20.29C). The otic vesicle is analogous to the lens of the eye and develops in a similar manner. During the third week, the midbrain induces the ectoderm to thicken and form the **otic placode,** a flat disk of ectoderm that quickly rounds up and becomes the otic vesicle. As does the lens vesicle, the otic vesicle pinches off from the ectoderm and becomes a hollow cavity surrounded by mesenchyme. The saccule and utricle develop from the central portion of the otic vesicle, while the cochlea coils outward from the ventral portion, and the semicircular canals develop from three lobes in the dorsal end of the otic vesicle.

- What structure develops into the inner ear?
- Explain why the internal and external surfaces of the tympanic membrane have different sources of innervation.

OLFACTORY EPITHELIUM AND OLFACTION

Our sense of smell detects molecules that flow in and out of the nasal cavity. Each breath circulates air past patches of **olfactory epithelium** high inside the nasal cavity, where molecules that bind with olfactory neurons cause impulses to be sent to the brain.

OLFACTORY EPITHELIUM

Olfaction is the only special sense in which neurons themselves respond directly to stimuli. The nasal cavity is lined with pseudostratified ciliated columnar epithelium, but only the olfactory epithelium, shown in Figure 20.30, contains **olfactory neurons** that detect odors. Special **Bowman's glands** secrete a film of mucus over the olfactory epithelium. This area covers the roof of the nasal cavity and extends about 1 cm down the median septum and the superior nasal concha.

Olfactory neurons are modified bipolar cells. The cell bodies extend dendrites to the surface of the epithelium, and the axons reach into the **olfactory bulbs** of the brain. Airborne **oderant molecules** (OH-der-ant) that dissolve in the mucous film trigger action potentials by binding to **oderant receptors,** or ion channels, on the cilia. A dozen or more modified **cilia** extend from a single **dendritic knob** at the tip of each dendrite. A 9 + 2 axoneme (see Cilia, Chapter 2) forms the base of each cilium, but the outer doublets disappear distally. Unmyelinated olfactory axons carry impulses from many bipolar cells through the **cribriform plate** (CRIH-brih-form) of the ethmoid bone to the olfactory bulbs of the cerebrum. Neurons in the olfactory bulbs distribute directly to olfactory areas of the cerebral cortex. Olfactory neurons in the nasal cavity replace themselves by cell division. This is the only neural tissue that regenerates itself.

Supporting cells surround each olfactory neuron with fields of microvilli that capture microorganisms and other airborne particles in the mucus. Small **basal cells** at the basement membrane supply new supporting cells by cell division. Although supporting cells may release materials into the mucus, **Bowman's glands** are the primary source of **olfactory mucus.** These alveolar glands are located in the lamina propria beneath the epithelium. They secrete polysaccharides, oderant-binding proteins, and enzymes through ducts into the nasal cavity. The polysaccharides make the mucus viscous and sticky, and the oderant-binding proteins enable oderants in the mucus to bind to olfactory neurons, while the enzymes clear the way for new oderants by degrading old ones.

OLFACTORY PATHWAY

The olfactory pathway conducts impulses directly from the olfactory epithelium to the primary olfactory cortex in the cerebrum. It is the only "sense" that is not relayed through the thalamus. The **olfactory nerve** conducts the unmyelinated fibers described above through the cribriform plate (Figure 20.30A) to the olfactory bulbs beneath the cerebral hemispheres. Actually part of the cerebral hemispheres, the **olfactory bulbs** begin to integrate olfactory signals for the brain, much as the retina does for the visual cortex. The olfactory bulbs contain six layers of cells and synapses that resemble the organization of the cerebral cortex. **Mitral** (MY-tral) and **tuft cells** in these layers are second order neurons, each of which distributes impulses from as many as 1000 olfactory neurons to the olfactory cortex in the **parahippocampal gyrus** (par-ah-HIP-oh-KAMP-al; see Chapter 19, Brain) on the medial surface of the temporal lobe and to the **medial olfactory area** anterior to the hypothalamus.

- Where is the olfactory epithelium located in the nasal cavity?
- To what parts of olfactory neurons do oderant molecules bind?
- Where is the olfactory cortex located in the cerebrum?

TASTE BUDS AND SENSE OF TASTE

Our sense of taste relies on taste buds that communicate to gustatory centers in the brain. Taste

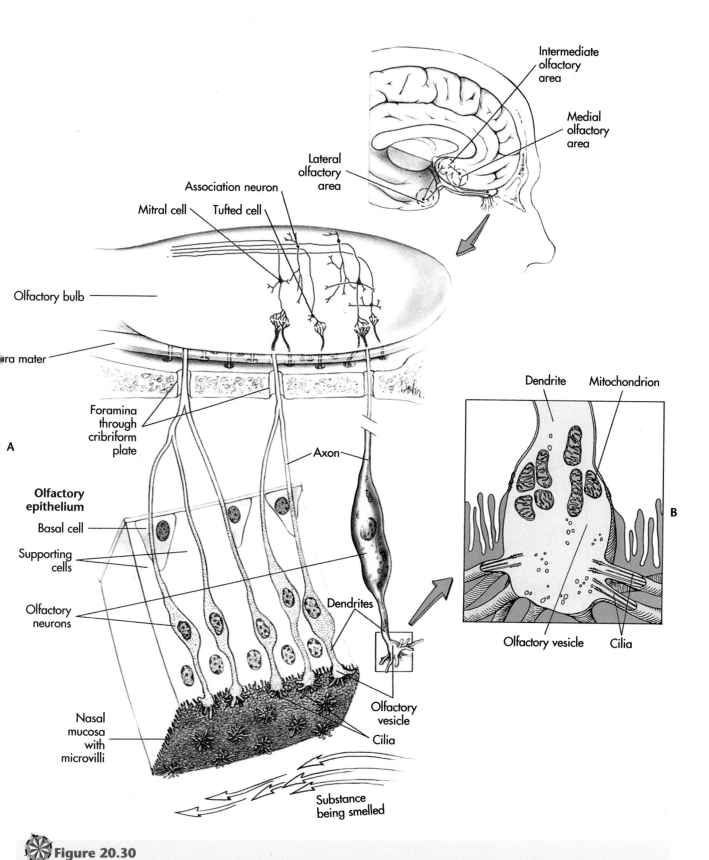

Intermediate olfactory area

Medial olfactory area

Lateral olfactory area

Association neuron

Mitral cell   Tufted cell

Olfactory bulb

Dura mater

Foramina through cribriform plate

A

Axon

Olfactory epithelium

Basal cell

Supporting cells

Olfactory neurons

Dendrites

Olfactory vesicle

Cilia

Nasal mucosa with microvilli

Substance being smelled

Dendrite   Mitochondrion

B

Olfactory vesicle   Cilia

✳ **Figure 20.30**

Olfactory epithelium and olfactory pathway. A, Passing materials stimulate olfactory neurons (bottom of figure), whose fibers connect through the cribriform plate to the olfactory bulbs of the brain. **B,** Detail of olfactory sensory ending.

buds are round clusters of about 80 taste cells each, embedded in the epithelium of the tongue, as well as in the soft palate and pharynx. The tongue alone contains as many as 7000 taste buds located in papillae (described in Chapter 23, Digestive System). The apex of the tongue is more sensitive to sweet and salt (Figure 20.31), whereas the sides and rear are more sensitive to sour and bitter. Do not think that apical taste buds detect only sweetness, however; all regions respond to all four stimuli. The brain interprets the various combinations of these stimuli, together with olfactory stimuli, as "flavors," but how taste cells re-

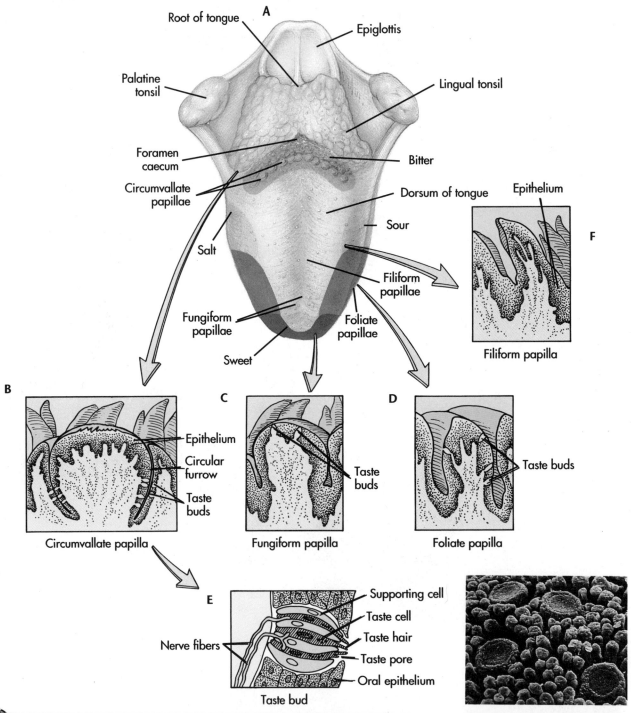

Figure 20.31

Taste buds and the tongue. Four sensory regions of the tongue, (A) with papillae that contain taste buds. The papillae are (B) circumvallate papillae, (C) fungiform papillae, and (D) foliate papillae. Structure of taste bud and taste cells (E). Filiform papillae (F) have no taste buds.

spond to stimuli and how the brain marshals the sense of taste from the information it receives are not nearly as well understood as sight and hearing.

TASTE CELLS AND TASTE BUDS

Most taste cells are tall, slender cells (Figure 20.31E) anchored to the basement membrane, with apical ends exposed to the oral cavity through a taste pore. Taste cells respond when molecules pass through the **taste pore** of a taste bud into the **taste chamber** and bind to the ends of the taste cells. Gated ion channels and membrane transduction pathways similar to those involved in hearing and sight propagate the signal along the membrane, where it flows across synapses to afferent neurons at the base of the taste bud. At one time it was thought that taste cells extended long **taste hairs** into the chamber for better reception, but modern observations with the electron microscope show only microvilli or an occasional cilium. Unlike the inner ear, no cells are specialized for support alone; all but one type of taste cell seem to be capable of both tasting and support.

Most taste buds contain four types of cells. Dark cells probably are most concerned with taste. Abundant, fine, cytoplasmic filaments give them their dark appearance, and microvilli decorate the apical ends. Dark cells also appear to secrete materials into the taste chamber to flush out old molecules. Light cells resemble dark cells, but have fewer cytoplasmic filaments. A third type of cell ends in a blunt knob instead of microvilli. In the bases of these cells are vesicles that resemble synaptic vesicles, which suggests that these cells transmit across synapses. A fourth type of cell, the basal cell, continually divides and produces the other three types at the base of the taste bud. As are other epithelial cells, taste cells are replaced about every 10 days. All four kinds of cells synapse with each other and with individual afferent neurons to the brain. Each taste bud thus represents a **"taste unit"** analogous to the motor units in skeletal muscles because one neuron receives from many cells.

GUSTATORY PATHWAY

Three cranial nerves conduct taste impulses to the brain (Figure 20.32). In keeping with the general rule that anterior cranial nerves serve anterior structures, branches of the **facial nerve** (VII; see Chapter 19, Cranial Nerves) serve the soft palate and anterior two thirds of the tongue. The **glossopharyngeal nerve** (IX; GLOSS-oh-fahr-IN-jee-al) conveys from the posterior third, and branches of the **vagus** (X; VAY-gus) deliver from the pharynx and esophagus. Second order neurons relay from synapses in the medulla to nuclei in the pons and thalamus. From the thalamus, axons of third order neurons project onto the **gustatory cortex** in the insula and operculum of the cerebrum, adjacent to the oral somatosensory cortex. Most impulses travel on the same side as the taste buds

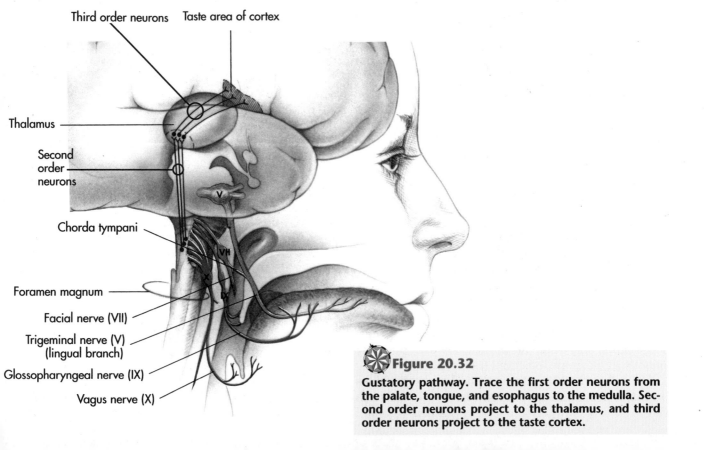

Third order neurons   Taste area of cortex

Thalamus

Second order neurons

Chorda tympani

Foramen magnum

Facial nerve (VII)

Trigeminal nerve (V) (lingual branch)

Glossopharyngeal nerve (IX)

Vagus nerve (X)

Figure 20.32

Gustatory pathway. Trace the first order neurons from the palate, tongue, and esophagus to the medulla. Second order neurons project to the thalamus, and third order neurons project to the taste cortex.

where they began, and only minor traffic crosses to the opposite side of the brain.

- Where are taste buds located?
- What are the two principal taste cells?
- Which cranial nerves conduct taste impulses to the brain, and where are the gustatory centers located?

SUMMARY

Communication to the brain begins with sensory receptors. These may be free or encapsulated nerve endings; both are concerned with the general and special senses. Sensory receptors detect chemical, mechanical, or light stimuli. Sensory receptors also are classified by the nature of the perceptions they evoke (thermoreceptor, nocieceptors) and by their location (interoceptors and exteroceptors). *(Page 601.)*

General sensory endings generate impulses that provide the sensations of heat and cold, touch, body sense, and pain. Free nerve endings detect temperature and pain. Pacinian corpuscles contribute to the sense of touch by detecting pressure and motion, and muscle spindles and Golgi tendon organs serve as the sources of body sense by detecting tension in muscles, tendons, and joints. Nocieceptors detect painful chemical and mechanical stimuli. *(Page 602.)*

Special sense organs provide vision, hearing, equilibrium, olfaction, and taste. Photoreceptor rods and cones support vision; mechanoreceptors in the inner ear provide hearing and equilibrium; and chemoreceptors initiate the signals that result in the sensations of smell and taste. *(Page 601.)*

The eyeball is composed of three tunics related to the meninges of the brain. The fibrous sclera covers the eyeball and admits light through the cornea. The vascular choroid supplies circulation and forms the iris and ciliary body. The rods and cones and ganglion cells of the neural tunic transduce light into neural impulses that the brain interprets. *(Page 605.)*

The eyes and brain cooperate in vision. The cornea and lens focus light rays onto the retina, which converts the stimuli to neural impulses that formulate visual images in the primary visual cortex of the brain. The brain obtains binocular vision by directing the eyes to the same subject and fusing the images from each eye into one three-dimensional image. The lens accommodates for near and far objects by changing shape just enough to focus light on the retina. The iris and pupillary light reflex govern how much light enters the retina. Conjunctiva, lacrimal apparatus, and eyelids lubricate the cornea, keep it optically smooth, and combat infection. *(Page 605.)*

The left and right primary visual cortex receives impulses from the corresponding halves of each retina by way of the optic nerve, optic chiasm, lateral geniculate nucleus, and optic radiations. *(Page 616.)*

The eyeball develops from the brain, epidermis, and mesenchyme. The optic vesicle forms the retina and induces the epidermis to form the lens. Choroid, sclera, and cornea develop from mesenchyme that condenses around the optic cup. The conjunctiva and eyelids derive from the epidermis. *(Page 618.)*

The ear consists of three divisions. The external ear collects sound waves and conducts them through the auditory canal to the tympanic membrane, which transmits them to the middle ear. The ossicles of the middle ear conduct to the oval window that opens into the inner ear, where the cochlea detects sound, and motion registers in the vestibule and semicircular canals. All are served by axons of the vestibulocochlear nerve. *(Page 620.)*

The organ of Corti converts sound waves to neural impulses. The cochlea distributes the different frequencies of sound waves that hair cells detect. Higher frequencies stimulate hair cells and neurons at the base of the organ of Corti, and lower frequencies travel deeper into the cochlea, where they stimulate hair cells tuned to these waves. Cochlear axons conduct to the auditory centers of the cerebral cortex. *(Page 622.)*

Hair cells in the semicircular canals and vestibule detect circular and linear motion. Signals from cristae ampullaris in three mutually perpendicular semicircular canals inform the brain of angular acceleration when the head turns, and the maculae sacculi and utriculi signal the brain about linear acceleration. *(Page 626.)*

Hair cells are mechanoreceptors that respond to motion from certain directions. Hair cells have ranks of stereocilia with ion channels at their tips that open when the hair bends toward the kinocilium, and close when it moves in the opposite direction. Because the hair cells of the cristae and maculae are oriented in specific directions, their signals inform the brain of the head's changing position. *(Page 628.)*

The ear develops from the pharyngeal grooves and pouches and from the otic vesicle. The external ear and auditory canal arise from the first pharyngeal groove and arches; the middle ear derives from the first pharyngeal pouch; and the tympanic membrane develops from the partition between pouch and groove. The tympanic cavity erodes around the ossicles and the osseous labyrinth erodes within the temporal bone. The cochlea, vestibule, and semicircular canals are formed by the otic vesicle. *(Page 630.)*

Taste and smell are derived from chemoreceptors in the mouth and nasal cavity. Chemical changes stimulate taste cells in taste buds on the tongue, palate, and pharynx, and olfactory neurons in the nasal cavity detect odors when molecules bind to receptors on their dendrites. *(Page 631.)*

STUDY QUESTIONS

1. Name the general and special senses. Give an example of the perception that each sense supports. Which of these examples involves free nerve endings, and which ones depend on encapsulated endings?   **1**

2. Which of the perceptions in question 1 is supported by chemoreceptors, which perception is supported by mechanoreceptors, and which perception is supported by photoreceptors? Are chemoreceptors free nerve endings or encapsulated?   **2**

3. Where in the retina and primary visual cortex will the image of the front bumper, rear bumper, and door of the car shown in Figure 20.16 be found? Name the neurons that connect these points on the retina with the corresponding locations in the visual cortex.   **3**

4. The conjunctiva and lacrimal apparatus have similar functions. What are these functions, and how are they similar? The cornea and lens focus light on the retina. What anatomical properties of these structures enable them to focus light?   **4**

5. Describe the three tunics of the eyeball. Which meninges do they represent, and what ocular structures are derived from them?   **5**

6. Which extrinsic ocular muscles elevate the eyeballs and keep them trained on the same object while they elevate?   **6**

7. Describe how lens accommodation and the pupillary light reflex operate. What anatomical structures are the sources of tension on the lens and pupil in each case?   **7, 8**

8. Which region of the organ of Corti absorbs high frequency sound waves? How do these waves reach this location, and what events transduce them into neural impulses? Why do low frequency waves not stimulate hair cells at this same location?   **9, 10, 11**

9. Describe the structure of a hair cell stereocilium. In which direction must the hair bend to stimulate vestibular or cochlear nerve fibers?   **11**

10. Which way do hair cells bend in the lateral semicircular canals when the head swings to the left? How does the pattern of signals from this motion differ from those received at rest, when the head is not turning?   **12, 13**

11. Why do maculae utriculi and sacculi appear to detect linear motion and the semicircular canals, rotational motion? What anatomical changes would have to take place for the opposite to be true?   **12, 13**

12. Diagram the olfactory nerve endings in the olfactory epithelium and their connections with the olfactory bulb. What events in the nasal cavity cause these neurons to stimulate mitral cells of the olfactory bulb?   **14**

13. Are taste buds free or encapsulated nerve endings? Describe the interconnections between taste cells and gustatory neurons.   **15**

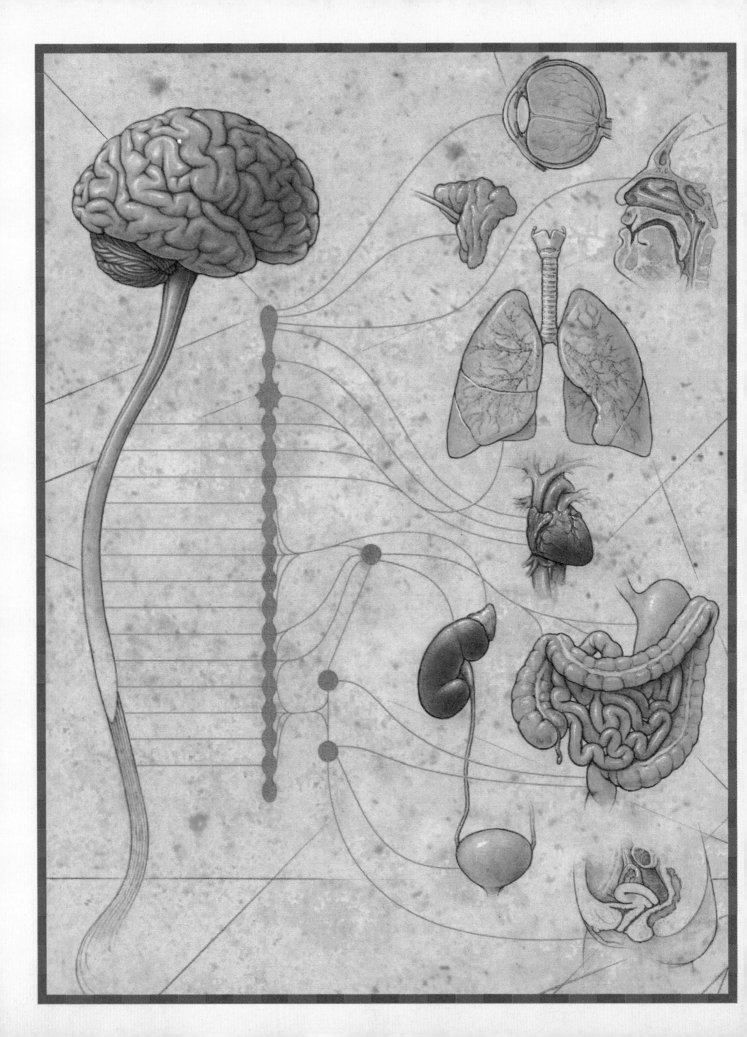

CHAPTER 21

Autonomic Nervous System

OBJECTIVES

After studying this chapter, you should be able to do the following:

1. Describe the principal actions of the autonomic nervous system.
2. Describe the anatomic features that define the autonomic nervous system.
3. Compare the anatomy of the autonomic nervous system with that of the somatic portion of the nervous system.
4. Describe the major anatomic features in the sympathetic and parasympathetic divisions, and point out differences and similarities between them.
5. Distinguish between thoracolumbar and craniosacral outflow.
6. Describe the effects of sympathetic and parasympathetic divisions on target organs.
7. Trace the sympathetic and parasympathetic pathways that regulate pupillary reflexes, heart rate, peristalsis, and vascular tone.
8. Describe the differences between white and gray rami communicantes and the axons in them.
9. Cite the differences between the autonomic nervous system and visceral afferent neurons.
10. Name the classes of ganglia, and list the pathways that axons can take through them.
11. Know the major classes of neurotransmitter substances used by the autonomic nervous system, and describe the meaning of the following terms: adrenergic, cholinergic, muscarinic, and nicotinic.
12. Outline the development of the autonomic nervous system.
13. Explain how injury to the superior cervical ganglion can result in Horner's syndrome.

VOCABULARY TOOLBOX

enteric	entero-, *G* the gut, intestine.	**prevertebral**	pre-, *L* before, in front. In front of the vertebral column.
ganglion, ganglia	gangli-, *G* a knot on a string, a swelling.	**rami communicantes**	rami, *L* branches. Communicating branches.
parasympathetic	para-, *G* beside. Beside the sympathetic system.	**splanchnic**	splanchn-, *G* the viscera.
paravertebral	para-, *G* beside. Beside the vertebral column.	**stellate**	stell-, *L* a star. Star-shaped.
		sympathetic	sym-, *G* with, path-, *G* suffering.
plexus, plexuses	plex-, *L* interwoven, network.		

RELATED TOPICS

- **Axons and synapses, Chapter 17**
- **Cell signaling, Chapter 2**
- **Divisions of the nervous system, Chapter 17**
- **Hypothalamus, Chapter 19**

- **Neural tube and neural crest, Chapters 4 and 17**
- **Pupillary light reflex, Chapter 1**
- **Spinal nerves, Chapter 18**

PERSPECTIVE

At the edges of our consciousness, another nervous system, the autonomic nervous system, attends to homeostasis in the visceral organs. Because it is not as dramatic as the thoughts and actions of the somatic system, we tend to take its steady physiological housekeeping for granted until its fine tuning goes wrong. The autonomic system is probably more important, medically speaking, than its somatic sister because it reaches deep into the fundamental cellular processes that make us function. Anatomy and physiology blend when autonomic impulses enter cells to spur secretion of hormones and digestive enzymes or to regulate heart rate and liver function. If we had to adjust these matters consciously, would there be time for anything else?

Normally, your pupils are the same size and both sides of your face perspire, but if an emergency room team finds you with dry, flushed skin on one side of the face and a constricted pupil behind a sagging eyelid, they suspect cervical damage on that side. You would be displaying the symptoms of Horner's syndrome (Figure 21.1), signs of injury to the **autonomic** (AW-toe-NOM-ik) **nervous system** (ANS).

Case study begins on p. 639

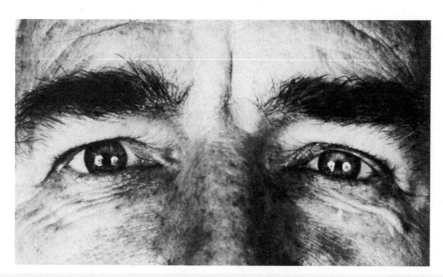

Figure 21.1

This person suffers Horner's syndrome on the left side. Although you cannot see that his skin is dry and flushed, the left pupil is constricted tightly (compare to right) and the left eyelid droops because the tarsal muscles that help support it do not contract.

BRIEF TOUR—AUTONOMIC NERVOUS SYSTEM

The symptoms of Horner's syndrome are the province of the autonomic nervous system—the "other" component of peripheral nerves (Figure 21.2) that attend to physiological homeostasis. The ANS controls all the glands and smooth muscles of the body, but it does not trigger skeletal muscles to contract. Its functions are diverse. It regulates perspiration, digestion, and pupillary reflexes, for example. It also controls blood pressure by regulating the smooth muscles that constrict or dilate the vessels, and it speeds or slows heart rate.

Unlike the somatic system, we usually do not think of the autonomic system as voluntary in function. Although we may be aware of such sensations as hunger or satiation, will alone cannot ordinarily turn them on or off. This inaccessibility results from spinal and cranial responses that continually balance physiological alternatives with little or no voluntary intervention from the cerebral cortex. Streams of impulses flow into the spinal cord and brain through **visceral afferent neurons** from sensory receptors throughout the body. With guidance from the hypothalamus and cerebral cortex, the spinal cord and brainstem can integrate this bodily information into outflow that travels through **visceral efferent neurons** back to visceral target organs, accelerating or decelerating processes within these parts of our bodies. These visceral, largely involuntary pathways constitute the autonomic nervous system. (Some specialists give this name only to the efferents, but this text considers both efferent and afferent pathways part of the autonomic system, just as somatic efferent and afferent pathways are considered together.) The following four unique features characterize the autonomic nervous system: (1) two opposing sets of efferent neural pathways for most organs, (2) two neurons in each efferent pathway, (3) autonomic ganglia where efferent neurons synapse outside the central nervous system, and (4) several neurotransmitters.

• What is the primary difference between the visceral nervous system and the autonomic nervous system?

TWO OPPOSING DIVISIONS

Opposing sympathetic and parasympathetic divisions bring about the right responses in the right places. These pathways act rather like antagonistic muscles, supplying dual sets of nerves to most visceral organs. As one set accelerates secretion or contraction, the other decelerates the process, and homeostasis rests in the balance between their effects. The pupillary muscles shown in Figure 21.3 are good examples. The iris receives separate sympathetic and parasympathetic nerves. Sympathetic neurons cause the pupil to dilate in dim light, and parasympathetic neurons stimulate the constrictor muscles when the light is bright. These opposing signals maintain homeostasis by continually balancing dilation and constriction as lighting conditions change. In Horner's syndrome, the balance has shifted permanently because the sympathetic neurons that dilate the pupil, constrict facial vessels, and promote sweating have been damaged. Without these pathways to

A

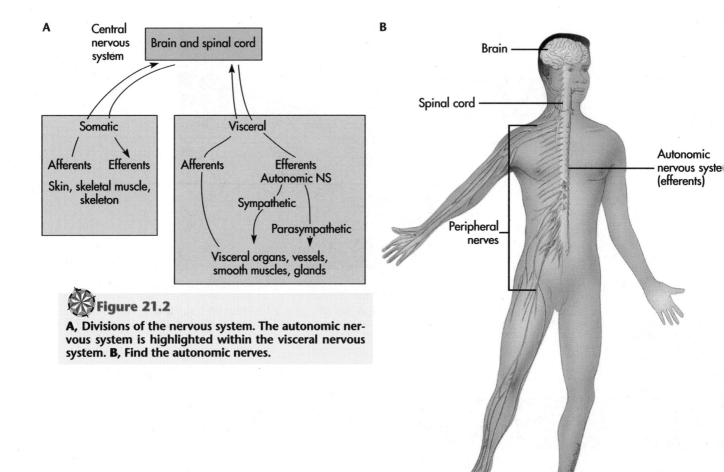

B

Figure 21.2

A, Divisions of the nervous system. The autonomic nervous system is highlighted within the visceral nervous system. B, Find the autonomic nerves.

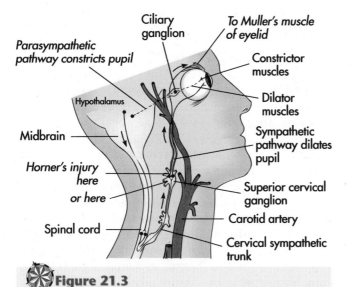

Figure 21.3

Neural pathways of Horner's syndrome. Sympathetic nerves descend the spinal cord from the hypothalamus and ascend the neck to the superior cervical ganglion and the head, but parasympathetic axons travel the oculomotor nerve from the midbrain.

oppose them, the parasympathetic neurons of the face (which lie entirely within the head) constrict the pupil and dilate the vessels. Not all sympathetic impulses are excitatory, however, nor are all parasympathetics inhibitory, as you will see.

TWO-NEURON PATHWAY

All autonomic efferent pathways consist of two neurons, as Figure 21.4 shows. Instead of innervating the target directly, **preganglionic** (pree-GANG-lee-on-ik) neurons connect with **postganglionic neurons** that end on glands or smooth muscles in the target organs. Synapses link these two neurons together in **autonomic ganglia** (GAN-glee-ah) found in the trunk and neck.

Although preganglionic neurons resemble somatic motor neurons in many respects, there are differences. Preganglionics are multipolar, with cell bodies in the lateral horns of the spinal gray matter or in the brainstem. Most of these cells have myelinated axons, and the axons may reach long distances to ganglia

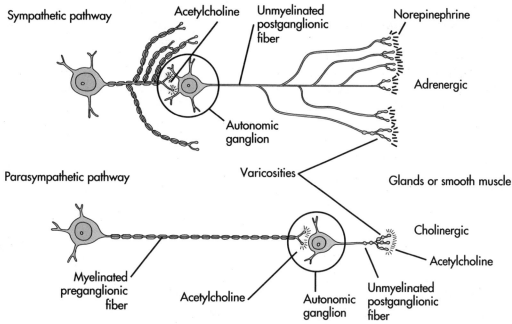

Figure 21.4

All autonomic pathways have two neurons. Preganglionic neurons originate in the spinal cord or brain and synapse with postganglionic neurons in ganglia that are close to the spinal cord (sympathetic) or visceral organs (parasympathetic). Acetylcholine transmits to postganglionic neurons in both systems, but epinephrine and norepinephrine substitute for acetylcholine at the nerve endings of sympathetic postganglionic neurons.

near target organs. Postganglionic neurons also are multipolar. With their cell bodies in ganglia near target organs, postganglionic axons usually are unmyelinated and short; they are especially short in the parasympathetic system. A single preganglionic neuron of the sympathetic division routinely innervates 20 postganglionic cells and may innervate up to 8000, thus spreading its effects widely to the correct destinations, as do the motor units of skeletal muscles. Postganglionic axons increase this branching effect with numerous **varicosities,** cytoplasmic swellings in the axon terminals (Figure 21.4) that mark synapses in contact with many muscle or glandular cells.

- Name the divisions of the autonomic nervous systems, and name the two types of autonomic neurons in these divisions.

AUTONOMIC GANGLIA AND PLEXUSES

Autonomic ganglia and plexuses distribute visceral efferents to appropriate target tissues and collect the appropriate visceral afferents from them. **Autonomic ganglia,** illustrated in Figure 21.5A, are the only structures outside the central nervous system that contain cell bodies of efferent neurons. Postganglionic neurons originate in autonomic ganglia, and their axons branch to their targets. Nerve **roots** carry preganglionic axons into the ganglia, and **branches** conduct postganglionic axons away from them. Because they contain synapses, autonomic ganglia are quite unlike dorsal root sensory ganglia that house

only cell bodies of afferent unipolar neurons, not synapses.

Autonomic ganglia fall into three classes, according to their locations. **Paravertebral ganglia** are located beside the vertebral column (Figure 21.5B), and **prevertebral ganglia** associate with the abdominal aorta. **Terminal ganglia** (not shown) are located directly in or on target organs themselves. One paravertebral ganglion, the superior cervical (Figure 21.3), often is implicated in Horner's syndrome.

Plexuses are networks of branches that distribute autonomic neurons and visceral afferents within organs. Axons converge and diverge in these networks, and microscopic ganglia within these plexuses contain the cell bodies and synapses. Plexuses associate with major arteries. The largest plexuses lace the descending aorta, and finer ones follow its branches into the visceral organs. Diffuse plexuses also cover the exterior of these organs, and sensory receptors there relay pain stimuli through afferent neurons back through the ganglia to the central nervous system.

- Name the three types of autonomic ganglia.
- In general, where are they located?

MULTIPLE NEUROTRANSMITTERS

The autonomic nervous system uses several important neurotransmitters. Preganglionic axons release acetylcholine (ah-SEET-il-KOE-leen) at synapses with postganglionic neurons in the sympathetic and parasympathetic divisions (Figure 21.4). Sympathetic postganglionic neurons use epinephrine (adrenalin; EP-in-EF-rin; ah-DREN-a-lin) or norepinephrine (noradrenalin) and therefore are called **adrenergic** (AD-ren-ER-jik) neurons. Parasympathetic pathways are **cholinergic** (KOE-lin-ER-jik) because preganglionic and postganglionic neurons use acetylcholine. Autonomic neurons and ganglia also use many other less common neurotransmitter substances (see Neurotransmitters and Receptors, this chapter).

- What types of neurotransmitters do sympathetic and parasympathetic systems use?
- What terms describe the pathways that use these neurotransmitters?

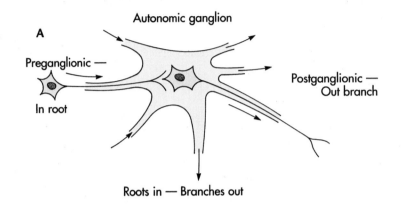

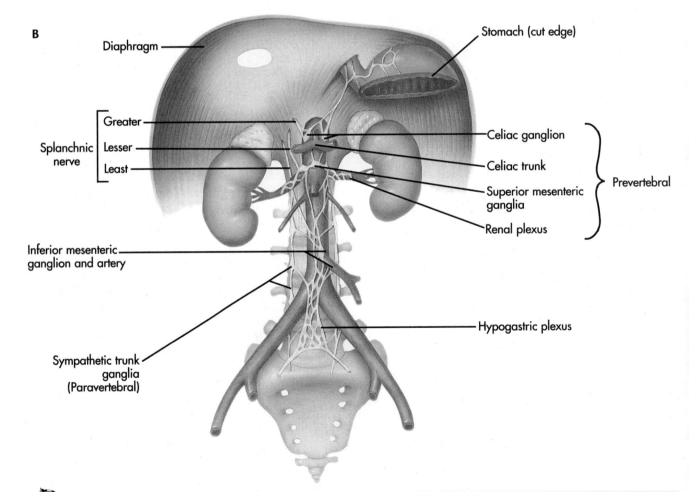

Figure 21.5

Autonomic ganglia of the abdomen. A, Roots enter and branches leave autonomic ganglia. **B,** Find the celiac ganglion and the superior and inferior mesenteric ganglia, all prevertebral ganglia surrounding the abdominal aorta. Sympathetic ganglia are paravertebral ganglia.

SYMPATHETIC AND PARASYMPATHETIC DIVISIONS

The sympathetic nervous system is most active in emergencies, while the parasympathetic system is most vigorous when the body is at rest. Sweaty hands and that tight feeling of apprehension are signs that the hypothalamus and sympathetic system are preparing for **"fight or flight"** (FF). Adrenalin surges from your adrenal glands, your blood pressure rises, your heart accelerates, and you feel jittery. Your blood glucose rises, your pupils dilate, and your spleen dumps more red blood cells into circulation to prepare for danger. Blood and oxygen flood into your muscles through dilated vessels. More air moves into the lungs through dilated bronchioles. With energy consumption rising, you are ready to go.

Compared with all this vigorous activity, the parasympathetic division promotes **"rest and relaxation"** (RR), when the emergency is over. For example, parasympathetic impulses lower your blood pressure and heart rate after a meal; your pupils constrict, and blood accumulates in your digestive tract. Peristalsis (gut movement) and absorption increase, your adrenals become quiescent, and your bladder and rectum may empty; all such processes are related to absorbing energy rather than expending it.

Table 21.1 summarizes the effects of both systems. To appreciate how the ANS works, you should learn the actions that control the heart and vessels, the gastrointestinal tract, the adrenal glands, and the sweat glands. One of the first things you will discover is that the sympathetic system is not exclusively excitatory, nor is the parasympathetic entirely inhibitory. The best way to keep their actions in mind is to visualize them in fight or flight and in rest and relaxation. It makes sense to divert energy into muscles and circulation at the expense of digestion for short periods, then to restore that energy when the emergency is over.

The anatomy of the sympathetic and parasympathetic systems reflects their actions. The sympathetic system reaches its targets by **thoracolumbar outflow** (thore-AK-oh-LUM-bar) through thoracic and lumbar spinal nerves, but **craniosacral outflow** (KRANE-ee-oh-SAY-cral) brings parasympathetic axons to these same target tissues from cranial and sacral nerves. Table 21.2 helps you compare these systems.

- Which system prepares for fight or flight? For rest and relaxation?
- Which system employs craniosacral outflow? Thoracolumbar outflow?

SYMPATHETIC DIVISION

SYMPATHETIC CHAIN

The sympathetic system communicates with the viscera by way of the sympathetic trunk (sympathetic chain; Figure 21.6). Extending along each side of the vertebral column from the head to the coccyx, two chains of ganglia (one shown) distribute thoracolumbar outflow to visceral organs in the head, neck, trunk, and extremities. Each chain contains 22 spindle-shaped or stellate (STELL-ate) paravertebral ganglia— 3 cervical, 11 thoracic, 4 lumbar, and 4 sacral—joined by narrow **cords** of nerve fibers. A twenty-third ganglion, the **ganglion impar** (IM-par), unites both chains at the coccyx. These numbers would correspond directly to the spinal nerves (8 cervical, 12 thoracic, 5 lumbar, 5 sacral, 1 coccygeal) if it were not for fusion or loss of ganglia during development.

The sympathetic chains (Figure 21.7) suggest supplementary spinal cords, lying lateral to the vertebral column. Connective tissue continuous with the epineurium of spinal nerves covers the ganglia and their roots and branches. Located at the angles of the ribs superiorly and then at the costovertebral joints, the trunks shift to the anterior surface of the lower lumbar vertebrae and pelvis before ending on the anterior surface of the coccyx.

RAMI COMMUNICANTES

Special branches of the spinal nerves deliver to the sympathetic trunk. Called **white rami communicantes** (RAM-eye ko-MUNE-ih-kahnt-ace), they connect preganglionic axons from all thoracic (T1 to T12) spinal nerves and the first two or three lumbar (L1 to L3) spinal nerves to corresponding ganglia. You can see which rami deliver to which organs in Table 21.3 and Figure 21.6. The white rami, usually one for each ganglion, glisten white in life because most of their axons are myelinated. Figure 21.8 shows that each white ramus (RAY-mus) branches from the anterior primary ramus of its spinal nerve (see Spinal Nerves, Chapter 18) after the spinal nerve has exited the intervertebral foramen. White rami also carry myelinated afferent axons from the viscera to the spinal cord (see Visceral Afferents, this chapter).

Another set of communicating branches, the **gray rami communicantes,** (not shown in Figure 21.6), conduct postganglionic axons to vessels in the body wall and extremities, as well as to sweat glands and smooth muscle fibers of the skin. Gray with unmyelinated axons, these branches join the anterior primary ramus of the spinal nerves proximal to the white rami (Figure 21.8). Axons in the gray rami return to the spinal nerve from the sympathetic trunk and thence out the anterior and posterior primary rami of the spinal nerve to their targets. Unlike the white rami,

Table 21.1   EFFECTS OF THE AUTONOMIC NERVOUS SYSTEM

Target Organ	Sympathetic Impulses (Adrenergic Receptors)	Parasympathetic Impulses
Eye		
Ciliary muscle	Relaxes, for far vision (β)*	Contracts, near vision
Iris	Dilator muscles contract (mydriasis) (α)*	Sphincter muscles contract (miosis)
Lacrimal gland	—	Secretes
Salivary glands	Sparse mucus secretion (α)	Profuse serous secretion
Thyroid gland	Stimulates (β)	
Lungs	Tracheal, bronchial muscles relax, bronchioles dilate (β)	Contract / Constrict
	Bronchial glands decrease (α) or increase secretion (β)	Secretion increases
Heart	Rate, output increase (β)	Decrease
Arteries		
Coronary	Dilate (α, β)	Constrict
Skeletal muscle	Dilate (β)	Dilate
Skin and mucosa	Constrict (α)	Dilate
Systemic veins	Constrict (α), dilate (β)	—
Esophagus	Peristalsis decreases (α, β)	Increases
Stomach	Peristalsis decreases (α, β)	Increases
	Sphincter constricts (α)	Dilates
	Secretion decreases (α)	Increases
Liver	Releases glucose into blood (β)	—
Gallbladder and bile ducts	Dilate (β)	Constrict
Pancreas	Acinar secretion decreases (α)	Increases
	Islet cell secretion, insulin decreases (α)	Increases
Intestines	Peristalsis decreases (α, β)	Increases
	Sphincters dilate (β)	Constrict
	Secretion decreases (α)	Increases
Adrenal medulla	Secretes epinephrine and norepinephrine (α)	—
Kidney	Arterioles constrict (α)	Dilate
Ureter	Tone decreases (α)	Increases
Urinary bladder	Detrusor muscle relaxes (β)	Contracts
	Trigone and sphincter contract (α)	Relax
Male genitalia	Ejaculation: seminal vesicles, vas deferens constrict (α)	Erects penis
Female genitalia	Vagina constricts	Erects clitoris
Uterus	In pregnancy: contraction (α), or relaxation (β)	Variable, depending on menstrual cycle
	Nonpregnant: relaxation (β)	
Skin	Arrector pili muscles contract (α)	—
	Palms and soles perspire (α)	Generalized perspiration
Fat cells	Fat molecules split, lipolysis (α, β)	—

From Bonica JJ, editor: The management of pain, vol 1, ed 2, Philadelphia, 1990, Lea and Febiger; Gilman, FG, Rall TW, Ries AS, and Taylor P, editors: Goodman and Gilman's the pharmacological basis of therapeutics, ed 8, New York, 1990, Pergamon Press.
See Neurotransmitters and Receptors, this chapter.

which are only thoracic or lumbar, all spinal nerves, including cervical and sacral, receive gray rami from the sympathetic trunk ganglia. After reentering the spinal nerves, these efferents follow a course through the branches, as do somatic motor axons.

- Where are the sympathetic chains located?
- What spinal segments deliver to the chain?
- What are the functions of the gray and white rami communicantes?
- How do preganglionic axons of neurons reach the sympathetic trunk?

PATHWAYS TO THE HEAD, NECK, AND TRUNK

Sympathetic impulses may ascend, descend, or pass directly through the sympathetic trunk toward target organs. Thoracic preganglionic axons enter the head and neck by ascending to higher ganglia in the sympathetic trunk where they synapse with postganglionic axons, illustrated in Figure 21.8A. Other thoracic preganglionic axons may (Figure 21.8B) synapse in a thoracic chain ganglion with postganglionic axons that run directly to heart and lungs.

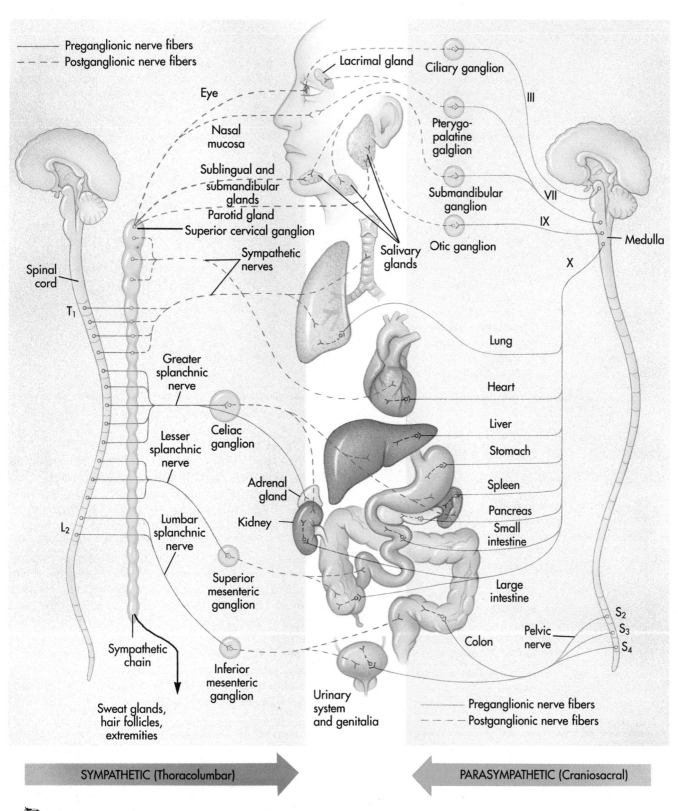

Figure 21.6

Sympathetic and parasympathetic pathways to the visceral organs. Preganglionic neurons are shown in solid lines, postganglionic neurons in dotted lines.

Table 21.2   SYMPATHETIC AND PARASYMPATHETIC DIVISIONS COMPARED

Item	Sympathetic	Parasympathetic
Effects	Fight or flight; energy consumptive	Rest and relaxation; energy productive
Outflow	Thoracolumbar: T1-L2 via sympathetic trunk to head, neck, thorax, abdomen, pelvis, body wall, and extremities	Craniosacral: via oculomotor, facial, glosso-pharyngeal, vagus cranial nerves, and sacral nerves 2-4 to head, neck, thorax, abdomen, and pelvis only; not to body wall or extremities
Ganglia	Sympathetic chain (paravertebral ganglia) beside vertebral column, prevertebral ganglia associated with abdominal aorta	Terminal ganglia only; located in or near target organs
Rami commun-icantes	White rami in T1-L2; gray rami in all segments of spinal cord and sympathetic trunk	None
Preganglionic neurons	Cell bodies in spinal cord, axons mye-linated, relatively short; synapses in sym-pathetic ganglia	Cell bodies in brainstem and spinal cord, axons myelinated, relatively long; synapses in terminal ganglia
Postganglionic neurons	Cell bodies in sympathetic ganglia, axons unmyelinated, relatively long distance to target	Cell bodies in terminal ganglia, axons un-myelinated, short distance to target
Neurotransmitters	Cholinergic in ganglia (acetylcholine), and adrenergic (epinephrine and nor-epinephrine) at end organs	Cholinergic; acetylcholine only
Receptors	Muscarinic and nicotinic at ganglia, alpha and beta adrenergic at end organs	Muscarinic and nicotinic

Table 21.3   AUTONOMIC SPINAL SEGMENTS

Location	Destination
T1-T4 (T5)*	Head and neck
T2-T8/9	Upper limbs: skin and vessels
T2-T8	Upper trunk: skin and vessels
T9-L2	Lower trunk: skin and vessels
T1-T5	Thoracic viscera: heart, lungs, esophagus
T5-L2	Abdominal viscera: gastrointestinal tract, ascending, transverse colon, liver, spleen, adrenal medulla, kidney, ureter
S2-S4	Pelvic viscera: descending colon, rectum, urinary bladder, external genitalia
T10-L2	Lower limbs: skin and vessels

From Bonica JJ, editor: The management of pain, vol 1, ed 2, Philadelphia, 1990, Lea and Febiger.

*Segments in parentheses may be missing.

Abbreviations: T—thoracic, L—lumbar, S—sacral.

Postganglionic axons (Figure 21.8C) reach the skin, vessels, body wall, and extremities through gray rami communicantes. In the abdominal cavity (Figure 21.8D), preganglionic axons run through the chain ganglia to synapses in prevertebral ganglia near the aorta and thence to abdominal organs. To reach the pelvis from lumbar regions of the spinal cord (Figure 21.8E), preganglionic axons descend the chain to synapse in ganglia at lower levels. It is not clear whether individual axons branch into more than one pathway.

CERVICAL

The sympathetic chain ascends the neck as three cervical ganglia with cords between them (Figure 21.6), along the carotid artery and the longus capitis (KAP-ih-tiss) and longus colli (KOAL-ee) muscles on each side of the neck (see Anterior and Lateral Muscles of the Neck, Chapter 10). Preganglionic axons from the first four or five thoracic rami communicantes (Table 21.3) but mostly numbers two and three, supply all three ganglia by way of the first pathway listed above. There are no cervical white rami, but eight gray rami return to the cervical nerves, one for each nerve.

The **superior cervical ganglion,** the largest of the ganglia (about 2-8 cm), supplies all sympathetic innervation to the head, shown in Figure 21.3. This ganglion is located below the carotid canal of the skull, posterior to the angle of the mandible at the level of the second cervical vertebra. Postganglionic axons branch from the ganglion primarily in the internal carotid nerve that follows the internal carotid artery through the canal, branching to the face, eyes, cranial cavity, cavernous sinuses, and hypophysis. Injuries to this nerve or ganglion cause Horner's syndrome. Caudal axons in the ganglion pass into the facial and glossopharyngeal cranial nerves, where they retard secretion in the salivary glands (see Table 21.1).

The **middle cervical ganglion** is the smallest of the three and communicates to the cardiac and thyroid plexuses of nerves. The **inferior cervical gan-**

glion often fuses with the first thoracic ganglion to form the **stellate ganglion;** it supplies branches to the vertebral nerve into the cranial cavity and also to the cardiac plexus. These nerves accelerate thyroxine secretion by the thyroid, as well as increase heart rate and output (see Table 21.1).

- Trace the sympathetic pathway from the spinal cord to the iris dilator muscles.

THORACIC

This portion of the sympathetic chain sends postganglionic axons from T1 through T5 to the heart, lungs, and bronchi, as well as to the esophagus and the thoracic aorta, generally increasing heart rate, dilating the pulmonary passages, and slowing movement by the esophagus (see Table 21.1 and Figure 21.6). Lower members of the chain contribute also to the esophagus, lungs, aorta, and splanchnic nerves of the abdomen. White and gray rami communicantes connect with all thoracic spinal nerves. Thoracic gan-

glia are found at the level of each intervertebral disk below the corresponding vertebra.

ABDOMINAL

The pattern of sympathetic innervation changes in the abdominal and pelvic cavities. There many preganglionic axons pass directly through the sympathetic trunk to synapses in three prevertebral ganglia—**celiac** (SEE-lee-ak), **superior mesenteric** (MEZ-en-TARE-ik), and **inferior mesenteric**—that surround the abdominal aorta. Only three lumbar ganglia receive white rami from the spinal cord. The celiac ganglion (Figures 21.5 and 21.6) associates with the celiac trunk and receives preganglionic axons from three **splanchnic nerves** (SPLANK-nik)— greater, lesser, and least—that branch from the lower ganglia of the thorax. Branches from the celiac ganglion distribute postganglionic axons to the liver, spleen, gallbladder, stomach, pancreas, small intestine, and ascending colon. The superior and inferior

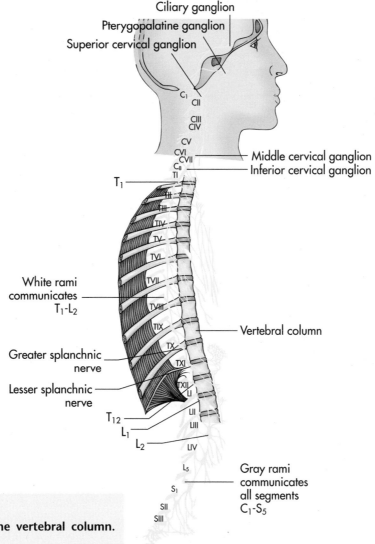

Figure 21.7

The sympathetic trunk follows the vertebral column. Right side view.

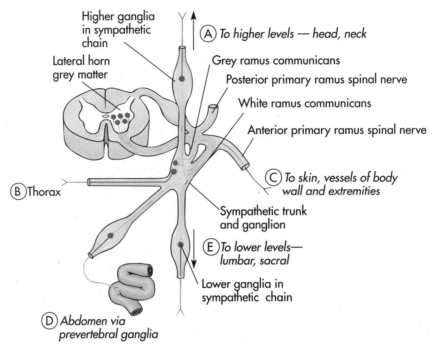

Higher ganglia
in sympathetic
chain

Lateral horn
grey matter

(A) *To higher levels — head, neck*

Grey ramus communicans

Posterior primary ramus spinal nerve

White ramus communicans

Anterior primary ramus spinal nerve

(C) *To skin, vessels of body
wall and extremities*

(B) Thorax

Sympathetic trunk
and ganglion

(E) *To lower levels—
lumbar, sacral*

Lower ganglia in
sympathetic chain

(D) *Abdomen via
prevertebral ganglia*

✳ Figure 21.8

Rami communicantes link spinal nerves and sympathetic trunk. Follow the pathways (find their letters) through the sympathetic ganglia.

mesenteric ganglia are located at the exits of arteries of the same name from the abdominal aorta, and they receive lumbar splanchnic nerves from the last thoracic and first three lumbar chain ganglia. Postganglionic axons follow branches of the abdominal aorta from the superior mesenteric ganglion to the remainder of the colon. The inferior mesenteric ganglion supplies the rectum, bladder, and external genitalia. In general these pathways slow the activites of the digestive system. They also raise blood glucose by stimulating its release from the liver (see Table 21.1).

The adrenal medulla is the only sympathetic target without a true ganglion. Preganglionic axons reach it by passing directly through the sympathetic trunk and celiac ganglion without synapsing until they end on **chromaffin cells** (KROME-ah-fin) that secrete epinephrine and norepinephrine when stimulated. Not neurons in the usual conductive sense, these secretory cells have the same relationships with preganglionic neurons as do postganglionic neurons, and the medulla itself may be viewed as a ganglion without postganglionic axons. Medullary cells and sympathetic ganglion cells both arise from the neural crest.

- Follow the flow of sympathetic stimuli from the spinal cord to the heart. What effect do these stimuli have there?
- What pathway do sympathetic impulses follow to the small intestine, and what effects do they produce?

SACRAL AND COCCYGEAL

The sympathetic chain extends into the pelvis as four sacral ganglia and one coccygeal ganglion (the

ganglion impar), but these ganglia have no direct link with the spinal cord in either region. In this sense they are comparable to the cervical ganglia at the superior end of the sympathetic chain. Outflow descends through these ganglia from lumbar levels to stimulate perspiration, contract arrector pili muscles of hair follicles, and dilate blood vessels of leg muscles and constrict those in the skin (see Table 21.1).

PARASYMPATHETIC DIVISION

The parasympathetic system is much simpler than the sympathetic. Outflow is cranial and sacral (craniosacral). There are no chain ganglia or communicating rami. Most outflow travels along four cranial nerves, and the rest travels through three sacral spinal nerves. Three quarters of total cranial traffic passes through one vast nerve, the vagus, to the heart, lungs, and abdominopelvic organs. The oculomotor, facial, and glossopharyngeal (GLOSS-oh-fahr-IN-jee-al) carry the remaining traffic to the head and neck. Nearly all parasympathetic ganglia are **terminal,** located in the walls or on the surface of organs as part of extensive plexuses. Preganglionic axons, of course, extend along these four nerves to the ganglia where synapses connect them with postganglionic cells that weave within the plexus to the targets. Parasympathetic outflow does not reach the body wall or extremities, so that peripheral vessels, sweat glands, and arrector pili muscles have no parasympathetic antagonists.

CRANIAL OUTFLOW
OCULOMOTOR NERVE
Parasympathetic axons of this nerve constrict the pupil and focus the lens (see Table 21.1 and Figure 21.6). Preganglionic axons originate in the tegmentum (TEG-ment-um) of the midbrain and follow the **oculomotor nerve** to the **ciliary ganglion** in the orbit, where they synapse with postganglionic neurons. After penetrating the sclera, some of these postganglionic axons end in the pupillary sphincter muscle and cause it to contract in bright light. Other parasympathetic axons ending in the ciliary muscle cause the lens to round up when the eye focuses on close objects.

FACIAL NERVE
The facial nerve distributes parasympathetic axons that stimulate glands of the nasal and oral cavity to secrete (see Table 21.1 and Figure 21.6). In addition to the lacrimal gland, these include mucous glands of the nasal cavity, palate, and tongue, as well as all salivary glands except the parotid. The **pterygopalatine ganglion** is the terminal ganglion for all but the tongue and salivary glands. The **submandibular ganglion** distributes postganglionic axons to the tongue and salivaries. Both sets of axons are espe-

cially active when you have a cold or are eating, but they are quiet in emergencies (sympathetic stimuli make mouths dry). Preganglionic axons originate in the reticular formation of the pons and pass through the facial nerve to the ganglia already named, where they synapse with unmyelinated postganglionic axons.

GLOSSOPHARYNGEAL
Parasympathetic axons in the glossopharyngeal nerve cause the parotid gland (pahr-AH-tid) to secrete more vigorously. Preganglionic axons from the reticular formation in the medulla oblongata pass into the **otic ganglion,** where they link with postganglionic neurons that end in the parotid gland. Some preganglionic axons pass through the otic ganglion and end directly on ganglion cells that stimulate secretion in the posterior third of the tongue.

- Which cranial nerve stimulates the lacrimal gland to secrete?
- Which nerves supply the submandibular, sublingual, and parotid gland?

VAGUS
The vagus (VAY-gus) is the largest cranial nerve; its branches reach from the posterior cranial fossa to the transverse colon (Figure 21.9). A **vagus nerve** descends on either side of the neck from the medulla oblongata through the jugular foramen, following the carotid arteries into the thoracic cavity and passing

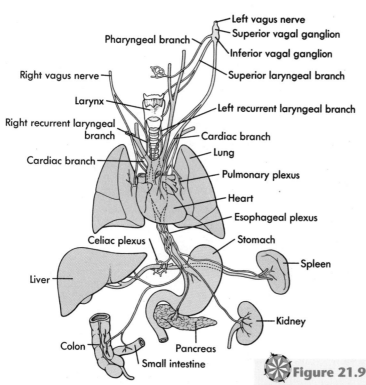

Figure 21.9

The vagus nerve sends branches to all abdominal organs.

the great vessels of the heart before converging as one nerve on the lower esophagus, sending off branches to these regions as it passes. The vagus then accompanies the esophagus through the diaphragm, and follows the lesser curvature of the stomach before distributing branches to all abdominal organs in Figure 21.6 and Table 21.1. With few exceptions, the parasympathetic axons of the vagus end in terminal ganglia located in plexuses on the target organs. Postganglionic axons mingle with sympathetic axons as they spread from these locations into the tissues.

Most superiorly, the pharyngeal branch of the vagus delivers special efferents to the voluntary muscles of the pharynx and soft palate and to the pharyngeal mucosa. Visceral afferents from the carotid bodies also travel in this branch. Muscles of the larynx and its mucosa receive from the laryngeal and recurrent branches, and the heart receives preganglionic axons through the cardiac branches. These latter axons inhibit heart rate; they enter through the **cardiac plexus** and end in the epicardium of the atria on postganglionic axons that target the sinoatrial and atrioventricular nodes and the atrioventricular bundle. Postganglionic axons in the walls of the bronchi constrict the respiratory passages and dilate bronchial vessels through connections from preganglionic axons in the pulmonary plexus. Similar circuitry supplies the esophagus.

In the abdomen, most branches of the vagus follow the arteries to their destinations in **myenteric ganglia** (MY-en-TARE-ik) of Auerbach (OUR-bak) and **submucosal ganglia** of Meissner (MICE-ner) in the plexuses of the gastrointestinal tract (Figure 21.10).

Preganglionic axons synapse at these locations with postganglionic axons that stimulate peristalsis and secretion through Auerbach's and Meissner's ganglia, respectively. The celiac and superior mesenteric plexuses of the sympathetic system carry these parasympathetic preganglionic axons to the small intestine, pancreas, liver, gall bladder, bile duct, and ascending and transverse colon.

- Name five organs supplied by the vagus nerve.

SACRAL OUTFLOW

Pelvic organs receive parasympathetic outflow from the third and fourth sacral nerves and sometimes from the second and fifth as well (see Table 21.1 and Figure 21.6). Preganglionic axons follow the sacral nerves and pelvic splanchnic branches into the **pelvic plexus,** where they mingle with sympathetic and visceral afferent axons. They synapse with postganglionic axons in **terminal ganglia** in the walls of the large intestine, urinary system, and genitalia. Impulses carried by these axons stimulate the bladder wall to contract and the prostate and seminal vesicles to secrete. Peristalsis and mucous secretion are intensified in the descending and sigmoid portions of the colon and rectum through the action of these parasympathetic axons, but the internal sphincter muscles of the anus and urethra relax. Complementary parasympathetic and sympathetic stimuli erect the penis and cause ejaculation, respectively. Parasympathetic stimuli contract the vagina and may cause the myometrium of the uterus to contract or relax, depending on pregnancy or the stage of the menstrual cycle (see Table 21.1).

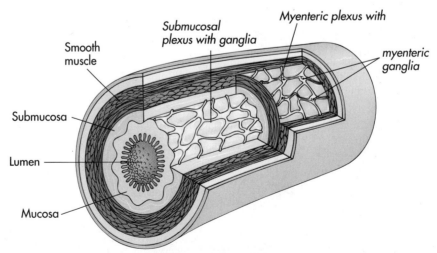

Smooth muscle

Submucosal plexus with ganglia

Myenteric plexus with

myenteric ganglia

Submucosa

Lumen

Mucosa

**Figure 21.10**

The enteric nervous system. Cutaway view shows the myenteric plexus in the smooth muscle layers of the gut tube and the submucosal plexus deeper in the wall of the digestive tract.

ENTERIC NERVOUS SYSTEM

The gut has its own nervous system. Somewhat as heartbeats originate from a pacemaker, the **enteric nervous system** (EN-ter-ik) spreads waves of contraction down the digestive tube (Figure 21.10). The stimuli come from within the gut itself, since the waves continue even when the sympathetic and parasympathetic nerves are disconnected. This remarkable ability lies in the myenteric and submucosal plexuses found throughout the digestive tube. These plexuses wrap the gut tube with interconnecting neural networks that cause the tube to dilate or constrict (myenteric plexus) and the lining to secrete (submucosal plexus). Many patients experience severe pain after a portion of the intestine has been removed because the peristaltic waves are not synchronized across the newly joined ends until enteric connections reestablish themselves.

The cell bodies of enteric neurons reside in **enteric ganglia,** the terminal myenteric and submucosal ganglia described above, that receive synapses from both the parasympathetic and the sympathetic systems. Furthermore, visceral afferent neurons establish **enteric reflexes** that influence peristalsis and secretion. These odd reflexes bypass the spinal cord. The neurons have cell bodies in enteric ganglia and synapses on postganglionic neurons in the celiac and mesenteric ganglia of the sympathetic system. Impulses return to the gut by way of postganglionic neurons. Such reflexes influence the action of sympathetic and parasympathetic control from the spinal cord and brain.

Occasional individuals lack enteric ganglia in the large intestine or rectum (one in 5000 to 8000) because neural crest cells failed to migrate to these locations or did not survive there (see Development, this Chapter). In this serious condition, known as **Hirschsprung's disease,** blockage expands the gut and abdomen above the section that no longer responds to autonomic signals, a condition called **megacolon.** Surgery to remove the defect is usually successful, but cautious diet and intravenous feeding are required for survival.

- Why do we say the gut has its own nervous system?
- What are the two principal anatomical structures of the enteric nervous system?

CENTRAL CONTROL

CEREBRAL CORTEX

Association areas in the cerebral cortex influence the action of the autonomic nervous system. Emotions, such as embarrassment that causes your face to flush or fright that sends goosebumps and shivers down your back, are examples of somatic and visceral linkage. Hunger before and satiety after meals, and sexual arousal are other examples. Descending pathways from centers in the frontal lobes link these sensations to the **hypothalamus.**

HYPOTHALAMUS

The hypothalamus controls numerous autonomic pathways through some 16 different nuclei (see Hypothalamus, Chapter 18). Two examples of hypothalamic action concern the **posterior hypothalamus** and **medial preoptic areas,** which raise and lower blood pressure, respectively, by modulating the vasomotor center located in the pons and medulla oblongata. The **vasomotor center** in turn is the focus of a reflex arc that continually monitors blood pressure by way of the glossopharyngeal and vagus nerves, and adjusts vascular tone by constricting blood vessel walls. The posterior hypothalamus intensifies these sympathetic signals, and the medial preoptic area inhibits them. The cerebral cortex and thalamus also regulate the vasomotor center.

BRAINSTEM

Other nuclei in the midbrain, pons, and medulla mediate visceral reflexes. Pretectal nuclei and **Edinger/Westphal nuclei** of the midbrain are parasympathetic links in the pupillary light reflex described in Chapters 1 and 20. These nuclei relay parasympathetic impulses that constrict the pupil in bright light. In the pons the **superior salivatory nucleus** mediates salivation. This nucleus relays impulses from receptors in the tongue to parasympathetic neurons in the facial nerve that stimulate the submandibular and sublingual glands to secrete saliva. The medulla is the site of numerous reflexes that regulate breathing, heart rate, coughing, and vomiting, by way of the glossopharyngeal and vagus nerves.

SPINAL CORD

Thoracolumbar and sacral outflow originate from columns of nuclei in the lateral horns of the spinal gray matter, summarized in Table 21.2.

- Cite an example of the action of the cerebral cortex, the hypothalamus, and the brainstem in the autonomic nervous system.

AUTONOMIC GANGLIA

The structure of the ganglia in the autonomic nervous system reveals little about the synapses within them. Embedded in the fascia between major organs and covered by connective tissue capsules, autonomic ganglia resemble lymph nodes in size and shape (Figure 21.11A). Ganglia range from round to irregular shapes, depending on the number of roots and branches. They may be nearly 3 cm long (1¼ inch) or as small as 1 mm in diameter. Plexuses surround many ganglia and make it difficult to tell where the ganglion ends and the plexus begins.

Internally, autonomic ganglia are strikingly homogeneous, as Figure 21.11B shows. The epineurial covering of autonomic nerves also coats autonomic ganglia, and connective tissue septa partition the ganglia into lobules that contain cell bodies of numerous **ganglion cells** (the neurons within the ganglion). Tangles of axons, dendrites, synapses, and cell bodies of gan-

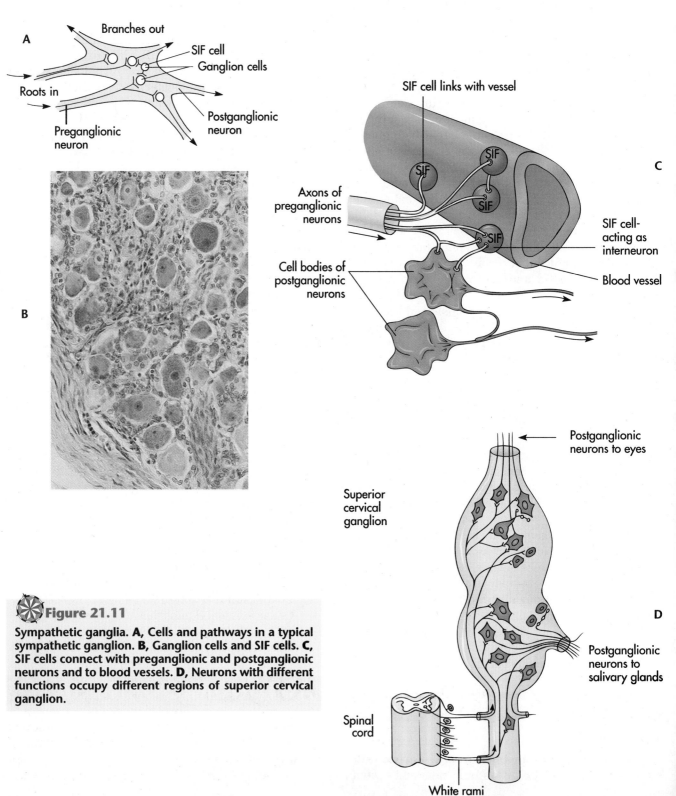

Figure 21.11

Sympathetic ganglia. **A**, Cells and pathways in a typical sympathetic ganglion. **B**, Ganglion cells and SIF cells. **C**, SIF cells connect with preganglionic and postganglionic neurons and to blood vessels. **D**, Neurons with different functions occupy different regions of superior cervical ganglion.

glion cells fill the lobules. Profuse capillary plexuses of vessels enter and leave on the roots and branches of the ganglia. Among this apparently disorderly mass are multipolar postganglionic neurons whose cell bodies are covered with satellite cells (see Neuroglia, Chapter 17). Other **small intensely fluorescent** (SIF) cells contain granules that fluoresce brilliantly in ultraviolet light.

SIF cells probably influence synaptic transmission and the flow of neural traffic through ganglia. More than two dozen neurotransmitter substances have been identified among SIF granules and elsewhere, and many are identical to peptide hormones released by endocrine tissues. The list of these substances is as complex as the ganglion anatomy itself. It includes **cholecystokinin,** the gastric hormone that regulates digestion in the stomach; **luteinizing hormone releasing factor,** the peptide that causes the pituitary gland to release luteinizing hormone; and **enkephalins** that inhibit pain by inhibiting synaptic transmission. Many SIF cells act as **interneurons** between pre- and postganglionic cells (Figure 21.11C) where their neurotransmitters may modulate synaptic transmission. Other SIF cells connect neurons with blood vessels, which suggests that blood-borne factors also regulate neural traffic.

Autonomic ganglia are not as disorganized as their spaghetti-like interiors suggest. Neurons with different functions have a tendency to occupy different locations. Figure 21.11D illustrates such a situation in the superior cervical ganglion, where cell bodies of neurons whose axons innervate the eye occupy the superior end, and cell bodies of neurons to the face occupy the inferior end.

- What are ganglion cells?
- Name two different types of ganglion cells.

NEUROTRANSMITTERS AND RECEPTORS

Autonomic synapses employ a wide variety of receptors, which gives the ANS a wide range of potential responses throughout the body. For example, the brain can suppress pain by releasing enkephalins that inhibit synaptic transmission of impulses from pain receptors.

Sympathetic pathways employ two classes of transmembrane receptor molecules—alpha and beta—to respond to adrenergic molecules flowing across synapses. **Beta adrenergic receptors** are members of the membrane transduction chain of proteins that stimulates or inhibits cell processes by activating **adenyl cyclase** (see Signal Transduction, Chapter 2). **Alpha adrenergic receptors** frequently elicit the opposite effects by inhibiting adenyl cyclase. In many cases the same neurotransmitter substance will stimulate cells by way of alpha receptors and inhibit them via betas. Because each class has many variants and different responses linked to them, it is not difficult to see how the ANS can control so many different metabolic processes.

Cholinergic parasympathetic pathways are equally diverse because of nicotinic and muscarinic receptors that respond differently to acetylcholine. Once again these differences concern the nature of receptor molecules in the signal transduction pathways to the interior of the cell. **Nicotinic receptors** (NIK-oh-TIN-ik) are sodium and potassium ion channel proteins gated by acetylcholine itself, so that the neurotransmitter directly stimulates the internal cellular machinery to respond. Though not ion channels, **muscarinic receptors** (MUS-kar-IN-ik) are members of the membrane transduction chain of proteins that can respond to ion channels by exciting or inhibiting cells.

- What classes of receptors serve sympathetic and parasympathetic synapses?
- Which receptors are ion channels?

VISCERAL AFFERENTS

Afferent neurons connect visceral organs to the central nervous system. Like somatic afferents from muscles and skin, **visceral afferents** are myelinated unipolar neurons with cell bodies in the **posterior root ganglia** of spinal nerves and **sensory ganglia** of the cranial nerves. Axons return through the same local plexuses, ganglia, and sympathetic and parasympathetic nerves that distribute outflow to the head, neck, and trunk. **Cranial nerves,** especially the vagus nerve (Figure 21.9), contribute heavily to this sensory inflow, as do the rami communicantes of spinal nerves. Visceral afferents return from the extremities and body wall by way of spinal nerves, so that somatic and visceral impulses travel the same nerves but not the same axons.

Visceral afferents relay impulses concerning the tone of vessels in the face and scalp to the **superior cervical ganglion** (sympathetic) by way of branches on the external carotid artery (see these regions in Figure 21.6). The facial nerve carries afferents from the nose, and together with the glossopharyngeal and vagus nerves also carries impulses from salivary glands and mucosa of the mouth and pharynx. The carotid bodies monitor oxygen tension in the carotid arteries through afferent axons that return to the medullary reflex centers by way of the vagus nerve. Impulses from the wall of the thorax communicate through the thoracic spinal nerves, but those from the heart and lungs travel the vagus nerve and the rami communicantes of the sympathetic trunk to the brain and spinal cord.

Abdominal organs depend heavily on sympathetic pathways through the celiac ganglion, the splanchnic nerves, and the sympathetic trunk. The stomach, liver, gallbladder, small intestine, pancreas, and adrenals all send afferent neurons to the spinal cord by way of the **celiac ganglion,** through the thoracic splanchnic nerves, to the sympathetic trunk and white rami communicantes. Afferent neurons communicate to the spinal cord from the rectum, bladder, and external genitalia, through the **pelvic plexus,** mesenteric ganglia, and lumbar splanchnic nerves. The **pudendal nerve** also carries visceral afferents from external genitals to the sacral spinal nerves.

- By what paths do visceral afferent neurons enter the spinal cord?
- Where are the cell bodies of these neurons located?

DEVELOPMENT

The autonomic nervous system develops from the neural tube and neural crest, shown in Figure 21.12. Preganglionic neurons develop from the **neural tube,** and the **neural crest** forms the postganglionic neurons, as well as the visceral afferents. As you might expect, cranial and sacral portions of the neural tube give rise to the parasympathetic system (Figure 21.12A), while the sympathetic system derives from thoracic and lumbar regions, along with contributions from cervical and sacral segments.

If you remember the discussion in Chapter 17, Neurons, Neuroglia, and Neural Pathways, you know that the neural plate in Figure 21.12B folds and becomes the neural tube, and the neural crest appears as two strips of cells beside the tube, during the third week of embryonic development. Preganglionic neurons sprout axons from the neural tube that grow ventrally during week 5, where they meet young postganglionic neurons that originate in the neural crest. The sympathetic chain begins to assemble during week 6. It links with rami communicantes by the end of week 8, so that in week 10 the heart and lungs begin to receive impulses from a few pioneering neurons.

Neural crest cells begin to migrate outward when the neural tube closes, and various fates await them at their destinations. Some of these cells follow ventral courses, where they form the sympathetic ganglia. After these same cells link with preganglionic neurons, postganglionic axons grow from the ganglia into the visceral tissues. Other neural crest cells that migrate deeper before encountering preganglionic neurons become the ganglion cells of either prevertebral sympathetic ganglia or the terminal ganglia of the parasympathetic system. The causes of Hirschprung's disease may prevent these neural crest cells from reaching the embryonic gut or from surviving there. Still other neural crest cells migrate into the medulla of the adrenal gland, where they become chromaffin cells that secrete epinephrine. Other neural crest cells migrate laterally and establish the dorsal root ganglia, where they become visceral or somatic afferent neurons by linking with sensory receptors peripherally and the spinal cord centrally.

- Which tissue, neural tube, or neural crest gives rise to sympathetic ganglia, preganglionic neurons, postganglionic neurons, chromaffin cells, and dorsal root ganglia?

SUMMARY

Table 21.2 summarizes and compares the sympathetic and parasympathetic divisions of the autonomic nervous system.

The autonomic nervous system (ANS) innervates all organs except the skeletal muscles. The autonomic system is efferent, visceral, and involuntary. Central control lies in the spinal cord and brainstem. Efferent neurons and autonomic ganglia distribute neural outflow to the glands and smooth muscles of the visceral organs, blood vessels, and skin. Visceral afferent neurons relay the effects of this traffic back to the spinal cord and brain. *(Page 639.)*

The autonomic nervous system maintains physiological homeostasis by controlling fundamental conditions and processes, such as blood pressure, heart rate, blood glucose, lens accommodation, peristalsis, and secretion. Homeostasis resides in balancing the opposing effects of sympathetic and parasympathetic divisions of the ANS. Sympathetic responses deal with emergencies, generally consume energy, and prepare for fight or flight. Parasympathetic effects are concerned with obtaining energy and are more active in rest and relaxation. *(Page 639.)*

Every autonomic pathway contains two neurons. A preganglionic neuron from the spinal cord or brainstem synapses in an autonomic ganglion with a postganglionic neuron whose axon ends on the target tissue. Sympathetic pathways originate from thoracic and lumbar segments (thoracolumbar outflow). Parasympathetic pathways are craniosacral, with contributions from cranial and sacral nerves. *(Page 640.)*

Sympathetic outflow proceeds through the white rami communicantes in thoracic and lumbar nerves, to the sympathetic chain in the abdomen, where postganglionic neurons distribute to target organs in the head, neck, and trunk. Gray rami communicantes conduct postganglionic axons from the sympathetic chain to the body wall and extremities. *(Page 643.)*

Parasympathetic outflow passes in cranial and sacral nerves through preganglionic neurons to terminal ganglia in the walls of target organs, where short postganglionic axons end on glands and smooth

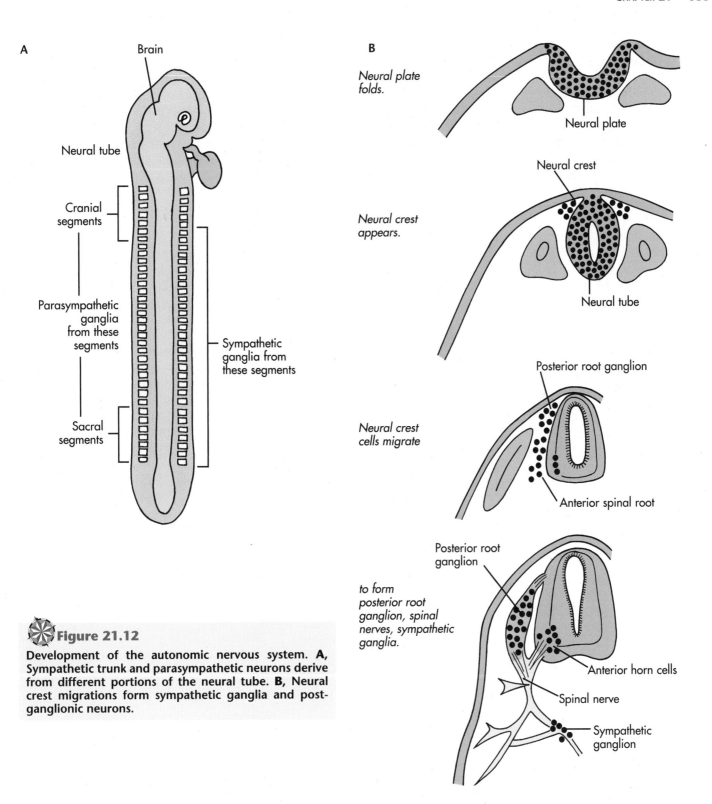

A
Brain
Neural tube
Cranial segments
Parasympathetic ganglia from these segments
Sacral segments
Sympathetic ganglia from these segments

B
Neural plate folds.
Neural plate
Neural crest appears.
Neural crest
Neural tube
Neural crest cells migrate
Posterior root ganglion
Anterior spinal root
to form posterior root ganglion, spinal nerves, sympathetic ganglia.
Posterior root ganglion
Anterior horn cells
Spinal nerve
Sympathetic ganglion

Figure 21.12

Development of the autonomic nervous system. A, Sympathetic trunk and parasympathetic neurons derive from different portions of the neural tube. B, Neural crest migrations form sympathetic ganglia and post-ganglionic neurons.

muscle fibers. Because parasympathetic axons do not reach peripheral blood vessels or the skin and extremities, these regions lack dual autonomic innervation. *(Page 643, 648.)*

Sympathetic impulses ascend the cervical chain. To dilate the iris, for example, sympathetic impulses ascend the cervical trunk to the superior sympathetic ganglion and the iris, while parasympathetic

circuitry follows the oculomotor nerve to the ciliary ganglion and constrictor muscles. *(Page 644, 649.)*

In the thorax, sympathetic neurons reach the cardiac plexus of the heart through the first five thoracic ganglia. Parasympathetic impulses counteract the stimulatory effects of these sypathetic neurons by descending through the vagus nerve and cardiac plexus to the epicardium. *(Page 647, 649.)*

The sympathetic system inhibits most of the gastrointestinal tract by way of preganglionic neurons that pass through the sympathetic trunk and splanchnic nerves directly to the prevertebral ganglia and postganglionic axons that end on glands and smooth muscle. Parasympathetic impulses oppose this action through branches of the vagus nerve and sacral outflow in the walls of the gut tube. The sacral sympathetic chain innervates vessels and skin of the lower extremities, but there is no parasympathetic outflow to these areas. *(Page 647, 649.)*

The ganglia and plexuses of the enteric nervous system distribute impulses that govern peristalsis and glandular secretion in the digestive system. Parasympathetic and sympathetic pathways control digestion at higher levels. *(Page 651.)*

Afferent neurons link visceral receptors with the spinal cord and brain by way of the spinal and cranial nerves. These unipolar neurons have cell bodies in dorsal root ganglia of the spinal nerves and in sensory ganglia of cranial nerves. The axons travel from the skin and extremites entirely within the spinal nerves, but visceral afferents connect through white rami communicantes. *(Page 653.)*

Autonomic ganglia contain synapses and cell bodies. The sympathetic system employs paravertebral, prevertebral, and terminal ganglia, but terminal ganglia are the only type in the parasympathetic system. In addition to the synaptic connections between pre- and postganglionic neurons, autonomic ganglia contain small intensely fluorescent (SIF) cells that store various neurotransmitter substances that probably modulate synaptic transmission. *(Page 651.)*

The autonomic nervous system employs multiple transmitter substances and receptors that receive them. Preganglionic (cholinergic) neurons release acetylcholine to muscarinic or nicotinic receptors in postganglionic neurons. The same is true of postganglionic parasympathetic neurons, but postganglionic sympathetic endings release epinephrine or norepinephrine (adrenergic) to alpha or beta receptors. Other peptide and catecholamine neurotransmitters and receptors are found throughout the autonomic nervous system. *(Page 653.)*

The autonomic nervous system develops from the neural tube and neural crest. Preganglionic axons sprout from the neural tube. The neural crest provides ganglia, ganglion cells, and postganglionic neurons, as well as visceral afferents. *(Page 654.)*

STUDY QUESTIONS   OBJECTIVES

1. Make a list of bodily functions controlled by the autonomic nervous system.   **1**
2. Why do anatomists separate the autonomic system from visceral afferents? What is unique about the anatomy of the ANS?   **2**
3. If the autonomic nervous system were wired in the same way as the somatic portion, what sympathetic pathways and neurons would innervate the stomach?   **3**
4. How are the sympathetic and parasympathetic divisions different? How are they similar? Consider their pathways, location of ganglia, neurotransmitters, and receptors.   **4**
5. Which divisions use thoracolumbar or craniosacral outflow? Which spinal and cranial segments are involved?   **5**
6. We have said the sympathetic system promotes energy consumption. What actions make this statement true? In contrast, what parasympathetic effects provide the body with energy?   **6**
7. Why does damage to the superior cervical ganglion result in Horner's syndrome? Find your answer by tracing the symptoms back to the ganglion through the relevant nerves.   **7, 13**
8. The sympathetic system goes into high gear in emergencies. It increases cardiac output, and the adrenal gland dumps epinephrine into the blood. What other changes take place? Trace the action through the sympathetic pathways to the heart and adrenal glands. Trace the pathways of the other effects.   **6, 7**
9. Parasympathetic pathways frequently reverse the effects of sympathetic stimulation. Trace the pathways that become active in the parasympathetic system after the emergency described in Question 8.   **6, 7**
10. What are the roles of the white and gray rami communicantes? Which axons travel in each ramus? Which segments of the spinal cord have gray rami? Which have white rami?   **8**
11. What efferent and afferent neural traffic do the rami communicantes carry at T5?   **8**
12. Visceral afferents resemble somatic afferents more than visceral efferents. What anatomic features make this so?   **9**
13. Autonomic ganglia fall into three classes based on their locations. What are these categories, and where are the ganglia located? To which systems do they belong?   **10**
14. Trace thoracolumbar outflow through the celiac ganglion to its destination. List the neurons and nerves involved.   **10**
15. Trace the pathways preganglionic neurons can take through sympathetic ganglia.   **10**
16. The parasympathetic system is cholinergic, and one of its cholinergic effects constricts the coronary arteries. What neurotransmitter and receptors are at work?   **11**
17. Beta-blocking drugs are used to control heart rate. What system do they affect, and what effects do you expect these drugs to have? What neurotransmitters are blocked?   **11**
18. What steps in development of the visceral nervous system account for the existence of neurons with cell bodies outside the spinal cord?   **12**

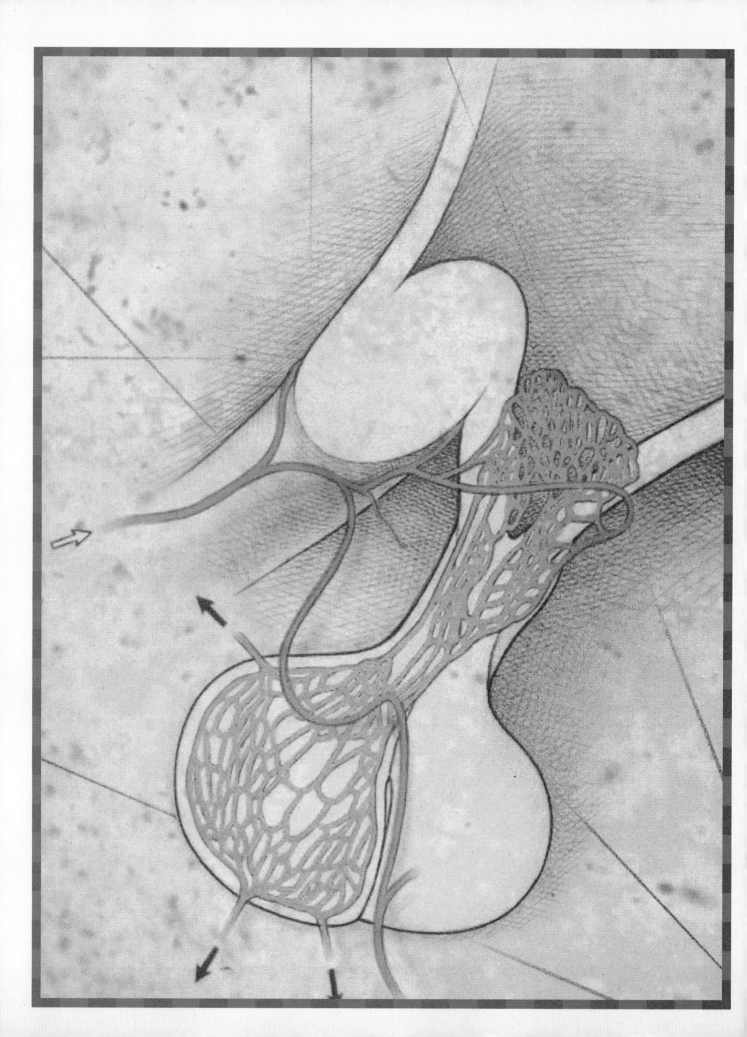

CHAPTER 22

Endocrine System

OBJECTIVES

After studying this chapter, you should be able to do the following:

1. Describe the anatomy and action of a typical endocrine gland. You should be able to name its hormone products, the targets the hormones act on, and what effect or effects each hormone produces.
2. Describe and trace the steps in the feedback control loops that regulate hormone production in endocrine glands.
3. Distinguish between exocrine and endocrine secretion, including anatomy, nature of products, and their effects.
4. List the pituitary gland hormones, the cells that secrete them, and the effects of these hormones on their target tissues.
5. Describe the anatomy of the pineal gland and some effects of its hormone, melatonin.
6. Describe the structure of the thyroid gland, the process of thyroxine production, and the action of this hormone on target tissues.
7. Describe the anatomy and action of the parathyroid glands. You should be able to identify the hormone these glands secrete and its effects on bone resorption and blood calcium.
8. Describe the islet cells of the pancreas and the hormones they secrete, as well as the action of these hormones on their target tissues.
9. Describe the action of insulin and glucagon, how glucose regulates the levels of these hormones, and how these hormones regulate the levels of glucose in the blood through negative feedback.
10. Describe the anatomy of the adrenal gland and relate the different cells of this gland to their products.
11. Discuss the adrenal hormones, their effects, and how the pituitary gland controls their secretion.
12. List several hormones produced by tissues outside the endocrine glands, and describe their effects.

VOCABULARY TOOLBOX

These terms describe endocrine glands:

adrenal	ad-, *L* toward; renalis, *L* kidneys. The adrenal glands lie atop the kidneys.	**pancreas**	pan, *G* all; kreas, *G* meat. The pancreas of animals is soft and meaty when eaten.
endocrine, exocrine	krinein, *G* to separate. Endocrine and exocrine glands secrete materials from the gland.	**pineal**	pinea, *L* a pine cone. The pineal gland is shaped like a pine cone.
hypophysis	hypo-, *G* under, beneath; phye, *G* to grow. The hypophysis is an outgrowth beneath the brain.	**pituitary**	pituit-, *L* a secretion of mucus, phlegm. The pituitary secretes hormones.
		thyroid	thyreoeides, *G* shield-shaped. The thyroid resembles a battle shield.

PERSPECTIVE

The nervous system and endocrine system are partners in communication. In the nervous system, long neurons secrete neurotransmitters across short distances, but in the endocrine system, small cells secrete hormones that travel long distances to other cells. The adrenal glands, for example, regulate heart rate by secreting epinephrine into the blood, whereas the brain sends impulses directly to the heart through the vagus nerve. It is difficult to say precisely where the nervous system ends and the endocrine system begins because both secrete many identical molecules and signal the same organs. Many neurotransmitter substances in the nervous system (epinephrine, for example) are hormones in the endocrine system, and vice versa. This redundancy seems to assure that many intercellular links fine-tune and tightly control metabolism, be they physical connections, such as synapses between neurons, or more ephemeral, hormonal links between glands and their target tissues.

Unlike the skeletal muscles, skeleton, and the nervous system, in which visible physical connections such as tendons, joints and synapses link the parts, the organs of the endocrine system are separate structures located in different body regions. Their hormones nevertheless control the action of other organs and tissues, linked to them by the major carrier of nutrition and oxygen—the blood.

ENDOCRINE SYSTEM

GLANDS, HORMONES, AND TARGETS

The endocrine system consists of glands and many scattered cells that regulate other tissues by secreting hormones into the blood. Hormones are molecules that stimulate gene action, growth, and synthesis in "target" tissues to which the hormone molecules bind. Figure 22.1 illustrates the principal **endocrine glands** and their locations in the body. In the head and neck are the **pituitary, pineal, thyroid,** and **parathyroid** glands. The **thymus** gland is located in the chest, and the **pancreas, adrenal glands, ovaries, testis,** and **placenta** are located in the abdominopelvic cavity. Other endocrine cells and tissues occur throughout the digestive and nervous systems. Chapter 26, Reproductive System, describes the ovaries, testes, and placenta, and Chapter 16, Lymphatic and Immune Systems, discusses the thymus.

ENDOCRINE AND EXOCRINE GLANDS

The body contains two classes of glands. Exocrine glands, such as the salivary glands, sweat glands, and various intestinal glands, all secrete their products through ducts, but **endocrine glands** are ductless. Figure 22.2 shows that the differences between these glands concern four important features.

1. Endocrine (END-oh-krin) glands have no ducts to discharge their products.
2. Instead, endocrine glands secrete directly to capillaries and circulating blood rather than outside of the body (sweat glands) or to internal cavities, such as the mouth and intestines (salivaries, digestive).

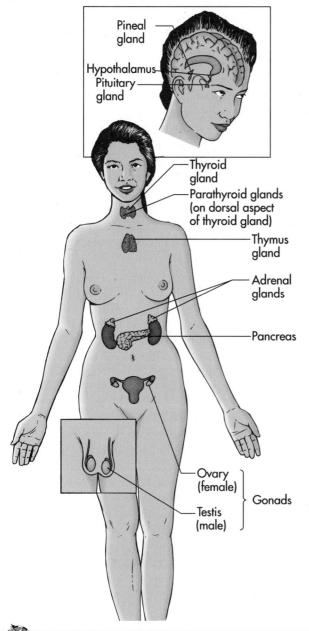

Figure 22.1

The principal endocrine glands.

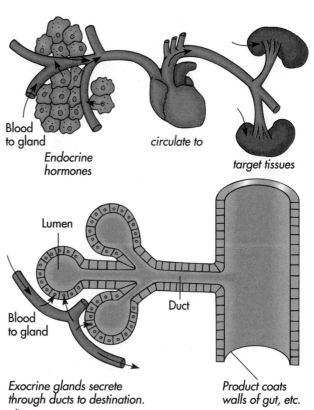

Exocrine glands secrete through ducts to destination.

Product coats walls of gut, etc.

Figure 22.2

Endocrine and exocrine glands. Endocrine glands secrete hormones directly into the blood, which conducts them to the target tissues. Exocrine glands secrete products into lumen and ducts that lead to cavities in the body or to the outside.

3. Endocrine glands release only small amounts of hormone, unlike the large quantities of perspiration, saliva, and gastric juice secreted from exocrine glands.

4. Furthermore, endocrine hormones stimulate synthesis, growth, and gene action in their target tissues, whereas the primary action of exocrine products is to cool the body, to lubricate tissues, or to digest other molecules.

TYPICAL ENDOCRINE GLAND—BRIEF TOUR

Look for five characteristics of the endocrine system illustrated in Figure 22.3.

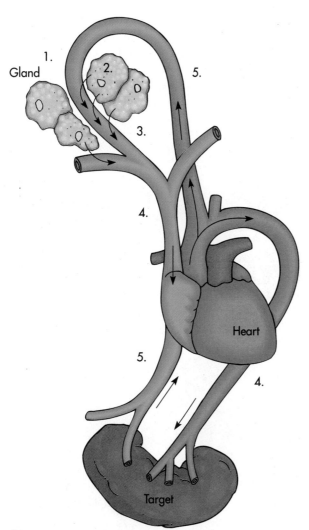

⚙ Figure 22.3

Characteristics of the endocrine system. (1) The gland secretes one or several hormones. (2) The cells secrete steroids, peptides, or amino acid derivatives directly into capillaries (3). (4) Veins carry hormones to the heart, and arteries carry the hormones to target organs. (5) Targets release regulatory molecules that circulate back through the heart to the gland.

1. Know which hormones the gland produces. Tables in each section of this chapter itemize the hormones of each gland and their effects on target tissues.

2. The anatomy of endocrine cells discloses how they secrete. Endocrine cells display abundant endoplasmic reticulum, smooth in the case of steroid hormones, rough for peptides. The number and size of mitochondria suggest how vigorously synthesis proceeds. The Golgi apparatus packages products into various secretion granules whose abundance also gives a clue about synthesis. The size, shape, and internal structure of these granules often indicate which hormone the granules contain.

3. Endocrine glands are highly vascular and contain little connective tissue. Trace the vessels into and out of the gland to see how they distribute blood. Look for the associations between secreting cells and capillaries that provide efficient access to the circulation.

4. Follow hormones through major vessels to the heart and target tissues. Each section of text identifies these vessels so that you can work out the pathway from source to target with the help of descriptions and diagrams in Chapter 14, Vessels.

5. Endocrine glands are themselves regulated by negative feedback loops illustrated in Figure 22.3. They are targets of the targets. Many target tissues release molecules that circulate back to the gland and inhibit secretion of the hormone. Look for this "back side" of the loop in each section, and trace the flow of these inhibitory molecules back to the gland to understand how the loop regulates hormone production.

• What is a hormone?
• What are the primary differences between endocrine and exocrine glands?
• What five points should you look for in every endocrine gland?

CLASSES OF HORMONES

Endocrine cells synthesize three classes of hormone molecules—peptides, steroids, and derivatives of **amino acids** (Figure 22.4). Peptide hormones from the pituitary or pancreatic islets, for example, are short chains of amino acids. Steroids (STEERoyds), such as the sex hormones from the ovary and testis and cortisol from the adrenal gland, are lipid molecules assembled from cholesterol. In addition the adrenal and thyroid glands derive the hormones epinephrine and thyroxine from amino acids. Hormones bind to specific receptor proteins on target cells, which enables cells to discriminate among the many kinds of circulating hormone molecules. Cells do not respond to hormones unless they have appropriate receptors to recognize the hormones in circulation. For example, bone especially responds to calci-

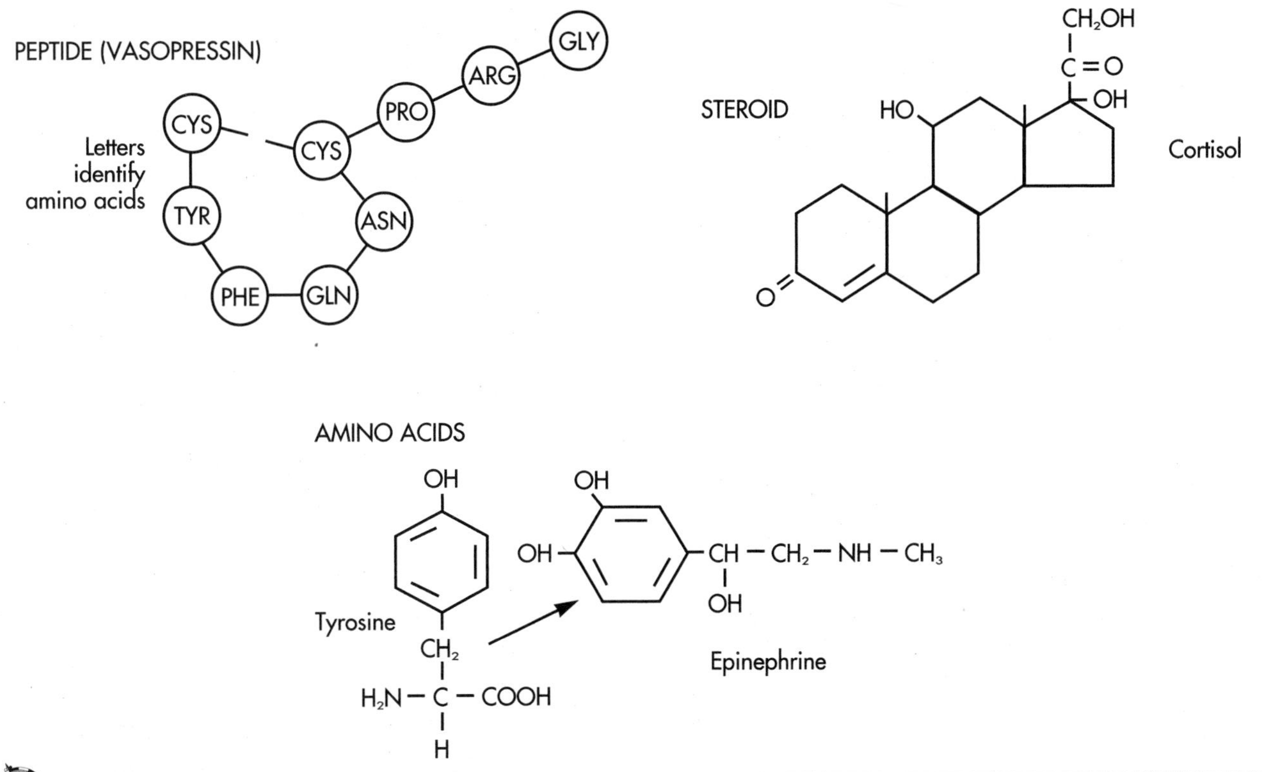

PEPTIDE (VASOPRESSIN)

Letters identify amino acids

CYS — CYS — PRO — ARG — GLY

TYR — ASN

PHE — GLN

STEROID

HO

CH_2OH
$C=O$
OH

Cortisol

O

AMINO ACIDS

OH

Tyrosine

CH_2

$H_2N - C - COOH$
H

OH
OH

$CH - CH_2 - NH - CH_3$
OH

Epinephrine

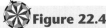

Figure 22.4

Three classes of hormones. Vasopressin, like all peptides, is a chain of amino acids linked by covalent bonds. Vasopressin contains nine amino acids. Cortisol is a steroid that contains five hydrocarbon rings, making this molecule a lipid. Tyrosine is an amino acid, and cells convert it to epinephrine. Epinephrine's hydrocarbon ring derives from tyrosine.

tonin because it is the property of receptors on bone cells to bind calcitonin molecules specifically. Cells lacking the same receptors do not respond to calcitonin.

LONG-DISTANCE COMMUNICATION

The endocrine system and nervous system each communicate over long distances but by different mechanisms. Instead of secreting molecules that travel short distances across synapses (Figure 22.5), the endocrines rely on the cardiovascular system to circulate their hormones to target tissues throughout the body. Despite this difference the systems are similar in many ways. Neurons secrete neurotransmitter molecules into the narrow gap between cells at synapses, and when the neurotransmitter binds to receptors on the postsynaptic cell, that cell responds. Endocrine glands release hormones into the circulation, which bridges the long distance to the target tissue, but at their destinations hormone molecules nevertheless bind to cellular receptors that bring about the response of the tissue to the hormone.

In fact, this superficial resemblance is much deeper, because some endocrine hormones are secreted by neurons and some neurotransmitter substances act as endocrine hormones, as you will see in

this chapter. For this reason, many anatomists and physiologists speak of the **neuroendocrine system** (NOOR-oh-END-oh krin), recognizing the profound overlaps and cooperation between the nervous and endocrine systems.

HORMONAL SIGNALING

Endocrine responses are slower than those of the nervous system. Endocrine glands signal their target tissues by secreting small "pulses" of hormone molecules to which the target responds. When an endocrine gland releases a burst of hormone the quantity in circulation rises rapidly and then declines rapidly, initiating a fresh response by the target tissues while the pulse is at its "peak." Minutes, hours, or days may pass before tissues respond, compared with small fractions of seconds in the nervous system. Because of their "pulsed release," most hormones have short half-lives in the blood. Half of the thyroxine your thyroid is secreting now will be removed from circulation within 6 minutes, and half of the remainder will be gone in another 6 minutes.

- What are the names of the three classes of hormone molecules?
- What is the principal difference between hormonal and neural communication between cells?

Endocrines

A

Circulation
to target

Neurons

B

Neuroendocrine

C

Circulation

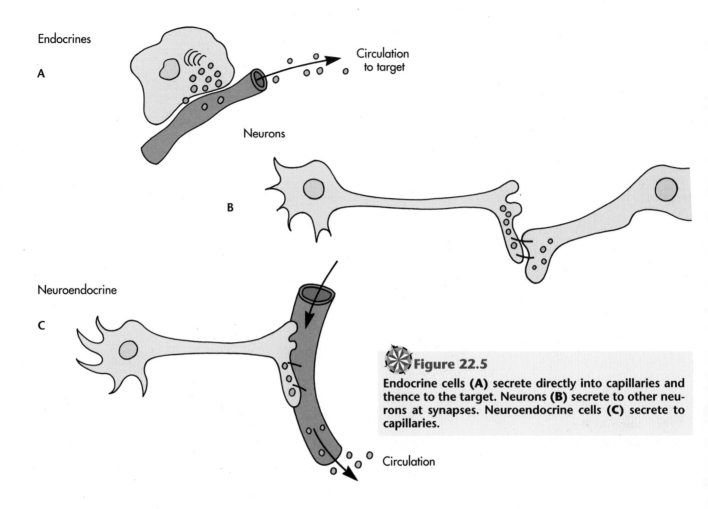

✳ **Figure 22.5**
Endocrine cells **(A)** secrete directly into capillaries and thence to the target. Neurons **(B)** secrete to other neurons at synapses. Neuroendocrine cells **(C)** secrete to capillaries.

THE PITUITARY GLAND

The pituitary gland is considered the "master gland" because it regulates many other glands. The thyroid, mammary glands, adrenals, ovaries, testes, and placenta are all under pituitary (pih-TOO-ih-TARE-ee) control. This pea-sized gland (about 1 cm diameter) extends from the hypothalamus beneath the brain, to rest in the sella turcica of the sphenoid bone on the floor of the cranial cavity (Figure 22.6A). The pituitary gland also is called the **hypophysis** (hy-POF-ih-sis); both names recognize the glandular nature of the pituitary and its broad physiological effects.

The anatomy of the pituitary reveals it as two closely related but different portions. Figure 22.6B shows that the **infundibulum** (in-fun-DIB-you-lum) connects the pituitary with the hypothalamus. The infundibulum is actually an extension of the hypothalamus that ends in the **posterior pituitary,** or **neurohypophysis** (NOOR-oh-hy-POF-ih-sis), at the rear of the gland. As the name indicates, the cells of this portion are neural in nature, and they secrete the neuropeptide hormones, oxytocin and vasopressin, described in Table 22.2 and below. The greater bulk of the

gland, the **anterior pituitary** or **adenohypophysis** (ah-DEEN-oh-hy-POF-ih-sis), secretes hormones that regulate the thyroid, the mammary glands, the adrenals, and the gonads (also in Table 22.2). A small sleeve of the adenohypophysis, known as the **pars tuberalis,** wraps around the infundibulum. The indistinct **pars intermedia** lies between the neurohypophysis and the adenohypophysis. Table 22.1 keeps track of these names; ask your instructor which ones to use.

The adenohypophysis and neurohypophysis have different origins that account for the presence of neural and glandular tissue. The neurohypophysis arises from the infundibulum (Figure 22.7). The adenohypophysis originates from a pocket of epithelium

Table 22.1	DIVISIONS OF THE PITUITARY GLAND	
Anterior pituitary	Adenohypophysis	Pars distalis Pars tuberalis Pars intermedia
Posterior pituitary	Neurohypophysis	Pars nervosa

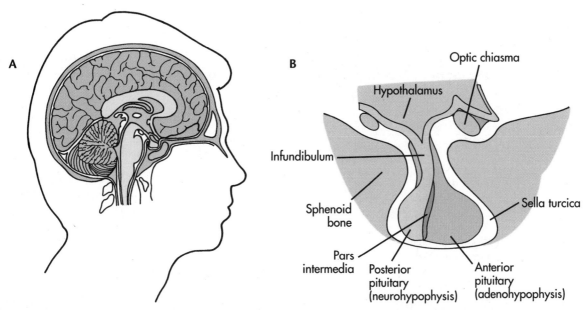

Figure 22.6

The pituitary gland. **A,** Midsagittal section reveals the pituitary gland beneath the hypothalamus, lying in the sella turcica. **B,** Divisions of the pituitary gland.

called **Rathke's** (RATH-keys) **pouch** that grows upward from the roof of the embryonic mouth and fuses with the posterior pituitary. Rathke's pouch pinches off from the roof of the mouth, but the hypophysis and infundibulum remain attached to the brain.

- What are the names of the different regions of the hypophysis?
- From what tissues do the infundibulum and adenohypophysis arise?

ADENOHYPOPHYSIS

At least five varieties of cells produce hormones in the **adenohypophysis** (Table 22.2). **Somatotrophs** synthesize **growth hormone** (GH), also called **somatotropin** (STH), that broadly affects body growth by regulating the growth of bones and muscle mass. Somatotrophs (so-MAT-oh-TROFES) are spherical cells that occupy fully 40% of the adenohypophysis. They contain prominent mitochondria and abundant secre-

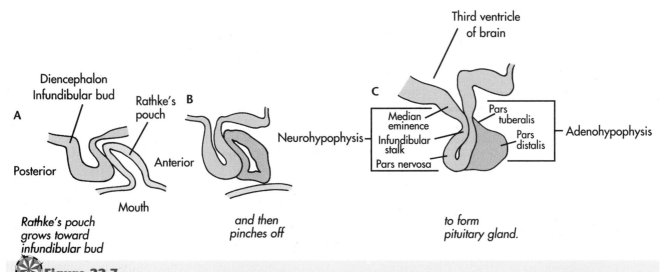

Figure 22.7

The pituitary develops from the infundibulum and Rathke's pouch. **A,** The infundibulum grows downward ,and Rathke's pouch grows upward. **B,** They join and Rathke's pouch pinches off. **C,** The infundibulum becomes the neurohypophysis, and Rathke's pouch becomes the adenohypophysis.

Table 22.2 PITUITARY HORMONES

Name	Hormone	Type	Action	Releasing Hormone
ADENOHYPOPHYSIS				
Somatotroph	Growth hormone (GH)	Protein	Stimulates growth in many tissues, especially cartilage in epiphyseal plates of long bones	Growth hormone-releasing hormone (GRH)
Mammotroph	Prolactin (PRL)	Protein	Promotes breast enlargement during pregnancy and milk production (lactation)	Prolactin-releasing hormone (PRH)
Thyrotroph	Thyroid-stimulating hormone (TSH)	Glycoprotein	Stimulates thyroid follicle cells to produce thyroxine	Thyroid stimulating-hormone releasing hormone (TRH)
Gonadotroph	Follicle-stimulating hormone (FSH)	Glycoprotein	Stimulates ovarian follicles to mature; spermatogenesis in testis	Gonadotropin-releasing hormone (GnRH)
	Luteinizing hormone (LH)	Glycoprotein	Stimulates ovulation; maturation of corpus luteum; testosterone production in testis	
Corticotroph	Adrenocorticotropic hormone (ACTH)	Peptide (39)*	Stimulates steroid production by adrenal cortex	Corticotropin-releasing hormone (CRH)
Pars intermedia	Melanocyte-stimulating hormone (MSH)	Peptide (13,22)	Unknown; darkening of melanocytes	—
NEUROHYPOPHYSIS				
Paraventricular nucleus	Oxytocin (OT)	Peptide (9)	Promotes contraction of myometrium during labor; causes breasts to eject milk (milk letdown) at nursing	—
Supraoptic nucleus	Vasopressin (antidiuretic hormone, ADH)	Peptide (9)	Promotes water resorption by kidneys	—

From Halmi NS: The hypophysis. In Weiss L, editor: Cell and tissue biology, *ed 6, Baltimore, 1988, Urban and Schwarzenberg; and Griffin JE and Ojeda SR:* Textbook of endocrine physiology, *ed 2, New York, 1992, Oxford University Press.*
*The number of amino acids in the peptide chain is given in parentheses.

tory granules filled with growth hormone. The hypothalamus regulates secretion of growth hormone by secreting its own growth hormone–releasing hormone (GHRH) that causes somatotrophs to release growth hormone. Overproduction of growth hormone can lead to giantism, and underproduction, to dwarfism.

Mammotrophs (MAM-oh-trofes) are small spindle-shaped cells that produce **prolactin** (PRL), the hormone that prepares the breasts to produce milk. During pregnancy, the secretion granules triple in size, and mammotrophs themselves proliferate. The hypothalamus promotes release of prolactin by secreting prolactin-releasing hormone (PRH).

Thyrotrophs (THY-ro-trofes) secrete **thyroid-stimulating hormone** (TSH), also known as thyrotropin, that stimulates the thyroid to produce and release thyroxine. These polyhedral thyrotroph cells store TSH in small secretion granules. The hypothalamus promotes this activity by secreting thyroid-releasing hormone (TRH).

Gonadotrophs (go-NAD-oh-trofes) are fusiform cells that produce **gonadotropins,** the peptides that promote the development and production of eggs and sperm. **Follicle stimulating hormone** (FSH) stimulates ovarian follicles to mature in females and promotes sperm production in males. **Luteinizing hor-**

mone (LH; LOOT-ee-in-eye-zing) initiates ovulation in females and testosterone production in males. Gonadotropin-releasing hormones (GnRH) from the hypothalamus stimulate gonadotrophs to release their hormones into the circulation.

Corticotrophs (KORE-tih-ko-TROFES) produce several hormones from a single, large polypeptide precursor molecule called **proopiomelanocortin** (POMC; pro-OH-pea-oh-MEL-an-oh-KORT-in), which is stored in small secretion granules before its release into circulation. This precursor cleaves into smaller fragments that act as **endorphins,** the central nervous system's own painkillers; **melanocyte-stimulating hormone** (MSH); and **adrenocorticotropic hormone** (ACTH; ah-DREE-no-KORE-tih-ko-TRO-pik), also known as corticotropin, a peptide that regulates the adrenal cortex.

NEUROHYPOPHYSIS

The posterior pituitary is a neuroendocrine tissue. The axons of neurons from several hypothalamic (HY-po-THAL-am-ik) nuclei shown in Figure 22.8 descend into the neurohypophysis, where the axon terminals release **oxytocin** (OT; oxy-TOE-sin) and **vasopressin** (VAY-zo-PRESS-in), much as neurotransmitters are secreted at synapses. No synapses receive

these peptide hormones, however, because the hormones enter capillaries of the posterior pituitary and escape into the systemic circulation. Axons from the **supraoptic nucleus** release vasopressin, also known as **antidiuretic hormone** (ADH; AN-tee-die-you-RET-ik), and neurons from the **paraventricular nucleus** release oxytocin. ADH conserves body fluids by causing the kidney tubules to reabsorb water from urine. ADH is also known as vasopressin because, in higher concentrations, it is a potent vasoconstrictor that raises blood pressure. Oxytocin is most active during childbirth, when it initiates uterine contraction.

- How does the adenohypophysis differ in location and function from the neurohypophysis?
- What class of hormone do both portions secrete?
- What are the names of cells in the adenohypophysis that secrete hormones?

VESSELS

The hypophysis is intensely vascularized. Hypophyseal vessels link the pituitary to the vessels that distribute blood to its target tissues and the tissues that regulate the pituitary. Pituitary hormones reach the blood rapidly through these vessels, but the vessels also bring to the interior of the gland other hormones that regulate the adenohypophysis and neuro-

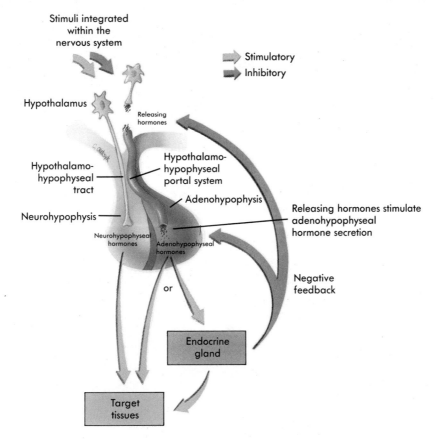

 **Figure 22.8**

Axons secrete into capillaries of the neurohypophysis. These axons follow the hypothalamo-hypophyseal tracts from the hypothalamus to the neurohypophysis.

hypophysis. Figure 22.9 shows that the **superior** and **inferior hypophyseal arteries** deliver blood to the pituitary from the internal carotid artery. The superior hypophyseal artery supplies capillaries in the adenohypophysis, and the inferior hypophyseal artery supplies those in the neurohypophysis. Several **hypophyseal veins** ultimately drain these capillaries and any hormones they have received into the internal jugular veins.

When you examine the figure carefully, you will see that the superior hypophyseal artery branches into capillaries in the infundibulum, from which **hypophyseal portal venules** drain into the adenohypophyseal capillaries. This is another example of a portal system, in which veins drain from one capillary plexus in the hypothalamus into a second plexus in the adenohypophysis (see Hepatic Portal Vein, Chapter 14, Vessels). This hypothalamo-hypophyseal

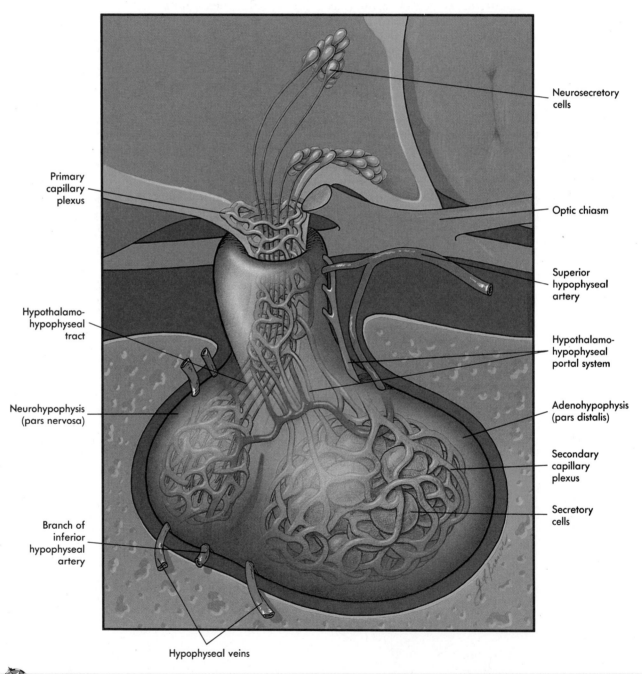

Primary capillary plexus

Hypothalamo-hypophyseal tract

Neurohypophysis (pars nervosa)

Branch of inferior hypophyseal artery

Hypophyseal veins

Neurosecretory cells

Optic chiasm

Superior hypophyseal artery

Hypothalamo-hypophyseal portal system

Adenohypophysis (pars distalis)

Secondary capillary plexus

Secretory cells

Figure 22.9

Hypophyseal circulation. The hypothalamo-hypophyseal portal system delivers from the thalamus and infundibulum to the adenohypophysis.

portal system delivers releasing hormones from the hypothalamus that in turn control the quantities of hormones released by the adenohypophysis.

RELEASING HORMONES AND CONTROL LOOPS

Hypothalamic-releasing hormones illustrate the physiological link between the hypothalamus and adenohypophysis provided by the hypophyseal portal system. Corticotropin-releasing hormone (CRH) governs how much ACTH corticotroph cells release from the adenohypophysis. The portal system conducts CRH from the hypothalamus to the adenohypophysis, where CRH binds to corticotrophs and accelerates release of ACTH. Different releasing hormones from the hypothalamus, listed in Table 22.2, control other cells of the adenohypophysis.

Negative feedback loops from the targets of pituitary hormones control the quantity of releasing hormones in circulation. Follow the control loop for ACTH in Figure 22.10 to determine what to look for in other endocrine glands. ACTH stimulates the adrenal gland to produce cortisol, a hormone that causes the liver and other tissues to increase the amount of glucose circulating in the blood. The cardiovascular system carries ACTH from the pituitary to the heart, and thence to the adrenal glands, which release cortisol into the blood. Cortisol in turn brings about its effects in the liver and elsewhere. Some ACTH and cortisol molecules circulate back through the heart and return to the hypothalamus and pituitary, where they suppress CRH and ACTH production. Cortisol inhibits the hypothalamus from releasing CRH, and cortisol also retards corticotrophs from releasing ACTH. ACTH also shortcuts this loop and directly reduces the amount of CRH elaborated from the hypothalamus. With so many inhibitory effects at work, what causes the brain to release CRH in the first place? The answer seems to be that stress induces the hypothalamus to release CRH and ACTH, and thereby to stimulate the adrenal glands.

- How do releasing hormones from the hypothalamus reach the adenohypophysis?
- How does cortisol return to the hypothalamus and pituitary, and what effect has it when it arrives?

THE PINEAL GLAND

The pineal (pie-NEE-al) **gland is one of several sites in the brain where melatonin is produced.** A dose of this peptide taken before and after a long flight can relieve the unpleasant effects of jet lag, especially when you are traveling east toward the new day. **Melatonin** (MEL-ah-TONE-in) may synchronize many cyclic processes in humans, from sleep to cell

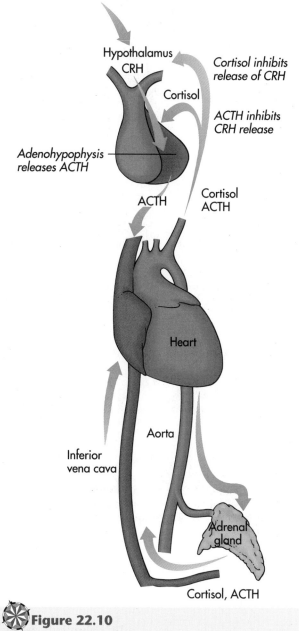

Figure 22.10

Flow of ACTH, a typical pituitary hormone. See text for description.

division, and the pineal gland may help control these events.

The pineal gland, an enlargement of the posterior wall of the third ventricle of the brain, overlooks the superior colliculus of the midbrain (Figure 22.11). Approximately 5 mm long by 4 mm wide, the gland terminates in a pineal stalk of similar dimensions that projects posteriorly from the roof of the **diencephalon** and the choroid plexus of the third ventricle. A delicate connective tissue capsule envelops the gland and is continuous with the pia mater and

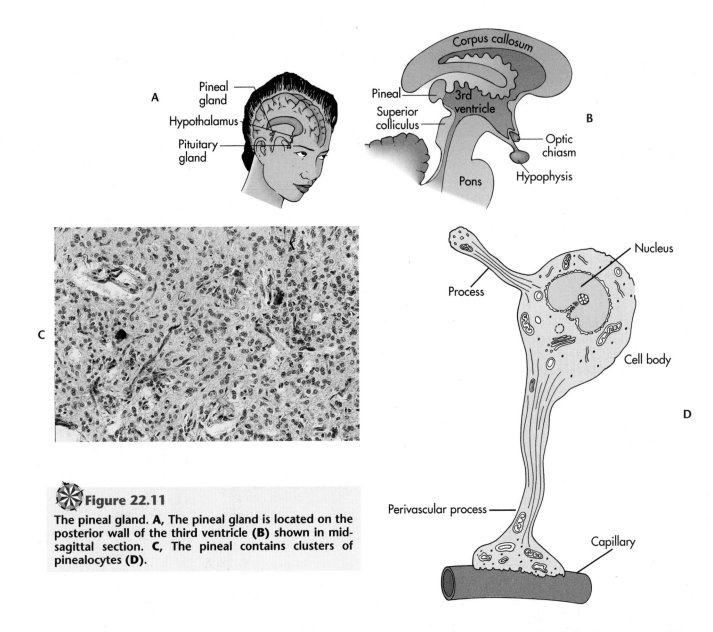

Figure 22.11

The pineal gland. A, The pineal gland is located on the posterior wall of the third ventricle (B) shown in mid-sagittal section. C, The pineal contains clusters of pinealocytes (D).

arachnoid of the brain. Internally the capsule extends septa and trabeculae that support lobules of clustered **pinealocytes,** the principal pineal cell and melatonin source. The posterior cerebral arteries supply blood to capillaries among the lobules, and tributaries of the great cerebral vein drain blood and pineal products from the gland into the systemic circulation.

That pinealocytes display large cell bodies and cellular processes reminiscent of neurons is not surprising, because both pinealocytes and neurons derive from the neural ectoderm. Melatonin storage granules fill these processes, which frequently end on capillaries, the endocrine connections that qualify the pineal as an endocrine organ. Sympathetic nerve fibers supply the pineal.

• What is the principal hormone of the pineal gland?
• Where is this gland located, and how do its products reach the systemic circulation?

THE THYROID

The **thyroid gland regulates the basic metabolic rate of the body** through its main secretion, **thyroxine** (Table 22.3). Figure 22.12A shows that the thyroid is situated below the larynx, in the *V* between the common carotid arteries at the level of the fifth and sixth tracheal cartilages. The **lobes** ($5 \times 3 \times 2$ cm, or $2 \times 1\frac{1}{4} \times \frac{3}{4}$ inches) of the gland, which resemble butterfly wings or shields, are joined by a narrow **isthmus** across the trachea and covered by a delicate connective tissue **capsule.** Each lobe contains numerous **thyroid follicles,** the functional units of the gland that produce thyroxine. The same connective tissue that covers the gland also joins the follicles to each other and provides access for circulation and nerves to each follicle. Crowding compresses the follicles into various polyhedral shapes. **Superior** and **infe-**

Table 22.3   THYROID AND PARATHYROID HORMONES

Cells	Hormone	Target	Action	Regulators
THYROID GLAND				
Follicle cells	Thyroxine (T4)	Various: liver, intestine, muscle, adipose	Stimulates cellular respiration	TSH (adenohypophysis) stimulates release
C cells (parafollicular cells)	Calcitonin (CT)	Osteoclasts	Reduces blood calcium ion levels; inhibits resorption of bone	Rising blood calcium ion concentration
PARATHYROID GLAND				
Chief cells	Parathormone (PTH)		Raises blood calcium ion levels	Decreasing blood calcium ion
		Osteoclasts	Stimulates resorption of bone	
		Kidneys	Stimulates calcium resorption	
		Intestine	Promotes uptake of calcium from diet	
Oxyphil cells	Unknown	—	—	—

From Griffin JE: The thyroid. In Griffin JE and Ojeda SR: Textbook of endocrine physiology, *ed 2, New York, 1992, Oxford University Press; and Gershon MD and Nunez EA: The thyroid gland. In Weiss L, editor:* Cell and tissue biology, *ed 6, Baltimore, 1988, Urban and Schwarzenberg.*

rior thyroid arteries supply rich beds of capillaries that receive secretions from each follicle. **Thyroid veins** drain these products to the internal jugular and brachiocephalic veins and thence to the heart. Lymphatics also share the spaces between follicles.

FOLLICLES

Each follicle is a hollow ball of cuboidal follicle epithelial cells that surround a large central **lumen** filled with **colloid** secretion from the follicle cells (Figure 22.12B). Like all epithelia, this epithelium is mounted on a basement membrane anchored to the connective tissue sheath around each follicle. Follicle cells produce thyroxine, and the cytoplasm reflects this activity. Extensive rough endoplasmic reticulum and Golgi apparatus surround the central nucleus of each cell, and numerous secretory vesicles occupy the apical cytoplasm beneath the plasma membrane. Microvilli and coated pits attest to vigorous exocytosis that releases materials into the lumen. Tight junctions prevent leakage from the lumen by sealing the apical borders of the cells together. Numerous folds in the basal membrane provide transport surfaces for thyroid hormone production.

Follicles also contain a few large **parafollicular cells** (about 1% of the total cells) that do not border on the lumen of the follicle. Parafollicular cells (C cells) release **calcitonin** (KAL-sih-TONE-in) from abundant small secretion vesicles held in the cytoplasm. Calcitonin promotes calcification of bone and teeth by withdrawing calcium from the blood (also discussed

in Chapter 5, Bone). Rising levels of calcium ions in the blood stimulate C cells to produce and release calcitonin. This peptide hormone and several closely related ones are the products of a single gene. Calcitonin itself inhibits osteoclasts from reabsorbing bone; the ruffled border of these cells rapidly withdraws and resorption slows in the presence of calcitonin. A closely related hormone named N-procalcitonin stimulates proliferation of osteoblasts, the bone-depositing cells, but calcitonin has no such accelerating effect. Calcitonin enters and leaves the blood quickly. Half of it is gone 5 minutes after entering the blood; it is removed and degraded by the kidney, liver, and possibly bone. Calcitonin also interacts with parathormone from the parathyroid glands to control blood calcium levels (see Parathyroids, p. 674).

THYROXINE PRODUCTION

Follicular epithelial cells act both as exocrine and endocrine glands. As exocrine cells they produce **thyroglobulin,** a glycoprotein storage form of thyroxine, and secrete it into the lumen of the follicle for storage, as you can see in Figure 22.13. In their role as endocrine cells, these same cells reabsorb thyroglobulin back into the cytoplasm and convert it into active hormone that then diffuses into the circulation outside the follicles. Synthesis of thyroxine molecules begins with the amino acid tyrosine and iodide ions. The basal membrane of the follicle cells transports these molecules into the cytoplasm, where the rough endoplasmic reticulum and Golgi apparatus as-

A

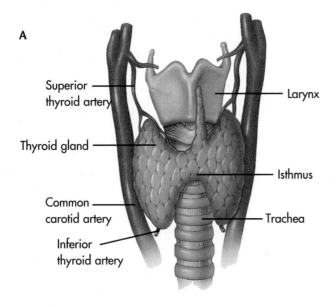

Superior thyroid artery

Larynx

Thyroid gland

Isthmus

Common carotid artery

Trachea

Inferior thyroid artery

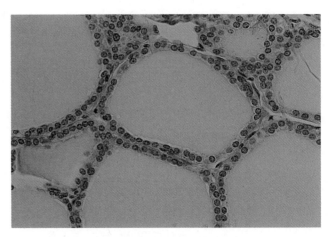

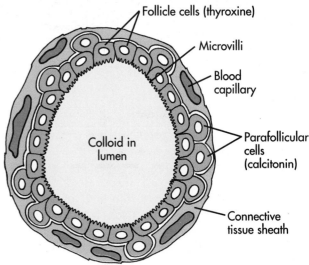

Follicle cells (thyroxine)

Microvilli

Blood capillary

Colloid in lumen

Parafollicular cells (calcitonin)

Connective tissue sheath

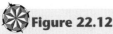

Figure 22.12

The thyroid gland. **A,** The thyroid lies in front of the trachea, below the larynx, between the common carotid arteries. **B,** Thyroid follicles fill the gland. **C,** Follicle cells surround a lumen in the center of the follicle. Parafollicular cells are also present.

semble tyrosine and polysaccharides into thyroglobulin. At the same time, the cells secrete into the colloid an enzyme, peroxidase, which oxidizes iodide ions to iodine molecules. The iodine molecules bind with thyroglobulin to form iodinated thyroglobulin that is now withdrawn from storage and transported back across the apical membrane into the cytoplasm. Here the triiodinated form of the hormone, triiodothyronine (T3; TRY-eye-oh-doe-THY-roe-neen), and the tetraiodinated form, thyroxine itself (T4), are stripped from their polysaccharide carriers to diffuse outside to the capillaries as active thyroid hormone molecules.

- Where is the thyroid gland located?
- Which cells are responsible for producing thyroxine and calcitonin?
- Where in thyroid follicles do iodine molecules become bound to thyroxine?
- Which aspect of thyroxine production is exocrine and which is endocrine?

THYROID CONTROL LOOP
ACTION OF THYROXINE

Thyroxine affects target cells in nearly all tissues and organs. The thyroid control loop shown in

Figure 22.14 begins when thyroxine circulates into these cells and binds to thyroxine receptors in their nuclei. This process stimulates cell metabolism and heat production by activating genes responsible for carbohydrate metabolism and protein synthesis. Thyroxine stimulates the intestine to absorb glucose and causes muscle fibers and adipose tissue to take up glucose more rapidly. Thyroxine is necessary for growth and development and must be present in order for growth hormone to act in bone and muscle.

THYROID RESPONSE

The thyroid is a target of the pituitary gland, which makes the effects of the thyroid gland part of a larger control loop from the pituitary. **Thyroid-stimulating hormone** (TSH) from the pituitary stimulates the thyroid to release thyroxine. Thyrotroph cells of the adenohypophysis secrete TSH that circulates to the thyroid gland and stimulates follicular cells to release thyroxine when TSH molecules bind to receptors on the follicular cell membranes.

The thyroid gland and nervous system balance how much TSH is released. Thyroxine itself is part of a short feedback loop that prevents excessive release of TSH. As thyroxine levels increase, this loop dimin-

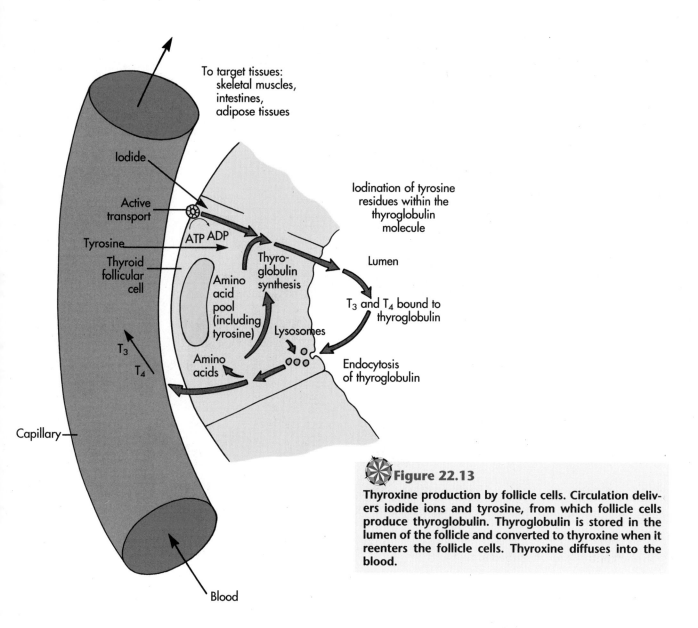

To target tissues:
skeletal muscles,
intestines,
adipose tissues

Iodide

Active
transport

ATP ADP

Tyrosine

Thyroid
follicular
cell

Amino
acid
pool
(including
tyrosine)

Thyro-
globulin
synthesis

Amino
acids

T₃

T₄

Capillary

Blood

Iodination of tyrosine
residues within the
thyroglobulin
molecule

Lumen

T₃ and T₄ bound to
thyroglobulin

Lysosomes

Endocytosis
of thyroglobulin

Figure 22.13

Thyroxine production by follicle cells. Circulation delivers iodide ions and tyrosine, from which follicle cells produce thyroglobulin. Thyroglobulin is stored in the lumen of the follicle and converted to thyroxine when it reenters the follicle cells. Thyroxine diffuses into the blood.

ishes the amount of TSH thyrotrophs release, which in turn causes fewer thyroxine molecules to enter circulation. The nervous system also assures that there will be no underproduction of thyroxine by producing **thyrotropin-releasing hormone,** TRH. This hormone enters the hypophyseal portal system and circulates into the adenohypophysis, where it causes thyrotroph cells to release TSH. It is not known whether thyroxine also regulates TRH and extends the back side of the control loop to the hypothalamus.

The consequences of overproduction and underproduction of thyroxine show how important normal thyroid function is. Overproduction is called **hyperthyroidism** and underproduction, **hypothyroidism.** Hyperthyroidism is usually caused by **Graves' disease,** an autoimmune condition that stimulates the thyroid to produce more than normal quantities of thyroxine. The thyroid enlarges and forms a **goiter** (GOY-ter), as the follicles increase in size; increased

metabolism stimulates appetite but reduces body weight. Patients sweat profusely and tolerate heat poorly.

In contrast, hypothyroidism almost always results from failure of the thyroid to produce adequate thyroxine. Only rarely does hypothyroidism result from the inability to respond to pituitary hormones. Goiters occur in hypothyroid patients when the gland is unable to increase production of thyroxine when stimulated by TSH, and instead grows larger in response to TSH. In infants, hypothyroidism leads to mental retardation and reduced growth; in adults, lethargy, slowed mental function, and heart failure often result.

- What is the action of thyroxine, TSH, and TRH?
- Which glands secrete these hormones?
- Which of these hormones influence the activity of the thyroid gland?
- What are the main effects of thyroxine on target tissues?

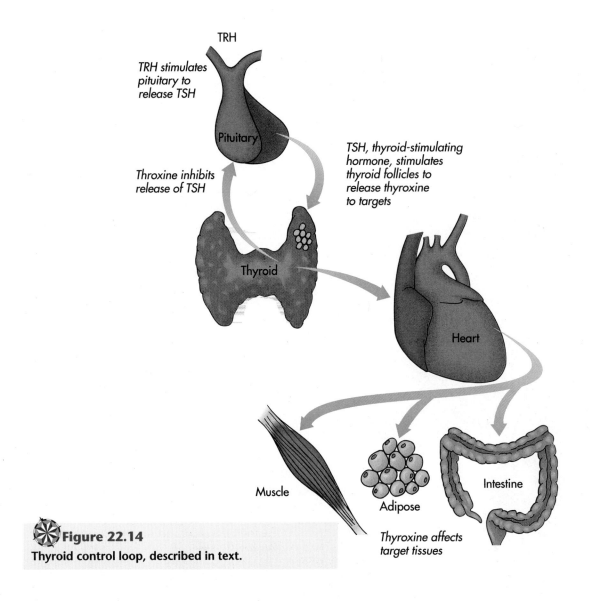

TRH

TRH stimulates pituitary to release TSH

Pituitary

Throxine inhibits release of TSH

TSH, thyroid-stimulating hormone, stimulates thyroid follicles to release thyroxine to targets

Thyroid

Heart

Muscle

Adipose

Intestine

Thyroxine affects target tissues

✵ **Figure 22.14**
Thyroid control loop, described in text.

THE PARATHYROIDS

PARATHORMONE

The parathyroid glands secrete parathormone **(PTH), which mobilizes calcium ions** from the skeleton into the blood. These reddish-brown, oval glands (Figure 22.15A), usually 4 in number and 5 mm in diameter, are embedded on the posterior surface of the thyroid gland. The parathyroids derive from the third and fourth pharyngeal pouches, paired outpocketings of the embryonic pharynx, which neatly explains why there are four glands (see Chapter 23, Digestive System). Delicate connective tissue capsules anchor one gland in the superior and inferior halves of each lobe. Instead of follicles, the units of structure and function are nests of cuboidal cells supported by a fine connective tissue network that branches throughout each gland, carrying capillaries from the **inferior thyroid artery** and nerve endings from the middle cervical ganglion. **Thyroid veins**

drain parathormone-laden blood to the heart for circulation around the body.

Two types of cells appear to be involved in parathormone production. The most abundant of these are **chief cells,** or principal cells, shown in Figure 22.15B. The characteristic rough endoplasmic reticulum, Golgi apparatus, and secretion granules of protein production are not so dramatically obvious in chief cells as in thyroid follicular cells. **Oxyphil cells** (OXEE-fill), the second type of parathyroid cells, are unusual because their cytoplasm is packed with little else than mitochondria. The function of oxyphil cells is unknown. Intermediate cell types suggest that oxyphil cells may derive from chief cells.

PARATHORMONE CONTROL LOOP

Parathormone is a polypeptide assembled in the rough endoplasmic reticulum and packaged in the Golgi apparatus for secretion from the cell membrane by exocytosis. Synthesis begins with a large precursor

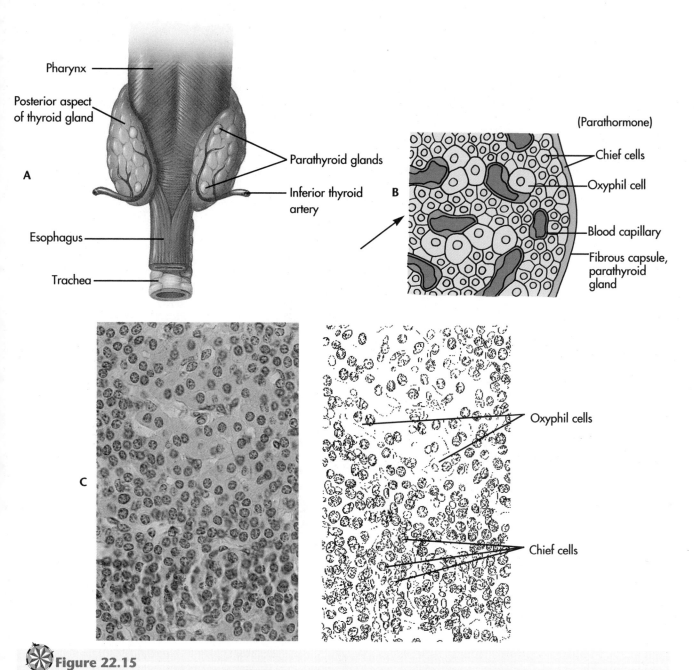

Figure 22.15

A, The parathyroid glands reside in the posterior substance of the thyroid glands. **B, C** and **D,** Section through the parathyroid gland reveals clusters of chief cells.

molecule from which short segments are clipped before the mature PTH molecule, 84 amino acid residues long, is released.

Parathormone raises blood calcium levels in three ways (Figure 22.16). First, the hormone appears to stimulate osteoclasts to reabsorb calcium more rapidly from bone. PTH also promotes absorption of calcium ions in the small intestines by activating the synthesis of vitamin D. Furthermore, PTH conserves blood calcium by reducing the amount excreted by the kidney. All of these actions are part of a far-reaching negative feedback system that controls blood cal-

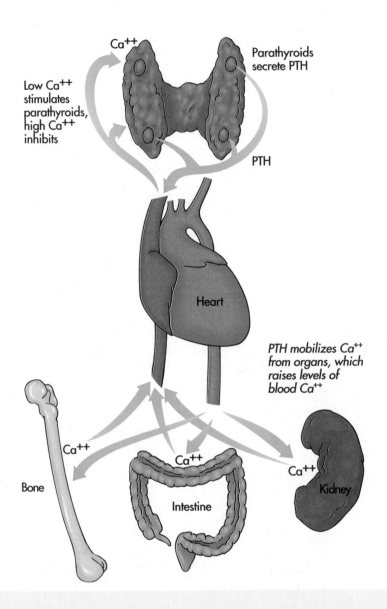

Ca⁺⁺

Low Ca⁺⁺ stimulates parathyroids, high Ca⁺⁺ inhibits

Parathyroids secrete PTH

PTH

Heart

PTH mobilizes Ca⁺⁺ from organs, which raises levels of blood Ca⁺⁺

Ca⁺⁺

Bone

Ca⁺⁺

Intestine

Ca⁺⁺

Kidney

Figure 22.16

Parathormone control loop. Low calcium concentration stimulates the parathyroids to release parathormone (PTH), which in turn raises the level of blood calcium and inhibits parathormone production.

cium levels. The parathyroids themselves are important links in this system. Low blood calcium levels stimulate PTH production and high levels inhibit production, precisely what a self-regulating feedback loop should do.

- What hormone does the parathyroid gland produce, and what is the effect of this hormone?
- Name the cells involved in PTH synthesis.
- Where are the parathyroid glands located?

INTERACTION WITH CALCITONIN

Together, parathormone and calcitonin control blood calcium. Whereas parathormone counteracts the tendency of calcium levels to fall, calcitonin counteracts increases. The result is a balance between the antagonistic actions of these hormones. Figure 22.17 illustrates how this interaction works (discussed also in Chapter 5, Bone). When circulating calcium increases, the additional calcium stimulates the thyroid to release calcitonin. Blood calcium declines when this hormone goes to work because it promotes calcification of bone by withdrawing calcium from the blood. Parathormone comes into play when calcium levels begin to fall. These conditions cause the parathyroid glands to release greater quantities of PTH, which stimulates osteoclasts to mobilize calcium from bone, the intestine to absorb more calcium from the diet, and the kidney to conserve what calcium the blood contains. When these actions restore blood calcium to normal levels, parathormone secretion declines.

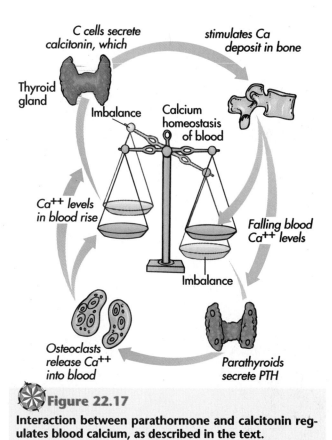

Figure 22.17

Interaction between parathormone and calcitonin regulates blood calcium, as described in the text.

- Why are parathormone and calcitonin said to be antagonistic hormones?

PANCREATIC ISLETS

The pancreas is both an endocrine and exocrine gland. Located beneath the greater curvature of the stomach, most of the pancreas (Figure 22.18) consists of exocrine glands called **acini** that secrete digestive enzymes to the small intestine for digestion (Chapter 23, Digestive System, supplies details). As a classic exocrine gland, the lumen of each acinus receives secretions from the glandular wall, and a duct drains these products from each acinus into the pancreatic duct and the duodenum. The **islets of the pancreas,** shown in Figure 22.19, are scattered among the closely packed acini throughout the pancreas. These clusters are also known as the **islets of Langerhans,** after Paul Langerhans, who discovered them as a medical student in 1868. Islets are clusters of several hundred cells that secrete peptide hormones into capillaries that intimately surround and invade each islet. Delicate reticular fibers also surround each islet and support it with a network that reaches every islet cell.

Blood enters the pancreas by way of the **splenic artery** and through branches and anastomoses of the

gastroduodenal (GAS-troe-do-oh-DEEN-al) and **superior mesenteric arteries.** Branches of these vessels distribute to pancreatic acini and islets before returning to the **hepatic portal vein** by way of the **splenic** and **superior mesenteric veins.** The liver, an important target of the islets, receives blood directly from the hepatic portal vein.

ISLET CELLS

Islets contain several types of cells—A, B, D, EC, and F—at least four of which synthesize and secrete different hormones (Figure 22.19 and Table 22.4). B-cells, or **beta cells,** are the most abundant, comprising about 70% of the total. Usually found in the center of each islet, they synthesize **insulin** (see Table 22.4) that is stored in dense crystalline cytoplasmic secretion granules until release. Insulin stimulates the liver, skeletal muscle, and adipose tissues to take up glucose from the blood and convert it to glycogen. Insulin (insula, L, island) is the principal hormone produced by the islets of Langerhans, as one might expect from such abundant cells. Insulin synthesis begins with a large **preproinsulin** protein from which two functional polypeptide chains of amino acids are cleaved.

Most A-cells, or **alpha cells,** are located in the periphery of the islets, where they display numerous dense, round secretion granules that contain **glucagon** and other peptide hormones listed in Table 22.4. Second only to beta cells in abundance, alpha cells represent approximately 20% of the total islet cells. Glucagon molecules are polypeptide chains that counteract the effects of insulin by promoting the breakdown of glycogen and fats into glucose. The balance between insulin and glucagon from the islets strongly influences the overall pattern of glucose metabolism by contributing to the storage of glucose as fats and glycogen on one hand (insulin) and to the mobilization of glucose from fats and glycogen on the other (glucagon).

Like alpha cells, D-cells, or **delta cells** (Figure 22.19), occupy the periphery of islets, but they make up only 5% of all islet cells. Delta cells synthesize and secrete **somatostatin** (so-MAT-oh-STAT-in) from round, moderately dense secretion granules. Somatostatin is a small polypeptide with many targets and effects. One of its main effects takes place in the pituitary, where it inhibits the release of growth hormone. Closer to home, somatostatin also acts directly on alpha and beta cells to inhibit release of glucagon and insulin.

Several other types of islet cells account for the final 5% of pancreatic endocrine cells. EC cells are **enterochromaffin cells** (EN-ter-oh-KROME-ah-fin) that release **serotonin** (SEAR-oh-TONE-in) from secretion granules of various sizes. Serotonin is an amino acid derivative with many effects and numerous sources in

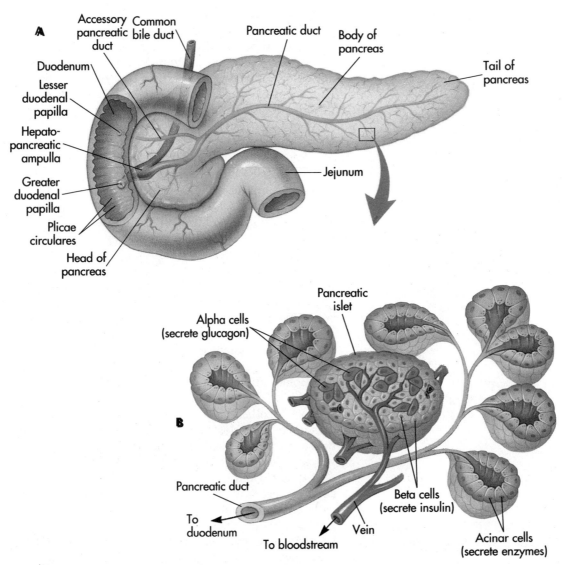

Figure 22.18

Pancreas and pancreatic islets. **A,** Gross anatomy of the pancreas. Find the arteries and veins that vascularize this gland. **B,** Islets are scattered among acini and lead to capillaries, rather than to the pancreatic duct.

the nervous system and blood. For example, basophils and mast cells promote tissue inflammation by releasing serotonin. **F-cells** secrete **pancreatic polypeptide** in small, granular secretory vesicles. This polypeptide stimulates the stomach to secrete pepsin, but it also inhibits peristalsis in the intestines and bile secretion from the gall bladder. F-cells also are found in acini and pancreatic ducts.

- How do pancreatic islets differ from acinar exocrine glands?
- Which hormones do alpha, beta, delta, and F-cells secrete?
- Which islet cells release serotonin?

INSULIN, GLUCAGON, AND DIABETES

Insulin and glucagon are two key hormones that regulate glucose metabolism, continually adjusting storage and release of glucose so that relatively constant amounts circulate in the blood. Insulin, you will recall, promotes uptake of glucose from circulation and its storage, while glucagon enhances release of glucose into circulation (see Figure 22.20). The islets maintain this balance by regulating how much insulin and glucagon enters circulation. When glucagon increases, insulin decreases, and vice versa. Confusing though it may sound, insulin and glucagon maintain this reciprocal relationship themselves by regulating each other's release. Insulin inhibits glucagon release. Insulin drains from the beta cells in the center of the islets past the alpha cells at the periphery, where the molecules bind to insulin receptors in the alpha cells and depress glucagon secretion. Glucagon has a similar feedback effect on beta cells, but to reach the beta cells, it must first circulate through the heart and re-

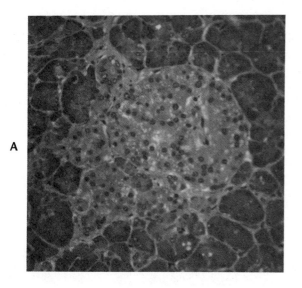

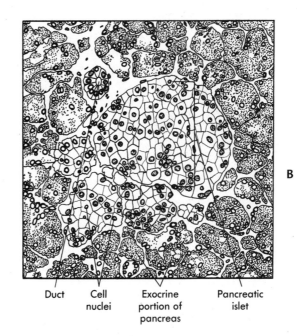

Duct   Cell   Exocrine   Pancreatic
        nuclei portion of   islet
                pancreas

Figure 22.19

Pancreatic islets. Section through one islet **(A)** reveals four types of islet cells. **(B)** Be sure you recognize the location of each type of cell in the islet. Are the cells central or peripheral?

Table 22.4   PANCREATIC ISLET CELLS AND HORMONES

Name	Abundance, Location	Secretion Granules*	Product	Action
Alpha, A-cells	15%-20%, islet periphery and core	Spherical; dense homogeneous matrix; 250 nm	Glucagon, gastric inhibitory peptide, cholescystokinin, ACTH/endorphin	Glucagon mobilizes glucose from glycogen and mobilizes fats
Beta, B-cells	70%, islet core	Irregular; dense crystalline matrix; 300 nm	Insulin	Promotes cellular uptake of glucose and conversion to glycogen
Delta, D-cells	5%, islet periphery	Spherical; moderately dense homogeneous matrix; 300-350 nm	Somatostatin	Inhibits release of insulin and glucagon in alpha and beta cells; inhibits hypothalamus and pituitary
Enterochromaffin, EC-cells	<5%	Irregular; dense homogeneous matrix; 175-400 nm	Serotonin	Enhances gastrointestinal motility
F-cells	<5%, islets and acini, ducts	Spherical; variable density, homogeneous granular matrix cores; 150-200 nm	Pancreatic polypeptide (PP)	Stimulates gastric secretion, inhibits intestinal peristalsis, bile secretion

From Bauer EG. In Weiss L, editor: Cell and tissue biology, *ed 6, Baltimore, 1988, Urban and Schwarzenberg.*
Abbreviations: nm, nanometers.

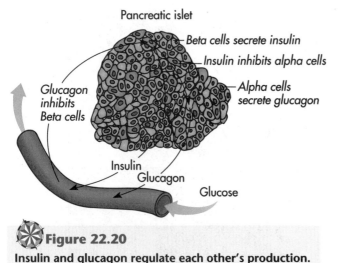

Pancreatic islet

Beta cells secrete insulin

Insulin inhibits alpha cells

Alpha cells secrete glucagon

Glucagon inhibits Beta cells

Insulin

Glucagon

Glucose

Figure 22.20

Insulin and glucagon regulate each other's production.

turn to the pancreas before slowing insulin release.

Diabetes mellitus (DIE-ah-BEET-ease MEL-ih-tus) is a common insulin deficiency disease that illustrates insulin's fundamental role in metabolism. This disease has two forms, one in which no insulin is produced (**insulin dependent diabetes mellitus,** IDDM) and the other in which the tissues become resistant to the action of insulin (**noninsulin dependent diabetes mellitus,** NIDDM). The first usually attacks children and adults less than 30 years old, while NIDDM usually appears in older age. In either disease, blood glucose rises drastically and the kidney excretes this excess, while patients lose weight from their inability to absorb glucose. Insulin dependent diabetes destroys the beta cells in an autoimmune process initiated by viral infections. Insulin replacement therapy is necessary for survival. Beta cells continue to produce normal quantities of insulin in the noninsulin dependent disease, but not enough to overcome the inability of the tissues to absorb glucose. Consequently, treatment carefully balances diet and glucose intake with guarded doses of insulin. Although genetics can predispose people to this form of the disease, no viral link is known.

- What are the effects of insulin and glucagon on blood glucose levels?
- Why does blood glucose increase in diabetes?
- What is the defect in IDDM and in NIDDM?

THE ADRENALS

The adrenal glands secrete hormones that have wide-ranging effects on glucose metabolism, electrolyte balance, sexual development, and stress. **Steroid hormones** affect the first three processes, and **epinephrine** enables the sympathetic nervous system to deal with fight or flight stresses (Table 22.5). As you might expect from the discussion of the pituitary and pancreas, different adrenal cells synthesize different hormones. Those in the **adrenal cortex** (ah-DREE-nal) synthesize all the steroids, and cells in the central medulla produce epinephrine and smaller quantities of norepinephrine. These differences reflect the mesodermal origin of the cortex and the neural ectodermal beginnings of the medulla.

Table 22.5   ADRENAL CELLS AND HORMONES

Name	Cellular Features	Product	Action
ADRENAL CORTEX			
Zona glomerulosa	Occasional lipid droplets	Aldosterone (mineralocorticoid)	Conserves sodium by enhancing reabsorption of sodium in kidney; balances potassium by stimulating secretion from the kidney
Zona fasciculata	Numerous lipid droplets	Cortisol (glucocorticoid)	Promotes glucose production from proteins, carbohydrates (gluconeogenesis); suppresses inflammation
Zona reticularis	Fewer lipid droplets than zona fasciculata	Androgens (sex hormone)	Promotes libido, axillary hair, pubic hair
ADRENAL MEDULLA			
Chromaffin cells	Catecholamine chromaffin secretion granules	Epinephrine and smaller amounts of norepinephrine (catecholamines)	Fight or flight responses

From Kaplan N: The adrenal glands. In Griffin JE and Ojeda SR: **Textbook of endocrine physiology,** *ed 2, New York, 1992, Oxford University Press; and Black VA: The adrenal gland. In Weiss L:* **Cell and tissue biology,** *ed 6, Baltimore, 1988, Urban and Schwarzenberg.*

The adrenal glands, illustrated in Figure 22.21, are situated above the kidneys in the adipose tissue that surrounds the fibrous capsule of each kidney. The adrenals are sometimes called **suprarenal glands** because of these locations. Shaped like a pyramid or crescent, each adrenal ($1 \times 5 \times 4$ cm, or $\frac{1}{2} \times 2 \times 1\frac{3}{4}$ inches) drapes over the superior pole of its kidney, moored by its own delicate connective tissue capsule. The cortex (Figure 22.22) is about 1 cm deep and covers the entire surface of the gland, whereas the medulla occupies the core, sharply distinguished from the cortex by a prominent **corticomedullary boundary.** Cholesterol precursors to the steroid hormones give the gland and its cortex a yellow appearance, but the medulla is colored reddish-brown by blood, which also is present in the cortex but is masked there by cholesterol.

Circulation arrives at the adrenals from the descending aorta by way of **adrenal arteries** that enter the margins of the glands. A profuse vascular tree and capillaries supply the adrenal cortex and medulla before exiting through the **central vein,** at the hilus on the inferior surface of each gland. This blood returns to the inferior vena cava directly (right adrenal) or by way of the left renal vein (left adrenal).

- Where are the adrenal glands located in the body?
- What are the two main divisions of the adrenal glands?
- How does blood reach the adrenal glands and where does it flow after leaving them?

ADRENAL CORTEX

The adrenal cortex consists of polyhedral parenchymal cells (pahr-EN-kim-al) that display all the features expected of steroid-secreting cells. Abundant smooth endoplasmic reticulum and prominent Golgi apparatus around the central nucleus testify to vigorous steroid synthesis. Many large mitochondria with tubular cristae supply energy, and numerous lipid droplets store cholesterol precursors that the endoplasmic reticulum converts to steroid hormones. Desmosomes and gap junctions, rather than tight junctions, are the principal intercellular connections. You will recall that gap junctions indicate that cells communicate with each other (Chapter 2, Cells).

The anatomy of the adrenal cortex resembles the cooling coils of your air conditioner, where warm air passes through a honeycomb of filaments and tubes that cool the passing air. In the adrenal cortex (Figure 22.22), radial capillaries circulate blood from the outer surface of the gland through the cortex to the medulla, past palisades of parenchymal cells that release hormones into the blood as it passes. Parenchymal cells form three layers—the **zona glomerulosa, zona fasciculata,** and **zona reticularis**—that synthesize different steroids. The zona glomerulosa receives blood first because it is the most superficial layer. Round clusters of glomerulosa cells synthesize **aldosterone** (al-DOS-ter-own) a steroid hormone that conserves sodium ions and stimulates secretion of potassium ions in the kidney. This hormone enters the capillaries that flow among the clusters of cells. Beneath the zona glomerulosa, tall columns of parenchymal cells in the zona fasciculata are surrounded by straight radial capillaries. These parenchymal cells are especially rich with lipid droplets, indicating that these cells vigorously synthesize **cortisol** (KORE-tih-

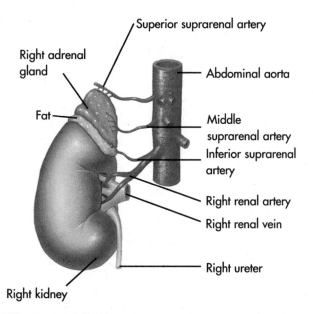

Superior suprarenal artery

Right adrenal gland

Abdominal aorta

Fat

Middle suprarenal artery

Inferior suprarenal artery

Right renal artery

Right renal vein

Right ureter

Right kidney

Figure 22.21

Adrenal glands cover the superior pole of each kidney.

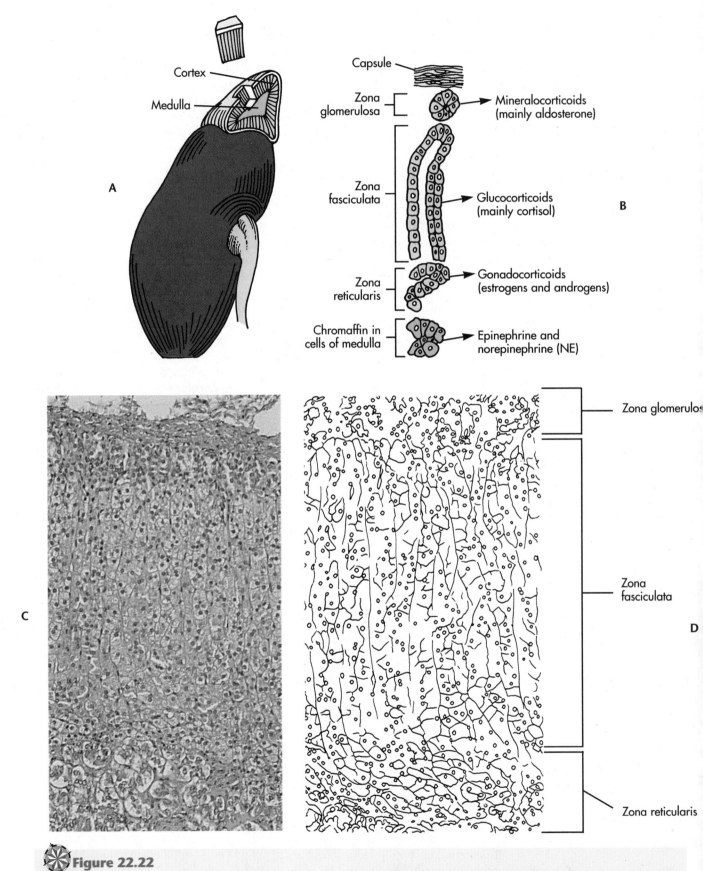

Figure 22.22

Adrenal cortex. **A,** A section from the adrenal cortex reveals the zona glomerulosa, zona fasciculata, and zona reticularis. **B,** Blood enters from the capsule of the gland and circulates past clusters and columns of cortex cells. **C** and **D,** Stained tissue section and interpretive diagram.

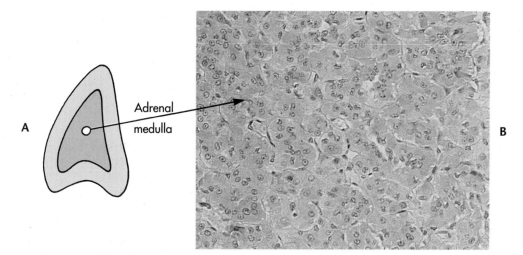

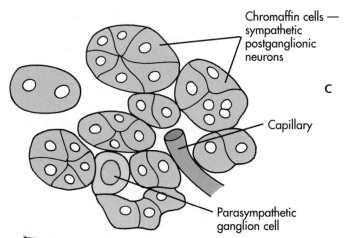

Figure 22.23

The adrenal medulla. **A,** The medulla occupies the center of the adrenal gland. **B,** It contains clusters of chromaffin cells and occasional parasympathetic ganglion cells illustrated in **C.**

sole), a hormone that has broad metabolic effects. Cortisol joins aldosterone from the zona glomerulosa and circulates to the zona reticularis, where irregular networks of cells release into the capillaries small quantities of **androgens** (AN-droe-jenz; male sex hormones) that promote secondary sex characteristics in both sexes. The capillaries of the adrenal cortex are fenestrated and covered with a continuous basement membrane in all three zones. **Perivascular spaces** intervene between the capillaries and many parenchymal cells, so that secreted hormones go to the capillaries directly or through the perivascular spaces before entering the capillaries. Thus charged with adrenal hormones, this blood will now enter the adrenal medulla.

- What are the names of the three zones of the adrenal cortex?
- Name an example of each type of steroid the cortex produces.
- Which zone is responsible for which hormone?

ADRENAL MEDULLA

The adrenal medulla differs considerably from the adrenal cortex in its anatomy, the products it elaborates, and its developmental origin (Figure 22.23). Clusters of **chromaffin cells** secrete **epinephrine** and **norepinephrine** into capillary networks and small venules that enter from the adrenal cortex. Chromaffin cells are **neuroendocrine cells** of the sympathetic nervous system, derived from the neural crest. They are innervated by preganglionic neurons, but chromaffin cells themselves have no axons or dendrites.

Chromaffin cells are round or polyhedral cells, named for the ability of their secretion granules to stain brown when treated with chromic acid. There appear to be two varieties of chromaffin cells, one in which dense secretion granules store norepinephrine and another that stores epinephrine in lighter granules. These granules also store opiate peptides, or

enkephalins (en-KEF-al-inz), intrinsic neural pain killers. Chromaffin cells display other standard endocrine features in addition to secretion granules. The Golgi apparatus is prominent; rough endoplasmic reticulum and cylindrical mitochondria are abundant; and desmosome, gap junctions, and tight junctions join the cells together.

- What is the name of the principal cells of the adrenal medulla?
- What products do they secrete?
- What is the major difference between these cells and those of the adrenal cortex?

CORTISOL CONTROL LOOP

The steroid hormones of the adrenal cortex have profound effects on glucose metabolism and electrolyte balance and rather lesser effects on sex characteristics. Those steroids affecting glucose are called **glucocorticoids,** the principal one being cortisol (hydrocortisone). **Mineralocorticoids** control electrolytes, and aldosterone is the primary member of this group. Fundamentally, cortisol promotes gluconeogenesis (GLUE-cone-ee-oh-JEN-eh-sis), a process that produces glucose by degrading proteins. Like glucagon, cortisol and gluconeogenesis tend to raise blood glucose, but glucagon stimulates degradation of glycogen into glucose rather than by degrading proteins. Cortisol is also well known for its ability to

suppress immune responses in tissues, especially inflammatory reactions. You probably have used topical hydrocortisone creams for mild skin irritations. Aldosterone causes the kidney to conserve sodium ions by reabsorbing them back into the blood and to secrete potassium ions into the filtrate from the blood.

Adrenocorticotropic hormone (ACTH) regulates synthesis and release of glucocorticoids and androgens, shown in Figure 22.24. ACTH circulates from the pituitary gland into the cortex, where it binds primarily with zona fasciculata cells and causes them to release cortisol that then circulates to primary targets, including the liver, muscles, skin, adipose tissue, and lymphatic tissue of the immune system. In the liver, cortisol binds with cortisol receptors in the cytoplasm and activates synthesis of the enzymes responsible for gluconeogenesis. Cortisol promotes breakdown of fats into fatty acids and glycerol in adipose tissue, which contributes to gluconeogenesis and also raises blood glucose. Furthermore, cortisol is necessary for proper vascular tone. Cortisol controls its own production by regulating ACTH; high levels of circulating cortisol suppress the release of ACTH by the adenohypophysis, thus completing the feedback control loop. Low levels of cortisol, on the other hand, enhance ACTH secretion so that cortisol and glucose levels tend to remain relatively constant. Cortisol deficiency can occur when the adrenals fail to secrete enough hormone **(Addison's disease)** or when there is too little ACTH from the adenohypophysis to stimulate the adrenal cortex. Failures in either organ can result in excess cortisol production, as when the adrenal itself overproduces or when the pituitary or tumors of the pituitary elaborate too much ACTH **(Cushing's syndrome).**

The production of aldosterone is controlled by its target organ, the kidney. This gland, or more specifically the juxtaglomerular apparatus (Chapter 25), produces two peptides—renin and angiotensin—that

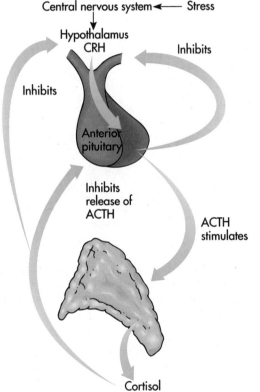

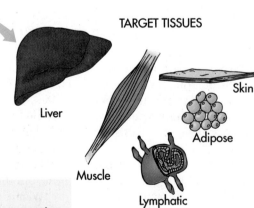

Figure 22.24

Cortisol control loop. Cortisol controls its own production by inhibiting release of CRH from the hypothalamus and ACTH from the adenohypophysis. ACTH also inhibits CRH release.

stimulate the adrenal cortex to produce aldosterone (see Chapter 25, Kidney).

- Which hormone controls cortisol production?
- What are the main effects of cortisol?
- What is the effect of low cortisol levels on ACTH production?

OTHER ENDOCRINE TISSUES

Endocrine glands are not the only tissues that secrete hormones. Numerous cells and tissues throughout the brain, gut, and cardiovascular system do the same, and this section selects a few examples to show how pervasive endocrine action is (Figure 22.25).

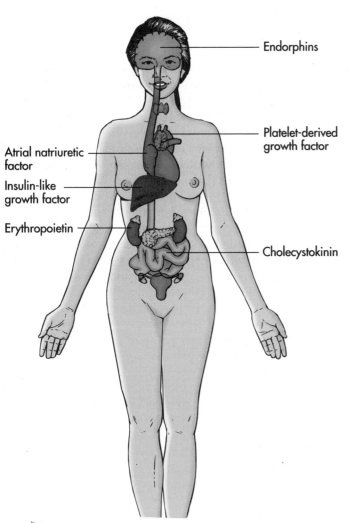

Figure 22.25

Endocrine tissues are located in the brain and spinal cord; they also occur in the intestine, the heart, the kidney, and the liver.

ENDORPHINS

These peptides belong to a family of peptide hormones that appear to have many different effects, most notably as the body's internal pain killers. Synthesized in the brain, especially in the adenohypophysis, they bind to receptors in the brain that increase the thresholds to pain. **Endorphins** (en-DOR-fins; orphe, *G*, melodious) are implicated in other effects, as well; they appear to enhance the release of thyroid releasing hormone from the hypothalamus and also to influence the neurosecretion of vasopressin, ACTH, and growth hormone.

CHOLECYSTOKININ

CCK cells in the mucosa of the intestine secrete cholecystokinin into the vascular system. The digestive system adjusts digestion to the amount of food in the gastrointestinal tract. Demand peaks at meals and falls between meals. **Cholecystokinin** (CCK; KOAL-eh-SIS-to-KINE-in) is one of several peptides (vasoactive intestinal peptide, gastrin, pancreatic polypeptide, somatostatin) that participate in these processes. Carried to the pancreas, CCK stimulates acinar cells to secrete digestive enzymes. CCK also causes the gall bladder to contract and release bile, and it inhibits secretion by the stomach.

ATRIAL NATRIURETIC FACTOR

The right atrium of the heart secretes atrial natriuretic factor (ANF). As blood returns to the heart, especially to the right side, it stretches the atrial and ventricular walls. Extensive atrial stretching increases the amount of urine produced by the kidney, and **atrial natriuretic factor** (NATE-ree-YOUR-et-ik; natrium, *L*, sodium; ure-, *L*, urine) is the link between the heart and kidney that removes excessive fluid from the body. As you can see, this is another negative feedback loop that reduces the amount of venous blood returning to the heart, by excreting excess fluid as urine. Muscle fibers in the atrial wall secrete ANF, a polypeptide, depending on how much the muscle fibers stretch. ANF circulates to the kidneys, where it causes them to release greater quantities of water and sodium ions into the urine until declining venous return diminishes the amount of circulating ANF.

ERYTHROPOIETIN

This glycoprotein is produced in the kidney and stimulates erythropoiesis in the bone marrow in response to oxygen demand. **Erythropoietin** (EPO; eh-RITH-roe-POY-et-in) accumulates in the blood of anemia sufferers whose marrow does not produce sufficient red blood cells.

INSULIN-LIKE GROWTH FACTOR

This hormone is the effective agent for growth hormone. The liver and many other tissues produce **in-**

sulin-like growth factor (IGF-I), which resembles the insulin molecule, in response to growth hormone. IGF-I in turn promotes local cell growth and division in the target tissues of insulin. For example, in cartilage, IGF-I stimulates matrix production; in adipose tissue, it stimulates the synthesis of lipids and glycogen; and in connective tissue, it stimulates proliferation of fibroblasts.

PLATELET-DERIVED GROWTH FACTOR

The body elaborates a large spectrum of growth factors that enable it to stimulate local cell populations, including tumors, to grow. Circulating platelets of the blood are in an especially favorable location to deliver such factors. **Platelet-derived growth factor** (PDGF) is a protein that binds to receptors on the surface of target cells and thereby promotes initial steps in the cell cycle. Other growth factors, including IGF and **epidermal growth factor** (EGF), are necessary for the cells to proliferate.

- Name several sources of hormones outside the regular endocrine glands.
- What hormone does each one of these sources produce?

SUMMARY

The endocrine system consists of glands and many scattered cells that regulate other tissues by secreting hormones into the blood. Hormones stimulate gene action, growth, and synthesis. Endocrine glands have no ducts but are highly vascular, secreting their products directly into the cardiovascular system that delivers to target tissues throughout the body. Endocrine glands secrete proteins, peptides, steroid hormones, and hormones derived from amino acids. Some of these molecules are identical to those secreted by the nervous system, but hormones are usually transported over long distances rather than short distances across synapses of the nervous system. Negative feedback loops control hormone production and are effective because relatively few molecules of any hormone circulate. *(Page 661.)*

The pituitary gland is the master gland that controls many other endocrine glands. This gland is located beneath the hypothalamus of the brain. The adenohypophysis governs the thyroid, mammary glands, adrenals, and gonads, whereas the neurohypophysis controls body fluid volume and uterine muscle. The hypophyseal portal system delivers releasing hormones from the hypothalamus to the adenohypophysis. These releasing hormones control the secretion of hormones from the anterior pituitary. Neurosecretion from axons of the hypothalamus elaborates the hormones of the posterior pituitary. The neurohypophysis and adenohypophysis differ in that

the former derives from the hypothalamus and the latter, from epithelium of the mouth. *(Page 664.)*

The pinealocytes of the pineal gland release melatonin, which appears to regulate diurnal rhythms. The pineal is an extension of the roof of the diencephalon. *(Page 669.)*

The thyroid gland promotes growth and stimulates carbohydrate metabolism and protein synthesis. The thyroid gland is located below the larynx in front of the trachea. Thyroid follicles produce thyroxine that affects nearly all body tissues by binding to thyroxine receptors and activating genes of carbohydrate metabolism. The hypothalamus and pituitary gland control thyroxine production by releasing TRH and TSH hormones, respectively, but overall thyroxine production is balanced by thyroxine itself. TRH stimulates the adenohypophysis to release TSH, and TSH stimulates thyroid follicles to release thyroxine. Thyroxine counteracts the stimulatory effects of TSH by inhibiting the follicle cells from releasing thyroxine. Thyroid glands also secrete calcitonin, which diminishes the concentration of calcium ions circulating in the blood. *(Page 670.)*

Parathormone from the four parathyroid glands promotes the mobilization of calcium ions into the blood. Located in the rear of the thyroid gland, parathyroid chief cells release parathormone in response to declining calcium levels in the blood. This peptide hormone raises blood calcium in three ways. (1) It stimulates osteoclasts to reabsorb calcium from bone. (2) It reduces the amount of calcium excreted by the kidneys, and (3) with vitamin D it stimulates the intestines to absorb calcium from the diet. Parathormone and calcitonin interact to maintain relatively constant amounts of circulating calcium ions. *(Page 674.)*

Pancreatic islets contain at least five types of endocrine cells. Scattered throughout the pancreas, clusters of alpha cells secrete glucagon, beta cells produce insulin, delta cells synthesize somatostatin, enterochromaffin cells secrete serotonin, and F-cells produce pancreatic polypeptide. Insulin and glucagon control glucose metabolism by promoting uptake and storage of glucose (insulin) and release of glucose from storage into the blood (glucagon). Glucose levels in turn regulate insulin and glucagon production. In diabetes, islet cells may cease insulin production, or the tissues may become resistant to insulin. Either deficit tends to raise the quantity of glucose in circulation. *(Page 677.)*

The adrenal glands have widespread effects on glucose metabolism, electrolyte balance, and sexual development through the action of steroid hormones from the adrenal cortex. The adrenal medulla secretes an entirely different hormone, epinephrine, that promotes the fight or flight behavior of the sympathetic nervous system. The adrenals form caps over the superior poles of the kidneys. Zona glomerulosa cells of the adrenal cortex produce aldosterone, which causes the kidneys to retain sodium ions and to secrete potassium ions. Zona fasciculata cells produce cortisol, which has profound effects on glucose and lipid metabolism and also suppresses inflammation. The zona reticularis produces androgens that promote secondary sex characteristics in both sexes. The pituitary gland controls the adrenal cortex. The adrenal medulla, which is composed of neuroendocrine cells, is very different from the cortex. Chromaffin cells of the medulla are postsynaptic cells that lack axons and dendrites. *(Page 680.)*

Other endocrine cells secrete peptide hormones from locations scattered throughout the body. For example, the nervous system produces endorphins, the gastrointestinal tract secretes cholecystokinin, the heart produces atrial natriuretic factor, the kidneys produce erythropoietin, the liver produces insulin-like growth factor, and platelets in the blood release platelet-derived growth factor. *(Page 685.)*

STUDY QUESTIONS
OBJECTIVES

1. Follow the synthesis and secretion of a hormone from a typical endocrine gland to its target. What cellular structures and vessels must it traverse from its formation in the gland cells to its receptor in the target cells? **1**

2. How do negative feedback loops regulate hormone production and secretion? Choose an example from any gland and describe the loop that controls that gland. **2**

3. What are the differences between endocrine and exocrine glands and their products? Give examples of each type. **3**

4. From memory, make a table of the cells of the adenohypophysis, the hormones they secrete, and any releasing hormones that regulate these cells. **4**

5. How does the hypophyseal portal system deliver releasing hormones to the adenohypophysis? **4**

6. Describe how the neurohypophysis secretes oxytocin and vasopressin. Why is this process called neurosecretion, and how does it differ from secretion in the adenohypophysis? **4**

7. What unique endocrine features set the pineal gland apart from all others? **5**

8. Describe how the anatomy of thyroid follicles is related to thyroxine production. Where is thyroglobulin produced, stored, and iodinated? Where do iodide and tyrosine enter the gland, and where does thyroxine leave? **6**

9. What is the effect of thyroxine on its target organs, and what vessels carry this hormone to its destinations? How are the symptoms of hypo- and hyperthyroidism related to thyroxine action? **6**

10. Diagram the thyroid control loop. Show the action of thyroxine on target cells and the effect of the pituitary and thyroxine itself on thyroxine production. **6**

11. What cells assemble parathormone, and how does this hormone interact with calcitonin from the thyroid gland to control blood calcium levels? **7**

12. Where are the islet cells located in the pancreas? How does the process of secretion in these and acinar glands differ? What are the different types of islet cells, and what hormones do they secrete? What target tissues do they affect, and how do these tissues respond to the hormone? **8**

13. What is the effect of insulin on glucose metabolism and the levels of glucose in the blood? How does diabetes mellitus interfere with insulin action? What is the effect of glucagon on glucose metabolism? What are the effects of blood glucose levels on production of glucagon and insulin? **9**

14. Where are the adrenal glands located? What are the anatomical and functional differences between the medulla and cortex of these glands? **10**

15. Which pituitary hormone controls cortisol secretion by the adrenal cortex, and how does cortisol affect the secretion of this hormone? **11**

16. What are the functions of endorphins, cholecystokinin, platelet-derived growth factor, erythropoietin, and atrial natriuretic factor? **12**

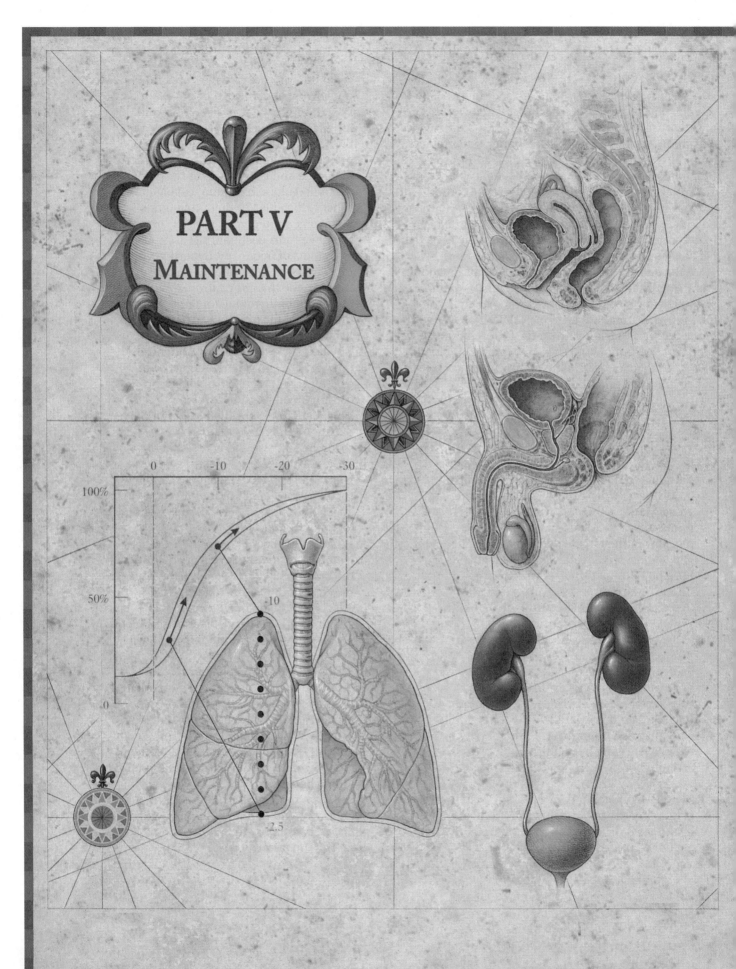

PART V
MAINTENANCE

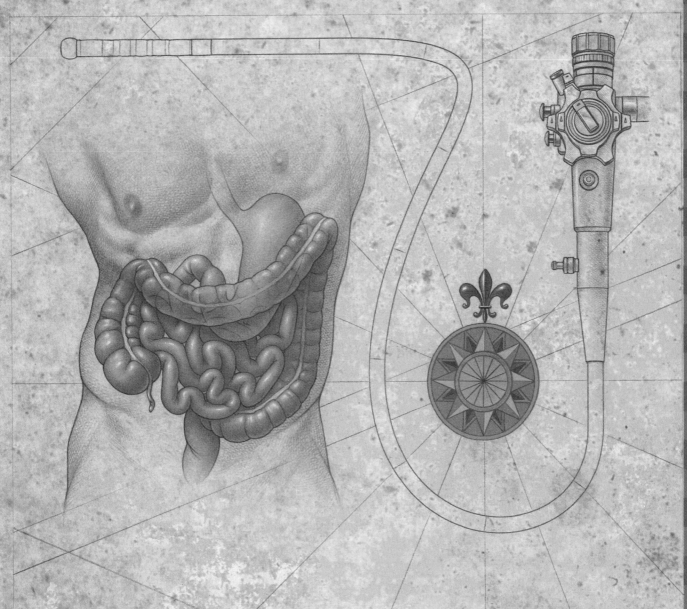

What features of the gut, lungs, kidneys and reproductive organs set them apart from all others? Maintenance and epithelia. These organ systems work in the background, supplying the fundamental needs of cell metabolism. The digestive system provides nutrients, the lungs supply oxygen and purge carbon dioxide, the kidneys excrete metabolic wastes and regulate the ionic composition of the body's fluids, and the reproductive system supplies the ultimate service necessary for continuance of the species—reproduction.

Anatomically, all four systems are visceral organs located in the various divisions of the thoracic and abdominopelvic cavities. Epithelium is the fundamental tissue in all of them. The epithelial lining of the digestive tube absorbs nutrients, and glands secrete copious mucus and digestive enzymes. The alveolar endothelium of the lungs exchanges oxygen and carbon dioxide with the blood, and the nasal passages and pulmonary passages also secrete mucus. The epithelium of the nephrons of the kidney filters blood and regulates urine production. Epithelia line the ducts of the male and female genitalia.

Part V further emphasizes the value of large epithelial surface areas for metabolic maintenance. **Chapter 23** introduces the anatomy of the digestive system to show how the pits and folds of the gut lining take on nutrients needed for cellular respiration and metabolism. Respiration also consumes oxygen and releases carbon dioxide; the respiratory system does this with the help of an immense tree of passageways in the lungs that exchanges these gases with blood, described in **Chapter 24**. In **Chapter 25,** you will find that kidney tubules regulate the composition of blood by extracting surplus substances and by conserving essential electrolytes and metabolites. Finally, **Chapter 26** considers the ultimate source of metabolism and individuals in the epithelial surfaces of the reproductive system.

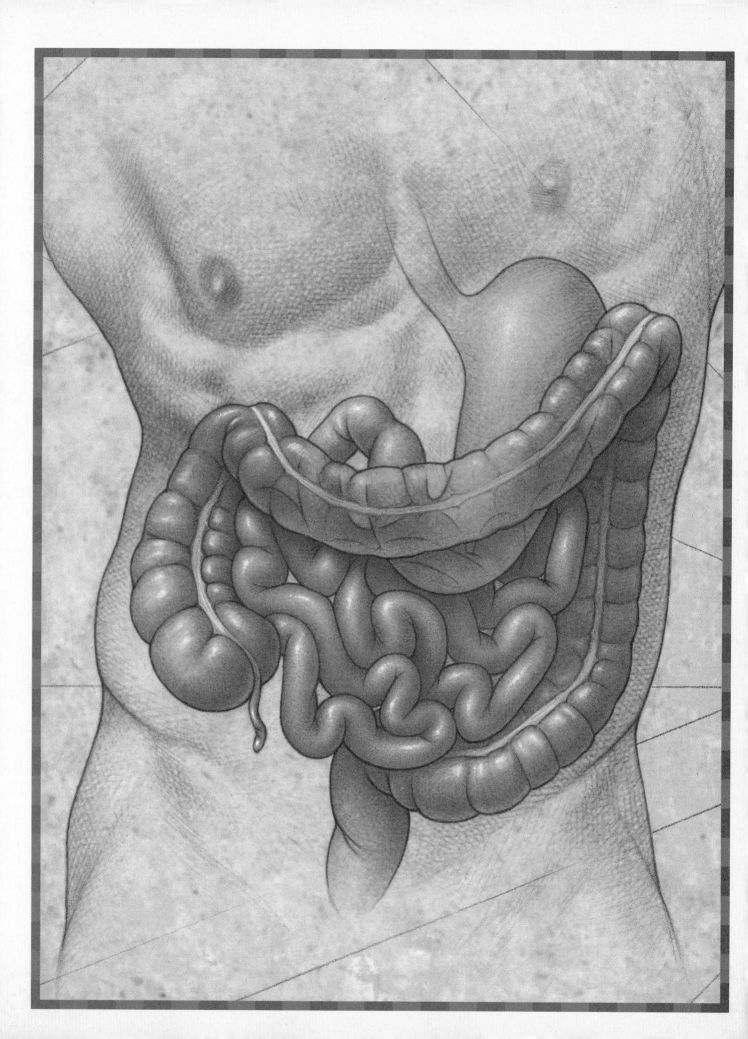

CHAPTER 23

Digestive System

OBJECTIVES

After studying this chapter, you should be able to do the following:

1. Describe the various steps in the process of digestion, and name the organs of the digestive system that carry out these steps.
2. Describe the four fundamental tissue layers of the digestive tube and the services each supplies to the process of digestion.
3. Describe the organs and passages of the digestive system and their functions (a very big objective; there are many organs).
4. Compare the structure of the mucosa in different regions of the digestive tube, and explain how these differences help each region function.
5. Describe the anatomy and functions of the salivary glands.
6. Describe the anatomy of the small intestine and its role in absorption and digestion.
7. Describe how the mucosa has increased the surface area of the intestinal epithelium.
8. Describe the anatomy and functions of the liver, pancreas, and gall bladder.
9. Describe the problem of autodigestion in the mucosa, and explain anatomical ways different regions of the digestive system solve this problem.
10. Describe the major events in development of the digestive system.

 VOCABULARY TOOLBOX

Principal names in the digestive tract.

biliary	bili-, *L* bile, anger.	**liver**	lifer, *Anglo-Saxon* the liver.
duodenum	duoden-, *L* twelve.	**oral**	ora, *L* the mouth.
esophagus	eso-, *G* inward; phag, *G* eat.	**palate**	palatum, *L* the roof of the mouth.
fauces	fauc-, *L* the throat.	**pancreas**	*G* pan, all; kreas, meat.
gall bladder	gealla, *Anglo-Saxon* bile, anything bitter.	**pharynx**	pharyn-, *G* the throat.
ileum	ileo, *G* twisted; *L* the intestine.	**stomach**	stomachus, *G* the stomach.
jejunum	jejun-, *L* hunger.	**tongue**	tunge, *Anglo-Saxon* the tongue.

RELATED TOPICS

- Enteric nervous system, Chapter 21
- Epithelia and glands, Chapter 3
- Lymphatics, lymph nodules, and nodes, Chapter
- Muscles of the pelvis, Chapter 10
- Taste buds, Chapter 20

PERSPECTIVE

Inside the body is another external surface, the epithelium that lines the digestive tract. Its surface area is larger than the skin and just as tough, in its own way. Virtually all nutrients must cross it to enter the body. Even though food may disappear down your throat and be inside your body in a visual sense, it does not enter functionally until the digestive system absorbs it from the lumen of the digestive tract. Defecation is proof that materials that do not cross this epithelial surface remain outside.

The digestive system is both external and internal because it is a tube-within-a-tube. The digestive tube is clearly within the body, connecting mouth and anus, but the contents of this inner tube and its epithelial lining are external.

BRIEF TOUR—ORGANS AND FUNCTIONS

The digestive system is a highly specialized tube with many different organs and glands that together perform all the various steps of eating and digesting a meal. Their overall function, of course, is to supply the body with energy and essential molecules and minerals needed for growth and maintenance. Glucose, other carbohydrates, and fats are the principal energy sources. To build proteins, the body requires certain amino acids that it cannot synthesize by itself; most metabolic pathways and many enzymes require vitamins and mineral ions in order to function. All of these molecules are absorbed into the body by the digestive system.

Figure 23.1 outlines the steps in the digestive process and the organs that accomplish them.

1. **Ingestion** is the act of taking food across the lips into the mouth, perhaps the most enjoyable service of the digestive system.
2. In **mastication,** mandible and teeth chew large pieces of food into smaller bits. Salivary glands and oral mucosa moisten and lubricate the food, and the tongue mixes the pieces together.
3. In **deglutition** (dee-glue-TIH-shun), the tongue and pharynx swallow the food and the esophagus propels it to the stomach.
4. In **digestion,** the stomach and small intestine break down food by secreting enzymes that chemically cleave the particles into nutrient molecules. The stomach stores food temporarily until the small intestine can process it, and the pancreas, liver, and gall bladder secrete enzymes or bile into the small intestine.
5. Through **absorption,** the wall of the small intestine takes up nutrients into the blood and lymph.
6. Finally, for **defecation,** the large intestine consolidates undigested materials into fecal matter and the rectum expels them through the anus.

The entire tube from mouth to anus is known as the **alimentary,** or **digestive, tract;** from the esophagus to the rectum it is the **gastrointestinal,** or **GI, tract.**

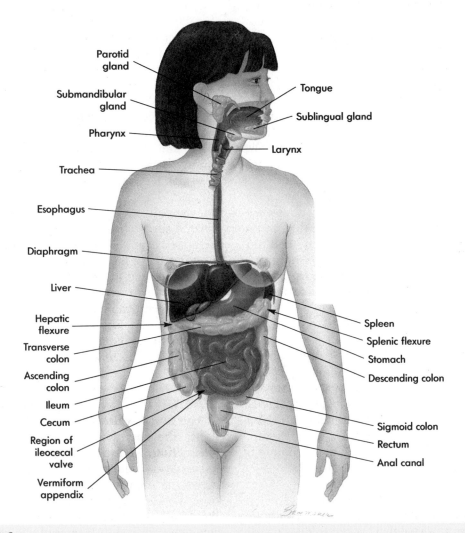

Parotid gland
Submandibular gland
Pharynx
Trachea
Esophagus
Diaphragm
Liver
Hepatic flexure
Transverse colon
Ascending colon
Ileum
Cecum
Region of ileocecal valve
Vermiform appendix
Tongue
Sublingual gland
Larynx
Spleen
Splenic flexure
Stomach
Descending colon
Sigmoid colon
Rectum
Anal canal

Figure 23.1
Digestive system and the digestive process. Follow the process through the organs of the system.

Most digestive organs form the wall of the tract, but special structures that project into the tube, such as teeth and tongue, are considered **accessory organs,** and glands outside of the tube, such as the salivaries, liver, and pancreas, are **accessory glands.**

PROBLEMS AND ANATOMICAL SOLUTIONS

The digestive system's benefits do not come without problems and solutions that are reflected in the system's anatomy.

SURFACE-VOLUME

The lining of the intestinal tract is not flat. Millions of folds, projections, pits, and glands convolute its surface into as much as 186 m² (2000 ft²) of area for absorption. If the lining were entirely smooth and flat, the intestines would have to be impossibly large to meet the demands of an adult requiring 2000 calories of energy a day. Folding allows the gut to serve a large body by packing a huge surface area into relatively small space. This chapter devotes considerable attention to this aspect of digestive anatomy; so should you. Learn the structures each region employs, and use the summary at the end of the chapter (see Surface Tactics) to help organize the information.

AUTODIGESTION

Digestive enzymes do not discriminate between food proteins and gut proteins. The same conditions that promote chemical digestion within the gut also threaten to digest its own lining. Two main anatomical tactics combat this continual dilemma. In one, the gut secretes a film of protective mucus that helps isolate the lining from its contents. Look for mucous-secreting cells throughout the digestive lining to see how this defense works. The second tactic continually renews the lining. Cells of the gut lining divide rapidly and replace it entirely in a matter of days. Look for specialized areas of the lining that provide these new cells.

FOREIGN CELLS AND SUBSTANCES

It is not surprising that the gut's well-developed lymphatic system protects it against infection. With so much surface exposed, it is inevitable that viruses, fungi, bacteria, and toxic molecules will enter the tissues. Scattered through the lining, lymph nodules recognize intruders and destroy them. Lymph nodes are close by, associated with the vasculature of most digestive organs. Look for both defenses.

SECRETION AND ABSORPTION

If you have read Chapter 12, Integument, you know that skin protects against dehydration; so does the digestive system, but in a different way. Even though the digestive system secretes almost 7 l of digestive fluids daily (Table 23.1), it recovers most of this potentially fatal loss by reabsorbing water, especially in the intestines. Look for structures in the epithelium of the gut that secrete and absorb water.

Table 23.1	DIGESTIVE SECRETION
Name	**Quantity (ml)**
Saliva	1000
Gastric juice	1500
Pancreatic juice	1000
Bile	1000
Small intestine	1800
Submucosal glands	200
Large intestine	200
TOTAL	6700

From Guyton AC: Textbook of medical physiology, *ed 7, Philadelphia, 1986, WB Saunders Co.*

- What is the primary function of the digestive system?
- Name the steps in digestion.
- Name two physiological problems that the digestive system solves.

FOUR-LAYERED TUBE

The digestive system is equipped with four layers of tissue—mucosa, submucosa, muscularis externa, and serosa—as shown in Figure 23.2. The innermost layer, the **mucosa,** is in contact with the gut contents. Beneath this layer is connective tissue, **submucosa.** Next are several layers of smooth muscle, known collectively as the **muscularis externa.** Finally, a **serosa** (seh-ROE-za) usually covers the external surface of the gut tube. Each region of the gut modifies these layers, especially the mucosa, according to that section's role in the digestive process. For example, the mucosa lubricates the esophagus as food particles pass, and it secretes digestive enzymes in the small intestine. You should be able to recognize each layer in the different digestive organs and describe how it differs in the esophagus, stomach, or any other region.

MUCOSA

The mucosa lines the digestive tract and establishes the chemical conditions for digestion and absorption. The mucosa itself is composed of three tissue layers. An **epithelium** lines the lumen and secretes the molecules needed for each step in digestion. Where little absorption or secretion occurs (as in the mouth and pharynx), the epithelium is stratified squamous; it is simple cuboidal in many glands, and in the stomach and intestines it becomes simple columnar for secretion and absorption. Various pits,

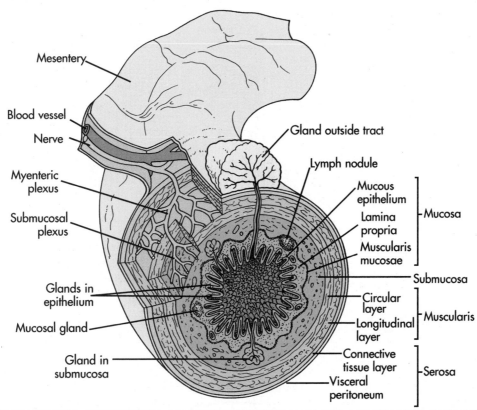

Mesentery

Blood vessel

Nerve

Myenteric plexus

Submucosal plexus

Glands in epithelium

Mucosal gland

Gland in submucosa

Gland outside tract

Lymph nodule

Mucous epithelium

Lamina propria

Muscularis mucosae

⎱ Mucosa

Submucosa

Circular layer

Longitudinal layer

⎱ Muscularis

Connective tissue layer

Visceral peritoneum

⎱ Serosa

✦**Figure 23.2**

The digestive system is a four-layered tube with glands. Find mucosa, submucosa, muscularis externa, and serosa. Glands are located in mucosa, submucosa, and outside the tube.

folds, projections, and glands increase its surface area.

A **basement membrane** attaches the epithelium to the **lamina propria** (LAM-in-ah PRO-pree-ah; near layer). The lamina propria is highly vascular. Its capillaries are the first to absorb nutrients from the epithelium and the last to supply water and electrolytes for secretion. Numerous nerve endings and lymphatic vessels populate the lamina propria. A thin layer of smooth muscle fibers, the **muscularis mucosae** (MUS-kue-LAR-iss mew-KOE-sa; muscles of the mucosa), identifies the boundary with the submucosa.

SUBMUCOSA

This layer is loose connective tissue equipped with vessels, nerves, and lymphatics that continue into the mucosa. **Submucosal plexuses** of the autonomic nervous system (see Enteric Nervous System, Chapter 21) lace both the submucosa and the next layer, muscularis externa.

MUSCULARIS EXTERNA

This layer usually is double; an inner layer of smooth muscle fibers encircles the tube, and an outer layer of fibers runs lengthwise, parallel to the axis of the tube. The muscularis is responsible for swallowing and peristalsis. The inner circular fibers constrict

the tube and longitudinal fibers bend and twist it. Some organs add an oblique third layer that helps with mixing and churning. **Myenteric plexuses** of the enteric nervous system stimulate these motions.

SEROSA

The serosa is the slick, smooth, slippery mesothelium that covers the exterior of the gut tube and allows the intestines to slip and slide past each other when the body moves and when peristalsis squeezes them. The **mesothelium,** named for its mesodermal origin, is several layers of squamous cells that secrete a thin film of serous fluid for lubrication. Beneath the mesothelium is a thin layer of connective tissue. Another covering, the **adventitia** (ad-ven-TISH-ya), replaces the serosa where adhesion rather than slipperiness is useful. Adventitia is fibrous connective tissue that anchors the esophagus in the neck and thoracic cavity, and retains the rectum in the wall of the pelvic cavity. You also will find adventitia in other locations. Occasionally, adventitia can displace patches of serosa on the intestines so that instead of sliding they adhere, causing considerable pain.

GLANDS

The same epithelium that lines the gut tube also lines the glands that secrete mucus and enzymes.

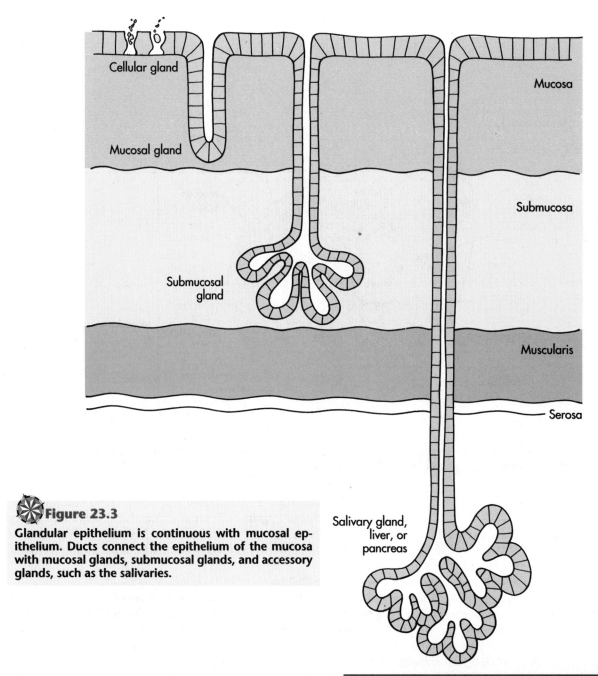

Figure 23.3

Glandular epithelium is continuous with mucosal epithelium. Ducts connect the epithelium of the mucosa with mucosal glands, submucosal glands, and accessory glands, such as the salivaries.

Many mucous glands are **simple cellular glands,** single goblet cells, shaped like a goblet in Figure 23.3, scattered on the surface of the epithelium. Larger tubules grow into the wall of the gut.

How far the tubules penetrate is important. **Mucosal glands** occupy only the lamina propria, and a duct connects each to the lumen of the intestine; but **submucosal glands** reach the submucosa, and their ducts pass through the mucosa. **Accessory glands,** such as the salivaries and pancreas, pass through the entire wall and reside outside the gut tube altogether, connected only by a duct. Different varieties of gland populate different regions of the gut. You should be able to compare the glands of different regions.

- Name the four tissue layers of the digestive system and three categories of its glands.

ORAL CAVITY

The oral cavity is the entrance to the digestive system; it is the space where teeth, tongue, and saliva begin to disassemble food. The **oral cavity** is really two spaces. Food passes the lips in Figure 23.4 (the **rima oris**) into the **vestibule,** which is the space between the jaws and the cheeks and lips. The vestibule opens to the mouth, or **oral cavity proper,** inside the arch of the mandible and maxilla. The **hard** and **soft palates** frame the roof of the mouth, and the mylohyoid muscles of the mandible form the floor from which arises the tongue. The oral cavity exits to the pharynx by way of the **fauces** (FOW-sees), a narrow constriction at the rear, and the teeth and mandible form the lateral walls of the oral cavity.

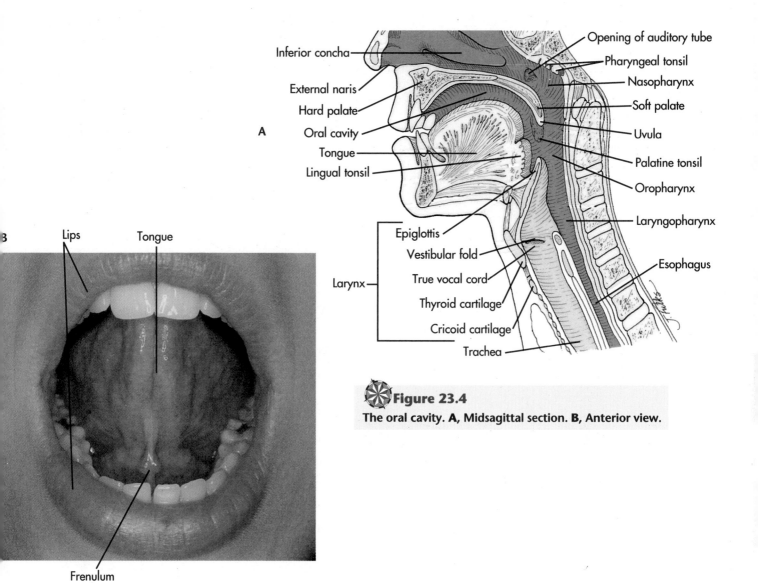

A

B

Inferior concha

External naris

Hard palate

Oral cavity

Tongue

Lingual tonsil

Larynx

Epiglottis

Vestibular fold

True vocal cord

Thyroid cartilage

Cricoid cartilage

Trachea

Opening of auditory tube

Pharyngeal tonsil

Nasopharynx

Soft palate

Uvula

Palatine tonsil

Oropharynx

Laryngopharynx

Esophagus

Lips   Tongue

Frenulum

Figure 23.4

The oral cavity. A, Midsagittal section. B, Anterior view.

Besides its roles in ingestion and deglutition, the oral cavity also assists breathing, and the brain marshals its lips and tongue for speech. Taste buds on the tongue provide our sense of taste. You can appreciate how valuable the oral cavity is when you consider the problems of people who have lost their mandible or tongue to disease or injury.

Only the mucosa and submucosa of the four gut tissues are present in the oral cavity. Mucosa lines the entire cavity, except for the surfaces of teeth, with **stratified squamous epithelium** (SKWAY-mus) supported by a basement membrane and the lamina propria. Because there is no muscularis mucosae, submucosa blends with the mucosa and joins directly to muscles (lips, cheeks, soft palate, floor of the cavity) or to periosteum of bones (hard palate, gums). Intense vascularization in the lamina propria and submucosa gives the lips, cheeks, gingivae (gums), and other oral tissues their characteristic red appearance.

Most of the epithelium is keratinized by an outer layer of cornified cells that resembles the epidermis

of the skin. Only the epithelia of the cheeks, soft palate, and the floor of the mouth do not keratinize, which means that no stratum corneum or keratin filaments form in these areas. Numerous **simple tubular glands** in the submucosa moisten the epithelium with mucus from ducts that open onto the surface of the palate, tongue, floor, and cheeks. Sublingual and submandibular **salivary glands** secrete into the floor of the mouth on either side of the **frenulum** (FRENyou-lum), the fibrous tendon that connects the underside of the tongue with the floor of the mouth. The parotid glands secrete into the vestibule at the level of the second and third upper molars.

Besides being a nice place to get a kiss, lips are interesting because they mark the boundary between interior and exterior at the **red zone.** The epidermis becomes thinner and less cornified as it enters the red zone from the exterior surface of the skin and no longer produces hairs and sebaceous glands. Keratinization stops altogether in the vestibule, and mucous glands now continually moisten the epithelium that lines the mouth. Lips themselves occasionally dry out because the red zone lacks these mucous glands. Dermis of the skin also becomes submucosa as it crosses the red zone. Dermal papillae disappear in the vestibule, but capillaries remain close to the surface and color the lips and mucosa.

- Name the two portions of the oral cavity.
- Which tissues of the digestive tube are present in the oral cavity?
- What are the names of the entry to and the exit from the oral cavity?

TONGUE

The tongue mixes food and helps us speak, and its nerve endings enhance the pleasure of eating by sensing texture and taste. In Figure 23.5 the free **apex** and **body** of the tongue originate from its **base,** or **root,** on the mandible and the hyoid bone, in front of the epiglottis. The **foramen cecum** (SEE-kum) marks the boundary between body and base of the tongue, as well as the origin of the thyroid gland (see Development, this chapter). Numerous papillae cover

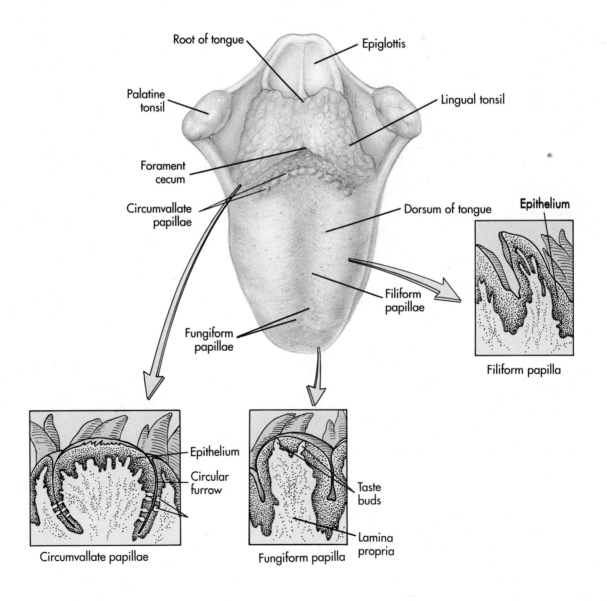

Figure 23.5

The dorsal surface of the tongue, with papillae.

the dorsal surface and sides of the tongue, and the shallow **median sulcus** divides the body of the tongue into left and right halves. The **lingual tonsils** lie between the root of the tongue and the **epiglottis** (ep-ih-GLOT-iss), and the **palatine tonsils** occupy the **palatoglossal arch** behind the fauces. The tongue receives blood from branches of the external carotid artery and drains to the internal jugular vein. The mandibular nerve (V) and the facial nerve (VII) supply the anterior two thirds of the tongue, and the glossopharyngeal nerve (IX) innervates the posterior one third and sides. Taste buds are innervated by the facial and glossopharyngeal nerves.

Stratified squamous epithelium covers the dorsal surface of the tongue, admitting numerous mucous glands from the submucosa beneath. The epithelium of the dorsal and lateral surfaces is keratinized and thrown into many projections, called papillae, that sense the texture and taste of food. Wedge-shaped **filiform papillae** (FILL-ih-form pah-PILL-ee) cover the anterior two thirds of the tongue. These papillae give this portion of the tongue a rough surface for slurping ice cream and popsicles, and they house tactile sensory endings, but no taste buds. Filiform papillae give way to **fungiform papillae** at the apex and sides of the tongue. Dome-shaped like mushrooms, as their name implies, fungiform papillae have taste buds on their lateral walls; so do **vallate papillae,** which resemble fungiform papillae but are larger and are separated from each other by folds of epithelium. About 12 to 18 vallate papillae populate a transverse band across the root of the tongue. (Chapter 20 discusses taste buds.) Lamina propria fills the core of all papillae, and submucosal glands moisten the surface of the tongue. **Mucous glands** supply the body and apex of the tongue, and **serous glands** (SEER-us; Von Ebner's glands) release their watery secretion onto the root of the tongue.

Two sets of muscles maneuver the tongue in speech and eating (described fully in Chapter 10). The tongue is mounted on the floor of the oral cavity by **extrinsic muscles** that originate outside the tongue on the mandible and hyoid bone. The geniohyoid (genio, *G* chin, jaw; mandible to tongue) muscle draws the body of the tongue forward and down, whereas the styloglossus (styloid process to tongue) and hyoglossus (hyoid to tongue) muscles raise and depress the tongue, respectively, to the rear. The origins and insertions of **intrinsic muscles** are entirely within the tongue. These muscles change the shape rather than the position of the tongue. When they contract, longitudinal muscles shorten and widen the tongue, horizontal muscles narrow and lengthen it, and vertical muscles flatten it.

- What are the functions of the tongue?
- Name the structures that enable the tongue to perform these functions.

- Which two layers of the digestive tube are present in the tongue?

TEETH

DENTITION

Teeth grip, slice, and crush food throughout a lifetime of chewing that prepares chunks of food for digestion. Figure 23.6A shows the full, adult armament of 32 permanent teeth mounted in the arch of the mandible and maxilla. A full set of teeth is known as the **dentition.** The permanent dentition includes four incisors, two canines, two premolars, and three molars in each jaw (Figure 23.6B). This set replaces the **deciduous** or **baby dentition** of 20 teeth that includes the same number of incisors and canines but has only two molars in each jaw. **Incisors** have sharp edges that grip and cut food as it enters the mouth; **canines** also grip with their conical crowns. **Premolars** and **molars** mesh with their counterparts in the opposing jaw and crush materials caught between their broad, shallow cusps. Roots of the teeth are enclosed by bony **alveoli** in the mandible and maxilla. Incisors, canines, and premolars have one root each, but molars have two or three.

STRUCTURE

Enamel and dentine are the hard, permanent portions of teeth illustrated in Figure 23.7A. The **enamel** is the smooth, white material that covers the **anatomical crown** of the tooth, a molar in this example, and the **roots** are the portions not covered by enamel. Crown and root meet at the **neck,** the boundary between these two regions. The part of a tooth you and your dentist actually see above the gums, however, is the **clinical crown,** whose lower margins of enamel are obscured by the **gingivae** (JIN-jih-vee). A narrow groove, the **gingival sulcus** (SUL-kus), between each tooth and its surrounding gingivae ends in a basement membrane firmly attached to the crown of the tooth. This barrier prevents infection, unless it is breached by bacteria accumulated in the sulcus. **Dentine** underlies enamel in the crown, and a thin sheath of **cementum** covers dentine of the root. The interior of every tooth contains a **pulp cavity,** a highly vascular space filled with nerves and connective tissue. This cavity and its contents extend down each root as the **root canal** and open through an **apical foramen** to the surrounding bone. Like nutrient foramina in bone, these openings are the points of entry for vessels from the external carotid artery and branches of the mandibular nerve (cranial nerve V). The **periodontal ligament** (PAIR-ee-oh-DON-tal) connects the root to the alveolar bone of the jaw. Large collagen fibers in this ligament orient in different directions and literally suspend a tooth within its alveolus. Because the

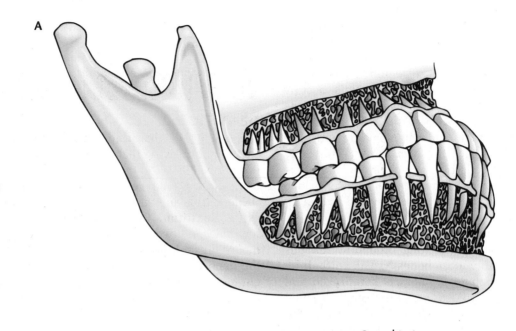

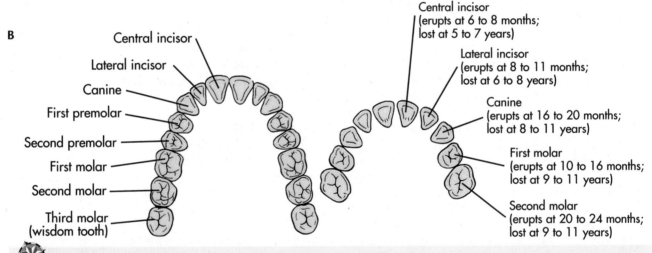

Central incisor
(erupts at 6 to 8 months;
lost at 5 to 7 years)

Lateral incisor
(erupts at 8 to 11 months;
lost at 6 to 8 years)

Canine
(erupts at 16 to 20 months;
lost at 8 to 11 years)

First molar
(erupts at 10 to 16 months;
lost at 9 to 11 years)

Second molar
(erupts at 20 to 24 months;
lost at 9 to 11 years)

Central incisor

Lateral incisor

Canine

First premolar

Second premolar

First molar

Second molar

Third molar
(wisdom tooth)

✹ Figure 23.6

The dentition. **A,** Right lateral view of teeth with roots exposed in alveolar bone. **B,** Permanent and deciduous dentition, upper jaw. Eruption and loss are listed for deciduous teeth.

connection is not entirely rigid, teeth move slightly as they mesh, or occlude, with their partners. To extract a tooth, the periodontal ligament must be severed.

DENTINE

Dentine is harder than bone but softer than enamel, the hardest substance in the body. Hardness comes from the **extracellular matrix** of dentine, which resembles that of bone in its general composition (see Cells and Matrix, Chapter 5, Bone). As with bony matrix, collagen fibers fill a glycoprotein gel, but more fibers and hydroxyapatite are present than in bone. Numerous **dentinal tubules** (DEN-tin-al) radiate through the matrix, like the canaliculi of bone, from the inner pulp boundary to the junction with enamel or cementum, as seen in Figure 23.7B. A single long cytoplasmic process of an odontoblast fills each tubule and maintains the matrix of the dentine. **Odontoblasts** (oh-DON-toe-blasts) are the cells that form dentine, and in this regard they are comparable to the osteoblasts of bone. The cell bodies of odontoblasts are located in the pulp just outside the dentine. A single sensory nerve ending follows each odontoblast process into its dentinal tubule. Odontoblasts slowly apply new dentine to the inner surface of the dentine, and consequently the pulp cavity and root canal tend to narrow in older age as the dentine builds up.

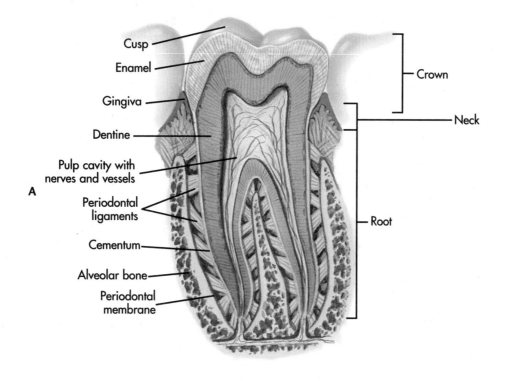

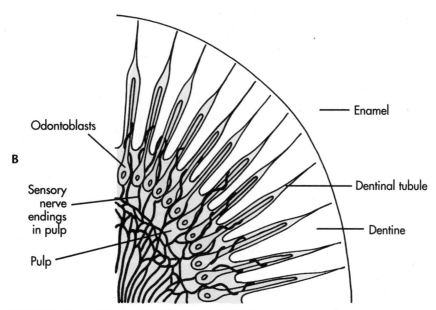

Figure 23.7

Tooth structure. **A,** Sagittal section through a molar reveals enamel, dentine, and pulp cavity. **B,** A portion of the dentine beneath the crown reveals dentinal tubules and odontoblasts.

ENAMEL

Enamel is the hardest material in the human body. The composition of enamel is unusual because it appears to mature from the time it is deposited in developing teeth. Its principal components are proteins called **enamelogenins** (ee-NAM-el-oh-JEEN-inz; in immature enamel) and **enamelins** (in the mature form). Glycoproteins and abundant hydroxyapatite are present, but there are no collagen fibers. This crystalline matrix organizes into **enamel rods** that extend inward from the outer surface to the junction with dentine. **Interrod matrix** cements the rods together. Enamel is formed by **ameloblasts** (ah-MEL-oh-BLASTS) while the tooth develops, but enamel is not replaced after teeth erupt because the ameloblasts degenerate.

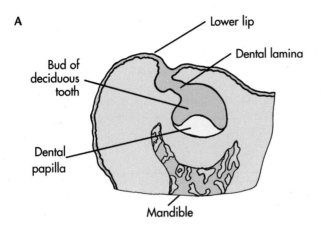

ERUPTION

Each tooth begins as a tooth bud in the tissue that will form the mandible and maxilla (Figure 23.8). Ectoderm grows downward in each bud as the **dental lamina,** and mesoderm grows upward toward the dental lamina as the **dental papilla** (Figure 23.8A). Ameloblasts form in the bulb at the end of the lamina, and this swelling becomes the **enamel organ** that produces enamel (Figure 23.8B). At the same time, odontoblasts in the dental papilla begin to form dentine. Enamel and dentine develop between the layers of ameloblasts and odontoblasts, and the boundary between the dentine and enamel persists in the crown of the mature tooth (see Figure 23.7). Even as the de-

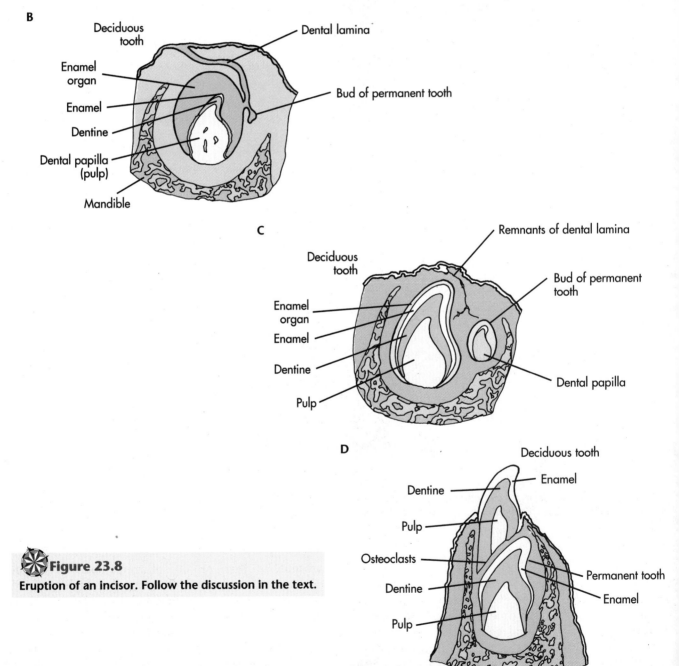

✳ **Figure 23.8**
Eruption of an incisor. Follow the discussion in the text.

ciduous tooth forms, a bud is set aside for its permanent replacement. The permanent tooth develops in the same way, although on a delayed schedule. Deciduous and permanent teeth both are beginning to develop in Figure 23.8C. The permanent tooth is well advanced when the deciduous tooth erupts through the gums (Figure 23.8D) and obliterates the enamel organ that laid down the enamel. Being internal structures, the odontoblasts remain through the life of the tooth and the dental papilla becomes the pulp cavity.

- Name the four types of teeth in your dentition.
- How do they differ from each other?
- What structures are found in the crowns and roots of teeth?
- Which cells form enamel and dentine?

SALIVARY GLANDS

Saliva is a complex fluid mixture of mucus, enzymes, ions, and water that moistens the mouth, helps suppress bacterial growth, and liquifies food. The mucous glands of the oral mucosa secrete this fluid more or less constantly, but three pairs of **parotid, submandibular,** and **sublingual glands**— the salivary glands—secrete saliva especially when the oral cavity contains food. All three are rather typical compound glands, as seen in Figure 23.9A. A connective tissue capsule covers each gland and subdivides it into lobules filled with tubules and alveoli that secrete to branched ducts (see Glands, Chapter 3, Tissues). Their products are viscous mucus and watery serous fluids. The principal enzymes in serous secretions are **alpha amylase** (AM-ill-ace), which attacks starch, and **lysozyme** (LIE-so-ZIME), whose bacteriocidal action lyses bacteria. The autonomic nervous system controls the salivary glands by way of the superior cervical ganglion (sympathetic) and the facial (VII) and hypoglossal nerves (XII) (parasympathetic). Anticipation of a good meal, pleasing taste and texture, or just the presence of food itself stimulates salivation. Facial and lingual branches of the external carotid arteries and tributaries of the external jugular veins supply the salivary glands with blood.

PAROTID GLANDS

The parotids (par-AH-tidz) are serous glands, the largest of the salivaries. They are located superficially, outside the wall of the oral cavity (Figure 23.9A) in the angle of the mandible, forward of the mastoid process below the zygomatic arch. The parotid glands mostly secrete from alveoli that drain into the **parotid (Stensen's) duct.** This duct passes forward across the masseter muscle, into the buccina-

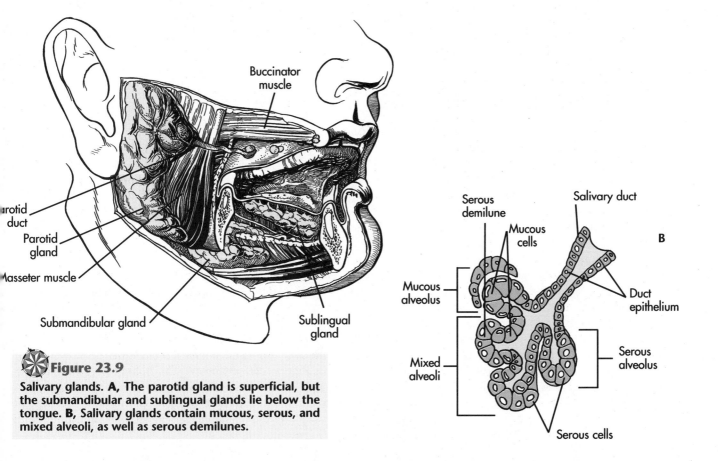

Figure 23.9
Salivary glands. **A,** The parotid gland is superficial, but the submandibular and sublingual glands lie below the tongue. **B,** Salivary glands contain mucous, serous, and mixed alveoli, as well as serous demilunes.

tor muscle of the cheek, before discharging into the vestibule, lateral to the second upper molar.

SUBMANDIBULAR GLANDS

Also known as the submaxillary gland, the submandibular gland lies between the base of the tongue and the submandibular fossa of the mandible. Most of its alveoli are serous, and serous demilunes (Figure 23.9B) cap its mucous alveoli, providing for an overall mixed seromucous secretion. Ducts from each lobule lead to the central **submandibular (Wharton's) duct** that opens 5 cm forward of the gland on each side of the frenulum.

SUBLINGUAL GLANDS

These glands also are located beneath the body of the tongue, on either side of the median septum. Most alveoli of the sublingual gland are mucous, just the converse of the submandibular, with some serous demilunes. Several ducts drain each gland to a row of openings located posterior to Wharton's duct.

To summarize, the parotid gland secretes into the vestibule, and the submandibular and sublingual glands discharge below the tongue. The parotid differs from the "subs" in other regards. Most of the secretory units in the parotid gland are alveolar, whereas tubuloalveolar units are the rule in the others. Submandibular and sublingual are mixed seromucous glands, whereas the parotid is serous.

- Name the three major salivary glands.
- Are the products of each one serous, mucous, or mixed?
- Where does each gland release its secretion?

PHARYNX

The pharynx is the first clearly tubular portion of the digestive system; in addition to linking the oral cavity with the esophagus, as illustrated by Figure 23.4, it also communicates with four more spaces. Air passes between the nasal cavity and the larynx (LAHR-inks), and air also enters the pair of auditory tubes that connect to the middle ears. The pharynx (FAHR-inks) itself consists of three regions. The **nasopharynx** leads from the choanae at the rear of the nasal cavity above the soft palate. The **oropharynx** receives from the mouth and nasopharynx and leads to the **laryngopharynx** (la-RIN-joe-FAHR-inks), which in turn is continuous with the esophagus posteriorly and the larynx anteriorly. The soft palate and epiglottis act as valves that ensure that food and air go to their correct destinations when you swallow and breathe (see Swallowing).

Only a medially located organ can provide all of these connections. The pharynx is located below the sphenoid bone and the basilar part of the occipital bone, in front of the first six cervical vertebrae, and between the carotid arteries and jugular veins of the neck. Muscles suspend the pharynx laterally from the styloid processes, pterygoid plates, and the mandible.

Three layers of the gut tube are present in the pharynx. The mucosa is covered with stratified squamous epithelium, except in the nasopharynx and larynx, where the epithelium becomes columnar. Lamina propria is distinctly fibrous, especially in the nasopharynx. Mucous glands populate the lamina propria and the pharyngeal muscle layer. The pharyngeal constrictor and longitudinal muscles, used in swallowing, represent the muscularis externa. The exterior surface of the pharynx is composed of fascia that connects to surrounding muscles, vessels, and bones.

SWALLOWING

Figure 23.10 shows the motions of swallowing (deglutition). Their effect is like squeezing toothpaste from a tube. As you press the bottom of the toothpaste tube, you mimic the actions of the tongue and pharynx as they propel food into the pharynx and then into the esophagus; but of course it is not this simple. The muscles of deglutition also prevent food from entering the nasopharynx and larynx and reopen these passages quickly so that breathing may continue.

Swallowing food involves three stages. The first is voluntary, and the last two are reflexive, governed by the swallowing center in the medulla oblongata and by fibers of the fifth, ninth, tenth, and twelfth cranial nerves to the pharynx and esophagus. The first step is the **buccal** or **voluntary phase,** in which the tongue presses against the hard palate (Figure 23.10A), progressing from the apex toward the root (like the toothpaste tube), squeezing the bolus of food to the fauces at the rear of the oral cavity.

The **pharyngeal phase** (fahr-IN-gee-al) takes over now, as receptors in the pharyngeal columns and root of the tongue detect food particles. In response to this stimulus, the levator veli palatini muscles (described in Chapter 10) raise the soft palate and jam it against the roof of the nasopharynx (Figure 23.10B), thereby blocking entry to the nasal cavity. At the same time, the space narrows between the palatopharyngeal folds on either side of the root of the tongue, blocking large unmasticated pieces from entering the pharynx. The slit between the vocal cords also narrows, preventing food from entering the trachea, and the larynx rises (a movement you can feel by palpating your larynx; Figure 23.10C and D), which depresses the epiglottis and further guards against food entering the trachea. The pharyngeal constrictor muscles (named in Figure 23.10B) propel the bolus toward the esophagus. These last motions also cause the pharyngoesophageal sphincter muscle to open, allowing the esophagus to receive the swiftly descending bolus (Figure 23.10E).

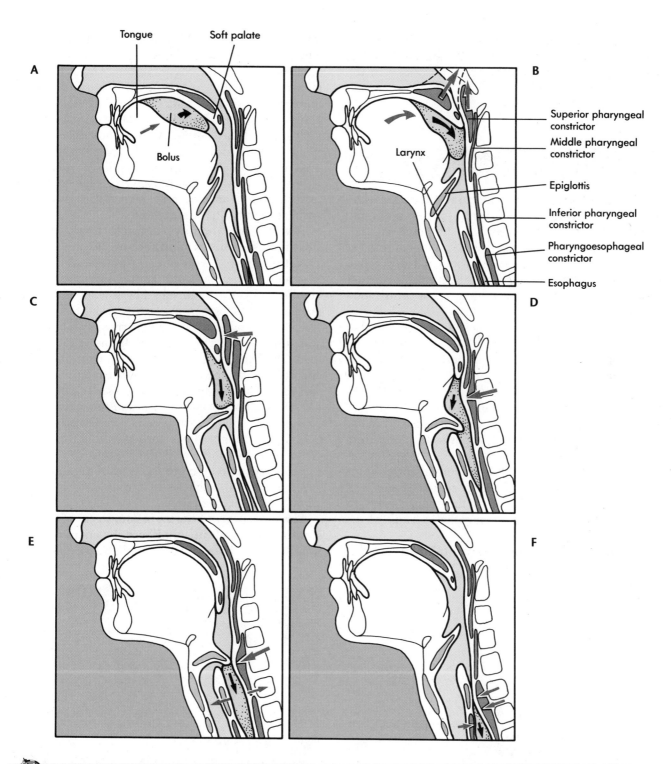

Tongue   Soft palate

A

B
Superior pharyngeal
constrictor

Middle pharyngeal
constrictor

Epiglottis

Inferior pharyngeal
constrictor

Pharyngoesophageal
constrictor

Esophagus

Bolus

Larynx

C

D

E

F

Figure 23.10

The three phases of swallowing (deglutition) are: buccal phase (**A**), pharyngeal phase (**B-E**), and esophageal phase
(**F**). Follow the description in the text.

In the **esophageal phase** (ee-so-FAGE-ee-al) the esophagus propels the bolus to the stomach by alternately dilating and constricting, as described in the next section and Figure 23.12.

- Name the three phases of swallowing.
- Cite an event that occurs in each phase.

ESOPHAGUS

The esophagus is the muscular tube that conducts food from the pharynx to the stomach. Fluids drain through the esophagus directly, as swallowing a cold drink tells you, but it takes about 2 seconds for solids to be squeezed through. The esophagus secretes only mucus for lubrication; no further steps in digestion or absorption occur until food reaches the stomach.

About 25 cm (10 inches) long and 2 cm (¾ inch) in diameter, the esophagus (Figure 23.11A) descends the neck from the level of the sixth cervical vertebra and the cricoid cartilage of the larynx, between the trachea and vertebral column, and then leads through the thoracic cavity, exiting through the esophageal hiatus of the diaphragm into the abdominal cavity and stomach. Arteries supply the esophagus with blood from the thyrocervical trunk, thoracic aorta, bronchial arteries, and celiac trunk of the abdominal aorta. Veins drain to the inferior thyroid, azygos, hemiazygos, and gastric veins. Innervation is by the vagus nerve (X) (parasympathetic) and the sympathetic trunk of the autonomic nervous system.

All four layers of the digestive tube are present for the first time in the esophagus (Figure 23.11B), which makes this a good basic organ to compare with other portions of the digestive tract. Anatomic landmarks begin in the mucosa with its **stratified squamous epithelium** for transporting food. **Submucosal glands** lubricate the epithelium with mucus, and cell division replaces epithelial cells that food scrapes away. Because no absorption takes place in the esophagus, it is not surprising that mucous glands are the only glands present. The **muscularis externa** propels food to the stomach, and **adventitia** retains the esophagus in the neck and thorax.

The mucosa displays a thick, nonkeratinized stratified squamous epithelium (Figure 23.11B). Twenty to thirty layers of cells are mounted upon a basement membrane and lamina propria. Delicate bundles of smooth muscle fibers belonging to the muscularis mucosae run lengthwise beside the lamina propria and mark the boundary between mucosa and submucosa. When the esophagus is not passing food, the mucosa collapses into longitudinal folds and accommodates to the space between the trachea and vertebral column, but food boli stretch these folds flat when they pass. Lubrication for this process comes from submucosal glands lined with cuboidal epithelium. Mucus flows through a duct from each gland onto the surface of the mucosa. Submucosal vessels distribute blood to the wall of the esophagus, and submucosal plexuses of the neurenteric system stimulate mucus secretion.

The muscularis externa is composed of two substantial layers of muscle fibers. Fibers of the **outer longitudinal layer** extend lengthwise along the

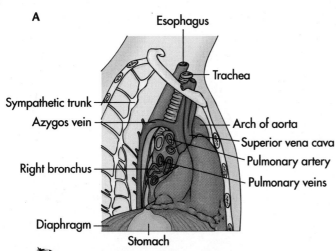

A

Esophagus

Trachea

Sympathetic trunk

Azygos vein

Arch of aorta

Superior vena cava

Pulmonary artery

Pulmonary veins

Right bronchus

Diaphragm

Stomach

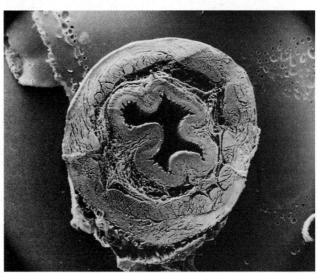

B

Figure 23.11

The esophagus connects pharynx and stomach. A, Midsagittal view of esophagus passing through the thoracic cavity. **B,** Transverse section through the esophagus shows mucosa, submucosa, muscularis externa, and adventitia. (Scanning electronmicrograph by permission of R.G. Kessel and R.H. Kardon, *Tissues and organs: A text-atlas of scanning electron microscopy*, 1979, W.H. Freeman and Co.)

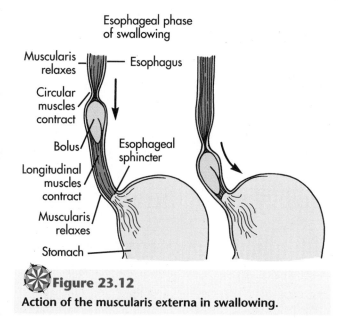

Esophageal phase
of swallowing

Muscularis
relaxes

Esophagus

Circular
muscles
contract

Bolus

Esophageal
sphincter

Longitudinal
muscles
contract

Muscularis
relaxes

Stomach

Figure 23.12
Action of the muscularis externa in swallowing.

esophagus, and the **inner circular fibers** wrap around the tube. Figure 23.12 illustrates the waves of contraction that propel food down the esophagus. Both layers dilate at the leading side of each wave and constrict behind, propelling the bolus to the stomach. Plexuses between these layers distribute more fibers of the neurenteric system to both sets of muscles and control the waves of contraction. The superior (cervical) third of the esophagus contains skeletal muscle fibers that continue into the esophagus from the pharynx. Smooth and skeletal fibers work together in the middle (thoracic) portion, and the inferior third gives way altogether to smooth muscle fibers.

Adventitia covers the cervical and thoracic portions of the esophagus, connecting it to the vertebral column posteriorly, to the vessels of the neck and thoracic cavity on either side, and to the trachea anteriorly (Figure 23.11A). Only the short abdominal portion below the diaphragm is covered by serosa.

- What is the function of the esophagus?
- Where is the esophagus located?
- Which four gut layers are present in the esophagus?

PERITONEAL CAVITY AND MESENTERIES

PERITONEAL CAVITY

The remaining organs of the digestive system take up most of the space within the abdominopelvic cavity, leaving only a small **peritoneal cavity** (PAIR-ih-tone-ee-al) crowded between them. The peritoneal cavity, illustrated in Figure 23.13, is lined by the peritoneum that covers the walls of the abdominopelvic cavity and the organs within this cavity. The peritoneum encloses a potential space

filled with only a few milliliters of serous fluid secreted by the peritoneum. The peritoneal cavity is rather like the space taken up by a balloon inflated in a jar of marbles. The balloon reaches between the marbles and it also lines the walls of the jar, coating everything it can reach with a thin sheet of rubber. Like the marbles, the abdominal organs lie inside the abdominopelvic cavity but outside the peritoneal cavity, covered with a thin peritoneum.

When the peritoneum becomes infected, a condition called **peritonitis,** the inflamed tissue secretes copious ascites fluid that contains peritoneal cells and microorganisms. Peritonitis can be life-threatening because the microorganisms can easily reach all the abdominal organs.

ABDOMINAL ORGANS

The abdominopelvic cavity is bounded superiorly by the diaphragm and inferiorly by the floor of the pelvic cavity. The vertebral column forms a prominent posterior ridge in the wall of the cavity. The peritoneum covers the organs and lines the walls; **parietal peritoneum** covers the walls, and the visceral organs are covered with **visceral peritoneum.** Visceral peritoneum is the serosa that covers the gut tube. There is no entrance or exit to the peritoneal cavity in males, but the oviducts and female reproductive tract open the cavity to potential, though infrequent, infection in women (see Chapter 26, Reproductive System).

Opening the peritoneal cavity exposes the abdominal organs for examination, as shown in Figure 23.16B. The liver occupies the dome beneath the diaphragm, and the stomach lies beneath the left lobe of the liver. The spleen tucks behind the stomach to its left, and the gall bladder peers out beneath the margin of the liver. The kidneys and duodenum lie out of sight behind the parietal peritoneum in the posterior wall of the abdomen, and the pancreas lies behind the peritoneum in the curve of the duodenum.

The greater omentum is a curtain of peritoneum that lies between the intestines and the anterior abdominal wall. It is a pouch that encloses an inner space called the omental bursa (Figure 23.13). When the greater omentum is folded upward over the stomach and liver, the ascending, transverse, and descending portions of the colon are seen framing the coils of jejunum and ileum of the small intestine (Figure 23.16B). The cecum and vermiform appendix mark the junction of the ileum with the colon. The sigmoid colon enters the pelvic cavity, passing behind the peritoneum on the way to rectum and anus. The bladder protrudes above the pubis.

MESENTERIES

Most digestive organs are tethered to the posterior wall of the abdominal cavity by **mesenteries.** Mesen-

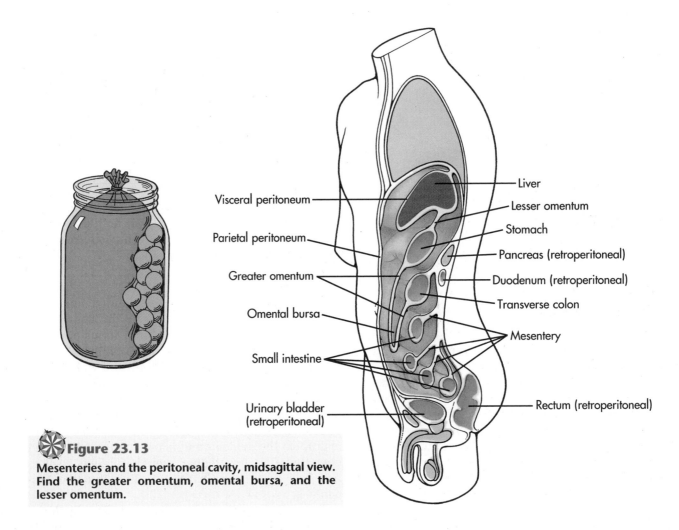

Visceral peritoneum

Parietal peritoneum

Greater omentum

Omental bursa

Small intestine

Urinary bladder
(retroperitoneal)

Liver

Lesser omentum

Stomach

Pancreas (retroperitoneal)

Duodenum (retroperitoneal)

Transverse colon

Mesentery

Rectum (retroperitoneal)

Figure 23.13

Mesenteries and the peritoneal cavity, midsagittal view. Find the greater omentum, omental bursa, and the lesser omentum.

teries (MEZ-en-TAIR-eez) are folds of visceral peritoneum, visible in Figure 23.13, that hang like curtains from the abdominal wall and envelop the digestive tube and its vessels and nerves. A short mesentery called the triangular ligament suspends the liver from the diaphragm, and the lesser omentum in turn suspends the stomach from the liver. The greater omentum is a fold of mesentery that continues from the stomach in front of the colon and small intestine. The **transverse mesocolon** suspends the transverse colon and thereby partitions the liver and stomach from the organs below. The pancreas, duodenum, urinary bladder, and rectum are **retroperitoneal** organs because they are embedded in the body wall behind the peritoneum. The jejunum and ileum are suspended by a fan-shaped mesentery that carries branches of the superior and inferior mesenteric arteries. How the small intestine becomes surrounded by large intestine is shown in Figure 23.14. The gut begins as a relatively straight tube suspended by a dorsal mesentery, but the small intestine loops inside the bend of the colon and the mesentery follows the intestine.

The greater omentum (oh-MEN-tum) is harder to understand, and you should read the development

section in this Chapter to see how it forms. An **omentum** is a mesentery that joins organs to other organs rather than to the abdominal wall. Figure 23.13 shows that the **greater omentum** connects the stomach with the transverse colon. The greater omentum is a pouch that encloses a portion of the peritoneal cavity called the **omental bursa.** The bursa reaches deep into the greater omentum and extends superiorly behind the liver and stomach. The omental bursa opens to the greater part of the peritoneal cavity through the **epiploic foramen** (ep-ee-PLOE-ik), an entrance (not shown in the figure) beneath the **lesser omentum** that connects the lesser curvature of the stomach to the liver.

The greater omentum contains numerous fat deposits considered to protect the abdominal organs, but the greater omentum is protective in another sense as well. Its abundant lymph nodes and nodules increase the surface available for immunological surveillance. More significantly, the greater omentum helps seal off infection from the peritoneal cavity by adhering to sites of inflammation, such as to the appendix when appendicitis occurs.

• What is the difference between the abdominal cavity and peritoneal cavity?

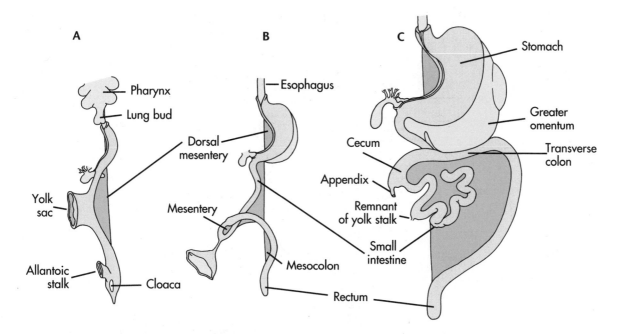

Figure 23.14

The small intestine begins (**A**) as a straight tube that (**B**) loops inside the large intestine and then (**C**) elongates extensively.

- Which cavity contains the digestive organs?
- What is the difference between a mesentery and an omentum?

STOMACH

The stomach is a flattened, J-shaped pouch that receives food from the esophagus at the **cardia** and releases it to the duodenum at the pyloric end (Figure 23.15). The **body** of the stomach expands the narrow digestive tube to about 7 to 10 cm. The **fundus** bulges above the cardia at the superior end, and the **pyloric portion** (pie-LOR-ik) tapers gradually to the **pyloric sphincter** that guards the outlet at the **pylorus** to the duodenum. The lesser curvature of the stomach is the inside, posterior border of the "J"; the greater curvature, the outside anterior border, suspends the greater omentum. A prominent notch on the lesser curvature, called the **incisure** (in-SIZE-your), marks the slightly enlarged **antrum.** The body of the stomach lies almost vertically between the left lobe of the liver and the spleen, and the pyloric portion curves to the right past the midline of the abdomen. The stomach retains all four of the gut's tissue layers.

The size and position of the stomach can change considerably. Empty of food, the stomach collapses, but it can expand readily to hold 2 or more liters. With only the lesser curvature and the cardiac and pyloric ends firmly attached to the lesser omentum and

the abdominal wall, your stomach is free to enlarge with food and to be displaced up and down by other organs and gravity when you sit or stand. In addition to these passive movements, periodic waves of contraction constrict the stomach into almost separate parts and thoroughly mix the stomach contents into liquid **chyme** (KIME).

The stomach receives blood from branches of the celiac trunk, and its veins are tributaries of the hepatic portal vein. Lymph drains from four separate fields in the stomach wall into lymph nodes associated with the abdominal aorta and thence to the thoracic duct. Sympathetic innervation arrives from the sympathetic trunk by way of the celiac ganglion, and the vagus nerve delivers parasympathetic fibers.

- Name the four regions of the stomach.
- Where is the stomach located in the abdominal cavity?

TISSUE LAYERS

The mucosa and muscularis externa are dramatically modified in the stomach. The stratified squamous epithelium of the esophagus abruptly becomes **simple columnar epithelium** below the diaphragm, and the mucosa of the stomach now begins to secrete strongly acidic **gastric juice,** as much as a liter a day. Millions of **gastric pits** and **gastric glands** in the mucosa increase the surface area of the epithelium, releasing hydrochloric acid and hydrolytic enzymes, and absorbing nutrients. Cell division and

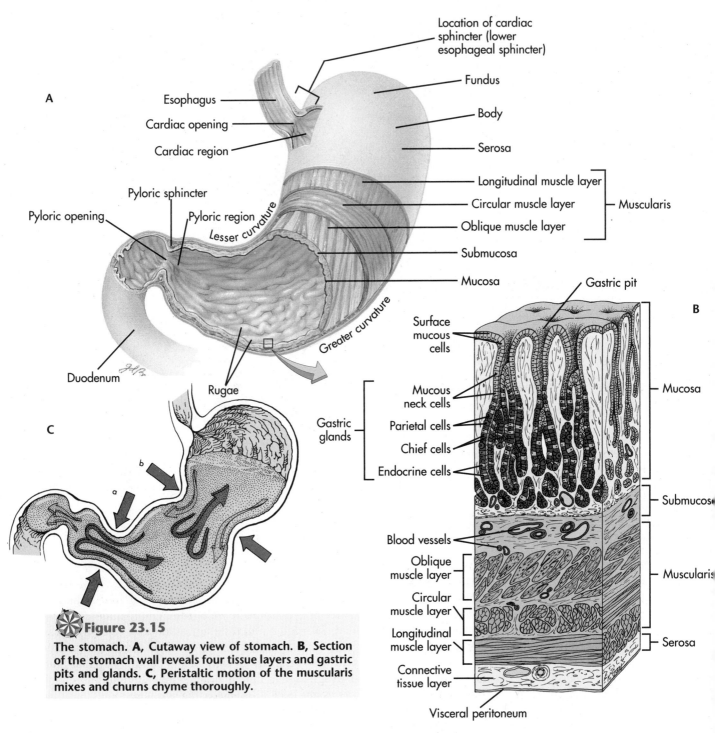

✵**Figure 23.15**

The stomach. A, Cutaway view of stomach. **B,** Section of the stomach wall reveals four tissue layers and gastric pits and glands. **C,** Peristaltic motion of the muscularis mixes and churns chyme thoroughly.

protective mucus minimize autodigestion, and lymph nodules in the lamina propria intercept antigens. (This general pattern remains throughout the lower portions of the gut, until stratified squamous epithelium recurs in the anal canal.) The muscularis acquires three layers that churn and mix the chyme, and it forms the pyloric sphincter. Numerous internal folds called **rugae** (ROO-gee) allow the stomach to enlarge and constrict as it fills and empties. Unlike gastric pits, rugae are not permanent; they take up the slack of an empty stomach and stretch flat when the stomach fills.

SEROSA

Serosa covers the exterior of the stomach; that is, the visceral peritoneum covers the anterior and posterior surfaces and is continuous with the greater omentum inferiorly and lesser omentum superiorly.

MUSCULARIS

The muscularis externa has added a third, oblique layer of muscle fibers deep to the familiar circular and longitudinal layers. Figure 23.15A cuts away these layers one at a time to show in which direction the fibers run. Outer fibers follow the curvature of the stomach longitudinally. Circular fibers appear next and oblique fibers are innermost. Oblique fibers are especially prominent at the cardiac end, and circular fibers concentrate in the pyloric sphincter. Neurenteric nervous system and myenteric plexuses are scattered among all three layers; they initiate waves of contraction, occurring about 12 seconds apart, that churn the contents of the stomach from the cardiac to the pyloric end.

SUBMUCOSA

Submucosa delivers vessels, nerves, and lymphatics to the stomach wall and extends into the rugae. It is loose connective tissue.

MUCOSA

The mucosa is the most highly modified tissue of the stomach. Three and a half million gastric pits and 15 million glands punctuate the mucosa, adding surface that secretes gastric juice into the lumen. The stomach mucosa also can absorb certain molecules such as ethanol and aspirin, which is why you can feel the effects of these drugs within several minutes of swallowing them.

The anatomy of gastric pits and glands (Figure 12.15B) illustrates how the mucosa increases the gastric surface area. The epithelium folds its luminal surface into the mucosa as gastric pits (tubular ducts) that lead to gastric glands deep in the mucosa. Gastric glands branch so that two or more may secrete into a single gastric pit. A narrow **neck,** or **isthmus,** connects each gland to a pit. A basement membrane surrounds every pit and gland, and lamina propria fills the narrow space crowded between the glands. Muscularis mucosae marks the boundary with the submucosa.

Pits and glands are lined by five types of columnar cells that contribute to gastric juice.

1. Deep in the glands, granular **chief** or **peptic cells** secrete **pepsinogen,** a precursor of pepsin, the hydrolytic enzyme that cleaves proteins into smaller peptide fragments.
2. **Parietal cells (oxyntic cells)** line the walls of the glands. These cells have clear cytoplasm and internal spaces (canaliculi) that secrete **hydrochloric acid** (0.1 normal HCl), which acidifies the chyme (pH 2 to pH 3) for digestion. Parietal cells also secrete **intrinsic factor,** an important glycoprotein that carries vitamin B_{12} across the intestinal epithelium when this vitamin is absorbed. Inability to secrete intrinsic factor means that the intestinal epithelium cannot absorb B_{12}, no matter how much B_{12} may be present. The consequences of this deficiency are serious; the lack of vitamin B_{12} leads to pernicious anemia because erythroid stem cells (and other dividing cells) require B_{12} to synthesize DNA.

Numerous mucous cells secrete a film of mucus that isolates the epithelium from these harsh conditions in the lumen.

3. **Neck mucous cells** line the upper reaches of the glands and secrete acidic mucus.
4. **Surface mucous cells** line the pits and luminal surface of the epithelium with films of neutral mucus. Mucous cells are called goblet cells because of their shape. A broad base and narrow stem lead to the bowl of the goblet that produces and stores mucin vesicles before secreting them from the lip of the goblet.
5. Back at the base of the glands, **enteroendocrine cells** secrete **gastrin** and several other peptide hormones that adjust the output of the glands to the amount of food in the stomach and duodenum. Most enteroendocrine cells are pyramidal in shape and contain distinctive secretion vesicles in the cytoplasm. These secretions go directly into the blood in endocrine fashion, not to the lumen, so that other regions of the stomach and intestine also respond to their presence. Secreted in the pylorus, gastrin prompts the stomach to discharge chyme into the duodenum and the fundus to secrete gastric juice (see Other Endocrine Tissues, Chapter 22).

Gastric glands differ among the regions of the stomach. Glands of the fundus and body are like those just described, but those of the cardia and pylorus secrete more mucus. Fewer parietal cells are present in the latter regions, and the glands branch less and coil more.

• What features of the gastric mucosa differ from the mucosa of the esophagus?

AUTODIGESTION

Secreting mucus is not the only tactic that minimizes autodigestion. Pepsin does not attack proteins until it becomes active outside the cells that secrete it. Pepsin is the active enzyme created when HCl cleaves off the terminal fragment from the inactive pepsinogen molecule. The enzyme thus is activated in the lumen of the stomach, away from the chief cells that secreted it.

Cell division is another, equally important, protection because it replaces the epithelium through epithelial cell proliferation in the necks of the glands. Maturing cells move up the walls of the pits and into the glands, while old cells are lost from the luminal surface and from the walls of the glands.

• Name three mechanisms that protect the stomach lining from autodigestion.

SMALL INTESTINE

The small intestine completes the digestive process by introducing more digestive enzymes and absorbing nutrients through a surface area increased by new folds and projections not seen in the stomach. All four tissue layers are present. The mucosa and submucosa secrete digestive enzymes from glands that resemble gastric glands; the pancreas and gall bladder deliver hydrolytic enzymes and bile through ducts into the lumen. The small intestine is 2 to 7 m (7 to 23 ft) long, depending on muscle tone of the muscularis externa, and its mucosa forms various circular folds and villi that vastly increase the absorptive surface. The small intestine continues to resolve problems of autodigestion by secreting mucus and renewing its epithelium.

DUODENUM

The small intestine is composed of three divisions (Figure 23.16)—the **duodenum, jejunum,** and **ileum.** The first two complete the digestive process, and the last two absorb its products. The duodenum (do-oh-DEE-num) receives chyme from the pylorus,

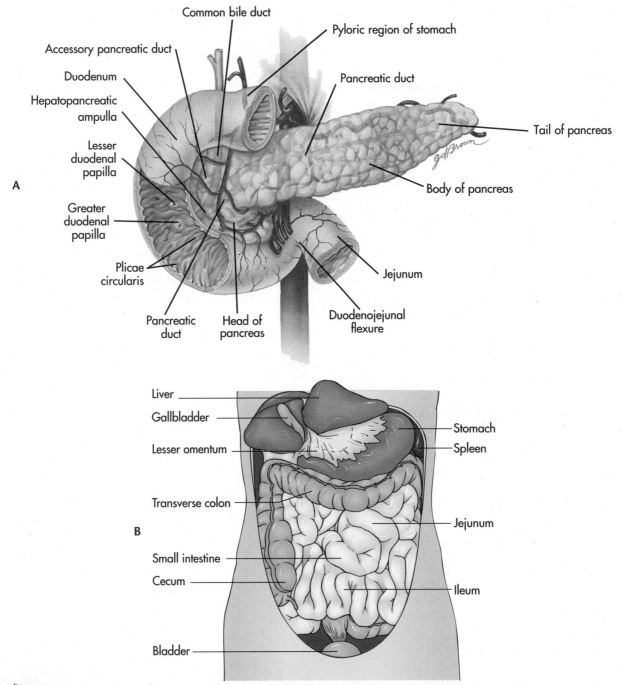

Figure 23.16

The small intestine. A, Duodenum with pancreas. B, Jejunum and ileum.

and the ileum (ILL-ee-um) discharges it through the ileocaecal valve into the ascending colon. Most of the duodenum is retroperitoneal, curving like the letter "C" for about 12 inches behind the parietal peritoneum of the posterior abdominal wall. Only the first part reaches into the abdominal cavity, where it receives the stomach. The duodenum crosses the vertebral column after completing the "C" and bends cranially to emerge once again into the abdominal cavity and become the jejunum (je-JUNE-um).

The duodenum receives numerous substances, including chyme from the pylorus that it delivers to the jejunum, hydrolytic enzymes and bicarbonate ions from the pancreas, and bile from the gall bladder by way of an opening called the **duodenal papilla.** The duodenum secretes its own intestinal enzymes and alkaline mucus from characteristic **submucosal glands,** also known as **Brunner's glands.** The alkaline mucus neutralizes the acidic chyme from the stomach, raising its pH to 8, the optimum for intestinal enzymes to act. Numerous circular folds (plicae circulares) and intestinal villi and mucosal glands augment the surface of the duodenal epithelium.

JEJUNUM AND ILEUM

The jejunum and ileum coil from their mesenteries below the transverse mesocolon within the folds of the ascending and descending colon (Figure 23.16B). The jejunum emerges from the duodenum at a sharp caudal bend below the transverse mesocolon and continues for about 12 feet before becoming the ileum. No external landmark shows where the jejunum ends and the ileum begins, but the mucosa makes clear that the superior end of the jejunum differs from the inferior end of the ileum (see below). The jejunum and ileum retain mucosal glands, but have no submucosal glands. The ileum extends for about 8 feet and then discharges into the caecum through the **ileocecal valve** (ILL-ee-oh-SEE-kal).

- Name the three regions of the small intestine.
- What are the functions of each region?
- Where is each located in relation to the colon?

MUCOSA

The mucosa of the small intestine exposes at least 186 m² (2000 ft²) of surface in an adult weighing 50 kg (110 lbs) or more. Because the dimensions of the small intestine (3 cm × 5 m) never remotely approach so large an area, the mucosa employs three anatomical structures—**plicae circulares, intestinal villi,** and **microvilli**—to make the difference. As shown in Figure 23.17A, hundreds of plicae circulares (PLEE-kee), or circular folds, in the mucosa and submucosa increase the internal surface of all three parts of the small intestine. The figure shows that folds are closely spaced and about 0.5 cm tall in the

duodenum and jejunum, but they are sparser and shorter in the ileum. Additional surface area comes from millions of intestinal villi that project from the surfaces of the plicae, as shown in the figure. Finally, billions of microvilli cover the luminal surfaces of the epithelial cells on the villi and plicae (Figure 23.17C). All of these structures pack more than 2000 ft² of surface into about 20 feet of tubing.

PLICAE CIRCULARES

Figure 23.17 shows a transverse section through a circular fold in the jejunum. A core of submucosa extends from the wall of the digestive tube into the plica and delivers vessels, lymphatics, and nerve fibers to the mucosa that covers the fold. All molecules absorbed or secreted will pass through these vessels or lymphatics. Intestinal villi sprout from the mucosa on all surfaces of the plica.

INTESTINAL VILLI

If submucosa is the core of plicae circulares, then lamina propria is the core of each villus. This loose connective tissue supports a **simple columnar epithelium** upon a basement membrane, all shown in Figure 23.18. A few **goblet cells** and numerous **enterocytes** make up the epithelium of intestinal villi. Goblet cells secrete a protective mucous film over the surface of the villus, and enterocytes absorb nutrients and secrete intestinal enzymes needed for digestion. Enterocytes are typical polarized epithelial cells (see Epithelia, Chapter 3). **Microvilli** extend from a terminal web of microfilaments at the luminal surface of each cell, to form the brush border of microvilli covering the villus. Bands of **tight junctions** join the cells below the luminal surface, and deep basal infoldings of the plasma membrane enhance the transport of nutrients into the lamina propria. Active transport and endocytosis take up nutrients that are transported across the basal surface and basement membrane into the capillaries of the lamina propria.

The lamina propria core of each villus contains capillaries, lymphatics, nerve endings, smooth muscle fibers, fibroblasts, and macrophages. Capillaries, fed by arterioles from the superior mesenteric artery capture nutrients, such as glucose, amino acids, and electrolytes, absorbed by the epithelium and transport them through the mesenteries to the liver by way of the hepatic portal vein. Lymphatic vessels form **lacteals** (LAK-tee-als) that drain the villus of lipids (see Chapter 16, Lymphatic and Immune Systems). These molecules enter the epithelium by endocytosis and become encapsulated in lipoprotein carrier molecules as **chylomicrons** (KYE-low-MY-krons). Lacteals take up chylomicrons and circulate them through the thoracic duct to the blood. Smooth muscle fibers shorten and relax the villi, rinsing fresh chyme over the surface of the villi.

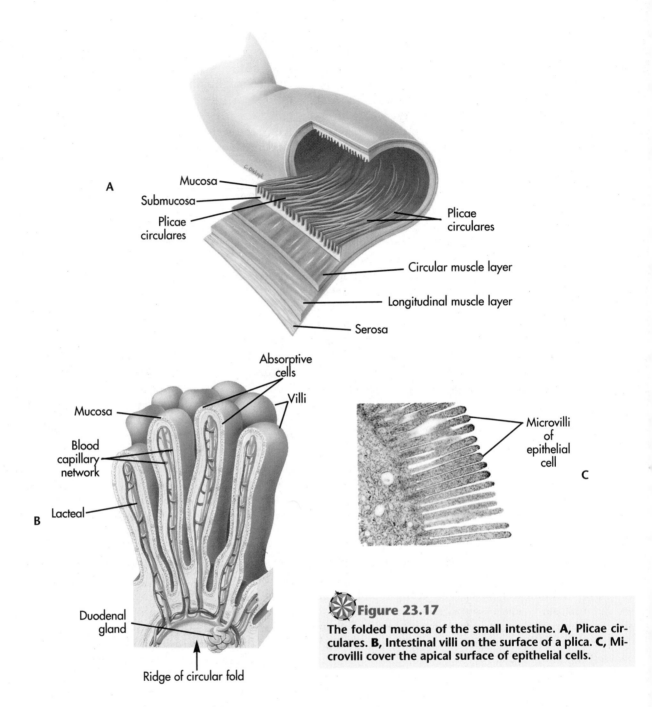

A

Mucosa
Submucosa
Plicae circulares
Plicae circulares
Circular muscle layer
Longitudinal muscle layer
Serosa

Absorptive cells
Villi
Mucosa
Blood capillary network
B
Lacteal
Microvilli of epithelial cell
C
Duodenal gland
Ridge of circular fold

✳ **Figure 23.17**

The folded mucosa of the small intestine. **A,** Plicae circulares. **B,** Intestinal villi on the surface of a plica. **C,** Microvilli cover the apical surface of epithelial cells.

INTESTINAL GLANDS

Intestinal glands, or **crypts of Lieberkuhn,** indent the mucosa between villi, as illustrated in Figure 23.18, and provide additional surface that secretes intestinal enzymes into the lumen. **Paneth cells** populate the base of the crypts, but their function is unclear. Paneth cells may control the bacterial flora of the intestine because lysozyme, an enzyme that attacks bacterial cell walls, is one of their secretory products.

Intestinal crypts also are important sources of new enterocytes and goblet cells that renew the epithelium. Intense cell division and growth in the walls of intestinal crypts supplies cells that move upward over the surface of the villi to the apex, where they are lost into the lumen and digested. Cell division and movement replace the epithelium approximately every 7 to 8 days.

• What features of the mucosa increase its surface area?

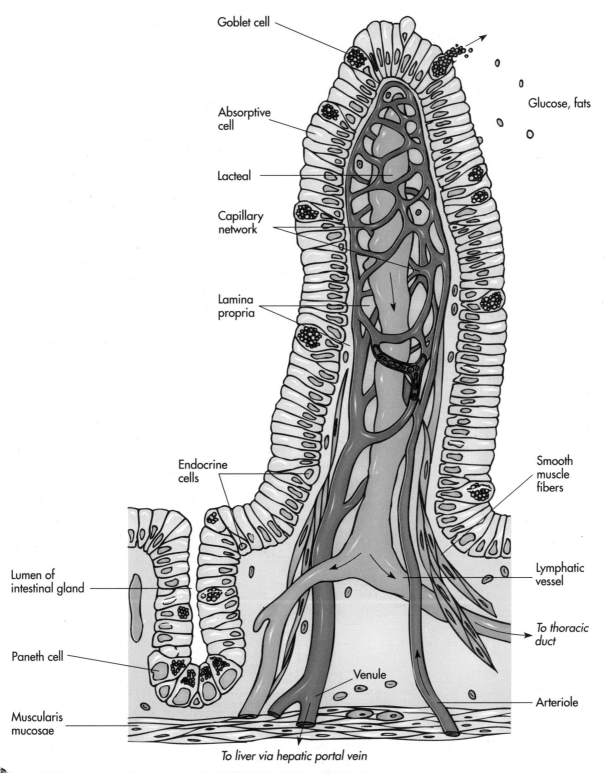

Goblet cell

Absorptive
cell

Lacteal

Capillary
network

Lamina
propria

Endocrine
cells

Lumen of
intestinal gland

Paneth cell

Muscularis
mucosae

Glucose, fats

Smooth
muscle
fibers

Lymphatic
vessel

To thoracic
duct

Venule

Arteriole

To liver via hepatic portal vein

Figure 23.18

Anatomy of an intestinal villus. Trace the structures that glucose and fats must negotiate in order to reach the liver (glucose) and the thoracic duct (fats). See text.

SUBMUCOSA

The submucosa supplies the usual vessels, lymphatics, and plexuses to the wall of the small intestine, but two other features make intestinal submucosa notable. The submucosa of the duodenum includes submucosal glands (**Brunner's glands)** that secrete **alkaline mucus** that neutralizes chyme from the stomach. Located in the submucosa, these glands are highly coiled compound tubular glands with ducts that discharge into the intestinal crypts. As you might expect, most submucosal glands are found where chyme enters the small intestine, in the superior portion of the duodenum, just below the pylorus and duodenal ampulla.

The second special feature is **Peyer's patches,** which are clusters of unencapsulated lymph nodules found in the mucosa and submucosa of the digestive tract, especially in the small intestine. Seen from the exterior, Peyer's patches are whitish aggregates of lymphoid tissue whose macrophages and lymphocytes patrol the mucosa for foreign cells and substances. The largest, 3 to 5 cm long, are found in the ileum.

MUSCULARIS EXTERNA AND SEROSA

The muscularis externa of the small intestine reverts to its fundamental inner circular and outer lon-gitudinal layers. The serosa is described in Peritoneal Cavity and Mesenteries, this chapter.

LIVER

METABOLIC MIDDLEMAN

The dark-red liver is the body's largest single gland (1 to 1.5 kg or 2.5 to 3.6 lbs). It is a "metabolic middleman" because it takes up and secretes more than 500 different kinds of molecules that are intimately involved in cellular function (Figure 23.19). The liver is also both an **endocrine** and **exocrine** gland. The liver secretes serum proteins, including fibrinogen (clotting) and albumins, directly into blood as part of its endocrine actions. Although these proteins are not hormones, the process of secretion is nevertheless typically endocrine, moving the proteins from hepatocytes, the liver cells, to capillaries. Hepatocytes also detoxify many drugs and convert nitrogen wastes to urea. Moreover the liver stores and releases glucose, keeping blood glucose levels relatively constant, and it synthesizes cholesterol and the lipoproteins that transport cholesterol in the blood, all fundamentally endocrine secretions. Its principal exocrine role is to secrete bile through bile ducts into the duodenum in typical exocrine fashion.

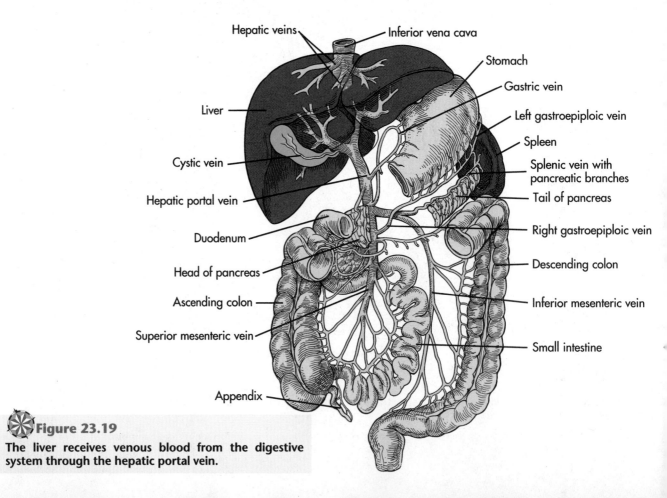

Hepatic veins — Inferior vena cava
Liver — Stomach
— Gastric vein
— Left gastroepiploic vein
Cystic vein — Spleen
Hepatic portal vein — Splenic vein with pancreatic branches
— Tail of pancreas
Duodenum — Right gastroepiploic vein
Head of pancreas — Descending colon
Ascending colon — Inferior mesenteric vein
Superior mesenteric vein — Small intestine
Appendix

✸ **Figure 23.19**

The liver receives venous blood from the digestive system through the hepatic portal vein.

The location of the liver reflects its middleman's role. The gland lies between the diaphragm above and the stomach and intestines below, as shown in Figure 23.19. In vascular terms the liver is immediately downstream from the digestive tract, where glucose and many other molecules enter the liver through the **hepatic portal vein,** and its products go directly through the **inferior vena cava** to the heart and lungs and then into the systemic circulation.

The liver takes its dome-like shape from the diaphragm, which covers its superior, diaphragmatic surface (Figure 23.20). The stomach, duodenum, colon, and right kidney also shape the inferior (visceral) surface. The **sagittal fossa** divides the liver into two great lobes, right and left. The right lobe is the larger and displays two smaller **quadrate** and **caudate lobes** on its visceral surface, defined by the gall bladder and the inferior vena cava, respectively.

Folds of visceral peritoneum enclose the liver and its vessels. The hepatic portal vein, hepatic artery, and common bile duct all enter at the **porta hepatis** (heh-PAT-iss) on the posterior visceral surface. The hepatic veins drain into the inferior vena cava on the posterior diaphragmatic surface. Visceral peritoneum binds the liver to the diaphragm and to the posterior wall of the abdomen, but there is an extensive **bare area** on the diaphragmatic surface where the peritoneum does not reach. Connective tissue attaches this area directly to the diaphragm. The sickle-shaped **falciform ligament** (FAL-si-form; falx, *L,* sickle) connects the ventral surface to the abdominal wall, and the **round ligament** that runs along the margin of the falciform ligament contains the fibrous remains of the umbilical veins that connected the placenta to the liver during fetal life.

- Where is the liver located in the abdominal cavity?
- Name the four lobes of the liver and the vessels that deliver to and remove blood from it.

HEPATIC LOBULES AND HEPATOCYTES

Hepatic lobules are structural units that carry out the liver's activities. Approximately 1 million of these miniature stubby cylinders, each about 2 mm long and 0.7 mm in diameter, are found throughout the

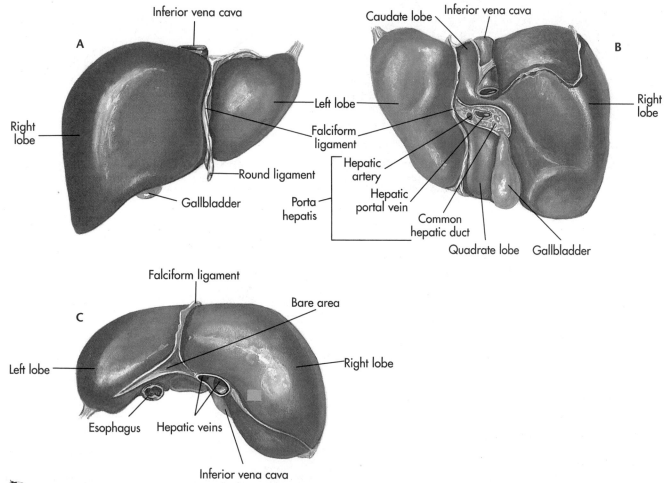

Figure 23.20

External anatomy of the liver. A, Anterior surface. B, Visceral surface. C, Diaphragmatic surface.

liver (Figure 23.21), crowded so closely together that they press each other into hexagonal shapes. Blood circulates over curtains of **hepatocytes** that radiate from the center of each lobule, taking up and releasing products as blood passes. Some lobules are covered with loose connective tissue capsules, but most merge together without these partitions. **Portal triads** communicate with lobules at the corners where several lobules meet. As the name indicates, triads contain three vessels. The large central vessel is a branch of the **hepatic portal vein,** flanked by a branch of the **hepatic artery** and an **interlobular branch** of the bile duct. Each triad delivers to capillaries called **sinusoids** (SIGN-you-soydz) located in the adjacent lobules. Portal and arterial blood mixes in the sinusoids and flows past hepatocytes, draining through a **central vein** from each lobule that leads ultimately to the hepatic veins. Bile from the lobules drains into the interlobular branches of the bile duct by way of **bile**

canaliculi (KAN-al-IK-you-lee).

Figure 23.22 shows how hepatic lobules act as endocrine and exocrine glands. In endocrine secretion, hepatocytes take up and secrete molecules into the sinusoids. A highly fenestrated capillary endothelium admits plasma but not blood cells into the **peri-sinusoidal space,** or **space of Disse** (DISS), where it bathes every hepatocyte. Hepatocytes take up glucose and other molecules that cross the endothelium into this space, and secretions are released into the perisinusoidal space and flushed away into the sinusoids. In this manner, newly synthesized fibrinogen and serum proteins enter the blood along with urea and other products of the liver.

- Name several functions of liver lobules.
- Which of these is endocrine in nature and which is exocrine?
- What vascular structures deliver blood and remove bile from liver lobules?

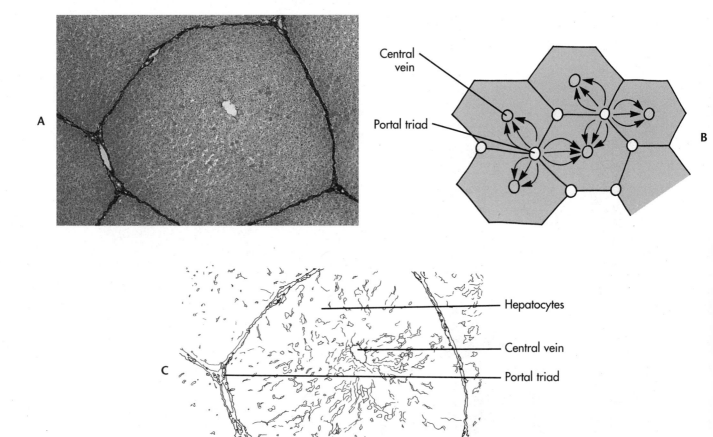

✸ Figure 23.21

Hepatic lobules. A, Transverse section through several tightly packed hepatic lobules. **B,** Portal triads deliver blood to portions of neighboring lobules, and this blood drains, along with blood from other portal triads, through the central vein of each lobule. Each lobule and its central vein receive blood from several portal triads, and a single portal triad serves several lobules.

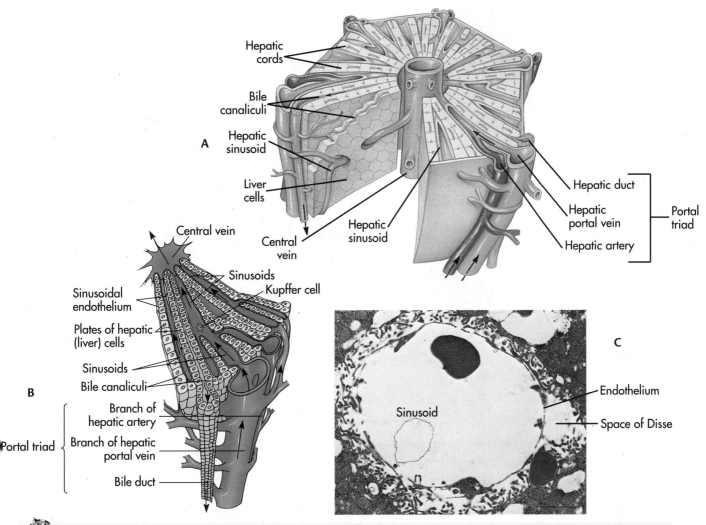

Figure 23.22

A and B, Structure of hepatic lobules and portal triads. **C,** Transverse section through a hepatic sinusoid shows that the space of Disse surrounds the endothelium of the sinusoid. (Electronmicrograph courtesy of C. Kelloes and M. Farmer, Center for Advanced Ultrastructural Research, University of Georgia.)

BILE AND BILE CANALICULI

Bile is the product of exocrine action. Bile contains bilirubin, bile salts, lipoproteins, lecithin, cholesterol, and water that together emulsify fats in the small intestine. Once hepatocytes release it, bile never comes in contact with the blood because it flows separately to the gall bladder and thence to the intestine. Production begins when macrophages in the spleen and Kupffer cells (specialized macrophages) in the sinusoidal endothelium of the liver ingest erythrocytes and break down their hemoglobin molecules into **bilirubin** that enters the blood. Hepatocytes take up this bilirubin (BILLY-rube-in) and secrete it along with other bile products into narrow **bile canaliculi** that resemble the lumens of tubular glands between the hepatocytes. These canaliculi lead to **bile ductules** and **interlobular branches** of the bile ducts in the portal triads and ultimately to the duodenum.

VASCULAR AND BILIARY TREES

Blood reaches hepatic lobules through a vascular tree, and a biliary tree drains bile to the gall bladder and duodenum, shown in Figure 23.23. Ninety percent of the blood enters through the **hepatic portal vein,** laden with nutrients from the intestines, and the rest arrives freshly oxygenated through the **hepatic artery,** not shown. These vessels follow each other as they branch into each lobe among the hepatic lobules. After it has circulated through the lobules, blood exits through central veins that follow a venous tree to the **hepatic veins.** In contrast, bile drains into the interlobular branches of the left and right **hepatic ducts** that unite to form the **common hepatic duct.** From this point, bile may pass through the **cystic duct** for temporary storage in the gall bladder, or it may leave the liver through the **common bile duct** to the duodenum (see Gall Bladder, next section).

• Why is bile secretion an exocrine function?

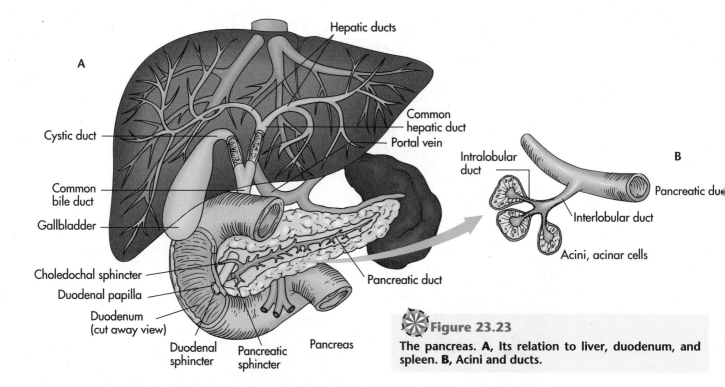

The pancreas. **A,** Its relation to liver, duodenum, and spleen. **B,** Acini and ducts.

GALL BLADDER

The **gall bladder** **stores and concentrates bile.** Because the liver continually produces bile but the duodenum uses it only intermittently, the gall bladder stores the excess until it is used. The gall bladder is the oval sac below the visceral surface of the liver, between the quadrate lobe and main body of the right lobe, shown in Figure 23.23. It accommodates up to 50 ml of bile that discharges when the bladder contracts.

Simple columnar epithelium lines the lumen of the gall bladder and absorbs enough water and ions across microvilli to concentrate the bile fivefold. As usual, lamina propria and basement membrane support this epithelium and, in this instance, allow it to fold into numerous rugae-like ridges when the bladder empties. The muscularis externa enables the gall bladder to contract and to stretch when full. There is no submucosa. Elastic fibers are interspersed among three layers of circular, oblique, and longitudinal muscle fibers. Externally a thick, collagenous serosa supplies vessels, nerves, and lymphatics, substituting for the missing submucosa.

Bile reaches the gall bladder through a tree of ducts inside (**intrahepatic**) and outside (**extrahepatic**) the liver. Pressured by continual secretion (about 0.5 ml/min), bile from hepatocytes enters the bile canaliculi and flows through the bile ductules into interlobular bile ducts of the portal triads. From the triads, tributaries feed the left and right hepatic ducts that leave the lobes of the liver at the porta he-

patis and become the common hepatic duct, the first extrahepatic duct in this sequence. The common hepatic duct runs for about 3 cm before joining the common bile duct and cystic duct. The cystic duct leads to the gall bladder, and the common bile duct empties into the duodenum at the **duodenal ampulla** (also known as the **ampulla of Vater**) along with the **pancreatic duct.**

The duodenum controls the filling and emptying of the gall bladder. Three sphincter muscles guard the common bile duct (**choledochal sphincter** [KOAL-eh-DOK-al]), pancreatic duct (**pancreatic sphincter**), and duodenal papilla. Bile enters the gall bladder when the choledochal sphincter closes and enters the duodenum when this and the sphincter for the duodenal ampulla are dilated. The duodenum adjusts flow, according to amino acids and fatty acids that enter from the stomach. These components stimulate enteroendocrine cells to release **cholecystokinin** (KOAL-eh-sis-toe-KINE-in), which causes the gall bladder to contract and the sphincters to open.

As you might expect, the gall bladder may become infected by bacteria that reflux up the cystic duct from the duodenum. Such infections frequently accompany **gallstones** that form when salts crystallize while the gall bladder removes water from the bile. Passing a stone is extremely painful if the stone lodges behind the sphincters in the cystic or common bile ducts.

- Trace the various ducts in the biliary tree from bile canaliculi to the common bile duct.

PANCREAS

Like the liver, the pancreas is both an exocrine and endocrine gland, with different cells for each function. Exocrine cells secrete batteries of hydrolytic enzymes involved in digestion into the duodenum. Endocrine cells, the pancreatic islets discussed in Chapter 22, supply insulin and glucagon that regulate glucose metabolism in the liver.

The **pancreas** is a delicate, pinkish-white gland (Figure 23.23) that lies transversely behind the peritoneum beneath the stomach, between the bend of the duodenum at the right and the spleen and kidney on the left. It is a lobular, elongated organ (12.5 to 15 cm or 5 to 6 inches long), with a broad **head** lying within the bend of the duodenum, a narrower **body,** and a tapering **tail** that ends behind the spleen. The **pancreatic duct** delivers alkaline pancreatic juice from the head of the pancreas to the duodenum by way of the duodenal papilla, which it shares with the common bile duct from the gall bladder.

Internally the pancreas consists of many **pancreatic acini** grouped into lobules that give the gland its lobular texture. **Pancreatic islets** are scattered within these lobules. A thin, delicate connective tissue capsule covers the entire gland and partitions the lobules from each other. Figure 23.23B shows that acini are densely packed within every lobule. Approximately 50 cuboidal **acinar cells** (see Acinar Cells, Chapter 2) in each acinus synthesize and release their enzymes into a small central lumen. Tight junctions seal the apical membranes of these cells into one secretory unit. Abundant mitochondria, rough endoplasmic reticulum, and zymogen granules indicate that considerable energy goes into synthesis of pancreatic enzymes. **Intralobular ducts** collect the enzymes from acini, and **interlobular ducts** drain each lobule into the pancreatic duct that extends along the entire pancreas. The cuboidal epithelium of these ducts adds alkaline mucus to the pancreatic juice. Activation mechanisms comparable to those in the stomach inhibit autodigestion by converting inactive pancreatic enzymes to active forms in the duodenum.

The pancreas receives its circulation from the splenic and superior mesenteric arteries, and corresponding veins drain to the hepatic portal vein. Even though the intestine controls pancreatic secretion hormonally, the pancreas nevertheless is innervated by the autonomic nervous system and visceral afferents. Lymphatics drain to local lymph nodes and ultimately to the thoracic duct.

- Where is the pancreas located in the abdominal cavity?
- What is the name of the cells that secrete pancreatic enzymes?
- Where does the pancreatic duct deliver these enzymes?

LARGE INTESTINE

The principal role of the large intestine is to reabsorb water from chyme and to consolidate the indigestible residue as semisolid feces for defecation (Figure 23.24). The mucosa therefore has intestinal glands and goblet cells but lacks villi and circular folds. Large pouches in the wall increase the surface area moderately.

The large intestine, or **colon,** is approximately 1.5 m in length and 7 to 8 cm in diameter. It receives from the **ileocecal valve** in the right iliac region of the abdomen and discharges through the **anus** in the floor of the pelvic cavity. External sacs or pouches called **haustrae** (HOUSE-tree), and longitudinal bands of muscularis externa, the **taeniae coli** (TANE-ee-ah KOAL-eye), distinguish the large intestine from the small intestine. On the exterior, fatty lobes, the **appendices epiploica** (ah-PEN-dih-sees EP-ee PLOE-ik-ah), also add to the difference.

The large intestine begins as the **cecum** (SEE-kum), a blind pouch below the ileocecal valve illustrated in Figure 23.24A. The narrow **vermiform appendix** (verm-, *L,* worm) extends from the cecum. Above the ileocecal valve the **ascending colon** rises along the right wall of the abdominal cavity and bends sharply **(right colic flexure)** across the abdomen at the level of the liver as the **transverse colon.** At the **left colic flexure** the **descending colon** follows the left abdominal wall, where it then takes an S-shaped course as the **sigmoid colon** to the posterior wall of the pelvic cavity. At this point it becomes the **rectum** and passes behind the peritoneum to the **anal canal** and anus.

Three strips of **taeniae coli** (*L,* bands of the colon) muscle follow all regions of the colon, one band anteriorly and two posteriorly, helping segment the large intestine into three rows of bucket-like pouches called the **haustrae** (*L,* bucket, to draw up), which are separated from each other by circular semilunar folds. Appendices epiploica depend from the taeniae coli between the haustrae. Internally, numerous **intestinal crypts** (Figure 23.24B) are lined with mucus-secreting goblet cells and absorptive cells that are responsible for gelling the feces. The large intestine contains as many as 500 different strains of bacteria that digest cellulose, synthesize vitamins, and provide micronutrients that the colon absorbs.

FOUR TISSUE LAYERS

The mucosa and muscularis externa are the most highly modified tissues of the large intestine. The muscularis diminishes to the taeniae coli bands, and circular fibers concentrate in the semilunar folds between the haustrae. Inner circular fibers become thicker in the rectum and form the internal anal sphincter. Serosa covers most of the large intestine, except for ad-

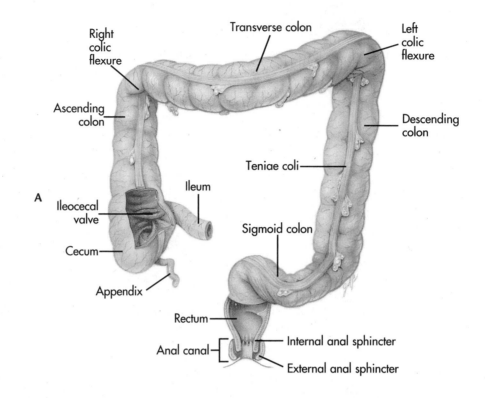

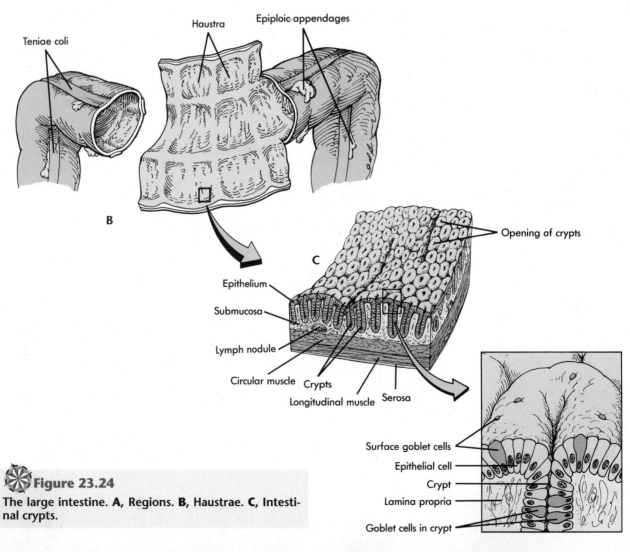

Figure 23.24

The large intestine. **A,** Regions. **B,** Haustrae. **C,** Intestinal crypts.

ventitia around the rectum and anal canal. Submucosa is little changed from previous sections of the gastrointestinal tract and contains numerous Peyer's patches.

MUCOSA

The mucosa of the cecum and colon lacks villi, but it displays numerous intestinal glands, seen in Figure 23.24 B and C. These simple tubular glands are packed closely together in hexagonal arrangements within the lamina propria, bottoming out about 0.5 mm deep on the muscularis mucosae. The two primary roles of the glands of the large intestine are reflected by the epithelial cells that line them. The most abundant cells are **columnar absorptive cells** that take up water and electrolytes from the chyme, as well as fats that the ileum has not absorbed. Microvilli on the apical surfaces improve absorption, and sodium-potassium ATPases on the basal membranes transport ions to the lamina propria. These absorptive cells develop in the base of the gland and mature as they move up the walls, so that they are functioning several days later when they reach the upper walls and luminal surface, where they slough off into the lumen. About one in every five epithelial cells is a **goblet cell** that secretes mucus into the lumen. **Enteroendocrine cells,** mostly EC cells that secrete serotonin, also populate the glands.

APPENDIX, RECTUM, AND ANAL CANAL

Tissues of these regions differ from those of the colon itself. Although the vermiform appendix is much narrower than the cecum and colon, it possesses intestinal glands backed by abundant lymphoid tissues in the lamina propria and submucosa. The glands produce little mucus, however, and fibrous connective tissue frequently replaces the mucosa and lymphoid tissue, especially after infection and inflammation. There are no taeniae coli or haustrae in the appendix; the muscularis externa spreads uniformly.

Taeniae coli also are absent in the rectum. The upper part of the rectum (Figure 23.25), above its passage through the floor of the pelvis, is lined with circular folds that resemble the plicae circulares of the small intestine. Intestinal glands also populate these structures, but the folds become **longitudinal anal columns** as the upper rectum gives way to the anal canal. Stratified squamous epithelium replaces the columnar cells about 2 cm above the anus. The lining is thereafter continuous with the skin at the anus. The muscularis externa forms an involuntary **internal anal sphincter** muscle, and the body wall provides a **voluntary external sphincter** (see Muscles of the Pelvis, Chapter 10).

- What are the primary functions of the large intestine?
- Name the various divisions of the large intestine.

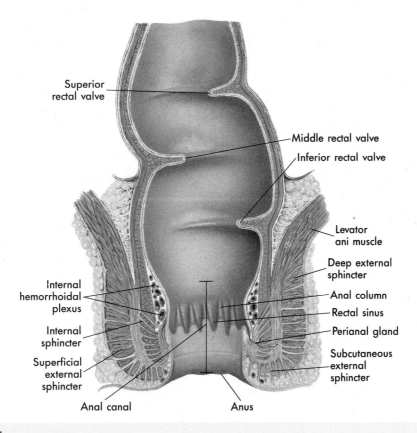

Figure 23.25
Rectum and anal canal, midsagittal view.

SURFACE TACTICS

Figure 23.26 summarizes the various folds that pack the enormous surface of the mucosa. As you can see, the small intestine has embellished its internal surface area most lavishly, and the oral cavity and anal canal are the least folded.

Columnar epithelium is the most folded tissue. Its folds either enter the mucosa (infold) or project from it (outfold). Infoldings line the entire gastrointestinal tract from stomach to large intestine, but only the small intestine has outward folds as well. Folding starts in the stomach where gastric pits and glands penetrate into the mucosa. The epithelial surface expands enormously in the small intestine when the mucosa sends plicae circulares and intestinal villi outward into the lumen, as well as intestinal glands into the mucosa. Beyond the areas where most digestion and absorption are completed, the outward projections disappear and the large intestine again shows the stomach-like, infolded intestinal glands. Submucosal glands appear in the oral cavity, pharynx, esophagus, and duodenum.

Stratified squamous epithelium has few permanent folds. The oral papillae, tonsils, and anal columns are the only significant permanently folded landmarks in the oral cavity and the anal canal.

• Which type of gut epithelium folds most profusely?

• Which regions of the gut have mucosal infoldings, projections, or both?

DEVELOPMENT

The organs of the digestive system develop from endoderm and mesoderm, two of the three germ layers described in Chapter 4, Development. The epithelium of the gut tube and the secretory epithelium of the accessory glands both derive from endoderm. The rest of the mucosa, the remaining layers of the gut tube, and the mesenteries, together with the peritoneal cavity (coelome), derive from mesoderm. Review these germ layers and chambers in Chapter 4 before continuing this section.

GUT TUBE FORMS

The gut tube begins during the third week of development as a flat sheet of endoderm on the underside of the embryo. During this week, gastrulation spreads a layer of mesoderm above the endoderm and the embryo begins to elongate. The tube begins to form when the endoderm, accompanied by mesoderm, grows rapidly forward as the **foregut** and also posteriorly as the **hindgut,** so that by the end of week 4, the gut resembles the letter *T,* as seen in Figures 23.27A and 4.5. The stem of the *T,* the **midgut,** re-

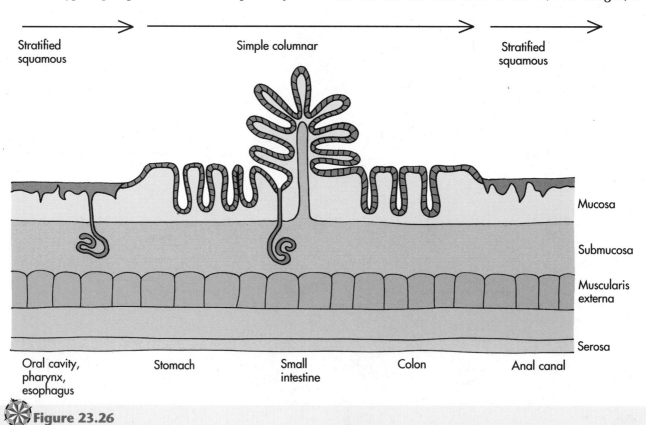

Stratified squamous   Simple columnar   Stratified squamous

Mucosa

Submucosa

Muscularis externa

Serosa

Oral cavity, pharynx, esophagus   Stomach   Small intestine   Colon   Anal canal

Figure 23.26

Modifications of the digestive epithelium. Follow description in the text.

mains open to the yolk sac beneath. The definitive gut tube will have formed when the midgut closes off from the yolk sac and the mouth and anus open. The digestive organs take form as accessory glands sprout from the wall of the tube, and the tissue layers begin to differentiate and function. The foregut forms all regions from the oral cavity and pharynx to the duodenum, including the liver, gall bladder and pancreas; the midgut derivatives extend from the duodenum to the transverse colon; and the hindgut produces the remainder of the intestinal tract, from the transverse colon to the anus.

PHARYNX

The oral cavity and pharynx begin as one large chamber at the head of the foregut, from which pharyngeal pouches grow outward between the **pharyngeal arches** in Figure 23.27B. Four of these paired arches, described in Chapter 4, develop together with the pharyngeal pouches. The first arch forms the maxilla and mandible, the second and third arches form the hyoid bone, and the fourth arch becomes the cartilages of the larynx. The **pharyngeal pouches** (Figure 23.27C) grow outward from the pharynx between the arches, toward the **pharyngeal grooves** that crease the exterior of the embryo (see Figure 4.7). Only the first pouch and groove meet; the pouch becomes the eustachian tube and the groove forms the auditory canal of the ear with the tympanic membrane between them. The remaining pharyngeal grooves never culminate in adult structures, but the pouches that might have met them give rise to the palatine tonsils, thymus, and parathyroid glands. Parathyroid C cells develop from a small fifth pair of pouches located behind the fourth pair. The thyroid gland develops from the floor of the second arch and migrates caudally beneath the larynx to meet the parathyroids, leaving only the foramen cecum in the root of the tongue to mark its origin. The tongue itself develops from the floor of the first and second pharyngeal arches.

ABDOMINAL ORGANS

The gastrointestinal tract begins at the embryonic esophagus, where the pharynx narrows and the **lung buds** mark the entrance to the larynx and trachea. A **dorsal mesentery** suspends the entire tube from this point caudally to the hindgut (Figure 23.27D), and its folds and loops will follow the organs that now begin

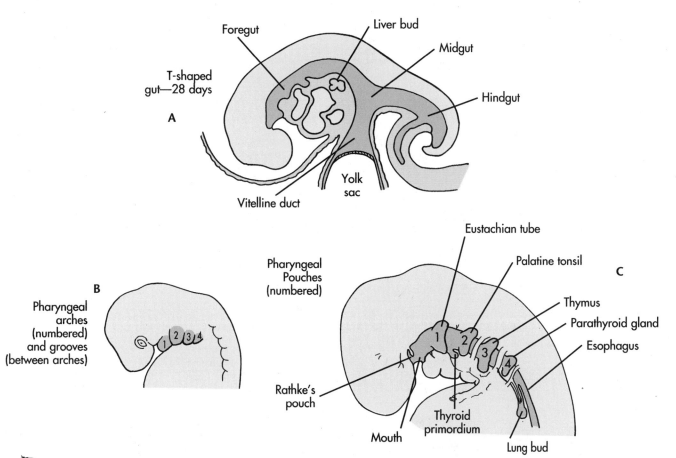

⚘ **Figure 23.27**

Development of the digestive system. Follow discussion in the text.   *Continued.*

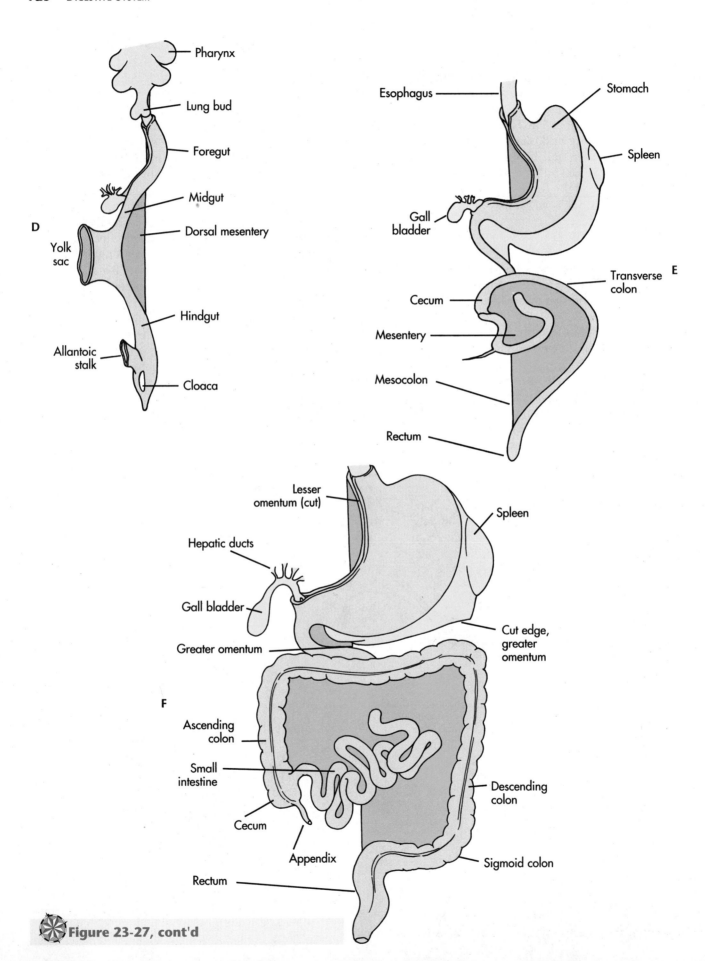

Figure 23-27, cont'd

to appear. The stomach and liver develop first, followed by the gall bladder and pancreas. During the fourth week, the stomach begins to dilate and rotate its dorsal margin to the left (Figure 23.27E) so that this margin forms the greater curvature and comes to rest more anteriorly. The portion of the dorsal mesentery that accompanies this motion becomes the greater omentum, and the space that was originally on the right side of this mesentery is now enclosed within the omental bursa. The original **ventral mesentery** of the stomach becomes the lesser omentum (Figure 23.27F).

While these changes are underway, the **liver** and **gall bladder diverticula** (DIE-ver-TIK-you-lah) grow outward from the gut tube caudal to the stomach, leaving the bile duct and cystic ducts as connections with the wall of the duodenum. At the same time, two more outgrowths develop from the duodenum. These are the **dorsal** and **ventral pancreatic diverticula** that fuse

with each other and grow and branch into the surrounding mesoderm as the pancreas. The pancreatic duct forms from the ventral diverticulum. The dorsal diverticulum occasionally persists as an accessory pancreatic duct.

The jejunum, ileum, cecum, ascending colon, and part of the transverse colon develop from the midgut. The anterior portion of the tube (jejunum and ileum) loops around the posterior (colon), and the small intestine grows rapidly, filling the abdominal cavity within the curve of the colon (Figure 23.27E, F). The hindgut ends in a temporary chamber called the **cloaca,** which divides into the anal canal and urogenital organs (see Chapters 25 and 26).

- Name a derivative from the foregut, midgut, and hindgut.
- What embryonic structures correspond to the foramen cecum, the auditory canal, pancreatic duct, and descending colon?

SUMMARY

The digestive system is a highly specialized tube that provides the body with nutrition. Food enters the mouth (ingestion), is chewed and liquified (mastication), and swallowed (deglutition) by the pharynx and esophagus. The gastrointestinal tract chemically breaks down food particles (digestion) and absorbs nutrients from them (absorption). Defecation removes what is not digested. *(Page 693.)*

The digestive tract is made of four layers of tissue. Mucosa, the innermost layer, digests and absorbs nutrients across a simple columnar epithelium supplemented with numerous glands in the wall of the tube and outside. Nutrients circulate through vessels in the submucosa, and neurons of this same tissue control secretion and peristalsis. Smooth muscle fibers of the muscularis externa propel food through the digestive tract. Serosa covers the exterior of the digestive tube and lines the peritoneal cavity as the visceral peritoneum. *(Page 694.)*

Numerous epithelial glands secrete mucus, various juices, digestive enzymes, and bile. These glands may be as simple as goblet cells or as complex as accessory glands, including the salivaries, liver, and pancreas. Between these extremes, mucosal and submucosal glands elaborate protective mucus, as well as digestive juices that control pH and promote chemical digestion. *(Page 695.)*

The oral cavity is the entrance to the digestive system; it is the space where teeth, tongue, and saliva begin to disassemble food. Mucosa and submucosa line this chamber with stratified squamous epithelium and submucosal mucous glands. Teeth cut and crush, and the tongue mixes food with saliva from the parotid, submandibular, and sublingual glands. Papillae and taste buds on the tongue also sense texture and taste. The muscular pharynx is a primary swallowing organ, along with the tongue and esophagus. Also lined with mucosa and submucosa, the pharynx directs food from the oral cavity to the esophagus. *(Page 696.)*

The esophagus is the first portion of the digestive tube that contains all four layers of tissue. Waves of contraction in the muscularis externa conduct food down this narrow tube, through the neck and thorax, to the stomach. The mucosal lining continues as stratified squamous epithelium with submucosal mucous glands that provide lubrication and protection. Unlike most of the organs below it, the esophagus is held in position by adventitia. *(Page 706.)*

The gastrointestinal tract fills the abdominal cavity, and the peritoneal cavity surrounds the GI tract with visceral peritoneum that allows the organs to move about. Mesenteries suspend these organs from the posterior abdominal wall and supply vessels and nerves to them, and omenta connect organs to each other. The lesser omentum joins the stomach with the liver, and the greater omentum encloses the omental bursa. Other organs, such as the duodenum and rectum, are retroperitoneal. *(Page 707.)*

The stomach is a pouch that stores food until the small intestine processes it. Lined with simple columnar epithelium and gastric pits and glands that increase the surface area, the stomach also begins to digest and absorb some food products. Pepsin and hydrochloric acid cleave proteins, and mucus protects the epithelium. Submucosa delivers vessels and visceral nerve fibers. Three layers of muscularis externa mix the stomach contents, and serosa covers the exterior surface. *(Page 709.)*

The small intestine completes digestion and absorption. The pancreas, gall bladder, and intestinal glands supply bile and hydrolytic enzymes for digestion. Plicae circulares, intestinal villi, and microvilli increase the absorptive surface. The duodenum receives chyme from the pylorus (the lowest portion of the stomach), neutralizes the chyme, and dips behind the peritoneum, to emerge as the folds of jejunum and ileum that are suspended by fan-shaped mesenteries within the embrace of the large intestine. *(Page 711.)*

The liver is a metabolic and anatomical "middleman" that takes up and secretes numerous products from the digestive system into the blood. The liver also secretes bile that the gall bladder stores before releasing it into the duodenum. Hepatic lobules filled with hepatocytes act as both endocrine and exocrine glands, secreting products into vascular trees that lead to the inferior vena cava and into a biliary tree that empties bile to the duodenum. The liver is shaped like a dome by the diaphragm above it and by the stomach and intestines below. *(Page 716.)*

The pancreas is also exocrine and endocrine, but has different cells for each function. Lobules and acini of the exocrine pancreas secrete hydrolytic enzymes into the duodenum, and the islet cells produce endocrine hormones. The pancreas is located transversely within the bend of the duodenum, into which the pancreatic duct empties its enzymes. *(Page 721.)*

The large intestine reabsorbs water and solidifies the feces before defecation. The colon and cecum receive chyme from the ileocecal valve. No villi or circular folds augment the mucosa of the colon, but numerous intestinal glands secrete mucus and reabsorb water and ions into the circulation. Muscularis externa concentrates in longitudinal bands and transverse strips that define the haustrae. Simple columnar epithelium becomes stratified squamous in the anal canal, and an enlarged muscularis externa forms the internal anal sphincter. *(Page 721.)*

The internal surface of the digestive system is multiplied by various folds that extend into the mucosa or out into the lumen. From this basic tactic arise circular folds, intestinal glands of various types, and intestinal villi and microvilli, together forming a remarkably enlarged surface packed into an efficiently small space. *(Page 724.)*

The digestive tube develops from endodermal tubes. The foregut forms the oral cavity, pharynx, esophagus, stomach, liver, gall bladder, and pancreas; the small intestine and part of the colon arise from the midgut; and the hindgut forms the remainder of the colon and the rectum. Beginning with a continuous curtain of dorsal mesentery, the small intestine loops inside the large intestine. The dorsal mesentery of the stomach expands while the stomach rotates and thereby forms the greater omentum and the omental bursa. *(Page 724.)*

STUDY QUESTIONS   OBJECTIVES

1. Which organs or spaces carry out the following steps in the digestive process: ingestion, deglutition, mastication, absorption, defecation, and digestion?   **1**

2. Which structures in the digestive tract swallow food?   **1, 2**

3. In which portions of the digestive system are all four tissue layers present? What layers are missing elsewhere?   **2**

4. Which organ in the digestive tract displays intestinal glands and strips of muscularis externa, but no villi?   **2**

5. Compare the anatomy of mucosa in the stomach and duodenum.   **2, 3**

6. Trace the path—the epithelia and tissue layers crossed, and vessels traveled—by which glucose reaches liver hepatocytes from the lumen of the jejunum.   **2, 3**

7. Which sections of the digestive tract does stratified squamous epithelium line?   **2, 3**

8. In what respects are the oral and esophageal mucosa alike? How are they different?   **2, 4**

9. Find one characteristic feature of the wall of the duodenum that differs in structure from the wall of the jejunum.   **2, 6**

10. Make a list of mucosal, submucosal, and accessory glands associated with each organ or section of the digestive tract. Do any sections lack such glands? Which sections possess all three types?   **3, 5**

11. Why are lips red but apt to dry out, compared with epidermis of the skin and oral mucosa?   **4**

12. How does the lining of the jejunum increase its internal surface, compared with the large intestine?   **4, 7**

13. Which regions of the digestive tract have three layers of muscularis externa? Which have two, and which have no muscularis externa?   **6**

14. How do the pancreas and liver function as endocrine glands? As exocrine glands? What anatomical differences and similarities can you find between them?   **8**

15. Trace the flow of bile from liver hepatocytes to the duodenum.   **8**

16. How do the stomach and esophagus protect their epithelial linings?   **9**

17. Trace the major events in the development of the pharynx, stomach, small intestine, and greater omentum.   **10**

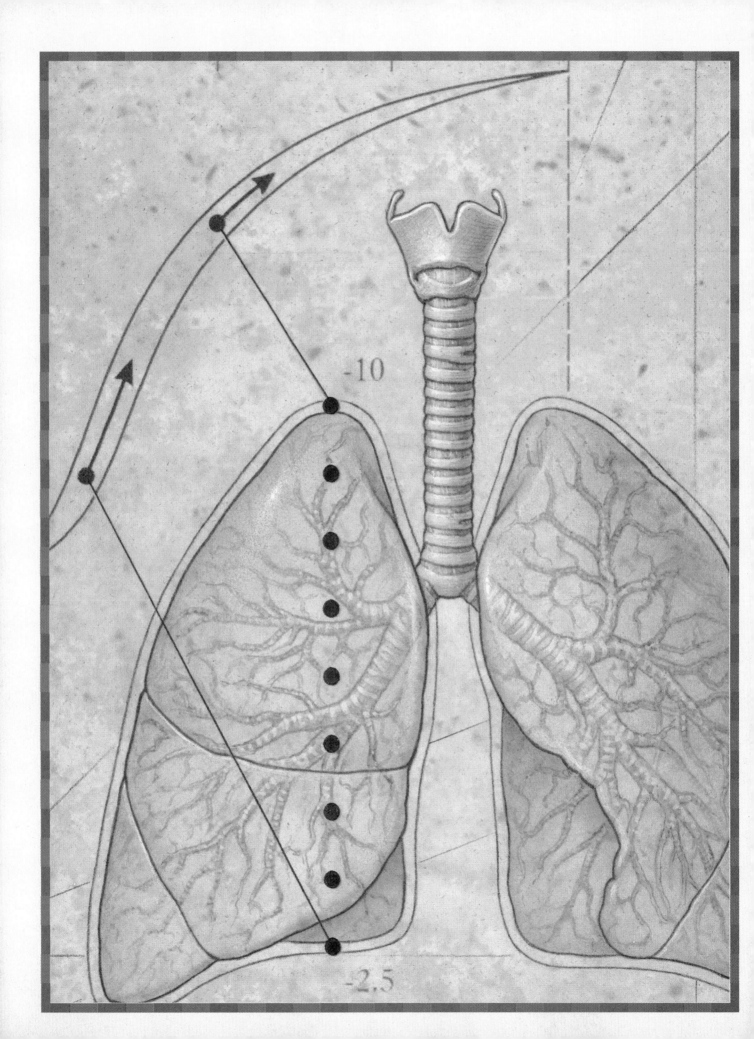

CHAPTER 24

Respiratory System

OBJECTIVES

After studying this chapter, you should be able to do the following:

1. Describe the major structures and functions of the respiratory system.
2. Describe the location of the lungs in the thoracic cavity and the structures that shape their external surfaces.
3. Name and describe the passages of the tracheobronchial tree.
4. Describe the various structural divisions of the lungs—lobes, bronchopulmonary segments, lobules, primary lobules, acini, alveoli—and the anatomical relationships between them.
5. Compare the structure of the passages in the respiratory and conducting zones in the tracheobronchial tree.
6. Describe the anatomical relationships among the thoracic cavity, pleural cavities, and the pulmonary spaces within the lungs.
7. Trace the flow of inhaled oxygen molecules through the respiratory system into the blood, and trace molecules of carbon dioxide in the opposite direction, from blood to the outside air.
8. Explain why alveoli are so small and numerous.
9. Describe the anatomical defects in cystic fibrosis, asthma, and emphysema.
10. Explain how the anatomy of the nasal cavity and its mucosa moistens, heats, and cleans air flowing into the respiratory tract.
11. State plausible functions for the paranasal sinuses.
12. Describe the chambers of the pharynx and their functions.
13. Describe the motions of the laryngeal cartilages during speech.
14. Describe the motions of thorax, diaphragm, abdominal muscles, and lungs during ventilation.
15. Describe the developmental events that form the tracheobronchial tree.

 **VOCABULARY TOOLBOX**

Terms for breathing and the lungs:

alveolus	*L* a cavity or pit.	**pleural**	pleur-, *G* the side, a rib.
bronchus, bronchial, bronchiole	bronch-, *G* the windpipe.	**pulmonary**	pulmo-, *L* the lung.
		respiratory, respiration	-spire or spir-, *L* breathe.
exhale, inhale	hal-, *L* breathe, breathing.	**trachea**	trache-, *L* the windpipe.
larynx	*G* the gullet, the larynx.	**ventilation**	vent-, *L* the wind.
lung	lungen, *Anglo-Saxon* the lung.		

RELATED TOPICS

- Bones of the paranasal sinuses, Chapter 6
- Larynx and swallowing, Chapter 23
- Muscles of larynx and of ventilation, Chapter 10
- Olfactory mucosa, Chapter 20
- Pulmonary circuit, Chapter 14
- Surface-volume relationships, Chapter 1

PERSPECTIVE

The respiratory system provides the body with oxygen for cellular respiration in the tissues and removes carbon dioxide that cellular respiration produces. The next few beats of your heart will spread blood thinly over an immense epithelial surface in the capillaries of your lungs. Less than a micron away from this flow, air is circulating through millions of small balloon-like alveoli covered by these capillaries. In a fraction of a second, as blood and air stream past each other, oxygen and carbon dioxide exchange places across this short distance. Blood rushes away charged with oxygen and the air, with carbon dioxide. In the tissues this exchange reverses; there, cellular respiration consumes oxygen from the blood and discharges carbon dioxide back into the circulation.

This chapter shows you the anatomy that delivers blood and air to the lungs, and it illustrates how disease interferes with exchange of oxygen and carbon dioxide.

BRIEF TOUR—RESPIRATORY SYSTEM

The respiratory system consists of the nasal cavity, pharynx, larynx, trachea, and lungs. When you take a breath, air enters the nose and nasal cavity (Figure 24.1A), passes through the pharynx and larynx into the **trachea,** and reaches the bronchi and bronchioles of the **lungs** that lead to 300 million alveoli, where oxygen and carbon dioxide diffuse across thin partitions between the air and blood (Figure 24.1B). Beginning with the trachea, these passages are known as the **tracheobronchial tree** (TRAKE-ee-oh-BRONK-ee-al), whose millions of branches reach thin, delicate pulmonary exchange surfaces in the alveoli. Exhaling propels the air back down the tree in the opposite direction and out through the nasal cavity or mouth. The larynx uses this outflow to make sounds that the tongue and lips mold into speech. The ribs and intercostal muscles and diaphragm ventilate the lungs, and cranial and spinal nerves mediate breathing and speech.

The nasal cavity, pharynx, and larynx constitute the **upper respiratory tract,** and the trachea and passages within the lungs make up the **lower respiratory tract.** Mucosa lines all these surfaces with a versatile epithelium that humidifies and heats air before it reaches the delicate surfaces in the lungs. This mucosa also traps foreign particles in a film of mucus and supplies numerous macrophages to suppress infections.

ANATOMY OF A COUGH

Coughing is a defense that drives out noxious materials that have entered the airways. Coughing calls all respiratory organs into action. Your own coughs probably start with an irritation in your larynx or trachea, where receptors detect particles or fluids going "down the wrong tube" into the trachea. Suddenly, you inhale deeply and air rushes into your lungs. Then the glottis of your larynx closes and your chest and abdominal muscles contract, building up pressure behind the glottis, which instantaneously releases an explosion of air that rattles the passages of the lungs and upper airways with the familiar coughing sound. Drainage from a cold may have caused you to cough,

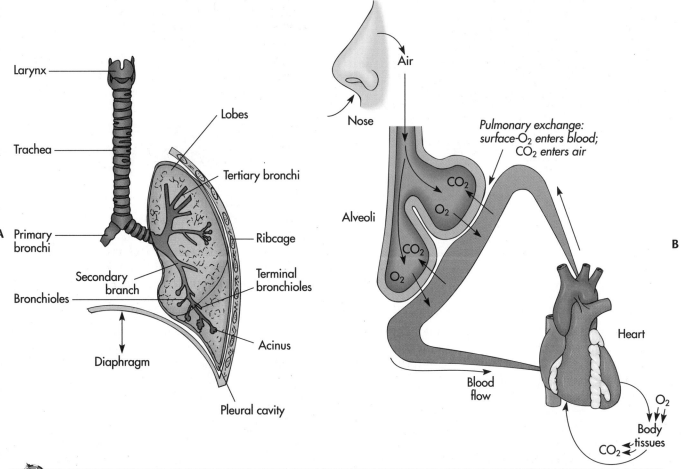

✴ **Figure 24.1**

A, Respiratory system and tracheobronchial tree. **B,** Pulmonary exchange surface. Blood takes up oxygen from the air and gives off carbon dioxide.

but tobacco smoke, allergies, and asthma also cause coughing. Occasionally, chronic coughing signals underlying disorders that deserve medical examination. Coughing also draws attention to yourself and may have psychological origins.

LUNGS AND PLEURAL CAVITIES

The lungs are the paired, conical, spongy organs of the thoracic cavity that represent the "business end" of the respiratory system. The lungs fill the pleural cavities (PLOOR-al) on either side of the mediastinum (MEE-dee-ah-STINE-um) and conform to the walls of these chambers (Figures 24.2 and 24.3). The

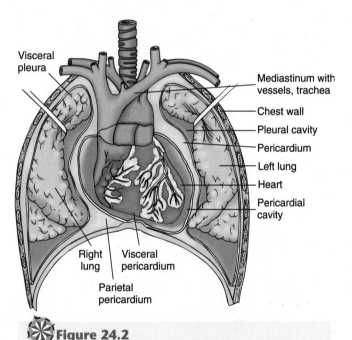

Figure 24.2

Lungs and pleural cavities. Frontal section through the thorax.

Labels on figure:
Visceral pleura
Mediastinum with vessels, trachea
Chest wall
Pleural cavity
Pericardium
Left lung
Heart
Pericardial cavity
Right lung
Visceral pericardium
Parietal pericardium

lateral surface of each lung is convex and tapers superiorly to an **apex** shaped by the ribcage; the diaphragm molds the concave **base** of each lung, and the **mediastinal surface** bears the imprints of the trachea, the esophagus, and the heart and its great vessels. The heart impresses the large **cardiac notch** in the smaller left lung. Except for a **root** that tethers it to the mediastinum, each lung is free to inflate and deflate within its pleural cavity. **Pneumothorax** (NEW-moe-THORE-ax) emphasizes how important this attachment is and how inflatable the lungs are. In pneumothorax, air enters the pleural cavity through injury or surgery in the wall of the thorax and deflates the lung to a bundle of

tissue crumpled around the root of the lung on the mediastinum. Thin **visceral pleura** covers the surface of each lung and the root. This sheet of tissue is continuous with the **parietal pleura** that lines the interior wall of the thoracic cavity and the superior surface of the diaphragm. Blood gives the lungs a reddish-gray color, and air spaces lend a spongy, lobular texture to the exterior surfaces of the lungs.

Deep **fissures** divide the lungs into **lobes** that correspond to major branches of the tracheobronchial tree. The visceral pleura follows these fissures and covers each lobe intimately, like a plastic bag clinging to its contents. Two fissures segment the right lung into three lobes (Figure 24.3). The **horizontal fissure** separates the **superior** from the **middle lobe,** and the long **oblique fissure** divides the middle from the **inferior lobe.** Because the heart intrudes into the left pleural cavity, the left lung has only **superior** and **inferior lobes** separated by the oblique fissure. The **root** enters each lung in a slight depression called the **hilus** (HY-lus) on the mediastinal surface of each lung. Each root carries a bronchus and branches of the vagus nerve and sympathetic trunk in the autonomic nervous system, in addition to the pulmonary and bronchial arteries and veins, and lymph vessels.

- What structures shape the lungs?
- Which lung is larger?
- Where is the root of the lung and what does this structure do?
- How many lobes overall do the lungs display?

TRACHEOBRONCHIAL TREE

The tracheobronchial (TB) tree distributes air throughout the lungs. Branching, you may remember from Chapter 1, is an anatomical tactic that creates large surfaces by dividing spaces into many smaller, compact units. Beginning with the **trachea** in Figure 24.4, first **bronchi,** then **bronchioles** branch pairwise; one branches into two, then these two into four, and four into eight, and so forth. **Respiratory bronchioles** and **alveolar ducts** continue branching in this binary pattern until they end in **alveolar sacs** and **alveoli.** From the single entrance through the trachea, air passes some 23 ranks or generations of branches to reach some 300 million alveoli and 65 m² (700 ft²) of surface at the tips of the tree. The larger passages (trachea, bronchi, and bronchioles) are known as the **conducting zone** because air flows in and out through them to the **respiratory zone** (respiratory bronchioles, alveolar ducts, sacs, and alveoli) where air and blood exchange oxygen and carbon dioxide. Sixteen or seventeen generations of branching occur in the conducting zone, and the rest occupy

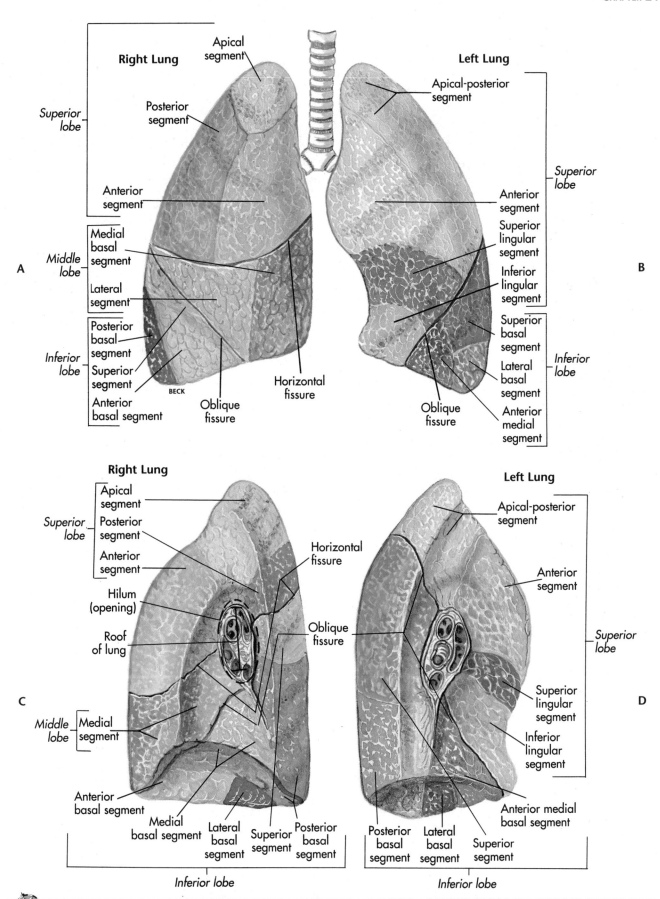

Right Lung

Apical segment

Posterior segment

Superior lobe

Anterior segment

Middle lobe

Medial basal segment

Lateral segment

Inferior lobe

Posterior basal segment

Superior segment

Anterior basal segment

BECK

Oblique fissure

Horizontal fissure

Left Lung

Apical-posterior segment

Superior lobe

Anterior segment

Superior lingular segment

Inferior lingular segment

Superior basal segment

Lateral basal segment

Inferior lobe

Anterior medial segment

Oblique fissure

A

B

Right Lung

Apical segment

Posterior segment

Superior lobe

Anterior segment

Hilum (opening)

Roof of lung

Middle lobe Medial segment

Anterior basal segment

Medial basal segment

Lateral basal segment

Superior segment

Posterior basal segment

Horizontal fissure

Oblique fissure

Left Lung

Apical-posterior segment

Anterior segment

Superior lobe

Superior lingular segment

Inferior lingular segment

Anterior medial basal segment

Posterior basal segment

Lateral basal segment

Superior segment

Inferior lobe

Inferior lobe

C

D

Figure 24.3

The lungs. A and B, Anterior view. C and D, Mediastinal surfaces.

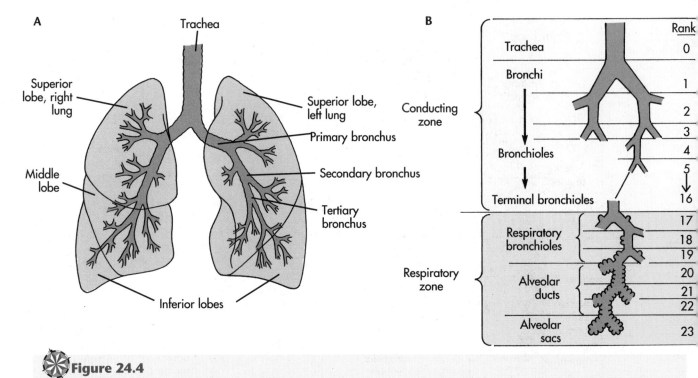

Figure 24.4

Tracheobronchial tree. **A,** Trachea, bronchi. **B,** Ranks of bronchi and bronchioles in the tracheobronchial tree.

the respiratory zone. **Chronic obstructive pulmonary disease** (COPD) is a group of conditions that interfere with the flow of air through the tracheobronchial tree and with gas exchange. Cystic fibrosis, asthma, and emphysema belong to this group and are discussed in the sections that follow.

Four tissue layers—mucosa, submucosa, muscularis, and adventitia—occur throughout the tracheobronchial tree (Figure 24.5). The **mucosa** lines all passages with an epithelium that becomes steadily thinner as branching proceeds. Beginning as tall columnar cells in the trachea and bronchi, the epithelium becomes cuboidal in the bronchioles, and finally thin squamous cells line the alveoli. The epithelium of the conducting zone secretes a mucous film that captures airborne particles, but because diffusion in the respiratory passages requires thin walls to be effective, respiratory passages do not secrete mucus. **Mucous glands** in the mucosa and submucosa supply mucus to the bronchi, and goblet cells release mucus throughout the conducting zone. Collagen fibers wrap around the branches of the tree in the **submucosa** to prevent them from bursting when the lungs inflate with air. Elastic fibers also allow stretching, but these fibers snap back when the lungs exhale, helping drive air out of the lungs. The **muscularis** appears as smooth muscle fibers, and in the **adventitia,** cartilage supports the trachea and bronchi. As the branches become smaller and more numerous, airflow through

them slows and the need for cartilage disappears.

Look for four different changes in the structure of the tracheobronchial tree as you read the sections below. You should be able to summarize and give an example of each change.

1. Passages become narrower and more numerous as branching progresses.
2. Cartilage supports large passages where rushing air tends to collapse them, but this need diminishes in smaller passages where air flows more slowly.
3. Without cartilages to restrict them, smooth muscle fibers constrict and dilate the bronchioles to regulate airflow.
4. The height of epithelial cells diminishes distally and production of mucus gives way to surfactant (see Pulmonary Exchange Surfaces, this chapter) in the respiratory passages.

- Name the passages in the tracheobronchial tree.
- Which ones are in the conducting zone? In the respiratory zone?
- How many ranks of branching are there in the TB tree?

VASCULAR TREE

Two vascular trees supply blood to the lungs. The **pulmonary tree** supplies deoxygenated blood to

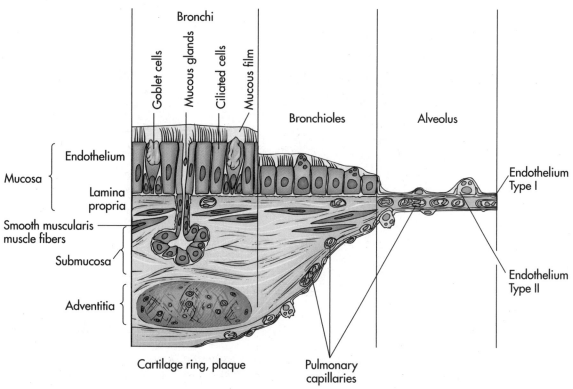

Bronchi

Goblet cells
Mucous glands
Ciliated cells
Mucous film

Bronchioles

Alveolus

Mucosa {
  Endothelium
  Lamina propria
}

Smooth muscularis muscle fibers

Submucosa

Adventitia

Cartilage ring, plaque

Pulmonary capillaries

Endothelium Type I

Endothelium Type II

✳ Figure 24.5

Tissue layers of the tracheobronchial tree, longitudinal section. Bronchi, bronchioles, and alveoli have been joined together as if they were a single tube. Air flows through the lumen above the mucosa.

the alveoli, where gas exchange takes place, and to the smaller bronchioles. A separate **bronchial tree** (BRONK-ee-al) supplies oxygenated blood to the larger conducting passages because they absorb little oxygen from the passing air. In Figure 24.6, deoxygenated blood from the right ventricle of the heart enters the pulmonary tree through the **pulmonary trunk** and pulmonary arteries. The **pulmonary arteries** branch into smaller arteries that follow the bronchi of the tracheobronchial tree, and blood flows from them into capillaries that cover the alveoli and respiratory passages (see Pulmonary Circulation, Chapter 14). Now oxygenated, this blood drains through tributaries of the **pulmonary veins** into the left atrium of the heart. The bronchial tree delivers freshly oxygenated blood to the lungs by way of **bronchial arteries** from the left ventricle of the heart. **Bronchial veins** drain some of this blood from the lungs, but most of it mixes with pulmonary blood and returns to the heart through the pulmonary veins.

The pulmonary arteries follow the tracheobronchial tree closely, branching wherever the bronchi or bronchioles branch. The pulmonary trunk divides into the left and right pulmonary arteries that accompany the primary bronchi through the root into each lung. Branches of these arteries in turn follow the secondary bronchi and the tertiary bronchi. This remarkable repetition continues to follow the bronchioles

deep into the respiratory zone of the lungs, so that each alveolar sac obtains its own individual branch of the artery. The pulmonary venous tree also follows this pattern, but not as faithfully.

- What is the main difference between the vessels of the pulmonary tree and the bronchial tree?

TRACHEA

The trachea conducts air between the larynx and the lungs. The trachea (Figure 24.7A) is a cylindrical tube, 2 cm × 11 cm long (¾ in × 4½ in), that descends the neck from the larynx, anterior to the esophagus and the sixth cervical vertebra, into the thorax and mediastinum, where it branches into two bronchi at the level of the fifth thoracic vertebra. Between 16 and 20 hyaline **cartilage rings** and bands support the trachea, which otherwise would collapse during inhalation as air flows rapidly through under low pressure. These rings actually are shaped like the letter *C,* with a gap on the posterior side that accommodates the esophagus between the trachea and vertebral column and allows the trachea to expand slightly. Thick collagen fibers spiral around the trachea, preventing it from rupturing when coughing pressurizes the lungs, and elastic fibers run lengthwise, allowing it to stretch and recoil as the lungs move.

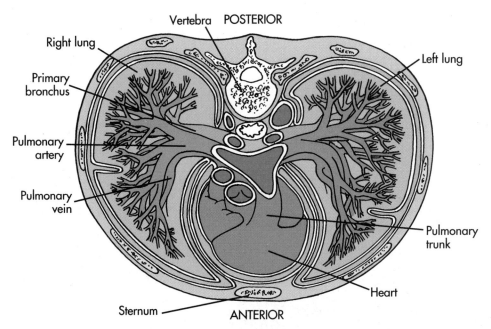

Vertebra POSTERIOR

Right lung

Primary bronchus

Pulmonary artery

Pulmonary vein

Left lung

Pulmonary trunk

Heart

Sternum ANTERIOR

✳️ **Figure 24.6**

Transverse section through the thorax at the level of the sixth thoracic vertebra, showing bronchi and pulmonary arteries and veins entering the lungs. Branches of the pulmonary arteries follow the tracheobronchial tree.

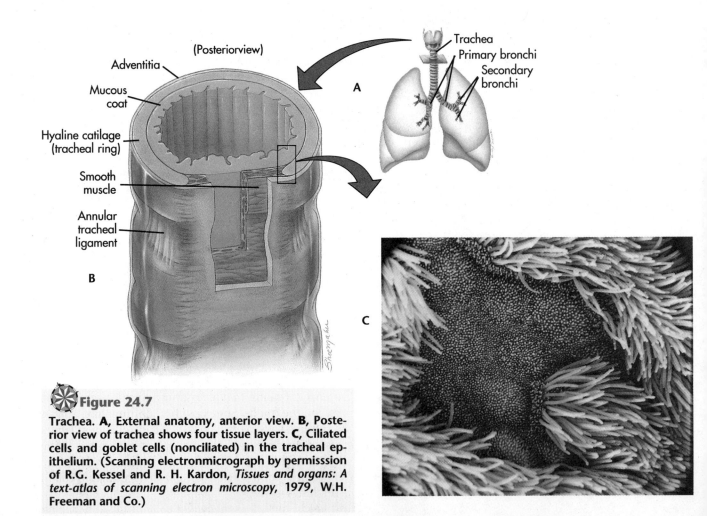

(Posterior view)

Adventitia

Mucous coat

Hyaline catilage (tracheal ring)

Smooth muscle

Annular tracheal ligament

B

Trachea

Primary bronchi

Secondary bronchi

A

C

✳️ **Figure 24.7**

Trachea. **A,** External anatomy, anterior view. **B,** Posterior view of trachea shows four tissue layers. **C,** Ciliated cells and goblet cells (nonciliated) in the tracheal epithelium. (Scanning electronmicrograph by permisssion of R.G. Kessel and R. H. Kardon, *Tissues and organs: A text-atlas of scanning electron microscopy,* 1979, W.H. Freeman and Co.)

The wall of the trachea consists of mucosa, submucosa, muscularis, and adventitia (Figures 24.7B and C and 24.5). **Pseudostratified columnar epithelium** lines the trachea with **goblet cells** and **ciliated cells** that catch particulates in the **mucus escalator,** a film of mucus carried upward by cilia into the larynx where it is swallowed or expelled by coughing. Hooks on the tips of the cilia grab the mucus on the upstroke but slide easily past on the downstroke. A fine net of collagen fibers in the lamina propria completes the **mucosa** of the trachea. This connective tissue blends with coarser fibers and larger vessels in the underlying **submucosa.** Vessels, nerves, and lymphatics populate both layers, and macrophages wander freely throughout in search of foreign matter. Numerous mucosal and **submucosal mucous glands** supplement the mucus formed by goblet cells. **Muscularis** and fibrous **adventitia** join the cartilage rings together, providing a strong, fibrous wall that links to the muscles of the neck and to the mediastinum.

- What services do cartilage rings, goblet cells, and ciliated cells provide to the trachea?

BRONCHI

Bronchi branch into each lung. Bronchi are the largest branches of the tracheobronchial tree, ranging from 1.5 cm in diameter down to 1 mm (Figure 24.8A). **Primary bronchi** (BRONK-eye), the largest of these airways, branch from the trachea at the **carina** and enter the lungs. The right primary bronchus is larger but shorter than the left (which passes beneath the aorta) and is more directly in line with the trachea, so that objects in the airway lodge here more often than in the left lung. As soon as it enters the lung, each primary bronchus branches into **secondary bronchi** that lead to each lobe of the lung. Within each lobe, **tertiary bronchi** lead to individual **bronchopulmonary segments** (BRONK-oh-PULL-mon-air-ee) of that lobe. Subsequent branches deliver to the alveoli and respiratory zone within that segment.

Bronchopulmonary segments are independent branches of the tracheobronchial tree. The right lung usually has ten of these segments and the left, nine. Along with its own air flow, each segment receives individual branches of the pulmonary artery and veins. Septa composed of connective tissue partition bronchopulmonary segments from their neighbors, isolating each segment. Each is named, and pulmonary surgeons know their locations thoroughly. Infections tend to localize in one or more bronchopulmonary segments, where they can be found by X-ray, and if necessary they can be surgically removed with little disturbance of other segments.

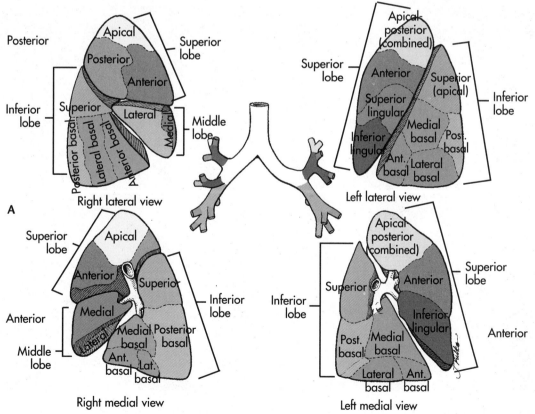

✵ Figure 24.8

A, Tertiary bronchi and bronchopulmonary segments.

Continued.

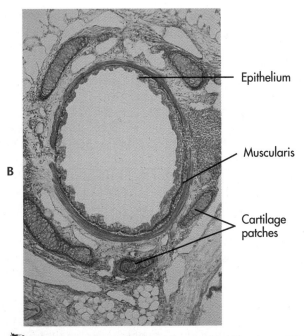

B

- Epithelium
- Muscularis
- Cartilage patches

Figure 24.8, cont'd

B, Transverse section through a bronchus.

The largest bronchi resemble the trachea, but as branching progresses and the bronchi narrow, they begin to resemble bronchioles (see Figures 24.5 and 24.8B). Initially, cartilage rings support the primary bronchi, but in tertiary bronchi the rings diminish to patches that disappear as bronchi become narrower. The muscularis becomes more regular and proportionately thicker as the cartilage diminishes. Pseudostratified ciliated epithelium lines all bronchi, but the cells are shorter in the smaller bronchi. This epithelium contains goblet cells that secrete the mucus that ciliated cells propel toward the larynx and esophagus. The lamina propria also becomes thinner, but the submucosa retains its mucous glands.

CYSTIC FIBROSIS

Cystic fibrosis is a genetic disease inherited through an abnormal gene on chromosome 7. The gene is quite common in American whites but is rare in Asian and African Americans. Although 1 in 20 whites carries the gene, far fewer individuals are affected (1 in 2000 newborns) because the mutant form is recessive. Even so, this incidence is high. Cystic fibrosis interferes with secretion by the exocrine glands of the body generally and the mucous glands of the lungs in particular (Figure 24.9). The lungs secrete an

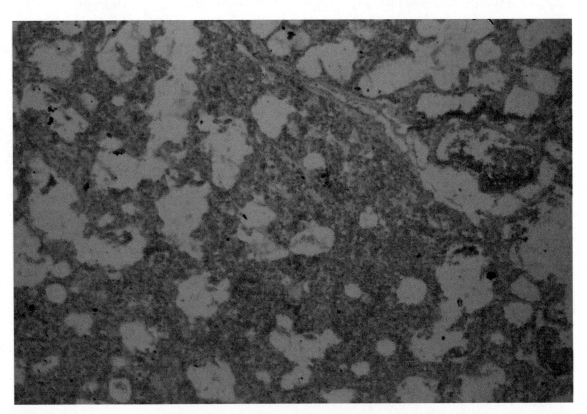

Figure 24.9

Cystic fibrosis. Sagittal section of whole lung reveals enlarged bronchi and relatively normal alveolar tissue near the surface.

especially viscous mucus that accumulates in the bronchi and overwhelms the cilia. Patients contract respiratory infections when this first line of defense no longer clears bacteria from the lungs. Inflammation causes the walls of the bronchioles to thicken, and these obstructions ultimately cause the larger bronchi to dilate, as you can see in the figure. Remarkably the acini and alveoli of the respiratory zone are relatively unaffected by cystic fibrosis. Treatment involves antibiotics to combat infection and therapy to remove mucus from the lungs. Genetic pedigrees and sweat tests for hypersaline perspiration have routinely been used to diagnose cystic fibrosis, but with the recent identification and isolation of the cystic fibrosis gene, patients can be tested for the gene directly.

- How do bronchi differ from the trachea?
- What are the usual symptoms of cystic fibrosis in the lungs?

BRONCHIOLES

Inhaled air flows from bronchi to the bronchioles. Bronchioles (BRONK-ee-olz) are less than 1 mm in diameter, profusely branched, and lack cartilage and mucous glands (Figure 24.10). As the diameter of bronchioles diminishes, the simple columnar epithelium becomes cuboidal; ciliated cells disappear; the muscularis becomes relatively thicker; and the lamina propria, submucosa, and adventitia merge as one connective tissue. Special secretory **Clara cells** replace goblet cells, but it is not clear what role their serous secretions play. Finally, the last rank of bronchioles, called **terminal bronchioles,** leads to the respiratory surfaces. From this point onward, inhaled air encounters the respiratory zone and its thin-walled alveoli.

ASTHMA

Asthma (AZ-thma) narrows the bronchioles and makes breathing difficult. Patients suffer episodes of wheezing and coughing when narrowed bronchioles resist the flow of air. Wheezing and shortness of breath occur when interstitial fluid accumulates from inflammation in the bronchial mucosa or when smooth muscle fibers constrict the bronchioles. The mucosa is especially reactive to irritation, and various factors, including allergens, exercise, emotional states, and air pollution, may initiate an episode. Treatment with bronchodilating agents that block contraction of the muscularis often brings temporary relief.

- What are the primary differences between bronchioles and bronchi?
- What is a terminal bronchiole?

PULMONARY LOBULES

Once in the respiratory zone, air enters lobules that contain the alveoli and passages leading to them. Also known as secondary lobules, these lobules

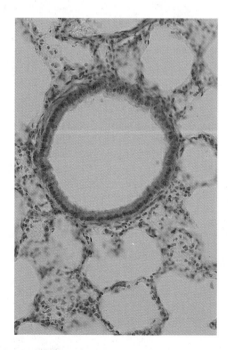
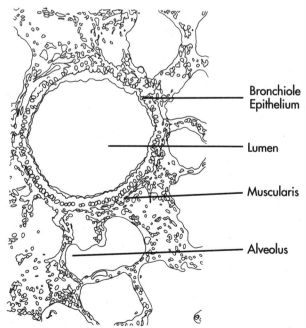

Bronchiole
Epithelium

Lumen

Muscularis

Alveolus

Figure 24.10
Transverse section through a large bronchiole.

appear on the exterior of the lungs as slightly convex bulges up to 2 cm (¾ in) across, separated by delicate fibrous septa (Figure 24.11). These external lobules are conical in shape, but they become irregular, smaller, and poorly defined deeper in the lungs. Although a single bronchiole delivers to each lobule, there is so much variation among secondary lobules that no single secondary lobule can serve as the stan-

dard for them all. There are smaller, more uniform units of pulmonary structure and function within secondary lobules, however.

A **primary lobule** (Figure 24.11B) consists of all the respiratory passages ventilated by a single terminal bronchiole. A typical primary lobule begins when a terminal bronchiole branches into two **respiratory bronchioles,** the first passages that display a few blis-

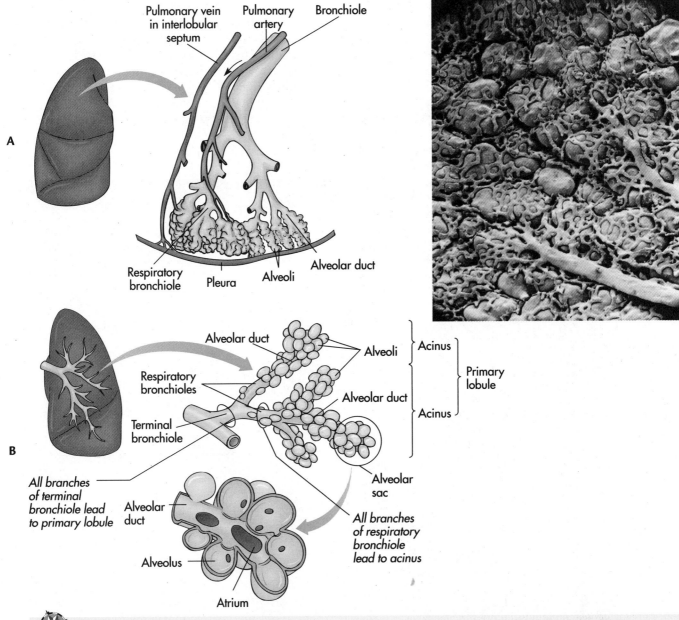

Figure 24.11

Structural units of the lungs. A, Secondary lobules are evident on the surface of the lungs. These lobules contain many terminal and respiratory bronchioles. **B,** Primary lobules include all the passages that branch from one terminal bronchiole. Acini are smaller; they contain all the branches from a respiratory bronchiole. **C,** Pulmonary capillaries cover alveoli.(Scanning electronmicrograph by permission of R.G. Kessel and R.H. Kardon, *Tissues and organs: A text-atlas of scanning electron microscopy,* 1979, W.H. Freeman and Co.)

ter-like **alveoli** (AL-vee-OH-lie), the respiratory surfaces of the lungs. Respiratory bronchioles lead in turn to one or more ranks of **alveolar ducts,** where alveoli begin to crowd the walls of the ducts with even more respiratory surfaces. Ducts end in **alveolar sacs,** which are clusters of alveoli that resemble clusters of grapes, grouped around a central lumen or **atrium.** A secondary lobule contains many primary lobules, since each primary lobule measures approximately 1 mm in diameter.

Physicians and researchers recognize an even smaller structural unit, the acinus. An **acinus** (ah-SIGN-us) is one half of a primary lobule. An acinus is considered to consist of all the alveolar ducts, sacs, and alveoli that lead from a single respiratory bronchiole. According to the pairwise pattern of branching, each primary lobule contains two acini, one for each respiratory bronchiole.

Pulmonary blood follows the respiratory branches of acini. Deoxygenated blood enters the primary lobule, shown in Figure 24.11C, through branches of **pulmonary arteries** that accompany the bronchioles and alveolar ducts to the alveoli. The blood becomes oxygenated in fine **pulmonary capillary beds** that cover each alveolus, and returns by way of tributaries to the **pulmonary veins.** Small **bronchiolar plexuses** and arterial branches supply the walls of bronchioles (and bronchi), but most of this bronchiolar blood mixes with pulmonary blood and returns through the pulmonary veins.

Except for their alveoli, the walls of respiratory bronchioles resemble those of other bronchioles. The epithelium consists of low cuboidal cells without cilia; Clara cells are present, and individual fibers of the muscularis now wrap around the bronchiole between the alveoli. Alveoli now cover more surface than the walls of the bronchioles themselves, and it is across these thin alveolar partitions that oxygen diffuses into the blood.

- Which airways are found within primary lobules? Within acini?
- What characteristic makes these passages respiratory?
- How do respiratory bronchioles differ from terminal bronchioles?

PULMONARY EXCHANGE SURFACES

Gases are exchanged in the alveolar septa. Alveoli and alveolar sacs trap capillaries and connective tissue between them in thin partitions called **alveolar septa** (SEP-tah). These walls contain the gas exchange surfaces, and they account for the spongy, elastic nature of the lungs because they are so numerous. Figure 24.12 shows the anatomy of gas exchange in an alveolar septum. Pulmonary capillaries reside in a thin matrix of connective tissue that is coated by thin squamous **alveolar epithelium.** Because the entire parti-

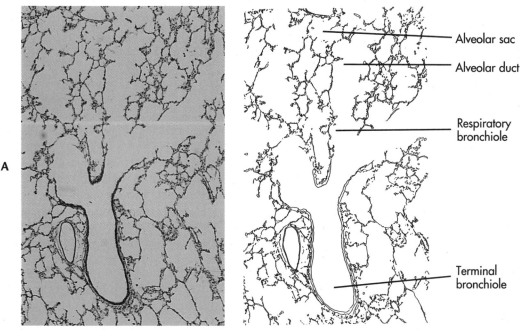

A

Figure 24.12

Alveolar septa and pulmonary exchange. **A,** Alveoli and alveolar septa. A terminal bronchiole leads to alveolar sacs and alveoli in human lung.

Alveolar sac

Alveolar duct

Respiratory bronchiole

Terminal bronchiole

Continued.

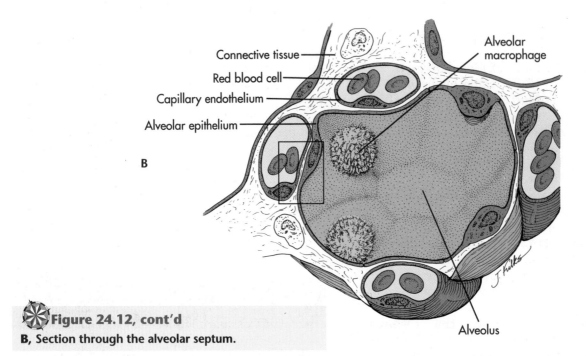

Figure 24.12, cont'd
B, Section through the alveolar septum.

tion is no more than 2.5 μm thick, gases readily diffuse between the capillaries and the epithelium on both sides.

Oxygen reaches the blood by diffusing across the alveolar epithelium and the capillary endothelium shown in the figure, usually less than 0.5 μm total distance from air to blood. Two principal types of cells make up the alveolar epithelium. **Alveolar epithelial cells,** or Type I pneumocytes, are more prominent. Very thin (0.2 μm across), except for nuclei that bulge into the airspace, these cells are the primary gas exchange cells of the alveoli. Type I pneumocytes send out flat cytoplasmic processes onto both surfaces of the alveolar septum and line more than one alveolus, capturing capillaries between the processes. Tight junctions seal the processes of neighboring cells together. Type II pneumocytes, or **septal cells,** secrete **surfactant** (surf-ACT-ant), a phospholipid detergent whose principal component is dipalmitoylphosphatidylcholine (DPPC; di-PALM-ih-TOE-il-FOS-fah-TIDE-il-KOE-leen). Forming a film on the surface of the alveoli, this fluid dissolves gases and dissipates surface tension that would otherwise collapse the alveoli. As air flushes about an alveolus, oxygen and other gases dissolve in the surfactant film and diffuse into the alveolar epithelium.

Once across the epithelium, the gas molecules pass the fused basement membranes that join the alveolar epithelium to the capillaries. **Capillary endothelial cells** are equally thin, and both oxygen and nitrogen cross into the blood plasma. Oxygen enters the **erythrocytes** and binds to **hemoglobin** molecules in exchange for carbon dioxide molecules that are released. Exchange is complete when carbon dioxide molecules diffuse back across the same pathway into the alveoli and are swept away by exhalation.

The connective tissue matrix is not a passive component of the alveolar septum. **Collagen fibers** in the matrix lend it strength, and **elastic fibers** provide the lungs' resilience by stretching at inhalation and recoiling during exhalation. Finer **reticular fibers** add to this flexible network. The matrix also is a pathway for **lymphatic drainage** and **alveolar macrophages,** or **dust cells,** that migrate freely within the septum and pass through the capillary endothelium into alveoli on immunological patrol.

EMPHYSEMA

Emphysema patients complain of shortness of breath because the alveolar spaces for exchange have enlarged, as you can see in Figure 24.13A and C. Enlarging the alveoli may sound like a good thing, but according to surface-volume rules, enlargement paradoxically means fewer such spaces and smaller exchange surface overall (see Chapter 1, Anatomic Basics). The situation resembles what happens when soap bubbles fuse and merge into larger ones that take up the same space but with much less surface. In one form of emphysema called **panacinar emphysema** (PAN-ah-sign-ar EM-fih-seem-ah), acini become blocked and capture space from their neighbors. Secondary lobules become single, large chambers without the original septa that made them so effective. Breakdown of the collagen and elastic fiber network apparently causes enlargement. According to the **protease-antiprotease hypothesis** (PRO-tee-ace), lung tissue releases proteases (hydrolytic enzymes that attack connective tissue proteins). Normal lungs hold this potential destruction in check by releasing inhibitors of proteases. Emphysema is considered to result when too little inhibitor or too much protease is present.

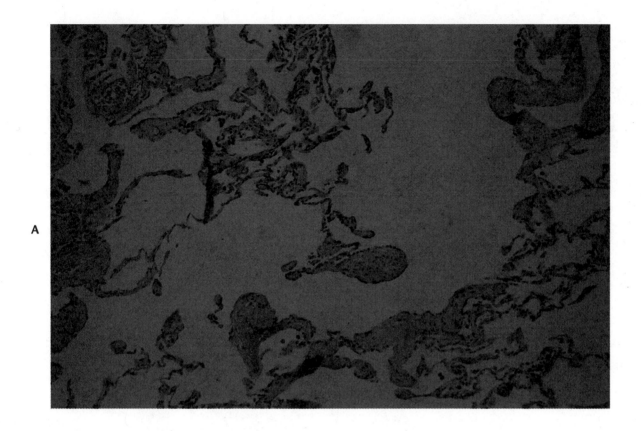

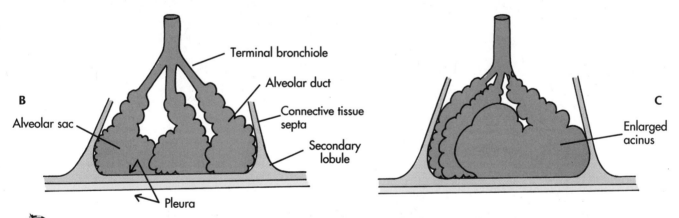

Figure 24.13

Emphysema. **A,** Sagittal section of lung with enlarged acini and lobules throughout the lung. **B,** Panacinar emphysema. **C,** When an acinus is blocked, it enlarges and obliterates adjacent acini.

- Name the cells that line the alveolar septa.
- What functions do these cells have?
- Why is surfactant necessary?
- What is the effect of emphysema on gas exchange?

NASAL CAVITY AND PARANASAL SINUSES

The bones of the face are filled with air spaces that reach deep into the skull. The **nasal cavity,** il-lustrated in Figure 24.14, leads from the nostrils (external nares) through the choanae (KOE-an-eye; internal nares) to the pharynx and the lungs. The **paranasal sinuses** lead into the skull from the nasal cavity through openings in the walls of this cavity, but the sinuses are blind chambers, and air simply circulates back out through the same openings. Ciliated columnar epithelium and mucous glands line all these internal surfaces, and branches of both the internal and external carotid arteries supply the highly vascular submucosa that connects the mucosa to the periosteum and perichondrium of the skull. The warm, moist,

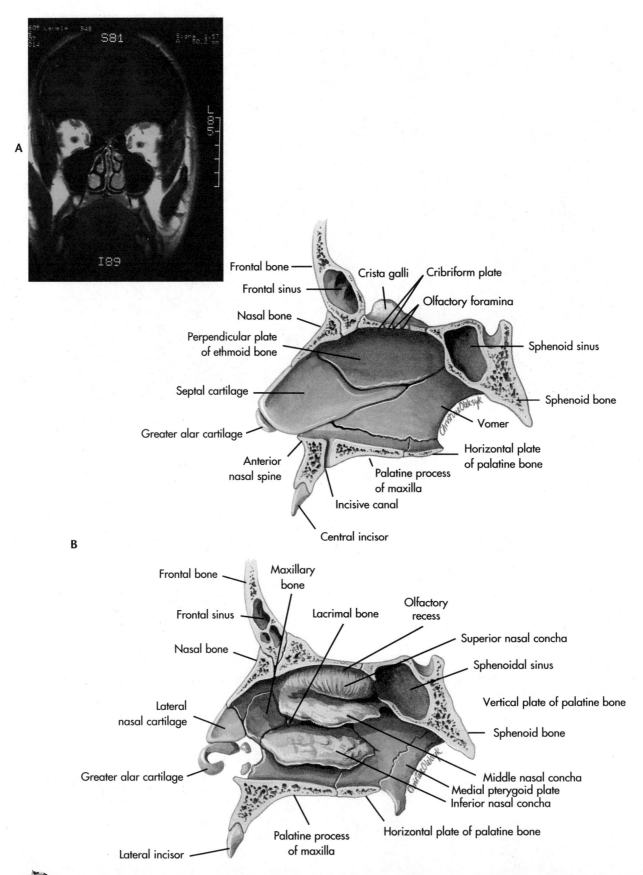

Figure 24.14

The nasal cavity. **A,** Frontal section through the nasal cavity, MRI scan. **B,** Nasal septum *(top);* nasal conchae on right lateral wall of nasal cavity *(bottom).*

sticky surfaces of the nasal cavity **heat, humidify,** and **clean** the air flowing into the lungs.

NASAL CAVITY

The nasal cavity is a pair of spaces in the facial bones of the skull that lie on either side of the nasal septum above the palate in the figure. Moist, highly vascular mucosa cleans the air and brings it to body temperature and humidity before the air reaches the delicate surfaces of the lungs. The **mucosa** also is responsible for that runny, stuffy feeling of common colds (**rhinitis;** rine-EYE-tis) and for **sinusitis** (SIGN-us-EYE-tis), inflammation of the paranasal sinuses. The roof of the nasal cavity is equipped with olfactory receptors and special mucous glands for olfaction, described in Chapter 20, Receptors and Special Sensory Organs.

Air enters the **nose** through the **nares** (NAIR-eez) and passes through the **vestibule** behind the nares before reaching the nasal cavity itself. Five **nasal cartilages,** named in Figure 24.14B, enclose the vestibule, shape the nose, and attach the nose to the skull. Skin and sebaceous glands line the vestibule. **Nasal hairs** guard access to the nasal cavity itself, but it is not clear what they may block. The epidermis becomes ciliated pseudostratified columnar **respiratory epithelium** posterior to the hairs.

The maxilla and palatine bones form the hard palate **floor of the nasal cavity,** and the vomer and ethmoid form the **nasal septum** that separates the left and right portions of the nasal cavity. The ethmoid and sphenoid bones form the **roof** of the cavity, and both bones contribute to the **lateral walls,** along with the nasal conchae, maxilla, palatine, and lacrimal bones (see Chapter 6, Axial Skeleton). In contrast with the flat nasal septum, three **nasal conchae** (KON-kee)—superior, middle, and inferior—curl into the nasal cavity from the lateral walls. Each concha encloses a round space called a **meatus** (me-AH-tus). The **inferior meatus,** guarded by the **inferior nasal concha,** is the largest space, and the **superior meatus** with its **superior nasal concha** is the smallest. A narrow groove in the **middle meatus,** called the **infundibulum,** opens to the frontal, ethmoid, and maxillary sinuses beneath the **middle concha.** The **sphenoethmoid recess** (SFEN-oh-ETH-moyd) lies above and lateral to the superior nasal concha and opens to the sphenoid sinuses. Nasal surgeons reach the paranasal sinuses with instruments and fiber optic viewing tubes that pass through the entrances to the sinuses.

Water and heat from the mucosa humidify and warm the air as it flows over the conchae and through the meati in Figure 24.15. The nasal conchae increase these air conditioning surfaces, as well as spinning the air and making it turbulent so that even small cellular particles 4 μm across collide with the mucous mat that covers the epithelium. Mucous glands produce this sticky film and cilia circulate the mucus to the choana and pharynx for swallowing. Secretion and flow renew the mucous film at least every 20 minutes.

- What structures increase the surface of the nasal cavity available for air conditioning?
- What are the major actions of this air conditioning?

PARANASAL SINUSES

Four pairs of paranasal sinuses lead into the skull from small openings or **ostia** (mouths) in the walls of the nasal cavity, as illustrated in Figure 24.16. The **maxillary sinuses** are the largest, being entirely enclosed by the maxilla below the orbit and zygoma. The **frontal sinuses** lie within the frontal bone, behind the glabella above the roof of the orbit, and below the anterior cranial fossa. Ostia connect both frontal and maxillary sinuses with the middle meatus. The **ethmoid sinuses** are a series of small air cells located in the ethmoid bone in the lateral walls of the nasal cavity behind the frontal sinuses. The more anterior cells open into the middle meatus, and the posterior cells communicate by way of the superior meatus. The most posterior chambers are the **sphenoid sinuses** that lie in the sphenoid bone beneath the sella turcica and open to the sphenoethmoid recess above the superior nasal concha. Ciliated columnar epithelium lines all these spaces, but it is thinner and there are fewer mucous glands and less connective tissue than in the nasal mucosa. In **chronic sinusitis** the sinuses inflame and secrete copious mucus. In severe cases the sinuses inflate painfully with mucus, pus, and microorganisms.

The nasal sinuses are excellent subjects for debating structure and function because no single function is compelling. Consider the following possibilities:

1. The sinuses lighten the bones of the skull. But at only about 70 ml total space, how advantageous is it for a 54 kg (120 lb) person to save 70 g (2.5 oz) of bone?
2. They are resonating chambers for the voice. That is certainly so, but the nasal conchae, larynx, and chest also help resonate.
3. They condition inhaled air. If this is a primary function, why are the ostia so narrow?
4. They are shock absorbing chambers, much like automobile air bags, that collapse in protection of deeper structures from trauma. Perhaps so.

- Name the four sets of paranasal sinuses and describe their locations.

PHARYNX

The pharynx connects the nasal cavity with the larynx and trachea. This multipurpose tube, shown

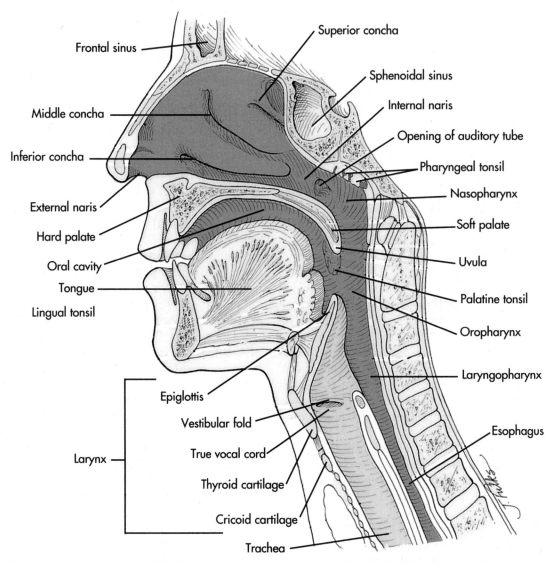

Figure 24.15

Midsagittal section through the nasal cavity and pharynx. Right side in view.

in Figure 24.15, has six openings, four of which are especially important in breathing. Most important for respiration, of course, is the passage that connects the nasal cavity with the trachea. Air from the nasal cavity flows through the **choanae** (internal nares), into the **nasopharynx,** the **oropharynx,** then the **laryngopharynx,** and finally enters the trachea through the **larynx.** Some of this air also enters the openings of the **auditory (eustachian) tubes** (you-STAY-kee-an) on each side of the nasopharynx, equalizing the pressure in the middle ear behind the tympanic membrane with that on the outside (see Chapter 20, Sensory Receptors and Special Sense Organs). The oropharynx also passes food from the fauces to the laryngopharynx and the esophagus (see Chapter 23, Digestive System).

Three regions of the pharynx, shown in Figure 24.15, direct air to the trachea and help prevent food and other materials from entering the airways. The first of these, the nasopharynx (NAY-zo-FAHR-inks), lies posterior to the choana and superior to the soft palate. The auditory tubes open into it, and the **pharyngeal tonsils** line the posterior wall. Respiratory infections and colds often enlarge these lymphoid glands, so that they are called **adenoids** (AD-en-oyds; aden-, *G,* gland). Surgical removal of tonsils (**lingual tonsils** of the oropharynx) and adenoids is a common treatment in childhood for excessive respiratory tract infections. The oropharynx (OR-oh-FAHR-inks), the second region of the pharynx, enters from the oral cavity below the soft palate and transports food and air to the laryngopharynx. Finally, the laryngopharynx (lar-IN-joe-

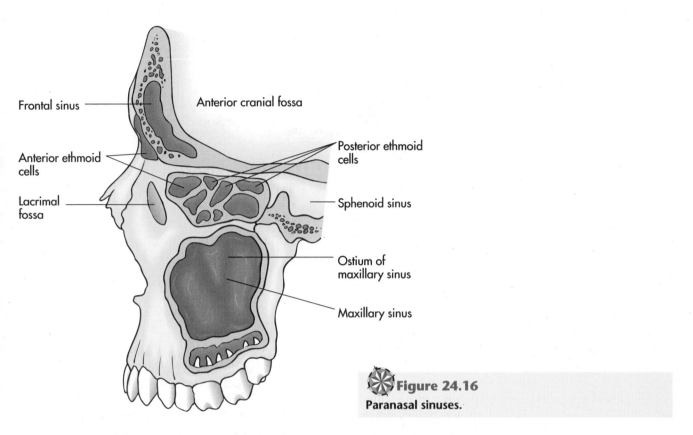

Frontal sinus

Anterior cranial fossa

Anterior ethmoid cells

Posterior ethmoid cells

Lacrimal fossa

Sphenoid sinus

Ostium of maxillary sinus

Maxillary sinus

Figure 24.16
Paranasal sinuses.

FAHR-inks) is located below the hyoid (HIGH-oyd) bone and above the openings to the esophagus and trachea; this cavity includes the larynx. Pseudostratified columnar ciliated respiratory epithelium, like that of the nasal cavity, lines the nasopharynx and the larynx. Stratified squamous epithelium lines the oropharynx and the posterior wall of the laryngopharynx, however, in keeping with the combined digestive and respiratory roles of these passages.

The soft palate directs airflow in the oral and nasal cavity. Closed during swallowing, the soft palate usually prevents food from entering the nasal cavity, but air may enter through the mouth or nasal cavity. In mouth breathing the palate rises against the wall of the nasopharynx to allow air into the oropharynx, but in nasal breathing the palate depresses and closes off the oral cavity. Coughing raises the soft palate, but sneezing depresses it so that the air rushes out through the nasal cavity. Receptors in the nasal mucosa initiate the sneezing reflex, and receptors in the larynx and trachea are responsible for coughing.

- Name the six openings of the pharynx.
- What position does the soft palate take in speech, sneezing, and nasal breathing?

LARYNX

The box-like larynx is the organ of sound. Its cartilages and ligaments support the trachea and guard its entrance, as illustrated in Figure 24.15, and its vocal folds produce sound when air passes between them. The hard, protruding **Adam's apple** identifies the larynx externally. Internally the larynx forms the forward and lateral walls of the laryngopharynx at the levels of the fourth to sixth cervical vertebrae. Seen from the posterior in Figure 24.17, the larynx resembles a cup with a lid, the **epiglottis** (ep-ih-GLOT-iss), which folds down over the cup, known as the **vestibule,** to prevent food from entering (see Chapter 23, Digestive System). **Aryepiglottic folds** (arry-EP-ih-GLOT-ik) surround the vestibule, and the **vocal folds** can be seen deep in the floor of the vestibule on either side of the **glottis** (GLOT-iss), the opening to the trachea. **Vestibular folds** surround the vocal folds. When the glottis is wide open, as shown in the figure, air simply enters and leaves, but when the folds are apposed closely, the glottis narrows and the vocal folds vibrate in the passing air, producing sound.

The mucosa of the larynx is **pseudostratified ciliated columnar epithelium** above and below the vocal folds, but **stratified squamous epithelium** covers the vocal folds themselves and patches of the vestibule. This distribution reflects wear and tear on tissues that come in contact with others. Stratified squamous epithelium also covers the portions of the epiglottis that fold against the tongue. Both epiglottis and vestibule are extremely sensitive to touch. Foreign material lodged here provokes violent coughing.

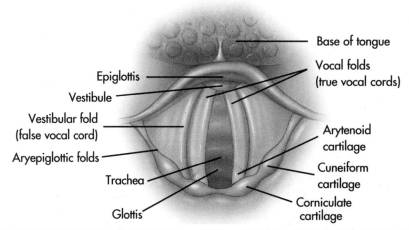

Base of tongue

Epiglottis

Vestibule

Vestibular fold
(false vocal cord)

Aryepiglottic folds

Trachea

Glottis

Vocal folds
(true vocal cords)

Arytenoid
cartilage

Cuneiform
cartilage

Corniculate
cartilage

**Figure 24.17**

The larynx viewed posteriorly from above.

CARTILAGES

Nine cartilages support the larynx, and the muscles and ligaments attached to them maneuver the vocal cords and epiglottis during speech and swallowing. Three large, unpaired cartilages lie in the anterior midline (Figure 24.18), and three small, paired cartilages help support the vocal folds and aryepiglottic folds.

UNPAIRED

The **thyroid cartilage** is the largest cartilage in the larynx. It protrudes anteriorly as the Adam's apple, and its broad **ali** (AIL-eye; *L,* wings) spread laterally around the larynx where they end in vertical posts called the **superior** and **inferior cornua** (KOR-new-ah; cornu, *L,* horn). Ligaments and muscles join the superior border of the thyroid cartilage to the hyoid bone, and the inferior border fits neatly around the cricoid cartilage below. The **cricoid cartilage** (KRY-koyd) resembles a class ring, the only true ring around the larynx; it is narrow anteriorly and broader posteriorly, where it fits neatly within the inferior cornua of the thyroid cartilage. The trachea attaches to the lower margin of the cricoid cartilage, and the paired cartilages articulate on the superior margin. The epiglottis is a patch of elastic cartilage that resembles a leaf, with its narrow stem hinged to the **thyroid notch** of the thyroid cartilage. The epiglottis spreads from this attachment point into the aryepiglottic folds and forms the trap door lid to the vestibule.

PAIRED

These smaller cartilages open and close the vocal folds and support the aryepiglottic folds. The vocal cords stretch posteriorly from the thyroid cartilage to the **arytenoid cartilages** (AR-ih-TEE-noyd) that mount upon the rear of the cricoid carti-

lage. The pyramidal shape of the arytenoids allows them to articulate firmly with the cricoid cartilage while they adduct and abduct the vocal cords. The superior apex of each arytenoid supports the **corniculate cartilages** (kore-NIK-you-late; they resemble cattle horns) that in turn suspend the aryepiglottic folds. The narrow **cuneiform cartilages** (KUE-nee-ih-form; cune, *L,* wedge; wedge-shaped) are outriggers from the corniculate cartilages that taper further along the aryepiglottic folds. The intrinsic muscles of the larynx, described in Chapter 10 and Figure 10.16, adjust tension and pitch of the vocal cords by swiveling the arytenoid cartilages and tilting the thyroid cartilage.

- Name the cartilages of the larynx.
- Which cartilages maneuver the vocal folds, and which ones support the aryepiglottic folds?

VENTILATION

Ventilation is the name of the process that moves air in and out of the lungs, in contrast to breathing, which refers to all aspects of lung function. The lungs themselves fill and empty passively, as the thoracic cavity alternately enlarges and diminishes the space available for them inside the pleural cavities, shown in Figure 24.19. Movements of the **ribs** and **diaphragm** bring these changes about. The ribs enlarge the thoracic cavity by swinging upward on their articulations laterally and anteriorly. This motion brings air into the lungs. Air flows out when the ribs depress, returning the thoracic cavity to its previous volume. The diaphragm is the primary muscle of ventilation; it increases the volume further by lowering the floor of the thoracic cavity; it decreases thoracic volume by raising the floor during exhalation.

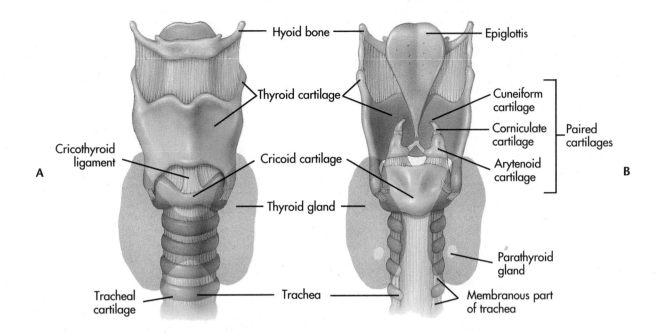

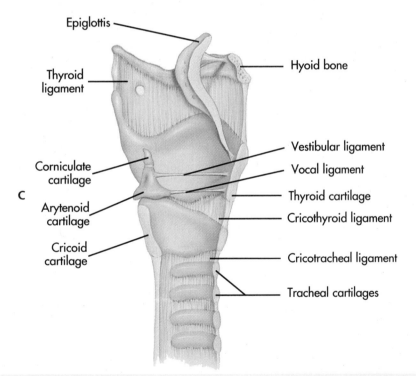

Figure 24.18
Three views of the larynx. A, Anterior. B, Posterior. C, Midsagittal section, right side.

Follow these motions in your next few breaths. When the ribs rise and the diaphragm flattens, air rushes into the lungs and equalizes the pressure that has been momentarily lowered when the thoracic cavity begins to enlarge. The lungs expand laterally and forward, becoming longer as they follow the diaphragm caudally. When you exhale, however, the pressure inside the cavity rises briefly as the walls of the thorax descend and abdominal organs press the diaphragm back up into the thorax. Now the air flows

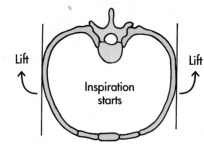

A

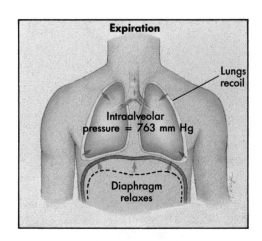

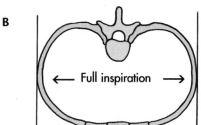

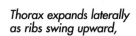

B

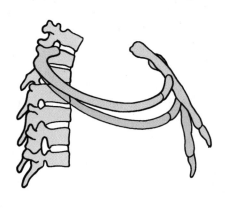

Thorax expands laterally as ribs swing upward,

while the rib cage also expands anteriorly.

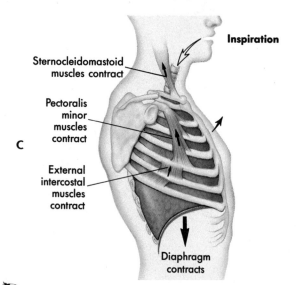

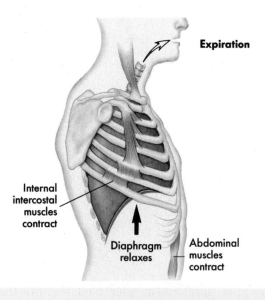

C

Figure 24.19

Ventilation. A, Anterior view of thorax. Deep inhalation elevates the rib cage and depresses the diaphragm. Deep exhalation does the reverse; it depresses the rib cage and elevates the diaphragm. **B,** Rib movements for lateral and anterior expansion. **C,** Muscle action.

outward until the pressure inside equals that outside, and the lungs return to their previous size. If you are like most people, you probably breathe 20 to 40 times a minute at rest, but your breaths become deeper and more rapid when you exercise and slower and shallower in sleep. You can regulate your own breathing voluntarily, but in the end, breathing is involuntary, under control of the autonomic nervous system and the levels of oxygen and carbon dioxide in the blood.

QUIET BREATHING

Three groups of muscles—thoracic, diaphragmatic, and abdominal—ventilate the lungs. Their overall effects are straightforward, even though their exact individual roles may not be so obvious. See Chapter 10 for details of these muscles and Chapters 6 and 8 for descriptions of ribs and their articulations. In quiet breathing the **external intercostal muscles** raise the ribs, but the **internal intercostal muscles** also are active during inhalation and exhalation. The internal intercostal muscles immediately lateral to the sternum contract during inhalation, but more laterally the internal intercostals are active in exhalation. Diaphragm muscles depress the **central tendon** of the diaphragm and compress the abdominal organs below it. Three **hiatuses** (HY-ate-us-ez; hiatus, *L*, a gap, opening) in the diaphragm allow the diaphragm to rise and fall around the esophagus, the descending aorta, and the inferior vena cava without permitting the abdominal organs to protrude through the gaps. Abdominal muscles have relatively little role in quiet inhalation. When you exhale, the external intercostal muscles relax, and gravity and the elasticity of ligaments depress the ribs. The tone of the abdominal muscles helps the abdominal organs to press the diaphragm upward into the thorax.

DEEP BREATHING

Deep breathing brings all muscle groups into vigorous action. Take a deep breath to feel these muscles at work. If you stand in front of a mirror, you will see the **sternocleidomastoid muscles** of your neck raise the sternum and clavicle. Beneath these muscles, the **scalenes** elevate the first two ribs, as shown in Figure 24.19, and both the **pectoralis minor** and the external intercostals raise ribs three, four, and five. The diaphragm depresses more fully. The abdominal muscles come into play during exhalation. The **external oblique muscles** of the abdomen, together with the **internal obliques, transversus abdominis,** and **rectus abdominis,** all compress the abdominal organs against the diaphragm and forcibly expel air from the lungs, while the internal intercostals compress the ribcage. Rib cage and abdominal muscles contract vigorously when you cough. Try it and see.

HEIMLICH MANEUVER

Shown in Figure 24.20, this motion expels objects that have lodged in the larynx, blocking inhalation. The **Heimlich maneuver** (HYME-lik) attempts to duplicate the motions of coughing that the obstruction has prevented. Embrace the patient from behind, just below the margin of the rib cage, and hug firmly and rapidly upward, as if the person's abdominal muscles were contracting violently.

- How does air flow into the lungs?
- What three groups of muscles enable this inflow?
- Which centers of the medulla control breathing?

DEVELOPMENT

Infants born prematurely much before 27 weeks of pregnancy (38 weeks is normal) have little chance of survival because their lungs are not ready to breathe. The respiratory zone is just beginning to form at this time. The steps that form these all-important surfaces begin during the fourth week of development and continue through a child's first year of life, as shown in Figure 24.21.

Pulmonary epithelium derives from endoderm, and mesoderm produces all other lung tissues, except neurons from the ectoderm (see Germ Layers, Chapter 4). About the twenty-sixth day of development, the respiratory diverticulum grows ventrally from the foregut (Figure 24.21A) and forms a **laryngotracheal groove** (lar-IN-joe-TRAKE-ee-al; Figure 24.21B) that becomes the larynx. During the next 34 weeks, three periods of development transform this bud into the tracheobronchial tree and lungs. The **pseudoglandular stage** (weeks 5 to 16) forms the bronchi, bronchioles, and mucous glands of the conducting zone. The lungs acquire respiratory passages during the **canalicular stage** (KAN-al-IK-you-lar; weeks 17 to 25), and in the **terminal sac stage** (week 26 to birth), alveolar sacs and more alveolar ducts form and begin to secrete surfactant. This period allows the lungs to function should birth come prematurely, but with no more than 5 million alveoli. Pulmonary capillaries and alveoli continue to develop during the first year of **postnatal development,** giving the lungs their full complement of 300 million alveoli.

PSEUDOGLANDULAR STAGE

The lungs resemble a highly branched exocrine gland during this period. The respiratory diverticulum elongates (Figure 24.21A through D) and branches into two **lung buds** that together become the trachea and primary bronchi of the tracheobronchial tree. Pairwise branching next produces secondary bronchi and lobes of the lungs, then tertiary bronchi and bronchopulmonary segments. Branching continues into the surrounding mesenchyme tissue

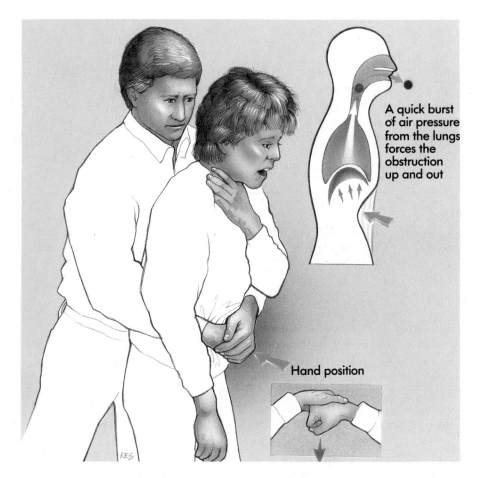

A quick burst of air pressure from the lungs forces the obstruction up and out

Hand position

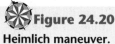**Figure 24.20**
Heimlich maneuver.

until terminal bronchioles form in the sixteenth week. Hyaline cartilage and smooth muscle from this mesenchyme spread through the tree as it enlarges. The epithelium is tall columnar in the main branches and cuboidal in the bronchioles. Goblet cells and ciliated epithelial cells appear throughout, and mucous glands begin to form during the twelfth week.

CANALICULAR STAGE

This period forms the passages of the respiratory zone. Acini develop, and the gas exchange epithelium appears and begins to secrete surfactant. Figure 24.21E shows that rapid bursts of branching form short respiratory bronchioles and alveolar ducts, which then elongate, and that the mesenchyme spreads thinly around them. Type I and II epithelial cells appear. Type I cells that come in contact with pulmonary capillaries flatten and spread out as they form immature respiratory exchange surfaces between them. Type II cells remain cuboidal and begin to release surfactant; they also are progenitors of both types of pneumocytes.

TERMINAL SAC STAGE

Acini are completed in this period, developing from sac-like structures, known as terminal sacs, at the ends of the respiratory tree (Figure 24.21F). **Terminal sacs** form the last few generations of alveolar ducts. Collagen fibers, elastic fibers, and smooth muscle begin to wrap around these parts of the tree, and the lungs begin to grow and expand rapidly.

POSTNATAL DEVELOPMENT

Alveoli and alveolar septa mature during the first year of life. Alveoli accumulate on the walls of alveolar ducts (Figure 24.21G), and alveolar sacs form at the ends of these ducts because elastic fibers draw them into shape. Fibroblasts that appeared in the mesenchyme around the ducts in the terminal sac stage now deposit substantial elastic fibers around the entrance to every alveolus. These fibers resemble purse strings, but they do not actually draw the alveolus closed. Instead the epithelial wall bulges outward from the elastic bands to form alveoli. As the alveolar walls spread, the alveolar septum narrows.

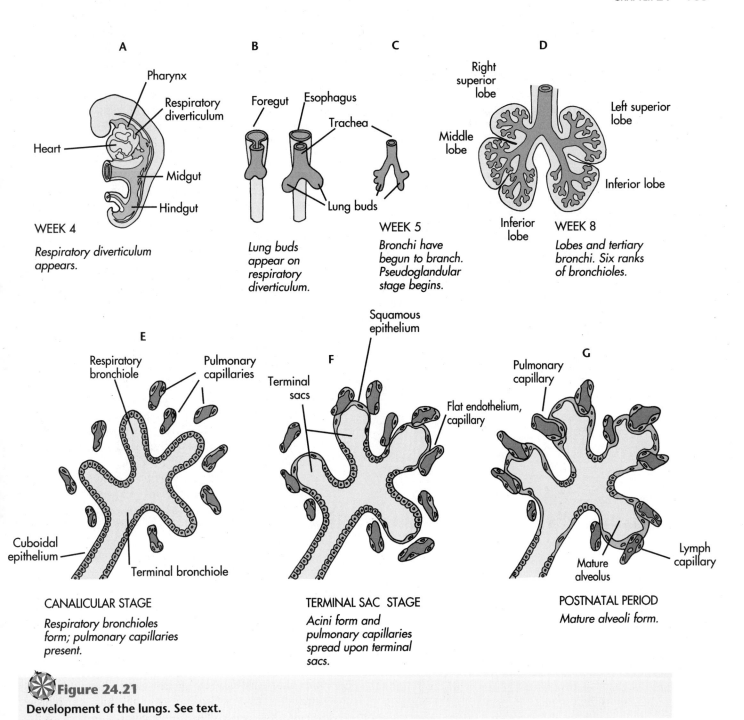

Figure 24.21

Development of the lungs. See text.

The capillaries that surrounded each alveolar duct now are squeezed thinly between both walls of the septum, and the mature gas exchange surface forms.

- Name the phases of lung development and the major events that occur in each of them.
- When is the lung capable of exchanging gases?

SUMMARY

The respiratory system exchanges two important gases with the blood and tissues. Cellular res-

piration in the tissues requires oxygen from the outside air and continually produces carbon dioxide that the system returns to the air. The lungs distribute air and blood over a large, thin exchange surface where oxygen and carbon dioxide diffuse past each other into the blood and air, respectively. Figure 24.22 illustrates that pulmonary diseases interfere with this exchange by reducing the exchange surface area (emphysema) or by limiting access to it (cystic fibrosis, asthma). *(Page 733.)*

The lungs are the principal organs of the respiratory system because they contain the gas ex-

change surfaces. Air enters the upper respiratory tract (nasal cavity, pharynx, and larynx) before reaching the trachea and lungs (lower respiratory tract). The tracheobronchial tree delivers air to 300 million alveoli, the gas exchange surfaces. *(Page 733.)*

The lungs reside in the pleural cavities, which in turn occupy the thoracic cavity on either side of the mediastinum. Air flows passively in and out the tracheobronchial tree when the thoracic cavity enlarges and becomes smaller. Slippery visceral pleura allows the lungs to accommodate these changes, and the roots of the lungs attach the lungs to the mediastinum. *(Page 734)*

The tracheobronchial tree consists of about 23 generations of branching tubes that deliver air to the alveoli. The trachea, bronchi, and bronchioles are the larger airways and constitute the conducting zone of the tree, through which air flows to the respiratory surfaces that exchange oxygen and carbon dioxide.

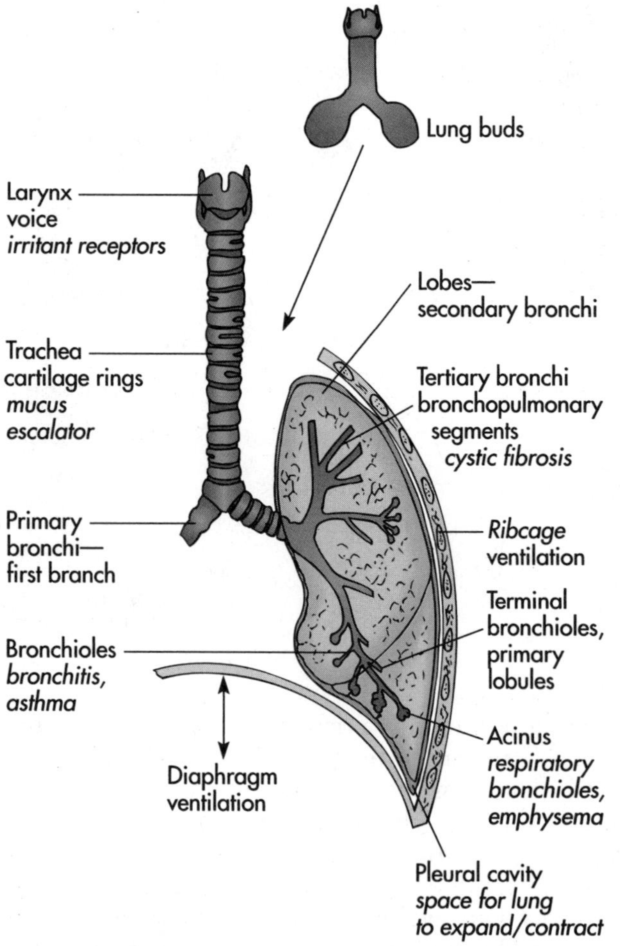

Figure 24.22

Anatomical summary of lung and tracheobronchial tree.

The epithelium of the conducting zone is columnar and cuboidal, with goblet and ciliated cells that trap particulates in the mucus escalator. Respiratory passages are thin-walled respiratory bronchioles, alveolar ducts, alveolar sacs, and alveoli. These surfaces secrete surfactant that keeps these alveoli open, but they do not secrete mucus. *(Page 734.)*

Two vascular trees that accompany the branches of the tracheobronchial tree deliver blood to the lungs. The arteries of the pulmonary tree deliver deoxygenated blood from the pulmonary trunk and right side of the heart to capillaries that cover the alveoli, and venous tributaries drain to the pulmonary veins and the left side of the heart. A smaller bronchial tree delivers oxygenated blood from the left side of the heart to the walls and connective tissues of the bronchi and larger bronchioles. Some of this blood returns to the right side of the heart by way of bronchial veins, but most mixes with oxygenated pulmonary blood. *(Page 736.)*

The lungs are divided into subunits. The largest divisions are lobes, which contain bronchopulmonary segments. The segments in turn contain secondary lobules that contain primary lobules and, within the latter, acini. Acini consist entirely of respiratory passages—all the passages that branch from one respiratory bronchiole. *(Page 741.)*

Pulmonary exchange surfaces consist of capillaries and a thin layer of connective tissue crowded between thin alveolar epithelial cells. These structures called alveolar septa usually are less than 2 microns across, thin enough for gases to diffuse from either side into the capillaries. The lungs contain approximately 700 ft² of alveolar septa. Type I pneumocytes form most of this surface, and Type II pneumocytes secrete surfactant that reduces surface tension between air and water on the surface of the epithelial cells. Elastic fibers permit inflation and deflation, and macrophages remove airborne microorganisms. *(Page 743.)*

Chronic obstructive pulmonary disease (COPD) diminishes the effectiveness of the respiratory zone of the lungs. Thick, congestive mucus in the conducting airways predisposes cystic fibrosis patients to respiratory infections because these passages cannot be cleared as readily as normal airways. The bronchi and bronchioles of asthma patients become sensitive to various stimuli that constrict and obstruct airflow to the acini. Emphysema usually is associated with chronic inflammation of the bronchi and results in enlarged alveoli when the fibrous support of respiratory surfaces collapses. *(Page 736, 740, 741, 744.)*

The nasal cavity conditions air as it is inhaled. Nasal conchae and a vascular mucosa moisten, heat, and clean air as it flows over the conchae and through the meati of the nasal cavity. The respiratory epithelium is pseudostratified ciliated columnar, with mucous glands and goblet cells that trap particulates in a film of mucus that is conducted toward the pharynx by cilia. Extensive capillary beds supply heat and fluid that condition air. *(Page 745.)*

The paranasal sinuses are chambers in the skull that are lined with the same type of epithelium as the nasal cavity, but the small openings that communicate with the nasal cavity do not appear to admit air rapidly enough for them to be major air conditioning structures. The paranasal sinuses appear to have no single compelling role in the respiratory system. *(Page 747.)*

The pharynx delivers air between the nasal cavity and larynx. The soft palate and epiglottis direct this flow and prevent food and foreign objects from entering the larynx. The nasopharynx communicates with the eustachian tubes, and the palatine tonsils line its walls. *(Page 747.)*

The larynx produces sound for speech. This cartilaginous box suspends the trachea below the hyoid bone. The vocal folds, stretched between the thyroid and arytenoid cartilages, produce sound when exhaled air causes them to vibrate. Pitch and tone are regulated by the tension and length of the cords. The epiglottis guards the entrance to the glottis, the opening between the vocal folds. *(Page 749.)*

Air flows into the lungs when the thoracic cavity expands and flows out when it becomes smaller. During inhalation, the intercostal muscles raise the rib cage, the diaphragm flattens, and the lungs expand downward and outward into the enlarged space. In exhalation the muscles of the rib cage relax or forcefully collapse the thoracic cavity and air flows outward as the abdominal organs, with help from muscles of the abdominal wall, press the diaphragm upward. Breathing centers in the medulla coordinate ventilation with oxygen and carbon dioxide levels in the blood. *(Page 750.)*

The tracheobronchial tree is the developmental core of the lungs. Beginning as pairwise buds and branches, the tree branches from the trachea progressively outward to the respiratory zone and alveoli. The conducting zone forms during the pseudoglandular stage, and the respiratory zone appears during the canalicular stage. The terminal sac stage adds more alveolar ducts, and alveoli begin to form and secrete surfactant, but most alveoli accumulate during the first year of life as pulmonary capillaries connect with them and thin, mature alveolar septa form. *(Page 753.)*

STUDY QUESTIONS
OBJECTIVES

1. What are the major functions of the organs and passageways of the respiratory system? Which organs perform these functions?   **1**

2. Which walls in the thoracic cavity shape the bases, apices, and medial surfaces of the lungs?   **2**

3. Name the passages of the tracheobronchial tree. How many generations of branching do these passages represent overall, on the average? How many generations are in the conducting zone and how many in the respiratory zone?   **3**

4. Describe the location of an acinus within a bronchopulmonary segment. If there are six ranks of branches in the respiratory zone of the lungs, how many more alveolar sacs than terminal bronchioles would you expect a lung to contain?   **4**

5. The epithelium becomes thinner the more distally on the tracheobronchial tree it is located. What other changes in the walls of airways accompany thinner epithelium?   **5**

6. In what way does the thoracic cavity contain the lungs and pleural cavities? Are the lungs themselves inside the pleural cavities?   **6**

7. Trace a molecule of oxygen from the nasal cavity to an alveolar sac in the first bronchopulmonary segment of the tracheobronchial tree. Follow a molecule of carbon dioxide from the blood across the gas exchange surface into an alveolar sac.   **7**

8. What physiological principle leads to small, numerous alveoli? Why are the enlarged acini of emphysema less effective gas exchangers?   **8**

9. Describe the general features of COPD. What characteristics classify cystic fibrosis, asthma, and emphysema among chronic obstructive pulmonary diseases?   **9**

10. How does the nasal cavity condition air?   **10**

11. Which are the four paranasal sinuses? Where are they located, and where do they communicate with the nasal cavity?   **11**

12. What is the position of your soft palate when you say the letter *K*? What is the position taken by the soft palate when you inhale through your nose, exhale through your mouth, or exhale through your nose?   **12**

13. How does the larynx prevent most foreign bodies from entering the trachea?   **13**

14. How does the cough reflex activate the muscles and motions of ventilation? What muscles are responsible for gentle inhaling, deep inhalation, and forcible exhalation?   **14**

15. What state of development and function have the lungs reached at 28 weeks of pregnancy? At birth?   **15**

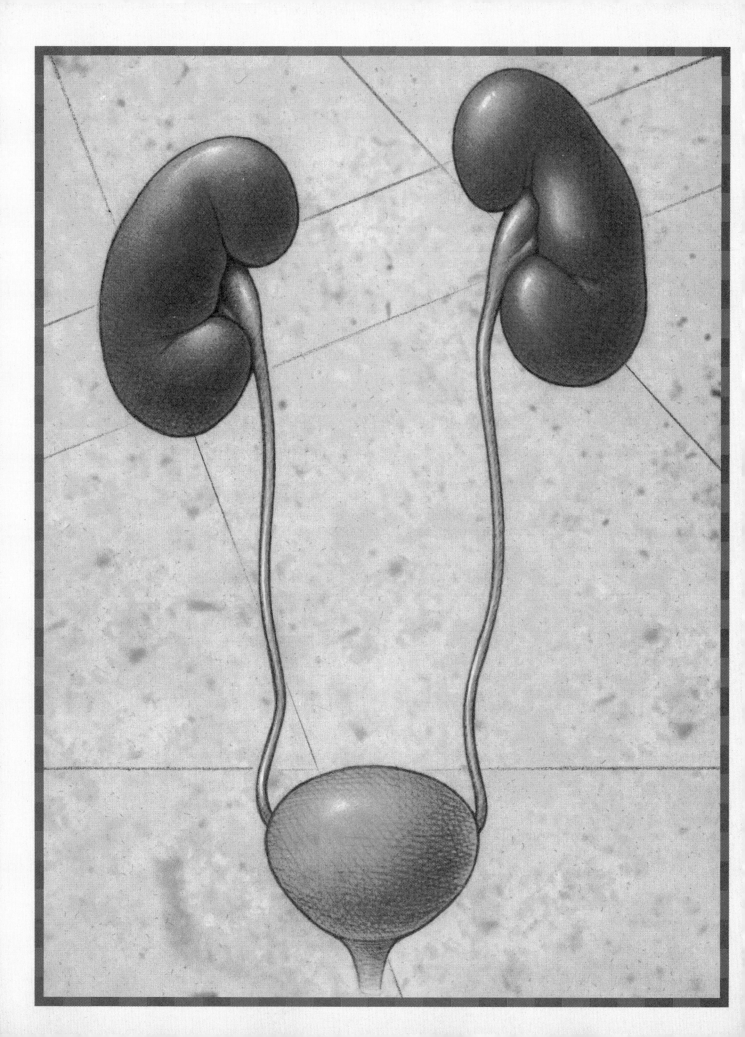

CHAPTER 25

Urinary System

OBJECTIVES

After studying this chapter, you should be able to do the following:

1. Name the organs of the urinary system and describe their functions.
2. Describe the flow of blood, filtrate, and urine through the kidneys, as well as the flow of urine through the ureters and bladder.
3. Describe the gross anatomy of the kidneys and locate them in the body.
4. Describe the internal organization of the kidney.
5. Trace the flow of blood through the renal arterial and venous trees.
6. Describe the various structures and functions of nephrons.
7. Describe the elements of the glomerular filtration barrier and the types of cells and molecules each level blocks from passing into the filtrate.
8. For each region of the tubules, name the molecules reabsorbed and the mechanisms of resorption.
9. Explain how kidney diseases interfere with filtration and reabsorption.
10. Describe how the juxtaglomerular apparatus responds when filtration is (1) too rapid or (2) too slow.
11. Describe the anatomy of the ureters, bladder, and urethra, and point out features that all three structures share.
12. Explain how the micturition reflex works.
13. Describe the connections that join donor kidneys to the recipient's circulation and urinary system.
14. Describe the development of nephrons and of collecting portions of the kidney.
15. Describe how developmental errors interfere with kidney anatomy.

 **VOCABULARY TOOLBOX**

Terms related to the urinary system.

calyx, calyces	calyx, *G* cup. The calyces form cups around the medullary pyramids.	**nephron**	nephr-, *G* the kidney.
cortex, cortical	cort-, *L* the bark, shell. The cortex is the outer tissue of the kidney.	**parenchyma, parenchymal**	par, *G* beside, near; enchyma, *G* an infusion. Parenchymal cells lie beside the branches of vessels and the ureter.
detrusor	trus, *L* push, thrust. This muscle squeezes urine from the bladder.		
glomerulus, glomerular	glomus, *L* a ball of yarn, a ball.	**pelvis**	pelv-, *L* a basin. The renal pelvis receives urine from many branches.
kidney	kiden or kydn, *Middle English;* the excretory organ of the urinary system.	**renal, renin**	ren-, *L* a kidney.
		trigone	trigon-, *G* triangle, triangular.
medulla, medullary	medull-, *L* marrow, pith. The inner layer of kidney tissue.	**urine, ureters, urethra**	ure, *G* the tail.

PERSPECTIVE

The urinary system regulates the composition of blood and body fluids. Virtually every organ in the body releases substances into the blood and removes others from it. The kidneys balance these changes by conserving the constituents needed for body functions and excreting the excesses. Kidneys also secrete hormones such as erythropoietin, which stimulates bone marrow to produce erythrocytes. As you might expect, different parts of the kidney are responsible for different functions, so that when diseased kidneys fail, the composition of urine often identifies the malfunctioning structures for the physician. This chapter shows you how the kidney's anatomy is related to urine production.

The urinary system consists of the kidneys, **ureters, urinary bladder, and urethra,** shown in Figure 25.1. **Homeostasis** of blood and tissue fluids is its primary action (see Homeostasis, Chapter 1). The kidneys continually adjust to changing conditions in the body by filtering blood and excreting excess substances in the form of urine. As blood circulates through the kidneys, excess water, urea, electrolytes, and drugs are removed, while essential quantities of water and electrolytes, as well as certain substances, such as serum proteins and glucose, remain in the blood. Conducted by the ureters to the urinary bladder, urine is stored until the urethra empties to the outside.

The kidneys also act as **endocrine glands** (see Other Endocrine Tissues, Chapter 22). Kidneys regulate arterial blood pressure by secreting renin into the blood; renin causes the walls of vessels to constrict, raising blood pressure. By secreting erythropoietin, the kidneys also stimulate bone marrow to produce erythrocytes, and they convert vitamin D into its active form as it circulates through the kidney. When the liver and small intestine do not supply adequate glucose, the kidney synthesizes it from circulating amino acids.

Nephrons are the fundamental structures of the kidney that perform these excretory and endocrine functions (Figure 25.2). Each kidney contains about 1.3 million nephrons, more than enough to maintain homeostasis. At any given moment when the body is at rest, the nephrons receive one quarter of the blood pumped by the heart, some 1900 l (475 gal) of blood per day; only the brain receives a larger share. Only 1 or 2 l of urine actually derive from this enormous flow of blood, however; the rest of the fluid returns to the body through a highly selective process that excretes the excess substances and returns the essential ones. Selection begins when about 10% of the blood fluid, about 190 l (48 gal) daily, filters into the nephrons and becomes filtrate. The nephrons reabsorb nearly all of this potential urine back to the blood, excepting urea and excess water. These and other substances that remain in the filtrate become urine.

- What is the primary role of the urinary system?
- What endocrine roles do the kidneys perform?
- What is a nephron?
- How much blood circulates daily through the kidneys, and how much of this fluid becomes urine?

Case studies begin on p. 780.

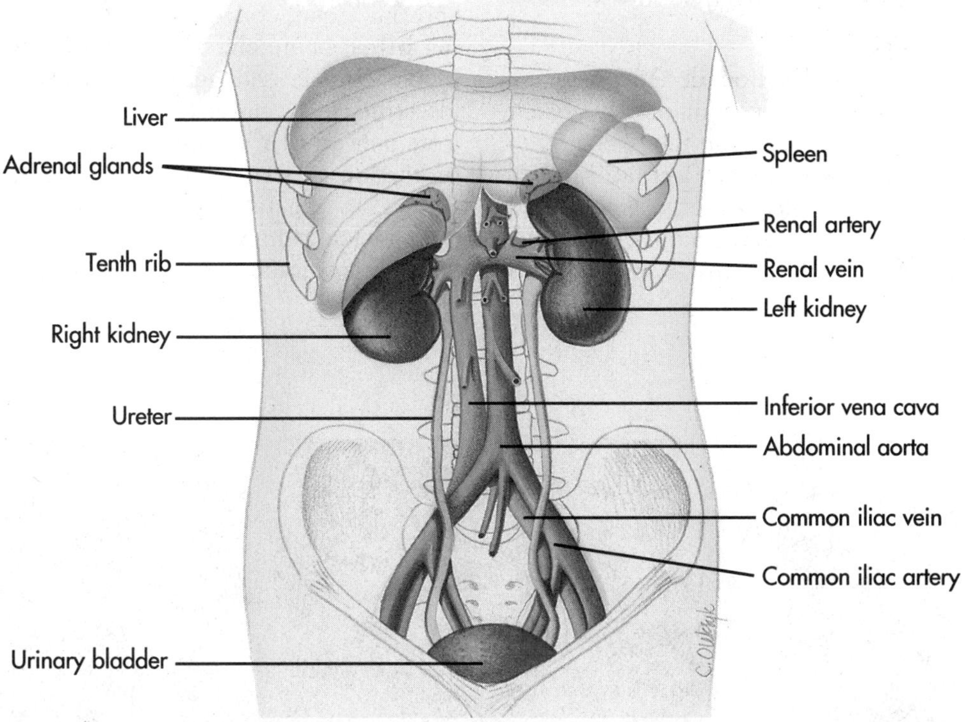

Figure 25.1
The urinary system in a female. The parietal peritoneum has been removed to show the kidneys, ureters, and vessels.

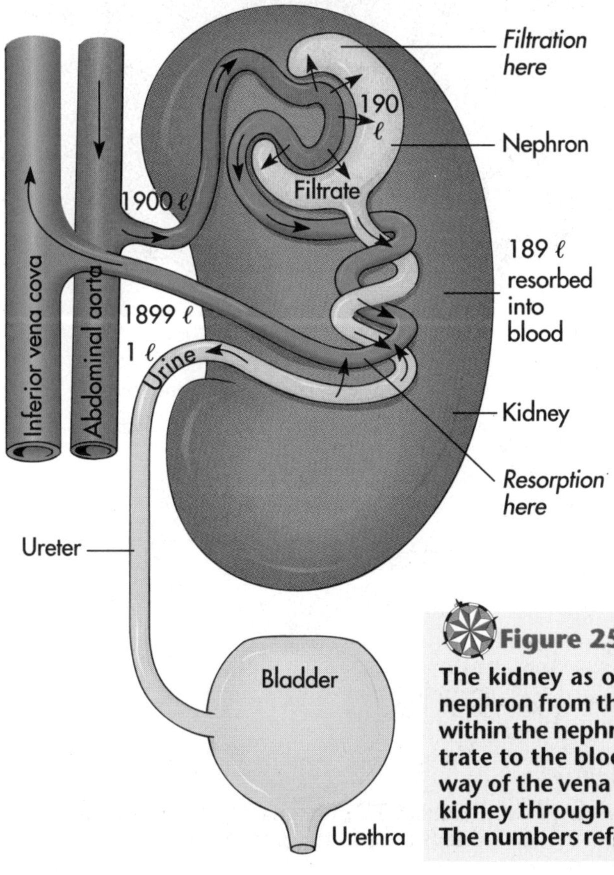

Figure 25.2
The kidney as one large nephron. Blood flows to the nephron from the aorta and some of it becomes filtrate within the nephron. The nephron returns most of the filtrate to the blood, which then returns to the body by way of the vena cava. A small amount of urine exits the kidney through the ureter to the bladder and outside. The numbers refer to liters of blood, filtrate, or urine.

KIDNEYS

EXTERNAL ANATOMY

The pair of kidneys is located in the posterior abdominal wall, behind the peritoneum, on each side of the vertebral column between the twelfth thoracic and third lumbar vertebrae. Each **kidney** is shaped like a bean and is nearly as large as your fist, about 12 cm long, 5 cm broad, and 2.5 cm thick (5 × 2 × 1 inch). The **superior pole** (Figure 25.3A) of each kidney is somewhat smaller than the **inferior pole,** and the broad, convex **lateral margin** contrasts with the **hilus,** or notch, on the **medial margin.** The left kidney is usually larger and higher than the right. The lowermost ribs protect both kidneys against damage from behind, while anteriorly the liver provides some protection to the right kidney and the stomach and spleen, to the left. A **renal capsule** (REEN-al) encloses each kidney in a thin fibrous covering (Figure 25.3B). The renal capsule in turn is cushioned by a

A

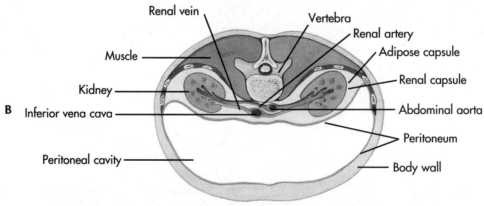

B

Renal vein
Muscle
Kidney
Inferior vena cava
Peritoneal cavity
Vertebra
Renal artery
Adipose capsule
Renal capsule
Abdominal aorta
Peritoneum
Body wall

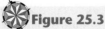

Figure 25.3

Gross anatomy of kidneys. A, Anterior view of right kidney. The fibrous capsule has been removed. **B,** Transverse section through the abdomen shows the location of kidneys behind the peritoneum.

shock-absorbing layer of fatty tissue, the **adipose capsule.** Finally, surrounding the adipose capsule, the **renal fascia** attaches the kidney to the body wall and the parietal peritoneum. In occasional individuals a mesentery suspends the kidney from the posterior abdominal wall, and the kidney then is said to float among the adjacent organs (floating kidney). An **adrenal gland** caps the superior pole of each kidney within the adipose capsule. Even though the adrenal glands regulate kidney activity, vessels do not connect them directly to the kidneys; adrenal hormones must circulate back to the heart before entering the kidneys (see Adrenal Glands, Chapter 22).

The **renal artery** enters the kidney at the hilus, and the **renal vein** and **ureter** exit from the same point. **Sympathetic neurons** from the splanchnic nerves and celiac plexus enter also through the hilus and end on internal blood vessels that stimuli cause to constrict. **Parasympathetic fibers** from the vagus nerve and celiac plexus also innervate the kidneys, and visceral afferents register pain. Four or five **lymph vessels** drain from the hilus into the thoracic duct.

MEDULLA AND CORTEX

The longitudinal section through the kidney shown in Figure 25.4A illustrates how blood reaches the

nephrons and how urine exits the kidneys. Reddish-brown kidney tissue called **parenchyma** (pahr-EN-kih-ma) contains the nephrons and displays two fundamental cellular layers—the outer **cortex** and inner **medulla.** Nephrons reside in the cortex, where numerous nephrons and blood-rich capillaries give the parenchyma a dark, granular appearance. Medullary parenchyma, however, appears striated because straight ducts and capillaries pass through it. Medullary parenchyma also is lighter in color than the cortex because less blood circulates through the medulla.

Together, the medulla and cortex subdivide into 11 more or less distinct **lobes.** Lobar organization is more obvious in the medulla, where each lobe appears as a **medullary pyramid** (MED-you-lair-ee) whose apex, or **renal papilla** (pah-PILL-ah), receives urine from **collecting ducts** and **papillary ducts** that drain from the nephrons. Deeper within the lobes, each collecting duct receives urine from about 100 nephrons. These clusters of nephrons are called **renal lobules.** Accordingly, each lobe contains perhaps 1000 lobules and 100,000 nephrons. Folds of parenchyma called **renal columns** extend into the medulla between the medullary pyramids, carrying vessels and nerves to the nephrons in the cortex. The cortical portions of lobes have no obvious boundaries between them.

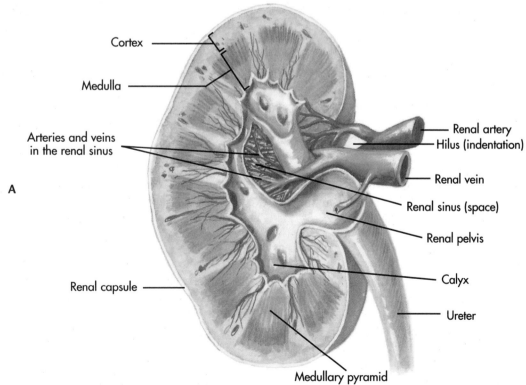

Cortex

Medulla

Arteries and veins in the renal sinus

A

Renal capsule

Renal artery

Hilus (indentation)

Renal vein

Renal sinus (space)

Renal pelvis

Calyx

Ureter

Medullary pyramid

Figure 25.4

Cortex, medulla, and renal pelvis, longitudinal section, right kidney. A, This diagram shows the relation between renal pelvis, medullary pyramids, and renal cortex.

Continued.

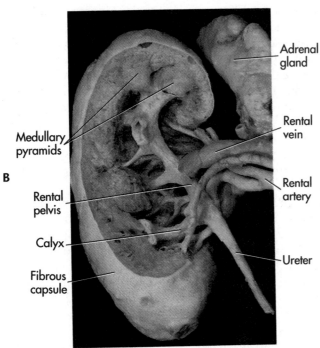

B

Adrenal
gland

Medullary
pyramids

Rental
vein

Rental
artery

Rental
pelvis

Calyx

Ureter

Fibrous
capsule

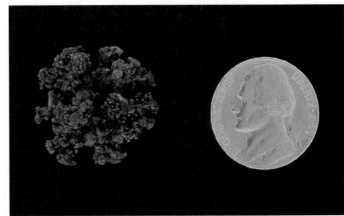

C

Figure 25.4, cont'd
B, Photograph of similar view. **C,** Renal calculus.

- Where are the kidneys located?
- What is renal parenchyma and how can it be identified?
- Where are the cortex and medulla located in the kidney?

RENAL PELVIS

The **renal pelvis** of the ureter is the large central chamber that receives urine from the **major calyces** (KALE-ih-sees), the main branches of the renal pelvis (Figure 25.4A and B). **Minor calyces** branch from the **major calyces;** each **minor calyx** covers the apex of a medullary pyramid. Urine drains from the pyramids into the minor calyces, and thence to the major calyces and the renal pelvis. Each minor calyx drains about 100,000 nephrons. Kidney stones, or **calculi** (KAL-kue-lee), may accumulate in the calyces and renal pelvis, where they sometimes block urine flow. Large calculi, like that shown in Figure 25.4C, form casts of the renal pelvis and its branches. To pass even a small stone through the ureter is excruciatingly painful. Many kidney stones can be broken into small fragments by **lithotripsy** (LITH-oh-trip-see), an ultrasound treatment that shatters stones into very small pieces that can pass through the ureter.

The **renal sinus** is a space that lies between the branches of the ureter and the parenchyma. Arteries, veins, lymphatics, and nerves travel into and out of the kidney through the renal sinus. Adipose tissue occupies the space between vessels and nerves in the renal sinus. The renal sinus forms (see Development, this chapter) when the ureter branches into the kidney.

RENAL VASCULAR TREES

Arterial and venous trees deliver blood to the kidneys. The **renal arterial tree** delivers blood from the aorta to all nephrons of each kidney (Figure 25.5A). The **renal venous tree** returns the reconditioned blood to the renal vein and inferior vena cava (see Vascular Trees, Chapter 14). The arterial tree begins branching when the **renal artery** separates into five **segmental arteries,** each with blood for a renal segment, as shown in Figure 25.5B and C. With their own largely independent blood supplies, **renal segments** are analogous to bronchopulmonary segments of the lungs. Surgeons prefer to enter the kidney between these segments to remove calculi because they risk less damage to major vessels.

Segmental arteries travel within the renal sinus, around the renal pelvis, into the renal columns, and then branch into **interlobar arteries** that carry blood between the lobes of the kidney. Next, the interlobars radiate into adjacent lobes as **arcuate arteries** (AR-kue-ate), following the boundary between cortex and medulla. Inside each lobe the arcuates send many **interlobular arteries** (note the spelling) into the cortex between lobules. Finally, **afferent arterioles** deliver blood to individual nephrons in each lobule. As a result of this organization, each renal segment receives

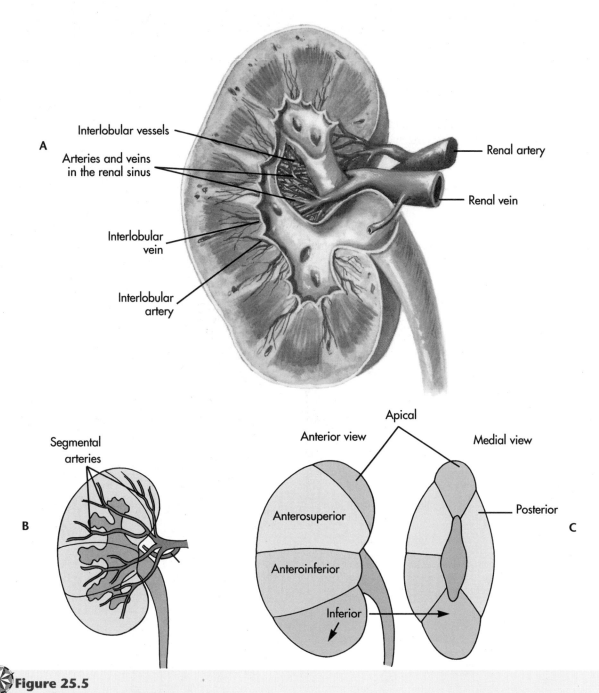

A

Interlobular vessels

Arteries and veins
in the renal sinus

Interlobular
vein

Interlobular
artery

Renal artery

Renal vein

B

Segmental
arteries

Apical

Anterior view

Medial view

Anterosuperior

Anteroinferior

Inferior

Posterior

C

✳ **Figure 25.5**

Renal vascular trees and segments of the kidney. A, Longitudinal section illustrates the branches of the arterial and venous trees. B, Segmental arteries. C, Segments of the kidney.

its own blood supply, but each renal lobe within the segment receives blood from several different interlobar arteries, and each lobule receives from several interlobular arteries.

The venous tree parallels the arterial tree for most of its length, draining reconditioned blood from nephrons into the renal vein. The most important difference between the trees occurs as blood leaves the nephrons into interlobular veins or more directly into arcuate veins. Beyond this point, however, arterial

and venous trees are similar. Arcuate veins deliver blood from several lobes to interlobar veins; these veins drain each renal segment into segmental veins that finally discharge into the renal vein and inferior vena cava.

Renal arteries and veins may vary considerably from one individual to the next, but these differences rarely interfere with kidney function. There may be two renal veins for each kidney, or vessels may be much longer or shorter than normal (Figure 25.5).

Short renal veins in a donor kidney are more difficult for renal transplant surgeons to connect to a recipient's circulation.

- Name the branches of the renal pelvis.
- Which branches receive urine directly from renal lobes?
- Which renal vessels receive blood from the abdominal aorta, and which ones discharge into the glomerulus?

NEPHRONS

Nephrons are structural units of excretion. They are long narrow tubules, some 2.5 cm (1 in) long by 0.5 mm in diameter, as illustrated in Figure 25.6. The cup-shaped apex, or **renal corpuscle** (KORE-puss-l), of each nephron filters the blood, and the rest of the tubule reabsorbs back into the blood constituents that have entered the filtrate inside the

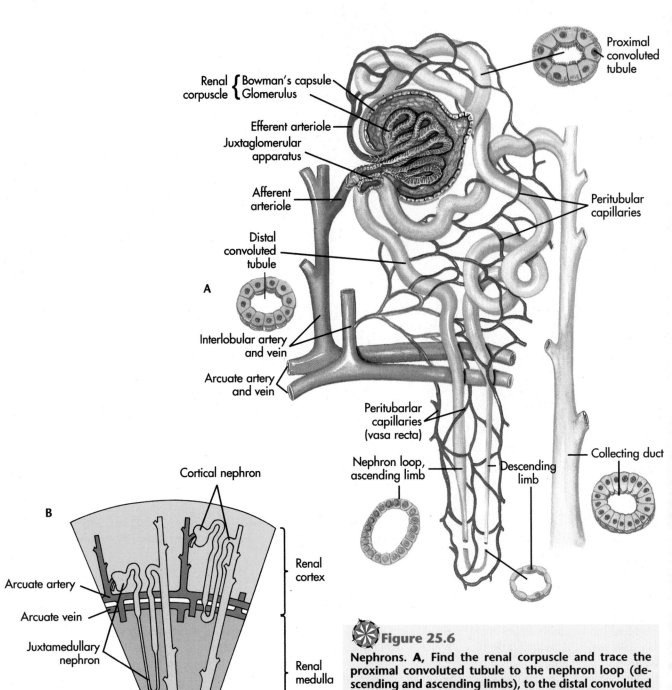

Figure 25.6

Nephrons. **A,** Find the renal corpuscle and trace the proximal convoluted tubule to the nephron loop (descending and ascending limbs), to the distal convoluted tubule, and to the collecting duct. Follow the flow of blood from the afferent arteriole into the glomerular capillaries, out the efferent arteriole, to the peritubular capillaries and vasa recta, to an interlobular and arcuate vein. **B,** Cortical and juxtamedullary nephrons.

nephron. A renal corpuscle consists of the **glomerular**, or **Bowman's, capsule** and the **glomerulus** (glow-MARE-you-luss), an arrangement that resembles a Thermos vacuum bottle. Like the bottle, the glomerular capsule is a double-walled chamber that encloses the glomerulus, a tuft of capillaries in the space where the bottle would store soup or cold drinks. These capillaries enter the **vascular pole** of the renal corpuscle, equivalent to the mouth of the bottle. Fluid filters from the glomerulus into the **capsular space,** which corresponds to the vacuum space between the walls of the bottle. The outer, or **parietal wall,** of the glomerular capsule encloses the capsular space, and the inner, or **visceral wall,** surrounds the glomerulus itself.

The fluid, now called **filtrate,** flows from the capsular space into the **proximal convoluted tubule** (PCT) and then down the descending limb of the **nephron loop,** or **Henle's loop** (HEN-leez), and up the ascending limb, where it passes near Bowman's capsule at the **juxtaglomerular apparatus** (JUKS-tah-glow-MARE-you-lar) that regulates filtration rate (see page 773). The **distal convoluted tubule** (DCT) carries filtrate to the end of the nephron. The filtrate flows into a common **collecting duct** that receives filtrate from adjacent nephrons. Collecting ducts drain through the medullary pyramids in individual **papillary ducts** (PAP-ill-air-ee) that in turn drain from the apex of each pyramid into the minor calyces and the ureter. Collecting ducts and papillary ducts contribute to the striated appearance of the medulla.

Blood arrives at the vascular pole of each glomerular capsule through an **afferent arteriole,** traverses the **glomerular capillaries,** and leaves through an **efferent arteriole.** A branching network of **peritubular capillaries** (PAIR-ih-TUBE-you-lar) carries blood from the efferent arteriole over the external surface of the proximal and distal convoluted tubules and collecting tubules, taking up fluid the nephron reabsorbs from the filtrate and receiving other regulatory products. This capillary network extends down Henle's loop and the collecting ducts as the **vasa recta** (VAH-sa REK-tah), upright vessels that also contribute to the striated appearance of the medulla. Most blood bypasses the vasa recta, however, which circulate only about 1% of the kidney's blood. Interlobular and arcuate veins return venous blood from the peritubular capillaries and vasa recta.

Kidneys contain cortical nephrons and juxtamedullary nephrons. The difference between these types, shown in Figure 25.6B, concerns how the kidney concentrates urea in the urine. Eighty-five percent of human nephrons are cortical; they reside entirely in the cortex, their short nephron loops descending only part of the way toward the medulla. About 15% are juxtamedullary nephrons that reside at the boundary between cortex and medulla and send very long nephron loops deep into the medulla. These loops help concentrate urea in the urine (see page 768), and they add to the striated appearance of the medullary pyramids.

- Name the regions of the nephron.
- Which portion contains the capsular space?
- Which section follows the proximal convoluted tubule?

GLOMERULAR FILTRATION BARRIER

THREE ELEMENTS

Renal corpuscles are responsible for filtering blood across a highly selective yet rapidly permeable **glomerular filtration barrier** that lies between the glomerular capillaries and the visceral wall of the glomerular capsule, shown in Figure 25.7A through C. The glomerular filtration barrier consists of three elements. Each element acts somewhat as a sieve, finer than the previous one, that prevents progressively smaller molecules from passing. The first element is the **endothelium** of the capillaries, composed of squamous endothelial cells punctured by numerous small openings or **fenestrae.** The **glomerular basement membrane** is the second part of the barrier. This modified basement membrane attaches the capillary endothelium to the third element—the **visceral endothelium** of the glomerular capsule—whose highly specialized **podocytes** (PO-do-SITES; Figure 25.7B and C) tightly grasp each capillary. Podocytes are unique cells that extend cellular processes called **trabeculae** around the glomerular capillaries. Numerous small **pedicels** (PED-ih-cells), or foot processes, extend like a fringe from the trabeculae and interdigitate with the pedicels from other trabeculae and podocytes. This arrangement creates a narrow **filtration slit** between the pedicels, through which molecules pass on their way into the capsular space. To enter the capsular space, components of the blood must (1) pass from the lumen of the glomerular capillaries through the fenestrae in the endothelium of the capillaries, (2) cross the glomerular basement membrane, and (3) finally, flow through the filtration slit. About 10% of the blood that enters the glomerulus passes through this barrier and becomes filtrate, amounting to about 190l (48 gal) of filtrate per day or 125mm/min in a normal individual.

Figure 25.8 details the elements of the filtration barrier. You are looking at a transverse section through a loop of glomerular capillaries; blood is flowing toward you in some capillaries and away in others. Capsular space and filtrate entirely surround this loop. Thin, squamous endothelial cells with nu-

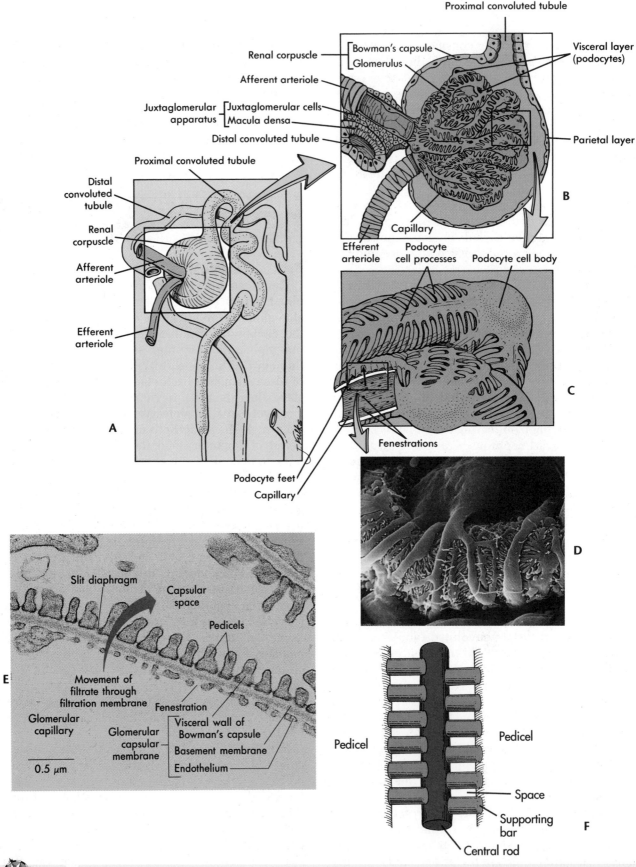

Figure 25.7

Renal corpuscle and glomerular filtration barrier. **A,** Renal corpuscle. **B,** Longitudinal section through the corpuscle, with glomerulus inside glomerular capsule. **C,** Podocytes wrap the glomerular capillaries with interdigitating pedicels. **D,** Scanning electron micrograph of podocytes. (Scanning electronmicrograph by permission of R.G. Kessel and R.H. Kardon, *Tissues and organs: A text-atlas of scanning electron microscopy,* 1979, W.H. Freeman and Co.) **E,** Transmission electron micrograph shows filtration slit and diaphragm between the pedicels of different podocytes. **F,** Structure of filtration slit diaphragm.

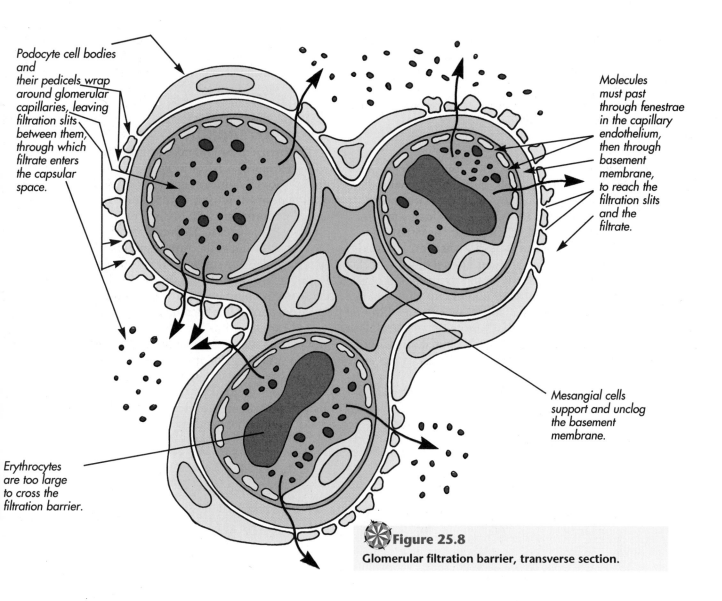

Podocyte cell bodies and their pedicels wrap around glomerular capillaries, leaving filtration slits between them, through which filtrate enters the capsular space.

Molecules must past through fenestrae in the capillary endothelium, then through basement membrane, to reach the filtration slits and the filtrate.

Mesangial cells support and unclog the basement membrane.

Erythrocytes are too large to cross the filtration barrier.

Figure 25.8
Glomerular filtration barrier, transverse section.

merous **fenestrae** (FEH-nes-tree; *L*, window), or pores, line the capillary walls, and podocytes cover the external surface of the entire capillary loop. A central core of **mesangial cells** (mezz-AN-jee-al) secretes a matrix that supports the loop. This matrix is continuous with the glomerular basement membrane between the capillary endothelium and the podocytes. Mesangial cells, capillary endothelium, and podocytes are thought to produce and maintain the glomerular basement membrane.

FILTRATION
STEP 1

At 70 nm in diameter, numerous open fenestrae in the capillary endothelium allow proteins and smaller molecules to pass rapidly from the lumen of the glomerular capillaries. Large macromolecular immunoglobin complexes, platelets, erythrocytes, and leucocytes are blocked, however, and remain in the blood. Unlike fenestrae in the choroid plexus of

the brain and in endocrine glands, these fenestrae have no closing diaphragms.

STEP 2

Water, smaller proteins, glucose, amino acids, electrolytes, drugs, vitamins, urea, and other molecules that have passed through the fenestrae now approach the glomerular basement membrane. This structure is a molecular sieve made of **proteoglycans** (PRO-tee-oh-GLY-kans), negatively charged macromolecular complexes of protein and polysaccharides (see Basement Membranes, Chapter 3). The molecular pores of the glomerular basement membrane exclude proteins larger than 7 nm in diameter, or about 65,000 Daltons molecular weight. Because the membrane is negatively charged, electrostatic repulsion also excludes many negatively charged molecules. As a result, most serum proteins, low density lipoproteins (LDL), high density lipoproteins (HDL), and very low density lipoproteins (VLDL) remain in the blood.

Small peptides, such as cholecystokinin (a chain of 11 amino acids), do pass through, along with water molecules, electrolytes, glucose, amino acids, drugs, urea, and other small molecules. Mesangial cells are thought to remove any material that clogs the spaces in the glomerular basement membrane.

STEP 3

The final elements of the filtration barrier are the filtration slits, approximately 20 to 30 nm wide, between the podocytes. The slits provide a large area for filtration, but they are too wide to exclude anything that has already passed through the basement membrane. A thin **filtration slit diaphragm,** approximately 7 nm thick can be seen connecting pedicels across the slit (Figure 25.7E and F), but it is not clear what substances if any this diaphragm may block. After passing through the filtration slit, the filtrate contains peptides and small proteins, electrolytes (principally Na^+, K^+, and Cl^-), and metabolites such as glucose, amino acids, vitamins, drugs, urea, and water.

- What structures constitute the glomerular filtration barrier?
- Which of these elements is made of podocytes?

- How much filtrate passes through the glomerular filtration barrier in a day?
- What materials and substances are retained by the capillary fenestrae?

NEPHROTIC SYNDROME AND PROTEINURIA

There are many forms of renal failure that occur when glomerular filtration and tubular reabsorption decline. Renal failure may manifest itself in the release of high quantities of protein in the urine, a condition called **proteinuria** (PRO-teen-YOUR-ee-ah). Healthy glomeruli release 2 to 2.5 gm of protein into the filtrate daily, but because the tubules reabsorb nearly all of this, normal urine contains only 150 mg of this protein. In **nephrotic syndrome** (nef-ROT-ik), however, the urine may contain up to 20 gm of proteins because the filtration barrier has become more permeable.

As you should expect, in nephrotic syndrome the glomerular basement membrane itself is more permeable to proteins. The molecular pores in the glomerular basement membrane increase in size, allowing more proteins to pass. Filtration slits and pedicels also are involved, as illustrated in Figure 25.9. The podocytes withdraw their pedicels, and the filtration

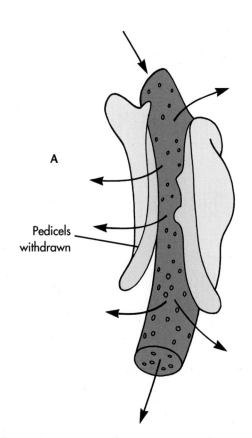

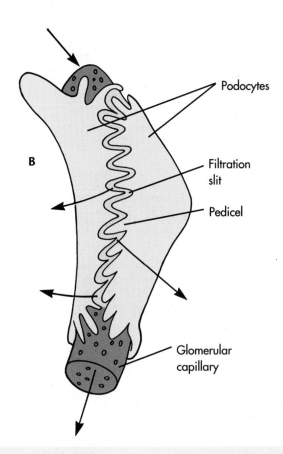

A — Pedicels withdrawn

B — Podocytes, Filtration slit, Pedicel, Glomerular capillary

✳ **Figure 25.9**

Defects in the glomerular filtration barrier of nephrotic syndrome. Nephrotic syndrome disrupts filtration slits (A), compared with (B) normal slits.

slits become wider as the margins of the cells flatten. Filtration slit diaphragms break into pieces, and the podocytes detach from the basement membrane. All of these effects may stem from malfunctioning podocytes that may fail to renew the basement membrane that attaches the cells and allows pedicels to form.

GLOMERULONEPHRITIS

In **glomerulonephritis** (glow-MARE-you-low-nef-RYE-tiss), inflammation of the glomerulus can lead to kidney failure by destroying nephrons. Antigen-antibody complexes lodge in the glomerular basement membrane as part of the inflammatory process, and collagen fibers begin to replace the proteoglycans of the membrane. In **pyelonephritis** (PIE-el-oh-NEF-rye-TISS), inflammation begins in the renal pelvis when bacterial infections reflux up the ureters. The inflammation can spread through the collecting system to the glomeruli and cause glomerulonephritis. **Hematuria** (HEEM-at-YOUR-ee-ah), or the appearance of blood in the urine, is a sign that glomerulonephritis or other source of damage has introduced erythrocytes into the filtrate and urine. Hematuria also can indicate that infection has ruptured capillaries in the wall of the bladder or ureter.

TUBULES AND REABSORPTION

The tubular portion of the nephron returns most of the components of the filtrate to the blood. The fluid that is not reabsorbed by the wall of the tubule becomes urine. The proximal convoluted tubule, nephron loop, distal tubule, and the collecting duct successively remove materials from the filtrate passing through them, shown in Figure 25.10A. Except for part of the nephron loop, the wall of the tubule is simple cuboidal epithelium surrounded by a basement membrane to which are bound interstitial cells and peritubular capillaries outside the tubule. The distal end of the nephron loop consists of simple squamous epithelium. Reabsorbed molecules pass from the lumen of the tubule, across the epithelium and basement membrane, into the peritubular capillaries. Tubular reabsorption is vital. The flow of filtrate is so great that all of your blood would be lost in 20 minutes, if nephrons did not return the filtrate to the blood.

The structure of the endothelium reflects the process of reabsorption. The endothelium selectively absorbs molecules from the filtrate by the action of sodium-potassium pumps in the cell membrane (Figure 25.10). Located on numerous infoldings of the plasma membrane at the base and sides of the epithelial cells, these ion pumps consume much of the kidney's energy as they pump sodium ions from the interior of the cells, out across the basement membrane, to the outside of the tubule. Sodium ions diffuse from the lumen of the tubule down this concentration gradient into the cells, across a brush border of **microvilli** on the luminal surface of the endothelium. Water follows the sodium ions by osmosis. Transport proteins in the membrane also cotransport metabolites, such as glucose and amino acids, from the tubules (see Activities of the Cell Membrane, Chapter 2). Mitochondria supply the ATP needed for transport. As you follow the filtrate through the tubule, follow also the changes in microvilli, mitochondria, and membrane infoldings that reflect the reabsorptive activities of each region.

Unlike glucose and amino acids, proteins do not return to the blood intact because the epithelial cells convert them to amino acids. Protein molecules enter the epithelium by endocytosis, and lysosomes degrade the proteins to amino acids in the cytoplasm. Amino acids then return to the peritubular capillaries.

PROXIMAL CONVOLUTED TUBULE

This longest portion of the nephron reabsorbs about 70% of water and electrolytes from the filtrate and all amino acids, glucose, vitamins, and proteins (Figure 25.10A). The anatomy of the endothelium reflects this vigorous work. A rich **brush border** of microvilli provides extra surface for resorption of water and ions and for endocytosis of proteins. Numerous **mitochondria** in the base of the cells provide energy for the sodium pumps, and abundant deep infoldings in the plasma membrane on the basal ends of the cells provide additional surface for more pumps that transport electrolytes and metabolites.

NEPHRON LOOP

The nephron loop concentrates salts in the medullary pyramids. The descending limb (Figure 25.10B) is freely permeable to water but relatively impermeable to sodium ions, so that water, rather than sodium ions, diffuses into the medullary pyramid and adjacent vasa recta. As the filtrate rounds the bend of the loop, permeability begins to change as the epithelium becomes more permeable to sodium and less permeable to water (Figure 25.10C). As a result, sodium ions begin to leave the loop as the filtrate rises through the ascending limb. Because the ascending limb becomes impermeable to water and pumps sodium and chloride ions from the filtrate into the interstitial spaces of the medulla, the salt concentration becomes four to five times higher in the medullary pyramids than in the cortex. There is relatively little blood in the vasa recta to dilute this salt because only about 1% of the blood in the kidney circulates through these vessels.

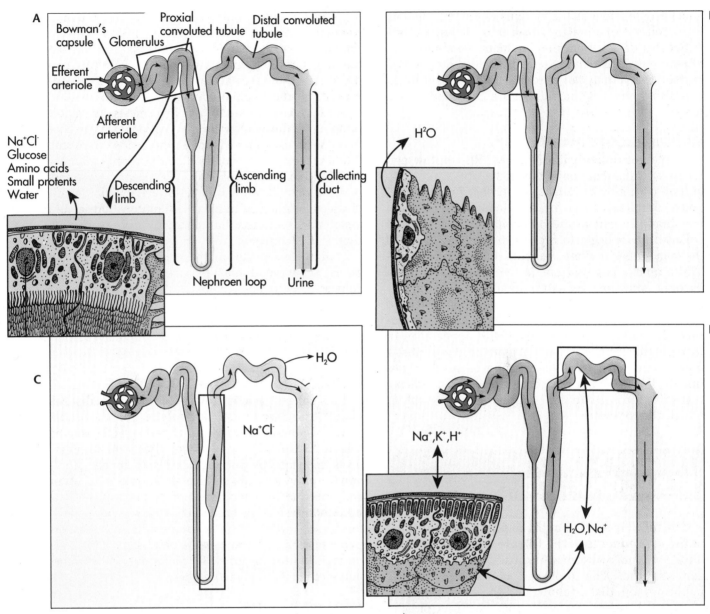

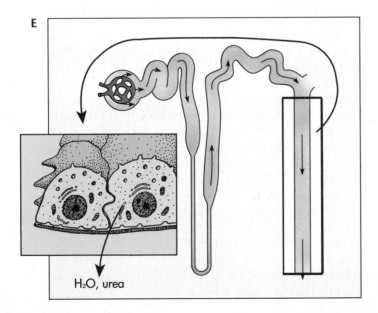

A

Bowman's capsule — Glomerulus — Proxial convoluted tubule — Distal convoluted tubule

Efferent arteriole

Afferent arteriole

Na⁺Cl⁻
Glucose
Amino acids
Small protents
Water

Descending limb

Ascending limb

Collecting duct

Nephroen loop

Urine

B H^2O

C H_2O

Na^+Cl^-

D Na^+, K^+, H^+

H_2O, Na^+

E H_2O, urea

✳ **Figure 25.10**

Anatomy of tubular reabsorption in a juxtaglomerular nephron. A, Proximal convoluted tubule reabsorbs electrolytes, water, glucose, proteins, amino acids, and vitamins from the filtrate. Cuboidal cells in the proximal tubule contain numerous microvilli, mitochondria, and basal infoldings. **B,** Descending limb of the nephron loop removes water but is impermeable to sodium ions. Cells are cuboidal in the proximal portion and squamous distally, where microtubules, mitochondria, and infoldings are sparse. **C,** Ascending limb of nephron loop is composed of squamous cells distally and cuboidal cells proximally. This limb reabsorbs sodium ions but not water. **D,** Distal convoluted tubule. Cuboidal cells with sparser brush border, fewer mitochondria, and infoldings transport less vigorously. The epithelium reabsorbs sodium and chloride ions. It secretes hydrogen ions, potassium ions, and ammonium ions, as well as urea. **E,** Collecting duct. Cuboidal cells with sparse microvilli, mitochondria, and membrane infoldings. Water and urea diffuse from the filtrate.

DISTAL CONVOLUTED TUBULE

This short portion of the nephron fine-tunes the composition of the filtrate by reabsorbing certain molecules and secreting others. Tubular reabsorption of electrolytes and water from the filtrate continues in the distal tubule. Microvilli are sparse and mitochondria fill the basal infoldings (Figure 25.10D). The adrenal gland promotes reabsorption of sodium ions and the secretion of potassium ions in the distal tubule by secreting **aldosterone.** To balance this return of sodium to the body fluids, **atrial natriuretic factor** (NATE-ree-you-RET-ik), a hormone secreted by the atrium of the heart, stimulates the distal tubule to secrete sodium and potassium into the filtrate, a process called **tubular secretion.** The distal tubule also regulates blood pH by secreting into the filtrate excess hydrogen ions produced by cellular metabolism.

COLLECTING DUCT

The collecting ducts conduct filtrate from the distal convoluted tubules through the medullary pyramids to the papillary ducts. As filtrate passes for a second time through the salty medulla (Figure 25.10E), urea diffuses from the collecting duct into the medulla and water also tends to follow. How much water diffuses from the filtrate into the medullary pyramids is controlled by **antidiuretic hormone** (ADH; AN-tee-die-you-RET-ik; also known as vasopressin) that is secreted by the posterior pituitary gland. High levels of ADH promote this return from the filtrate, called **antidiuresis,** by raising the permeability of the endothelial cells to water, but low ADH levels decrease their permeability so that the filtrate now becomes dilute, abundant urine **(diuresis).**

TWO-SOLUTE HYPOTHESIS

The medullary pyramids concentrate urea in the urine by removing proportionately more water than urea, as the filtrate flows through the nephron loop and the collecting duct. Water exits the filtrate because water diffuses readily into the salty environment of the medulla. Two solutes—sodium ions and urea—establish these favorable conditions, according to the **two-solute hypothesis.** The nephron loop concentrates sodium ions by pumping them into the medulla, and the vasa recta remove water. Urea itself also contributes to these conditions by diffusing from the collecting ducts. As a result, enough water leaves the filtrate and enough urea remains in the filtrate to concentrate urea 70 times higher in the urine than in the glomerular filtrate.

TUBULAR DISORDERS

Relatively few disorders affect the renal tubules. **Fanconi syndrome** is an example of diseases that interfere with tubular resorption. In this disease the proximal convoluted tubule fails to reabsorb amino acids, glucose, and phosphate from the filtrate. **Cystinuria** (SIS-tin-YOUR-ee-ah) is a similar condition in which the amino acid cystine accumulates in urine because the proximal tubule does not reabsorb it. Disorders of tubular secretion also occur when the endothelium secretes too few or too many molecules into the filtrate. Blood becomes acidic in **renal tubular acidosis** because the distal tubule does not secrete excess hydrogen ions into the filtrate.

* On what function does the kidney expend most of its energy?
* What materials are reabsorbed by the proximal convoluted tubule?
* Which sections of the tubule make the medulla salty?
* Name disorders of tubular secretion and tubular reabsorption.

JUXTAGLOMERULAR APPARATUS

The juxtaglomerular apparatus regulates filtration rate. The juxtaglomerular apparatus (JGA),

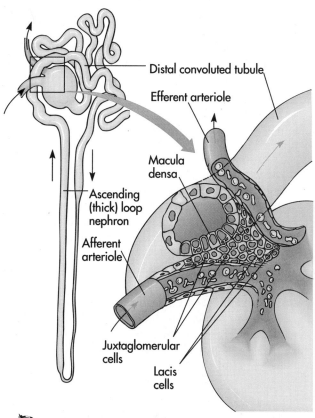

⁂ Figure 25.11
Juxtaglomerular apparatus.

shown in Figure 25.11, is an association between the glomerulus and the tubule, which adjusts the rate of filtration in the glomerulus. The JGA consists of three parts.

1. The **macula densa,** a patch of tall epithelial cells located at the beginning of the distal convoluted tubule.
2. The **juxtaglomerular cells,** a collar of modified smooth muscle fibers in the tunica media of the afferent arteriole that brings blood into the glomerulus.
3. **Lacis cells** (LAKE-iss), which appear to be modified mesangial cells that lie between the juxtaglomerular cells and macula densa.

The juxtaglomerular apparatus is an elegant negative feedback structure that apparently enables the kidney to regulate its own rate of filtration. The filtrate returning from the nephron loop apparently is used to regulate filtration upstream in the glomerulus. The macula densa evidently monitors the composition of the filtrate. When the filtration rate falls, the macula densa causes the juxtaglomerular cells to release the enzyme **renin** (REEN-in) into the glomerulus, where it is thought to increase the filtration rate by constricting the efferent arterioles. When the macula densa detects the filtration rate rising, it causes the juxtaglomerular cells to secrete less renin. As a result the rate of filtration is balanced by its effects on the composition of the filtrate.

What about the Lacis cells? They are agents of the sympathetic nervous system that increase blood pressure by also causing the juxtaglomerular cells to release renin (fight or flight; see Sympathetic Nervous System, Chapter 21).

The release of renin begins a cascade of events that affects blood pressure both inside the kidney and throughout the body. Renin converts **angiotensinogen** (AN-jee-oh-ten-SIN-oh-jen), a peptide, into **angiotensin I** (AN-jee-oh-TEN-sin). The cardiovascular endothelium, especially in the lungs, converts this peptide into **angiotensin II.** Angiotensin II is a potent vasoconstrictor that causes smooth muscle fibers in arterioles to constrict, but this constriction is brief because angiotensin II is rapidly destroyed in the blood. Nevertheless, angiotensin II has a longer lasting effect through the adrenal glands, which it stimulates to produce **aldosterone** (AL-doe-STEER-own). A steroid hormone, aldosterone raises blood pressure and blood volume by enhancing the reabsorption of sodium and chloride ions and water from the distal convoluted tubule into the blood.

- Name the three constituent cells of the juxtaglomerular apparatus.
- What function does each type perform?

URETERS, BLADDER, AND URETHRA

Ureters conduct urine to the urinary bladder, where the urine is stored until it is evacuated through the urethra.

URETERS

The muscular ureters pump urine into the bladder. Contrary to popular impression, urine does not simply drip down the ureters into the bladder. Ureters (you-REET-ers) actively pump urine into the bladder for the same reason that one has to blow air to inflate a balloon. The ureters work as effectively upside down or in weightlessness, which they could not do if they relied on gravity to drain them.

From the minor calyces to its entrance into the bladder, each ureter is a continuous structure (Figures 25.12A and 25.1). The renal portions and the ureter proper all derive from the same origin (see Development, this chapter). Each ureter narrows from about 10 mm diameter as it leaves the hilus of the kidney. From there it extends about 25 cm (10 in) behind the peritoneum to the bladder. The **abdominal portion** of the ureter passes in the fascia along the psoas muscle and crosses the common iliac arteries and veins onto the posterior wall of the pelvic cavity, still behind the peritoneum. The **pelvic portions** of the ureters bend medially and enter the base of the bladder through one-way valves that prevent backflow from the bladder.

Three tissue layers—an adventitia, muscularis, and mucosa—line all portions of the ureter and bladder, making their fundamental structures quite similar. The fibrous **adventitia,** the external covering shown in Figure 25.12B and C, is continuous with the renal capsule and the adventitia of the bladder. The adventitia supplies the ureters with blood and lymph vessels, as well as axons of sympathetic and parasympathetic neurons. Visceral afferents to the lower thoracic and first lumbar segments register acute stretching pain when a kidney stone passes. The **muscularis,** the middle layer of tissue, is composed of inner longitudinal and outer circular layers of smooth muscle fibers, precisely reversing the arrangement in the digestive tract. A third outer longitudinal layer of fibers supplements the first two in the lower third of each ureter. These muscle fibers propel urine toward the bladder by peristalsis. The **mucosa** lines the ureters and bladder with **transitional epithelium.** A basement membrane joins this layer to a fibrous **lamina propria** beneath. The mucosa folds deeply into the star-shaped pattern shown in Figure 25-12B and C when the ureters and bladder are empty and expands fully when these structures hold urine.

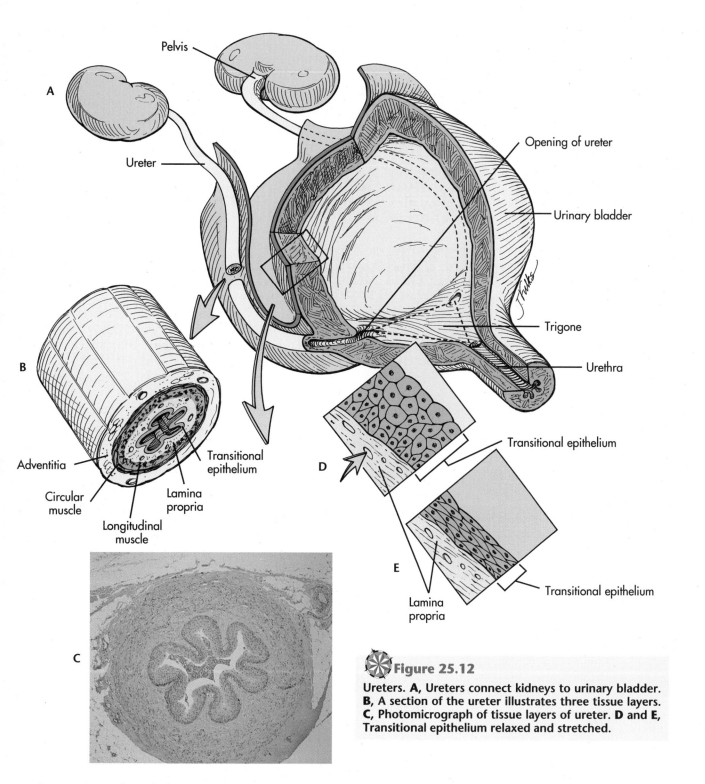

Figure 25.12

Ureters. A, Ureters connect kidneys to urinary bladder. **B,** A section of the ureter illustrates three tissue layers. **C,** Photomicrograph of tissue layers of ureter. **D** and **E,** Transitional epithelium relaxed and stretched.

The transitional epithelium consists of several layers of cells (Figure 25-12D and E), beginning with cuboidal cells at the basement membrane and progressing to squamous cells at the luminal surface. The epithelium thickens as the ureters descend to the bladder. Beginning in the minor calyces with two or three layers, the epithelium progresses to four and five layers in the ureters proper and to six and more layers in the bladder. Regardless of the layer, all cells

are thought to attach to the basement membrane. Stretching draws the cells flat, but when the wall relaxes, some cells lie over others (see Epithelia, Chapter 3).

- Name the renal and extrarenal regions of the ureter.
- Which three layers of tissue make up the walls of the ureter?
- What is transitional epithelium?

BLADDER

The bladder stores urine until the micturition reflex releases it. The bladder (Figure 25.13A) lies in the floor of the pelvic cavity, anterior to the rectum and vagina (in females) and posterior to the pubic bone. A **fibroelastic adventitia** attaches its inferior and lateral walls, while the peritoneum covers the superior third with serosa. Capable of inflating with as much as 800 ml of urine, the bladder expands into the abdominopelvic cavity by virtue of a transitional epithelium, a fibroelastic lamina propria, and a muscularis called the **detrusor muscle** (DEE-true-sore). The layers of smooth muscle fibers in the detrusor are not as distinct as those in the esophagus, for instance, but they repeat the inner longitudinal, middle circular, outer longitudinal pattern of the ureters. An empty bladder collapses into ruga-like folds (see Stomach, Chapter 23) that flatten and stretch when it fills. The ureters enter the posterior floor of the bladder at the corners of a permanently smooth, flat, triangular area called the **trigone** (TRY-goan), and the urethra exits at its anterior corner.

The bladder is innervated by the autonomic nervous system. Visceral afferents from the bladder wall and urethra reach the sacral spinal cord by way of pelvic nerves. These same nerves carry parasympathetic fibers back to the bladder and urethra. Sympathetic pathways innervate these same targets by way of the pelvic splanchnic nerves.

URETHRA

The urethra drains urine from the bladder outside the body. Urethral structure resembles that of the bladder and ureters, with a **mucosa** and **muscularis**, but an outer **spongy layer** replaces the adventitia. The urethra (you-REE-thra) is 3 to 5 cm long (1 to 2 in) in females and about 20 cm (8 in) long in males, where the extra distance is taken up by the prostate gland and penis. In both sexes the urethra exits the trigone of the bladder through an involuntary smooth muscular **internal sphincter** and passes through the pelvic diaphragm, where it is guarded by the **urethral sphincter,** a voluntary skeletal muscle (see Table 10.14, Muscles of the Perineum) stimulated by axons of somatic motor neurons in the pudendal nerve.

FEMALE

The urethra opens to the vestibule, located between the vagina and clitoris in Figure 25.14A. Stratified squamous epithelium, continuous with that of the vestibule, lines the urethra for most of its length. In Figure 25.14B, highly vascular, fibrous lamina propria supports the epithelium, and smooth muscle fibers of inner longitudinal and outer circular layers are continuous with the bladder. The outer circular fibers thicken at the bladder to form the internal urethral sphincter. The lumen folds into a U-shaped trough around a central ridge.

MALE

Three separate portions of the urethra conduct urine and semen in males (Figure 25.14C). All consist of mucosa and muscularis. The **prostatic urethra** (pro-STAT-ik) is approximately 3 cm (1½ in) long and passes through the prostate gland (PRAH-state), directly beneath the exit from the bladder and its in-

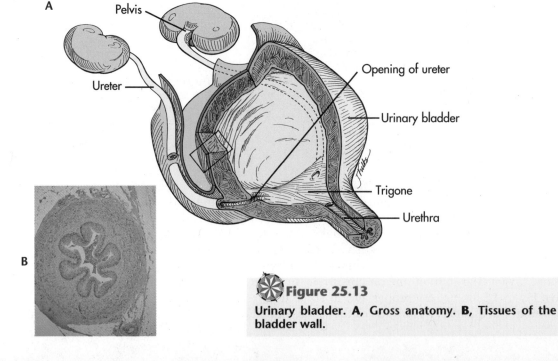

A Pelvis
Ureter
Opening of ureter
Urinary bladder
Trigone
Urethra
B

✳ **Figure 25.13**

Urinary bladder. **A,** Gross anatomy. **B,** Tissues of the bladder wall.

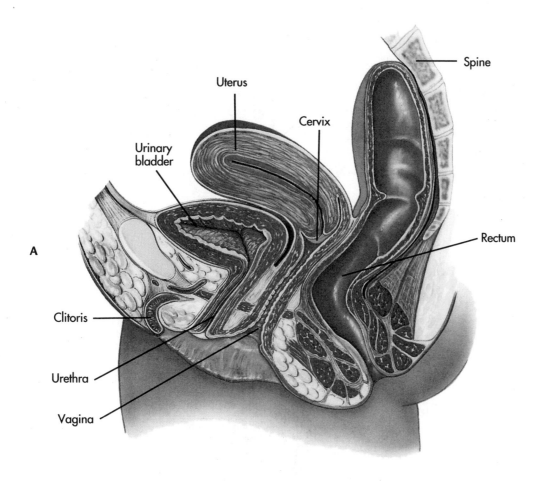

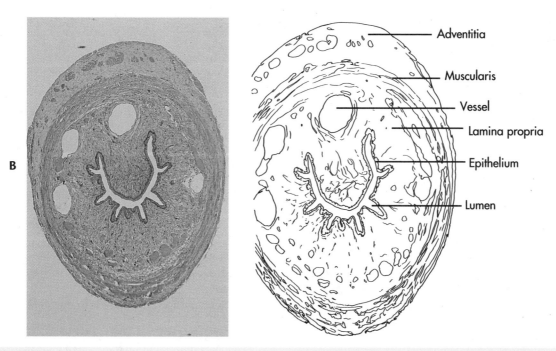

Figure 25.14

Urethra. A, Female, midsagittal section through the pelvis. B, Transverse section through female urethra.

Continued.

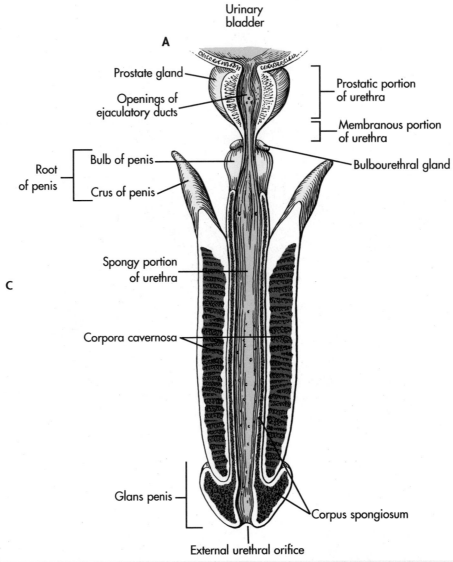

Urinary
bladder

A

Prostate gland

Openings of
ejaculatory ducts

Prostatic portion
of urethra

Membranous portion
of urethra

Root
of penis

Bulb of penis

Crus of penis

Bulbourethral gland

C

Spongy portion
of urethra

Corpora cavernosa

Glans penis

Corpus spongiosum

External urethral orifice

Figure 25.14, cont'd
C, Male urethra.

ternal sphincter. This portion receives secretions from the prostate gland through many small prostatic ducts. The ejaculatory ducts from the testis also join the urethra within the prostate. The **membranous urethra,** the shortest section (only 1.5 cm long), penetrates the pelvic diaphragm and is entirely surrounded by the urethral sphincter muscle. This portion corresponds to the female urethra in location and structure. In the remaining 15 cm, the **spongy portion,** or **penile urethra** (PEEN-ile), extends from the bulb of the penis to the glans. It receives from the bulbourethral glands proximally and from numerous mucous glands (urethral glands) in the body of the penis. All these numerous openings can catch the narrow tip of a catheter inserted into the urethra to withdraw urine. The narrowest diameter of the urethra, about 3 mm, occurs in the membranous portion. **Stratified**

columnar epithelium lines all portions of the urethra, except for transitional epithelium at the bladder and stratified squamous epithelium within the glans.

- Which portion of the bladder receives the ureters and urethra?
- Where are the sphincters of the urethra located?
- What are the names of the three portions of the male urethra?
- Which portion corresponds to the female urethra?

MICTURITION
The micturition reflex voids urine from the bladder. As the bladder inflates with urine, the bladder wall stimulates the urethral sphincter to open and the detrusor muscle to contract, thereby driving out the urine (Figure 25.15). These stimuli cease as the

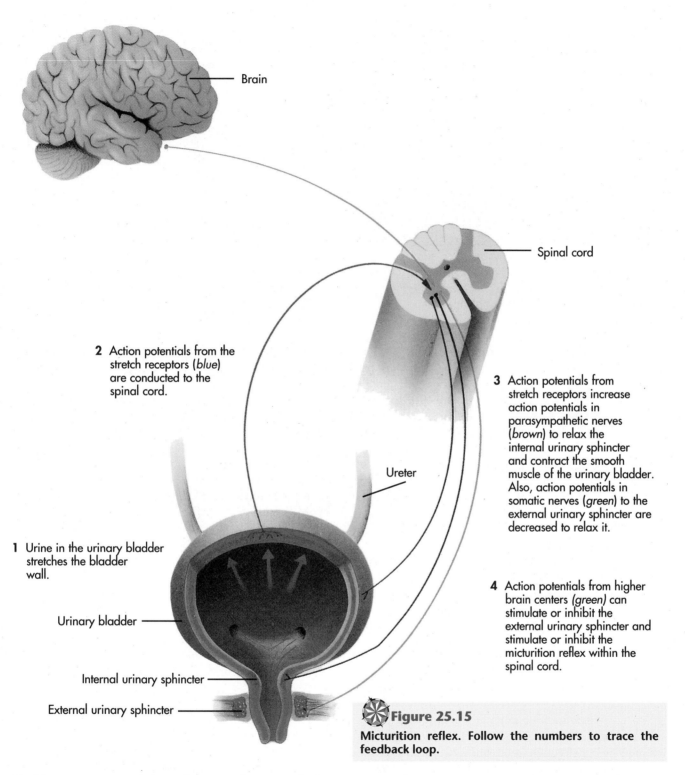

2 Action potentials from the stretch receptors (*blue*) are conducted to the spinal cord.

1 Urine in the urinary bladder stretches the bladder wall.

Urinary bladder

Internal urinary sphincter

External urinary sphincter

Brain

Spinal cord

3 Action potentials from stretch receptors increase action potentials in parasympathetic nerves (*brown*) to relax the internal urinary sphincter and contract the smooth muscle of the urinary bladder. Also, action potentials in somatic nerves (*green*) to the external urinary sphincter are decreased to relax it.

Ureter

4 Action potentials from higher brain centers (*green*) can stimulate or inhibit the external urinary sphincter and stimulate or inhibit the micturition reflex within the spinal cord.

Figure 25.15

Micturition reflex. Follow the numbers to trace the feedback loop.

bladder empties, and the sphincter closes until the bladder signals it to open again. You can temporarily override these stimuli and delay micturition (MIK-tshur-IH-shun), and you also can urinate voluntarily even though the bladder is not full.

This control is possible because involuntary and voluntary sphincters guard the urethra. The involuntary sphincter is a collar of smooth muscle fibers at the exit from the bladder, and the voluntary urethral sphincter muscle is controlled by stimuli from the pu-

dendal nerve. As the bladder stretches and fills with about 300 ml of urine, **stretch receptors** in the detrusor muscle and internal sphincter initiate waves of reflexive contraction that press the urine against this sphincter. Simultaneously, visceral afferents from these same receptors inhibit contraction of the external sphincter and make the central nervous system aware of impending micturition. At this point, one can relax the external sphincter and micturition will proceed. Alternatively the micturition reflex decays and pressure declines until the reflex recovers. Its intensity increases and builds more rapidly next time as the bladder continues to fill. Eventually, urine pressure will force the involuntary sphincter open and, overriding the owner's control, will relax the external sphincter as well.

- What stimuli cause micturition?
- Which sphincter is under voluntary control, and how do micturition reflexes open it?

KIDNEY TRANSPLANTATION

▲ **The anatomy of kidney transplantation is interesting** because the donor kidney is implanted in a retroperitoneal position resembling that of the patient's kidneys, but in a different location. The new kidney is implanted in the iliac fossa, its renal vessels are connected to the iliac artery and vein, and its ureter is connected to the bladder, but the patient's kidneys usually are not removed.

Up to 24,000 new patients are expected to begin treatment for **end-stage renal disease** (irreversible kidney failure) each year in the United States. Many will elect **dialysis therapy** (use of an artificial kidney), and others will receive a transplanted kidney from a living donor or from a cadaver donor. Approximately 80% of dialysis patients and of those who receive a cadaver kidney are alive after 5 years of treatment. The survival rate is slightly better, 90%, for recipients of kidneys from living, related donors.

Transplantations from living, closely related family members are most successful. Careful matching of human leucocyte antigens between donor and recipient tissue is necessary, and the persons receiving the implant must remain permanently on immunosuppressive drugs, usually cyclosporine. Although cadaver kidneys can be preserved for up to 72 hours and still function, those from a living donor of close genetic relationship to the recipient are preferred for immediate transfer of an actively functioning kidney. A living kidney is removed, together with as much of its vessels, ureter, and surrounding fatty capsule as possible, through an incision over the twelfth rib, after the rib itself is removed. The surgeon is careful not to enter the peritoneal cavity. The kidney is dissected free, the ureter is cut to prove by its flow of urine that the kidney functions, and finally the renal artery and vein are disconnected from the aorta and inferior vena cava. Cadaver kidneys, however, are usually removed by entering the abdominal cavity, and portions of the aorta and inferior vena cava may be removed with them.

The preferred site of implantation is the right iliac fossa in the wall of the abdominal cavity. The fascia between the peritoneum and the iliopsoas muscle offers a ready site where the vessels of the donor kidney can be attached to the iliac artery and vein of the patient. This site is also readily accessible should repairs be necessary later. The right side is preferred because the sigmoid colon interferes with dissection on the left. The renal artery usually is sutured end-to-end to the internal iliac artery or onto one side of the common iliac artery. Next, the renal vein is sutured to the side of the common iliac vein. With vascular supply reestablished, the ureter is then connected to the bladder by dissecting a passage for it in the bladder wall and suturing it in place from the interior of the bladder. This procedure establishes a third entrance to the bladder, but alternatively the patient's own ureter may be attached to the donor ureter or to the renal pelvis.

- Where is the donor kidney implanted?
- Which recipient vessels are connected to the donor kidney?

KIDNEY DEVELOPMENT

The kidneys are the last of three different excretory organs to appear during the course of development and evolution. All are associated with the intermediate mesoderm, a strip of tissue between the somite mesoderm and the lateral plate that was described in Embryonic Disk—Mesoderm, Chapter 4. Like the somites, this mesoderm becomes segmented, beginning from the cephalic end (Figure 25.16).

PRONEPHROS

In humans these segments develop into the **pronephros** (pro-NEF-ross) during the fourth week but quickly regress without functioning (Figure 25.16A). A permanent **pronephric duct** grows caudally toward the hindgut and attaches temporarily to the segments of the pronephros. Most vertebrates replace the pronephros with the mesonephros, the major excretory organ of amphibians, reptiles, and other fish.

MESONEPHROS

The **mesonephros** (MEZ-oh-NEF-ross) develops from caudal segments that form while the pronephros is degenerating (Figure 25.16B), and it functions briefly during the fifth week. Each segment forms a glomerulus and a nephron that connects to the pronephric duct, now called the **mesonephric duct.** The mesonephric duct discharges its waste into the hindgut. Kidney development stops here in amphibians and reptiles but goes one step further in birds and mammals.

METANEPHROS

Permanent metanephric kidneys (MET-ah-NEF-rik) **begin to develop in the fifth week,** when a bud

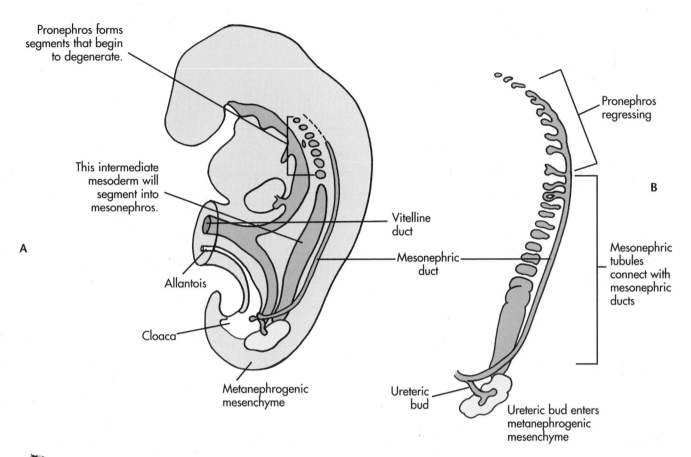

Pronephros forms segments that begin to degenerate.

This intermediate mesoderm will segment into mesonephros.

Allantois

Cloaca

Metanephrogenic mesenchyme

Vitelline duct

Mesonephric duct

Ureteric bud

A

Pronephros regressing

B

Mesonephric tubules connect with mesonephric ducts

Ureteric bud enters metanephrogenic mesenchyme

Figure 25.16
Three phases of renal development—pronephros, mesonephros, and metanephric kidney. A, Week 4. B, Week 5.

of tissue known as the **ureteric bud** grows from the mesonephric duct into the most caudal section of intermediate mesoderm. The ureteric bud (YOU-ree-TARE-ik) branches into the **metanephrogenic mesenchyme,** and these branches become the collecting system of the kidney. They induce the mesenchyme to form nephrons and glomeruli. As a result the cortex (nephrons and glomeruli) and medullary pyramids (collecting and papillary ducts) surround the calyces, the renal pelvis, and the ureter at the hilus of the kidney.

As Figure 25.17A shows, the ureteric bud branches into two, then four, then eight, and more branches, as each generation of binary branching adds another rank of branches. The overall process resembles branching in the tracheobronchial tree (see Chapter 23, Respiratory System). There are at least 15 generations of branches; the ureter and calyces represent the first generations, and the collecting ducts represent the final generations.

Branches induce nephrons as the branches invade the mesenchyme (Figure 25.17B). Each new branch induces the mesenchyme surrounding it to form a nephron, so that each round of branching forms two new nephrons. A glomerulus forms at the distal end of each nephron, and the nephron connects with the

branch that induced it. Branching stops in the fourteenth week, but branches continue to induce nephrons as the branches grow longer. This results in arcades of nephrons, first one nephron, then the next, until about 10 have appeared on each arcade, with the most recent nephron located superficially in the cortex. At 36 weeks, 2 weeks before term, the kidney has achieved its adult complement of more than 1 million nephrons, and no more nephrons form.

MIGRATION OF THE KIDNEYS

If the kidneys remained where they developed, they would be located in the pelvic cavity, not beneath the diaphragm. The kidneys appear to move cranially, but in actuality the lumbar and sacral regions grow rapidly longer while the kidneys remain in place. Related changes cause the vertebral column to become longer than the spinal cord (see Chapter 18, Spinal Cord and Spinal Nerves). As the pelvis tows the ureters caudally, new arteries supply the kidneys from progressively higher levels as the trunk elongates, so that the renal arteries represent the highest level of a series that first arose in the pelvis.

• What are the names of the three types of kidney?
• Which structures does the ureteric bud form, and which structures does it induce to form?

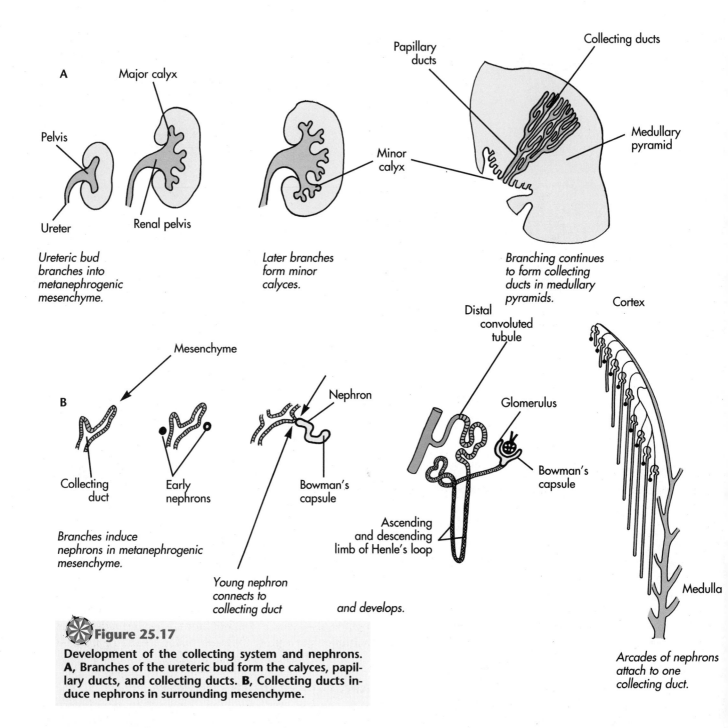

Figure 25.17

Development of the collecting system and nephrons. **A,** Branches of the ureteric bud form the calyces, papillary ducts, and collecting ducts. **B,** Collecting ducts induce nephrons in surrounding mesenchyme.

DEVELOPMENTAL ANOMALIES

Developmental errors cause rather striking abnormalities in kidneys (Figure 25.18). One kidney (Figure 25.18A) may fail altogether to develop **(unilateral renal agenesis).** This condition is not life-threatening when the remaining kidney has enough nephrons, but **bilateral renal agenesis** (ay-JEN-ih-sis) is fatal before or shortly after birth. One obvious explanation of this condition is that the ureteric bud fails to develop and induce nephrons to form. In **polycystic kidneys** (POL-ee-SIS-tik), nephrons fail to connect with collecting ducts. Filtrate inflates affected nephrons, and the kidneys resemble clusters of grapes. Kidneys also can fail to migrate from the pelvis. In **pelvic kidney,** one kidney or both may remain in the pelvis (Figure 25.18B), blocked from further movement by the common iliac artery. Alternatively, migrating kidneys may fuse into a **horseshoe kidney** (Figure 25.18C) that remains low in the abdomen, caught by the inferior mesenteric artery; such kidneys function normally. Function is abnormal in **oligonephronic hypoplasia** (OL-ih-go-NEF-ron-ik HIGH-po-PLAY-zha), however. Fewer nephrons and medullary pyramids form in this condition, and the glomeruli and tubules enlarge as much as tenfold. Larger endothelial cells line these tubules. As you

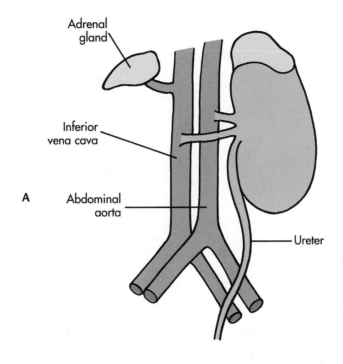

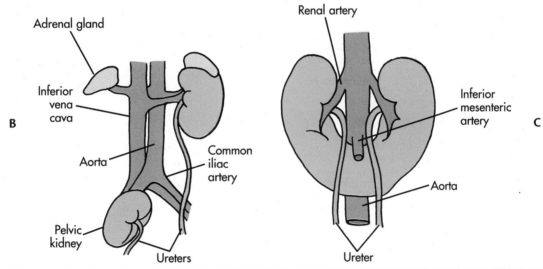

Figure 25.18

Developmental anomalies of the kidneys. A, Unilateral renal agenesis. B, Pelvic kidney. C, Horseshoe kidney.

might expect from surface-volume considerations, such glomeruli filter less effectively, and the kidneys eventually fail.

- Name conditions caused by failure of kidneys to migrate or failure of ureteric buds to develop.

SUMMARY

See Figure 25.19 for a summary of the structure and functions of the urinary system.

The urinary system is a homeostatic system; it regulates the composition of blood and body fluids.

The kidneys excrete excess components in the urine, and the ureters, bladder, and urethra conduct this fluid to the outside of the body. *(Page 760.)*

Nephrons are the structural units of kidney function. They consist of a renal corpuscle and tubule, surrounded with capillaries. The glomerulus filters blood through the glomerular filtration barrier into the filtrate of the glomerular capsule. The tubule selectively reabsorbs most of these constituents into the peritubular capillaries and vasa recta, but those substances that remain in the tubule become urine. *(Page 760, 766.)*

The kidneys are paired, bean-shaped organs lo-

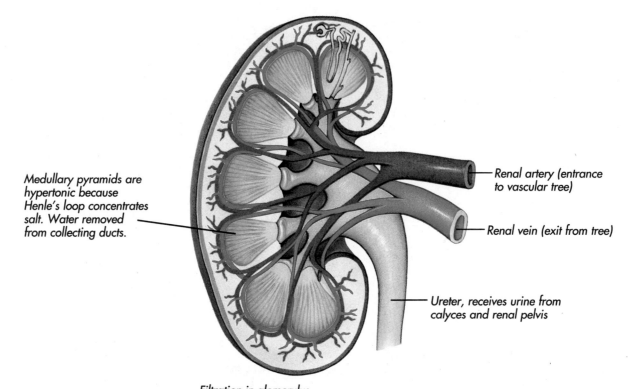

Medullary pyramids are
hypertonic because
Henle's loop concentrates
salt. Water removed
from collecting ducts.

Renal artery (entrance
to vascular tree)

Renal vein (exit from tree)

Ureter, receives urine from
calyces and renal pelvis

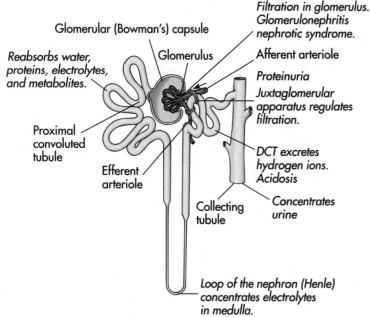

Glomerular (Bowman's) capsule

Glomerulus

Reabsorbs water,
proteins, electrolytes,
and metabolites.

Proximal
convoluted
tubule

Efferent
arteriole

Collecting
tubule

Loop of the nephron (Henle)
concentrates electrolytes
in medulla.

Filtration in glomerulus.
Glomerulonephritis
nephrotic syndrome.

Afferent arteriole

Proteinuria

Juxtaglomerular
apparatus regulates
filtration.

DCT excretes
hydrogen ions.
Acidosis

Concentrates
urine

✷ **Figure 25.19**
Summary of kidney structure and function.

cated within the posterior wall of the peritoneal cavity. Nephrons are located in the cortex and medulla, and the conducting system of calyces, renal pelvis, and ureter drains from the hilus. A renal arterial tree delivers blood to the nephrons, and a venous tree drains the peritubular capillaries and vasa recta. *(Page 762.)*

The glomerular filtration barrier selectively filters constituents from the blood into the filtrate. Three levels of the barrier are responsible for selectivity. The first level is the fenestrated capillary endothelium, which prevents all cells and large macromolecular immune complexes from reaching the next level. The glomerular filtration barrier is a thick glomerular basement membrane whose molecular pores keep most proteins from reaching the third level. Only small proteins, peptides, and metabolites pass. The filtration slits between the podocytes of the glomerular capsule are the third element of the barrier, but it is not clear whether the filtration slit diaphragm prevents any molecules from entering the filtrate. *(Page 767.)*

The tubule selectively reabsorbs certain substances from the filtrate, but not urea and products of medication. Active transport via sodium-potassium ion pumps in the cell membrane is the primary means of resorption. The proximal convoluted tubule reabsorbs electrolytes, glucose, amino acid, proteins, and water. The nephron loop concentrates sodium and chloride ions in the medullary pyramids, making this region of the kidney salty. The distal convoluted tubule reabsorbs sodium ions and secretes hydrogen ions and potassium ions into the filtrate. As collecting ducts carry urine from the distal convoluted tubules through the medulla to the papillary ducts and minor calyces, water is extracted from the urine so that the wastes become more concentrated. *(Page 771.)*

The juxtaglomerular apparatus regulates filtration by measuring how fast filtration proceeds and by adjusting the rate of filtration in the glomerulus. The

JGA is a negative feedback structure. The macula densa measures filtration rate, and the juxtaglomerular cells adjust filtration rate by secreting renin into the glomerulus. Sympathetic neurons also cause the lacis cells to stimulate the juxtaglomerular cells to secrete renin. *(Page 773.)*

The ureter branches into the kidney as the renal pelvis and the major and minor calyces. These tributaries drain urine through the ureters to the bladder, where it is stored until micturition releases it through the urethra. All three structures consist of mucosa, muscularis, and an external covering of adventitia or other tissue. Transitional epithelium lines most of the calyces, pelvis, ureter, and bladder, but the urethra is lined mostly with stratified columnar epithelium. Two and three layers of smooth muscle fibers in the muscularis of the ureters pump urine to the bladder. Similar layers in the detrusor muscle of the bladder expel urine at micturition, which is controlled by the micturition reflex. *(Page 774.)*

Kidney transplantation is the preferred treatment for end-stage renal disease. This operation inserts a kidney from a live donor or cadaver into a pocket formed in the iliac fossa of the recipient. The renal vessels are connected to the iliac vessels, artery-to-artery, vein-to-vein, and the ureter is joined to the bladder. *(Page 780.)*

Kidneys develop through three successive forms—the pronephros, mesonephros, and metanephros. The metanephric kidney is the permanent form in humans, and it develops when the ureteric bud branches into the surrounding metanephrogenic mesenchyme and induces nephrons to form. The calyces, papillary ducts, and collecting ducts represent successive rounds of binary branching. These branches induce nephrons to develop in the cortex and medulla. Glomeruli form at the distal ends of nephrons, and the proximal ends join to the collecting tubules. *(Page 780.)*

Developmental errors change the anatomy of kidneys. Kidneys may fail to form when the ureteric bud itself does not form or does not induce nephrons. Kidneys that fail to migrate remain in the pelvis. Others that fuse during migration become horseshoe kidneys, while still others become polycystic when their nephrons do not connect to collecting ducts. *(Page 782.)*

STUDY QUESTIONS   OBJECTIVES

1. Describe the functional and anatomical relationships between the organs of the urinary system.   **1**
2. Trace the course of erythrocytes, molecules of glucose, and urea through the urinary system, beginning at the renal artery.   **2, 5**
3. Where are the kidneys located in adults, and in the fetus? Describe the capsules that cover them and the vessels (including lymphatics) and nerves that communicate with them.   **3, 14**
4. What tissues and structures are found in the cortex, medulla, renal pelvis, and renal sinus?   **4**
5. Describe the structures of a nephron. What are the anatomical and functional differences between the renal corpuscle and renal tubule?   **6**
6. What is the function of the nephron loop? How do form and function differ in cortical and juxtamedullary nephrons?   **6**
7. What cells and structures are found in renal parenchyma? Where is parenchyma located in the kidneys?   **6**
8. Describe the three elements of the glomerular filtration barrier. Which molecules and cells does each level allow to pass, and which does it block? Describe the passage of amino acid molecules through this barrier from blood plasma into the filtrate.   **7**
9. Make a table that shows the regions of the tubule and the substances that they reabsorb (or secrete) from the filtrate.   **8**
10. What forms of damage to the glomerulus allow either proteins (proteinuria) or erythrocytes (hematuria) to appear in the urine?   **9**
11. Describe the anatomy of the juxtaglomerular apparatus, and explain its response to low and high filtration rates.   **10**
12. Describe the tissue layers of the walls of the ureters, bladder, and urethra. Which layer ensures the flow of urine to the outside?   **11**
13. How does the nervous system recognize the need to urinate, and how do voluntary responses and the micturition reflex respond to this stimulus?   **12**
14. Describe the implantation site and the systemic connections made in kidney transplantation.   **13**
15. What are the developmental and evolutionary differences between pronephric, mesonephric, and metanephric kidneys?   **14**
16. Describe the events that establish the collecting system and nephrons in the kidney. How do these changes explain the location of nephrons in the cortex?   **14**
17. What developmental errors account for pelvic kidneys, horseshoe kidneys, renal agenesis, and polycystic disease?   **15**

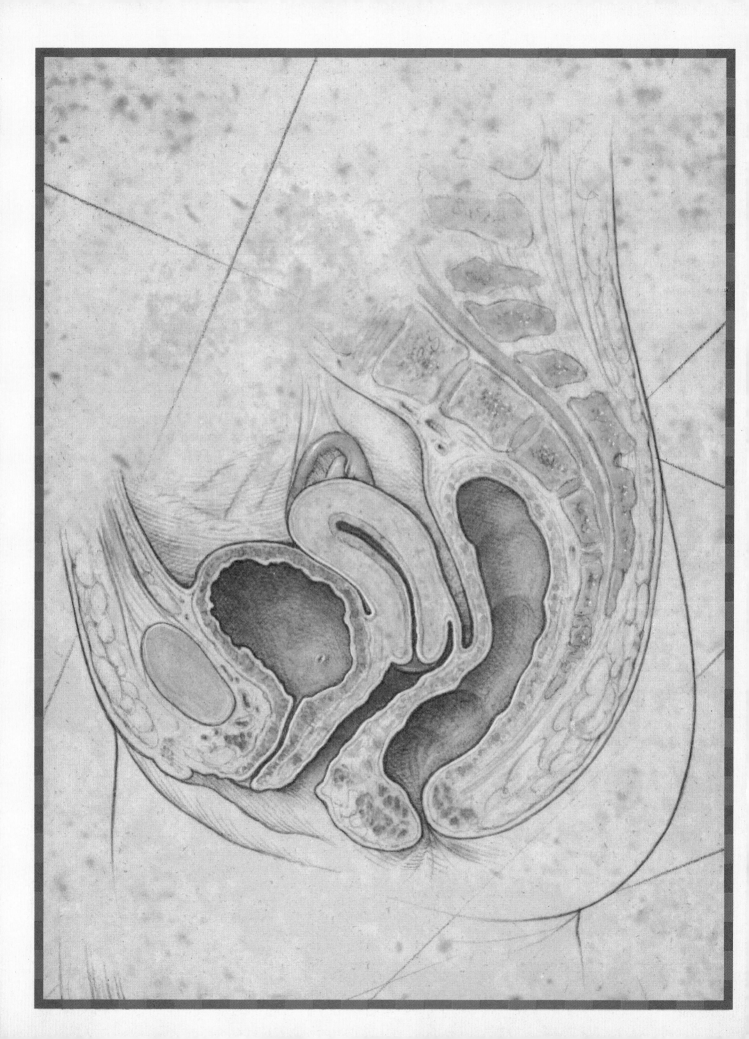

CHAPTER 26

Reproductive System

OBJECTIVES

After studying this chapter, you should be able to do the following:

1. Describe the organs, ducts, and glands, together with their functions, of the male and female reproductive systems.
2. Follow the movement of sperm and eggs through the glands and ducts of the male and female genitalia.
3. Diagram the branching ducts of the internal genitalia in males and females, and describe the changes in the mucosa and muscularis of these ducts as you follow them to the exterior openings.
4. Describe the sequence of events in spermatogenesis and oogenesis that produce sperm and eggs, respectively.
5. Outline the events of the female reproductive cycle.
6. Name the components of indifferent external genitalia and the structures that each forms in males and females.
7. Name the genital structures of one sex that are homologous in the other sex.
8. Name the internal genitalia that develop from mesonephric ducts or from Mullerian ducts in males and females.
9. Explain what is meant by chromosomal, gonadal, and somatic sex characteristics, and what is meant by gender identity.
10. Trace the major anatomical and functional changes that occur in a developing fetus, trimester by trimester.
11. Describe the anatomy of the placental barrier.
12. Describe the anatomical relationships between the placenta and the fetal membranes.
13. Describe the changes in uterus and breasts that occur during pregnancy and at birth.

 VOCABULARY TOOLBOX

Genital terms.

clitoris	clito-, *G* closed.	**scrotum**	*L* a pouch.
corpus	corp-, *L* a body.	**sperm**	sperm-, *G* seed, semen, sperm.
egg	eggja, *Old Norse* egg.	**testis**	testi-, *L* a testicle.
epididymis	didym, *L* double, twin, the testes.	**uterus**	uter-, *L* the womb.
	Epididymis, outside the testes.	**vagina**	vagin-, *L* a sheath.
ovary	ovari-, *L* an ovary.	**vas**	vas-, *L* a vessel or duct.
penis	peni-, *L* the penis.	**vulva**	vulv-, *L* a covering, the vulva.
placenta	placent-, *L* a round, flat cake.		

RELATED TOPICS

- **Embryonic development, Chapter 4**
- **Glands, Chapter 3**
- **Meiosis, Chapter 2**
- **Urethra, migration of kidneys, Chapter 25**

PERSPECTIVE

Of all the body's organs, only the reproductive system requires two versions to complete its work. The skeleton and muscles of men and women certainly differ, but these and other systems function by themselves without the need of a second sex. The complementary anatomy of male and female reproductive systems develops from the same tissues, which become male or female, depending on the sex chromosomes inherited from the parents. As a result, many aspects of male and female genitalia are homologous, the clitoris and the penis, for example. This chapter describes the anatomy of male and female reproductive systems; it traces the homologies between them, and it illustrates cases in which an individual's chromosomal and somatic sex differ. Most important, this chapter follows the reproductive system through a pregnancy to the birth of a new individual.

The reproductive system sustains the existence of the human species by producing new individuals. Figure 26.1 outlines the **glands** and **ducts** of the **internal genitalia** that communicate with the **external genitalia**. In females the ovaries periodically release an egg into the uterine tubes. In males the testes continually produce sperm that accumulate in the epididymis and vas deferens until they and seminal fluid from accessory glands are ejaculated through the penis. Introduced into the vagina, sperm pass through the cervix, into the uterus, and on to the uterine tubes, where fertilization may occur. An embryo passes into the uterus and begins to develop in the lining. The placenta links mother and fetus throughout pregnancy until 38 weeks after conception, when labor delivers the fetus and discards the placenta.

The external genitalia—scrotum and penis in males, vulva in females—are largely concerned with copulation, while the internal genitalia produce gametes, bring them together for fertilization, and nurture an embryo to birth. The ovaries and testes are exocrine and endocrine glands combined. They release eggs and sperm through ducts, and they secrete endocrine hormones through the blood, which promote oogenesis and spermatogenesis, the processes that produce eggs and sperm. The placenta also is an endocrine gland, but all other reproductive glands are exocrine glands that secrete through ducts. The prostate and seminal vesicles of males secrete seminal fluids, and the uterine and vestibular glands of females accommodate embryos in the womb and lubricate penises, respectively.

- Name several glands and ducts in the male and female reproductive systems.
- Name the parts of male and female external genitalia.

Case study begins on p. 807.

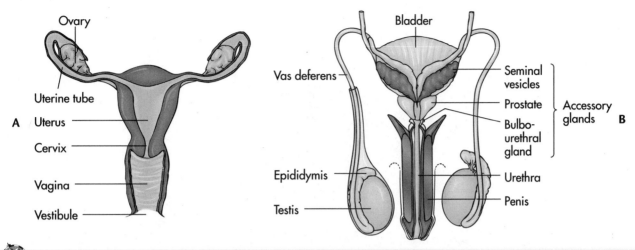

Figure 26.1
Reproductive systems. **A**, Female. **B**, Male.

MALE

BRIEF TOUR

Follow sperm through the **male genitalia** (JEN-ih-TAIL-ee-ah) in Figure 26.2A. Testosterone continually stimulates sperm to develop in the **testes,** which are suspended in the **scrotum** at a slightly lower temperature (35° versus 37° C or 95° versus 98.6° F) necessary for testosterone synthesis. Sperm develop in the **seminiferous tubules,** then pass to the **epididymis** and **vas deferens** (also known as ductus deferens) for storage until ejaculation ejects them out of the body. Autonomic reflexes stimulate waves of contraction in the walls of the vas deferens. These contractions sweep sperm back through the **inguinal canal** to the **prostatic urethra,** buried within the **prostate gland** beneath the bladder, where the sperm mix with seminal fluid from the **seminal vesicles** and prostate. The **membranous urethra** carries the semen through the floor of the pelvis into the base of the **penis,** where **bulbourethral glands** add their products, and finally to the outside. The **corpora cavernosa** and **corpus spongiosum** erect the penis by filling with blood during intercourse.

PERINEUM

The **perineum (PAIR-ih-NEE-um) is the diamond-shaped area between the anus and scrotum** and the ischial tuberosities of the pelvis (Figure 26.2B). Dark, wrinkled skin covers the scrotum and penis. An irregular ridge on the underside of both structures, called the **median raphe** (RAH-fay), marks the location where skin from both sides fuses during development (see Male-Female Homologies, this chapter). This line also marks the **median sep-**

tum that divides the scrotum into two sacs—left and right—that hold the testes.

SCROTUM

Each side of the scrotum suspends a testicle in a small sac lined by tunica vaginalis (VA-jin-al-iss; Figure 26.3). The **parietal tunic** lines the scrotal wall, and the **visceral tunic** covers the testis. The left testicle usually is lower than the right. The testicle is relatively free to move about this space, suspended from the posterior wall by the **spermatic cord,** which brings the vas deferens and testicular vessels and nerves to the testis. Occasionally a testicle twists on its moorings, with painful and potentially serious results. **Torsion,** as this condition is known, threatens the life of the testicle when a twisted testicular artery or vas deferens blocks blood from entering or sperm from leaving.

TUNICS

Figure 26.3 also shows how the tunics of the scrotum are continuous with the abdominal wall. Beginning from the interior, the **spermatic cord** contains the vas deferens and the vessels and nerves that followed the testis into the scrotum. Accompanying the vas is the **testicular artery** and **vein;** the latter forms the **pampiniform plexus** (PAM-pin-ih-form) around the artery, so that cool venous blood, returning from the testis, lowers the temperature of the arterial blood before it reaches the testis. Surrounding these vessels are the remains of the peritoneum, then an **inner layer of fascia** that is continuous with the inner fascia of the abdominal wall, followed by the **cremaster muscle** (CREE-mas-ter) from the internal oblique muscles of the abdomen, and then by **external fascia**

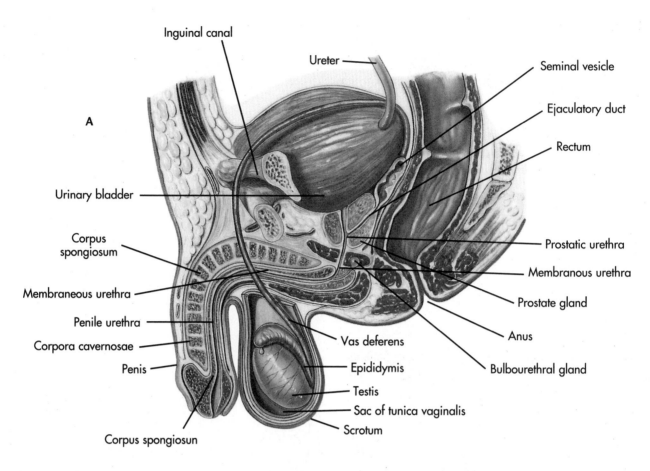

A

Inguinal canal

Ureter

Seminal vesicle

Ejaculatory duct

Rectum

Urinary bladder

Corpus spongiosum

Membraneous urethra

Penile urethra

Corpora cavernosae

Penis

Corpus spongiosun

Vas deferens

Epididymis

Testis

Sac of tunica vaginalis

Scrotum

Prostatic urethra

Membranous urethra

Prostate gland

Anus

Bulbourethral gland

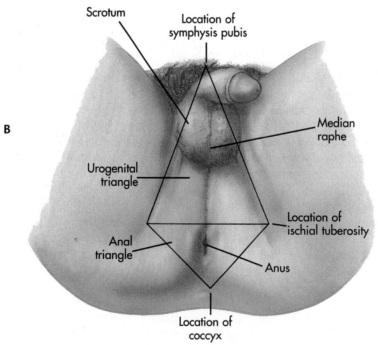

B

Scrotum

Location of symphysis pubis

Median raphe

Urogenital triangle

Anal triangle

Location of ischial tuberosity

Anus

Location of coccyx

✸ **Figure 26.2**

Male reproductive organs. A, Sagittal section through pelvis. **B,** Perineum.

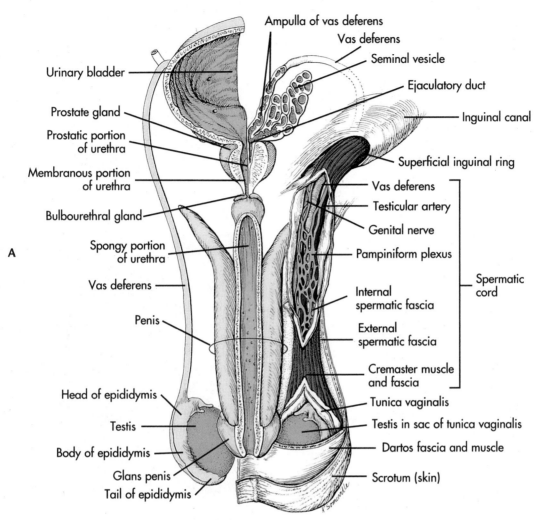

Ampulla of vas deferens
Vas deferens
Seminal vesicle
Ejaculatory duct
Inguinal canal
Superficial inguinal ring
Vas deferens
Testicular artery
Genital nerve
Pampiniform plexus
Internal spermatic fascia
External spermatic fascia
Cremaster muscle and fascia
Tunica vaginalis
Testis in sac of tunica vaginalis
Dartos fascia and muscle
Scrotum (skin)

Spermatic cord

Urinary bladder
Prostate gland
Prostatic portion of urethra
Membranous portion of urethra
Bulbourethral gland
Spongy portion of urethra
Vas deferens
Penis
Head of epididymis
Testis
Body of epididymis
Glans penis
Tail of epididymis

A

⊕Figure 26.3

The testis and scrotum. A, The testicles reside in the scrotum, suspended by the spermatic cord in the cavity of the tunica vaginalis.   *Continued.*

from the external oblique muscles. Finally, just beneath the skin is a thin net of smooth muscle fibers, the **dartos tunic** (DAR-tose), derived from cutaneous smooth muscle.

DESCENT

Descent of the testes is an important process in male genital development because it brings the testes to a cooler environment that permits testosterone production and spermatogenesis to proceed (Figure 26.3B). Indeed the scrotum has no insulating fat to retard cooling. If his testes fail to descend into his scrotum during fetal life, a boy will remain sexually immature. Each testis, attached to the posterior wall of the abdominal cavity by mesentery and the testicular artery and nerve, descends through the **inguinal canal** into its side of the scrotum, apparently guided by the **gubernaculum** (GOO-ber-NAK-you-lum), a fold of fibrous connective tissue that leads from the

young testis within the pelvic wall to the scrotum. You may recall from Chapter 25 that the kidneys migrate from the pelvis by retaining their original position while the pelvis and abdomen elongate away from them. The testes do the opposite; they follow the pelvis away from the kidneys. Each testis carries with it a portion of the abdominal cavity and its peritoneal lining into the scrotum. This lining remains in the scrotum as the **tunica vaginalis** when the inguinal canal closes after the testis' passage. The testes remain attached to the posterior wall of the scrotum and receive the testicular arteries, veins, and nerves as they did in their original location. Occasionally a loop of intestine enters the tunica vaginalis when the inguinal canal remains patent (open) or when weak musculature allows the canal to open. **Inguinal hernias** (IN-gwin-al HER-nee-as) such as these are more common in men than women (where the inguinal canal is smaller) and usually are repaired surgically.

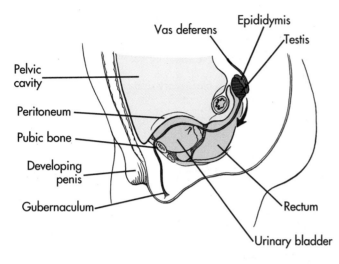

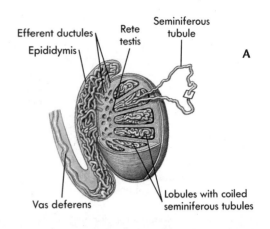

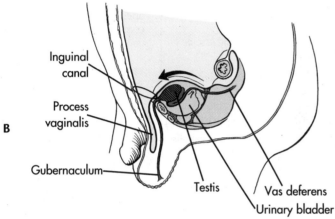

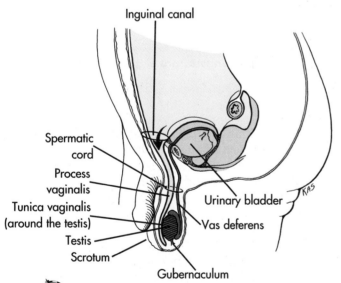

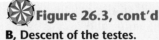
Figure 26.3, cont'd
B, Descent of the testes.

- How is the tunica vaginalis a remnant of the peritoneal cavity?
- What is the spermatic cord?
- What is the relationship between the coverings of the scrotum and the abdominal wall?

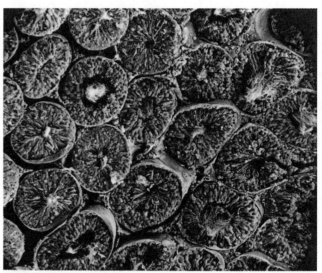

Figure 26.4

The testis. **A,** Internal anatomy; seminiferous tubules deliver to the epididymis and vas deferens. **B,** Interstitial tissue lies between the seminiferous tubules. (Scanning electronmicrograph by permission of R.G. Kessel and R.H. Kardon, *Tissues and organs: A text-atlas of scanning electron microscopy,* 1979, W.H. Freeman and Co.)

TESTES

Each **testis** is ovoid in shape, approximately 4.5 cm long, 2.5 cm wide and 3 cm in depth (1½ × 1 × 1⅕ in), and is covered with the slick, visceral portion of the tunica vaginalis shown in Figure 26.4A. Just beneath is the tough, fibrous **tunica albuginea** (AL-byou-jin-ee-ah; alb-, *L*, white) that surrounds the entire testis and makes it firm.

Fibrous septa subdivide the testis into several hundred narrow, conical lobules. Each lobule (Figure 26.4A) contains up to four tightly coiled loops of **seminiferous tubules,** and altogether some 250 tubules occupy each testis. The tubules empty sperm into the **rete testis** (REE-tee; ret-, *L*, a net), a tubular network in the posterior region of the testis. From the rete testis, **efferent ductules** lead sperm outside the testis to the epididymis. **Interstitial cells** (also known as Leydig cells) that produce testosterone lie in the loose connective tissue between the seminiferous tubules, along with testicular vessels, nerves, and lymphatics (Figure 26.4B). This location makes testosterone easily accessible to spermatogenesis inside the seminiferous tubules.

SEMINIFEROUS TUBULES

Seminiferous tubules are lined with **germinal epithelium** attached to a basement membrane, as well as a thin layer of **myoid cells** (MY-oyd) and lymphatic vessels, shown in Figure 26.5 A and B. Myoid cells resemble smooth muscle cells of the tunica intima of ar-

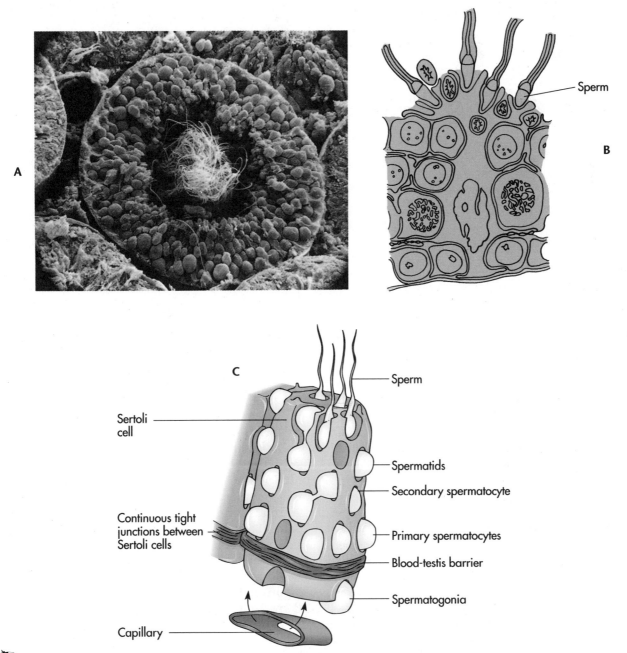

Figure 26.5

Spermatogenesis. A, Scanning electronmicrograph displays the various cells of spermatogenesis. Sperm line the lumen of the seminiferous tubule. (Scanning electronmicrograph by permission of R.G. Kessel and R.H. Kardon, *Tissues and organs: A text-atlas of scanning electron microscopy,* 1979, W.H. Freeman and Co.) **B**, Spermatogenesis in the germinal epithelium. **C**, Sertoli cells bind spermatocytes and form the blood-testis barrier.   *Continued.*

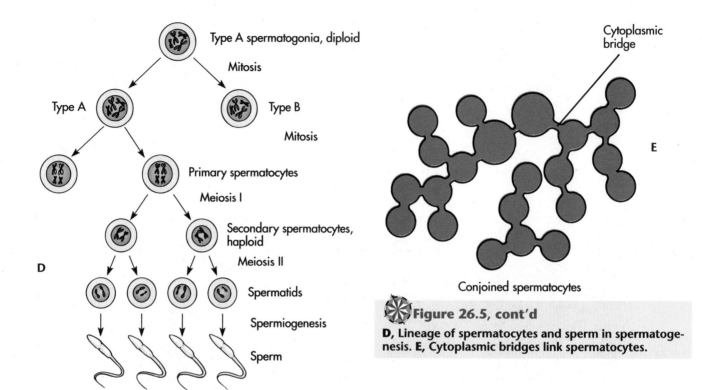

Figure 26.5, cont'd

D, Lineage of spermatocytes and sperm in spermatogenesis. **E,** Cytoplasmic bridges link spermatocytes.

teries and probably help squeeze sperm from the tubules. The germinal epithelium establishes the environment for **spermatogenesis** (sper-MAT-oh-JEN-eh-sis). First of all, it provides altogether 2800 cm² (3 ft²) of surface where sperm develop. Each tubule is about 75 cm (30 in) long and 0.2 mm in diameter. Placed end to end the tubules of both testes would span four football fields. Second, the germinal epithelium is stratified, so that spermatogenesis begins in the basal layers and releases mature sperm from the apical layers into the lumen of the tubule, somewhat as keratinocytes are released from the epidermis of the skin. Third, spermatogenesis is supported by **sustentacular cells of Sertoli** (SIR-toll-ee) that extend from the basement membrane to the lumen. **Sertoli cells** promote spermatogenesis by transporting nutrients and regulatory factors across plasma membrane connections between themselves and the developing spermatocytes. Sertoli cells contain abundant **androgen binding protein** (ABP) that concentrates testosterone for spermatogenesis, as well as **inhibin** (IN-hih-bin) that regulates the rate of spermatogenesis. These cells also form a **blood-testis barrier** (Figure 26.5C) that protects sperm from immunological attack. Remember, sperm are potentially foreign cells with different genotypes than the man who produces them (see Spermatogenesis).

Belted tight junctions join Sertoli cells near their bases and apparently block lymphocytes from the interior of the tubules.

• What are the functions of seminiferous tubules and Sertoli cells?

SPERMATOGENESIS

Spermatogenesis transforms diploid spermatogonia into haploid sperm by meiosis. Spermatogonia (sper-MAT-oh-GOAN-ee-ah) are diploid cells that remain quiescent until puberty (see Gonads and Primordial Germ Cells, this chapter). Table 26.1 compares the steps of spermatogenesis to those of oogenesis in females. **Type A spermatogonia** are the stem cells that maintain the testis' ability to produce sperm by dividing into more Type A cells, as illustrated in Figure 26.5D. Type A cells also produce B cells. **Type B spermatogonia** become sperm.

Type B spermatogonia divide mitotically, producing clones of **primary spermatocytes** that are conjoined by the narrow **cytoplasmic bridges** and docked onto Sertoli cells, as illustrated in Figure 26.5E. As many as 16 primary spermatocytes may derive from a single Type B spermatogonium. The process by which each primary spermatocyte develops into sperm begins with meiosis. The first meiotic division (Meiosis I;

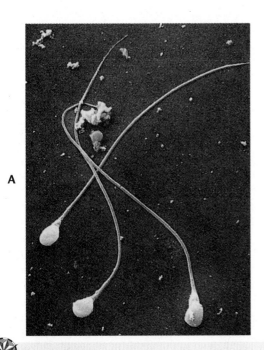

Figure 26.6

Steps in spermiogenesis. **A,** Mature sperm, the product of spermiogenesis. **B,** Spermiogenesis transforms spermatids into sperm.

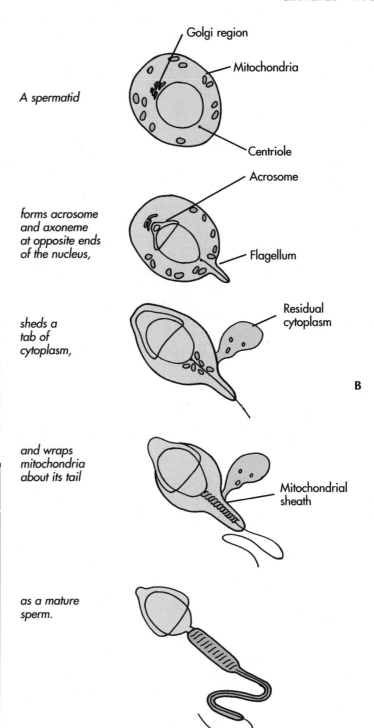

A spermatid

Golgi region
Mitochondria
Centriole

forms acrosome and axoneme at opposite ends of the nucleus,

Acrosome
Flagellum

sheds a tab of cytoplasm,

Residual cytoplasm

B

and wraps mitochondria about its tail

Mitochondrial sheath

as a mature sperm.

Table 26.1	SPERMATOGENESIS AND OOGENESIS	
Stages	**Spermatogenesis**	**Oogenesis**
Diploid stem cell	Spermatogonia	Oogonia
Mitotic division produces	Primary spermatocyte	Primary oocyte
First meiotic division produces	Secondary spermatocyte	Secondary oocyte, first polar body
Second meiotic division produces	Spermatid	Ootid, second polar body
Maturation produces	Sperm	Ovum, egg

see Figure 26.5D) starts with 23 pairs of doubled chromosomes in each spermatocyte nucleus. This division cleaves each primary spermatocyte into two smaller haploid **secondary spermatocytes** that the Sertoli cells carry deeper into the tubule. Each nucleus now contains 23 doubled chromosomes as described in Meiosis, Chapter 2. Next, the second meiotic division delivers 23 single chromosomes to each of two haploid **spermatids** (SPERM-ah-tids). Overall, these meiotic divisions produce four haploid spermatids from each diploid primary spermatocyte.

In the final steps of spermatogenesis, **spermiogen-** **esis** (SPER-me-oh-JEN-eh-sis) transforms spermatids into sperm by forming the **head,** the **midpiece,** and the **tail.** Figure 26.6 shows these changes. Chromatin condenses in the **nucleus,** and the Golgi apparatus spreads the **acrosome** over the anterior surface of the nucleus. Two or three **mitochondria** complete the midpiece by wrapping themselves around the base of the lengthening **axoneme,** which has formed from the centriole at the posterior end of the nucleus. Sertoli cells take up excess lobes of cytoplasm from transforming spermatids.

From Type B spermatogonium to mature sperm,

spermatogenesis takes about 74 days, and the sperm require another 14 days to enter the vas deferens. Some 200 million sperm develop daily. Typical ejaculates contain at least 20 million sperm/ml, the minimum considered necessary for fertility. Of these sperm, at least 50% must be motile and 60% must appear normal for an adequate chance that fertilization will occur.

- Name the stages of spermatogenesis and events that produce them.
- How many sperm does a man typically produce every day?

MALE GENITAL DUCTS

Sperm pass through the genital ducts (Figure 26.4) that collect, store, activate, and finally deliver sperm through the penis. As you might expect, the epithelial lining of the ducts reflects these various functions. Simple cuboidal epithelium of the rete testis and efferent ductules is supplanted by tall columnar secretory cells in the epididymis and finally by stratified columnar epithelium in the urethra. The ducts also become more muscular. At first only a basement membrane and fibrous connective tissue support the rete testis, but further along, the epididymis, urethra, and especially the vas deferens, acquire a thick muscularis whose smooth muscle fibers propel sperm during ejaculation.

RETE TESTIS

This network of delicate tubules collects sperm from the seminiferous tubules and sends them out of the testis through the efferent ductules. Embedded in highly vascular connective tissue, the ducts of the **rete testis** are lined with simple cuboidal cells. Each cell sprouts numerous microvilli to absorb fluid from the seminiferous tubules and a single cilium to move the fluid toward the efferent ductules.

EFFERENT DUCTULES

Sperm are conducted from each testis to the epididymis by 10 to 15 narrow **efferent ductules.** Patches of ciliated cells beat toward the epididymis, and numerous nonciliated cells have extensive microvilli that continue to absorb seminiferous fluid.

EPIDIDYMIS

Sperm are stored and activated on the postlateral surface of each testis by 4 to 5 cm (2 in) of highly coiled **epididymis** (ep-ih-DID-ih-mis). Immobile sperm enter the **head of the epididymis,** collect in the **body,** and emerge fully mobile from the **tail of the epididymis** into the vas deferens. The actual cumulative length of this passage is 5 m (16 ft). The epididymis is the first passage with a true mucosa and muscularis. The pseudostratified epithelium of the mucosa contains tall and short microvillar cells that absorb the final fragments

of cytoplasm from spermiogenesis. A vascular lamina propria underlies this epithelium, and the muscularis becomes more substantial as it approaches the vas deferens.

VAS DEFERENS

The vas deferens conducts sperm from the epididymis of each testis to the ejaculatory duct and urethra. The **vas deferens** (DEF-er-enz) is a substantial tube with a thick muscularis, a mucosa, and an adventitia, easily palpated as the firm tube in each side of the scrotum. The vas is about 4 mm in diameter and 45 cm (18 in) long, and it reaches through the inguinal canal to the prostate. The vas enlarges into the **ampulla** as it nears the seminal vesicles. The pseudo-stratified columnar epithelium of the vas resembles that of the epididymis. Numerous longitudinal folds of epithelium flatten as sperm pass during ejaculation, and reappear afterward. Compared with the thin, smooth muscle in the epididymis, the muscularis has become a multilayered tissue, with inner and outer longitudinal layers separated by middle circular fibers.

EJACULATORY DUCT

These short tubes continue to the urethra from the seminal vesicles. Only 1 to 2 cm long, the **ejaculatory ducts** run entirely within the prostate gland. Their pseudostratified columnar epithelium resembles that of the ampullae.

URETHRA

The **urethra** (YOU-reeth-rah; also described in Chapter 25) conducts urine or semen through the penis and consists of mucosa and muscularis. It is divided into three sections. First, the **prostatic urethra** (pro-STAT-ik; about 3 cm long) conducts urine from the bladder and joins the ejaculatory ducts within the prostate gland. Secretions from the prostate gland also enter through many small prostatic ducts. (Ejaculation prevents urine from mixing with semen by closing the internal sphincter muscle of the bladder.) Second, the **membranous urethra,** the narrowest and shortest segment, only 3 mm × 1.5 cm, penetrates the pelvic diaphragm and the external urethral sphincter muscle. Finally, the **spongy portion,** or **penile urethra** (PEEN-ile), extends about 15 cm from the bulb of the penis to the glans, where it opens as a vertical slit. Bulbourethral glands secrete into the proximal end of the penile urethra, and numerous mucous glands (urethral glands) release into the body of the penis. **Stratified columnar epithelium** lines all portions of the urethra, except for transitional epithelium at the bladder and stratified squamous epithelium within the glans.

- Trace sperm through the internal ducts.
- Which tubes are muscular, and which are lined with columnar epithelium?

ACCESSORY GLANDS

The prostate gland, seminal vesicles, and bulbourethral glands add seminal fluid to the sperm. All three are compound exocrine glands with branching ducts or passages that lead from tubuloalveolar secretory chambers (see Glands, Chapter 3). A typical ejaculate gathers altogether 4 to 6 ml of whitish viscous fluid from these glands.

PROSTATE

The prostate gland is really three glands in one that together secrete clear, colorless prostatic fluid, rich in proteolytic enzymes, into the urethra (Figure 26.7A). The size of a walnut, the prostate (PRAH-state) is embedded in a fibroelastic capsule beneath the bladder, where the ejaculatory ducts join the prostatic urethra. The prostatic utricle is a small diverticulum that opens to the urethra from the posterior, between the ejaculatory ducts. The glands of the prostate surround the urethra and provide about one fifth of the total ejaculate. The innermost and smallest are the mucosal glands that secrete through numerous small openings in the anterior wall of the urethra. Next in size and location are the submucosal glands that secrete into the lateral wall of the urethra, and fi-

nally a large opening in the posterior wall receives the products of the main prostate gland, the outermost and largest. Most of the prostatic fluid is provided by 30 to 50 tubuloalveolar glands. All three glands are lined with tall columnar epithelium and numerous shorter cells and are enveloped by fibromuscular tissue that appears to expel fluid from the glands during ejaculation. Prostatic fluid is especially rich in fibrinolysin, which liquifies semen, and acid phosphatase, as well as citric acid and zinc ions.

Prostate glands often enlarge in older men and may then interfere with micturition. The gland may become as large as an orange and encroach upon the bladder's space for urine. Enlargement may be benign, through hyperplasia of the mucosal glands deep in the prostate, or possibly malignant, if the main prostate enlarges. Middle-aged men in the United States have a 10% chance of developing prostate cancer, about the same risk that women have of developing breast cancer.

Telling the difference between benign and cancerous enlargement calls for several tests and decisions about treatment. Physicians can palpate enlarged prostates through the wall of the rectum, but because tumors grow slowly and may be small (1 to 3 cc vol-

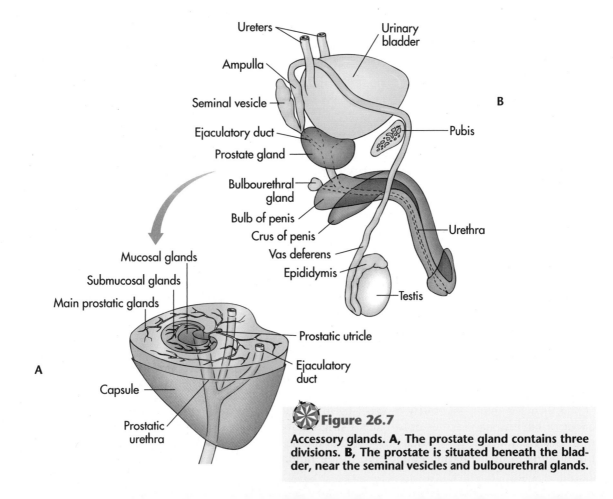

Figure 26.7

Accessory glands. **A,** The prostate gland contains three divisions. **B,** The prostate is situated beneath the bladder, near the seminal vesicles and bulbourethral glands.

ume) when they metastasize, rectal exams miss many early cancers. **Prostate-specific antigen** (PSA) tests are more effective. These tests detect a blood protein made only in the prostate. Elevated PSA levels do not necessarily mean that a prostate is cancerous, but 90% of dangerous prostate tumors are associated with higher than normal levels of PSA.

Up to two thirds of men with elevated PSA have prostate tumors that can be removed surgically, suppressed by radiation, or carefully monitored if the tumor is small and growing slowly. Older men currently are encouraged to watch and wait because the risks of surgery may outweigh the risks from a slowly growing tumor. Surgery is recommended for younger men in their fifties or sixties who face a longer time in which prostate cancer can develop. Surgery risks impotence, however, because the same autonomic nerves that erect the penis also pass through the prostate.

- What products does the prostate gland produce?
- Why is prostate cancer difficult to detect?

SEMINAL VESICLES

These long (5 to 10 cm) lobular glands adjoin the vas deferens at the entrance to the prostate gland, as shown in Figure 26.7B. The interior mucosa forms a maze of chambers lined with pseudostratified tall columnar epithelium that secretes 60% of the ejaculate fluid. Rough endoplasmic reticulum and dense secretory granules enable the cytoplasm to secrete proteins that are contained in the viscous, yellowish product, rich in prostaglandins and fructose, that enters the ejaculatory ducts. Testosterone stimulates all this activity.

BULBOURETHRAL GLANDS

Also known as **Cowper's glands,** the pea-sized **bulbourethral glands** lie on either side of the urethral bulb in Figure 26.7B, where their mucoid secretion precedes ejaculation into the spongy portion of the urethra, neutralizing urine and perhaps lubricating the urethra.

PENIS

The penis discharges sperm or urine through the urethra, as shown in Figure 26.8A. The thin skin that covers the entire penis is loosely attached to the corpora by connective tissue, except at the glans, where the connection with the corpus spongiosum is quite firm. As in the scrotum, there is no adipose tissue in the penis. The epidermis that covers the glans is continuous with the **foreskin,** a fold of skin that ensheathes the glans but is removed in circumcised males.

The penis is given its ability to erect by a pair of **corpora cavernosa** (CAV-er-nos-a) and a single **corpus spongiosum** that each reach the length of the penis (Figure 26.8A). Fibrous connective tissue binds

the two corpora cavernosa together on the dorsal side of the penis, and the corpus spongiosum contains the urethra. A fibrous sheath wraps all three corpora into a somewhat triangular shaft of the penis. The proximal ends of the corpora cavernosa end in tapering **crura** (crus, singular) that curve laterally and are ensheathed by the **ischiocavernosus muscle** (ISS-she-oh-KAV-ern-oh-sus) that anchors the penis firmly to the ramus of the ischium and pubis (Figure 26.8B). By contrast the corpus spongiosum has a bulbous end that accepts the urethra and attaches to the pelvic diaphragm by way of the **bulbospongiosus muscle** (BUL-bo-SPON-jee-oh-sus). Distally the corpus spongiosum gives the **glans** the characteristic mushroom-cap shape that shields the ends of the corpora cavernosa. The urethra opens at a vertical slit in the glans. Fibrous **trabeculae** (Figure 26.8C) reach in all directions within the corpora, leaving spaces between them, lined with vascular epithelium, that inflate with arterial blood during sexual arousal.

Blood reaches the cavernous sinuses from the internal iliac artery through branches of the **internal pudendal artery** (PEW-den-dal), returning through the prominent **deep dorsal vein** and the **prostatic plexus** of veins proximal to the corpus spongiosum. A pair of **dorsal arteries** lies on either side of the dorsal vein between the corpora cavernosa, as shown in Figure 26.8C.

ERECTION AND EJACULATION

Erection and **ejaculation** are controlled through the autonomic nervous system, but sensations from the skin are somatic. Psychological and tactile stimulation cause the parasympathetic nervous system to erect the penis. Neurons of the sacral outflow from the spinal cord release the neurotransmitter, **nitric oxide** (NO), that dilates the corpora of the penis and allows blood to inflate them as rigid rods. Sympathetic neurons from the lumbar spinal cord and the hypogastric plexus stimulate ejaculation. These impulses cause the seminal vesicles and prostate to release their fluids and also cause the vas deferens to constrict. As the seminal fluid begins to flow, the bulbospongiosus and ischiocavernosus muscles squeeze the bulbourethral glands and the bases of the corpora, ejecting the semen through the penis.

- Which corpus or corpora enclose the urethra? Attach to the ischial tuberosity? Form the glans?

FEMALE

BRIEF TOUR

The anatomy of the female genitalia is simpler than that of males in many respects, as you can see in Figure 26.9. Unlike that of males, the female pelvic cavity is divided into two pouches by the broad liga-

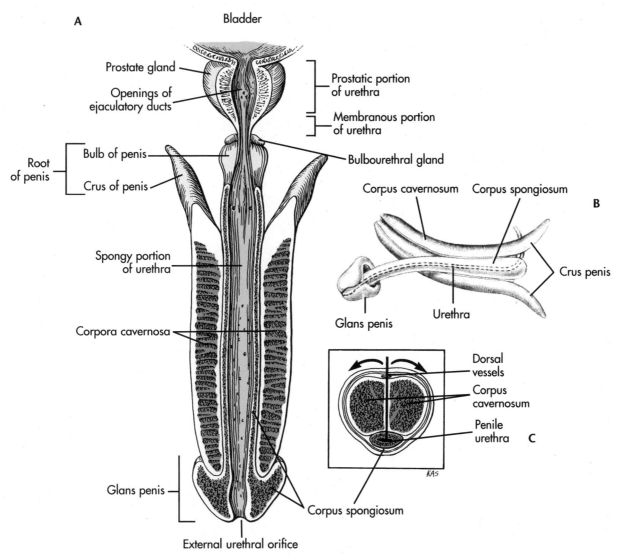

A

Bladder

Prostate gland

Openings of
ejaculatory ducts

Prostatic portion
of urethra

Membranous portion
of urethra

Root
of penis

Bulb of penis

Crus of penis

Bulbourethral gland

Corpus cavernosum   Corpus spongiosum

B

Spongy portion
of urethra

Crus penis

Corpora cavernosa

Glans penis   Urethra

Dorsal
vessels

Corpus
cavernosum

Penile
urethra   C

Glans penis

KAS

Corpus spongiosum

External urethral orifice

Figure 26.8

Penis. A, Corpora cavernosa and corpus spongiosum, frontal section. B, The corpora attach to the ramus of the ischium. C, Transverse section reveals dorsal vessels and two corpora cavernosa. The corpus spongiosum surrounds the penile urethra.

ment, a fold of mesentery that reaches up from the floor of the pelvic cavity between the bladder and rectum. This ligament is important because it envelops the uterus, vagina, and uterine tubes that reach laterally to the ovaries, like the crossbar of the letter *T*. All vessels and nerves supplying these structures enter through the broad ligament. Figure 26.10 shows the **uterovesical** (YOU-ter-oh-VEH-sih-kal) pouch between the uterus and bladder and the **rectouterine** (REK-toh-YOU-ter-in) pouch between the uterus and rectum. The ovaries are the principal sources of estrogen and progesterone, the endocrine hormones that promote oogenesis, the uterine cycle, and secondary sex characteristics. The ovaries release eggs for fertilization in the uterine tubes. The uterine tubes sweep an embryo into the uterus, where it can develop in the uterine lining. The uterus leads through the cervix to

the vagina, which receives the penis and sperm at intercourse. Sperm move from deep in the vagina, through the cervix and uterus, and into the uterine tubes where fertilization takes place. In pregnancy, mother and fetus establish circulation through the placenta. Labor dilates the cervix and the vagina to deliver the fetus.

OVARIES

The ovaries are smooth, oval, orange organs that produce eggs and hormones. As shown in Figure 26.9, each ovary is approximately 4 cm long, 2 cm wide, and slightly less than 1 cm thick (1½ × ⅘ × ⅖ in). A single mesovarium suspends each ovary, with its vessels and nerves, from the broad ligament near the anterior wall of the pelvis. Ovarian vessels and nerves enter at the **hilus** of the ovary. Thin columnar

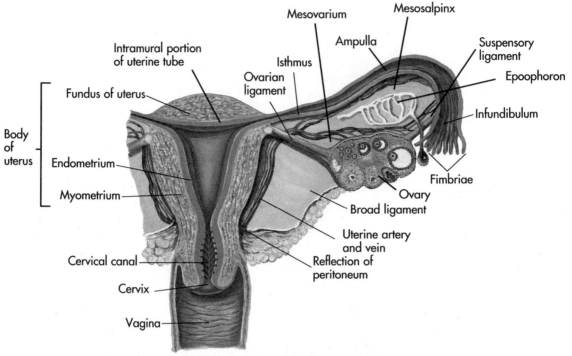

Figure 26.9
Female genitalia in the broad ligament.

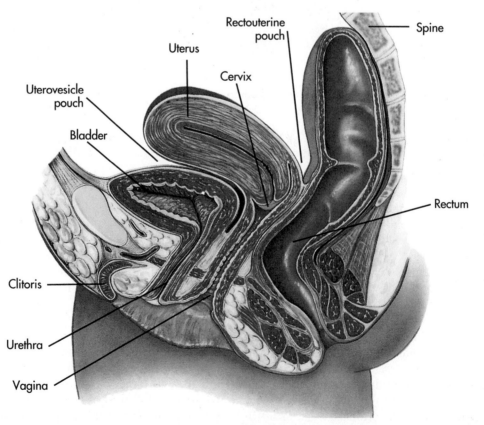

Figure 26.10
Female genitalia in sagittal section.

germinal epithelium covers the surface of each ovary and is continuous with the visceral peritoneum of the broad ligament. Immediately beneath the germinal epithelium lies the whitish **tunica albuginea** (Figure 26.11), a substantial fibrous connective tissue that supports the numerous ovarian follicles and loose connective tissue stroma within the ovary. Deep in the ovary the highly vascular **medulla** distributes branches of the **ovarian artery** and **veins** and divisions of the **ovarian nerves** into the stroma and the epithelium. No ovarian follicles are found in the medulla.

OOGENESIS AND OVARIAN FOLLICLES

Oogenesis is the process that produces eggs. Homologous to spermatogenesis, **oogenesis** (oh-oh-JEN-eh-sis; see Table 26.1) transforms diploid **oogonia** (oh-oh-GOAN-ee-ah) into haploid eggs, but there are fundamental differences between the two processes, aside from their products. First, unlike spermatogonia, which continually replace themselves throughout a man's reproductive life, the supply of oogonia does not renew itself. Oogonia proliferate, become **primary oocytes** in the fetus, and remain **arrested** in the first meiosis until ovulation begins at puberty. Some oocytes thus may wait as much as 50 years before they are ovulated. A girl is born with about 2 million primary oocytes, all she will ever have, but only

about 400 ever mature as eggs. Most of the rest will become **atretic** (ah-TREH-tik), that is, degenerate and die. Secondly, meiotic divisions are **unequal,** so that one oocyte produces only one large egg cell and several small polar bodies (described later), rather than four small cells as in spermatogenesis. Finally, oogenesis and meiosis are completed only after a sperm fertilizes the egg.

Whereas sperm develop within seminiferous tubules, oogenesis proceeds in ovarian follicles, which are clusters of epithelial cells (Figure 26.11A) that surround individual oocytes (read Male-Female Homologies, this chapter, to see how seminiferous tubules and ovarian follicles are comparable). Follicles and oocytes progress through several stages of development. These steps begin with **primordial follicles,** in each of which a single layer of squamous follicle cells surrounds a primary oocyte, and a basement membrane links the whole follicle with surrounding ovarian stroma. Follicle cells derive from the same germinal epithelium as Sertoli cells, and they have a similar supporting role, described later. Primordial follicles cluster beneath the epithelium of the ovary, but they do not all mature simultaneously. Beginning at puberty, every 28 days follicle stimulating hormone (FSH) and the ovarian cycle (described in Figure 26.12) cause a group of follicles and oocytes to enlarge and mature, but only one dominant follicle

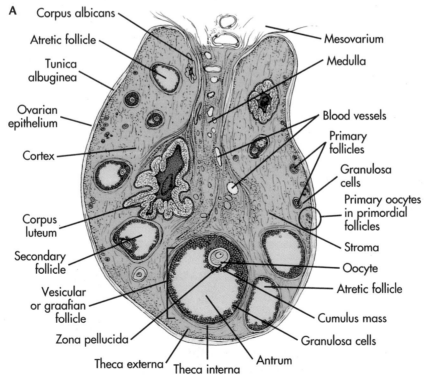

Figure 26.11

Ovary. **A,** Transverse section through the ovary, showing ovarian follicles.

Continued.

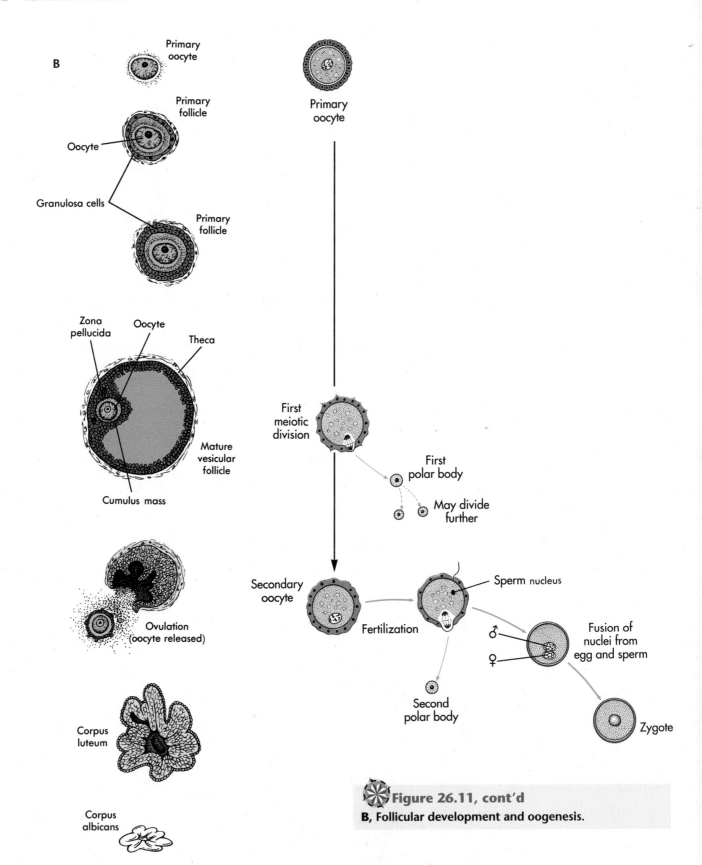

Figure 26.11, cont'd
B, Follicular development and oogenesis.

(occasionally more) releases its oocyte for fertilization during each cycle. The rest of the group become atretic.

Primordial follicles become primary follicles. The oocyte itself and its follicle cells enlarge, and a

thick, gelatinous **zona pellucida** (PEL-loose-ih-da) forms between the follicle cells and the oocyte. The oocyte takes on materials from the follicle cells through cytoplasmic extensions across the zona pellucida. Several layers of follicle cells accumulate in

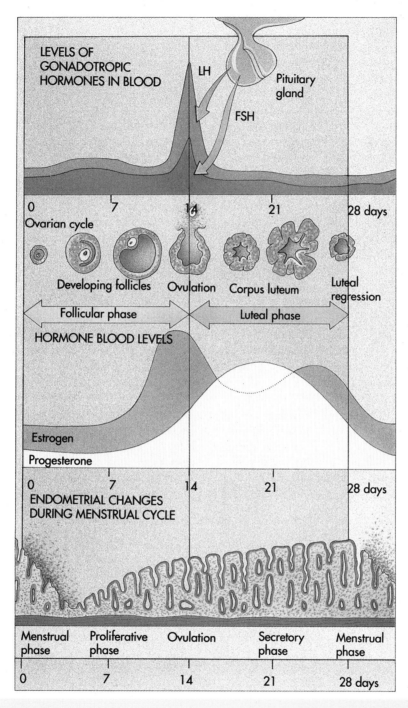

⚘ Figure 26.12

Ovarian and uterine cycles. The ovarian cycle produces eggs, while the uterine cycle prepares the endometrium to receive an embryo should an egg be fertilized. Both cycles run simultaneously, typically 28 days, counting from the beginning of menstruation. During the first 2 weeks *(red)* the ovary is in the follicular phase, preparing follicles to ovulate, while menstruation sheds the uterine lining and the proliferative phase replaces it. Follicle stimulating hormone promotes follicular maturation in the ovary, and estrogens from the follicles stimulate the endometrium to regenerate. A surge of luteinizing hormone from the pituitary causes ovulation on day 14. The last 2 weeks *(yellow)* find the ovary in its luteal phase, producing a corpus luteum, while the uterus enters the secretory phase, as the secretions of its new glands prepare for implantation. Luteinizing hormone promotes the luteal phase, and estrogens and progesterone from the corpus luteum stimulate uterine glands to secrete. The cycles end with menstruation.

secondary follicles (Figure 26.11A) and begin to secrete estrogens. The follicle is now a small endocrine gland. The follicle cells are known as **granulosa cells** because they contain cytoplasmic granules that release estrogen. Spindle-shaped **thecal cells** (THEE-kal) accumulate from the ovarian stroma around the follicle and its basement membrane, where they provide substrates for estrogen synthesis. The well-vascularized inner **theca interna** that provides these materials blends with the fibrous **theca externa** on the outside of the follicle. When enough estrogen accumulates to form spaces between the granulosa cells, the structure is now a young **vesicular (Graafian) follicle.** These small spaces merge into one huge **antrum** cavity in mature, pea-sized vesicular follicles. The oocyte and its follicle cells are nearly isolated in this estrogen lake, except for a small neck of follicle cells called the **cumulus oophorus** (KUME-you-lus oh-oh-FOR-us) that connects them with the wall of the follicle. (The oocyte and granulosa cells die in atretic vesicular follicles, but the thecal cells return to the stroma from which they originated.) Activated by **luteinizing hormone** (LH; LOOT-ee-in-EYE-zing), the vesicular follicle ruptures and releases from the ovary the oocyte, which is surrounded by a crown of granulosa cells, the **corona radiata.**

The primary oocyte completes the **first meiotic division** in the vesicular follicle just before ovulation (Figure 26.11B). Each nucleus receives one chromosome from each homologous pair of chromosomes, but cytoplasmic division is unequal. The first division discards one nucleus, along with a small piece of cytoplasm, into the **first polar body.** The other nucleus (chance determines which) remains in the large **secondary oocyte** and quickly begins the **second meiotic division** but arrests again at metaphase until fertilization causes the process to resume. Both oocyte and polar body can be seen inside the zona pellucida in the figure. Though technically still a secondary oocyte, but universally called an egg or ovum, this cell survives only one or two days in the uterine tube unless it is fertilized. Entry of a sperm causes the nucleus to complete the second meiotic division. Again, unequal cleavage sheds a **second polar body.** The remaining egg pronucleus, now within the true mature haploid egg, or **ootid** (OH-oh-tid), fuses with the sperm pronucleus to form the zygote nucleus, and embryonic development begins.

CORPUS LUTEUM

The corpus luteum is the remains of the vesicular follicle (Figure 26.11A and B). After ovulation, blood forms a large clot in the antrum, and the granulosa and thecal cells transform under the influence of LH into large, yellow **lutein cells** (LOOT-ee-in) that continue secreting progesterone and estrogens during the final 2 weeks of the uterine cycle.

Thereafter this short-lived endocrine gland degenerates, and its cells are replaced with white, fibrous scar tissue known as the **corpus albicans** (AL-bih-kans; alb-, *L*, white).

Pregnancy rescues the corpus luteum, but only temporarily. After an egg is fertilized and the embryo implants into the uterus, the corpus luteum enlarges (from 2 cm to 5 cm) and continues to produce hormones as the **corpus luteum of pregnancy.** The placenta stimulates this output with **human chorionic gonadotropin** (HCG), and the corpus luteum sustains the uterine lining until week 9 or 10, when the placenta takes over the production of estrogen and progesterone until delivery. The corpus luteum then becomes a corpus albicans.

- Name the stages of follicular development, and describe the characteristics of each stage.
- Which phase of oogenesis occurs in each follicular stage?
- What is the corpus luteum?

UTERINE TUBES

Two uterine tubes establish the chemical environment for fertilization and conduct fertilized eggs to the uterus. Also known as **oviducts** or **Fallopian tubes,** the uterine tubes run transversely in the free border of the broad ligament (Figure 26.9), known as the **mesosalpinx** (mezz-oh-SAL-pinks), superior to each ovary. Each tube is approximately 12 cm (5 in) long and consists of four regions through which fertilized eggs pass into the uterus. The funnel-like **infundibulum** captures ovulated eggs within its tentacular folds, or **fimbriae** (FIM-bree-ee; fimbri-, *L*, a fringe or fibers). Cilia on the inner surface conduct the egg to the **ampulla** (*L*, a flask), a long swollen section that is the site of fertilization. Next, a short narrow muscular **isthmus** leads to the uterus, where the **intramural portion** of the uterine tube penetrates the thick wall of the uterus and releases the embryo into the lumen. Careful examination of the mesovarium reveals the **epoophoron** (EP-oh-oh-FOR-on) and **Gartner's duct,** the remnants of mesonephric ducts (see Male-Female Homologies, this chapter).

Three tissue layers form the wall of the uterine tube (Figure 26.13). Great folds of **mucosa** nearly fill the lumen so that sperm and egg must travel deep between the folds, driven by cilia that beat toward the uterus or by peristaltic contractions from the smooth muscle fibers of the **muscularis.** Typical inner circular and outer longitudinal layers of fibers thicken at the isthmus and finally become continuous with the myometrium of the uterus. Thin **serosa,** continuous with that of the mesosalpinx and broad ligament, covers the superficial surface of the uterine tubes.

- Which three layers of tissue make up the walls of the uterine tubes?
- Name the four regions of a uterine tube.

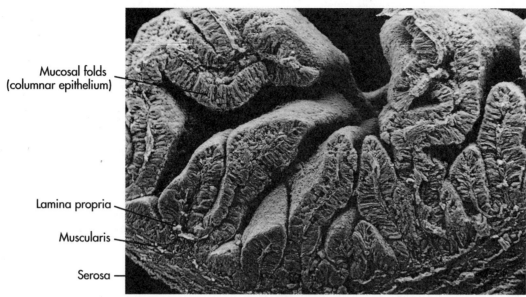

Mucosal folds (columnar epithelium)

Lamina propria

Muscularis

Serosa

Figure 26.13

Uterine tube. Transverse section through the ampulla, where fertilization occurs. (Scanning electronmicrograph by permission of R.G. Kessel and R.H. Kardon, *Tissues and organs: A text-atlas of scanning electron microscopy*, 1979, W.H. Freeman and Co.)

UTERUS AND UTERINE CYCLE

The uterus carries fetuses to term. The uterus (YOU-ter-us) is the size and shape of a small pear (7.5 × 5 × 3 cm) located in the broad ligament between the bladder and rectum (see Figures 26.9 and 26.14A). The uterus opens through the **cervix** (SIR-viks) to the vagina at its narrow end, and the large **body** of the uterus drapes forward over the bladder when a woman stands erect (shown in Figure 26.10). Uterine tubes exit to either side in the folds of the broad ligament, marking the **fundus** as the base of the pear between the uterine tubes. Inside, the **lumen** of the uterus is a triangular potential space flattened between the thick anterior and posterior walls of the uterus, with angles reaching the uterine tubes and cervix. The entrance to the cervix is called the **internal os;** it leads through the narrow **cervical canal** and opens at the **external os** into the vagina. Fibrous tissue makes the cervix firm, but the body and fundus of the uterus are more flaccid.

The same three tissues that line the uterine tubes also line the uterus, but with important differences that you will see. The **serosa,** which covers externally, is continuous with the peritoneum of the uterine tubes and broad ligament. The serosa houses sympathetic ganglia and plexuses, and uterine arteries branch through it into the myometrium and endometrium beneath.

The myometrium is the thickened muscularis that delivers the baby through the birth canal. Smooth muscle fibers sweep around the uterus in poorly defined deep longitudinal, middle circular, and superficial oblique layers. Figure 26.14B shows the loops in a resting and a pregnant uterus. The fibers are better defined in the more fibrous cervix, however, where the deep and superficial layers are clearly longitudinal and the middle remains circular. Coiled branches of uterine and ovarian arteries and veins pass through the middle layer of fibers into the endometrium.

The endometrium secretes glycogen-rich material into the lumen of the uterus in preparation for pregnancy (Figure 26.12). Simple columnar epithelium lines the lumen with ciliated cells and microvillar cells and forms simple tubular **endometrial glands** that are underlain by **endometrial stroma,** the uterine version of lamina propria.

As you know, the endometrium (end-oh-MEET-reeum) does not remain continually in this state of biological anticipation. The **ovarian cycle,** described in Figure 26.12, promotes development and secretion by the endometrial glands; but unless pregnancy begins, the surface of the endometrium, known as the **functional layer** (Figure 26.12), is shed by **menstruation** (MEN-strew-AYE-shun) and new tissue regenerates from the thinner **basilar layer.** Ischemia (lack of blood) is the immediate cause of menstruation; the spiral arteries of the myometrium and lamina propria constrict and shut off blood supply to the functional layer, which then degenerates and is shed.

The endometrium also lines the cervix but is not lost during menstruation. Simple columnar epithelium and elaborate mucous glands fill the cervical canal with mucus that sperm must negotiate when entering the uterus. The epithelium becomes stratified squamous as it nears the external os.

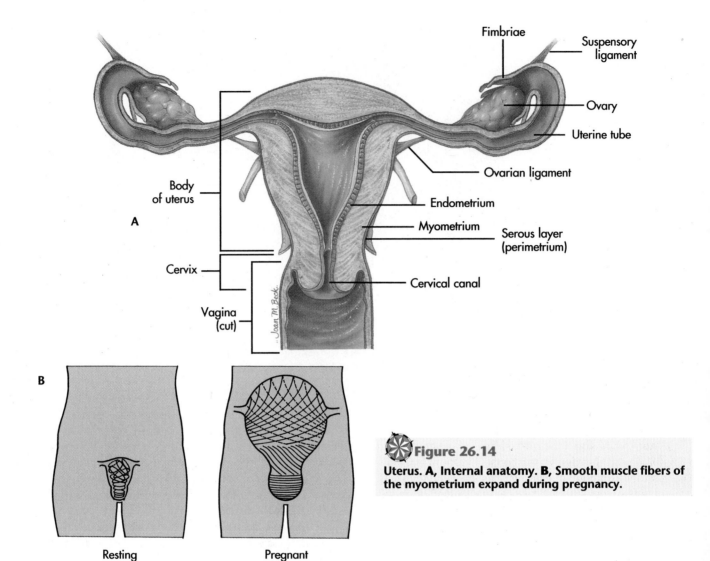

Figure 26.14

Uterus. A, Internal anatomy. **B,** Smooth muscle fibers of the myometrium expand during pregnancy.

- Describe the location of the uterus in the pelvic cavity.
- What are the functions of the endometrium, myometrium, and serosa of the uterus?

VAGINA

The vagina lies between the bladder and rectum. Opening through the pelvic diaphragm, it connects the vestibule to the uterus (Figure 26.9). About 9 cm (4 in) long, it receives the penis at intercourse and becomes the **birth canal** in delivery. The cervix opens at the deepest apex of the vagina at the **fornix** on the anterior wall and bends forward at a right angle to the vagina. **Mucosa** and **muscularis** line the walls of the vagina, but these tissues differ considerably from those of the uterus. First, the vaginal wall is much thinner and neighboring organs collapse it into lengthwise folds. Second, stratified squamous epithelium lines the lumen of the vagina. In the vagina, many transverse **rugae** engorge with blood and help

stimulate the penis during sexual intercourse. Inner circular and outer longitudinal layers of muscularis surround the mucosa, and **adventitia** anchors the vagina in the pelvic floor, except for the small portion of the vagina that protrudes into the pelvic cavity and is covered by serosa.

Vaginal smears of epithelial cells can readily detect the phases of the ovarian cycle because estrogens thicken the vaginal epithelium and promote the release of glycogen during the cycle. Vaginal bacteria convert glycogen to lactic acid that combats infection by reducing the pH of the vagina.

VULVA AND PERINEUM

The perineum of females is shaped like a diamond, as in males. The urethra and vagina open to the **vulva,** the female external genitalia (Figure 26.15). The **labia minora** guard the **vestibule** that receives these tubes and also receives lubricating secretions from the **vestibular glands.** Posterior to these glands

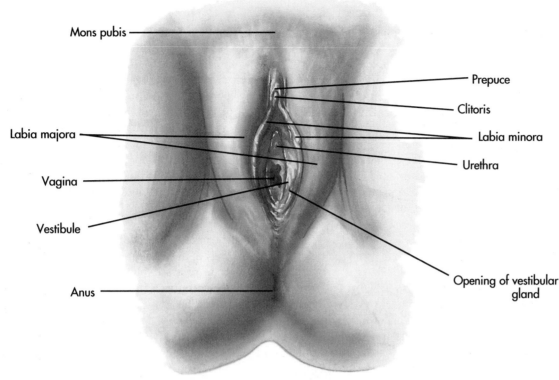

Mons pubis

Prepuce

Clitoris

Labia majora

Labia minora

Urethra

Vagina

Vestibule

Anus

Opening of vestibular gland

Figure 26.15
Vulva.

lie the **bulbs of the vestibule,** erectile tissues homologous with the bulb of the penis and the corpus spongiosum. The **hymen** (HIGH-men) is a thin sheet of tissue that partially covers the entrance to the vagina until accident or the first intercourse ruptures it. An erectile **clitoris** (KLIT-or-iss) marks the anterior end of the vestibule, where the labia minora meet as a fold of tissue called the **prepuce** (PREE-pyuse) that envelops the clitoris. The **labia majora** are the outer folds of skin and adipose tissue that border the vulva laterally. The labia majora join posteriorly at the **perineal body** anterior to the anus, and they merge anteriorly in the **mons pubis,** a fatty pad located over the pubis.

- Into which space in the vulva do the vagina and urethra open?

MALE-FEMALE HOMOLOGIES

Kristin and Tom thought the interview had gone quite well. Their first steps toward becoming parents were under way. Unlike many couples who discover their infertility after marriage, they had planned from the start to adopt children. Low sperm counts or blocked uterine tubes rarely reveal themselves until couples fail to achieve pregnancy, but Kristin's expe-

rience was different, and that was the reason she and Tom were seeing the adoption agency. Kristin has inherited androgen insensitivity syndrome, a condition that interferes with and even removes the receptors that take up testosterone in her tissues. Her cells do not respond to testosterone, which all women produce in small quantities. She first became aware of her condition when her menstrual periods failed to start. You can imagine her anguish when tests revealed that she was born without a uterus or ovaries, and that in fact her cells were chromosomally male, XY, rather than female, XX.

Androgen insensitivity syndrome is a relatively common disparity between chromosomal and somatic sex, brought about by a recessive gene for defective testosterone receptors, inherited from both parents, that blocks the response of cells to testosterone. Unable to use testosterone, Kristin developed female external genitalia and secondary sex characteristics, but axillary and pubic hair did not form, the clue that led her doctor to recommend chromosomal testing. Unless signaled otherwise by the Y chromosome, genitalia and secondary characteristics tend to develop as female. In Kristin's case the necessary male-determining genes of the Y chromosome were present, but her cells lacked the ability to respond to testosterone from the testes. As a result, dysfunctional testes developed and descended into her labia majora, the tissue that in males would become the scrotum. Internal female genitalia failed to appear, except for a blind, shortened vagina.

The sex of an individual expresses itself in at least four different levels. **Chromosomal sex,** XX in females and XY in males, dictates the gonadal and somatic sex and perhaps also gender identity. **Gonadal sex** refers to the presence of ovaries or testes, and **somatic sex** expresses an individual's secondary sex characteristics, female or male. **Gender identity** is a person's psychological sense of herself or himself as feminine or masculine. Kristin's experience is one of many variations in which one or more levels of sexual expression do not match.

COMMON ORIGINS, COMPLEMENTARY ORGANS

Internal and external genitalia originate from indifferent tissues that are capable of developing in either male or female directions. Consequently, male and female genitalia have **homologous parts,** summarized in Table 26.2, that develop from the same embryonic tissue. For example, scrotum and labia majora have the same origins, which is why testes were found in the labia in Kristin's case.

EXTERNAL GENITALIA

At 7 weeks of development (Figure 26.16) the external genitalia are ready to become scrotum or vulva. A broad **genital eminence** is surrounded by a pair of **genital swellings** and topped by a **genital tubercle.** The **urogenital sinus** opens in the midline between the **genital folds,** and the anus opens posteriorly.

In females the genital tubercle becomes the **clitoris,** and the **labia majora** develop from the genital swellings and the **labia minora,** from the genital folds. The **urogenital sinus** splits into the urethra and vagina, which open into the deepened **vestibule** between the labia minora. The clitoris and **glans** recede to the anterior end of the vestibule, covered by the **prepuce.**

By contrast, structures that remain open in females close in males. The genital tubercle becomes the **glans of the penis,** and the genital eminence elongates as the **penis** itself, incorporating the opening of the urogenital sinus into a **urogenital groove** on the underside. The genital folds, the borders of this groove, fuse except for an orifice that migrates to the glans. Destined to form the vestibule in females, this tissue becomes the **penile urethra** in males. The genital swellings sweep posteriorly around the penis and fuse in the midline to form the **median raphe.** Occasionally the urethral groove fails to fuse entirely in males and leaves one or more openings on the underside of the penis, a condition known as **hypospadias** (HIGH-po-SPADE-ee-as) that is easily corrected by surgery. A complementary situation

Table 26.2	GENITAL HOMOLOGIES	
Male	**Indifferent**	**Female**
Testis	Indifferent gonad	Ovary
Spermatogonia, spermatocytes, sperm	Primordial germ cells	Oogonia, oocytes, eggs
Seminiferous tubules	Sex cords	Ovarian follicles
Sertoli cells	Germinal epithelium	Granulosa cells
Interstitial cells	Mesenchyme	Stromal cells, thecal cells
Gubernaculum testis	Gubernaculum	Round ligament
Efferent ductules	Mesonephric tubules	Epoophoron
Epididymus	Mesonephric duct	Epoophoron
Vas deferens	—	Gartner's duct
Appendix testis	Mullerian duct	Uterine tubes, uterus
Urinary bladder, urethra	Urogenital sinus	Urinary bladder, urethra, vestibule
Prostatic utricle	—	Vagina
Prostate gland	—	Urethral glands
Bulbourethral glands	—	Vestibular glands
Penis	Genital tubercle	Clitoris
Glans	—	Glans
Corpora cavernosa	—	Corpora cavernosa
Corpus spongiosum	—	Bulb of the vestibule
Ventral penis	Urogenital folds	Labia minora
Scrotum	Urogenital swellings	Labia majora

From Moore KL: The developing human, *ed 3, Philadelphia, 1982, WB Saunders Co.*

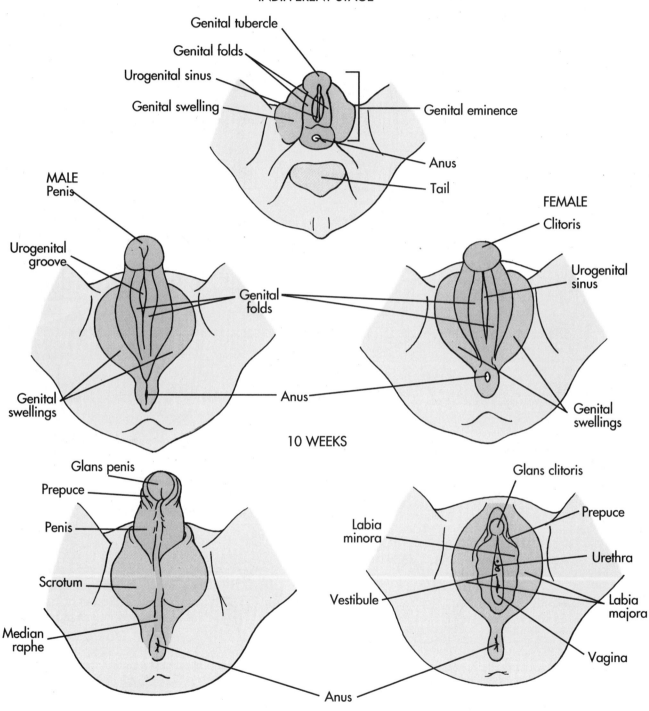

INDIFFERENT STAGE

Genital tubercle
Genital folds
Urogenital sinus
Genital swelling
Genital eminence
Anus
Tail

MALE
Penis
Urogenital groove
Genital swellings

FEMALE
Clitoris
Urogenital sinus
Genital folds
Anus
Genital swellings

10 WEEKS

Glans penis
Prepuce
Penis
Scrotum
Median raphe
Anus

Glans clitoris
Prepuce
Labia minora
Urethra
Vestibule
Labia majora
Vagina

NEAR TERM

Figure 26.16
External genitalia. Transformation of indifferent genitalia into scrotum and penis or vulva.

of urogenital fusion has been reported in at least one female who menstruated through an enlarged clitoris.

- What are the homologs of the clitoris, penile urethra, and labia majora?

DUCTS

Genital ducts also begin indifferently as a dual system of potentially male and potentially female passages illustrated by Figure 26.17. The duct system prepares for development by forming a set of **mesone-**

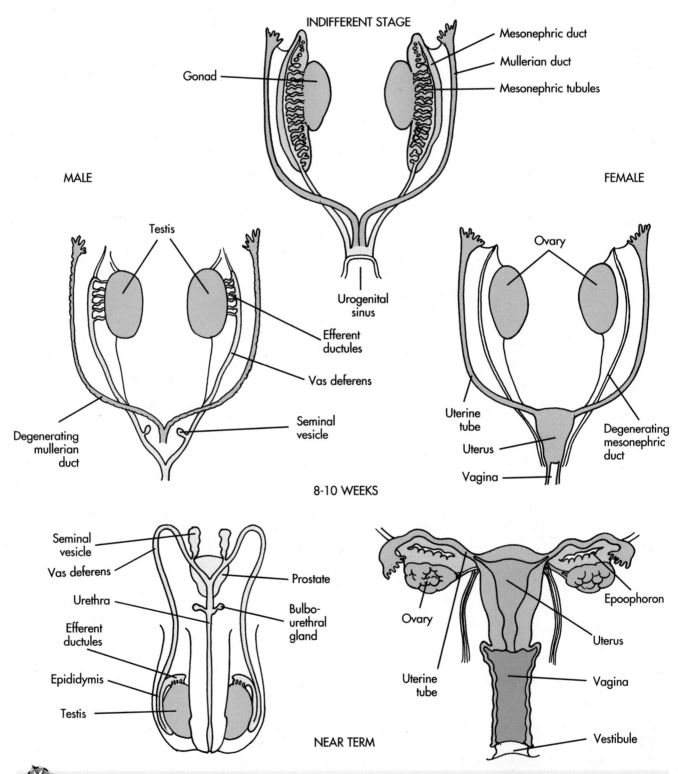

Figure 26.17

Internal genitalia transformed into male and female patterns.

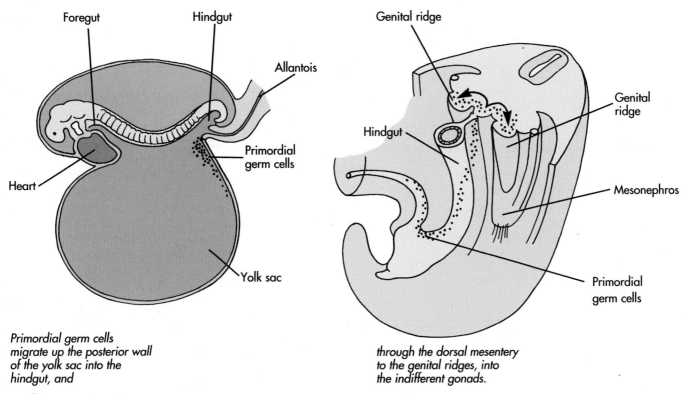

Foregut   Hindgut

Allantois

Primordial germ cells

Heart

Yolk sac

Genital ridge

Genital ridge

Hindgut

Mesonephros

Primordial germ cells

Primordial germ cells migrate up the posterior wall of the yolk sac into the hindgut, and

through the dorsal mesentery to the genital ridges, into the indifferent gonads.

Figure 26.18
Migration of primordial germ cells into the gonadal ridges.

phric ducts and another set of **Mullerian ducts** (mew-LAIR-ee-an) that lead to the urogenital sinus. In males the mesonephric duct develops and the Mullerian duct degenerates, but just the opposite takes place in females. Mesonephric ducts serve the mesonephric kidney temporarily in both sexes (See Chapter 25, Urinary System), and they become the epididymis and vas deferens in males. The mesonephric tubules that remain become the efferent ductules of the testis. The testis secretes a **Mullerian inhibiting factor** that causes the Mullerian ducts to regress. The **prostatic utricle** (YOU-trih-kl; Figure 26.7A) represents the remains of the Mullerian duct system in males. In females the Mullerian ducts become uterine tubes and also fuse as the uterus and superior portion of the vagina. Remnants of the mesonephric duct and tubules (see Chapter 25) remain as the **epoophoron** of the ovarian ligament, precisely in the position they would have taken had the ovary been a testis.

• What are the homologs of the mesonephric ducts, the Mullerian ducts, and the uterus?

GONADS AND PRIMORDIAL GERM CELLS

As you have seen, the gonads contain germ cells committed to gametogenesis and somatic cells (Sertoli cells and follicle cells) that promote gametogenesis but remain diploid and do not become eggs or sperm. Gonads establish the conditions for sperm and eggs to develop, but they are not the primary source of gametes, as students often assume. Eggs and sperm instead trace their origins to **primordial germ cells** (PGCs) that originate in the embryonic yolk sac and move through the dorsal mesentery into the **gonadal ridges** during the fifth week of development, as Figure 26.18 shows. Once arrived, they proliferate into clones of spermatogonia or oogonia, depending on the chromosomal sex of the individual.

Indifferent gonadal ridges consist of mesenchyme covered by peritoneal epithelium. As primordial germ cells enter the gonadal ridge, they cause epithelial cells to enclose them in clusters called **sex cords. Testis-determining factor** from the Y chromosome causes these cords to become seminiferous tubules and the rete testis. Primordial germ cells become spermatogonia, and the epithelial cells develop as Sertoli cells, while the **mesenchyme** produces interstitial cells and the tunica albuginea. Some of the mesonephric tubules connect the testis to the mesonephric duct, but excess tubules remain at the head and tail of the epididymis as the **aberrant ductules.**

In the absence of testis-determining factor, the PGCs and epithelial cells of the sex cords proliferate and supply the ovary with follicles. The sex cords fragment into **primordial follicles,** and the mesenchyme becomes the **ovarian stroma.** Epithelial cells become **granulosa cells** of the follicles, but **the-**

cal cells derive from the stroma. Without Mullerian inhibiting factor to interfere, the caudal portions of the Mullerian ducts form the cervix, body, and fundus of the uterus, and the uterine tubes cloak the ovaries.

- From what tissues do PGCs, Sertoli cells, follicle cells, and interstitial cells develop?

PREGNANCY AND BIRTH

The reproductive system culminates in pregnancy and birth. As the fetus develops, its mother's body changes as well. The placenta quickly becomes the fetus' initial respiratory, excretory, and digestive systems. The mother's uterus expands to accommodate the growing fetus, and her breasts prepare to lactate. This section follows the anatomy of these events.

FETUS

Predicting the date of birth is an important part of monitoring the health of the mother and her fetus as pregnancy proceeds. Delivery is expected at the end of 38 weeks after fertilization, or 40 weeks counting from the **last menstrual period** (LMP). Knowing these times within a margin of several days allows an obstetrician to detect trouble. Trouble is surprisingly frequent. An estimated 75% to 80% of conceptions fail. Most of these failures occur for unknown reasons in the first 2 weeks, before a woman may know she is pregnant, when the embryo is establishing connections with the uterus and the germ layers are forming. The period of next greatest risk lasts until 9 weeks, while all the fetus' organs are being established. Once these organ systems become established, most developmental problems interfere with individual organs rather than with entire systems or the entire fetus. **Prenatal medicine** is a rapidly expanding field that attempts to diagnose and treat fetal disorders that arise from heredity, inadequate nutrition, substance abuse, disease, and general environmental effects that reach the fetus through its mother. Table 26.3 lists major events of fetal development. See Chapter 4, Developmental Basics, to help with the next sections.

Table 26.3 FETAL DEVELOPMENT

System	First Trimester	Second Trimester	Third Trimester
Musculoskeletal	Somites, limb buds, outgrowth, trunk musculature; upper limbs, definitive proportions	Ossification shows in maternal X-rays at 16W; mother begins to feel quickening	—
Integumentary	Epidermis, dermis, nails, hair follicles form	Vernix caseosa covers the skin with fetal sebum and epithelial cells; lanugo by 20W, fine hair; eyebrows and scalp hair at 20W	Scalp and lanugo hair abundant; subcutaneous fat fills out wrinkled skin; white fat increases to about 3.5% of body weight
Cardiovascular	Heart, great vessels form, beating at 3W; circulation at 4W, basic fetal circulation at 8W; erythropoiesis in yolk sac at 2W, liver at 6W, transfers largely to spleen at 12W; fetal hemoglobin; ultrasound detects heartbeat at 5W	Vascular proliferation	Erythropoiesis transfers from spleen to bone marrow at 28W; pulmonary circulation takes over from umbilical and placental at birth; newborn begins to produce adult hemoglobin
Lymphatic and immune	T lymphocytes circulate at 18W, reach adult numbers 32W; B lymphocytes appear in liver, marrow	IgG transfers passive immunity beginning 20-22W; no active immunity IgM unless infection (rubella, syphilis); B lymphocytes circulate in blood 15W	Complement system active in neonate
Central nervous system	Brain, spinal cord 4W; Neuronal proliferation; sucking reflex at 12W	Neural proliferation ends, migration to definitive locations most active; neural connections, synapses, organization begins; myelination begins, continues beyond 1 year	Neural connections, organization underway; CNS can produce breathing reflexes and control body temperature; pupillary light reflex at 30W; cognition circuits probably not active until birth Eyes reopen
Sensory	—	—	
Endocrine	—	Pancreatic islets 3M, insulin at 5M	—

Continued.

Table 26.3   FETAL DEVELOPMENT—cont'd

System	First Trimester	Second Trimester	Third Trimester
Digestive	All digestive organs laid down and take definitive positions; liver occupies most of abdominal cavity 9-10W; midgut herniates into extra-embryonic coelome, then returns to abdomen at 10W; villi begin 7W, taste buds 8W, parietal cells in stomach, bile secretion begins 12W	Gastric, intestinal, hepatic, pancreatic secretions begin; peristalsis 13W; swallowing reflexes 16-17W, Peyers patches 20W; swallows about 450 ml amniotic fluid/day	Digestive secretion is mature; HCl found in stomach 32W; sucking and swallowing coordinated 34W, peristaltic waves in small intestine 34W
Respiratory	Placenta is fetal respiratory organ; lung buds, conducting zone forms trachea, bronchi, bronchioles; pseudoglandular stage; upper tract, nasal cavity, larynx by 5W	Canalicular stage—respiratory zone forms	Terminal sac stage; alveolar cells of lung begin to make surfactant at 24W; fetus survives premature delivery because lungs are functional
Urinary	Metanephric kidney forms 5W, migrate at 9W; urine starts to form 9-12W, excreted into amniotic fluid; ureters connect to bladder, 9W; nephrons begin forming	Nephrons continue to develop	Nephrogenesis complete 36W
Genital	External genitalia assume male or female form	Primordial follicles in ovaries at 16W; uterus formed in females and vagina begins to canalize; testes begin descent in males at 20W	Testes approach scrotum at 28W and descend by 32W
General	Ultrasound detects embryo at 2W transvaginally, 5W transabdominally	—	Substantial weight gain; fetus becomes plump, 7%-8% of body weight is white fat; 26W fetuses survive premature birth

From Little GA: Fetal growth and development. In Eden RD and Boehm FH, editors: Assessment and care of the fetus, Norwalk, 1990, Appleton and Lange; and Moore KL: The developing human, ed 3, Philadelphia, 1982, WB Saunders Co.

Abbreviations: M, months of development after conception; W, weeks of development after conception.

FIRST TRIMESTER—0 TO 13 WEEKS

The embryo forms its fundamental tissues and organs and becomes a fetus during the first 13 weeks of development, the first **trimester** (TRY-mess-ter). Germ layers transform into rudiments of organs—somites, neural plate, foregut, yolk sac, limb buds—during the embryonic period of the first 9 weeks. The heart begins beating at 3 weeks, blood circulates at 4 weeks, and basic fetal vessels are in place at 8 weeks. **Transvaginal ultrasound** can show the heart beating at 5 weeks. Body cavities have closed around the umbilical cord, and the peritoneal cavity has become separate thoracic and abdominopelvic cavities. The first neurons begin to proliferate and migrate from the neural tube, which then forms the spinal cord and brain. By 10 weeks, the facial features become obviously human, digits and nails have begun to form on the limbs, and the genitalia have assumed male or fe-

male form. Reliable sex identification by ultrasound awaits the third trimester, however.

After 9 weeks, the embryo is considered a fetus. Its tissues have a delicacy and transparency rarely preserved in the anatomy lab. With very little fiber to support them, first trimester tissues and organs are more fragile than an adult brain. Nevertheless the fetus kicks at the uterus and swallows amniotic fluid. By the thirteenth week, the fetus is 10 cm (4 in) long, crown to rump, and weighs approximately 72 gm.

SECOND TRIMESTER—14 TO 26 WEEKS

Now the fetus grows rapidly, and the mother begins to "show" as the uterus pushes toward the navel. She can feel her fetus moving and knows when it is most active. The fetus' external features are exquisite. Skin is wrinkled, but rich with pink capillaries, and the epidermis begins to keratinize. Scalp hair and fine body hair known as **lanugo** (la-NEW-go) appear, and creamy **vernix caseosa** (VER-niks KASE-ee-oh-sah) covers the epidermis with sebum and epithelial cells from the skin and the amnion, described below. Ears stand out from the head, legs have elongated, and nails are growing.

Capillary beds proliferate throughout most tissues and ossification begins. Neural proliferation subsides as the newly formed neurons begin to establish synapses and to integrate themselves into neural pathways and reflexes. Myelin sheaths begin to cover axons. Stomach, intestines, pancreas, and liver begin secreting, and the oral cavity and esophagus begin swallowing amniotic fluid from the amniotic cavity that collects as a viscous gel, **meconium** (meh-KONE-ee-um), in the large intestine to be defecated with the first bowel movement. The lungs begin to form respiratory bronchioles and alveolar ducts, and nephrons continue to accumulate in the kidneys. The ovaries are equipped with primordial follicles early in this trimester, and the uterus and vagina form in a female. By the end of this period, testes begin to descend to the scrotum in a male. The fetus is 25 cm (10 in) long and weighs approximately 1000 gm (2 lbs), more than a 10-fold weight gain.

THIRD TRIMESTER—27 TO 39 WEEKS

The fetus becomes capable of living outside the uterus because at 26 weeks its lungs have enough alveoli to function and have been making surfactant since 24 weeks. Delivery now is legally a birth. The nervous system now can regulate breathing reflexes, and it can control body temperature through neural pathways that continue to organize for several years after birth. The digestive system is ready for food, and the kidneys have completed nephrogenesis. Both mother and fetus gain weight. A skinny second trimester fetus becomes chubby and outright plump at term, storing 7% to 8% of its body weight in energy-rich fat reserves. The mother may gain as much as 30 lbs, including fetus, placenta, and enlargement of her own tissues. The uterus reaches the xyphoid process of the sternum and is pushing intestines and other organs aside. The fetus' head has **engaged** in the pelvic cavity. It is time to deliver. The fetus is 36 cm (14 in) long, and weighs approximately 3400 gm (7.5 lbs).

- Name events that take place in each of the trimesters.
- When does the fetus become fat?
- When does the heart begin to beat?
- When do the intestines begin secreting hydrolytic enzymes?

PLACENTA AND FETAL MEMBRANES

The placenta connects the fetus with the mother. It is the fetus' lungs, kidneys, liver, and intestines all rolled into one, a vascular exchange surface and endocrine gland attached to the body of the uterus, as shown in Figure 26.19A. The mature placenta (plah-SENT-ah), a pad of mostly fetal tissue that has spread within the uterine wall, weighs about 1 lb at birth and is about 18 cm in diameter and 2 cm thick (7 × ¾ in). The placenta is a temporary organ; it exists only during pregnancy and disconnects from the uterus at birth. Fetal blood takes up oxygen, nutrients, and hormones from maternal blood, and returns hormones, carbon dioxide, urea, and other wastes to the maternal blood. These components cross a barrier made of fetal tissues that separates maternal from fetal blood. The plasma and cells of fetal blood do not mix with maternal blood.

The placenta forms within the endometrium, where the embryo has implanted itself. As the embryo develops, an extra-embryonic tissue called the **chorion** (KORE-ee-on) surrounds the embryo itself and spreads into the maternal endometrium, where it establishes the vascular connection with the mother. Figure 26.19B outlines the anatomy of this junction. Bush-like projections from the chorion, called **chorionic villi,** extend into **intervillous spaces** eroded into the endometrium by the growing placenta. Two **umbilical arteries** carry fetal blood through the **umbilical cord** to the placenta, and one **umbilical vein** returns this blood to the fetus. From the center of the placenta, these vessels branch toward the chorionic villi and circulate blood among the branches of the villi. Coiled **helical arteries** squirt maternal blood from the uterine and ovarian arteries into the intervillous spaces, past the branches of the villi. The lining of the intervillous spaces is fetal tissue, however, not maternal vascular endothelium. As the mother's blood washes over the surface of the villi, oxygen and other molecules from the maternal blood cross the surface of the villi into the fetal blood, and fetal products cross into the intervillous spaces. Maternal venules drain the maternal blood from the intervillous spaces to the uterine veins. Because maternal blood directly bathes the chorionic villi, this arrangement is known as a **hemochorial placenta** (HE-mo-KORE-ee-al); that is, maternal bloopd (hemo-) is directly in contact with the chorion (-chorial).

Exchange takes place across four tissues of the chorionic villi, shown in Figure 26.19C. The most important of these is the **syntrophoblast** (sin-TROE-fo-blast) that covers the exterior of the chorionic villi. The syntrophoblast is a syncytium (sin-SISH-ee-um), a highly unusual tissue that consists of many nuclei in a common cytoplasm with no plasma membrane partitions between the nuclei. This tissue lines the intervillous spaces and, as the endocrine center of the placenta, secretes estrogens, progesterone, and chorionic gonadotropin. The **cytotrophoblast** (SITE-oh-TROE-fo-blast), the cellular source of the syntro-

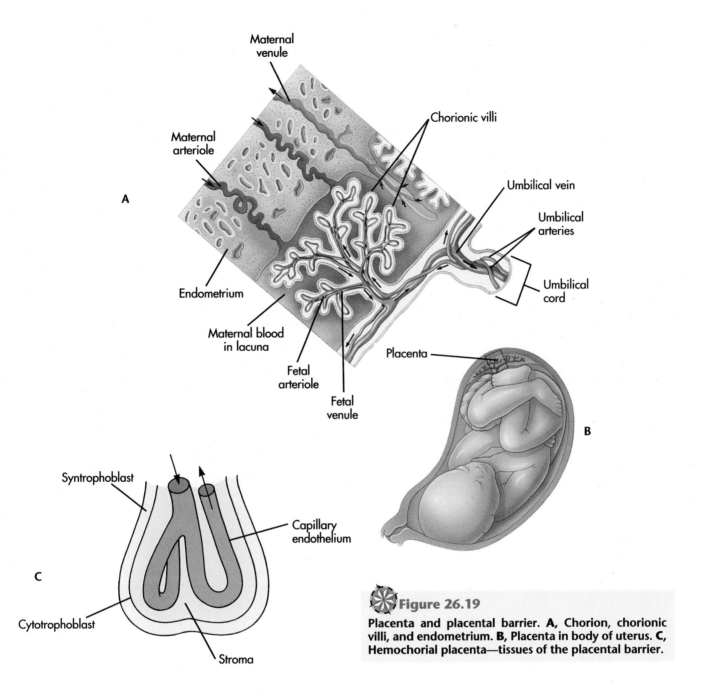

Figure 26.19

Placenta and placental barrier. A, Chorion, chorionic villi, and endometrium. B, Placenta in body of uterus. C, Hemochorial placenta—tissues of the placental barrier.

phoblast, lies directly beneath the syntrophoblast. Loose connective tissue **stroma,** and finally the **capillary endothelium,** complete the placental barrier that separates maternal and fetal blood.

Although small molecules do traverse the placental barrier, proteins and other macromolecules do not cross. Maternal antibodies and viruses are exceptions, however, because the placenta transports antibodies from the mother, and the fetus can contract viral infections from her blood. Amino acids are actively transported from the mother, but oxygen crosses passively by facilitated diffusion. Many other small molecules cross as well, including alcohol, nicotine, drugs, and environmental chemicals that may be toxic

to the fetus.

- Which two tissues compose the placenta?
- Name the four tissues of the placental barrier.
- Why is this barrier considered to be hemochorial?
- What fetal functions does the placenta perform?

FETAL MEMBRANES

Several fetal membranes protect the fetus as it develops within the uterus. One of these membranes contributes the fetal tissues of the placenta, and all of the membranes are withdrawn, along with uterine tissue, from the uterus at birth, as the **afterbirth.** Beginning with the embryo itself (Figure 26.20A) at 5 weeks, the **amnion** encloses the embryo within the

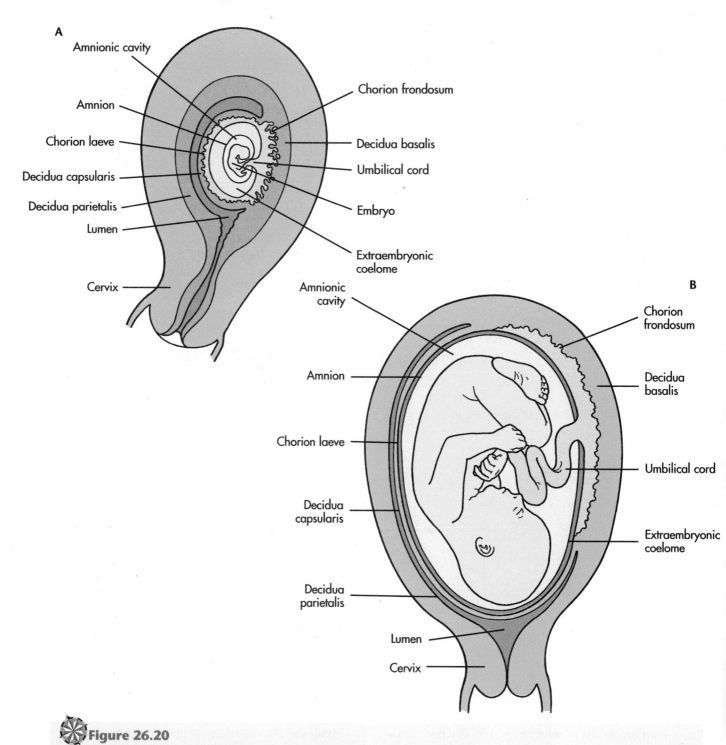

A

- Amnionic cavity
- Amnion
- Chorion laeve
- Decidua capsularis
- Decidua parietalis
- Lumen
- Cervix
- Chorion frondosum
- Decidua basalis
- Umbilical cord
- Embryo
- Extraembryonic coelome

B

- Amnionic cavity
- Amnion
- Chorion laeve
- Decidua capsularis
- Decidua parietalis
- Lumen
- Cervix
- Chorion frondosum
- Decidua basalis
- Umbilical cord
- Extraembryonic coelome

Figure 26.20

Fetal membranes. A, During the first trimester. B, Near term.

amniotic cavity, a space filled with protective amniotic fluid. The **umbilical cord** connects the embryo to the **chorion,** which surrounds the amnion and embryo with a second fluid-filled chamber known as the extra-embryonic coelome. Chorionic villi cover the exterior of the chorion and are especially large in the placenta, where the embryo is attached to the uterine wall. This region of the chorion is known as the **chorion frondosum** (fron-DOH-sum). Elsewhere, it is

called the **chorion laeve** (LEV-ay; *L,* smooth) where the chorionic villi become short and later regress. The amnion, chorion, and their respective chambers are fetal in origin, derived from the fertilized egg.

As you examine Figure 26.20A, remember that the embryo is implanted within the endometrium and that the fetus pushes part of the uterine wall into the lumen of the uterus as the fetus grows. The endometrium that accompanies these expanding fetal membranes is known as the **decidua** (de-SID-you-ah) because it will be discarded as part of the afterbirth. The **decidua basalis** is associated with the placenta; the **decidua capsularis** covers the chorion laeve, and the **decidua parietalis** is the endometrial lining of the rest of the uterus. By the fifth month of pregnancy (Figure 26.22B), the fetus has so enlarged that the amnion, chorion laeve, and decidua capsularis press against the decidua parietalis on the opposite wall of the uterus and nearly obliterate the lumen of the uterus. Delivery ruptures the decidua capsularis and chorion, as well as the amnion ("the water breaks").

• Name the tissues from which the chorion and decidua basalis derive.

UTERUS

The uterus enlarges as it keeps pace with the growing fetus (Figure 26.21). The muscle fibers of the myometrium increase in number and enlarge from 0.25 mm to as much as 5 mm long during the first half of pregnancy. Fiber growth diminishes thereafter, and the myometrium becomes thinner as it stretches and prepares for delivery. The **first stage of labor** begins when rhythmic contractions press the fetus firmly

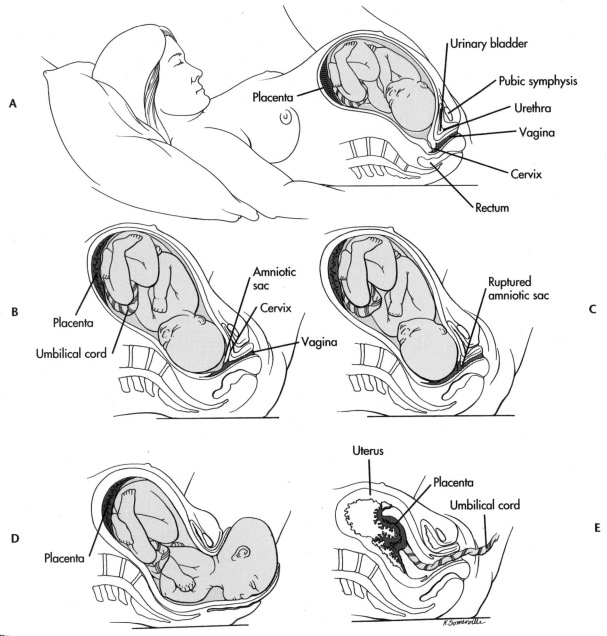

⚘ **Figure 26.21**
Labor, described in text.

down into the cervix (Figure 26.21A, B, and C). Squeezed between the fetus and the pelvis, the amnion and decidua rupture, releasing the amniotic fluid, and the cervix begins to dilate, which sets off a reflex that releases **oxytocin** from the neurohypophysis. This hormone increases the power and frequency and often the pain of contractions, while the cervix continues dilating. The first stage may last 12 hours or more, especially on the first delivery, and ends with the cervix fully dilated 10 cm (4 in) or more, so that the uterus and vagina are now one **birth canal.**

Stage two delivers the baby. Now the mother, overwhelmed by the urge to bear down, pushes the baby through the canal (Figure 26.21D). The baby crowns (its scalp appears at the vaginal opening), the head emerges, and then a new slippery, slithery, squalling baby arrives, all in about 50 minutes or even less in later births.

Finally, **stage three** delivers the placenta and its membranes. The uterus continues contracting after delivery (Figure 26.21E), which squeezes out the placenta and helps reduce loss of blood by constricting the helical arteries.

- What happens to the fetus and its membranes at each stage of labor?

BREASTS

The breasts contain the mammary glands. Although they are anatomic derivatives of the skin, their ability to nurse an infant places the breasts here, in our discussion of pregnancy. Each nearly hemispherical breast contains 15 to 20 glandular **lobes** (Figure 26.22) between the skin and pectoralis major muscles of the chest, supported by subcutaneous adipose and fibrous connective tissue. Each lobe contains a highly branched, compound tubular **mammary gland** whose duct opens at the **nipple,** a raised apical tissue centered in the **areola** (AHR-ee-oh-la). Both nipple and areola are hairless and darker with melanocytes than the surrounding epidermis. They are equipped with sebaceous glands and sensory nerve endings, as well as smooth muscle fibers that cold, touch, or sexual arousal erect. Estrogens are responsible for enlargement of the breasts at puberty, but estrogens stimulate only the duct system and adipose tissue to develop, not the glands themselves that will secrete milk after birth. Each lobe contains many **lobules,** blind ductules that lead to a single **lactiferous sinus** (lak-TIF-er-us) behind the orifice of each main duct at the nipple. Adipose tissue fills out between lobules, and fibrous **suspensory ligaments** support the lobes from the deep fascia at the base of the breast. These ligaments and lobes can be seen in mammogram X-rays.

It is rather common for breasts to enlarge temporarily in teenage boys. Such **gynecomastia** (GUY-neh-ko-MAST-ee-ah) is usually a sign of unexplained excess estrogens, but gynecomastia in later life may indicate that an estrogen-producing tumor has developed.

Not until her first pregnancy do secretory portions of the mother's mammary glands develop. Then, under the influence of progesterone and estrogen,

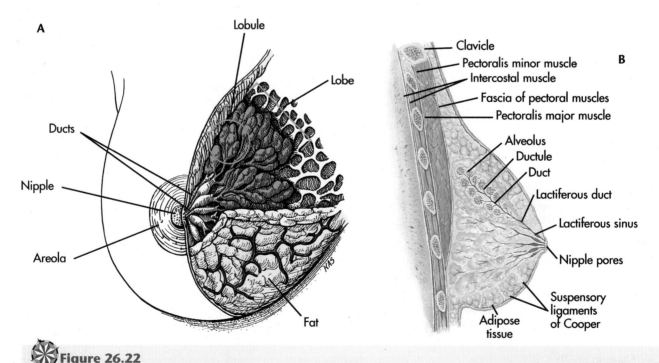

⚜ **Figure 26.22**

The breast and mammary glands. A, Frontal section. **B,** Sagittal section.

cuboidal epithelial cells form clusters of **alveoli** at the ends of the ductules, and beginning in the second trimester, **prolactin** and **somatomammotropin** from the adenohypophysis promote secretion. The cells become columnar, and **colostrum** (ko-LOSS-trum), a low-fat, protein-rich, colorless fluid, accumulates in the lobules and lactiferous sinuses as the breast enlarges. **Lactation** begins several days after birth and delivers milk rich in fats, lactose, vitamins, the proteins casein and lactalbumin, and antibodies that confer passive immunity to the baby. Breasts release milk reflexively **(milk let-down)** when the child suckles, a reaction mediated by the neurohypophyseal hormone, oxytocin. Lobules regress when lactation ends.

- In what respects do the breasts and mammary glands differ anatomically?

SUMMARY

The reproductive systems of males and females consist of glands, ducts, and external genitalia. Ovaries and testes are both endocrine glands that produce steroid sex hormones and exocrine glands that release eggs and sperm, respectively, into ducts that deliver them for fertilization. Complementary external genitalia enable males to deliver sperm to the female and allow females to give birth. *(Page 788.)*

The external genitalia of males consists of scrotum and penis and, in females, of the vulva. The female vestibule corresponds to the penile urethra of males. *(Page 788.)*

In males, spermatogenesis transforms diploid spermatogonia into mature haploid sperm in the seminiferous tubules of the testis. Sperm are stored in the epididymis, ejaculated through the vas deferens and urethra, and mixed with seminal fluids from the accessory glands. The corpora cavernosa and corpus spongiosum erect the penis. *(Page 789.)*

Oogenesis transforms diploid oogonia into haploid eggs within follicles of the ovary. Ovulation releases an egg into the uterine tubes, and if it is fertilized there, the embryo implants into the endometrium lining the uterus. The placenta nourishes the fetus until labor delivers the baby. *(Page 799.)*

The internal genitalia of males and females consist of ducts that carry sperm or eggs. Sperm develop in seminiferous tubules, are stored in the epididymis, and are delivered to the urethra by the vas deferens and ejaculatory ducts. Ductwork is simpler in females and consists of uterine tubes that deliver to the uterus and vagina. Most passages consist of mucosa, muscularis, and serosa or adventitia. *(Page 789, 799.)*

Spermatogenesis occurs in the germinal epithelium of seminiferous tubules. Diploid spermatogonia multiply into primary spermatocytes that then divide by meiosis into haploid secondary spermatocytes and spermatids. Spermiogenesis transforms spermatids into sperm. Sertoli cells promote these processes by attaching to the developing cells and forming a blood-testis barrier. Interstitial cells promote spermatogenesis by synthesizing testosterone. *(Page 794.)*

Oogenesis takes place within ovarian follicles, which are homologous to seminiferous tubules. The stages of oogenesis are also homologous with those of spermatogenesis. Diploid oogonia proliferate in fetal ovaries and become primary oocytes, arrested in the first meiotic division and held within primordial follicles. Oocytes mature under the influence of follicle-stimulating hormone when primordial follicles mature as Graafian follicles, containing haploid secondary oocytes that have completed the first meiotic division. The final meiotic division takes place after fertilization. Granulosa cells and thecal cells synthesize estrogens and progesterone. *(Page 801.)*

External genitalia develop from indifferent structures that assume homologous male or female forms. The genital tubercle becomes the clitoris in females and the penis in males. The genital swellings become the labia majora of females and the scrotum of males, and the genital folds develop into labia minora or penile urethra, in the respective sexes. *(Page 808.)*

Primordial germ cells migrate into the gonads and establish the cells that become eggs or sperm. Ovary and testis provide the environment for gametes to develop, but they are not the ultimate source of gametes. *(Page 811.)*

Two indifferent sets of ducts are the source of passages in the internal genitalia. The mesonephric ducts develop as the vas deferens in males, and the Mullerian ducts remain rudimentary. Females reverse this pattern; the Mullerian ducts form the uterine tubes and uterus, and the mesonephric ducts degenerate. Androgen insensitivity syndrome illustrates that genitalia tend to develop as female unless switched to male by the Y chromosome. *(Page 810.)*

A fetus progresses through embryonic and fetal stages during pregnancy. The embryonic stage establishes the organ systems during the first 9 weeks. This period is especially risky for the embryo; most conceptions end during this time, and major damage can occur to the embryo when organ systems develop abnormally or fail to form altogether. Organs begin to function during the fetal period, which lasts from the tenth week until term. Relatively few conceptions end during the fetal period, and developmental abnormalities usually concern parts of organs rather than whole systems. *(Page 812.)*

The placenta serves as the fetal lungs, digestive system, kidneys, and liver. It is a temporary organ formed by fetal chorion and maternal endometrium. The hemochorial placenta exposes maternal blood to fetal chorionic villi that exchange small molecules with the mother, but generally exclude proteins and other macromolecules. The umbilical cord joins the placenta to the fetus. The placenta also is an endocrine gland that secretes hormones that maintain the uterine lining during pregnancy. *(Page 814.)*

Fetal membranes are associated with the placenta. The amnion retains the fetus within the amniotic cavity; the chorion surrounds the amnion, and the deciduae are endometrial tissues shed from the uterine lining with the afterbirth. *(Page 815.)*

Labor delivers the fetus and the placenta. The first stage of labor dilates the cervix and forms the birth canal. The second stage delivers the baby, and in the third the uterus releases the placenta and its membranes. *(Page 817.)*

The breasts contain the mammary glands. *(Page 818.)*

STUDY QUESTIONS

1. Name the glands, ducts, and structures of the external genitalia in males and females.   **1**
2. Describe the anatomy of a testis and an ovary, and cite major points of difference between these organs.   **1**
3. How does the tunica vaginalis come to exist in the scrotum? Why were Kristin's sterile testes found in her labia majora?   **1**
4. Name the accessory glands of males and females and the products they secrete.   **1**
5. Explain how the corpus spongiosum and the corpora cavernosa erect the penis.   **1**
6. In what respects do the mucosa, muscularis, and serosa differ among the uterine tubes, uterus, cervix, and vagina?   **1, 3**
7. Outline the ducts of male and female internal genitalia, and follow the passage of sperm and eggs through them.   **2, 3**
8. Describe the layers of cells in a seminiferous tubule and the steps of spermatogenesis that they represent.   **4**
9. Follow the steps of oogenesis through the phases of follicular development in an ovary.   **4, 5**
10. Which cells in testis, ovary, and placenta produce hormones?   **4, 7**
11. Which structures do the genital tubercle, genital swellings, genital folds, and urogenital sinus become in females and males?   **6, 7**
12. Which cells in seminiferous tubules and ovarian follicles are homologous? What functions do these homologous parts perform?   **7**
13. Where do primordial germ cells originate, and how are they related to sperm and eggs?   **8**
14. How does androgen insensitivity syndrome illustrate the relationships between chromosomal, gonadal, and somatic sex? What is gender identity?   **9**
15. What major developmental events occur in each trimester of pregnancy?   **10**
16. When do reflexes begin to organize in the nervous system? When do the lungs begin to produce surfactant? When does the heart begin beating? When do the external genitalia take on male or female form?   **10**
17. Describe the anatomy of the maternal and fetal tissues of the placenta. Trace the exchange of oxygen and carbon dioxide between mother and fetus across the tissues of the placental barrier. What parts of the placenta and endometrium are removed from the uterus after delivery?   **11**
18. What are the anatomical relationships between the amnion, chorion, and decidua membranes?   **12**
19. When does a uterus begin to enlarge in pregnancy? How large does it become? What changes in the anatomy of mammary glands during pregnancy prepare the breasts to lactate?   **13**
20. Where are the following coverings located: tunica vaginalis, tunica albuginea, corpus albicans, decidua basalis, mucosa, syntrophoblast, germinal epithelium, theca interna, and corona radiata? What function does each covering perform?   **Various**

MINI-ATLAS OF HUMAN ANATOMY

The following photographs of actual human dissections show the relationships among body systems in various regions of the body, from head to foot. Within each region the photographs range from superficial to deep structures.

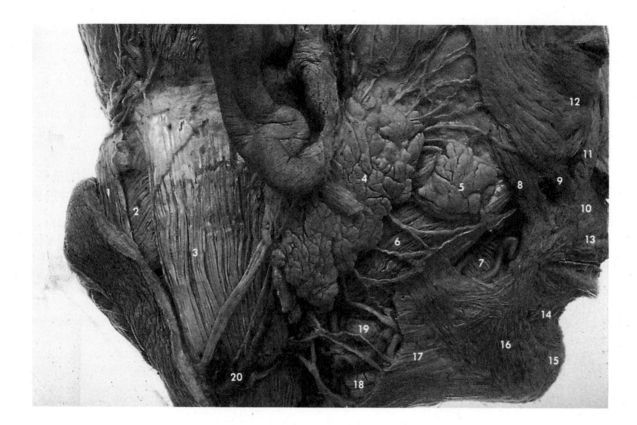

1 Trapezius
2 Splenius capitis
3 Sternocleidomastoid
4 Parotid gland and facial nerve branches at anterior border
5 Accessory parotid gland
6 Masseter
7 Buccinator
8 Zygomaticus major
9 Zygomaticus minor
10 Levator labii superioris

11 Levator labii superioris alaeque nasi
12 Orbicularis oculi
13 Orbicularis oris
14 Depressor labii inferioris
15 Mentalis
16 Depressor anguli oris
17 Platysma
18 Submandibular gland
19 Submandibular lymph nodes
20 External jugular vein

 Figure A.1

Right lower face and upper neck.

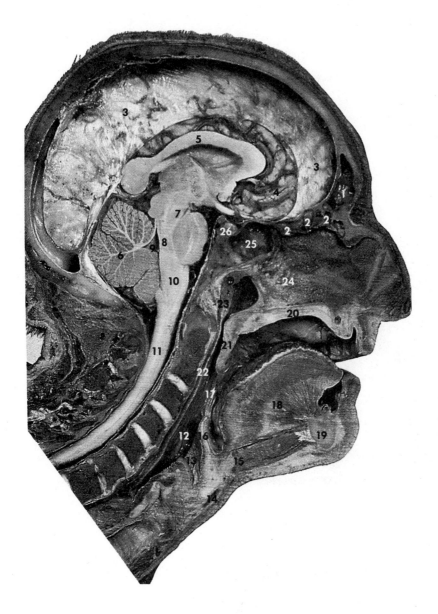

1 Frontal sinus	**14** Thyroid cartilage
2 Ethmoidal air cells	**15** Hyoid bone
3 Falx cerebri	**16** Epiglottis
4 Medial surface of cerebral hemisphere	**17** Oropharynx
5 Corpus callosum	**18** Tongue
6 Cerebellum	**19** Mandible
7 Midbrain	**20** Hard palate
8 Pons	**21** Soft palate
9 Fourth ventricle	**22** Nasopharynx
10 Medulla oblongata	**23** Pharyngeal tonsil
11 Spinal cord	**24** Nasal septum
12 Laryngopharynx	**25** Sphenoidal sinus
13 Opening into larynx	**26** Pituitary gland

 Figure A.2

Right half of the head, in sagittal section.

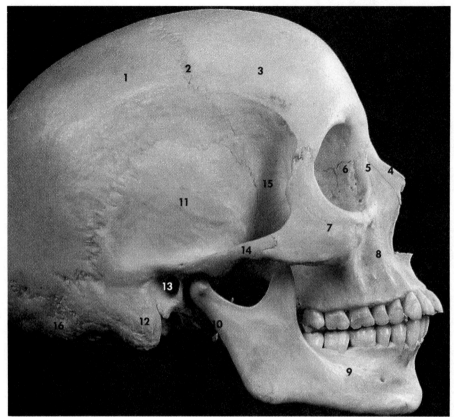

1 Parietal bone
2 Coronal suture
3 Frontal bone
4 Nasal bone
5 Frontal process of maxilla
6 Lacrimal bone
7 Zygomatic bone
8 Maxilla
9 Body of mandible
10 Styloid process
11 Temporal bone
12 Mastoid process
13 External auditory meatus
14 Zygomatic arch
15 Greater wing of sphenoid
16 Occipital bone

Figure A.3
The skull, from the right.

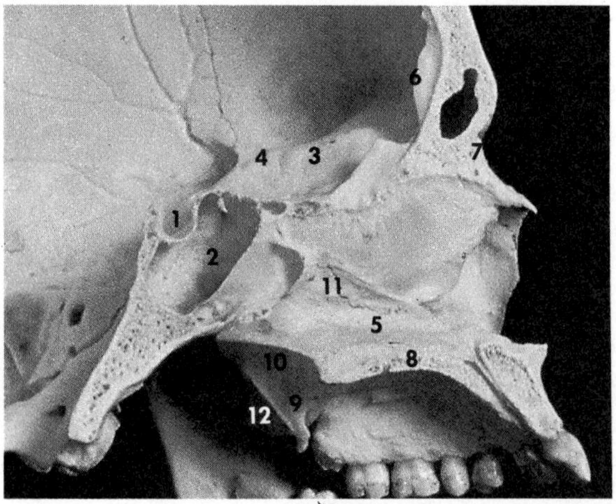

1 Pituitary fossa (sella turcica)
2 Sphenoidal sinus
3 Cribriform plate of ethmoid bone
4 Air cells of ethmoidal sinus
5 Inferior nasal concha
6 Frontal sinus
7 Nasal bone
8 Palatine process of maxilla
9 Pterygoid hamulus
10 Medial pterygoid plate
11 Perpendicular plate of palatine bone
12 Lateral pterygoid plate

Figure A.4
Lateral wall of the skull, midline sagittal section.

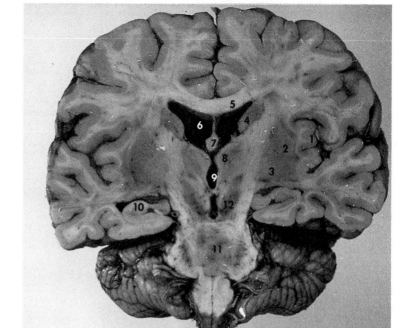

1 Insula
2 Putamen
3 Globus pallidus
4 Body of caudate nucleus
5 Corpus callosum
6 Lateral ventricle
7 Fornix
8 Thalamus
9 Third ventricle
10 Hippocampus
11 Pons
12 Substantia nigra

Figure A.5
The brain, coronal section.

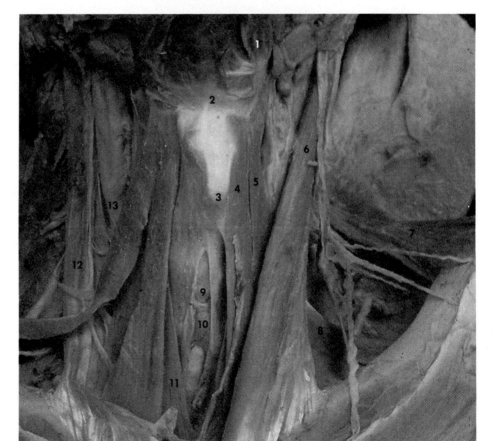

1 Anterior belly of digastric
2 Hyoid bone
3 Laryngeal prominence
   (Adam's apple)
4 Sternohyoid
5 Superior belly of omohyoid
6 Sternocleidomastoid
7 Trapezius
8 Inferior belly of omohyoid
9 Cricoid cartilage
10 Isthmus of thyroid gland
11 Sternothyroid
12 Internal jugular vein
13 Common carotid artery

Figure A.6
Front of the neck, superficial.

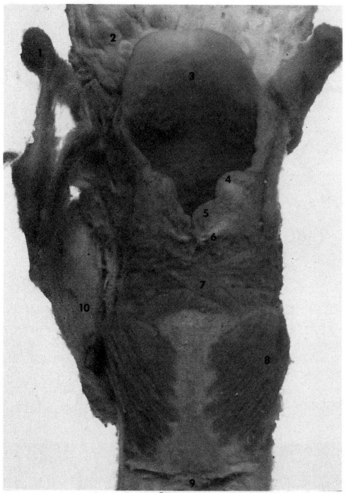

1 Greater horn of hyoid bone
2 Tongue
3 Epiglottis
4 Cuneiform cartilage
5 Corniculate cartilage
6 Transverse arytenoid muscle
7 Oblique arytenoid muscle
8 Posterior cricoarytenoid muscle
9 Trachea
10 Thyroid cartilage

Figure A.7
Larynx, posterior view.

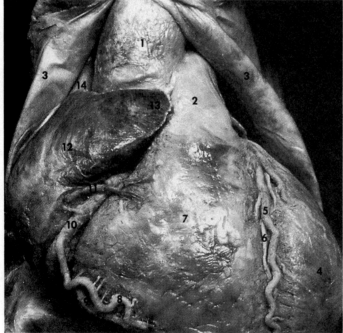

1 Ascending aorta
2 Pulmonary trunk
3 Pericardium (turned laterally)
4 Left ventricle
5 Anterior interventricular branch
   of left coronary artery
6 Great cardiac vein
7 Right ventricle
8 Marginal branch of right
   coronary artery
9 Small cardiac vein
10 Right coronary artery
11 Anterior cardiac vein
12 Right atrium
13 Auricle of right atrium
14 Superior vena cava

Figure A.8
The heart.

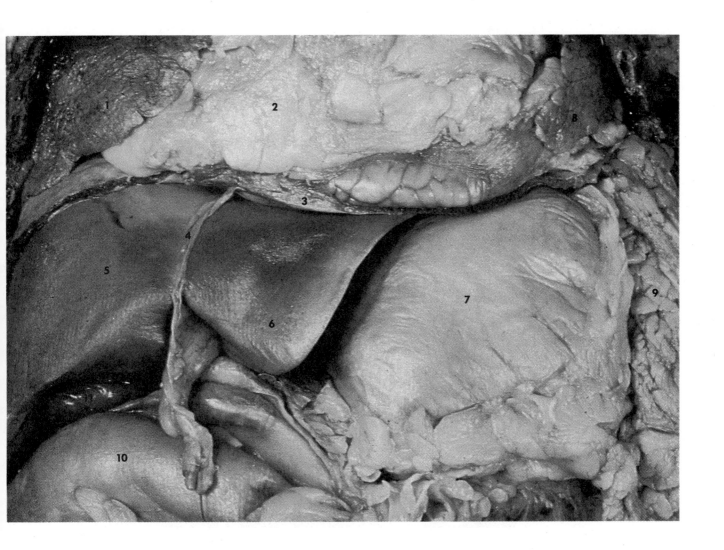

1 Inferior lobe of right lung
2 Pericardial fat
3 Diaphragm
4 Falciform ligament
5 Right lobe of liver
6 Left lobe of liver

7 Stomach
8 Inferior lobe of left lung
9 Greater omentum
10 Transverse colon
11 Gallbladder

Figure A.9

Upper abdomen, anterior view.

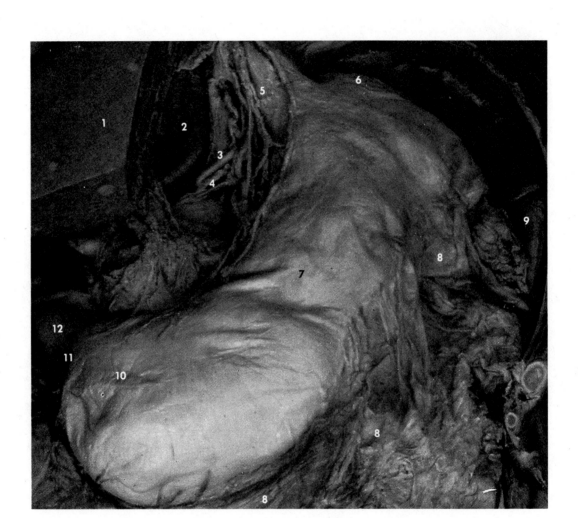

1 Right lobe of liver (cut)	**7** Body of stomach
2 Caudate lobe of liver	**8** Greater omentum
3 Left gastric artery	**9** Spleen
4 Left gastric vein	**10** Pyloric part of stomach
5 Esophagus	**11** Pylorus
6 Fundus of stomach	**12** Duodenum

**Figure A.10**

Stomach, anterior view.

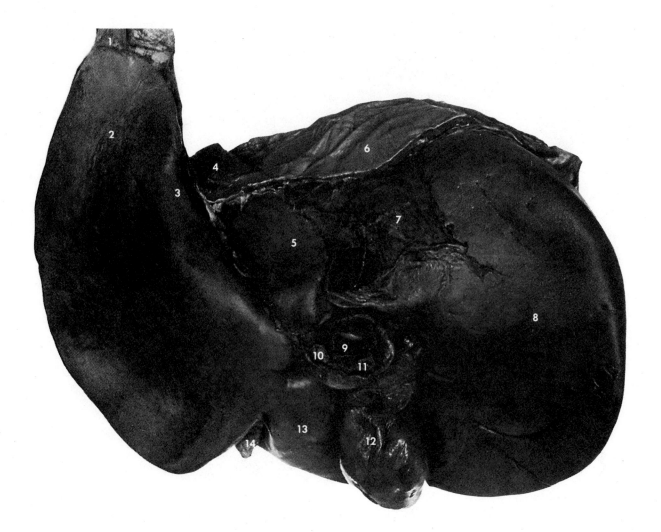

1 Part of coronary ligament
2 Left lobe
3 Esophageal groove
4 Inferior vena cava
5 Caudate lobe
6 Diaphragm on part of bare area
7 Bare area
8 Right lobe

9 Hepatic portal vein
10 Hepatic artery
11 Common hepatic duct
12 Gallbladder
13 Quadrate lobe
14 Ligamentum teres and
   falciform ligament

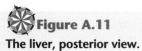
Figure A.11

The liver, posterior view.

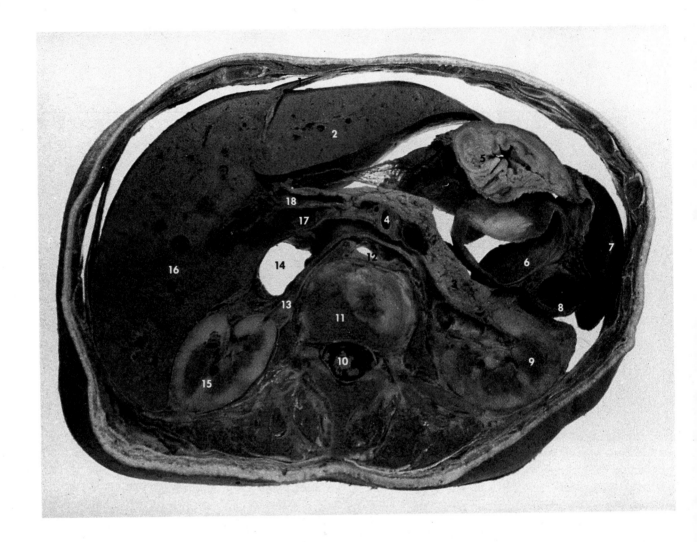

1	Falciform ligament	11	Body of first lumbar vertebra
2	Left lobe of liver	12	Abdominal aorta
3	Pancreas	13	Right renal artery
4	Superior mesenteric artery	14	Inferior vena cava
5	Stomach	15	Right kidney
6	Transverse colon	16	Right lobe of liver
7	Spleen	17	Hepatic portal vein
8	Descending colon	18	Hepatic artery
9	Left kidney		

 **Figure A.12**

Upper abdomen, transverse section.

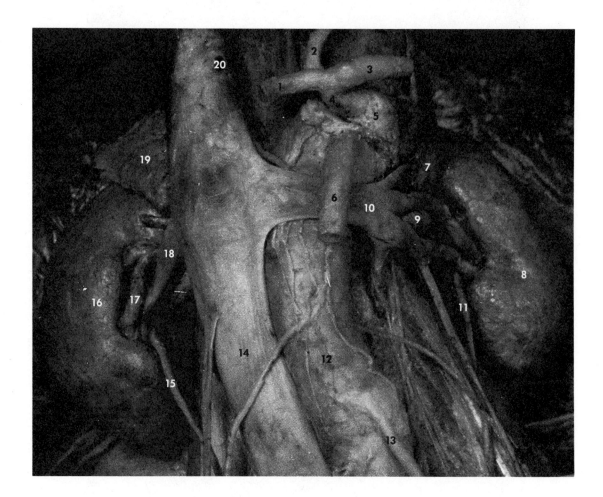

1 Common hepatic artery	11 Left ureter
2 Left gastric artery	12 Abdominal aorta
3 Splenic artery	13 Inferior mesenteric artery
4 Celiac trunk	14 Inferior vena cava
5 Celiac ganglion	15 Right ureter
6 Superior mesenteric artery	16 Right kidney
7 Left adrenal gland	17 Right renal artery
8 Left kidney	18 Right renal vein
9 Left renal artery	19 Right adrenal gland
10 Left renal vein	20 A hepatic vein

 **Figure A.13**

Kidneys and adrenal glands.

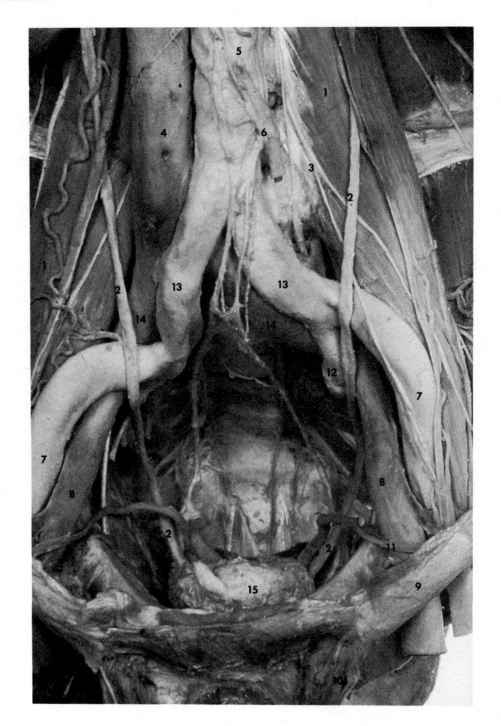

1 Psoas major	**7** External iliac artery	**13** Common iliac artery
2 Ureter	**8** External iliac vein	**14** Common iliac vein
3 Genitofemoral nerve	**9** Inguinal ligament	**15** Urinary bladder
4 Inferior vena cava	**10** Spermatic cord	
5 Aorta	**11** Ductus deferens	
6 Inferior mesenteric artery	**12** Internal iliac artery	

Figure A.14

Posterior abdominal wall and pelvis.

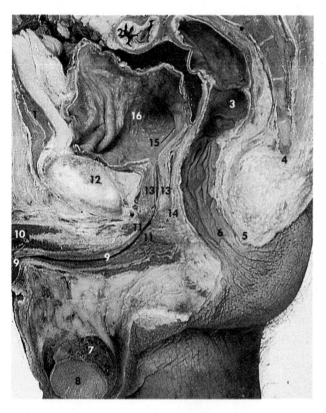

1 Rectus abdominis
2 Sigmoid colon
3 Rectum
4 Coccyx
5 External anal sphincter
6 Anal canal
7 Epididymis
8 Testis
9 Spongy urethra and corpus spongiosum
10 Corpus cavernosum
11 Sphincter urethrae
12 Symphysis pubis
13 Prostate
14 Ejaculatory duct
15 Internal urethral orifice
16 Urinary bladder

Figure A.15

Male pelvis, midline sagittal section.

1 Ureter
2 Ovary
3 Uterine tube
4 Fundus of uterus
5 Body of uterus
6 Cervix of uterus
7 Urinary bladder
8 Urethra
9 Symphysis pubis
10 Labium minus
11 Vagina
12 Rectum
13 Anal canal
14 External anal sphincter

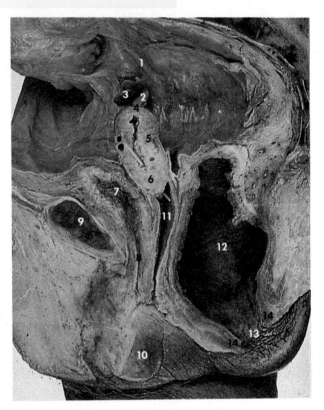

Figure A.16

Female pelvis, midline sagittal section.

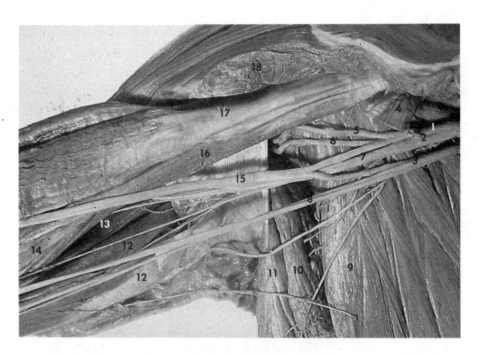

1 Lateral cord
2 Posterior cord
3 Medial cord
4 Pectoralis minor (cut)
5 Musculocutaneous nerve
6 Axillary nerve
7 Radial nerve
8 Ulnar nerve
9 Subscapularis
10 Teres major
11 Latissimus dorsi
12 Long head of triceps
13 Lateral head of triceps
14 Medial head of triceps
15 Median nerve
16 Coracobrachialis
17 Biceps brachii
18 Deltoid (cut)

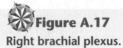

Figure A.17
Right brachial plexus.

1 Brachioradialis
2 Extensor carpi radialis longus
3 Extensor carpi radialis brevis
4 Abductor pollicis longus
5 Extensor pollicis brevis
6 Extensor pollicis longus
7 Extensor digitorum
8 Extensor digiti minimi
9 Extensor indicis
10 Extensor carpi ulnaris
11 Abductor digiti minimi
12 Extensor retinaculum

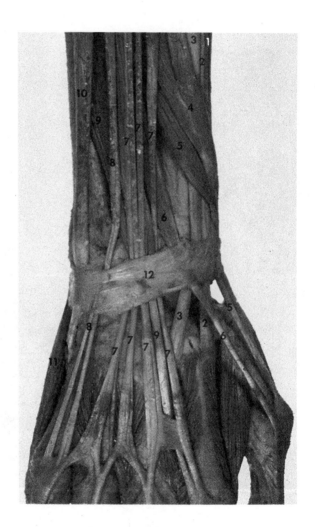

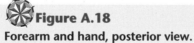Figure A.18
Forearm and hand, posterior view.

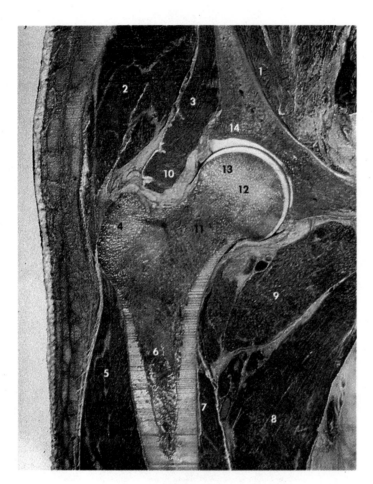

1 Iliacus
2 Gluteus medius
3 Gluteus minimus
4 Greater trochanter
5 Vastus lateralis
6 Shaft of femur
7 Vastus medialis
8 Adductor longus
9 Pectineus
10 Capsule of hip joint
11 Neck of femur
12 Head of femur
13 Hyaline cartilage of head
14 Hyaline cartilage of acetabulum

Figure A.19
Hip joint, coronal section.

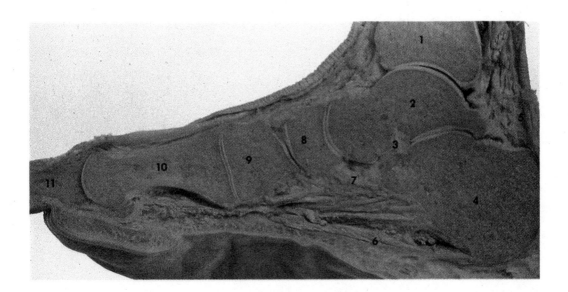

1 Tibia	5 Calcaneal tendon (Achilles' tendon)	9 Medial cuneiform
2 Talus	6 Flexor digitorum brevis	10 First metatarsal
3 Ligament	7 Plantar calcaneonavicular ligament	11 Proximal phalanx of great toe
4 Calcaneus	8 Navicular	

Figure A.20
Foot, sagittal section.

MINI-ATLAS OF SURFACE ANATOMY

Although the skin conceals most internal organs, it reveals many structures indirectly. The photographs in this mini-atlas indentify skeletal, muscular, and visceral structures evident on the surface of the skin.

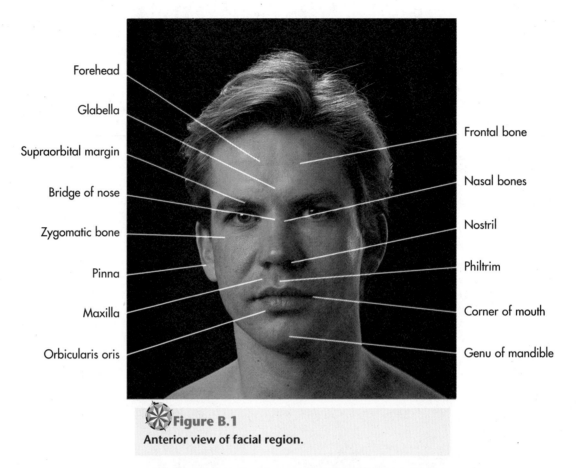

Forehead

Glabella

Supraorbital margin

Bridge of nose

Zygomatic bone

Pinna

Maxilla

Orbicularis oris

Frontal bone

Nasal bones

Nostril

Philtrim

Corner of mouth

Genu of mandible

Figure B.1
Anterior view of facial region.

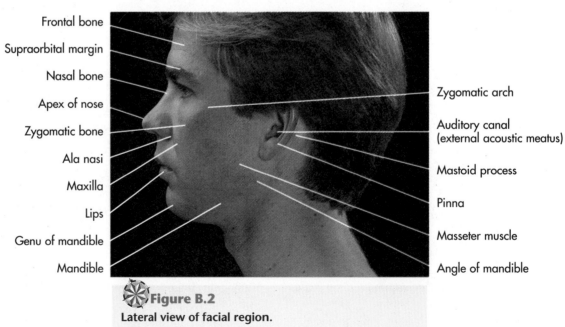

Frontal bone

Supraorbital margin

Nasal bone

Apex of nose

Zygomatic bone

Ala nasi

Maxilla

Lips

Genu of mandible

Mandible

Zygomatic arch

Auditory canal
(external acoustic meatus)

Mastoid process

Pinna

Masseter muscle

Angle of mandible

Figure B.2
Lateral view of facial region.

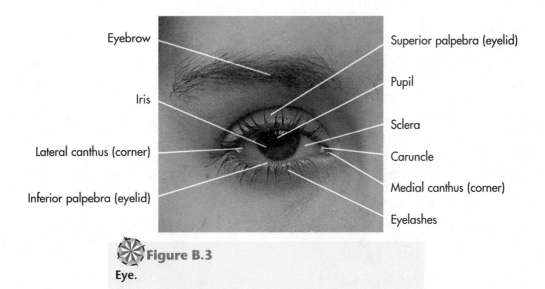

Eyebrow

Iris

Lateral canthus (corner)

Inferior palpebra (eyelid)

Superior palpebra (eyelid)

Pupil

Sclera

Caruncle

Medial canthus (corner)

Eyelashes

Figure B.3
Eye.

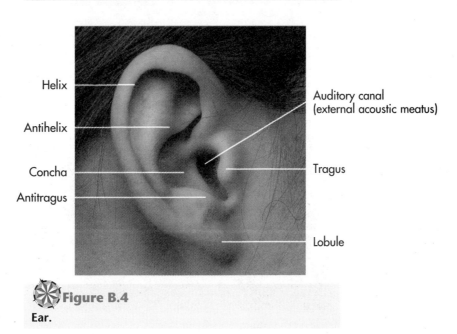

Helix

Antihelix

Concha

Antitragus

Auditory canal
(external acoustic meatus)

Tragus

Lobule

Figure B.4
Ear.

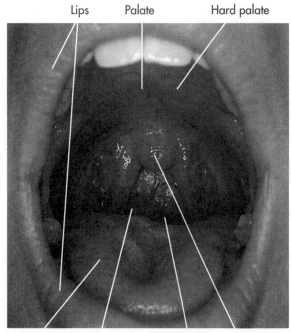

Lips   Palate   Hard palate

Tongue   Palatopharyngeal   Oropharynx   Uvula
         arch

Figure B.5

Oral cavity, tongue depressed.

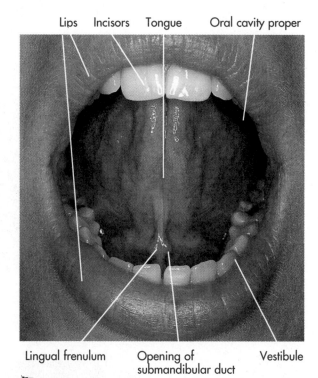

Lips   Incisors   Tongue   Oral cavity proper

Lingual frenulum   Opening of   Vestibule
                    submandibular duct

Figure B.6

Oral cavity, tongue elevated.

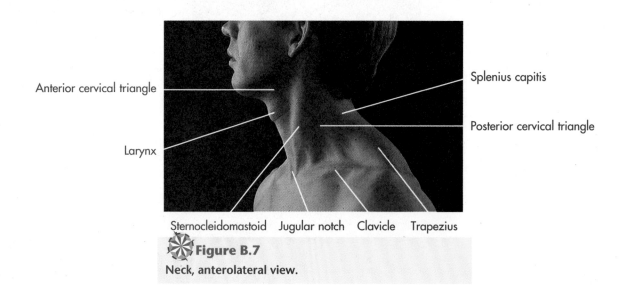

Anterior cervical triangle   Splenius capitis

Larynx   Posterior cervical triangle

Sternocleidomastoid   Jugular notch   Clavicle   Trapezius

Figure B.7

Neck, anterolateral view.

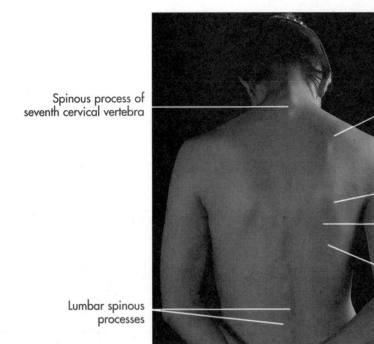

Spinous process of
seventh cervical vertebra

Superior border
of scapula

Acromion process

Scapula

Medial border
of scapula

Inferior angle
of scapula

Lumbar spinous
processes

Figure B.8
Back, female.

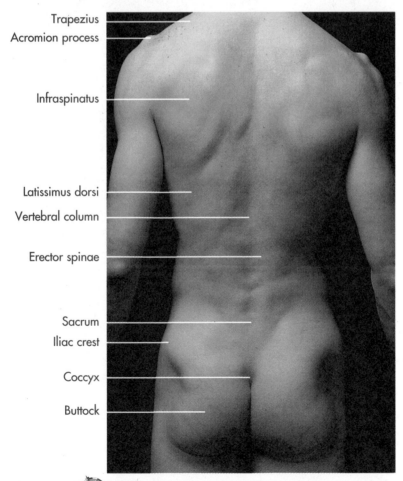

Trapezius
Acromion process

Infraspinatus

Latissimus dorsi

Vertebral column

Erector spinae

Sacrum

Iliac crest

Coccyx

Buttock

Figure B.9
Posterior view of male trunk.

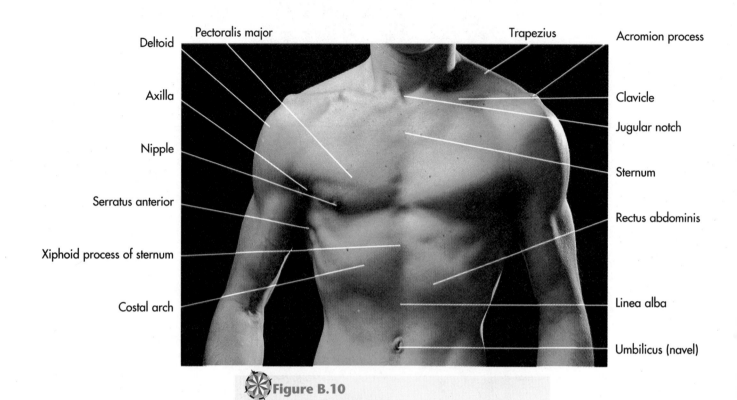

Deltoid

Pectoralis major

Trapezius

Acromion process

Axilla

Clavicle

Nipple

Jugular notch

Serratus anterior

Sternum

Xiphoid process of sternum

Rectus abdominis

Costal arch

Linea alba

Umbilicus (navel)

Figure B.10

Anterior view of male trunk.

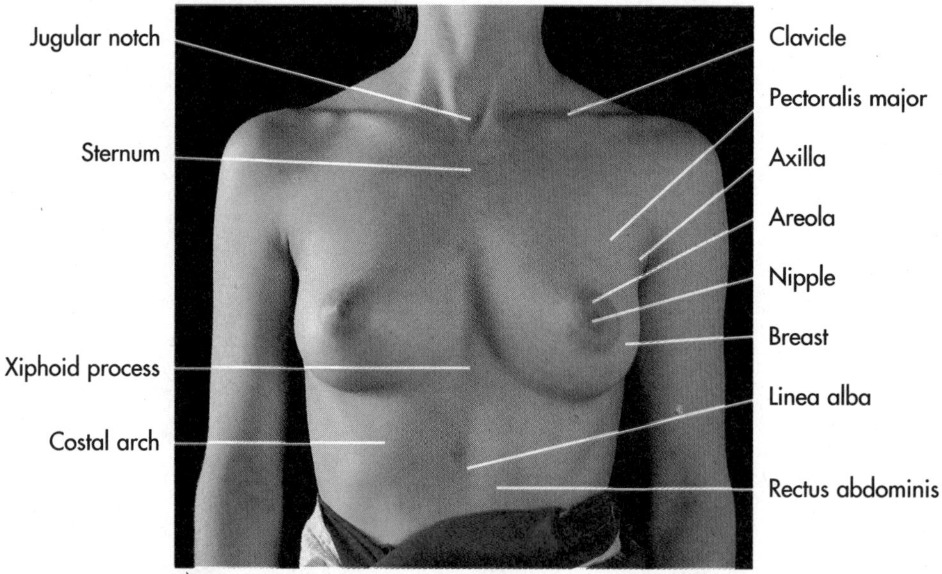

Jugular notch

Clavicle

Sternum

Pectoralis major

Axilla

Areola

Nipple

Breast

Xiphoid process

Linea alba

Costal arch

Rectus abdominis

Figure B.11

Anterior view of female thorax.

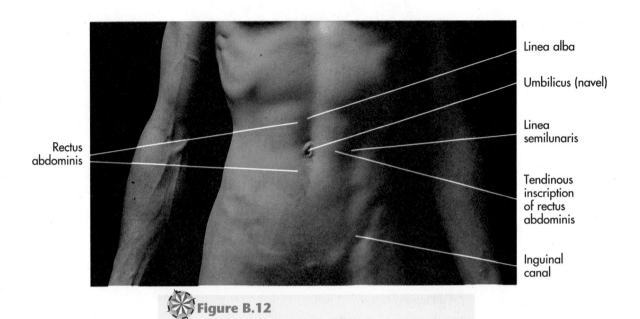

Linea alba

Umbilicus (navel)

Linea semilunaris

Tendinous inscription of rectus abdominis

Inguinal canal

Rectus abdominis

Figure B.12
Anterior view of male abdomen.

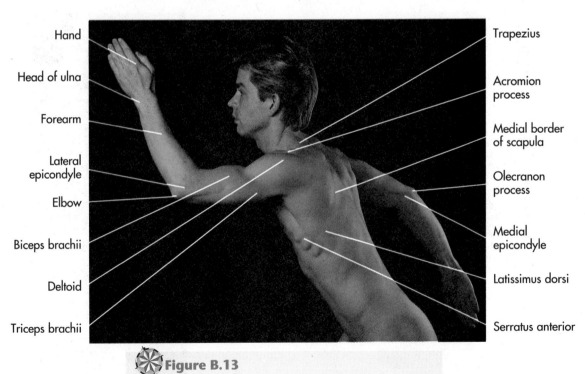

Hand

Head of ulna

Forearm

Lateral epicondyle

Elbow

Biceps brachii

Deltoid

Triceps brachii

Trapezius

Acromion process

Medial border of scapula

Olecranon process

Medial epicondyle

Latissimus dorsi

Serratus anterior

Figure B.13
Lateral view of trunk and upper extremeties.

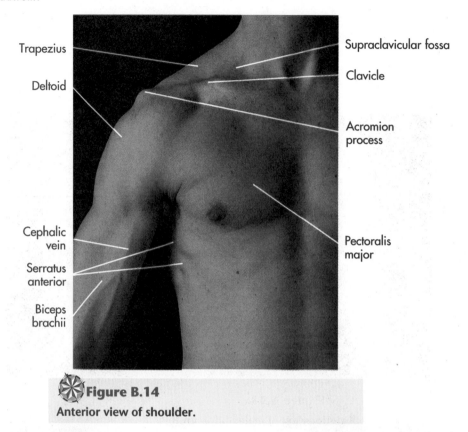

Trapezius

Deltoid

Cephalic vein

Serratus anterior

Biceps brachii

Supraclavicular fossa

Clavicle

Acromion process

Pectoralis major

Figure B.14
Anterior view of shoulder.

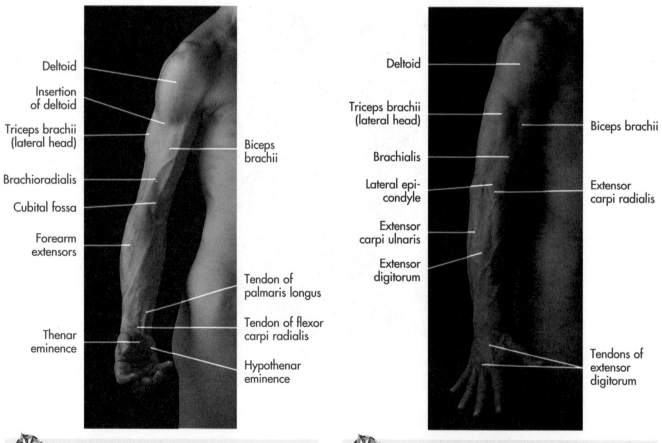

Deltoid

Insertion of deltoid

Triceps brachii (lateral head)

Brachioradialis

Cubital fossa

Forearm extensors

Thenar eminence

Biceps brachii

Tendon of palmaris longus

Tendon of flexor carpi radialis

Hypothenar eminence

Figure B.15
Anterior view of upper extremity, digits flexed.

Deltoid

Triceps brachii (lateral head)

Brachialis

Lateral epicondyle

Extensor carpi ulnaris

Extensor digitorum

Biceps brachii

Extensor carpi radialis

Tendons of extensor digitorum

Figure B.16
Lateral view of upper extremity, digits extended.

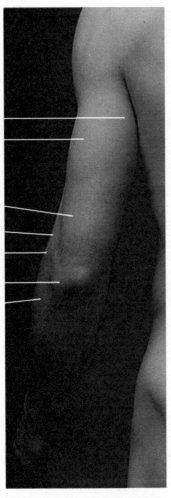

Triceps brachii, medial head

Triceps brachii, long head

Triceps brachii, lateral head

Brachioradialis

Lateral epicondyle of humerus

Olecranon process

Extensor carpi radialis longus

 **Figure B.17**
Posterior view of elbow.

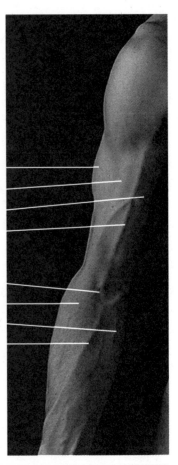

Triceps brachii, lateral head

Biceps brachii

Basilic vein

Cephalic vein

Cubital fossa

Brachioradialis

Median antebrachial vein

Median cubital vein

Figure B.18
Anterior view of elbow.

Distal interphalangeal joints

Proximal interphalangeal joints

Metacarpophalangeal joints

Tendons of extensor digitorum

Dorsal metacarpal veins

Tendons of extensor pollicis longus

Anatomical snuff box

Styloid process of ulna

Styloid process of radius

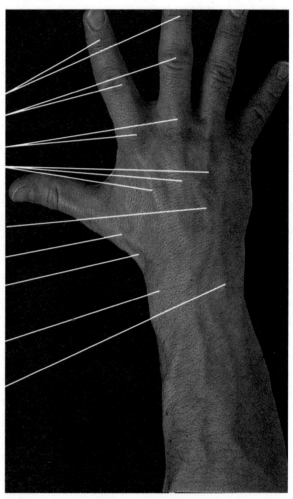

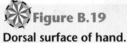

Figure B.19
Dorsal surface of hand.

Tendons of extensor
digitorum

Tendon of
extensor digiti
minimi

Hypothenar
eminence

Tendon of extensor
pollicis longus

Tendon of flexor
carpi ulnaris

Styloid process
of radius

Styloid process
of ulna

Thenar eminence

Tendon of flexor
digitorum superficialis

Tendon of flexor
carpi radialis

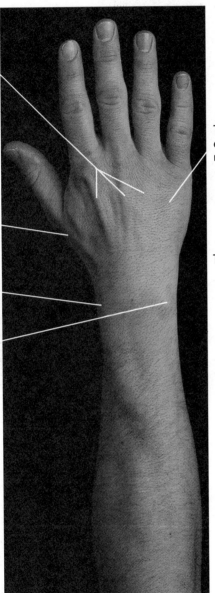

Figure B.20
Anterior surface of hand and forearm.

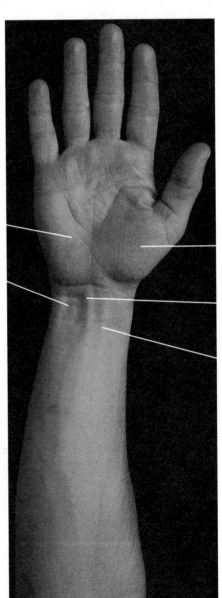

Figure B.21
Posterior surface of hand and forearm.

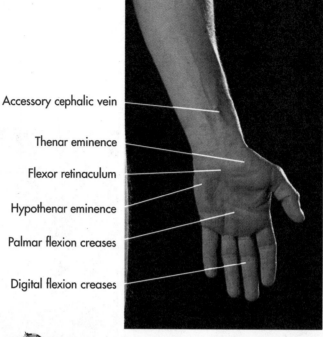

Accessory cephalic vein

Thenar eminence

Flexor retinaculum

Hypothenar eminence

Palmar flexion creases

Digital flexion creases

Figure B.22
Anterior surface of hand and wrist, digits extended.

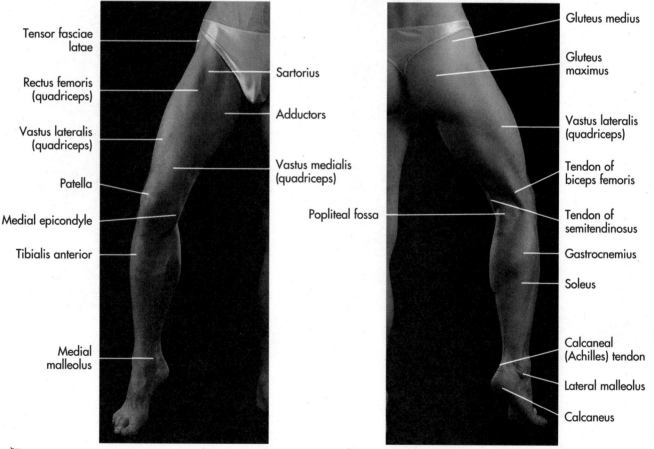

Tensor fasciae latae

Rectus femoris (quadriceps)

Vastus lateralis (quadriceps)

Patella

Medial epicondyle

Tibialis anterior

Medial malleolus

Sartorius

Adductors

Vastus medialis (quadriceps)

Popliteal fossa

Gluteus medius

Gluteus maximus

Vastus lateralis (quadriceps)

Tendon of biceps femoris

Tendon of semitendinosus

Gastrocnemius

Soleus

Calcaneal (Achilles) tendon

Lateral malleolus

Calcaneus

Figure B.23
Anterior view of lower extremity.

Figure B.24
Posterior view of lower extremity.

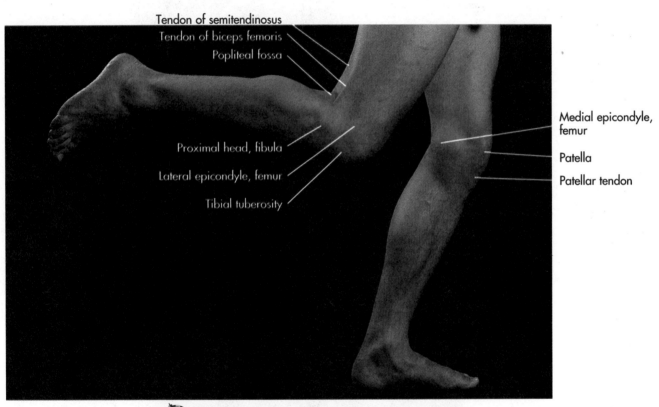

Tendon of semitendinosus
Tendon of biceps femoris
Popliteal fossa

Medial epicondyle, femur

Patella

Patellar tendon

Proximal head, fibula
Lateral epicondyle, femur
Tibial tuberosity

Figure B.25
Lateral and medial view of thigh and knee.

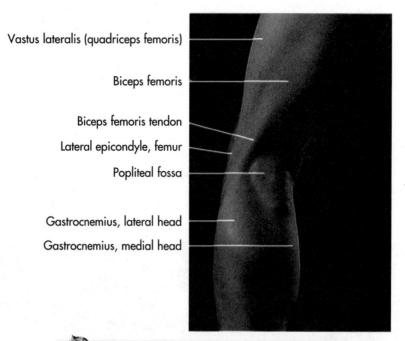

Vastus lateralis (quadriceps femoris)

Biceps femoris

Biceps femoris tendon

Lateral epicondyle, femur

Popliteal fossa

Gastrocnemius, lateral head

Gastrocnemius, medial head

Figure B.26
Posterior view of thigh and knee.

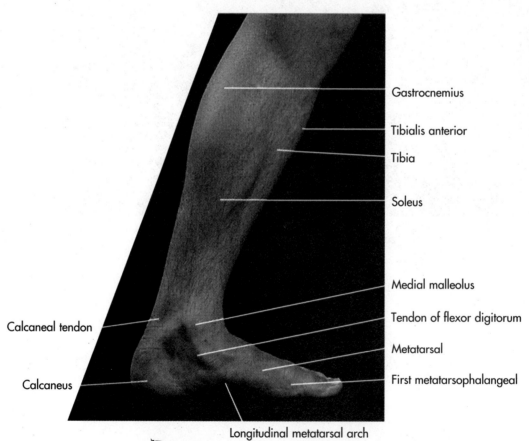

Gastrocnemius

Tibialis anterior

Tibia

Soleus

Medial malleolus

Tendon of flexor digitorum

Metatarsal

First metatarsophalangeal

Calcaneal tendon

Calcaneus

Longitudinal metatarsal arch

Figure B.27
Medial view of leg, ankle, and foot.

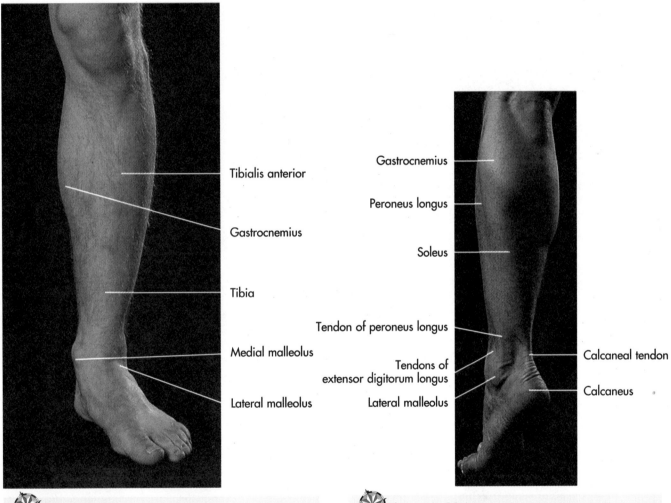

Tibialis anterior

Gastrocnemius

Tibia

Medial malleolus

Lateral malleolus

Gastrocnemius

Peroneus longus

Soleus

Tendon of peroneus longus

Tendons of
extensor digitorum longus

Lateral malleolus

Calcaneal tendon

Calcaneus

Figure B.28
Anterior view of leg, ankle, and foot.

Figure B.29
Posterolateral view of leg, ankle, and foot.

A-band The main, central band of a sarcomere that contains myosin (thick) myofilaments.

abdomen Belly; anterior region of the trunk between the thorax and the pelvis.

abdominal aorta The portion of the aorta in the abdomen. Leads blood from the thoracic aorta and terminates in the common iliac arteries.

abdominal cavity The body cavity of the abdomen; bounded by the diaphragm superiorly, the abdominal cavity is continuous with the pelvic cavity, and contains the liver, stomach, and intestines.

abdominopelvic cavity (ab-DAH-min-oh-PEL-vik) The cavity of the abdomen and pelvis; consists of separate but continuous abdominal and pelvic cavities.

abduction Movement of a body part away from the midline of the trunk or other part.

abductor Muscles that draw a body part away from the midline. The deltoid muscle is an abductor that draws the upper extremity away from the trunk.

aberrant ductule Remnants in the epididymis of males of mesonephric tubules that connected the mesonephric kidney to the mesonephric duct. Some mesonephric tubules remain as functional efferent ductules.

accommodation Change in shape of the lens in focusing an image on the retina. The lens becomes thicker and more convex as it accommodates to near objects, thinner and flatter to accommodate to distant objects.

acetabulum (ah-seh-TAB-yoo-lum) Cup-shaped depression on the external surface of the coxal bone into which the head of the femur fits at the hip joint.

acetylcholine (ACh) (AS-eh-til-KOE-leen) Neurotransmitter substance re-leased from motor neurons, all preganglionic neurons of the parasympathetic and sympathetic divisions, all postganglionic neurons of the parasympathetic division, some postganglionic neurons of the sympathetic division, and some central nervous system neurons.

acetylcholinesterase (AChase) Enzyme found in the synaptic cleft of synapses and motor endplates that degrades acetylcholine to acetic acid and choline, thus limiting the stimulatory effect of acetylcholine.

Achilles tendon See *calcaneal tendon*.

achondroplasia An inherited form of dwarfism that interferes with growth of the extremities; often inherited as a dominant gene.

acinus, *pl.* acini (AS-ih-nus, -eye) Grape-shaped secretory portion of a gland. The terms *acinus* and *alveolus* are sometimes used interchangeably. Some authorities differentiate the terms: acini have a constricted opening into the excretory duct, whereas alveoli have an enlarged opening. The term *acinus* is traditionally used for the pancreas. See *alveolus*.

acne vulgaris (AK-nee vul-GAR-iss) The condition common in teenagers in which increasing levels of testosterone cause sebaceous glands to be blocked by excessive quantities of sebum, and skin to become inflamed.

acromegaly Disorder marked by progressive enlargement of the head and face, hands and feet, and thorax, caused by excessive secretion of growth hormone by the adenohypophysis.

acromion (ah-KROM-ee-ahn) The point of the shoulder blade.

acrosome Cap on the head of the spermatozoon, containing hydrolytic enzymes that help the spermatozoon to penetrate the ovum.

actin A major class of contractile proteins found in muscle fibers and in the cytoskeleton of most cells.

actin microfilament Microfilaments are a class of cytoplasmic filaments less than 7 nm in diameter. Actin microfilaments are contractile filaments of actin that comprise the cytoskeleton of most cells and are responsible for cell movement and changes in shape.

actin myofilament Thin myofilament within the I-bands of sarcomeres; composed of actin molecules, tropomyosin, and troponin. Actin myofilaments slide along myosin myofilaments during muscle contraction.

actinic keratosis, *pl.* keratoses (ak-TIN-ik KARE-ah-TOE-sis, -sees) Small patches of sun-induced keratinization on the skin.

action potential Change in membrane potential in an excitable tissue, especially in the axons of neurons, that acts as an electrical signal and is propagated in an all-or-none fashion.

active transport Process in which membrane carrier proteins move molecules or ions across cell membranes against a concentration gradient; consumes ATP energy.

acute appendicitis Inflammation of the vermiform appendix. See *vermiform appendix*.

acute contagious conjunctivitis Acute inflammation of the conjunctiva.

acute myositis (MY-oh-SITE-iss) Irritation and inflammation resulting from torn muscle fibers and muscle connective tissue. See *RICE program*.

Adam's apple The prominence of the larynx, specifically the thyroid cartilage.

Addison's disease A condition characterized by fluid and electrolyte imbalance and by low blood sugar that results from insufficient cortisol production by the adrenal cortex.

adduction Movement of a body part toward the midline of the trunk or other body part.

adductors Any skeletal muscles that adduct a body part.

adenohypophysis (ah-DEEN-oh-hy-POF-ih-sis) The anterior pituitary, the portion of the hypophysis derived from the oral ectoderm.

adenoid (AD-en-oyd) Enlarged pharyngeal tonsil resulting from chronic inflammation.

adenosine diphosphate (ADP) This molecule consists of a ribose sugar to which is bonded the organic base adenine and a chain of two phosphate groups. Adenosine diphosphate adds a third phosphate group to the chain to form adenosine triphosphate.

adenosine triphosphate (ATP) This molecule consists of a ribose sugar to which is bonded the organic base adenine and a chain of three phosphate groups. Energy stored in the phosphate bonds is used in nearly all cell reactions that require energy.

adhesive plaque A variety of junctional complexes that join epithelial cells to the basement membrane.

adipocyte Round cells filled with fat.

adipose cell (ADD-ih-pos) See *adipocyte.*

adipose tissue Fatty variety of connective tissue proper, containing adipocytes, found in skin and mesenteries. See *brown adipose tissue* and *white adipose tissue.*

ADP See *adenosine diphosphate.*

adrenal gland Also called the suprarenal gland. Located near the superior pole of each kidney, the medulla of the adrenal is a highly modified sympathetic ganglion that secretes epinephrine and norepinephrine; the cortex secretes aldosterone and cortisol.

adrenaline See *epinephrine.*

adrenergic (ah-DREN-er-JIK) Refers to nerve fibers of the autonomic nervous system that secrete norepinephrine, or to drugs that mimic the actions of the sympathetic nervous system.

adrenergic receptor Receptor molecule that binds to adrenergic agents, such as epinephrine and norepinephrine, and which can initiate cellular responses.

adrenocorticotropic hormone (ACTH) (ah-DREE-no-KORE-tih-ko-TRO-pik) Also known as corticotropin. Hormone of the adenohypophysis that governs the nutrition and growth of the adrenal cortex, and stimulates the cortex to secrete cortisol.

adventitia (ad-ven-TISH-ya) Outermost covering of organs or structures that binds to other organs, vessels, or to the body wall. Adventitia develops from other tissue, usually mesoderm, rather than from the organ itself.

aerobic respiration Breakdown of glucose in the presence of oxygen to produce carbon dioxide, water, and ATP molecules. See *respiration.*

afferent arteriole Branch of an interlobular artery of the kidney that conveys blood to the glomerulus.

afferent neuron Neurons whose axons conduct impulses in spinal and cranial nerves from peripheral receptors to the central nervous system, that is, the spinal cord and brain.

agonist Skeletal muscles whose motion is opposed by antagonistic muscles. See *antagonistic pair.*

agranulocyte Nongranular leukocyte (monocyte or lymphocyte).

ala, *pl.* **alae** (AIL-ah; -eye) Wing-shaped structure, such as the wings of the sphenoid bone of the skull.

alar plate The dorsolateral walls of the neural tube that become the dorsal half of the spinal cord; nerve fibers in this region conduct sensory impulses.

albinism (AL-bin-ism) Inherited inability to produce melanin, resulting in unpigmented skin and hair.

albumin (al-BYOO-min) Major class of blood plasma proteins. Instrumental in movement of water and electrolytes between the blood and interstitial fluid.

aldosterone (al-DOSS-ter-own) Steroid hormone produced by the adrenal cortex that facilitates sodium reabsorption and potassium and hydrogen secretion in the distal renal tubule.

alimentary tract The digestive tract; the esophagus, stomach, and intestines.

all-or-none When a stimulus is applied, a neuron either produces an action potential or does not. Skeletal muscle cells either contract to the maximum extent possible or do not contract. No intermediate response.

allantois Tube extending from the hindgut of the embryo into the umbilical cord; forms the urinary bladder.

allergen Antigen that causes an allergic response. An allergic reaction is caused by an allergen, such as pollen, food, dander, and insect venom.

alveolar duct Ductlike respiratory passages of the lungs, covered with alveoli, that branch from respiratory bronchioles and lead air to alveolar sacs and alveoli.

alveolar gland Gland in which the secretory unit has a saclike form and an obvious lumen.

alveolar sac The chamber formed by two or more alveoli of the lungs that share a common opening. See *alveolus.*

alveolar septum, *pl.* **septa** (SEP-tum, -tah) The thin partitions between alveoli of the lungs, containing pulmonary capillaries that exchange carbon dioxide and oxygen with the air.

alveolus, *pl.* **alveoli** (al-VEE-oh-LUS, -LEE) A cavity or pit. Examples include the bony sockets into which roots of teeth fit, the small respiratory sacs of the lungs, and the secretory portions of glands.

amacrine cell (AM-ah-krin) Cells of the neural retina that lack long cytoplasmic processes.

ameloblast (ah-MEL-oh-BLAST) The cells that produce enamel and the crown of teeth.

amnion The extraembryonic membrane that surrounds the embryo and encloses amniotic fluid in the amniotic cavity.

amniotic cavity Protective, fluid-filled cavity surrounding the developing embryo.

amorphous (a-MORE-fuss) Without obvious form of its own, formless.

amorphous ground substance The proteoglycan gel of connective tissue.

amphiarthrosis, *pl.* **amphiarthroses** Slightly movable joints composed of cartilaginous (intervertebral disks) or fibrous connective tissue (interosseous ligaments).

ampulla Saclike dilation of a tube. In the semicircular canals the ampulla contains the crista ampul- laris. The name also is applied to the wide portion of the uterine tube between the infundibulum and the isthmus. The ampulla of Vater receives the common bile duct and pancreatic duct before entering the duodenum.

amygdala Component of the limbic system of the brain that coordinates appropriate behavior.

amylase (AM-ill-ace) A group of starch-splitting enzymes that cleave starch, glycogen, and related polysaccharides.

anabolic steroids Steroids, such as corticoids, that stimulate metabolism and growth. Anabolism refers to metabolic synthesis of proteins and polysaccharides that can result in growth.

anaerobic respiration Breakdown of glucose in the absence of oxygen to produce lactic acid and ATP.

anagen (AN-ah-gen) A hair follicle's period of hair production, which may last several months or years before the follicle becomes dormant. See *catagen* and *telogen*.

anal canal Terminal portion of the digestive tract between the rectum and anal sphincter.

anal sphincter The muscular rings that close the anus.

analgesic Drug, such as aspirin, that relieves pain. Pain-relieving action of such a drug.

anaphase Period during cell division when chromatids separate (or in the case of first meiosis, when homologous chromosome pairs separate).

anastomosis, *pl.* **anastomoses** (ah-NAS-toe-MO-sis, -sees) Interconnecting junctions between vessels that give organs or tissues alternative sources of blood. Convergence; the reverse of branching.

anatomical position Position in which a person is standing erect facing the observer, arms at the sides, and the palms of the hands facing forward.

anatomy Scientific discipline that investigates the structure of the body.

anchoring filament Thin filaments that connect the edges of endothelial cells of lymphatic capillaries to surrounding tissues. Anchoring filaments are thought to assist the entry of materials into these capillaries by pulling open clefts between the edges of endothelial cells.

androgen (AN-droe-jen) A group of male sex hormones. A notable androgen is testosterone.

androgen binding protein To affect their target cells, androgens must bind to androgen binding proteins, receptors in the plasma membrane that initiate the cells' response to these hormones.

androgen insensitivity syndrome A heritable condition in which target cells do not respond to androgens, for lack of effective androgen binding proteins. Formerly called testicular feminization, in males this condition produces feminine external genitalia and secondary sex characteristics.

anemia Any condition in which the number of red blood cells per m³ the amount of hemoglobin in 100 ml of blood, or the volume of packed red blood cells per 100 ml of blood is less than normal.

anencephaly Defective development of the brain and absence of the bones of the cranial vault. Only a rudimentary brainstem and some trace of basal ganglia are present.

aneurysm Dilated portion of an artery caused by weakness in the tunica media.

angina pectoris Severe constricting pain in the chest, often radiating to the left shoulder and down the arm; caused by ischemia of the heart muscle, usually because of coronary artery disease that occludes coronary vessels.

angioblast The precursor cells that form the vascular endothelium.

angiogenesis The process that forms new vessels by sprouting from old vessels, in contrast with vasculogenesis, wherein vessels coalesce from small clusters of cells.

angiogenin A protein, often produced by tumors, that stimulates capillary development.

angioplasty Operation to enlarge the narrowed lumen of a coronary artery by introducing a balloon-tipped catheter and dilating the lumen by withdrawing the inflated tip past the obstruction.

angiotensin I (AN-jee-oh-TEN-sin) Peptide produced by the action of renin on angiotensinogen.

angiotensin II Peptide derived from angiotensin I; stimulates vasoconstriction and aldosterone secretion.

angiotensinogen (AN-jee-oh-ten-SIN-oh-jen) Precursor molecule to angiotensin I; produced by the liver.

anion Ion carrying a negative charge, such as chloride ion, Cl^-.

anisotropic (an-EYE-so-TROP-ik) Nonrandom organization of filaments, particularly in A-bands of sarcomeres. See *isotropic*.

ankle The structure between the leg and foot. The joint between the talus (ankle bone) and tibia and fibula of the leg.

ankylosing spondylitis Inflammation and immobilization of the cartilaginous joints of the vertebral column and pelvis.

ankylosis (ANK-ee-LOW-sis) The immobilizing of joints by fibrous or bone tissue.

annular ligament This ring-shaped ligament retains the circular end of the radius in a sleeve that allows the radius to rotate upon the ulna when the forearm pronates and supinates.

annulospiral ending (ANN-yoo-loe-SPY-ral) Also called primary sensory endings. Annulospiral endings in muscle spindles detect muscle contraction.

anococcygeal ligament (AY-no-kok-SID-jee-al) A ligament of the pelvic diaphragm that connects the anus to the coccyx.

antagonist Muscle that works in opposition to another (agonist) muscle.

antagonistic pair Pairs of muscles that produce opposite motions.

anterior The front; the side of the body that goes first when a person walks forward.

anterior chamber of eye Chamber in the anterior portion of the eyeball between the cornea and the iris.

anterior interventricular sulcus Groove on the anterior surface of the heart, marking the location of the septum between the ventricles.

anterior pituitary See *adenohypophysis*.

anterograde amnesia Inability to remember events after an injury or illness, even though memory of prior events is unaffected.

anterograde degeneration When an axon of a neuron is cut, as in an injury, the neuron may die. In anterograde degeneration, neurons also degenerate in the direction that impulses would flow normally from the degenerating neuron. See *retrograde degeneration*.

antibody (AN-tih-BOD-ee) Proteins found in the plasma of blood that bind specifically to antigens.

antibody-mediated immunity Immunity due to B-cells and the production of antibodies.

anticholinesterase Molecules that inhibit acetylcholinesterase, the enzyme that degrades acetylcholine at synapses and motor endplates. Anticholinesterases allow more acetylcholine to accumulate at synapses and can thereby enhance the ability of damaged synapses to function normally. Anticholinesterases are used to boost the action of defective acetylcholine receptors in myasthenia gravis.

anticoagulant Agent that prevents coagulation. Heparin is an anticoagulant that suppresses clotting of blood.

antidiuresis (AN-tee-die-you-REE-sis) The inhibition or slowing of urine production by the kidney. The opposite of diuresis.

antidiuretic hormone (ADH) (ANT-ee-die-you-RET-ik) Hormone secreted from the neurohypophysis that causes the kidney to reduce the output of urine; also called vasopressin because it causes vasoconstriction.

antigen (AN-tih-JEN) Any substance that induces a state of sensitivity and/or resistance to infection or toxic substances after a latent period; substance that stimulates the immune system.

antigen-presenting cell Cells that present fragments of virus or other antigens proliferating within them for recognition by T-lymphocytes and subsequent removal of the infected cells.

antiport Membrane transport protein that transports two molecules across the membrane in opposite directions at the same time. As sodium ions enter, for example, potassium ions leave the sodium/potassium ATPase.

antrum Cavity of an ovarian follicle filled with fluid that contains estrogen.

anulus fibrosus Fibrous material forming the outer portion of an intervertebral disk. Not to be confused with annular ligament.

anus Lower opening of the digestive tract through which fecal matter is extruded.

aorta (ay-OR-tah) The main trunk of the systemic arterial tree; this large, elastic artery carries blood from the left ventricle of the heart through the thorax and abdomen.

aortic body One of the smallest bilateral structures, similar to the carotid bodies, attached to a small branch of the aorta near its arch; contains chemoreceptors that respond primarily to decreases in blood oxygen; less sensitive to decreases in blood pH or increases in carbon dioxide.

apex The tapered, often superior, end of an organ or structure. The apex of the heart is the rounded tip, which is directed anteriorly and slightly inferiorly.

aphasia Speechlessness; impaired or absent speech.

apical domain The portion of the plasma membrane of epithelial cells that faces the lumen. Separated by tight junctions from the basal domain, this portion of the membrane secretes cytoplasmic products into the lumen and takes up materials from the lumen into the cytoplasm.

apical ectodermal ridge (AER) Layer of surface ectodermal cells at the distal margin of the embryonic limb bud that stimulates outgrowth of the limb.

apical foramen (tooth) Opening at the apex of the root of a tooth that gives passage to the nerve and blood vessels.

aplastic anemia Anemia characterized by greatly decreased formation of erythrocytes and hemoglobin by bone marrow.

apocrine gland (AP-o-krin) Gland whose cells release cytoplasm with its secretion (*e.g.,* mammary glands). See *holocrine gland* and *merocrine gland.*

apocrine sweat gland These glands, associated with hair follicles in the axillae, genital, and anal regions of the skin, release milky secretions into the necks of hair follicles. See *apocrine gland.*

aponeurosis (APO-noor-OH-sis) A broad, flat, tendon sheet. The external oblique muscle inserts by way of an aponeurosis onto the linea alba of the abdomen.

apoprotein (AP-oh-PRO-teen) A small protein that enables another, usually larger, protein to act, by binding to it. Apoproteins enable low density lipoproteins to bind to receptors on liver cells.

appendices epiploica (ah-PEN-dih-sees EP-ee-PLOE-ik-ah) Fatty appendages of the large intestine.

appendicular Relating to an appendix or appendage, such as the limbs and their associated girdles.

appendicular skeleton The portion of the skeleton consisting of the upper limbs and the lower limbs and their girdles.

appositional growth A pattern of growth, especially in bone and cartilage, that adds new layers of material onto existing layers.

apraxia Disorder of motor coordination characterized by partial or complete inability to perform purposeful movements.

aqueous humor Watery, clear solution that fills the anterior and posterior chambers of the eye.

arachnoid Thin, cobweblike meningeal layer surrounding the brain; the middle of the three layers of the meninges.

arachnoid villi Tufts of arachnoid tissue that return cerebrospinal fluid to the blood in the superior sagittal sinus of the brain.

arch of the aorta Curve between the ascending and descending portions of the aorta in the thorax.

areola Circular pigmented area surrounding the nipple; its surface is dotted with little projections from areolar glands beneath the skin.

areolar Relating to tissue with small areas or spaces within it.

areolar connective tissue Also known as loose connective tissue, the matrix of this connective tissue is soft and watery. Found in dermis, mesenteries, and the digestive tract.

arm That part of the upper limb between the shoulder and the elbow; the brachium.

arrector pili, *pl.* **arrectores pilorum** Smooth muscle attached to the hair follicle and dermis that erects the hair.

arrhythmia (ay-RYTH-mee-ah) See *dysrhythmia.*

arterial (ar-TEER-ee-al) Concerning the arteries, or the arterial system.

arteriole (ar-TEER-ee-ole) Minute artery that conducts blood to capillaries. Smallest artery having all three vascular tunics.

arteriosclerosis Hardening of the arteries that results from deposits of calcium and collagen in the tunica intima.

arteriovenous anastomosis Vessel that shunts blood directly from an arteriole to a venule, without passing through capillaries.

artery Blood vessel that carries blood away from the heart.

arthrodesis Surgical procedure that fuses together and immobilizes bones of a diseased, degenerated joint.

arthrodial joint Also known as gliding or plane joints, these synovial joints involve flat surfaces that slide upon each other. The temporomandibular joint and articular facets of vertebrae are arthrodial joints.

arthroplasty Arthroplastic surgery remodels or replaces damaged joints. Total hip replacement is a form of arthroplasty.

arthroscopic surgery Surgery of synovial joints that employs an arthroscope to enter the synovial capsule and visualize the interior of the joint.

arthrotomy Surgery that enters the synovial cavity by cutting the synovial capsule. Arthroscopic surgery is a form of arthrotomy.

articular cartilage Hyaline cartilage covering the ends of bones within a synovial joint.

aryepiglottic fold (arry-EP-ih-GLOT-ik) Laryngeal folds that surround the opening of the glottis.

arytenoid cartilage (AR-ih-TEE-noyd) The vocal cords mount upon these cartilages, which adduct and abduct the vocal cords.

ascending aorta Part of the aorta immediately leaving the left ventricle, from which the coronary arteries arise.

ascending colon Portion of the colon between the small intestine and the right colic flexure.

ascending tract Bundles of nerve fibers that conduct impulses up the spinal cord to the brain. The fibers within particular tracts have particular origins and destinations. For example, those of the spinothalamic tract arise in the spinal cord and terminate in the thalamus.

association area These centers in the cerebral cortex interpret visual, auditory, gustatory, somatic sensory, olfactory, and vestibular inputs.

association neuron Neurons that connect different areas of the cerebral cortex in the same hemisphere, in contrast with commissural neurons that connect between the hemispheres.

asthma (AZ-thma) Condition of the lungs in which narrowing of airways is caused by contraction of smooth muscle, edema of the mucosa, and mucus in the lumen of the bronchi and bronchioles.

astigmatism Condition of the lens or cornea of the eye in which rays of light do not focus on a single point on the retina, because the lens or cornea does not curve uniformly.

astrocyte Star-shaped neuroglia cell involved with the blood-brain barrier and other neural structures.

atheroma Patches of lipid and collagen deposits in the tunica intima of large and medium-sized arteries.

atherosclerosis (ATH-er-oh-sklair-OH-sis) A condition characterized by irregularly distributed patches of lipid and collagen deposits in the tunica intima of large and medium-sized arteries.

athetosis Involuntary writhing motions of the limbs, particularly the upper limbs, associated with injury to the basal ganglia.

atopy Localized immediate hypersensitivity, such as asthma and hives.

ATP See *adenosine triphosphate*.

atretic (ah-TREH-tik) Degenerating. Atretic ovarian follicles fail to develop and are reabsorbed by the ovary.

atrial diastole (di-AH-sto-lee) The period during the cardiac cycle when the atria are filling with blood; not contracting. See *atrial systole*.

atrial natriuretic factor (NATE-ree-you-RET-ik) Peptide released from the atria when atrial blood pressure increases; lowers blood pressure by increasing the rate of urine production, thus reducing blood volume.

atrial septal defect Perforation of the interatrial septum, the partition between the left and right atria of the heart, that allows blood to flow between the atria.

atrial systole (SIS-tow-lee) Contraction of the atria. See *atrial diastole*.

atrioventricular bundle Bundle of modified cardiac muscle fibers that projects from the AV node through the interventricular septum.

atrioventricular (AV) node Small node of specialized cardiac muscle fibers that gives rise to the atrioventricular bundle of the conduction system of the heart.

atrioventricular (AV) valve Valve that closes the openings between the atria and ventricles.

atrium, *pl.* **atria** The thin-walled chambers of the heart that receive blood from the body or from the lungs and deliver it to the ventricles.

atrophy Wasting of tissues, as from decreased cell size or number.

auditory canal The pair of tubes that lead from the external ears to the tympanic membranes.

auditory cortex Portion of the cerebral cortex that is responsible for the conscious sensation of sound; located in the dorsal portion of the temporal lobe within the lateral fissure and on the superolateral surface of the temporal lobe.

auditory ossicle Bone of the middle ear; the malleus, incus, and stapes are ossicles.

auditory (eustachian) tube (you-STAY-kee-an) The pair of tubes that connect the nasopharynx with the middle ears.

auricle Part of the external ear that protrudes from the side of the head. Also, the small earlike pouch projecting from the superior, anterior portion of each atrium of the heart.

auricular Relating to the ear.

auricular cartilage The elastic cartilage that supports and shapes the external ear.

auscultation (aws-CUL-tay-shun) Listening for internal sounds of the body, especially the heart, lungs, or digestive tract. May be done directly or with the help of a stethoscope.

autodigestion Chemical process in the lumen of the digestive tract that tends to digest the lining of the tract.

autoimmune disease Disease resulting from a specific immune system reaction against self-antigens.

autonomic Refers to efferent neurons that innervate cardiac muscle, smooth muscle, and glands; characterized by having two neurons in series between the central nervous system and effector organs.

autonomic ganglia Ganglia containing the neuron cell bodies of the autonomic nervous system.

autonomic nervous system (ANS) Composed of nerve fibers that send impulses from the central nervous system to smooth muscle, cardiac muscle, and glands.

autophagia (aw-toe-FAY-jee-ah) Process in which the endomembrane system disposes of a cell's own organelles. Self-eating.

autophagic vesicle (aw-toe-FAY-jik) Cytoplasmic vesicles that capture spent ribosomes and mitochondria for processing by the endomembrane system.

autosome Any chromosome other than a sex chromosome; autosomes normally occur in pairs in somatic cells and singly in gametes.

axial Head, neck, and trunk, as distinguished from the extremities.

axial myopia Myopia or nearsightedness due to elongation of the globe of the eye.

0axial segmentation The concept that the trunk of the body is segmented from head to pelvis along the central body axis; based on the segmental features of the vertebrae, ribs, and spinal nerves.

axial skeleton Skull, vertebral column, and rib cage.

axilla Armpit.

axis of symmetry The central line of symmetry that can be passed through the entire body or a part.

axon (AK-son) Single peripheral process of a neuron that conducts action potentials away from the cell body toward synapses with other neurons.

axon collateral Branch of an axon. An axon can have many collaterals.

axon hillock The tapering portion of an neuron cell body that leads to an axon.

axonal transport Rapid movement of vesicles along microtubules within axons.

axoneme (AX-o-neem) The central 9 + 2 arrangement of microtubules in cilia and flagella that enables these organelles to propel cells and fluid.

axoplasm Cytoplasm of the axon.

axoplasmic flow Slow movement of axon cytoplasm along an axon.

B-lymphocyte Type of lymphocyte responsible for detecting antigens on the exterior of cells or dissolved in body fluids.

ball-and-socket joint A variety of synovial joint, such as the hip joint, which enables movement in three directions simultaneously. Named for the spherical head of the femur that inserts into the concave socket of the coxal bone. See *acetabulum.*

basal body The centrosomelike structure that anchors cilia and flagella into the cytoplasm. Basal bodies contain nine triplet microtubules and one central microtubule, a 9 + 1 arrangement.

basal cell carcinoma A benign tumor of the epidermis.

basal domain The portion of the plasma membrane of epithelial cells that attaches the base of the cell to the basement membrane. Separated by tight junctions from the apical domain of the membrane, the basal domain transports different products into and out of the cell than does the apical domain.

basal ganglia Nuclei at the base of the cerebrum involved in motor functions.

basal layer of epidermis The deepest cell layer of the epidermis. Also known as the stratum basale or stratum germinativum, the basal layer is the source of dividing cells that form all other layers of the epidermis.

basal plate The ventrolateral walls of the neural tube that become the ventral half of the spinal cord, carrying motor impulses.

base The broad, usually inferior, end on which an organ or structure rests or is attached to another structure. The basal ends of epithelial cells are attached to the basement membrane. The base of the heart is the flatter portion, directed posteriorly and superiorly, that is suspended by the great vessels.

basement membrane Specialized extracellular layer located at the base of epithelial cells that attaches them to the underlying connective tissue.

basilar membrane Wall of the membranous labyrinth bordering the scala tympani; supports the spiral organ (of Corti).

basophil (BAYS-oh-fil) White blood cell-containing granules that stain specifically with basic dyes; promotes inflammation.

Bell's palsy Paralysis of facial muscles and senses caused by injury to the facial nerve (cranial nerve VII).

belly of muscle Largest portion of a skeletal muscle located between the origin and insertion.

belt desmosome A variety of junctional complex that employs a belt of desmosome fibers to join together the membranes of adjacent cells.

benign Indicating the mild or harmless nature of an illness or tumor.

biaxial motion Motion of synovial joints in two axes simultaneously. The radius bone experiences biaxial motion when the forearm simultaneously flexes at the elbow and pronates.

bicuspid valve Valve closing the orifice between left atrium and left ventricle of the heart. One of two atrioventricular valves, the other being the tricuspid valve. Also called mitral valve.

bifid Separated into two parts.

bilateral symmetry The property of having mirror-image right and left sides.

bile Fluid secreted from the liver into the duodenum; consists of bile salts, bile pigments, bicarbonate ions, cholesterol, fats, fat-soluble hormones, and lecithin. Emulsifies fats in digestion.

bile canaliculus (KAN-al-IK-you-lus) A type of intercellular channel approximately 1 μm or less in diameter that occurs between liver cells and into which bile is secreted; empties bile into the hepatic ducts and ultimately the gallbladder and duodenum.

bile salt Organic salt secreted by the liver that functions as an emulsifying agent.

bilirubin (BIL-ee-ROO-bin) Bile pigment derived from hemoglobin during destruction of erythrocytes.

binocular vision Vision using two eyes at the same time; responsible for depth perception when the visual fields of both eyes converge on the same objects.

bipennate muscle Skeletal muscles in which short muscle fibers are arranged in chevronlike patterns that give more power, but less motion, than parallel muscles of similar size.

bipolar neuron Small neuron of the central nervous system that extends two processes, an axon and a dendrite, from the cell body.

Birbeck granule Tennis racket–shaped membranous organelles of unknown function in Langerhans cells of the epidermis.

birth canal The dilated passage formed by the cervix and vagina through which a fetus passes toward birth.

blastocoele Cavity in the blastocyst stage of embryonic development.

blastocyst Stage of mammalian embryos that consists of the inner cell mass and a thin trophoblast layer enclosing the blastocoele.

blind spot Point in the retina where the optic nerve penetrates the fibrous tunic; contains no rods or cones and therefore does not respond to light. See *optic disk.*

blood The fluid and cells that circulate through the heart, arteries, capillaries, and veins; means by which oxygen and nutritive materials are transported to the tissues and carbon dioxide and various metabolic products are removed for excretion.

blood clot Coagulated form of blood.

blood doping Technique that temporarily increases the oxygen capacity of a patient's blood by transfusing additional erythrocytes stored earlier from the patient's own blood.

blood island Aggregation of mesodermal cells in the embryonic yolk sac that forms vascular endothelium and primitive blood cells. See *vasculogenesis.*

blood pressure Pressure of the blood within the blood vessels; commonly expressed in units of millimeters of mercury supported in a glass column by the pressure of blood.

blood smear A technique for spreading a small droplet of blood thinly across a microscope slide for staining and observation of blood cells.

blood-brain barrier Permeability barrier controlling the passage of most large-molecular compounds from the blood to the cerebrospinal fluid and brain tissue; consists of capillary endothelium and may include astrocytes.

blood-testis barrier Comparable to the blood-brain barrier, the blood-testis barrier prohibits passage of certain blood components into sperm developing in the seminiferous tubules of the testis.

blood-thymic barrier Layer of reticular cells that separates capillaries from thymic tissue in the cortex of the thymus gland; prevents large molecules from leaving the blood and entering the cortex.

body fold The folds of the embryonic body wall that grow ventrally and medially, and fuse to form the trunk and enclose the ventral body cavity.

body of a bone The central mass of a bone from which extend various processes. The shaft of a long bone, or squama of a flat bone.

bone A specialized connective tissue, together with cartilage and blood, that is characterized by its hard, rigid extracellular matrix containing hydroxyapatite. Forms the skeleton.

bony labyrinth Part of the inner ear; contains the membranous labyrinth that forms the cochlea, vestibule, and semicircular canals.

Bowman's capsule See *glomerular capsule.*

brachial Relating to the arm, the brachium.

brachial plexus Distributes axons of cervical and thoracic spinal nerves (C4 through T1) into the nerves of the upper extremities.

brachium The arm; the portion of the upper extremity between the elbow and the shoulder.

bradycardia (BRAY-dee-KAR-dee-ah) Slow heart rate; a form of cardiac dysrhythmia.

brainstem Portion of the brain consisting of the midbrain, pons, and medulla oblongata.

branched gland Endocrine glands in which the secretory portion of the gland branches. Not to be confused with compound glands, in which the ducts branch.

branchial arch Typically, six arches in vertebrates; in the lower vertebrates branchial arches bear gills, but they appear transiently in embryos of higher vertebrates and give rise to pharyngeal structures in the head and neck.

breast The breasts contain the mammary glands, derivatives of the epidermis.

bricks and mortar model of epidermis The concept that the epidermal barrier consists of keratinocytes (bricks) sealed together with waterproof intercellular lipids (mortar).

broad ligament Peritoneal fold passing from the lateral margin of the uterus to the wall of the pelvis on either side.

Broca's speech area An association area in the frontal lobe of the cerebral cortex that enables one to formulate words into speech.

Brodmann area Korbinian Brodmann found that the microanatomy of the cerebral cortex varies, and mapped more than 40 different regions on the surface of the cerebrum.

bronchial associated lymphoid tissues (BALT) Permanent lymphatic nodules associated with the bronchial passages of the lungs.

bronchial tree (BRONK-ee-al) The treelike branched passages of the lungs that lead from the single large trachea progressively to numerous small alveoli.

bronchiole (BRONK-ee-ole) The finer branches of the bronchial tree, less than 1 mm in diameter; have no cartilage in the wall, but do have relatively more smooth muscle and elastic fibers.

bronchopulmonary segments (BRONK-oh-PULL-mo-NAIR-ee) Largely independent segments of the lungs that receive air and blood separately from other segments. Air is delivered by tertiary bronchi.

bronchus, *pl.* bronchi (BRONK-us, -ee) Large, cartilage-supported passages of the lungs that lead from the trachea to the bronchioles.

brown adipose tissue Together with white adipose tissue, brown adipose tissue is a variety of connective tissue that stores fat in adipose cells. Numerous small fat droplets provide energy for heat in infants. The brown color is due to blood supply, not to the color of fat.

Brunner's gland Submucosal glands of the duodenum that secrete alkaline mucus. Also known as duodenal glands.

brush border Epithelial surface consisting of microvilli.

bulb of the penis Expanded posterior part of the corpus spongiosum of the penis.

bulb of the vestibule Mass of erectile tissue on either side of the vagina.

bulbar conjunctiva (BUL-bar) Conjunctiva that covers the anterior surface of the eyeball, in contrast with the palpebral conjunctiva that lines the inner surface of the eyelids.

bulbar septum The septum that separates the pulmonary trunk and ascending aorta in the conus arteriosus of the embryonic heart.

bulbourethral gland Two small compound glands in males that produce a mucoid secretion that discharges through a small duct into the spongy urethra.

bundle branch block Delayed conduction of depolarizations from the AV node through the bundle branches of specialized conducting fibers in the wall of the ventricles of the heart.

bursa, *pl.* bursae Closed sac or pocket that contains synovial fluid and usually occurs near moving joints and surfaces, where it reduces friction from tendons, ligaments, skin, and bone.

bursitis Inflammation of a bursa.

calcaneal tendon Also known as Achilles tendon. Common tendon of the gastrocnemius, soleus, and plataris muscles that inserts onto the calcaneus (heel bone).

calcitonin Hormone released from parafollicular cells of the parathyroid gland that causes a decrease in blood levels of calcium ions.

calcium ion signaling pathway A major membrane signal transduction pathway that enables cells to respond to external changes. Minute flows of calcium ions across the membrane trigger internal responses.

calculus, *pl.* calculi (KAL-kue-luss, -lee) Hard accretions, or stones, such as kidney and gallstones, that can block internal passages.

callus Thickening of the stratum corneum of the skin in response to pressure or friction. The hard bonelike substance that develops at the site of a broken bone.

calmodulin Protein receptor for calcium ions that plays a role in many Ca++-regulated processes such as smooth muscle contraction.

calyx, *pl.* calyces (KAIL-iks, -ih-sees) Flower-shaped or funnel-shaped structure; specifically, one of the branches or recesses of the renal pelvis into which the tips of the renal pyramids project.

canal of Schlemm Series of veins at the base of the cornea that drain excess aqueous humor from the anterior chamber of the eye; also known as scleral venous sinus.

canalicular stage (KAN-al-IK-you-lar) Stage of lung development at 17 to 25 weeks that forms the passages of the respiratory zone.

canaliculus, *pl.* **canaliculi** (KAN-al-IK-you-lus, -lee) Microscopic canals in bone through which osteocytes communicate by way of narrow cell processes.

cancellous bone Bone with a lattice-like appearance; spongy bone.

canine Referring to the cuspid tooth.

canthus, *pl.* **canthi** Angle between the eyelids at the medial and lateral margins of the eye.

capacitation Process whereby spermatozoa acquire the ability to fertilize ova; occurs in the female genital tract.

capillary (KAP-ill-air-ee) Minute blood vessel consisting only of simple squamous epithelium; major site for the exchange of substances between the blood and tissues.

capillary exchange Process in which materials cross the endothelial wall of capillaries between the blood and interstitial fluid of surrounding tissues.

capitulum Head-shaped structure; the rounded surface of the humerus that articulates with the radius bone at the elbow.

carbaminohemoglobin Form of hemoglobin bound to carbon dioxide molecules by means of reactive amino groups on the hemoglobin molecule.

carbohydrate A monosaccharide, such as glucose, or molecules, such as glycogen, composed of monosaccharides. Carbohydrates contain atoms of carbon, hydrogen, and oxygen.

carbonic anhydrase Enzyme that catalyzes the reaction between carbon dioxide and water to form carbonic acid.

carcinoma Tumor of epithelial tissues that most commonly occurs in skin, uterus, breast, intestine, and tongue.

cardiac Related to the heart.

cardiac cycle Cyclic behavior of the beating heart. Consists of the periods of systole and diastole.

cardiac muscle One of three major varieties of muscle fibers. Characterized by involuntary contraction, cross-

striations, aerobic respiration, and intercalated disks. Found only in the heart.

cardiac nerve Nerve in the autonomic nervous system that extends from the sympathetic chain ganglia to the heart.

cardiac output Volume of blood in milliliters pumped by the heart per minute.

cardiac region Region of the stomach near the heart and the entrance of the esophagus.

cardiac reserve Work that the heart is able to perform beyond that required during ordinary circumstances of daily life.

carina The triangular piece of hyaline cartilage that supports the junction of the trachea with the bronchi. A modified tracheal ring.

carotene (KAR-oh-teen) A class of yellow or orange plant pigment used as a source of vitamin A. Found in foods such as carrots, sweet potatoes, and egg yolks.

carotid body One of the small organs near the carotid sinuses; contains chemoreceptors that respond primarily to decreases in blood oxygen; less sensitive to decreases in blood pH or increases in carbon dioxide.

carotid sinus Enlargement of the internal carotid artery near the point where the internal carotid artery branches from the common carotid artery; contains baroreceptors (pressure receptors).

carpal Bone of the wrist.

carpal tunnel syndrome Inflammation of the common flexor sheath of the wrist caused by repetitive overuse.

carpus The wrist.

carrier Person in apparent good health whose chromosomes contain a heterozygous pathologic gene that may be transmitted to his or her children.

carrier protein Membrane protein that transports molecules across the plasma membrane of cells. See *antiport, symport,* and *uniport.*

cartilage Firm, smooth, resilient nonvascular connective tissue of joints and the skeleton.

cartilaginous Pertaining to cartilage.

cartilaginous joint Bones connected by cartilage; includes synchondroses and symphyses.

caruncle Raised mound of tissue at the medial angle or canthus of the eye.

catabolism All of the decomposition reactions that occur in the body; releases energy.

catagen (KAT-ah-jen) The phase of the hair follicle growth cycle when growth shuts down and the follicle regresses. See *anagen, telogen.*

catalyst Substance that increases the rate at which a chemical reaction proceeds without itself being changed permanently. Enzymes.

cataract Loss of transparency of the lens or capsule of the eye due to a protein buildup.

cauda equina Bundle of spinal nerves arising from the caudal end of the spinal cord below the first lumbar vertebra, and extending through the subarachnoid space in the vertebral canal.

caudal Concerning the tail. Structures in the trunk near the coccyx are caudal.

caudate nucleus Slender C-shaped portion of the basal ganglia of the brain, involved in general muscle tone.

caveola, *pl.* **caveolae** Shallow invagination in the membranes of smooth muscle cells that may perform the calcium-storing function similar to both the T-tubules and sarcoplasmic reticulum of skeletal muscle.

cecum (SEE-kum) Cul-de-sac forming the first portion of the large intestine into which empties the ileum. Also spelled cecum.

cell Basic living unit of all plants and animals.

cell body The central portion of a neuron containing the cell nucleus and synthetic organelles, from which extend one axon and one or more dendrites. Also known as perikaryon or soma. Other dendritic cells, such as glial cells and podocytes, contain cell bodies.

cell cycle The lifecycle of cells in which cells grow and then divide.

cell-mediated immunity Immunity due to the actions of T-lymphocytes.

cementum Layer of modified bone covering the dentine of the root and neck of a tooth; blends with the fibers of the periodontal membrane.

central nervous system (CNS) Major subdivision of the nervous system, consisting of the brain and spinal cord.

centriole A small, cylindrical organelle composed of nine triplet microtubules. Two make a centriole pair, usually lying perpendicular to each other in the centrosome.

centromere (SEN-tro-meer) The central portion of a chromosome that attaches to the spindle during mitosis and meiosis.

centrosome Specialized zone of cytoplasm, close to the nucleus, that contains two centrioles.

cephalic (seh-FAL-ik) Concerning the head. Cephalic structures are in or near the head.

cerebellum Portion of the brain attached to the brainstem at the pons; important in coordinating movement, maintaining muscle tone, and balance.

cerebral aqueduct Narrow canal in the midbrain that connects the third and fourth ventricles.

cerebral cortex The superficial gray matter of the cerebral hemispheres.

cerebral hemispheres The cerebrum of the brain is divided into left and right cerebral hemispheres.

cerebrospinal fluid The fluid of the spinal canal and ventricles of the brain and of the meninges that surround the brain and spinal cord.

cerebrum Portion of the brain derived from the telencephalon; the cerebral hemispheres.

ceruminous gland Modified sebaceous glands in the external auditory meatus that produce cerumen (earwax).

cervical canal Canal extending from the isthmus of the uterus to the opening of the uterus into the vagina.

cervical plexus Redistributes axons of cervical spinal nerves (C1 through C4) into the nerves of the head, neck, and shoulders.

cervix (SIR-viks) Lower part of the uterus extending from the isthmus of the uterus into the vagina.

chalazion Inflammation in the meibomian (tarsal) glands of the eyelid; also called a meibomian cyst.

cheek Side of the face forming the lateral wall of the mouth.

chemical synapse In this common type of synapse, axons release neurotransmitter molecules, such as acetylcholine, that stimulate dendrites or cell bodies of other neurons.

chemistry Science dealing with the atomic composition of substances and the reactions they undergo.

chemoreceptor Sensory cell that is stimulated to produce action potentials by a change in the concentration of chemicals. Examples of chemoreceptors include taste receptors, olfactory receptors, and carotid bodies.

chiasmata (kye-AS-mah-tah) Segments of chromosomes are exchanged when homologous chromosomes pair during meiosis. Chiasmata are the junctions between chromosomes where these exchanges occur. See *crossing over.*

chief cell Cells of the parathyroid gland that secrete parathyroid hormone. Cells of gastric glands, also called peptic cells, that secrete pepsinogen.

chloride Chlorine ion, Cl^-.

choana, *pl.* **choanae** See *internal naris.*

cholecystokinin (CCK) (KOAL-eh-sis-toe-KINE-in) Hormone liberated by the upper intestinal mucosa on contact with gastric contents; stimulates contraction of the gallbladder and secretion of pancreatic juice.

choledochal sphincter (KOAL-ee-DOK-kal) This sphincter guards the opening of the common bile duct into the duodenal ampulla. Opened, the sphincter allows bile to exit to the duodenum.

cholesterol (ko-LES-ter-ol) Major component of plasma membranes, and precursor molecule in synthesis of steroids. Synthesized in liver and absorbed from the diet, cholesterol is carried by lipoproteins in the blood to its target tissues and can accumulate in the tunica intima of vessels as atherosclerotic plaques.

cholinergic Referring to nerve fibers that secrete acetylcholine, or to drugs that bind to and activate cholinergic receptor sites.

cholinergic neuron Refers to nerve fibers that secrete acetylcholine as a neurotransmitter substance.

chondroblast Cartilage-producing cell.

chondrocyte (KON-droe-site) Mature cartilage cell.

chorda tympani Branch of the facial nerve that conveys taste sensation from the front two thirds of the tongue.

chordae tendineae (KORD-ee TEN-din-ee) Tendinous cords running from the papillary muscles to the cusps of the atrioventricular valves in the heart.

chorion (KORE-ee-on) The outer extraembryonic membrane of embryos that invades the endometrium of the uterus and establishes the placenta.

chorion frondosum (fron-DOH-sum) The portion of the chorion that is the placenta and contains chorionic villi that exchange components between fetal and maternal blood.

chorion laeve (LEV-aye) Smooth nonvillous, nonplacental portion of the chorion.

chorionic villi Projections from the surface of the placenta that exchange blood components between fetus and mother.

choroid Portion of the vascular tunic associated with the sclera of the eye. Iris and ciliary body are the other components of the vascular tunic.

choroid plexus Specialized plexus located within the ventricles of the brain that secretes cerebrospinal fluid.

chromaffin cell Chromaffin cells of the adrenal medulla are postsynaptic neurons that lack axons and dendrites.

chromatid Chromosomes consist of two identical chromatids after the chromosomes have replicated. See *mitosis.*

chromatin The form taken by DNA and chromosomes in the nucleus during interphase. See *cell cycle.*

chromatolysis Early cytoplasmic steps in neuron regeneration.

chromosomal sex A person's sex chromosome composition; XX is female and XY is male.

chromosome Colored body in the nucleus, composed of DNA and proteins, and containing the primary genetic information of the cell; 23 pairs in humans.

chronic Long-term condition with little change.

chronic obstructive pulmonary disease (COPD) A group of conditions that interfere with flow of air through the tracheobronchial tree, and with gas exchange in the alveoli. Includes cystic fibrosis, asthma, and emphysema.

chronic sinusitis Sustained inflammation of the nasal mucosa.

chyle Milky-colored lymph with a high content of fat absorbed from the small intestine.

chylomicron (KY-low-MY-kron) Microscopic particle of lipid surrounded by protein, occurring in chyle and in blood.

chyme (KIME) Semifluid mass of partly digested food passed from the stomach into the small intestine.

chymotrypsin Proteolytic enzyme formed in the small intestine from the pancreatic precursor protein, chymotrypsinogen.

ciliary body Structure, continuous with the anterior margin of the choroid layer, that focuses the lens and secretes aqueous humor.

ciliary gland Modified sweat gland that opens into the follicle of an eyelash, keeping it lubricated.

ciliary muscle Smooth muscle in the ciliary body of the eye that accommodates the lens. See *accommodation.*

ciliary process Highly vascular structures of the ciliary body of the eye that produce aqueous humor.

ciliary ring Portion of the ciliary body of the eye that contains ciliary muscle fibers.

cilium, *pl.* **cilia** Short, motile whiplike organelle containing nine doublet microtubules and a central pair of microtubules that enables cells to move fluids. See *axoneme, flagellum.*

circle of Willis Circle of interconnected arteries at the base of the brain, named for Thomas Willis, seventeenth-century English anatomist.

circumcision Operation in which part or all of the prepuce of the penis is removed.

circumduction Movement of a body part, notably the limbs, in a circular motion.

circumvallate papilla Type of papilla located on the surface of the tongue and surrounded by a circular groove and fold of epithelium.

cis face The side of Golgi membranes nearer the nucleus that receives proteins for sorting according to their destination inside or outside the cytoplasm. Sorted proteins leave the opposite face, the trans face.

cisterna, *pl.* **cisternae** (sis-TER-nah, -nee) Enlarged, fluid-filled internal space. Interior space of the endoplasmic reticulum. See *terminal cisterna.*

cisterna chyli (sis-TER-nah KYE-lee) Enlarged inferior end of the thoracic duct that receives chyle from the intestine.

Clara cell Special epithelial secretory cells of the bronchioles.

clathrin A major variety of protein that coats the cytoplasmic surfaces of coated pits for receptor-mediated endocytosis.

clavicle The collarbone of the pectoral girdle. Articulates with the sternum and scapula.

cleavage Inward pinching of the plasma membrane that divides a cell into two cells during cytokinesis.

cleft palate Failure of the embryonic palate to fuse along the midline, resulting in an opening through the roof of the mouth into the nasal cavity.

clitoris Small cylindrical, erectile body, rarely exceeding 2 cm in length, situated at the most anterior portion of the vulva.

clivus (KLY-vus) The slope of the sphenoid bone that accommodates the medulla oblongata anterior to the foramen magnum.

cloaca The endodermally lined chamber into which the hindgut and allantois empty in embryos. Divides into rectum and urethra in both sexes, and the vestibule of the vagina in females.

clone A population of genetically identical cells derived by proliferation of a single precursor cell.

clotting factor Proteins and other molecular components that are necessary for the clotting of blood. Hemophilia, a disorder that interferes with clotting, is due to an inherited absence of clotting factors.

coagulation Process of changing from liquid to solid, especially of blood; formation of a blood clot.

coated pit Depressions in the plasma membrane that capture materials by receptor-mediated endocytosis and transfer these materials to the endomembrane system of the cell.

coccygeal plexus Distributes axons of sacral and coccygeal spinal nerves (S4 through Cx1) to the skin and muscles of the coccyx.

coccyx (KOX-iks) The fused terminal bones of the vertebral column that curve forward and help support the floor of the pelvic cavity.

cochlea The coiled auditory portion of the inner ear that contains the spiral organ that detects the stimuli called sound.

cochlear duct Interior of the membranous labyrinth of the cochlea; cochlear canal.

cochlear nerve Nerve that carries sensory impulses from the spiral organ of Corti to the vestibulocochlear nerve.

cochlear prosthesis Implanted device that can partially replace a damaged spiral organ of the ear.

coelome Embryonic cavity of the trunk, from which the pericardial, pleural, and peritoneal cavities derive.

colestipol (ko-LES-tih-pole) A drug that can reduce the level of blood cholesterol.

collagen Major ropelike fibrous protein of the extracellular matrix of connective tissue.

collagen fiber Bundles of collagen molecules in the extracellular matrix that provide connective tissue with strength and resilience.

collagenase A class of enzymes that attack collagen and collagen fibers.

collateral ganglia Sympathetic ganglia that are found at the origin of large abdominal arteries, including the celiac, superior, and inferior mesenteric arteries.

collecting duct Straight tubule that extends from nephrons of the kidney into the medullary pyramids. Filtrate from the distal convoluted tubules enters the collecting duct and is carried to the calyces by papillary ducts.

colon Division of the large intestine that extends from the cecum to the rectum.

color blindness Inability to distinguish between certain colors due to a deficiency of one or more visual pigments in the cones of the eye.

colostrum (ko-LOSS-trum) Thin, white fluid; the first milk secreted by the breast after pregnancy and childbirth; contains less fat and lactose than milk secreted later.

columnar (kol-UM-nar) Shaped like a column; columnar epithelial cells.

comedo (KOM-ee-doh) Blackheads; caused by sebum accumulating in hair follicle openings.

commissural axon Commissural axons cross between opposite sides of the brain and spinal cord.

commissure Connection of nerve fibers between the cerebral hemispheres or from one side of the spinal cord to the other.

common bile duct Duct formed by the union of the common hepatic and cystic ducts; it empties into the duodenum.

common hepatic duct Part of the biliary duct system that is formed by the junction of the right and left hepatic ducts.

compact bone Bone that has fewer internal spaces than cancellous bone; composed of osteons.

complement Group of serum proteins that stimulates phagocytosis and inflammation.

complicated plaque Advanced atherosclerotic plaques that can obstruct blood flow by causing clots to form or by releasing fragments of the arterial wall into the blood flow.

compound gland Exocrine gland in which the ducts branch, as opposed to simple glands with ducts that do not branch.

concentration gradient A difference in concentration among dissolved molecules at different locations in a cell. Nonuniform distribution of dissolved molecules.

concha, *pl.* **conchae** (KON-kah, -kee) The three bony shell-like ridges on the lateral wall of the nasal cavity; the inferior nasal concha, for example.

conducting artery Large elastic arteries, such as the aorta, that conduct blood to the different regions of the body where smaller arteries distribute to organs. See *distributing artery.*

conducting zone The bronchi and bronchioles that conduct air in the lungs between the trachea and the respiratory surfaces that exchange oxygen and carbon dioxide. See *respiratory zone.*

condyle (KON-dile) Rounded articulating surface of a joint.

condyloid joint A variety of synovial joint, such as the metacarpophalangeal joints, characterized by oval articulating surfaces that allow biaxial motion. Also known as ellipsoid joint.

cone Photoreceptor in the retina of the eye; responsible for color vision.

congenital Occurring at birth; may be genetic or due to external influences (drugs) during development.

congestive heart failure Condition in which the heart, for various reasons, becomes unable to pump adequate quantities of blood and, consequently, body fluid accumulates in the lungs and extremities.

conjunctiva Mucous membrane covering the anterior surface of the eyeball and lining the lids.

connective tissue One of the four major tissue types, characterized by abundant extracellular matrix; the supporting or framework tissue of the body; derived from mesoderm.

constipation Condition in which bowel movements are infrequent or incomplete.

continuous capillary Capillary in which pores are absent; less permeable to large molecules than discontinuous capillaries.

contractile ring An element of the cytoskeleton, composed of actin microfilaments, that cleaves the cytoplasm of cells during cytokinesis.

contralateral Structures on opposite sides, such as the left and right arms, are contralateral.

conus arteriosus (KONE-us) The tapering tube that receives blood pumped from the ventricles of the em-

bryonic heart. Separates into pulmonary trunk and ascending aorta.

conus medullaris The tapering inferior end of the spinal cord from which lumbar and sacral spinal nerves exit.

coracoid process (KOR-ah-koyd) The process on the scapula that resembles the beak of a crow.

cord Cluster of cells. Ovarian cords are clusters of primordial follicles in the ovarian stroma.

corn Thickening of the stratum corneum of the skin over a bony projection in response to friction or pressure.

cornea Transparent portion of the sclera that comprises the outer wall of the anterior portion of the eye.

corners of the heart These locations on the chest are best for hearing the sounds of each valve of the heart.

corniculate cartilage (kore-NIK-you-late) Conical nodule of elastic cartilage, resembling cattle horns, that surmounts the apex of each arytenoid cartilage.

cornified epithelia Epithelia in which the keratinocytes become flattened and dehydrated.

cornu, *pl.* **cornua** (KOR-new, -new-ah) Horn-shaped projections, notably from the body of the hyoid bone.

corona radiata Innermost cells of the cumulus oophorus that form a radiating crown of cells around the oocyte.

coronal Plane separating the body or any part of the body into anterior and posterior portions; frontal section.

coronary Resembling a crown; encircling.

coronary artery One of two arteries that arise from the base of the aorta and carry blood to the muscle of the heart.

coronary artery disease Atherosclerosis of the coronary arteries and their branches that serve the heart.

coronary sinus (SINE-us) Short trunk that receives most of the veins of the heart and empties into the right atrium.

coronary sulcus (SUL-kus) Groove on the outer surface of the heart marking the division between the atria and the ventricles.

coronoid process The process on the ramus of the mandible that resembles a crow's beak.

corpus Any body or mass; the main part of an organ or structure.

corpus albicans (AL-bih-kans) Atrophied corpus luteum that remains as a connective tissue scar in the ovary.

corpus callosum Largest commissure of the brain, connecting the cerebral hemispheres.

corpus cavernosum, *pl.* **corpora cavernosa** (KAH-ver-NO-sum, -sah) Two parallel columns of tissue that erect the penis or the body of the clitoris when engorged with blood.

corpus luteum Yellow endocrine body formed in the ovary in the site of a ruptured vesicular follicle immediately after ovulation; secretes progesterone and estrogen.

corpus luteum of pregnancy Large corpus luteum in the ovary of a pregnant female; secretes large amounts of progesterone and estrogen.

corpus spongiosum Median column of erectile tissue located between and ventral to the two corpora cavernosa in the penis; posteriorly it forms the bulb of the penis, and anteriorly it terminates as the glans penis; it is traversed by the urethra. In females it forms the bulb of the vestibule.

corpus striatum Collective term for the caudate nucleus, putamen, and globus pallidus of the basal ganglia; so named because of the striations caused by intermixing of gray and white matter that results from the number of tracts crossing the anterior portion of the corpus striatum.

cortex Outer portion of an organ or structure, such as the cortex of lymph nodes, adrenal gland, and kidney.

cortical nephron The entire nephron is located in the cortex; its short nephron loop does not extend into the medulla of the kidney. See *juxtamedullary nephron.*

corticotroph (KORE-tih-co-TROFE) The cells of the anterior pituitary that produce ACTH.

corticotropin releasing hormone Hormone from the hypothalamus that stimulates the adenohypophysis to release adrenocorticotropic hormone.

cortisol (KORE-tih-sol) Steroid hormone released by the zona fasciculata of the adrenal cortex; increases blood glucose and inhibits inflammation.

costal Related to ribs.

costal cartilage The ribs terminate in costal cartilages that connect most ribs to the sternum.

cotransport Simultaneous movement of two substances in the same direction by a carrier protein across the plasma membrane.

Cowper's gland See *bulbourethral gland.*

cranial cavity The rounded chamber of the skull that houses the brain.

cranial fossa The brain rests in three depressions, or cranial fossas, in the base of the cranium.

cranial nerve Nerve that originates from within the brain; there are twelve pairs of cranial nerves.

cranial vault Eight skull bones that surround and protect the brain; brain case.

craniosacral outflow Efferent impulses in the parasympathetic division of the autonomic nervous system that flow from cranial and sacral levels of the brainstem and spinal cord.

craniosynostosis (KRAY-nee-oh-SIN-os-TOE-sis) Facial and skull deformity caused by insufficient bone growth and premature ossification of sutures (synostosis) between cranial and facial bones.

cranium The brain case.

cremaster muscle (kree-MAS-ter) Extension of abdominal muscles from the internal oblique muscles; in males, raises the testicles; in females, envelops the round ligament of the uterus.

cribriform plate (KRIB-rih-form) Sievelike, perforated portion of the ethmoid bone that receives olfactory neurons from the nasal cavity.

cricoid cartilage (KRY-koyd) The most inferior laryngeal cartilage.

cricothyrotomy Incision through the skin and cricothyroid membrane of the larynx that allows air to bypass an obstructed upper respiratory tract.

crista, *pl.* cristae (KRIS-tah, -tee) Shelflike infoldings of the inner membrane of a mitochondrion. See also *crista ampullaris.*

crista ampullaris Fold on the inner surface of the ampulla of each semicircular canal that detects acceleration.

crista galli (KRIS-tah GAL-lee) A landmark structure of the anterior cranial fossa that anchors the anterior end of the falx cerebri to the base of the cranium.

crossing over When homologous chromosomes pair during meiosis, chromatids cross over each other and exchange segments of chromosomes at crossing points called chiasmata. See *chiasmata.*

Crouzon's syndrome A variety of craniosynostoses in which the orbits, maxilla, and nasal cavity are too small.

crown That part of a tooth that is covered with enamel and extends above the gums.

cruciate ligament Ligaments of the knee that resemble a cross.

crus, *pl.* crura (KRUSS, KROOR-ah) The portion of the lower extremity between the knee and ankle. Any leglike structure.

crus of the penis Posterior portion of the corpus cavernosum of the penis attached to the ischiopubic ramus of the pelvis.

crypt Glands of the intestinal mucosa; formerly known as crypts of Lieberkuhn.

cryptorchidism Failure of the testis to descend.

crystalline protein Transparent proteins that fill the epithelial cells of the lens in the eye.

cuboidal (kue-BOYD-al) Structures that resemble a cube; especially cuboidal epithelial cells.

cumulus oophorus (KUME-you-lus oh-oh-FOR-us) Mass of epithelial cells surrounding the oocyte in a vesicular follicle; also called the cumulus mass.

cuneiform cartilage (KUE-nee-ih-form) Small, tapered rod of elastic cartilage above the corniculate cartilages in the larynx.

cupula Gelatinous mass that overlies the hair cells of the cristae ampullares of the semicircular canals.

curare Plant extract that binds to acetylcholine receptors and prevents the normal action of acetylcholine at synapses and motor endplates.

Cushing's syndrome Overproduction of cortisol by the adrenal cortex in response to excessive production of ACTH in the anterior pituitary.

cusp Pointed or rounded projections or edges of a structure, especially of teeth and cardiac valves.

cutaneous (kue-TAIN-ee-us) Pertains to the skin. Cutaneous receptors detect stimuli interpreted as pain, touch, or heat or cold.

cuticle Outer thin covering layer of epidermis, usually horny; the growth of the epidermis onto the nail, or the outer covering of a hair.

cutis laxa (KUE-tiss LAX-ah) An inherited connective tissue disorder affecting the elastic fibers of the dermis.

cyanotic (SY-an-OT-ik) Blue coloration of the skin and mucous mem-

branes due to insufficient oxygenation of blood.

cystic duct Duct leading from the gallbladder; joins the common hepatic duct to form the common bile duct.

cystic fibrosis (SIS-tik fi-BRO-sis) An inherited mucosal disease that interferes with secretion by mucosa. Particularly viscous mucus accumulates in respiratory passages of the lungs and pancreatic ducts.

cystinuria (SIS-tin-YOUR-ee-ah) A disorder of nephron tubules in which the amino acid, cystine, accumulates in urine because the proximal tubule is unable to reabsorb this substance.

cytokeratin A family of proteins that compose the intermediate filaments of the cytoskeleton. These proteins are characteristic of epidermis.

cytokinesis Division of the cytoplasm during cell division.

cytology Study of anatomy, physiology, pathology, and chemistry of the cell.

cytoplasm (SITE-oh-plasm) Protoplasm of the cell.

cytoplasmic bridge Narrow cytoplasmic connections, notably between developing spermatocytes, that join cells in a shared cytoplasm.

cytoplasmic keratin See *cytokeratin.*

cytoskeleton A cytoplasmic network of protein filaments and fibers that support cells, give them shape, and move organelles about the cytoplasm.

cytosol The soluble, nonmembranous components of the cytoplasm.

cytotrophoblast (SITE-oh-TROE-fo-BLAST) Inner cellular layer of the trophoblast.

dartos muscle (DAR-tose) Layer of smooth muscle in the skin of the scrotum; contracts in response to lower temperature and relaxes in response to higher temperatures; raises and lowers testes in the scrotum.

decidua (de-SID-you-ah) The portion of the endometrium discarded as part of the afterbirth. The d. basalis is associated with the placenta, the d. capsularis covers the chorion and fetus, and the d. parietalis derives from the opposite wall of the uterus.

deciduous dentition The first set of teeth; primary teeth.

decussate To cross. Used primarily to describe the crossing of axons between opposite sides of the brain and spinal cord.

deep Tissues and organs that are beneath the surfaces of others are deep, in contrast to superficial structures.

deep fascia (FASS-shya) The sheets of dense connective tissue that sheathe the limbs and trunk, and which surround the skeletal muscles and bone.

deep inguinal ring Opening in the transverse fascia through which the spermatic cord (or round ligament in females) enters the inguinal canal.

defecation Discharge of feces from the rectum and anus.

deglutition (dee-glue-TIH-shun) Act of swallowing.

dendrite (DEN-drite) Peripheral processes of neurons that extend from the cell body and receive signals from synapses. Conducts impulses toward the cell body.

dendritic cell Any cell having a branching, treelike appearance; neurons, neuroglial cells, podocytes.

dendritic spine (den-DRIH-tik) Small projections of plasma membrane on the surface of dendrites that increase the surface area available for junctions with synapses.

dense connective tissue Tendons, ligaments, cornea, and dermis are dense connective tissues in which large bundles of collagen fibers and relatively little amorphous ground substance crowd the extracellular matrix.

dental arch Curved maxillary or mandibular arch in which the teeth are located.

dentine Bony material forming the mass of the tooth beneath the enamel crown.

deoxyhemoglobin Hemoglobin without oxygen bound to it.

deoxyribonucleic acid (DNA) Type of nucleic acid containing deoxyribose as the sugar component, found principally in the nuclei of cells; genetic material of cells; the template from which mRNA and other types of RNA are synthesized by transcription.

depolarization Decrease or reversal of a cell's membrane potential. The cytoplasmic face of the membrane becomes less negative or even more positive than the external face. Phase of the action potential in which the membrane potential approaches zero, or becomes positive.

depression Linear movement of a structure in an inferior direction.

depressor Any skeletal muscle that depresses a body part.

dermal papilla, *pl.* **papillae** (pah-PILL-ah, -ee) The boundary between the epidermis and dermis of the skin undulates, consisting of numerous folds and projections of the dermis, called dermal papillae.

dermal ridge Ridges of dermis that underlie the epidermal ridges of the feet and hands.

dermatitis Inflammation of the skin.

dermatoglyphic pattern (DER-mat-oh-GLIF-ik) Fingerprint. The looped and whorled patterns of epidermal ridges on the hands and feet.

dermatome (DERM-ah-tome) Area of skin whose sensory neurons all travel in an individual spinal nerve.

dermis Dense, irregular connective tissue that forms the deep layer of the skin.

Descemet's membrane (des-eh-MAYS) The innermost basement membrane of the cornea.

descending aorta Part of the aorta, further divided into the thoracic aorta and abdominal aorta, in which blood descends to the abdomen.

descending colon Part of the colon extending from the left colonic flexure to the sigmoid colon.

desmosome A variety of junctional complex that adheres patches of plasma membrane between cells. Each desmosome contains a dense plate at the point of adhesion, and fibers between the cells.

desquamation Peeling or scaling off of the superficial cells of the epidermis.

detrusor muscle (deh-TRUE-sore) The layer of smooth muscle fibers in the wall of the bladder. Tone and contraction of this muscle evacuates urine from the bladder.

diabetes insipidus Chronic excretion of large amounts of urine accompanied by extreme thirst; results from inadequate output of antidiuretic hormone.

diabetes mellitus (DIE-ah-BEET-eez MEL-ih-tus) Metabolic disease in which carbohydrate use is reduced and that of lipid and protein enhanced; caused by deficiency of insulin or an inability to respond to insulin. See *insulin-dependent* and *non–insulin-dependent diabetes mellitus*.

dialysis therapy Treatment for end-stage kidney disease that circulates a patient's blood over a dialysis mem-

brane to remove wastes. Artificial kidney.

diapedesis Passage of blood or any of its formed elements through the intact walls of blood vessels.

diaphragm (DIE-ah-fram) Musculomembranous partition between the abdominal and thoracic cavities.

diaphysis (dye-AH-fih-sis) Narrow shaft of a long bone.

diarrhea Abnormally frequent discharge of more or less fluid fecal matter from the bowel.

diarthrosis A class of freely movable joints; commonly, synovial joints.

diastole (die-AS-tow-lee) Relaxation of the heart chambers during which they fill with blood; usually refers to ventricular relaxation, but also refers to atrial diastole.

diencephalon Second portion of the embryonic brain; located deep in the adult cerebrum atop the brainstem.

differential blood cell count A clinical technique that identifies and counts the various leukocytes, platelets, and erythrocytes of the blood.

diffusion Tendency for solute molecules to move from an area of high concentration to an area of low concentration in solution; due to the constant random motion of all atoms, molecules, and ions in solution.

digestion The process that converts food into nutrient molecules that are absorbed by the digestive tract.

digestive tract Esophagus, stomach, small intestine, and large intestine.

digit Finger, thumb, or toe.

dilator pupillae Radial smooth muscle fibers of the iris diaphragm that cause the pupil of the eye to dilate.

dipalmitoylphosphatidlycholine (DPPC) (di-PALM-ih-TOE-il-FOS-fah-TIDE-il-KOE-leen) The major component of surfactant, the secretion that reduces surface tension in alveoli of the lungs and prevents the alveoli from collapsing.

diploid Normal number of chromosomes in somatic cells, 46 chromosomes in humans.

diplopia Condition in which a single object is perceived as two. Both eyes do not focus on the same object.

discontinuous capillary A variety of capillary with large gaps between the endothelial cells and in the basement membrane that allow direct access to interstitial spaces. Found in liver sinusoids and bone marrow. See *continuous capillary* and *fenestrated capillary.*

dislocation Dislocation, also called luxation, separates the articulating surfaces of a joint from each other. Most dislocations involve synovial joints.

distal Distant, away from the center, or from the base of an extremity, or from another point of reference. Used particularly for extremities; the wrist is distal to the forearm, and the arm is distal to the shoulder. Opposite of proximal.

distal convoluted tubule Convoluted tubule of the nephron that extends from the ascending limb of the nephron loop and ends in a collecting duct.

distal phalanx The distal segment of a finger or toe.

distributing artery Medium-sized artery with tunica media composed principally of smooth muscle; regulates blood flow to different regions of the body by constricting or dilating this muscle tissue.

diuresis Increased production of urine by the kidneys.

DNA See *deoxyribonucleic acid.*

dorsal Concerning the back.

dorsal cavity The body cavity that contains the brain and spinal cord within the skull and vertebral column.

dorsal mesentery Sheet of tissue that suspends the embryonic gut and lungs from the dorsal midline. Becomes the mesenteries of the digestive tract.

dorsal root Sensory (afferent) root of a spinal nerve.

dorsal root ganglion Collection of sensory neuron cell bodies within the dorsal root of a spinal nerve.

dorsiflexion Movement of the wrist or ankle so that the dorsal surface of the hand or foot moves superiorly. Opposite of palmar and plantar flexion.

Duchenne muscular dystrophy (DO-shenn; DIS-troe-fee) An inherited disease that causes skeletal muscles to atrophy. Cause has been traced to absence of the protein dystrophin in the sarcoplasmic reticulum of skeletal muscle fibers.

duct A narrow tube. This term and variations of it refer to the thoracic duct of the lymphatic system, the ductus arteriosus, the cerebral aqueduct, and many other tubular structures. In particular, ducts conduct secretions from exocrine glands. Such ducts may empty into the lumen of the digestive tract, nasal and respiratory passages and other cavities, and onto the surface of the skin.

ductus arteriosus Fetal vessel connecting the left pulmonary artery with the descending aorta; closes at birth.

ductus deferens Duct of the testicle, running from the epididymis to the ejaculatory duct; also called vas deferens.

ductus venosus (DUK-tus veh-NO-sus) The continuation of the umbilical vein from the placenta through the liver to the inferior vena cava in the fetus.

duodenal gland Small submucosal gland that opens into the base of intestinal glands; secretes a mucoid alkaline fluid.

duodenal papilla The combined opening of the common bile duct and pancreatic duct into the duodenum.

duodenum (do-oh-DEE-num) First division of the small intestine; connects the stomach to the jejunum.

dura mater Tough, fibrous membrane forming the outer covering of the brain and spinal cord.

dust cell A class of alveolar macrophages in the alveolar septa of the lungs that detect and remove foreign particles by phagocytosis.

dysrhythmia (dis-RITH-mee-ah) Abnormal heartbeat; may be rapid (tachycardia) or slower (bradycardia) or irregular.

E

eardrum Tympanic membrane; tissue membrane that separates the external from the middle ear; vibrates in response to sound waves.

eccrine Refers to water-producing sweat glands.

ectoderm Outermost of the three germ layers of an embryo; gives rise to nervous system and epidermis.

eczema Inflammation of the skin, typically with vesicles that often break open.

edema Excessive accumulation of fluid, usually causing swelling, especially of extremities.

efferent arteriole Vessel that carries blood from the glomerulus to the peritubular capillaries of the kidneys.

efferent ductule Small ducts leading sperm from the testis to the head of the epididymis; derived from mesonephric tubules that drained the mesonephros into the mesonephric duct. See *aberrant ductule.*

efferent neuron Neuron whose axon conducts action potentials away from the brain or spinal cord toward muscles and glands in the body.

Ehlers-Danlos syndrome An inherited connective tissue disease of collagen fibers. Defective cross-linking between collagen fibers leads to unusually flexible joints and skin, slowly healing injuries, and abdominal hernias.

ejaculation Reflexive expulsion of semen from the penis.

ejaculatory duct Short duct formed by the union of the vas deferens and the excretory duct of the seminal vesicle; opens into the prostatic urethra.

elastic artery Large artery, such as the aorta, characterized by elastic fibers in the tunica media.

elastic cartilage A variety of cartilage characterized by elastic fibers. Elastic cartilage supports the external ear.

elastic fiber Connective tissue fibers made of the stretchy protein, elastin.

elastin Major connective tissue protein of elastic tissue that has a structure like a coiled spring.

electrical synapse Transmission across this variety of synapse is by flow of ionic current, often through gap junctions that join the neurons together.

electrocardiogram (ECG) Graphic record of the changing electrical currents during the cardiac cycle, obtained with the electrocardiograph.

electrolyte Ions in solution; named for the ability of such solutions to conduct an electrical current.

electromyography A technique that analyzes the action of skeletal muscles by recording their electrical activity from electrodes inserted into the muscle.

elevation Movement of a structure in a superior (upward) direction.

ellipsoid joint See *condyloid joint.*

embolus, *pl.* **emboli** (EM-bo-luss, -lee) Obstruction or occlusion of a vessel by a circulating clot, a mass of bacteria, or other foreign material.

embryo Developing human from conception to the eighth week of development.

embryonic disk Flat disk of tissue in which the embryo forms.

embryonic period The first 8 weeks of development, in which the major organ systems establish themselves.

enamel Hard substance covering the exposed portion (crown) of the tooth.

enamel organ The patch of ameloblasts that secretes the enamel of each tooth.

enamelin The principal protein of mature enamel in teeth.

enamelogenin (ee-NAM-el-oh-JEEN-in) The principal proteins of immature enamel in teeth.

encapsulated nerve ending A class of sensory nerve endings that are covered with a capsule of fibrous connective tissue.

end-stage renal disease Irreversible kidney failure that results from a variety of conditions that destroy the nephrons of the kidney. Necessitates dialysis treatment or kidney transplantation for survival.

endocardial cushion Connective tissue of the embryonic heart that divides the opening between the atrium and ventricle into the right and left atrioventricular orifices.

endocardium Innermost layer of the heart, including endothelium and connective tissue; tunica intima of the heart.

endochondral ossification Bone formation that occurs by the formation and growth of a cartilage template, which is then replaced by bone.

endocrine gland A class of gland without ducts that secretes hormones internally, usually into the circulation.

endocytosis Uptake of material into endocytotic vesicles in the cytoplasm through the cell membrane.

endocytotic vesicle The process of endocytosis takes up substances from outside the cell membrane by capturing them in endocytotic vesicles, membranous sacs that transport their contents to the endosome for distribution.

endoderm Innermost of the three germ layers of an embryo; becomes lining of the digestive tract, the trachea, and the lungs.

endolymph Fluid found within the membranous labyrinth of the inner ear. See *perilymph*.

endolymphatic sac This pouch beneath the brain stores endolymph for the cochlear duct of the spiral organ of Corti, within the osseous labyrinth of the ear.

endomembrane system The system of internal membranes that distributes materials within the cytoplasm of cells. The Golgi apparatus, endosome, lysosomes, and endocytotic vesicles are major components.

endometrial glands The glands of the endometrium of the uterus.

endometrium Mucous membrane comprising the inner layer of the uterine wall; consists of a simple columnar epithelium and a lamina propria that contains simple tubular glands.

endomysium The delicate, innermost layer of connective tissue that surrounds skeletal muscle fibers in a skeletal muscle.

endoneurium (EN-do-NOOR-ee-um) The innermost layer of delicate connective tissue that surrounds individual nerve fibers in peripheral nerves.

endoplasmic reticulum (ER) Double-walled membranous network inside the cytoplasm; rough ER has ribosomes attached to the surface; smooth ER has no ribosomes.

endorphin (en-DOR-fin) Opiatelike polypeptide found in the brain and other parts of the body; binds in the brain to the same receptors that bind opiates.

endosteal dura The dura mater covering of the brain has two fibrous layers, the outer endosteal dura that adheres to the skull, and an inner meningeal dura.

endosteum (en-DOSS-tee-um) Membranous lining of the medullary cavity of long bones and the cavities of spongy bone.

endothelial injury hypothesis This hypothesis proposes that atherosclerosis progresses because of injury to the endothelial lining of vessels.

endothelium Sleeve of flat cells lining blood and lymphatic vessels and the chambers of the heart.

enkephalin (en-KEF-al-in) A class of peptides found in the brain; binds to specific receptor sites, some of which may be pain-related opiate receptors.

enteric ganglion, *pl.* **enteric ganglia** Ganglia of the enteric nervous system that distribute autonomic impulses to myenteric and submucosal plexuses of the gut wall.

enteric nervous system The network of neurons and ganglia that causes waves of peristaltic contractions to spread along the digestive tube, from esophagus to rectum. Stimuli for this periodic action arise in the wall of the gut tube itself, not from the central nervous system.

enterochromaffin (EC) cell (EN-ter-oh-KROME-ah-fin) Serotonin-releasing cells of the islets of the pancreas.

enterocyte The epithelial cells of the intestine.

enzyme Protein that acts as a catalyst.

eosinophil (EE-oh-SIN-oh-fils) Granular leukocytes that stain orange and red with acidic dyes; inhibits inflammation.

ependymal cell A variety of neuroglial cell that lines the central canal of the spinal cord and ventricles of the brain.

epiblast The upper of two cell layers in the embryonic disk (the hypoblast is the lower) present during the second week of development, before gastrulation.

epicardium Serous membrane covering the surface of the heart. Also called the visceral pericardium.

epicondyle Bony projection, on or at the side of a condyle, onto which ligaments and tendons insert.

epidermal melanin unit (EMU) The epidermal cells that receive melanin granules from a single melanocyte are considered an epidermal melanin unit.

epidermis (EP-ih-DER-miss) Outer layer of the skin, formed of epithelial tissue that covers the dermis.

epidermolysis bullosa simplex (EBS) (bull-OH-sah) A group of inherited disorders of the epidermis that interfere with differentiation of keratinocytes.

epididymis (ep-ih-DID-ih-miss) Convoluted tube connected to the posterior surface of the testis, consisting of a head, body, and tail; site of storage and maturation of spermatozoa.

epidural space The space that contains a fatty cushion between the meningeal dura and endosteal dura of the spinal cord.

epiglottis (ep-ih-GLOT-iss) Plate of elastic cartilage covered with mucous membrane; serves as a valve over the glottis of the larynx during swallowing.

epimyocardium The outer tissue layer that covers the heart and lines the pericardial cavity.

epimysium (ep-ih-MISS-ee-um) Fibrous envelope that surrounds a skeletal muscle.

epinephrine (EP-ih-NEF-rin) The major hormone produced by the adrenal medulla. Epinephrine raises cardiac output and blood glucose level. Chemically related to norepinephrine, a neurotransmitter of the sympathetic division of the autonomic nervous system.

epineurium (EP-ih-NOOR-ee-um) Connective tissue sheath surrounding a nerve.

epiphyseal line Dense plate of bone that indicates the former site of the epiphyseal plate in a growing long bone.

epiphyseal plate Site at which long bones grow; located between the epiphysis and diaphysis of a long bone; plate of hyaline cartilage where growth is followed by endochondral ossification; also called the metaphysis or growth plate.

epiphysis, *pl.* **epiphyses** (eh-PIF-ih-sis, -sees) Portion of a bone developed from a secondary ossification center and joined to the remainder of the bone by the epiphyseal plate.

epiploic foramen (ep-ee-PLOE-ik) The entrance to the bursa of the greater omentum.

epithelial Having to do with epithelium; epithelial tissue.

epithelium, *pl.* **epithelia** (EP-ih-THEEL-ee-um, -ah) One of the four primary types of tissues; it covers and lines organs with sheets of cells underlain by a basement membrane.

eponychium (EH-po-NEEK-ee-um) Outgrowth of the skin that covers the proximal and lateral borders of the nail; cuticle.

epoophoron (EP-oh-oh-FOR-on) Remnants in the broad ligament of females of the mesonephric tubules that joined the transient mesonephric kidney with the mesonephric duct.

erection Rigid, firm condition of erectile tissue when inflated with blood; becomes hard and unyielding; refers to the penis and clitoris.

erythema (air-ih-THEEM-ah) A condition in which the skin reddens from proliferation of capillaries, usually from infection or inflammation.

erythroblastosis fetalis Destruction of erythrocytes in the fetus or newborn caused by antibodies produced in the Rh-negative mother acting on Rh-positive blood of the fetus or newborn.

erythrocyte (eh-RITH-row-site) Disk-shaped red blood cells that contain hemoglobin, which transports oxygen in the blood.

erythropoiesis Production of erythrocytes.

erythropoietin (EPO) (eh-RITH-roe-POY-et-in) Protein secreted by the kidneys that stimulates erythropoiesis by causing proerythroblasts to form and by releasing reticulocytes from bone marrow.

esophageal sphincter Ring of muscle that regulates the passage of materials into or out of the esophagus. The upper esophageal sphincter is at the superior opening of the esophagus, and the lower esophageal sphincter is at the inferior end.

esophagus Portion of the digestive tract between the pharynx and stomach.

estrogen Steroid hormone secreted primarily by the ovaries; involved in maintenance and development of female reproductive organs, secondary sexual characteristics, and the menstrual cycle.

eumelanin (you-MEL-ah-nin) Brown or black melanin pigment of the skin, in contrast to phaeomelanin, which is yellow.

eustachian tube Auditory canal; connects the middle ear to the nasopharynx.

evagination Protrusion of some part or organ.

eversion Turning outward, particularly of the foot.

exocrine gland Class of gland that secretes products through a duct to the lumen of a cavity or onto the surface of the skin.

exocytosis Elimination of bulk material from a cell by way of vesicles; opposite direction to endocytosis.

extension To stretch out. Drawing a body part outward; a straightening motion that contrasts with flexion and bending.

extensor Skeletal muscles that extend body parts.

external Structures that lie outside another structure, rather than inside, are external.

external anal sphincter Ring of skeletal muscle fibers surrounding the anus.

external auditory meatus Short canal that terminates on the eardrum and opens to the exterior; part of the external ear.

external ear Portion of the ear that includes the auricle and external auditory meatus; terminates at the eardrum.

external elastic lamina The outermost layer of the tunica media in arteries.

external genitalia The external reproductive organs, including the vulva, penis, and scrotum.

external naris, *pl.* **nares** Nostril; anterior or external opening of the nasal cavity.

external os The external opening of the cervix into the vagina.

exteroceptor Class of sensory receptor in the skin or mucous membranes that responds to stimulation by external agents or physical forces.

extracellular Outside the cell.

extracellular matrix Noncellular material located between connective tissue cells.

extrafusal fiber Skeletal muscle fibers; as opposed to intrafusal fibers, the modified skeletal muscle fibers of muscle spindles. See *muscle spindle*.

extrinsic muscle Muscle located outside the structure being moved. Extrinsic muscles originating from the hyoid bone move the tongue, and extrinsic ocular muscles move the eyeballs.

eyelid Palpebra; the movable fold of skin that covers the eyeball.

F-actin Fibrous actin molecule that is composed of a series of globular actin molecules (G-actin).

facet Literally, a small face. A small, smooth, often flat articular surface between bones. See *arthrodial joint*.

facilitated diffusion This process moves molecules across plasma membranes into or out of cells from a high to low concentration; mediated by carrier proteins in the membrane and does not consume ATP energy.

falciform ligament (FAL-sih-form) Sickle-shaped fold of peritoneum extending to the surface of the liver from the diaphragm and anterior abdominal wall. See *falx*.

fallopian tube See *uterine tube*.

false pelvis Portion of the pelvis superior to the pelvic brim; composed of the bone on the posterior and lateral sides and of muscle on the anterior side.

false rib These ribs (eighth through tenth) either attach indirectly to the sternum through the cartilage of the seventh rib, or do not attach to the sternum at all (eleventh through twelveth). See *vertebrochondral* and *vertebral rib*.

falx (FALKS) Sickle-shaped curtain of tissue. Falx cerebelli: fold of dura between the cerebellar hemispheres; falx cerebri: between the cerebral hemispheres; falciform ligament: the ventral mesentery of the liver.

Fanconi syndrome This disorder of tubular reabsorption in the kidney interferes with reabsorption of glucose, amino acids, and phosphate by the proximal convoluted tubule.

farsightedness See *hyperopia*.

fascia Loose areolar connective tissue found beneath the skin (hypodermis), or dense connective tissue that encloses and separates muscles.

fascicle, fasciculus Bundle of fibers in a tendon or muscle fibers in a muscle, or nerve fibers in a peripheral nerve, bound together by connective tissue.

fat Greasy semisolid material found in animal tissues and many plants; composed of glycerol and fatty acids bound together by covalent bonds.

fatty acid Any organic acid composed of a long chain of carbon atoms with an acidic group at one end.

fatty streak The first stage of atheroma formation in large arteries. Streaks of fatty tissue appear in the tunica intima and may become sclerotic plaques.

fauces (FOW-sees) Opening between the mouth and the pharynx.

feces Matter discharged from the bowel during defecation, consisting of the undigested residue of the food, epithelium, intestinal mucus, bacteria, and waste material.

female pronucleus Nucleus of the ovum formed after a sperm has penetrated the ovum. Each pronucleus carries the haploid number of chromosomes (23).

fenestra, *pl.* **fenestrae** (feh-NES-trah, -tree) An opening in the wall of a structure, often in an epithelium.

fenestrated capillary (FEN-es-TRATE-ed) In contrast with more typical continuous capillaries, the endothelium of fenestrated capillaries has numerous fenestrae, or pores, that allow substances to cross the endothelium more readily. A diaphragm stretches across most pores. See *continuous capillary* and *discontinuous capillary*.

fertilization Fusion of egg and sperm. Fertilization starts embryonic development; begins with the penetration of the secondary oocyte by the spermatozoon and is completed with the fusion of the male and female pronuclei.

fetal period The last 7 months of development, during which the organ systems grow and become functional.

fetus Developing human from the ninth week of development until birth.

fibrin The fibrous protein that coagulates blood.

fibrinogen (fye-BRIN-oh-jen) The protein precursor to fibrin.

fibroblast (FYE-bro-blast) Spindle-shaped or stellate cells that form connective tissue.

fibrocartilage A variety of cartilage characterized by large, dense bundles of collagen fibers. Fibrocartilage is found in intervertebral disks and symphyses.

fibroma A benign tumor derived from fibroblasts.

fibrosarcoma A malignant tumor derived from fibroblasts.

fibrous Pertaining to a tissue or structure that contains fibers.

fibrous capsule The kidney is covered and protected with a fibrous connective tissue capsule.

fibrous connective tissue Fibrous connective tissue contains large, densely-packed bundles of collagen fibers, but little amorphous ground substance.

fibrous joint Bones connected by fibrous tissue with no joint cavity; includes sutures, syndesmoses, and gomphoses.

fibrous tunic Outer layer of the eye; composed of the sclera and the cornea.

fibula The lateral, slender bone of the leg.

filiform papilla (FILL-ih-form pah-PILL-ah) Filament-shaped papilla on the surface of the tongue.

filopodium, *pl.* **filopodia** Long cellular process containing actin microfilaments that attaches to the substrate and draws cells forward.

filtrate Liquid that has passed through a filter; fluid that enters the nephron through the filtration membrane of the glomerulus in the kidney.

filtration Movement, due to a pressure difference, of a liquid through a filter that prevents some or all of the substances in the liquid from passing through.

filtration slit The third element of the glomerular filtration barrier in the nephrons of the kidney. The filtration slit lies between foot processes of podocytes.

filtration slit diaphragm The thin diaphragm that stretches across the filtration slit of the glomerular filtration barrier.

filum terminale Terminal thread; cord of pia mater tethering the end of the spinal cord to the end of the vertebral canal.

fimbria, *pl.* **fimbriae** (FIM-bree-ah, -ee) Fringelike border of the ostium of the uterine tube that envelops the ovary.

first class lever Few bones act as first class levers, in which the fulcrum lies between the load and the effort.

first degree burn Burns that damage the epidermis but do not remove it. See *second, third degree burn*.

first meiotic division First of two cell divisions that reduce the chromosome number in half (from diploid to haploid) during development of eggs and sperm. Homologous pairs of chromosomes separate during this division.

fixator Muscle that stabilizes the origin of a prime mover.

flaccid paralysis Inability to initiate contractions in muscle. The muscle remains without tone and does not contract.

flagellum, *pl.* **flagella** Long, whiplike organelle consisting of nine doublet microtubules and two central single microtubules that enables cells, notably sperm, to move. See *axoneme, cilium*.

flat muscle These skeletal muscles are wide, relatively thin muscles, such as the external oblique and rectus abdominis muscles of the abdomen; a variety of parallel muscles.

flatus Gas in the gastrointestinal tract that may be expelled through the anus.

flexion Bending motion.

flexor A skeletal muscle that flexes a body member.

floating kidney In occasional individuals, a mesentery suspends a kidney in the peritoneal cavity, outside the normal location in the fibrous capsule behind the peritoneum.

floor plate The narrow, thin, anterior median portion of the neural tube that contributes to the isthmus of the spinal cord. See *roof plate*.

flower spray ending A variety of sensory endings found in muscle spindles that detect contraction; also known as secondary sensory endings.

foam cell Modified smooth muscle fibers of the subendothelial connective tissue of vessels that accumulate cholesterol in atherosclerotic plaques.

foliate papilla Leaf-shaped papilla on the lateral surface of the tongue.

follicle-stimulating hormone (FSH) Hormone of the adenohypophysis that, in females, stimulates the vesicular (graafian) follicles of the ovary, and assists in follicular maturation and the secretion of estrogen; in males, stimulates the epithelium of the seminiferous tubules and is partially responsible for inducing spermatogenesis.

follicular phase Time between the end of menses and ovulation, characterized by rapid division of endometrial cells and development of follicles in the ovary; the proliferative phase.

fontanelle (fon-tah-NEL) Membranous sheet of unossified connective tissue between the cranial bones of infants.

foot process Also known as pedicel. Small, numerous cytoplasmic processes of podocytes in the glomerular filtration barrier of the kidney that form the margin of the filtration slit by interdigitating with foot processes of adjacent podocytes.

foramen, *pl.* foramina (foe-RAM-en, -in-ah) A hole, often in a bone, that admits nerves or vessels.

foramen cecum Median pit on the dorsum of the posterior part of the tongue; the site of the origin of the thyroid gland in the embryo.

foramen ovale (oh-VAL-ee) The oval opening in the interatrial septum of the heart during fetal life; closes postnatally, forming the fossa ovale.

force That which produces motion.

forearm Portion of the upper limb between the elbow and wrist; the antebrachium.

foregut Cephalic portion of the primitive digestive tube in the embryo.

foreskin See *prepuce.*

formed elements Cells (erythrocytes and leukocytes) and cell fragments (platelets) of blood.

fornix Recess at the cervical end of the vagina; also the recess deep to each eyelid where the palpebral and bulbar conjunctivae meet.

fossa, *pl.* fossae (FOSS-ah, -ee) A depression in a bone or a partition into which some other structure may fit. The brain rests in cranial fossae in the floor of the cranial cavity.

fossa ovalis (FOSS-ah oh-VAL-iss) Small depression in the interatrial septum of the heart that marks the location of the fetal foramen ovale. The fossa ovalis is formed by the closure of the foramen ovale following birth.

fovea centralis Depression in the middle of the macula of the retina where there are only cones and no blood vessels.

freckle Small patches or points of skin in which clusters of melanocytes have produced excess mela- nin pigment.

free nerve ending A class of sensory nerve endings that are not covered with a fibrous capsule.

frenulum (FREN-you-lum) Membranous fold extending from the floor of the mouth to the midline of the undersurface of the tongue.

frontal section A plane that passes through the body or a body part parallel to its central axis, and that separates anterior structures from posterior. Also called coronal section.

FSH surge Increase in plasma follicle-stimulating hormone (FSH) levels before ovulation.

fulcrum Pivot point of a lever that enables the lever to exert force on the load.

full-thickness burn Burn that destroys the epidermis and the dermis of the skin and sometimes the underlying tissue as well; also called a third degree burn.

fundus Bottom or rounded end of a hollow organ; the fundus of the stomach or uterus.

fungiform papilla Mushroom-shaped papilla on the surface of the tongue.

funiculus, *pl.* funiculi Bundle of nerve fibers in the spinal cord.

fusiform (FEW-sih-form) Spindle-shaped. Structures such as the biceps brachii muscle of the arm are fusiform; their ends taper.

gallbladder Pear-shaped receptacle on the inferior surface of the liver; serves as a storage reservoir for bile.

gallstone Stones (calculi) that collect in the bladder and biliary tree. Associated with excess cholesterol.

gamete Ovum or spermatozoon.

gamma-aminobutyric acid (GABA) An inhibitory neurotransmitter molecule.

ganglion, *pl.* ganglia (GANG-glee-ahn, -ah) Any cluster of nerve cell bodies in the peripheral nervous system.

ganglion cell Used variously to refer to nerve cells within a ganglion. Ganglion cells of the retina extend axons that form the optic nerve. Ganglion cells of autonomic ganglia are postganglionic neurons.

gangrene Necrosis of tissue due to obstruction, loss, or diminution of blood supply.

gap junction Small channel between cells that allows the passage of ions and small molecules between cells; provides means of intercellular communication.

Gartner's duct This duct is the remnant in females of the mesonephric duct that becomes the vas deferens in males.

gastric gland Gland located in the mucosa of the fundus and body of the stomach.

gastric inhibitory polypeptide Hormone secreted by the duodenum that inhibits gastric acid secretion.

gastric juice The acidic fluid secreted by the mucosa of the stomach.

gastric pit Numerous pits in the mucous membrane of the stomach, at the bottom of which are the gastric glands that secrete mucus, hydrochloric acid, intrinsic factor, pepsinogen, and hormones.

gastrin Hormone secreted in the mucosa of the stomach and duodenum that stimulates secretion of hydrochloric acid by the parietal cells of the gastric glands.

gastrocolic reflex Local reflex resulting in mass movement of the contents of the colon that occurs after the entrance of food into the stomach.

gastroenteric reflex Reflex initiated by stretch of the duodenal wall, or by the presence of irritating substances in the duodenum, that causes a reduction in gastric secretions.

gastroesophageal opening Opening of the esophagus into the stomach.

gastrointestinal Referring to the stomach and intestines; GI tract.

gastrulation The developmental process that rearranges tissue layers of the embryonic disk into three definitive germ layers.

gate theory of pain The concept that perception of pain is modified by interactions among neurons that sensitize to pain (open the gate) or inhibit (close the gate).

gated ion channel Ion channel protein molecules that can be opened or closed by certain conditions on the plasma membrane. Sodium and potassium ion channels are gated channels governed by the membrane potential.

gender identity Sexual identity based on a person's sense of masculine or feminine psychological identity.

gene Functional unit of heredity. Each gene occupies a specific place or locus on a chromosome, is capable of reproducing itself exactly at each cell division, and is capable of directing the formation of an enzyme or other protein.

general sense organ A class of receptors that respond to stimuli that are perceived as touch, pain, heat and cold, and body sense.

genetics Branch of science that investigates heredity.

genital tubercle Median elevation just cephalic to the urogenital orifice of a fetus; gives rise to the penis of males or the clitoris of females.

genitalia (JEN-ih-TAIL-ee-ah) The reproductive organs and structures.

genotype Genetic makeup of an individual.

genu, *pl.* **genua** (JEH-noo, -noo-ah) The knee.

genu valgus Knock-knee.

genu varus Bowleg.

germ cell Spermatozoon or ovum.

germ layer Any of three layers in the embryo (ectoderm, mesoderm, and endoderm) from which organs and tissues arise.

germinal center Lighter-staining center of a lymphatic nodule; area of rapid lymphocyte division.

gingiva (JIN-jih-vah) The gums. Dense fibrous oral mucosa that surrounds the teeth in the mandible and maxilla.

ginglymus See *hinge joint.*

girdle Belt or zone; the bony region where the limbs attach to the trunk.

glabella (glah-BELL-ah) The slight depression in the forehead between the orbits.

glabrous (GLA-brus) Hairless skin, as on the palms of the hands and soles of the feet.

gland Secretory organ from which secretions may be released into the blood, into a cavity, or onto a surface. See *endocrine* and *exocrine glands.*

glans penis Conical expansion of the corpus spongiosum that forms the head of the penis.

glaucoma Disease of the eye involving increased intraocular pressure caused by decreased drainage of the aqueous humor into the canal of Schlemm; causes optic disk degeneration, nerve fiber damage, and defective vision.

glenoid Socket of a joint. Glenoid fossa of the shoulder.

glial cell A major class of neural cells that support and nourish the neurons of the nervous system, but unlike neurons, do not conduct action potentials; synonymous with neuroglia.

gliding joint See *arthrodial joint.*

globin Protein portion of hemoglobin molecules.

globulin (GLOB-yoo-lin) A major class, together with albumins, of plasma proteins in the blood.

glomerular basement membrane The second element of the glomerular filtration barrier in nephrons of the kidney. Molecular pores in the membrane, and its amnionic composition, exclude most proteins and many negatively charged molecules from passing through the membrane into the filtrate of the nephron.

glomerular capsule Apex of a renal tubule that expands around a tuft of glomerular capillaries in the kidney and collects renal filtrate. Also known as Bowman's capsule.

glomerular filtration barrier Membrane formed by the glomerular capillary endothelium, the basement membrane, and the podocytes of the glomerular (Bowman's) capsule.

glomerular filtration rate Amount of plasma (filtrate) that filters into the glomerular (Bowman's) capsules of the kidneys per minute; normally about 125 ml per minute.

glomerulonephritis (glow-MARE-you -LOW-nef-RYE-tiss) Inflammation of the glomerulus; can lead to kidney failure by destroying nephrons.

glomerulus (glow-MARE-you-luss) Mass of capillary loops at the apex of each nephron, surrounded by the glomerular (Bowman's) capsule.

glottis The opening into the larynx from the pharynx.

glucagon Hormone secreted by the islets of Langerhans of the pancreas; acts primarily on the liver to release glucose into the circulatory system.

glucocorticoid Steroid hormone (cortisol) released by zonula fasciculata of the adrenal cortex; increases blood glucose and inhibits inflammation.

gluconeogenesis (GLUE-ko-NEE-oh-JEN-eh-sis) Formation of glucose from proteins (amino acids) or lipids (glycerol).

glucose Six-carbon monosaccharide; dextrose or grape sugar.

glycocalyx (GLY-ko-CALE-iks) Polysaccharide-protein layer that covers the external surface of cells.

glycolysis Anaerobic process during which glucose is converted to pyruvic acid for net production of two ATP molecules for every glucose molecule converted.

goblet cell Mucus-producing epithelial cell with an apical end distended with mucin; resembles a goblet.

goiter An enlarged thyroid gland.

Golgi apparatus Named for Camillo Golgi, Italian histologist and Nobel laureate, 1843–1926. Specialized endoplasmic reticulum that concentrates and packages materials for secretion from the cell and distribution within the cell.

Golgi tendon organ (GOAL-jee) Proprioceptive nerve ending in a tendon; detects tension in the tendon and contributes to body sense.

gomphosis (GOM-foe-sis) Fibrous joint in which a peglike process fits into a hole; specifically, roots of teeth in alveoli of the mandible and maxilla.

gonad Organ that produces sex cells; testis and ovary.

gonadal ridge Ridge of intermediate mesoderm in the embryo that receives primordial germ cells and develops into testis or ovary.

gonadal sex Sexual identity based on the appearance of the gonads: ovaries or testes.

gonadotroph (go-NAD-oh-trofe) Cells of the anterior pituitary that produce gonadotropin hormones, including follicle-stimulating hormone and luteinizing hormone.

gonadotropin Hormone capable of promoting gonadal growth and function. Two major gonadotropins are luteinizing hormone (LH) and follicle-stimulating hormone (FSH).

gonadotropin-releasing hormone (GnRH) Hypothalamic-releasing hormone that stimulates the secretion of gonadotropins (LH and FSH) from the adenohypophysis.

gout Metabolic disease characterized by deposits of uric acid crystals in the joints.

gouty arthritis A form of arthritis associated with deposition of uric acid crystals in synovial joints.

graafian follicle See *vesicular follicle.*

granular layer of epidermis An intermediate layer of the epidermis, also called the stratum granulosum, in which the cytoplasm of keratinocytes contains keratohyalin granules.

granulocyte (GRAN-you-low-site) Mature leukocyte (neutrophil, basophil, or eosinophil) that contains prominent cytoplasmic granules.

granulopoiesis The process that produces granular leukocytes in the bone marrow. Analogous to erythropoiesis.

granulosa cell The granular cells that secrete estrogens in ovarian follicles.

Graves' disease An autoimmune condition that causes the thyroid to produce excessive quantities of thyroxine.

gray matter Collections of nerve cell bodies, their dendritic processes, and associated neuroglial cells, within the central nervous system.

gray ramus communicans, *pl.* **rami communicantes** Connection between spinal nerves and sympathetic chain ganglia through which unmyelinated postganglionic axons project.

greater duodenal papilla Point of opening of the common bile duct and pancreatic duct into the duodenum.

greater omentum Peritoneal fold passing from the greater curvature of the stomach to the transverse colon, hanging like an apron in front of the intestines.

greater vestibular gland Two mucus-secreting glands on either side of the lower part of the vagina. The equivalent of bulbourethral glands in males.

growth cone The growing end of an axon.

growth hormone Somatotropin; stimulates general growth of the individual; stimulates cellular amino acid uptake and protein synthesis.

gubernaculum (GOO-ber-NAK-you-lum) Column of tissue that connects the fetal testis to the developing scrotum; involved in testicular descent.

gustatory Associated with the sense of taste.

gustatory hair Microvillus of gustatory cell in a taste bud.

gut associated lymphoid tissues (GALT) Permanent lymphatic nodules associated with the wall of the digestive tract; Peyer's patches.

gynecomastia Excessive development of the male mammary glands, which sometimes secrete milk.

gyrus, *pl.* **gyri** Rounded folds of the cerebral cortex.

H-zone Area in the center of the A-band into which no actin myofilaments slide; contains only myosin myofilaments.

hair The long, slender structures, composed of keratinized epithelial cells, that protrude from hair follicles on the skin.

hair bulb Knoblike base of a hair follicle.

hair cell Sensory cells located in the organ of Corti in the inner ear.

hair follicle Invagination of the epidermis into the dermis; produces a hair and receives the ducts of sebaceous and apocrine glands.

hair placode (PLA-kode) Thickened patch of epidermal and mesodermal cells of the embryonic skin from which a hair follicle and hair develops.

hair shaft Stem of a hair.

half-life The time it takes for one half of a substance or a population of cells to be lost through biological processes.

hallux valgus Angling of the big toe toward the other toes of the foot at the metatarsal phalangeal joint.

hand (manus) Distal extremity of the forearm, consisting of the wrist, metacarpus, fingers, and thumb.

hand plate The flattened, distal portion of the upper limb bud that becomes the hand.

haploid Having only one set of chromosomes, in contrast to diploid, two sets; characteristic of gametes.

hard keratin The keratins of hair and nails, in contrast to the soft keratins of epithelium.

hard palate Floor of the nasal cavity that separates the nasal cavity from the oral cavity; formed by the palatine processes of the maxillary bones and the horizontal plates of the palatine bones.

Hassall's corpuscle Concentric arrangement of small bodies of keratinized epithelial cells around clusters of degenerating leukocytes in the medulla of the thymus.

haustra, *pl.* **haustrae** (HOUSE-tra, -tree) Sacs of the colon, caused by contraction of the taeniae coli, which are slightly shorter than the gut, so that the latter is thrown into pouches.

haversian canal Canal containing blood vessels, nerves, and loose connective tissue and running in the center of osteons, the units of compact bone. Also known as central canal. Named for a seventeenth-century English anatomist, Clopton Havers.

haversian system See *osteon.*

head Uppermost portion of the human body.

head of the epididymis The larger, upper portion of the epididymis, consisting of fifteen to twenty coiled tubules that receive sperm from the efferent ductules of the testis.

heart Hollow-chambered organ composed of cardiac muscle that receives blood from the veins and pumps it into the arteries.

heart attack Sudden blockage of coronary circulation to the muscle of the ventricles. See *infarct* and *myocardial infarction.*

heart failure Inability of the heart to meet the body's circulatory needs.

heart murmur Any abnormal heart sound; often caused by malfunctioning valves.

heart skeleton Fibrous connective tissue that provides a point of attachment for cardiac muscle cells, electrically insulates the atria from the ventricles, and forms the fibrous rings around the valves.

heat The sensation produced by fire or a lamp, as opposed to the sensation of cold. The basis of heat is the flow of kinetic energy of atoms, which becomes zero at absolute zero.

heel strike The phase of walking and running in which the calcaneus or heel bone of the foot being put forward contacts the ground.

Heimlich maneuver (HIME-lik) Action that can expel an obstruction from the larynx or trachea, in which a person stands behind the victim, wraps his arms around the victim, and suddenly thrusts the fist into the abdomen between the navel and the rib cage to force air up the trachea and dislodge the obstruction.

helicotrema Opening at the apex of the cochlea through which the scala vestibuli and the scala tympani of the cochlea connect.

hematocrit Percentage of blood volume occupied by erythrocytes.

hematopoiesis (hee-MAT-oh-POY-ee-sis) Development of blood cells, *i.e.,* erythrocytes, leukocytes, and thrombocytes, in the bone marrow.

hematopoietic inductive microenvironment Areas of bone marrow that induce production of blood cells.

hematopoietic tissue Blood-forming tissue.

hematuria (HEE-mat-YOUR-ee-ah) Condition characterized by blood or blood cells in the urine.

heme Oxygen-carrying, red-colored iron porphyrin portion of the hemoglobin molecule.

hemidesmosome Similar to half a desmosome, this junctional complex attaches epithelial cells to the basement membrane, rather than to other epithelial cells.

hemochorial placenta (HE-mo-KORE-ee-al) The type of placenta in humans, in which the chorion is in direct contact with maternal blood.

hemocytoblast Progenitor stem cell from which the different lines of blood cells develop.

hemoglobin Red respiratory protein of erythrocytes that transports oxygen and carbon dioxide.

hemoglobinopathy (HEE-mo-GLOW-bin-OP-ath-ee) A variety of blood disorders characterized by defective hemoglobin molecules.

hemolysis Rupture of red blood cells and release of hemoglobin.

hemolytic anemia Any anemia resulting from abnormal destruction of erythrocytes in the body.

hemophilia Inherited defect in the clotting mechanism that causes hemorrhage.

hemorrhage Loss of blood from a ruptured vessel.

hemorrhagic anemia Anemia resulting directly from loss of blood.

hemostasis (HEE-mo-stay-sis) Arrest of bleeding; interruption of flow of blood to a site of injury.

Henle's loop (HEN-leez) See *nephron loop.*

heparin Anticoagulant that prevents platelet agglutination and thus prevents thrombus formation.

hepatic artery Branch of the aorta that delivers blood to the liver.

hepatic cord Plate of liver cells that radiates away from the central vein of a hepatic lobule.

hepatic duct The left and right hepatic ducts drain bile from the liver and join to form the common hepatic duct.

hepatic lobule The structural and functional unit of the liver; composed of hepatocytes, a central vein, sinusoids, and portal triads.

hepatic portal system System of portal veins that carry blood from the intestines, stomach, spleen, and pancreas to the liver. See *portal system.*

hepatic portal vein Portal vein formed by the superior mesenteric and splenic veins that enters the liver. See *portal system.*

hepatic sinusoid Terminal blood vessel having an irregular and larger caliber than an ordinary capillary within hepatic lobules.

hepatic vein Vein that drains blood from the liver into the inferior vena cava.

hepatocyte (heh-PAT-o-site) Liver cell.

hepatopancreatic ampulla Dilation within the major duodenal papilla that normally receives both the common bile duct and the main pancreatic duct.

hepatopancreatic ampullar sphincter Smooth muscle sphincter of the hepatopancreatic ampulla; sphincter of Oddi.

hernia (HER-nee-uh) Protrusion of any organ or body structure through an abnormal opening in the tissues containing it.

heterozygous A gene is heterozygous when it consists of two different alleles, or alternative forms, of the same gene.

hiatus (HIGH-ate-us) An opening.

high density lipoprotein (HDL) Any plasma lipid-protein complex with a density of 1.06 to 1.21 g per ml.

high endothelial venule (HEV) Small veins in the lymphatic system with high-walled endothelium through which blood lymphocytes migrate.

hilum or hilus Indented surface of many organs, such as lungs and kidney, which serves as a point where nerves and vessels enter or leave.

hindgut Caudal or terminal part of the embryonic gut.

hinge joint A synovial joint that allows back-and-forth movement in one plane, like a door hinge. Also known as ginglymus.

hippocampus S-shaped fold of gray matter involved in short-term memory; located on the floor of the inferior horn of the lateral ventricle in the temporal lobe of the brain.

Hirschprung's disease Absence of peristalsis, accumulation of feces, and enlargement of the colon that results from congenital absence of parasympathetic nerves in the smooth muscle wall of the colon.

histamine Amine released by mast cells and basophils that promotes inflammation.

histologist Scientist who studies the microscopic anatomy of tissues.

histology The science that deals with the microscopic structure of cells, tissues, and organs in relation to their function.

holocrine gland (HOL-oh-krin) Gland whose secretion is formed by the disintegration of entire cells, for example, sebaceous glands. See *apocrine gland* and *merocrine gland.*

homeostasis State of equilibrium or constancy in function and form of the body's metabolism, cells, and tissues.

homologous Alike in structure or origin.

homologous pair of chromosomes Two chromosomes in a diploid cell that are identical in size, shape, and genes. One homolog is received from the mother and one from the father.

homozygous Homozygous genes consist of two identical alleles.

homunculus, *pl.* homunculi Literally, a small man. An image of a man used to show the location of motor and sensory functions from different parts of the body that project onto the cerebral cortex.

horizontal canal Vascular canal in osteons of compact bone that runs perpendicular to the long axis of the bone and to the central canals; connects central canals with each other and with the exterior circulation.

horizontal section A cut made along a transverse plane through the body or a body part.

hormone Substance secreted by endocrine tissues into the blood that acts on a target tissue to produce a specific response.

hormone receptor Protein or glycoprotein molecule of cells that specifically binds to hormones and produces a specific response.

horn Subdivision of gray matter in the spinal cord. The axons of sensory neurons synapse with neurons in the posterior horn, the cell bodies of motor neurons are located in the anterior horn, and the cell bodies of autonomic neurons are found in the lateral horn.

horseshoe kidney A single, anomalous horseshoe-shaped kidney formed by the fusion of the lower portion of two kidneys.

housemaid's knee Inflammation of the prepatellar bursa of the knee caused by excessive kneeling on hard surfaces.

human chorionic gonadotropin (HCG) Hormone produced by the placenta; stimulates secretion of testosterone by the fetus; during the first trimester HCG stimulates ovarian secretion from the corpus luteum of the estrogen and progesterone required for the maintenance of the placenta. In a male fetus, stimulates secretion of testosterone by the fetal testis.

humerus The bone of the arm that articulates with the scapula at the shoulder, and at the elbow with the radius and ulna.

humoral immunity Immunity due to antibodies circulating in the plasma of the blood.

hyaline cartilage Gelatinous, glossy, bluish cartilage tissue consisting of cartilage cells and their matrix; contains chondrocytes, collagen, proteoglycans, and water.

hyaline membrane disease Respiratory distress syndrome. Seen especially in premature neonates; associated with reduced amounts of lung surfactant.

hydrocephalus A condition of excessive fluid within the ventricles of the brain that tends to swell the brain and thin the cerebral cortex.

hydrogen bond Hydrogen atoms bound covalently to either nitrogen or oxygen atoms have a small positive charge that weakly attracts the small negative charges of other oxygen or nitrogen atoms; can occur within a molecule or between different molecules. Influences the shape of protein and polysaccharide molecules.

hydrophilic Polar molecular functional groups that react with, or dissolve readily in water; water-loving.

hydrophobic Polar molecular functional groups that react with or dissolve readily in nonpolar solvents; water-fearing.

hydroxyapatite Mineral with the empiric formula 3 $Ca_3(PO_4)_2$ $Ca(OH)_2$; the main mineral of bone and teeth.

hymen Thin, membranous fold partly occluding the vaginal orifice; normally disrupted by sexual intercourse.

hyoid (HY-oid) U-shaped bone between the mandible and larynx that receives the muscles of swallowing and supports the larynx.

hypercalcemia Abnormally high levels of calcium in the blood.

hyperopia Farsightedness, due to an error in either refraction of the cornea and lens or to shortening of the globe of the eye, in which parallel rays of light are focused behind the retina.

hyperpolarization Increase in the charge difference, or membrane potential, across the cell membrane.

hypertension Persistent high blood pressure; usually defined as systolic pressure greater than 140 mm of mercury and diastolic pressure greater than 90 mm.

hyperthyroidism Abnormally high secretion of thyroid hormone by the thy-

roid gland, resulting in increased metabolic activity and weight loss.

hypertonic Solution that causes cells to shrink; concentration of solutes exceeds that in the cytoplasm of cells or body fluids.

hypertrophy Increase in bulk or size that is not due to an increase in number of cells.

hypoblast The lower of two tissues of the embryonic disk prior to gastrulation. See *epiblast.*

hypocalcemia Abnormally low levels of calcium in the blood.

hypochromic anemia Anemia characterized by a decrease of hemoglobin in erythrocytes.

hypodermis Loose areolar connective tissue, found deep in the dermis, that connects the skin to muscle or bone.

hyponychium (HY-po-NEEK-ee-um) Thickened portion of the stratum corneum of the epidermis under the free edge of the nail.

hypophysis (hy-POF-ih-sis) Endocrine gland attached to the hypothalamus by the infundibulum. Also called the pituitary gland.

hypopolarization Decrease in the electrical charge difference, or membrane potential, across the cell membrane.

hypospadias (HY-po-SPADE-ee-as) Developmental anomaly in the wall of the urethra in which the tube opens on the undersurface of the penis; also, a similar defect in females in which the urethra opens into the vagina. Correctible by surgery.

hypothalamohypophyseal portal system Series of blood vessels that conduct blood from the hypothalamus to the adenohypophysis; vessels originate from capillary beds in the hypothalamus and terminate as a capillary bed in the adenohypophysis. See *portal system.*

hypothalamohypophyseal tract Nerve tract, consisting of the axons of neurosecretory cells, that extends from the hypothalamus into the neurohypophysis. Hormones produced by these cells are transported through the hypothalamohypophyseal tract to the neurohypophysis, where they are stored for later release.

hypothalamus (HY-po-THAL-ah-muss) Important autonomic and neuroendocrine control center inferior to the thalamus.

hypothenar Fleshy mass of tissue on the medial side of the palm; contains

muscles responsible for moving the little finger.

hypothyroidism Abnormally low secretion of thyroid hormone by the thyroid gland that results in lowered metabolic rates and lethargy.

hypotonic Solution that causes cells to swell; concentration of solutes is lower than that of cytoplasm or body fluids.

hypoxia Lower than normal levels of oxygen in the blood or tissues.

I-band Area between the ends of adjacent A-bands within a myofibril; the Z-line divides the I-band into two equal parts.

ichthyosis (IK-thee-OH-sis) Congenital skin disease in which excessive keratinization results in dry epidermis that peels away in scales and sheets.

ileocecal sphincter Thickening of circular smooth muscle between the ileum and the cecum that forms the ileocecal valve.

ileocecal valve (ILL-ee-oh-SEE-kal) Valve formed by the ileocecal sphincter between the ileum and the cecum.

ileum (ILL-ee-um) Third portion of the small intestine, extending from the jejunum to the ileocecal opening into the large intestine.

iliac Pertaining to the ilium.

ilium Lateral, rounded, winglike portion of the os coxa, or hip bone.

immune competent cell B- and T-lymphocytes that are capable of recognizing and removing foreign cells and antigens after being exposed to antigens.

immune complex Antibody bound to an antigen.

immune surveillance Concept that circulating cells of the immune system recognize and remove foreign or malignant cells from virtually all body tissues.

immunity The condition of possessing resistance to infectious disease and harmful substances.

immunization Process by which a patient is rendered immune by deliberately introducing an anti-gen or antibody.

immunodeficiency Failure of some component of the immune system to operate, to recognize cells or substances as foreign, or to remove them.

immunoglobulin Antibody found in the serum portion of plasma.

implantation Attachment of the blastocyst to the endometrium of the uterus; occurs 6 or 7 days after fertilization of the ovum.

impotence Inability to accomplish the male sexual act; caused by psychological or physical factors, or a combination of the two.

incisor The anterior cutting teeth.

incus Middle of the three ossicles in the middle ear. See *malleus* and *stapes.*

indifferent gonad Primitive gonad in the embryo that has not differentiated into an ovary or testis.

infarct (IN-farkt) Area of necrosis resulting from a sudden insufficiency of arterial or venous blood supply, especially in the myocardium of the heart.

infectious mononucleosis Viral infection that causes lymphocytes, many of which resemble monocytes, to increase.

inferior Lower, in reference to the anatomical position.

inferior colliculus The lower of two rounded eminences on the roof of the midbrain; involved with hearing.

inferior vena cava (VEE-nah KAY-vah) Largest vein of the body. Returns blood from the lower limbs and the greater part of the pelvic and abdominal organs to the right atrium.

infertility Inability to produce offspring; does not necessarily indicate sterility, in which gonads produce no functional eggs or sperm.

inflammation See *inflammatory response.*

inflammatory response Complex sequence of events involving chemicals and immune cells that results in the isolation and destruction of antigens and tissues near the antigens.

infundibulum (in-fun-DIB-you-lum) Funnel-shaped structure; the infundibulum attaches the hypophysis to the hypothalamus; the funnel-like expansion of the uterine tube near the ovary.

ingestion Intake of solids or liquids into the body by way of the mouth.

inguinal canal Passage through the lower abdominal wall that transmits the spermatic cord in males and the round ligament in females.

inguinal hernia (ING-gwin-al HER-nee-a) Hernia of the inguinal (groin) region. See *hernia.*

inguinal ligament Fibrous band extending from the anterior superior spine of the ilium to the pubic tubercle.

inhibin Hormone secreted by Sertoli cells of the seminiferous tubules that suppresses the release of follicle-stimulating hormone (FSH).

initial segment The initial segment of an axon, located near the axonal hillock, initiates action potentials.

inner cell mass Group of cells at one side of the blastocyst, part of which forms the body of the embryo.

inner chamber of mitochondrion The innermost compartment of mitochondria.

inner ear Contains the bony and membranous labyrinths, in which reside the sensory organs for hearing and balance.

inner membrane of mitochondrion The second membrane layer of the mitochondrion, folded into many cristae; surrounds inner chamber.

insertion More movable attachment point of a muscle; usually the lateral or distal tendon of a limb muscle.

insula Oval region of the cerebral cortex buried deep in the lateral fissure.

insulin Protein hormone secreted from the pancreas that increases the uptake of glucose and amino acids by most tissues.

insulin-dependent diabetes mellitus (IDDM) Severe diabetes mellitus, characterized by little or no insulin production. See *non–insulin-dependent diabetes mellitus.*

interatrial septum Partition between the atria of the heart.

intercalated disk Cell-to-cell attachment with gap junctions between cardiac muscle cells.

intercalated duct Minute duct of glands such as the salivary gland and the pancreas; leads from the acini to the interlobular ducts.

intercellular Between cells.

intercellular cleft Openings between epithelial cells of blood capillaries and lymph capillaries that allow monocytes and macrophages to pass to or from the tissues.

intercostal Referring to structures between ribs.

intercostal muscle Muscles between the ribs that participate in inhalation and exhalation.

interferon Protein that prevents viral replication.

interleukin Protein produced by macrophages and T-cells that activates T- and B-cells.

interlobular duct Any duct leading from a lobule of a gland.

intermediate filament See *intermediate filament (IF) protein.*

intermediate filament (IF) protein A class of pro-tein filaments measuring 8 to 10 nm in thickness that comprise the cytoskeleton in eukaryotic cells.

intermediate mesoderm In the embryo, a band of mesoderm that lies between the paraxial and lateral plate mesoderm; develops into the pronephros and mesonephros.

intermediate olfactory area Part of the olfactory cortex responsible for modulating olfactory sensations.

intermuscular septum Sheet of fascia that separates muscles.

internal Located or occurring in the interior of a body or organ.

internal anal sphincter Smooth muscle ring located at the upper end of the anal canal.

internal elastic lamina Innermost fenestrated layer of elastic tissue in the wall of an artery.

internal genitalia Internal reproductive organs; in females, the vagina, uterus, uterine tubes, and ovaries; in males, the prostate, seminal vesicles, vas deferens, and testes.

internal naris, *pl.* **nares** Opening from the nasal cavity into the nasopharynx.

internal os The internal entrance of the cervix into the lumen of the uterus.

internal urinary sphincter A sphincter composed of the thickened middle smooth muscle layer of the bladder around the urethral opening.

interneuron Any neuron in a neural pathway between a primary afferent neuron and a motor neuron.

internode Myelinated segment of an axon between two nodes of Ranvier.

interoceptor Sensory nerve ending located in visceral tissues.

interosseous membrane Thin layer of tissue lying between the bones of the forearm and of the leg.

interpapillary peg A downgrowth of epidermis that divides dermal ridges into secondary ridges.

interphase Period of the cell cycle when cell growth and DNA replication occur.

interstitial Space within tissue. Interstitial growth means growth from within.

interstitial cell Cells located between the seminiferous tubules of the testes; secretes testosterone. Also known as Leydig cell.

interventricular foramen Two small passages that connect the third ventricle of the brain with the lateral ventricles.

interventricular groove Shallow sulcus on the anterior and posterior sides of the heart that indicates the location of the interventricular septum.

interventricular septum Partition between the ventricles of the heart.

intervertebral disk Fibrocartilage disk that lies between the vertebrae of the spinal column.

intervillous space Spaces between placental villi that contain maternal blood.

intestinal crypt See *intestinal gland.*

intestinal gland Tubular glands in the mucosa of the small and large intestines.

intestinal villi Fingerlike projections of the inner lining of the small intestine that increase surface area available for absorption.

intracellular Inside a cell; in the cytoplasm.

intrafusal fiber (intra-FUSE-al) Specialized muscle fiber in a muscle spindle.

intramembranous ossification Bone formation occurring within a connective tissue membrane.

intramural plexus Combined submucosal and myenteric plexuses of the digestive tract.

intramuscular septum Connective tissue partitions that divide skeletal muscles into fascicles, and which provide pathways for vessels and nerves to the muscle fibers within the fascicles.

intrinsic factor Factor secreted by the parietal cells of gastric glands; required for adequate absorption of vitamin B$_{12}$.

intrinsic muscle Muscles located within the structure being moved, such as the intrinsic muscles of the tongue, in which vertical, transverse, and longitudinal fibers control the shape of the tongue.

invagination Infolding or inpocketing of an epithelium or organ.

inversion Turning inward, medially, as inversion of the foot.

iodopsin Visual pigment found in the cones of the retina.

ion Atom or group of atoms carrying a charge of electricity by virtue of having gained or lost one or more electrons.

ion channel Pore in the cell membrane through which ions, *e.g.,* sodium and potassium, move.

ion channel protein A variety of membrane transport protein that transports ions through channels in the protein molecule between the cytoplasmic and external sides of the plasma membrane. See *voltage-gated ion channel.*

ipsilateral Structures that are on the same side of the body are ipsilateral. The left upper and lower limbs are ipsilateral. See *contralateral.*

iris Specialized portion of the vascular tunic; the colored portion of the eye that can be seen through the cornea.

iron deficiency anemia Anemia due to dietary lack of iron or to iron loss as a result of chronic bleeding.

irregular connective tissue Connective tissue in which the collagen bundles are oriented in many different directions, instead of preferential orientation, as in regular connective tissue.

ischemia Reduced blood supply to an area of the body.

ischemic (iss-KEM-ik) Relating to ischemia.

ischium (ISS-kee-um) Posterior and inferior portion of the os coxa, or hip bone.

islet of Langerhans See *pancreatic islet.*

isometric contraction Muscle contraction in which the length of the muscle does not change but the tension produced increases.

isotonic contraction Muscle contraction in which the tension produced by the muscle stays the same but the muscle becomes shorter.

isotropic Mostly random organization of proteins in I-bands of skeletal muscle fibers. See *anisotropic.*

isthmus Constriction connecting two larger parts of an organ; the constriction between the body and the cervix of the uterus, or the portion of the uterine tube between the ampulla and the uterus.

jaundice Yellowish color of the integument, sclerae, and other tissues that results from excessive bile pigments in blood.

jejunum Second portion of the small intestine; located between the duodenum and the ileum.

jugular Relating to the throat or neck.

junctional complex Cellular structures that join cells to other cells or to the basement membrane and other structures. See *adhesive plaque, belt*

desmosome, desmosome, gap junction, hemidesmosome, and tight junction.

juxtaglomerular apparatus (JGA) (JUKS-tah-glow-MARE-you-lar) Structure consisting of juxtaglomerular cells of the afferent arteriole and macula densa cells of the distal convoluted tubule, located on the renal corpuscle of nephrons; secretes renin and regulates glomerular filtration in the kidney.

juxtaglomerular cell Modified smooth muscle fibers, located in the afferent arteriole of the glomeruli of kidneys, that secrete renin.

juxtamedullary nephron Nephron located near the junction of the renal cortex and medulla, which extends a long nephron loop into the medulla.

K

Kartagener's syndrome (kar-TAH-jeners) Hereditary condition characterized by inversion of visceral organs, chronic dilation of the bronchi and bronchioles of the lungs, and sinusitis. Related to defective ciliary axonemes.

keratin Fibrous protein complex found in the stratum corneum, hair, and nails; provides protection against abrasion.

keratin filament (KARE-ah-tin) Class of intermediate filaments that form a network in epithelial cells, giving strength to tissues.

keratinization (KARE-ah-tin-ih-ZAY-shun) Production of keratin and differentiation of epithelial cells as they move to the skin surface.

keratinized Referring to a structure that contains keratin, a protein found in skin, hair, and nails.

keratinizing epithelium Keratinizing epithelium becomes dry and tough; epidermis of skin.

keratinocyte (KARE-ah-TIN-oh-site) Epidermal cell that produces keratin.

keratogenous zone (KARE-ah-TOD-jen-us) The region near the base of the hair bulb where epithelial cells synthesize the keratins of the hair shaft.

keratohyalin granule (KARE-ah-toe-HY-ah-lin) Non–membrane-bound protein granules in the cytoplasm of stratum granulosum cells of the epidermis.

keratohyalin matrix The material, derived from keratohyalin granules and fibers, that fills cornified cells of the epidermis.

kidney The organ that excretes urine. The kidneys are two bean-shaped organs, each approximately 11 cm long, 5 cm wide, and 3 cm thick, lying on either side of the spinal column, posterior to the peritoneum, approximately opposite the twelfth thoracic and first three lumbar vertebrae.

kinetochore (kin-NET-oh-kore) Centromere; the location on a chromosome that attaches to the mitotic spindle. See *kinetochore fiber.*

kinetochore fiber Spindle fibers that link the kinetochores of mitotic chromosomes to polar spindle fibers, enabling chromosomes to move toward the poles of the spindle during mitosis.

kinin (KINE-in) Serum protein that causes vasodilation and increases vascular permeability.

kinocilium Motile cilium, in contrast to immobile stereocilia, containing nine peripheral doublet microtubules and two single central microtubules.

knee Genu; the joint of the lower extremity between the thigh and leg, where the femur, tibia, and patella articulate.

kyphosis (ky-FOE-sis) A forward curvature of the vertebral column, especially of the thoracic region. Humpback.

labium majus, *pl.* **labia majora** In female external genitalia, one of two rounded folds of skin surrounding the labia minora and vestibule; homologue of the scrotum in males.

labium minus, *pl.* **labia minora** In female external genitalia, one of two narrow longitudinal folds of mucous membrane enclosed by the labia majora and bounding the vestibule; they unite anteriorly to form the prepuce.

labor Process of expulsion of the fetus and the placenta from the uterus.

labrum Lip or lip-shaped structure.

labyrinth Intricate structure consisting of winding passageways; the bony and membranous labyrinths of the inner ear.

lacis cell (LAKE-iss) A variety of modified smooth muscle cells located in the juxtaglomerular apparatus of the kidney.

lacrimal apparatus Composed of a lacrimal or tear gland in the superolateral corner of the orbit of the eye, and a duct system that extends from the eye to the nasal cavity.

lacrimal duct Canal that carries excess tears away from the eye; located in the

medial canthus and opening on a small projection called the lacrimal papilla.

lacrimal gland Tear gland located in the superolateral corner of the orbit.

lacrimal papilla Small projection of tissue in the medial canthus or corner of the eye; the lacrimal canal opens within the lacrimal papilla.

lacrimal sac Enlargement in the lacrimal canal that leads into the nasolacrimal duct.

lactation The secretion of milk from the breasts after childbirth; also, the period during which milk is secreted and the infant is nursed.

lacteal (LAK-tee-al) Lymphatic vessel in the wall of the small intestine that carries chyle from the intestine and absorbs fat.

lactic acid Three-carbon molecule derived from pyruvic acid as a product of anaerobic respiration. $NADH_2$ (reduced nicotine adenine dinucleotide) reacts with pyruvic acid to form lactic acid and NAD^+ (oxidized nicotine adenine dinucleotide).

lactiferous duct One of fifteen or twenty ducts that drain the lobes of the mammary gland and open onto the surface of the nipple.

lactiferous sinus Dilation of the lactiferous duct just before it enters the nipple.

lactoferrin (LAK-toe-FAIR-in) An iron-binding protein found in neutrophils and in many tissue secretions, such as milk, tears, saliva, and bile.

lactose Disaccharide present in mammalian milk; composed of glucose and galactose.

lacuna, *pl.* **lacunae** (lah-KU-nah, -nee) Small space or cavity; space within the matrix of bone or car-tilage occupied by an osteocyte or chondrocytes; space containing maternal blood within the placenta.

lambdoidal suture (LAM-doid-al) Immobile articulation between the occipital and parietal bones of the skull.

lamella, *pl.* **lamellae** (lah-MELL-ah, -ee) Thin sheet or layer of bone.

lamellar bone (lah-MELL-ar) Type of bone tissue structure found in adult mammals, composed of parallel or concentric plates.

lamellated corpuscle Pacinian corpuscle; oval receptor found in the deep dermis or hypodermis (responsible for transmitting stimuli of deep cutaneous pressure and vibration) and in tendons (responsible for proprioception).

lamellipodia (la-MELL-ih-POE-dee-ah) Spreading extensions of cytoplasm that precede a moving cell.

lamin protein fiber Lattice proteins of the nuclear envelope.

lamina Thin plate; the thinner, transverse portion of the vertebral arches of vertebrae.

lamina densa Dense layer of the basement membrane. Thicker, deeper, denser layer of the basement membrane beneath the lamina rara; binds to underlying connective tissue by way of the lamina reticularis.

lamina propria (LAM-in-ah PRO-pree-ah) Layer of connective tissue underlying the epithelium of a mucous membrane.

lamina rara Literally, light layer. Lighter, thinner, superficial layer of the basement membrane that binds directly to epithelial cells.

lamina reticularis Network layer of the basement membrane. A diffuse network of collagen fibers characterizes this layer beneath basement membranes; the fibers extend into the extracellular matrix of the connective tissue, joining this tissue to the basement membrane.

laminar flow Smooth, parallel flow of layers of a fluid along smooth, parallel paths; in contrast with turbulent flow.

Langerhans cell (LANG-er-hans) Nonpigmented granular dendrocytes of the epithelium. Considered to be antigen-presenting cells; named after Paul Langerhans, nineteenth-century German anatomist.

lanugo (lah-NEW-go) Fine, soft, unpigmented fetal hair.

large intestine Portion of the digestive tract extending from the small intestine to the anus.

laryngitis Inflammation of the mucous membrane of the larynx.

laryngopharynx (lah-RIN-joe-FAHR-inks) That part of the pharynx lying posterior to the larynx.

laryngotracheal groove Depression in the posterior portion of the pharynx in the embryo, which develops into the larynx and the trachea.

larynx (LAHR-inks) The organ that produces sound when air vibrates its vocal cords. Located between the pharynx and the trachea, the larynx consists of a framework of cartilages and elastic membranes housing the vocal folds and the muscles that control the position and tension of these elements.

lateral To the left or right side of the trunk or other body part.

lateral geniculate nucleus Nucleus of the thalamus where fibers from the optic tract terminate.

lateral plate mesoderm A lateral portion of the mesoderm of the embryonic disk.

LDL receptor A protein receptor that binds with low-density lipoproteins.

left colic flexure Curved junction of the transverse and descending colon.

leg That part of the lower limb between the knee and ankle.

lens The transparent spherical structure, located between the iris and the vitreous humor, that focuses light rays upon the retina.

lens accommodation Change in thickness of the lens that focuses an object on the retina; the lens becomes rounder for near objects, flatter for distant objects.

lens fiber The variety of epithelial cells that compose the lens of the eye.

lesser duodenal papilla The opening of the accessory pancreatic duct into the duodenum.

lesser omentum Peritoneal fold that passes from the liver to the lesser curvature of the stomach and to the upper border of the duodenum for a distance of approximately 2 cm beyond the pylorus.

lesser vestibular gland A number of these minute mucous glands open on the surface of the vestibule between the openings of the vagina and urethra.

leukemia Malignant proliferation of abnormal leukocytes in the blood.

leukocyte (LOO-ko-site) White blood cell.

leukocytosis Abnormally large number of leukocytes in the blood; greater than 10,000 per cubic mm.

leukotriene Specific class of physiologically active fatty acid derivatives present in many tissues.

levels of organization Refers to the concept that biological structures reveal themselves at various levels of organization, from molecule to organism.

lever Rigid shaft capable of turning about a fulcrum or pivot point. Bones act as levers in the skeleton, powered by muscles.

Leydig cell See *interstitial cell.*

LH surge Increase in plasma luteinizing hormone (LH) levels occurring before ovulation and responsible for initiating ovulation.

lidocaine Local anesthetic used in treatment of cardiac dysrhythmias.

ligament Band or sheet of dense connective tissue that connects two or more bones, cartilages, or other structures across a joint; a mesentery supporting an abdominal organ.

ligamentum arteriosum (LIG-ah-MEN-tum ar-TEER-ee-OH-sum) Fibrous remains of the ductus arteriosus between the pulmonary trunk and aorta.

ligamentum flavum, *pl.* **ligamenta flava** Paired bands of yellow elastic tissue that connect the laminae of adjacent vertebrae.

ligamentum teres (TEAR-eez) The fibrous remains of the fetal umbilical vein, found on the anterior abdominal wall.

ligamentum venosum (veh-NO-sum) Fibrous cord within the liver that is the remnant of the ductus venosus, the major vein of the fetal liver that received blood from the umbilical vein.

limb Arm, forearm, wrist, and hand; or thigh, leg, ankle, and foot, considered as a whole.

limb bud Protrusion of ectoderm and mesoderm from the body wall of the embryo that develops into a hindlimb or forelimb.

limb placode Thickening in the epithelium of the embryo where the hind and forelimbs develop.

limbic system Parts of the brain involved with emotions and olfaction; includes the cingulate gyrus, hippocampus, habenular nuclei, parts of the basal ganglia, and the hypothalamus (especially the mammillary bodies, the olfactory cortex, and various nerve tracts [e.g., fornix]).

limbus The junction of the cornea with the sclera of the eyeball. The transparent cornea becomes opaque sclera at this point.

linea alba (LIN-ee-ah AL-bah) The white line; a tendinous band running in the midline of the anterior abdominal wall between the xiphoid process and the symphysis pubis, onto which insert the oblique and transverse abdominal muscles.

linea aspera A long, prominent ridge on the posterior surface of the shaft of the femur that receives intermuscular septa and insertions of adductor muscles.

lingual tonsil Collection of lymphoid tissue on the posterior portion of the dorsum of the tongue.

lip Anterior border of the mouth; the pair of muscular folds with an outer mucosa covered by stratified squamous epithelium.

lipase Any fat-splitting enzyme.

lipid Substance composed principally of carbon, oxygen, and hydrogen; contains a lower ratio of oxygen to carbon and is less polar than carbohydrates; generally soluble in nonpolar solvents.

lipid bilayer Double layer of lipid molecules forming the plasma membrane and other cellular membranes.

lipid hypothesis A hypothesis that proposes that atherosclerosis is caused by abnormal lipid metabolism.

lithotripsy (LITH-oh-trip-see) Procedure employing sound waves to pulverize kidney stones.

liver Largest gland of the body, lying in the upper-right quadrant of the abdomen just inferior to the diaphragm; secretes bile and is of great importance in carbohydrate and protein metabolism and in detoxifying chemicals.

liver diverticulum A sac or pouch formed in the endoderm of the midgut of the embryo that gives rise to the liver.

lobe A rounded, projecting part of an organ; the lobe of a lung, the liver, or a gland. Major subdivision of organs such as the lungs and cerebral hemispheres. Lobes usually are separated from each other by prominent fissures or grooves.

lobule Small lobe or a subdivision of a lobe, *e.g.,* a lobule of the lung or a gland; hepatic lobule.

local anesthetic Injectable or topical agent used to produce a loss of sensation in a particular part of the body.

locus Specific place or location of a structure; the locus of a gene on a chromosome.

longitudinal axis The central line passing through the length of the body or a body part.

loop of Henle See *nephron loop.*

loose connective tissue A soft, watery form of connective tissue found in mesenteries and the walls of digestive organs. Also known as areolar connective tissue.

lordosis (lor-DOE-sis) An exaggerated dorsal curvature of the lumbar region of the vertebral column that is sometimes related to loss of vertebral arches.

lovastatin (LOW-va-STA-tin) A cholesterol-lowering drug.

low density lipoprotein (LDL) Lipoproteins in the plasma, with a density of 1.019 to 1.063 gm/ml, that carry cholesterol molecules.

lower respiratory tract The larynx, trachea, and lungs.

lucid layer of epidermis Fourth layer of the epidermis, composed of transparent cells without nuclei.

lumbar plexus Network of nerves formed by branches of the lumbar nerves located in or dorsal to the psoas major muscle.

lumbosacral plexus The interconnections between ventral primary branches of the lumbar, sacral, and coccygeal nerves.

lumen The interior of a tubular structure, such as an intestine or blood vessel.

luminal end The surfaces of epithelial cells that face the lumen of a gland or other internal organ.

lunate Shaped like a half- or crescent moon.

lung bud Outgrowth of the endoderm from the pharynx of the embryo that gives rise to the epithelial lining of the respiratory tract.

lunula (LOON-you-la) White, crescent-shaped portion of the nail matrix visible through the proximal end of the nail.

luteal phase That portion of the menstrual cycle after ovulation, from formation of the corpus luteum to menstruation; usually of 14 days duration; coincides with the secretory phase of the uterine lining.

lutein cell (LOO-tee-in) A variety of cells in the corpus luteum of the ovary that secrete progesterone and estrogen.

luteinizing hormone (LH) (LOO-tee-in-eyes-ing) In females, LH stimulates the final maturation of follicles and secretion of progesterone by them, and conversion of the ruptured follicle into the corpus luteum; in males, LH stimulates the secretion of testosterone by the testes.

luteinizing hormone-releasing hormone (LHRH) See *gonadotropin-releasing hormone.*

luxation Complete dislocation, usually of a bone. See *subluxation.*

lymph Clear, yellowish fluid of lymph vessels that is derived from interstitial fluid.

lymph capillary Beginning of the lymphatic system of vessels; lined with flattened endothelium lacking a basement membrane.

lymph node Encapsulated mass of lymph tissue found among lymph vessels.

lymph nodule Small, sometimes permanent accumulation of lymph tissue that lacks a distinct boundary.

lymph sinus Channels in the lymph node supported by a reticulum of cells and fibers.

lymph vessel One of the system of vessels carrying lymph from the lymph capillaries to the veins.

lymphatic (lim-FAT-ik) Referring to lymph or the lymphatic system. A lymph vessel.

lymphatic capillary See *lymph capillary.*

lymphatic nodule Spherical masses of lymphoid cells.

lymphatic organ Large masses of lymphatic tissue, especially the spleen, tonsils, and thymus.

lymphatic pump Tissue motion and unidirectional valves move lymph fluid through lymph vessels in a manner similar to that of the musculovenous pumps of veins.

lymphatic tissue Network of reticular fibers, lymphocytes, and other cells.

lymphatic trunk Large collecting lymph vessel formed by the union of smaller vessels.

lymphatic valve Structure located in lymph vessels that prevents the backflow of lymph; fundamental component of lymphatic pumps.

lymphoblast Cell that matures into a lymphocyte.

lymphocyte (LIMF-oh-site) Nongranulocytic white blood cell formed in lymphoid tissue.

lymphocytic leukemia Leukemia characterized by enlargement of lymphoid tissue.

lymphoid tissue Lymphocytes and other cells distributed within a reticular framework; found in lymph nodules, nodes, thymus, and spleen.

lymphokine Chemical produced by lymphocytes that activates macrophages, attracts neutrophils, and promotes inflammation.

lymphopoiesis Formation of lymphocytes.

lysis Breaking apart; process by which a cell swells, ruptures, and releases its cytoplasmic contents.

lysosome (LIE-so-some) Membrane-bound vesicle containing hydrolytic enzymes that function as intracellular digestive enzymes.

lysozyme (LIE-so-zime) Enzyme that digests the cell walls of certain bacteria and causes the cells to lyse; present in tears and some other fluids of the body, including digestive fluids.

M-band Line in the center of the H-zone, composed of delicate filaments, that holds the myosin myofilaments in place in the sarcomeres of striated muscle fibers.

M-cell Specialized cells that admit antigens to Peyer's patches in the mucosa of the digestive tract.

macrophage (MAK-row-fayje) Any large mononuclear phagocytic cell.

macula, *pl.* **maculae** (MAK-you-lah, -lee) A patch or plaque of cells. See *maculae densa, lutea, sacculi,* or *utriculi.*

macula densa Cells of the distal convoluted tubule, located at the renal corpuscle, that form part of the juxtaglomerular apparatus.

macula lutea Small yellowish spot in the retina directly behind the lens in which cones are densely packed.

macula sacculi Sensory area on the anterior wall of the saccule, containing hair cells and a gelatinous mass embedded with otoliths that are sensitive to gravity and acceleration.

macula utriculi Sensory area on the lateral wall of the utricle, similar to the macula sacculi, that is sensitive to linear acceleration and gravity.

macular degeneration Partial degeneration of the macula lutea that leads to blindness; common in elderly people.

main prostate gland The prostate gland has three subdivisions consisting in turn of deep mucosal glands and submucosal glands, and the outermost, largest division, containing the main prostate glands.

major calyx, *pl.* **calyces** (KAIL-iks, -ih-sees) Primary branch of the renal pelvis; there are usually two or three in each kidney.

major histocompatibility complex (MHC) Group of genes that control the production of major histocompatibility complex proteins, which are glycoproteins found on the surfaces of cells. The major histocompatibility proteins serve as self-markers for the immune system and are used by antigen-presenting cells to present antigens to lymphocytes.

male pattern alopecia (AL-o-PEE-shya) Genetically determined, androgen-stimulated hair loss occurring primarily in males.

male pronucleus Haploid nucleus, derived from sperm, within the fertilized egg.

malignant tumor A progressive abnormal growth that invades surrounding tissues and organs.

malleus Largest of the three auditory ossicles; attached to the tympanic membrane. See *incus* and *stapes*.

mamma, *pl.* **mammae** Breast. The organ of milk secretion in females; one of two hemispheric projections situated in the subcutaneous fascia over the pectoralis major muscle on either side of the chest; rudimentary in males.

mammary gland The glands of the breast that secrete milk. See *mamma*.

mammary ligament Cooper's ligaments; well-developed ligaments that extend from the overlying skin to the fibrous stroma of the mammary gland and support the breast.

mammillary body Nipple-shaped structures at the base of the hypothalamus.

mammotroph (MAM-oh-trofe) A variety of cell in the adenohypophysis that secretes prolactin.

mandible (MAN-dih-ble) The bone of the lower jaw that anchors the lower teeth.

manipulative motion Muscular motions, performed by distal muscles of the limbs, that maneuver the hands and feet. See *positioning motion*.

manubrium Part of a bone resembling a handle; the manubrium of the sternum represents the handle of a sword.

manus The hand.

Marfan's syndrome A hereditary disorder of connective tissue characterized by myopia, scoliosis, disproportionately long extremities, and heart defects.

marrow Soft tissue in the center of bones that produces erythrocytes and leukocytes, and stores adipose tissue.

masseter Facial muscle originating on the zygomatic arch and inserting on the mandible; elevates the mandible in chewing.

mast cell Connective tissue cell that contains basic staining granules; promotes inflammation.

mastication Process of chewing.

mastoid (MASS-toid) Resembling a breast; the mastoid process of the temporal bone is a tuberosity that receives the sternocleidomastoid muscle.

mastoid air cell Spaces within the mastoid process of the temporal bone, connected to the middle ear by ducts.

matrix Noncellular substance surrounding the cells of connective tissue.

matrix cell Basal cells of sebaceous glands; or cells of the hair matrix.

maxilla (max-ILL-ah) The bone of the upper jaw.

meatus, *pl.* **meati** (me-AY-tus, -tee) Passageway or tunnel; curved space beneath the nasal conchae of the nasal cavity. See *external auditory meatus*.

mechanical advantage The relative advantage in force and motion that a bone or lever has, depending on the distance between the force and fulcrum and between the fulcrum and load. Levers have high mechanical advantage when a small force moves a large object.

mechanoreceptor A sensory receptor that responds to mechanical pressure. Examples are pressure receptors in the carotid sinus or touch receptors in the skin. See *lamellated corpuscle*.

meconium (meh-KONE-ee-um) First intestinal discharges of the newborn infant, greenish in color and consisting of epithelial cells, mucus, and bile.

medial Toward the midline of the trunk or other body part.

median The middle of the trunk or other body part.

median plane The sagittal plane that passes through the middle of the trunk or other structure.

median raphe Fibrous line that marks the fusion of body folds in the floor of the pelvis and other body parts.

mediastinum (MEE-dee-ah-STY-num) The middle partition of the thorax that contains the heart, great vessels, trachea, and esophagus.

mediastinum testis Fibrous tissue continuous with the tunica albuginea that projects into the testis near the posterior border.

medulla Central, inner part of an organ or structure, as in the ovary, testis, kidney, brain, and bones.

medulla oblongata Inferior portion of the brainstem that connects the spinal cord to the brain and contains autonomic centers controlling such functions as heart rate, respiration, and swallowing.

medullary (MED-you-LAIR-ee) Referring to the medulla of organs.

medullary cavity Large, marrow-filled cavity in the diaphysis of a long bone.

medullary pyramid Pyramidal masses of the kidney that project into the renal pelvis, and that contain collecting and papillary ducts. Wedge-shaped tracts of the medulla oblongata lying on either side of the anterior median sulcus.

medullary ray Extension of the nephrons into the medulla of the kidney, consisting of collecting ducts and nephron loops (of Henle).

medullary sinus Channel carrying lymph through the lymph node toward the efferent lymphatic vessel.

megacolon Abnormal enlargement of the colon. See *Hirschsprung's disease*.

megakaryoblast A variety of proliferating cell that gives rise to megakaryocytes.

megakaryocyte (MEG-a-KAR-ee-oh-site) Large nucleated cell that produces blood platelets.

meibomian cyst See *chalazion*.

meibomian gland Sebaceous gland near the inner margins of the eyelid; secretes a thin film of sebum that lubricates the eyelid and retains tears. See *tarsal gland*.

meiosis Process of cell division that results in the formation of gametes. Consists of two divisions of a parental cell that results in one egg or four sperm, each of which contains one half the number of chromosomes of the parent cell.

Meissner's corpuscle (MICE-nerz) Named for George Meissner, nineteenth-century German histologist. See *tactile corpuscle*.

melanin (MEL-ah-nin) Brown, black, or yellow pigment that is responsible for skin and hair color.

melanocyte (MEL-ah-no-site) Cell found mainly in the stratum basale of the epidermis that produces melanin pigment.

melanocyte-stimulating hormone (MSH) Peptide hormone secreted by the adenohypophysis; increases melanin production by melanocytes, making the skin darker in color.

melanoma (MEL-ah-NO-ma) A tumor arising from melanocytes.

melanosome Membranous organelle containing the pigment melanin.

melatonin (MEL-ah-TONE-in) Hormone secreted by the pineal body; inhibits secretion of gonadotropin-releasing hormone from the hypothalamus.

membrane-bound receptor Receptor molecule, such as a hormone receptor protein, that is bound to the cell membrane of the target cell.

membrane-coating granule Membrane-bound granule with a diameter of 100 to 500 nm, located in the epidermis and other keratinized stratified squamous epithelia.

membrane receptor protein See *membrane-bound receptor.*

membranous labyrinth Membranous structure within the inner ear, consisting of the cochlea, vestibule, and semicircular canals.

membranous urethra Portion of the male urethra, approximately 1 cm in length, extending from the prostate gland to the beginning of the penile urethra.

memory cell Small lymphocytes derived from B-cells or T-cells that rapidly respond to a subsequent exposure to the same antigen.

menarche Establishment of menstrual function; the time of the first menstrual period.

meningeal dura Deep layer of the dura mater associated with meninges, in contrast with the more superficial periosteal dura that is associated with bone.

meninx, *pl.* meninges Connective tissue membranes surrounding the brain. See *dura mater, arachnoid,* and *pia mater.*

meniscus, *pl.* menisci (men-ISS-kus, -kee) Crescent-shaped extension of fibrocartilage from the synovial capsule between the articulating surfaces of bones in a synovial joint.

menopause Permanent cessation of the menstrual cycle.

menses Periodic hemorrhage from the uterine mucous membrane, occurring at approximately 28-day intervals.

menstrual cycle Series of changes that occur in sex-ually mature, nonpregnant women and result in menses. Specifically refers to the uterine cycle but is often used to include both the uterine and ovarian cycles.

menstruation (MEN-strew-AYE-shun) Periodic shedding of blood and tissue from the endometrium of the uterus.

Merkel cell Named for Friedrich Merkel, nineteenth-century German anatomist. See *tactile disk.*

Merkel's disk See *tactile disk.*

merocrine gland Gland that secretes products without loss of cellular cytoplasm; an example is water-producing sweat glands. See *apocrine gland* and *holocrine gland.*

mesangial cell (mezz-AN-jee-al) Phagocytic cell found in the glomerular filtration barrier between the capillaries of the renal glomerulus.

mesaxon The plasma membrane of Schwann cells that grow concentrically, enfolding an axon.

mesencephalon Midbrain in both the embryo and adult; consists of the cerebral peduncle and the corpora quadrigemini.

mesenchyme Loose embryonic connective tissue consisting of mesenchymal cells, supported in a loose, fluid, homogeneous ground substance.

mesenchyme cell Undifferentiated, stellate mesodermal cell of the early embryo. Derived from mesoderm or neural crest.

mesentery (MEZ-en-tair-ee) Double layer of peritoneum that supports abdominal organs by extending from the abdominal wall to the abdominal viscera, and by conveying vessels and nerves.

mesoderm Middle of the three germ layers of an embryo. Gives rise to connective tissues, skeleton, muscle, vascular system, kidneys, and gonads.

mesonephric duct A pair of ducts in the embryo that drain the mesonephric tubules; becomes the vas deferens in males; remains vestigial in females.

mesonephros (MEZ-oh-NEF-ross) One of three excretory organs appearing during embryonic development; forms caudal to the pronephros as the pronephros disappears. It is well developed and functions before the metanephros gives rise to the kidney; the mesonephros regresses, but males retain its duct system as the efferent ductules and epididymis.

mesosalpinx (mez-oh-SAL-pinks) Part of the broad ligament supporting the uterine tube.

mesothelium (mez-oh-THEEL-ee-um) The squamous epithelium that lines serous cavities, such as the pleural and peritoneal cavities.

mesovarium Short peritoneal fold connecting the ovary with the broad ligament of the uterus.

messenger RNA (mRNA) Type of RNA that moves out of the nucleus and into the cytoplasm, where it serves as a template that specifies the structure of proteins.

metabolic rate Total amount of energy produced and used by the body per unit of time.

metabolism Sum of all the chemical reactions that take place in the body, consisting of anabolism and catabolism. Cellular metabolism refers to the chemical reactions within cells.

metacarpal The long bones of the hand between the carpus and the phalanges.

metanephros (MET-ah-NEF-ross) Metanephric kidney. Most caudally located of the three excretory organs appearing during embryonic development; becomes the permanent kidney of mammals. It develops caudal to the mesonephros as the mesonephros regresses.

metaphase Phase of mitosis when the chromosomes line up along the equator of the spindle.

metaphase plate The equator of the mitotic spindle where chromosomes line up during metaphase.

metarteriole Small, peripheral blood vessels that contain scattered smooth muscle fibers in their walls; located between the arterioles and the true capillaries.

metastasis (meh-TAS-ta-sis) Movement of tumor cells from one location in the body to another.

metastasize (meh-TAS-ta-size) To accomplish metastasis.

metatarsal Five long bones of the foot, located between the tarsus and phalanges.

metencephalon Second-most posterior division of the embryonic brain; becomes the pons and cerebellum in the adult.

metopic suture (meh-TOPE-ik) A suture, or immovable joint, that is sometimes discernible down the midline of the frontal bone of the skull. Ossifies prematurely.

micelle Droplets of lipid surrounded by bile salts in the small intestine.

microfilament Small fibrillar networks in the cytoskeleton of cells; provide structure and motion to the cytoplasm and mechanical support for microvilli and stereocilia.

microglia Small neuroglial cells that become phagocytic and mobile in response to inflammation; considered to be macrophages within the central nervous system.

microtubule Hollow molecular tube composed of tubulin, measuring approximately 25 nm in diameter and usually several micrometers long. Supports the cytoskeleton of the cell and is a component of centrioles, spindle fibers, cilia, and flagella.

microvillus, pl. microvilli Minute projections on the luminal surface of the cell membrane that greatly increase the area available for membrane transport.

micturition (MIK-tshur-IH-shun) Urination.

micturition reflex Contraction of the bladder stimulated by stretching of the bladder wall; empties the bladder.

middle ear Air-filled space, containing the ossicles within the temporal bone, between the external and internal ear.

middle phalanx Phalanges of the fingers and toes found between the proximal phalanx and the distal phalanx of each finger or toe, except the thumb and great toe.

midgut The middle segment of the embryonic intestine that gives rise to the stomach, small intestine, and colon.

midpiece of sperm Region of the sperm between the head and the tail.

midsagittal Sagittal plane dividing the body or any part of the body into equal left and right halves; median plane.

milk letdown Expulsion of milk from the alveoli of the mammary glands; stimulated by oxytocin.

mineral Inorganic nutrient necessary for normal metabolic functions.

mineralocorticoid Steroid hormone (*e.g.*, aldo-sterone) produced by the zona glomerulosa of the adrenal cortex; facilitates exchange of potassium for sodium in the distal convoluted tubule; causes sodium reabsorption and potassium and hydrogen ion secretion.

minor calyx, pl. calyces (KAIL-iks, -ih-sees) Branches of the major calyx that receive urine from the renal papillae in kidneys; the total number varies from seven to thirteen in each kidney.

miosis (my-OH-sis) Excessive constriction of the pupil.

mitochondrial DNA Circular DNA molecules in the mitochondria of eukaryotic cells, which encode several mitochondrial enzymes.

mitochondrion, pl. mitochondria (my-toe-CON-dree-on, -ah) Small, spherical, rod-shaped or thin filamentous structure in the cytoplasm of cells that is a site of ATP production.

mitosis Cell division resulting in two daughter cells with exactly the same number and type of chromosomes as the mother cell.

mitotic spindle Microtubules that radiate from the centrioles and attach to the centromeres of chromosomes during metaphase of cell division.

mitral valve (MY-tral) Bicuspid valve; valve found between the left atrium and the left ventricle of the heart; left atrioventricular valve.

modiolus Central core of spongy bone about which turns the spiral canal of the cochlea.

molar Tricuspid tooth; the three posterior teeth of each dental arch.

mole Aggregation of melanocytes in the epidermis or dermis of the skin.

molecule Two or more atoms of the same or different elements joined by a chemical bond.

monocyte Nongranulocytic, relatively large mononuclear leukocyte normally found in lymph nodes, spleen, bone marrow, and loose connective tissue.

mononuclear phagocytic system Phagocytic cells, each with a single nucleus; derived from monocytes.

monosaccharide Simple sugar carbohydrate that cannot form any simpler sugar by hydrolysis. Glucose.

mons pubis Fatty prominence over the symphysis pubis in females.

morphogenesis (MORF-oh-JEN-eh-sis) Development of the shape and structure of a tissue, an organ, or an organism.

morula (MOR-you-lah) Embryonic mass of twelve or more cells produced by the early cleavage divisions of the zygote.

motor endplate Synaptic connection between a motor neuron and a striated muscle fiber.

motor neuron Neuron that innervates skeletal, smooth, or cardiac muscle fibers.

motor nuclei A collection of motor neuron cell bodies in the central nervous system that give origin to efferent axons of peripheral nerves.

motor unit A single neuron and the muscle fibers it innervates.

mucocutaneous ending A class of general sensory receptor (Krause's end-bulbs) that is located beneath the papillary layer of dermis in the skin at the anal, vaginal, oral, and nasal orifices.

mucosa The layer of connective tissue and epithelium that lines internal surfaces of the respiratory, urinary, and digestive systems. Mucous membranes are mucosa.

mucosal gland See *mucous gland*.

mucous gland Small glands of the mucosa and submucosa that secrete mucus.

mucous membrane Thin sheet consisting of epithelium and connective tissue (lamina propria) that lines cavities which open to the outside of the body; many contain mucous glands that secrete mucus.

mucous neck cell A variety of mucus-secreting cells in the neck of a gastric gland.

mucus Viscous secretion produced by and covering mucous membranes; lubricates mucous membranes and traps foreign substances.

Muller cell Fibers of the retina that run through the thickness of the retina from the internal limiting membrane to the bases of the rods and cones.

Muller's muscle Rudimentary smooth muscle of the eyelid.

Mullerian duct A pair of embryonic ducts that give rise to the vagina, uterus, and uterine tubes in females, and which are rudimentary in males.

Mullerian inhibiting factor Substance that inhibits the development of the Mullerian ducts in males and promotes the development of the vas deferens.

multicellular gland A gland composed of many cells, in contrast with unicellular glands such as goblet cells. Salivary glands, pancreas, and gastric glands are multicellular. See *gland*.

multipennate muscle A muscle in which fiber bundles converge on many tendons. See *bipennate muscle*.

multiple sclerosis (MS) Chronic neurological disease in which the myelin sheath surrounding the axons is removed, inhibiting conduction of electrical impulses.

multipolar neuron One of three anatomical categories of neurons, consisting of a cell body, an axon, and two or more dendrites.

mumps Inflammation of the parotid gland.

murmur Soft sound heard on auscultation of the heart, lung, or blood vessels.

muscarine Alkaloid compound that is found in certain mushrooms; binds to and activates muscarinic receptors.

muscarinic receptor Class of cholinergic receptor in the autonomic nervous system that is specifically activated by muscarine or acetylcholine.

muscle Muscle tissue is one of the four primary types of tissues, and is characterized by its contractile abilities.

muscle compartments Skeletal muscles group around the skeleton of the limbs in flexor, extensor, abductor, and adductor compartments that bring about these motions because they are located on the anterior or posterior, or medial or lateral sides of the limb, respectively.

muscle fascicle Bundle of muscle fibers surrounded by perimysium within a muscle.

muscle fiber Muscle cell.

muscle group A group of skeletal muscles concerned with a particular type of motion. The forearm flexors draw the hand and wrist toward the body.

muscle spindle A structure composed of three to ten specialized muscle fibers supplied by gamma motor neurons and wrapped in sensory nerve endings; detects stretch of the muscle and is involved in maintaining muscle tone. See *annulospiral ending, flower spray ending, nuclear bag fiber,* and *nuclear chain fiber.*

muscle tone Relatively constant tension produced by a muscle for long periods of time as a result of asynchronous contraction of motor units. The resilient property of muscle tissue that enables it to contract.

muscle twitch Contraction of a whole muscle in response to a stimulus that causes an action potential in one or more muscle fibers.

muscular artery An artery with a tunica media composed of concentrically arranged smooth muscle fibers.

muscular fatigue Fatigue due to a depletion of ATP within the muscle fibers.

muscular tissue One of the four primary tissue types; characterized by its contractile abilities.

muscularis Muscular coat of a hollow organ or tubular structure.

muscularis externa The visceral muscle fiber layer of the digestive tube. One of the four primary layers of the wall of the digestive tract.

muscularis mucosae (MUS-kue-LAR-iss mew-KOE-see) Thin layer of smooth muscle found in most parts of the digestive tube; located between the lamina propria and the submucosa.

musculi pectinati Prominent ridges of atrial myocardium located on the inner surface of the right atrium and both auricles.

musculovenous pump The combination of venous valves and motions of skeletal muscles that pumps blood through the veins.

myasthenia gravis (MY-as-THEE-nee-ah GRAV-is) Muscular weakness and atrophy caused by destruction of acetylcholine receptors in motor endplates.

myelencephalon Most caudal portion of the embryonic brain; becomes the medulla oblongata.

myelin Lipoprotein surrounding the axons of some nerve fibers.

myelin sheath Envelope surrounding most axons; formed by Schwann cell membranes wrapped around the axon.

myelinated axon Axon having a myelin sheath.

myeloblast Stem cell from which granulocytes develop.

myelocytic leukemia A form of granulocytic leukemia characterized by excessive myeloblasts in the blood and organs.

myeloid tissue (MY-eh-loyd) Bone marrow consisting of developing erythrocytes, granulocytes, and megakarocytes in a network of reticular cells and fibers.

myeloperoxidase (MY-eh-lo-per-OX-ih-days) A peroxidase, found in phagocytic cells, that can oxidize halogens.

myenteric ganglia An aggregation of nerve cells in the muscular coat of the intestine.

myenteric plexus Plexus of unmyelinated fibers and postganglionic autonomic cell bodies lying in the muscular coat of the esophagus, stomach, and intestine; communicates with the submucosal plexuses.

myoblast (MY-oh-blast) Undifferentiated, dividing cell that can develop into a skeletal muscle fiber.

myocardial infarction Sudden necrosis of cardiac muscle due to an infarct. See *infarct.*

myocardium Middle layer of the heart, consisting of cardiac muscle.

myoepithelial cell (MY-o-ep-ih-THEEL-ee-al) Cell of the muscle epithelium located around sweat, mammary, lacrimal, and salivary glands; thought to squeeze secretions from these glands.

myofibril (MY-oh-FYE-bril) Fine longitudinal fibril within a muscle fiber; composed of thick and thin myofilaments in sarcomeres.

myofibroblast A fibroblast with contractile characteristics of a smooth muscle fiber.

myofilament Extremely fine molecular fiber of the myofibrils of muscle; thick myofilaments consist of myosin, and thin myofilaments, of actin.

myoid cell Cell of the inner part of the inner segment of rods and cones.

myometrium Muscular wall of the uterus; composed of smooth muscle.

myopia Nearsightedness; condition due to an error in either refraction of lens or cornea, or to elongation of the globe of the eye in which light rays are focused in front of the retina. See *hyperopia.*

myosin (MY-oh-sin) A globulin present in muscle tissue; binding of myosin with actin produces muscle contraction.

myosin myofilament Thick myofilament of muscle fibrils; composed of myosin molecules.

myositis Inflammation of a muscle.

myositis ossificans (MY-ose-EYE-tiss OSS-ih-fih-KANS) Ossifying muscle inflammation.

myotome (MY-oh-tome) Muscle plate of an embryo that develops into skeletal muscle; also a muscle group innervated by a single spinal nerve.

N

Na+/K+ ATPase Sodium-potassium ATPase, the sodium-potassium pump. The membrane enzyme that transports sodium and potassium ions in opposite directions across the plasma mebrane, using energy derived by split-ting ATP. Establishes resting membrane potential of cells, especially neurons.

nail Fingernails and toenails. Consist of several layers of dead epithelial cells containing hard keratin that form a tough plate on the ends of the digits.

nail bed The epidermis beneath the nails.

nail groove The fold of epidermis, from which fingernails and toenails grow, that extends into the dermis at the base of the nail.

nail matrix Portion of the nail groove from which the cells of the nail form.

nail plate The portion of the nail that grows out of the nail groove. Rests upon the nail bed.

naris, *pl.* **nares** (NAIR-iss, -eez) See *internal* and *external naris.*

nasal bone (NAY-zal) The pair of small bones that forms the ridge of the nose.

nasal cartilage The hyaline cartilage that supports and shapes the tip of the nose.

nasal cavity Cavity between the external nares and the nasopharynx. The nasal cavity is divided into two chambers by the nasal septum and is bounded inferiorly by the hard and soft palates.

nasal concha, *pl.* **conchae** (KON-ka, -kee) Three folds of bone and mucosa that extend into the nasal cavity from the lateral wall, increasing the surface area for conditioning air. Consist of superior, middle, and inferior nasal conchae.

nasal septum Bony partition that separates the nasal cavity into left and right portions; composed of the vomer, the perpendicular plate of the ethmoid bone, and hyaline cartilage.

nasolacrimal duct Duct that leads from the lacrimal sac to the nasal cavity.

nasopharynx Part of the pharynx that lies above the soft palate; anteriorly, it opens into the nasal cavity, and posteriorly, to the oropharynx.

navicular (nah-VIK-yoo-lar) The tarsal bone that articulates with the talus, the cuneiform bones, and the cuboid bone of the foot.

nearsightedness See *myopia.*

neck (tooth) Slightly constricted part of a tooth, between the crown and the root.

neck mucus cells Mucus-secreting cells located in the neck of gastric glands of the stomach.

necrosis Death of cells or tissues.

necrotic Pertaining to or affected by necrosis. Dying tissues or cells.

negative feedback Mechanism by which any deviation from a normal value is resisted or minimized.

neoplasm Abnormal growth; benign or malignant tumor.

nephron Functional unit of the kidney, consisting of the renal corpuscle, the proximal convoluted tubule, the nephron loop (loop of Henle), and the distal convoluted tubule.

nephron loop U-shaped portion of the nephron between the proximal and distal convoluted tubules; consists of ascending and descending limbs. In juxtamedullary nephrons, extends into medullary pyramids. Also known as loop of Henle, or Henle's loop.

nephrotic syndrome (nef-ROT-ik) A form of renal failure in nephrons in which the glomerular filtration barrier permits large quantities of proteins to enter the urine.

nerve Bundle of nerve fibers and nerve fascicles accompanied by connective tissue, and located outside of the central nervous system.

nerve ending A general term that refers to ends of neurons and the structures with which they associate. Includes sensory receptors, motor endplates, and axon terminals.

nerve fiber An axon and its surrounding sheath of Schwann cells and myelin.

nervous tissue One of the four primary types of tissue; characterized by its ability to conduct action potentials.

neural (NOOR-al) Relating to any structure composed of nerve cells.

neural canal The lumen enclosed by the neural tube. Becomes the central, or spinal, canal of the spinal cord.

neural crest Strip of tissue found at the edge of the neural plate as it rises to meet at the midline to form the neural tube.

neural crest cell Cells derived from the margins of the neural plate in the embryo; together with mesoderm, neural crest cells form the mesenchyme of the embryo; neural crest cells give rise to part of the skull, the teeth, melanocytes, sensory neurons, and autonomic neurons.

neural fold The lateral portions of the neural plate that grow upward and medially, and fuse together as the neural tube.

neural groove The slot or depression between the neural folds that becomes the neural canal after the neural folds have closed and formed the neural tube.

neural plate Region of the dorsal surface of the embryo that is transformed into the neural tube.

neural retina The light sensitive layer of the retina, containing rods and cones, in contrast with the pigmented retina.

neural tube Tube formed from the neuroectoderm by the closure of the neural groove. Develops into the spinal cord and brain.

neurilemma The sheath of Schwann cells that covers axons in peripheral nerves.

neurilemmacyte Schwann cells and oligodendrocytes, the neuroglial cells that myelinate axons in the peripheral

and central nervous systems, respectively.

neuroectoderm That portion of the ectoderm of an embryo that gives rise to the brain and spinal cord.

neuroendocrine cell Neurons that secrete directly into capillaries, like cells of endocrine glands.

neuroendocrine system (NOOR-oh-END-oh-krin) A term that recognizes the relationships between the cells of the nervous and endocrine systems, and consolidates them into one larger system.

neurofilament Intermediate filaments, elements of the cytoskeleton of axons.

neuroglia (noor-OH-glee-ah) Cells in the nervous system other than the neurons; includes astrocytes, ependymal cells, microglia, oligodendrocytes, satellite cells, and Schwann cells.

neurohypophysis (NOOR-oh-hy-POF-ih-sis) Portion of the hypophysis derived from the brain; commonly called the posterior pituitary. Major secretions include antidiuretic hormone and oxytocin.

neuromuscular junction Specialized synapse between a motor neuron and a muscle fiber. See *motor endplate.*

neuron (NOOR-on) The conductive cells of the nervous system are neurons. In a typical neuron, dendrites and one axon extend from the central cell body that contains the cell nucleus. Morphologic and functional unit of the nervous system.

neurotransmitter Specific substances, such as acetylcholine, released by axon terminals, that diffuse across synapses onto receptors on the membranes of dendrites that stimulate or inhibit responses in the postsynaptic neuron.

neutropenia (NEW-tro-PEE-nee-ah) Low quantities of neutrophils in the blood.

neutrophil (NEW-tro-fil) Type of white blood cell; small phagocytic white blood cell with a lobed nucleus and small granules in the cytoplasm.

neutrophilia Accumulation of neutrophils in the blood.

nexin link Links involving the protein nexin that join doublet microtubules to each other in axonemes of cilia and flagella, and enable motion.

nicotine Alkaloid compound found in tobacco that binds to and activates nicotinic receptors.

nicotinic receptor Class of cholinergic receptor molecule that is specifically activated by nicotine and by acetylcholine.

9 + 2 arrangement of microtubules The characteristic arrangement of microtubules in axonemes of cilia and flagella. Consists of nine outer doublet microtubules and two separate single cen-tral microtubules. See *axoneme.*

nipple Projection at the apex of the breast, on the surface of which the lactiferous ducts open; surrounded by a circular pigmented area, the areola.

Nissl body The rough endoplasmic reticulum found in cell bodies of neurons.

nitric oxide Chemical symbol NO (nitrogen and oxygen); this gas is a vasodilator that allows blood into the cavernous sinuses to erect the penis. Among an increasing array of functions, NO may also have a role in memory by enhancing neurotransmission at synapses.

nociceptor A sensory receptor that detects painful or injurious stimuli.

node of Ranvier Short gap in the myelin sheath of a nerve fiber between adjacent Schwann cells.

noise-induced hearing loss (NIHL) Deafness arising from chronic exposure to loud noise.

non–insulin-dependent diabetes mellitus (NIDDM) The form of diabetes mellitus in which the tissues become resistant to the action of insulin, in contrast with insulin-dependent diabetes, in which the pancreas produces insufficient insulin.

nonkeratinizing epithelia Nonkeratinizing epithelia remain moist and soft, as in the lining of the oral cavity.

nonpigmented granular dendrocyte Also known as Langerhans cells, these epidermal cells are considered to present antibodies to the immune system.

norepinephrine Neurotransmitter substance released from most of the postganglionic neurons of the sympathetic division; hormone released from the adrenal cortex that increases cardiac output and blood glucose levels.

nose Visible structure that forms a prominent feature of the face; can also refer to the nasal cavity.

notochord (NO-toe-cord) Small axial rod of tissue lying ventral to the neural tube in embryos. A characteristic of all vertebrates, in humans it becomes the nucleus pulposus of the intervertebral disks.

notochord process The strip of mesoderm cells that advances cranially during gastrulation. Forms the notochord and becomes incorporated into the vertebral column.

nuchal Pertaining to the nape of the neck.

nuclear bag fiber Modified muscle fibers of muscle spindles in which nuclei cluster in the center of the cell.

nuclear chain fiber A modified muscle fiber in muscle spindles that contains strings of nuclei in the center of the cell.

nuclear envelope Double membrane structure surrounding and enclosing the nucleus.

nuclear pore Porelike openings in the nuclear envelope where the inner and outer membranes fuse.

nucleic acid DNA and RNA, a family of nucleotide polymers that stores genes of cells, controls protein synthesis, and governs the phenotypes of cells.

nucleolus, *pl.* **nucleoli** Rounded, dense, well-defined nuclear body with no surrounding membrane; assembles ribosomes.

nucleotide Basic building block of nucleic acids, consisting of a sugar (either ribose or deoxyribose) and one of several types of organic bases (adenine, cytosine, guanine, and thymine). The sequence of nucleotides in DNA and RNA encodes genetic information.

nucleus, *pl.* **nuclei** Cell organelle containing most of the genetic material of the cell; collection of nerve cell bodies within the central nervous system; center of an atom consisting of protons and neutrons.

nucleus pulposus Soft central portion of intervertebral disks.

nutrient Chemicals taken into the body that are used to produce energy, provide building blocks for new molecules, or function in other chemical reactions.

nutrient foramina Foramina on the surfaces of bones through which vessels communicate with the interior of the bone.

nutrition Process by which nutrients are obtained and used by the body.

nystagmus Rhythmical oscillation of the eyeballs as they focus on objects.

oblique Structure or motion oriented at an angle to the perpendicular or horizontal.

obturator (OB-tyur-AY-tor) A structure that closes an opening. The obturator foramen of the os coxa is blocked by muscle and membrane.

occipital (ok-SIP-ih-tal) Inferior, posterior portion of the head. The back of the skull.

ocular dominance column Alternating strips of visual cortex that receive axons from the retina of the left or right eye.

oderant molecule Molecules that bind to oderant receptors in the nasal cavity and stimulate responses that the brain interprets as odors.

oderant receptor Receptors of the olfactory pathway that detect odor.

odontoblast (oh-DON-toe-blast) A variety of cells that produce dentine of the teeth.

olecranon fossa A depression on the posterior distal surface of the humerus into which the olecranon process of the ulna fits when the forearm extends.

olecranon process The process on the proximal end of the ulna that projects past the trochlear notch.

olfaction Sense of smell; the process that senses smell.

olfactory bulb Ganglionlike enlargement of the rostral end of the olfactory tract that lies over the cribriform plate; receives the olfactory nerves from the nasal cavity.

olfactory cortex Termination of the olfactory tract in the cerebral cortex within the lateral fissure of the cerebrum.

olfactory epithelium Epithelium of the olfactory recess, containing olfactory receptors.

olfactory receptor Sensory receptors that respond to chemicals suspended in air. Located in the olfactory recess of the nasal cavity. Receives stimuli linked to the sense of smell.

olfactory recess Extreme superior region of the nasal cavity that contains olfactory epithelium.

olfactory tract Nerve tract that projects from the olfactory bulb to the olfactory cortex.

oligodendrocyte Neuroglial cell whose cytoplasmic extensions form myelin sheaths around axons in the central nervous system.

oligonephronic hypoplasia A condition in which kidneys have insufficient quantites of nephrons. The existing nephrons enlarge in compensation for their reduced numbers.

omental bursa The pouch enclosed by the folds of the greater omentum in the abdominal cavity.

omentum Fold of peritoneum passing from the stomach to another organ.

oncogene Numerous genes regulate normal development and differentiation of cells and tissues. When such genes mutate and function abnormally, those that cause cancer are known as oncogenes.

oncology Study and treatment of cancer.

oocyte Diploid, immature ovum.

oogenesis Formation and development of a secondary oocyte or ovum.

oogonium, *pl.* **oogonia** Stem cell from which oocytes are derived.

ootid (OH-oh-tid) Mature ovum, having a haploid nucleus.

opioid Synthetic opiumlike narcotics.

opposition Movement of the thumb and little finger toward each other; movement of the thumb toward any of the fingers.

opsin Protein portion of the rhodopsin molecule. A class of proteins that bind to the substance retinal to form the visual pigments of the rods and cones of the eye.

optic Related to vision.

optic chiasma Point of crossing of optic nerve fibers in the optic neural pathway.

optic cup Embryonic outgrowth from the brain that is the primordium of the eye and becomes the retina.

optic disk Point in the retina at which axons of ganglion cells of the retina converge to form the optic nerve, which then penetrates through the fibrous tunic of the eyeball. See *blind spot.*

optic nerve Nerve carrying visual signals from the eye to the optic chiasma.

optic stalk Constricted proximal portion of the optic vesicle in the embryo; develops into the optic nerve.

optic tract Nerve tract that extends from the optic chiasma to the lateral geniculate nucleus of the thalamus.

optic vesicle Paired evaginations from the walls of the embryonic forebrain, from which the optic cup develops.

ora serrata The pouchlike projections of the retina at its anterior border around the ciliary body of the eyeball.

oral Relating to the mouth.

oral cavity The mouth; the space surrounded by the lips, cheeks, teeth, and palate; the rima oris is its entrance, the fauces, its posterior exit.

orbit Eye socket; formed by seven skull bones that surround and protect the eye.

organ Part of the body composed of two or more tissue types and exercising one or more specific functions.

organ of Corti Spiral organ; rests on the basilar membrane and supports the hair cells that detect sounds.

organ rudiment Embryonic tissue precursors, or primordia of organs and tissues. The pancreas, for example, develops from the pancreatic rudiment.

organ system Group of organs classified as a unit because they perform a common function or set of functions.

organelle Specialized part of a cell serving one or more specific individual functions.

organism Any living thing considered as a whole, whether composed of one cell or many cells.

organogenesis Organ development.

orgasm Climax of the sexual act, associated with a pleasurable sensation.

origin Less movable attachment point of a muscle; usually the medial or proximal end of a muscle associated with the limbs.

oropharynx (OR-oh-FAHR-inks) Portion of the pharynx that lies posterior to the oral cavity; it is continuous above with the nasopharynx and below with the laryngopharynx.

osmosis Diffusion of solvent (water) through a membrane from a less concentrated solute to a more concentrated solute.

osmotic pressure Force required to prevent the movement of water across a selectively permeable membrane. A measure of the concentration of dissolved molecules (solutes).

osseous labyrinth The bony chambers in the temporal bone of the skull that house the membranous labyrinth; the vestibule, semicircular canals, and spiral organ of Corti.

ossicle One of the three small bones of the middle ear: the malleus, incus, and stapes.

ossification Bone formation.

osteoarthritis Form of arthritis that inflames the synovial capsule and erodes the articular cartilages.

osteoblast Bone-forming cell.

osteoclast Large multinucleated cell that absorbs bone.

osteocyte (OSS-tee-oh-SITE) Mature bone cell surrounded by bone matrix.

osteoid (OSS-tee-oid) The organic components of bone matrix, principally proteoglycan ground substance and collagen fibers. In contrast, hydroxyapatite is the principal mineral component of bone.

osteomalacia (OSS-tee-oh-mah-LAY-she-ah) Softening of bones due to calcium depletion. Adult rickets.

osteon (OSS-tee-ahn) Structural unit of compact bone. A single central canal, with its vessels, and the hard concentric lamellae and osteocytes surrounding it; also called a haversian system.

osteopenia (OSS-tee-oh-PEE-nee-ah) Loss of bone mass in aging, in which more bone is removed than is replaced.

osteopetrosis (OSS-tee-oh-pe-TROE-sis) A rare human genetic disorder that features excessive ossification of bone due to lack of osteoclasts.

osteophyte (OSS-tee-oh-fite) Excessive growths of bone at articular cartilage of synovial joints. Painful and immobilizing consequences of osteoarthritis.

osteoporosis (OSS-tee-oh-pore-OH-sis) Reduction in bone mass, resulting in larger spaces and thinner trabeculae in cancellous bone.

osteotomy Surgical cutting of a bone.

ostium, *pl.* **ostia** Small opening; the opening of the uterine tube near the ovary or the opening of the uterus into the vagina, for example.

otic vesicle The organ primordium from which the osseous labyrinth of the ear develops.

otitis Inflammation of the ear. Called otitis media for the middle ear; otitis externa, for the external ear.

otolith Crystalline particles of calcium carbonate and protein embedded in the maculae, sacculi, and ampullae of the inner ear.

otosclerosis Formation of spongy bone about the stapes and oval window; results in deafness.

otospongiosus Degenerative process that forms cancellous bone in the middle and inner ear. In one form of otospongiosus the ossicles of the middle ear fuse together and become immobile. See *otosclerosis.*

outer chamber of mitochondrion The space between the inner and outer membranes of mitochondria.

outer membrane of mitochondrion The outermost membrane that covers mitochondria.

oval window Membranous structure to which the stapes attaches; transmits vibrations to the inner ear.

ovarian cycle Series of events that occur regularly in the ovaries of sexually mature, nonpregnant females; results in ovulation and the production of the hormones estrogen and progesterone.

ovarian epithelium Peritoneal covering of the ovary.

ovarian follicle Spherical cell cluster in the ovary containing an oocyte.

ovarian ligament Bundle of fibers passing to the uterus from the ovary.

ovarian stroma The connective tissue of the ovary that supports ovarian follicles.

ovary The pair of female reproductive glands located in the pelvic cavity; produces ova, estrogen, and progesterone.

oviduct See *uterine tube.*

ovulation Release of an ovum, or secondary oocyte, from the vesicular follicle.

ovum, *pl.* **ova** Female gamete or sex cell that has completed both meiotic divisions; contains the genetic material transmitted from the female.

oxidation Loss of one or more electrons from a molecule.

oxygenated blood Blood whose hemoglobin is charged with oxygen from the lungs.

oxyhemoglobin Oxygenated hemoglobin.

oxyntic cell See *parietal cell.*

oxyphil cell (OX-ee-fill) Oxyphil cells, a major type of cell in the parathyroid gland, contain abundant mitochondria but have no known function.

oxytocin (OT) (oxy-TOE-sin) Peptide hormone secreted by the neurohypophysis; increases uterine contraction and stimulates milk ejection from the mammary glands.

P-wave First set of peaks and valleys in the electrocardiogram; represents depolarization of the atria.

pacinian corpuscle (pa-SIN-ee-an KOR-pus-el) See *lamellated corpuscle.*

palate (PAL-at) Roof of the mouth.

palatine Concerning the roof of the mouth.

palatine tonsil The pair of large oval masses of lymphoid tissue embedded in the lateral wall of the oropharynx.

palmar flexion The flexing motion of the wrist that draws the palm of the hand toward the forearm. The opposite motion, dorsiflexion, draws the dorsal surface of the hand toward the forearm.

palpate (PAL-pate) To examine by touch.

palpebra, *pl.* **palpebrae** An eyelid.

palpebral conjunctiva Portion of the conjunctiva that covers the inner surface of the eyelids.

palpebral fissure Opening between the upper and lower eyelids.

pampiniform plexus (pam-PIN-ih-form) The network of coiled veins that surrounds the testicular artery. Blood flowing through this network from the scrotum is thought to cool incoming blood in the testicular artery. See also *venae comitantes.*

panacinar emphysema (PAN-ah-SINE-ar EM-fih-SEE-mah) A form of emphysema characterized by degeneration of the alveolar septum. In this condition, numerous small alveoli fuse into fewer large, less effective alveoli.

pancreas Abdominal gland that secretes pancreatic juice into the intestine, and insulin and glucagon from the pancreatic islets into the bloodstream.

pancreatic acinus The cup-shaped secretory portions of the pancreatic exocrine glands; lined by pancreatic acinar cells.

pancreatic duct Excretory duct of the pancreas that extends through the gland from tail to head, where it empties into the duodenum at the greater duodenal papilla.

pancreatic islet Islets of Langerhans; the endocrine glands of the pancreas; cellular masses varying from a few to hundreds of cells lying in the interstitial tissue of the pancreas; composed of different cell types that secrete insulin and glucagon.

pancreatic juice External secretion of the pancreas; clear, alkaline fluid containing hydrolytic enzymes.

pancreatic somatostatin Somatostatin released from the pancreas.

pancreatic sphincter This ring of smooth muscle fibers guards the exit of the pancreatic duct from the pancreas.

Paneth cell Prominent, nondividing secretory cells found at the base of intestinal crypts. Extensive endoplasmic reticulum and secretory granules suggest secretion, but despite the presence of lysozyme in lysosomes, the function of these cells is unclear.

pannus Intrusion of granular tissue from the capsule of synovial joints that spreads over the surface of articular cartilages and eventually immobilizes the joint. More generally, any granular or scaly intruding tissue.

papilla A small, nipplelike process. Projections of the dermis, containing blood vessels and nerves, into the epidermis. Projections on the surface of the tongue.

papillary duct These straight ducts descend through the medulla of the kidney to empty urine into the minor calyces at the apex of the renal pyramids.

papillary muscle (PAP-ill-air-ee) Nipplelike, conical projection of myocardium within the ventricle; the chordae tendineae are attached to the apex of the papillary muscle.

papillary plexus The fine network of capillaries found in the papillary layer of dermis, directly beneath the epidermis.

paracortical layer The zone of tissue between the cortex and medulla in lymph nodes. Characterized by high endothelial vessels.

parafollicular cell Endocrine cells scattered throughout the thyroid gland that secrete the hormone calcitonin.

parahormone Substance secreted by a wide variety of tissues; not a true hormone, but with hormonelike localized effects.

parakeratosis (PARA-kare-ah-TOE-sis) Unusual form of keratinization in which nucleated keratinocytes enter the cornified layer of the epidermis. Occurs in actinic keratoses.

parallel muscle A class of skeletal muscles in which the muscle fibers and fascicles run parallel to each other between origin and insertion. Includes strap, flat, and fusiform muscles.

paramesonephric duct Two embryonic tubes extending along the mesonephros and emptying into the cloaca; form the uterine tubes, uterus, and part of the vagina in females. Degenerates in males. Also known as Mullerian duct.

paranasal sinus Air-filled cavities within skull bones that connect to the nasal cavity; located in the frontal, maxillary, sphenoid, and ethmoid bones.

parasagittal Sagittal plane located on one side of the midline.

parasympathetic Subdivision of the autonomic nervous system; characterized by craniosacral outflow through axons from the brainstem and sacral region of the spinal cord; usually stimulates vegetative functions such as digestion, defecation, and urination.

parathormone (PTH) The hormone of the parathyroid glands that raises blood calcium levels and mobilizes calcium from bone.

parathyroid gland Four glandular masses imbedded in the posterior surface of the thyroid gland; secretes parathormone and calcitonin.

paraxial mesoderm This mesoderm lies immediately lateral to the notochord in embryos; it is the source of somites.

parenchyma (pahr-EN-kih-mah) A general term for the essential internal functioning tissues of organs, in contrast to the supportive stroma.

parietal (pahr-EYE-et-al) Relating to the walls of cavities.

parietal cell Gastric gland cell that secretes hydrochloric acid.

parietal pericardium Serous membrane lining the fibrous portion of the pericardial sac.

parietal peritoneum Layer of peritoneum lining the abdominal and pelvic walls.

parietal pleura Serous membrane that lines the walls of the pleural cavity.

Parkinson's disease A degenerative disease of the substantia nigra that results in muscular tremors and immobility, shuffling gait, and masklike facial expression.

parotid duct The duct that delivers secretions from the parotid gland to the vestibule of the oral cavity. Also known as Stensen's duct.

parotid gland Largest of the salivary glands; situated anterior to each ear.

paroxysmal atrial tachycardia (PAT) A common form of cardiac dysrhythmia featuring episodes of rapid heartbeat (tachycardia).

partial pressure Pressure exerted by a single gas in a mixture of gases.

partial-thickness burn Burn that damages only the epidermis (first degree burn) or the epidermis and part of the dermis (second degree burn).

parturition Childbirth.

patella Kneecap. Bone of the kneecap.

patent (PAY-tent) Open.

patent ductus arteriosus Though the ductus arteriosus normally closes after birth, a patent ductus arteriosus remains open and admits deoxygenated blood from the pulmonary trunk to the aorta.

pathologist Scientist who studies the causes and symptoms of disease.

pectoral Relating to the chest.

pectoral girdle Site of attachment of the upper limb to the trunk; consists of the scapula and the clavicle.

pedicel Also called foot processes, these small cytoplasmic processes form the slit pores of the glomerular capsule in the kidneys.

pedicle Stalk or base of a structure, *e.g.*, the pedicle of the vertebral arch.

pedigree Ancestral line of descent, especially as diagrammed on a chart.

peduncle Stalk or stem connecting portions of the brainstem to the cerebrum or cerebellum.

pelvic Relating to the pelvis.

pelvic brim Imaginary plane passing from the sacral promontory to the pubic crest of the pelvis.

pelvic cavity The cavity that lies within the frame of the pelvis.

pelvic diaphragm The internal layer of muscles (levator ani and coccygeus) that spans the pelvic outlet.

pelvic girdle Bone that attaches the lower limb to the trunk; ring of bone formed by the sacrum and the ossa coxae.

pelvic inlet Superior opening of the true pelvis.

pelvic kidney An anomaly of kidney development in which one or both kidneys fail to ascend into the abdomen and remain in the pelvis.

pelvic nerve Parasympathetic nerve that arises from the sacral region of the spinal cord.

pelvic outlet Inferior opening of the true pelvis.

pelvis Any basin-shaped structure; cup-shaped ring of bone at the lower end of the trunk, formed from the ossa coxae, sacrum, and coccyx.

penile urethra That portion of the male urethra that extends through the penis.

penis Male external genital organ for urination and intercourse.

pennate muscle Muscles with fasciculi arranged like the barbs of a feather along a common tendon.

pepsin Principal digestive enzyme of gastric juice, formed from pepsinogen; digests proteins into smaller peptide chains.

pepsinogen Precursor of pepsin, formed and secreted by the chief cells of the gastric mucosa; the acidity of the gastric juice and pepsin itself cleaves pepsinogen into pepsin.

peptidase An enzyme capable of hydrolyzing peptide links of peptides.

peptide A short chain of amino acids, from ten to twenty amino acids in length.

peptide bond Chemical bond that links chains of amino acids together as proteins.

perforating canal Canal containing blood vessels and nerves that penetrates compact bone perpendicular to the central canals.

perforating fiber Connective tissue fiber from a tendon or ligament that penetrates the periosteum of a bone and anchors the periosteum and tendon or ligament to the bone.

periarterial sheath Dense accumulations of lymphocytes (white pulp) surrounding arteries within the spleen.

pericapillary cell Slender connective tissue cells outside of capillary walls; relatively undifferentiated; may become a fibroblast, macrophage, or smooth muscle cell.

pericardial Around the heart.

pericardial cavity (PEAR-ih-KAR-dee-al) Space within the mediastinum in which the heart is located.

pericardial coelome (SEE-loam) Embryonic coelome that surrounds the heart and becomes the pericardial cavity.

pericardial fluid Viscous fluid contained within the pericardial cavity between the visceral and parietal pericardium; functions as a lubricant for pericardium of the beating heart.

pericarditis Inflammation of the pericardium.

pericardium Serous membrane covering the heart.

perichondrium Double-layered connective tissue sheath surrounding cartilage.

periderm (PAIR-ih-derm) The outer, initial layer of epidermis in embryo and fetus that precedes keratinization in the second trimester.

perikaryon (PAIR-ih-KARRY-ahn) The central portion of a neuron, containing the cell nucleus and synthetic organelles, from which extend axon and dendrites. Also known as cell body or soma.

perilymph Fluid contained within the bony labyrinth of the inner ear. See *endolymph*.

perimetrium Outer serous coat of the uterus.

perimysium Fibrous sheath enveloping a bundle of skeletal muscle fibers in a muscle fascicle.

perineal body The fibrous disk that is the central hub of muscles of the perineum.

perineum (PAIR-ih-NEE-um) Area inferior to the pelvic diaphragm between the thighs; extends from the coccyx to the pubis.

perineurium Connective tissue sheath surrounding a nerve fascicle.

periodontal ligament (PAIR-ee-oh-DON-tal) Connective tissue that surrounds the tooth root and attaches it to its bony socket.

periosteum (PAIR-ee-OSS-tee-um) Thick, double-layered connective tissue sheath covering the entire surface of a bone except the articular cartilage.

peripheral circulatory system All vessels of the circulatory system outside of the heart.

peripheral nervous system (PNS) Major subdivision of the nervous system, consisting of spinal and cranial nerves and their ganglia.

perisinusoidal space Space between the endothelium of sinusoids and hepatocytes in the liver. Also known as the space of Disse.

peristaltic wave Wave of relaxation and contraction that progresses along the digestive tube or other tubular structures, such as the ureters.

peritoneal cavity The cavity between organs in the abdominopelvic cavity enclosed by peritoneum.

peritoneum Serous membrane that lines the peritoneal cavity and covers the abdominal viscera.

peritonitis Potentially life-threatening inflammation of the peritoneum.

peritubular capillary (PAIR-ih-TUBE-you-lar) The capillary network located in the cortex of the kidney; associated with the distal and proximal convoluted tubules.

permanent tooth One of the 32 teeth belonging to the second or permanent dentition.

pernicious anemia Anemia resulting from inadequate intake or absorption of vitamin B_{12}.

peroneal Associated with the fibula.

peroxisome (per-OX-ih-soam) Membrane-bound body similar to a lysosome in appearance but often smaller and irregular in shape; contains enzymes that either decompose or synthesize hydrogen peroxide.

perspiration The cooling, watery fluid secreted by sweat glands through the skin. The process of secreting this fluid.

petechiae (pe-TEEK-ee-ee) Minute red skin spots produced by hemorrhage of dermal capillaries.

Peyer's patch (PIE-ers) Lymph nodules found in the lower half of the small intestine and the appendix.

phaeomelanin (FEE-oh-MEL-ah-nin) Brown or black melanin pigment of skin and hair.

phagocyte Cell that ingests bacteria, foreign particles, and other cells.

phagocytosis Process by which cells ingest large particles such as other cells, bacteria, and dead tissue.

phagosome (FAG-oh-soam) Vesicle of the endomembrane system of cells that takes up external materials by phagocytosis.

phalanx, *pl.* **phalanges** (FAY-lanks; fah-LAN-jes) Bone of the fingers or toes.

pharyngeal (fahr-IN-gee-al) Concerning the pharynx.

pharyngeal arch The mesodermal supporting arches that surround the embryonic pharynx.

pharyngeal groove The paired invaginations of external embryonic ectoderm between the pharyngeal arches. The first pharyngeal groove forms the auditory canal of the external ear.

pharyngeal pouch Paired evaginations of embryonic internal pharyngeal endoderm between the pharyngeal arches that give rise to the auditory tube, tonsils, thyroid gland, parathyroid glands, and thymus.

pharyngeal tonsil Two aggregates of lymphoid nodules on the posterior wall of the nasopharynx. Also called adenoids when inflamed and enlarged. See *lingual tonsil*.

pharynx (FAHR-inks) Upper expanded portion of the digestive tube between the esophagus and the oral and nasal cavities.

phenotype Anatomical and physiological characteristics expressed by genes.

phlebitis Inflammation of veins.

phlebosclerosis (FLEE-bo-sklair-OH-sis) Fibrous thickening or hardening of the veins, as opposed to arteriosclerosis, hardening of the arteries.

phosphatase (FOSS-fah-tase) Alkaline phosphatase has an important role in supplying enough phosphate ions for calcium and phosphate to crystallize as hydroxyapatite.

phospholipid (FOSS-fo-LIP-id) Phosphate-containing lipid molecule with a polar end and a nonpolar end; main component of the lipid bilayer.

phospholipid bilayer The fundamental structure of cellular membranes.

phosphorylation Addition of phosphate to an organic compound. Protein kinases are enzymes that phosphorylate proteins.

photopigment The light-sensitive pigments of rods (rhodopsin) and cones (iodopsin) in the retina.

photoreceptor A sensory receptor that is sensitive to light. Examples are rods and cones of the retina.

phrenic nerve Nerve derived from spinal nerves C3 through C5; supplies the diaphragm.

physiology Scientific discipline that deals with the vital processes or functions of living things.

pia mater Literally, tender mother; delicate vascularized membrane that is the inner covering of the brain and spinal cord. See *meninx*.

piezoelectric (PIE-eh-zoh-electric) The electronic ability of certain crystalline materials, including bone, to produce minute electric currents when subjected to force, and conversely, to deform slightly when subjected to electric current.

pigmented retina Pigmented portion of the retina containing melanocytes that absorb stray light.

pilosebaceous apparatus Specialized epidermal structure composed of a hair follicle, sebaceous gland, and arrector pili muscle.

pineal gland A small pine cone-shaped structure that projects from the epiphysis of the diencephalon; produces melatonin.

pinkeye Acute contagious conjunctivitis; inflammation of the conjunctiva.

Pinkus corpuscle See *tactile disk*.

pinna See *auricle*.

pinocytosis Cell drinking; cellular uptake of liquid by endocytosis.

pioneer neurons Neurons that establish neural pathways during development.

pituitary gland (pih-TOO-ih-TARE-ee) See *hypophysis*.

pivot joint A class of synovial joint that permits rotation about the longitudinal axis of a bone. The radioulnar joint of the elbow is a pivot joint.

placenta The transient, largely fetal organ that transfers materials between maternal and fetal circulation.

placental barrier The epithelial partition between maternal and fetal blood in the placenta.

placode A thickened patch of embryonic cells that is precursor to adult organs and tissues. The lens develops from the lens placode in the embryonic ectoderm of the head.

plane An imaginary surface formed by extension through any axis or two points. Examples include midsagittal planes, coronal planes, and transverse planes.

plane joint See *arthrodial joint*.

plantar flexion Motion that draws the foot inferiorly in the anatomical position.

plasma Fluid portion of blood.

plasma cell Cell derived from B-lymphocytes; produces circulating antibodies.

plasma membrane Cell membrane; outermost component of the cell, surrounding the cytoplasm.

plasma profile An analysis of the components of blood plasma that discloses the levels of blood electrolytes, metabolites, proteins, and enzymes. This information can monitor normal body processes and disclose disease.

plasmin Enzyme derived from plasminogen; dissolves clots by converting fibrin into soluble products.

platelet Disk-shaped cell fragment found in blood; contains granules and clear cytoplasm but has no definite nucleus; activates clotting process.

platelet aggregation The process by which platelets (thrombocytes) seal off small breaks in vessels.

platelet-derived growth factor (PDGF) A protein derived from platelets that stimulates growth of vascular endothelial and smooth muscle cells.

pleura Serous membrane covering the lungs and lining the thoracic wall, diaphragm, and mediastinum.

pleural cavity (PLOOR-al) Potential space between the parietal and visceral layers of the pleura.

pleural fluid Serous fluid found in the pleural cavity; helps to reduce friction when the pleural membranes rub together.

plexus, *pl.* **plexuses** Intertwining network of nerves or blood vessels.

plicae circulares (PLEE-kee) Numerous circular folds of the mucous membrane of the small intestine.

pluripotent Pluripotent cells are able to differentiate into several different types of cells.

pluripotential stem cell Stem cells that have the ability to differentiate into several different types of cells.

pneumothorax (NEW-moe-THORE-ax) Equalization of pressure in the pleural cavity with that of atmospheric air.

podocyte Epithelial cell of glomerular capsules attached to the outer surface of the glomerular capillary basement membrane by cytoplasmic foot processes. Forms filtration slit.

polar body During oogenesis, the cytoplasm is delivered unequally to daughter cells at each meiotic division. The egg cell receives almost all cytoplasm, with only a small tab of cytoplasm delivered to the polar body formed at both divisions. Polar bodies degenerate and do not normally participate in fertilization.

polar spindle fiber Fibers of the mitotic spindle that draw chromosomes toward the opposite poles of the spindle.

polarity The property of being different at opposite ends. Epithelial polarity concerns the different functions of the luminal and basal membranes of epithelial cells. Polar molecules distribute their electronic charges nonuniformly, so that some regions are more negative than others. Polar molecules dissolve readily in polar solvents, such as water, and nonpolar molecules dissolve in nonpolar solvents, such as oils and lipids.

pollicis Associated with the thumb; literally, of the thumb.

polycystic kidney (POL-ee-SIS-tik) A hereditary condition of the kidneys in which the nephrons inflate with fluid, and the kidney fills with cysts because collecting ducts fail to connect to the nephrons.

polycythemia Increased number of erythrocytes in blood.

polydactyly (pol-ee-DAK-till-ee) Condition caused by a recessive gene in which extra cartilaginous models of the phalanges result in formation of additional fingers or toes.

polymorphonuclear leukocyte (POLY-mor-fo-NUKE-lee-ar) The granular leukocytes that possess multilobed nuclei; eosinophils, neutrophils, and basophils.

polysaccharide Carbohydrate containing a large number of monosaccharide molecules.

pons The portion of the brainstem between the me-dulla and midbrain.

popliteal Posterior region of the knee.

port wine stain An intensely vascular patch of skin that appears dark red.

porta hepatis (heh-PAT-iss) The point on the inferior surface of the liver where the hepatic artery and hepatic portal vein enter and the hepatic ducts exit.

portal system System of veins in which blood, after passing through one capillary bed, flows into a second capillary network. The hepatic portal system is an example.

portal triad Branches of the portal vein, hepatic artery, and hepatic duct that supply liver lobules and that are bound together in the connective tissue between liver lobules.

positioning motion The gross movements of extremities controlled by the proximal muscles that position the hands and feet.

positive feedback Mechanism by which any deviation from a normal value is made greater.

posterior That which follows. In humans, toward the back. Behind.

posterior chamber of the eye Chamber of the eye between the iris and the lens, filled with aqueous humor.

posterior columns Ridges on the posterior surface of the spinal cord and medulla oblongata that carry ascending tracts in the fasciculi gracilis and cuneatus.

posterior horn The posterior gray matter of the spinal cord.

posterior interventricular sulcus Groove on the diaphragmatic surface of the heart, marking the location of the septum between the two ventricles.

posterior mesentery Also known as the dorsal mesentery, this mesentery attaches the gut tube to the posterior body wall in embryonic development.

posterior pituitary See *neurohypophysis*.

posterior root Also called dorsal root; all afferent axons of spinal nerves enter the dorsal surface of the spinal cord by way of posterior roots of the spinal nerves.

posterior root ganglion Also called dorsal root ganglion, this is a cluster of cell bodies of afferent neurons in the dorsal roots of spinal nerves.

postganglionic neuron The second neuron in autonomic neural pathways. With its cell body and dendrites in an autonomic ganglion, a postganglionic neuron receives stimuli from the first, or preganglionic, neuron and sends impulses along its axon to a target organ.

postmenopausal period The period of time following menopause.

postsynaptic neuron The neuron that receives transmissions across a synapse.

potassium ion channel Integral proteins of plasma membranes that specifically transport potassium ions through channels across the membrane.

potential difference Difference in electrical charge, measured as the charge difference across the cell membrane.

precapillary sphincter Smooth muscle sphincter that regulates blood flow through a capillary.

preganglionic neuron Autonomic neuron whose cell body is located within the central nervous system and sends its axon to an autonomic ganglion where it synapses with postganglionic neurons.

premenstrual syndrome (PMS) In some women of reproductive age, the regular monthly experience of physiological and emotional distress, usually during the several days preceding menses, typically involving fatigue, edema, irritability, tension, anxiety, and depression.

premolar Bicuspid tooth; two cusps on the crown.

prenatal medicine The field of medicine that treats pregnant women and their fetuses before delivery.

prepuce In males, the free fold of skin that more or less completely covers the glans penis; the foreskin. In females, the external fold of the labia minora that covers the clitoris.

presbycusis Hearing loss that occurs with age.

presbyopia The inability of the lens to accommodate, or to focus upon, near objects. Commonly acquired in middle age when the ciliary muscles begin to fail and the lens becomes less elastic.

presynaptic neuron A neuron that sends transmissions across a synapse.

primary bronchus Two tubes arising at the inferior end of the trachea; each primary bronchus enters a lung.

primary follicle Ovarian follicle before the appearance of an antrum; contains the primary oocyte.

primary lobule A structural and functional respiratory unit of the lungs, which consists of all passages that branch from a single terminal bronchiole.

primary nodule Primary nodules are a form of lymph nodule, the simplest lymphatic tissue. Primary nodules are dense clusters of small lymphocytes. See *secondary nodule.*

primary oocyte Oocyte before completion of the first meiotic division (arrested at prophase I).

primary spermatocyte Diploid spermatocyte that arises from a spermatogonium; gives rise to secondary spermatocytes after the first meiotic division.

prime mover Muscle that plays a major role in accomplishing a movement.

primitive groove The depression that marks the primitive streak and the embryonic axis on the embryonic disk in early development.

primitive node The gastrulation center of the primitive streak.

primitive ridge The pair of ridges beside the primitive groove of the primitive streak.

primitive streak Median structure in the embryonic disk from which arises the mesoderm by inward and then lateral migration of cells.

primordial follicle Immature ovarian follicles.

primordial germ cell Progenitor cells of the sex cells, found initially outside the gonad in the wall of the yolk sac.

procaine A topical local anaesthetic.

process A portion of a bone that projects outward from the central body.

progeria Severe retardation of growth after the first year of life, accompanied by a senile appearance and death at an early age.

progesterone Steroid hormone secreted by the ovaries; necessary for uterine and mammary gland development and function.

prognathia (prog-NATH-ee-ah) Protruding jaw.

progress zone The zone of mesoderm and ectoderm at the tip of limb buds that promotes growth and differentiation of limbs.

projection neuron A neuron whose long axon extends from one level or region of the brain or spinal cord to a different location, as if it were a projectile shot from a cannon to its target.

prolactin Hormone of the adenohypophysis that stimulates the production of milk.

prolactin-releasing hormone (PRH) Neurohormone released from the hypothalamus that stimulates the adenohypophysis to release prolactin.

prolapse Downward or outward displacement of an internal body part. The rectal mucosa protrudes through the anus in prolapse of the rectum.

proliferative phase See *follicular phase.*

pronation (pro-NAY-shun) Rotation of the forearm so that the anterior surface is down (prone) or posterior.

pronator Any muscle that pronates the forearm.

prone Lying face down.

pronephric duct The temporary duct that connects pronephric tubules in developing kidneys. Precursor to the mesonephric duct.

pronephros (pro-NEF-ross) Primitive kidney. In the embryos of higher vertebrates, a series of tubules emptying into the coelomic cavity. It is a temporary structure in the human embryo, followed by the mesonephros and still later by the metanephros, which gives rise to the kidney.

pronormoblast Early stage of erythrocyte development, in which the nucleus becomes functional.

pronucleus The haploid nuclei, derived from egg and sperm nuclei, that are capable of fusing together in a fertilized egg as a diploid zygote nucleus with which embryonic development begins.

proopiomelanocortin (POMC) (pro-OH-pea-oh-MEL-an-oh-KORT-in) The corticotrophic cells of the adenohypophysis (anterior pituitary) produce a single large precursor molecule, proopiomelanocortin, that is cleaved into endorphin, melanocyte-stimulating hormone, and adrenocorticotropic hormone.

prophase First stage in mitosis, in which chromatin strands condense to form chromosomes.

proprioception Information about the position of the body and its various parts.

proprioceptor Sensory receptor, associated with joints and tendons, that detects stimuli interpreted as body sense.

prostaglandin Class of physiologically active substances present in many tissues; among effects are vasodilation, stimulation and contraction of uterine smooth muscle, and promotion of inflammation and pain.

prostate cancer Cancer of the prostate gland.

prostate gland (PRAH-state) Gland that surrounds the beginning of the urethra in males. The gland secretes a milky fluid that is discharged by 20 to 30 excretory ducts into the prostatic urethra as part of the semen.

prostate specific antigen (PSA) A protein characteristic of only the prostate gland. Oversecretion and accumulation of elevated levels of PSA in the blood often indicate prostate cancer.

prostatic fluid (pro-STAT-ik) The fluid products of the prostate gland.

prostatic plexus Network of veins that returns blood from the cavernous sinuses of the penis.

prostatic urethra Part of the male urethra, approximately 2.5 cm in length, that passes through the prostate gland.

prostatic utricle Remnant of the degenerated Mulle-rian ducts in the prostate gland of males.

protease Enzyme that cleaves protein molecules into peptide fragments.

protease/antiprotease hypothesis (PRO-tee-ace) This hypothesis attempts to account for emphysema by supposing that excessive protease enzymes, or insufficient inhibitors to block the action of these enzymes, cause the alveolar septa of lungs to break down.

protein Macromolecule consisting of long chains of amino acids linked together by peptide bonds.

protein kinases A class of important enzymes in cell-signaling pathways that activate other enzymes by attaching phosphate molecules to them (phosphorylating them).

proteinuria (PRO-teen-YOUR-ee-ah) Excessive protein in the urine.

proteoglycan (PRO-tee-oh-GLY-can) The highly hydrated polysaccharide/protein molecules of the extracellular ground substance in connective tissue.

prothrombin Glycoprotein present in blood that is converted to thrombin in the presence of prothrombin activator. Major component of the blood clotting pathway.

proton Positively charged subatomic particle in the nuclei of atoms.

proto-oncogene Normal genes that are known candidates to mutate as oncogenes. See *oncogene*.

protoplasm Living matter of which cells are formed.

protoplasmic astrocyte A variety of astrocytes, themselves a class of neuroglial cells of the nervous system. Protoplasmic astrocytes extend rather thick cytoplasmic processes from their central cell bodies.

protraction (pro-TRAK-shun) Movement of a body part forward, or anteriorly, in a straight line.

proximal Near to the reference point, especially to the midline of the body, or to the base of the limbs. See *distal*.

proximal convoluted tubule (PCT) Part of the nephron that extends from the glomerulus to the descending limb of the nephron loop.

proximal phalanx The segment of each digit that articulates with the metacarpals of the hand or the metatarsals of the foot.

pseudoglandular stage The phase of lung development at 5 to 16 weeks of development that forms the passages of the conducting zone of the lungs.

pseudostratified epithelium (SUE-doe-) Epithelium consisting of a single layer of columnar cells but appearing as several layers. May be ciliated.

pseudounipolar neuron See *unipolar neuron*.

psoriasis (sor-EYE-ah-sis) Condition of unknown origin that produces large silvery scales over reddish bumps in the skin that bleed readily when scratched.

pterygoid Wing-shaped structure.

pterygoid process (TEAR-ih-goid) Winglike processes that extend from the body of the sphenoid bone of the skull downward on each side of the choanae.

ptosis Falling down of an organ; drooping of the upper eyelid.

puberty Series of events that transform a child into a sexually mature adult.

pubic symphysis The fibrocartilaginous joint that joins the pubic bones of the pelvis.

pubis The anteriormost of the three bones of the pelvis.

pudendal (pew-DEN-dal) Concerning the external genitalia, particularly of females.

pudendal cleft Cleft between the labia majora.

pudendum See *vulva*.

pulmonary artery The arteries that conduct deoxygenated blood from the pulmonary trunk to the right and left lungs.

pulmonary semilunar valve The heart valve that admits blood from the right ventricle into the pulmonary trunk, and prevents backflow.

pulmonary trunk Large, elastic artery that carries blood from the right ventri-

cle of the heart to the right and left pulmonary arteries.

pulmonary vascular tree The vascular tree that distributes blood between the alveoli of the lungs and the heart. Returns oxygenated blood to the heart.

pulmonary vein Four veins that carry blood from the lungs to the left atrium of the heart.

pulp The tissue that resides in the pulp cavity of a tooth, or in the spleen.

pulp cavity of tooth Central hollow chamber of a tooth consisting of the crown cavity and the root canal.

pulp vein Veins that drain blood from the red pulp of the spleen.

pulse Rhythmical dilation of an artery produced by the increased volume of blood ejected into the arteries by the left ventricle.

punctum Small opening of the lacrimal canaliculus.

pupil Circular opening in the iris through which light enters the eye.

Purkinje cell Large neurons of the cerebral cortex with elaborately branched dendrites.

Purkinje fiber Modified cardiac muscle cells found beneath the endocardium of the ventricles. Specialized to conduct action potentials.

pus Fluid product of inflammation; contains leukocytes, debris of dead cells, and tissue elements liquefied by enzymes.

putamen The lateral portion of the lentiform nucleus of the cerebral hemispheres.

pyelonephritis (PIE-el-oh-NEF-rye-TISS) Inflammation of the kidney and renal pelvis.

pyloric opening Opening connecting the stomach with the superior part of the duodenum.

pyloric sphincter Thickening of the circular layer of the gastric musculature encircling the junction between the stomach and duodenum.

pylorus The inferior, tapering region of the stomach.

pyramidal cell A variety of neurons found in the cerebral cortex, characterized by pyramid-shaped cell bodies.

pyramidal tract Tracts that descend through the pyramids, swellings on the anterior surface of the medulla oblongata.

pyruvic acid End product of glycolysis. Two three-carbon pyruvic acid molecules are produced from a six-carbon molecule of glucose.

Q

QRS complex Principal deflection in the electrocardiogram, representing ventricular depolarization.

quadratus Four-sided; square.

quadriceps femoris The four-headed muscle in the anterior compartment of the thigh that extends the leg and flexes the thigh.

R

radial keratotomy (RK) A microsurgical procedure that adjusts the focus of the eye by making small cuts in the cornea.

radial pulse Pulse detected in the radial artery.

radial spoke protein A variety of protein found in the radial spokes of axonemes in cilia and flagella.

radius The lateral bone of the forearm.

ramus, *pl.* **rami** A branch, usually of a peripheral nerve. The part of a bone that forms an angle with the main body of the bone. See *median raphe*.

ramus communicans, *pl.* **rami communicantes** Branches of thoracic and lumbar spinal nerves that communicate with the sympathetic chain of the autonomic nervous system. See *white ramus communicans* and *gray ramus communicans*.

raphe (RAY-fee) Median ridge or line representing fusion of lateral structures; the central line running over the scrotum from the anus to the root of the penis.

Rathke's pouch (RATH-keys) A diverticulum in the roof of the oral opening that meets the infundibulum evaginating from the floor of the diencephalon and forms the adenohypophysis.

receptor-mediated endocytosis Cellular uptake of materials that bind to receptors in the plasma membrane.

rectouterine pouch (REK-toh-YOU-ter-in) The space in the pelvic cavity of females between the broad ligament, which supports the uterus and ovaries, and the rectum posteriorly.

rectum Portion of the digestive tract that extends from the sigmoid colon to the anal canal.

red muscle fiber Also called slow-twitch muscle fibers, these skeletal muscle fibers contain large quantities of the oxygen-carrying red protein, myoglobin. Red fibers are smaller, contract slower, and develop less power, but resist fatigue better than white fast-twitch muscle fibers.

red pulp Reddish-brown substance of the spleen, consisting of venous sinuses and the tissues intervening between them, called pulp cords.

reduced hemoglobin Form of hemoglobin in red blood cells after the oxygen of oxyhemoglobin is released in the tissues.

referred pain Visceral pain perceived to be from different regions of the body than the actual source.

reflex Automatic response to a stimulus that occurs without conscious thought; produced by a reflex arc.

reflex arc Smallest neural pathway of the nervous system that is capable of receiving a stimulus and producing a response; composed of a receptor, afferent neuron, interneuron, efferent neuron, and effector organ.

refraction Bending of a light ray when it passes from one medium into another of different density. This physical property enables the cornea and lens to focus light on the retina.

regional anatomy Anatomy studied and organized by regions of the body, such as the head, neck, and axilla. See *systemic anatomy*.

regular dense connective tissue A variety of con-nective tissue characterized by large collagen fibers, densely packed and oriented in a particular direction. Tendon and ligaments are such connective tissue, in contrast with loose connective tissue and dense irregular connective tissue.

regurgitation (ree-GUR-jih-TAY-shun) Backflow. In vomiting, stomach contents regurgitate through the esophagus and oral cavity. Blood regurgitates through leaky valves in the heart.

renal agenesis (ay-JEN-ih-sis) Failure of one or both kidneys to develop.

renal arterial tree The branches of the renal artery that deliver blood to glomeruli of the kidney.

renal artery The artery that originates from the aorta and delivers blood to the kidney.

renal capsule (REEN-al) The fibrous capsule that encloses the kidney.

renal column Cortical substance separating the renal pyramids.

renal corpuscle (KORE-pussl) Structure in the kidney composed of the glomerulus and the glomerular (Bowman's) capsule.

renal cortex Outer part of the kidney, consisting of the renal corpuscle and the proximal and distal convoluted tubules; also the renal columns, which are extensions inward between the pyramids.

renal failure Failure of nephrons to filter and/or to reabsorb filtrate.

renal fascia Connective tissue surrounding the kidney that forms a sheath or capsule for the organ.

renal fat pad Fat layer between the renal capsule and fibrous capsule of the kidney that functions as a shock-absorbing material.

renal lobule Clusters of nephrons in the cortex of the kidney.

renal medulla Inner portion of the kidney, consisting of the renal pyramids and the medullary rays that extend into the cortex.

renal papilla (pah-PILL-ah) The apex of a medullary pyramid in the kidney, which releases urine.

renal pelvis Funnel-shaped expansion of the upper end of the ureter that receives the calyces.

renal pyramid See *medullary pyramid*.

renal sinus The fat-filled space between the renal pelvis and the medulla of the kidney.

renal tubular acidosis Accumulation of excess acid in the blood due to failure of nephrons to reabsorb bicarbonate ions, which can neutralize the acid, from the filtrate.

renal venous tree The tributaries of the renal veins that receive blood from the nephrons of the kidney.

renin (REEN-in) Enzyme secreted by the juxtaglomerular apparatus that converts angiotensinogen to angiotensin I.

replication Duplication of DNA molecules.

repolarization Phase of the action potential in which the membrane potential returns to the resting membrane potential.

respiration Cellular process in which oxygen is used to oxidize organic molecules, providing a source of energy, carbon dioxide, and water. Also refers to movement of air into and out of the lungs, the exchange of oxygen and carbon dioxide with blood, the transportation of these gases in the blood, and exchange of them between the blood and the tissues.

respiratory bronchiole Smallest bronchiole (0.5 mm in diameter) that connects terminal bronchioles to alveolar ducts.

respiratory epithelium The pseudo-stratified ciliated columnar epithelium of the nasal passages.

respiratory membrane Membrane in the lungs composed of alveoli and capillaries that exchanges oxygen and carbon dioxide between the air and blood.

respiratory zone Passages in the lung that exchange oxygen and carbon dioxide between blood and air across a lining of thin squamous epithelium.

resting potential The difference in electrical charge on the cytoplasmic surface of a cell membrane relative to the charge at the exterior surface of the membrane.

rete testis (REE-tee) Network of passages that receive sperm from the seminiferous tubules and tubuli recti in the testes.

reticular (reh-TIK-you-lar) Relating to a fine network of cells or collagen fibers.

reticular activating system A network of nuclei in the medulla that alerts the cerebral hemispheres to consciousness.

reticular cell Cell whose cytoplasmic processes form a cellular network with those processes of similar cells; along with a network of reticular fibers, reticular cells form the framework of bone marrow and lymphatic tissues.

reticular fiber Delicate collagen fibers of connective tissue that form supportive networks of cells in the liver and lymph nodes.

reticular formation The nuclei and fibers of the reticular activating system in the medulla. See *reticular activating system*.

reticulocyte (reh-TIK-you-low-site) Young nucleated red blood cells that synthesize hemoglobin molecules in the endoplasmic reticulum; become erythrocytes after completing synthesis.

reticuloendothelial system See *mononuclear phagocytic system*.

retina Light-sensitive neural tunic of the eyeball.

retinaculum Dense, regular connective tissue sheath holding down the tendons at the wrist, ankle, and other sites.

retinal Vitamin A derivative (an aldehyde of retinol) that binds to opsin to form rhodopsin.

retinal detachment Separation of the sensory retina from the pigmented retina; rods and cones in the sensory retina degenerate because of a lack of nutrition from the vascular choroid layer behind the pigmented retina.

retinal pigment epithelium The pigmented layer of the retina that backs the rods and cones.

retraction (ree-TRAK-shun) Linear movement of a body part in the posterior direction. To retract the shoulders is to draw them backward, bracing to attention.

retrograde degeneration When an axon is cut, the neuron may die. In retrograde degeneration, neurons upstream, opposite the direction that impulses flow normally into the degenerating neuron, also degenerate. See *anterograde degeneration* and *Wallerian degeneration*.

retroperitoneal Structures, such as the kidneys, that lie behind the peritoneum in the body wall, outside the peritoneal cavity, are retroperiotoneal.

rheumatic fever (roo-MAT-ik) A bacterial disease that can cause tender joints, high fever, and skin rash. Permanent damage to heart valves can result when the inflamed inner lining of the heart, the endocardium, scars and the cusps of valves do not seal.

rheumatoid arthritis (ROO-mah-toid arth-RYE-tis) Painful, disfiguring autoimmune form of inflammation of the joints or bones.

rheumatoid nodule (ROO-mat-oyd) In rheumatoid arthritis patients, a type of watery nodule, or sac, that forms around a joint and leads to its immobilization.

rhinitis (rine-EYE-tis) Inflammation of the nasal mucosa.

rhodopsin Light-sensitive substance found in the rods of the retina; composed of opsin loosely bound to retinal.

ribonuclease Enzyme that splits RNA into its component nucleotides.

ribonucleic acid (RNA) Macromolecules formed by transcription from DNA; directs protein synthesis. Found in nuclei and cytoplasm of all cells. See *messenger RNA, ribosomal RNA,* and *transfer RNA*.

ribosomal RNA (rRNA) RNA that is associated with ribosomal proteins to form ribosomes.

ribosome (RYE-bo-soam) Small, spherical, cytoplasmic organelle where protein synthesis occurs.

RICE program A therapeutic treatment for pulled muscles (myositis). Involves Rest, Icing the muscle, Compression bandaging, and Elevating the injured limb.

right lymphatic duct Lymphatic duct that empties into the right subclavian vein; drains the right side of the head and neck, the right upper thorax, and the right upper limb.

rigor mortis Increased rigidity of muscle after death due to cross-bridge formation between actin and myosin as calcium ions leak from the sarcoplasmic reticulum.

rima oris The lips that form the boundaries of the entrance to the mouth.

rod Photoreceptor in the retina of the eye; responsible for black-and-white vision in low-intensity light.

roof plate The thin portion of the neural tube, posterior to the neural canal, which becomes the dorsal portion of the isthmus of the spinal cord. See *floor plate*.

root Slender connections between larger body parts. The roots of the spinal nerves join the spinal nerves with the spinal cord, and the narrow root of each lung distributes nerves, vessels, and air passages between the lung and the mediastinum.

root canal The extension of the pulp cavity of a tooth down its root as a canal that opens through the tip of the root to the adjacent bone.

root of lung The connection between the lung and the wall of the mediastinum, containing the bronchus and vessels to each lung.

root of penis Proximal attached part of the penis, including the two crura and the bulb.

root of tooth That part below the neck of a tooth covered by cementum rather than enamel and attached by the periodontal ligament to the alveolar bone.

rotation Movement of a structure about its axis.

rotator cuff muscle A group of four deep muscles that attach the humerus to the scapula; supraspinatus, infraspinatus, teres minor, and subscapularis muscles.

rough ER Endoplasmic reticulum with ribosomes on its exterior that synthesizes proteins.

round ligament Fibromuscular band attached to the uterus on either side, in front of and below the opening of the uterine tube; it passes through the inguinal canal to the labium majus.

round window Membranous structure separating the scala tympani of the inner ear from the middle ear.

rubrospinal tract Descending spinal tract that conducts motor control from the red nucleus, the largest nucleus of the midbrain, down the brainstem to the anterior horn of the spinal gray matter.

Ruffini corpuscle Receptor located deep in the dermis that responds to continuous touch or pressure; named for Angelo Ruffini, Italian histologist (1864–1929).

ruga, *pl.* **rugae** (ROO-gah, -gee) Fold or ridge; fold of the mucous membrane of the stomach when the organ is contracted; transverse ridge in the mucous membrane of the vagina.

rule of nines The empirical rule that calculates the coverage of burn injuries by dividing the skin into multiples of nine percent of the total. The entire head is nine percent, the upper extremity nine percent, and the trunk eighteen percent, for example.

S phase The period of interphase when DNA synthesis occurs and the chromosomes duplicate.

saccadic motion The rapid, jerky, repetitive motions of the eyes as they read a page or search for objects.

saccule Part of the membranous labyrinth of the ear; contains a sensory structure, the macula, that detects static equilibrium.

sacral (SAY-kral) Concerning the sacrum.

sacral plexus The interconnecting network of nerve fibers derived from lumbar nerve L4 through sacral nerve S3 that lead to the lower extremity.

sacroiliac joint The slightly mobile synovial joint that joins the ilium of the os coxa to the sacrum.

sacrum The triangular bone at the base of the vertebral column that attaches the pelvis to the vertebral column.

saddle block A clinical procedure that administers spinal anaesthetics into the lumbar region of the vertebral canal.

saddle joint A variety of synovial joint with articular surfaces shaped like saddles. Motion may be in either of two directions. The carpometacarpal joint of the thumb is a saddle joint.

sagittal In the line of an arrow shot from a bow; plane running vertically through the body and dividing it into right and left portions.

sagittal section A section through the body that parallels the body axis and separates structures at the left from those at right.

sagittal suture (SAJ-ih-tal) The suture that joins the parietal bones of the calvarium to each other at the midline.

salivary amylase Enzyme secreted in the saliva that breaks down starch to maltose and isomaltose.

salivary gland Gland that produces and secretes saliva into the oral cavity. The three major pairs of salivary glands are the parotid, submandibular, and sublingual glands.

saltatory conduction Conduction in which action potentials jump from one node of Ranvier to the next node along myelinated axons.

sarcolemma (SAR-ko-LEM-mah) Plasma membrane of a muscle fiber.

sarcoma A variety of cancer derived from the epidermis.

sarcomere Part of a myofibril between adjacent Z-lines.

sarcoplasm Cytoplasm of a muscle fiber, excluding the myofilaments.

sarcoplasmic reticulum (SAR-ko-plaz-mik reh-TIK-yoo-lum) Modified endoplasmic reticulum of muscle fibers.

satellite cell Specialized neuroglial cell that surrounds the cell bodies of neurons within ganglia.

satiety The condition of being fulfilled or fully gratified, with regard to beauty, sex, hunger, or thirst.

scala The strips of tissue in the membranous labyrinth of the cochlea that partition the chambers of the spiral canal.

scala tympani Division of the spiral canal of the cochlea lying below the spiral lamina and basilar membrane.

scala vestibuli Division of the cochlea lying above the spiral lamina and vestibular membrane.

scapula The posterior bone of the shoulder that receives the humerus.

scar tissue Fibrous tissue that replaces normal tissue, usually after injury or inflammation.

schindylesis (skin-dih-LEE-sis) A type of suture in which a V-shaped margin of a bone receives a wedge-shaped border of another bone.

Schwann cell A variety of neuroglial cell that forms a myelin sheath around the axons of the peripheral nerves.

Schwann tube Schwann cells of regenerating axons form tubes that guide the outgrowth of new axons.

sciatic Relating to the hip.

sciatic nerve Tibial and common peroneal nerves bound together into the lower extremities.

sclera White of the eye; white, opaque portion of the fibrous tunic of the eye.

scleral venous sinus See *canal of Schlemm.*

sclerotic plaque (sklair-OT-ik) The second stage in development of atherosclerotic plaques, in which fatty streaks thicken and begin to narrow arteries as collagen capsules form around foam cells in the tunica media. See *fatty streak* and *complicated plaque.*

sclerotome (SKLAIR-oh-tome) The cluster of cells derived from a somite that with other sclerotomes gives rise to the axial skeleton. Somites subdivide into dermatomes, sclerotomes, and myotomes.

scoliosis (SKO-lee-OH-sis) Lateral curvature of the vertebral column.

scrotum Musculocutaneous sac containing the testes.

sebaceous gland Gland of the skin, usually associated with a hair follicle, that produces sebum.

sebum (SEE-bum) Oily, white, fatty substance produced by the sebaceous glands.

second class lever One of three classes of lever, such as a wheelbarrow, in which the load is between the muscular force upon a bone and the joint. The action of the calcaneal tendon on the ankle is an example of a second class lever.

second degree burn These burns remove the epidermis and penetrate into the dermis, and usually require skin grafting to close the wounds.

second heart sound Sound ("dupp") made by closure of the semilunar valves.

second meiotic division The final division in haploid cells at meiosis that separates sister chromatids.

second messenger Molecule that is produced in a cell in which the first messenger interacts with a membrane-bound receptor molecule; the second messenger then acts as a signal and carries information to a site within the cell; *e.g.,* cyclic AMP.

second polar body The small rudimentary fragment of cytoplasm and nucleus discarded after the second meiotic division of the oocyte.

secondary bronchus Branch from a primary bronchus that conducts air to each lobe of the lungs. There are two branches from the primary bronchus in the left lung and three branches in the right lung.

secondary follicle Follicle in which the secondary oocyte is surrounded by granulosa cells at the periphery of the fluid-filled antrum.

secondary nodule Small lymphocytes surround an inner core of large lymphocytes in secondary nodules. See *primary nodule.*

secondary oocyte Oocyte in which the second meiotic division stops at metaphase II unless fertilization occurs.

secondary palate Roof of the mouth in the early embryo that gives rise to the hard and the soft palates.

secondary (memory) response Immune response that occurs when the immune system is exposed to an antigen against which it has already produced a primary response.

secondary spermatocyte Spermatocyte derived from a primary spermatocyte by the first meiotic division; in the second meiotic division, each secondary spermatocyte divides into two spermatids.

secretin Hormone formed by the epithelial cells of the duodenum; stimulates secretion of pancreatic juice with high concentrations of bicarbonate ions.

secretion General term for a substance produced inside a cell and released from the cell across its plasma membrane; also, the process of such production and release.

secretory phase Refers to the secretory activity of endometrial glands of the uterus during the second half of the menstrual cycle. See *luteal phase.*

semen Thick, yellowish-white, viscous fluid containing spermatozoa and secretions of the testes, seminal vesicles, prostate, and bulbourethral glands, released by ejaculation from the penis.

semicircular canal Canal in the petrous portion of the temporal bone that contains sensory organs that detect dynamic equilibrium. Each inner ear contains three semicircular canals.

semilunar valve (SEM-ee-LOON-ar) Prevents regurgitation of blood from the aorta or pulmonary trunk back into

the ventricle; consists of three half-moon–shaped cusps that blood pressure seals firmly together.

seminal fluid The fluid portion of semen.

seminal vesicle Paired glands that secrete components of seminal fluid into the ejaculatory ducts.

seminiferous tubule Tubule in the testis in which spermatozoa develop.

sensory ganglion Collection of sensory neuron cell bodies within a cranial nerve.

sensory retina See *neural retina.*

septal cell Alveolar epithelial cells that line the alveoli of the lungs with surfactant.

septum, *pl.* **septa** Thin wall dividing two cavities or masses of soft tissue.

septum pellucidum One of two thin plates of brain tissue that separate the left and right ventricles.

septum primum First septum in the embryonic heart that arises on the wall of the atrium, separating that chamber into right and left atria.

septum secundum Second membrane that partitions the atrium into left and right chambers; arises after the septum primum and to its right.

seromucous gland A variety of exocrine gland, composed of serous and mucous structures, that secretes watery serous fluid and viscous mucus.

serosa (seh-ROE-zah) Outermost covering of an organ or structure that lies in a body cavity. See *adventitia.*

serotonin (sear-oh-TONE-in) Vasoconstrictor substance released by blood platelets.

serous (SEER-us) Relating to or producing a watery substance.

serous fluid Watery fluid similar to lymph that is produced by and covers serous membranes; lubricates serous membranes.

serous gland Exocrine glands that secrete watery fluids, in contrast with viscous secretions from mucous glands.

serous membrane Thin sheet composed of epithelial and connective tissues; lines peritoneal, pleural, and pericardial cavities.

serous pericardium Serous membrane lining the pericardial sac.

serrated Saw-toothed edge. The serratus anterior and serratus posterior muscles have serrated origins on the rib cage.

Sertoli cell (SIR-toll-ee) Elongated cell in the wall of the seminiferous tubules

to which spermatids attach during spermatogenesis.

serum Fluid portion of blood remaining after removal of fibrin and blood cells.

sesamoid bone Seed-shaped bone found within tendons; the patella within the quadriceps tendon.

sex chromosome Pair of chromosomes responsible for sex determination; XX in female and XY in male.

sex cord Cluster of primordial germ cells in the indifferent gonadal ridges of embryos. Sex cords become seminiferous tubules in males, ovarian follicles in females.

sex determining region of the Y (SRY) Gene on the Y chromosome that encodes the testis determining factor, which in turn directs genitalia to develop as male.

Sharpey's fiber See *perforating fiber.*

shivering Rapid, rhythmic, involuntary contractions of skeletal muscle; initiated by the nervous system as body temperature falls.

sickle cell anemia Anemia characterized by the presence of crescent-shaped erythrocytes and excessive hemolysis; inherited as a recessive allele.

sigmoid colon Part of the colon between the descending colon and the rectum.

sigmoid mesocolon Fold of peritoneum attaching the sigmoid colon to the posterior abdominal wall.

signal transduction Cellular process that enables cells to respond internally to external stimuli on the cell membrane.

simple columnar epithelium A variety of epithelium, associated with secretion and absorption, that consists of a single layer of columnar cells.

simple cuboidal epithelium A variety of epithelium that consists of a single layer of cube-shaped cells; associated with secretion and absorption.

simple epithelium Epithelium consisting of a single layer of cells.

simple gland A variety of gland that consists of a secretory portion at the end of an unbranched duct.

sinoatrial (SA) node (SINE-oh-AYE-tree-al) Mass of specialized cardiac muscle fibers; acts as the pacemaker of the cardiac conduction system.

sinus (SINE-us) Hollow in a bone or other tissue; enlarged channel for blood or lymph. See *paranasal sinus.*

sinus venosus Region of the embryonic heart that conducts blood into the atrium; becomes a portion of the right atrium, including the SA node.

sinusitis (SINE-us-EYE-tis) Inflammation of the paranasal sinuses.

sinusoid (SINE-you-soyd) Capillary with larger diameter (10 to 20 μm or more) than ordinary capillaries; lined with fenestrated endothelium. Hepatic sinusoids.

situs inversus Reversed symmetry of the visceral organs: heart on right, liver on left, spleen on right, appendix on left. Inherited as a recessive allele. See *Kartagener's syndrome.*

skeletal muscle One of three classes of muscles, skeletal muscles move the skeleton. Skeletal muscle fibers are striated, multinucleated, and are controlled voluntarily, in contrast with cardiac and visceral (smooth) muscle fibers.

skeleton of the heart The fibrous rings that support the valves of the heart and provide origins and insertions for cardiac muscle.

sliding filament hypothesis Theory describing the mechanism by which actin and myosin myofilaments slide over one another during muscle contraction.

small intensely fluorescent (SIF) cells These cells secrete neurotransmitter substances and hormones in autonomic ganglia, storing their products in secretion granules that fluoresce in ultraviolet light. SIFs probably regulate synaptic transmission in these ganglia.

small intestine Portion of the digestive tube between the stomach and the cecum; consists of the duodenum, jejunum, and ileum.

smooth endoplasmic reticulum Endoplasmic reticulum that lacks ribosomes. The site of polysaccharide and lipid synthesis in the cytoplasm.

smooth muscle The class of muscle characterized by lack of cross-striation, by involuntary control, and by the ability to sustain tension over long periods. Known also as visceral muscle for its location in visceral organs, glands, and vessels.

smooth muscle fiber An individual smooth muscle cell.

sodium ion channel A membrane transport protein that transports sodium ions through a molecular pore in the protein molecule.

sodium-potassium ATPase Another name for sodium-potassium pumps. Used to recognize the ability of these membrane transport proteins to liber-

ate chemical energy enzymatically by cleaving molecules of ATP into ADP and phosphate.

sodium-potassium pump Biochemical mechanism that uses energy derived from ATP to achieve the active transport of potassium ions, opposite to that of sodium ions.

soft palate Posterior muscular portion of the palate, forming an incomplete septum between the mouth and the oropharynx and between the oropharynx and the nasopharynx.

solute The name given to any kind of molecule that dissolves in a solvent, usually water.

solution Homogeneous mixture formed when a solute dissolves in a solvent.

solvent Any liquid, usually water, that holds another substance in solution.

soma (SOE-mah) The central, enlarged portion of a neuron containing the cell nucleus and synthetic organelles, from which extend axon and dendrites. Also known as perikaryon or cell body.

somatic (so-MAT-ik) Relating to the body; the cells of the body, except the reproductive cells.

somatic mesoderm Mesoderm, derived from the lateral plate mesoderm, that forms the parietal walls of the coelome in association with ectoderm. See also *splanchnic mesoderm.*

somatic nervous system Composed of nerve fibers that send impulses from the central nervous system to skeletal muscle, skeleton, and skin.

somatic sex Masculinity or feminity based on secondary sex characteristics of the body.

somatomammotropin Hormone of the adenohypophysis that, with prolactin, stimulates the mammary glands to secrete colostrum and milk.

somatomotor center Center in the cerebral cortex anterior to the central sulcus that initiates voluntary action of the skeletal muscles.

somatopleura The composite tissue of the embryonic body wall that consists of somatic mesoderm and ectoderm. See also *splanchnopleura.*

somatosensory center Center in the cerebral cortex posterior to the central sulcus that receives sensory impulses from the body.

somatostatin (so-MAT-oh-STAT-in) Hypothalamic hormone capable of in-

hibiting the release of growth hormone by the adenohypophysis.

somatotroph (so-MAT-oh-TROFE) A variety of cell in the adenohypophysis that secretes somatotropin, growth hormone.

somatotropin Protein hormone of the adenohypophysis; it promotes body growth, fat mobilization, and inhibition of glucose utilization. Also known as growth hormone.

somesthetic Body sensations consciously perceived.

somite (SO-mite) Paired segments consisting of cell masses formed in the early embryonic mesoderm on either side of the neural tube; produce sclerotomes, dermatomes, and myotomes.

space of Disse See *perisinusoidal space.*

special sense Sight, hearing, balance, taste, and smell.

special sense organ The eyes, ears, tongue, and nose; that is, the structures that supply the special senses—sight, hearing, balance, taste, and smell.

specialized conducting fiber See *Purkinje fiber.*

spectrin Contractile protein on the cytoplasmic face of erythrocyte cell membranes that helps maintain the disklike shape of these cells.

speech Use of the voice in conveying ideas.

spermatic cord Cord formed by the vas deferens and associated vessels and nerves; extends through the inguinal canal into the scrotum.

spermatid (SPER-mah-tid) Cell derived from the secondary spermatocyte; gives rise to a spermatozoon.

spermatocyte Cell arising from a spermatogonium destined to become spermatozoon.

spermatogenesis Formation and development of spermatozoa.

spermatogonium, *pl.* **spermatogonia** (sper-MAT-oh-GOAN-ee-um, -ee-ah) Cell that divides by mitosis to form primary spermatocytes.

spermatozoon, *pl.* **spermatozoa** Sperm cell. Male gamete or sex cell, composed of a head and a tail. The spermatozoon contains the genetic information transmitted by males.

spermiogenesis The process that converts spermatids into sperm. The flagellum forms, the nucleus and acrosome form, and the spermatid loses most of its cytoplasm.

sphenoid (SFEE-noid) Wedge-shaped; in the skull, the sphenoid bone articulates with bones of the cranial cavity and the face.

sphincter (SFINK-ter) Circular muscle, usually visceral muscle, that closes or constricts body openings, such as the eyelids, pylorus, and anus.

sphincter pupillae Circular smooth muscle fibers of the iris diaphragm that constrict the pupil of the eye.

sphingomyelin The principal phospholipid of the myelin sheath of nerve fibers.

sphygmomanometer (SFIG-mo-ma-NOM-eh-ter) Instrument used with a pressure cuff to measure blood pressure.

spina bifida Absence of the vertebral arches of vertebrae, leaving gaps through which the spinal meninges may protrude.

spinal canal The central canal that carries cerebrospinal fluid down the spinal cord.

spinal cord The extension of the brain that descends the vertebral canal of the vertebral column.

spinal nerve Thirty-one pairs of nerves formed by the joining of the dorsal and ventral roots that arise from the spinal cord.

spinal tap A clinical procedure, similar to saddle block, that withdraws samples of cerebrospinal fluid from the spinal meninges through a syringe that enters between the lumbar vertebrae.

spinal tract Clusters of ascending or descending nerve fibers in the spinal cord, comparable to fascicles of spinal nerves.

spindle fiber Specialized microtubule that develops from each centrosome and extends toward the chromosomes during cell division.

spinous layer of epidermis The second layer of cells in the epidermis. Derived from cells of the basal layer, cells of the spinous layer begin to differentiate as keratinocytes, acquiring desmosomes and internal networks of keratin intermediate filaments. Further differentiation carries them into the granular layer.

spinous process (SPY-nus) The median bony processes that extend from the vertebral arches of vertebrae.

spiral artery One of the corkscrewlike arteries in the premenstrual endometrium; most obvious during the secretory phase of the uterine cycle.

spiral ganglion Contains cell bodies of sensory neurons that innervate hair cells of the organ of Corti.

spiral lamina This thin leaf of bone, attached to the modiolus, supports the basilar and vestibular membranes of the spiral organ of Corti.

spiral ligament Structure attaching the basilar membrane to the lateral wall of the bony labyrinth.

spiral organ Organ of Corti; rests on the basilar membrane and consists of the hair cells that detect sound.

splanchnic mesoderm Mesoderm that lines the visceral wall of the coelome. Associated with endoderm. See also *somatic mesoderm*.

splanchnic nerve Sympathetic nerve formed by preganglionic fibers that pass through the sympathetic chain ganglia to abdominal organs before synapsing.

splanchnopleura The composite tissue formed of endoderm and splanchnic mesoderm. See also *somatopleura*.

spleen Large lymphatic organ in the upper part of the abdominal cavity on the left side between the stomach and diaphragm, composed of white and red pulp. It responds to foreign substances in the blood, destroys worn out erythrocytes, and stores blood cells.

splenic cord A variety of cellular partition in the red pulp of the spleen that filters platelets, macrophages, and erythrocytes from the blood.

spongy urethra Portion of the male urethra, approximately 15 cm in length, that traverses the corpus spongiosum of the penis.

sprain A condition that stretches the capsule that encloses synovial joints. Severe strain may tear the capsule and ligaments of the joint.

squama (SKWAY-ma) The scalelike body of flat bones of the cranium.

squamous (SKWAY-mus) Scalelike, flat.

squamous cell carcinoma A metastatic cancer of the epidermis that derives from squamous epithelial cells.

squamous epithelium (SKWAY-mus) A variety of epithelium that consists of thin, scalelike cells; may be a single layer (simple squamous) or many-layered (stratified squamous).

squamous suture A type of suture that joins bones with overlapping slanted edges.

stance phase of gait The phase of walking in which each lower extremity in turn supports the weight of the body.

stapedius Small skeletal muscle attached to the stapes.

stapes Smallest of the three auditory ossicles; attached to the oval window.

statoconia Dense calcium carbonate granules of the vestibule and semicircular canals that detect motion and position.

stellate (STELL-ate) Star-shaped. Cells that extend numerous cytoplasmic processes from the central cell bodies are stellate cells, such as fibroblasts, astrocytes, and macrophages.

stem cell A progenitor cell from which tissues develop and are maintained by cell division.

stenosis (STEN-oh-sis) Narrowing of vessels and cardiac valves that reduces blood flow.

stepping reflex Reflexes of the lower extremities that move opposite limbs in walking and running.

stereocilium Elongated nonmobile microvillus that characterizes hair cells of the inner ear.

sternebra, *pl.* sternebrae (STER-nee-brah, -bree) Four cartilaginous models that fuse together during development to form the sternum.

sternum The breastbone.

steroid (STEER-oyd) Large family of lipids, including some reproductive hormones, vitamins, and cholesterol.

steroid hormone Lipid hormones, such as cortisol, estrogen, and testosterone, that are derived chemically from cholesterol.

stethoscope The instrument used to hear heart and lung sounds. Two tubes conduct sound from a diaphragm placed on the chest, to the ears of a clinician.

stomach Large sac between the esophagus and the small intestine, lying just beneath the diaphragm.

strabismus Lack of parallelism between the visual axes of the eyes, in which the eyes do not focus on the same object.

strap muscle A long, narrow variety of skeletal muscle with parallel muscle fibers that resembles a strap; sartorius muscle.

stratified epithelium Epithelium consisting of more than one layer of cells.

stratum Layer of tissue.

stratum basale (bay-SAL-eh) Basal or deepest layer of the epidermis; the layer of the epidermis where cells reproduce by mitosis.

stratum corneum Most superficial layer of the epidermis, consisting of flat, keratinized, dead cells.

stratum germinativum See *stratum basale.*

stratum granulosum Layer of cells in the epidermis filled with granules of keratohyalin.

stratum lucidum Clear layer of the epidermis found in the thick skin between the stratum granulosum and the stratum corneum.

stratum spinosum Layer of many-sided cells in the epidermis with intercellular connections (desmosomes) that give the cells a spiny appearance.

stress fiber Clusters and bunches of actin microfilaments connected with the plasma membrane in the cytoplasm of stretching cells.

stretch reflex The reflex that increases a muscle's tension when it is subjected to increased load. Stretch reflexes accommodate limbs to greater or lesser loads.

stria, *pl.* striae Line or streak in the skin that is a different texture or color from the surrounding skin. Stretch mark.

striated Striped; marked by stripes or bands. Striated muscle fibers display cross-bands.

striation A stripe or band.

stride The cyclic motions of walking, beginning with heel-strike or toe-off, that bring the limbs to the same position before repeating.

stroke Term denoting a sudden neurological affliction, usually related to blockage or hemorrhage of cerebral vessels.

stroke volume Volume of blood pumped out of one ventricle of the heart in a single beat, approximately 70 ml.

styloid process The slender, bony process that suspends muscles of the tongue and pharynx from the base of the skull.

subcutaneous Under the skin; same tissue as the hypodermis.

subcutaneous bursa Bursae located between the skin and tendons or bone. See *bursa.*

subcutaneous fascia Sheets or bands of fibrous connective tissue that connect the skin to underlying muscles and bones.

subcutaneous plexus Deep network of vessels of the skin that communicates with superficial layers of the skin and deeper body structures.

subdural hematoma Bleeding that causes blood to accumulate in the subdural space.

subdural space The potential space formed between the arachnoid and dura mater of the brain and spinal cord.

subendothelial connective tissue The connective tissue of the tunica intima that supports the endothelium of vessels.

sublingual gland Pair of salivary glands in the floor of the mouth inferior to the tongue.

subluxation Partial dislocation of a synovial joint, in which the articular surfaces of the bones have not completely separated.

submandibular duct This duct delivers secretions from the submandibular gland to the floor of the oral cavity; also known as Wharton's duct.

submandibular gland Salivary glands in the neck, located in the space bounded by the two bellies of the digastric muscle and the angle of the mandible.

submucosa Layer of connective tissue beneath a mucous membrane. The submucosa of the gut wall lies deep to the mucosa.

submucosal gland Exocrine glands in the submucosa of the intestinal wall.

submucosal mucous gland Glands of the submucosa of the duodenum that secrete alkaline mucus.

submucosal plexus Ganglionated plexus of unmyelinated nerve fibers in the intestinal submucosa.

submuscular bursa Sac of synovial fluid between certain muscles and bones or ligaments that dissipates friction. See *bursa.*

substantia nigra Black nuclear mass in the midbrain; involved in coordinating movement and maintaining muscle tone.

subtendinous bursa Sac of synovial fluid between certain tendons and surrounding tissues that dissipates friction. See *bursa.*

subthreshold stimulus Stimulus resulting in a local potential so small that it does not reach threshold and produce an action potential.

sulcus, *pl.* sulci Furrow or groove on the surface of the brain between the gyri.

superficial Near the surface of an organ or tissue.

superficial inguinal ring Slitlike opening in the aponeurosis of the external oblique muscle of the abdominal wall through which the spermatic cord

(round ligament in females) emerges from the inguinal canal.

superior Up, or higher, with reference to the anatomical position.

superior colliculus Two rounded eminences on the roof of the midbrain that aid in coordinating eye movements.

superior vena cava Vein that returns blood from the head and neck, upper limbs, and thorax to the right atrium.

supination (SOO-pin-AYE-shun) Rotation of the forearm (when the forearm is parallel to the ground) so that its anterior surface is up (supine).

supinator The muscle in the posterior compartment of the forearm that supinates the forearm.

supine Lying on one's back, face up.

surface mucous cells A variety of mucus-secreting cells that line the stomach and other digestive organs, in contrast to mucous cells that line gastric pits and intestinal glands.

surface/volume rule The proposition stating that the surface area of a body increases slower than its volume, and thus limits the size of cells and tissues.

surfactant (surf-ACT-ant) Lipoproteins forming a monomolecular layer over pulmonary alveolar surfaces; stabilizes alveolar volume by reducing surface tension and the tendency for the alveoli to collapse. Produced by septal cells.

suspension Liquid through which a solid is dispersed and from which the solid separates unless the liquid is kept in motion.

suspensory ligament Band of peritoneum that extends from the ovary to the body wall; contains the ovarian vessels and nerves. Also, small ligament, attached to the margin of the lens in the eye and the ciliary body, that holds the lens in place.

sustentacular cells of Sertoli See *Sertoli cell.*

sutural bone Small accessory bones incorporated into the lambdoidal and sagittal sutures of the calvarium.

suture (SOO-tshur) Junction between bones of the skull.

sweat Perspiration; secretions produced by the sweat glands of the skin.

sweat gland Variety of gland that produces a watery secretion called sweat. Some sweat glands produce viscous organic secretions. See *apocrine sweat gland.*

sweat pore The orifice of a duct that delivers perspiration from a sweat gland to the surface of the skin.

swing phase of gait The phase of walking and running when each limb leaves the ground and swings forward for the next step.

sympathetic chain ganglion Collection of autonomic ganglia that are connected to each other to form a chain along both sides of the vertebral column.

sympathetic division Subdivision of the autonomic division of the nervous system characterized by having the cell bodies of its preganglionic neurons located in the thoracic and upper lumbar regions of the spinal cord (thoracolumbar division); most active in preparing the body for physical activity; fight-or-flight response.

symphysis, *pl.* **symphyses** (SIM-fih-sis, -sees) Fibrocartilage joint between two bones; pubic symphysis.

symport A variety of membrane transport protein that transports two different molecules simultaneously across the plasma membrane in the same direction.

synapse (SIN-aps) Junction between neurons that transmits from axon to dendrite or cell body. Syn-apses also connect neurons to skeletal muscle fibers, gland cells, or sensory receptors. See *chemical synapse* and *electrical synapse.*

synapsis The process by which homologous chromosomes pair during prophase of meiosis I.

synaptic cleft (SIN-ap-tik) A narrow space across which neurotransmitter molecules diffuse at chemical synapses between the axon terminal of a presynaptic neuron and dendrite or cell body of a postsynaptic neuron.

synaptic vesicle Small vesicles that release neurotransmitter molecules from the axon terminal at chemical synapses.

synaptonemal complex The protein structure that binds homologous chromosomes together during synapsis in meiosis.

synarthrosis, *pl.* **synarthroses** An immovable joint.

synchondrosis, *pl.* **synchondroses** (SIN-kond-ROE-sis, -sees) Union between two bones formed by hyaline cartilage.

syncytiotrophoblast Outer layer of the trophoblast, composed of a syncytium; also called syntrophoblast.

syncytium A multinucleated colony sharing a common cytoplasm produced when many cells fuse. A func-

tional syncytium behaves like a true syncytium, but the cells remain separate. Numerous gap junctions allow visceral smooth muscle cells and cardiac cells to function as a syncytium.

syndactyly (sin-DAK-till-ee) A recessive genetic condition in which cartilaginous models of bone fuse during development, reducing the number of digits.

syndesmophyte Cartilaginous growth that can accumulate in cartilaginous joints and immobilize them. See *osteophyte.*

syndesmosis, *pl.* **syndesmoses** (SIN-dez-mo-sis, -sees) Form of fibrous joint in which opposing surfaces that are some distance apart are united by ligaments.

synergist Muscle that works with other muscles to cause a movement.

synostosis, *pl.* **synostoses** (SIN-oss-TOE-sis, -sees) Ossified joint, usually an ossified suture.

synovial Relating to synovial joints.

synovial capsule The fibrous sleeve that surrounds synovial joints.

synovial cavity The internal space that holds synovial fluid between the articulating surfaces of bones in synovial joints.

synovial fluid Slippery, lubricating fluid found inside synovial joints, tendon sheaths, and bursae; produced by the synovial membranes.

synovial joint Joint in which the ends of the bones are covered with articular cartilage and are enclosed in a joint cavity filled with synovial fluid. The bones are held together by the joint capsule.

synovial membrane The inner connective tissue lining of the synovial capsule that secretes synovial fluid.

synovium See *synovial membrane.*

syntrophoblast (sin-TROE-fo-blast) The superficial, syncytial layer of the placenta that invades the endometrial wall of the uterus. See *syncytium.*

system Complex whole composed of related parts; an organ system consists of anatomically and functionally related organs.

systemic anatomy The study of anatomy by organ system rather than by body region. See *regional anatomy.*

systole (SIS-tow-lee) Contraction of the heart chambers during which blood leaves the chambers; usually refers to ventricular contraction, but also applies to atrial contraction.

T-cell Thymus-derived lymphocyte responsible for cell-mediated immunity.

T-tubule Tubelike invagination of the sarcolemma that conducts action potentials to the sarcoplasmic reticulum of skeletal and cardiac muscle fibers.

T-wave Deflection in the electrocardiogram following the QRS complex; represents ventricular repolarization.

tachycardia (TAK-ih-KAR-dee-ah) Rapid heart beat. See *dysrhythmia.*

tactile corpuscle Oval receptor found in the papillae of the dermis; responsible for fine, discriminative touch; also known as Meissner's corpuscle.

tactile disk Cuplike receptor found in the epidermis; responsible for light touch and superficial pressure; Merkel's disk. These encapsulated sensory endings each contain about 50 Merkel cells.

taenia colus, *pl.* **taeniae coli** (TANE-ee-ah KOAL-us, -eye) Three longitudinal bands of smooth muscle fibers extending along the wall of the colon.

talus A bone of the ankle (a tarsal).

target tissue Tissue on which a hormone acts.

tarsal Each of the seven ankle bones.

tarsal gland Meibomian gland; sebaceous glands of the eyelid that secrete an oily film over the surface of the eyeball that lubricates and helps prevent evaporation of the tear film from the conjunctiva.

tarsal plate Crescent-shaped layer of connective tissue that helps maintain the shape of the eyelid.

taste Sensation produced when a chemical stimulus is applied to the taste receptors in the tongue.

taste bud Sensory structure, mostly on the tongue, that functions as a taste receptor.

taste pore Small opening in a taste bud that admits molecules to the bud where they are detected.

Tay-Sachs disease An inherited (recessive, chromosome 15), progressive, lethal neural disorder that results from the absence of the enzyme hexosaminidase. This deficiency leads to accumulation of phospholipids in the brain, dementia, blindness, and death in early childhood.

tectorial membrane A membrane attached to the spiral lamina that extends over the hair cells of the spiral organ of Corti; kinocilia extend from the apical surface of the hair cells to the tectorial membrane.

tectum Roof of the midbrain.

tegmentum Floor of the midbrain.

telencephalon Anterior division of the embryonic brain, from which the cerebral hemispheres develop.

telodendron, *pl.* **telodendria** A terminal branch of an axon. See *terminal arborization.*

telogen (TELL-o-jen) The dormant phase of the hair growth cycle.

telophase Telophase is the last phase of mitosis, in which two daughter nuclei are established from the chromatids that have reached opposite ends of the mitotic spindle. Cytokinesis cleaves the cytoplasm into two daughter cells, and interphase resumes.

tendon Band or cord of dense connective tissue that connects a muscle to a bone or other structure.

tendon reflex The response that inhibits muscle contraction and thereby limits the maximum tension a muscle produces. This reflex is mediated by Golgi tendon organs in the tendons of skeletal muscles. See *Golgi tendon organ.*

tennis elbow Painful inflammation of the tendons of the forearm extensor muscles at their origins on the lateral epicondyle of the humerus.

Tenon's capsule The thin connective tissue sheath that covers the posterior portion of the eyeball, from the ciliary border to the optic nerve, and allows the eyeball to move within the orbit.

tenosynovitis Inflammation of the synovial tendon sheath.

tensor tympani The small skeletal muscle that inserts on the malleus and adjusts the tension of the tympanic membrane to the intensity of sound.

teres Round, smooth, and cylindrical. The teres major muscle of the shoulder is smooth and cylindrical.

terminal arborization All the end branches, or telodendria, of an axon.

terminal bouton Enlarged axon terminal or presynaptic terminal that forms synapses with other neurons.

terminal bronchiole Ultimate airway of the conducting zone of the lungs; lined by simple columnar or cuboidal epithelium without mucous goblet cells. Most of the cells are ciliated, but a few nonciliated, serous-secreting cells occur. Leads to respiratory bronchioles and respiratory zone.

terminal cisterna Enlarged ends of the sarcoplasmic reticulum, associated with T-tubules.

terminal ganglia Autonomic ganglia that reside in tissues and organs.

terminal hair Long, coarse, usually pigmented hair found in the scalp, eyebrows, and eyelid; replaces vellus hair.

terminal sac stage The final period of lung development in the fetus, from 26 weeks' gestation to birth, in which the acini of the lungs form and the lungs become functional.

terminal sulcus V-shaped groove on the surface of the tongue at its posterior margin.

terminal web The network of intermediate filaments that supports microvilli and other membranous structures at the apical ends of epithelial cells.

tertiary bronchus A branch of secondary bronchi that enters an individual bronchopulmonary segment of the lungs.

testicle The testis.

testis, *pl.* **testes** Paired male reproductive glands located in the scrotum; produce spermatozoa, testosterone, and inhibin.

testis determining factor The putative protein, encoded on the sex-determining region of the Y chromosome, that switches genital development to male.

testosterone Steroid hormone secreted primarily by the testes; promotes spermatogenesis, maintenance and development of male reproductive organs, secondary sexual characteristics, and sexual behavior.

tetrad A structure formed during synapsis in meiosis I when two homologous chromosomes, each with two chromatids, pair, thereby forming the four-stranded structure called a tetrad.

tetraiodothyronine One of the iodine-containing thyroid hormones. Also called thyroxine, it contains four iodine atoms.

tetralogy of Fallot The "blue baby" syndrome. Four congenital defects of the heart and great vessels lead deoxygenated blood into the aorta and body tissues. Obstruction of pulmonary circulation with hypertrophy of the right ventricle, ventricular septal defect, and partial transposition of the aorta all enable deoxygenated blood to enter the aorta instead of the pulmonary trunk.

thalamus Large mass of gray matter that forms the larger dorsal subdivi-

sion of the diencephalon. "Gateway" to the cerebrum. Ascending tracts terminate in thalamic nuclei on synapses with neurons that project to the cerebral cortex.

thalassemia (THAL-ah-SEEM-ee-ah) A group of inherited disorders of hemoglobin metabolism in which there is a decrease in synthesis of a particular globin chain without change in the structure of that chain.

theca A sheath or capsule.

theca externa External fibrous layer of the theca or capsule of a vesicular follicle.

theca interna Inner vascular layer of the theca or capsule of the vesicular follicle; produces estrogen and contributes to the formation of the corpus luteum after ovulation.

thecal cell The variety of cells composing the theca interna and externa of vesicular follicles of the ovary.

thenar Fleshy mass of tissue at the base of the thumb; contains muscles responsible for moving the thumb.

thermoreceptor The class of receptors that responds to temperature differences; stimuli received by thermoreceptors contribute to the perception of heat and cold.

thick myofilament The myosin-containing filaments of the A-bands of sarcomeres in skeletal and cardiac muscle fibers. Crosslinks between thick and thin myofilaments enable muscle fibers to contract.

thigh That part of the lower limb between the hip and knee.

thin myofilament The actin-containing filaments of the I-bands of sarcomeres in skeletal and cardiac muscle fibers. See *thick myofilament.*

third class lever A bone such as the radius moves as a third class lever when the biceps flexes the radius. In third class levers the effort (insertion of the biceps on the radius) is between the fulcrum (the elbow) and the load (the wrist and hand).

third degree burn Full thickness burn; severe burn that removes the epidermis and dermis of the skin, exposing fascia, bone, and muscle. Third degree burns require skin grafts to cover the wound.

third heart sound Sound sometimes heard at, and corresponding with, the first phase of rapid ventricular filling.

thoracic (thore-ASS-ic) Having to do with the thorax, or chest.

thoracic aorta The portion of the aorta passing through the thoracic cavity to the diaphragm.

thoracic cavity Space within the thoracic walls, bounded below by the diaphragm and above by the neck.

thoracic duct Largest lymph vessel in the body, beginning at the cisterna chyli and emptying into the left subclavian vein; drains the left side of the head and neck, the left upper thorax, the left upper limb, and the inferior half of the body.

thoracolumbar division Synonym for the sympathetic division of the autonomic nervous system.

thoracolumbar outflow This term characterizes the sympathetic division of the autonomic nervous system in which efferent pathways from the thoracic segments and the first two lumbar segments of the spinal cord reach visceral organs. See *craniosacral outflow.*

thorax Chest; upper part of the trunk between the neck and abdomen.

thoroughfare channel Central capillary channel that conducts blood through a capillary bed from an arteriole to a venule.

threshold potential The minimum membrane potential at which an action potential can begin. See *threshold stimulus.*

threshold stimulus Stimulus just large enough to produce an action potential in a neuron.

thrombin Enzyme, formed in blood, that converts fibrinogen into fibrin.

thrombocyte Platelet.

thrombocytopenia (THROM-bo-site-oh-PEE-nee-ah) Condition in which there is an abnormally small number of platelets in the blood. Low platelet count, less than 150,000 per cubic mm.

thrombocytosis Condition in which excessive platelets circulate in the blood. High platelet count, greater than 450,000 per cubic mm. Can lead to thrombosis.

thrombopoiesis The bone marrow process that produces thrombocytes.

thrombopoietin A glycoprotein that stimulates thrombopoiesis.

thrombosis Clotting within a blood vessel that may cause infarction of tissues supplied by the vessel.

thrombospondin A clotting protein that binds fibrin fibers to the surfaces of platelets.

thrombus Clot in the cardiovascular system formed from constituents of blood; may block a vessel or attach to the vessel or heart wall without obstructing the lumen.

thymopoietin (THIGH-mo-POY-et-in) Thymus growth factor that stimulates T-lymphocytes to proliferate in the cortex of the thymus.

thymosin The hormone of the thymus gland that stimulates T-lymphocytes to express antigen receptors on their plasma membranes.

thymus gland (THIGH-muss) Bi-lobed lymph organ located in the inferior neck and superior mediastinum; produces T-lymphocytes.

thyroglobulin Thyroid hormone-containing protein, stored in the colloid within the thyroid follicles.

thyroid The thyroid gland, or pertaining to this gland.

thyroid cartilage Largest laryngeal cartilage. It forms the laryngeal prominence, or Adam's apple.

thyroid gland Endocrine gland located inferior to the larynx and consisting of two lobes connected by an isthmus; secretes the thyroid hormones triiodothyronine and tetraiodothyronine.

thyroid-stimulating hormone (TSH) Glycoprotein hormone released from the hypothalamus; stimulates thyroid hormone secretion from the thyroid gland.

thyrotroph (THY-ro-trofe) A variety of cell in the adenohypophysis that produces thyroid stimulating hormone.

thyrotropin See *thyroid-stimulating hormone.*

thyroxine See *tetraiodothyronine.*

tibia The larger, medial bone of the leg.

tight junction A variety of junctional complex that fuses the membranes of adjacent cells together.

tinnitus Spontaneous sensation of noise without sound stimuli.

tissue Collection of similar cells and the substances between them.

tissue repair Substitution of viable cells for damaged or dead cells by regeneration or replacement.

toe-off phase of gait The instant during walking and running when each toe pushes off from the floor.

tolerance Failure of the specific immune system to respond to an antigen.

tongue Muscular organ occupying most of the oral cavity when the mouth is closed; facilitates mastication, swallowing, and speech.

tonsil Refers to large collections of lymphatic tissue beneath the mucous membrane of the oral cavity and pharynx; lingual, pharyngeal, and palatine tonsils.

tooth Hard, conical structure set in the alveoli of the upper or lower jaws; used in mastication and assists in speech.

tophi (TOE-fee) Fibrous nodules that encapsulate crystals of uric acid in joints affected by gouty arthritis. Tophi can lead to osteoarthritis.

torsion of testicle Twisting of the spermatic cord that suspends the testis in the scrotal sac and delivers arteries and veins to the testis. A painful and potentially serious disorder that can destroy the testis.

trabecula, *pl.* **trabeculae** (tra-BEK-you-lah, -lee) Rods or fibers forming a network that supports tissues; the plates of cancellous bone, or the fibrous trabeculae that support the internal tissues of lymph nodes.

trabeculae carnae (tra-BEK-you-lee KAR-nee) Muscular bundles lining the walls of the ventricles of the heart.

trachea Air tube extending from the larynx into the thorax, where it divides to form the bronchi; composed of 16 to 20 rings of hyaline cartilage.

tracheobronchial tree (TRAKE-ee-oh-BRONK-ee-al) The branching, treelike passages of the respiratory system that conduct air to and from the alveoli of the lungs. The trachea forms the trunk and the bronchi and bronchioles, its branches.

tracheostomy Incision into the trachea, creating an opening into which a tube can be inserted to facilitate the passage of air.

tract A bundle of axons in the spinal cord and brain, comparable to a nerve fascicle in spinal nerves. See *spinal tract.*

trans face, Golgi apparatus Cytoplasmic vesicles leave the trans face of the Golgi membranes; the trans face usually faces away from the nucleus. See *cis face.*

transcription Cellular process that synthesizes RNA from DNA templates in the genes.

transcytosis The cellular transport process in which transport vesicles and their contents shuttle back and forth through the cytoplasm between the apical and basal plasma membranes of epithelial cells. Endocytosis takes up materials from outside, and exocytosis releases contents to the outside.

transdermal drug delivery Administration of drugs in low concentrations over long periods, through the skin, by means of an adhesive patch.

transfer RNA (tRNA) RNA that attaches to individual amino acids and transports them to the ribosomes, where they are connected to form a protein polypeptide chain.

transfusion Transfer of immunologically matching blood from one person to another.

transient pore When transcytosis occurs in thin epithelial cells, the cell is so thin that a vesicle leaving the outer surface may immediately reach the inner surface, as if it were briefly an open, transient, pore through the entire cell.

transitional epithelium Stratified epithelium that may be either cuboidal or squamous; appears squamous when stretched, cuboidal when relaxed.

translation Synthesis of polypeptide chains on ribosomes, directed by information contained in messenger RNA molecules.

transport vesicle A variety of membranous cytoplasmic vesicles that transports its contents through the cytoplasm.

transvaginal ultrasound An ultrasonic procedure that visualizes an early fetus in the uterus by means of a sound transducer inserted into the vagina.

transverse Across the central axis of the body or a part of the body.

transverse colon Part of the colon between the right and left colic flexures.

transverse mesocolon Fold of peritoneum attaching the transverse colon to the posterior abdominal wall.

transverse section Plane separating the body or any part of the body into superior and inferior portions; a cross-section. Transverse sections cut a tissue or organ across its longitudinal axis.

transverse tubule Tubule that extends from the sarcolemma to the sarcoplasmic reticulum of striated muscle fibers. See *T-tubule*.

triad A structure composed of three parts. Two terminal cisternae of the sarcoplasmic reticulum and a T-tubule between them. See *portal triad*.

triaxial motion of synovial joints Motion of the bones of a joint about three axes that pass through the joint.

tricuspid valve (try-KUS-pid) Valve closing the orifice between the right atrium and right ventricle of the heart.

triglyceride Glycerol with three attached fatty acids.

trigone (TRY-goan) Triangular smooth area at the base of the bladder, between the openings of the two ureters and that of the urethra.

triiodothyronine One of the thyroid hormones; contains three iodine atoms.

trimester (TRY-mess-ter) One third of the period of pregnancy and fetal development, which is divided into three 13-week trimesters.

trochanter (troe-KANT-er) Greater and lesser trochanters of the femur are processes that receive tendons of pelvic muscles.

trochlea (TROKE-lee-ah) Structure shaped like or serving as a pulley or spool; the trochlea of the elbow and of the superior oblique muscle of the eyeball.

trochoid joint (TROE-koyd) Also known as pivot joints, trochoid joints are a class of synovial joints that enable rotation. The radius rotates in a trochoid joint on the capitulum of the humerus, for example. See *pivot joint.*

trophoblast Cell layer forming the outer layer of the blastocyst, which invades the uterine mucosa during implantation; the trophoblast does not become part of the embryo but contributes to the formation of the placenta.

tropomyosin (TROE-poe-MYO-sin) Fibrous protein component of actin myofilaments in muscle fibers.

troponin (TRO-poe-nin) Globular protein component of the actin myofilament.

true pelvis Portion of the pelvis inferior to the pelvic brim.

true rib See *vertebrosternal rib.*

truncus arteriosus (TRUNK-us) The truncus arteriosus receives blood from the ventricle of the embryonic heart and subsequently divides into the ascending aorta and the pulmonary trunk.

trunk The central axial portion of the body that contains the visceral organs. Head, neck, and extremities extend from the trunk.

trypsin Proteolytic enzyme formed in the small intestine from the inactive pancreatic precursor trypsinogen.

tubercle A small projection that receives tendons or ligaments on a bone.

tuberosity A large projection that receives tendons or ligaments on a bone.

tubular gland (TOOB-you-lar) A variety of gland in which the secretory portion is tubular rather than cup-shaped (alveolar), such as gastric glands.

tubular reabsorption Movement of materials, by means of diffusion, active transport, or co-transport, from the filtrate within a nephron to the blood.

tubular secretion Movement of materials by means of active transport from the blood into the filtrate of a nephron.

tubulin The principal structural protein of microtubules in the cytoskeleton.

tunic The enveloping layers of a body part; three tunics of vessels; the fibrous, vascular, and neural tunics of the eyeball, and four tunics of the digestive tube.

tunica adventitia (TUNE-ik-ah) Outermost fibrous coat of a vessel or an organ that is derived from the surrounding connective tissue.

tunica albuginea (AL-byou-JIN-ee-ah) Dense, white, collagenous tunic surrounding the testis.

tunica intima (IN-tih-mah) Innermost coat of a blood vessel; consists of endothelium, subendothelial connective tissue, and an inner elastic membrane.

tunica media Middle, usually muscular, coat of an artery or other tubular structure.

tunica vaginalis (VA-jin-AL-iss) Closed sac, derived from the peritoneum, that contains the testis and epididymis. It forms from an outpocket of the abdominal cavity, the processus vaginalis.

tunnel of Corti The channel between the inner and outer rows of hair cells in the spiral organ of Corti of the cochlea.

two-solute hypothesis The hypothesis that proposes that kidneys concentrate urea in the urine through the use of two solutes: urea itself and sodium chloride ions.

tympanic cavity The portion of the middle ear that houses the ossicles and tympanic membrane; connects to the auditory tube and inner ear.

tympanic membrane Eardrum; cellular membrane that separates the external from the middle ear; vibrates in response to sound waves.

ulna The medial bone of the forearm.

umbilical cord The cord that connects the fetus to the placenta, and carries two umbilical arteries and one umbilical vein.

unequal cleavage A process in which cytokinesis delivers unequal amounts of cytoplasm to each daughter cell.

uniaxial motion In uniaxial motion of synovial joints, a body part moves in one axis only. Flexion of the elbow is uniaxial motion.

unicellular gland A single-celled gland.

unipennate muscle (YOU-nih-PEN-ate) Skeletal muscles that have one set of oblique fibers that originate from one tendon and insert on another.

unipolar neuron One of the three anatomic categories of neurons, consisting of a nerve cell body with a single axon projecting from it; also called a pseudounipolar neuron.

uniport Membrane transport proteins that transport one molecule or ion across the plasma membrane at a time.

unmyelinated axon Axons that lack a myelin sheath.

unsaturated Carbon chain of a fatty acid that possesses one or more double or triple bonds.

upper respiratory tract The nasal cavity, pharynx, and associated structures.

ureter (you-REET-er) Tube conducting urine from the kidney to the urinary bladder.

ureteric bud (YOU-ree-TARE-ik) An early structure of developing kidneys; the ureter buds from the mesonephric duct into surrounding mesenchyme tissue.

urethra (you-REE-thrah) The tube that discharges urine from the bladder.

urethral gland One of numerous mucous glands in the wall of the spongy urethra in males.

urethral sphincter The sphincter muscles that guard the exit of urine from the bladder through the urethra. The internal sphincter is visceral muscle, and the external sphincter is skeletal muscle.

urogenital diaphragm The floor of the pelvic cavity beneath the bladder, penetrated by the anus, urethra, and, in females, the vagina.

urogenital fold Paired longitudinal ridges developing in the embryo on either side of the urogenital orifice. In males, they form part of the penis; in females, they form the labia minora.

urogenital groove The space between the urogenital folds of developing external genitalia.

urogenital sinus Derived from the cloaca, this chamber becomes the bladder and urethra in both sexes and also contributes to the vestibule of the vagina in females. See *cloaca*.

uterine cavity Space within the uterus extending from the cervical canal to the openings of the uterine tubes.

uterine cycle Series of events that occur in a regular fashion in the uterus of sexually mature, nonpregnant females; prepares the uterine lining for implantation of the embryo.

uterine part Portion of the uterine tube that passes through the wall of the uterus.

uterine tube The pair of tubes leading on either side from the uterus to the ovary; consists of the infundibulum, ampulla, isthmus, and uterine parts; also called the fallopian tube or oviduct.

uterovesical pouch (YOU-ter-oh-VEH-sih-kal) The space in the pelvic cavity of females between the broad ligament, which supports the uterus and the bladder.

uterus Hollow muscular organ in which the fertilized ovum develops into a fetus.

utricle (YOU-trih-kl) Part of the membranous labyrinth; contains the sensory structure, the macula, that detects static equilibrium.

uveal tract Also known as the vascular tunic of the eye, the uveal tract consists of the choroid, ciliary body, and iris.

uvula (YOU-view-la) Small, grapelike appendage at the posterior margin of the soft palate.

vagina Genital canal in females, extending from the uterus to the vulva.

vaginal smear Procedure that collects epithelial cells from the vagina to monitor the menstrual cycle.

vallate papillae The variety of papillae on the tongue that are surrounded by a groove and a fold of epithelium. Also known as circumvallate papillae.

varicose vein Permanent dilation and tortuosity of veins as a result of incompetent valves (valves that do not close).

varicosity A local cytoplasmic swelling in axons of postganglionic neurons that makes synapses with glands and visceral muscles in the autonomic nervous system.

vas deferens (DEF-er-enz) See *ductus deferens*.

vasa nervorum Vessels of the spinal nerves.

vasa recta (VAH-sa REK-tah) Specialized capillaries that follow the nephron loop from the cortex of the kidney into the medulla and back to the cortex.

vasa vasorum (VAH-sa vah-ZOR-um) Small vessels distributed to the outer and middle coats of larger blood vessels.

vascular tree The branching pattern formed by arteries and veins.

vascular tunic Middle layer of the eye; contains many blood vessels. See *choroid* or *uveal tract*.

vasculogenesis The process in which embryonic blood vessels develop by coalescing from smaller clusters of mesenchymal cells. See *angiogenesis*.

vasoconstriction (VAH-zo-) Decreased diameter of blood vessels; restricts blood flow.

vasodilation Increased diameter of blood vessels; increases blood flow.

vasomotor center Area within the medulla oblongata that regulates the diameter of blood vessels by way of the sympathetic nervous system.

vasopressin (VAY-zo-PRESS-in) Hormone secreted from the neurohypophysis that causes vasoconstriction and acts on the kidney to reduce urinary volume; also called antidiuretic hormone.

vein Blood vessel that carries blood toward the heart; after birth, all veins except the pulmonary veins carry unoxygenated blood.

vellus Short, fine, usually unpigmented hair that covers the body except for the scalp, eyebrows, and eyelids. Much of the vellus is replaced at puberty by terminal hairs.

vena cava (VEE-nah KAVE-ah) The largest veins of the body, which return blood to the right atrium of the heart. The superior vena cava drains the head, neck, chest, and upper extremities; the inferior vena cava drains from below the diaphragm.

venae comitantes (VEEN-ee KOM-ih-TAN-tes) Venous anastomoses that surround major arteries. The pulsing artery pumps blood through the veins, past venous valves.

venous sinus Endothelium-lined venous channel in the dura mater that receives cerebrospinal fluid from the arachnoid granulations.

venous tree (VEEN-ous) The branching pattern of smaller tributary veins that collect blood from the tissues and deliver it to larger veins.

ventilation Movement of gases into and out of the lungs.

ventral aorta The vessel that conducts blood from the conus arteriosus of the embryonic heart to the aortic arches.

ventral cavity Together with the dorsal cavity, the ventral cavity is a feature of the human body plan. The ventral cavity consists of the thoracic and abdominopelvic cavities.

ventral mesentery The sheet of tissue that attaches the anterior surface of certain organs to the anterior body wall.

ventral root Motor (efferent) root of a spinal nerve.

ventricle Chambers of the heart that pump blood into arteries. In the brain, the spaces filled with cerebrospinal fluid.

ventricular diastole Dilation and refilling of the ventricles of the heart.

ventricular fold See *vestibular fold*.

ventricular septal defect A congenital opening in the interventricular septum of the heart that allows blood to flow between the ventricles and to bypass the pulmonary circuit.

ventricular systole Contraction of the ventricles.

venule (VEN-yool) Minute vein, consisting of endothelium and a few scattered smooth muscles, that carries blood away from capillaries.

vermiform appendix Wormlike sac extending from the blind end of the cecum.

vermis The highly folded medial cortex of the cerebellum.

vernix caseosa (VER-niks KASE-ee-oh-sah) Fatty, cheeselike substance, composed of sloughed epithelial cells and secretions, that covers the fetal skin.

vertebral (VER-TEH-bral) Concerning a vertebra or the vertebrae.

vertebral canal The canal, formed by the vertebral column, that encloses the spinal cord.

vertebral column The 26 vertebrae considered together; bears the weight of the trunk, protects the spinal cord,

is the site of exit of the spinal nerves, and provides attachment sites for muscles.

vertebral rib Floating rib; the eleventh and twelfth ribs articulate only with the vertebral column, not with the sternum.

vertebrochondral rib The eighth, ninth, and tenth ribs articulate with the vertebral column and indirectly with the sternum, through their costal cartilages and the costal cartilage of the seventh rib.

vertebrosternal rib The first seven ribs; these ribs articulate with the vertebral column and, through costal cartilages, directly with the sternum.

vertigo The false, disorienting sense of motion associated with defective vestibular organs and nerves.

very low density lipoprotein (VLDL) Proteins that carry large numbers of cholesterol molecules in the blood.

vesical (VEH-sih-kal) Vesical refers to the bladder. Not to be confused with vesicle.

vesicle Small sac containing a liquid or gas, *e.g.*, a blister in the skin or an intracellular, membrane-bound sac.

vesicular follicle Graafian follicle; ovarian follicle in which the oocyte attains its full size and is surrounded by granulosa cells at the periphery of the fluid-filled atrium; the follicular cells proliferate, and the theca develops into internal and external layers. See *theca interna* and *theca externa*.

vestibular fold False vocal cord. One of two folds of mucous membrane stretching across the laryngeal cavity from the angle of the thyroid cartilage to the arytenoid cartilage superior to the vocal cords; helps close the glottis.

vestibular gland See *greater vestibular gland*.

vestibular membrane Membrane separating the cochlear duct and the scala vestibuli.

vestibule A space or cavity occuring at the entrance to another structure, as in the anterior part of the nasal cavity just inside the external nares that is enclosed by cartilage; space between the lips and the alveolar processes and teeth; middle region of the inner ear containing the utricle and saccule; space behind the labia minora containing the openings of the vagina, urethra, and vestibular glands.

vestibulocochlear nerve Cranial nerve composed of vestibular and cochlear nerve fibers that conduct equilibrium

and auditory impulses, respectively, to the brain.

villus, *pl.* villi Projections of the mucous membrane of the intestine; they are leaf-shaped in the duodenum and become shorter, more finger-shaped, and sparser in the ileum.

viscera The organs of the thoracic and abdominopelvic cavities.

visceral (VISS-er-al) Relating to the viscera.

visceral afferent neuron A neuron of the autonomic nervous system that conducts impulses toward the brain and spinal cord.

visceral efferent neuron A neuron of the autonomic nervous system that conducts impulses away from the brain and spinal cord.

visceral nervous system The autonomic nervous system; the nerve fibers that serve the visceral organs, vessels, sweat glands, and arrector pili muscles.

visceral pericardium Serous membrane covering the surface of the heart. Also called the epicardium.

visceral peritoneum Layer of peritoneum covering the abdominal organs.

visceral pleura (PLOOR-ah) Serous membrane that invests the lungs and extends into the fissures between the lobes.

visceroreceptor Sensory receptor associated with the visceral organs.

visual association area The area of the cerebral cortex immediately surrounding the primary visual cortex, which helps interpret visual images.

visual cortex Area in the occipital lobe of the cerebral cortex that integrates visual information and produces the sensation of vision.

visual field The area of vision of each eye.

visual radiation Nerve fibers that project from the lateral geniculate body to the visual cortex of the brain. Also known as optic radiation.

vitamin A group of organic substances, present in minute amounts in natural foodstuffs, that are essential to normal metabolism; insufficient amounts in the diet may cause deficiency diseases.

vitiligo (VIT-ih-LYE-go) White patches of skin caused by loss of melanin pigment.

vitreous humor Transparent jellylike material that fills the globe of the eyeball between the lens and the retina.

vocal cord True vocal cord. A pair of fibrous folds of elastic ligaments, covered by mucous membrane, stretching from the thyroid cartilage to the arytenoid cartilage; vibration of the vocal cords produces sound for speech.

vocal fold See *vocal cord.*

Volkmann's canal See *horizontal canal.*

voltage-gated ion channel Ion channel proteins in which the channel opens and closes, depending on the local membrane potential.

vomer (VOE-mer) Resembles a plowshare, which gives this bone its name. Forms nasal septum.

vomit To eject matter from the stomach or small intestine through the mouth.

Von Ebner's gland Serous glands of the tongue.

vulva External genitalia of females, composed of the mons pubis, the labia majora and minora, the clitoris, the vestibule of the vagina and its glands, and the opening of the urethra and of the vagina; the pudendum.

Wallerian degeneration The process by which axons degenerate when severed. See *anterograde* and *retrograde degeneration.*

Wernicke's speech area Area of cortex in the parietal lobe of the cerebrum that enables one to formulate ideas into words.

Wharton's duct See *submandibular duct.*

white adipose tissue Fatty connective tissue characteristic of adults; white adipose cells contain one large lipid droplet, in contrast with brown adipose cells that contain many smaller lipid droplets; found subcutaneously, in mesenteries and retroperitoneally.

white matter Bundles of myelinated axons in the brain and spinal cord.

white muscle fiber Fast-twitch skeletal muscle fibers; contract rapidly, anaerobically, with low resistance to fatigue; have low myoglobin content.

white pulp That part of the spleen consisting of lymphatic nodules and diffuse lymphatic tissue; associated with arteries.

white ramus communicans, *pl.* **rami communicantes** Connection between spinal nerves and sympathetic chain ganglia through which myelinated preganglionic axons project. See *gray ramus communicans.*

wisdom tooth Third molar tooth on each side in each jaw.

Wormian bone See *sutural bone.*

wrist Also known as the carpus, the wrist consists of eight short bones that allow the hand to flex and extend on the forearm.

xiphoid (ZIF-oid) Sword shaped, particularly the tip of the sword; the inferior part of the sternum.

xiphoid process The fibrocartilage tip of the sternum, which receives the vertebrochondral ribs.

yolk sac Highly vascular sac that contains yolk in bird, fish, and reptile embryos. Rudimentary in humans.

Z-line Delicate membranelike structure, found at ends of sarcomeres, to which the actin myofilaments attach. Also known as Z-disks.

zona fasciculata Middle layer of the adrenal cortex that secretes cortisol.

zona glomerulosa Superficial layer of the adrenal cortex that secretes aldosterone.

zona pellucida (PEL-loose-ih-dah) Layer of gel surrounding the oocyte.

zona reticularis Deep layer of the adrenal cortex that secretes androgens and estrogens.

zonula adherens Adherent band beneath tight junctions of epithelial cells.

zonula occludens Junctional complex between cells in which the cell membranes may be fused; occludes or blocks off the space between the cells. Also called tight junction.

zonular filament The group of filaments that suspend the lens from the ciliary body.

zygomatic arch (ZY-go-MAT-ik) Bony arch created by the junction of the zygomatic and temporal bones.

zygote Diploid cell resulting from fertilization of an ovum by a sperm cell. Embryonic development begins with the zygote.

zymogen granule (ZIME-o-jen) Enzyme-packed granules secreted by pancreatic acinar and other cells.

Cerebral aqueduct, 559, 567
Cerebral circulation, *432*, 433-434
Cerebral cortex, *572*
    anatomy of
        internal, 576-577, *577*
        micro, 575-576, *576*
    areas of, 573-575, *574*, *575*
    association centers of, 569
    auditory association area of, 574, *574*
    autonomic nervous system control by, 653
    Broca's speech area of, 575
    functions of, 572
    homunculi of, 574, *575*
    limbic system of, 583-584, *584*
    olfactory centers of, 574, *574*
    sensory and motor areas of, 574, *574*
    sensory and motor centers of, 596
    sensory centers of, 569
    taste area of, 574, *574*
    visual centers of, 574, *574*
    Wernicke's speech area of, 574-575
Cerebrospinal fluid, 562
    assessment of, 533, *533*
    blocked circulation of, 587-588
    circulation of, 586-588, *587*
    functions of, 559, 584
    renewal of, 586
Cerebrum
    basal ganglia of, *582*, 582-583
    functions of, 571-572
    hemispheres of, 571-577
        fissures, lobes, sulci, and gyri of, *572*, 572-573
    peduncles of, 569
Ceruminous gland(s), 621
Cervical canal, 807
Cervical muscle(s), 304
    anterior, 282, *282*
    origin, insertion, action, and innervation of, 283t, 284t
Cervical nerve(s), 534
Cervical plexus, 339, *339*
Cervical triangle(s), *280*, 280-281, *281*, 304
Cervix, 790, 807
    dilation of, 801
    during labor, 820
Chemoreceptor(s), 601
Chiasmatic groove, 152
Chief cell(s), function of, 676
Chloride, plasma profile of, 458t
Cholecystokinin, actions of, 687, 722
Choledochal sphincter, 722
Cholesterol
    dietary, pathway of, 471-472, *472*
    increased levels of, 474
        causes of, 472-473
        drugs for reducing, 473
    plasma profile of, 458, 458t
    serum levels of, 471-473

Cholesterol—cont'd
    synthesis of, 471
Cholinergic neuron(s), 644
Chondroblast(s), function of, 125
Chondrocalcin, function of, 127
Chondrocyte(s), *126*
    function of, 125
Chondronectin, function of, 127
Chordae tendineae, 388
Chorion, 816
Chorion frondosum, 818
Chorion laeve, 818
Choroid, 605, 611-612, *612*
Choroid plexus, 559
    anterior, 569
    of cerebral cortex, 576
    cerebrospinal fluid production by, 586
Chromaffin cell(s), 650
    hormone secretion by, 685
Chromatid(s)
    in meiosis, 53
    in mitosis, 51-52
Chromatin fiber(s), 33
Chromatolysis, 521
Chromosomal sex, 810
Chromosome(s), *33*
    homologous pairs of, 53
    in meiosis, 52-53
    in mitotis, 51-52
Chronic obstructive pulmonary disease, 738, 758-759
Chylomicron(s), 715
    characteristics of, 485-486
    function of, 471-472
Chyme, 711
Cilia
    epithelial, 68
    function of, 58
    properties of, 43
Ciliary body, 605, 613-614
Cingulate sulcus, 573
Circle of Willis, 433
Circulation
    cerebral, *432*, 433-434
    collateral, *423*
    coronary; see Coronary circulation
    fetal, 447, *448*
    postnatal, 447, *448*
    pulmonary, 445, *446*, 447
    thoracic, 434
Circulatory system, 8; see also Cardiovascular system; Immune system; Lymphatic system
Circumduction, characteristics of, 221
Circumflex femoral artery(ies), 444
Cisternae of skeletal muscle, 249
Clara cell(s), 743
Clathrin, 46
Clavicle, 191, 193, *313*
Cleavage, 101, 109

Clitoris, 809
    development of, 810
Clivus, 152
Cloaca, 729
Clone(s) in cell cycle, 50-51
Coagulation
    granules in, 468-469
    hemostatic plug formation in, *469*
    liver in, 456
    platelets in, 467-468
Coarctation of aorta, *404*, 405
Cocaine, axonal effects of, 522
Coccygeal nerve, 534
Coccygeal plexus, 336, *542*, 543
Coccygeus muscle(s), origin, insertion, action, and innervation of, 300t
Coccyx, 171
    sympathetic pathways to, 650
Cochlea, 620, 622
Cochlear duct, 622-624, *623*
Cochlear nerve, 625
Cochlear prostheses, 626
Codeine, neural effects of, 522-523
Colestipol for reducing serum cholesterol, 473
Colic artery(ies), *439*
Collagen
    alveolar, 746
    in cartilage, 125, 127
    in dense connective tissue, 83
    diseases affecting, 81
    in extracellular matrix, 81, *81*
    in fibrocartilage, 91-92
    gene transcription in, 35
    in lamellar bone, 121
    production of, 28
    in proteoglycans, 120
Colliculi, superior, 569
Colon, 723, *724*, 725, 730
    descending, 710, 723
    epithelium of, 72
    functions of, 723
    mucosa of, 725
    sigmoid, 723
    tissue layers of, 723, 725
    transverse, 723
Colon cancer
    development of, 53, 55
    susceptibility gene in, 55
Colostrum, 821
Coma, cause of, 583
Comedones, 361
Computed tomography, 21, *21*
Concentration gradients, 507
Concha, description and example of, 145t
Condylar process, 164
Condyle(s), 118
    description and example of, 145t
    of knee, 187
    tibial, *188*

20A, 19-29 Scott Bodell. 19-6C, 19-7C, 19-8, 19-10, 19-14, 19-20B, 19-29B Jack Tandy. 19-15 R. T. Hutchings. 19-17 after William Ober. 19-27A,B R.S.R. Associates.

Chapter 20

20-1 to 20-3, 20-7, 20-8A, 20-10, 20-13, 20-18, 20-19, 20-29 R.S.R. Associates. 20-4B, 20-12, 20-20, 20-31, 20-32 Marsha J. Dohrmann. 20-5, 20-16 Christine Oleksyk. 20-6 John V. Hagen. 20-8B, 20-14A, 20-15, 20-22, 20-23, 20-27A Barbara Cousins. 20-14B Victor Eroschenko. 20-21 Rolin Graphics. 20-27C Carlyn Iverson.

Chapter 21

21-1 Trevor-Roper PD and Curran PV: *The eye and its disorders*, ed 2, Oxford, 1984, Blackwell Scientific Publications, Ltd. 21-2A, 21-3, 21-7 to 21-10, 21-11C,D Rolin Graphics. 21-2B, 21-8 Christine Oleksyk. 21-4, 21-5A, 21-11A, 21-12 R.S.R. Associates. 21-5B Carlyn Iverson. 21-6 Barbara Cousins. 21-11A-D Victor Eroschenko.

Chapter 22

22-1 to 22-3, 22-10, 22-11A,B, 22-12C, 22-14, 22-16, 22-17, 22-18A, 22-20, 22-24, 22-25 Rolin Graphics. 22-5, 22-6, 22-7, 22-11D, 22-13, 22-15C,D, 22-18B, 22-23 R.S.R. Associates. 22-8 Christine Oleksyk. 22-9 G. David Brown. 22-11C, 22-12B, 22-15B, 22-19A, 22-22, 22-23B Victor Eroschenko. 22-12A, 22-15A, 22-21 Carlyn Iverson. 22-18 Barbara Cousins.

Chapter 23

23-1 Lisa Shoemaker. 23-2, 23-13 Jack Tandy. 23-3, 23-7B, 23-8, 23-11C, 23-12, 23-13 (inset), 23-14, 23-21B, 23-22B, 23-26, 23-27, 23-29, 23-30 R.S.R. Associates. 23-4A Jody Fulks. 23-4B Trent Stephens. 23-5 Marsha J. Dohrmann. 23-6A, 23-11A, 23-16B, 23-18, 23-22B, 23-23B Rolin Graphics. 23-6B Kathryn Born. 23-7A, 23-9A, 23-20, 23-23A David J. Mascaro and Associates. 23-9B, 23-15B Kathy Mitchell Grey. 23-15A, 23-16A, 23-24A G. David Brown. 23-17A,B Christine Oleksyk. 23-19 David J. Mascaro and Associates. 23-22A Victor Eroschenko. 23-23A Barbara Cousins.

Chapter 24

24-1B, 24-2, 24-5, 24-11A,B, 24-16, 24-22 Rolin Graphics. 24-3 Ernest W. Beck. 24-4, 24-6, 24-12C, 24-13B, 24-19B, 24-21 R.S.R. Associates.

24-7A Lisa Shoemaker. 24-8A, 24-12B, 24-15, 24-19A Jody L. Fulks. 24-8B, 24-10A, 24-12A, 24-15A Victor Eroschenko. 24-9, 24-13A Newland. 24-14A Harrington. 24-14B, 24-17 Christine Oleksyk. 24-18 Carlyn Iverson. 24-20 Kate Sweeney.

Chapter 25

25-1, 25-3B, 25-15 Christine Oleksyk. 25-2, 25-11, 25-19B Rolin Graphics. 25-3A SIU Biomedical Comm./Custom Medical Stock Photo. 25-4A, 25-5A David J. Mascaro and Associates. 25-5B, 25-6B, 25-7F, 25-8, 25-9, 25-16, 25-17, 25-18 R.S.R. Associates. 25-6A Ernest W. Beck. 25-7A-D, 25-10 Pagecrafters. 25-12A,B, D,E, 25-13 Jody L. Fulks. 25-12C, 25-14B Victor Eroschenko. 25-14A,C Kevin A. Somerville.

Chapter 26

26-1, 26-5C,D, 26-7 Rolin Graphics. 26-2A, 26-10 Ronald J. Ervin. 26-2B, 26-15 David J. Mascaro and Associates. 26-3, 26-8, 26-11B, 26-12, 26-14A, 26-19A, 26-21, 26-22A Kevin Somerville. 26-4A William Ober. 26-5E, 26-6B, 26-14B, 26-16, 26-17, 26-18 R.S.R. Associates. 26-11A Carlyn Iverson. 26-20 Jack Tandy.